Springer Collected Works in Mathematics

For further volumes:
http://www.springer.com/series/11104

Marcel Riesz

Marcel Riesz

Collected Papers

Editors

Lars Gårding · Lars Hörmander

Reprint of the 1988 Edition

Springer

Author
Marcel Riesz (1886, Györ, Hungary –
1969, Lund, Sweden)
Lund University
Lund, Sweden

Editors
Lars Gårding
Lars Hörmander
Lund University
Lund, Sweden

ISSN 2194-9875
ISBN 978-3-642-34603-3 (Softcover)
978-3-540-18115-6 (Hardcover)
DOI 10.1007/978-3-642-37535-4
Springer Heidelberg New York Dordrecht London

Library of Congress Control Number: 2012954381

Mathematics Subject Classification (1980): 01 A 75

Printed on acid-free paper

Springer is part of Springer Science+Business Media (www.springer.com)

Preface

The work of Marcel Riesz spans a large area of analysis, from Fourier analysis to the theory of partial differential equations. Much of it is not available in print any longer, which has made the publication of his collected works important.

This volume contains almost all his published papers. The Cambridge Tract volume [16] with G.H. Hardy as coauthor, and the survey article [31] in Enc. d. Math. Wiss. with E. Hilb as coauthor have not been included though. We have also omitted some papers published in Hungarian or Swedish [1, 3, 15, 44], two purely didactic papers [51, 54], and a late informally published manuscript [57]. The thesis [2] is included in an English translation made for this volume by Professor John Horváth.

The numbers in brackets refer to the bibliography which is essentially taken from the memorial article on Marcel Riesz published by one of us in Acta Mathematica *124* (1970) and included as an introduction here. Another very detailed report on the mathematical work of Marcel Riesz has been published in Hungarian by J. Horváth in Matematikai Lapok *26* (1975), 11–37 and *28* (1980), 65–100. A French translation is available in Cahiers du Sém. d'Histoire des Math. *3* (1982), 83–121 and *4* (1983), 1–59.

The publication of this volume has been made possible by grants from the Swedish Natural Science Research Council, Återförsäkringsaktiebolaget Sverige, Ivar Bendixson's foundation at the University of Stockholm, and the Filip Lundberg foundation. We would also like to thank Professors Béla Szökefalvi-Nagy and Károly Tandori for their help in the preparation of the manuscript, Professor John Horváth for his excellent translation of Marcel Riesz' thesis, and Springer-Verlag for its superb technical work.

Lars Gårding
Lars Hörmander

Table of contents

Marcel Riesz in memoriam

Lars Gårding

Acta mathematica *124* (1970), I–XI

Marcel Riesz was born in Györ, Hungary, November 16, 1886 and died in Lund, Sweden, on September 4, 1969. He studied in Budapest, Göttingen and Paris. In 1911 Mittag-Leffler invited him to come to Sweden where he taught at Stockholms Högskola. In 1926 he was appointed professor of mathematics at the university of Lund. After retiring from this position in 1952 he went to the United States where he was visiting research professor at the universities of Maryland and Chicago and other places. He returned to Lund in 1962. He was a member of the Swedish Academy of Sciences, the Physiographical Society in Lund, Videnskabernes Selskab in Copenhagen, had honorary degrees from the universities of Copenhagen and Lund and was an honorary member of the Swedish mathematical society.

Marcel Riesz was the youngest member of a generation of brilliant Hungarian mathematicians that included among others Leopold Fejér, Riesz's elder brother Frederick Riesz and Alfred Haar. His first paper ([1], 1906) is an exposition in Hungarian of a subject of current interest at the time, namely various summation methods for Taylor series of analytic functions. One of these methods, due to Mittag-Leffler, sums the series in a star-shaped region bounded by singular points, the Mittag-Leffler star. Common interest in these matters was the beginning of the association with Mittag-Leffler. They seem to have met for the first time in Stockholm in 1908.

In the period from 1908 to 1916, Riesz did very successful work in the summability theory of power series, trigonometric series and Dirichlet series. The starting point was Fejér's discovery that the Fourier series of a function f is summable by arithmetic means to $f(x)$ at every point of continuity. Lebesgue had shown that the same holds at all Lebesgue points. The current scales of summation methods were those of Hölder and Cesàro, recently shown to be equivalent. The Cesàro mean $s_n^{(\alpha)}$ of order $\alpha > -1$ of a sequence $\{s_n\}_0^\infty$, $s_0 = 0$, or a series with partial sums s_n is by definition the quotient of the coefficients of the nth terms in the formal power series for $(1-x)^{-\alpha}\sum_0^\infty s_n x^n$ and $(1-x)^{-\alpha-1}$. The Cesàro limit is $\lim s_n^{(\alpha)}$ for $n \to \infty$. When $\alpha = 1$, this procedure sums the series by arithmetic means and the result of Lebesgue applies. One of the classical results in Fourier series, namely that Fejér's (and Lebesgue's) theorem holds for summation of order $\alpha > 0$ is essentially due to Riesz. He announced it in 1909 [7] and published his proof much later [25]. In the meantime there were proofs by Chapman in 1910–11 and by Hardy in 1913. Riesz's main contribution, however, was in another direction. He observed that Cesàro's summation is equivalent to another one which has the advantage of being applicable also to functions $s(x)$ of locally bounded variation for $x \geq 0$ such that $s(0) = s(0+) = 0$. Its means are

$$s_\alpha(x) = x^{-\alpha}\int_0^x (x-t)^\alpha\, ds(t) \tag{1}$$

and the limit is defined accordingly. In case of a series we put $s(x) = s_n$ when $n \leq x < n+1$. The right-hand side of (1) appears also in the classical Riemann–Liouville integral of a function f,

$$I^\alpha f(x) = \frac{1}{\Gamma(\alpha)}\int_0^x (x-t)^{\alpha-1} f(t)\, dt. \tag{2}$$

The semi-group properties of the operator $\alpha \to I^\alpha$, namely that $I^\alpha I^\beta = I^{\alpha+\beta}$ and $I^0 =$ identity has immediate application to the means (1). The corresponding summation method applies to generalized Dirichlet series

$$f(z) = \int_0^\infty e^{-zt}\, ds(t),$$

where z is a complex number. The means are

$$x^{-\alpha}\int_0^x e^{-zt}(x-t)^\alpha\, ds(t)$$

and the α-sum of $f(z)$ is obtained by letting $x \to \infty$. These integrals were introduced by Riesz in 1909 [7] under the name of typical means. They had an instant success and led to a Cambridge Tract on Dirichlet series 1915 written jointly by Hardy and Riesz and reprinted in 1952.

Some ten years later, Riesz had established himself as an expert in the theory of trigonometric series and together with E. Hilb wrote the article Neuere Untersuchungen über trigonometrische Reihen in Encyklopädie der Mathematischen Wissenschaften. It would carry us too far to penetrate in detail Riesz's contributions to this intricate and much studied subject, but at least three more things stand out. His interpolation formula [14] for trigonometric polynomials has a permanent place in interpolation theory and provides simple proofs of, e.g., the well-known inequalities of Bernstein and Markoff with the best constants. One recurrent theme in Riesz's work is a theorem by Fatou that says that the power-series of an analytic function converges at every point of the boundary of the circle of convergence where the function is regular and the coefficients of the series tend to zero. Riesz gave a beautiful proof [17] of this result that also showed that the convergence is locally uniform on every arc of regularity. This proof is available in Landau's little booklet *Darstellung und Begründung einiger neuerer Ergebnisse der Funktionentheorie*. Riesz also extended Fatou's theorem to Dirichlet series and to summability situations. His best known result in Fourier series, however, lies in another direction. Every trigonometric series with real coefficients can be written as

$$u(e^{i\theta}) = \sum_0^\infty \operatorname{Re} c_n e^{in\theta}. \tag{3}$$

Formally, this is the real part of the power series

$$f(z) = \sum_0^\infty c_n z^n, \quad z = re^{i}$$

for $r = 1$. Its imaginary part

$$v(e^{i\theta}) = \sum_0^\infty \operatorname{Im} c_n e^{in\theta}$$

is called the conjugate series to (3). By Parseval's formula

$$\int_0^{2\pi} u(e^{i\theta})^2 d\theta = \pi \sum |c_n|^2 = \int_0^{2\pi} v(e^{i\theta})^2 d\theta, \tag{4}$$

provided we normalize so that $c_0 = 0$. Hence $u \to v$ is an isometry $L^2 \to L^2$ where L^2 refers to the interval $0, 2\pi$. The connection between conjugate series was an early subject of study. In particular, as early as 1912, W. H. and G. C. Young had considered the question whether the map $u \to v$ might be, if not an isometry, then at least a homeomorphism $L^p \to L^p$,

$1<p<\infty$, but they had some indirect evidence for a negative answer. Here was an outstanding unsolved problem and when Riesz wrote to Hardy in 1923 that he had found a proof (see [33]), Hardy wrote back "some months ago you said 'j'ai démontré que 2 sér. trig. conjuguées sont toujours en même temps les séries de F de fonctions de la classe $L^p(p>1)\ldots$'. *I want the proof*. Both I and my pupil Titchmarsh have tried in vain to prove it ...". And in the next letter "Very many thanks—you supply all that is essential, I have sent on your letter to Titchmarsh. Most elegant and beautiful. Of course p. 2 is the real point. It is amazing that none of us should have seen it before (even for $p=4$!)". What Hardy is refering to is Riesz's use of Cauchy's formula which gives $\int_0^{2\pi} f(e^{i\theta})^p d\theta=0$ when p is an integer. Taking the real part and putting $p=2$ proves (4) while, if $p=4$,

$$\int_0^{2\pi} (u^4-6\,u^2v^2+v^4)\,d\theta=0.$$

This identity and a moment's reflection show that $\int_0^{2\pi} v^p d\theta \leqslant c_p \int_0^{2\pi} u^p d\theta$ when $p=4$ and, more generally, when p is an even integer. The remaining arguments of the proof were more technical but they eventually lead Riesz to a basic result in analysis, his convexity theorem [32]. It combines elementary form with great power and runs as follows: the logarithm of the maximum $M_{\alpha\beta}$ of the absolute value of a bilinear form

$$A=\sum_{j=1}^{m}\sum_{k=1}^{n} a_{jk}\,x_j\,y_k$$

under the conditions

$$\sum_{j=1}^{m}|x_j|^{1/\alpha}\leqslant 1,\quad \sum_{k=1}^{n}|y_k|^{1/\beta}\leqslant 1$$

is a convex function of α,β in the triangle

$$\alpha\leqslant 1,\ \beta\leqslant 1,\ \alpha+\beta\geqslant 1.$$

A little later, Riesz's student Olof Thorin proved convexity when $\alpha\geqslant 0$, $\beta\geqslant 0$ provided the variables x and y are allowed to be complex. The convexity has many applications, one of them to Riesz's theorem about conjugate functions. Let c_p be the best constant in the inequality $\|v\|_p\leqslant c_p\|u\|_p$ where $\|f\|_p=(\int_0^{2\pi}|f(e^{i\theta})|^p d\theta)^{1/p}$. Then, by Hölder's inequality, c_p is the supremum of the absolute value of the bilinear form $A(u,w)=\int_0^{2\pi} u(e^{i\theta})w(e^{i\theta})\,d\theta$ when $\|u\|_p\leqslant 1$ and $\|w\|_q\leqslant 1$ where $p^{-1}+q^{-1}=1$. Further, $c_p=c_q$. Hence, since $c_2=1$ and, by the argument given above, $c_p<\infty$ when p is an even integer, it follows that $c_p<\infty$ when $1<p<\infty$. Riesz's convexity theorem has been the model for similar results, in recent times also for the theory of interpolation spaces.

Before Riesz moved to Lund he also worked with the moment problem ([24]), in particular the Hamburger moment problem: find a real measure $d\mu \geqslant 0$ on the real line with given moments

$$c_n = \int_{-\infty}^{+\infty} x^n \, d\mu(x).$$

Hans Hamburger's almost complete analysis of this problem in 1920 attracted the attention of many mathematicians, among them Riesz and Torsten Carleman. Here Carleman, who could fit the problem into his theory of singular integral equations, was perhaps more successful. Riesz was prepared for the subject through his friend Erik Stridsberg who took a lively interest in Stieltjes's classical paper from 1894 about the moment problem on the half-axis $x \geqslant 0$.

The meat of the moment problem is the analysis of the set of solutions and the problem of uniqueness. Existence of a solution μ is a relatively simple matter. It is obviously necessary that, for every real polynomial P,

$$P(x) = \sum a_n x^n \geqslant 0 \Rightarrow \sum a_n c_n \geqslant 0 \tag{5}$$

and it is easy to see that this is equivalent to the condition that all the quadratic forms $\sum_0^n \sum_0^n c_{j+k} y_j y_k$ are $\geqslant 0$. To prove that (5) is also sufficient one extends the non-negative linear functional $\varphi(P) = \Sigma a_n c_n$ from polynomials to the space of real continuous functions of at most polynomial growth. This procedure is now familiar to every student of analysis who knows the Hahn–Banach theorem, but it is interesting to note that Riesz used it in 1918 (see [24], the third note p. 2) in a lecture to the Stockholm Mathematical Society, well before Banach (1923, 1929) and Hahn (1926).

Riesz's work after he moved to Lund marks a break with the past. He acquired new interests, starting work in potential theory and wave propagation including Dirac's equation of the electron and relativity theory. He also took a continuing interest in elementary number theory. His most important contributions are in potential theory and wave propagation. In both cases he invented new multi-dimensional analogues of the Riemann-Liouville integral (2). The one used in potential theory is the fractional potential of order α of a mass distribution $f dx$,

$$I^\alpha f(x) = H_m(\alpha)^{-1} \int |x-y|^{\alpha-m} f(y) \, dy. \tag{6}$$

Here $|x-y|$ is the euclidean distance in R^m, $0<\alpha<m$, and

$$H_m(\alpha) = \pi^{m/2} \Gamma\left(\frac{\alpha}{2}\right) \Big/ \Gamma\left(\frac{m-\alpha}{2}\right)$$

is chosen so that $I^\alpha I^\beta = I^{\alpha+\beta}$ when $\alpha > 0$, $\beta > 0$, $\alpha+\beta < m$. By a passage to the limit, I^0 is the identity. In particular, if

$$U^\alpha \mu(x) = H_m(\alpha)^{-1} \int |x-y|^{\alpha-m} d\mu(x)$$

is the α-potential of a mass distribution μ one has

$$\int U^\alpha \mu(x) U^\beta \nu(x)\, dx = H_m(\alpha+\beta)^{-1} \int |x-y|^{\alpha+\beta-m} d\mu(x)\, d\nu(x)$$

when $\alpha > 0, \beta > 0, \alpha+\beta < m$. Hence

$$\int U^\alpha \mu(x)\, d\mu(x) = H_m(\alpha) \int (U^{\alpha/2} \mu(x))^2\, dx \tag{6}$$

is non-negative. When $\alpha = 2 < m$, this represents the self-energy of the mass distribution μ with respect to the Newtonian potential. In this case the equilibrium distribution μ_F on a compact set F had been characterized by Gauss as having minimal energy in the class of mass distributions μ supported by F and having a given total mass. After his student Otto Frostman had put this principle on a strict basis, Riesz noted that the formula (6) indicates that the equilibrium distribution with respect to an α-potential could be defined by this principle of Gauss. In the case $0 < \alpha \leqslant 2$, which is the condition for the kernel $|x-y|^{\alpha-m}$ to be subharmonic, this program met with complete success in Frostman's well-known thesis. The equilibrium distribution μ_F exists, it is unique, its potential on F is essentially constant, there is a notion of α-capacity and there are generalizations of the balayage process and Green's function. After Frostman, Riesz himself wrote on the subject [42]. Viewed in perspective, these contributions opened the way to the present-day variety of generalizations and refinements of classical potential theory.

At some points in his work on α-potentials, Riesz uses the fact that if f is smooth enough, (6) is a regular analytic function of α when $c < \operatorname{Re} \alpha < m$ with a given c and that $\Delta I^\alpha = I^{\alpha-2}$ where $\Delta = (\partial/\partial x_1)^2 + \ldots + (\partial/\partial x_m)^2$ is Laplace's operator. Analyticity properties became very important in the integral

$$I^\alpha f(x) = H_m(\alpha)^{-1} \int_{x-C} r(x-y)^{\alpha-m} f(y)\, dy \tag{7}$$

associated with the wave operator

$$\Delta = (\partial/\partial x_1)^2 - \ldots - (\partial/\partial x_m)^2.$$

Here $r^2(x) = x_1^2 - \ldots - x_m^2$ is the square of the Lorentz distance and $C: r^2(x) \geqslant 0$, $x_1 > 0$ is the forward light-cone so that $x-C$ is the retrograde light-cone with its vertex at x. Finally,

$$H_m(\alpha)=\pi^{(m-2)/2}\,2^{\alpha-1}\Gamma\left(\frac{\alpha}{2}\right)\Gamma\left(\frac{\alpha+2-m}{2}\right). \tag{8}$$

Clearly, (7) is an analytic function of α when the integral converges, i.e. when Re $\alpha>m-2$. The factor H_m is chosen so that

$$\Delta I^{\alpha}=I^{\alpha-2},\ I^{\alpha}I^{\beta}=I^{\alpha+\beta} \tag{9}$$

for large α and β. The main point is now that, if f is smooth enough, the right side of (7) is analytic in the halfplane Re $\alpha>-c$ with any given c, that (9) holds under corresponding circumstances and that $I^0=1$. The real explanation for this is perhaps the following. Let f have compact support and let

$$F(\xi+i\eta)=\int e^{-x(\xi+i\eta)}f(x)\,dx$$

be the Fourier–Laplace transform of f. Then

$$I^{\alpha}f(x)=(2\pi)^{-m}\int e^{x(\xi+i\eta)}\,r(\xi+i\eta)^{-\alpha}F(\xi+i\eta)\,d\eta,$$

where the integral is independent of the choice of $\xi\in C$ and the right side is indeed entire analytic in α when f is infinitely differentiable. Although Riesz was conscious of this, he always preferred direct proofs avoiding the use of the Fourier–Laplace transform. In his last published paper [59] he gives a variant of an earlier direct proof that I^0 is the identity. This point of view is connected with his intense interest in special relativity and questions of invariance under the Lorentz group.

It is probable that Riesz conceived the integral (7) during a series of seminars in the beginning of the thirties conducted by his colleague Nils Zeilon. The subject was Hadamard's theory of finite parts of integrals as exposed in his well-known book on the Cauchy problem. Hadamard gives explicit formulas for the solution of Cauchy's problem with data on a space-like surface for the wave equation with variable coefficients. In 1938 Riesz was ready with a beautiful variant of these formulas based on an operation I^{α} analogous to (7) and with similar properties relative to the wave operator with variable coefficients. He published an account in a note added to a lecture [43] given a year earlier where he treats the case of constant coefficients. In the lecture he remarked among other things that the fact that $1/H_m(2)=0$ when $m>2$ is even reflects Huygens' principle and he finished by giving an invariant formula for the solution of Cauchy's problem when $m=4$ and analyzing it geometrically. In the fifties, during his stay in the United States, Riesz devoted much thought to this formula and extended it to characteristic surfaces in an intriguing paper [55].

The short paper [43] contains the main body of Riesz's work on the wave equation and it still makes impressive reading. The detailed account came in 1949 in a massive paper [47] in the Acta Mathematica. Considering that Riesz was a natural perfectionist and, in his later years, a slow worker, this paper is the result of an enormous effort on his part. It was made possible only through collaboration with his friend K. G. Hagstroem who graciously offered his hospitality and his services as a secretary during many summers in the forties.

Riesz's work on the wave equation did not contribute to the existence theory of Cauchy's problem and one of its main goals, the proper interpretation of divergent integrals, has since been attained in a much larger framework through Laurent Schwartz's theory of distributions. But the method of analytical continuation which makes sense also in this theory, is a lasting contribution.

Around 1930, the teaching of quantum mechanics and relativity theory became a necessity at every self-respecting university, but at the time not many physicists had the mathematical training necessary for these subjects. For many years Riesz lectured periodically on tensors and matrices to a motley crew of physicists and mathematicians. He also started research of his own and published his first paper in relativistic quantum theory in 1946. His main interest was the Dirac equation and the Clifford algebra. In [53] he introduced what amounts to a Clifford structure on a Lorentz manifold. Riesz also applied his work on the wave equation to the classical relativistic theory of the electron [49, 50]. The popular lecture [44] on the models of non-euclidean geometry, a masterpiece in its kind, is a byproduct of his interest in the geometry of relativity.

Riesz wrote clearly and well and paid much attention to form. His favourite language was French and his style, steeped in the classical tradition, sometimes borders on the precious. Mathematical research always involves competition for fame and a place in the hierarchy, but he made it seem a gentleman's game. He was of course no stranger to ambition and had to assert himself both in Sweden and in the cosmopolitan world he came from. He admired his illustrious brother Frederick and they had cordial relations. They wrote one paper [22] together, but otherwise there is a clear distance in content between their work, perhaps a result of a conscious effort on Marcel's part. Seen together they had much physical resemblance but very different temperaments, Frederick calm with great poise and Marcel quick and restless in comparison. Marcel Riesz knew an astonishing number of mathematicians and over the years made and kept many friends among them.

Mittag-Leffler had made Stockholms Högskola a center of mathematical research. It had a peak of activity before and around the turn of the century and its other mathematicians, Bendixson, von Koch and Fredholm were famous names. Some ten years later, however, their scientific activity was on the wane for various reasons. Riesz filled a mathematical vacuum. He learnt Swedish quickly and he was very active in the local mathematical society where he soon became the dominating figure. He was lively, accessible, an enthusiastic teacher and a good lecturer with a thorough knowledge of his field. His charming expository lecture [15] from 1913, written in Swedish, has a distinct personal touch reflecting these qualities. In 1923, Riesz lost a competition for a chair in Lund to Carleman. Shortly afterwards von Koch died and Bendixson, Fredholm and Phragmén made a move to appoint Riesz as von Koch's successor. The move failed and the call went to Carleman. Shortly afterwards, Riesz got a position in Lund. At least in the beginning, he must have felt his stay in Lund as an exile. He had been very successful in Stockholm where, among others, Harald Cramér and Einar Hille had been his personal students.

Lund did not have much of a mathematical tradition but Riesz's arrival meant a change of atmosphere. He was now an international star, active with his own research and he also had the time and incentive to broaden his interests. Frostman's thesis was a success and there were others after him. Lars Hörmander was one of Riesz's last personal students in Sweden. Riesz's work on fractional potentials was the origin of the contributions from Lund to the theory of partial differential operators.

I met Marcel Riesz in 1937, my first year at the university. He then had a small circle of graduate students. Each one got personal attention. Riesz loved to talk about mathematics and he appreciated having listeners. He could go on for hours and when he was in good form, his grip on the listener never slackened. Riesz lived alone and these personal lectures took place sometimes in his home, sometimes in his favourite café and sometimes over the telphone. The subjects were mostly those that occupied him at the time, fractional potentials, relativity theory and the Clifford algebra. He was not inclined to give away thesis subjects, but if somebody had a promising idea he was a very stimulating partner and spared no effort. He worked constantly, often at late hours and periodically with great intensity. These habits did not change much with advancing age and eventually took their toll. After about ten years in the United States he had a breakdown that forced him to return to Sweden and begin a more sedate life.

Although Riesz's interest in administrative matters was strictly limited to those concerning his own subject, he had great influence in the faculty merely by his good sense, his wit and his charm. He was a shrewd judge of personalities and human affairs and had an uncanny knack for the right word at the right moment. His memory was extraordinary and he was a fascinating teller of stories. Unfortunately he never wrote his memoirs. He had a long and active life and bore the burden of his last illness with great courage.

The original bibliography for this article (pp: X–XI) is omitted here and appears in an edited version at the end of this volume.

2.

Summable trigonometric series and summable power series[1]

Mathematikai és Physikai Lapok *19* (1910), 1–56

Introduction

Since the investigations of Du Bois-Reymond we know that the points of divergence of the Fourier series of a continuous function may form an everywhere dense set. Thus not even continuous functions have the property that their Fourier series converges.

The recent tendency in the theory of functions is not exhausted by merely looking for convergence criteria but searches for other connections between the function and its Fourier series. Fejér[2] was the first to apply consciously this far-reaching point of view in the theory of these series. He recognized that the method of summation, applied by Cesàro[3] and Frobenius[4] to other questions (which, instead of the limit of the partial sums, investigates the limit of their arithmetic means), is of capital importance for the Fourier series.

The divergent series which are summable by means of this method belong at any rate to the divergent series which are the easiest to handle and therefore the most useful. A series of this kind will be said to be *summable* in brief, and the limit of the arithmetic means will be called the *Cesàro-sum* of the series.

Fejér's main result is the following:

The Fourier series of any finite and integrable function $f(x)$ is summable at every point where the function is continuous or has a discontinuity of the first kind, the value of the Cesàro-sum is $f(x)$ at the points of continuity and $\frac{1}{2}[f(x+0)+f(x-0)]$ at the points of discontinuity of the first kind.

This penetrating theorem was significantly generalized later from different points of view by Fejér himself and by Lebesgue[5].

It is also an important property of Fejér's summation that the arithmetic means converge uniformly to the values of the function in any interval of continuity of the function, while, as it was recently shown by Lebesgue[6], the Fourier series of an everywhere continuous function may converge everywhere without being uniformly convergent.

---

[1] Parts of the results of this article have already appeared in the *Comptes Rendus* (7. octobre 1907.)

[2] Fejér: Investigations concerning Fourier series. Doctoral thesis (in Hungarian). Math. és Phys. Lapok 1902.

— Untersuchungen über Fourier'sche Reihen. Math. Ann., vol. 58.

[3] Cesàro: Sur la multiplication des séries. Bull. Sci. Math., vol. 14 (1890).

[4] Frobenius: Über die Leibnitz'sche Reihe. Crelle's Journal, vol. 89.

[5] Lebesgue: Sur la convergence des séries de Fourier. Math. Ann., vol. 61.

[6] Lebesgue: Comptes Rendus, 17. novembre 1905.

It is known that the Fourier series are special trigonometric series, whose coefficients are given by the familiar Fourier integral formulas. Thus when we consider these series we always start from a given function. The most important theorems in the theory of convergent trigonometric series given a priori were given by Riemann[7], Cantor[8] and Du Bois-Reymond[9]. Their results will occur during our discussions. Now I single out only one of the most typical among them, Cantor's theorem. This reads as follows:

If a trigonometric series converges in the interval $(0,2\pi)$ *(except possibly for a reducible set of points*[10]*) and its sum vanishes, then also all its coefficients vanish.*

Cantor's proof is based on theorems of Riemann[7] and Schwarz.

We saw above on the example of Fourier series that the method of summation by arithmetic means suits the nature of trigonometric series better than forming the ordinary sum. Thus the question arises naturally whether Cantor's theorem can be generalized to divergent but summable trigonometric series. In his doctoral dissertation, Fejér is still skeptical about this question, since the Cesàro-sum of the divergent series

$$\tfrac{1}{2}+\cos x+\cos 2x+\ldots+\cos nx+\ldots$$

is zero at every point of the interval $(0,2\pi)$, except at the points 0 and 2π, nevertheless not all the coefficients of this series do vanish. In the "Mathematische Annalen", however, Fejér already extends Riemann's theorem used by Cantor thus he so-to-say prepares the answer to be given to this question.

During our common stay in Göttingen, Mr. Fejér himself was kind enough to call my attention to this subject, for which it is my pleasant duty here to thank him.

This work is divided into two parts. In the first part I establish among others the following theorems:

If the series

$$\sum_{n=1}^{\infty}\frac{|a_n|+|b_n|}{n^2}$$

formed with the coefficients of some trigonometric series is convergent, then the Cesàro-sum of this trigonometric series can vanish at every point of the interval $(0,2\pi)$ *only if all its coefficients vanish.*

If the coefficients of a trigonometric series tend to zero, and if in the interval $(0,2\pi)$, *with the exception of a reducible set, the series is summable and its Cesàro-sum is zero, then all its coefficients vanish.*

For the proof of our theorem we had to generalize Schwarz's mentioned theorem in a suitable way. Our result reads as follows:

---

[7] Riemann: Über die Darstellbarkeit einer Funktion durch eine trigonometrische Reihe. Gesammelte Werke (1876) p. 213.

[8] Cantor: Über die Ausdehnung eines Satzes aus der Theorie der trigonometrischen Reihen. Math. Ann., vol. 5.

[9] Du Bois-Reymond: Beweis dass die Coefficienten etc. Abh. bayer. Akad., vol. 12 (1875).

[10] A set of points is said to be reducible if one of its derived sets is empty.

Every continuous function, whose generalized second derivative is continuous, and whose generalized fourth derivative is zero, is a polynomial of degree three.[11]

This result, together with Fejér's generalization of Riemann's theorem, and furthermore the solution of a functional equation, lead immediately to our first theorem.

To prove our statement concerning series with coefficients tending to zero, we had to deduce an additional simple lemma.

As a consequence of these investigations, we present then the analogue of Du Bois-Reymond's fundamental theorem, which we prove with the help of a mean-value theorem, interesting in itself.

In the second part of our article, we investigate the behavior of some power series on their circle of convergence, completely independently of the first part.

Our investigations enable us to form strongly divergent summable series, and on the other hand they indicate the applicability of the iterated summation process. Namely, we present certain classes of power series which are summable by the n times iterated summation process, while a number less than n does not achieve the goal. As far as we know, this has not yet been done.

Further on we extend our results to the case when the order of summation is not an integer. At this point our investigations are related to some results of Hadamard.

I. Cantor's theorem

We mentioned in the introduction that Cantor proved his theorem on the basis of Riemann's and Schwarz's theorems. Riemann's theorem is the following:

Let

$$a_0+a_1 \cos x+b_1 \sin x+\ldots+a_n \cos nx+b_n \sin nx+\ldots \tag{1}$$

be a trigonometric series, whose coefficients tend to zero. The uniformly convergent series

$$F(x)=\frac{a_0}{2}x^2-A_1-\frac{A_2}{4}-\ldots-\frac{A_n}{n^2}-\ldots \tag{2}$$

obtained by integrating twice, where

$$A_0=a_0, \qquad A_n=a_n \cos nx+b_n \sin nx,$$

represents of course a continuous function. Now *if the series* (1) *converges at the point x and its sum is $f(x)$, then the limit*

[11] We remark here that so far only trivial results concerning generalized derivatives of higher than the second order have been obtained. Also Harnack's results, which are simple applications of well-known properties of uniformly convergent sequences, are obvious (Math. Ann., vols. 23 and 24).

$$\lim_{h=0} \frac{F(x+h)-2F(x)+F(x-h)}{h^2} =$$

$$= \lim_{h=0} \left[A_0 + A_1 \left(\frac{\sin \frac{h}{2}}{\frac{h}{2}} \right)^2 + \ldots + A_n \left(\frac{\sin \frac{nh}{2}}{\frac{nh}{2}} \right)^2 + \ldots \right] \tag{3}$$

exists and its value is $f(x)$.[12]

Furthermore, for all values of x

$$\lim_{h=0} \frac{F(x+h)-2F(x)+F(x-h)}{h} = 0. \tag{4}$$

This theorem is then combined with the theorem of Schwarz, according to which:

If at every point of the interval (a, b) *the second generalized derivative of some continuous function* $F(x)$ *satisfies*

$$\lim_{h=0} \frac{\Delta^2 F(x,h)}{h^2} = \lim_{h=0} \frac{F(x+h)-2F(x)+F(x-h)}{h^2} = 0,$$

then $F(x)$ *is a linear function of* x.

On the other hand, Cantor[13] shows that *the coefficients of a trigonometric series converging in an arbitrarily small interval tend to zero.* Comparing these results with Riemann's theorems, we obtain easily Cantor's theorem.[14]

Let us suppose, namely, that the series (1) converges with the exception of a finite number of points, and that its sum is zero. By virtue of the two quoted theorems, it is clear that the function $F(x)$ is linear between two consecutive exceptional points. But, as we have said, the function is continuous also at these points, hence the function $F(x)$ represents a continuous polygonal line, whose vertices are the possible existing exceptional points. However, even at such exceptional points we have

---

[12] The expression (3) is called the generalized second differential quotient or the generalized second derivative of $F(x)$. The way of forming generalized derivatives of higher order is clear. Their main interest is that they are defined by only one passage to the limit, and that they exist whenever the corresponding ordinary derivative exists and then the two values coincide. In our article an important role will be given mainly to the fourth generalized derivative.

[13] Cantor: Über einen die trigonometrischen Reihen betreffenden Lehrsatz. Crelle's Journal, vol. 72. — Über trigonometrische Reihen, Math. Ann., vol. 4. Cantor proves here also the more general theorem that, if

$$\lim_{n=\infty} (a_n \cos nx + b_n \sin nx) = 0$$

holds everywhere in some interval, then necessarily

$$\lim_{n=\infty} a_n = 0 \quad \text{and} \quad \lim_{n=\infty} b_n = 0.$$

[14] Kronecker shows in a witty way that this result can be omitted from the proof of the theorem. Beweis dass eine für jeden Werth etc. Crelle's Journal, vol. 72.

$$\lim_{h=0} \frac{F(x+h)-2F(x)+F(x+h)}{h} = 0,$$

$$\lim_{h=0} \frac{F(x+h)-F(x)}{h} = \lim_{h=0} \frac{F(x-h)-F(x)}{-h},$$

thus the two lines meeting at the point with abscissa x are each other's prolongation, and so the graph of $F(x)$ is a straight line. Let the equation of this straight line be

$$y = cx + c',$$

then for all values of x we have

$$A_0 \frac{x^2}{2} - A_1 - \ldots - \frac{A_n}{n^2} - \ldots = cx + c',$$

$$A_0 \frac{x^2}{2} - cx - c' = A_1 + \frac{A_2}{2^2} + \ldots + \frac{A_n}{n^2} + \ldots.$$

The left hand side can be periodic only if

$$A_0 = c = 0.$$

However, we know that the series on the right hand side converges uniformly and so its coefficients are given by the Fourier integral formulas, i.e.,

$$a_n = 0, \quad b_n = 0.$$

If the trigonometric series possesses infinitely many exceptional points, and these form a reducible set, then the same method can be applied.

Generalization of Cantor's theorem to summable series

We have already mentioned that our investigations concern such possibly divergent series

$$c_0 + c_1 + c_2 + \ldots + c_n + \ldots$$

that the limit

$$\lim_{n=\infty} S_n = \lim_{n=\infty} \frac{s_0 + s_1 + \ldots + s_n}{n+1},$$

formed with the partial sums s_n, exists. It is to be observed that the limit exists whenever the series converges, and then it coincides with the sum of the series. Fejér's results about Fourier series showed, however, that this method of summation has a much wider range of applicability than ordinary convergence. In the present article we shall also construct trigonometric series which are summable everywhere with the indicated method but whose convergence is in general out of question.

For the time being we want to emphasize an important property of summable series, which we shall apply permanently in the sequel.

If the limit $\lim_{n=\infty} S_n$ exists, it is clear that

$$\lim_{n=\infty} (S_n - S_{n-1}) = 0,$$

similarly

$$\lim_{n=\infty} \left(\frac{n+1}{n} S_n - S_{n-1}\right) = 0,$$

hence

$$\lim_{n=\infty} \frac{s_n}{n} = 0 \tag{5}$$

and

$$\lim_{n=\infty} \frac{c_n}{n} = \lim_{n=\infty} \frac{s_n - s_{n-1}}{n} = 0. \tag{6}$$

Thus these last two conditions are necessary for the summability of a series.[15]

In our introduction we mentioned reasons which made it desirable to investigate whether Cantor's theorem can be extended to summable series. At the same time, we announced the following two theorems:

If the series

$$\sum_{n=1}^{\infty} \frac{|a_n| + |b_n|}{n^2} \tag{7}$$

formed with the coefficients of some trigonometric series is convergent, then the Cesàro-sum of this trigonometric series can vanish at all points of the interval $(0, 2\pi)$ *only if all the coefficients of the series vanish.*

If the coefficients of the series tend to zero with increasing n, then we may allow a reducible set of exceptional points in the statement of our theorem.

These two theorems shed new light on Cantor's original result. There, even in the presence of exceptional points, one did not have to assume that the coefficients become infinitely small, because this followed already from the convergence of the series.

It is to be noted that the series always satisfies (7) if there exist two numbers c and α ($\alpha < 1$) such that

$$|a_n| < cn^\alpha \quad \text{and} \quad |b_n| < cn^\alpha.$$

On the other hand, from what we said above, it follows that if our series is summable at the point x, then

$$\lim_{n=\infty} \frac{a_n \cos nx + b_n \sin nx}{n} = 0.$$

But we assumed summability in the whole interval and so it follows from Cantor's mentioned theorem[13] that we always have

$$\lim_{n=\infty} \frac{a_n}{n} = \lim_{n=\infty} \frac{b_n}{n} = 0.$$

[15] Fejér l. c.

Thus we see to which series our theorem cannot be applied.

Fejér's generalization of Riemann's theorem will serve as the starting point for the proofs of our theorems:

Let

$$a_0 + \sum_{n=1}^{\infty} a_n \cos nx + b_n \sin nx$$

be such a trigonometric series which is summable at the point x by the method of arithmetic means, and yields the value $f(x)$. Integrating termwise four times, we obtain the uniformly convergent[16] *series*

$$f(x) = a_0 \frac{x^4}{24} + \sum_{n=1}^{\infty} \frac{a_n \cos nx + b_n \sin nx}{n^4}$$

satisfying

$$\lim_{h=0} \frac{F(x+2h) - 4F(x+h) + 6F(x) - 4F(x-h) + F(x-2h)}{h^4} =$$

$$= \lim_{h=0} \left[a_0 + \sum_{n=1}^{\infty}{}' (a_n \cos nx + b_n \sin nx) \left(\frac{\sin \frac{nh}{2}}{\frac{nh}{2}} \right)^4 \right] = f(x).$$

Thus we can see that we only need an appropriate generalization of Schwarz's theorem in order to obtain the required analogue of Cantor's theorem. We shall deduce a direct generalization of Schwarz's theorem from the following theorem:

Let $F(x)$ be a continuous function whose generalized second derivative is continuous[17]*, and which has a generalized fourth derivative*

$$\Phi(x) = \lim_{h=0} \frac{\Delta^4 F(x)}{h^4} =$$

$$= \lim_{h=0} \frac{F(x+2h) - 4F(x+h) + 6F(x) - 4F(x-h) + F(x-2h)}{h^4},$$

at each point of the interval considered. Then the quantity $\frac{\Delta^4 F(x_0)}{h^4}$ lies between the greatest lower bound and least upper bound of the values taken by the function Φ in the interval $(x_0 - 2h, x_0 + 2h)$[18] (I).

This mean-value theorem, which plays a fundamental role in our ulterior investigations, reminds us somewhat of Rolle's theorem. In its proof we shall be helped by the following lemma.

---

[16] Provided that we assume that

$$|a_n| \text{ and } |b_n| < kn^{3-\varepsilon}, \quad \varepsilon > 0.$$

[17] It is easy to show that it then coincides with the ordinary second derivative.

[18] Of course, the value of h is so chosen that $x_0 - h$ and $x_0 + h$ are in the interior of the given interval.

The generalized fourth derivative of a function satisfying our hypothesis is positive or zero at each point where the second derivative has a minimum[19], *and is negative or zero at every point which corresponds to a maximum.*

This theorem is plausible if the first four derivatives of the function exist in the ordinary sense, because then the fourth derivative of the function is again the second derivative of its second derivative. However, in our investigations the generalized fourth derivative is defined by a single passage to the limit, and has no direct connection with the second derivative. Thus the above indicated property is all the more interesting.

A theorem of Hölder[20], which is completely analogous to our theorem (I), will serve to prove our lemma. Namely:

If the generalized second derivative of a continuous function $F(x)$ exists at every point of some interval, and the infimum and supremum of this derivative in the interval (x_0-h, x_0+h) are m and M, respectively, then the quantity

$$\frac{\Delta^2 F(x_0)}{h^2}=\frac{F(x_0+h)-2F(x_0)+F(x_0-h)}{h^2}$$

lies between the values m and M.

The proof of our lemma follows immediately from this theorem. Indeed, suppose that e.g., at the point x_0, where the second derivative has a minimum, the generalized fourth derivative would be negative. In this case we can find a positive quantity h' such that

$$\frac{F(x_0+2h)-4F(x_0+h)+6F(x_0)-4F(x_0-h)+F(x_0-2h)}{h^4}<0,$$

whenever

$$h<h'.$$

From here we obtain by a small transformation

$$\frac{F(x_0+2h)-2F(x_0)+F(x_0-2h)}{4h^2}<\frac{F(x_0+h)-2F(x_0)+F(x_0-h)}{h^2}.$$

Thus we can get a sequence of inequalities:

$$\frac{F(x_0+h)-2F(x_0)+F(x_0-h)}{h^2}<$$

$$<\frac{F\left(x_0+\frac{h}{2}\right)-2F(x_0)+F\left(x_0-\frac{h}{2}\right)}{\frac{1}{4}h^2}<$$

---

[19] Since by our assumption this second derivative is continuous, it has a maximum and minimum.

[20] Hölder: Zur Theorie der trigonometrischen Reihen. Math. Ann., vol. 24 (p. 183). See also Lebesgue: Sur les séries trigonométriques. Ann. l'École Norm. Sup., 1903 (p. 458), and Leçons sur les séries trigonométriques (p. 6).

$$< \frac{F\left(x_0+\frac{h}{4}\right)-2F(x_0)+F\left(x_0-\frac{h}{4}\right)}{\frac{1}{16}h^2} < \ldots <$$

$$< \frac{F\left(x_0+\frac{h}{2^n}\right)-2F(x_0)+F\left(x_0-\frac{h}{2^n}\right)}{\frac{1}{4^n}h^2} < \ldots < F''(x_0).$$

But x_0 admits an h''-neighborhood where $F''(x)$ does not assume values less than $F''(x_0)$, and so by Hölder's theorem for sufficiently large n the relations

$$\frac{F\left(x_0+\frac{h}{2^n}\right)-2F(x_0)+F\left(x_0-\frac{h}{2^n}\right)}{\frac{1}{4^n}h^2} \geqq F''(x_0)$$

should hold, which contradicts our above conclusion.

Having proved our lemma, one can justify theorem (I) as follows:

Let us examine the function $F(x)$ at the points x_0-2h, x_0-h, x_0, x_0+h, x_0+2h. We can always find a polynomial $P(x)$ of degree four such that at these five points the equality $F(x)-P(x)=0$ holds. An elementary calculation[21] yields that $P(x)$ has the form

$$P(x)=\frac{\Delta^4 F(x_0)}{24h^4}x^4+ax^3+bx^2+cx+d.$$

Since the function $F(x)-P(x)$ vanishes at the endpoints and in at least three inner points of the interval (x_0-2h, x_0+2h), it has at least two maxima and two minima in the interior of the interval. Furthermore, any two points which yield a maximal value of the function are separated by a point corresponding to a minimum, and vice versa. At a point corresponding to a maximum the function $F''(x)-P''(x)$ is negative or zero, while at a point which yields a minimum this value is positive or zero. Thus this second derivative admits at least one maximum and one minimum in the interior of the interval (x_0-2h, x_0+2h).

Let e.g., x_1 be a point where the difference $F''(x)-P''(x)$ assumes a minimal value. According to our lemma, at this point the generalized fourth derivative of the function $F(x)-P(x)$ is positive or zero.

[21] $F(x)$ and $P(x)$ coincide at the five points considered, hence their fourth divided differences with respect to these points are also equal. However, the divided n-th differences of a polynomial of degree n are the same for any choice of x and h, and they are equal to the leading coefficient multiplied by $n!$. From this remark the above form of $P(x)$ follows directly.

$$\lim_{t=0} \frac{\Delta^4[F(x_1)-P(x_1)]}{t^4} = \lim_{t=0} \frac{\Delta^4 F(x_1)}{t^4} - \lim_{t=0} \frac{\Delta^4 P(x_1)}{t^4} =$$

$$= \lim_{t=0} \frac{\Delta^4 F(x_1)}{t^4} - \frac{\Delta^4 F(x_0)}{h^4} \geqq 0.$$

Thus

$$\frac{\Delta^4 F(x_0)}{h^4} \leqq \lim_{t=0} \frac{\Delta^4 F(x_1)}{t^4} = \Phi(x_1).$$

If on the other hand the difference $F''(x)-P''(x)$ assumes a maximal value at the point x_2, then

$$\frac{\Delta^4 F(x_0)}{h^4} \geqq \Phi(x_2)$$

and so the proof of our mean-value theorem (I) is complete.

From this result we could obtain immediately the Du Bois-Reymond type theorem for summable series, from where the Cantor type theorem would follow as a special case, we consider, however, more instructive to deduce first from what has been said so far the direct generalization of Schwarz's theorem, which then entails the statement announced at the beginning of our work.

The required generalization reads as follows:

Any continuous function having a continuous second derivative, whose generalized fourth derivative vanishes, is a polynomial of degree three.

It follows namely from our theorem that for arbitrary values of x_0 and h,

$$F(x_0+2h)-4F(x_0+h)+6F(x_0)-4F(x_0-h)+F(x_0-2h)=0. \tag{8}$$

Thus there exists a polynomial $P_3(x)$ of degree three such that

$$K(x)=F(x)-P_3(x)=0$$

at the five places which figure above.

We shall show that this equality holds identically.

On the one hand, from the above interpolation formula it is immediately clear that

$$K(x_0+nh)=0 \qquad (n=0,\ \pm 1,\ \pm 2,\ldots) \tag{9}$$

since also P_3 satisfies (8) and so we have as well

$$K(x_0+2h)-4K(x_0+h)+6K(x_0)-4K(x_0-h)+K(x_0-2h)=0. \tag{10}$$

Thus it is enough to prove our assertion for the interval $(x_0-2h,\ x_0+2h)$; we have exactly the same situation in the other intervals.

It is evident that

$$K(x_0+2h)-4K\left(x_0+\frac{3h}{2}\right)+6K(x_0+h)-4K\left(x_0+\frac{h}{2}\right)+K(x_0)=0.$$

Comparing this with (9):

$$K\left(x_0+\frac{3h}{2}\right)=-K\left(x_0+\frac{h}{2}\right).$$

With a similar argument:

$$-K\left(x_0+\frac{h}{2}\right)=K\left(x_0-\frac{h}{2}\right)=-K\left(x_0-\frac{3h}{2}\right). \tag{11}$$

Applying (10) again

$$K\left(x_0+\frac{3h}{2}\right)-4K(x_0+h)+6K\left(x_0+\frac{h}{2}\right)-4K(x_0)+K\left(x_0-\frac{h}{2}\right)=0. \tag{12}$$

(11) and (12) can be valid at the same time only if

$$K\left(x_0+\frac{3h}{2}\right)=K\left(x_0+\frac{h}{2}\right)=K\left(x_0-\frac{h}{2}\right)=K\left(x_0-\frac{3h}{2}\right)=0. \tag{13}$$

Continuing the halving, we see that K vanishes on a dense set of points, hence by its continuity everywhere.

After these preliminary theorems, we can return to the trigonometric series.

Let the trigonometric series

$$\tfrac{1}{2}a_0+a_1\cos x+b_1\sin x+a_2\cos 2x+b_2\sin 2x+\dots$$

be given, which satisfies the condition (7), and whose Cesàro-sum vanishes at each point of the interval $(0,2\pi)$ without exception. After integrating termwise four times, we obtain a function $F(x)$ whose generalized fourth derivative is everywhere zero by Fejér's quoted theorem. Since furthermore the second derivative of the function $F(x)$ is continuous by condition (7), it follows from our preceding theorem that $F(x)$ is a polynomial of degree three and therefore

$$-\frac{1}{48}a_0x^4+C_0x^3+C_1x^2+C_2x+C=$$

$$=a_1\cos x+b_1\sin x+\dots+\frac{a_n\cos nx+b_n\sin nx}{n^4}+\dots.$$

The argument can follow completely the pattern of that used for Cantor's theorem. The non-periodic part on the left hand side must vanish, from where vanishing of the individual coefficients follows.

Now we would like to apply what has been said so far to an example. Let us examine why our considerations cannot be applied e.g. to the series

$$\tfrac{1}{2}+\cos x+\dots+\cos nx+\dots$$

playing an important role in Fejér's investigations, whose Cesàro-sum vanishes with the exception of the points $x=2k\pi$ $(k=\pm 1, \pm 2, \dots)$, as we have already mentioned.

According to our theorem, the series

$$\frac{1}{48}x^4 + \sum_{n=1}^{\infty} \frac{\cos nx}{n^4}$$

obtained by integrating termwise four times is the Fourier series of some polynomial of degree three in the interval $(0,2\pi)$.[22] (Our theorem may be applied to the whole interval $(0,2\pi)$ because there is no exceptional point in the interior of this interval.) Thus

$$P(x) = -\frac{1}{48}x^4 + C_0x^3 + C_1x^2 + C_2x + C_3 = \sum_{n=1}^{\infty} \frac{\cos nx}{n^4}. \tag{14}$$

We can also compute easily the values of the coefficients C and so we may obtain the values of some series from our function-theoretical considerations. In fact the Fourier series of $P(x)$, $P'(x)$ and $P''(x)$ are uniformly convergent and thus these continuous functions coincide with the values of their Fourier series on the whole interval $(0,2\pi)$, to wit

$$P(2\pi) = P(0), \quad P'(2\pi) = P'(0), \quad P''(2\pi) = P''(0). \tag{15}$$

Hence

$$C_0 = \frac{1}{12}\pi, \quad C_1 = -\frac{1}{12}\pi^2, \quad C_2 = 0.$$

Taking into account the equality

$$P''(0) = 2C_1,$$

we obtain as a byproduct the known identity

$$\sum_{n=1}^{\infty} \frac{1}{n^2} = \frac{1}{6}\pi^2.$$

Knowing the values of C_0, C_1, C_2, the well-known identity

$$\sum_{n=1}^{\infty} \frac{1}{n^4} = \frac{1}{90}\pi^4 = C_3$$

can be deduced easily, again from equation (14).

It is now visible that the equality

$$-\frac{1}{48}x^4 + \frac{\pi}{12}x^3 - \frac{\pi^2}{12}x^2 + \frac{\pi^4}{90} = \sum_{n=1}^{\infty} \frac{\cos nx}{n^4}$$

is valid only in the interval $(0,2\pi)$. For instance in the interval $(2\pi, 4\pi)$ the value of the series

$$\sum_{n=1}^{\infty} \frac{\cos nx}{n^4}$$

is

$$-\frac{1}{48}(x-2\pi)^4 + \frac{\pi}{12}(x-2\pi)^3 - \frac{\pi^2}{12}(x-2\pi)^2 + \frac{\pi^4}{90} \neq P(x).$$

---

[22] Of course, this can also be seen directly.

Thus one can see that our previous reasoning cannot be applied, because the original polynomial of degree three is continued by different polynomials of degree three when passing through the points $x \equiv 0 \pmod{2\pi}$.

The situation is, however, different if the coefficients of the examined trigonometric series become infinitely small. Then we can show that the series obtained by integrating four times is continued necessarily by the *same* polynomial of degree three beyond the point 2π, and in fact on the whole real axis; thus the implications of page 20 can be applied here.

We shall prove that if the coefficients of a trigonometric series

$$a_0 + a_1 \cos x + b_1 \sin x + \ldots + a_n \cos nx + b_n \sin nx + \ldots$$

tend to zero with increasing n, then the series $F(x)$ obtained by integrating termwise four times satisfies

$$\lim_{h=0} \frac{\Delta^4 F(x)}{h^3} = \lim_{h=0} \frac{F(x+2h) - 4F(x+h) + 6F(x) - 4F(x-h) + F(x-2h)}{h^3} = 0, \quad (16)$$

for all values x. What is important for us is precisely the fact that this equality is completely independent of whether the original series is summable at the point considered.

Let us suppose for a moment that we have already proved this theorem. Let x_0 be an isolated exceptional point of our in general summable and vanishing trigonometric series. According to what has been said on page 20, to the right of x_0 (in a certain neighborhood) the function $F(x)$ coincides with a polynomial of the form

$$a_1 x^3 + b_1 x^2 + c_1 x + d_1,$$

and to the left with another

$$a_2 x^3 + b_2 x^2 + c_2 x + d_2.$$

Applying relation (16) to this function $F(x)$, we obtain that

$$a_1 = a_2.$$

On the other hand, the series

$$F(x) = \frac{a_0 x^4}{24} + \sum_{n=1}^{\infty} \frac{a_n \cos nx + b_n \sin nx}{n^4}$$

as well as its first and second derivatives are continuous everywhere, from where the equalities

$$b_1 = b_2, \qquad c_1 = c_2, \qquad d_1 = d_2$$

then follow.

Repeating this procedure, the same result can be obtained also for non-isolated singular points by which our theorem, stated on page 15, concerning trigonometric series with coefficients tending to zero will be proved in its entirety.

We must, however, first prove the equality (16). To this end we only need to modify slightly the argument followed by Riemann in the proof of the analogous theorem (p. 13). (Our assertion also follows easily from this theorem.)

We know that

$$\frac{F(x+4h)-4F(x+2h)+6F(x)-4F(x-2h)+F(x-4h)}{16h^4}=$$

$$=A_0+A_1\left(\frac{\sin h}{h}\right)^4+\ldots+A_n\left(\frac{\sin nh}{nh}\right)^4+\ldots$$

where

$$A_n=a_n \cos nx+b_n \sin nx .$$

Let us divide this series into three parts. Let the first part contain all those terms of the series whose index is less than m. We choose this number m so that beyond it a_n and b_n are less than ε. Let us assign to the second part those terms for which $nh \leq$ than a certain constant number c. Finally put into the third part the remaining terms. It is clear that if h approaches zero, then the first part remains bounded and so is less than a constant number G, the second part is less than $\varepsilon \frac{c}{h}$, and the third part is

$$<\sum_{c<nh}\frac{1}{\frac{1}{16}n^4h^4}<\varepsilon\left(\frac{16}{c^4}+\frac{16}{3hc^3}\right).^{23}$$

Thus

$$2h\left[A_0+A_1\left(\frac{\sin h}{h}\right)^4+\ldots+A_n\left(\frac{\sin nh}{nh}\right)^4+\ldots\right]<2h\left(G+\frac{\varepsilon c}{h}+\frac{16\varepsilon}{c^4}+\frac{16\varepsilon}{3hc^3}\right),$$

i.e., indeed

$$\lim_{h=0}\frac{\Delta^4F(x)}{h^3}=0.$$

Lebesgue's concept of integral

We shall now proceed to the theorem of Du Bois-Reymond which infers from the convergence of a trigonometric series that we are dealing with a Fourier series. We shall present an analogous theorem for summable series. The full significance of Du Bois-Reymond's fundamental theorem becomes clear, however, only with the introduction of Lebesgue's concept of the integral. We shall describe briefly this concept of integral.

Measurable sets. The problem of measuring sets was investigated first by Cantor and then by Jordan. The completely general and precise definition of

---

[23] We obtain this from the inequality

$$\sum_{k=n+1}^{\infty}\frac{1}{n^\alpha}<\int_n^\infty\frac{1}{x^\alpha}dx=\frac{1}{(\alpha-1)n^{\alpha-1}}\qquad(\alpha>1).$$

the concept of measure is, however, largely the merit of Borel.[24] In the case of e.g. sets of dimension one he postulates the following:[25]

To every bounded one-dimensional set let there belong a number, the measure of the set, which is positive or zero, and which shall possess the following properties:

1°. *The measures of two congruent sets are equal.*

2°. *If a set is the disjoint union of finitely or countably many sets, then the measure of the former set shall equal the sum of the measures of the latter ones.*

3°. *The interval* (0,1) *has measure* 1.

It is evident that in the case of sets of several dimensions, only 3° must be replaced appropriately.

In his thesis,[26] Lebesgue presents a procedure starting from these postulates, which can be applied to determine the measure of sets of arbitrary dimension. For the sake of simplicity we describe the method for a one-dimensional set; the reasoning can be generalized directly.

Let E be a set of points which is contained entirely in an interval I. Let us include all the points of this set in countably many intervals. Determine the sum of the measures[27] of these intervals, and let α denote the greatest lower bound of these sums when we choose the intervals in all possible ways. In this manner we assign a definite number to each set. Let β be the number which we assign in an analogous manner to the points of the interval I lying outside E. If $\alpha+\beta$ is equal to the length of the interval I, then we call E a measurable set, and α the measure of the set E. It is easy to show that the concept of measure so defined satisfies Borel's postulates in every respect.

From both Borel's postulates and Lebesgue's constructive definition follows the existence of sets of measure zero (ensemble de mesure nulle). Thus any bounded countable set, e.g. the set of rational numbers in an interval, has measure zero. But such is also Cantor's famous nowhere dense perfect set, although its cardinality is the continuum.

A *function* f of one or several variables is called *measurable* if for arbitrary numbers a, b the set of all those points where $a<f<b$ is measurable. It is clear that the continuous and the monotone functions, and hence the functions of bounded variation (fonction à variation bornée) are all measurable. *The limit function* — whenever it exists — *of a sequence of measurable functions is also measurable.*

The definition of the integral. Let the measurable function f be given; divide the real interval ranging from $-\infty$ to $+\infty$ into countably many parts with the aid of increasing numbers l_i (i ranges over the integers from $-\infty$ to $+\infty$). We choose the intercalated numbers so that $l_{i+1}-l_i$ remains less than η. Denote by e_i the measure of those points in which

[24] Borel: Leçons sur la théorie des fonctions. (Paris, Gauthier-Villars, 1898.)

[25] Lebesgue: Leçons sur l'intégration et la recherche des fonctions primitives. (Paris, Gauthier-Villars, 1904.)

[26] Intégrale, longueur, aire. Annali di matematica 1902.

[27] It is immediately clear from Borel's postulates that the measure on any interval is given by its length.

$$l_i \leqq f < l_{i+1}.$$

Then let us form the infinite series

$$\sum_{-\infty}^{\infty} l_i e_i = A.$$

This series will in general not be unconditionally convergent, it will, however, be so if the function f is bounded, because then we are dealing with a finite sum. It can possibly be convergent also for other functions.

Assume that our series is unconditionally convergent and let us insert further numbers between the numbers l_i. So we replaced the series A by another series A_1, for which it is easy to verify that it is also unconditionally convergent. We continue this procedure so that the differences of the neighboring inserted numbers decrease to zero uniformly. The sums $A, A_1, A_2, \ldots$ thus obtained constitute an increasing sequence but they all remain below the same finite bound. (They are less than $A + \eta I$, where I denotes the length of the interval I.) We call the limit of these sums A the *integral* of the function f over I. As we have seen, this always exists whenever some series A is unconditionally convergent. All those functions, for which this condition is satisfied, are said to be *integrable in the sense of Lebesgue.*[28]

Now we quickly list without proof the most important properties of the Lebesgue integral.

If the Riemann-integral of some bounded function exists, then the Lebesgue one also exists and coincides with the former. For the sake of brevity, we will use henceforth the term integrable always in the Lebesgue sense.

If f is integrable and φ is integrable and bounded, then the product $f\varphi$ is integrable. Thus the Fourier coefficients of any integrable function can be defined. It is clear that the addition theorems of integrals also hold.

If the limit of a sequence of integrable functions is bounded, then it is also integrable. We think it is not necessary to detail the importance of this theorem. But we can assert even more.

Let the sequence of measurable functions $f_1, f_2, \ldots, f_n, \ldots$, whose absolute values are less than a given number K, be given. If these functions converge in the interval I to a function f, then its integral is the limit of the integrals of f_n.

It is clear that this theorem, which we will apply often, holds even if in some set of measure zero the functions f_n do not converge to f.

Du Bois-Reymond's theorem

According to Cantor's theorem, a function can have at most *one* convergent trigonometric expansion. By virtue of this theorem this expansion is always that of Fourier if the Fourier series of the function converges and represents the function, but does there exist in other cases such a trigonometric series which converges everywhere and represents the function?

---

[28] It is evident that the value of the integral does not depend on the manner of partitioning.

Dini, Ascoli and P. Du Bois-Reymond investigated this question. The last one advanced the problem by a big step. Harnack and Hölder perfected the proof of his result in some respect.

We state Du Bois-Reymond's theorem immediately in Lebesgue's version. This nice version was made possible by the introduction of the concept of integral explained above.

Namely, if a trigonometric series is *convergent* then the function represented by it is *measurable*, and if we suppose further that it is *bounded*, then it is also *integrable*. The theorem mentioned is now the following:

Every trigonometric series which converges in the interval $(0,2\pi)$ *(with at most the exception of a reducible set), and which represents in it a bounded function, is the Fourier of this function.*

From the point of view of Du Bois-Reymond, the existence of the Fourier coefficients had to be supposed separately. As Lebesgue remarks, *there exists* indeed *a bounded function which is not integrable in the sense of Riemann*, and which is *represented* everywhere by a convergent trigonometric series. Visibly the introduction of the new integral concept unifies the theory of trigonometric series and makes it more general.

In what follows we present the analogue of Du Bois-Reymond's theorem for the case of summable series.

Let a trigonometric series be given, such that the series

$$\sum_{n=1}^{\infty} \frac{|a_n|+|b_n|}{n^2} \tag{17}$$

formed from its coefficients a_n, b_n *is convergent. If the trigonometric series is summable everywhere and its Cesàro-sum is a bounded function, then our series is the Fourier series of this function.*

If the coefficients of the series tend to zero with increasing n, then our assertion remains valid even if we allow the existence of exceptional points forming a reducible set.

Our first theorem is a simple consequence of the mean-value theorem stated on page 16.

Indeed, let a power series of the form

$$\sum c_n z^n = \sum (a_n + i b_n) z^n,$$

whose real or imaginary part is summable on the circumference of the circle with unit radius, be given. It is easy to prove that then the power series converges inside the circle. Namely, as a consequence of the summability we have

$$\lim_{n=\infty} \frac{c_n}{n} = 0.$$

Suppose for instance that the real part is summable. On the circumference of the unit circle, with the substitution $z=e^{i\varphi}$, this will have the form

$$\sum_{n=0}^{\infty} a_n \cos n\varphi - b_n \sin n\varphi .$$

Let $f(\varphi)$ denote the Cesàro-sum of this series. Let us integrate termwise this series four times and denote by $F(\varphi)$ the sum of the uniformly convergent series thus obtained:

$$F(\varphi)=\frac{a_0\varphi^4}{24}+\sum_1^\infty\frac{a_n\cos n\varphi-b_n\sin n\varphi}{n^4}.$$

We introduce furthermore the following notation:

$$f(r,\varphi)=a_0+\sum_1^\infty r^n(a_n\cos n\varphi-b_n\sin n\varphi),$$

$$F(r,\varphi)=\frac{a_0\varphi^4}{24}+\sum_1^\infty r^n\,\frac{(a_n\cos n\varphi-b_n\sin n\varphi)}{n^4}.$$

We see that

$$f(1,\varphi)=f(\varphi)\quad\text{and}\quad F(1,\varphi)=F(\varphi).$$

As we know, the generalized fourth derivative of the series $F(\varphi)$ is everywhere the corresponding value $f(\varphi)$. Since the second derivative of $F(\varphi)$ is continuous, in view of theorem (I) of page 16, the quantity

$$\frac{\Delta^4F(\varphi,h)}{h^4}=\frac{F(\varphi+2h)-4F(\varphi+h)+6F(\varphi)-4F(\varphi-h)+F(\varphi-2h)}{h^4}\tag{18}$$

lies for all values of φ and h between the greatest lower bound and the least upper bound of the boundary function $f(\varphi)$. However, for a fixed value h, the range of the harmonic function

$$\frac{\Delta^4F(r,\varphi,h)}{h^4}=\frac{F(r,\varphi+2h)-4F(r,\varphi+h)+6F(r,\varphi)-4F(r,\varphi-h)+F(r,\varphi-2h)}{h^4}$$

lies between the greatest lower bound and the least upper bound of its boundary values, i.e., of the quantities $\frac{\Delta^4F(\varphi,h)}{h^4}$, and hence, by the preceding, between the infimum and the supremum of the function $f(\varphi)$.

On the other hand it is clear that choosing fixed values r and φ,

$$\lim_{h=0}\frac{\Delta^4F(r,\varphi,h)}{h^4}=\frac{\partial^4F(r,\varphi)}{\partial\varphi^4}=f(r,\varphi).$$

Thus also all the values $f(r,\varphi)$ are included between the greatest lower bound and the least upper bound of the boundary function $f(\varphi)$.

If we now assume that $f(\varphi)$ is bounded, then also $f(r,\varphi)$ is such for all values of r and φ. Thus then, by one of Lebesgue's quoted theorems,

$$\frac{1}{\pi}\int_0^{2\pi}f(\varphi)\cos n\varphi\,d\varphi=\lim_{r=1}\frac{1}{\pi}\int_0^{2\pi}f(r,\varphi)\cos n\varphi\,d\varphi=\lim_{r=1}a_nr^n.$$

That is

$$a_0 = \frac{1}{2\pi}\int_0^{2\pi} f(\varphi)\,d\varphi, \qquad a_n = \frac{1}{\pi}\int_0^{2\pi} f(\varphi)\cos n\varphi\,d\varphi, \qquad b_n = \frac{1}{\pi}\int_0^{2\pi} f(\varphi)\sin n\varphi\,d\varphi.$$

We give also another proof which then can be generalized to functions with exceptional points. For the moment we assume that such exceptional points do not exist. We shall establish the following theorem:[29]

If the generalized second derivative F of a continuous function Φ is also continuous, and its generalized fourth derivative exists everywhere and is bounded, then

$$F(x) = \int_0^x \int_0^\theta f(t)\,dt\,d\theta + Ax + B,$$

where A and B are suitable constants.

Since

$$\lim_{h=0} \frac{\Delta^4 \Phi(x,h)}{h^4} = f(x),$$

and thus $\dfrac{\Delta^4 \Phi(x,h)}{h^4}$ is bounded for all x and h, denoting by Φ_1 an indefinite integral of Φ, and by Φ_2 an indefinite integral of Φ_1, we have

$$\lim_{h=0} \frac{\Delta^4[\Phi_1(x,h) - \Phi_1(0,h)]}{h^4} = \int_0^x f(t)\,dt,$$

$$\lim_{h=0} \left[\frac{\Delta^4 \Phi_2(x,h)}{h^4} - \frac{\Delta^4 \Phi_2(0,h)}{h^4} - \frac{\Delta^4 \Phi_1(0,h)}{h^4} x\right] = \int_0^x \int_0^\theta f(t)\,dt\,d\theta.$$

Clearly, the first two limits exist because Φ_2 being the twice iterated integral of a two times differentiable function, possesses a fourth derivative in the usual sense. Also the integral on the right side exists, thus the third limit on the left exists too.

On the other hand it is evident that

$$\lim_{h=0} \frac{\Delta^4 \Phi_2(x,h)}{h^4} = \Phi''(x) = F(x)$$

and thus indeed

$$F(x) = \int_0^x \int_0^\theta f(t)\,dt\,d\theta + Ax + B.$$

Now let us return to the original meaning of $f(x)$. Let $f(x)$ be a trigonometric series which is summable by means of the Fejér process:

$$f(x) = a_0 + a_1\cos x + b_1 \sin x + \ldots + a_n \cos nx + b_n \sin nx + \ldots$$

[29] The proof follows essentially the reasoning which Lebesgue used in the proof of an analogous theorem.

and

$$\Phi(x)=\frac{a_0x^4}{24}+\ldots+\frac{a_n\cos nx+b_n\sin nx}{n^4}+\ldots,$$

then the series

$$F(x)=\Phi''(x)=\frac{a_0x^2}{2}-\ldots-\frac{a_n\cos nx+b_n\sin nx}{n^2}+\ldots$$

is also uniformly convergent by the condition (17). To determine its coefficients, let us use the method of Fourier. We get:

$$-\frac{a_n}{n^2}=\frac{1}{\pi}\int_0^{2\pi}\int_0^x\int_0^\theta f(t)\cos nx\,dt\,d\theta\,dx-\frac{2a_0}{n^2},$$

$$-\frac{b_n}{n^2}=\frac{1}{\pi}\int_0^{2\pi}\int_0^x\int_0^\theta f(t)\sin nx\,dt\,d\theta\,dx+\frac{2a_0\pi}{n}\qquad(n>0).$$

It is easy to show that we can interchange the order of integration and obtain thus

$$a_n=\frac{1}{\pi}\int_0^{2\pi}f(t)\cos nt\,dt+2\left[a_0-\frac{1}{2\pi}\int_0^{2\pi}f(t)\,dt\right],$$

$$b_n=\frac{1}{\pi}\int_0^{2\pi}f(t)\sin nt\,dt-n\left[2\pi a_0-\frac{1}{\pi}\int_0^{2\pi}f(t)(2\pi-t)\,dt\right].$$

We know that from the summability of our series the equalities

$$\lim_{n=\infty}\frac{a_n}{n}=0\quad\text{and}\quad\lim_{n=\infty}\frac{b_n}{n}=0$$

follow.

Since the first integral which figures in the expression of b_n tends to zero, it is clear from what we have just said that

$$2\pi a_0-\frac{1}{\pi}\int_0^{2\pi}f(t)(2\pi-t)\,dt=0.$$

On the other hand, naturally the first integral which figures in the expression of a_n also tends to zero. Thus if the term in brackets in the expression of a_n is not zero, then from a sufficiently large n onward a_n has a constant sign and its absolute value is larger than a certain positive number, and so our series could not be summable at the point zero. Thus, finally,

$$a_0=\frac{1}{2\pi}\int_0^{2\pi}f(t)\,dt,$$

$$a_n=\frac{1}{\pi}\int_0^{2\pi}f(t)\cos nt\,dt,\tag{20}$$

$$b_n=\frac{1}{\pi}\int_0^{2\pi}f(t)\sin nt\,dt.$$

If the coefficients of the series tend to zero and the series is not summable in a reducible set in the interval $(0,2\pi)$, or its Cesàro-sum does not remain between finite bounds at these points, then in every interval which does not contain such a point the function $\Phi(x)$ differs from the series $\frac{a_0x^4}{24}+\sum\frac{a_n\cos nx+b_n\sin nx}{n^4}$ by a polynomial of degree at most three.

However, it follows from the lemma $\lim\limits_{h\to 0}\frac{\Delta^4\Phi(x)}{h^3}=0$ that these polynomials of degree three are the same, and then the procedure already known can be applied.

Some remarks

In what follows, we shall scrutinize the generality of our results.

I. As is well known, if the series

$$a_0+a_1\cos x+b_1\sin x+\ldots+a_n\cos nx+b_n\sin nx+\ldots$$

is summable in some interval, then the often used relations

$$\lim_{n\to\infty}\frac{a_n}{n}=0 \quad\text{and}\quad \lim_{n\to\infty}\frac{b_n}{n}=0$$

subsist. We mostly imposed, however, also the restriction that the series

$$\sum_{n=1}^{\infty}\frac{|a_n|+|b_n|}{n^2}, \tag{21}$$

formed from the coefficients of our series, is convergent. It is evident that the last restriction is at any rate a sufficient but not a necessary condition in our proof. In our investigations we only need that the function represented by the series

$$\sum_{n=1}^{\infty}\frac{a_n\cos nx+b_n\sin nx}{n^4} \tag{22}$$

be bounded and admit a continuous second derivative. In any case, the second derivative is a "measurable function", thus it is integrable if it is bounded and so has a Fourier series. Using integration by parts we obtain that the coefficients of this series are

$$\alpha_n=-\frac{a_n}{n^2} \quad\text{and}\quad \beta_n=-\frac{b_n}{n^2}. \tag{22'}$$

On the other hand, the second derivative of the function (22) can be written as

$$-\lim_{h\to 0}\Bigg[(\alpha_1\cos x+\beta_1\sin x)\left(\frac{\sin h}{h}\right)^2+\ldots+ \\ +(\alpha_n\cos nx+\beta_n\sin nx)\left(\frac{\sin nh}{nh}\right)^2+\ldots\Bigg], \tag{23}$$

where α_n and β_n have the meaning under (22'). Thus quite generally we can raise the following question: What hypothesis do the coefficients of a Fourier series have to satisfy, so that summing by Riemann's method it should yield a continuous function?

Simple considerations show that the following question is completely equivalent with this:

Given a sequence of numbers $a_0, a_1, b_1, \ldots, a_n, b_n, \ldots$, what is the necessary and sufficient condition that they shall be the Fourier coefficients of some continuous function?

The answer is seemingly easy. It follows for instance from Hölder's theorem (page 17) that a necessary and sufficient condition is that as h decreases to zero (23) shall converge to some value, uniformly at all points x. Or, on the basis of what has been said in the introduction, we see that a necessary and sufficient condition is the uniform convergence of the Fejér sums.

However, both assertions contain essentially infinitely many conditions. For the time being, it is a very interesting but extremely difficult problem to construct conditions of a simple form which involve only the coefficients and not the variable x.

II. The restrictions in (21) could be eliminated also by showing that for our investigations we do not need the continuity of the (generalized) second derivative. Thus for instance we could replace the analogue of Schwarz's theorem by the following theorem: Every continuous function, whose generalized fourth derivative vanishes identically, is a polynomial of degree three.

The function which is

$$=x^2 \quad \text{if} \quad x \geqq 0$$

and

$$=-x^2 \quad \text{if} \quad x \leqq 0$$

is a good conterexample. The generalized fourth derivative of this continuous function vanishes everywhere, moreover its first derivative is also continuous; its generalized second derivative exists everywhere but has a jump at the point $x=0$.

This example shows clearly that our restriction concerning the continuity of the generalized second derivative was necessary.

Similar statements follow for other derivatives but we do not go into details because this would lead very far from our subject proper.

III. So far we mentioned the Riemann and Fejér methods of summation. We examined what are the consequences regarding the coefficients if any one of these processes yield zero as the sum of the series. We did not discuss in this sense the so-called Poisson method of summation. We saw earlier that for the trigonometric series

$$f(\varphi)=a_0+a_1\cos\varphi+b_1\sin\varphi+\ldots+a_n\cos n\varphi+b_n\sin n\varphi+\ldots \tag{24}$$

we can form the series[30]

$$f(r,\varphi)=a_0+(a_1\cos\varphi+b_1\sin\varphi)r+\ldots+(a_n\cos n\varphi+b_n\sin n\varphi)r^n+\ldots \qquad (25)$$

which converges for all values $r<1$ whenever for instance our series is the Fourier series of some function $f(\varphi)$.

Or it is e.g. convergent if we can find a (real) number k such that

$$\lim_{n=\infty}\frac{a_n}{n^k}=\lim_{n=\infty}\frac{b_n}{n^k}=0.$$

If the trigonometric series converges at a point φ and r tends to 1 from below, then by Abel's classical theorem we have

$$\lim_{r=1} f(r,\varphi)=f(\varphi). \qquad (26)$$

As we know, the limit on the left side may, however, exist even if the series is divergent. (So e.g., as it was shown by Frobenius, this limit exists whenever we have summability with the method of arithmetic means.) In every such case Poisson, who investigated the series (25) in connection with the Fourier series, considered the limit in (26) the sum of the series $f(\varphi)$. We shall call this summation process along a radius the *Poisson summation.*

It would be interesting to examine our problems thoroughly also in this light.

From our previous investigations the following statement results directly:

If the harmonic function

$$a_0+(a_1\cos\varphi+b_1\sin\varphi)r+\ldots+(a_n\cos n\varphi+b_n\sin n\varphi)r^n+\ldots$$

belonging to $f(\varphi)$ is bounded in the interior of the unit circle and its radial limits on the circumference are zero with the exception of at most a set of measure zero, then also the coefficients a_n, b_n vanish and so the function is identically zero.

This theorem includes (in the case of a circle), one of the fundamental theorems of harmonic functions, according to which, if a harmonic function vanishes on the boundary, then it also vanishes inside the curve. Namely, the boundedness of the function follows simply from the fact that we can approach any point of the boundary on an arbitrary path.

The proof proceeds on the basis of Lebesgue's integral theorems like that of the theorem on page 26.

As we know, the real part

$$\tfrac{1}{2}+r\cos\varphi+\ldots+r^n\cos n\varphi+\ldots \qquad (z=re^{i\varphi}),$$

of the function

$$\frac{1}{1-z}-\tfrac{1}{2}=\tfrac{1}{2}+z+z^2+\ldots$$

[30] It is known that such a series $f(r,\varphi)$ represents a harmonic function and that conversely every harmonic function can be written in this form.

has the limit zero everywhere on the circumference of the unit circle with the exception of the point $z=1$, but, as we see, the function does not vanish identically.

One can even construct a non-vanishing function which is harmonic in the interior of a circle and whose radial limits are zero *without exception*.

Such a function is for instance

$$u=\frac{y}{(x-1)^2+y^2}\cdot\frac{x^2+y^2-1}{(x-1)^2+y^2}$$

in the interior of the unit circle

$$x^2+y^2-1=0.$$

(One sees that in the neighborhood of the point

$$x=1, \quad y=0$$

this function remains bounded only along the real axis.) I thank Mr. Holmgren for the oral communication of this example.

Replacing the coordinates x,y by polar coordinates r,φ we obtain a trigonometric series whose *sum by the Poisson method is zero without exception*, but the coefficients of the series do nevertheless not all vanish.

This example shows that if we want to extend Cantor's theorem to Poisson summation then we also must impose restrictions on the coefficients of our series. As we have seen, we obtain a suitable restriction if we assume that the harmonic function associated with the series is bounded. As Fatou notes, for this it is necessary that the sum of the squares of the coefficients

$$\sum a_n^2+b_n^2$$

be convergent. It is easy to see that, dropping the boundedness, even this latter restriction is suitable to generalize Cantor's theorem. As it is shown by F. Riesz,[31] this latter condition holds if and only if our trigonometric series is the Fourier series of some integrable function with integrable square.

II. Summability of power series on the circle of convergence

It is known that there exist power series for which every point of the circle of convergence is singular, i.e. for which this circle is a natural boundary, but which nevertheless converge on the whole circumference of this circle. On the other hand, for instance the function

$$\frac{1}{1-z}=1+z+z^2+\ldots+z^n+\ldots$$

has on the unit circle (and in the whole plane) the unique singularity $z=1$ and its series is divergent on the whole circle. Thus we see how in some cases the

[31] F. Riesz: C.R. 11 mars 1907. — Götting. Nachr. 1907.

singularities on the circle influence the convergence of the series even at points of regularity.

Among the variety of cases a new point of view was introduced by an interesting theorem of Fatou:[32]

Every power series

$$c_0+c_1z+c_2z^2+\dots,$$

whose coefficients tend to zero and whose circle of convergence is the unit circle, converges at every point of regularity of the circumference of this circle.

On the other hand, it is easy to see that Fatou's condition is also necessary for the convergence of the series at all points of regularity on the unit circle (if such do at all exist).[33]

Our requirement that the radius of the circle of convergence is the unit does, of course, not restrict the applicability of the theorem. For if the coefficients tend to zero, then the radius of convergence is $\geqq 1$. In general, if this radius is finite and $\neq 0$ then by a simple transformation we obtain a series to which Fatou's theorem can be applied directly.

In the sequel we shall always assume that the *radius of convergence* of the power series considered *is the unit.*

If we differentiate the series

$$c_0+c_1z+c_2z^2+\dots$$

which figures in the theorem above, we obtain the series

$$\sum n c_n z^{n-1} = \sum C_n z^n$$

whose characteristic property is that

$$\lim_{n=\infty} \frac{C_n}{n} = 0.$$

To such series Fatou's theorem can in general no longer be applied, and this series may be divergent even at every point of regularity of the circle of convergence. *Such series are, however, summable by the method of arithmetic means at every point of regularity of the circle of convergence.* Namely, we shall prove the following general theorem.

The equality

$$\lim_{n=\infty} \frac{C_n}{n^k} = 0 \qquad (k=1,2,3,\dots) \tag{27}$$

[32] Fatou: Séries trigonométriques et séries de Taylor. Acta Math., vol. 30 (p. 389).

[33] Moreover, if this condition is not fulfilled, the power series cannot converge at any point of the circumference.

is a necessary and sufficient condition[34] *that the series*

$$C_0 + C_1 z + \ldots + C_n z^n + \ldots$$

shall be summable at each point of regularity of the circle of convergence by the k-fold application of the method of arithmetic means. By theorems of Frobenius[4] and Hölder,[35] this iterated Cesàro-sum then yields the value of the function at the respective point.

In our proof we will rely on Fatou's above theorem.

Assume first $k=1$, i.e., that in the power series

$$C_0 + C_1 z + C_2 z^2 + \ldots + C_n z^n + \ldots$$

we have

$$\lim_{n=\infty} \frac{C_n}{n} = 0. \tag{28}$$

Let for instance $z=1$ be a point of regularity of the function. Though our reasoning will be general, for the sake of simplicity we perform it only for this point.

We form the direct continuation of the power series by expanding around inner points of the circle of convergence, and thus obtain new elements of the analytic function defined by the power series. The circle of convergence of these elements cover a domain at whose points the various function elements obtained from the direct analytic continuation assume the same value. We denote by $f(z)$ the single-valued analytic function thus defined in our domain.

For the sake of simplicity we assume also that

$$f(1) = f'(1) = 0. \tag{29}$$

This can always be achieved by subtracting a suitable linear expression, while it is clear that changing the initial terms of the series does not have an effect on its summability. This said, a simple calculation leads to the goal.

The partial sums of the series at $z=1$ are

$$s_0 = C_0, \quad s_1 = C_0 + C_1, \ldots, s_n = C_0 + \ldots + C_n, \ldots.$$

Their arithmetic means are

$$S_0 = s_0, \ldots, S_n = \frac{s_0 + s_1 + \ldots + s_n}{n+1}, \ldots.$$

---

[34] If the equality in (27) does not hold, but the quantities $\frac{C_n}{n^k}$ remain between finite bounds, then from the investigations below it follows easily that the arithmetic means of k-th order also remain between finite bounds.

[35] Hölder: Grenzwerthe von Reihen an der Convergenzgrenze. Math. Ann., vol. 20.

One can deduce easily the identities

$$
\begin{aligned}
&s_0+s_1z+\ldots+s_nz^n+\ldots=\frac{f(z)}{1-z},\\
&s_0+(s_0+s_1)z+\ldots+(s_0+s_1+\ldots+s_n)z^n+\ldots=\frac{f(z)}{(1-z)^2},\\
&S_0z+S_1z^2+\ldots+S_nz^{n+1}+\ldots=\int_0^z\frac{f(z)}{(1-z)^2}\,dz,\\
&S_0z+(S_1-S_0)z^2+\ldots+(S_n-S_{n-1})z^{n+1}+\ldots=(1-z)\int_0^z\frac{f(z)}{(1-z)^2}\,dz
\end{aligned}
\tag{30}
$$

where

$$|z|<1$$

and where we integrated inside the unit circle. On the other hand, by our condition (29) in a neighborhood of the point $z=1$ the function $f(z)$ can be expanded in a series thus:

$$f(z)=\frac{f''(1)}{2!}(z-1)^2+\frac{f'''(1)}{3!}(z-1)^3+\ldots.$$

Therefore the function

$$(1-z)\int_0^z\frac{f(z)}{(1-z)^2}\,dz$$

(integrating over an appropriate admissible path) also behaves regularly in a neighborhood of the point $z=1$. Thus if we show that the equality

$$\lim_{n=\infty}(S_n-S_{n-1})=0$$

subsists, then Fatou's theorem can be applied to the power series (30), and so this series converges at the point $z=1$. In other words this means that the value

$$S_0+(S_1-S_0)+(S_2-S_1)+\ldots+(S_n-S_{n-1})+\ldots=\lim_{n=\infty}S_n$$

exists, i.e., our series is indeed summable at the point $z=1$.

However, the relation

$$\lim_{n=\infty}(S_n-S_{n-1})=0 \tag{31}$$

follows simply from the equality

$$\lim_{n=\infty}\frac{s_n}{n}=0. \tag{32}$$

Namely

$$|S_n - S_{n-1}| = \left| \frac{s_0 + \ldots + s_{n-1} + s_n}{n+1} - \frac{s_0 + \ldots + s_{n-1}}{n} \right| =$$

$$= \left| -\frac{s_0 + \ldots + s_{n-1}}{n(n+1)} + \frac{s_n}{n+1} \right| < \frac{|s_0| + \frac{|s_1|}{1} + \ldots + \frac{|s_{n-1}|}{n-1}}{n} + \frac{|s_n|}{n+1}.$$

But the first quantity is the arithmetic mean of the values

$$|s_0|, \frac{|s_1|}{1}, \frac{|s_2|}{2}, \ldots$$

whence it can be seen from a well-known theorem that (31) indeed follows from (32). Thus we only have to verify the equality

$$\lim_{n=\infty} \frac{s_n}{n} = 0.$$

The points of regularity of the power series

$$\varphi(z) = \sum_1^\infty \frac{C_n}{n} z^n = \sum_1^\infty c_n z^n = \int_0^z \frac{f(z) - C_0}{z} dz$$

are the same as those of the function $f(z)$, hence $z=1$ is also a point of regularity of the function $\varphi(z)$. On the other hand, from the condition

$$\lim_{n=\infty} c_n = \lim_{n=\infty} \frac{C_n}{n} = 0$$

in (28) it follows, applying Fatou's theorem again, that the power series $\varphi(z)$ converges at the point $z=1$. As we know, every convergent series is summable by the method of arithmetic means, and the Cesàro-sum is equal to the sum of the series.

Let

$$\sigma_n = c_0 + c_1 + \ldots + c_n,$$

then the limit

$$\lim_{n=\infty} \frac{\sigma_0 + \sigma_1 + \ldots + \sigma_{n-1}}{n} = \lim_{n=\infty} \left(c_0 + c_1 + \ldots + c_n - \frac{c_1 + 2c_2 + \ldots + nc_n}{n} \right)$$

exists and, according to what was said before, it coincides with the value

$$\lim_{n=\infty} (c_0 + c_1 + \ldots + c_n).$$

Thus

$$\lim_{n=\infty} \frac{(c_0 + c_1 + \ldots + nc_n)}{n} = \lim_{n=\infty} \frac{(C_0 + C_1 + \ldots + C_n)}{n} = 0,$$

i.e., (32) holds indeed. Of course we have this equality not only at the point $z=1$ but for all points of regularity of the circle of convergence.

If $k>1$, than we can prove our statement the way already indicated by passing from k to $k+1$ and applying the following lemma:

If

$$\lim_{n=\infty} \frac{C_n}{n^k} = 0 \qquad (k \text{ a positive integer})$$

for some power series

$$C_0 + C_1 z + C_2 z^2 + \ldots + C_n z^n \ldots,$$

then, forming the partial sums s_n of the series, at any point of regularity of the circle of convergence one has

$$\lim_{n=\infty} \frac{s_n}{n^k} = 0. \tag{33}$$

The condition

$$\lim_{n=\infty} \frac{C_n}{n^k} = 0$$

is at the same time necessary for the summability by the k-fold iteration of the method of arithmetic means at at least one point of the circle of convergence. We have namely the following theorems:

If a numerical series

$$c_0 + c_1 + c_2 + \ldots + c_n + \ldots$$

is summable by the k-fold iteration of the method of arithmetic means, then the partial sums σ_n of this series necessarily satisfy the equality

$$\lim_{n=\infty} \frac{\sigma_n}{n^k} = 0. \tag{34}$$

From here it follows that *also the relation*

$$\lim_{n=\infty} \frac{c_n}{n^k} = 0$$

is a necessary condition for the said summability. Of course relation (34) holds also for arithmetic means of various orders of the partial sums.

An easy consequence of the mentioned facts is the following theorem, which for a special case was already given at the beginning of our article (page 15):

If a trigonometric series

$$a_0 + a_1 \cos x + b_1 \sin x + \ldots + a_n \cos nx + b_n \sin nx + \ldots$$

is summable in some interval by arithmetic means of order k, then

$$\lim_{n=\infty} \frac{a_n}{n^k} = \lim_{n=\infty} \frac{b_n}{n^k} = 0. \tag{35}$$

Hence one sees that the condition $\lim \frac{C_n}{n^k} = 0$ is indeed necessary in our theorem concerning power series.

Summation in the case of an arbitrary real exponent

So far in our investigations we always assumed that the number k was a positive integer, or if it was not, we had to replace it by a larger integer.

The sequel will be concerned with arbitrary real numbers k. Since the absolute value of $\frac{1}{n^{i\lambda}}$ is equal to 1 for any real λ, the introduction of complex exponents is of no interest. We can raise the following question: *Can we, similarly to the methods so far, define a method of summation corresponding to arbitrary real numbers k, and will it lead to a goal in an appropriate case?*

It is clear that the concept of an arithmetic mean does not have a direct generalization. However, another method of summation, which in many respects shows an analogy with forming arithmetic means, can be generalized.

Let the series

$$a_0+a_1+\ldots+a_n+\ldots$$

be given. We introduce the following notations:

$$a_0+a_1+\ldots+a_n=S_n^{(0)},$$

$$S_0^{(0)}+S_1^{(0)}+\ldots+S_n^{(0)}=S_n^{(1)}$$

and generally

$$S_0^{(\gamma-1)}+S_1^{(\gamma-1)}+\ldots+S_n^{(\gamma-1)}=S_n^{(\gamma)}. \tag{36}$$

If we denote furthermore the arithmetic means of the partial sums $S_n^{(0)}$ by $s_n^{(1)}$, their arithmetic means by $s_n^{(2)}$, and in general we use the notations

$$s_n^{(\gamma)}=\frac{s_0^{(\gamma-1)}+s_1^{(\gamma-1)}+\ldots+s_n^{(\gamma-1)}}{n}, \tag{37}$$

then, as Knopp[36] has shown, *the existence of the limit*

$$\lim_{n=\infty} s_n^{(\gamma)} \tag{38}$$

implies always the existence of the limit

$$\lim_{n=\infty} \frac{\gamma!}{n^\gamma} S_n^{(\gamma)} \tag{39}$$

(but not conversely). *This value then coincides with the previous one.*

Clearly, the quantities $S_n^{(\gamma)}$ can also be defined by the following equality

$$\sum_{n=0}^{\infty} S_n^{(\gamma)}x^n=\left(\frac{1}{1-x}\right)^{\gamma+1}\sum_{n=0}^{\infty} a_n x^n, \qquad |x|<1. \tag{40}$$

[36] Knopp: Grenzwerthe von Reihen bei der Annäherung an die Convergenzgrenze. Inauguraldissertation. Berlin, 1907.

This can raise the idea to define these sums for an arbitrary value of γ. (On page 36 we gave an integral representation which is suitable to define the first arithmetic mean. With a suitable transformation of this, we could form also the generalized Cesàro-sum, but its handling is very complicated.)

In the case of an arbitrary γ we define $S_n^{(\gamma)}$ also by the above identity

$$\sum_{n=0}^{\infty} S_n^{(\gamma)} x^n = \left(\frac{1}{1-x}\right)^{\gamma+1} \sum_{n=0}^{\infty} a_n x^n, \tag{41}$$

whence it immediately follows that

$$S_n^{(\gamma)} = \sum_{i=0}^{n} B_{n-i}^{(\gamma+1)} a_i = \sum_{i=0}^{n} \frac{\Gamma(\gamma+n-i+1)}{\Gamma(\gamma+1)\Gamma(n-i+1)} a_i, \tag{42}$$

where $B_i^{(\gamma+1)}$ denotes the respective binomial coefficient.

Knopp shows[37] that if for an arbitrary value $\gamma > -1$

$$\lim_{n=\infty} \frac{\Gamma(\gamma+1)}{n^\gamma} S_n^{(\gamma)} \tag{43}$$

exists, then this limit also exists for all larger values of γ and coincides with the former. This theorem enables us to define precisely the degree of indeterminacy of a series. Namely if the above limit exists for some value $\gamma > -1$, then one can find a $\gamma_0 \geqq -1$ such that (43) exists for all $\gamma > \gamma_0$ and never exists if $-1 \leqq \gamma < \gamma_0$. Knopp called this number the *"Oscillationsgrad"* of the series

$$a_0 + a_1 + \ldots + a_n + \ldots,$$

and generalized for it a well known theorem of Cesàro.

We shall now establish the following theorem:

The equality

$$\lim_{n=\infty} \frac{C_n}{n^k} = 0$$

is a necessary and sufficient condition for the existence of the limit

$$\lim_{n=\infty} \frac{\Gamma(k+1)}{n^k} S_n^{(k)} \qquad (k \neq -1, -2, \ldots)$$

at every point of regularity of the circle of convergence of the power series

$$C_0 + C_1 z + C_2 z^2 + \ldots + C_n z^n + \ldots.$$

First we shall show the sufficiency of our condition. Indeed, let a power series

$$F(z) = C_0 + C_1 z + C_2 z^2 + \ldots + C_n z^n + \ldots \tag{44}$$

[37] Knopp: Multiplication divergenter Reihen. Sitzungsberichte der Berl. Math. Gesellsch. 1907 (p. 4). (Archiv der Math. u. Phys., vol. 12).

be given for which

$$\lim_{n=\infty} \frac{C_n}{n^k} = 0.$$ [38]

Assume for a moment that

$$k>0.$$

Since furthermore our theorem is already proved for all integers, we may suppose that k is not an integer, and thus let

$$\mu-2<k<\mu-1, \tag{45}$$

where μ denotes an integer. Let us integrate the series termwise μ times. The coefficients of the series

$$\Phi(z)=D_0 z^{\mu}+D_1 z^{\mu+1}+\ldots+D_n z^{n+\mu}+\ldots \tag{46}$$

thus obtained satisfy the equality

$$\lim_{n=\infty} n^{\mu-k} D_n = \lim_{n=\infty} \frac{C_n}{n^k} = 0, \tag{46'}$$

and since $\mu-k>1$, this series converges unconditionally and uniformly on the circumference of the circle of convergence. It follows from this that the function $\Phi(z)$ is uniformly continuous in both the interior and on the circumference of the circle. Let us examine the values of the two series at an arbitrary point of regularity x. Denoting by $\sigma_n^{(k)}(x)$ the sum of order k formed by the first n terms of the series Φ at this point, we can write

$$\sum_{n=0}^{\infty} \sigma_n^{(k)}(x) z^n = \left(\frac{1}{1-z}\right)^{k+1} \sum_{n=0}^{\infty} D_n x^{n+\mu} z^n.$$

Hence an application of Cauchy's integral theorem yields easily:

$$\sigma_n^{(k)}(x) = \frac{1}{2\pi i} \int_C \frac{\Phi(xz)}{z^{\mu}} \left(\frac{1}{1-z}\right)^{k+1} \frac{1}{z^{n+1}}\, dz, \tag{47}$$

where C denotes the circumference of the circle of convergence. This statement is only valid, i.e., the integration can be carried out without obstacle only if the integrand is finite or integrably infinite on the circumference. This can always be achieved by assuming that

$$\Phi(x)=\Phi'(x)=\ldots=\Phi^{(\mu-1)}(x)=0, \tag{48}$$

where, as we know, we mean by $\mu-1$ the integer immediately following k. Then the integrand vanishes at the point $z=x$. The validity of the equalities (48) can, however, always be secured by subtracting a polynomial of degree at most $\mu-1$. It is easy to show also for this method of summation that changing the initial terms preserves summability, and the value of

[38] From here it follows clearly that the power series converges inside the unit circle. Furthermore we assume now also that this circle is at the same time the circle of convergence of the power series.

$$\lim_{n=\infty} \frac{\Gamma(k+1)}{n^k} S_n^{(k)} \quad (k \neq -1, -2, \ldots),$$

changes in the same manner as if the series were convergent, or would consist only of finitely many terms. With a slight transformation we obtain further:

$$\sigma^{(k)}(x) = \frac{1}{2\pi i} \int_C \frac{\Phi(z)}{z^{n+\mu+1}} \left(\frac{1}{x-z}\right)^{k+1} x^{n+k+\mu+1} dz. \tag{49}$$

The sum of order k formed from the first n terms of the series $F(x)$ will be:

$$S_n^{(k)}(x) = \frac{1}{2\pi i} \int_{C'} \frac{\Phi(z)}{z^{n+\mu+1}} \frac{d^\mu}{dx^\mu} \left[\left(\frac{1}{x-z}\right)^{k+1} x^{n+k+\mu+1}\right] dz, \tag{50}$$

where we obtain the curve C' by describing around the point x a circle c with arbitrarily small radius which contains an arc of the circle C. Omitting this arc, the remaining part of C and the arc of the circle c inside the circle C constitute the curve C'. We assume furthermore that

$$\Phi^{(\mu)}(x) = \ldots = \Phi^{(2\mu-1)}(x) = 0. \tag{51}$$

We shall show that

$$\lim_{n=\infty} \frac{1}{n^k} S_n^{(k)}(x) = 0. \tag{52}$$

Let us perform the differentiation first and then, to simplify the later calculations, suppose that

$$x = 1,$$

and thus by (48) and (51), of course,

$$\Phi(1) = \Phi'(1) = \ldots = \Phi^{(2\mu-1)}(1) = 0. \tag{53}$$

The quantity obtained by differentiation will be:

$$\sum_{\lambda=0}^{\mu} A_\lambda \left(\frac{1}{1-z}\right)^{k+1+\lambda}, \tag{54}$$

where the A_λ are numerical coefficients such that the quantity

$$\frac{A_\lambda}{n^\mu} \tag{55}$$

moves within finite bounds. Thus if, for an arbitrary integer $0 \leqq \lambda \leqq \mu$, we show that

$$\begin{aligned} &\lim_{n=\infty} \frac{1}{n^k} \int_C n^\mu \frac{\Phi(z)}{z^{n+\mu+1}} \left(\frac{1}{1-z}\right)^{k+1+\lambda} dz = \\ &= \lim_{n=\infty} n^{\mu-k} \int_C \frac{\Phi(z)}{z^{n+\mu+1}} \left(\frac{1}{1-z}\right)^{k+1+\lambda} dz = 0, \end{aligned} \tag{56}$$

then, since λ can assume only finitely many distinct values, at the same time (52) will also be proved.[39]

Let

$$k+1+\lambda=\nu, \tag{57}$$

then

$$\nu<2\mu. \tag{58}$$

In any case, the point $z=1$ admits a neighborhood on the circumference where the function $\Phi(z)$ is still regular. Let z_1 and z_2 be two points in this neighborhood such that going around in the positive sense the order of the three points shall be z_1, 1, z_2. Let C_1 denote the arc $z_1 z_2$ containing 1 and C_2 the remaining arc $z_2 z_1$.

We construct a polynomial (in general of degree 5) such that the quantity

$$H(z)=\left(\frac{1}{1-z}\right)^{\nu}-P(z), \tag{59}$$

as well as its first and second derivatives vanish at the points z_1 and z_2.[40]

By (46)

$$\Phi(z)=D_0 z^{\mu}+\ldots+D_n z^{n+\mu}+\ldots$$

and by (46′)

$$\lim_{n=\infty} n^{\mu-k} D_n=0,$$

hence

$$\lim_{n=\infty} n^{\mu-k} \int_C \frac{\Phi(z)}{z^{n+\mu+1}} P(z)\,dz=0, \tag{60}$$

for, by the already applied Cauchy theorem, the integral gives the coefficients of the power series of $\frac{\Phi(z)}{z^{\mu}} P(z)$ (up to a constant factor).

On the other hand, from the inequality

$$\mu-2<k<\mu-1$$

the inequality

$$\mu-k<2 \tag{61}$$

follows, while based on (59) it is easy to show integrating by parts that

$$\lim_{n=\infty} n^2 \int_{C_2}\left[\left(\frac{1}{1-z}\right)^{\nu}-P(z)\right]\frac{1}{z^{n+\mu+1}}\,dz=0 \tag{62}$$

and this relation holds all the more if we put $\mu-k$ instead of 2 in the exponent.

[39] Namely, by (53) the integration along C' can be replaced by an integration along C.

[40] We choose e.g., a definite value of our multi-valued function at the point z_1, and then we continue the function along the circle.

Let us now apply those formulas which from the coefficients of the Fourier series of two functions yield the Fourier coefficients of the product.

If the Fourier series of $f(x)$ is

$$\tfrac{1}{2} a_0 + \sum_{p=1}^{\infty} (a_p \cos px + b_p \sin px),$$

that of $\varphi(x)$ is

$$\tfrac{1}{2} \alpha_0 + \sum_{p=1}^{\infty} (\alpha_p \cos px + \beta_p \sin px),$$

and that of their product $F(x)$ is

$$\tfrac{1}{2} A_0 + \sum_{p=1}^{\infty} (A_p \cos px + B_p \sin px),$$

then by introducing the notation $a_{-k} = a_k$, $b_{-k} = b_k$, we have

$$\begin{aligned} A_0 &= \tfrac{1}{2} a_0 \alpha_0 + \tfrac{1}{2} \sum_{p=1}^{\infty} (a_p \alpha_p + b_p \beta_p), \\ A_n &= \tfrac{1}{2} a_0 \alpha_n + \tfrac{1}{2} \sum_{p=1}^{\infty} [a_p (\alpha_{p+n} + \alpha_{p-n}) + b_p (\beta_{p+n} + \beta_{p-n})], \\ B_n &= \tfrac{1}{2} a_0 \beta_n + \tfrac{1}{2} \sum_{p=1}^{\infty} [a_p (\beta_{p+n} - \beta_{p-n}) - b_p (\alpha_{p+n} - \alpha_{p-n})]. \end{aligned} \tag{63}$$

These equalities, which have a completely general validity,[41] can be verified very easily in our case. Namely the series occurring in our applications all converge unconditionally and uniformly.

Now we shall show that if we have

$$\lim_{n=\infty} n^s a_n = \lim_{n=\infty} n^s b_n = 0 \tag{64}$$

and

$$\lim_{n=\infty} n^s \alpha_n = \lim_{n=\infty} n^s \beta_n = 0 \qquad (s > 1),$$

then also for the coefficients of the product

$$\lim_{n=\infty} n^s A_n = \lim_{n=\infty} n^s B_n = 0. \tag{65}$$

It is sufficient to prove for instance that

$$\lim_{n=\infty} n^s \sum_{p=1}^{\infty} a_p (\alpha_{p+n} + \alpha_{p-n}) = 0.$$

Let

$$\sum_{p=0}^{\infty} |a_p| = M_1$$

and

[41] Hurwitz, Lebesgue, Fatou l.c.

$$\sum_{p=0}^{\infty} |\alpha_p| = M_2 .$$

Let n_0 denote a number such that

$$n^s |a_n| < \varepsilon$$

and (66)

$$n^s |\alpha_n| < \varepsilon ,$$

whenever

$$n > n_0 .$$

It is immediately clear that if

$$2n+1 \geqq N \geqq 2n ,$$

then

$$N^s \sum_{p=1}^{\infty} |a_p| |\alpha_{p+N}| < \varepsilon M_1 ,$$

and on the other hand

$$\left| N^s \sum_{p=1}^{\infty} a_p \alpha_{p-N} \right| < \left| N^s \sum_{p=1}^{n} a_p \alpha_{p-N} \right| + \left| N^s \sum_{p=n+1}^{\infty} a_p \alpha_{p-N} \right| <$$

$$< 2^s \varepsilon M_1 + 2^{s+1} \varepsilon M_2 .$$

Visibly, (65) does indeed hold.

The following theorem, which can be considered as a generalization of a theorem of Riemann, follows easily from what has just been said:

Let a function $f(x)$ be given whose Fourier series fulfils the condition in (64), *and let $\varphi(x)$ be a function satisfying*

$$\lim_{n=\infty} n^s \int_a^b \varphi(x) \cos nx \, dx = 0 ,$$

and (67)

$$\lim_{n=\infty} n^s \int_a^b \varphi(x) \sin nx \, dx = 0 \qquad (s>1) ,$$

where a and b are two fixed numbers. Then we have

$$\lim_{n=\infty} n^s \int_a^b f(x) \varphi(x) \cos nx \, dx = 0 \tag{68}$$

and

$$\lim_{n=\infty} n^s \int_a^b f(x) \varphi(x) \sin nx \, dx = 0 . \tag{69}$$

Now we can return to the investigations of the integral

$$\lim_{n=\infty} n^{\mu-k} \int_{C_2} \Phi(z) \left[\left(\frac{1}{1-z} \right)^{\nu} - P(z) \right] \frac{1}{z^{n+\mu+1}} \, dz . \tag{70}$$

If we decompose the integrand into its real and imaginary parts, and compare the equalities in (46′) and (62) with the theorem just deduced, then we obtain directly that the limit of the quantity in (70) is zero.[42]

When integrating along C_1 we apply integration by parts twice. It is clear from (59) that the totally integrated parts vanish. The integration by parts can be carried out for all values

$$\mu-1 \leqq \nu = k+\lambda+1 \leqq 2\mu$$

if in addition to (53) we assume also that

$$\Phi^{(2\mu)}(1) = \Phi^{(2\mu+1)}(1) = 0. \tag{71}$$

After the integration by parts the factor

$$\frac{n^{\mu-k}}{(n+\mu)(n+\mu-1)},$$

whose limit is zero for $n=\infty$ by the inequality $\mu-k<2$, will stand in front of the integral sign. On the other hand the expression which is multiplied by $\frac{1}{z^{n+\mu-1}}$ is finite and hence, by a classical theorem of Riemann,

$$\lim_{n=\infty} n^{\mu-k} \int_{C_1} \Phi(z)\left[\left(\frac{1}{1-z}\right)^{\nu} - P(z)\right] \frac{1}{z^{n+\mu+1}}\,dz = 0. \tag{72}$$

Thus this relation is valid also when integrating along the whole circle, and thus taking (60) into account, we finally established that if the condition

$$\lim_{n=\infty} \frac{C_n}{n^k} = 0$$

is satisfied, at the point of regularity $z=1$ of the power series

$$C_0 + C_1 z + \ldots + C_n z^n + \ldots$$

the limit

$$\lim_{n=\infty} \frac{\Gamma(k+1)}{n^k} S_n^{(k)}$$

exists and, of course, yields the value of the function.

In our proof we assumed that $k>0$. It is clear that this restriction is not essential and that *with the procedure used we can prove the existence of the above limit for any k* ($\neq -1, -2, \ldots$) *if only the condition*

$$\lim_{n=\infty} \frac{C_n}{n^k} = 0$$

---

[42] In the proof we also have to use the fact that both the real and the imaginary part of

$$\left(\frac{1}{1-z}\right)^{\nu} - P(z)$$

have the properties of the function φ figuring in (67), if now by s we mean $\mu-k$.

is fulfilled. Thus we obtain Fatou's theorem as a special case. If, conversely, we would want to deduce from this theorem the general one, then we would obtain interesting results concerning the Riemann-Liouville generalized derivatives.[43] However, we cannot go into details now.

We have mentioned that the condition $\lim\limits_{n=\infty} \frac{C_n}{n^k} = 0$ *is also necessary* for the power series to be summable by the described method at at least one point of the circle of convergence. This statement can be checked easily by reversing our arguments.

From an interesting theorem of Knopp[44] it follows furthermore that if

$$\lim_{n=\infty} \frac{C_n}{n^k} = 0 \quad \text{and} \quad k > 1,$$

then also in this case, at every point of regularity the partial sums of the series satisfy the equality

$$\lim_{n=\infty} \frac{s_n}{n^k} = 0.$$

After the completion of this research, through Knopp's dissertation I became aware of some results of Hadamard which in many respects are similar to the ones obtained here.[45] Namely, when investigating in his "thèse" the nature of the singular points lying on the circle of convergence, he introduces the concept of order of a power series. He defines this important concept first with the help of the already mentioned Riemann-Liouville generalized derivatives, and then obtains by deep arguments that *the order of the power series*

$$f(x) = a_0 + a_1 x + a_2 x^2 + \ldots + a_n x^n + \ldots$$

on its circle of convergence is always

$$\omega = 1 + \limsup_{n=\infty} \frac{\log |a_n|}{\log n}. \tag{73}$$

Thereafter he states the following theorem:[46]

Let a power series, whose order ω is finite, be given, and let $\omega' > 0$ be a number larger than ω (and not equal to it); let furthermore x be a point of regularity of the circle of convergence, then we have

$$f(x) = \lim_{n=\infty} \frac{a_0 B_n^{(\omega')} + a_1 B_{n-1}^{(\omega')} x + \ldots + a_n x^n}{B_n^{(\omega')}}, \tag{74}$$

where $B_n^{(\omega')}$ is the coefficient of x^n among the binomial coefficients which figure in the series expansion of $(1-x)^{-\omega'}$.

---

[43] These quantities define the derivatives of order k for the case of a non-integer k.

[44] Knopp: Eine notwendige und hinreichende Konvergenzbedingung. Rend. Circ. Math. Palermo, vol. 25 (1908).

[45] Hadamard: Essai sur l'étude des fonctions données par leur développement de Taylor. Thèse. J. Math. 1892 (p. 182).

[46] La série de Taylor et son prolongement analytique (Scientia. Paris, 1901). (pp. 47 and 52).

On the other hand, this limit exists at all points of regularity only if $\omega' > \omega$.

The limit (74) coincides essentially with the limit

$$\lim_{n=\infty} \frac{\Gamma(\omega')}{n^{\omega'-1}} S_n^{(\omega'-1)} \tag{75}$$

used often so far. Indeed the quantity under the symbol "lim" in (74) is equal to

$$\frac{1}{B_n^{(\omega')}} \frac{n^{\omega'-1}}{\Gamma(\omega')} \left[\frac{\Gamma(\omega')}{n^{\omega'-1}} S_n^{(\omega'-1)}\right].$$

However,

$$\lim_{n=\infty} \frac{n^{\omega'-1}}{B_n^{(\omega')} \cdot \Gamma(\omega')} = 1 \tag{76}$$

except for the points $\omega' = 0, -1, -2, \ldots$, where this expression is indeterminate. This equality follows namely in a simple manner from the product representation of the Γ-function.

Thus from the existence of one of these two limits the existence of the other one follows and the two values always coincide.

From here it is, however, obvious that Hadamard is mistaken when giving the necessary condition, and in my opinion the otherwise very beautiful method used by him is not even suitable for yielding such a condition.

We have namely proved that the equality

$$\lim_{n=\infty} \frac{a_n}{n^{\omega'-1}} = 0 \tag{77}$$

is a necessary and sufficient condition for the existence of the limit (74) due to Hadamard at all points of regularity of the circle of convergence. But this does not coincide in every case with Hadamard's condition. If e.g., the coefficients a_n have the form

$$a_n = \frac{n^k}{\varphi(n)},$$

where the absolute value of $\varphi(n)$ grows beyond all bounds with increasing n but

$$\lim_{n=\infty} \frac{\varphi(n)}{n^{\varepsilon}} = 0,$$

however small the positive number ε is, then naturally the order of the power series

$$a_0 + a_1 x + \ldots + a_n x^n + \ldots$$

is

$$\omega = k + 1.$$

Now, according to Hadamard's theorem, summation of order $\omega - 1 = k$ does not lead yet to the goal, while according to our results this method furnishes the value of the function at every point of regularity of the circle of convergence.

Thus, for instance, by Hadamard's condition the series

$$\sum_{n=2}^{\infty} \frac{x^n}{\log n}$$

(its order is 1) cannot be convergent, and the series

$$\sum_{n=2}^{\infty} \frac{n x^n}{\log n}$$

(its order is 2) cannot be summable by the method of simple arithmetic means at the points of regularity of the unit circle, although according to Fatou's theorem and our results these series have these properties.

It is furthermore remarkable that our method of summation works even if k is a negative number (not an integer).

Thus for instance in the case of the series

$$\sum \frac{x^n}{n^\alpha} \qquad (\alpha > 0)$$

the sums

$$\frac{\Gamma(-k+1)}{n^{-k}} S_n^{(-k)} \qquad (0 < k < \alpha) \tag{78}$$

oscillate much stronger than the partial sums of the series. However, the limit of (78) exists nevertheless.

It is also clear that for our investigations it is not necessary that the point considered be a point of regularity. *It is for instance sufficient if the boundary function of the power series has at this point and in its neighborhood a certain number of derivatives (taken with respect to the arc). I have shown elsewhere that also this condition can be generalized substantially.*

Application to trigonometric series

Our results provide a simple method for constructing trigonometric series whose coefficients become infinitely large and which in spite of this are everywhere summable by Fejér's method. Such series had not yet been known. (We observe, however, that series with an analogous character can also be obtained in other ways.)

Namely one can verify easily with Lindelöf's perfected methods of the calculus of residues[47] that the series already mentioned

$$\sum_{n=2}^{\infty} \frac{z^n}{\log n}, \qquad \sum_{n=2}^{\infty} \frac{n z^n}{\log n}$$

have the unique singular point $z = 1$ on the unit circle.

---

[47] Lindelöf: Calcul des résidus etc. (Gauthier-Villars, Paris, 1905). (See e.g. p. 109.)

Thus among the series

$$\sum_{n=2}^{\infty} \frac{n \sin n\varphi}{\log n} \quad \text{and} \quad \sum_{n=2}^{\infty} \frac{n \cos n\varphi}{\log n}$$

formed from the latter one, the first one is summable with the method of Fejér everywhere, while the second with the exception of the points $\varphi \equiv 0 \pmod{2\pi}$.

For instance the series

$$\sum_{n=0}^{\infty} n^{\nu} \sin n\varphi \quad \text{and} \quad \sum_{n=0}^{\infty} n^{\nu} \cos n\varphi \qquad (0 \leqq \nu < 1)$$

have similar properties.

We observe that none of the functions represented by the everywhere summable series

$$\sum_{n=2}^{\infty} \frac{n \sin n\varphi}{\log n} \quad \text{and} \quad \sum_{n=0}^{\infty} n^{\nu} \sin n\varphi \qquad (0 \leqq \nu < 1)$$

is bounded, which, at least for the second series, also follows from the (generalized) Du Bois-Reymond theorem proved in the course of our article.

4.

Sur les séries trigonométriques

C. R. Acad. Sci. Paris *145* (1907), 583–586

M. Cantor a démontré le théorème suivant :

Une série trigonométrique convergeant en tout point de l'intervalle $(0, 2\pi)$ (*sauf peut-être en un ensemble réductible de points*) *vers la valeur nulle, a tous ses coefficients nuls.*

La démonstration repose sur un théorème de Riemann et sur un théorème de M. Schwarz, d'après lequel *une fonction continue* $f(x)$, *dont la dérivée seconde généralisée*

$$\lim_{h=0} \frac{\Delta^2 f(x)}{h^2} = \lim_{h=0} \frac{f(x+h) - 2f(x) + f(x-h)}{h^2}$$

s'annule identiquement, est une fonction linéaire.

La question suivante se pose :

Lorsqu'une série trigonométrique divergente est sommable par une autre méthode (*par exemple par la méthode bien connue de M. Fejér*) *et sa* somme *s'annule dans tout l'intervalle* $(0, 2\pi)$ (*sauf peut-être en certains points exceptionnels*), *peut-on en conclure encore que tous ses coefficients s'annulent?*

Au premier moment il paraît que la réponse sera négative. En effet, M. Fejér a donné un exemple simple, la série

$$\frac{1}{2} + \cos x + \cos 2x + \ldots$$

dont la *somme* est nulle en chaque point, excepté les points 0 et 2π [1].

---

[1] C'était M. Fejér lui-même qui a bien voulu attirer mon attention sur ce sujet. *R.*

On a pourtant les théorèmes :

Une série trigonométrique, dont les coefficients a_n et b_n sont tels que la série

$$\text{(A)} \qquad \sum_{n=1}^{\infty} \frac{|a_n| + |b_n|}{n^2}$$

converge [1] *et dont la* somme *s'annule sans exception en tous les points de l'intervalle* (0, 2π), *a ses coefficients nuls.*

Les coefficients de la série tendant vers 0 *avec* $\frac{1}{n}$, *le théorème reste encore exact lorsqu'on admet un ensemble réductible de points exceptionnels.*

M. Fejér a généralisé le théorème de Riemann par le théorème suivant [2] :

Soit $\frac{1}{2}a_0 + \Sigma a_n \cos nx + b_n \sin nx$ *une série trigonométrique qui est sommable au point* x *par le procédé de la moyenne arithmétique et y donne la valeur* $f(x)$. *En intégrant quatre fois terme à terme, on parvient à une série uniformément convergente* $F(x)$ *pour laquelle on a*

$$\lim_{h=0} \frac{F(x+4h) - 4F(x+2h) + 6F(x) - 4F(x-2h) + F(x-4h)}{16h^4}$$
$$= \lim_{h=0} \left[\frac{1}{2} a_0 + \sum_{n=1}^{\infty} (a_n \cos nx + b_n \sin nx) \left(\frac{\sin nh}{nh} \right)^4 \right] = f(x).$$

Nous allons généraliser maintenant le théorème de M. Schwarz et cette généralisation, jointe à ce théorème de M. Fejér, fournira la démonstration de notre énoncé :

Une fonction continue $F(x)$ *admettant en tout point d'un intervalle une*

---

(1) Donc, notre théorème subsiste s'il existe deux constantes c et α (où $\alpha < 1$), telles que

$$|a_n| \quad \text{et} \quad |b_n| < cn^{\alpha}.$$

D'ailleurs, il résulte d'une remarque de M. Fejér [*Untersuchungen über trigonom. Reihen* (*Math. Ann.*, t. LVIII, p. 63)] que pour les coefficients d'une série trigonométrique sommable, dans un intervalle, on a

$$\lim_{n=\infty} \frac{a_n}{n} = 0, \qquad \lim_{n=\infty} \frac{b_n}{n} = 0.$$

On voit donc les séries pour lesquelles notre théorème est douteux.

(2) Fejér, *loc. cit.*, p. 68.

dérivée seconde généralisée continue (¹) *et une dérivée quatrième généralisée*

$$\Phi(x) = \lim_{h=0} \frac{\Delta^4 F(x)}{h^4}$$
$$= \lim_{h=0} \frac{F(x+2h) - 4F(x+h) + 6F(x) - 4F(x-h) + F(x-2h)}{h^4},$$

la quantité $\frac{\Delta^4 F(x_0)}{h^4}$ *est comprise entre les limites inférieure et supérieure de* Φ *dans l'intervalle* $(x_0 - 2h, x_0 + 2h)$ (I).

On démontrera d'abord le lemme suivant :

La dérivée quatrième généralisée d'une fonction satisfaisant aux conditions énoncées est positive ou nulle en tout point où la dérivée seconde a un minimum, et négative ou nulle en tout point correspondant à un maximum.

Pour la démonstration on se servira du théorème suivant de M. *Hölder* (²) :

Si une fonction continue $f(x)$ *a en tout point une dérivée seconde généralisée* $\varphi(x)$, *la quantité*

$$\frac{\Delta^2 f(x_0)}{h^2} = \frac{f(x_0+h) - 2f(x_0) + f(x_0-h)}{h^2}$$

sera comprise entre les limites inférieure et supérieure de φ *en* $(x_0 - h, x_0 + h)$.

Ayant démontré notre lemme nous considérons la fonction $F(x)$ aux points $x_0 - 2h$, $x_0 - h$, x_0, $x_0 + h$, $x_0 + 2h$. Alors il existe un polynome de quatrième degré $P(x)$ tel que $F(x) - P(x) = 0$ pour les cinq points considérés. Un calcul élémentaire donne que $P(x)$ doit être de la forme

$$P(x) = \frac{\Delta^4 F(x_0)}{24 h^4} x^4 + a x^3 + b x^2 + c x + d.$$

La différence $F(x) - P(x)$ a au moins deux maximums et deux minimums entre $x_0 - 2h$ et $x_0 + 2h$; de plus deux points correspondant à un maximum sont toujours séparés par un point qui donne un minimum et inversement. A un maximum correspond une valeur négative ou nulle et à un minimum une valeur positive ou nulle de la dérivée seconde $F''(x) - P''(x)$. Donc cette dérivée a au moins un maximum et un minimum à l'intérieur de l'intervalle $(x_0 - 2h, x_0 + 2h)$ et de ce fait joint à notre lemme on conclut aisément notre théorème I.

On a comme cas particulier une généralisation immédiate du théorème de M. Schwarz.

Toute fonction continue elle-même et admettant une dérivée seconde con-

---

(¹) De la continuité il suit que c'est une dérivée seconde au sens ordinaire.

(²) Hölder, *Zur Theorie der trigonometrischen Reihen* (*Math. Ann.*, t. XXIV, p. 183). — Voir aussi : Lebesgue, *Sur les séries trigonométriques* (*Ann. de l'École Norm.*, 1903, p. 458) et *Leçons sur les séries trigonométriques*, p. 6.

tinue [1] *et une dérivée quatrième généralisée partout nulle est un polynome de troisième degré.*

De plus des théorèmes analogues peuvent se démontrer pour d'autres dérivées généralisées.

De notre énoncé on déduit, comme le fait M. Lebesgue dans un cas analogue, les conséquences suivantes :

Si, pour les coefficients $a_n + ib_n$ *d'une série de Taylor la somme* $\sum \frac{|a_n + ib_n|}{n^2}$ *converge et la partie réelle (ou imaginaire) de cette série est sommable sur son cercle de convergence par la méthode de M. Fejér, la partie réelle (ou imaginaire) de la serie à l'intérieur du cercle reste comprise entre les limites inférieure et supérieure des valeurs prises sur la circonférence.*

Lorsqu'une série trigonométrique toujours satisfaisante à notre condition (A) *est sommable en tout point de l'intervalle* $(0, 2\pi)$ *et y donne une fonction bornée* $f(x)$, *elle est série de Fourier de la fonction* $f(x)$ *(la notion d'intégrale étant prise dans le sens de M. Lebesgue).*

Ce théorème, *généralisation du théorème de Dubois-Reymond, comprend comme cas particulier notre généralisation du théorème de M. Cantor.*

Le cas des coefficients tendant vers zéro, où l'on peut admettre des points exceptionnels, ne comporte aucune difficulté, car en tout point on a $\lim \frac{1}{h^3}\Delta^4 F(x) = 0$.

---

(1) La fonction égale à x^2 pour des valeurs $x \geqq 0$ et à $-x^2$ pour des valeurs $x < 0$ montre très nettement que notre restriction concernant la continuité de la dérivée seconde est nécessaire.

(7 octobre 1907.)

5.

Sur les séries de Dirichlet

C. R. Acad. Sci. Paris *148* (1909), 1658–1660

1° Nous allons montrer, dans les lignes suivantes, comme il est utile d'appliquer certaines méthodes de sommation aux séries de Dirichlet. D'abord nous énoncerons les théorèmes suivants :

I. *La série de Dirichlet* $f(s) = \sum a_n n^{-s}$ *étant sommable par les moyennes arithmétiques d'ordre* k ($k = 1, 2, 3, \ldots$) *en un point* s_0 *est uniformément sommable par les moyennes d'ordre* $k-1$ *sur toute portion finie de la droite* $R(s) = R(s_0) + 1$, $R(s)$ *désignant la partie réelle de* s. *Il en résulte que la sommabilité est uniforme dans toute bande définie par les inégalités*

$$R(s) \geqq R(s_0) + 1, \qquad |t| < T,$$

t désignant la partie imaginaire de $s = \sigma + it$ *et* T *désignant un nombre positif quelconque* [1].

L'intérêt des résultats de ce genre est bien mis en évidence par le théorème qui suit et qui se démontre aisément :

II. *La série* $f(s) = \sum a_n n^{-s}$ *étant sommable d'ordre* k ($k = 0, 1, 2, \ldots$) *en un point* s_0 *et* ε *désignant un nombre positif quelconque, on a, dans le domaine*

$$R(s) \geqq R(s_0) + \varepsilon,$$

l'égalité

$$\lim_{|s|=\infty} \frac{f(s)}{|s|^{k+1}} = 0,$$

et cela uniformément avec $|s|$.

Si l'on introduit les moyennes arithmétiques d'ordre non entier, on peut démontrer de même qu'en agrandissant l'abscisse de s d'une quantité positive, l'ordre de sommation pourra être abaissé de la même quantité. L'existence d'une droite de sommabilité ainsi qu'une formule correspondant à celle de M. Cahen pour l'abscisse de con-

---

(1) Dans ce théorème que je possède depuis environ une année je me suis rencontré avec M. Bohr qui a publié ses résultats dans une Note très intéressante insérée dans ces *Comptes rendus*, le 11 janvier 1909. Il conclut, de la sommabilité d'ordre k pour des valeurs telles que $R(s) > \sigma_0$, la sommabilité d'ordre $k-1$ pour des valeurs telles que $R(s) > \sigma_0 + 1$. Notre énoncé comprend celui de M. Bohr.

vergence s'établissent d'une manière très simple. La formule explicite se trouve (pour k entier) dans la Note citée de M. Bohr.

La généralisation du théorème II, pour k non entier, me paraît très probable, mais je n'ai pas encore surmonté toutes les difficultés de la démonstration.

2° Dans un autre ordre d'idées j'ai démontré les théorèmes suivants :

III. *Lorsque la fonction définie dans une portion du plan par la série convergente* $f(s) = \Sigma a_n n^{-s}$ *est régulière* (¹) *aux points du demi-plan*

$$R(s) \geqq c,$$

et y satisfait à la condition

$$\lim_{|s|=\infty} \frac{f(s)}{|s|^k} = 0, \tag{1}$$

uniformément avec $|s|$, *la série* $\Sigma a_n n^{-s}$ *est sommable par les moyennes arithmétiques d'ordre* k' (k' *désignant un nombre quelconque* $> k$) *en tout point de la droite* $R(s) = c$ *et du demi-plan* $R(s) > c$. *La sommabilité est uniforme en toute bande du domaine* $R(s) \geqq c$ *qui est parallèle à l'axe réel.*

IV. *La droite* $R(s) = c$ *contenant des pôles ou des points critiques algébriques et, parmi ceux-ci, certains d'un ordre d'infinitude* $\geqq 1$, *aussi que la condition* (1) *étant remplie pour un certain nombre* k, *la série n'est sommable par aucune moyenne arithmétique en aucun point de la droite.*

La démonstration de ces deux théorèmes se fonde sur une application convenable de la formule de Cahen établie en toute rigueur par MM. Hada-

---

(¹) Le théorème subsiste pour les points réguliers de la droite $R(s) = c$ si l'on admet sur elle des points singuliers (en nombre fini) au voisinage desquels $|f(s)|$ est intégrable. En effet, tout revient à la question si l'intégrale

$$\int_{T_1}^{T_2} f(c+ti)\, e^{wti}\, dt$$

tend ou non vers zéro avec $\frac{1}{w}$ ($c + T_1 i$ et $c + T_2 i$ étant deux points qui entourent le point singulier). On voit de cette remarque que même des singularités de la forme $\frac{1}{z \log z}$ peuvent être admises. Il en résulte de plus que, dans le théorème IV qui suit, on peut donner des singularités d'une nature beaucoup plus générale que celles que j'y caractérise, détruisant la sommabilité de la même manière que celles-ci. Je me réserve de revenir sur ces questions dans un travail plus détaillé.

mard et Perron (par le dernier même pour des séries qui ne possèdent pas de domaine de convergence absolue).

On voit du théorème IV qu'en particulier *la série de Dirichlet qui représente la fonction* $\zeta(s)$ *n'est pas sommable* (1) *par aucune moyenne arithmétique en aucun point de la droite* $\Re(s)=1$. Avant que j'eusse trouvé mon théorème IV, M. Féjer me communiqua que, dans ce cas particulier intéressant, ses calculs sur la valeur asymptotique de certaines suites lui avaient indiqué le résultat négatif ci-dessus. Notre théorème nous montre que de même *les séries qui représentent les fonctions* $[\zeta(s)]^\lambda$ *à un exposant* >1 *ne sont sommables par aucune moyenne arithmétique sur la droite* $\Re(s)=1$.

Par contre, *les séries de Dirichlet qui représentent les fonctions* $[\zeta(s)]^\lambda$, λ *désignant un nombre réel* <1, *sont convergentes sur toute la droite* $\Re(s)=1$ (*sauf le point* $s=1$ *pour* $\lambda>0$).

Ce théorème se démontre par le théorème plus général qui suit :

Supposons qu'il existe pour la fonction $f(s)$ *représentée dans une certaine portion du plan par la série convergente* $\sum a_n n^{-s}$ *un nombre* k *tel que*

$$|f(s)|<c|s|^k$$

pour

$$\Re(s)\geqq 1,$$

et des $|s|$ *assez grands. Nous supposons, en outre, que la fonction est régulière dans ce domaine, sauf tout au plus en des points isolés de la droite* $\Re(s)=1$, *dans le voisinage desquels* $|f(s)|$ *est intégrable et, de plus, qu'on a*

$$\lim_{n=\infty}\frac{a_1+a_2+\ldots+a_n}{n}=0.$$

Dans ces conditions la série converge en tout point régulier (2) *de la droite* $\Re(s)=1$.

Par une petite généralisation d'un théorème de M. Landau (3), nous voyons très nettement que les fonctions $[\zeta(s)]^\lambda$ $(\lambda<1)$ sont de l'espèce ci-dessus.

---

(1) Je viens de constater que ceci résulte aussi, d'une manière très simple, d'une formule due à M. Kinkelin.

(2) La convergence subsiste pour une grande classe des points singuliers.

(3) E. Landau, *Beiträge zur analytischen Zahlentheorie* (*Rendiconti del Circolo mat. di Palermo,* t. XXVI, p. 228).

Dans une Note prochaine nous introduirons une méthode générale nouvelle applicable à des séries de la forme générale $\sum a_n e^{-\lambda_n s}$ qui se prête heureusement à sommer les séries de $\zeta(s)$ et de ses puissances et qui fournira des résultats utiles sur le produit de deux séries de Dirichlet.

(21 juin 1909.)

6.

Sur la sommation des séries de Dirichlet

C. R. Acad. Sci. Paris *149* (1909), 18–21

Soit donnée la série

$$\sum_1^\infty a_n e^{-\lambda_n s},$$

les λ désignant une suite de constantes positives croissant vers l'infini. Nous posons

$$a_n e^{-\lambda_n s} = c_n, \qquad \sum_1^n c_k = s_n \qquad \text{et} \qquad \sigma(\lambda) = s_{n-1} \quad \text{pour} \qquad \lambda_{n-1} < \lambda \leqq \lambda_n;$$

$$(1)\quad \mathbf{S}_n^{(1)} = c_1 + c_2 + \ldots + c_n - \frac{\lambda_1 c_1 + \ldots + \lambda_n c_n}{\lambda_n}$$
$$= \frac{s_1(\lambda_2 - \lambda_1) + s_2(\lambda_3 - \lambda_2) + \ldots + s_{n-1}(\lambda_n - \lambda_{n-1})}{\lambda_n} = \frac{1}{\lambda_n}\int_0^{\lambda_n} \sigma(\lambda)\, d\lambda,$$

et, dans un ordre d'idées général,

$$(2)\qquad \mathbf{S}_n^{(k)} = c_1\left(1 - \frac{\lambda_1}{\lambda_n}\right)^k + c_2\left(1 - \frac{\lambda_2}{\lambda_n}\right)^k + \ldots + c_n\left(1 - \frac{\lambda_n}{\lambda_n}\right)^k.$$

Nous appellerons la sommation par ces moyennes la *sommation typique d'ordre k des séries aux exposants* $\lambda_1, \lambda_2, \ldots, \lambda_n, \ldots$.

Dans cette Note nous devons nous contenter de signaler quelques propriétés remarquables de ces moyennes.

Dans le cas $\lambda_n = n$, c'est-à-dire dans le cas qui provient d'une simple transformation de la *série entière,* notre méthode nous conduit à des moyennes entièrement équivalentes aux *moyennes arithmétiques.* (La forme est différente, excepté $k = 0$ et 1.)

Au cas $\lambda_n = \log n$, qui est des plus intéressants, nous appellerons la sommation typique *sommation logarithmique.* Un calcul élémentaire montre que la limite qu'on obtient de (1) en y posant $\lambda_n = \log n$ et la limite

$$(3)\qquad \lim_{n=\infty} \frac{\frac{s_1}{1} + \frac{s_2}{2} + \ldots + \frac{s_n}{n}}{\log n}$$

existent en même temps et sont égales. De plus, on a les théorèmes :

I. *Une série quelconque étant sommable par les moyennes arithmétiques d'ordre k est sommable par les moyennes logarithmiques du même ordre.*

II. *Supposons que la fonction $f(s)$ représentée dans une partie du plan par la série convergente $\sum a_n e^{-\lambda_n s}$ est régulière dans le domaine $R(s) > c$ et n'admet sur la droite $R(s) = c$ d'autres points singuliers que des pôles et des points critiques algébriques (en nombre fini) d'un ordre d'infinitude $\leqq m$* (¹). *Supposons, de plus, qu'il existe un nombre m' tel que la fonction $f(s)$ satisfait, dans le demi-plan $R(s) \geqq c$, à la condition*

$$|f(s)| < C|s|^{m'}$$

(pour des $|s|$ assez grandes). Alors la série est sommable en tout point régulier de la droite $R(s) = c$ par les moyennes typiques d'ordre k, k désignant un nombre positif tel que

$$k > m' \qquad \text{et} \qquad k > m - 1.$$

La sommation est uniforme sur toute portion finie de la droite qui ne contient que des points réguliers.

Par exemple, la fonction $\zeta(s)$ rentre bien dans cette classe. *Sa série est sommable par les moyennes logarithmiques d'ordre* 1 *sur toute la droite*

$$R(s) = 1,$$

sauf le point $s = 1$. Ce fait résulte aussi immédiatement de la formule de M. Kinkelin, en tenant compte de la forme (3) de nos moyennes.

Le résultat le plus remarquable que nous avons obtenu par les moyennes considérées se rapporte à la multiplication des séries de Dirichlet. Soient $\sum a_n e^{-\lambda_n s}$ et $\sum b_n e^{-\mu_n s}$ deux séries de cette espèce. Leur produit formel donne de nouveau une série de Dirichlet qu'on obtient en ordonnant les quantités $\lambda_n + \mu_m$ d'après leur grandeur. Nous écrivons l'égalité formelle

$$\sum_1^\infty a_n e^{-\lambda_n s} \sum_1^\infty b_n e^{-\mu_n s} = \sum_1^\infty c_n e^{-\nu_n s}.$$

Le théorème que j'ai en vue (généralisation du théorème de Cesàro sur la multiplication de Cauchy) est le suivant :

III. *Les séries $\sum_1^\infty a_n$* et *$\sum_1^\infty b_n$ étant convergentes la série $\sum_1^\infty c_n$ est sommable par les moyennes typiques d'ordre* 1 (formées avec les exposants ν) *et la somme ainsi définie est égale au produit $\sum_1^\infty a_n \sum_1^\infty b_n$.*

(¹) Dans un travail détaillé, nous démontrerons ce théorème dans une forme beaucoup plus générale. Cf. la remarque faite au sujet du théorème III de notre Note intitulée : *Sur les séries de Dirichlet* (21 juin 1909).

M. Landau avait démontré le théorème d'Abel pour le produit des séries de Dirichlet dont au moins l'une possède un domaine de convergence absolue ([1]). Pour le cas général il regarde la question comme *ouverte*. *Par notre résultat cette question est résolue par l'affirmative.*

A cette occasion nous ne nous arrêtons pas à la question de la multiplication des séries plusieurs fois indéterminées, l'indétermination étant définie par nos moyennes typiques. Nous préférons envisager un peu plus attentivement le cas particulier $\lambda_n = \log n$. Nos séries de la forme $\sum a_n n^{-s}$ et $\sum b_n n^{-s}$ étant convergentes, nous avons vu que leur produit est sommable par *les moyennes logarithmiques d'ordre* 1 ([2]).

Par contre on peut donner des séries de Dirichlet convergentes dont le produit n'est sommable par aucune moyenne arithmétique ([3]). Telles sont par exemple aux points $1 + ti\,(t \neq 0)$ les séries qui représentent les fonctions $[\zeta(s)]^\lambda$ et $[\zeta(s)]^\mu$, λ et μ désignant deux nombres positifs < 1 et tels que $\lambda + \mu \geqq 1$. Ce fait gagnera en intérêt, si nous ajoutons que la sommation par des moyennes logarithmiques d'ordre 1 n'est aucunement plus génerale que l'ensemble des méthodes des moyennes arithmétiques. Par exemple les séries $\sum n \log n \cos n\varphi$ et $\sum n \log n \sin n\varphi\,[\varphi \not\equiv 0 \pmod{2\pi}]$ sont sommables par toute moyenne arithmétique dont l'ordre (entier on non) est > 1. On voit immédiatement qu'elles ne le sont pas par les moyennes logarithmiques d'ordre 1.

([1]) E. Landau, *Ueber die Multiplikation Dirichlet'scher Richen* (*Rendiconti del Circolo mat. Palermo*, t. XXIV; p. 133).

([2]) On obtient un résultat intéressant en joignant ce fait au théorème suivant : *La série $\sum c_n$ étant sommable par les moyennes logarithmiques d'ordre* 1, *la série $\sum \frac{c_n}{\log n}$ est sommable par les moyennes arithmétiques du même ordre. Il en résulte que la série $\sum c_n n^{-s}$ et ses dérivées sont sommables par les mêmes moyennes pour toute valeur de s telle que* $\mathrm{R}(s) > 0$.

([3]) Voir le théorème IV de notre Note citée.

(5 juillet 1909.)

7.

Sur les séries de Dirichlet et les séries entières

C. R. Acad. Sci. Paris *149* (1909), 909–912

1° Un beau théorème de la théorie des séries entières dû à M. Fatou se fonde sur un théorème de Riemann concernant les séries trigonométriques. La méthode ingénieuse de Riemann nous a servi de modèle pour former une méthode conduisant au théorème qui suit :

I. *La condition*

$$\lim_{n=\infty} \frac{a_1+a_2+\ldots+a_n}{e^{\lambda_n c}} = 0 \qquad (c>0) \tag{1}$$

supposée remplie, la série de Dirichlet $\Sigma a_n e^{-\lambda_n s}$ *est convergente en tout point régulier de la droite* $R(s)=c$. *De plus, la convergence est uniforme dans tout intervalle de régularité. On sait que la même condition* (1) *est nécessaire pour que la droite contienne au moins un point de convergence.*

Ce théorème comprend comme cas particulier celui de M. Fatou.

2° Nous posons [1]

$$a_n e^{-\lambda_n s} = c_n, \qquad c_1+c_2+\ldots+c_n = s_n, \qquad \sigma(\lambda) = s_n \quad \text{pour} \quad \lambda_n < \lambda \leqq \lambda_{n+1},$$

$$\Omega^{(k)}(\omega) = \sum_{\lambda_1}^{\lambda_n<\omega} c_n\left(1-\frac{\lambda_n}{\omega}\right)^k = \frac{k}{\omega^k}\int_0^{\omega} \sigma(\lambda)(\omega-\lambda)^{k-1}\,d\lambda, \tag{2}$$

k étant $\geqq 0$ (entier ou non).

Nous appelons les expressions (2) *moyennes typiques* (*d'ordre* k) relatives aux séries de la forme $\Sigma a_n e^{-\lambda_n s}$.

La formule (2) fait voir que les domaines de sommabilité correspondant aux diverses méthodes sont des demi-plans et que la sommabilité est uniforme dans tout domaine intérieur au demi-plan correspondant [2].

---

(1) Voir deux Notes insérées aux *Comptes rendus*, qui portent respectivement les titres : 1° *Sur les séries de Dirichlet* (21 juin 1909); 2° *Sur la sommation des séries de Dirichlet* (5 juillet 1909). On peut aussi, dans le cas général, définir des moyennes qui jouent le même rôle que les moyennes arithmétiques jouaient dans le cas particulier $\lambda_n = \log n$.

(2) Remarquons que k désignant un nombre négatif quelconque, le théorème II de notre première Note citée subsiste pour des séries quelconques, en y remplaçant les mots « moyennes arithmétiques » par les mots « moyennes typiques ».

3° En remplaçant dans la condition (1) les sommes partielles de la série

$$a_1 + a_2 + \ldots + a_n + \ldots$$

par les moyennes typiques de cette série, on obtient un théorème, généralisation de I, qui comprend comme cas particulier le théorème suivant, démontré dans ma Thèse :

II. *La série entière* $\Sigma c_n z^n$ *est sommable par les moyennes arithmétiques d'ordre k en tout point régulier de la circonférence de rayon* 1, *si la condition* $\lim c_n n^{-k} = 0$ *est remplie. La même condition est nécessaire pour qu'il y ait au moins un point de sommabilité d'ordre k sur la circonférence.*

4° La régularité n'est aucunement une condition nécessaire pour que le théorème I et sa généralisation soient valables. Pour le moment nous nous bornerons au cas particulier $\lambda_n = n$ (¹). Considérons la fonction

$$F(r, \varphi) = \sum (a_n \cos n\varphi + b_n \sin n\varphi) r^n,$$

où

$$\lim a_n n^{-k} = \lim b_n n^{-k} = 0.$$

Les théorèmes que nous trouverons pourront s'appliquer immédiatement aux séries entières (²). Nous ferons les hypothèses suivantes :

1° $\lim_{r=1} F(r, \varphi) = F(\varphi)$ est bien déterminé pour $\varphi_1 \leqq \varphi \leqq \varphi_2$, sauf peut-être sur un ensemble de mesure nulle; 2° la fonction $F(\varphi)$ est intégrable au sens de M. Lebesgue; 3° on a uniformément

$$\lim_{r=1} \int_{\varphi_1}^{\varphi} F(r, \varphi)\, d\varphi = \int_{\varphi_1}^{\varphi} F(\varphi)\, d\varphi \qquad (\varphi_1 \leqq \varphi \leqq \varphi_2).$$

Convenons de définir $F(\varphi) = 0$ sur les points de la circonférence extérieurs à l'arc (φ_1, φ_2). Cela posé, la condition nécessaire et suffisante pour que la méthode des moyennes arithmétiques d'ordre k s'applique à la série

$$\sum a_n \cos n\varphi + b_n \sin n\varphi$$

au point φ' $(\varphi_1 < \varphi' < \varphi_2)$, c'est que cette même méthode s'applique à la série de Fourier de $F(\varphi)$ au point φ'. *On réussit donc aux points de continuité ou de discontinuité de première espèce de la fonction* $F(\varphi)$. Pour $k \geqq 1$ cela est évident d'après un théorème célèbre de M. Fejér. J'ai dû montrer que *pour* $0 < k < 1$ *les résultats principaux de M. Fejér subsistent encore.*

(¹) La méthode que nous suivons s'étend aisément au cas général.

(²) M. Dienes a eu l'idée de combiner un théorème de M. Hadamard : 1° avec le théorème cité de M. Fatou, 2° avec notre théorème II, 3° avec des théorèmes de M. Borel et de M. Lindelöf. Par cette idée, il parvint à quatre théorèmes nouveaux. La combinaison des méthodes, au lieu de celle des théorèmes, l'aurait conduit à des résultats plus généraux.

Dans le même ordre d'idées, j'ai démontré le résultat qui suit. Considérons une méthode de sommation qui ait les deux propriétés suivantes : 1° elle fournit en tout point régulier du cercle de convergence d'une série entière la valeur de cette série; 2° elle est au moins aussi générale que l'une des méthodes des moyennes arithmétiques d'ordre positif. Supposons d'autre part que la série entière

$$\Phi(r, \varphi) = \sum c_n r^n e^{ni\varphi}$$

ait un rayon de convergence fini R et que les fonctions $\Phi(r, \varphi)$ et $\Phi(\varphi) = \lim_{r=R} \Phi(r, \varphi)$ vérifient les conditions (1), (2) et (3) énoncées tout à l'heure pour les fonctions $F(r, \varphi)$ et $F(\varphi)$. Dans ces conditions, *la méthode de sommation* dont nous venons de parler, *appliquée à la série*

$$\sum c_n R^n e^{ni\varphi},$$

donne la valeur

$$\frac{1}{2}[\Phi(\varphi + 0) + \Phi(\varphi - 0)]$$

en chaque point du cercle de convergence, où la fonction $\Phi(\varphi)$ *admet une discontinuité de première espèce.*

Plusieurs méthodes de M. Mittag-Leffler, ainsi qu'une méthode due à M. Lindelöf, rentrent dans la catégorie ci-dessus. On peut obtenir un théorème analogue pour les points frontières de l'étoile de M. Mittag-Leffler.

5° J'ai trouvé des théorèmes analogues à I et à ses généralisations pour les intégrales

$$\int_0^\infty \varphi(t)\, e^{-tx}\, dt.$$

Les moyennes typiques s'y définissent aussi par les intégrales de Riemann-Liouville.

(22 novembre 1909.)

8.

Sur un problème d'Abel

Rend. Circ. Mat. Pal. *30* (1910), 339–345

(Extrait de deux lettres à M. G. MITTAG-LEFFLER).

I.

. . . .

Je me hâte de vous communiquer que le texte français d'un passage de votre conférence de Rome, sur lequel je me suis permis une légère remarque, peut être maintenu presqu'entièrement. Le texte qui se trouve dans votre lettre et qui correspond à l'original suédois comporte une restriction qui n'est pas nécessaire: c'est la condition de LIPSCHITZ. En effet je viens de démontrer le théorème suivant (en parlant toujours, pour fixer les idées, de la méthode de votre troisième Note):

Désignons par α *et* $\alpha'(<\alpha)$ *deux quantités positives. Si la branche fonctionnelle* $FA^{(\alpha)}(x)$ *s'approche indéfiniment d'une valeur déterminée, quand* x, *de l'intérieur de la figure* $V^{(\alpha)}$ *décrite autour de la demi-droite entre l'origine et un sommet, tend vers ce sommet, l'expression* $c_0 + \sum_{\mu=1}^{\infty} G_\mu(x\,|\,\alpha')$ *reste convergente si l'on y introduit l'affixe du sommet et elle exprime par conséquent la valeur de la branche fonctionnelle en ce point.*

Le théorème que je viens d'énoncer, suppose essentiellement $\alpha' < \alpha$. Dans le cas $\alpha' = \alpha$ je peux donner des exemples qui le mettent en défaut.

Pour démontrer cet énoncé je vais d'abord établir le théorème qui suit.

Soit $f(z)$ *une fonction de la variable complexe, continue dans le domaine* D

$$|z| \leqq R \qquad (R > 1),$$

$$\vartheta \leqq \arg(z-1) \leqq 2\pi - \vartheta \qquad \left(0 < \vartheta < \frac{\pi}{2}\right),$$

et sur son contour, qui soit régulière dans le domaine D *excepté au point* $z = 1$. *Dans ces conditions, le développement taylorien de la fonction autour du point* $z = 0$

$$a_0 + a_1 z + \cdots + a_n z^n + \cdots$$

est convergent pour $z = 1$. *De plus, il converge uniformément sur la circonférence* $|z| = 1$.

Désignons par Γ une courbe fermée, renfermant le point $z = 0$, composée d'un arc de cercle de centre $z = 0$, de rayon r $(1 < r < R)$, et des deux demi-droites

$$\arg(z-1) = \pm\vartheta. \tag{1}$$

On a

$$a_0 + a_1 + \cdots + a_n = \frac{1}{2\pi i}\int_{\Gamma'} \frac{f(z)}{1-z}\frac{1}{z^{n+1}}\,dz,$$

Γ' désignant une courbe quelconque renfermant le point $z = 0$ et intérieure à la courbe Γ. En tenant compte de la continuité de $f(z)$ au point $z = 1$ on conclut

$$(2)\qquad a_0 + a_1 + \cdots + a_n = \frac{1}{2\pi i}\int_{\Gamma} \frac{f(z)}{1-z}\left(\frac{1}{z^{n+1}} - 1\right)dz.$$

Nous pouvons supposer, sans restreindre la généralité,

$$f(1) = 0.$$

Pour démontrer notre théorème, il nous suffira de montrer que l'intégrale (2) tend vers zéro avec $\frac{1}{n}$.

Désignons les points d'intersection du cercle de rayon r et des demi-droites (1) par z_1 et $\bar{z}_1$. Étant donnée une quantité positive ε, aussi petite qu'elle soit, nous pouvons choisir le rayon $r > 1$, assez rapproché de 1 pourqu'on ait

$$(3)\qquad |f(z)| < \varepsilon$$

pour les points de notre domaine D satisfaisant à l'inégalité

$$|z - 1| \leqq |z_1 - 1|.$$

Ayant fixé r, on peut choisir ν assez grand pour qu'on ait, pour toute valeur de $n > \nu$, sur l'arc de rayon r

$$(4)\qquad \left|\int_{z_1}^{\bar{z}_1} \frac{f(z)}{1-z}\frac{1}{z^{n+1}}\,dz\right| < \varepsilon',$$

ε' étant donné d'avance.

n désignant maintenant un nombre déterminé $> \nu$, soit z_0 le point du segment $(1, z_1)$ tel que

$$|z_0 - 1| = \frac{1}{n}.$$

On a

$$(5)\qquad \left\{\begin{aligned} &\left|\int_{z_0}^{z_1} \frac{f(z)}{1-z}\frac{1}{z^{n+1}}\,dz\right| < \varepsilon n \int_{z_0}^{z_1} \frac{1}{|z|^{n+1}}\,|dz| \\ &= \varepsilon n \int_{z_0}^{z_1} \frac{1}{|z|^{n+1}}\frac{|dz|}{d|z|}\,d|z| < \frac{\varepsilon n}{\cos\vartheta}\int_{z_0}^{z_1} \frac{1}{|z|^{n+1}}\,d|z| \\ &= \frac{\varepsilon n}{\cos\vartheta}\cdot\frac{1}{n}\left(\frac{1}{|z_0|^n} - \frac{1}{|z_1|^n}\right) < \frac{\varepsilon}{\cos\vartheta}. \end{aligned}\right.$$

De même,

$$(6)\qquad \left|\int_{z_0}^{\bar{z}_1} \frac{f(z)}{1-z}\frac{1}{z^{n+1}}\,dz\right| < \frac{\varepsilon}{\cos\vartheta}.$$

D'ailleurs l'intégrale

$$\int_{z_0}^{\bar{z}_0} \frac{f(z)}{1-z}\,dz,$$

prise sur Γ dans le sens positif, est plus petite que $2\pi\varepsilon$, car elle est égale à l'intégrale prise sur l'arc de cercle intérieur au domaine D, de centre $z = 1$, passant par les points z_0 et $\bar{z}_0$ et sur cet arc

$$(7) \qquad \left|\int \frac{f(z)}{1-z}dz\right| < \varepsilon \int \left|\frac{dz}{1-z}\right| < 2\pi\varepsilon.$$

Les inégalités (3), (4), (5), (6) et (7) montrent, qu'aussi petite que soit la quantité positive η, on peut choisir n assez grand pour qu'on ait

$$\left|\int_{z_0}^{\bar{z}_0} \frac{f(z)}{1-z}\left(\frac{1}{z^{n+1}} - 1\right)dz\right| < \eta,$$

l'intégrale étant prise sur Γ dans le sens positif. Passons maintenant à l'intégrale rectiligne prise sur le segment $(1, z_0)$

$$\left|\int_1^{z_0} \frac{f(z)}{1-z}\left(\frac{1}{z^{n+1}} - 1\right)dz\right|$$

$$= \left|\int_1^{z_0} f(z)\left(\frac{1}{z^{n+1}} + \cdots + \frac{1}{z}\right)dz\right| < (n+1)\varepsilon\int_1^{z_0}|dz| < \frac{n+1}{n}\varepsilon < 2\varepsilon.$$

De même

$$\left|\int_1^{\bar{z}_0} \frac{f(z)}{1-z}\left(\frac{1}{z^{n+1}} - 1\right)dz\right| < 2\varepsilon;$$

donc enfin nous avons établi que la quantité

$$|a_0 + \cdots + a_n| = \left|\frac{1}{2\pi i}\int_\Gamma \frac{f(z)}{1-z}\left(\frac{1}{z^{n+1}} - 1\right)dz\right|$$

tend vers zéro avec $\frac{1}{n}$.

[Le même raisonnement montre que la convergence de la série

$$a_0 + a_1 z + \cdots + a_n z^n + \cdots$$

est uniforme sur le cercle de rayon 1].

Cela posé, le théorème que j'énonce au commencement de ma lettre sur la convergence de vos séries de polynômes se démontre facilement grâce à la forme que vous avez donnée au reste de ces séries. Cette forme montre très nettement qu'à tout théorème concernant la convergence ou la divergence des séries entières, correspond un théorème concernant la convergence ou la divergence de vos séries de polynômes.

Pour faire la démonstration, supposons que le sommet en question soit le point $x = 1$. Effectuons la transformation conforme du plan de la variable u, sur le plan de la variable x, par la fonction génératrice

$$x = f(u\,|\,\alpha')$$

de votre troisième Note. La figure $V^{(\alpha')}$ correspond au cercle de centre o et de rayon 1 du plan des u. Comme la fonction $FA^{(\alpha)}(x)$ est régulière à l'intérieur et sur le contour de la figure génératrice $V^{(\alpha)}$, excepté au point $x = 1$, la fonction $\Phi(u) = F[f(u\,|\,\alpha')]$ est régulière dans tout le cercle

$$|u| \leqq 1$$

excepté aux points $u=1$ et $u=-1$. [Ce dernier point est singulier parce que $u=-1$ est un point singulier de la fonction $f(u\,|\,x')$]. Cependant, la fonction $\Phi(u)$ est finie et continue dans tout le cercle et sur tout son contour, y compris les points $u=\pm 1$. De plus, la fonction $FA^{(\alpha)}(x)$ est supposée régulière aussi à l'intérieur de la figure génératrice $V^{(\alpha)}$, où $x > x'$, et elle est continue aussi lorsque le point x tend vers $x=1$ en restant à l'intérieur de cette figure. La fonction $\Phi(u)$ est donc certainement régulière dans un voisinage de $u=1$ délimité par des demi-droites formant avec l'axe réel les angles $\pm\vartheta$ $\left(0<\vartheta<\frac{\pi}{2}\right)$. De plus, elle est continue lorsque la variable u tend dans ce voisinage vers $u=1$. Donc, enfin, on pourra tracer une courbe fermée Ω composée des segments de droite ci-dessus et d'une courbe ouverte touchant extérieurement le cercle de centre 0 et de rayon 1 au point $u=-1$, n'ayant que ce point commun avec le cercle et dans laquelle $\Phi(u)$ ait les propriétés suivantes:

Elle est finie et continue dans tout l'intérieur de la courbe et y est régulière, excepté aux points $u=\pm 1$. La forme du reste de votre série de polynômes sera donc

$$-\frac{1}{2\pi i}\int_{\Omega}\frac{\Phi(u)-\Phi(1)}{1-u}\left(\frac{1}{u^{n+1}}-1\right)du.$$

Une très légère modification du raisonnement que nous avons fait pour la courbe Γ, prouve que cette intégrale tend aussi vers zéro avec $\frac{1}{n}$.

On voit, dans quel cas très général le problème d'ABEL est résoluble par votre méthode.

Je remarque encore que la condition de continuité peut être remplacée par des conditions beaucoup plus générales, comme je l'ai indiqué dans ma Note: *Sur les séries de* DIRICHLET *et les séries entières* [1]).

Paris, 24 Mai 1910.

MARCEL RIESZ.

II.

. . . .

Je vous suis très reconnaissant de m'avoir observé qu'un passage de ma lettre (uniformité de la convergence) vous paraissait obscur. J'avais raison de dire que dans les conditions indiquées la convergence est uniforme, mais j'avais bien tort de ne pas en donner la démonstration détaillée. Mon excuse est que dans cette publication c'est l'application que j'avais principalement en vue. Comme le raisonnement (que je m'imaginais

[1]) Comptes rendus hebdomadaires des séances de l'Académie des Sciences (Paris), t. CXLIX (2e semestre 1909), pp. 909-912.

beaucoup plus simple, ne l'ayant jamais mis sur papier) me semble assez délicat, je me permets de vous le donner dans cette lettre. Dans un prochain travail je vais réunir mes deux raisonnements.

. . . .

J'ajoute qu'il n'était pas inutile de remarquer que je parlais de la méthode de votre troisième Note. Il est tout à fait essentiel que les deux arcs de la figure génératrice forment au point 1 un angle $(\alpha\pi)$ qui diminue avec α. Ce n'est pas le cas pour les méthodes basées sur l'intégrale de LAPLACE-ABEL dépendant d'un paramètre α. Et en effet je peux construire une fonction continue dans tout le domaine $|z| \leqq 1$ régulière en tout point, excepté le point $z = 1$, dont le développement taylorien n'est sommable au point $z = 1$ par aucune des méthodes de LAPLACE-ABEL, aussi petit que soit le paramètre α.

. . . .

En reprenant les notations de ma dernière lettre, on a

$$(1)\qquad f(x) - (a_0 + a_1 x + \cdots + a_n x^n) = \frac{1}{2\pi i}\int_\Gamma f(z)\frac{\left(\frac{x}{z}\right)^{n+1}}{z - x}\,dz = I_n(x),$$

x désignant un point quelconque de la circonférence

$$|x| = 1,$$

différent de $x = 1$. Nous allons démontrer que cette intégrale tend uniformément vers zéro avec $\frac{1}{n}$. Il suffit évidemment d'effectuer la démonstration pour les points situés sur la demi-circonférence supérieure.

Nous désignons les droites

$$\arg(z - 1) = \pm\,\vartheta$$

par d et $\bar{d}$. Nous supposons de nouveau

$$(2)\qquad f(1) = 0.$$

Nous fixons de nouveau le point z_1 sur la droite d, de manière que l'on ait

$$(3)\qquad |f(z)| \leqq \varepsilon$$

pour les points de notre domaine satisfaisant à l'inégalité

$$(4)\qquad |z - 1| \leqq |z_1 - 1|.$$

Après avoir fixé le point z_1 et par conséquent le rayon r, nous déterminons le nombre n de façon à satisfaire aux trois conditions suivantes:

$$A)\qquad \frac{1}{n} < r - 1.$$

$$B)\qquad \left|\frac{f(z)}{z - x}\left(\frac{x}{z}\right)^{n+1}\right| < \varepsilon',$$

z désignant un point quelconque de l'arc de rayon r, et x un point quelconque de la circonférence de rayon 1.

C) Désignons en outre par x_n le point de la circonférence $|x| = 1$ qui est à la distance $\frac{1}{n}$ de la demi-droite *d* et imposons enfin à *n* la condition suivante:

$$|x_n - 1| + \frac{1}{n} \leqq |z_1 - 1|.$$

Dans cette condition tout cercle décrit avec le rayon $\frac{1}{n}$ autour d'un point quelconque de l'arc $(x_n, 1)$ est intérieur au cercle qui est décrit autour de $x = 1$ et qui passe par z_1. Donc on a, en vertu de (3),

$$|f(z)| \leqq \varepsilon \tag{5}$$

dans tous ces divers cercles. (Évidemment, nous ne considérons que la partie de ces cercles qui est intérieure au domaine Γ).

Les conditions *A*), *B*) et *C*) étant remplies pour un certain nombre ν, le sont pour tout nombre

$$n \geqq \nu.$$

n étant choisi de cette manière, envisageons l'intégrale $I_n(x)$, *x* désignant un point quelconque ($\neq 1$) de la demi-circonférence supérieure. Pour les points *x* situés sur l'arc $(-1, x_n)$ on a

$$|I_n(x)| = \left|\frac{1}{2\pi i}\int_{z_1}^{\bar z_1} \frac{f(z)}{z-x}\left(\frac{x}{z}\right)^{n+1} dz + \frac{1}{2\pi i}\int_{\bar z_1}^{1} \frac{f(z)}{z-x}\left(\frac{x}{z}\right)^{n+1} dz + \frac{1}{2\pi i}\int_{1}^{z_1} \frac{f(z)}{z-x}\left(\frac{x}{z}\right)^{n+1} dz\right|$$

$$\leqq r\varepsilon' + \frac{\varepsilon n}{2\pi}\left(\int_{\bar z_1}^{1} \frac{1}{|z|^{n+1}}|dz| + \int_{1}^{z_1} \frac{1}{|z|^{n+1}}|dz|\right).$$

Or, d'après le raisonnement de ma lettre la somme de ces intégrales est $< \frac{2}{n \cos \vartheta}$. Donc on a, pour tous ces points *x*,

$$|I_n(x)| < \frac{2\varepsilon}{\cos \vartheta} + r\varepsilon' < \eta_n, \tag{6}$$

η_n tendant vers zéro avec $\frac{1}{n}$.

Passons maintenant aux points *x* situés sur l'arc $(x_n, 1)$ (excepté les extrémités de cet arc). Traçons autour d'un point quelconque de cet arc un cercle de rayon $\frac{1}{n}$ qui va couper le contour Γ en deux points ξ_n et ξ'_n. Ces points seront situés [2]) sur le segment $(1, z_1)$, ou bien le point ξ_n sera situé sur ce segment, tandis que ξ'_n sera situé sur le segment $(1, \bar z_1)$. (Dans le premier cas c'est le point supérieur que nous désignons par ξ_n). On a, en tout cas,

$$|I_n(x)| = \left|\frac{1}{2\pi i}\int_{\xi_n}^{\xi'_n} \frac{f(z)}{z-x}\left(\frac{x}{z}\right)^{n+1} dz + \frac{1}{2\pi i}\int_{\xi'_n}^{\xi_n} \frac{f(z)}{z-x}\left(\frac{x}{z}\right)^{n+1} dz\right|$$

$$\leqq \left|\frac{1}{2\pi i}\int_{\xi_n}^{\xi'_n} \frac{f(z)}{z-x}\left(\frac{x}{z}\right)^{n+1} dz\right| + \left|\frac{1}{2\pi i}\int_{\xi'_n}^{\xi_n} \frac{f(z)}{z-x}\left(\frac{x}{z}\right)^{n+1} dz\right|,$$

[2]) En vertu de *A*) aucun des cercles ne coupe l'arc de rayon *r*.

ce dernier chemin d'intégration se composant d'un ou de deux segments de droites, et le premier, de l'arc de cercle de rayon r et des segments de droites. La première intégrale est, d'après le raisonnement que nous venons de faire,

$$< r\varepsilon' + \frac{2\varepsilon}{\cos\vartheta};$$

donc il ne nous reste qu'à évaluer la seconde intégrale. On a de nouveau

$$\left|\frac{1}{2\pi i}\int_{\xi_n'}^{\xi_n}\frac{f(z)}{z-x}\left(\frac{x}{z}\right)^{n+1}dz\right| = \left|\frac{1}{2\pi i}\int_{\xi_n}^{\xi_n'} f(z)\frac{\left(\frac{x}{z}\right)^{n+1}-1}{z-x}dz + \frac{1}{2\pi i}\int_{\xi_n}^{\xi_n'}\frac{f(z)}{z-x}dz\right|$$

$$\leq \left|\frac{1}{2\pi i}\int_{\xi_n'}^{\xi_n}\frac{f(z)}{z}\left[\left(\frac{x}{z}\right)^{n}+\cdots+\left(\frac{x}{z}\right)+1\right]dz\right| + \left|\frac{1}{2\pi i}\int_{\xi_n'}^{\xi_n}\frac{f(z)}{z-x}dz\right|.$$

La longueur du chemin d'intégration de ξ_n à ξ_n' est

$$< \frac{4}{n}.$$

Par conséquent, la première intégrale est

$$< \frac{2\varepsilon}{\pi}\frac{n+1}{n} < 2\varepsilon.$$

D'ailleurs, on a

$$\left|\frac{1}{2\pi i}\int_{\xi_n'}^{\xi_n}\frac{f(z)}{z-x}dz\right| = \left|f(x) - \frac{1}{2\pi i}\int_C\frac{f(z)}{z-x}dz\right| \leq |f(x)| + \left|\frac{1}{2\pi i}\int_C\frac{f(z)}{z-x}dz\right|,$$

en désignant par C l'arc de cercle de rayon $\frac{1}{n}$ décrit autour du point x, joignant les points ξ_n et ξ_n' et situé à l'intérieur du domaine Γ. Or on a, d'après (3),

$$|f(x)| < \varepsilon$$

et d'après (5)

$$|f(z)| < \varepsilon$$

le long de l'arc C. Il en résulte

$$\left|\frac{1}{2\pi i}\int_C\frac{f(z)}{z-x}dz\right| < \frac{\varepsilon}{2\pi}\int_C\frac{|dz|}{|z-x|} < \varepsilon,$$

donc enfin

$$\left|\frac{1}{2\pi i}\int_{\xi_n'}^{\xi_n}\frac{f(z)}{z-x}\left(\frac{x}{z}\right)^{n+1}dz\right| < 3\varepsilon.$$

Paris, 12 Juin 1910.

MARCEL RIESZ.

9.

Une méthode de sommation équivalente à la méthode des moyennes arithmétiques

C. R. Acad. Sci. Paris *152* (1911), 1651–1654

L'équivalence des méthodes de sommation de Hölder et de Cesàro a été établie par MM. Knopp et Schnee. Dans cette Note, je me propose d'introduire une nouvelle méthode, équivalente à celles-ci et utilisable pour l'étude des séries de la forme

$$\sum_{n=0}^{\infty} a_n n^{-s}.$$

Considérons la série

$$(1) \qquad a_0 + a_1 + \ldots + a_n + \ldots,$$

et un nombre $k \geqq 0$ qui sera l'ordre de la sommation. Posons avec Cesàro

$$(2) \qquad (1-x)^{-(k+1)} = \sum_{n=0} C_n^{(k)} x^n,$$

$$(3) \qquad (1-x)^{-(k+1)} \sum_{n=0}^{\infty} a_n x^n = \sum_{n=0} S_n^{(k)} x^n.$$

Si la limite

$$(4) \qquad \lim_{n=\infty} \frac{S_n^{(k)}}{C_n^{(k)}} = S$$

existe, nous disons que la série (1) est sommable par les moyennes de Cesàro d'ordre k et nous appelons S sa somme.

Posons d'autre part, ω désignant un nombre positif quelconque,

$$(5) \qquad \sigma_k(\omega) = \sum_{r=0}^{<\omega} a_r (\omega - r)^k.$$

Nous allons indiquer la démonstration du théorème suivant :

La limite (4) *et la limite*

(6)
$$\lim_{\omega=\infty} \frac{\sigma_k(\omega)}{\omega^k}$$

existent en même temps et sont égales (¹).

La démonstration de notre théorème se compose de la démonstration des trois lemmes suivants :

I. *Lorsque la limite* (4) *existe, on a*

(7)
$$\lim_{n=\infty} \frac{S_n^{(l)}}{n^k} = 0,$$

l désignant un nombre quelconque $< k$.

II. *Lorsque la limite* (6) *existe, on a pour tout nombre entier* $E < k$

(8)
$$\lim_{n=\infty} \frac{S_n^{(E)}}{n^k} = 0.$$

III. *L'égalité* (8) *étant remplie pour tout nombre entier* $E < k$, *on aura*

(9)
$$\lim_{n=\infty} \left(\frac{\sigma_k(\omega)}{\omega^k} - \frac{S_n^{(k)}}{C_n^{(k)}} \right) = 0 \qquad (n \leqq \omega \leqq n+1).$$

La démonstration des lemmes I et III ne présentant aucune difficulté, je me borne à indiquer la démonstration du lemme II. Celle-ci devient aussi très simple, quand k est entier. En effet, la $k^{\text{ième}}$ différence des quantités $\sigma_k(n)$, $\sigma_k\left(n+\frac{1}{k}\right)$, $\sigma_k\left(n+\frac{2}{k}\right)$, ..., $\sigma_k(n+1)$ fournit $S_n^{(0)}$ à un facteur constant près, donc la relation (8) est vérifiée pour $E=0$. En formant ensuite les différences d'ordre $(k-1)$, $(k-2)$, ..., 1, on démontre la même relation pour $E = 1, 2, \ldots, (k-1)$.

Soit maintenant k non entier et, pour simplifier le raisonnement, supposons

---

(¹) Ce théorème a servi de base à la plupart de mes recherches sur l'application des moyennes arithmétiques. Dans une Note insérée dans ces *Comptes rendus* (5 juillet 1909), j'ai pensé pouvoir l'énoncer sous une forme plus simple, en ne faisant prendre à ω que des valeurs entières. Je me suis bientôt aperçu que cet énoncé était inexact. Je tiens à remercier M. Knopp qui me communiqua récemment des remarques intéressantes sur ce point. Les moyennes (6) forment d'ailleurs un cas particulier des expressions que j'ai appelées *moyennes typiques* et dont la forme définitive se trouve dans ma Note du 22 novembre 1909.

d'abord

(10) $$0 < k < 1.$$

Nous n'avons donc qu'à démontrer l'égalité

(11) $$\lim_{n=\infty} \frac{S_n^{(0)}}{C_n^{(k)}} = \lim_{n=\infty} \frac{S_n^{(0)}}{n^k} = 0.$$

Nous pouvons supposer, sans restreindre la généralité,

(12) $$\lim_{\omega=\infty} \frac{\sigma_k(\omega)}{\omega^k} = 0.$$

Posons

$$S(t) = S_r^{(0)}, \qquad \text{pour} \qquad r < t \leq r+1$$

et

$$\sigma'_k(u) = k[a_0(u-0)^{k-1} + a_1(u-1)^{k-1} + \dots + a_r(u-r)^{k-1}], \quad \text{pour} \quad r < u \leq r+1,$$

ce qui nous donne

(13) $$S(t) = \mathcal{G} \int_0^t \sigma'_k(u)(t-u)^{-k}\, du,$$

$\mathcal{G}$ désignant une constante facile à calculer. On a d'autre part

$$\sigma_k(n) = \sum_{r=0}^{n-1} a_r(n-r)^k = S_{n-1}^{(0)} + \sum_{r=0}^{n-2} S_r^{(0)}[(n-r)^k - (n-r-1)^k]$$

$$= S_{n-1}^{(0)} + k\int_0^{n-1} S(t)(n-t)^{k-1}\, dt$$

$$= S_{n-1}^{(0)} + k\mathcal{G}\int_0^{n-1} (n-t)^{k-1}\, dt \int_0^t \sigma'_k(u)(t-u)^{-k}\, du$$

$$= S_{n-1}^{(0)} + k\mathcal{G}\int_0^{n-1} \sigma'_k(u)\, du \int_u^{n-1} (n-t)^{k-1}(t-u)^{-k}\, dt$$

$$= S_{n-1}^{(0)} + k\mathcal{G}\int_0^{n-1} \sigma'_k(u)\, M(u)\, du,$$

où l'on a posé

$$M(u) = \int_u^{n-1} (n-t)^{k-1}(t-u)^{-k}\, dt < \int_u^n (n-t)^{k-1}(t-u)^{-k}\, dt = M,$$

M désignant un nombre indépendant de u et de n. De plus, $M(u)$ étant décroissant entre o et $n-1$, on a, en appliquant le second théorème de la moyenne,

$$|\sigma_k(n) - S_{n-1}^{(0)}| = |k\mathcal{G}\,M(0)\sigma_k(\tau)| < |k\mathcal{G}M\sigma_k(\tau)|,$$
$$0 < \tau < n-1,$$

ce qui, en vertu de la relation (12), entraîne l'égalité (11).

Désignons dans le cas général par K le plus grand nombre entier, contenu dans k et par $\omega_1, \omega_2, \ldots, \omega_{K+1}$ $(K+1)$ nombres distincts, compris entre n et $(n+1)$ tels que les différences $(\omega_1 - n), \ldots, (\omega_{K+1} - n)$ restent constantes, quand n tend vers l'infini. L'un de ces nombres étant ω, on a, en intégrant par partie,

$$\sigma_k(\omega) = \frac{1}{k(k-1)\ldots(k-K)} \int_0^\omega \sigma_K(t)(\omega - t)^{k-K-1}\,dt$$
$$= \frac{1}{k(k-1)\ldots(k-K)} \left(\int_0^n + \int_n^\omega\right).$$

On voit comme ci-dessus que la première intégrale, divisée par n^k, tend vers zéro avec $\frac{1}{n}$. En refaisant dans la seconde intégrale les intégrations par partie, on obtient $(K+1)$ expressions linéaires et indépendantes de $\sigma_0(n), \ldots, \sigma_K(n)$ qui, divisées par n^k, tendent vers zéro avec $\frac{1}{n}$. Donc, c'est aussi le cas pour les quantités $\sigma_0(n), \ldots, \sigma_K(n)$, et par conséquent aussi pour les quantités $S_n^{(0)}, \ldots, S_n^{(K)}$.

10.

Über einen Satz des Herrn Fatou

J. reine angew. Math. *140* (1911), 89–99

Herr *Fatou* hat einen äußerst bemerkenswerten Satz über Potenzreihen aufgestellt*), den wir unten wiedergeben. Der Satz ist von sehr einfacher Form, jedoch beruht der *Fatou*sche Beweis auf einem tiefliegenden und schwierigen *Riemann*schen Satze über die Darstellbarkeit einer Funktion durch eine konvergente trigonometrische Reihe.

Diesen *Fatou*schen Satz habe ich nach verschiedenen Richtungen erweitert. Einerseits habe ich in meiner Inauguraldissertation**), wo ich die Summierbarkeit durch arithmetische Mittel untersuche, eine notwendige und hinreichende Bedingung für die Summierbarkeit einer Potenzreihe in jedem regulären Punkte des Einheitskreises entwickelt. Andererseits habe ich in einer vorläufigen Mitteilung***) einen Satz über *Dirichlet*sche Reihen aufgestellt, der auch den *Fatou*schen Satz umfaßt. Zum Beweise dieses Satzes bedarf ich nur elementarer Hilfsmittel aus der Funktionentheorie; ich mache von den *Riemann*schen Resultaten keinen Gebrauch und wende von seinen Kunstgriffen nur solche an, die mir im Beweisgange unumgänglich scheinen.

Ich teile hier den Beweis nur für den speziellen Fall der Potenzreihen mit, wo er sich besonders einfach gestaltet. Unsere Abschätzungen gestatten auch den *Fatou*schen Satz dahin zu ergänzen, daß die darin

---

*) *P. Fatou*: Séries trigonométriques et séries de *Taylor*. Acta Mathematica. Bd. 30 (S. 389).

**) Summierbare trigonometrische Reihen und Potenzreihen. (Ungarisch.) Budapest 1908.

***) Sur les séries de *Dirichlet* et les séries entières. Comptes Rendus de l'Acad. des Sciences. Paris. (22 novembre 1909).

behauptete Konvergenz der Potenzreihe an den regulären Stellen des Einheitskreises, auf jedem Regularitätsbogen*) eine gleichmäßige ist.

Zugleich gebe ich die oben angedeutete Verallgemeinerung für arithmetische Mittel wieder.

Der *Fatou*sche Satz mit unserer Ergänzung lautet:

Satz I. Genügen die Koeffizienten einer Potenzreihe

$$f(z) = a_0 + a_1 z + \cdots + a_n z^n + \cdots$$

der Bedingung

$$\lim_{n=\infty} a_n = 0, \tag{1.}$$

so konvergiert die Potenzreihe an jeder regulären Stelle des Einheitskreises und zwar gleichmäßig an den Stellen eines Regularitätsbogens. Natürlich ist die Bedingung (1.) *auch notwendig, damit auf dem Einheitskreise wenigstens ein Konvergenzpunkt liegt.*

Wir setzen voraus, daß $z = 1$ eine reguläre Stelle der Funktion ist und beweisen zunächst, daß die Reihe für $z = 1$ konvergiert. (Die Gleichmäßigkeit der Konvergenz wird sich nachher leicht ergeben.)

Es seien z_1 und z_2 die Endpunkte eines Regularitätsbogens, der die Stelle $z = 1$ enthält, und zwar liege z_1 in der oberen und z_2 in der unteren Halbebene. Es bedeute ferner C_1 einen außerhalb des Einheitskreises und C_2 einen innerhalb dieses Kreises gelegenen Kurvenzug, beide mit den Endpunkten z_1 und z_2. Sie seien noch so beschaffen, daß die geschlossene Kurve $C_1 + C_2$ die Stellen $z = 1$ und $z = 0$ einschließt und daß die Funktion $f(z)$ auf dieser Kurve $C_1 + C_2$ und in ihrem Innern regulär ist.

Man hat dann in der Umgebung des Nullpunktes die Entwickelung

$$\frac{f(z)}{1-z} = \sum_{\varkappa=0}^{\infty} (a_0 + a_1 + \cdots + a_\varkappa) z^\varkappa$$

und somit nach dem *Cauchy*schen Satze

$$a_0 + a_1 + \cdots + a_n - f(1) = \frac{1}{2\pi i} \int_{C_1+C_2} \frac{f(z)}{1-z} \frac{1}{z^{n+1}} dz. \tag{2.}$$

Wir beweisen zuerst die Gleichung

$$\lim_{n=\infty} \int_{C_1} \frac{f(z)}{(1-z) z^{n+1}} dz = 0. \tag{3.}$$

*) Ein Bogen des Einheitskreises heiße ein *Regularitätsbogen* der Funktion $f(z)$, wenn in allen seinen Punkten (die Endpunkte inbegriffen) die Funktion regulär ist.

Durch partielle Integration erhält man

$$(4.)\quad \int\limits_{C_1} \frac{f(z)}{(1-z)z^{n+1}}\,dz = \frac{1}{n}\left(\frac{f(z_2)}{(1-z_2)z_2^n} - \frac{f(z_1)}{(1-z_1)z_1^n} + \int\limits_{C_1} \frac{d}{dz}\left(\frac{f(z)}{1-z}\right)\frac{1}{z^n}\,dz\right).$$

Der absolute Betrag des Klammerausdruckes bleibt für jedes n unter einer endlichen Schranke, folglich konvergiert (4.) mit wachsendem n wirklich gegen Null.

Wir wollen nun auch die Gleichung

$$(5.)\quad \lim_{n=\infty} \int\limits_{C_2} \frac{f(z)}{(1-z)z^{n+1}}\,dz = 0$$

begründen. Dazu verwandeln wir dieses Integral in eine unendliche Reihe von Integralen, die über einen Bogen des Einheitskreises erstreckt sind, und schätzen dann diese Reihe mittels der Bedingung $\lim\limits_{n=\infty} a_n = 0$ ab.

Man kann nämlich immer eine lineare Funktion $c_0 + c_1 z$ finden derart, daß die Funktion

$$H(z) = \frac{1}{1-z} + c_0 + c_1 z$$

an den Stellen z_1 und z_2 verschwindet. Es besteht ferner die Gleichung

$$(6.)\quad \frac{1}{2\pi i} \int\limits_{C_1+C_2} \frac{f(z)(c_0 + c_1 z)}{z^{n+1}}\,dz = a_{n-1}c_1 + a_n c_0, \qquad (n \geqq 1)$$

wo die rechte Seite infolge von (1.) mit wachsendem n gegen Null konvergiert. Andererseits folgert man, daß ebenso wie (3.) auch

$$(7.)\quad \lim_{n=\infty} \frac{1}{2\pi i} \int\limits_{C_1} \frac{f(z)(c_0 + c_1 z)}{z^{n+1}}\,dz = 0$$

ist. Durch Subtraktion der Gleichung (7.) von (6.) ergibt sich, daß ebenfalls

$$(8.)\quad \lim_{n=\infty} \frac{1}{2\pi i} \int\limits_{C_2} \frac{f(z)(c_0 + c_1 z)}{z^{n+1}}\,dz = \lim_{n=\infty} \int\limits_{C_2} \frac{f(z)(c_0 + c_1 z)}{z^{n+1}}\,dz = 0$$

ausfällt. Zeigen wir also, daß

$$(9.)\quad \lim_{n=\infty} \int\limits_{C_2} \frac{f(z)}{z^{n+1}}\left(\frac{1}{1-z} + c_0 + c_1 z\right)dz = \lim_{n=\infty} \int\limits_{C_2} \frac{f(z)H(z)}{z^{n+1}}\,dz = 0$$

ist, so wird damit auch (5.) bewiesen.

Wir betrachten nun die Potenzreihe

$$F(z) = \sum_{\varkappa=0}^{\infty} \frac{a_\varkappa}{(\varkappa+1)(\varkappa+2)} z^{\varkappa+2}.$$

Diese Reihe konvergiert im Innern des Einheitskreises und am Rande unbedingt und *gleichmäßig.* Ferner besteht im Innern des Kreises und auch an den Stellen z_1 und z_2 die Gleichung

$$F''(z) = f(z).$$

Man erhält also, da $H(z)$ an den Stellen z_1 und z_2 verschwindet, durch wiederholte partielle Integration die Umformung*)

$$(10.)\quad \int_{C_2} f(z)\frac{H(z)}{z^{n+1}}dz = \int_{C_2} F(z)\frac{d^2}{dz^2}\left(\frac{H(z)}{z^{n+1}}\right)dz - \left[F(z)\frac{d}{dz}\left(\frac{H(z)}{z^{n+1}}\right)\right]_{z_2}^{z_1}$$

$$= \sum_{\varkappa=0}^{\infty} a_\varkappa\left(\int_{C_2}\frac{z^{\varkappa+2}}{(\varkappa+1)(\varkappa+2)}\frac{d^2}{dz^2}\left(\frac{H(z)}{z^{n+1}}\right)dz - \left[\frac{z^{\varkappa+2}}{(\varkappa+1)(\varkappa+2)}\frac{d}{dz}\left(\frac{H(z)}{z^{n+1}}\right)\right]_{z_2}^{z_1}\right)$$

$$= \sum_{\varkappa=0}^{\infty} a_\varkappa \int_{C_2} z^\varkappa \frac{H(z)}{z^{n+1}}dz = \sum_{\varkappa=0}^{\infty} a_\varkappa \int_{C_2} z^{\varkappa-n-1}H(z)\,dz.$$

Statt die einzelnen Integrationen längs C_2 auszuführen, kann man auch nach dem *Cauchy*schen Satze auf jenem Bogen $(z_1 z_2)$ des Einheitskreises, der die Stelle $z = 1$ nicht enthält, integrieren, d. h.

$$(11.)\quad \int_{C_2} f(z)\frac{H(z)}{z^{n+1}}dz = \sum_{\varkappa=0}^{\infty} a_\varkappa \int_{z_1}^{z_2} z^{\varkappa-n-1}H(z)\,dz.$$

Damit sind wir aber gleich am Ziele. Denn wir haben zunächst für alle Werte von $\varkappa$ die Ungleichung

$$(12.)\quad \left|\int_{z_1}^{z_2} z^{\varkappa-n-1}H(z)\,dz\right| < G$$

und durch zweimalige partielle Integration die Ungleichung

$$(13.)\quad \left|\int_{z_1}^{z_2} z^{\varkappa-n-1}H(z)\,dz\right| = \left|\int_{z_1}^{z_2}\frac{z^{\varkappa-n}}{\varkappa-n}H'(z)\,dz\right|$$

$$= \left|\left[\frac{z^{\varkappa-n+1}}{(\varkappa-n)(\varkappa-n+1)}H'(z)\right]_{z_1}^{z_2} - \int_{z_1}^{z_2}\frac{z^{\varkappa-n+1}}{(\varkappa-n)(\varkappa-n+1)}H''(z)\,dz\right| < \frac{G}{|\varkappa-n|\,|\varkappa-n+1|},$$

wo G eine nur von den Stellen z_1 und z_2 abhängige, also von n und $\varkappa$ unabhängige Zahl bedeutet. Für $\varkappa = n-1$ und $\varkappa = n$ wenden wir die erste, für alle übrigen Zahlen die zweite Ungleichung an.

Wie klein auch die positive Größe ε gewählt sei, können wir laut (1.) eine Zahl m so angeben, daß für jedes $\varkappa > m$

$$(14.)\quad |a_\varkappa| < \varepsilon$$

ist. Ist dann noch

$$n - 2 > m,$$

so haben wir infolge der Ungleichungen (12.), (13.) und (14.) die neue Ungleichung

$$\left|\sum_{\varkappa=0}^{\infty} a_\varkappa \int_{z_1}^{z_2} z^{\varkappa-n-1}H(z)\,dz\right| < G\left(\sum_{\varkappa=0}^{m}\frac{|a_\varkappa|}{(\varkappa-n)(\varkappa-n+1)} + \varepsilon\sum_{\varkappa=m+1}^{n-2}\frac{1}{(\varkappa-n)(\varkappa-n+1)}\right.$$

*) Dieses Verfahren kann man in verschiedenen Fragen über trigonometrische Reihen nutzbar machen. Es kann z. B. zum Beweise des von *Fatou* verwendeten *Riemann*schen Satzes dienen.

$$+ 2\varepsilon + \varepsilon \sum_{\varkappa=n+1}^{\infty} \frac{1}{(\varkappa - n)(\varkappa - n + 1)}\Big).$$

Die erste Summe ist aber

$$\leqq \frac{\sum_{\varkappa=0}^{m} |a_\varkappa|}{(n-m)(n-m+1)}$$

und wird mithin, wenn n genügend groß gewählt wird, auch $< \varepsilon$. Die beiden anderen Summen sind resp. $\leqq 1$. Also wird

$$(15.)\qquad \left| \int_{C_2} \frac{f(z)\,H(z)}{z^{n+1}} dz \right| = \left| \sum_{\varkappa=0}^{\infty} a_\varkappa \int_{z_1}^{z_2} z^{\varkappa-n-1} H(z)\,dz \right| < 5G\varepsilon,$$

wenn nur n eine gewisse, von ε abhängige Größe überschreitet. In dieser Ungleichung liegt eben der Beweis von (5.). Aus (2.), (3.) und (5.) folgt dann unmittelbar

$$\lim_{n=\infty} (a_0 + a_1 + \cdots + a_n) = f(1).$$

Wir wollen noch die Gleichmäßigkeit der Konvergenz auf einem beliebigen Regularitätsbogen beweisen. Es sei $(z_1' z_2')$ ein Bogen, der im Bogen $(z_1 z_2)$ enthalten ist. Es genügt dann zu zeigen, daß die Reihe an den Stellen des Bogens $(z_1' z_2')$ gleichmäßig konvergiert, da jeder Regularitätsbogen in einem größeren enthalten ist.

Es bedeute nun x eine beliebige Stelle des Bogens $(z_1' z_2')$. Wir haben dann die Gleichung

$$\frac{1}{2\pi i} \int_{C_1+C_2} \frac{f(z)}{(x-z)} \left(\frac{x}{z}\right)^{n+1} dz = a_0 + a_1 x + \cdots + a_n x^n - f(x).$$

Das Integral über C_1 konvergiert, wie leicht ersichtlich, mit wachsendem n gleichmäßig gegen Null für jede Stelle x des Bogens $(z_1' z_2')$. Ferner bestimmen wir die lineare Funktion $c_0 + c_1 z$ so, daß

$$H(z) = \frac{1}{x-z} + c_0 + c_1 z$$

an den Stellen z_1 und z_2 verschwindet. Man hat die Werte

$$c_0 = \frac{z_1 + z_2 - x}{(z_1 - x)(z_2 - x)}, \quad c_1 = \frac{1}{(z_1 - x)(z_2 - x)},$$

die zwischen endlichen Grenzen bleiben, wenn x sich auf dem Bogen $(z_1' z_2')$ bewegt. Folglich bleibt die der Zahl G [in (12.) und (13.)] entsprechende Zahl auch zwischen endlichen Grenzen, womit unsere Behauptung über die Gleichmäßigkeit der Konvergenz völlig bewiesen ist.

Satz II. Genügen die Koeffizienten einer Potenzreihe

$$(16.)\qquad f(z) = a_0 + a_1 z + \cdots + a_n z^n + \cdots$$

der Bedingung

(17.) $$\lim_{n=\infty} \frac{a_n}{n^\gamma} = 0 \quad (\gamma \geqq 0),$$

so ist die Reihe an jeder regulären Stelle x des Einheitskreises durch arithmetische Mittel γ-ter Ordnung summierbar und zwar gleichmäßig an den Stellen eines Regularitätsbogens. Die Bedingung (17.) *ist auch notwendig, damit auf dem Einheitskreise wenigstens eine Stelle liegt, wo die Reihe durch arithmetische Mittel γ-ter Ordnung summierbar ist.*

Der Beweis der Notwendigkeit der Bedingung (17.) ist sehr leicht zu führen, so daß wir auf ihn gar nicht eingehen. Daß diese Bedingung auch hinreichend ist, kann, wenn γ eine ganze Zahl ist, vollständig ähnlich bewiesen werden, wie oben die Konvergenz. Der einzige Unterschied besteht in der Bildung von $H(z)$, indem man zu $\frac{1}{x-z}$ statt eines linearen Ausdruckes ein Polynom $(2\gamma+1)$-ter Ordnung addiert, womit man bewirken kann, daß an den Stellen z_1 und z_2 nicht nur $H(z)$, sondern auch seine Ableitungen bis zur γ-ten Ordnung verschwinden.

Ist aber γ keine ganze Zahl, so darf man, wie aus der Formel (22.) unmittelbar ersichtlich, nur über solche Kurven integrieren, welche die Stelle $z = x$ nicht einschließen. Ein Teil unseres obigen Beweises muß also durch einen neuen Beweisgang ersetzt werden, welcher aber uns gleich erlaubt, statt Satz II. den folgenden allgemeineren Satz zu begründen.

Satz III. Es sei

(18.) $$f(z) = a_0 + a_1 z + \cdots + a_n z^n + \cdots$$

eine Potenzreihe, deren Koeffizienten der Bedingung

(19.) $$\lim_{n=\infty} \frac{a_n}{n^\gamma} = 0 \quad (\gamma > 0)\ ^{*})$$

---

*) Bekanntlich wurde die Tatsache, daß unter gewissen Stetigkeitsbedingungen die ersten arithmetischen Mittel der *Fourier*schen Partialsummen einer Funktion konvergieren, von *Fejér* erkannt. Unser obiges Ergebnis, auf trigonometrische Reihen übertragen, kann auch als eine Erweiterung des *Fejér*schen Resultates erkannt werden.

Im Falle $\gamma = 0$, also im Falle der gewöhnlichen Konvergenz, wurde die Frage nach der Richtigkeit unseres Satzes III schon von *Pringsheim* gestellt. *Fejér* gab darauf eine negative Antwort, indem er eine im Innern und am Rande des Einheitskreises beschränkte und stetige Funktion angab, deren *Maclaurin*sche Entwicklung auf dem Rande in einer überall dichten Punktmenge divergiert (Über gewisse Potenzreihen an der Konvergenzgrenze. Sitzungsber. d. math.-phys. Kl. d. Kgl. Bayr. Akad. 1910). Will man bei $\gamma = 0$ die Regularität der betreffenden Stelle fallen lassen und dennoch die Konvergenz sichern, so muß man statt Stetigkeit vielmehr solche Einschränkungen treffen, wie in der Konvergenztheorie der *Fourier*schen Reihen.

genügen. Es sei ferner $(z_1 z_2)$ ein Bogen des Einheitskreises von der Beschaffenheit, daß die Funktion $f(z)$ gleichmäßig einer Grenzfunktion zustrebt, wenn z sich vom Innern des Kreises längs der entsprechenden Radien den Stellen des Bogens $(z_1 z_2)$ nähert. Die Reihe (18.) ist unter diesen Bedingungen an jeder inneren Stelle des Bogens $(z_1 z_2)$ durch arithmetische Mittel γ-ter Ordnung summierbar und zwar gleichmäßig auf jedem Bogen $(z_1' z_2')$, der in $(z_1 z_2)$ enthalten ist.

Der Einfachheit halber setzen wir wieder voraus, daß $z = 1$ ein innerer Punkt des Bogens $(z_1 z_2)$ sei, und beweisen dann, daß die Reihe

$$a_0 + a_1 + \cdots + a_n + \cdots$$

durch die arithmetischen Mittel γ-ter Ordnung summierbar ist.

Die Definition dieser Mittel lautet bekanntlich nach *Cesàro* folgendermaßen: Man setze

$$\left(\frac{1}{1-y}\right)^{\gamma+1} = \sum_{n=0}^{\infty} C_n^{(\gamma)} y^n$$

und

$$\left(\frac{1}{1-y}\right)^{\gamma+1} \sum_{n=0}^{\infty} a_n y^n = \sum_{n=0}^{\infty} S_n^{(\gamma)} y^n \tag{20.}$$

und bilde ferner den Grenzwert

$$\lim_{n=\infty} \frac{S_n^{(\gamma)}}{C_n^{(\gamma)}}. \tag{21.}$$

Existiert dieser Grenzwert, so nennen wir die Reihe

$$a_0 + a_1 + \cdots + a_n + \cdots$$

durch arithmetische Mittel γ-ter Ordnung summierbar und bezeichnen (21.) als ihre Summe*).

Es sei C_1' ein Kurvenzug, der die Punkte z_1 und z_2 im Innern des Einheitskreises verbindend in der Nachbarschaft des Kreisbogens $(z_2\ 1\ z_1)$ verläuft. Der Kurvenzug C_2 sei wieder wie oben definiert. Dann haben wir infolge der Formel (20.)

$$\frac{S_n^{(\gamma)}}{C_n^{(\gamma)}} = \frac{1}{C_n^{(\gamma)} 2\pi i} \int_{C_1'+C_2} \frac{f(z)}{(1-z)^{\gamma+1} z^{n+1}} dz. \tag{22.}$$

Jetzt beweisen wir zuerst, daß

*) Für nicht ganze Ordnungszahl tritt meines Wissens diese Definition zuerst bei *Hadamard* auf (Essai sur l'étude des fonctions données par leur développement de *Taylor*. Thèse, Paris 1892, p. 82). Vgl. auch: *Knopp*, Grenzwerte von Reihen bei der Annäherung an die Konvergenzgrenze. Inauguraldiss. (Berlin 1907), S. 17 u. ff.

$$(23.)\qquad \lim_{n=\infty} \frac{1}{C_n^{(\gamma)}} \int_{C_2} \frac{f(z)}{(1-z)^{\gamma+1} z^{n+1}} dz = 0$$

ist. Der Gedankengang ist mit dem früheren im wesentlichen identisch. Es sei z. B.

$$(24.)\qquad 0 < \gamma \leqq 1.$$

Wir bestimmen dann ein Polynom dritten Grades

$$P(z) = c_0 + c_1 z + c_2 z^2 + c_3 z^3$$

derart, daß die Funktion

$$H(z) = \frac{1}{(1-z)^{\gamma+1}} + c_0 + c_1 z + c_2 z^2 + c_3 z^3$$

in den Punkten z_1 und z_2 samt ihrer ersten Ableitung verschwindet. Diese Bedingung ergibt für die Werte c vier lineare Gleichungen, die, wie leicht ersichtlich, einander nicht widersprechen und die Werte c eindeutig bestimmen. Man beweist dann genau wie vorher, daß

$$(25.)\qquad \lim_{n=\infty} \frac{1}{C_n^{(\gamma)}} \int_{C_2} \frac{f(z)}{z^{n+1}} (c_0 + c_1 z + c_2 z^2 + c_3 z^3)\, dz = 0$$

ausfällt. Zeigen wir also noch, daß

$$(26.)\qquad \lim_{n=\infty} \frac{1}{C_n^{(\gamma)}} \int_{C_2} \frac{f(z)}{z^{n+1}} H(z)\, dz = 0$$

ist, so wird damit infolge von (24.) auch (23.) bewiesen.

Wir können aber mittels der Reihe

$$F(z) = \sum_{\varkappa=0}^{\infty} a_\varkappa \frac{z^{\varkappa+3}}{(\varkappa+1)(\varkappa+2)(\varkappa+3)},$$

die laut (19.) und (24.) im Innern und am Rande des Einheitskreises gleichmäßig konvergiert, wieder die Umformung

$$(27.)\qquad \int_{C_2} f(z) \frac{H(z)}{z^{n+1}} dz = \sum_{\varkappa=0}^{\infty} a_\varkappa \int_{C_2} z^\varkappa \frac{H(z)}{z^{n+1}} dz = \sum_{\varkappa=0}^{\infty} a_\varkappa \int_{C_2} z^{\varkappa-n-1} H(z)\, dz$$

rechtfertigen. In diesen Integrationen kann aber der Kurvenzug C_2 wieder durch jenen Kreisbogen $(z_1 z_2)$, der den Punkt $z = 1$ nicht enthält, ersetzt werden, d. h.

$$(28.)\qquad \int_{C_2} f(z) \frac{H(z)}{z^{n+1}} dz = \sum_{\varkappa=0}^{\infty} a_\varkappa \int_{z_1}^{z_2} z^{\varkappa-n-1} H(z)\, dz.$$

Wir haben wieder zunächst

$$(29.)\qquad \left| \int_{z_1}^{z_2} z^{\varkappa-n-1} H(z)\, dz \right| < G,$$

ferner durch dreimalige partielle Integration

$$(30.)\quad \left|\int_{z_1}^{z_2} z^{\varkappa-n-1} H(z)\,dz\right| < \frac{G}{|\varkappa-n|\,|\varkappa-n+1|\,|\varkappa-n+2|},$$

wo G wieder von $\varkappa$ und n unabhängig ist. Die erste Ungleichung wenden wir für $\varkappa = n-2$, $\varkappa = n-1$ und $\varkappa = n$ an, die zweite für alle übrigen Zahlen.

Wir bestimmen nun die positive ganze Zahl m so, daß für jedes $\varkappa > m$

$$(31.)\quad \left|\frac{a_\varkappa}{\varkappa^\gamma}\right| < \varepsilon,$$

wo ε eine beliebig klein gewählte positive Größe ist. Wenn noch

$$n-3 > m,$$

so haben wir zufolge der Ungleichungen (29.), (30.) und (31.) die neue Ungleichung

$$(32.)\quad \left|\frac{1}{C_n^{(\gamma)}}\sum_{\varkappa=0}^{\infty} a_\varkappa \int_{z_1}^{z_2} z^{\varkappa-n-1} H(z)\,dz\right| < \frac{G}{C_n^{(\gamma)}}\Bigg(\sum_{\varkappa=0}^{m}\frac{|a_\varkappa|}{(n-\varkappa)(n-1-\varkappa)(n-2-\varkappa)}$$
$$+\ \varepsilon n^\gamma \sum_{\varkappa=m+1}^{n-3}\frac{1}{(n-\varkappa)(n-1-\varkappa)(n-2-\varkappa)}$$
$$+\ 3\varepsilon n^\gamma + \varepsilon \sum_{\varkappa=n+1}^{2n}\frac{\varkappa^\gamma}{(\varkappa-n)(\varkappa-n+1)(\varkappa-n+2)}$$
$$+\ \varepsilon \sum_{\varkappa=2n+1}^{\infty}\frac{\varkappa^\gamma}{(\varkappa-n)(\varkappa-n+1)(\varkappa-n+2)}\Bigg).$$

Die erste Summe wird aber, da sie aus einer endlichen Anzahl von Gliedern besteht, mit genügend großem n kleiner als ε. Die zweite Summe ist kleiner als $\frac{1}{4}$. Die dritte Summe ist kleiner als $\frac{(2n)^\gamma}{4}$. Die vierte Summe ist, da $\gamma \leqq 1$ und mithin in ihr $\frac{\varkappa^\gamma}{(\varkappa-n+2)} < 2$ ausfällt, kleiner als

$$2\sum_{\varkappa=2n+1}^{\infty}\frac{1}{(\varkappa-n)(\varkappa-n+1)} < 2.$$

Also wird endlich bei genügend großem n

$$(33.)\quad \frac{1}{C_n^{(\gamma)}}\left|\int_{C_2}\frac{f(z)\,H(z)}{z^{n+1}}\,dz\right| = \left|\frac{1}{C_n^{(\gamma)}}\sum_{\varkappa=0}^{\infty} a_\varkappa \int_{z_1}^{z_2} z^{\varkappa-n-1} H(z)\,dz\right|$$
$$< \frac{G}{C_n^{(\gamma)}}\left(\varepsilon + \frac{\varepsilon n^\gamma}{4} + 3\varepsilon n^\gamma + \varepsilon n^\gamma + 2\varepsilon\right).$$

Da ferner

$$\lim_{n=\infty}\frac{n^\gamma}{C_n^{(\gamma)}} = \Gamma(\gamma+1)$$

eine endliche Zahl ist, so wird der letzte Ausdruck bei genügend großem n kleiner als eine beliebig klein vorgeschriebene positive Zahl, womit (23.) vollständig bewiesen ist.

Wir müssen noch die Gleichung

$$\lim_{n=\infty} \frac{1}{C_n^{(\gamma)} 2\pi i} \int\limits_{C_1'} \frac{f(z)}{(1-z)^{\gamma+1} z^{n+1}} dz = f(1) \tag{34.}$$

darlegen. Dazu bestimmen wir auf dem Bogen $(z_2\,1)$ den Punkt ξ_2 und auf dem Bogen $(1\,z_1)$ den Punkt ξ_1 derart, daß auf dem Rande*) und im Innern des Einheitskreises

$$|f(z) - f(1)| < \varepsilon, \tag{35.}$$

sobald

$$|z-1| < |\xi_1 - 1| = |\xi_2 - 1|,$$

wo ε eine beliebige positive Größe bedeutet. Man kann ferner bei genügend großem n auf dem Bogen $(\xi_2\,1)$ bzw. auf dem Bogen $(1\,\xi_1)$ die Punkte ν_2 und ν_1 so bestimmen, daß

$$|\nu_1 - 1| = |\nu_2 - 1| = \frac{1}{n}. \tag{36.}$$

Wir beschreiben dann um den Mittelpunkt $z=1$ mit dem Radius $\frac{1}{n}$ einen Kreisbogen B, der im Innern des Einheitskreises die Punkte $z = \nu_1$ und $z = \nu_2$ verbindet, und bezeichnen den aus den Kreisbögen $(z_2 \xi_2 \nu_2)$, B und $(\nu_1 \xi_1 z_1)$ gebildeten Kurvenzug mit C_1''. Offenbar kann das Integral über C_1' durch das Integral über C_1'' ersetzt werden. D. h.

$$\begin{aligned} \frac{1}{C_n^{(\gamma)} 2\pi i} \int\limits_{C_1'} \frac{f(z)}{(1-z)^{\gamma+1} z^{n+1}} dz &= \frac{1}{C_n^{(\gamma)} 2\pi i} \int\limits_{C_1''} \frac{f(z)}{(1-z)^{\gamma+1} z^{n+1}} dz \\ &= \frac{1}{C_n^{(\gamma)} 2\pi i} \int\limits_{C_1''} \frac{f(z)-f(1)}{(1-z)^{\gamma+1} z^{n+1}} dz + \frac{f(1)}{C_n^{(\gamma)} 2\pi i} \int\limits_{C_1''} \frac{1}{(1-z)^{\gamma+1} z^{n+1}} dz \\ &= \frac{f(1)}{C_n^{(\gamma)} 2\pi i} \int\limits_{C_1''} \frac{1}{(1-z)^{\gamma+1} z^{n+1}} dz + \frac{1}{C_n^{(\gamma)} 2\pi i} \int\limits_{z_2}^{\nu_2} \frac{f(z)-f(1)}{(1-z)^{\gamma+1} z^{n+1}} dz \\ &+ \frac{1}{C_n^{(\gamma)} 2\pi i} \int\limits_{B} \frac{f(z)-f(1)}{(1-z)^{\gamma+1} z^{n+1}} dz + \frac{1}{C_n^{(\gamma)} 2\pi i} \int\limits_{\nu_1}^{z_1} \frac{f(z)-f(1)}{(1-z)^{\gamma+1} z^{n+1}} dz. \end{aligned} \tag{37.}$$

Es ist aber

$$\begin{aligned} \frac{1}{C_n^{(\gamma)} 2\pi i} \int\limits_{C_1''} \frac{1}{(1-z)^{\gamma+1} z^{n+1}} dz &= \frac{1}{C_n^{(\gamma)} 2\pi i} \int\limits_{C_1''+C_2} \frac{1}{(1-z)^{\gamma+1} z^{n+1}} dz \\ &- \frac{1}{C_n^{(\gamma)} 2\pi i} \int\limits_{C_2} \frac{1}{(1-z)^{\gamma+1} z^{n+1}} dz = 1 - \frac{1}{C_n^{(\gamma)} 2\pi i} \int\limits_{C_2} \frac{1}{(1-z)^{\gamma+1} z^{n+1}} dz. \end{aligned}$$

Der letzte Ausdruck konvergiert, wie leicht ersichtlich, mit wachsendem n gegen Null, mithin ist

*) Wir bezeichnen auch die Randfunktion, wo sie existiert, mit $f(z)$.

$$(38.)\qquad \lim_{n=\infty} \frac{f(1)}{C_n^{(\gamma)} 2\pi i} \int_{C_1''} \frac{1}{(1-z)^{\gamma+1} z^{n+1}} dz = f(1).$$

Andererseits ist

$$\left| \frac{1}{C_n^{(\gamma)} 2\pi i} \int_{z_2}^{\nu_2} \frac{f(z)-f(1)}{(1-z)^{\gamma+1} z^{n+1}} dz \right| \leqq \left| \frac{1}{C_n^{(\gamma)} 2\pi i} \int_{z_2}^{\xi_2} \frac{f(z)-f(1)}{(1-z)^{\gamma+1} z^{n+1}} dz \right|$$
$$+ \left| \frac{1}{C_n^{(\gamma)} 2\pi i} \int_{\xi_2}^{\nu_2} \frac{f(z)-f(1)}{(1-z)^{\gamma+1} z^{n+1}} dz \right|.$$

Ferner hat man, da die Integrationsgrenzen von n unabhängig sind,

$$\lim_{n=\infty} \left| \frac{1}{C_n^{(\gamma)} 2\pi i} \int_{z_2}^{\xi_2} \frac{f(z)-f(1)}{(1-z)^{\gamma+1} z^{n+1}} dz \right| = 0.$$

Um das zweite Integral abzuschätzen, setzen wir

$$z = e^{i\varphi},\ \xi_2 = e^{i\tau},\ \nu_2 = e^{i\theta}.$$

Dann ist

$$|1-z| = \left|2 \sin \frac{\varphi}{2}\right| > \left|\frac{\varphi}{2}\right|,$$

also infolge von (35.)

$$(40.)\qquad \left| \frac{1}{C_n^{(\gamma)} 2\pi i} \int_{\xi_2}^{\nu_2} \frac{f(z)-f(1)}{(1-z)^{\gamma+1} z^{n+1}} dz \right| < \frac{\varepsilon}{2\, C_n^{(\gamma)} \pi} \int_{\tau}^{\theta} \frac{d\varphi}{\left|\frac{\varphi}{2}\right|^{\gamma+1}}$$
$$< \frac{\varepsilon 2^{\gamma+1}}{\gamma 2 C_n^{(\gamma)} \pi} \frac{1}{|\theta|^{\gamma}} < \frac{\varepsilon 2^{\gamma+1}}{2\gamma\, C_n^{(\gamma)} \pi} n^{\gamma} < g\varepsilon,$$

wo g eine von n unabhängige endliche positive Zahl bedeutet. Ähnliche Abschätzungen wie (39.) und (40.) bestehen auch für das von ν_1 bis z_1 erstreckte Integral.

Endlich ist

$$\left| \frac{1}{C_n^{(\gamma)} 2\pi i} \int_{B} \frac{f(z)-f(1)}{(1-z)^{\gamma+1} z^{n+1}} dz \right| < \frac{\varepsilon}{C_n^{(\gamma)} 2\pi} n^{\gamma+1} \frac{1}{\left(1-\frac{1}{n}\right)^{n+1}} \frac{2\pi}{n}$$
$$< \frac{8\varepsilon n^{\gamma+1} 2\pi}{C_n^{(\gamma)} 2\pi n} = \frac{8\varepsilon n^{\gamma}}{C_n^{(\gamma)}} < 8g\varepsilon.$$

Diese Abschätzungen und die Gleichungen (37.) und (38.) ergeben aber die Richtigkeit von (34).

Die in unserem Satze behauptete *gleichmäßige Summierbarkeit* beruht auf der Tatsache, daß jede in einer abgeschlossenen Punktmenge stetige Funktion daselbst *gleichmäßig stetig* ist.

Eine leichte Umformung unseres Beweises ergibt statt Satz III einen allgemeineren Satz, der sich in meiner zitierten „Comptes Rendus"-Mitteilung findet.

11.

Über summierbare trigonometrische Reihen

Math. Ann. *71* (1911), 54–75

Inhaltsübersicht.

Einleitung.*)

Aus den Arbeiten von Fejér ist es wohl bekannt, welche hervorragende Rolle die Mittelwerte der Teilsummen in verschiedenen Fragen der Fourierschen Reihen spielen.

Riemann stellte der Theorie der Fourierschen Reihen die Theorie der durch ihre Koeffizienten formal gegebenen, trigonometrischen Reihen an die Seite. Die Konvergenztheorie dieser Reihen knüpft an die Namen Riemann, Cantor und Du Bois-Reymond an.

Die Fejérschen Resultate über Fouriersche Reihen legen es nahe zu untersuchen, inwieweit die Sätze der oben genannten Forscher ihre Gültigkeit behalten, wenn man in der Theorie der formal gegebenen trigonometrischen Reihen statt Konvergenz nur die Summierbarkeit durch Mittelwerte der Teilsummen voraussetzt.

Es handelt sich zunächst um die Eindeutigkeitsfrage: *Kann eine Funktion durch zwei verschiedene summierbare trigonometrische Reihen dargestellt werden?*

---

*) Diese Arbeit ist ein wesentlich umgearbeiteter und erweiterter Teil meiner Inauguraldissertation: Összegezhető trigonometrikus sorok és összegezhető hatványsorok, Budapest 1908. Ein kurzer Bericht über die Resultate erschien in den Paris C. R.: Sur les séries trigonométriques (7 octobre 1907).

Wie bekannt, wurde die Eindeutigkeitsfrage der konvergenten trigonometrischen Reihen durch den Cantorschen Satz erledigt, der sich im wesentlichen auf zwei Sätze von Riemann und einen Satz von Schwarz stützt.

Die Frage nach dem Analogon des Cantorschen Satzes tritt zuerst in der Dissertation von Fejér auf. Er stellt sich hier der Frage skeptisch gegenüber, da doch die bekannte Reihe

$$\frac{1}{2} + \cos x + \cdots + \cos nx + \cdots$$

an jeder Stelle $x \not\equiv 0 \pmod{2\pi}$ durch sein Verfahren summierbar ist und den Wert Null hat. In seiner Annalenarbeit gibt er jedoch schon ein tiefreichendes Analogon des Riemannschen Satzes, wodurch die Lösung des Problems angebahnt wurde.

Eine diesbezügliche Unterhaltung mit Herrn Fejér, im Sommer 1907, gab mir die Veranlassung, das Problem aufzunehmen. Ich habe bewiesen, daß tatsächlich im allgemeinen *Eindeutigkeit* der Entwicklung stattfindet. Mein Beweis stützt sich nebst dem obengenannten Fejérschen Satze auf einen neuen Mittelwertsatz, aus dem dann ein Analogon des Schwarzschen Satzes folgt. Um diese Sätze anwenden zu können, mußte ich immerhin die untersuchten trigonometrischen Reihen einer gewissen *Einschränkung* unterwerfen.

Die Sonderstellung des obigen Fejérschen Beispieles erklärt sich dann dadurch, daß in unserem Falle, im Gegensatz zu dem ursprünglichen Cantorschen Theorem, *Ausnahmestellen* im allgemeinen *nicht* zugelassen werden können.

Ähnlich, wie der Cantorsche, läßt sich auch der Du Bois-Reymondsche Satz auf summierbare trigonometrische Reihen übertragen.

Wir werden am Ende unserer Arbeit auf arithmetische Mittel beliebiger — ganzzahliger oder nicht ganzzahliger — Ordnung eingehen. Dabei wird sich die Ordnungszahl 1 in gewisser Hinsicht als Grenzfall darbieten. Während nämlich die zwei genannten klassischen Sätze sich übertragen lassen, wenn die Ordnungszahl der Summation $k \leqq 1$ ist (für $k = 1$ mit der angedeuteten Einschränkung), ist dies bei $k > 1$ durchaus nicht der Fall.

Die Übereinstimmungen bezw. Gegensätze, die sich ergeben, bieten auch für die ursprünglichen Sätze manche neue Gesichtspunkte.

§ 1.

Die Sätze von Cantor, Riemann, Schwarz, Du Bois-Reymond und Hölder.

Der Satz von Cantor*) lautet:

Ist eine trigonometrische Reihe mit eventueller Ausnahme einer reduktiblen Punktmenge im Intervall $(0, 2\pi)$ *konvergent und hat daselbst den Wert Null, so sind alle ihre Koeffizienten gleich Null.*

Die Riemannschen Sätze**), auf welche sich der Cantorsche Beweis stützt, können folgendermaßen ausgesprochen werden:

Die Koeffizienten der trigonometrischen Reihe

$$(1)\qquad a_0 + a_1 \cos x + b_1 \sin x + \cdots + a_n \cos nx + b_n \sin nx + \cdots$$

mögen mit wachsendem n gegen Null konvergieren. Durch eine zweimalige gliedweise Integration entsteht aus der Reihe (1) die gleichmäßig konvergente Reihe

$$F(x) = \frac{a_0 x^2}{2} - \sum_{n=1}^{\infty} \frac{a_n \cos nx + b_n \sin nx}{n^2},$$

die natürlich eine stetige Funktion darstellt. Der erste Riemannsche Satz besagt nun:

Ist die Reihe (1) *an einer Stelle* x_0 *konvergent und hat daselbst den Wert* $f(x_0)$, *so existiert der Grenzwert*

$$\lim_{h=0} \frac{F(x_0-h) - 2F(x_0) + F(x_0+h)}{h^2}$$

und ist gleich $f(x_0)$.

Der zweite Riemannsche Satz besagt:

An jeder Stelle x *ist*

$$\lim_{h=0} \frac{F(x-h) - 2F(x) + F(x+h)}{h} = 0.$$

Dieser letzte Satz gründet sich allein auf die Voraussetzung, daß $\lim_{n=\infty} a_n = \lim_{n=\infty} b_n = 0$ ist und gilt völlig unabhängig von der Beschaffenheit der betreffenden Stelle.

---

*) Cantor, Beweis, daß eine für jeden reellen Wert von x durch eine trigonometrische Reihe gegebene Funktion $f(x)$ sich nur auf eine einzige Weise in dieser Form darstellen läßt, J. f. Math. 72 (1870); Über die Ausdehnung eines Satzes aus der Theorie der trigonometrischen Reihen, Math. Ann. 5 (1872).

**) Riemann, Über die Darstellbarkeit einer Funktion durch eine trigonometrische Reihe, Gesammelte Werke (1892), S. 227.

Der Satz von Schwarz lautet:

Verschwindet die verallgemeinerte zweite Ableitung

$$\lim_{h=0} \frac{F(x+h) - 2F(x) + F(x-h)}{h^2}$$

der stetigen Funktion $F(x)$ *in einem Intervalle überall, so ist die Funktion in diesem Intervalle linear.*

Cantor gebraucht bei dem Beweise seines Satzes noch das folgende Lemma*):

Ist in jedem Punkte eines Intervalles

$$\lim_{n=\infty} (a_n \cos nx + b_n \sin nx) = 0,$$

so ist auch

$$\lim_{n=\infty} a_n = \lim_{n=\infty} b_n = 0.$$

Endlich sprechen wir den Du Bois-Reymondschen**) Satz, der den Cantorschen umfaßt, gleich in der schönen Fassung von Lebesgue***) aus, die erst durch die Einführung seines *Integralbegriffes* ermöglicht wurde:

Jede trigonometrische Reihe, die mit eventueller Ausnahme einer reduktiblen Menge von Stellen, im Intervall $(0, 2\pi)$ *konvergiert und daselbst eine beschränkte Funktion darstellt, ist die Fouriersche Reihe dieser Funktion.*

Der Beweis dieses Satzes gründet sich nebst den obigen Riemannschen Sätzen am besten auf den folgenden Satz von Hölder†), der seinerseits wieder den Schwarzschen Satz umfaßt:

---

*) Cantor, Über einen die trigonometrischen Reihen betreffenden Lehrsatz, J. f. Math. 72. Einen viel einfacheren Beweis gibt er in der Arbeit: Über trigonometrische Reihen, Math. Ann. 4 (1871).

Kürzlich hat Osgood einen sehr einfachen Beweis gegeben: On Cantor's theorem concerning the coefficients of a convergent trigonometric series with generalizations, Amer. Math. Soc. Trans. 10. Für den Beweis vgl. auch Lebesgue, Leçons sur les séries trigonométriques, S. 110—111.

Cantor teilt in seiner Notiz zu dem Aufsatze „Beweis, daß eine etc." J. f. Math. 73 (1871), eine mündliche Bemerkung von Kronecker mit, die sehr einfach zeigt, daß das Lemma bei dem Beweise des Eindeutigkeitssatzes entbehrt werden kann. Wie leicht ersichtlich, ist das auch in unseren Untersuchungen der Fall. Wir werden dennoch das Cantorsche Lemma benutzen, denn durch seine Anwendung sieht man am natürlichsten, welchen Einschränkungen die Koeffizienten der summierbaren trigonometrischen Reihen unterliegen.

**) Du Bois-Reymond, Abh. d. bayer. Akad. 12 (1875).

***) Lebesgue, Sur les séries trigonométriques", Ann. éc. norm. 1903. Vgl. auch für all diese Fragen sein vorzügliches Werk: Leçons sur les séries trigonométriques Paris, Gauthier-Villars 1906, S. 110 u. ff.

†) Hölder, Zur Theorie der trigonometrischen Reihen, Math. Ann. 24, S. 183.

Wenn die stetige Funktion $F(x)$ an jeder Stelle eines Intervalles eine verallgemeinerte zweite Ableitung besitzt, deren Werte im Intervall $(x_0 - h, x_0 + h)$ zwischen m und M liegen, so liegt auch der Wert

$$\Delta^2 F(x_0, h) = \frac{F(x_0+h) - 2F(x_0) + F(x_0-h)}{h^2}$$

zwischen m und M.

§ 2.

Die Analoga der Sätze von Riemann, Schwarz und Hölder.

Vorab müssen wir einige Bemerkungen über summierbare Reihen machen.

Gegeben sei die Reihe

$$(2) \qquad c_0 + c_1 + \cdots + c_n + \cdots.$$

Wir setzen

$$s_0 = c_0, \quad s_1 = c_0 + c_1, \cdots, s_n = c_0 + c_1 + \cdots + c_n, \cdots$$

$$\sigma_0 = \frac{s_0}{1}, \quad \sigma_1 = \frac{s_0 + s_1}{2}, \cdots, \sigma_n = \frac{s_0 + s_1 + \cdots + s_n}{n+1}, \cdots.$$

Existiert der Grenzwert

$$(3) \qquad \lim_{n=\infty} \sigma_n = s,$$

so sagen wir, daß die Reihe (2) durch das Verfahren der ersten arithmetischen Mittel summierbar ist und nennen s ihre Summe. Wir wollen bis auf weiteres diese Reihen schlechthin *„summierbar"* nennen. Später werden wir dieses Wort in einem allgemeineren Sinne gebrauchen. Wie bekannt, ist die Existenz des Grenzwertes $\lim\limits_{n=\infty} s_n$ hinreichend, aber nicht notwendig für die Existenz des Grenzwertes $\lim\limits_{n=\infty} \sigma_n$. Existiert aber dieser letzte Grenzwert, so ist, wie zuerst Fejér*) hervorgehoben hat, *notwendigerweise*

$$(4) \qquad \lim_{n=\infty} \frac{s_n}{n} = \lim_{n=\infty} \frac{(n+1)\sigma_n - n\sigma_{n-1}}{n} = \lim_{n=\infty} (\sigma_n - \sigma_{n-1}) = 0,$$

also auch

$$(5) \qquad \lim_{n=\infty} \frac{c_n}{n} = \lim_{n=\infty} \frac{s_n - s_{n-1}}{n} = 0.$$

Diese Bemerkung läßt sich sofort auf die trigonometrischen Reihen anwenden. Es ergibt sich nämlich daraus mit Rücksicht auf das Cantorsche Lemma der folgende Satz:

Ist eine trigonometrische Reihe

$$a_0 + a_1 \cos x + b_1 \sin x + \cdots + a_n \cos nx + b_n \sin nx + \cdots$$

*) Fejér, Untersuchungen über Fouriersche Reihen, Math. Ann. 58, S. 63.

in einem Intervalle summierbar, so ist

$$\lim_{n=\infty} \frac{a_n}{n} = \lim_{n=\infty} \frac{b_n}{n} = 0. \tag{6}$$

Dies vorausgeschickt, können wir zur Besprechung der im Titel angedeuteten Sätze übergehen. Wie gesagt, liegt das Analogon des (ersten) Riemannschen Satzes bei Fejér fertig vor.

Es sei wiederum eine trigonometrische Reihe

$$a_0 + a_1 \cos x + b_1 \sin x + \cdots + a_n \cos nx + b_n \sin nx + \cdots \tag{7}$$

gegeben, deren Koeffizienten der Bedingung (5) genügen mögen.*) Eine viermalige gliedweise Integration ergibt die überall gleichmäßig konvergente Reihe

$$F(x) = \frac{a_0 x^4}{24} + \sum_{n=1}^{\infty} \frac{a_n \cos nx + b_n \sin nx}{n^4},$$

und der sogenannte vierte mittlere Differenzenquotient von $F(x)$ hat die Form

$$\frac{\Delta^4 F(x,h)}{h^4} = \frac{F(x+2h) - 4F(x+h) + 6F(x) - 4F(x-h) + F(x-2h)}{h^4}$$

$$= a_0 + \sum_{n=1}^{\infty} (a_n \cos nx + b_n \sin nx) \left(\frac{\sin \frac{nh}{2}}{\frac{nh}{2}} \right)^4 .$$

Der Fejérsche Satz besagt nun:

Ist die Reihe (7) *an einer Stelle* x_0 *summierbar mit dem Werte* $f(x_0)$, *so ist auch*

$$\lim_{h=0} \frac{\Delta^4 F(x_0, h)}{h^4} = f(x_0). \tag{8}$$

Um diesen Satz bei der geplanten Verallgemeinerung der Cantorschen und Du Bois-Reymondschen Sätze verwerten zu können, müssen wir noch die entsprechenden Analoga der Schwarzschen und Hölderschen Sätze entwickeln. Wir beweisen zuerst einen Mittelwertsatz, der dem Hölderschen entspricht und folgern aus ihm das Analogon des Schwarzschen Satzes.

Es sei $F(x)$ *eine stetige Funktion mit stetiger zweiter Ableitung, die an jeder Stelle eines Intervalles eine verallgemeinerte vierte Ableitung*

$$\Phi(x) = \lim_{h=0} \frac{\Delta^4 F(x,h)}{h^4}$$

$$= \lim_{h=0} \frac{F(x_0+2h) - 4F(x_0+h) + 6F(x_0) - 4F(x_0-h) + F(x_0-2h)}{h^4}$$

*) Fejér, l. c. S. 68—69. In der Fejérschen Fassung steht statt (6) eine leichtere Einschränkung der Koeffizienten, wir gebrauchen aber den Satz nur in obigem Umfange.

besitzt. Dann liegt der Wert $\frac{\Delta^4 F(x_0, h)}{h^4}$ *zwischen der oberen und unteren Grenze der Werte, die* Φ *im Intervalle* $(x_0 - 2h, x_0 + 2h)$ *annimmt.**)

Dieser Mittelwertsatz spielt in unseren Untersuchungen eine grundlegende Rolle. Wir beweisen ihn durch folgenden

Hilfssatz: *Die verallgemeinerte vierte Ableitung einer unseren obigen Bedingungen***) *genügenden Funktion ist positiv oder Null an jeder Stelle, wo die zweite Ableitung ein Minimum hat, negativ oder Null an jeder Stelle, die einem Maximum entspricht.*

Unser Hilfssatz folgt leicht aus dem zitierten Hölderschen Satze. Es habe nämlich an einer Stelle x_0 die zweite Ableitung ein Minimum. Wäre die verallgemeinerte vierte Ableitung dort negativ, so wäre ein h' zu finden, sodaß für jedes $h < h'$

$$\frac{F(x_0+2h) - 4F(x_0+h) + 6F(x_0) - 4F(x_0-h) + F(x_0-2h)}{h^4} < 0$$

ausfiele. Also wäre

$$\frac{F(x_0+2h) - 2F(x_0) + F(x_0-2h)}{4h^2} < \frac{F(x_0+h) - 2F(x_0) + F(x_0-h)}{h^2}.$$

So bekämen wir eine Folge von Ungleichungen

$$\frac{F(x_0+h) - 2F(x_0) + F(x_0-h)}{h^2} < \frac{F\left(x_0+\frac{h}{2}\right) - 2F(x_0) + F\left(x_0-\frac{h}{2}\right)}{\left(\frac{h}{2}\right)^2} < \cdots$$

$$< \frac{F\left(x_0+\frac{h}{2^n}\right) - 2F(x_0) + F\left(x_0-\frac{h}{2^n}\right)}{\left(\frac{h}{2^n}\right)^2} < \cdots < F''(x_0).$$

Andererseits hat nach unserer Voraussetzung $F''(x)$ in x_0 ein Minimum, also hat x_0 gewiß eine Umgebung h'', in welcher $F''(x)$ keinen kleineren Wert als $F''(x_0)$ annimmt. Dann müßte aber bei genügend großem n zufolge des Satzes von Hölder die Ungleichung

$$\frac{F\left(x_0+\frac{h}{2^n}\right) - 2F(x_0) + F\left(x_0-\frac{h}{2^n}\right)}{\left(\frac{h}{2^n}\right)^2} \geqq F''(x_0)$$

bestehen, was unserer obigen Ungleichung widerspricht. Aus unserem Hilfssatz können wir unseren Mittelwertsatz auf folgende Art beweisen:

*) h ist natürlich so gewählt, daß $x_0 - 2h$ und $x_0 + 2h$ im Inneren des gegebenen Intervalles liegen.

**) In diesem Satze kann man die Stetigkeit der zweiten Ableitung auch fallen lassen.

Wir betrachten die Werte der Funktion $F(x)$ an den Stellen x_0-2h, x_0-h, x_0, x_0+h, x_0+2h. Es gibt immer ein Polynom $P_4(x)$ vierten Grades von der Eigenschaft, daß an den obigen fünf Stellen die Gleichung $F(x)-P_4(x)=0$ besteht. Wir finden unmittelbar*), daß $P_4(x)$ die Form hat

$$P_4(x)=\frac{\Delta^4 F(x_0,h)}{24h^4}x^4+ax^3+bx^2+cx+d.$$

Da die stetige Funktion $F(x)-P_4(x)$ an den zwei Endpunkten und mindestens drei inneren Punkten des Intervalles $(x_0-2h,\ x_0+2h)$ verschwindet, hat sie im Inneren des Intervalles wenigstens zwei Maxima und zwei Minima. Überdies sind zwei Punkte, die einem Maximalwerte entsprechen, durch einen Punkt getrennt, der einem Minimum entspricht, und umgekehrt. Bei einem Maximum ist $F''(x)-P_4''(x)$ negativ oder Null, bei einem Minimum positiv oder Null. Also hat diese für stetig vorausgesetzte zweite Derivierte im Innern des Intervalles $(x_0-2h,\ x_0+2h)$ mindestens ein Maximum und ein Minimum.

Es sei z. B. x_1 eine solche Stelle, wo $F''(x)-P_4''(x)$ ein Minimum hat. Nach unserem Hilfssatze ist an dieser Stelle die verallgemeinerte vierte Derivierte von $F(x)-P_4(x)$ positiv oder Null. Das heißt

$$\begin{aligned}\lim_{t=0}\frac{\Delta^4[F(x_1,t)-P_4(x_1,t)]}{t^4}&=\lim_{t=0}\frac{\Delta^4 F(x_1,t)}{t^4}-\lim_{t=0}\frac{\Delta^4 P_4(x_1,t)}{t^4}\\&=\lim_{t=0}\frac{\Delta^4 F(x_1,t)}{t^4}-\frac{\Delta^4 F(x_0,h)}{h^4}\geqq 0.\end{aligned}$$

Also ist

$$\frac{\Delta^4 F(x_0,h)}{h^4}\leqq\lim_{t=0}\frac{\Delta^4 F(x_1,t)}{t^4}=\Phi(x_1). \tag{10}$$

Ist ferner x_2 eine Stelle, wo $F''(x)-P_4''(x)$ ein Maximum hat, so ist

$$\frac{\Delta^4 F(x_2,h)}{h^4}\geqq\Phi(x_2). \tag{11}$$

Durch die zwei Ungleichungen (10) und (11) ist unser Mittelwertsatz vollständig bewiesen.

Daraus leiten wir nun das gewünschte Analogon des Schwarzschen Satzes her, das folgendermaßen lautet: *Jede Funktion $F(x)$, die in einem*

---

*) In den untersuchten fünf Punkten haben $F(x)$ und $P_4(x)$ denselben Wert, also sind die Werte ihrer vierten Differenzenquotienten (die mit dem Mittelpunkt x_0 und mit der Differenz h gebildet sind) auch gleich. Aber der n^{te} Differenzenquotient eines Polynoms n^{ten} Grades ist konstant für jedes x und für jedes h und ist gleich dem Produkte aus $n!$ und dem Koeffizienten des höchsten Gliedes. Aus dieser Tatsache ersieht man sogleich die obige Form von $P_4(x)$, die sich natürlich auch durch Rechnung leicht verifizieren läßt.

Intervalle samt ihrer zweiten Ableitung) stetig ist und deren verallgemeinerte vierte Ableitung identisch verschwindet, ist in diesem Intervalle ein Polynom dritten Grades.*

Es folgt nämlich aus unserem Mittelwertsatze, daß für jeden Wert von x und h**)

$$F(x+2h)-4F(x+h)+6F(x)-4F(x-h)+F(x-2h)=0. \tag{12}$$

Wir behaupten nun, daß jede stetige Funktion, die dieser Gleichung genügt, ein Polynom dritten Grades ist. Denn es ist für zwei spezielle Werte x_0 und h_0

$$F(x_0+2h_0)-4F(x_0+h_0)+6F(x_0)-4F(x_0-h_0)+F(x_0-2h_0)=0.$$

Es gibt also ein Polynom dritten Grades $P_3(x)$ von der Eigenschaft, daß

$$K(x)=F(x)-P_3(x)=0 \tag{13}$$

an den fünf Stellen x_0+2h_0, x_0+h_0 usw. Wir behaupten nun, daß die Gleichung (13) identisch besteht. Es besteht nämlich identisch die Gleichung

$$P_3(x+2h)-4P_3(x+h)+6P_3(x)-4P_3(x-h)+P_3(x-2h)=0,$$

woraus dann in Verbindung mit (12) auch identisch

$$K(x+2h)-4K(x+h)+6K(x)-4K(x-h)+K(x-2h)=0. \tag{14}$$

Durch wiederholte Anwendung von (14) folgt aus (13), daß auch

$$K(x_0\pm nh_0)=0 \qquad (n=0,1,2,\cdots). \tag{15}$$

Es genügt also, den Beweis des identischen Bestehens von (13) für das Intervall $(x_0-2h_0,\ x_0+2h_0)$ zu führen. Dieselbe Beweisführung kann dann auf ein beliebiges Intervall $(x_0+n_1h_0,\ x_0+n_2h_0)$ angewandt werden.

Es ist aber auch

$$K(x_0+2h_0)-4K\left(x_0+\frac{3h_0}{2}\right)+6K(x_0+h_0)-4K\left(x_0+\frac{h_0}{2}\right)+K(x_0)=0.$$

---

*) Es wäre der späteren Anwendung wegen wünschenswert, unseren Satz zu beweisen, ohne die zweite Ableitung stetig vorauszusetzen. Die Funktion

$$f(x)=\begin{cases} = \ \ x^2, & \text{wenn } x\geqq 0,\\ = -x^2, & \text{„} \ \ x\leqq 0,\end{cases}$$

liefert aber ein *deutliches Gegenbeispiel.* Die vierte verallgemeinerte Ableitung dieser stetigen Funktion verschwindet überall, ihre erste Ableitung ist überall stetig, die verallgemeinerte zweite Ableitung existiert auch überall, sie erleidet aber an der Stelle $x=0$ einen Sprung.

**) Die Werte x und h sind natürlich so gewählt, daß alle Argumente in dem gegebenen Intervalle liegen.

Indem man dies mit (13) vergleicht, erhält man:

$$K\left(x_0+\frac{3h_0}{2}\right)=-K\left(x_0+\frac{h_0}{2}\right). \tag{16}$$

Dieselbe Schlußweise ergibt

$$-K\left(x_0+\frac{h_0}{2}\right)=K\left(x_0-\frac{h_0}{2}\right) \tag{17}$$

und

$$K\left(x_0-\frac{h_0}{2}\right)=-K\left(x_0-\frac{3h_0}{2}\right). \tag{18}$$

Ferner erhält man ebenfalls aus (14)

$$\begin{aligned} K\left(x_0+\frac{3h_0}{2}\right)-4K(x_0+h_0)+6K\left(x_0-\frac{h_0}{2}\right) \\ -4K(x_0)+K\left(x_0+\frac{h_0}{2}\right)=0. \end{aligned} \tag{19}$$

Die Gleichungen (16), (17), (18) und (19) können nur dann gleichzeitig bestehen, wenn

$$K\left(x_0+\frac{3h_0}{2}\right)=K\left(x_0+\frac{h_0}{2}\right)=K\left(x_0-\frac{h_0}{2}\right)=K\left(x_0-\frac{3h_0}{2}\right)=0.$$

Indem man die Halbierung fortsetzt, ersieht man, daß die Funktion $K(x)$ in Punkten einer überall dichten Menge, also infolge ihrer Stetigkeit überall verschwindet, womit das Analogon des Schwarzschen Satzes dargelegt ist.

Nach diesen vorbereitenden Sätzen können wir uns endlich zu unseren trigonometrischen Reihen wenden.

§ 3.

Der verallgemeinerte Cantorsche Satz.

I. Hauptsatz. *Es sei die trigonometrische Reihe*

$$a_0+a_1\cos x+b_1\sin x+\cdots+a_n\cos nx+b_n\sin nx+\cdots \tag{20}$$

überall summierbar und die aus ihr durch zweimalige gliedweise Integration gebildete Reihe

$$\frac{a_0x^2}{2}-\sum_{n=1}^{\infty}\frac{a_n\cos nx+b_n\sin nx}{n^2} \tag{21}$$

*gleichmäßig konvergent.**) *Verschwindet dann die Summe der Reihe* (20) *im ganzen Intervall* $(0,\,2\pi)$, *so verschwinden auch alle ihre Koeffizienten.*

*) Diese Bedingung ist offenbar erfüllt, wenn man zwei Konstanten C und k $(k<1)$ angeben kann, sodaß

$$|a_n|<Cn^k \quad\text{und}\quad |b_n|<Cn^k. \tag{22}$$

Andererseits folgen aber laut (6) aus der Summierbarkeit die Limesgleichungen

II. Hauptsatz. *Konvergieren die Koeffizienten der Reihe mit wachsendem n gegen Null, so bleibt der Satz auch dann noch richtig, wenn man eine reduktible Menge von Ausnahmestellen zuläßt,* d. h. solche Stellen, in welchen die Reihe entweder nicht summierbar, oder ihre Summe von Null verschieden ist.

Der Beweis unseres ersten Hauptsatzes folgt nun genau dem Cantorschen Verfahren. Nach dem Fejérschen Analogon des Riemannschen Satzes erfüllt die aus (20) durch viermalige gliedweise Integration erhaltene Reihe die Bedingungen, die unserem Analogon des Schwarzschen Satzes zugrunde liegen. Diese Reihe stellt also ein Polynom dritten Grades dar:

$$C_0x^3 + C_1x^2 + C_2x + C_3 = \frac{a_0}{24}x^4 + \sum_{n=1}^{\infty} \frac{a_n \cos nx + b_n \sin nx}{n^4}.$$

Damit aber diese Gleichung überall bestehe, müssen in ihr die nichtperiodischen Glieder verschwinden. Daraus folgt zunächst

$$a_0 = 0.$$

---

(23) $$\lim_{n=\infty} \frac{a_n}{n} = \lim_{n=\infty} \frac{b_n}{n} = 0.$$

Man ersieht aus (22) und (23), welcher Natur die Koeffizientenfolgen sind, auf welche unser Satz nicht anwendbar ist.

Wir bemerken noch, daß aus der Summierbarkeit der Reihe (20) an einer Stelle x_0 leicht die Konvergenz der Reihe $\sum_{n=1}^{\infty} \frac{a_n \cos nx_0 + b_n \sin nx_0}{n}$, also auch die Konvergenz von $\sum_{n=1}^{\infty} \frac{a_n \cos nx_0 + b_n \sin nx_0}{n^2}$ folgt. In der Voraussetzung der Konvergenz von (21) liegt also keine Einschränkung. Eine wesentliche Einschränkung scheint aber in der vorausgesetzten Gleichmäßigkeit zu liegen. Vgl. für all dies § 5 der vorliegenden Arbeit.

Übrigens ist die Bedingung (21) mit der folgenden gleichwertig: Die Konstanten

$$\alpha_n = \frac{a_n}{n^2} \quad \text{und} \quad \beta_n = \frac{b_n}{n^2}$$

seien die Fourierschen Konstanten einer stetigen Funktion. Man hat nämlich infolge von (23)

$$\lim_{n=\infty} n\alpha_n = \lim_{n=\infty} n\beta_n = 0.$$

Es folgt aber leicht (z. B. aus den Fejérschen Sätzen über Fouriersche Reihen), daß Größen von dieser Beschaffenheit dann und nur dann die Fourierschen Konstanten einer stetigen Funktion sein können, wenn die Reihe

$$\sum_{n=1}^{\infty} \alpha_n \cos nx + \beta_n \sin nx$$

gleichmäßig konvergiert.

Da ferner die trigonometrische Reihe auf der rechten Seite gleichmäßig konvergiert, sind die Werte

$$\frac{a_1}{1^4}, \frac{b_1}{1^4}, \ldots, \frac{a_n}{n^4}, \frac{b_n}{n^4}, \ldots$$

die Fourier-Koeffizienten einer durchweg konstanten Funktion, d. h.

$$a_1 = b_1 = \cdots = a_n = b_n = \cdots = 0, \text{ w. z. b. w.}$$

Wir wollen unseren Satz durch ein Beispiel erläutern. Wir untersuchen, warum sich unsere Methode auf die Reihe

$$\frac{1}{2} + \cos x + \cdots + \cos nx + \cdots$$

nicht anwenden läßt. Es liegt dies, kurz gesagt, an den Ausnahmestellen $x \equiv 0 \pmod{2\pi}$. Im Innern des Intervalles $(0, 2\pi)$ nämlich, wo die Reihe überall summierbar ist und den Wert Null hat, lassen sich unsere Schlüsse ohne weiteres anwenden. Folglich stellt die durch viermalige gliedweise Integration erzeugte Reihe

$$F(x) = \frac{1}{48} x^4 + \sum_{n=1}^{\infty} \frac{\cos nx}{n^4}$$

ein Polynom dritten Grades dar. Dieses Polynom ist, wie man auch direkt leicht berechnet:

$$P(x) = \frac{\pi}{12} x^3 - \frac{\pi^2}{12} x^2 + \frac{\pi^4}{90}.$$

Man hat also im Intervall $(0, 2\pi)$:

$$F(x) = \frac{1}{48} x^4 + \sum_{n=1}^{\infty} \frac{\cos nx}{n^4} = P(x).$$

Diese Beziehung besteht aber nur im Intervalle $(0, 2\pi)$. Im Intervalle $(2\pi, 4\pi)$ z. B. hat man:

$$F(x) = \frac{1}{48} x^4 - \frac{1}{48}(x - 2\pi)^4 + \frac{\pi}{12}(x - 2\pi)^3 - \frac{\pi^2}{12}(x - 2\pi)^2 + \frac{\pi^2}{90} \neq P(x).$$

Unser Verfahren scheitert also an dem Umstande, daß das ursprüngliche Polynom dritten Grades sich bei dem Überschreiten der Punkte $x \equiv 0 \pmod{2\pi}$ je in einem anderen Polynome fortsetzt.

Diese Bemerkung leitet uns über zu unserem zweiten Hauptsatze. Wenn nämlich die Koeffizienten der trigonometrischen Reihe gegen Null konvergieren (was bei dem ursprünglichen Cantorschen Satze immer der Fall ist), so setzt sich — wie wir sehen werden — die Reihe $F(x)$ bei der Überschreitung eventueller Ausnahmestellen in demselben Polynom dritten Grades fort. Der Gedankengang der vorigen Seite wird somit anwendbar.

Es sei nun unter dieser Voraussetzung x_0 eine isolierte Ausnahmestelle unserer trigonometrischen Reihe, die im allgemeinen summierbar ist und den Wert Null hat. Nach dem zitierten Fejérschen Satze und dem Analogon des Schwarzschen Satzes stimmt die Funktion $F(x)$ rechts von x_0 (in einer gewissen Umgebung) mit einem Polynom

$$\alpha_1 x^3 + \beta_1 x^2 + \gamma_1 x + \delta_1,$$

links von x_0 mit einem Polynom

$$\alpha_2 x^3 + \beta_2 x^2 + \gamma_2 x + \delta_2$$

überein. Diese zwei Polynome sind identisch. Denn infolge des Unendlichkleinwerdens der Koeffizienten unserer trigonometrischen Reihe läßt sich der zweite Riemannsche Satz auf die Reihe

$$F''(x) = \frac{a_0 x^2}{2} - \sum_{n=1}^{\infty} \frac{a_n \cos nx + b_n \sin nx}{n^2}$$

anwenden, das heißt, man hat auch für die Stelle x_0:

$$\lim_{h=0} \frac{F''(x_0+h) - 2F''(x_0) + F''(x_0-h)}{h} = 0.$$

Daraus und aus der Stetigkeit von $F(x)$, $F'(x)$, $F''(x)$ folgt sofort

$$\alpha_1 = \alpha_2, \quad \beta_1 = \beta_2, \quad \gamma_1 = \gamma_2, \quad \delta_1 = \delta_2.$$

Durch Fortsetzung dieser Schlußweise kann man dasselbe Resultat auch für nichtisolierte Ausnahmestellen beweisen, wenn nur diese eine reduktible Menge bilden. Damit ist unser zweiter Hauptsatz in seinem vollen Umfange bewiesen.

§ 4.

Der verallgemeinerte Du Bois-Reymondsche Satz.

Wir hoben schon im § 2 die Verdienste von Lebesgue um den Du Bois-Reymondschen Satz hervor. Auf dem Standpunkte von Du Bois-Reymond (Riemannscher Integralbegriff) mußte man, außer der durchgängigen Konvergenz der Reihe, die Integrierbarkeit der dargestellten Funktion voraussetzen. Lebesgue gibt nun Beispiele von beschränkten Funktionen an, die durch überall konvergente trigonometrische Reihen dargestellt werden, obgleich sie im Riemannschen Sinne nicht integrierbar sind.*) Man sieht, wie die Einführung des Lebesgueschen Integralbegriffes in der Theorie der trigonometrischen Reihen einen einheitlichen Gesichts-

*) Leçons sur les séries trigonométriques, S. 68. Vgl. für alle diese Bemerkungen Fatou, Séries trigonométriques et séries de Taylor (Thèse), Acta Math. 30, S. 336—337.

punkt gewährt. Im folgenden gebrauchen wir nun die Worte „Integral" und „integrierbar" im Lebesgueschen Sinne.

Unsere Verallgemeinerung des Du Bois-Reymondschen Satzes lautet:

III. Hauptsatz: *Es sei wiederum die trigonometrische Reihe*

$$f(x) = a_0 + a_1 \cos x + b_1 \sin x + \cdots + a_n \cos nx + b_n \sin nx + \cdots \tag{24}$$

überall summierbar und die Reihe

$$\sum_{n=1}^{\infty} \frac{a_n \cos nx + b_n \sin nx}{n^2} \tag{25}$$

gleichmäßig konvergent. Ist dann die Funktion $f(x)$ beschränkt, so ist (24) *ihre Fouriersche Reihe, d. h. es ist*

$$a_0 = \frac{1}{2\pi}\int_0^{2\pi} f(x)\,dx;\quad a_n = \frac{1}{\pi}\int_0^{2\pi} f(x)\cos nx\,dx;\quad b_n = \frac{1}{\pi}\int_0^{2\pi} f(x)\sin nx\,dx. \tag{26}$$

Der Beweis kann nach Heranziehung unserer Hilfssätze dem Lebesgueschen Beweisgange analog geführt werden.*)

Wir wollen zunächst den folgenden Satz begründen:

Hat eine Funktion $\Phi(x)$ überall eine verallgemeinerte vierte Ableitung $f(x)$, die zwischen endlichen Grenzen liegt, und außerdem eine überall stetige zweite Ableitung $F(x)$, so ist

$$F(x) = \int_0^x\int_0^\theta f(t)\,dt\,d\theta + Ax + B, \tag{27}$$

wo A und B Konstante bedeuten.

Da nämlich

$$\lim_{h=0} \frac{\Delta^4 \Phi(x,h)}{h^4} = f(x)$$

ist, so liegen auch die Größen $\frac{\Delta^4 \Phi(x,h)}{h^4}$ zwischen denselben endlichen Grenzen für jeden in Betracht kommenden Wert von x und h. Bezeichnen wir also ein unbestimmtes Integral von $\Phi(x)$ mit $\Phi_1(x)$ und ein unbestimmtes Integral von $\Phi_1(x)$ mit $\Phi_2(x)$, so kann man nach einem be-

*) Man kann auch, wie es Lebesgue im Falle konvergenter trigonometrischer Reihen durchgeführt hat, beweisen, daß unter unseren Bedingungen, die Werte der harmonischen Funktion

$$f(r,x) = \sum_{n=0}^{\infty} (a_n r^n \cos nx + b_n r^n \sin nx) \qquad (r < 1)$$

zwischen denselben Schranken liegen wie $f(x)$. Dadurch erhält man leicht einen zweiten Beweis unseres III. Hauptsatzes.

kannten Lebesgueschen Satze unter dem Limes-Zeichen integrieren und erhält so:

$$(28)\quad \begin{aligned} &\lim_{h=0} \frac{\Delta^4 [\Phi_1(x,h) - \Phi_1(0,h)]}{h^4} = \int_0^x f(t)\,dt; \\ &\lim_{h=0} \left[\frac{\Delta^4 \Phi_2(x,h)}{h^4} - \frac{\Delta^4 \Phi_2(0,h)}{h^4} - \frac{\Delta^4 \Phi_1(0,h)\,x}{h^4}\right] = \int_0^x \int_0^\theta f(t)\,dt\,d\theta. \end{aligned}$$

Es ist klar, daß die ersten zwei Ausdrücke in (28) für $h = 0$ einen Grenzwert haben, denn Φ_2 hat als zweimaliges Integral einer zweimal differenzierbaren Funktion auch im gewöhnlichen Sinne eine vierte Ableitung. Ferner ist die Integration auf der rechten Seite ausführbar, also muß der dritte Ausdruck unter dem „Limes"-Zeichen auch einen Grenzwert haben.

Da aber

$$\lim_{h=0} \frac{\Delta^4 \Phi_2(x,h)}{h^4} = F(x)$$

ist, so können wir nach (28) setzen

$$F(x) = \int_0^x \int_0^\theta f(t)\,dt\,d\theta + Ax + B.$$

Wenden wir nun dieses Resultat auf unsere trigonometrische Reihen an. Es sei

$$f(x) = a_0 + a_1 \cos x + b_1 \sin x + \cdots + a_n \cos nx + b_n \sin nx + \cdots$$

eine trigonometrische Reihe, die den Bedingungen unseres III. Hauptsatzes genügt. Dann sind die Reihen

$$\Phi(x) = \frac{a_0 x^4}{24} + \sum_{n=1}^{\infty} \frac{a_n \cos nx + b_n \sin nx}{n^4}$$

und

$$F(x) = \Phi''(x) = \frac{a_0 x^2}{2} - \sum_{n=1}^{\infty} \frac{a_n \cos nx + b_n \sin nx}{n^2}$$

gleichmäßig konvergent. Wir können also ihre Koeffizienten durch das Fouriersche Verfahren bestimmen:

$$-\frac{a_n}{n^2} = \frac{1}{\pi} \int_0^{2\pi} \int_0^x \int_0^\theta f(t) \cos nx\,dt\,d\theta\,dx - \frac{2a_0}{n^2},$$

$$-\frac{b_n}{n^2} = \frac{1}{\pi} \int_0^{2\pi} \int_0^x \int_0^\theta f(t) \sin nx\,dt\,d\theta\,dx + \frac{2a_0\pi}{n} - \frac{2A}{n}$$

$$(n > 0).$$

Die Integrationen können aber in beliebiger Reihenfolge ausgeführt werden*) und somit hat man:

$$-\frac{a_n}{n^2}=\frac{1}{\pi}\int_0^{2\pi}\cos nx\,dx\int_0^x f(t)\,dt\int_t^x d\theta-\frac{2a_0}{n^2}$$

$$=\frac{1}{\pi}\int_0^{2\pi}\cos nx\,dx\int_0^x f(t)(x-t)\,dt-\frac{2a_0}{n^2}$$

$$=\frac{1}{\pi}\int_0^{2\pi}f(t)\,dt\int_t^{2\pi}(x-t)\cos nx\,dx-\frac{2a_0}{n^2},$$

schließlich

$$a_n=\frac{1}{\pi}\int_0^{2\pi}f(t)\cos nt\,dt+2\left[a_0-\frac{1}{2\pi}\int_0^{2\pi}f(t)\,dt\right];$$

ähnlich:

$$b_n=\frac{1}{\pi}\int_0^{2\pi}f(t)\sin nt\,dt-n\left[2\pi a_0-2A-\frac{1}{\pi}\int_0^{2\pi}f(t)(2\pi-t)\,dt\right]$$

$$(n>0).$$

Aus der Summierbarkeit unserer trigonometrischen Reihe folgen aber die Gleichungen

$$\lim_{n=\infty}\frac{a_n}{n}=\lim_{n=\infty}\frac{b_n}{n}=0. \tag{29}$$

In dem Ausdrucke für b_n konvergiert nach einem bekannten Riemannschen Satze**) das erste Integral mit wachsendem n gegen Null. Aus (29) folgt also unmittelbar

$$2\pi a_0-2A-\frac{1}{\pi}\int^{2\pi}f(t)(2\pi-t)\,dt=0.$$

Ferner konvergiert auch in der Formel für a_n das erste Integral gegen Null. Wäre daher der Klammerausdruck von Null verschieden, so hätte a_n von einem genügend großen n an ein festes Vorzeichen und sein Betrag wäre größer als eine feste Zahl. Mithin könnte unsere Reihe (24) an der Stelle $x=0$ nicht summierbar sein.

---

*) Leçons sur les séries trig. S. 12 und 123.

**) S. 60—61 des Lebesgueschen Werkes findet man die Verallgemeinerung dieses Riemannschen Satzes auf Funktionen, die im Lebesgueschen Sinne integrierbar sind.

Also ist schließlich

$$a_0 = \frac{1}{2\pi}\int_0^{2\pi} f(t)\,dt,$$

$$a_n = \frac{1}{\pi}\int_0^{2\pi} f(t)\cos nt\,dt,$$

$$b_n = \frac{1}{\pi}\int_0^{2\pi} f(t)\sin nt\,dt.$$

Unser Satz findet gerade in diesen Formeln seinen Ausdruck.

Es sei nun unsere Reihe so beschaffen, daß ihre Koeffizienten gegen Null konvergieren, sie selbst aber im Interall $(0, 2\pi)$ eine reduktible Menge von Ausnahmestellen hat, in denen sie entweder nicht summierbar ist, oder aber ihre Summe nicht zwischen endlichen Grenzen bleibt. Wir können dann in jedem Intervalle (a, b), das keine Ausnahmestellen enthält, setzen

$$F(x) = \int_a^x\int_a^\theta f(t)\,dt\,d\theta + cx + d = \int_0^x\int_0^\theta f(t)\,dt\,d\theta + A'x + B'.$$

In einem anderen solchen Intervall muß dann $F(x)$ die Form haben

$$F(x) = \int_0^x\int_0^\theta f(t)\,dt\,d\theta + A''x + B''.$$

Da aber die Koeffizienten der trigonometrischen Reihe gegen Null konvergieren, haben wir nach dem zweiten Riemannschen Satze für jeden Wert von x:

$$\lim_{h=0} \frac{F(x+h) - 2F(x) + F(x-h)}{h} = 0,$$

woraus folgt, daß der lineare Ausdruck immer derselbe bleibt. Also hat man im ganzen Intervall $(0, 2\pi)$:

$$F(x) = \int_0^x\int_0^\theta f(t)\,dt\,d\theta + Ax + B$$

und somit kann unsere frühere Schlußweise angewendet werden.

§ 5.

Untersuchung der Gültigkeit der Cantorschen und Du Bois-Reymondschen Sätze für arithmetische Mittel beliebiger Ordnung.

Wir schicken wieder einige Bemerkungen über diese Mittel voraus. Es sei zunächst eine numerische Reihe

$$c_0 + c_1 + \cdots + c_n + \cdots \tag{29}$$

gegeben. Wir setzen mit Cesàro*)

$$\left(\frac{1}{1-y}\right)^{k+1} = \sum_{n=0}^{\infty} C_n^{(k)} y^n \tag{30}$$

und $(k \geqq 0)$

$$\left(\frac{1}{1-y}\right)^{k+1} \sum_{n=0}^{\infty} c_n y^n = \sum_{n=0}^{\infty} S_n^{(k)} y^n. \tag{31}$$

Existiert dann der Grenzwert

$$\lim_{n=\infty} \frac{S_n^{(k)}}{C_n^{(k)}} = \lim_{n=\infty} \frac{\Gamma(k+1)\, S_n^{(k)}}{n^k}, \tag{32}$$

so nennen wir die Reihe (29) *durch arithmetische Mittel k^{ter} Ordnung summierbar***) und bezeichnen (32) als ihre Summe.

Es bestehen nun die zwei leicht beweisbaren Sätze, von welchen der erste von Knopp***) herrührt:

Satz A. *Existiert der Grenzwert* (32) *für ein gewisses (ganzzahliges oder nichtganzzahliges) $k \geqq 0$, so existiert er auch für jedes $k' > k$, und diese Grenzwerte sind alle gleich.*

Satz B. *Existiert der Grenzwert für ein gewisses $k \geqq 0$, so ist*

$$\lim_{n=\infty} \frac{c_n}{n^k} = 0. \tag{33}$$

Aus diesem letzten Satze folgt nach Heranziehung des Cantorschen Lemma sofort der

---

*) Cesàro, Sur la multiplication des séries, Bull. Sc. Math. 14 (1890).

**) So viel ich weiß, wurden arithmetische Mittel nicht ganzer Ordnungszahl zuerst von Hadamard in Betracht gezogen: Essai sur l'étude des fonctions données par leur développement de Taylor (Thèse), J. de math. (1892) S. 182.

***) Knopp, Grenzwerte von Reihen bei der Annäherung an die Konvergenzgrenze (Inauguraldissertation), Berlin 1907. In dieser Arbeit werden eine Reihe wichtiger Sätze über arithmetische Mittel beliebiger Ordnungszahl dargelegt.

Satz C. *Ist eine trigonometrische Reihe*

$$a_0 + a_1 \cos x + b_1 \sin x + \cdots + a_n \cos nx + b_n \sin nx + \cdots$$

in einem Intervalle durch arithmetische Mittel k^{ter} Ordnung summierbar, so ist

(34) $$\lim_{n=\infty} \frac{a_n}{n^k} = \lim_{n=\infty} \frac{b_n}{n^k} = 0.$$

Nach diesen Vorbemerkungen können wir uns zur folgenden Frage wenden, welche einen guten Überblick über unsere bisherigen Betrachtungen gestattet:

Es sei die trigonometrische Reihe

(35) $$a_0 + a_1 \cos x + b_1 \sin x + \cdots + a_n \cos nx + b_n \sin nx + \cdots$$

durch die arithmetischen Mittel k^{ter} Ordnung im Intervall $(0, 2\pi)$ summierbar und habe überall den Wert Null. Für welche Werte von $k \geqq 0$ folgt daraus das Verschwinden ihrer Koeffizienten?*)

I. Es sei zunächst

$$0 \leqq k < 1.$$

Aus der Beziehung (34) folgt dann unmittelbar, daß die Reihe

$$\sum_{n=1}^{\infty} \frac{a_n \cos nx + b_n \sin nx}{n^2}$$

gleichmäßig konvergiert. Andererseits folgt aus dem zitierten Satz von Knopp, daß unsere trigonometrische Reihe auch durch arithmetische Mittel erster Ordnung überall summierbar ist und auch so summiert überall verschwindet. Es sind also alle Bedingungen unseres ersten Hauptsatzes erfüllt, folglich haben wir den Satz:

Ist die Reihe (34) *durch arithmetische Mittel kleinerer als erster Ordnung an jeder Stelle des Intervalles* $(0, 2\pi)$, *mit dem Werte Null summierbar, so verschwinden alle ihre Koeffizienten.*

Dieser Satz ist eine volle Verallgemeinerung des Cantorschen Satzes. Ähnlich läßt sich der Du Bois-Reymondsche Satz auf arithmetische Mittel kleinerer als erster Ordnung übertragen.

II. Es sei nun

$$k > 1.$$

Da zeigt das Beispiel

(36) $$\sin x + 2 \sin 2x + \cdots + n \sin nx + \cdots,$$

*) Die Reihe $\frac{1}{2} + \cos x + \cdots + \cos nx + \cdots$ ist in jedem Punkte $x \not\equiv 0 \pmod{2\pi}$ durch arithmetische Mittel beliebiger *positiver* Ordnung mit dem Werte Null summierbar. Will man also Ausnahmestellen zulassen, so hört die Gültigkeit des Cantorschen Satzes für $k > 0$ auf. Wir können uns daher im Text darauf beschränken, den Cantorschen Satz ohne Ausnahmestellen zu untersuchen.

daß *für arithmetische Mittel höherer als erster Ordnung der Cantorsche Satz und mithin auch der Du Bois-Reymondsche Satz nicht gelten. Diese Reihe ist,* wie leicht ersichtlich, *durch irgend welche arithmetische Mittel höherer als erster Ordnung überall summierbar und hat überall den Wert Null.**) (Dabei kann die Ordnung der Summation ganzzahlig oder nichtganzzahlig sein.)

Dies zeigt, daß für arithmetische Mittel höherer als erster Ordnungszahl unsere obige Frage negativ zu beantworten ist.

III. Es sei endlich

$$k = 1.$$

Dieser Fall ist es, dem eigentlich die ganze Arbeit gewidmet ist. Dennoch können wir in diesem Falle die in diesem Paragraphen gestellte Frage nicht befriedigend beantworten. Vielmehr müssen wir sie in einem wesentlichen Punkte als *offen* bezeichnen. In unserem ersten Hauptsatze

---

*) Wir wollen über diese Reihe noch einige Bemerkungen machen. Sie ist die formale Ableitung der Reihe

$$-\left(\frac{1}{2} + \cos x + \cdots + \cos nx + \cdots\right)$$

die, wie wir wissen, mit Ausnahme der Stellen $x \equiv 0 \pmod{2\pi}$ verschwindet, was auch die ähnliche Eigenschaft der Reihe (36) begründen mag. (Bei dieser letzteren bilden aber natürlich auch die erwähnten Stellen keine Ausnahme.) Setzen wir

$$re^{ix} = z = u + iv,$$

so sehen wir, daß die assoziierte Reihe

$$\sum_{n=1}^{\infty} nr^n \sin nx \qquad (r < 1),$$

der imaginäre Bestandteil der Funktion

$$\frac{z}{(1-z)^2}$$

ist. Sie ist mithin gleich der *harmonischen* Funktion

$$w(u, v) = -\frac{v}{(u-1)^2 + v^2} \cdot \frac{u^2 + v^2 - 1}{(u-1)^2 + v^2}.$$

Von dieser Funktion sieht man auch direkt, daß sie die folgende merkwürdige Eigenschaft hat: Sie ist harmonisch und regulär im Innern des Einheitskreises. Nähert sich der Punkt (u, v) längs eines beliebigen Radius dem Rande dieses Kreises, so nähert sich die Funktion dem Werte Null. (Natürlich ist diese Annäherung keine gleichmäßige.)

Dieses Beispiel einer solchen Funktion wurde meines Wissens zuerst von Herrn Petrini gefunden. Ich verdanke es der mündlichen Mitteilung von Herrn Holmgren. Ich bin durch diese Funktion auf die Reihe (36) gekommen.

(Während der Korrektur erschien eine Arbeit von Burkhardt: Zur Geschichte der divergenten Reihen von 1750—1860, Math. Ann. 70, die auf S. 371—372, unter anderen, sehr bemerkenswerte geschichtliche Angaben über die letzt erwähnten trigonometrischen Reihen enthält.)

haben wir ja außer der Summierbarkeit der Reihe (20) auch die gleichmäßige Konvergenz der Reihe (21) zur Voraussetzung. Aus der Summierbarkeit durch Mittel erster Ordnung folgen aber nur die Gleichungen

$$\lim_{n=\infty} \frac{a_n}{n} = \lim_{n=\infty} \frac{b_n}{n} = 0,$$

woraus keineswegs (wie im Falle $k < 1$) die gleichmäßige Konvergenz der Reihe (21) folgt. Dazu will ich noch folgendes hinzufügen:

In der Fußnote zu S. 64 bemerkten wir gelegentlich, daß aus der *Summierbarkeit einer Reihe*

$$c_0 + c_1 + \cdots + c_n + \cdots$$

durch einfache arithmetische Mittel die Konvergenz der Reihe

$$\frac{c_1}{1} + \frac{c_2}{2} + \cdots + \frac{c_n}{n} + \cdots$$

folgt. (Dies läßt sich durch zweimalige partielle Summation leicht beweisen und ist übrigens als Spezialfall in einem Satze enthalten, den ich über Dirichletsche Reihen*) ausgesprochen habe.) Unsere Reihe

$$\sum_{n=1}^{\infty} \frac{a_n \cos nx + b_n \sin nx}{n^2} = \sum_{n=1}^{\infty} (\alpha_n \cos nx + \beta_n \sin nx)$$

hat also allein infolge der durchgängigen Summierbarkeit der Reihe

$$\sum_{n=0}^{\infty} (a_n \cos nx + b_n \sin nx)$$

die Eigenschaft:

An jeder Stelle ist nicht nur die Reihe

$$\sum_{n=1}^{\infty} (\alpha_n \cos nx + \beta_n \sin nx), \tag{37}$$

sondern auch die Reihe

$$\sum_{n=1}^{\infty} (n\alpha_n \cos nx + n\beta_n \sin nx) \tag{38}$$

konvergent. Ob daraus die gleichmäßige Konvergenz der Reihe (37) folgt, kann ich nicht entscheiden.**) Damit bleibt aber eben die in diesem Paragraphen behandelte Frage, für $k = 1$, unbeantwortet.

---

*) Sur les séries de Dirichlet (Théorème I), Paris C. R , 21 juin 1909.

**) Wäre die Antwort bejahend, so könnten wir in unserem ersten und dritten Hauptsatze die einschränkende Bedingung fallen lassen, und die Sätze von Cantor und Du Bois-Reymond wären durch unsere Untersuchungen auch im Falle $k = 1$ in voller Allgemeinheit bestätigt. Aus einer negativen Antwort würde die Ungültigkeit der beiden Sätze natürlich nicht folgen.

Man könnte nun auch im Falle $k > 1$ nach geeigneten Einschränkungen über die Koeffizienten der Reihe suchen, welche das Bestehen der Sätze von Cantor und Du Bois-Reymond sichern. Solche bieten sich ziemlich leicht. Auch unser Mittelwertsatz ist, wenn man zu höheren Ableitungen übergeht, einer Verallgemeinerung fähig.

Während aber unser angeführtes Beispiel zeigt, daß in diesen Fällen Einschränkungen nötig sind, zeigt es keineswegs, daß man solche auch bei Mitteln erster Ordnung heranziehen muß. Wie gesagt, diese Frage bleibt auch nach der vorliegenden Arbeit unentschieden.

12.

Sur la représentation analytique des fonctions définies par des séries de Dirichlet

Acta. Math. *35* (1911), 253–270

(Lettre à M. MITTAG-LEFFLER.)

Je me permets de vous adresser un résumé sommaire des recherches que je vous avais communiquées dans mes lettres du 28 décembre 1909 et du 12 juin 1910.

Considérons une série de DIRICHLET

$$\sum_{n=0}^{\infty} a_n e^{-\lambda_n s}, \tag{1}$$

$$(0 \leq \lambda_n < \lambda_{n+1};\ \lim_{n=\infty} \lambda_n = \infty) \tag{2}$$

qui ait la droite de convergence $R(s)=d$, d désignant une quantité réelle, finie. Désignons par $f(s)$ la fonction analytique, représentée par la série dans le demi-plan $R(s)>d$. Comme nous apprend la théorie des séries entières, la droite $R(s)=d_1$ étant la droite de convergence de la série

$$f_1(s) = \sum_{n=0}^{\infty} a_n e^{-ns},$$

il y a nécessairement des points singuliers de la fonction $f_1(s)$ sur cette droite. On sait que cela n'est plus vrai en général, quand les λ_n désignent des nombres quelconques (satisfaisant toujours à la condition (2)). P. e. la série

$$\sum_{n=1}^{\infty} \frac{(-1)^n}{n^s} \tag{3}$$

a la droite de convergence $R(s) = o$. Néanmoins, la fonction qu'elle représente dans le demi-plan $R(s) > o$, est une fonction entière.

Nous sommes forcément conduits à conclure que dans ce cas et dans des cas pareils, c'est une singularité située à l'infini qui détruit la convergence au-delà de la droite de convergence.

Dans deux Notes des »Comptes Rendus» j'ai introduit une méthode de sommation (généralisation de la méthode des moyennes arithmétiques) que j'ai appelée méthode des *moyennes typiques* (5 juillet et 22 novembre 1909). Les expressions fournies par cette méthode convergent aussi ou dans le plan tout entier, ou dans un demi-plan, ou enfin elles divergent partout. C'est le second cas qui nous intéresse surtout. Le demi-plan de sommabilité peut être beaucoup plus étendu que le demi-plan de convergence, mais, pour la droite de sommabilité se présente le même fait qui se présentait pour la droite de convergence. *Elle ne contient pas nécessairement de point singulier sur sa partie finie.*[1]

Je vais montrer que par une généralisation convenable des méthodes fondées sur l'emploi de l'intégrale de LAPLACE-ABEL, on arrive à des expressions limites qui n'ont aucun inconvénient à cet égard. Lorsque la fonction est entière, nos expressions limites la représentent dans tout le plan. Au cas contraire, *la frontière du domaine de convergence contient au moins un point singulier fini.* De plus, par votre artifice d'introduire dans les expressions un certain paramètre tendant vers zéro, *on obtient des expressions limites multiples qui convergent à l'intérieur de l'étoile principale de la fonction et divergent à son extérieur.* Du reste, *la notion d'étoile que vous avez introduite en Analyse, joue ici le même rôle fondamental que dans la théorie des séries entières.*

La recherche des singularités d'une fonction définie par une série de TAYLOR, commence par la détermination du rayon de convergence. Il y correspondrait, pour les séries de DIRICHLET, la détermination de leur abscisse de convergence. Or, comme nous avons dit, la connaissance de cette abscisse et même celle des abscisses de sommabilité, ne nous apprend en général rien sur les singularités situées dans la partie finie du plan. Par contre, les résultats que je viens d'indiquer, nous fourniront immédiatement une méthode générale pour la recherche de ces singularités. Il me semble donc que les méthodes fondées sur l'intégrale de LAPLACE-ABEL ont même une plus grande importance dans le cas général que dans le cas particulier des séries entières.

*

[1] Ce fait a été déjà signalé (à un autre point de vue) par M. BOHR qui appliquait les moyennes arithmétiques entre autres à la série (3).

Permettez-moi de rappeler brièvement les résultats sur les séries entières que je vais généraliser dans la suite.

Envisageons la série entière

$$F(x) = \sum_{n=0}^{\infty} a_n x^n \tag{4}$$

qui ait le rayon de convergence fini $R \neq 0$, et formons la fonction

$$\Phi_1(x) = \sum_{n=0}^{\infty} \frac{a_n}{\underline{|n}} x^n, \tag{5}$$

qui, en vertu de notre supposition $R \neq 0$, est une fonction entière en x. Du reste, cette fonction et l'intégrale qui va suivre, interviennent déjà dans des recherches de LAPLACE et dans celles d'ABEL.

M. BOREL a démontré que l'on a

$$F(x) = \int_0^{\infty} e^{-\omega} \Phi_1(\omega x)\, d\omega \tag{6}$$

dans un domaine qu'il appelait *»polygone de sommabilité», l'intégrale convergeant uniformément dans tout domaine fini, intérieur à celui-ci.*

M. PHRAGMÉN a établi à son tour que *l'intégrale ne peut converger à l'extérieur du domaine de* M. BOREL, *ce domaine est* donc bien une *»étoile de convergence»* de l'expression (6) dans le sens que vous avez prêté à ce mot.

Le domaine, fixé par M. BOREL, se définit de la manière suivante: Menons par chaque point singulier ξ de la fonction une droite, perpendiculaire au segment $(0, \xi)$. Les points du plan qui sont du même côté de chacune de ces droites que l'origine, constituent l'intérieur du domaine BOREL.

On voit que ces points intérieurs x sont aussi caractérisés par le fait qu'on peut décrire autour de $\frac{x}{2}$ un cercle avec un rayon supérieur à $\frac{|x|}{2}$, tel que la fonction $F(x)$ soit régulière à son intérieur et sur sa périphérie.

Vous avez repris la question dans vos quatrième et cinquième Notes, et aussi dans votre conférence de Rome, *en démontrant que la généralisation de l'intégrale ci-dessus permet de représenter la fonction $F(x)$ dans toute son étoile principale.*

Formons la fonction entière

$$\Phi_\alpha(x) = \sum_{n=0}^{\infty} \frac{a_n}{\underline{|\alpha n}} x^n; \qquad (0 < \alpha;\ \underline{|\alpha n} = \Gamma(\alpha n + 1)). \tag{7}$$

Vous avez établi l'égalité

$$(8)\qquad F(x)=\int_0^\infty e^{-\omega^{\frac{1}{a}}}\Phi_a(\omega x)\,d\omega^{\frac{1}{a}}=\int_0^\infty e^{-\omega}\Phi(\omega^a x)\,d\omega,$$

les intégrales du second membre possédant une certaine étoile de convergence $B^{(a)}$ et la convergence étant uniforme dans tout domaine fini, intérieur à cette étoile. Les étoiles $B^{(a)}$ ont en outre la double propriété extrêmement remarquable qu'elles s'approchent indéfiniment de l'étoile principale de la fonction, quand α tend vers zéro, et du cercle de convergence de la série, quand α croît indéfiniment.

Enfin dans *votre conférence*, vous avez donné le résultat à la fois simple et élégant que voici:

On a uniformément, dans tout domaine fini, intérieur à l'étoile principale de la fonction, l'égalité:[1]

$$(9)\qquad F(x)=\lim_{a=0}\left(a_0+\frac{a_1}{\lfloor\alpha\,.\,1}x+\frac{a_2}{\lfloor\alpha\,.\,2}x^2+\cdots+\frac{a_n}{\lfloor\alpha\,.\,n}x^n+\cdots\right)=\lim_{a=0}\Phi_a(x).$$

Cela posé, revenons maintenant à l'étude des séries de Dirichlet. Posons

$$(10)\qquad e^{-s}=x,$$

$$(11)\qquad f(s)=\sum_{n=0}^{\infty}a_n e^{-\lambda_n s}=F(x)=\sum_{n=0}^{\infty}a_n x^{\lambda_n}.$$

Nous supposons de nouveau que notre série de Dirichlet converge dans un certain demi-plan.

Quant à la fonction $F(x)$, selon les formules (10) et (11), elle doit être étudiée (en général), sur *la surface de* Riemann de la fonction $\log x$. Dans un certain voisinage de l'origine, $F(x)$ est définie sur tous les feuillets par le développement ci-dessus. On forme ensuite votre étoile principale à l'aide des points singuliers qu'on rencontre, en formant le prolongement analytique de la fonction ainsi définie, le long des demi-droites issues de l'origine.

Envisageons maintenant le plan de la variable s. *Excluons de ce plan les demi-droites parallèles à l'axe réel négatif, allant des points singuliers de la fonction $f(s)$ à l'infini. Le domaine a, qui reste, est la transformée de votre étoile principale par la substitution $s=-\log x$*

La définition de l'étoile $B^{(a)}$ sur la surface de Riemann est essentiellement

[1] En vertu d'un théorème de M. Borel, généralisé par M. Phragmén, l'étoile principale n'est pas nécessairement le domaine total de convergence de l'expressionlimite (9).

la même que celle que vous avez donnée pour le plan. Faisons parcourir à x_0 tous les points à l'intérieur de l'étoile A, pour lesquels la courbe

$$R\left(\left(\frac{x_0}{x}\right)^{\frac{1}{\alpha}}\right) = 1 \qquad (12)$$
$$-\alpha\frac{\pi}{2} < \operatorname{Arg}\left(\frac{x_0}{x}\right) < \alpha\frac{\pi}{2}$$

est située à l'intérieur de l'étoile A, et appelons $B^{(\alpha)}$ l'étoile auxiliaire, obtenue de cette manière.

En ce qui concerne le plan de la variable $s = \sigma + it$, considérons le contour défini par les relations

$$(13) \qquad \sigma = \alpha \log\left(\cos\frac{t}{\alpha}\right), \qquad \left(-\alpha\frac{\pi}{2} < t < \alpha\frac{\pi}{2}\right)$$

et le domaine qu'il renferme

$$(13') \qquad \sigma < \alpha \log\left(\cos\frac{t}{\alpha}\right), \qquad \left(-\alpha\frac{\pi}{2} < t < \alpha\frac{\pi}{2}\right).$$

Le contour passe par l'origine, il est symétrique par rapport à l'axe réel négatif et il a deux arcs allant à l'infini. Il s'approche indéfiniment de l'axe réel négatif, lorsque α tend vers zéro.

Soit ξ un point singulier quelconque de la fonction $f(s)$, et excluons du plan tous les points s tels que le point ayant pour affixe la différence $s - \xi$, tombe à l'intérieur du domaine (13'). Le contour limitant le domaine exclu est évidemment congruent au contour (13) et passe par le point ξ. En reprenant cette opération pour tous les points singuliers de la fonction $f(s)$, désignons par $b^{(\alpha)}$ le domaine restant. Nos formules (12) et (13) montrent que le domaine $b^{(\alpha)}$ dérive de l'étoile $B^{(\alpha)}$ au moyen de la substitution $s = -\log x$. Lorsque α tend vers zéro, le domaine $b^{(\alpha)}$ s'approche indéfiniment du domaine a.

Écrivons encore

$$(14) \qquad \Phi_\alpha(x) = \frac{a_0}{\underline{|\alpha\lambda_0}} x^{\lambda_0} + \frac{a_1}{\underline{|\alpha\lambda_1}} x^{\lambda_1} + \cdots + \frac{a_n}{\underline{|\alpha\lambda_n}} x^{\lambda_n} + \cdots; \quad \underline{|\alpha\lambda_n} = \Gamma(\alpha\lambda_n + 1)$$

et

$$(15) \qquad \varphi_\alpha(s) = \frac{a_0}{\underline{|\alpha\lambda_0}} e^{-\lambda_0 s} + \frac{a_1}{\underline{|\alpha\lambda_1}} e^{-\lambda_1 s} + \cdots + \frac{a_n}{\underline{|\alpha\lambda_n}} e^{-\lambda_n s} + \cdots$$

On peut facilement établir que *la série $\varphi_\alpha(s)$ converge partout* et qu'elle *représente* par conséquent une *fonction entière,* lorsque la série $\sum a_n e^{-\lambda_n s}$ a un domaine de convergence (plan entier ou demi-plan).

Ces préliminaires posés, je vais démontrer les théorèmes suivants:

I. *Le point s étant intérieur au domaine* $b^{(\alpha)}$ *appartenant à notre fonction f, on a l'égalité*

$$(16)\qquad f(s) = F(x) = \int_0^{\infty} e^{-\omega}\, \Phi_\alpha(\omega^\alpha x)\, d\omega = \int_{-\infty}^{+\infty} e^{-e^v}\, \varphi_\alpha(s - \alpha v)\, e^v\, dv,$$

la seconde intégrale convergeant uniformément pour les points s d'un domaine fini[1] *quelconque, intérieur à* $b^{(\alpha)}$ *et divergeant à l'extérieur de* $b^{(\alpha)}$. *La première intégrale a les mêmes propriétés par rapport à l'étoile* $B^{(\alpha)}$.

II. *On a uniformément dans toute région finie*[1] *intérieure à l'étoile principale A, respectivement dans la région correspondante du domaine a*

$$(17)\qquad F(x) = \lim_{\alpha=0}\left(\frac{a_0}{\lfloor\alpha\lambda_0} x^{\lambda_0} + \frac{a_1}{\lfloor\alpha\lambda_1} x^{\lambda_1} + \cdots + \frac{a_n}{\lfloor\alpha\lambda_n} x^{\lambda_n} + \cdots\right) = \lim_{\alpha=0} \Phi_\alpha(x),$$

$$(18)\qquad f(s) = \lim_{\alpha=0}\left(\frac{a_0}{\lfloor\alpha\lambda_0} e^{-\lambda_0 s} + \frac{a_1}{\lfloor\alpha\lambda_1} e^{-\lambda_1 s} + \cdots + \frac{a_n}{\lfloor\alpha\lambda_n} e^{-\lambda_n s} + \cdots\right) = \lim_{\alpha=0} \varphi_\alpha(s).$$

Démonstration. La démonstration de nos énoncés se fonde essentiellement sur un corollaire immédiat du théorème connu que voici:

Supposons que *la série de* Dirichlet

$$f(s) = \sum_{n=0}^{\infty} a_n e^{-\lambda_n s}$$

converge au point s_0. Il en résulte qu'elle *converge uniformément dans le domaine infini*

$$-\vartheta \leqq \operatorname{Arg}(s - s_0) \leqq \vartheta; \quad 0 \leqq \vartheta < \frac{\pi}{2}.$$

Nous n'avons besoin de ce théorème que dans la forme plus restreinte qui suit.

Sous la condition indiquée, *la série converge uniformément dans toute demi-bande infinie, définie par les inégalités:*

$$R(s - s_0) \geqq \delta > 0$$

et

$$|I(s)| \leqq \tau$$

τ désignant un nombre fini, positif quelconque.

[1] La marche de notre démonstration va montrer qu'on peut aussi admettre certains domaines infinis.

Pour simplifier la démonstration de (16), supposons d'abord $\alpha = 1$ et considérons l'intégrale

$$\int_0^{\omega} e^{-\omega} \Phi_1(\omega x)\, d\omega.$$

Les raisonnements qu'on trouve dans les travaux de M. Borel et dans votre quatrième Note, reposent sur le fait que la différence

$$F(x) - \int_0^{\omega} e^{-\omega} \Phi_1(\omega x)\, d\omega \tag{20}$$

est égale à une intégrale complexe curviligne (dépendant de ω), dans laquelle intervient la fonction $F(z)$, le chemin d'intégration entourant l'origine et le point x.

Or, dans notre cas actuel, *l'origine est en général un point très singulier de la fonction*

$$F(z) = \sum_{n=0}^{\infty} a_n z^{\lambda_n}$$

et les théorèmes de Cauchy n'étant plus valables à son voisinage, la différence ci-dessus ne peut se mettre sous la forme mentionnée.

Toutefois, la difficulté n'est qu'apparente. En effet, grâce au théorème cité sur la convergence uniforme des séries de Dirichlet, nous pouvons montrer facilement que la différence ci-dessus s'exprime par la même intégrale curviligne que dans le cas des séries entières, si l'on applique un chemin d'intégration convenable qui passe par l'origine, au lieu de l'entourer.

Décrivons en effet autour de l'origine comme centre un cercle quelconque et traçons deux rayons de ce cercle qui forment avec l'axe réel positif les angles $(1 + \varepsilon)\frac{\pi}{2}$ et $-(1+\varepsilon)\frac{\pi}{2}$; $(0 < \varepsilon \leq 1)$. Désignons par U la courbe fermée, composée de ces deux rayons et du plus grand des arcs qu'ils interceptent sur la circonférence. On obtient de la formule de Hankel par un simple changement de variable la formule suivante:

$$\frac{1}{\underline{|\gamma}} = \frac{1}{2\pi i} \int_U e^u u^{\frac{1}{\gamma}} \frac{du}{u} \tag{21}$$

l'intégrale étant prise dans le sens positif.

Cette formule que vous avez déjà utilisée dans la théorie du prolongement analytique, me sert aussi de point de départ.

En effet, x étant un point donné quelconque, on peut choisir le rayon r du cercle considéré assez petit, pour que rx soit intérieur au cercle de convergence du développement

$$F(z) = \sum_{n=0}^{\infty} a_n z^{\lambda_n}.$$

Dans cette hypothèse, la série

$$F(xu) = \sum_{n=0}^{\infty} a_n (xu)^{\lambda_n} \tag{22}$$

est d'après le théorème cité[1] sur les séries de Dirichlet, uniformément convergente pour les points u de la courbe U. On peut par conséquent intégrer cette série terme à terme le long de cette courbe et écrire

$$\Phi_1(x) = \frac{1}{2\pi i} \int_U e^{\frac{1}{u}} F(xu) \frac{du}{u} \tag{23}$$

ou par un changement de variable

$$\Phi_1(x) = \frac{1}{2\pi i} \int_{U_x} e^{\frac{x}{z}} F(z) \frac{dz}{z}. \tag{24}$$

U_x désignant la courbe qu'on obtient en multipliant la courbe U par x. Les deux demi-droites dont les segments font partie de la courbe U_x, forment avec le segment $(0, x)$ les angles $\pm (1+\varepsilon)\frac{\pi}{2}$. $(0 < \varepsilon \leqq 1)$.

Soit maintenant s un point intérieur au domaine $b^{(1)}$ correspondant à la fonction $f(s)$ et posons de nouveau

$$x = e^{-s}.$$

Dans notre condition, x sera intérieur à l'étoile $B^{(1)}$, on pourrait par conséquent, dans le cas des séries entières, remplacer la courbe U_x par un cercle de centre $\frac{x}{2}$ et de rayon supérieur à $\frac{|x|}{2}$, dans lequel la fonction $F(z)$ serait encore régulière. Nous avons déjà dit que ce cercle ne peut nous servir dans le cas général (le

[1] En effet, si l'on écrit (22) sous la forme $\Sigma a_n e^{-\lambda_n s}$, les points s qui correspondent aux points xu, sont tous situés dans une demi-bande (de largeur $(1+\varepsilon)\pi \leqq 2\pi$).

point $z = 0$ étant aussi singulier) que dans une forme modifiée. Reprenons à cet effet les deux vecteurs, issus de l'origine, formant avec le segment $(0, x)$ les angles $\pm (1 + \varepsilon)\frac{\pi}{2}$, et désignons par D et D' leurs points d'intersection avec la circonférence ci-dessus, décrite autour de $\frac{x}{2}$. La courbe H qui va nous servir, sera composée des segments entre l'origine et les points D et D' et du plus grand arc de la circonférence. Comme nous avons supposé que le point x était à l'intérieur de l'étoile $B^{(1)}$, nous pouvons choisir le rayon de la circonférence assez rapproché de $\frac{|x|}{2}$ pour qu'il n'y ait de points singuliers ni sur la courbe H, ni à son intérieur (sauf peut-être le point $z = 0$). On peut par conséquent appliquer le théorème de Cauchy au domaine entre les courbes U_x et H et écrire

(25) $$\Phi_1(x) = \frac{1}{2\pi i}\int_H e^{\frac{x}{z}} F(z) \frac{dz}{z}$$

aussi bien que

(26) $$\Phi_1(\omega x) = \frac{1}{2\pi i}\int_H e^{\frac{\omega x}{z}} F(z) \frac{dz}{z}.$$

Il s'ensuit en intervertissant l'ordre des intégrations:

$$\int_0^\omega e^{-\omega}\Phi_1(\omega x)\, d\omega = \frac{1}{2\pi i}\int_H e^{\omega\left(\frac{x}{z}-1\right)} \frac{F(z)}{\frac{x}{z}-1}\frac{dz}{z} - \frac{1}{2\pi i}\int_H \frac{F(z)}{\frac{x}{z}-1}\frac{dz}{z}$$

c'est à dire

(27) $$F(x) - \int_0^\omega e^{-\omega}\Phi_1(\omega x)\, d\omega = \frac{1}{2\pi i}\int_H e^{\omega\left(\frac{x}{z}-1\right)} \frac{F(z)}{z-x} dz.$$

D'autre part, on a sur la courbe H

(28) $$R\left(\frac{x}{z} - 1\right) < \mu < 0$$

μ désignant une quantité facile à calculer. Par suite

(29) $$\left|e^{\left(\frac{x}{z}-1\right)}\right| < e^{\mu} < 1$$

d'où il résulte immédiatement que la dernière intégrale tend vers zéro avec $\frac{1}{\omega}$, c'est à dire

$$F(x) = \int_0^\infty e^{-\omega} \Phi_1(\omega x)\, d\omega \tag{30}$$

et

$$f(s) = \int_{-\infty}^{+\infty} e^{-e^v} \varphi_1(s - v)\, e^v\, dv. \tag{31}$$

On voit aussi que la première intégrale converge uniformément dans tout domaine fini, intérieur à l'étoile $B^{(1)}$ et situé sur un nombre fini de feuillets de notre surface de RIEMANN. Nos remarques préliminaires montrent suffisamment que la convergence est uniforme même dans des domaines tels que les points à l'infini du domaine correspondant du plan des s tombent dans un angle

$$-\vartheta \leq \operatorname{Arg} s \leq \vartheta; \left[0 \leq \vartheta < \frac{\pi}{2}\right]. \tag{32}$$

Le même raisonnement que M. PHRAGMÉN appliquait aux séries entières, prouve aussi dans notre cas général que les expressions (30) et (31) divergent nécessairement à l'extérieur des domaines correspondants $B^{(1)}$ et $b^{(1)}$.

Soit maintenant α un nombre positif quelconque. La démonstration que vous avez donnée pour vos théorèmes, repose sur les propriétés fondamentales de vos fonctions $E_\alpha(x)$ dont la découverte a inauguré tant de recherches nouvelles. Ici, je vais suivre une voie directe qui peut avoir un certain intérêt aussi au point de vue des séries entières.

La formule de HANKEL nous donne de nouveau la relation

$$\Phi_\alpha(x) = \sum_{n=0}^\infty \frac{a_n}{\underline{|\alpha\lambda_n}} x^{\lambda_n} = \frac{1}{2\pi i} \int_U e^{\frac{1}{u}} F(x u^\alpha) \frac{du}{u}. \tag{33}$$

qu'on trouve dans votre cinquième Note (pour les séries entières.)

Supposons maintenant que le point s soit intérieur au domaine $b^{(\alpha)}$. On peut alors trouver un contour $H^{(\alpha)}$ passant par l'origine, renfermant le point $x = e^{-s}$ tel que $F(z)$ n'ait de point singulier ni sur ce contour, ni à son intérieur, et de plus que l'on ait

$$R\left(\left(\frac{x}{z}\right)^{\frac{1}{\alpha}} - 1\right) < \mu < 0, \tag{34}$$

pour tous les points z du contour $H^{(\alpha)}$.[1]

Puis, on a par un simple changement de variable

$$(35)\qquad \Phi_\alpha(\omega^\alpha x) = \frac{1}{2\alpha\pi i}\int_{H^{(\alpha)}} e^{\omega\left(\frac{x}{z}\right)^{\frac{1}{\alpha}}} F(z)\frac{dz}{z}$$

et

$$(36)\qquad \int_0^\omega e^{-\omega}\Phi(\omega^\alpha x)\,d\omega = \frac{1}{\alpha 2\pi i}\int_{H^{(\alpha)}} e^{\omega\left(\left(\frac{x}{z}\right)^{\frac{1}{\alpha}}-1\right)} \frac{F(z)}{\left(\frac{x}{z}\right)^{\frac{1}{\alpha}}-1}\frac{dz}{z} - \frac{1}{\alpha 2\pi i}\int_{H^{(\alpha)}} \frac{F(z)}{\left(\frac{x}{z}\right)^{\frac{1}{\alpha}}-1}\frac{dz}{z}.$$

Le point $z = x$ étant un pôle simple, avec le résidu $-\alpha x$, de la fonction

$$\frac{1}{\left(\frac{x}{z}\right)^{\frac{1}{\alpha}}-1}$$

la dernière intégrale est égale à $F(x)$. On peut donc écrire

$$(37)\qquad F(x) - \int_0^\omega e^{-\omega}\Phi_\alpha(\omega^\alpha x)\,d\omega = -\frac{1}{\alpha 2\pi i}\int_{H^{(\alpha)}} e^{\omega\left(\left(\frac{x}{z}\right)^{\frac{1}{\alpha}}-1\right)} \frac{F(z)}{\left(\frac{x}{z}\right)^{\frac{1}{\alpha}}-1}\frac{dz}{z}.$$

L'inégalité (34) montre très nettement que le second membre tend vers zéro avec $\frac{1}{\omega}$, de sorte que l'on a

$$F(x) = \int_0^\infty e^{-\omega}\Phi(\omega^\alpha x)\,d\omega.$$

Les remarques que nous avons faites dans le cas $\alpha = 1$, s'appliquent sans difficulté à la convergence uniforme (dans les domaines caractérisés, intérieurs à $B^{(\alpha)}$), et à la divergence (aux points extérieurs à $B^{(\alpha)}$) de notre expression limite.

On voit que l'intégrale

$$(38)\qquad \int_0^\infty e^{-\omega}\left|\Phi_\alpha(\omega^\alpha x)\right|d\omega$$

[1] Cette inégalité montre qu'on peut choisir la courbe $H^{(\alpha)}$ de plus en plus aplatie et la faire tendre vers le segment $(0, x)$ lorsque α tend vers zéro.

converge aussi uniformément dans les domaines caractérisés, fait déjà observé par M. Borel pour le cas des séries entières et $\alpha = 1$.

Ces résultats nous permettent de *déterminer* par un calcul simple *les sommets de votre étoile principale* appartenant à la fonction $F(x)$.

Considérons une demi-droite issue de l'origine, formant avec l'axe réel positif l'angle t (cet angle pouvant varier entre $-\infty$ et $+\infty$). Déterminons le point d'intersection x_a de cette demi-droite et de la frontière de $B^{(a)}$. Posons

$$x = r e^{it}, \quad x_a = r_a e^{it}$$

et

$$(39) \qquad \int_0^\infty e^{-\omega} \,|\, \Phi_a(\omega^a x) \,|\, d\omega = \int_0^\infty e^{-\omega} \,|\, \Phi_a(\omega^a r e^{it}) \,|\, d\omega = \frac{1}{r^{\frac{1}{a}}} \int_0^\infty e^{-\frac{\omega}{r^{\frac{1}{a}}}} \,|\, \Phi_a(\omega^a e^{it}) \,|\, d\omega.$$

La valeur $\frac{1}{r_a^{\frac{1}{a}}}$ est égale à l'abscisse de convergence de la dernière intégrale. Cette abscisse se calcule moyennant une formule de M. Landau, analogue à la formule de M. Cahen pour l'abscisse de convergence d'une série de Dirichlet. On a

$$(40) \qquad \frac{1}{r_a^{\frac{1}{a}}} = \overline{\lim_{\omega = \infty}} \frac{\log \int_0^\omega |\, \Phi_a(v^a e^{it}) \,|\, dv}{\omega}$$

$$(41) \qquad \frac{1}{r_a} = \overline{\lim_{\omega = \infty}} \left(\frac{\log \int_0^\omega |\, \Phi_a(v^a e^{it}) \,|\, dv}{\omega} \right)^a$$

Le sommet $x_0 = R e^{it}$ de l'étoile principale, situé sur la demi-droite considérée, est évidemment donné par la formule

$$(42) \qquad \frac{1}{R} = \lim_{a=0} \overline{\lim_{\omega = \infty}} \left(\frac{\log \int_0^\omega |\, \Phi_a(v^a e^{it}) \,|\, dv}{\omega} \right)^a.$$

En reprenant un raisonnement de M. Borel,[1] donné au sujet des séries entières (dans le cas $\alpha = 1$), on arrive à une formule plus simple.

En effet les formules (34) et (35) montrent très nettement que pour tout point $x = r e^{it}$ intérieur à l'étoile $B^{(a)}$, on a

[1] Borel: Leçons sur les séries divergentes, Paris, 1901 (pp. 138—141).

$$\lim_{\omega=\infty} e^{-\omega} |\Phi_a(\omega^a x)| = \lim_{\omega=\infty} e^{-\omega} |\Phi(\omega^a r e^{it})| = 0$$

donc on a aussi

$$\lim_{\omega=\infty} e^{-\left(\frac{\omega}{r}\right)^{\frac{1}{a}}} |\Phi_a(\omega e^{it})| = 0$$

et par conséquent à partir d'une certaine valeur de ω, on peut écrire

$$|\Phi_a(\omega e^{it})| < e^{\left(\frac{\omega}{r}\right)^{\frac{1}{a}}}.$$

Il en résulte

$$\frac{\log|\Phi_a(\omega e^{it})|}{\omega^{\frac{1}{a}}} < \frac{1}{r^{\frac{1}{a}}}. \tag{43}$$

D'autre part, si le point $x = re^{it}$ est extérieur à l'étoile $B^{(a)}$, l'expression

$$e^{-\omega}|\Phi_a(\omega^a r e^{it})|$$

ne reste pas bornée, car en cas contraire, l'intégrale (16) convergerait en tout point

$$x = r' e^{it}$$

tel que

$$r' < r,$$

c'est à dire $x = re^{it}$ ne serait pas un point *extérieur* à l'étoile $B^{(a)}$.

La quantité r_a a donc la double propriété que pour $r < r_a$, on a à partir d'une certaine valeur de ω l'inégalité (43) et que pour $r > r_a$, on peut trouver des valeurs de ω croissant indéfiniment, pour lesquelles l'inégalité n'est pas remplie. Cela s'exprime par la formule

$$\frac{1}{r_a^{\frac{1}{a}}} = \overline{\lim_{\omega=\infty}} \frac{\log|\Phi_a(\omega e^{it})|}{\omega^{\frac{1}{a}}}, \tag{44}$$

ou

$$\frac{1}{r_a} = \overline{\lim_{\omega=\infty}} \frac{\log^a|\Phi_a(\omega e^{it})|}{\omega}.$$

On en conclut

$$\frac{1}{R} = \lim_{a=0} \overline{\lim_{\omega=\infty}} \frac{\log^a|\Phi_a(\omega e^{it})|}{\omega}. \tag{45}$$

Une transformation simple permet de déterminer les points singuliers correspondants de la fonction $f(s)$. Envisageons en effet la droite, parallèle à l'axe réel, avec l'ordonnée t. L'abscisse σ du premier point singulier qu'on rencontre, en prolongeant la fonction $f(s)$ à gauche, le long de cette droite, est donnée par la formule

(46) $$\sigma = \lim_{a=0} \overline{\lim_{\omega=\infty}} (a \log\log |\varphi_a(-\omega + it)| - \omega).$$

Remarquons que (44) et par conséquent les formules qui en découlent, peuvent être en défaut, quand r_a est infini. Dans ce cas en effet, la formule (44) peut donner au lieu de la valeur zéro, une valeur négative. Mais quand x_a est finie, elle fournit une valeur positive bien déterminée et inversement, si elle fournit une valeur positive, cette valeur est égale à $\frac{1}{r_a^{\frac{1}{a}}}$.

Ayant fini la démonstration et l'application du théorème I, il ne me reste qu'à démontrer le théorème II, c. à d. la relation

(18) $$f(s) = \lim_{a=0} \left(\frac{a_0}{\underline{|a\lambda_0}} e^{-\lambda_0 s} + \frac{a_1}{\underline{|a\lambda_1}} e^{-\lambda_1 s} + \cdots + \frac{a_n}{\underline{|a\lambda_n}} e^{-\lambda_n s} + \cdots \right) = \lim_{a=0} \varphi_a(s),$$

ou la relation équivalente

(17) $$F(x) = \lim_{a=0} \Phi_a(x)$$

$x = e^{-s}$ désignant un point intérieur[1] à l'étoile principale de la fonction F.

Considérons un contour fermé Γ quelconque, passant par l'origine et composé au voisinage de ce point de deux segments de droites, formant de nouveau avec l'axe réel positif les angles $\pm (1 + \varepsilon)\frac{\pi}{2}$, $(0 < \varepsilon \leq 1)$. Le point x étant supposé intérieur à l'étoile principale, on peut choisir a assez petit pour que les points xu^a soient aussi tous intérieurs à l'étoile, u désignant un point quelconque de la courbe Γ. De cette façon, la fonction $F(xu^a)$ sera régulière en u sur la courbe Γ et à son intérieur (excepté évidemment $u = 0$). On a par conséquent

(47) $$\Phi_a(x) = \frac{1}{2\pi i} \int_\Gamma e^{\frac{1}{u}} F(xu^a) \frac{du}{u},$$

[1] Pendant l'impression de cette lettre, M. Mittag-Leffler m'apprit que la démonstration qui suit, est essentiellement la même que celle qu'il avait donnée pour les séries entières, dans ses cours de 1905—06. Il avait aussi montré que le théorème reste valable en tout sommet ξ de l'étoile pour lequel il existe un angle de sommet ξ, renfermant le vecteur $(0, \xi)$ où la fonction est continue. Le lecteur verra sans difficulté que cette extension subsiste aussi dans notre cas général. (M. R.)

d'ailleurs, il est évident que

$$F(x) = \frac{1}{2\pi i}\int\limits_{\Gamma} e^{\frac{1}{u}} F(x) \frac{du}{u},$$

par suite

$$(48) \qquad F(x) - \Phi_a(x) = \frac{1}{2\pi i}\int\limits_{\Gamma} [F(x) - F(xu^a)]\, e^{\frac{1}{u}} \frac{du}{u}.$$

Désignons par P et P' deux points de la courbe Γ, tous les deux étant situés au voisinage de l'origine, le premier au-dessus de l'axe réel et le second au-dessous de cet axe. L'égalité ci-dessus peut s'écrire

$$(49) \quad F(x) - \Phi_a(x) = \frac{1}{2\pi i}\int\limits_{P}^{P'} [F(x) - F(xu^a)]\, e^{\frac{1}{u}} \frac{du}{u} + \frac{1}{2\pi i}\int\limits_{P'}^{P} [F(x) - F(xu^a)]\, e^{\frac{1}{u}} \frac{du}{u}.$$

Le premier chemin d'intégration, au voisinage de l'origine, devient arbitrairement petit, lorsque P et P' s'approchent suffisamment de l'origine, donc la valeur absolue de la première intégrale peut devenir $< \eta$, indépendamment de α, quelque petite que soit la quantité positive η. Quant aux points u qui forment le second chemin d'intégration, on pourra, après avoir fixé P et P', choisir α assez petit afin qu'on ait pour tous ces points:

$$|x - xu^a| < \delta$$

et

$$|F(x) - F(xu^a)| < \delta',$$

les quantités positives δ et δ' étant arbitrairement petites. Donc pour α assez petit, on aura aussi

$$\left| \frac{1}{2\pi i}\int\limits_{P'}^{P} [F(x) - F(xu^a)]\, e^{\frac{1}{u}} \frac{du}{u} \right| < \eta,$$

c'est à dire

$$|F(x) - \Phi_a(x)| < 2\eta,$$

ou

$$\lim_{a=0} \Phi_a(x) = F(x).$$

Notre raisonnement montre aussi que la convergence est uniforme dans tout domaine fini, intérieur à l'étoile principale et situé sur un nombre fini de feuillets de notre surface de RIEMANN. On peut, aussi pour cette expression limite, admettre des domaines tels que les points à l'infini du domaine correspondant du plan des s tombent dans un angle tel que (32).

Notons encore la formule intéressante:

$$\Phi_a(x) = \frac{1}{2\pi i}\int_{\Gamma} e^{\frac{1}{u}} F(xu^a)\frac{du}{u} = \frac{1}{2\pi i}\int_{\Gamma} e^{\frac{1}{u}}\frac{du}{u}\cdot\frac{1}{2\pi i}\int_{\Gamma'}\frac{F(z)}{z-xu^a}dz =$$

$$(50)\qquad = \frac{1}{2\pi i}\int_{\Gamma'} F(z)\,dz\,\frac{1}{2\pi i}\int_{\Gamma} e^{\frac{1}{u}}\frac{1}{z-xu^a}\frac{du}{u} = \frac{1}{2\pi i}\int_{\Gamma'} F(z)\frac{dz}{z}\,\frac{1}{2\pi i}\int_{\Gamma} e^{\frac{1}{u}}\frac{1}{1-\frac{x}{z}u^a}\frac{du}{u} =$$

$$= \frac{1}{2\pi i}\int_{\Gamma'} F(z)\,E_a\left(\frac{x}{z}\right)\frac{dz}{z},$$

où E_a désigne votre fonction entière bien connue, et Γ' désigne une courbe intérieure à l'étoile principale qui passe par l'origine et renferme la courbe Γ'', décrite par xu^a, lorsque u décrit la courbe Γ. On peut tirer facilement de cette formule une démonstration indirecte du théorème que je viens de démontrer.

* * *

J'ai terminé l'indication de mes recherches sur la représentation du prolongement analytique des séries de DIRICHLET. Permettez-moi encore de reprendre quelques remarques, publiées déjà en partie, qui rentrent plutôt dans l'ordre d'idées du problème d'ABEL.

Ces remarques, concernant la théorie des séries entières, se rattachent à votre conférence de Rome, où vous avez le premier étudié le problème. Elles nous fourniront des indications utiles aussi pour les séries de DIRICHLET.

Parmi les expressions limites diverses que vous avez étudiées dans vos travaux, *les séries de polynomes de votre troisième Note sont les plus efficaces au point de vue du problème* d'ABEL.[1]

D'abord j'ai étudié le problème opposé. J'ai examiné, si l'on peut trouver des fonctions telles que votre série de polynomes diverge dans un sommet de l'étoile correspondante quoique la fonction satisfasse en ce sommet à certaines conditions de continuité et de régularité.

Ainsi, j'ai construit *une fonction, continue dans la figure génératrice* $V^{(a)}$ *(contour y compris) appartenant au point* x *qui, de plus, était régulière dans toute cette*

[1] M. RIESZ: Sur un problème d'ABEL. (Extrait d'une lettre à M. MITTAG-LEFFLER, 24 mai 1910) Rendiconti del Circolo Matematico di Palermo. T. XXX. 1910 (Séance de 10 juillet 1910).

figure, sauf au point x et pourtant la série de polynomes correspondante

$$G_0(x/\alpha) + G_1(x/\alpha) + \cdots + G_n(x/\alpha) + \cdots$$

était divergente.

Je suis arrivé à construire cet exemple et d'autres exemples de divergence, concernant d'autres expressions-limites, en développant un *principe très fécond, dû à* M. Fejér qu'il vient d'appliquer dans une série de travaux fort remarquables.

Toutefois, j'ai pu vous communiquer dans ma lettre du 28 décembre 1909 qu'il est facile de transformer votre série de polynomes en une autre qui a la même étoile de convergence, mais qui converge aussi aux sommets de la nature indiquée ci-dessus. Ce sont les moyennes arithmétiques des sommes partielles de votre série de polynomes que j'avais en vue. En effet, en appliquant un théorème bien connu de M. Fejér, on montre immédiatement que ces expressions convergent même dans des conditions plus générales.[1] Cependant, j'ai trouvé aussitôt que, *grâce à une propriété heureuse de votre figure génératrice, on peut résoudre la question complètement sans former les moyennes arithmétiques, mais en faisant diminuer le paramètre* α dont dépend l'expression limite. C'est le résultat que j'avais publié dans la lettre que je vous avais adressée, sans faire aucune allusion aux moyennes arithmétiques de vos séries de polynomes.

Dans la même lettre j'avais aussi remarqué que les expressions limites fondées sur l'intégrale de Laplace-Abel ne sont pas aussi efficaces que vos séries de polynomes, quand il s'agit de déterminer la valeur limite de la fonction aux sommets, où elle est continue, la continuité étant toujours définie à l'aide de la figure génératrice correspondante. Néanmoins, on peut (comme je vous avais communiqué dans ma lettre du 28 décembre 1909) déduire aussi dans ce cas des expressions limites nouvelles, bien utilisables en ces points qui se forment d'une manière analogue aux moyennes arithmétiques. Le procédé consiste dans la formule:

$$\lim_{\Omega=\infty} \frac{1}{\Omega} \int_0^{\Omega} d\omega \int_0^{\omega} e^{-v} \Phi_a(v^a x)\, dv. \tag{51}$$

D'ailleurs, les avantages qu'on peut tirer pour les séries entières de l'application des figures génératrices, formant au point $x = 1$ un angle qui diminue

[1] Ce résultat a été aussi trouvé par M. Dienes qui l'avait publié avec d'autres résultats concernant les moyennes arithmétiques de vos séries de polynomes, ces autres résultats étant aussi des conséquences immédiates de vos théorèmes et de ceux de M. Fejér. (Comptes rendus, 25 juillet 1910.) Je dois observer que le raisonnement de ma lettre insérée aux Rendiconti fait voir que dans les énoncés de M. Dienes, l'introduction des moyennes arithmétiques pourrait être évitée.

avec α, suggèrent l'idée d'appliquer de telles figures aussi pour les séries de la forme $\Sigma a_n x^{\lambda_n}$. Pourtant, cette figure devant passer par l'origine à cause du caractère singulier de ce point, les expressions limites ainsi obtenues seront nécessairement, comme l'expression (16), des expressions limites doubles.

* * *

Tous les résultats que je viens d'indiquer dans cet exposé rapide, s'étendent facilement aux intégrales de la forme

$$F(x) = \int_0^\infty a(t)\, x^t\, dt = f(s) = \int_0^\infty a(t)\, e^{-ts}\, dt \tag{52}$$

en posant

$$\Phi_a(x) = \int_0^\infty \frac{a(t)}{\lfloor \alpha t} x^t\, dt = \varphi_a(s) = \int_0^\infty \frac{a(t)}{\lfloor \alpha t} e^{-ts}\, dt. \tag{53}$$

D'ailleurs, les séries $\sum_{n=0}^{\infty} a_n x^{\lambda_n}$ et les intégrales $\int_0^\infty a(t)\, x^t\, dt$ se ramènent facilement à la forme commune

$$\log x \int_0^\infty A(t)\, x^t\, dt.$$

Grâce à cette circonstance, les résultats ci-dessus sont bien utilisables pour la représentation d'une fonction analytique quelconque, au voisinage d'une classe étendue de points de non-uniformité.

Évidemment, l'expression (16) peut converger aussi dans des cas où la série de Dirichlet, ou l'intégrale $\int_0^\infty a(t)\, e^{-ts}\, dt$ correspondante ne converge nulle part.

Györ (Hongrie), 6 octobre 1910.

Marcel Riesz.

13.

Formule d'interpolation pour la dérivée d'un polynome trigonométrique

C. R. Acad. Sci. Paris *158* (1914), 1152–1154

1° I. *Considérons un polynome trigonométrique d'ordre n au plus*

$$F(\varphi) = a_n \cos n\varphi + b_n \sin n\varphi + \ldots + a_1 \cos \varphi + b_1 \sin \varphi + a_0;$$

posons

$$\varphi_1 = \frac{1}{2n}\pi, \qquad \varphi_2 = \frac{3}{2n}\pi, \qquad \ldots, \qquad \varphi_r = \frac{2r-1}{2n}\pi, \qquad \ldots, \qquad \varphi_{2n} = \frac{4n-1}{2n}\pi.$$

On a alors

$$F'(\varphi) = \frac{1}{2n}\sum_{r=1}^{2n} F(\varphi+\varphi_r)\frac{(-1)^{r+1}}{2\sin^2\frac{\varphi_r}{2}}. \tag{1}$$

2° Cette formule conduit immédiatement au beau théorème suivant, dû à M. Serge Bernstein :

Étant donné un polynome trigonométrique d'ordre n au plus tel que

$$|F(\varphi)| \leqq M$$

pour toutes les valeurs de φ, *on a pour toutes ces valeurs*

$$|F'(\varphi)| \leqq Mn \tag{2}$$

le signe d'égalité étant exclu dans la dernière formule, sauf dans le cas où $F(\varphi)$ *est de la forme* $F(\varphi) = M \sin n(\varphi - \omega) = M \cos n\left(\varphi - \frac{\pi}{2n} - \omega\right)$ (¹).

En effet, posons en particulier au lieu de F

$$G(\theta) = \sin n(\theta - \varphi);$$

on aura évidemment

$$G'(\varphi) = n.$$

En substituant d'autre part $G(\theta)$ dans la formule (1) et en observant

(¹) En réalité, M. Bernstein ne démontre la relation (2) que dans le cas où le polynome ne contient que des cosinus ou des sinus. Il en déduit pour le cas général $|F'(\varphi)| < 2Mn$. Cependant, M. Fejér, dans un travail où il donne une application très importante du théorème de M. Bernstein, avait déjà observé que le résultat de M. Bernstein concernant un polynome de sinus conduisait facilement à l'inégalité (2) aussi dans le cas général [*Ueber konjugierte trigonometrische Reihen* (*Journal de Crelle*, t. **144**, p. 50)].

que pour les valeurs $\varphi + \varphi_r$ cette fonction est égale à ± 1 avec des signes alternés, on obtient

$$\frac{1}{2n}\sum_{r=1}^{2n}\frac{1}{2\sin^2\frac{\varphi_r}{2}} = n.$$

Cette relation montre que les inégalités $|F(\varphi + \varphi_r)| \leqq M$ étant satisfaites pour une valeur particulière de φ, on a $|F'(\varphi)| < Mn$, sauf dans le cas où $F(\varphi + \varphi_r) = \pm M$ avec des signes alternés lorsque r varie de 1 à $2n$, le signe d'inégalité devant être remplacé par celui de l'égalité dans ce dernier cas. Si l'on suppose en outre que $|F(\theta)| \leqq M$ pour toutes les valeurs de θ, les relations $F(\varphi + \varphi_r) = \pm M$ (avec des signes alternés) entraînent évidemment que $F(\theta) = \pm M \sin n(\theta - \varphi)$.

3° Le résultat ci-dessus établissant une relation entre l'ordre de grandeur des valeurs $F(\varphi + \varphi_r)$ et $F'(\varphi)$ est manifestement, sous un certain rapport, plus précis que le théorème de M. Bernstein. En le combinant avec la méthode que M. Fejér applique dans son travail cité, nous obtenons le théorème suivant, analogue à un théorème de M. Fejér (Th. V, *loc. cit.*) :

II. *Supposons que la fonction représentée par la série entière $\Sigma a_n z^n$ à l'intérieur de la circonférence $|z| = 1$ soit continue sur cette circonférence. Supposons de plus que, par exemple, la partie réelle de la série converge uniformément dans toutes les racines de l'unité. Alors sa partie imaginaire convergera aussi uniformément pour toutes ces valeurs.*

En combinant d'autre part notre théorème I et la méthode de M. Fejér avec un théorème bien connu de M. G. Cantor sur les coefficients d'une série trigonométrique convergeant dans un intervalle et un théorème bien connu de M. Tauber, on arrive à la généralisation suivante du théorème cité de M. Fejér :

III. *Supposons que la fonction représentée par la série entière $\Sigma a_n z^n$ à l'intérieur de la circonférence $|z| = 1$ soit continue sur un arc de cette circonférence et que, par exemple, la partie réelle de la série converge uniformément sur tout arc intérieur à cet arc. Alors la partie imaginaire aura la même propriété.*

4° Je n'insisterai pas davantage sur les applications de la formule d'interpolation, e préfère ajouter quelques mots sur la vérification de cette formule. Elle se déduit immédiatement des identités suivantes :

$$\frac{1}{2n}\sum_{r=1}^{2n}\frac{(-1)^{r+1}\sin m\varphi_r}{2\sin^2\frac{\varphi_r}{2}} = m, \quad \frac{1}{2n}\sum_{r=1}^{2n}\frac{(-1)^{r+1}\cos m\varphi_r}{2\sin^2\frac{\varphi_r}{2}} = 0 \qquad (m \leqq n)$$

qu'on vérifie facilement en appliquant la formule d'interpolation de Lagrange à la fonction $\frac{x^{n-m}}{x^{2n}+1}$, en la dérivant par rapport à x et en y posant $x=1$. D'ailleurs, la seconde identité est évidente, les termes s'y détruisent deux à deux.

On a encore la formule suivante :

$$F(\varphi)=a_n \cos n\varphi+\frac{\cos n\varphi}{2n}\sum_{r=1}^{2n} F(\varphi_r)(-1)^{r+1}\cot\frac{\varphi_r-\varphi}{2}.$$

En dérivant cette dernière formule pour $\varphi=0$, on obtient par l'introduction d'une nouvelle variable notre formule (1).

De mon côté, je fus amené à la formule (1) par les considérations suivantes : on a la relation évidente

$$F'(\varphi)=\frac{1}{\pi}\int_0^{2\pi} F(\varphi+\theta)\left(\sum_{k=0}^{n} k\sin k\theta\right)d\theta; \tag{3}$$

d'autre part, on vérifie facilement les identités

$$\sum_{r=1}^{2n}\sin k\varphi_r \cos m\varphi_r=0 \qquad (k\leqq n,\ m\leqq n); \qquad \sum_{r=1}^{2n}\sin k\varphi_r \sin m\varphi_r=0$$
$$(k\leqq n,\ m\leqq n,\ k\neq m);$$

$$\sum_{r=1}^{2n}\sin^2 m\varphi_r=n \qquad (m\leqq n-1); \qquad \sum_{r=1}^{2n}\sin^2 n\varphi_r=2n.$$

Ces identités montrent qu'en essayant de remplacer l'intégrale (3) par une valeur moyenne, on doit remplacer la fonction $\sum_{k=1}^{n} k\sin k\theta$ par les valeurs

$$n\sin n\varphi_r+2\sum_{k=1}^{n-1} k\sin k\varphi_r=\frac{(-1)^{r+1}}{2\sin^2\frac{\varphi_r}{2}}.$$

Pour les dérivées d'ordre supérieur, on obtient des formules analogues quoique moins simples. Pour les dérivées d'ordre pair, on introduira les valeurs $\psi_r=\frac{r\pi}{n}$ au lieu des valeurs φ_r.

14.

Eine trigonometrische Interpolationsformel und einige Ungleichungen für Polynome

Jahresber. Deutsch. Math. Ver. *23* (1914), 354–368

Ich habe kürzlich eine trigonometrische Interpolationsformel mitgeteilt[1]), die ich hier herleiten und näher betrachten will.

Es sei $f(\varphi)$ ein trigonometrisches Polynom von höchstens n-tem Grade

$$f(\varphi) = a_n \cos n\varphi + b_n \sin n\varphi + \cdots + a_1 \cos \varphi + b_1 \sin \varphi + a_0.$$

Wir setzen

$$(1) \qquad \varphi_1 = \frac{\pi}{2n}, \; \varphi_2 = \frac{3\pi}{2n}, \; \ldots, \; \varphi_r = \frac{(2r-1)\pi}{2n}, \; \ldots, \; \varphi_{2n} = \frac{(4n-1)\pi}{2n}.$$

Es gilt dann die Formel

$$(2) \qquad f(\varphi) = a_n \cos n\varphi + \frac{\cos n\varphi}{2n} \sum_{r=1}^{2n} f(\varphi_r)\,(-1)^r \cot \frac{\varphi - \varphi_r}{2}.$$

In § 1 gebe ich zwei Beweise für diese Formel. In § 2 leite ich — wie ich das schon in der zitierten Arbeit getan habe, aber der Vollständigkeit halber auch hier ausführe — aus (2) durch Differentiation eine Formel für die Ableitung eines trigonometrischen Polynoms ab, die sofort den Satz von Herrn S. Bernstein über eine solche Ableitung ergibt. In § 3 spezialisiere ich die Formel (2) insbesondere für Sinuspolynome; ich erhalte für diesen Fall (und auch für Kosinuspolynome) einen Ausdruck, der lediglich mit einem Spezialfall der Lagrangeschen Interpolationformel identisch ist. Aus diesem Ausdrucke und dem Bernsteinschen Satze ergibt sich sofort der Satz von Herrn A. Markoff über die Ableitung gewöhnlicher Polynome. In § 4 gebe ich durch Wurzelabzählung einen neuen, sehr einfachen Beweis für den Bernsteinschen Satz und gewinne mit demselben Verfahren eine Verallgemeinerung dieses Satzes und einige andere verwandte Resultate. In § 5 benutze ich die Gelegenheit, eine Verallgemeinerung eines Satzes von Herrn Fejér über den Zusammenhang zwischen den Du-Bois-Reymondschen und Lebesgueschen Singularitäten, die ich auch in der genannten Arbeit ausgesprochen habe, darzulegen.

---

1) Formule d'interpolation pour la dérivée d'un polynome trigonométrique. Comptes Rendus. Paris. t. 158, p. 1154, 27 avril 1914.

§ 1. *Erster Beweis der Formel* (2). Wir können schreiben

$$(3)\qquad (-1)^r \cos n\varphi \cot \frac{\varphi-\varphi_r}{2} = \sin n\,(\varphi-\varphi_r) \cot \frac{\varphi-\varphi_r}{2}$$

$$= \frac{\sin (n+\frac{1}{2})(\varphi-\varphi_r)}{\sin \frac{\varphi-\varphi_r}{2}} - \cos n\,(\varphi-\varphi_r)$$

$$= 2\left(\tfrac{1}{2} + \cos(\varphi-\varphi_r) + \cos 2(\varphi-\varphi_r) + \cdots + \cos n(\varphi-\varphi_r)\right) - \cos n\,(\varphi-\varphi_r).$$

Schreibt man dann $f(\varphi_r) = \sum_{m=0}^{n} (a_m \cos m\varphi_r + b_m \sin m\varphi_r)$ und führt in (2) die mehrfache Summation zuerst nach r und dann nach m aus, so erhält man infolge bekannter Eigenschaften der Einheitswurzeln als Endresultat $f(\varphi)$.

Zweiter Beweis. In unserer Formel hat unter den Koeffizienten des trigonometrischen Polynoms a_n eine Sonderstellung. Dies erklärt sich z. B. dadurch, daß a_n in die Werte $f(\varphi_r)$ gar nicht hineinspielt, da $\cos n\varphi$ für alle φ_r gleich Null ist. Man sieht auch leicht, daß das Polynom durch die Angabe der Werte $f(\varphi_r)$ und des Wertes a_n nicht überbestimmt ist; diese $(2n+1)$ Werte bestimmen die $(2n+1)$ Koeffizienten.

Wir bezeichnen die rechte Seite von (2) mit $\bar{f}(\varphi)$ und wollen nun beweisen, daß identisch $\bar{f}(\varphi) = f(\varphi)$ ist.

Nach (3) ist $\cos n\varphi \cot \frac{\varphi-\varphi_r}{2}$ ein trigonometrisches Polynom von φ, in dem $\cos n\varphi$ nicht vorkommt, da $\cos n(\varphi-\varphi_r) = (-1)^{r+1} \sin n\varphi$ ist. Also ist

$$\bar{f}(\varphi) = a_n \cos n\varphi + \bar{b}_n \sin n\varphi + \bar{a}_{n-1} \cos(n-1)\varphi + \bar{b}_{n-1} \sin(n-1)\varphi + \ldots + \bar{a}_0.$$

Nun ersieht man aber aus (2), daß, für alle Werte von φ_r, $\bar{f}(\varphi_r) = f(\varphi_r)$ ist. Mithin

$$(4)\qquad f(\varphi_r) - \bar{f}(\varphi_r) = 0,$$

ebenso

$$(5)\qquad f(-\varphi_r) - \bar{f}(-\varphi_r) = 0.$$

Addiert man diese beiden Relationen, so folgt, daß das Kosinuspolynom

$$(6)\qquad (a_{n-1} - \bar{a}_{n-1}) \cos(n-1)\varphi + \ldots + (a_1 - \bar{a}_1) \cos\varphi + (a_0 - \bar{a}_0)$$

in mindestens $2n$ Punkten, nämlich in allen Punkten φ_r verschwindet. Subtrahiert man dagegen (5) aus (4), so folgt, daß das Sinuspolynom

$$(7)\qquad (b_n - \bar{b}_n) \sin n\varphi + (b_{n-1} - \bar{b}_{n-1}) \sin(n-1)\varphi + \ldots + (b_1 - \bar{b}_1) \sin\varphi$$

in mindestens $(2n+2)$ Punkten, nämlich in den Punkten 0 und π, und in den $2n$ Punkten φ_r verschwindet.

Nun gilt aber der folgende bekannte Satz, welchen man durch Einführung der Hilfsveränderlichen $\operatorname{tg}\frac{\varphi}{2}$ oder $e^{i\varphi}$ leicht beweist.

Beschreibt φ den Einheitskreis, so kann das trigonometrische Polynom von n-tem Grade $f(\varphi)$ in höchstens $2n$ inkongruenten Punkten verschwinden. Oder genauer, *das Polynom hat $2n$ nach 2π inkongruente (reelle oder imaginäre) Nullstellen, jede Nullstelle mit entsprechender Multiplizität gezählt.*

Hieraus folgt das Verschwinden aller Koeffizienten in (6) und (7), also die identische Gleichung $\bar{f}(\varphi) = f(\varphi)$.

§ 2. Aus (2) folgt durch Differentiation

$$f'(0) = \frac{1}{2n} \sum_{r=1}^{2n} f(\varphi_r) \frac{(-1)^{r+1}}{2 \sin^2 \frac{\varphi_r}{2}} \tag{8}$$

und hieraus allgemeiner[1])

$$f'(\varphi) = \frac{1}{2n} \sum_{r=1}^{2n} f(\varphi + \varphi_r) \frac{(-1)^{r+1}}{2 \sin^2 \frac{\varphi_r}{2}}. \tag{9}$$

Setzen wir in (8) $f(\varphi) = \sin n\varphi$ und berücksichtigen, daß $\sin n\varphi_r = (-1)^{r+1}$ ist, so erhalten wir

$$\frac{1}{2n} \sum_{r=1}^{2n} \frac{1}{2 \sin^2 \frac{\varphi_r}{2}} = n. \tag{10}$$

Wenn also die positive Zahl M so beschaffen ist, daß $|f(\varphi + \varphi_r)| \leqq M$ für einen bestimmten Wert von φ und alle φ_r, so ist für diesen Wert von φ

$$|f'(\varphi)| \leqq Mn,$$

das Gleichheitszeichen ausgeschlossen, wenn nicht $f(\varphi + \varphi_r)$ mit wechselndem Vorzeichen gleich $\pm M$ ist. Wenn aber außerdem das trigonometrische Polynom $f(\theta)$ auch für alle anderen Werte von θ dem absoluten Betrage nach $\leqq M$ ist, so hat dieses Polynom in allen Punkten $\varphi + \varphi_r$ abwechselnd ein relatives Maximum $+M$ oder ein relatives Minimum $-M$. Diese Eigenschaft besitzt aber auch die Funktion $M \sin n(\theta - \varphi)$, mithin hat eine der Funktionen $f(\theta) \pm M \sin n(\theta - \varphi)$ wenigstens $2n$ doppeltzählende Nullstellen und verschwindet somit nach einem angeführten Satze identisch. Hieraus folgt der Satz von Herrn S. Bernstein[2]):

1) Für zwei andere Beweise dieser Formeln siehe meine zitierte Arbeit.

2) S. Bernstein, Sur l'ordre de la meilleure approximation des fonctions continues par des polynomes de degré donné. Mémoires publiés par la Classe des Sciences de l'Académie de Belgique, t. 4. 1912. Herr Bernstein beweist den obigen Satz nur für reine Kosinus- und Sinuspolynome und erhält daraus für den allgemeinen Fall die Abschätzung $|f'(\theta)| < 2Mn$. Daß die Abschätzung $|f'(\theta)| \leq Mn$ auch im allgemeinen Falle gilt, wurde zuerst von Herrn Fejér gefunden, der in einer später zu zitierenden Arbeit bemerkt, daß dies sich aus dem Bernsteinschen Resultate über Sinuspolynome leicht ergibt.

I. *Ist das trigonometrische Polynom von höchstens n-tem Grade* $f(\theta)$ *dem absoluten Betrage nach überall* $\leqq M$, *so ist*

$$|f'(\theta)| \leqq Mn,$$

das Gleichheitszeichen ausgeschlossen, wenn $f(\theta)$ *nicht von der Form* $M \sin n(\theta - \varphi) = M \cos n\left(\theta - \varphi - \frac{\pi}{2n}\right)$ *ist.*

Aus der Interpolationsformel (9) ergibt sich leicht auch der folgende Satz:

I'. *Ist das Polynom mit beliebigen (komplexen) Koeffizienten und von höchstens n-tem Grade*

$$F(z) = c_n z^n + \cdots + c_1 z + c_0$$

auf dem Einheitskreise dem absoluten Betrage nach überall $\leqq M$, *so ist daselbst* $|F'(z)| \leqq Mn$, *das Gleichheitszeichen ausgeschlossen, wenn* $F(z)$ *nicht von der Form* $Me^{i\theta} z^n$ *ist.*

Der folgende Satz umfaßt sowohl diesen wie auch den Bernsteinschen Satz.

I''. *Ist das Laurentsche Polynom*

$$L(z) = c_n z^n + \cdots + c_1 z + c_0 + c_{-1} z^{-1} + \cdots + c_{-n} z^{-n}$$

auf dem Einheitskreise dem absoluten Betrage nach $\leqq M$, *so ist daselbst* $|L'(z)| \leqq Mn$, *das Gleichheitszeichen ausgeschlossen, sofern* $L(z)$ *nicht von der Form* $\alpha z^n + \beta z^{-n}$ *und* $|\alpha| + |\beta| = M$ *ist.*

Der erste Teil dieser beiden Sätze folgt sofort aus der Formel (9) und aus der Identität $\left|\frac{df(\varphi)}{de^{i\varphi}}\right| = \left|\frac{df(\varphi)}{d\varphi}\right|$. Die Formel zeigt auch, daß der Wert Mn nur dann erreicht werden kann, wenn $F(z)$ bzw. $L(z)$ in $2n$ äquidistanten Punkten des Einheitskreises abwechselnd gleich $\pm Me^{i\theta}$ ist. Hieraus folgt unmittelbar der vollständige Satz I' und auch der zweite Teil von I'' folgt leicht durch eine nochmalige Heranziehung der Ungleichung $|L(z)| \leqq M$.

§ 3. Wir spezialisieren nun die Formel (2) für ein reines Sinuspolynom

$$g(\varphi) = b_n \sin n\varphi + b_{n-1} \sin(n-1)\varphi + \cdots + b_1 \sin\varphi.$$

Nun ist aber

$$g(\varphi_r)(-1)^r = g(\varphi_{2n-r+1})(-1)^{2n-r+1}$$

und

$$\cot\frac{\varphi - \varphi_r}{2} + \cot\frac{\varphi - \varphi_{2n-r+1}}{2} = \frac{\sin\frac{2\varphi - \varphi_r - \varphi_{2n-r+1}}{2}}{\sin\frac{\varphi - \varphi_r}{2}\sin\frac{\varphi - \varphi_{2n-r+1}}{2}}.$$

Hier ist der Zähler wegen $\varphi_r + \varphi_{2n-r+1} = 2\pi$ gleich $-\sin\varphi$, und der Nenner gleich

$$\tfrac{1}{2}\left(\cos\frac{\varphi_{2n-r+1}-\varphi_r}{2} - \cos\frac{2\varphi-\varphi_r-\varphi_{2n-r+1}}{2}\right)$$

$$= \tfrac{1}{2}\left(\cos\frac{2\pi-2\varphi_r}{2} - \cos\frac{2\varphi-2\pi}{2}\right) = \tfrac{1}{2}(\cos\varphi - \cos\varphi_r).$$

Folglich

$$(11)\qquad g(\varphi) = \frac{\sin\varphi\cos n\varphi}{n}\sum_{r=1}^{n} g(\varphi_r)\frac{(-1)^{r+1}}{\cos\varphi-\cos\varphi_r}.\text{[1]}$$

Diese Formel kann man übrigens folgendermaßen als Spezialfall der Lagrangeschen Interpolationsformel herleiten:

$\cos n\varphi$ ist ein Polynom n-ten Grades in $\cos\varphi$, das für die n verschiedenen Werte $\cos\varphi = \cos\varphi_r$ $(r = 1, 2, \ldots, n)$ verschwindet. Hieraus folgt die bekannte Darstellung

$$(12)\quad \cos n\varphi = C(\cos\varphi - \cos\varphi_1)(\cos\varphi-\cos\varphi_2)\ldots(\cos\varphi-\cos\varphi_n),$$

(wo übrigens, wie bekannt $C = 2^{n-1}$). Ferner ist

$$\left(\frac{d\cos n\varphi}{d\cos\varphi}\right)_{\varphi=\varphi_r} = \frac{n\sin n\varphi_r}{\sin\varphi_r} = \frac{(-1)^{r+1}n}{\sin\varphi_r}.$$

Da nun andererseits $\frac{g(\varphi)}{\sin\varphi}$ ein Kosinuspolynom von höchstens $(n-1)$-tem Grade ist, ergibt die Partialbruchentwicklung des Quotienten

$$\frac{\frac{g(\varphi)}{\sin\varphi}}{\cos n\varphi}$$

den Ausdruck

$$\frac{g(\varphi)}{\sin\varphi\cos n\varphi} = \frac{1}{n}\sum_{r=1}^{n}\frac{g(\varphi_r)}{\cos\varphi-\cos\varphi_r}(-1)^{r+1}.\ \text{W. z. b. w.}$$

Aus (11) folgt unmittelbar der folgende bemerkenswerte Satz:

II. *Ist das Sinuspolynom von höchstens n-tem Grade $g(\varphi)$ dem absoluten Betrage nach überall $\leqq M$, so ist in den Intervallen*

$$-\frac{\pi}{2n} < \varphi < 0,\ 0 < \varphi < \frac{\pi}{2n},\quad \pi - \frac{\pi}{2n} < \varphi < \pi,\ \pi < \varphi < \pi + \frac{\pi}{2n}$$

$$(13)\qquad |g(\varphi)| \leqq M|\sin n\varphi|$$

das Gleichheitszeichen ausgeschlossen, sofern $g(\varphi)$ selbst nicht von der Form $\pm M\sin n\varphi$ ist.

---

[1] Für ein Kosinuspolynom $h(\varphi)$ erhält man durch Spezialisierung von (2)

$$h(\varphi) = a_n\cos n\varphi + \frac{\cos n\varphi}{n}\sum_{r=1}^{n}\frac{h(\varphi_r)\sin\varphi_r}{\cos\varphi-\cos\varphi_r}(-1)^{r+1}.$$

Ist $a_n = 0$, so ist auch dies ein Spezialfall der Lagrangeschen Interpolationsformel.

Es genügt, diesen Satz z. B. für das Intervall $0 < \varphi < \frac{\pi}{2n}$ zu beweisen. In diesem Intervalle hat aber der Nenner in allen Gliedern von (11) das gleiche (positive) Vorzeichen, außerdem ist in diesem Intervalle auch $\sin\varphi \cos n\varphi$ positiv, folglich ist für dieses Intervall

$$(14) \qquad |g(\varphi)| \leq \frac{M \sin\varphi \cos n\varphi}{n} \sum_{r=1}^{n} \frac{1}{\cos\varphi - \cos\varphi_r},$$

das Gleichheitszeichen ausgeschlossen, wenn $g(\varphi_r)$ nicht mit wechselndem Vorzeichen gleich $\pm M$, also $g(\varphi)$ identisch gleich $\pm M \sin n\varphi$ ist. Setzen wir aber in (11) statt $g(\varphi)$ die Funktion $\sin n\varphi$ ein, so erhalten wir die bekannte Formel[1]) (die natürlich für beliebiges φ gilt)

$$(15) \qquad \sin n\varphi = \frac{\sin\varphi \cos n\varphi}{n} \sum_{r=1}^{n} \frac{1}{\cos\varphi - \cos\varphi_r}.$$

Der Vergleich zwischen (14) und (15) ergibt unseren Satz.

Derselbe Gedankengang ergibt auch folgenden Satz, der für $\varphi = 0$ und $\varphi = \pi$ etwas mehr besagt:

III. *Ist das Sinuspolynom von höchstens n-tem Grade $g(\varphi)$ dem absoluten Betrage nach überall $\leq M$, so ist für*

$$-\frac{\pi}{2n} < \varphi < \frac{\pi}{2n} \quad \text{und} \quad \pi - \frac{\pi}{2n} < \varphi < \pi + \frac{\pi}{2n}$$

$$(16) \qquad \left|\frac{g(\varphi)}{\sin\varphi}\right| \leq M \left|\frac{\sin n\varphi}{\sin\varphi}\right|,$$

das Gleichheitszeichen ausgeschlossen, wenn $g(\varphi)$ nicht selbst von der Form $\pm M \sin n\varphi$ ist.[2])

Aus diesem Satze und dem Bernsteinschen ergibt sich unmittelbar der folgende Satz von Herrn A. Markoff[3]):

IV. *Ist im Intervalle $(-1, +1)$ das Polynom mit reellen*[4]) *Koeffizienten und von höchstens n-tem Grade*

$$H(x) = \alpha_n x^n + \alpha_{n-1} x^{n-1} + \cdots + \alpha_0$$

dem absoluten Betrage nach überall $\leq M$, so ist in diesem Intervalle

$$(17) \qquad |H'(x)| \leq M n^2,$$

---

1) Diese Partialbruchentwicklung für $\frac{\operatorname{tg} n\varphi}{\sin\varphi}$ erhält man natürlich auch durch logarithmische Differentiation von (12).

2) Im Beweise der beiden letzten Sätze benutzten wir nur die Voraussetzung, daß $|g(\varphi_r)| \leq M$ ist für alle φ_r, und nicht, daß $|g(\varphi)| \leq M$ für beliebiges φ. Mit dieser letzteren Voraussetzung kann man etwas mehr beweisen. Vgl. Satz II' S. 362.

3) A. Markoff, Über eine Mendeleieffsche Frage (russisch). St.-Petersburg Akad. Abh. (8°.) Bd. 62 (1889).

4) Werden auch komplexe Koeffizienten in den Sätzen II, III und IV zugelassen, so ist in diesen nur $Me^{i\theta}$ statt $\pm M$ zu setzen.

das Gleichheitszeichen ausgeschlossen, wenn $H(x)$ nicht von der Form $\pm M \cos(n \arccos x)$ *ist.*

Setzen wir nämlich in $H(x)$ $x = \cos\varphi$, so verwandelt sich, wie bekannt, $H(x)$ in ein Kosinuspolynom von höchstens n-tem Grade

$$H(x) = h(\varphi) = a_n \cos n\varphi + \cdots + a_1 \cos\varphi + a_0,$$

und es wird

$$H'(x) = -\frac{h'(\varphi)}{\sin\varphi},$$

wo nach dem Bernsteinschen Satze

$$|h'(\varphi)| \leqq Mn$$

ist. Also einerseits im Intervalle $\frac{\pi}{2n} \leqq \varphi \leqq \frac{\pi}{2}$

$$|H'(x)| \leqq \frac{Mn}{\sin\varphi} \leqq \frac{Mn}{\sin\frac{\pi}{2n}} < Mn^2.\text{[1]}$$

Andererseits ist nach Satz III im Intervalle $0 \leqq \varphi < \frac{\pi}{2n}$

$$|H'(x)| \leqq \frac{Mn \sin n\varphi}{\sin\varphi} \leqq Mn^2,\text{[2]}$$

und in diesen beiden Ungleichungen ist das Gleichheitszeichen ausgeschlossen, sofern $h'(\varphi)$ nicht von der Form $\pm Mn \sin n\varphi$ ist. Dann ist aber $h(\varphi) = \mp M \cos n\varphi + \text{Konst.}$, und diese Konstante muß wegen $|h(\varphi)| \leqq M$ gleich Null sein. — In den übrigen Quadranten geht natürlich der Beweis ebenso.

§ 4. Der Bernsteinsche Satz kann auch durch Wurzelabzählung leicht bewiesen werden.

---

1) $\sin\frac{\pi}{2n} > \frac{1}{n}$ folgt z. B. aus der bekannten Ungleichung

$$\sin\varphi > \frac{2}{\pi}\varphi \quad \left(0 < \varphi < \frac{\pi}{2}\right),$$

die, wie bekannt, aus der Tatsache folgt, daß $\frac{\sin\varphi}{\varphi}$ von 0 bis $\frac{\pi}{2}$ abnimmt.

2) Wir bedienen uns hierbei der bekannten Ungleichung

$$|\sin n\varphi| < n \sin\varphi.$$

Diese Ungleichung gilt, wie man durch vollständige Induktion leicht beweist, für ganzzahlige n im Intervalle $0 < \varphi < \pi$. Für nicht ganzzahlige $n > 1$ gilt sie auch in einem bestimmten Intervall $0 < \varphi < \alpha$, wo $\alpha > \frac{\pi}{2}$ sich durch eine geometrische Konstruktion leicht ergibt. Insbesondere für $0 < \varphi \leqq \frac{\pi}{2n}$ folgt die Ungleichung auch aus $\frac{\sin n\varphi}{\sin\varphi} = n\frac{\cos n\theta\varphi}{\cos\theta\varphi}$ $(0 < \theta < 1)$.

Wir bezeichnen die respektiven Maxima von $|f(\varphi)|$ und $|f'(\varphi)|$ mit M und M' und beweisen, daß

$$M' \leqq Mn \tag{18}$$

ist, das Gleichheitszeichen ausgeschlossen, wenn $f(\varphi)$ nicht von der Form $M \sin n(\varphi - \omega)$ ist.

Wir können ohne Einschränkung der Allgemeinheit

$$f'(0) = M'$$

annehmen. Wir bilden nun die Funktion

$$G(\varphi) = \frac{M'}{n} \sin n\varphi - f(\varphi).$$

Wäre nun in Gegensatz zu (18) $M' \geqq Mn$ und $G(\varphi)$ nicht identisch Null, so wäre $G(\varphi_1) \geqq 0$, $G(\varphi_2) \leqq 0$, $G(\varphi_3) \geqq 0$ usw. Es gäbe dann *zwischen* φ_r und φ_{r+1} eine Zahl ξ_r von der Beschaffenheit, daß $G'(\xi_r) < 0$, wenn r ungerade, und $G'(\xi_r) > 0$, wenn r gerade ist.[1]) Folglich hätte die Funktion $G'(\varphi)$ mindestens $2n$ getrennte Nullstellen, und da die Funktion $G'(\varphi)$ höchtens $2n$ Nullstellen haben kann, hat sie überhaupt keine mehrfachen Nullstellen. Nun ist aber $G'(0) = 0$, und da die Ableitungen von sowohl $f(\varphi)$ wie $\frac{M'}{n} \sin n\varphi$ in $\varphi = 0$ ihr Maximum erreichen, müßte $G'(\varphi)$ in $\varphi = 0$ mindestens von der zweiten Ordnung verschwinden. Somit führen die Annahmen $M' \geqq Mn$, $f'(0) = M'$ und $G(\varphi)$ nicht identisch gleich Null, auf einen Widerspruch.

Derselbe Gedankengang führt auch zu folgendem, allgemeinerem Satze den man übrigens auch mittels der Interpolationsformel (9) leicht beweisen kann.

V. *Ist* $\left|f\left(\varphi + \frac{\pi}{n}\right) - f(\varphi)\right| \leqq 2M$ *für alle Werte von* φ, *so ist*

$$|f'(\varphi)| \leqq Mn,$$

das Gleichheitszeichen ausgeschlossen, sofern $f(\varphi)$ *nicht von der Form* $M \sin n(\varphi - \omega) + C$ ist.

Wir bezeichnen wieder das Maximum von $|f'(\varphi)|$ mit M' und nehmen an, daß $f'(0) = M'$ ist. Wir bilden wieder die Funktion $G(\varphi) = \frac{M'}{n} \sin n\varphi - f(\varphi)$. Wäre nun $M' \geqq Mn$, so wäre

$$G(\varphi_1) \geqq G(\varphi_2) \leqq G(\varphi_3) \geqq \cdots \leqq G(\varphi_{2n-1}) \geqq G(\varphi_{2n}) \leqq G(\varphi_1).$$

Würde dann $G'(\varphi)$ nicht identisch verschwinden, so gäbe es *zwischen* φ_r und φ_{r+1} wieder einen Punkt ξ_r von der Beschaffenheit, daß $G'(\xi_r) < 0$,

1) Das gilt offenbar auch von dem Bogen $(\varphi_{2n}, \varphi_1)$, und im folgenden wird auch dieser Bogen mitgezählt.

wenn r ungerade, und $G'(\xi_r) > 0$, wenn r gerade ist. Folglich hätte $G'(\varphi)$ wieder $2n$ getrennte Nullstellen und überhaupt keine mehrfachen Nullstellen. Dies widerspricht aber der Tatsache, daß $G'(\varphi)$ für $\varphi = 0$ mindestens von der zweiten Ordnung verschwinden muß.

Wir beweisen mit diesem abzählenden Verfahren auch die Sätze II und III und sogar etwas mehr, indem wir zeigen, daß unter der dortigen Annahme die Ungleichung $|g(\varphi)| < M|\sin n\varphi|$ auch in den Endpunkten $-\frac{\pi}{2n}, \frac{\pi}{2n}, \pi - \frac{\pi}{2n}, \pi + \frac{\pi}{2n}$ gilt. Es handelt sich zunächst um den Satz:

II'. *Ist das Sinuspolynom von höchstens n-tem Grade $g(\varphi)$ dem absoluten Betrage nach überall $\leqq M$, so ist in den Intervallen*

$$(19)\qquad -\frac{\pi}{2n} \leqq \varphi < 0, \quad 0 < \varphi \leqq \frac{\pi}{2n}, \quad \pi - \frac{\pi}{2n} \leqq \varphi < \pi, \quad \pi < \varphi \leqq \pi + \frac{\pi}{2n}$$

$$|g(\varphi)| < M|\sin n\varphi|,$$

sofern nicht $g(\varphi)$ selbst von der Form $\pm M \sin n\varphi$ ist.

Wir bilden die Hilsfunktionen $G(\varphi) = M \sin n\varphi - g(\varphi)$ und $G_\beta(\varphi) = M \sin n\varphi - \beta g(\varphi)$, wo $0 < \beta < 1$. Nach unserer Voraussetzung ist $G_\beta(\varphi_r) > 0$, wenn r ungerade, und $G_\beta(\varphi_r) < 0$, wenn r gerade ist. Mithin hat $G_\beta(\varphi)$ *zwischen* φ_1 und φ_n wenigstens $(n-1)$ Nullstellen, und dasselbe gilt *zwischen* φ_{n+1} und φ_{2n}. Unsere Funktion verschwindet aber auch für $\varphi = 0$ und $\varphi = \pi$, und das gibt $2n$ Nullstellen. Mithin hat die Funktion $G_\beta(\varphi)$ je eine einfache Nullstelle zwischen zwei aufeinander folgenden Werten von φ_r, die Nullstellen zwischen φ_n und φ_{n+1} bzw. φ_{2n} und φ_1 sind $\varphi = \pi$ und $\varphi = 0$, und die Funktion hat überhaupt keine anderen, also auch keine komplexen Nullstellen. Hieraus folgt, daß die Funktion $G_\beta(\varphi)$ insbesondere für $0 < \varphi \leqq \frac{\pi}{2n} = \varphi_1$ nicht verschwindet, also dasselbe positive Vorzeichen wie $G_\beta\left(\frac{\pi}{2n}\right)$ besitzt. Nun ist aber gleichmäßig

$$G(\varphi) = \lim_{\beta \to 1} G_\beta(\varphi).$$

Verschwindet also die Funktion $G(\varphi)$ nicht identisch, so verschwindet sie nach einem bekannten Satze von Cauchy im Intervalle $0 < \varphi < \frac{\pi}{2n}$ überhaupt nicht und ist daselbst wie die Funktionen $G_\beta(\varphi)$ positiv.

Man zeigt auch leicht, daß $\frac{G_\beta(\varphi)}{\sin\varphi}$ in $\varphi = 0$ nicht verschwindet, also nach dem obigen positiv ist. Diese Funktion ist ein Kosinuspolynom von höchstens $(n-1)$-tem Grade, das *zwischen* φ_1 und φ_n einerseits, φ_{n+1}

und φ_{2n} andererseits, je $(n-1)$, also zusammen $(2n-2)$ Nullstellen und folglich keine anderen Nullstellen hat. Mithin ist $\frac{G_\beta(\varphi)}{\sin\varphi}$ *zwischen* $-\frac{\pi}{2n}$ und $\frac{\pi}{2n}$ positiv, somit gilt nach dem Cauchyschen Satze dasselbe auch für $\frac{G(\varphi)}{\sin\varphi}$. Also ist diese Funktion insbesondere für $\varphi = 0$ positiv.

All dies gilt natürlich auch für die Funktion $\overline{G}(\varphi) = M\sin n\varphi + g(\varphi)$ und auch — mutatis mutandis — in den übrigen Intervallen (19). Dadurch haben wir für die Sätze II und III einen neuen Beweis geliefert. Wir wenden uns nun zum Beweise der im Satze II′ gegebenen Ergänzung des Satzes II.

Bisher haben wir nur die Ungleichungen $|g(\varphi_\nu)| \leqq M$ benutzt. (Wie ich schon in der Fußnote 2) zu S. 359 bemerkt habe, genügen diese Ungleichungen vollständig zu den Sätzen II und III.) Nun setzen wir voraus, daß $|g(\varphi)| \leqq M$ für alle Werte von φ, und beweisen, daß unter dieser Voraussetzung auch

$$\left|g\left(\frac{\pi}{2n}\right)\right| < M\sin\left(n\frac{\pi}{2n}\right) = M$$

ist. Hätte nämlich z.B. $G(\varphi)$ in $\varphi = \frac{\pi}{2n}$ eine Nullstelle, so müßte diese mindestens von der zweiten Ordnung sein, da dann sowohl $|g(\varphi)|$ wie $M\sin n\varphi$ in $\varphi = \frac{\pi}{2n}$ ihr Maximum M erreichten. Die Funktion $G_\beta(\varphi)$ hat aber nur eine Nullstelle, die gegen $\frac{\pi}{2n}$ konvergieren könnte, nämlich diejenige *zwischen* $\varphi_1 = \frac{\pi}{2n}$ und $\varphi_2 = \frac{3\pi}{2n}$. Mithin könnte $G(\varphi)$ in $\varphi = \frac{\pi}{2n}$ höchstens von der ersten Ordnung verschwinden, also ist, unter der Voraussetzung $|g(\varphi)| \leqq M$, $\varphi = \frac{\pi}{2n}$ überhaupt keine Nullstelle der Funktion $G(\varphi)$. Dasselbe gilt natürlich von der Funktion $\overline{G}(\varphi)$ und für die Punkte $-\frac{\pi}{2n}$, $\pi - \frac{\pi}{2n}$ und $\pi + \frac{\pi}{2n}$.[1])

Der folgende Satz, der *scheinbar* einen entgegengesetzten Charakter hat, geht aus einem ähnlichen Gedankengange hervor:

VI. *Ist das trigonometrische Polynom von höchstens n-tem Grade $f(\varphi)$ dem absoluten Betrage nach überall $\leqq M$, besteht ferner für den speziellen*

---

1) Unser Beweisgang läßt die Richtigkeit des folgenden Satzes erkennen: Das einzige Sinuspolynom von höchstens n-tem Grade, das die Bedingungen

$$|g(\varphi_1)| > |g(\varphi_2)|,\ |g(\varphi_1)| \geq |g(\varphi_3)|,\ \cdots,\ |g(\varphi_1)| \geq |g(\varphi_n)|$$

und $g'(\varphi_1) = 0$ erfüllt, ist $C\sin n\varphi$.

Wert θ die Gleichung $f(\theta) = M$, so ist, wenn $f(\varphi)$ nicht identisch gleich $M \cos n(\varphi - \theta)$ ist, für

$$\theta - \frac{\pi}{n} \leqq \varphi \leqq \theta + \frac{\pi}{n} \quad (\varphi \neq \theta)$$

$$f(\varphi) > M \cos n(\varphi - \theta).$$

Wir können ohne Einschränkung der Allgemeinheit $\theta = 0$ annehmen. Es sei wiederum $0 < \beta < 1$. Wir bilden die Funktionen

$$G(\varphi) = M \cos n\varphi - f(\varphi)$$

und

$$G_\beta(\varphi) = M \cos n\varphi - \beta f(\varphi).\text{[1]}$$

Nun ist $G_\beta\left(\frac{r\pi}{n}\right) > 0$, wenn r gerade, und $G_\beta\left(\frac{r\pi}{n}\right) < 0$, wenn r ungerade ist. Mithin hat $G_\beta(\varphi)$ *zwischen* $\frac{r\pi}{n}$ und $\frac{(r+1)\pi}{n}$ je *eine* einfache Nullstelle und überhaupt keine anderen, also auch keine komplexen Nullstellen. Es ist nun gleichmäßig

$$\lim_{\beta \to 1} G_\beta(\varphi) = G(\varphi).$$

Das Polynom $G(\varphi)$ verschwindet aber in $\varphi = 0$ wenigstens von der zweiten Ordnung. Mithin müssen nach dem soeben erwähnten Satze von Cauchy die zwei Nullstellen zwischen $-\frac{\pi}{n} \equiv \frac{(2n-1)\pi}{n}$ und 0 bzw. zwischen 0 und $\frac{\pi}{n}$ gegen $\varphi = 0$ konvergieren. Folglich hat nach demselben Satze von Cauchy $G(\varphi)$ für $-\frac{\pi}{n} < \varphi < 0$ und $0 < \varphi < \frac{\pi}{n}$ keine Nullstellen. Es können aber auch $-\frac{\pi}{n}$ und $\frac{\pi}{n}$ keine Nullstellen von $G(\varphi)$ sein. Denn wäre z. B. $\frac{\pi}{n}$ eine Nullstelle, so müßte sie mindestens von der zweiten Ordnung sein. Es müßten also wenigstens zwei Nullstellen der Funktion $G_\beta(\varphi)$ gegen $\frac{\pi}{n}$ konvergieren. Da nun aber die Nullstelle zwischen 0 und $\frac{\pi}{n}$ gegen 0 konvergiert, so könnte höchstens diejenige zwischen $\frac{\pi}{n}$ und $\frac{2\pi}{n}$ gegen $\frac{\pi}{n}$ konvergieren, und da wäre $\frac{\pi}{n}$ nur eine einfache Null-

---

1) Man kann die Sätze II' und VI auch ohne Zuhilfenahme der Funktionen $G_\beta(\varphi)$ und des Cauchyschen Satzes beweisen, indem man, wie im zuletzt gegebenen Beweise des Bernsteinschen Satzes, die Nullstellen von $G'(\varphi)$ abzählt. Diese Beweise bieten aber nichts von besonderem Interesse.

Der Satz VI gestattet durch die Einführung von $x = h \cos(\varphi - \theta)$ eine Übersetzung für gewöhnliche Polynome und gibt in dieser Form eine recht interessante Ergänzung zu einem Satze von Herrn Bernstein (a. a. O. S. 14. Nr. 7 und Nr. 8). Auch dieser Bernsteinsche Satz läßt sich leicht durch das obige abzählende Verfahren beweisen.

stelle. Mithin sind $-\frac{\pi}{n}$ und $\frac{\pi}{n}$ keine Nullstellen. Dann ist aber $G\left(-\frac{\pi}{n}\right) < 0$, $G\left(\frac{\pi}{n}\right) < 0$, also

$$G(\varphi) < 0 \quad \text{für} \quad -\frac{\pi}{n} \leqq \varphi < 0 \quad \text{und} \quad 0 < \varphi \leqq \frac{\pi}{n}. \quad \text{W. z. b. w.}$$

Aus unserem letzten Satze ergibt sich leicht die folgende Behauptung, die den Satz V gewissermaßen präzisiert:

VII. *Ist das absolute Maximum der Ableitung eines trigonometrischen Polynoms von höchstens n-tem Grade* $f(\varphi)$ *dem absoluten Betrage nach gleich* M' *und ist* $f'(\theta) = \pm M'$, *so ist*

$$\left| f\left(\theta + \frac{\pi}{2n}\right) - f\left(\theta - \frac{\pi}{2n}\right) \right| \geqq \frac{2M'}{n},$$

das Gleichheitszeichen ausgeschlossen, wenn $f(\varphi)$ *nicht von der Form* $\pm \frac{M'}{n} \sin(\varphi - \theta) + C$ *ist.*

Wir können ohne Einschränkung der Allgemeinheit $\theta = 0$ und $f'(0) = +M'$ annehmen. Wir bilden die Funktion $G(\varphi) = \frac{M'}{n} \sin n\varphi - f(\varphi)$. Nach dem obigen Satze ist dann

$$f'(\varphi) > M' \cos n\varphi \quad \text{für} \quad -\frac{\pi}{n} \leqq \varphi < 0 \quad \text{und} \quad 0 < \varphi \leqq \frac{\pi}{n};$$

also ist für dieselben Werte $G'(\varphi) < 0$, mithin ist $G(\varphi)$ in dem Intervalle $\left(-\frac{\pi}{n}, +\frac{\pi}{n}\right)$ abnehmend, folglich

$$G\left(-\frac{\pi}{2n}\right) > G\left(\frac{\pi}{2n}\right),$$

d. h.

$$-\frac{M'}{n} - f\left(-\frac{\pi}{2n}\right) > \frac{M'}{n} - f\left(\frac{\pi}{2n}\right),$$

$$f\left(\frac{\pi}{2n}\right) - f\left(-\frac{\pi}{2n}\right) > \frac{2M'}{n}.$$

§ 5. Herr Fejér hat neulich einen unvermuteten Zusammenhang zwischen den Konvergenzeigenschaften der reellen und der imaginären Komponente einer Potenzreihe entdeckt, indem er auf Grund des Bernsteinschen Satzes das Folgende bewies:

Ist die Potenzreihe $f(z) = c_0 + c_1 z + \cdots + c_n z^n + \cdots$ *für* $|z| < 1$ *konvergent, und ist ihre Summe für* $|z| < 1$ *stetig für* $|z| \leqq 1$, *so folgt aus der gleichmäßigen Konvergenz ihrer reellen Komponente für* $|z| = 1$ *die gleichmäßige Konvergenz ihrer imaginären Komponente (und umgekehrt).*[1])

1) Fejér, Über konjugierte trigonometrische Reihen. J. f. Math. Bd. 144 (1914) S. 52. Theorem IV.

Ich habe in meiner zitierten Arbeit[1]) eine Verallgemeinerung dieses Satzes ausgesprochen, die ich hier mittels der Interpolationsformel (9) beweisen werde. Es handelt sich um den folgenden Satz:

VIII. *Sei die Funktion $f(z)$, die im Innern des Einheitskreises durch die Potenzreihe $c_0 + c_1 z + c_2 z^2 + \cdots + c_n z^n + \cdots$ dargestellt wird, in dem Kreissektor $\theta_1 \leq \operatorname{Arg} z \leq \theta_2$, $|z| \leq 1$ stetig, und es sei außerdem die eine Komponente der Reihe gleichmäßig konvergent auf jedem Bogen des Einheitskreises, der in dem Bogen $\theta_1 \leq \operatorname{Arg} z \leq \theta_2$ enthalten ist. Dann hat die andere Komponente dieselbe Eigenschaft.*

Wir schreiben formal

$$\sum_{n=0}^{\infty} c_n e^{ni\varphi} = \sum_{n=0}^{\infty} (a_n + ib_n) e^{ni\varphi}$$
$$= \sum_{n=0}^{\infty} ((a_n \cos n\varphi - b_n \sin n\varphi) + i(a_n \sin n\varphi + b_n \cos n\varphi)).$$

Wir nehmen an, daß z. B. die reelle Komponente auf jedem Bogen $\theta_1 + \delta \leqq \varphi \leqq \theta_2 - \delta$ (wo δ eine beliebig kleine positive Zahl bedeutet) gleichmäßig konvergiere. Wir setzen

$$s_n(\varphi) = \sum_{k=1}^{n} (a_k \cos k\varphi - b_k \sin k\varphi).$$

Dann ist nach der Voraussetzung, bei genügend großer Wahl von m, und für ein beliebiges $n \geqq m$,

$$|s_n(\varphi) - s_m(\varphi)| \leqq \varepsilon \quad \text{für} \quad \theta_1 + \frac{\delta}{2} \leqq \varphi \leqq \theta_2 - \frac{\delta}{2}, \tag{20}$$

wo ε eine beliebig kleine positive Zahl bedeutet.

Andererseits folgt aus der Konvergenz der reellen Komponente nach einem bekannten Satze von Herrn G. Cantor[2]), daß $a_n \to 0$ und $b_n \to 0$, d. h.

$$|a_n| < \varepsilon, \quad |b_n| < \varepsilon \quad \text{für } n \geqq m,$$

1) In derselben Arbeit gab ich auch ein Analogon des zitierten Fejérschen Satzes, in welchem ich, statt der Voraussetzung, daß die eine Komponente überall konvergiert, nur voraussetze, daß dies in allen Einheitswurzeln stattfindet, und ziehe den Schluß, daß unter dieser Voraussetzung die andere Komponente dieselbe Eigenschaft besitzt. Wenn aber eine Reihe von stetigen Funktionen in einer überall dichten Punktmenge eines Intervalls gleichmäßig konvergiert, so konvergiert sie gleichmäßig im ganzen Intervall. Folglich enthält der Satz II meiner zitierten Arbeit gar *nichts Neues*, denn seine Prämissen und seine Konklusion sind identisch mit denjenigen des Fejérschen Satzes.

2) Cantor, Über einen die trigonometrischen Reihen betreffenden Lehrsatz. J. f. Math. Bd. 72 (1870); Über trigonometrische Reihen. Math. Ann. Bd. 4 (1871). Siehe auch: Lebesgue, Leçons sur les séries trigonométriques S. 110—111, und Osgood, On Cantor's theorem concerning the coefficients of a convergent trigonometric series with generalisations. Amer. Math. Soc. Trans. Bd. 10 (1909).

also für alle Werte von φ

(21) $$|s_n(\varphi) - s_m(\varphi)| \leqq 2\varepsilon(n-m) < 2n\varepsilon.$$

Es sei nun θ ein beliebiger Punkt zwischen $\theta_1 + \delta$ und $\theta_2 - \delta$. Wir wenden auf die Ableitung von $(s_n(\theta) - s_m(\theta))$ die Interpolationsformel (9) an. Dies ergibt

$$s_n'(\theta) - s_m'(\theta) = \frac{1}{2n}\sum_{r=1}^{2n}(s_n(\theta+\varphi_r) - s_m(\theta+\varphi_r))\frac{(-1)^{r+1}}{2\sin^2\frac{\varphi_r}{2}}.$$

Wir zerlegen diese Summe in zwei Teile Σ' und Σ''. In Σ' kommen alle Glieder, für welche $\theta + \varphi_r$ zwischen $\theta_1 + \frac{\delta}{2}$ und $\theta_2 - \frac{\delta}{2}$ liegt, und in Σ'' die übrigen. In Σ' wenden wir die Ungleichung (20), in Σ'' die Ungleichung (21) an. Es ist dann

(22) $$\left|\frac{1}{2n}\sum{}'\right| \leqq \frac{\varepsilon}{2n}\sum_{r=1}^{2n}\frac{1}{2\sin^2\frac{\varphi_r}{2}} = \varepsilon n.$$

Σ'' enthält höchstens $2n$ Glieder, und da θ zwischen $\theta_1 + \delta$ und $\theta_2 - \delta$, $\theta + \varphi_r$ aber außerhalb des Bogens $\left(\theta_1 + \frac{\delta}{2}, \theta_2 - \frac{\delta}{2}\right)$ liegt, ist in allen diesen Gliedern $\sin^2\frac{\varphi_r}{2} \geqq \sin^2\frac{\delta}{4}$. Mithin

(23) $$\frac{1}{2n}\left|\sum{}''\right| < \frac{2n\cdot 2\varepsilon n}{2n\cdot 2\sin^2\frac{\delta}{4}} = \frac{\varepsilon n}{\sin^2\frac{\delta}{4}}.$$

Aus (22) und (23) folgt

$$|s_n'(\theta) - s_m'(\theta)| < C\varepsilon n,$$

wo C nur von δ abhängt. Da aber m eine bestimmte Zahl bedeutet, so ist bei genügend großem n auch

$$|s_m'(\theta)| < C\varepsilon n,$$

also

(24) $$|s_n'(\theta)| < 2C\varepsilon n \quad \text{für} \quad \theta_1 + \delta \leqq \theta \leqq \theta_2 - \delta.$$

Wir wenden nun folgenden Satz von Herrn Tauber an[1]):

Es existiere der Grenzwert

(25) $$\lim_{r\to 1}\sum_{k=0}^{\infty} u_k r^k = U, \qquad (r < 1)$$

---

1) Tauber, Ein Satz aus der Theorie der unendlichen Reihen. Monatsh. f. Math. Bd. 8 (1897).

und es sei

$$\lim_{n\to\infty} \frac{u_1 + 2u_2 + \cdots + nu_n}{n} = 0, \tag{26}$$

dann konvergiert die Reihe $\sum\limits_{k=0}^{\infty} u_k$, *und ihre Summe ist gleich* U. *Hängen die Werte* u_k *von einem Parameter* θ *ab, und sind die beiden obigen Bedingungen gleichmäßig erfüllt, so ist auch die Konvergenz der Reihe* $\Sigma u_k(\theta)$ *gleichmäßig.* [1])

Nach unserer Voraussetzung über die Stetigkeit unserer Potenzreihe konvergiert für $\theta_1 \leqq \theta \leqq \theta_2$, also auch für $\theta_1 + \delta \leqq \theta \leqq \theta_2 - \delta$ der imaginäre Teil

$$\sum_{k=0}^{\infty} (b_k \cos k\theta + a_k \sin k\theta) r^k$$

gleichmäßig gegen eine Grenzfunktion, wenn $r < 1$ gegen 1 konvergiert. Außerdem ist nach (24) für $\theta_1 + \delta \leqq \theta \leqq \theta_2 - \delta$ gleichmäßig

$$\lim_{n\to\infty} \frac{\sum\limits_{k=1}^{n} k(b_k \cos k\theta + a_k \sin k\theta)}{n} = 0,$$

also konvergiert für $\theta_1 + \delta \leqq \theta \leqq \theta_2 - \delta$ die Reihe

$$\sum_{k=0}^{\infty} (b_k \cos k\theta + a_k \sin k\theta)$$

gleichmäßig.

Zum Schlusse möchte ich noch bemerken, daß es interessant wäre, die in dieser Arbeit gegebenen Interpolationsformeln statt auf trigonometrische Polynome auch auf beliebige Funktionen anzuwenden und zu untersuchen, wie sich diese Formeln bei unendlich wachsendem n verhalten. Diese Frage ist bekanntlich für eine Anzahl Interpolationsformeln eingehend behandelt worden. [2])

---

1) Die Bedingungen (25) und (26) sind bekanntlich auch notwendig für die Konvergenz von Σu_k (bzw. für die gleichmäßige Konvergenz von $\Sigma u_k(\theta)$).

2) Für die Literatur vgl. D. Jackson, On the accuracy of trigonometric interpolation. Amer. Math. Soc. Trans. Bd. 14 (1913) und: A formula of trigonometric interpolation. Rend. d. Circ. Mat. di Palermo, Bd. 37 (1914).

17.

Neuer Beweis des Fatouschen Satzes

Gött. Nachr. (1916), 62–65

In seiner Habilitationsschrift, die er im Jahre 1854 zur Erlangung der „venia legendi" der philosophischen Fakultät der Universität Göttingen einreichte, und die zuerst in den Abhandlungen der Königlichen Gesellschaft der Wissenschaften zu Göttingen erschien, hatte Riemann einen Satz über trigonometrische Reihen dargelegt, der von überaus großer Tragweite ist[1]. Aus diesem Satze hatte später Fatou[2] einen Satz über Potenzreihen hergeleitet, den ich im Folgenden wiedergebe.

Für den Fatouschen Satz habe ich[3] schon in einer früheren Arbeit einen ziemlich einfachen Beweis gegeben, indem ich den Riemannschen Gedankengang dem speziellen Zwecke angepaßt habe. Ich habe den Satz auch in verschiedenen Richtungen erweitert[4]. Unter anderem habe ich für Dirichletsche Reihen einen Satz gegeben, der den Fatouschen Satz als Spezialfall umfaßt.

In der vorliegenden Mitteilung gebe ich einen neuen, sehr einfachen Beweis für den Fatouschen Satz, der auf ganz anderen

---

1) Über die Darstellbarkeit einer Function durch eine trigonometrische Reihe. Abhandlungen der Kgl. Gesellsch. d. Wissensch. zu Göttingen. Bd. 13. Vgl. Werke 1. Aufl., S. 237—239 und für die vorbereitenden Sätze S. 234—237.

2) Séries trigonométriques et séries de Taylor. Acta Math., Bd. 30. Vgl. S. 389—391.

3) Über einen Satz des Herrn Fatou. J. f. Math., Bd. 140.

4) A. a. O. und: Sur les séries de Dirichlet et les séries entières. Comptes Rendus, Bd. 149.

Prinzipien beruht, wie mein vorhergehender Beweis. Ich will auch beiläufig bemerken, daß dieser neue Beweis sich auch für alle oben erwähnten Verallgemeinerungen fast ebenso einfach gestaltet, jedoch gehe ich auf dieselben hier nicht ein.

Der Grundgedanke von Riemann wird auch in dem vorliegenden Beweise verwertet, obzwar in einer von der Riemannschen völlig verschiedenen Weise. Dieser Grundgedanke besteht darin, daß das Verhalten der Reihe, deren Konvergenz in einem gewissen Punkte untersucht wird, durch Multiplikation mit einer geeigneten Funktion in zwei Hilfspunkten verbessert wird, wodurch die Reihe im Allgemeinen leichter zugänglich wird. Durch die Multiplikation werden die Konvergenzeigenschaften der Reihe im untersuchten Punkte nicht wesentlich beeinflußt, also weder verbessert noch verschlimmert.

Der Fatousche Satz lautet:

Genügen die Koeffizienten einer Potenzreihe

$$f(z) = a_0 + a_1 z + \cdots + a_n z^n + \cdots$$

der Bedingung

$$\lim_{n=\infty} a_n = 0,$$

so konvergiert die Potenzreihe an jeder regulären Stelle des Einheitskreises und zwar gleichmäßig an den Stellen eines Regularitätsbogens[1]).

Jeder Regularitätsbogen (x_1, x_2) des Einheitskreises ist in einem größeren Regularitätsbogen (z_1, z_2) enthalten. Wir verlängern dann die Halbmesser $(0, z_1)$ und $(0, z_2)$ über die Punkte z_1 und z_2 hinaus und wählen die Zahl $R > 1$ so nahe an 1, daß die Funktion $f(z)$ im Innern und auf der Begrenzung des Kreissektors, der durch die Halbmesser $(0, z_1)$, $(0, z_2)$ bezw. durch deren Verlängerungen und einen Bogen mit dem Halbmesser R begrenzt wird, noch immer regulär sei. Die verlängerten Halbmesser $(0, z_1)$ und $(0, z_2)$ mögen den Bogen mit dem Halbmesser R in den Punkten z_1' und z_2' schneiden.

Wir bilden nun die Funktionen

$$g_n(z) = \frac{(f(z) - (a_0 + a_1 z + \cdots + a_n z^n))(z - z_1)(z - z_2)}{z^{n+1}}$$

und werden beweisen, daß, wie klein auch die positive Zahl ε sein mag, bei hinreichend großem n, auf der ganzen Begrenzung des Kreissektors

1) Ein Bogen des Einheitskreises heiße ein Regularitätsbogen, wenn in allen seinen Punkten, mit Einschluß der Endpunkte, die Funktion $f(z)$ regulär ist.

$$(1) \qquad |g_n(z)| < \varepsilon$$

wird. Da nun der absolute Betrag einer regulär analytischen Funktion sein Maximum auf der Begrenzung erreicht, so folgt aus (1) für einen beliebigen Punkt x des Bogens (z_1, z_2)

$$|g_n(x)| < \varepsilon$$

oder

$$|f(x) - (a_0 + a_1 x + \cdots + a_n x^n)| < \frac{\varepsilon}{|x - z_1|\,|x - z_2|},$$

d. h. unser obiger Satz.

Wir müssen also nur noch (1) für die Begrenzung des Kreissektors darlegen.

Laut unserer Voraussetzung können wir, wie klein auch die positive Zahl δ sein mag, eine positive ganze Zahl ν so bestimmen, daß

$$|a_m| < \delta \text{ für } m > \nu.$$

Setzen wir dann $|z| = r$ und wählen $n > \nu$, so ist für $|z| = r < 1$

$$|f(z) - (a_0 + a_1 z + \cdots + a_n z^n)|$$
$$= |a_{n+1} z^{n+1} + a_{n+2} z^{n+2} + \cdots| < \delta(r^{n+1} + r^{n+2} + \cdots) = \delta \frac{r^{n+1}}{1-r}.$$

Für die Punkte der Strecke $(0, z_1)$ ist $z = z_1 r$, $|z - z_1| = 1 - r$, $|z - z_2| \leqq 2$, folglich

$$(2) \qquad |g_n(z)| < \delta \frac{r^{n+1}}{1-r} \frac{2(1-r)}{r^{n+1}} = 2\delta.$$

Dieselbe Ungleichung gilt natürlich aus Symmetriegründen auch für die Punkte der Strecke $(0, z_2)$.

Es sei jetzt $1 \leqq r \leqq R$. Bezeichnen wir dann das Maximum von $|f(z)|$ im Kreissektor mit M und die größte der Zahlen $|a_0|$, $|a_1|, \ldots |a_\nu|$ mit A, so ist

$$|f(z) - (a_0 + a_1 z + \cdots + a_n z^n)|$$
$$< M + A(1 + r + \cdots + r^\nu) + \delta(r^{\nu+1} + r^{\nu+2} + \cdots + r^n)$$
$$< M + (\nu+1) A r^\nu + \delta \frac{r^{n+1}}{r-1} \leqq M + (\nu+1) A R^\nu + \delta \frac{r^{n+1}}{r-1}.$$

Für die Punkte der Strecke (z_1, z_1') ist $z = z_1 r$, $|z - z_1| = r - 1$, $|z - z_2| \leqq 2R$, mithin ist für diese Punkte

$$|g_n(z)| < \frac{r-1}{r^{n+1}} 2R(M + (\nu+1) A R^\nu) + 2R\delta.$$

Es ist nun für alle Werte $r \geqq 1$

$$\frac{r-1}{r^{n+1}} < \frac{r-1}{r^{n+1}-1} = \frac{1}{1+r+\cdots+r^n} < \frac{1}{n},$$

folglich ist bei hinreichend großem n

$$\frac{(r-1)}{r^{n+1}} 2R(M+(\nu+1)AR^\nu) < \delta,$$

also für alle Punkte der Strecke (z_1, z_1') und mithin auch für alle Punkte der Strecke (z_2, z_2')

$$|g_n(z)| < (2R+1)\delta. \tag{3}$$

Liegt endlich z auf dem Bogen (z_1', z_2'), so ist $|z-z_1| \leqq 2R$, $|z-z_2| \leqq 2R$, also nach den obigen Abschätzungen

$$|g_n(z)| < \frac{(2R)^2}{R^{n+1}}\left(M+(\nu+1)AR^\nu+\delta\frac{R^{n+1}}{R-1}\right)$$
$$= \frac{4R^2}{R^{n+1}}(M+(\nu+1)AR^\nu)+\frac{4R^2}{R-1}\delta.$$

Für hinreichend große Werte von n ist das erste Glied des letzten Ausdruckes kleiner als das zweite, folglich

$$|g_n(z)| < \frac{8R^2}{R-1}\delta. \tag{4}$$

Wählt man δ genügend klein, so ergeben die Ungleichungen (2), (3) und (4) die Ungleichung (1), also, nach dem obigen, den Fatouschen Satz.

18.

Sätze über Potenzreihen

Ark. Mat. Astr. Fys. *11*, no. 12 (1916), 1–16

Das Hauptziel der vorliegenden Arbeit ist einen neuen, wie es mir scheint, überaus einfachen Beweis für zwei Sätze von FATOU[1] zu geben. In einer früheren Arbeit[2] gab ich schon einen ziemlich einfachen Beweis für diese Sätze, indem ich den RIEMANNschen[3] Gedankengang, aus welchem FATOU seine Sätze hergeleitet hatte, dem speziellen Zwecke anpasste. In dieser letztgenannten Arbeit gab ich auch für die FATOUschen Sätze gewisse Verallgemeinerungen, die sich auf Summierung durch arithmetische Mittel beziehen. Auch für diese Verallgemeinerungen gebe ich in der vorliegenden Arbeit einen neuen, ganz elementaren Beweis, der aber verhältnismässig viel Rechenarbeit fordert.

Alle diese Sätze beziehen sich auf Potenzreihen. Ich habe aber auch für DIRICHLETsche Reihen einen Satz angegeben, der den FATOUschen Satz umfasst.[4] Mein neuer Beweis gestaltet sich nun für diese Reihen und auch für

---

[1] P. FATOU: Séries trigonométriques et séries de TAYLOR. — Acta Math. Bd. 30; vgl. S. 389—391.

[2] M. RIESZ: Über einen Satz des Herrn FATOU. — J. f. Math. Bd. 140. Vgl. auch meine Inauguraldissertation: Summierbare trigonometrische Reihen und Potenzreihen. (Ungarisch.) — Budapest 1908.

[3] B. RIEMANN: Ueber die Darstellbarkeit einer Function durch eine trigonometrische Reihe. — Vgl. Werke, 1. Aufl. (1876). S. 237—239.

[4] M. RIESZ: Sur les séries de DIRICHLET et les séries entières. — Comptes rendus, Paris. Bd. 149.

Integrale $\int_0^\infty a(v)e^{-vz}dv$ oder $\int_0^\infty e^{-vz}dA(v)$ fast ebenso einfach wie für Potenzreihen. Dies gilt auch betreffs der Sätze, die sich auf Summierung durch typische Mittel[1] beziehen. In dieser Arbeit beschränke ich mich aber auf Potenzreihen und komme in einer anderen Arbeit auf DIRICHLETsche Reihen und verwandte Ausdrücke zurück.

Ich werde voraussetzen, dass die betrachteten Funktionen in den untersuchten Punkten regulär sind. Bekanntlich ist die Regularität der Funktionen keineswegs eine notwendige Bedingung für die Gültigkeit dieser Sätze. Alle anderen Fälle können aber auf diesen Fall leicht zurückgeführt werden. Auch auf diese Frage gehe ich hier nicht ein.[2]

1°. Die Sätze von FATOU, die ich hier beweisen werde, lauten folgendermassen:

I. *Sind die Koeffizienten einer Potenzreihe*

$$f(z) = a_0 + a_1 z + \cdots + a_n z^n + \cdots$$

beschränkt, so sind auch die Teilsummen beschränkt in jedem Punkte des Einheitskreises, wo die Funktion $f(z)$ regulär ist. Die Teilsummen sind gleichmässig beschränkt auf jedem Regularitätsbogen[3] *des Einheitskreises.*

II.[4] *Genügen die Koeffizienten der Potenzreihe*

$$f(z) = a_0 + a_1 z + \cdots + a_n z^n + \cdots$$

der Bedingung

$$\lim_{n=\infty} a_n = 0,$$

so konvergiert die Potenzreihe an jeder regulären Stelle des Einheitskreises und zwar gleichmässig an den Stellen eines Regularitätsbogens.

---

[1] Vgl. für diesen Begriff: G. H. HARDY and M. RIESZ: The general theory of DIRICHLET's series. — Cambridge 1915.

[2] Vgl. meine oben angeführten Arbeiten.

[3] Ein Bogen des Einheitskreises heisse ein *Regularitätsbogen*, wenn in allen seinen Punkten (mit Einschluss der Endpunkte) die Funktion $f(z)$ regulär ist.

[4] Der Beweis des Satzes II erscheint auch in den Göttinger Nachrichten, 1916.

2°. Wir beweisen zunächst den Satz I um den Grundgedanken unseres Beweises klar hervortreten zu lassen.

Es sei der Bogen (x_1, x_2) ein Regularitätsbogen in dem oben angegebenen Sinne. Dann gibt es immer einen grösseren Regularitätsbogen (z_1, z_2), der den Bogen (x_1, x_2) in seinem Innern enthält. Wir verlängern dann die Halbmesser $(0, z_1)$ und $(0, z_2)$ über die Punkte z_1 und z_2 hinaus. Wir können immer einen Kreissektor finden, der durch die Halbmesser $(0, z_1)$, $(0, z_2)$ bzw. deren Verlängerungen und einen Bogen mit dem Halbmesser $R > 1$ begrenzt wird und so beschaffen ist, dass die Funktion $f(z)$ im Innern und auf dem Umfange dieses Kreissektors noch regulär sei. Die Halbmesser $(0, z_1)$ und $(0, z_2)$ mögen den Bogen mit dem Halbmesser R in den Punkten z'_1 bzw. z'_2 schneiden.

Wir bilden dann die Funktionen

$$g_n(z) = \frac{(f(z) - (a_0 + a_1 z + \cdots + a_n z^n))(z - z_1)(z - z_2)}{z^{n+1}}$$

und werden zeigen, dass unter der im Satze I gemachten Voraussetzung, alle diese Funktionen auf dem Umfange des Kreissektors, dem absoluten Betrage nach unter einer gemeinschaftlichen endlichen Schranke liegen.

Wir bezeichnen mit M das Maximum von $|f(z)|$ im Kreissektor. Ferner gibt es nach unserer Voraussetzung eine Zahl G derart, dass $|a_m| < G$ für alle Werte von m.

Es sei zunächst $r = |z| < 1$. Dann ist

$$|f(z) - (a_0 + a_1 z + \cdots + a_n z^n)| = |a_{n+1} z^{n+1} + a_{n+2} z^{n+2} + \cdots| \leqq$$

$$\leqq G(r^{n+1} + r^{n+2} + \cdots) = G\frac{r^{n+1}}{1-r}.$$

Ist dann z ein Punkt der Strecke $(0, z_1)$, d. h. ist $z = z_1 r$, so ist $|z - z_1| = 1 - r$, $|z - z_2| < 2$, mithin

$$(1) \qquad |g_n(z)| < G\frac{r^{n+1}}{1-r}\,\frac{2(1-r)}{r^{n+1}} = 2G.$$

Dieselbe Ungleichung gilt aus Symmetriegründen für die Strecke $(0, z_2)$.

Es sei jetzt $1 \leqq r = |z| \leqq R$. Es ist

$$|f(z) - (a_0 + a_1 z + \cdots + a_n z^n)| < M + G(1 + r + \cdots + r^n) <$$
$$< M + G \frac{r^{n+1}}{r-1}.$$

Ist dann z ein Punkt der Strecke (z_1, z'_1), so ist $|z - z_1| = r - 1$, $|z - z_2| < 2R$, folglich

$$(2) \quad |g_n(z)| < \left(M + G\frac{r^{n+1}}{r-1}\right)\frac{2R(r-1)}{r^{n+1}} <$$
$$< 2MR + 2GR = 2(M+G)R.$$

Natürlich gilt dieselbe Ungleichung auch für die Punkte der Strecke (z_2, z'_2).

Bedeutet z endlich einen Punkt des Bogens (z'_1, z'_2), so ist $|z - z_1| < 2R$, $|z - z_2| < 2R$, also nach der obigen Abschätzung

$$(3) \quad |g_n(z)| < \left(M + G\frac{R^{n+1}}{R-1}\right)\frac{(2R)^2}{R^{n+1}} < 4R^2\left(M + \frac{G}{R-1}\right) = M'.$$

Diese Zahl M' ist grösser als $2R(M+G)$ und $2G$, also ist nach (1), (2) und (3) auf der ganzen Begrenzung des Kreissektors

$$|g_n(z)| < M'.$$

Es sei jetzt x ein beliebiger Punkt des Bogens (z_1, z_2). Da die Funktionen $g_n(z)$ in dem Kreissektor und auf seiner Begrenzung regulär sind, erreicht ihr absoluter Betrag sein Maximum auf der Begrenzung. Folglich ist

$$|g_n(x)| < M'$$

d. h.

$$|a_0 + a_1 x + \cdots + a_n x^n| < \frac{M'}{|x - z_1|\,|x - z_2|} + M.$$

Bewegt sich x auf einem Bogen (x_1, x_2), der ganz im Innern des Bogens (z_1, z_2) enthalten ist, so bleibt der Ausdruck auf der rechten Seite gleichmässig beschränkt, und damit ist der Satz I vollständig dargelegt.

3°. Wir wenden uns jetzt zum Beweise des Satzes II, d. h. wir setzen

$$\lim_{n=\infty} a_n = 0$$

voraus. Wir bilden wieder die Funktionen $g_n(z)$ und werden zeigen, dass unter dieser Voraussetzung, bei hinreichend grossem n auf der ganzen Begrenzung des Kreissektors

(1) $$|g_n(z)| < \varepsilon$$

wird, wie klein auch die positive Zahl ε sein mag. Daraus folgt aber dann für jeden Punkt x des Bogens (z_1, z_2)

$$|g_n(x)| < \varepsilon,$$

oder

$$|f(x) - (a_0 + a_1 x + \cdots + a_n x^n)| < \frac{\varepsilon}{|x - z_1||x - z_2|}$$

d. h. der Satz II. Wir müssen also nur noch (1) für die Begrenzung des Kreissektors darlegen.

Wie klein auch die positive Zahl δ sein mag, so können wir immer eine Zahl ν so bestimmen, dass

$$|a_m| < \delta \text{ für } m > \nu.$$

Es sei wiederum zuerst $r = |z| < 1$. Dann ist, wenn $n \geqq \nu$ gewählt wird

$$|f(z) - (a_0 + a_1 z + \cdots + a_n z^n)| = |a_{n+1} z^{n+1} + a_{n+2} z^{n+2} + \cdots| \leqq$$
$$\leqq \delta (r^{n+1} + r^{n+2} + \cdots) = \delta \frac{r^{n+1}}{1 - r}.$$

Ist dann z ein Punkt der Strecke $(0, z_1)$ oder $(0, z_2)$, so folgt wie oben

(2) $$|g_n(z)| < 2\delta.$$

Sei jetzt $1 \leqq r = |z| \leqq R$. Bezeichnen wir dann das Maximum von $|f(z)|$ im Kreissektor wieder mit M und die grösste der Zahlen $|a_0|, |a_1|, \ldots |a_\nu|$ mit A, so ist

$$|f(z) - (a_0 + a_1 z + \cdots + a_n z^n)| < M + A(1 + r + \cdots + r^\nu) +$$
$$+ \delta(r^{\nu+1} + \cdots + r^n) < M + (\nu+1) A r^\nu +$$
$$+ \delta \frac{r^{n+1}}{r-1} \leqq M + (\nu+1) A R^\nu + \delta \frac{r^{n+1}}{r-1} \cdot$$

Liegt dann z auf der Strecke (z_1, z'_1), so ist wieder $|z - z_1| = r - 1$, $|z - z_2| < 2R$, mithin

$$|g_n(z)| < \frac{r-1}{r^{n+1}} 2R \left(M + (\nu+1) A R^\nu + \delta \frac{r^{n+1}}{r-1}\right) =$$
$$= \frac{r-1}{r^{n+1}} 2R(M + (\nu+1) A R^\nu) + 2R\delta.$$

Es ist nun für alle Werte $r \geqq 1$

$$\frac{r-1}{r^{n+1}} < \frac{r-1}{r^{n+1}-1} = \frac{1}{1 + r + \cdots + r^n} < \frac{1}{n},$$

folglich ist bei hinreichend grossem n

$$\frac{r-1}{r^{n+1}} 2R(M + (\nu+1) A R^\nu) < \delta,$$

also für alle Punkte der Strecke (z_1, z'_1) und ebenso auch für diejenigen der Strecke (z_2, z'_2)

$$|g_n(z)| < (2R+1)\delta. \tag{3}$$

Liegt endlich z auf dem Bogen (z'_1, z'_2), so ist wieder $|z - z_1| < 2R$, $|z - z_2| < 2R$, also nach der obigen Abschätzung

$$g_n(z) < \frac{(2R)^2}{R^{n+1}} \left(M + (\nu+1) A R^\nu + \delta \frac{R^{n+1}}{R-1}\right) =$$
$$= \frac{4R^2}{R^{n+1}} (M + (\nu+1) A R^\nu) + \frac{4R^2}{R-1} \delta.$$

Für hinreichend grosse Werte von n wird das erste Glied des letzten Ausdruckes kleiner als das zweite, folglich

$$(4) \qquad |g_n(z)| < \frac{8R^2}{R-1}\delta .$$

Bei genügend kleinem δ ergeben die Ungleichungen (2), (3) und (4) die Ungleichung (1), also nach den obigen Ausführungen den Satz II.

4°. Wir werden jetzt folgende Verallgemeinerung der FATOUschen Sätze darlegen:

III. *Genügen die Koeffizienten der Potenzreihe*

$$f(z) = a_0 + a_1 z + \cdots + a_n z^n + \cdots$$

der Bedingung

$$(1) \qquad \lim_{n=\infty} \frac{a_n}{n^k} = 0, \qquad (k \geqq 0),$$ [1]

so konvergieren die arithmetischen Mittel k-ter Ordnung der Reihe an allen Stellen des Einheitskreises, wo die Funktion $f(z)$ *regulär ist gegen den betreffenden Funktionswert und zwar gleichmässig an den Stellen eines Regularitätsbogens. Bekanntlich ist die Bedingung (1) auch notwendig, damit die Reihe an wenigstens einer Stelle des Einheitskreises durch arithmetische Mittel k-ter Ordnung summierbar sei.*

Wird nur vorausgesetzt, dass die Werte $\frac{a_n}{n^k}$ *dem absoluten Betrage nach unter einer endlichen Schranke liegen, so besitzen die arithmetischen Mittel k-ter Ordnung der Reihe an jeder regulären Stelle des Einheitskreises dieselbe Eigenschaft und zwar gleichmässig auf jedem Regularitätsbogen. Die letztgenannte Bedingung ist bekanntlich auch notwendig, damit die arithmetischen Mittel k-ter Ordnung der Reihe an wenigstens einem Punkte des Einheitskreises beschränkt bleiben.*

---

[1] Wir brauchen den Beweis natürlich nur für $k > 0$ führen. Unsere Sätze und auch das Resultat (15) sind sogar für $k > -1$ gültig, aber der Beweis würde sich unnötig komplizieren, wenn wir auch diese Fälle mit in Betracht ziehen würden.

Es genügt offenbar die erste Hälfte des Satzes (Bedingung (1)) darzulegen.

Die arithmetischen Mittel definieren wir auf üblicher Weise nach CESÀRO. Es sei eine beliebige Reihe $\sum_{n=0}^{\infty} b_n$ gegeben. Wir setzen

$$(2) \qquad (1-y)^{-(k+1)} = \sum_{n=0}^{\infty} C_n^{(k)} y^n,$$

$$(3) \qquad \sum_{n=0}^{\infty} S_n^{(k)} y^n = (1-y)^{-(k+1)} \sum_{n=0}^{\infty} b_n y^n.$$

Hieraus folgt

$$(4) \qquad S_n^{(k)} = \sum_{m=0}^{n} C_{n-m}^{(k)} b_m.$$

Wir nennen die Werte

$$\frac{S_n^{(k)}}{C_n^{(k)}}$$

die arithmetischen Mittel k-ter Ordnung der Reihe $\sum_{n=0}^{\infty} b_n$. Existiert der Grenzwert

$$(5) \qquad \lim_{n=\infty} \frac{S_n^{(k)}}{C_n^k}, \qquad (k \geqq 0),$$

so nennen wir die Reihe $\sum_{n=0}^{\infty} b_n$ durch arithmetische Mittel k-ter Ordnung summierbar und bezeichnen (5) als ihre Summe.[1]

[1] Für nicht ganze Ordnungszahl tritt meines Wissens diese Definition zuerst bei HADAMARD auf; vgl. Essai sur l'étude des fonctions définies par leur développement de TAYLOR. — Thèse, Paris 1892, p. 82. Vgl. auch KNOPP: Grenzwerte von Reihen bei der Annäherung an die Konvergenzgrenze. — Inauguraldissertation, Berlin 1907. Es lässt sich sogar für $k > -1$ eine vernünftige Theorie dieser Mittel aufbauen. Für $k \geqq 0$ habe ich eine Mittelbildung eingeführt, die mit der obigen gleichwertig ist und sich in vielen Fällen vorteilhafter anwenden lässt. Vgl. Une méthode de sommation équivalente à la méthode des moyennes arithmétiques. — Comptes rendus, Paris. Bd. 152.

Nach der obigen Definition ist

$$(6)\qquad C_n^{(k)} = \frac{(k+1)(k+2)\dots(k+n)}{\underline{|n}} = \frac{\Gamma(k+n+1)}{\Gamma(k+1)\underline{|n}}.$$

Aus diesen Formeln folgt nach GAUSS

$$(7)\qquad \lim_{n=\infty} \frac{C_n^{(k)}}{n^k} = \frac{1}{\Gamma(k+1)},$$

und daraus folgt dann wieder, dass der Grenzwert (5) und der Grenzwert

$$\lim_{n=\infty} \frac{\Gamma(k+1)\,S_n^{(k)}}{n^k}$$

gleichzeitig existieren und einander gleich sind.

Es folgt auch aus (7) und aus $C_n^{(k)} > 0$, dass zwei Zahlen B_1 und B_2 existieren derart, dass

$$(8)\qquad C_n^{(k)} < B_1\,(n+1)^k \qquad (k>0)$$

und

$$(9)\qquad (n+1)^k < B_2\,C_n^{(k)}\,{}^1 \qquad (k>0).$$

Aus den Definitionsformeln (2) und (3) folgt für beliebige Werte k und l

$$(10)\qquad C_n^{(k+l+1)} = \sum_{m=0}^{n} C_m^{(k)}\,C_{n-m}^{(l)} \qquad (k \gtreqless 0,\ l \gtreqless 0)$$

und

$$(11)\qquad S_n^{(k+l+1)} = \sum_{m=0}^{n} S_m^{(k)}\,C_{n-m}^{(l)} \qquad (k \gtreqless 0,\ l \gtreqless 0).$$

Es sei jetzt j eine positive ganze Zahl.[2] Dann ist

---

[1] Die Zahlen B_1, B_2, B_3 u. s. w. sind positive Grössen, bei denen es nur darauf ankommt, dass sie endlich bleiben. Die Ungleichungen gelten auch für $k > -1$.

[2] Für ganze Ordnungszahlen sind bekanntlich die CESÀROschen Grenzwerte mit den HÖLDERschen Grenzwerten vollständig gleichwertig.

$$(12)\qquad C_n^{(j)} = \frac{(j+1)(j+2)\dots(j+n)}{\lfloor n} = \frac{\lfloor n+j}{\lfloor n \lfloor j} = \frac{(n+1)(n+2)\dots(n+j)}{\lfloor j}.$$

Die arithmetischen Mittel j-ter Ordnung der Reihe $\sum_{n=0}^{\infty} a_n z^n$ haben den (natürlich auch von n abhängigen) Zähler

$$\sum_{m=0}^{n} C_{n-m}^{(j)} a_m z^m.$$

Unsere erste Aufgabe ist nun eine aus der Funktion $f(z)$ und ihren Ableitungen zusammengesetzte Funktion zu finden, deren Potenzreihenentwickelung mit diesem Polynom beginnt. Es ist leicht zu zeigen, dass die Funktion

$$(13)\quad F_{n,j}(z) = \frac{(-1)^j}{\lfloor j} z^{n+j+1} \frac{d^j}{dz^j}\left(\frac{f(z)}{z^{n+1}}\right) = \sum_{h=0}^{n} \frac{\lfloor n+j-h}{\lfloor n \lfloor h \lfloor j-h} f^{(h)}(z)(-z)^h = \sum_{h=0}^{n} \frac{C_n^{(j-h)}}{\lfloor h} f^{(h)}(z)(-z)^h$$

die gewünschte Eigenschaft besitzt. Dies ersieht man am leichtesten aus der Formel (12). Dieselbe Formel ergibt auch

$$(14)\quad F_{n,j}(z) = \sum_{m=0}^{n} C_{n-m}^{(j)} a_m z^m + (-1)^j \sum_{m=n+j+1}^{\infty} C_{m-n-j-1}^{(j)} a_m z^m = \sum_{m=0}^{n} C_{n-m}^{(j)} a_m z^m + (-1)^j z^{n+j+1} \sum_{m=0}^{\infty} C_m^{(j)} a_{m+n+j+1} z^m.$$

Es sei nun $k > 0$ eine beliebige, ganze oder nicht ganze Zahl, und es sei für diese Zahl k die Bedingung (1) erfüllt. Wir definieren dann die Zahlen z_1, z_2, R, z'_1, z'_2, M und betrachten denselben Kreissektor wie oben. Wir bilden ferner die Funktionen

$$g_{n,j,k}(z) = \frac{\left(F_{n,j}(z) - \sum_{m=0}^{n} C_{n-m}^{(j)} a_m z^m\right)(z-z_1)^{j+k+1}(z-z_2)^{j+k+1}}{n^k z^n}$$

und werden zeigen, dass auf der Begrenzung des Kreissektors für hinreichend grosse Werte von n

$$(15) \qquad |g_{n,j,k}(z)| < \varepsilon$$

wird, wie klein auch die positive Zahl ε sein mag,

Nach unserer Voraussetzung gibt es eine Zahl ν derart, dass

$$(16) \qquad |a_m| < \delta m^k \quad \text{für } m > \nu,$$

wo δ eine beliebig kleine positive Zahl bedeutet. Wählen wir also $n > \nu$, so ist nach (14) und (16) für $r = |z| < 1$

$$(17) \quad \left|F_{n,j}(z) - \sum_{m=0}^{n} C_{n-m}^{(j)} a_m z^m\right| < r^{n+j+1} \delta \sum_{m=0}^{\infty} (m+n+j+1)^k C_m^{(j)} r^m .$$

Da nun sicher

$$(m+n+j+1) < 2(n+j)(m+1)$$

ist, und in (8) und (9) statt k natürlich auch j oder $j+k$ geschrieben werden kann, so ist

$$\begin{aligned}(m+n+j+1)^k C_m^{(j)} &< 2^k (n+j)^k (m+1)^k C_m^{(j)} < \\ &< 2^k (n+j)^k (m+1)^k B_1 (m+1)^j = 2^k B_1 (n+j)^k (m+1)^{k+j} < \\ &< 2^k B_1 B_2 (n+j)^k C_m^{(k+j)} < B_3 n^k C_m^{(k+j)} .\end{aligned}$$

Daraus ergibt sich, dass (17) kleiner ist als

$$B_3 \delta r^n n^k (1-r)^{-(k+j+1)} .$$

Da auf der Strecke $(0, z_1)$ wieder $|z - z_1| = (1-r)$, $|z - z_2|$

< 2, so folgt, dass auf dieser Strecke und aus denselben Gründen auch auf der Strecke $(0, z_2)$

$$(18) \qquad |g_{n,j,k}(z)| < 2^{k+j+1} B_3 \delta = B_4 \delta .$$

Es sei jetzt $1 \leq r = |z| \leq R$. Es folgt aus der expliziten Form (13) von $F_{n,j}(z)$, dass es eine Zahl M_j gibt derart, dass in dem ganzen Kreissektor

$$|F_{n,j}(z)| < M_j n^j$$

ist. Ausserdem folgt aus (16), dass für $k > 0$ und hinreichend grosse Werte von n

$$|a_m| < \delta n^k$$

für alle $m = 0, 1, \ldots n$. Daraus ergibt sich

$$(19) \qquad |F_{n,j}(z) - \sum_{m=0}^{n} C_{n-m}^{(j)} z^m| < M_j n^j + \delta n^k \sum_{m=0}^{n} C_{n-m}^{(j)} r^m .$$

Da

$$\sum_{m=0}^{n} C_{n-m}^{(j)} r^m = r^n \sum_{m=0}^{n} C_m^{(j)} r^{-m} <$$

$$< r^n \sum_{m=0}^{\infty} C_m^{(j)} r^{-m} = r^n \left(1 - \frac{1}{r}\right)^{-(j+1)} = r^{n+j+1} (r-1)^{-(j+1)},$$

so ist (19) kleiner als

$$M_j n^j + \delta n^k r^{n+j+1} (r-1)^{-(j+1)} .$$

Es ist also auf der Strecke (z_1, z'_1), wo $|z - z_1| = r - 1$, $|z - z_2| < 2R$ ist, und ebenso auch auf der Strecke (z_2, z'_2)

$$(20) \qquad |g_{n,j,k}(z)| <$$

$$< \frac{(2R)^{j+k+1} (r-1)^{j+k+1} (M_j n^j + \delta n^k r^{n+j+1} (r-1)^{-(j+1)})}{n^k r^n} <$$

$$< \frac{B_5 (r-1)^{j+k+1} n^{j-k}}{r^n} + B_6 r^{j+1} (r-1)^k \delta .$$

Wir werden jetzt $\frac{r-1}{r^\mu}$ auch für nicht ganzzahlige Werte von μ abschätzen. Es sei $\mu \geqq 1$ und $r \geqq 1$; dann ist

$$\frac{r-1}{r^\mu} < \frac{r-1}{r^\mu - 1}.$$

Man überzeugt sich durch Differentiation leicht, dass dieser letzte Ausdruck abnimmt, wenn r von 1 bis ∞ wächst. Mithin ist

$$\frac{r-1}{r^\mu} < \frac{r-1}{r^\mu - 1} \leqq \frac{1}{\mu},$$

also für $1 \leqq r \leqq R$

$$\frac{B_5 (r-1)^{j+k+1} n^{j-k}}{r^n} < B_7 n^{j-k} \frac{(r-1)^j}{r^n} = B_7 n^{j-k} \left(\frac{r-1}{r^{\frac{n}{j}}}\right)^j <$$

$$< B_7 n^{j-k} \left(\frac{j}{n}\right)^j < B_8 n^{-k}.$$

Bei hinreichend grossem n wird der letzte Ausdruck beliebig klein, mithin folgt aus (20), dass auf den betrachteten Strecken (z_1, z'_1) und (z_2, z'_2) bei hinreichend grossem n

(21) $$|g_{n,j,k}(z)| < B_9 \delta$$

wird.

Auf dem Bogen (z'_1, z'_2) ist endlich nach den obigen Abschätzungen

$$|g_{n,j,k}(z)| < B_{10} n^{j-k} R^{-n} + B_{11} \delta.$$

Dies ergibt für hinreichend grosse Werte von n die Abschätzung

(22) $$|g_{n,j,k}(z)| < B_{12} \delta.$$

Bei hinreichend kleinem δ folgt schliesslich aus den Ungleichungen (18), (21) und (22) die Ungleichung (15). Dieselbe ergibt dann für einen beliebigen Punkt x des Bogens (z_1, z_2)

$$|F_{n,j}(x) - \sum_{m=0}^{n} C_{n-m}^{(k)} a_m x^m| < \frac{\varepsilon n^k}{|x - z_1|^{j+k+1} |x - z_2|^{j+k+1}}.$$

Wenn sich also x auf einem Bogen bewegt, der in (z_1, z_2) ganz enthalten ist, so ist für hinreichend grosses n und beliebig kleines, positives η

$$(23) \qquad |F_{n,j}(x) - \sum_{m=0}^{n} C_{n-m}^{(k)} a_m x^m| < \eta n^k.$$

Es sei jetzt zunächst k eine ganze Zahl und $j = k$. Dann ist nach dem obigen

$$|F_{n,k}(x) - \sum_{m=0}^{n} C_{n-m}^{(k)} a_m x^m| < \eta n^k.$$

Aus dieser Ungleichung und aus der expliziten Form (13) von $F_{n,k}(x)$ folgt dann, dass für hinreichend grosse n

$$|C_n^{(k)} f(x) - \sum_{m=0}^{n} C_{n-m}^{(k)} a_m x^m| < \eta' n^k,$$

wo η' eine beliebig Zahl $> \eta$ bedeutet. Auch η' kann also beliebig klein gemacht werden. Die letzte Formel ergibt also

$$\lim_{n=\infty} \frac{\sum_{m=0}^{n} C_{n-m}^{(k)} a_m x^m}{C_n^k} = f(x)$$

und zwar gleichmässig auf jedem Regularitätsbogen. Damit ist der Satz III für ganzzahlige Werte von k erledigt.

Es sei jetzt k keine ganze Zahl. Wir setzen dann $j = [k] + 1$. Aus (23) und aus der expliziten Form (13) von $F_{n,j}(z)$ ersehen wir, dass bei hinreichend grossem n

(24) $$\left| C_n^{(j)} f(x) - \sum_{m=0}^{n} C_{n-m}^{(j)} a_m x^m \right| < \eta' n^k,$$

wo η' wieder eine beliebige Zahl $> \eta$ bedeutet. Es ist nun nach (11)

$$\sum_{m=0}^{n} C_{n-m}^{(k)} a_m x^m = \sum_{p=0}^{n} C_{n-p}^{(k-j-1)} \sum_{\mu=0}^{p} C_{p-m}^{(j)} a_\mu x^\mu =$$

$$= \sum_{p=0}^{n} C_{n-p}^{(k-j-1)} \left(\sum_{\mu=0}^{p} C_{p-\mu}^{(j)} a_\mu x^\mu - C_p^{(j)} f(x) \right) + f(x) \sum_{p=0}^{n} C_{n-p}^{(k-j-1)} C_p^{(j)}.$$

Die letzte Summe ist nach (10) gleich $C_n^{(k)}$. Wir erhalten also

(25) $$\left| C_n^{(k)} f(x) - \sum_{m=0}^{n} C_{n-m}^{(k)} a_m x^m \right| =$$

$$= \left| \sum_{p=0}^{n} C_{n-p}^{(k-j-1)} \left(C_p^{(j)} f(x) - \sum_{\mu=0}^{p} C_{p-\mu}^{(j)} a_\mu x^\mu \right) \right|.$$

Die Grössenordnung von $C_p^{(k-j-1)}$ ist $p^{k-j-1} = p^{k-[k]-2}$. Folglich ist die Reihe

(26) $$\sum_{p=0}^{\infty} \left| C_p^{(k-j-1)} \right| = N$$

konvergent. Ferner ersieht man leicht aus (24), dass für hinreichend grosses n und beliebiges $0 \leq p \leq n$

$$\left| C_p^{(j)} f(x) - \sum_{\mu=0}^{p} C_{p-\mu}^{(j)} a_\mu x^\mu \right| < \eta' n^k.$$

Aus dieser Ungleichung, aus (25) und aus (26) folgt dann

$$\left| C_n^{(k)} f(x) - \sum_{m=0}^{n} C_{n-m}^{(k)} a_m x^m \right| < \eta' N n^k,$$

also schliesslich auch für nicht ganzzahlige k

$$\lim_{n=\infty} \frac{\sum_{m=0}^{n} C_{n-m}^{(k)} a_m x^m}{C_n^k} = f(x)$$

und zwar gleichmässig auf jedem Regularitätsbogen.

19.

Sur l'hypothèse de Riemann

Acta Math. *40* (1916), 185–190

1°. Dans cette Note, je me propose de représenter la fonction $\frac{\Gamma\left(1-\frac{s}{2}\right)}{\zeta(s)}$ par une intégrale de la forme

$$\int_0^\infty F(x)\, x^{-\left(\frac{s}{2}+1\right)} dx,$$

$F(x)$ désignant une fonction entière donnée explicitement. Comme on sait, l'hypothèse de Riemann consiste en ce que la fonction $\zeta(s)$ n'a pas de zéros dans le demi-plan $R(s) > \frac{1}{2}$. Notre représentation fournira une condition nécessaire et suffisante pour la validité de cette hypothèse de Riemann. Je ne sais pas encore décider si cette condition facilitera la vérification de l'hypothèse.

2°. Formons les quantités

$$C_k = \sum_{n=1}^{\infty} n^{-2k} \qquad (k = 1, 2, \ldots)$$

et la fonction entière

$$F(x) = \sum_{k=1}^{\infty} \frac{(-1)^{k+1}}{C_k \Gamma(k)} x^k.$$

D'après le calcul des résidus, on a

$$F(x) = \frac{i}{2} \int_{a-i\infty}^{a+i\infty} \frac{x^z}{\zeta(2z)\Gamma(z) \sin \pi z} dz, \qquad \left(\frac{1}{2} \leq a < 1\right)^{1} \qquad (1)$$

---

[1] En ce qui concerne l'ordre de grandeur de $\frac{1}{\zeta(z)}$ *cf.* Landau: Handbuch der Lehre der Verteilung der Primzahlen (Teubner, 1909) t. I p. 177–178.

l'intégrale étant absolument et uniformément convergente pour

$$-\frac{\pi}{2}+\delta \leqq \operatorname{Arg} x \leqq \frac{\pi}{2}-\delta \qquad (\delta > 0) \tag{2}$$

Il en résulte immédiatement, en posant $a=\frac{1}{2}$, que l'on a uniformément

$$\lim_{x=\infty} \frac{F(x)}{x^{\frac{1}{2}}}=0,$$

ou en adoptant une notation due à M. Landau

$$F(x)=o(x^{\frac{1}{2}}),^{1} \tag{3}$$

x tendant vers l'infini à l'intérieur d'un angle de la forme (2).

3°. Formons maintenant l'intégrale

$$\varphi(s)=\int_0^\infty x^{-\left(\frac{s}{2}+1\right)} F(x)dx, \tag{4}$$

le chemin d'intégration étant l'axe réel positif. En vertu de (3), cette intégrale converge absolument et *uniformément* dans toute bande de la forme

$$1+\delta \leqq R(s) \leqq 2-\delta \qquad (\delta>0),$$

et elle y représente donc une fonction analytique. D'autre part, on peut écrire

$$n^{-s}\varphi(s)=\int_0^\infty x^{-\left(\frac{s}{2}+1\right)} F\left(\frac{x}{n^2}\right)dx$$

$$\zeta(s)\varphi(s)=\sum_{n=1}^{\infty}\int_0^\infty x^{-\left(\frac{s}{2}+1\right)} F\left(\frac{x}{n^2}\right)dx.$$

Il est légitime d'intervertir l'ordre de la sommation et de l'intégration. Cela est évident pour l'intégrale prise entre 0 et 1 p. ex., les termes de la série étant de l'ordre de grandeur n^{-2}. Pour l'intégrale prise entre 1 et ∞, la formule (1) nous donne

[1] Comme d'ordinaire, la notation $\Phi(x)=o(x^a)$ signifie que $\frac{\Phi(x)}{x^a}$ tend vers zéro et la notation $\Phi(x)=O(x^a)$ signifie que $\frac{\Phi(x)}{x^a}$ reste fini, quand x tend vers l'infini.

$$\left|F\left(\frac{x}{n^2}\right)\right| < M\left(\frac{x}{n^2}\right)^a,$$

M désignant un nombre positif ne dépendant que de a. En choisissant $\frac{1}{2} < a < \frac{R(s)}{2}$, on voit que l'interversion en question est légitime.

Or, la série

$$\sum_{n=1}^{\infty} F\left(\frac{x}{n^2}\right)$$

étant uniformément convergente dans tout domaine borné et ayant par conséquent la valeur

$$xe^{-x},$$

on a finalement

$$\zeta(s)\varphi(s) = \int_0^\infty x^{-\left(\frac{s}{2}+1\right)} x e^{-x} dx = \Gamma\left(1 - \frac{s}{2}\right)$$

c.-à-d.

$$\frac{\Gamma\left(1-\frac{s}{2}\right)}{\zeta(s)} = \int_0^\infty x^{-\left(\frac{s}{2}+1\right)} F(x) dx. \tag{5}$$

4°. Si l'on avait, x tendant vers l'infini sur l'axe réel positif,

$$F(x) = O\left(x^{\frac{1}{4}+\delta}\right) \tag{6}$$

pour tout nombre positif δ, l'intégrale (5) convergerait absolument et uniformément dans toute bande de la forme

$$\frac{1}{2} + \delta \leqq R(s) \leqq 2 - \delta.$$

Donc, moyennant l'hypothèse (6), la fonction $\zeta(s)$ de Riemann n'a pas de zéros dans le demi-plan $R(s) > \frac{1}{2}$.

D'autre part, en utilisant un théorème de M. Carathéodory, M. Landau avait démontré que moyennant l'hypothèse de Riemann, on a

$$\frac{1}{\zeta(s)} = O(|s|^r)^1 \text{ pour } R(s) \geqq \frac{1}{2} + \delta,$$

[1] Landau: *loc. cit.* t. II p. 870. M. Littlewood a démontré que moyennant la même hypothèse, on a même $\frac{1}{\zeta(s)} = O(|s|^\varepsilon)$, ε désignant un nombre positif arbitrairement petit. (Quelques conséquences de l'hypothèse que la fonction $\zeta(s)$ de Riemann n'a pas de zéros dans le demi-plan $R(s) > \frac{1}{2}$. Comptes rendus, 29 janvier 1912.)

r désignant un nombre positif ne dépendant que de δ. Donc, en admettant que l'hypothèse de Riemann soit vraie, on pourrait mettre dans (1) $a = \frac{1}{4} + \delta$, ce qui entraînerait (6).

On en conclut que (6) *est la condition nécessaire et suffisante pour la validité de l'hypothèse de Riemann.*

Remarquons que la formule (1) met en évidence que la relation (6) est remplie dans tout angle (2), dès qu'elle l'est sur l'axe réel positif.

Inversement, on voit sans peine, en combinant (3) avec un théorème connu de MM. PHRAGMÉN et LINDELÖF, que (6) étant remplie sur une certaine demi-droite située dans l'angle (2), elle le sera sur l'axe réel positif.

On pourrait remplacer $F(x)$ par d'autres fonctions de structure analogue. On voit que la méthode que nous avons suivie permet aussi d'exprimer la valeur réciproque d'une fonction quelconque représentée par une série de Dirichlet.

Il est facile de vérifier l'identité

$$F(x) = \sum_{n=1}^{\infty} \mu(n) \frac{x}{n^2} e^{-\frac{x}{n^2}},$$

les $\mu(n)$ désignant les nombres de MÖBIUS.

5°. Nous ajouterons quelques remarques sur la distribution des zéros de la fonction $F(x)$.

On a, pour x réel et positif, $\lim\limits_{x=\infty} \frac{e^{-x} F(-x)}{x} = -1$. D'autre part, pour un x quelconque $\neq 0$, $F(x)$ satisfait à l'inégalité $|F(x)| < |x| e^{|x|}$. De ces deux faits il résulte d'après des théorèmes connus de POINCARÉ[1] et de M. HADAMARD que $F(x)$ est de *genre un*.

Désignons maintenant par γ_ν les zéros de $\frac{F(x)}{x}$ $(\nu = 1, 2, \ldots)$. De l'inégalité ci-dessus, on conclut immédiatement moyennant le théorème en question de M. HADAMARD que la série $\sum\limits_{\nu=1}^{\infty} \frac{1}{|\gamma_\nu|^{1+\varepsilon}}$ est convergente, ε désignant un nombre positif arbitrairement petit. Nous allons démontrer que la série $\sum\limits_{\nu=1}^{\infty} \frac{1}{|\gamma_\nu|}$ est divergente.

En effet, si elle était convergente on aurait d'après les faits précédents et

[1] De la convergence de $\sum\limits_{1}^{\infty} \frac{1}{|\gamma_\nu|}$ il s'ensuit $\left|\prod\limits_{1}^{\infty} \left(1 - \frac{x}{\gamma_\nu}\right)\right| = O(e^{\varepsilon |x|})$, ε désignant un nombre positif arbitrairement petit.

le théorème cité de Poincaré $F(x) = xe^{-x}\prod_{1}^{\infty}\left(1 - \frac{x}{\gamma_\nu}\right)$. Observons maintenant que la relation $F(x) = O(x^{\frac{1}{4}-\delta})\ (\delta > 0)$, x tendant vers l'infini sur l'axe réel positif, est inadmissible. Elle entraînerait en effet le résultat manifestement inexact que la fonction $\zeta(s)$ ne possède pas de zéros imaginaires. Mais si $F(x)$ était de la forme ci-dessus, il résulterait immédiatement du dit théorème de Poincaré $F(x) = O(e^{-x(1-\varepsilon)}) = O(x^{\frac{1}{4}-\delta})$. Nous avons donc démontré que la série $\sum \frac{1}{|\gamma_\nu|}$ est divergente.

Ainsi nous avons obtenu pour $F(x)$ la forme canonique suivante:

$$F(x) = xe^{\gamma x}\prod_{\nu=1}^{\infty}\left(1 - \frac{x}{\gamma_\nu}\right)e^{\frac{x}{\gamma_\nu}}. \tag{7}$$

Le nombre γ est évidemment réel.

Nous allons maintenant démontrer que $F(x)$ *a une infinité de zéros imaginaires.* Ceci résultera d'un théorème important de M. Pólya qu'il a eu l'amitié de me communiquer lors d'un entretien à Göttingen l'été 1913.[1]

Le théorème en question peut s'énoncer de la manière suivante:

Supposons que la fonction entière $\Phi(x) = \sum_{n=0}^{\infty} a_n x^n$ *admette une représentation de la forme*

$$\Phi(x) = c\,x^r e^{-\gamma x^2 + \delta x}\prod_{\mu=1}^{\tau}(1 + \beta_\mu x)(1 + \bar\beta_\mu x)\prod_{\nu=1}^{\infty}(1 + \delta_\nu x)e^{-\delta_\nu x}, \tag{8}$$

γ étant $\geqq 0$, β_μ et $\bar\beta_\mu$ désignant des quantités imaginaires conjuguées, c, δ et δ_ν des quantités réelles, et $\sum_{\nu=1}^{\infty}\delta_\nu^2$ étant convergent. Alors, pour que la fonction $\Phi(x)$ n'admette pas en même temps une représentation de la forme

$$\Phi(x) = c\,x^r e^{\gamma x}\prod_{\mu=1}^{\tau}(1 + \beta_\mu x)(1 + \bar\beta_\mu x)\prod_{\nu=1}^{\infty}(1 + \delta_\nu x), \tag{9}$$

γ et tous les δ_ν étant des nombres réels de même signe et le produit $\prod_{\nu=1}^{\infty}(1 + \delta_\nu x)$ étant

[1] Le théorème se trouve établi dans le travail suivant de M. Pólya, paru pendant l'impression du travail présent: Algebraische Untersuchungen über ganze Funktionen vom Geschlechte Null und Eins. J. f. Math (145) 1915; *cf.* p. 247—249.

absolument convergent, il faut et il suffit que chacune des deux suites $a_0, a_1, \ldots a_n, \ldots$ *et* $a_0, -a_1, \ldots (-1)^n a_n, \ldots$ *présente une infinité de changements de signe.*

En ce qui concerne notre fonction $F(x)$, il est manifeste, d'après ce que nous venons de démontrer, qu'elle ne peut admettre une représentation de la forme (9). D'autre part, il résulte de la formule (7) que si la fonction $F(x)$ n'avait qu'un nombre fini de zéros imaginaires, elle admettrait une représentation de la forme (8). La suite $\frac{1}{C_1 \Gamma(1)}, \frac{1}{C_2 \Gamma(2)}, \ldots$ ne présentant aucun changement de signe, il s'ensuivrait d'après le théorème de M. Pólya que $F(x)$ admet en même temps une représentation de la forme (9). Ceci étant impossible, il se trouve établi que $F(x)$ possède vraiment une infinité de zéros imaginaires.

D'autre part, de la relation

$$\sum_{n=1}^{\infty} F\left(\frac{x}{n^2}\right) = x e^{-x}$$

et de l'inadmissibilité de la relation $F(x) = O(x^{\frac{1}{4}-\delta})$, il résulte immédiatement que $F(x)$ change de signe au moins une fois sur l'axe réel positif.

20.

Über einen Satz des Herrn Serge Bernstein

Acta Math. *40* (1916), 337–347

(Aus zwei Briefen an Herrn G. MITTAG-LEFFLER.)

I.

...Neulich haben wir uns über einen Satz von Herrn S. BERNSTEIN unterhalten. Ich kann Ihnen jetzt für denselben einen ganz elementaren Beweis mitteilen, der auf der LAGRANGEschen Interpolationsformel beruht.

Ich werde zunächst den zu beweisenden Satz folgendermassen aussprechen. *Es sei $f(x)$ ein Polynom von höchstens n-tem Grade mit beliebigen komplexen Koeffizienten*

$$f(x) = a_0 + a_1 x + \cdots + a_n x^n.$$

Es bedeute L das Maximum des absoluten Betrages von $f(x)$ auf der Strecke $(-1, +1)$. Bezeichnen wir dann mit ξ einen beliebigen Punkt, der ausserhalb[1] *der Strecke $(-1, +1)$ liegt, so ist*

$$|f(\xi)| \leq L(A+B)^n, \tag{1}$$

wo A und B die Halbachsen der Ellipse bedeuten, die durch den Punkt ξ geht und die Punkte -1 und $+1$ zu Brennpunkten hat.

Herr BERNSTEIN[2] beweist für Polynome mit reellen Koeffizienten

$$|f(\xi)| < L(A+B)^n \qquad (n > 0) \tag{2}$$

[1] Der Satz bleibt richtig, wenn ξ auf der Strecke $(-1, +1)$ gelegen ist, wird aber in diesem Falle trivial, da jetzt die Ellipse in die (doppelte) Strecke $(-1, +1)$ übergeht und $A+B=1$ wird.

[2] S. BERNSTEIN: (I), Sur l'ordre de la meilleure approximation des fonctions continues par des polynomes de degré donné, Mémoires publiés par la Classe des Sciences de l'Académie de Belgique, Bd. 4, 1912, S. 13–15. In dieser Arbeit wird der Satz nur für reelle ξ bewiesen. Der allgemeine Beweis befindet sich in der Arbeit (II): Sur une propriété des polynomes, Bulletin de la Soc. Math. de Kharkof, Bd. 14.

und für Polynome mit beliebigen komplexen Koeffizienten

$$|f(\xi)| < 2L(A+B)^n.$$

Ich werde Ihnen demnächst einen auf der Theorie der analytischen Funktionen fussenden Beweis mitteilen, durch welchen ich zeigen kann, dass (2) auch für Polynome mit beliebigen komplexen Koeffizienten richtig ist. Mein elementares Verfahren ergibt aber für beliebige ξ nur die weniger besagende Ungleichung (1). (Wie ich am Ende dieses Briefes zeigen werde, ergibt z. B. für *reelle* ξ die LAGRANGEsche Interpolationsformel sogar mehr als (2), aber nicht für ein beliebiges ξ).

Um die Interpolationsformel anwenden zu können, bestimmen wir zunächst ein Polynom $(n+1)$-ten Grades $g(x)$, das in den Punkten

$$x_0 = 1,\ x_1 = \cos\frac{\pi}{n}, \ldots x_r = \cos\frac{r\pi}{n}, \ldots x_n = -1$$

verschwindet. Zu diesem Ende setzen wir für $-1 \leqq x \leqq 1$, $x = \cos\varphi$. Verändert sich nun φ von 0 bis π, so beschreibt x die Strecke $(+1, -1)$. Hieraus erhellt, dass das fragliche Polynom

$$g(x) = \sin(\arccos x)\sin(n \arccos x) = G(\varphi) = \sin\varphi \sin n\varphi$$

gesetzt werden kann.[1]

Nach der LAGRANGEschen Interpolationsformel ist nun

$$f(x) = g(x)\sum_{r=0}^{n}\frac{f(x_r)}{(x-x_r)g'(x_r)}.$$

Um diesen Ausdruck explizit zu berechnen, schreiben wir

$$g'(x_r) = \left[\frac{d}{d\varphi}(\sin n\varphi \sin\varphi)\frac{d\varphi}{dx}\right]_{\varphi=\varphi_r} = -\left[\frac{n\cos n\varphi\sin\varphi + \cos\varphi\sin n\varphi}{\sin\varphi}\right]_{\varphi=\varphi_r}.$$

Hieraus erhalten wir

$$g'(x_r) = (-1)^{r+1}n \qquad (0 < r < n)$$

und

---

[1] Dass $g(x)$ wirklich ein Polynom $(n+1)$-ten Grades in $x = \cos\varphi$ ist, folgt z. B. aus der Formel

$$g(x) = \frac{\sqrt{1-x^2}}{2i}\left[(x+i\sqrt{1-x^2})^n - (x-i\sqrt{1-x^2})^n\right].$$

Hier kann der Quadratwurzel irgend einer der beiden möglichen Werte beigelegt werden, der aber dann an allen drei Stellen derselbe sein muss.

$$g'(1) = -2n,$$

$$g'(-1) = (-1)^{n+1}\,2n.$$

Mithin

$$(3)\qquad f(x) = \frac{g(x)}{n}\left(-\frac{f(1)}{2(x-1)} + (-1)^{n+1}\frac{f(-1)}{2(x+1)} + \sum_{r=1}^{n-1}(-1)^{r+1}\frac{f(x_r)}{x-x_r}\right).$$

Es sei nun ξ ein beliebiger spezieller Wert von x, der ausserhalb der Strecke $(-1, +1)$ liegt. Wir können dann nach dem Vorgehen von Herrn Bernstein

$$(4)\qquad \xi = \cos(a+bi) = \frac{e^b+e^{-b}}{2}\cos a - i\frac{e^b-e^{-b}}{2}\sin a$$

setzen. Da Kosinus eine gerade Funktion ist, können wir hierbei $b > 0$ wählen. Dann zeigt aber die Formel (4) folgendes. Ist $a + bi$ einer der Werte von arc cos ξ und ist $b > 0$, so ist ξ auf einer Ellipse belegen, die die Punkte -1 und $+1$ zu Brennpunkten hat und deren Halbachsen $\frac{e^b+e^{-b}}{2}$ und $\frac{e^b-e^{-b}}{2}$ sind. Die Summe dieser Halbachsen ist natürlich e^b.

Wir schätzen nun $|\sin(a+bi)|$ ab.

$$|\sin(a+bi)| = \left|\frac{e^b+e^{-b}}{2}\sin a + i\frac{e^b-e^{-b}}{2}\cos a\right| =$$

$$= \frac{1}{2}\sqrt{e^{2b}+e^{-2b}-2\cos 2a} \leqq \frac{e^b+e^{-b}}{2} < e^b;$$

ebenso

$$|\sin n(a+bi)| \leqq \frac{e^{nb}+e^{-nb}}{2} < e^{nb}.$$

Mithin

$$(5)\quad |g(\xi)| = |\sin(n \arccos \xi)\sin(\arccos \xi)| = |\sin n(a+bi)\sin(a+bi)| < e^{(n+1)b}.$$

Wir bezeichnen nun mit δ das Minimum des Abstandes von ξ zu den Punkten der Strecke $(-1, +1)$. Dann erhalten wir aus (3) und aus (5) zunächst folgende sehr grobe Abschätzung

$$|f(\xi)| \leqq \frac{e^{(n+1)b}}{n}\frac{Ln}{\delta} = L\frac{e^{(n+1)b}}{\delta}.$$

Dieses Resultat wenden wir jetzt auf das Polynom $(f(x))^m$ an, wo m eine beliebige positive ganze Zahl bedeutet. Wir erhalten

$$|f(\xi)|^m \leqq L^m \frac{e^{(mn+1)b}}{\delta},$$

also

$$|f(\xi)| \leqq L e^{nb} \sqrt[m]{\frac{e^b}{\delta}}.$$

Lassen wir nun m ins Unendliche wachsen, so ergibt sich

$$|f(\xi)| \leqq L e^{nb}.$$ [1] W. z. b. w.

Für reelle Werte von ξ ergibt die Interpolationsformel folgende Verallgemeinerung eines Satzes von TCHEBYCHEFF.[2]

Unter den am Anfange dieses Briefes gemachten Voraussetzungen ist für jeden reellen Wert ξ, der ausserhalb der Strecke (— 1, + 1) *liegt,*

(6) $$|f(\xi)| \leqq L\,|\cos n\,(\text{arc}\cos \xi)|,$$ [3]

das Gleichheitszeichen ausgeschlossen, wenn das Polynom $f(x)$ nicht von der Form $Le^{i\Theta}\cos(n \text{ arc} \cos x)$ *ist, wobei Θ eine beliebige reelle Zahl bedeutet.*

In der Tat ergibt die Formel (3) für ein beliebiges x

(7) $$|f(x)| \leqq \frac{L\,|g(x)|}{n}\left(\frac{1}{2\,|x-1|} + \frac{1}{2\,|x+1|} + \sum_{r=1}^{n-1} \frac{1}{|x-x_r|}\right).$$

---

[1] Man kann einen ähnlichen Beweis und dasselbe Resultat erhalten, wenn man in $f(x)$ $x=\cos\varphi$ setzend, $f(\cos\varphi)$ in eine endliche FOURIERsche Reihe entwickelt, d. h.

$$f(\cos\varphi) = \sum_{\varkappa=0}^{n} a_\varkappa \cos \varkappa\varphi$$

schreibt und dann die Koeffizienten $a_\varkappa$ abschätzt; man erhält für $|f(\xi)|$ zunächst eine grobe Abschätzung, die dann, wie oben, verfeinert werden kann.

Den Kunstgriff, eine grobe Abschätzung auf eine beliebig hohe Potenz einer Funktion anzuwenden, um daraus eine genauere Abschätzung für die Funktion selbst herzuleiten, entnehme ich einem Schreiben von Herrn LANDAU. Herr LANDAU wendet daselbst den genannten Kunstgriff auf ein von dem unsrigen völlig verschiedenes Problem an.

[2] TCHEBYCHEFF: Sur les fonctions qui s'écartent peu de zéro pour certaines valeurs de la variable, Werke, Bd. 2, S. 335—356. Vergl. auch BERNSTEIN (I), S. 13—14. TSCHEBYCHEFF und BERNSTEIN beweisen diesen Satz nur für Polynome mit reellen Koeffizienten und zeigen in diesem Falle, dass das Gleichheitszeichen nur für das Polynom $f(x) = \pm L \cos(n \text{ arc} \cos x)$ bestehen kann. In seiner späteren Arbeit (II) zeigt Herr BERNSTEIN, dass für Polynome mit reellen Koeffizienten die Ungleichung (6) für beliebige komplexe ξ gilt, die der Bedingung $|\xi| \geqq 1$ genügen.

[3] Dass diese Abschätzung genauer ist als (1), folgt z. B. aus der Ungleichung $|\cos n(a+bi)| = \frac{1}{2}\sqrt{e^{2nb} + e^{-2nb} + 2\cos 2na} \leqq \frac{e^{nb}+e^{-nb}}{2} < e^{nb}$. Vergl. BERNSTEIN (I), S. 14.

Setzen wir andererseits in (3) statt $f(x)$ die Funktion $\cos(n \arccos x)$ ein und beachten, dass $\cos(n \arccos x_r) = (-1)^r$ ist, so erhalten wir

$$(8) \qquad \cos(n \arccos x) = -\frac{g(x)}{n}\left(\frac{1}{2(x-1)} + \frac{1}{2(x+1)} + \sum_{r=1}^{n-1} \frac{1}{x-x_r}\right).$$

Bedeutet nun ξ einen beliebigen reellen Wert, der entweder > 1 oder < -1 ist, und setzen wir diesen Wert in (7) und (8) ein, so haben alle Glieder der Klammer in (8) dasselbe Vorzeichen, mithin ist der absolute Betrag dieser Klammer gleich dem Werte der Klammer in (7), folglich ist die rechte Seite von (7) gleich $L|\cos(n \arccos \xi)|$. Also endlich

$$(6) \qquad |f(\xi)| \leqq L|\cos(n \arccos \xi)|.$$

Man sieht auch unmittelbar (da $g(\xi)$ für die in Betracht kommenden Werte nicht verschwindet), dass das Gleichheitszeichen in (7), also auch in (6), nur dann bestehen kann, wenn die Werte $f(x_r)$ abwechselnd gleich $\pm e^{i\Theta}L$ sind. Dann ist aber, wie leicht ersichtlich,

$$f(x) = \pm e^{i\Theta} L \cos(n \arccos x).$$

Es ist einleuchtend, dass es in dem zuletzt bewiesenen Satze genügt, statt $|f(x)| \leqq L$, für $-1 \leqq x \leqq 1$, nur $|f(x_r)| \leqq L\ (r = 0, 1, \ldots n)$ vorauszusetzen.

Islinge, den 23. Mai 1915.

II.

... Wie ich schon in meinem vorigen Briefe angedeutet habe, lässt sich, für $n > 0$, die Ungleichung (1) jenes Briefes durch die schärfere Ungleichung (2) ersetzen; d. h. unter den dortigen Voraussetzungen gilt für einen beliebigen Punkt ξ, der ausserhalb der Strecke $(-1, +1)$ liegt, die Ungleichung

$$|f(\xi)| < L(A+B)^n.$$

Der Beweis beruht auf einfachen funktionentheoretischen Betrachtungen, welche die wahre Grundlage des Satzes zu bilden scheinen.

Man denke sich die Ebene längs der Strecke $(-1, +1)$ aufgeschnitten und bilde in dieser aufgeschnittenen Ebene die Funktion

$$\psi(z) = z + \sqrt{z^2 - 1}.$$

Es wird derjenige Zweig der Quadratwurzel genommen, der für alle reelle $z > 1$ positiv ist. Dadurch ist $\psi(z)$ in der aufgeschnittenen Ebene eindeutig bestimmt.

Wir beweisen nun den folgenden (übrigens als bekannt zu betrachtenden) Hilfssatz:

Auf jeder Ellipse, mit den Brennpunkten -1 *und* $+1$ *und den Halbachsen* A *und* B, *ist* $|\psi(z)|$ *konstant und ist gleich* $A+B$. *Speziell ist auf der doppelten Strecke* $(-1, +1)$ $|\psi(z)| = 1$.

Bedeutet z einen Punkt der genannten Ellipse, so kann $z = A\cos\varphi + iB\sin\varphi$ gesetzt werden. Ausserdem ist $A^2 - B^2 = 1$. Also

$$\psi(z) = A\cos\varphi + iB\sin\varphi + \sqrt{A^2\cos^2\varphi + 2iAB\cos\varphi\sin\varphi - B^2\sin^2\varphi - 1}$$

$$= A\cos\varphi + iB\sin\varphi + \sqrt{A^2\cos^2\varphi + 2iAB\cos\varphi\sin\varphi - B^2\sin^2\varphi - A^2 + B^2}$$

$$= A\cos\varphi + iB\sin\varphi + \sqrt{B^2\cos^2\varphi + 2iAB\cos\varphi\sin\varphi - A^2\sin^2\varphi}.$$

Der Wert der Quadratwurzel ist $\pm(B\cos\varphi + iA\sin\varphi)$. Da demjenigen Punkte der Ellipse, der auf der positiven reellen Achse liegt, der Wert $\varphi = 0$ entspricht, muss das Vorzeichen $+$ genommen werden. Mithin ist

$$\psi(z) = (A+B)e^{i\varphi}$$

und

$$|\psi(z)| = A+B.$$

Die Doppelstrecke $(-1, +1)$ kann als diejenige Ellipse aufgefasst werden, für welche $A = 1$ und $B = 0$ wird, folglich übergeht hier die obige Gleichung in $|\psi(z)| = 1$. Dies folgt übrigens auch daraus, dass auf dieser Strecke z reell und $\sqrt{z^2 - 1} = \pm i\sqrt{1 - z^2}$ rein imaginär ist.

Wir betrachten nun in der aufgeschnittenen Ebene die Funktion

$$g(z) = \frac{f(z)}{(\psi(z))^n}.$$

Diese Funktion ist in jedem inneren Punkte der aufgeschnittenen Ebene (auch im Unendlichen) regulär analytisch, mithin erreicht ihr absoluter Betrag sein Maximum auf der Grenze, d. h. auf den beiden Rändern der Doppelstrecke $(-1, +1)$. Hier ist aber der absolute Betrag des Nenners gleich 1, und der absolute Betrag des Zählers ist nach unserer Voraussetzung $\leqq L$. Hieraus folgt schon, dass in der ganzen Ebene $|g(z)| \leqq L$ ist. Da aber der Zähler von $g(z)$ ein Polynom, und der Nenner, für $n > 0$, kein Polynom ist, ist $g(z)$ keine Konstante. Mithin ist in jedem Punkte, der ausserhalb der Strecke $(-1, +1)$ liegt,

$$|g(z)| = \left|\frac{f(z)}{(\psi(z))^n}\right| < L,$$

d. h.

$$|f(z)| < L(A+B)^n. \quad \text{W. z. b. w.}$$[1]

Unser Beweis lässt sich auch leicht auf den Fall übertragen, wo $f(z)$ ein Polynom in z und $\sqrt{z^2-1}$ vom Gesamtgrade $\leqq n$ ist. Hierdurch erhält man, wenn man noch $z = \cos x$ setzt, einen Satz über trigonometrische Polynome, auf welchen ich aber hier nicht näher eingehen will, da er auch aus unseren späteren Betrachtungen folgt.

Ich sagte oben, dass der angewandte Hilfssatz als bekannt zu betrachten ist. In der Tat bildet die Funktion $x = \psi(z)$ einen Zweig der Umkehrung der Funktion

(9) $$z = \frac{1}{2}\left(x + \frac{1}{x}\right).$$

Die Funktion (9) und ihre Umkehrung werden z. B. im zweiten Bande des *Traité d'Analyse* von Herrn Picard eingehend behandelt. (Vergl. 2. Aufl. S. 314—317.) Die Funktion (9) bildet jeden um den Anfangspunkt beschriebenen Kreis, dessen Halbmesser $r > 1$ ist, auf eine Ellipse ab, deren Halbachsen A und B der Gleichung $A + B = r$ genügen, und deren Brennpunkte -1 und $+1$ sind. Der Kreis mit dem Halbmesser 1 wird auf die doppelte Strecke $(-1, +1)$ abgebildet. In diesen Tatsachen ist der obige Hilfssatz enthalten.

Jeder Kreis, dessen Halbmesser $r < 1$ ist, wird durch die Funktion (9) wieder auf eine Ellipse abgebildet, deren Halbachsen die Gleichung $A - B = r$ erfüllen und deren Brennpunkte ebenfalls -1 und $+1$ sind. Die x-Ebene wird also durch die Funktion (9) auf die doppelte z-Ebene abgebildet. Die Umkehrfunktion dieser Abbildung hat folglich zwei Werte, die durch die Formeln

$$x_1 = z + \sqrt{z^2 - 1}$$

und

$$x_2 = z - \sqrt{z^2 - 1}$$

dargestellt werden. Natürlich ist $x_1 x_2 = 1$.

[1] Dieser Beweis, den mir Herr Marcel Riesz seiner Zeit mitteilte, wurde von mir in zwei kleineren Aufsätzen »Über einen Satz des Herrn Serge Bernstein, Sitzungsberichte der Kgl. Bayer. Akad. d. Wissensch., Jahrg. 1915, S. 419—424« und »Sur un théorème de M. Serge Bernstein, The Tohôku Math. Journal, Vol. 9, Nos. 1, 2, 1916, S. 1—6« ohne andere Bemerkung wiedergegeben, als mit einem Hinweis auf den obigen Brief. Durch unvorhergesehene Umstände wurde jedoch das Drucken desselben derart verzögert, dass meine Aufsätze früher als der Brief erschienen. Hierdurch könnte das Missverständnis aufkommen, der Beweis stamme von mir. Dies ist aber, wie ich hier ausdrücklich betonen will, durchaus nicht der Fall.

G. Mittag-Leffler.

Um das grosse Interesse der betrachteten Abbildung zu zeigen, will ich hier einen Satz aussprechen, der aus den angeführten Betrachtungen bei PICARD unmittelbar folgt.

Jede Funktion, die ausserhalb der Strecke $(-1, +1)$ regulär analytisch ist, kann in eine Reihe der Form

$$\sum_{n=0}^{\infty} a_n (z - \sqrt{z^2 - 1})^n = \sum_{n=0}^{\infty} a_n (z + \sqrt{z^2 - 1})^{-n} \tag{10}$$

entwickelt werden.

Dieser Satz beruht eben auf der Tatsache, dass die Funktion $z - \sqrt{z^2 - 1} = \frac{1}{z + \sqrt{z^2 - 1}}$ in der aufgeschnittenen Ebene regulär analytisch ist, und diese Ebene auf das Innere des Einheitskreises abbildet.

Der zuletzt ausgesprochene Satz gestattet eine recht interessante Anwendung. Herr HAAR hat nämlich den folgenden merkwürdigen Entwicklungssatz gefunden. (Über analytische Funktionen mit singulärer Linie, Göttinger Nachrichten, 1914, Heft 1, S. 115–123.)

Sei $F(z)$ eine Funktion der komplexen Veränderlichen z, die ausserhalb des doppelpunktlosen analytischen Kurvenstückes k überall regulär analytisch ist. Allsdann existieren unendlich viele, längs der Kurve k definierte und nur von dieser Kurve abhängige, stetige Funktionen $f_1(z), f_2(z), \cdots, f_n(z), \cdots$ so, dass für jede Stelle z, die ausserhalb der Kurve liegt, die unendliche Reihe

$$F(\infty) + a_1 \int_{(k)} \frac{df_1(\zeta)}{\zeta - z} + a_2 \int_{(k)} \frac{df_2(\zeta)}{(\zeta - z)^2} + \cdots + a_n \int_{(k)} \frac{df_n(\zeta)}{(\zeta - z)^n} + \cdots \tag{11}$$

konvergiert und die Funktion $F(z)$ darstellt; hierbei bedeuten die a_n Konstanten, und die Integrale sind im STIELTJESschen Sinne genommen.

Herr HAAR beweist dies durch eine konforme Abbildung auf Grund eines C. NEUMANNschen Theorems über Entwicklungen nach Kugelfunktionen zweiter Art. Es ist nun leicht zu zeigen, dass, wenn man statt dieses Entwicklungssatzes den obigen anwendet, die STIELTJESschen Integrale vermieden werden können. Es sei in der Tat $\varphi(z)$ die von Herrn HAAR angewandte Funktion, welche die durch die Kurve k aufgeschnittene Ebene auf eine durch die Strecke $(-1, +1)$ aufgeschnittene Ebene konform abbildet, wobei die unendlich fernen Punkte der beiden Ebenen einander entsprechen mögen. Dann ist nach dem obigen Satz, wenn wir

$$\beta(z) = \varphi(z) - \sqrt{(\varphi(z))^2 - 1}$$

setzen,

(12)
$$F(z) = F(\infty) + \sum_{n=1}^{\infty} a_n (\beta(z))^n .$$

Da nun die Funktion $\beta(z)$ in der ganzen durch die Kurve k aufgeschnittenen Ebene stetig ist und im Unendlichen verschwindet, so können wir schreiben

$$(\beta(z))^n = \frac{1}{2i\pi} \int \frac{(\beta(\zeta))^n}{\zeta - z} d\zeta ,$$

wobei das Integral über die beiden Ränder der Kurve k erstreckt wird und zwar so, dass die aufgeschnittene Ebene zur linken liegt. Integriert man dann nach dem Vorgehen von Herrn HAAR $(n-1)$-mal partiell, so müssen die total integrierten Glieder verschwinden, da die Funktion $(\beta(z))^n$ im Unendlichen von der n-ten Ordnung verschwindet. Mithin erhält man

$$(\beta(z))^n = \int \frac{\gamma_n(\zeta)}{(\zeta - z)^n} d\zeta .$$

Jedem Punkte ζ der Kurve k (ausser den Endpunkten) entsprechen in der aufgeschnittenen Ebene zwei Punkte ζ' und ζ''. Setzen wir $g_n(\zeta) = \gamma_n(\zeta') - \gamma_n(\zeta'')$ und in den Endpunkten $g_n(\zeta) = 0$, so erhalten wir

(13)
$$(\beta(z))^n = \int_{(k)} \frac{g_n(\zeta)}{(\zeta - z)^n} d\zeta .$$

Aus (12) und (13) folgt nun die Darstellung

$$F(z) = F(\infty) + \sum_{n=1}^{\infty} a_n \int_{(k)} \frac{g_n(\zeta)}{(\zeta - z)^n} d\zeta ,$$

wobei die Funktionen $g_n(\zeta)$ auf der Kurve k stetig (und ausser den Endpunkten sogar regulär analytisch) sind.

Es sei noch bemerkt, dass die Funktionen $f_n(\zeta)$ in der Entwicklungsformel von Herrn HAAR *von der zweiten angefangen* eine stetige Ableitung besitzen, wie das aus dem Beweise von Herrn HAAR hervorgeht. Man könnte also auch in der Formel (11) für $n = 2, 3, \ldots$

$$\int_{(k)} \frac{d f_n(\zeta)}{(\zeta - z)^n} = \int_{(k)} \frac{f'_n(\zeta)}{(\zeta - z)^n} d\zeta = \int_{(k)} \frac{g_n(\zeta)}{(\zeta - z)^n} d\zeta$$

schreiben. Die Einheitlichkeit der Darstellung würde aber dann verloren gehen, da $f_1(\zeta)$ keine stetige Ableitung besitzt.

Zuletzt will ich noch Ihre Aufmerksamkeit auf einige Sätze lenken, die in denselben Gedankenkreis gehören, wie der BERNSTEINsche Satz.

Aus dem, was wir oben über die Abbildung $z = \frac{1}{2}\left(x + \frac{1}{x}\right)$ sagten, erhellt sofort, dass der BERNSTEINsche Satz mit dem folgenden Satze gleichwertig ist:

Es sei $h(x) = \sum_{k=0}^{n} a_k \left(x + \frac{1}{x}\right)^k$ auf dem Einheitskreise dem absoluten Betrage nach $\leqq L$; dann ist im Inneren des Einheitskreises $|h(x)| < L|x|^{-n}$ und im Äusseren des Einheitskreises $|h(x)| < L|x|^n$.

Dieser Satz bedarf ja kaum eines Beweises. Er folgt unmittelbar aus der Tatsache, dass $h(x)x^n$ innerhalb und $h(x)x^{-n}$ ausserhalb des Einheitskreises regulär analytisch sind.

Derselbe Gedankengang führt zu dem folgenden allgemeineren Satze:

Ist $h(x) = \sum_{k=-m}^{n} a_k x^k$ auf dem Einheitskreise dem absoluten Betrage nach $\leqq L$, so ist innerhalb des Einheitskreises $|h(x)| \leqq L|x|^{-m}$ und ausserhalb des Einheitskreises $|h(x)| \leqq L|x|^n$. Das Gleichheitszeichen ist ausgeschlossen, wenn $h(x)$ nicht $= Le^{i\Theta}x^{-m}$ bzw. $= Le^{i\Theta}x^n$ ist, hierbei bedeutet Θ eine beliebige reelle Zahl. (Man kann selbstredend den Satz dadurch verallgemeinern, dass man die Anzahl der Glieder in irgendeiner der beiden Richtungen ins Unendliche wachsen lässt, vorausgesetzt dass die unendliche Reihe auf dem Einheitskreise z. B. gleichmässig konvergiert.)

Setzt man in diesem Satze $x = e^s$, so erhält man das folgende Ergebnis:

Ist $H(s) = \sum_{k=-m}^{n} a_k e^{ks}$ auf der imaginären Achse dem absoluten Betrage nach $\leqq L$, so ist für $R(s) > 0$ (wo $R(s)$ den reellen Teil von s bedeutet) $|H(s)| \leqq Le^{nR(s)}$ und für $R(s) < 0$, $|H(s)| \leqq Le^{-mR(s)}$ u. s. w.[1]

All dies kann auch auf Ausdrücke von der Form $H(s) = \sum_{k=0}^{n} a_k e^{\lambda_k s}$ übertragen werden. Hierbei bedeuten die λ_k reelle Grössen, die den Ungleichungen

---

[1] Setzt man in diesem Satze $s = i\alpha$, so kann man $H(s) = H(i\alpha)$ in der Form $\sum_{k=0}^{N}(A_k \cos k\alpha + B_k \sin k\alpha)$ schreiben. Das obige kann also auch als ein Satz über trigonometrische Polynome ausgesprochen werden.

$-\infty < \lambda_0 < \lambda_1 < \ldots < \lambda_n < +\infty$ genügen. Setzt man dann voraus, dass auf der imaginären Achse $|H(s)| \leq L$ ist, so folgt auf Grund eines bekannten PHRAGMÉN-LINDELÖFschen Satzes, dass für $R(s) > 0$, $|H(s)| \leq L e^{\lambda_n R(s)}$ und für $R(s) < 0$, $|H(s)| \leq L e^{\lambda_0 R(s)}$ ist. Hierbei ist auch das Gleichheitszeichen auszuschliessen, wenn $H(s)$ nicht $= L e^{i\Theta} e^{\lambda_n s}$ bzw. $= L e^{i\Theta} e^{\lambda_0 s}$ ist, wobei Θ wieder eine beliebige reelle Konstante bedeutet.

Für Integrale $\int_u^v a(\tau) e^{\tau s} d\tau$, oder allgemeiner für Ausdrücke von der Form $\int_u^v e^{\tau s} d A(\tau)$, wo $A(\tau)$ eine beliebige Funktion beschränkter Schwankung bedeuten kann, gilt ein ähnlicher Satz. (Dieser letzte Ausdruck umfasst natürlich die beiden vorangehenden.)

Diese Sätze enthalten ja eigentlich nichts neues; ich wollte eben nur ihren Zusammenhang mit dem schönen Satze von Herrn BERNSTEIN betonen.

Islinge, den 25. Mai 1915.

Marcel Riesz.

21.

Ein Konvergenzsatz für Dirichletsche Reihen

Acta Math. *40* (1916), 349–361

Vor mehr als sechs Jahren habe ich in einer kurzen Mitteilung[1] einen Satz über DIRICHLETsche Reihen

$$a_0 e^{-\lambda_0 z} + \cdots + a_n e^{-\lambda_n z} + \cdots \qquad (0 \leqq \lambda_0 < \cdots < \lambda_n < \cdots; \lim_{n=\infty} \lambda_n = \infty)$$

ausgesprochen, dessen Beweis (für den allgemeinen Fall) ich erst jetzt veröffentliche. Ich hegte nämlich immer die Hoffnung meinen ursprünglichen, recht verwickelten Beweis wesentlich vereinfachen zu können. Für den speziellen Fall der gewöhnlichen DIRICHLETschen Reihen $\Sigma a_n n^{-s}$ hat Herr LANDAU meinen damaligen Beweis mit meiner Zustimmung veröffentlicht.[2] Gleichzeitig gab er eine Reihe wichtiger Anwendungen meines Satzes auf gewisse zahlentheoretische Funktionen. In einer anderen Arbeit veröffentlichte er einen zweiten, auch von mir herrührenden, verwandten Satz, nebst Beweis und wandte denselben auf eine idealtheoretische Funktion an.[3]

Unser Satz behauptet, dass, wenn die Koeffizienten der DIRICHLETschen Reihe eine gewisse Bedingung erfüllen, die Reihe an allen Regularitätsstellen ihrer Konvergenzgeraden konvergiert. Ich habe schon in der erwähnten, kurzen Mit-

[1] M. RIESZ: Sur les séries de DIRICHLET et les séries entières, Comptes rendus, Paris, Bd. 149 (1909, 2).

[2] E. LANDAU: Über die Bedeutung einiger neuen Grenzwertsätze der Herren HARDY und AXER, Prace matematyczno-fizyczne, Warszawa, Bd. 21 (1910). Vergl. S. 151—167.

[3] E. LANDAU: Über eine idealtheoretische Funktion, Transactions of the Amer. Math. Soc., Bd. 13 (1912). Vergl. S. 18—19. Für andere Anwendungen dieser Sätze vergl. D. CAUER: Neue Anwendung der PFEIFFERschen Methode zur Abschätzung zahlentheoretischer Funktionen, Inauguraldissertation, Göttingen (1914) und G. H. HARDY: On the expression of a number as the sum of two squares, Quarterly Journal of Mathematics, no. 183 (1915).

teilung hervorgehoben, dass unser allgemeiner Satz den bekannten FATOUschen Satz über Potenzreihen als Spezialfall umfasst. Es ist mir nun neuerdings gelungen, für diesen letztgenannten Satz einen überaus einfachen Beweis zu finden, der an anderer Stelle veröffentlicht wurde.[1] Derselbe gestaltet sich für unseren allgemeinen Satz über DIRICHLETsche Reihen fast ebenso einfach wie für Potenzreihen. In Abschnitt 1 der vorliegenden Arbeit soll nun der Beweis für den genannten allgemeinen Satz entwickelt werden. Im Abschnitte 2 wird unter anderem gezeigt, dass unser Satz den FATOUschen Satz enthält. In Abschnitt 3 wird die Bedingung (1) von Abschnitt 1 durch eine gleichwertige ersetzt, und daraus werden dann ein Spezialfall des Satzes und einige Sätze für gewisse spezielle Exponentenfolgen hergeleitet. Im Abschnitte 4 wird endlich auseinandergesetzt, wie unsere Ergebnisse auf verwandte Integralausdrücke übertragen werden können.

In der vorliegenden Arbeit wird überall vorausgesetzt, dass die betrachtete Funktion an den fraglichen Stellen regulär ist. Ich gedenke in einer späteren Arbeit auszuführen, wie die Regularität durch weniger einschränkende Bedingungen ersetzt werden kann.[2] Auch auf die Ausdehnung der vorliegenden Ergebnisse auf *typische Mittel*[3] komme ich bei einer späteren Gelegenheit zurück.

1. Satz I. *Erfüllen die Koeffizienten einer* DIRICHLET*schen Reihe*

$$f(z) = a_0 e^{-\lambda_0 z} + \cdots + a_n e^{-\lambda_n z} + \cdots$$

die Bedingung

$$(1) \qquad \lim_{n \to \infty} \frac{a_0 + a_1 + \cdots + a_n}{e^{\lambda_n c}} = 0, \qquad (c > 0),^4$$

und ist die infolge der Bedingung (1) *für* $R(z) > c$ *reguläre Funktion* $f(z)$ *auch in gewissen Punkten der Geraden* $R(z) = c$ *regulär, so konvergiert die Reihe in diesen Punkten.*[5] *Die Konvergenz ist eine gleichmässige auf jeder abgeschlossenen Strecke der*

---

[1] M. RIESZ: Neuer Beweis des FATOUschen Satzes, Göttinger Nachrichten (1916); Sätze über Potenzreihen. Arkiv för Mat., Astr. och Fys. (1916).

[2] Vergl. M. RIESZ: Comptes rendus, a. a. O. und Über einen Satz des Herrn FATOU, Journal für Mathematik, Bd. 140 (1911).

[3] Für die ausführliche Darstellung der Eigenschaften dieser Mittel vergl. G. H. HARDY and M. RIESZ: The general theory of DIRICHLET's series, Cambridge Tract in Mathematics etc. (1915).

[4] Die Bedingung ist auch bei $c = 0$ hinreichend aber nicht mehr notwendig. Bleibt der Ausdruck auf der linken Seite von (1) nur beschränkt, so gilt dasselbe für die Teilsummen der Reihe an jeder Regularitätsstelle der Geraden $R(z) = c$. Ähnliches gilt, wenn in den späteren Bedingungen die Voraussetzung, dass gewisse Ausdrücke gegen Null konvergieren, ersetzt wird durch die Voraussetzung, dass diese Ausdrücke beschränkt bleiben, oder in der Ausdrucksweise von Herrn LANDAU, wenn o durch O ersetzt wird.

[5] Wie gewöhnlich, setzen wir $R(a + bi) = a$ und $I(a + bi) = b$.

Geraden, welche ausschliesslich aus Regularitätsstellen besteht. Wie bekannt,[1] *ist die obige Bedingung* (1) *auch notwendig, damit die Gerade* $R(z) = c$ *wenigstens eine Konvergenzstelle der Reihe enthalte.*

Wir setzen

$$A_n = a_0 + \cdots + a_n$$

und

$$A(v) = A_n, \qquad \lambda_n < v \leqq \lambda_{n+1}$$

$$A(v) = 0, \qquad 0 \leqq v \leqq \lambda_0.^{2}$$

Dann ist bekanntlich nach der Abelschen Transformation

$$(2) \qquad a_0 e^{-\lambda_0 z} + a_1 e^{-\lambda_1 z} + \cdots + a_n e^{-\lambda_n z} =$$

$$= A_0 (e^{-\lambda_0 z} - e^{-\lambda_1 z}) + A_1 (e^{-\lambda_1 z} - e^{-\lambda_2 z}) + \cdots + A_{n-1} (e^{-\lambda_{n-1} z} - e^{-\lambda_n z}) +$$

$$+ A_n e^{-\lambda_n z} = z \int_0^{\lambda_n} A(v) e^{-vz} dv + A_n e^{-\lambda_n z}.$$

Die Bedingung (1) kann auch in der Gestalt

$$(3) \qquad \lim_{v=\infty} A(v) e^{-vc} = 0$$

geschrieben werden. Dann folgt aber aus (2), dass die Reihe $f(z) = \sum_{n=0}^{\infty} a_n e^{-\lambda_n z}$ in jedem samt Begrenzung in der Halbebene $R(z) > c$ gelegenen, endlichen Gebiete gleichmässig konvergiert und daselbst eine regulär analytische Funktion darstellt. Es ist auch für $R(z) > c$

$$(4) \qquad f(z) = z \int_0^{\infty} A(v) e^{-vz} dv.$$

Es möge nun die Funktion $f(z)$ auch auf der Geraden $R(z) = c$ Regularitätsstellen besitzen und $x = c + \tau i$ sei eine solche Stelle. Wir bestimmen zwei Punkte $z_1 = c + \tau_1 i$ und $z_2 = c + \tau_2 i$ derart, dass $\tau_1 < \tau < \tau_2$ sei, und die Funktion $f(z)$ auf der ganzen Strecke z_1, z_2 (mit Einschluss der Endpunkte) regulär sei. Wir können dann die Zahl $b < c$ so nahe an c wählen, dass die Funktion $f(z)$ auch

[1] J. L. W. V. Jensen: Sur une généralisation d'un théorème de Cauchy, Comptes rendus, Paris, Bd. 106 (1888, 1). Diese Behauptung wird übrigens auch in Abschnitt 2 der vorliegenden Arbeit bewiesen.

[2] Natürlich kann λ_0 auch 0 sein.

noch im Rechteck z_1, z_2, $b + \tau_2 i$, $b + \tau_1 i$ dieselbe Eigenschaft besitze. Wir setzen der Kürze halber $z_1' = b + \tau_1 i$, $z_2' = b + \tau_2 i$. Wir setzen endlich $z_1'' = d + \tau_1 i$, $z_2'' = d + \tau_2 i$, wo d eine beliebige, aber feste Zahl $> c$ bedeutet.

Wir betrachten jetzt die Funktionen

$$g_n(z) = e^{\lambda_n (z-c)}(z - z_1)(z - z_2)(f(z) - (a_0 e^{-\lambda_0 z} + \cdots + a_n e^{-\lambda_n z}))$$

und werden zeigen, dass bei hinreichend grossem n auf der ganzen Begrenzung des Rechtecks z_1', z_1'', z_2'', z_2'

$$(5) \qquad |g_n(z)| < \varepsilon$$

wird, wie klein auch die positive Zahl ε sein mag.[1]

Wegen (1) können wir für ein beliebig kleines, positives δ eine Zahl ν finden derart, dass

$$(6) \qquad |A_n e^{-\lambda_n c}| < \delta \text{ für } n > \nu$$

und daraus folgt, dass

$$(7) \qquad |A(v) e^{-vc}| < \delta \text{ für } v > \lambda_\nu.$$

Wir setzen der Kürze halber noch $\sigma = R(z)$. Für $\sigma > c$ ist dann wegen (2) und (4)

$$(8) \qquad |f(z) - (a_0 e^{-\lambda_0 z} + \cdots + a_n e^{-\lambda_n z})| = \left| z \int_{\lambda_n}^{\infty} A(v) e^{-vz} dv - A_n e^{-\lambda_n z} \right|.^{2}$$

Wählen wir jetzt $n > v$, so folgt aus (6), (7) und (8)

$$(9) \qquad |f(z) - (a_0 e^{-\lambda_0 z} + \cdots + a_n e^{-\lambda_n z})| < \delta |z| \int_{\lambda_n}^{\infty} e^{-v(\sigma - c)} dv + \delta e^{-\lambda_n (\sigma - c)} =$$

$$= \delta e^{-\lambda_n(\sigma - c)} \left(1 + \frac{|z|}{\sigma - c}\right) \leqq \delta e^{-\lambda_n (\sigma - c)} \left(1 + \frac{K}{\sigma - c}\right),$$

wo K das Maximum von $|z|$ im Rechtecke z_1', z_1'', z_2'', z_2' bedeutet.

[1] In $g_n(z)$ könnte auch statt $e^{\lambda_n (z-c)}$ die Funktion $e^{\lambda (z-c)}$ stehen, wo λ eine beliebige Zahl bedeutet, die der Bedingung $\lambda_n \leqq \lambda \leqq \lambda_{n+1}$ genügt.

[2] Dieser Ausdruck kann auch in der Form $z \int_{\lambda_{n+1}}^{\infty} A(v) e^{-vz} dv - A_n e^{-\lambda_{n+1} z}$ geschrieben werden.

Es möge noch H das Maximum der Werte $|z-z_1|$ und $|z-z_2|$ für alle Punkte z des genannten Rechtecks bedeuten.

Ist nun z ein beliebiger Punkt der Strecke z_1, z_1'', so ist $|z-z_1| = \sigma - c$. Also ist nach (9) auf dieser Strecke

$$(10)\quad |g_n(z)| < e^{\lambda_n(\sigma-c)}(\sigma-c)\,H\,\delta\,e^{-\lambda_n(\sigma-c)}\left(1+\frac{K}{\sigma-c}\right) =$$

$$= \delta(\sigma-c)H + \delta H K \leqq \delta H(d-c+K).$$

Dieselbe Ungleichung gilt aus Symmetriegründen auch für die Strecke z_2, z_2''.

Es sei jetzt z ein Punkt der auf der Geraden $\sigma = d$ gelegenen Strecke z_1'', z_2''; dann ist wegen (9)

$$(11)\qquad |g_n(z)| < e^{\lambda_n(d-c)}H^2\,\delta\,e^{-\lambda_n(d-c)}\left(1+\frac{K}{d-c}\right) = \delta H^2\left(1+\frac{K}{d-c}\right).$$

Um $|g_n(z)|$ auch auf den anderen Teilen der Begrenzung des Rechtecks z_1', z_1'', z_2'', z_2' abzuschätzen, bezeichnen wir mit M das Maximum von $|f(z)|$ in diesem Rechtecke und mit A die grösste der Zahlen $|A_0|, |A_1|, \cdots |A_\nu|$. Dann ist mit Rücksicht auf (2), (6) und (7) für $b \leqq \sigma \leqq c$, $\tau_1 \leqq I(z) \leqq \tau_2$

$$(12)\quad |f(z)-(a_0e^{-\lambda_0 z}+\cdots+a_ne^{-\lambda_n z})| < M + \delta e^{-\lambda_n(c-\sigma)} + KA\int_0^{\lambda_\nu} e^{-(v-c)}dv +$$

$$+ K\delta\int_{\lambda_\nu}^{\lambda_n} e^{-v(\sigma-c)}dv < M + \delta e^{\lambda_n(c-\sigma)} + KA\lambda_\nu e^{\lambda_\nu(c-\sigma)} + \frac{K\delta}{c-\sigma}e^{\lambda_n(c-\sigma)}.$$

Auf der Strecke z_1', z_1 ist $|z-z_1| = c-\sigma$, folglich

$$|e^{\lambda_n(z-c)}(z-z_1)(z-z_2)| \leqq e^{-\lambda_n(c-\sigma)}(c-\sigma)H$$

also mit Benutzung von (12) und auf Grund von $c-\sigma \leqq c-b$

$$(13)\qquad |g_n(z)| < (c-\sigma)e^{-\lambda_n(c-\sigma)}H(M+KA\lambda_\nu e^{\lambda_\nu(c-b)}) + \delta H(c-b+K).$$

Durch Differentiation folgt unmittelbar, dass der Ausdruck $u\,e^{-\lambda_n u}$, als Funktion von $u \geqq 0$ betrachtet, sein Maximum für $u = \frac{1}{\lambda_n}$ erreicht. Daher ist

$$(c-\sigma)e^{-\lambda_n(c-\sigma)} \leqq \frac{1}{\lambda_n} \qquad (\sigma \leqq c).$$

Daraus folgt, dass das erste Glied auf der rechten Seite von (13) bei hinreichend grossem n kleiner als δ wird und somit ist dargelegt, dass bei hinrechend grossem n auf der ganzen Strecke z_1', z_1 und ebenso auch auf der Strecke z_2', z_2

$$|g_n(z)| < \delta(1 + H(c-b+K)) \tag{14}$$

ausfällt.

Auf der Strecke z_1', z_2' ist endlich wegen (12) und $c-\sigma = c-b$

$$|g_n(z)| < e^{-\lambda_n(c-b)}H^2\left(M + \delta e^{\lambda_n(c-b)} + KA\lambda_\nu e^{\lambda_\nu(c-b)} + \frac{K\delta}{c-b}e^{\lambda_n(c-b)}\right) =$$

$$= e^{-\lambda_n(c-b)}H^2\left(M + KA\lambda_\nu e^{\lambda_\nu(c-b)}\right) + \delta H^2\left(1 + \frac{K}{c-b}\right).$$

Für hinreichend grosses n ist das erste Glied des letzten Ausdruckes kleiner als das zweite, folglich

$$|g_n(z)| < 2\delta H^2\left(1 + \frac{K}{c-b}\right). \tag{15}$$

Bei hinreichend kleinem δ ergeben die Ungleichungen (10), (11), (14), (15) die Ungleichung (5) für die ganze Begrenzung des Rechtecks z_1', z_1'', z_2'', z_2'.

Damit sind wir aber gleich am Ziele. Denn die Funktionen $g_n(z)$ sind im betrachteten Rechtecke regulär, mithin erreicht ihr absoluter Betrag sein Maximum auf der Begrenzung. Daraus ergibt sich, bei hinreichend grossem n, für einen beliebigen inneren Punkt x der Strecke z_1, z_2

$$|g_n(x)| < \varepsilon,$$

oder

$$|f(x) - (a_0 e^{-\lambda_0 x} + \cdots + a_n e^{-\lambda_n x})| < \frac{\varepsilon}{|x-z_1||x-z_2|}, \tag{16}$$

d. h. die behauptete Konvergenz unserer Reihe.

Auf die gleichmässige Konvergenz kann man nun folgendermassen schliessen. Besteht die Strecke x_1, x_2 der Geraden $R(z) = c$ ausschliesslich aus Regularitätsstellen, so ist sie in einer grösseren Strecke enthalten, welche dieselbe Eigenschaft besitzt. Bezeichnen wir die Endpunkte dieser grösseren Strecke wiederum mit z_1 und z_2, so folgt aus (16) die gleichmässige Konvergenz für die Punkte der Strecke x_1, x_2.

2. In den folgenden Abschnitten geben wir unserem Satze verschiedene neue Formen, aus denen dann einige interessante Spezialfälle des Satzes folgen.

Durch die lineare Substitution $z' = z - c$ geht Satz I in den folgenden über.

Satz I'. *Es sei c eine beliebige, aber feste, positive, reelle Zahl. Erfüllen dann die Koeffizienten der* DIRICHLET*schen Reihe*

$$f(z) = a_0 e^{-\lambda_0 z} + \cdots + a_n e^{-\lambda_n z} + \cdots$$

die Bedingung

$$\lim_{n=\infty} \frac{a_0 e^{\lambda_0 c} + \cdots + a_n e^{\lambda_n c}}{e^{\lambda_n c}} = 0, \tag{17}$$

und ist die infolge dieser Bedingung für $R(z) > 0$ reguläre Funktion $f(z)$ auch in gewissen Punkten der imaginären Achse regulär, so konvergiert die Reihe in diesen Punkten. Die Konvergenz ist gleichmässig auf jeder Strecke der imaginären Achse, die ausschliesslich aus Regularitätsstellen besteht. Die Bedingung (17) ist auch notwendig, damit die imaginäre Achse wenigstens einen Konvergenzpunkt der Reihe enthalte.

Dieser Satz geht durch die Transformation $e^{-z} = x$ für $\lambda_n = n$ in den genannten, folgendermassen lautenden Satz von FATOU[1] über.

Erfüllen die Zahlen a_n die Bedingung

$$\lim_{n=\infty} a_n = 0, \tag{18}$$

so konvergiert die Potenzreihe

$$F(x) = a_0 + a_1 x + \cdots + a_n x^n + \cdots$$

an jeder Regularitätsstelle des Einheitskreises. Die Konvergenz ist auf jedem Regularitätsbogen eine gleichmässige.

In der Tat nimmt, für $\lambda_n = n$, die Bedingung (17) die folgende Gestalt an:

$$\lim_{n=\infty} \frac{a_0 + a_1 R + \cdots + a_n R^n}{R^n} = 0, \qquad (R > 1) \tag{19}$$

wo R eine feste Zahl bedeutet. Wir haben also nur zu zeigen, dass die Bedingungen (18) und (19) vollständig gleichwertig sind.

Es sei zuerst (18) erfüllt. Dann lässt sich bei beliebig kleinem ε die Zahl ν derart bestimmen, dass für alle $m > \nu$

$$|a_m| < \varepsilon$$

wird. Mithin ist für $n > \nu$

[1] P. FATOU: Séries trigonométriques et séries de TAYLOR, Acta Math., Bd. 30 (1906). Vergl. S. 389—391.

$$\frac{|a_0 + a_1 R + \cdots + a_n R^n|}{R^n} < \frac{|a_0 + \cdots + a_\nu R^\nu| + \varepsilon(R^{\nu+1} + \cdots + R^n)}{R^n} <$$

$$< \frac{|a_0 + \cdots + a_\nu R^\nu|}{R^n} + \varepsilon \frac{R}{R-1}.$$

Für hinreichend grosses n ist dieser Ausdruck $< \frac{2R}{R-1}\varepsilon$, d. h. (19) folgt wirklich aus (18).

Es sei nun umgekehrt (19) erfüllt. Dann ist bei genügend grossem n

$$\frac{|a_0 + \cdots + a_{n-1}R^{n-1} + a_n R^n|}{R^n} < \varepsilon$$

und

$$\frac{|a_0 + \cdots + a_{n-1}R^{n-1}|}{R^n} < \frac{|a_0 + \cdots + a_{n-1}R^{n-1}|}{R^{n-1}} < \varepsilon.$$

Daraus folgt durch Subtraktion $|a_n| < 2\varepsilon$. Mithin hat auch die Bedingung (19) die Bedingung (18) zur Folge.

Wir kehren jetzt zu dem allgemeinen Satze I' zurück. In demselben spielt augenscheinlich die positive Konstante c eine rein zufällige Rolle. Die Vermutung liegt also nahe, dass die Beziehung (17) für alle positiven Zahlen besteht, wenn sie für eine einzige solche Zahl besteht. Dies ergibt sich in der Tat aus dem folgenden, übrigens bekannten Satze.

Ist

$$\lim_{n=\infty} \frac{a_0 e^{\lambda_0 \alpha} + \cdots + a_n e^{\lambda_n \alpha}}{e^{\lambda_n \alpha}} = 0, \tag{20}$$

wo α eine beliebige, feste, komplexe Zahl bedeutet, so ist auch

$$\lim_{n=\infty} \frac{a_0 e^{\lambda_0 \beta} + \cdots + a_n e^{\lambda_n \beta}}{e^{\lambda_n \beta}} = 0, \tag{21}$$

wo β eine beliebige komplexe Zahl sein kann, deren reeller Teil positiv ist.

Unsere Behauptung folgt leicht aus der Abelschen Transformation. Wir setzen

$$A_{\alpha,n} = a_0 e^{\lambda_0 \alpha} + \cdots + a_n e^{\lambda_n \alpha}$$

$$A_\alpha(v) = A_{\alpha,n} \text{ für } \lambda_n < v \leqq \lambda_{n+1} \text{ und } = 0 \text{ für } 0 \leqq v \leqq \lambda_0.$$

Dann ist nach (2)

$$A_{\beta,n} = a_0 e^{\lambda_0 \beta} + \cdots + a_n e^{\lambda_n \beta} = (\alpha - \beta) \int_0^{\lambda_n} A_\alpha(v) e^{v(\beta-\alpha)} dv + A_{\alpha,n} e^{\lambda_n(\beta-\alpha)}. \tag{22}$$

Nach (20) kann man eine Zahl ν bestimmen derart, dass

$$|A_{\alpha,n}| < \varepsilon\, e^{\lambda_n R(\alpha)} \text{ für } n > \nu$$

und also auch

$$|A_\alpha(v)| < \varepsilon\, e^{v R(\alpha)} \text{ für } v > \lambda_\nu.$$

Aus (22) ergibt sich nun mit Rücksicht auf die letzten Ungleichungen

$$|A_{\beta,n}| < |\alpha-\beta| \left| \int_0^{\lambda_\nu} A_\alpha(v)\, e^{v(\beta-\alpha)}\, dv \right| + |\alpha-\beta|\, \varepsilon \int_{\lambda_\nu}^{\lambda_n} e^{v R(\beta)}\, dv + \varepsilon\, e^{\lambda_n R(\beta)} <$$

$$< |\alpha-\beta| \left| \int_0^{\lambda_\nu} A_\alpha(v)\, e^{v(\beta-\alpha)}\, dv \right| + \varepsilon\, e^{\lambda_n R(\beta)} \left(\frac{|\alpha-\beta|}{R(\beta)} + 1 \right).$$

Bei hinreichend grossem n ist das erste (von n unabhängige) Glied des letzten Ausdruckes kleiner als das zweite und daraus erhält man unsere obige Behauptung.

Aus diesem letzten Ergebnisse ersieht man auch leicht, dass die Bedingung (17) (bzw. die Bedingung (1)) auch notwendig ist, damit die imaginäre Achse (bzw. die Gerade $R(z) = c$) wenigstens einen Konvergenzpunkt enthalte.

Es sei in der Tat die Reihe $\Sigma a_n e^{-\lambda_n z}$ in einem Punkte $z = ti$ der imaginären Achse konvergent. Man kann dann ohne Beschränkung der Allgemeinheit voraussetzen, dass die Reihe gegen den Wert Null konvergiert, da dies immer durch Veränderung von z. B. a_0 erreicht werden kann, was den Grenzwert (17) nicht beeinflusst. Dann ist aber (20) für $\alpha = -ti$, also auch (21) für $\beta = c$ erfüllt. Dies besagt aber eben, dass auch (17) erfüllt ist.

3. In diesem Abschnitte werden wir (17) durch eine gleichwertige Bedingung ersetzen, die folgendermassen lautet.

Es gebe eine positive Konstante K derart, dass bei beliebig kleinem ε und hinreichend grossem n

(23) $$|a_n + \cdots + a_{n+p}| < \varepsilon, \text{ sobald } 0 \leq \lambda_{n+p} - \lambda_n \leq K$$

ist.[1]

[1] Die Bedingungen (1) und (17) erinnern an den Ausdruck von Herrn Cahen für die Konvergenzabszisse der Dirichletschen Reihen. Dieser Ausdruck ist bekanntlich im Allgemeinen nur dann gültig, wenn die Konvergenzabszisse positiv ist. Die Bedingung (23) erinnert nun an einen Ausdruck für dieselbe Konvergenzabszisse, der immer gültig ist, und den, von einander unabhängig, die Herren Lindh und Kojima und in einem Spezialfalle auch Herr Knopp gefunden haben. Die Bedingung (23) ist auch mit

$$\lim_{n=\infty} e^{\lambda_n c}(a_n e^{-\lambda_n c} + a_{n+1} e^{-\lambda_{n+1} c} + \cdots) = 0 \qquad (c > 0)$$

Man hat sicher mit Rücksicht auf das bekannte ABELsche Lemma (das auch aus der oben angewandten ABELschen Transformation folgt)

$$(24)\qquad |a_n + \cdots + a_{n+p}| = |e^{-\lambda_n c} a_n e^{\lambda_n c} + \cdots + e^{-\lambda_{n+p} c} a_{n+p} e^{\lambda_{n+p} c}| \leqq$$

$$\leqq e^{-\lambda_n c} \operatorname{Max} |a_n e^{\lambda_n c} + \cdots + a_{n+r} e^{\lambda_{n+r} c}|, \quad r = 0, 1, \cdots p.$$

Ist nun (17) erfüllt und ist n genügend gross, so ist

$$|a_0 e^{\lambda_0 c} + \cdots + a_n e^{\lambda_n c} + \cdots + a_{n+r} e^{\lambda_{n+r} c}| < \frac{\varepsilon}{2 e^K} e^{\lambda_{n+r} c}, \quad r = 0, 1, \cdots$$

folglich, wenn $\lambda_{n+r} - \lambda_n \leqq \lambda_{n+p} - \lambda_n \leqq K$ ist,

$$|a_n e^{\lambda_n c} + \cdots + a_{n+r} e^{\lambda_{n+r} c}| < \frac{\varepsilon}{e^K} e^{\lambda_{n+r} c} < \varepsilon e^{\lambda_n c},$$

also nach (24)

$$|a_n + \cdots + a_{n+p}| < \varepsilon, \qquad 0 \leqq \lambda_{n+p} - \lambda_n \leqq K,$$

d. h. die Beziehung (23).

Es sei jetzt umgekehrt (23) erfüllt. Aus dem ABELschen Lemma folgt

$$|a_m e^{\lambda_m c} + \cdots + a_{m+p} e^{\lambda_{m+p} c}| \leqq e^{\lambda_{m+p} c} \operatorname{Max} |a_m + \cdots + a_{m+r}|, \quad r = 0, 1, \cdots p,$$

also nach (13) für jedes $m > \nu$, wo ν eine genügend grosse Zahl bedeutet,

$$(25)\qquad |a_m e^{\lambda_m c} + \cdots + a_{m+p} e^{\lambda_{m+p} c}| < \varepsilon e^{\lambda_{m+p} c}, \quad 0 \leqq \lambda_{m+r} - \lambda_m \leqq K.$$

Wir teilen jetzt die Exponenten $\lambda_{\nu+1}, \cdots \lambda_n$ folgendermassen in Gruppen ein

$$\lambda_{\nu+1}, \lambda_{\nu+2}, \cdots \lambda_{p_1};\ \lambda_{p_1+1}, \lambda_{p_1+2}, \cdots \lambda_{p_2};\ \cdots;\ \lambda_{p_i+1}, \lambda_{p_i+2}, \cdots, \lambda_{p_{\bar{i}+1}};\ \cdots;\ \lambda_{p_j+1}, \lambda_{p_j+2}, \cdots \lambda_n;$$

wo

---

gleichwertig. Diese letzte Bedingung erinnert wieder an eine Formel der Herren PINCHERLE und SCHNEE für negative Konvergenzabszissen. (Für Literaturangaben betreffs dieser verschiedenen Formeln vergl. HARDY and RIESZ, a. a. O., S. 8.)

Es ist einleuchtend, dass aus (23) eine analoge Bedingung für jeden anderen festen Wert K' folgt. Denn besteht (23) für einen festen Werten K, so besteht es naturgemäss für jeden anderen festen Wert, der kleiner als K ist. Andererseits zieht (23) eine analoge Bedingung für alle ganzzahlige Multipla von K mit sich. Aus diesen beiden Tatsachen folgt unsere Behauptung, und wir könnten auch ohne Beschränkung der Allgemeinheit $K = 1$ voraussetzen. Daraus folgt dann leicht, dass die Bedingung (23) mit der folgenden gleichwertig ist:

Bei beliebig kleinem ε und hinreichend grossem n ist

$$|a_n + \cdots + a_{n+p}| < \varepsilon(\lambda_{n+p} - \lambda_n + 1) \qquad (p \geqq 0);$$

oder auch mit der folgenden: bei beliebig kleinem ε und hinreichend grossem n ist

$$|a_{n-p} + a_{n-p+1} + \cdots + a_n| < \varepsilon(\lambda_n - \lambda_{n-p} + 1) \qquad (p \geqq 0).$$

$$\lambda_n - \lambda_{p_j} > K, \qquad 0 < \lambda_n - \lambda_{p_j+1} \leqq K,$$

$$\cdots\cdots \qquad \cdots\cdots$$

$$\lambda_{p_i+1} - \lambda_{p_i} > K, \qquad 0 < \lambda_{p_{i+1}} - \lambda_{p_i+1} \leqq K,$$

$$\cdots\cdots$$

$$0 \leqq \lambda_{p_1} - \lambda_{\nu+1} \leqq K.$$

Dann ist

$$\lambda_{p_i+1} < \lambda_n - (j-i)K. \tag{26}$$

Wir setzen jetzt

$$\sum_{h=\nu+1}^{p_1} a_h e^{\lambda_h c} = A_1, \ \sum_{h=p_1+1}^{p_2} a_h e^{\lambda_h c} = A_2, \cdots \sum_{h=p_i+1}^{p_{i+1}} a_h e^{\lambda_h c} = A_{i+1}, \cdots \sum_{h=p_j+1}^{n} a_h e^{\lambda_h c} = A_{j+1},$$

so ist nach (25) und (26)

$$|A_{i+1}| < \varepsilon\, e^{(\lambda_n - (j-i)K)c}.$$

Es ist mithin

$$|a_{\nu+1} e^{\lambda_{\nu+1} c} + \cdots + a_n e^{\lambda_n c}| = |A_{j+1} + A_j + \cdots + A_{i+1} + \cdots + A_1| <$$

$$< \varepsilon e^{\lambda_n c} + \varepsilon e^{(\lambda_n - K)c} + \varepsilon e^{(\lambda_n - 2K)c} + \cdots = \frac{\varepsilon e^{\lambda_n c}}{1 - e^{-Kc}}.$$

Folglich ist bei genügend grossem n

$$|a_0 e^{\lambda_0 c} + \cdots + a_n e^{\lambda_n c}| \leqq |a_0 e^{\lambda_0 c} + \cdots + a_\nu e^{\lambda_\nu c}| + |a_{\nu+1} e^{\lambda_{\nu+1} c} + \cdots + a_n e^{\lambda_n c}|$$

$$< \frac{2\varepsilon e^{\lambda_n c}}{1 - e^{-Kc}}$$

d. h. die Beziehung (17). Also sind die Bedingungen (17) und (23) wirklich gleichwertig.

Wir wollen nun aus Satz I′ mit Hilfe von (23) einige speziellere Resultate herleiten.

Ist $\lambda_n = n$ und wählt man $K < 1$, so geht (23) in $|a_n| < \varepsilon$ über, wir erhalten also wieder den Fatouschen Satz.

Die Bedingung (23) ist sicher erfüllt, wenn a_n den folgenden zwei Bedingungen genügt

$$\lim_{n=\infty} a_n = 0; \qquad \lim_{n=\infty} \frac{a_n}{\lambda_n - \lambda_{n-1}} = 0. \tag{27}$$

Dies ergibt natürlich wieder den FATOUschen Satz, aber auch folgenden Satz über *gewöhnliche* DIRICHLETsche Reihen.

Ist $\lim_{n=\infty} n a_n = 0$, *so konvergiert die Reihe* $\sum_{n=1}^{\infty} a_n n^{-s}$ *an jeder Regularitätsstelle der imaginären Achse.*

Wie ich in der erstgenannten LANDAUschen Arbeit gezeigt habe, folgt für *gewöhnliche* DIRICHLETsche Reihen der Satz I leicht aus diesem spezielleren Satze.[1]

Ich will schliesslich bemerken, dass für alle solche Exponentenfolgen, für welche die Differenz $\lambda_n - \lambda_{n-1}$ beschränkt bleibt, die beiden Bedingungen (27) offenbar durch die eine Bedingung $\lim_{n=\infty} \frac{a_n}{\lambda_n - \lambda_{n-1}} = 0$ ersetzt werden können.[2] Ist dagegen die Differenz $\lambda_n - \lambda_{n-1}$ immer grösser als eine feste positive Zahl, so können die Bedingungen (27) durch die Bedingung $\lim_{n=\infty} a_n = 0$ ersetzt werden.

4. Unsere Ergebnisse lassen sich leicht auf Integrale von der Form

$$\int_0^{\infty} a(v) e^{-vz} dv \tag{28}$$

oder STIELTJESsche Integrale von der Form

$$\int_0^{\infty} e^{-vz} d A(v) \tag{29}$$

übertragen. In (28) bedeutet $a(v)$ eine Funktion, die über jede endliche Strecke integriert werden kann, und in (29) bedeutet $A(v)$ eine Funktion, die auf jeder endlichen Strecke von beschränkter Schwankung ist. Der Ausdruck (29) umfasst natürlich sowohl die Integrale (28) als auch die DIRICHLETschen Reihen.

Für die Ausdrücke (28) und (29) ist die Bedingung (1) naturgemäss durch

$$\lim_{\omega=\infty} e^{-\omega c} \int_0^{\omega} a(v) dv = 0$$

bzw. durch

$$\lim_{\omega=\infty} e^{-\omega c} A(\omega) = 0$$

zu ersetzen. Die Bedingung (23) geht entsprechend in

---

[1] Vergl. a. a. O. S. 156—158.

[2] Für eine andere Herleitung dieses Ergebnisses vergl. HARDY and RIESZ, a. a. O. S. 47.

$$\left|\int_{\omega}^{\omega+K'} a(v)\,dv\right| < \varepsilon, \qquad \omega > \mu,\ K' \leqq K$$

bzw.

$$|A(\omega + K') - A(\omega)| < \varepsilon, \qquad \omega > \mu,\ K' \leqq K$$

über.

Alle Beweise laufen genau so wie oben, nur ist statt der Abelschen Transformation stets eine partielle Integration vorzunehmen.

Islinge, den 11. März 1916.

22.

(with F. Riesz)
Über die Randwerte einer analytischen Funktion

4e Congr. des Math. Scand. Stockholm 1916 (1920), 27–44

I.

Nach einem klassischen Satze aus der Funktionentheorie ist eine analytische Funktion $f(z)$, die längs eines Kurvenbogens regulär und gleich Null ist, überall gleich Null. Bekanntlich besteht der ähnlich lautende Satz auch für den Fall, wo das reguläre Verhalten der Funktion nicht mehr auf dem Kurvenbogen selbst, sondern nur auf der einen Seite in der Nähe desselben vorausgesetzt wird, während man längs des Kurvenbogens nur stetigen Anschluss an den Randwert Null verlangt. Speziell kann also eine Funktion, die im Inneren eines Kreises regulär und nicht identisch gleich Null ist, auf dem Kreise selbst nicht längs eines noch so kleinen Bogens stetig verschwinden; oder, was scheinbar etwas allgemeiner lautet, eine im Inneren des Kreises reguläre und nicht überall konstante Funktion kann sich nicht längs eines Bogens stetig an einen konstanten Randwert anschliessen. Nun hat Herr Fatou in seiner Pariser Thèse[1] viel mehr bewiesen; er hat nämlich gezeigt, dass jene Punkte des Bogens, für welche die im Kreise reguläre, nicht identisch verschwindende, in der Nähe des Bogens beschränkte (jedoch auf dem Bogen nicht notwendig stetige) Funktion $f(z)$ bei radialer Annäherung gegen Null strebt, nicht nur den Bogen nie ganz ausfüllen, sondern ihre Komplementärmenge sogar einen von Null verschiedenen

[1] P. Fatou, *Séries trigonométriques et séries de Taylor*, Acta mathematica, 30 (1906), p. 335—400; vgl. speziell p. 394/5.

LEBESGUE'schen Inhaltsmass besitzt. Durch Anwendung dieses Satzes auf das Produkt $(f(z)-A_1)\ldots(f(z)-A_n)$ zeigt er ferner für den Fall, wo $f(z)$ nicht konstant ist, dass auch jene Stellen, für welche der Grenzwert gleich einer beliebigen unter den vorgegebenen Zahlen $A_1,\ldots,A_n$ ist, niemals den Bogen ausfüllen, sondern dass ihre Komplementärmenge noch immer einen von Null verschiedenen Inhaltsmass besitzt; hieraus folgt speziell, dass die Funktion längs des Bogens unendlich viele verschiedene Randwerte annimmt.

Was nun den Inhaltsmass jener Punktmenge selbst betrifft, für welche die Randwerte verschwinden, so wird hierüber durch FATOU's Resultat nur so viel ausgesagt, dass derselbe kleiner ist als die Länge des Bogens. Nun aber legt die Tatsache, dass $f(z)$ nicht längs eines noch so kleinen Bogens verschwinden kann, die Vermutung nahe, dass jene Menge eine *Nullmenge* ist d. i. den Inhaltsmass Null besitzt. Diese Vermutung hat auch Herr FATOU ausgesprochen, jedoch ohne Beweis; über den Beweis sagte er nur, er halte ihn für recht schwierig. Herr FATOU hat in diesem Punkte die Tragweite seiner Methoden unterschätzt; wir werden zeigen, dass die Richtigkeit der Vermutung leicht durch einen Gedankengang folgt, der gewissen Ausführungen seiner Arbeit nahe verwandt ist und sich unmittelbar auf seine Resultate stützt.

Die zunächst auferlegte Einschränkung auf Kreisbogen resp. Kreisgebiet ist nicht von Belang; später, nachdem wir das gegenseitige Entsprechen der Ränder bei konformer Abbildung eines durch eine rektifizierbare Kurve umgrenzten Gebietes auf das Kreisinnere näher untersucht haben, wird sich unter anderm auch die Richtigkeit der Vermutung für beliebige, rektifizierbare Randkurven ergeben.

Die zunächst zu beweisende speziellere Vermutung sprechen wir in folgendem Satze aus:

Die Gesamtheit der Nullstellen der Randfunktion einer innerhalb eines Kreises regulär analytischen und beschränkten Funktion $f(z)$ ist vom Masse Null.

Dazu bemerken wir, dass es uns freisteht, unter Randwerten jene bei radialer Annäherung zu verstehen, oder aber alle Wege zuzulassen, die nur unter von Null verschiedenem Winkel münden, oder allgemeiner, zwischen zwei im betreffenden Punkte endenden Sehnen verlaufen. Nach einem Satze von FATOU gibt es nämlich fast überall, d. i. höchstens mit Ausnahme einer Nullmenge, wohlbestimmte, vom Wege unabhängige Randwerte[1]; eine solche Ausnahmsmenge kommt aber für uns nicht in Betracht. Jene Punkte, in denen der Randwert gleich Null ist, bilden eine messbare Menge M vom Masse m, und der zu beweisende Satz behauptet, dass $m = 0$ ist. Nehmen wir an, es wäre $m \neq 0$. Wir dürfen ferner annehmen, es handle sich um den Einheitskreis $|z| = 1$. Nun führen wir eine Hilfsfunktion $g(z)$ ein, wie folgt. Wir definieren zunächst die harmonische Funktion $u(x, y)$, indem wir derselben auf der Menge M den konstanten Randwert $\frac{A}{m}$, auf der Komplementärmenge M^* den Randwert $\frac{A}{m - 2\pi}$ vorschreiben; A bedeutet eine positive Grösse, über die wir noch verfügen. Nach Fatou erfüllt die aus den vorgegebenen Randwerten durch das Poisson'sche Integral hergestellte harmonische Funktion $u(x, y)$ die gestellten Forderungen fast überall d. h. höchstens mit Ausnahme einer Nullmenge.[2] Da ferner das längs des Randes erstreckte Integral der Randwerte gleich $\frac{A}{m} m + \frac{A}{m - 2\pi}(2\pi - m) = 0$ ist, so ist auch $u(0, 0) = 0$. Es sei nun $v(x, y)$ jene zu $u(x, y)$ konjugierte harmonische Funktion, für welche $v(0, 0) = 0$ ist. Dann ist die Funktion

$$g(z) = g(x + iy) = e^{u(x,y) + iv(x,y)}$$

im Einheitskreise regulär und beschränkt; für dieselbe ist $g(0) = e^0 = 1$ und der absolute Wert $|g(z)| = e^{u(x,y)}$ hat auf der Menge M fast überall den Randwert $e^{\frac{A}{m}}$, auf der Komplementärmenge M^* fast überall den Randwert $e^{\frac{A}{m-2\pi}}$.

Bilden wir jetzt die Funktion $f(z) g(z)$; diese ist ebenfalls regulär für $|z| < 1$ und beschränkt für $|z| \leq 1$; daher gelten für dieselbe die Cauchy'schen Integralformeln nicht nur längs der Kreise $|z| = r < 1$, sondern auch noch auf dem Einheitskreise selbst, da bekanntlich nach Lebesgue bei beschränkten Folgen Grenzübergang und Integration vertauschbar sind.[3] Speziell ist

[1] Fatou, l. c., p. 366/7; vgl. auch C. Carathéodory, *Über die gegenseitige Beziehung der Ränder* etc., Math. Annalen, 73, p. 305—320, Kap. I.

[2] Fatou, l. c., p. 348/9 und 357.

[3] H. Lebesgue, *Leçons sur l'intégration et la recherche des fonctions primitives* (Paris, Gauthier-Villars, 1914), p. 114.

$$f(0) = f(0)\,g(0) = \frac{1}{2\,i\pi}\int\limits_{|z|=1}\frac{f(z)\,g(z)}{z}\,dz = \frac{1}{2\,i\pi}\int\limits_{M^*}\frac{f(z)\,g(z)}{z}\,dz,$$

da ja der Integrand auf der Menge M fast überall verschwindet. Nun aber ist

$$\left|\int\limits_{M^*}\frac{f(z)\,g(z)}{z}\,dz\right| \leq e^{\frac{A}{m-2\pi}}\int\limits_{M^*}|f(z)|\,|dz| = e^{\frac{A}{m-2\pi}}\int\limits_0^{2\pi}|f(e^{i\theta})|\,d\theta,$$

und da das rechtstehende Integral einen bestimmten endlichen Wert besitzt, während der Faktor $e^{\frac{A}{m-2\pi}}$, in welchem wir über A noch frei verfügen, für genügend grosse Werte von A beliebig klein wird, so folgt

$$f(0) = 0.$$

Da nun $f(0) = 0$ ist, so ist auch die Funktion $\frac{f(z)}{z}$ regulär; sie ist ferner ebenfalls beschränkt und ihre Randwerte stimmen dem absoluten Werte nach mit jenen von $f(z)$ überein; durch dieselbe Schlussweise folgt daher, dass auch $\frac{f(z)}{z}$ im Punkte $z = 0$ verschwindet. Dasselbe ergibt sich sukzessive für $\frac{f(z)}{z^2}$, $\frac{f(z)}{z^3}$, u. s. f., woraus dann bekanntlich das identische Verschwinden von $f(z)$ folgt.

Es mag noch bemerkt werden, dass es im obigen Beweise keineswegs auf die spezielle Struktur unserer Hilfsfunktion $g(z)$, sondern nur auf folgende fundamentale Eigenschaften derselben ankommt: $g(z)$ ist beschränkt, $|g(0)| = 1$, auf der Komplementärmenge M^* ist $|g(z)|$ kleiner als eine vorgegebene Zahl, die beliebig klein gewählt werden kann. Die obige Konstruktion ergibt natürlich auch leicht andere Funktionen allgemeinerer Struktur und mit denselben charakteristischen Eigenschaften.

II.

In dem soeben geführten Beweise haben wir die vorausgesetzte Beschränktheit von $f(z)$ nur insofern ausgenützt, indem daraus erstens die Existenz — fast überall — der Randwerte $f(e^{i\theta})$, zweitens die Integrierbarkeit von $|f(e^{i\theta})|$ und drittens die Gültigkeit der CAUCHY'schen Integralformeln bei Integration längs des Einheitskreises für das Produkt $f(z)\,g(z)$ folgte. Nun ist aber für diese Folgerungen die Beschränktheit keinesfalls eine notwendige Bedingung. So z. B. bleiben sämtliche Folgerungen bestehen, wenn man über die Funktion

$$f(z) = a_0 + a_1 z + a_2 z^2 + \cdots$$

nur quadratische Integrierbarkeit voraussetzt, d. h. wenn man voraussetzt, dass das Integral

$$\int\limits_{|z|=r} |f(z)|^2\,|dz| = r\int\limits_0^{2\pi} |f(re^{i\theta})|^2 d\theta = 2r\pi\,(|a_0|^2 + |a_1|^2 r^2, \ldots)$$

für $r < 1$ unterhalb einer endlichen Schranke liege; bekanntlich ist diese Voraussetzung gleichwertig mit der Konvergenz der Reihe

$$|a_0|^2 + |a_1|^2 + |a_2|^2 + \cdots$$

Unter dieser Voraussetzung gibt es nach dem sogenannten RIESZ-FISCHER'schen Satze eine nebst dem Quadrate ihres absoluten Wertes integrierbare Funktion $f(e^{i\theta})$, deren formal gebildete FOURIER'sche Entwicklung, nach den Potenzen $e^{ni\theta} = \cos n\theta + i \sin n\theta$ von $e^{i\theta} = \cos\theta + i\sin\theta$ geordnet, gleich

$$a_0 + a_1 e^{i\theta} + a_2 e^{2i\theta} + \cdots$$

wird.[1] Die so definierte Funktion $f(e^{i\theta})$ ist nach LEBESGUE fast überall gleich dem Differenzialquotienten ihres Integrals[2] und daher nach FATOU zugleich Randwert der Funktion $f(z)$.[3]

---

[1] F. RIESZ, *Sur les systèmes orthogonaux de fonctions*, Comptes Rendus, Paris, 18 mars 1907; *Über orthogonale Funktionensysteme*, Nachrichten d. k. Ges. d. Wiss., Göttingen 1907, p. 116—122; E. FISCHER, *Sur la convergence en moyenne*, Comptes rendus, Paris, 12 mai 1907.

[2] LEBESGUE, l. c., p. 124/5.

[3] FATOU, l. c, p. 348 u. 357.

Ferner ist auch $|f(e^{i\theta})|$ integrierbar und endlich gelten die CAUCHY'schen Integralformeln für das Produkt $f(z)g(z)$ — wie durch Anwendung des sogenannten PARSEVAL'schen Satzes[1] unmittelbar hervorgeht —, wenn man nur über $g(z)$ das gleiche wie über $f(z)$ d. i. quadratische Integrierbarkeit voraussetzt, also jedenfalls in dem hier betrachteten Spezialfalle, wo doch $g(z)$ sogar beschränkt ist. Somit besteht der Satz:

Die Gesamtheit der Nullstellen der Randfunktion einer innerhalb des Einheitskreises regulär analytischen und quadratisch integrierbaren Funktion ist vom Masse Null.

Die spezielle Formulierung dieses Satzes wie auch der folgenden Betrachtungen für den Einheitskreis anstatt eines beliebigen Kreises ist unwesentlich und dient nur zur bequemeren Ausdrucksweise.

Der Satz bleibt auch bestehen, wenn man darin $|f(z)|^2$ durch irgendeine Potenz $|f(z)|^p$ ersetzt, deren Exponent p grösser als 1 ist; der Beweis hierfür bietet keine Schwierigkeit, wenn man sich auf jene Resultate stützt, die der eine von uns über Funktionen, deren p-te Potenz integrierbar ist, vor einigen Jahren entwickelt hat.[2] Doch bietet diese Vorallgemeinerung für sich noch nicht viel Interessantes; wir haben dieselbe nur angeführt, weil sie auf den tiefer liegenden und die übrigen Resultate umfassenden Grenzfall $p=1$ hinweist. Wir werden zu diesem Grenzfalle auch durch eine Fragestellung geführt, die sich uns sofort aufdrängt, sobald wir die bisher entwickelten Resultate geometrisch deuten.

III.

Betrachten wir neben der Funktion

$$f(z)=a_0+a_1z+a_2z^2+\cdots$$

ihre Integralfunktion

$$F(z)=a_0z+\frac{a_1z^2}{2}+\frac{a_2z^3}{3}+\cdots$$

[1] FATOU, l. c., p. 351/2 und 378/9.

[2] F. RIESZ, *Untersuchungen über Systeme integrierbarer Funktionen*, Math. Annalen, 69, p. 449—497.

Sind für $f(z)$ und für den Einheitskreis die Voraussetzungen irgend eines der beiden Sätze oder die entsprechenden Voraussetzungen für irgend ein $p > 1$ erfüllt, so ist $F(z)$ daselbst beschränkt, auf dem Kreise selbst noch stetig und von beschränkter Schwankung. Daher leistet $v = F(z)$ die konforme Abbildung des Einheitskreises auf ein gewisses, über die v-Ebene ausgebreitetes, endliches RIEMANN'sches Flächenstück, dessen Berandung durch die *rektifizierbare* Kurve

$$v = F(e^{i\theta}) \qquad (0 \leqq \theta \leqq 2\pi)$$

gebildet wird. Nun lässt sich bekanntlich — was übrigens auch unmittelbar einleuchtet — für Punktmengen auf rektifizierbaren Kurven der LEBESGUE'sche Inhaltsmass durch Heranziehung der Bogenlänge ganz analog wie für die Gerade oder für den Kreis definieren. In unserem Falle entspricht dann jeder messbaren Punktmenge M auf dem Einheitskreise eine messbare Menge auf der Bildkurve, und der Inhaltsmass dieser Bildmenge ist gleich dem Integral

$$\int\limits_M |dF(z)| = \int\limits_M |F'(z)||dz| = \int\limits_M |f(e^{i\theta})|d\theta.$$

Verschwindet speziell der Randwert $f(e^{i\theta})$ für die Menge M, so ist also die Bildmenge vom Masse Null; ist dagegen $f(e^{i\theta})$ fast überall von Null verschieden und ist M selbst nicht vom Masse Null, dann kann auch das Integral nicht verschwinden und somit ist auch der Inhaltsmass der Bildmenge von Null verschieden. Unsere Sätze sagen also aus, dass unter den gemachten Voraussetzungen die Bildmenge nur dann vom Masse Null sein kann, wenn auch die ursprüngliche Menge es ist. Ist andererseits die ursprüngliche Menge vom Masse Null, so ist es auch die Bildmenge, da ja der Wert eines über eine Nullmenge erstreckten Integrals gleich Null ist. *Ursprüngliche Menge und Bildmenge sind somit gleichzeitig entweder beide oder keine vom Masse Null.*

Lassen wir jetzt die gemachten Voraussetzungen fallen und betrachten wir eine beliebige innerhalb des Einheitskreises reguläre und beschränkte Funktion $F(z)$, für welche die Randwerte $F(e^{i\theta})$ existieren und eine Funktion von beschränkter Schwankung ausmachen. Die Existenz der Randwerte verlangen wir zunächst nur z. B. im Sinne radialer Annäherung;

wir bemerken jedoch sofort, dass eine solche Präzisierung eigentlich überflüssig ist, da die Funktion $F(z)$ nicht nur innerhalb des Einheitskreises, sondern auch mit Einschluss des Randes überall stetig ist. Die Randfunktion $F(e^{i\theta})$ ist nämlich, da sie von beschränkter Schwankung ist, im allgemeinen jedenfalls stetig und könnte nur noch höchstens eine abzählbare Menge von einfacher Sprungstellen aufweisen. Andererseits aber gilt nach einem wenig bekannten Satze des Herrn PRINGSHEIM, der später unabhängig auch durch Herrn FATOU, wie auch durch Herrn LINDELÖF entdeckt wurde, dass die Randwerte einer beschränkten, regulär analytischen Funktion keine einfache Sprungstellen aufweisen.[1] Daher ist also $F(z)$ mit Einschluss des Randes stetig.

Die Gleichung $v = F(z)$ leistet auch hier die konforme Abbildung des Einheitskreises auf ein endliches RIEMANN'sches Flächenstück, das durch die rektifizierbare Kurve $v = F(e^{i\theta})$ begrenzt wird. Auch umgekehrt führt die konforme Abbildung des Einheitskreises auf ein beliebiges endliches, einfach zusammenhängendes, von einer rektifizierbaren Kurve umgrenztes RIEMANN'sches Flächenstück, wie aus dem OSGOOD-CARATHÉODORY'schen Satze über das gegenseitige stetige Entsprechen der Ränder bei konformer Abbildung[2] hervorgeht, zu einer Funktion $F(z)$, die auch noch auf dem Kreise selbst stetig ist und für welche der Randwert $F(e^{i\theta})$, da $v = F(e^{i\theta})$ die Gleichung der rektifizierbaren Randkurve darstellt, sicher von beschränkter Schwankung ist. Die Frage, die wir nun stellen, ist die, *ob auch in diesem allgemeinen Falle das gegenseitige Entsprechen der Nullmengen bestehen bleibt?*

Gelingt es uns, diese Frage zu bejahen, so folgt sofort auch dieselbe Erscheinung für die konforme Abbildung beliebiger durch rektifizierbare Kurven begrenzter Gebiete aufeinander, da man ja durch geeignete Querschnitte und Heranziehen des Einheitskreises als Zwischenglied bei der konformen Abbildung immer auf den hier behandelten Fall zurückgehen kann.

---

[1] A. PRINGSHEIM, *Über das Verhalten von Potenzreihen auf dem Convergenzkreise*, Sitzungsberichte, München 1900, Heft. 1., p. 96—98; FATOU, l. c., p. 363; E. LINDELÖF, *Sur un principe général de l'Analyse* etc., Acta Soc. Sc. Fennicae, 46, nº 4 (1915), p. 7.

[2] CARATHÉODORY, l. c. Einen besonders schönen und einfachen Beweis dieses Satzes gab E. LINDELÖF, *Sur la représentation conforme*, Comptes Rendus, Paris, 26 janvier 1914; die scheinbare Einschränkung auf schlichte Gebiete ist unwesentlich. Vgl. auch die vorher zitierte Arbeit von LINDELÖF, wo man weitere Literaturangaben findet.

IV.

Die soeben aufgeworfene Frage spalten wir in zwei Teile: 1) *Entsprechen den Nullmengen auf dem Kreise auch Nullmengen auf der Bildkurve* $v = F(e^{i\theta})$? 2) *Ist das Original einer Nullmenge ebenfalls eine Nullmenge?*

Wir beginnen mit der ersten Frage. Es genügt, dieselbe nur für abgeschlossene Nullmengen zu prüfen; die in diesem Falle erhaltene Antwort wird für beliebige Nullmengen gelten. In der Tat kann man die beliebige Nullmenge M auf dem Kreise derart in Bogensysteme einschliessen, dass deren Gesamtlänge beliebig klein ist; jedem solchen Bogensystem entspricht im Bilde ein Bogensystem, das die Bildmenge M' einschliesst und dessen Gesamtlänge daher wenigstens so gross ist wie der Inhaltsmass resp. — um auch mit der eventuellen Nichtmessbarkeit der Bildmenge zu rechnen — wie der äussere Inhaltsmass von M'. Greifen wir aus der Gesamtheit der die Menge M einschliessenden Bogensysteme eine unendliche Folge heraus, für welche die Gesamtlängen gegen 0 konvergieren, und es sei M_1 die Menge jener Punkte, die in sämtlichen Systemen der Folge enthalten sind. Dann sind M_1 und ihre Bildmenge M'_1 als Durchschnitt von abzählbar unendlich vielen messbaren Mengen, nämlich der Bogensysteme, ebenfalls messbar; ferner ist M_1 wieder eine Nullmenge, da sie ja in Bogensystemen von beliebig kleiner Gesamtlänge enthalten ist; dagegen ist die Menge M'_1, da sie die Menge M' enthält, sicher keine Nullmenge, wenn M' es nicht ist. Nun enthält aber jede messbare Menge, die keine Nullmenge ist, solche abgeschlossene Teilmengen, die es auch nicht sind. Sei M'_2 eine solche abgeschlossene Teilmenge von M'_1; dann ist auch das Original M_2 abgeschlossen, und damit ist die ursprüngliche Fragestellung tatsächlich auf den Spezialfall abgeschlossener Mengen zurückgeführt.

Um das Problem noch weiter zu vereinfachen, führen wir als neue Veränderliche die von irgendeinem Punkte aus gemessene Bogenlänge s der Bildkurve $v = F(e^{i\theta})$ ein. Die Funktion $F(e^{i\theta})$ übergeht hiedurch in die Funktion $v(s) = x(s) + iy(s)$, wo $x(s)$ und $y(s)$ die Koordinaten der Kurvenpunkte als Funktionen der Bogenlänge darstellen. Daher existiert fast überall auf der Kurve der Differentialquotient $v'(s) = x'(s) + iy'(s)$ und es ist fast überall $|v'| = \sqrt{x'^2 + y'^2} = 1$. Fassen wir nun alle Teilmengen M'_3 von M'_2 ins Auge, die aus der letzteren Menge durch je einen Kurvenbogen ausgeschnitten werden. Das Integral von $v'(s)$ über M'_3 kann in einzelnen Fällen verschwinden; für alle Mengen M'_3 verschwindet es jedoch dann und nur dann, wenn $v'(s)$ selbst auf der Menge M'_2 fast überall verschwindet; und da andererseits $|v'|$ fast überall gleich 1 ist, so findet dies dann und nur dann statt, wenn M'_2 selbst eine Nullmenge ist. Die Frage, ob M'_2 eine Nullmenge ist, wird also entschieden sein, wenn wir das Integral von $v'(s)$ über die Mengen M'_3 berechnen.

Hiemit ist die anfangs gestellte Frage auf die folgende zurückgeführt worden: Es sei M_3 eine beliebige abgeschlossene Nullmenge auf dem Einheitskreise; M'_3 ihre Bildmenge; verschwindet das Integral von $v'(s)$ über M'_3 immer, oder gibt es Mengen M'_3, für welche das Integral einen von Null verschiedenen Wert hat? Wir werden sehen, dass das Integral immer verschwindet, und damit wird auch die anfangs gestellte Frage bejahend beantwortet. Zu diesem Zwecke führen wir eine Hilfsfunktion $h(z)$ ein, die folgende Eigenschaften besitzt: *sie ist innerhalb des Einheitkreises regulär, auf demselben noch stetig, für alle Punkte der Menge M_3 gleich 1 und für die Komplementärmenge dem absoluten Werte nach überall kleiner als 1.* Eine solche Funktion $h(z)$ erhält man leicht aus dem Fatou'schen Beispiel einer im Kreise regulären, auf dem Kreise noch stetigen Funktion, die in allen Punkten von M_3 und nur dort verschwindet.[1] Wegen der Wichtigkeit der Funktion $h(z)$ für unsern Beweis wiederholen wir die Fatou'sche Konstruktion, entsprechend modifiziert, indem wir sie sofort unseren Zwecken anpassen.

Wir definieren zuerst entlang des Kreises eine Funktion $P(\theta)$, die für alle Punkte der Komplementärmenge von M_3 positiv und regulär analytisch ist, bei Annäherung an jeden beliebigen Punkt von M_3 ins Unendliche wächst, dabei aber längs des ganzen Kreises integrierbar ist. Dafür bemerken wir zunächst, dass die Menge M_3 ihre Komplementärmenge in endlich oder abzählbar unendlich viele Kreisbögen $a_n < \theta < b_n$ zerlegt. Der erste Fall ist trivial, da dann M_3 und damit auch M'_3 aus einer endlichen Anzahl von Punkten besteht. Im

[1] Fatou, l. c., p. 393. Die von Fatou ausserdem gestellte Forderung, dass der Einheitskreis eine natürliche Grenze der Funktion sei, kommt für uns nicht in Betracht.

zweiten Falle ordnen wir jedem Bogen $a_n < \theta < b_n$ eine Funktion $\varphi_n(\theta)$ zu, die längs des Bogens positiv und regulär analytisch ist, bei Annäherung an a_n und b_n ins Unendliche wächst, jedoch längs des Bogens ein endliches Integral I_n besitzt. Eine solche Funktion ist z. B. $\varphi_n(\theta) = (\theta - a_n)^{-\frac{1}{2}} + (b_n - \theta)^{-\frac{1}{2}}$. Dann wählen wir die positiven Konstanten c_n so, dass $\Sigma c_n I_n$ konvergiert. Endlich lässt sich nach einem bekannten Satze, da die Reihe mit positiven Gliedern $\Sigma(b_n - a_n)$ konvergiert, eine Folge von ins Unendliche wachsenden positiven Konstanten d_n derart bestimmen, dass auch noch die Reihe $\Sigma d_n(b_n - a_n)$ konvergiert. Setzen wir nun für jeden Bogen $a_n b_n$

$$P(\theta) = c_n \varphi_n(\theta) + d_n,$$

so erfüllt die Funktion P sämtliche Anforderungen; sie ist auf jedem Bogen $a_n < \theta < b_n$ positiv und regulär analytisch; ihr Integral ist gleich $\Sigma c_n I_n + \Sigma d_n(b_n - a_n)$, und schliesslich, da die Funktion auf $a_n b_n$ grösser als d_n ist und da die d_n ins Unendliche wachsen, so wird die Funktion für alle Punkte von M_3 stetig unendlich.

Aus der Funktion $P(\theta)$ wird mittels des Poisson'schen Integrals eine positive, im Innern des Einheitskreises harmonische Funktion $P(x, y)$ gebildet, die sich stetig an die Randwerte $P(\theta)$ anschliesst, was für die Punkte von M_3 so zu verstehen ist, dass dort der Grenzwert positiv unendlich wird. Die konjugiert harmonische Funktion $Q(x, y)$, bei der wir die verfügbare additive Konstante nach Belieben wählen, besitzt in den Punkten der Komplementärmenge ebenfalls stetige (ja sogar analytische) Randwerte; auf das Verhalten derselben in den Punkten von M_3 kommt es uns nicht weiter an. Nun setzen wir

$$h(z) = h(x + iy) = \frac{P(x, y) + iQ(x, y)}{1 + P(x, y) + iQ(x, y)};$$

Zähler und Nenner sind innerhalb des Einheitskreises regulär analytische Funktionen von $z = x + iy$; dasselbe gilt auch für $h(z)$, da $P(x, y)$ positiv ist und daher der Nenner nicht verschwinden kann. Bei Annäherung an einen Punkt von M_3 wächst $P(x, y)$, also auch $|P(x, y) + iQ(x, y)|$ ins Unendliche; daher ist der Grenzwert von $h(z)$ gleich 1. Bei Annäherung an einen Punkt der Komplementärmenge strebt $P + iQ$ gegen einen Grenzwert, dessen reeller Teil positiv ist; somit ist der Grenzwert von $h(z)$ dem absoluten Werte nach kleiner als 1. Endlich ist die Funktion $h(z)$ mit Einschluss des Randes stetig.

Betrachten wir nun das Integral

$$\int\limits_{|z|=1} (h(z))^n \, dF(z) = \int\limits_0^{2\pi} (h(e^{i\theta}))^n dF(e^{i\theta}).$$

Wir behaupten, dass auch für dieses Integral, das von der STIELTJES'schen Form ist, entsprechend der CAUCHY'sche Integralsatz gilt, und dass also das Integral für jedes positive, ganzzahlige n verschwindet. Dafür genügt es zu bemerken, dass $(h(z))^n$ eine im Einheitskreise reguläre, einschliesslich des Randes stetige Funktion ist und dass sie somit daselbst durch ganze rationale Funktionen, so z. B. nach FEJÉR durch die arithmetischen Mittel der Partialsummen ihrer Potenzreihe, gleichmässig angenähert werden kann. Es genügt also zu zeigen, dass die entsprechenden, mit den Annäherungsfunktionen gebildeten Integrale verschwinden, was schliesslich auf das Verschwinden der Integrale

$$\int\limits_{|z|=1} z^k dF(z) \qquad (k = 0, 1, 2, \ldots)$$

hinauskommt. Nun aber ist

$$\int\limits_{|z|=1} z^k dF(z) = -k \int\limits_{|z|=1} z^{k-1} F(z) dz = 0$$

nämlich für $k = 0$ wegen des Faktors k und für $k > 0$ deshalb, weil der Integrand auf dem Kreise stetig und im Innern regulär ist. Also ist auch für alle $n = 1, 2, \ldots$

$$\int\limits_{|z|=1} (h(z))^n dF(z) = \int\limits_0^{2\pi} (h(e^{i\theta}))^n dF(e^{i\theta}) = 0.$$

Führen wir schliesslich die Bogenlänge s der Kurve $v = F(e^{i\theta})$ als Veränderliche ein, so übergeht $F(e^{i\theta})$ in $v(s)$, $h(e^{i\theta})$ in eine Funktion $u(s)$ und die soeben erhaltene Identität in

$$\int_0^L (u(s))^n v'(s) ds = 0 \qquad (n = 1, 2, \ldots),$$

wo L die Länge der Bildkurve ist. Lassen wir nun n ins Unendliche wachsen, so strebt $(u(s))^n$ auf der Menge M'_3 gegen 1, sonst gegen 0; dabei bleibt der Integrand, in welchem ja beide Faktoren dem absoluten Werte nach $\leqq 1$ sind, beschränkt. Das Integral strebt somit gegen das Integral von $v'(s)$ über M'_3; daher ist auch dieses Integral gleich 0, was ja zu beweisen war.

Damit ist gezeigt, dass *jeder Nullmenge auf dem Kreise eine Nullmenge auf der Bildkurve entspricht.*

V.

Auf die zweite Frage, *ob auch jeder Nullmenge auf der Bildkurve eine Nullmenge auf dem Kreise entspricht,* ist die Antwort jetzt schon ausserordentlich leicht. Die soeben erkannte Eigenschaft der Abbildung ist nämlich zugleich eine notwendige und hinreichende Bedingung dafür, damit die Funktion $F(e^{i\theta})$ *absolut stetig* d. i. ein Integral ihres — fast überall existierenden — Differentialquotienten sei.[1] Andererseits ist nach FATOU die Funktion

$$f(e^{i\theta}) = \frac{dF(e^{i\theta})}{de^{i\theta}} = -ie^{-i\theta}\frac{dF(e^{i\theta})}{d\theta}$$

Randwert der Funktion $f(z) = F'(z)$,[2] und somit folgen aus den CAUCHY'schen Integralformeln für $F(z)$ durch partielle Integration die entsprechenden Formeln für $f(z)$. Speziell gelten die klassischen Integralformeln für das Integral von $f(z)z^k$ bei Integration über den Einheitskreis für alle ganzzahligen, positiven oder nicht-positiven Werte von k. Somit gelten die entsprechenden Formeln auch für die Produkte $f(z)P(z)z^k$, wo $P(z)$ irgend eine ganze rationale Funktion von z ist. Da schliesslich jede innerhalb des Einheitskreises reguläre und beschränkte Funktion $g(z)$ durch beschränkte Folgen ganzer rationaler Funktionen z. B. durch die FEJÉR'schen Mittel ihrer Potenzreihe, dar-

[1] Vgl. H. LEBESGUE, *Sur les intégrales singulières*, Annales de la Fac. des Sc. de Toulouse, 3e série, I, p. 43, wo man auch weitere Literaturangaben findet.

[2] FATOU, l. c., p. 345 u. 357.

gestellt werden kann,[1] so gelten nach einem grundlegenden LEBESGUE'schen Satze über Integration von Funktionenfolgen[2] die CAUCHY'schen Formeln auch für die Produkte $f(z)g(z)z^k$. Wählt man nun speziell für $g(z)$ die in I eigenführte Hilfsfunktion, so folgt durch genau dieselbe Schlussweise wie in I, dass *die Funktion* $f(e^{i\theta})$ *nur auf einer Nullmenge verschwinden kann.* Für die Funktion $F(e^{i\theta})$ aber bedeutet dieses Ergebnis nichts anderes, als dass $F'(e^{i\theta})$ auch nur auf einer Nullmenge verschwindet und dass somit das Integral von $|F'(e^{i\theta})|$ für jede andere messbare Menge von Null verschieden ist, d. h. dass einer solchen Menge in der Abbildung sicher keine Nullmenge entspricht.

Damit wurde auch die zweite Frage im allgemeinsten Falle bejahend beantwortet.

VI.

Zum Schlusse seien hier noch einige verschiedenartige Sätze angeführt, die sich durch Verschmelzung unserer Resultate mit bekannten Sätzen ergeben.

Der erste ist eine Ausdehnung des in I zitierten FATOU'schen Satzes über die Existenz der Randwerte einer innerhalb eines Kreises regulären und beschränkten Funktion auf beliebige, durch rektifizierbare Kurven begrenzte Gebiete und ergibt sich aus demselben durch Zerschneiden und konforme Abbildung. Er lautet: *Ist die Funktion $f(z)$ in einem durch eine oder mehrere rektifizierbare Kurven begrenzten Gebiete regulär und beschränkt, so besitzt sie auf dem Rande fast überall Randwerte, die vom Wege unabhängig sind, wenn nur dieser unter einem von Null verschiedenen Winkel mündet.*[3]

Der zweite Satz ist eine entsprechende Verallgemeinerung des in I beweisenen Satzes und besagt, dass *unter den soeben*

[1] L. FEJÉR, *Über gewisse durch die Fouriersche und Laplacesche Reihe definierten Mittelkurven und Mittelflächen*, Rendiconti, Palermo, 38 (1914), p. 79—97. Die angeführte Tatsache ist übrigens im Grunde schon in den bekannten Resultaten von Fejér und Lebesgue über die arithmetischen Mittel der Fourier'schen Reihe enthalten.

[2] H. LEBESGUE, *Sur la méthode de M. Goursat pour la résolution de l'équation de Fredholm*, Bulletin de la Soc. math. de France, 36 (1908), p. 3—19, n° 7. Der Satz besagt, dass eine Folge gliedweise integriert werden darf, wenn sie eine integrierbare Majorantenfunktion besitzt d. i. eine integrierbare Funktion, die sämtliche Glieder dem absoluten Werte nach übersteigt.

[3] Der Beweis folgt leicht aus unserem Satze, dass bei der konformen Abbildung Nullmengen einander gegenseitig entsprechen.

gemachten Voraussetzungen die Randwerte der Funktion höchstens für eine Nullmenge verschwinden.

In II haben wir unser in I erwickeltes Ergebnis in anderer Richtung erweitert. Wir haben gezeigt, dass die Voraussetzung über die Beschränktheit von $f(z)$ auch durch die Bedingung ersetzt werden kann, dass das Integral von $|f(z)|^p$ längs konzentrischen Kreisen beschränkt bleibt. Dabei wurde $p > 1$ vorausgesetzt. Den Fall $p = 1$ haben wir nur gestreift. Jetzt lässt sich auch dieser Fall vollständig erledigen. Es besteht nämlich folgender Satz:

Es sei $f(z)$ eine innerhalb des Einheitskreises regulär analytische Funktion, und es sei für $r < 1$

$$I(r) = \int\limits_{|z|=r} |f(z)|\,|dz|$$

beschränkt, Dann existiert bei radialer Annäherung (oder auch längs Wegen, die unter einem von Null verschiedenen Winkel münden) für $r \to 1$ die Randfunktion $\lim f(z)$ *fast überall, sie ist im* Lebesgue'*schen Sinne integrierbar und fast überall von Null verschieden.*

Unsere Voraussetzung deckt sich nämlich, indem man $f(z) = F'(z)$ setzt, vollständig mit den in IV und V über $F(z)$ gemachten Annahmen. In der Tat folgt die Beschränktheit von $I(r)$, sobald $F(z)$ auf dem Einheitskreise von beschränkter Schwankung ist, aus der Poisson'schen Integraldarstellung von $F(z)$ mittels ihrer Randwerte, bzw. aus der daraus durch Differentiation hergeleiteten Formel für $f(z)$. Umgekehrt folgt aus der Beschränktheit von $I(r)$, das ja zugleich die totale Schwankung von $F(z)$ längs das Kreises $|z| = r$ darstellt, einerseits die Beschränktheit von $F(z)$ und damit die Existenz — zunächst nur fast überall — der Randwerte $F(e^{i\theta})$ dann aber folgt auch, dass $F(e^{i\theta})$ für die Menge jener θ, für welche sie als Grenzwert definiert ist, eine Funktion beschränkter Schwankung darstellt. Setzt man diese Randfunktion in das Poisson'sche Integral ein, so erhält man wieder die Funktion $F(z)$. Nun aber hat das Poisson'sche Integral an allen Stetigkeitsstellen, wie auch an allen einfachen Sprungstellen der Randfunktion bestimmte Randwerte; da eine Funktion beschränkter Schwankung nur solche Stellen besitzt, existieren also in unserem Falle die Randwerte ohne Ausnahme. Schliesslich werden die Sprungstellen durch den obigen PRINGSHEIM'schen Satz wieder ausgeschlossen. Durch diese Betrachtungen ist nun unser Satz auf die Ergebnisse von IV und V zurückgeführt.

Besonders durchsichtig erscheint die Bedingung, dass $I(r)$ beschränkt ist, wenn man mit Herrn HARDY bemerkt, dass $I(r)$ (und auch noch der Integralmittelwert $\frac{1}{2\pi r} I(r)$) monoton wachsende Funktionen von r sind,[1] so dass die Bedingung erfüllt oder nicht erfüllt ist, je nachdem $I(r)$ für $r \to 1$ einem endlichen Grenzwerte zustrebt oder aber ins Unendliche wächst; eine dritte Möglichkeit gibt es nicht.

Von besonderem Interesse dürfte der folgende Satz sein, der die Frage nach den Gültigkeitsgrenzen der CAUCHY'schen Integralformeln wenigstens für den Kreisfall zu einem gewissen Abschluss bringt.

Damit es zu einer innerhalb des Einheitskreises regulär analytischen Funktion $f(z) = a_0 + a_1 z + \cdots + a_n z^n + \cdots$ *eine längs des Einheitskreises erklärte im* LEBESGUE'*schen Sinne integrierbare Funktion* $\varphi(z)$ *gebe derart, dass*

$$\frac{1}{2i\pi} \int\limits_{|z|=1} \frac{\varphi(z)}{z^{n+1}} dz = a_n$$

und

$$\frac{1}{2i\pi} \int\limits_{|z|=1} \varphi(z) z^n dz = 0 \quad (n = 0, 1, 2 \ldots)$$

seien, ist es notwendig und hinreichend, dass das Integral

$$I(r) = \int\limits_{|z|=r} |f(z)|\,|dz| = r \int\limits_0^{2\pi} |f(re^{i\theta})|\, d\theta$$

für alle $r < 1$ *unterhalb einer endlichen, von* r *unabhängigen Schranke bleibe. Ist irgendeine dieser beiden gleichwertigen Bedingungen erfüllt, so besitzt* $f(z)$ *fast überall einen Randwert; dieser ist im* LEBESGUE'*schen Sinne integrierbar und unterscheidet sich von* $\varphi(z)$ *höchstens in einer Nullmenge. Bezeichnet man den Randwert auch mit* $f(z)$, *so gelten natürlich die Formeln*

---

[1] G. H. HARDY: The Mean Value of the Modulus of an Analytic Function, Proc. of the London Math. Soc., ser. 2, vol. 14 (1915), p. 269—277.

$$a_n = \frac{1}{2i\pi} \int\limits_{|z|=1} \frac{f(z)}{z^{n+1}} dz$$

und

$$0 = \frac{1}{2i\pi} \int\limits_{|z|=1} f(z) z^n dz \quad (n = 0, 1, 2, \ldots),$$

die mit den Formeln

$$f(x) = \frac{1}{2i\pi} \int\limits_{|z|=1} \frac{f(z)}{z-x} dz \qquad (|x| < 1)$$

und

$$0 = \frac{1}{2i\pi} \int\limits_{|z|=1} \frac{f(z)}{z-x} dz \qquad (|x| > 1)$$

gleichbedeutend sind.

Ist in der Tat die erste unserer Bedingungen erfüllt, so lässt $f(z)$ vermittels $\varphi(z)$ eine Poisson'sche Integraldarstellung zu, und hieraus folgt die Beschränktheit von $I(r)$. Die Umkehrung ergibt sich unmittelbar aus unseren obigen Ausführungen.

Aus unserem letzten Resultate und einem bekannten Riemann-Lebesgue'schen Satze folgt nun leicht:

Ist für eine innerhalb des Einheitskreises regulär analytische Funktion $f(z) = \Sigma a_n z^n$

$$I(r) = \int\limits_{|z|=r} |f(z)|\, dz \qquad (r < 1)$$

beschränkt, so ist

$$\lim_{n=\infty} a_n = 0.$$

Der entsprechende Satz gilt für harmonische Funktionen $\Sigma(a_n \cos n\theta + b_n \sin n\theta) r^n$ *nicht,* wie das bekannte Beispiel

$$g(r,\theta) = \frac{1 - r^2}{2(1 + r^2 - 2r\cos\theta)} = \frac{1}{2} + \sum_{n=1}^{\infty} r^n \cos n\theta$$

zeigt. Es ist für $r < 1$

$$\int_0^{2\pi} |g(r,\theta)|\, d\theta = \pi$$

und dennoch streben die Koeffizienten nicht gegen Null.

Zuletzt sei noch das folgende verwandte Ergebnis hervorgehoben, welches die Grössenordnung der FOURIER-Koeffizienten einer Funktion beschränkter Schwankung betrifft. Bekanntlich gilt für diese der Satz, dass $n|a_n|$ und $n|b_n|$ unter einer endlichen Schranke liegen. Dagegen gelten nicht notwendigerweise die Limesgleichungen $na_n \to 0$, $nb_n \to 0$.[1] Aus der oben bewiesenen absoluten Stetigkeit von $F(e^{i\theta})$, wodurch dieselbe als Integralfunktion erscheint und aus einem schon oben erwähnten RIEMANN-LEBESGUE'schen Satze ergibt sich, da doch jetzt nb_n und $-na_n$ die FOURIER-Koeffizienten von $\frac{dF(e^{i\theta})}{d\theta}$ sind, dass für $F(e^{i\theta})$ die Limesgleichungen immer gelten, was auch so ausgesprochen werden kann:

Sind die mit einander konjugierten trigonometrischen Reihen

$$\frac{a_0}{2} + (a_1 \cos x + b_1 \sin x) + \cdots + (a_n \cos nx + b_n \sin nx) + \cdots$$

und

$$-\frac{b_0}{2} + (-b_1 \cos x + a_1 \sin x) + \cdots + (-b_n \cos nx + a_n \sin nx) + \cdots$$

beide zugleich FOURIER'*sche Reihen je einer Funktion beschränkter Schwankung, so gelten die Limesgleichungen*

$$na_n \to 0, \quad nb_n \to 0.$$

[1] Während der Drucklegung dieses Vortrages hat F. RIESZ sogar ein Beispiel einer periodischen, überall stetigen Funktion beschränkter Schwankung gegeben, für welche die Limesgleichungen nicht gelten: *Über die Fourierkoeffizienten einer stetigen Funktion von beschränkter Schwankung*, Math. Zeitschrift, 2 (1918), p. 312—315.

23.

Sur le principe de Phragmén-Lindelöf

Proc. Camb. Phil. Soc. *20* (1920–21), 205–207; *21* (1922–23), 6

(Extrait d'une lettre adressée à M. G. H. Hardy.)

1. La lecture attentive de votre travail récemment paru dans les *Acta Mathematica** et de ceux de MM. Phragmén et Lindelöf†, P. Persson‡, F. Carlson¶, etc., me fait croire que le théorème suivant n'a jamais été observé.

Si la fonction $g(z)$ holomorphe dans le demi-plan $R(z) \geqq 0$ (sauf peut-être à l'infini) y satisfait à l'inégalité

$$(1) \qquad |g(z)| < C_1 e^{lr} \qquad (r = |z|)$$

et sur l'axe imaginaire à celle-ci

$$(2) \qquad |g(z)| < C_2 e^{-kr},$$

C_1, C_2, k et l désignant des constantes positives, cette fonction s'annulera identiquement.

Ce théorème rend la première démonstration dans votre travail beaucoup plus simple qu'elle ne l'était. Vous le verrez dans un instant. Mais démontrons d'abord le théorème lui-même. Il découlera immédiatement des principes généraux de MM. Phragmén et Lindelöf.

Formons la fonction

$$\psi(z) = e^{\omega z} g(z),$$

ω désignant un nombre positif. Désignons par C le plus grand des nombres C_1 et C_2. Sur l'axe réel positif on aura d'après (1)

$$|\psi(z)| < Ce^{(\omega+l)r}$$

et sur l'axe imaginaire

$$|\psi(z)| < Ce^{-kr}.$$

Il s'ensuit d'après un théorème connu de Phragmén-Lindelöf ‖,

* G. H. Hardy, "On two theorems of F. Carlson and S. Wigert," *Acta Mathematica*, t. 42, 1920, pp. 327–339.

† E. Phragmén et Ernst Lindelöf, "Sur une extension d'un principe classique de l'Analyse et sur quelques propriétés des fonctions monogènes dans le voisinage d'un point singulier," *Acta Mathematica*, t. 31, 1908, pp. 381–406.

‡ Paul Persson, "Recherches sur une classe de fonctions entières," *Thèse pour le doctorat*, Upsal, 1908.

¶ F. Carlson, "Sur une classe de séries de Taylor," *Thèse pour le doctorat*, Upsal, 1914.

‖ *Loc. cit.*, pp. 393–394; an explicit proof of the particular result required is given by H. Cramér, "Sur une classe de séries de Dirichlet," *Thèse pour le doctorat* (Stockholm), Upsal, 1917, pp. 34–36.

ϕ désignant un nombre arbitraire situé entre $-\frac{\pi}{2}$ et $\frac{\pi}{2}$,

$$|\psi(z)| < Ce^{(-k|\sin\phi| + (\omega+l)\cos\phi)r}.$$

On aura donc, δ étant choisi assez petit, pour $\phi = \pm\left(\frac{\pi}{2} - \delta\right)$,

(3) $$\psi(z) < C,$$

inégalité valable aussi sur l'axe imaginaire. En appliquant encore une fois le théorème en question de Phragmén-Lindelöf aux angles

$$\left(-\frac{\pi}{2}, -\frac{\pi}{2}+\delta\right), \quad \left(-\frac{\pi}{2}+\delta, \frac{\pi}{2}-\delta\right) \quad \text{et} \quad \left(\frac{\pi}{2}-\delta, \frac{\pi}{2}\right)$$

tous moindres que π, on voit que (3) subsiste dans tout le demi-plan. C'est à dire, dans tout ce demi-plan,

(4) $$|g(z)\, e^{\omega z}| < C.$$

En tout point intérieur au demi-plan, on a $|e^z| > 1$. On conclut donc de (4), en faisant tendre ω vers l'infini,

$$g(z) \equiv 0.$$

2. Appliquons ce résultat à démontrer le théorème de M. Carlson.

Si $f(z) = f(re^{i\theta})$ *est holomorphe dans l'angle* $-\alpha \leqq \theta \leqq \alpha$, *où* $\alpha \geqq \frac{\pi}{2}$, *et y satisfait à l'inégalité*

$$|f(z)| < Ae^{pr} \qquad (p < \pi),$$

et $$f(n) = 0 \qquad (n = 0, 1, 2, \ldots),$$

alors $f(z)$ *s'annule identiquement.*

En effet, la fonction $g(z) = \frac{f(z)}{\sin \pi z}$ satisfait aux hypothèses du numéro précédent, par suite elle est identiquement nulle. Quant à la condition (2), cela est évident. En ce qui concerne (1), on trouve les éléments d'une démonstration rigoureuse dans votre travail ci-mentionné (p. 329). (En réalité, votre démonstration en question du théorème de M. Carlson présente une petite lacune. Pour pouvoir appliquer le théorème de Phragmén-Lindelöf, vous auriez dû démontrer que pour tout angle $-\alpha + \delta \leqq \theta \leqq \alpha - \delta$, il y a une majorante (une certaine fonction exponentielle) qui dépend seulement de r. Vous démontrez seulement que $+ 1(\theta) \leqq k$ sur chaque vecteur issu de l'origine, sans vous préoccuper de l'uniformité.)

Voilà une autre application de ce théorème. M. Cramér a déduit d'un beau théorème qui lui est dû, le corollaire qui suit*.

* H. Cramér, "Un théorème sur les séries de Dirichlet et son application," *Arkiv för Matematik, Astronomi och Fysik*, t. 13, No. 22, 1918, pp. 1–14 (p. 12).

Si le module de la fonction entière $\phi(z)$ finit par rester inférieur à $e^{k|z|}$, où $k > 0$, il y a sur chaque vecteur issu de l'origine une infinité de points aux modules indéfiniment croissants pour lesquels

$$|\phi(z)| > e^{-(k+\epsilon)|z|}$$

et cela a lieu pour tout $\epsilon > 0$.

L'application du théorème démontré au début à la fonction $g(z) = \phi(z)\phi(-z)$ donne de suite le résultat de M. Cramér. D'ailleurs il est facile de déduire ce résultat directement du théorème fondamental de Phragmén-Lindelöf que nous venons d'appliquer.

3. Un corollaire immédiat de notre théorème est encore le suivant.

Une fonction $g(z)$ qui est holomorphe dans un angle d'étendue $\geqq \pi$ et y satisfait à l'inégalité

$$|g(z)| < Ce^{-kr} \qquad (k > 0)$$

s'annule identiquement.

Par une substitution de variable on en obtient immédiatement:

Soit $\Phi(x)$ une fonction analytique de la variable complexe $x = re^{i\phi}$ qui jouit des propriétés suivantes:

1°. *Elle est holomorphe à l'intérieur et sur le contour d'un domaine T renfermant l'angle $-\frac{\pi}{2a} \leqq \phi - \phi_0 \leqq \frac{\pi}{2a}$, sauf peut-être certaines parties de cet angle situées à l'intérieur d'un cercle autour de l'origine.*

2°. *A l'intérieur et par suite au contour de ce domaine, on a*

$$|\Phi(x)| < Ce^{-kr^a}, \tag{5}$$

C et k étant des constantes positives.

Dans ces conditions, on a identiquement $\Phi(x) = 0$.

Le meilleur théorème de ce genre fut jusqu'ici, à ce qu'il semble, un théorème de M. Paul Persson démontré dans sa thèse (*l.c.* p. 8) et cité *in extenso* dans la thèse de M. Carlson (*l.c.* p. 36). Chez M. Persson, la condition

$$|\Phi(x)| < e^{-v(r)r^a}, \quad \lim_{r=\infty} v(r) = \infty$$

figure au lieu de (5).

24.I

Sur le problème des moments

Première Note

Ark. Mat. Astr. Fys. *16*, no. 12 (1921), 1–23

1. Comme problème des moments, nous désignerons ici le problème suivant:

Trouver les conditions nécessaires et suffisantes pour que les équations

$$c_n = \int_{-\infty}^{+\infty} t^n \, d\varphi(t) \qquad (n = 0, 1, 2, \ldots)$$

admettent une solution $\varphi(t)$ jamais décroissante. Deux solutions qui coïncident en leurs points de continuité à une constante additive près étant considérées comme identiques, trouver, en plus, les conditions nécessaires et suffisantes pour que la solution du problème soit unique.

Dans un mémoire élaboré qui va paraître prochainement, nous donnons un exposé complet de la question, en nous étendant même au cas où l'on remplace l'intervalle $(-\infty, +\infty)$ par un intervalle quelconque. Ici, nous nous restreindrons à donner un aperçu sommaire[1] de nos recherches, en insistant surtout sur quelques inégalités qui semblent mériter une attention particulière.

---

[1] Dans ce résumé sommaire, nous ne tenons pas à donner des renseignements détaillés quant à l'origine de certains théorèmes empruntés à différents auteurs. Nous comblerons cette lacune dans ledit mémoire.

On sait que, pour l'intervalle $(0, \infty)$, le problème des moments fut posé et résolu d'une façon magistrale par STIELTJES. Dans un travail important, récemment paru, M. HAMBURGER[1] a complètement résolu le problème posé pour l'intervalle $(-\infty, +\infty)$. Ses résultats principaux seront exposés dans la suite.

2. On a le théorème que voici:

Pour qu'il existe une solution $\varphi(t)$ jamais décroissante des équations en nombre infini

$$c_n = \int_{-\infty}^{+\infty} t^n d\varphi(t) \qquad (n = 0, 1, 2, \ldots)$$

c'est-à-dire pour qu'il existe une distribution de masses positives $d\varphi(t)$ dont les moments sont $c_0, c_1, c_2, \ldots$, il faut et il suffit que les formes quadratiques

$$\sum_{i,k=0}^{n} c_{i+k} x_i x_k \qquad (n = 0, 1, 2, \ldots) \tag{1}$$

soient toutes définies positives ou semi-définies positives. Pour que la solution admette une infinité de points de croissance, il faut et il suffit que lesdites formes soient définies positives.

La nécessité des conditions ci-dessus est manifeste, nous les admettrons donc dans la suite. Plus loin, nous indiquerons la démonstration de la réciproque. Nous laisserons, toutefois, dans toutes nos considérations de côté le cas presqu'évident des formes semi-définies, c'est-à-dire, nous admettrons que *les formes* (1) *sont définies positives.* Cette hypothèse s'exprime par les inégalités

$$|c_{i+k}|_0^n > 0 \qquad (n = 0, 1, 2, \ldots) \tag{2}$$

où nous posons, pour abréger,

$$\begin{vmatrix} c_0 & c_1 & \ldots & c_n \\ c_1 & c_2 & \ldots & c_{n+1} \\ \cdot & \cdot & \cdot & \cdot \\ c_n & c_{n+1} & \ldots & c_{2n} \end{vmatrix} = |c_{i+k}|_0^n.$$

Ces inégalités étant remplies, nous dirons que les nombres $c_0, c_1, c_2, \ldots$ forment une *suite définie positive.*

[1] Über eine Erweiterung des Stieltjesschen Momentenproblems, Math. Ann., I, t. 81, p. 235—319, II, t. 82, p. 120—164, III, t. 82, p. 168—187.

Préliminaires algébriques.

3. Etant donné un polynome

$$f(t) = x_0 + x_1 t + \cdots + x_n t^n,$$

nous poserons avec M. STRIDSBERG

$$f(c) = x_0 c_0 + x_1 c_1 + \cdots + x_n c_n.$$

$g(t)$ étant un autre polynome, nous entendrons par $f(c)\,g(c)$ ou $f^2(c)$, ce que deviennent le produit $f(t)\,g(t)$ ou le carré $f^2(t)$, lorsque, développés suivant les puissances de t, on y remplace chaque puissance t^j par le nombre correspondant c_j. En particulier

$$c^k f(c) = x_0 c_k + x_1 c_{k+1} + \cdots + x_n c_{k+n}.$$

Par contre $[f(c)][g(c)]$ ou $[f(c)]^2$ signifient respectivement le produit des valeurs $f(c)$ et $g(c)$ ou le carré de la valeur $f(c)$.

Par un théorème d'algèbre connu, on voit sans peine que notre hypothèse sur les formes quadratiques est équivalente à celle-ci, où $f(t)$ désigne un polynome arbitraire:

L'hypothèse $f(t) \geq 0$ *et non identiquement nul, entraîne* $f(c) > 0$.

4. Posons maintenant, dans cet ordre d'idées, le problème de minimum qui suit:

α désignant un nombre réel donné, trouver parmi tous les polynomes non négatifs $g(t)$, de degré $2n$ au plus et satisfaisant à l'égalité $g(\alpha) = 1$, celui qui rend l'expression $g(c)$ minimum.

Un calcul simple montre que la solution est unique et qu'elle est fournie par le *carré* du polynome

$$(3) \qquad f_n(t \mid \alpha) = \frac{\begin{vmatrix} 0 & 1 & \alpha & \dots & \alpha^n \\ 1 & c_0 & c_1 & \dots & c_n \\ t & c_1 & c_2 & \dots & c_{n+1} \\ \cdot & \cdot & \cdot & \cdot & \cdot \\ t^n & c_n & c_{n+1} & \dots & c_{2n} \end{vmatrix}}{\begin{vmatrix} 0 & 1 & \alpha & \dots & \alpha^n \\ 1 & c_0 & c_1 & \dots & c_n \\ \alpha & c_1 & c_2 & \dots & c_{n+1} \\ \cdot & \cdot & \cdot & \cdot & \cdot \\ \alpha^n & c_n & c_{n+1} & \dots & c_{2n} \end{vmatrix}} = \frac{\begin{vmatrix} 0 & \alpha^k \\ t^i & c_{i+k} \end{vmatrix}_0^n}{\begin{vmatrix} 0 & \alpha^k \\ \alpha^i & c_{i+k} \end{vmatrix}_0^n},$$

et que le minimum en question est

$$(3') \qquad \varrho_n(\alpha) = f_n^2(c \mid \alpha) = \frac{|c_{i+k}|_0^n}{-\begin{vmatrix} 0 & \alpha^k \\ \alpha^i & c_{i+k} \end{vmatrix}_0^n} > 0.$$

En posant

$$(4) \qquad h_n(t \mid \alpha) = h_n(\alpha \mid t) = \frac{-\begin{vmatrix} 0 & \alpha^k \\ t^i & c_{i+k} \end{vmatrix}_0^n}{|c_{i+k}|_0^n},$$

on pourra encore écrire

$$(4') \qquad \varrho_n(\alpha) = \frac{1}{h_n(\alpha \mid \alpha)}$$

et

$$(4'') \qquad f_n(t \mid \alpha) = \frac{h_n(t \mid \alpha)}{h_n(\alpha \mid \alpha)} = \varrho_n(\alpha) h_n(t \mid \alpha).$$

Evidemment le nombre $\varrho_n(\alpha)$ peut en même temps être considéré comme le minimum de la forme quadratique $\sum_{i,k=0}^{n} c_{i+k} x_i x_k$ sous la condition $x_0 + x_1 \alpha + \cdots + x_n \alpha^n = 1$.

Ecrivons encore quelques relations importantes qui forment l'origine des formules qui précèdent. On a par la propriété minimum de $f_n(t \mid \alpha)$,

$$(c - \alpha)^k f_n(c \mid \alpha) = 0 \qquad (k = 1, 2, \ldots n).$$

De ces formules, on conclut facilement que le polynome $f_n(t \mid \alpha)$, considéré comme fonction de t, a tous ses zéros réels et distincts et que son degré qui, en général, est n, est au moins $n - 1$. Si ce dernier cas intervient, nous dirons que n est un indice exceptionnel par rapport à α. On voit aussi, par la propriété minimum de notre polynome que, dans ce dernier cas, on a $f_n(t \mid \alpha) = f_{n-1}(t \mid \alpha)$, et que, n étant un indice exceptionnel, $n - 1$ et $n + 1$ ne pourront l'être. Dans la suite, nous supposerons, en général, que les indices dont il sera question, ne sont pas exceptionnels, sans nous arrêter aux modifications sans importance qu'il faut apporter à nos énoncés dans le cas d'exception.

On a évidemment

$$(5)\qquad 0 < \varrho_n(\alpha) \leqq \varrho_{n-1}(\alpha),$$

le signe d'égalité étant exclu toutes les fois que n n'est pas un indice exceptionnel. Par conséquent, il existe la limite

$$(6)\qquad \lim_{n\to\infty} \varrho_n(\alpha) = \varrho(\alpha).$$

5. Désignons les zéros de $f_n(t\,|\,\alpha)$ (considéré comme polynome de t) dans un ordre quelconque, par $\alpha_{n,1}, \alpha_{n,2}, \ldots \alpha_{n,n}$. En posant encore $\alpha_{n,0} = \alpha$, formons le polynome

$$(7)\qquad F_n(t\,|\,\alpha) = (t-\alpha) f_n(t\,|\,\alpha)$$

dont les zéros sont $\alpha_{n,0} = \alpha$, $\alpha_{n,1}, \ldots \alpha_{n,n}$. Ce polynome est entièrement caractérisé par les propriétés suivantes: Il est de degré $n+1$ au plus, et on a

$$\begin{aligned} c^k F_n(c\,|\,\alpha) &= 0, \qquad \text{pour } k = 0, 1, \ldots n-1, \\ F_n(\alpha\,|\,\alpha) &= 0, \\ F'_n(\alpha\,|\,\alpha) &= 1. \end{aligned}$$

On démontre sans peine que le polynome

$$(8)\qquad f_n(t\,|\,\alpha_{n,j}) = \frac{F_n(t\,|\,\alpha)}{(t-\alpha_{n,j})\,F'_n(\alpha_{n,j}\,|\,\alpha)} \qquad (j = 1, 2, \ldots n),$$

a les mêmes propriétés par rapport au point $\alpha_{n,j}$, que $f_n(t|\alpha)$ avait par rapport à α, ce qui justifie la notation ci-dessus. La formule (8) subsiste évidemment pour $j = 0$, c.-à-d. pour $\alpha_{n,0} = \alpha$. Aux points $\alpha_{n,j}$, il correspond des nombres positifs $\varrho_n(\alpha_{n,j})$, de même façon que $\varrho_n(\alpha)$ correspondait à α. On a une formule analogue à (3') et en outre

$$(9)\qquad \varrho_n(\alpha_{n,j}) = f_n^2(c\,|\,\alpha_{n,j}) = f_n(c\,|\,\alpha_{n,j}) > 0.$$

Par la formule d'interpolation de Lagrange, on établit facilement que, $g(t)$ désignant un polynome quelconque, de degré

$2n$ au plus, on a

$$g(c) = \sum_{j=0}^{n} \varrho_n(\alpha_{n,j})\, g(\alpha_{n,j}), \tag{10}$$

et notamment

$$c_k = \sum_{j=0}^{n} \varrho_n(\alpha_{n,j})\, \alpha_{n,j}^k \qquad (k = 0, 1, \ldots 2n). \tag{10'}$$

Posons, en particulier, $\alpha = 0$ et $f_n(t \mid 0) = f_n(t)$ ainsi que $F_n(t \mid 0) = F_n(t)$. On aura

$$f_n(t \mid 0) = f_n(t) = \frac{\begin{vmatrix} 1 & t & \ldots & t^n \\ c_1 & c_2 & \ldots & c_{n+1} \\ \cdot & \cdot & \cdot & \cdot \\ c_n & c_{n+1} & \ldots & c_{2n} \end{vmatrix}}{| c_{i+k} |_1^n} \tag{11}$$

et

$$\varrho_n(0) = \frac{| c_{i+k} |_0^n}{| c_{i+k} |_1^n}, \tag{12}$$

$$\lim_{n \to \infty} \varrho_n(0) = \varrho(0). \tag{12'}$$

Nous désignons les zéros de $F_n(t) = t f_n(t)$ par $\theta_{n,0} = 0$, $\theta_{n,1}, \ldots \theta_{n,n}$.

6. En même temps que la suite $c_0, c_1, c_2, \ldots$ est définie positive, les suites $c_{2p}, c_{2p+1}, \ldots$ $(p = 1, 2, \ldots)$ le sont aussi. Pour ces suites, on pourra donc poser des problèmes analogues à ceux qu'on vient de résoudre pour la suite originale. Nous rassemblons ici quelques formules dont nous aurons besoin dans la suite.

Le minimum de la forme quadratique $\sum_{i,k=0}^{n-p} c_{2p+i+k}\, x_i x_k$, sous la condition $x_0 + x_1\alpha + \cdots + x_{n-p}\alpha^{n-p} = 1$, est

$$\varrho_n^{(2p)}(\alpha) = \frac{| c_{i+k} |_p^n}{-\begin{vmatrix} 0 & \alpha^{k-p} \\ \alpha^{i-p} & c_{i+k} \end{vmatrix}_p^n}. \tag{13}$$

Nous posons

$$(13') \qquad \lim_{n\to\infty} \varrho_n^{(2p)}(\alpha) = \varrho^{(2p)}(\alpha).$$

En particulier, pour $\alpha = 0$:

Le minimum de $\sum_{i,k=0}^{n-p} c_{2p+i+k} x_i x_k$, sous la condition $x_0 = 1$, est

$$(13'') \qquad \varrho_n^{(2p)}(0) = \frac{|c_{i+k}|_p^n}{|c_{i+k}|_{p+1}^n}.$$

Un problème équivalent au dernier est le suivant: Parmi tous les polynomes non négatifs

$$g(t) = t^{2p} + \cdots + y_{2n} t^{2n}$$

de degré $2n$ au plus, trouver celui qui rend $g(c)$ minimum.

Le minimum en question est fourni par le polynome

$$(t^p f_n^{(2p)}(t))^2$$

où

$$(14) \qquad f_n^{(2p)}(t) = \frac{1}{|c_{i+k}|_{p+1}^n} \begin{vmatrix} 1 & t & \ldots\ldots & t^{n-p} \\ c_{2p+1} & c_{2p+2} & \ldots & c_{p+n+1} \\ \cdot & \cdot & \cdot\cdot\cdot & \cdot \\ c_{p+n} & c_{p+n+1} & \ldots & c_{2n} \end{vmatrix}.$$

La valeur du minimum est encore $\varrho_n^{(2p)}(0)$, donnée par la formule (13''). Evidemment, les propriétés des derniers polynomes sont analogues à celles des polynomes $f_n(t)$. En particulier, ces polynomes ont des zéros réels et distincts.

7. Soient maintenant α et β deux nombres réels quelconques. Nous allons exprimer le polynome $h_n(t\,|\,\alpha)$ et la valeur minimée $\varrho_n(\alpha)$ par les polynomes et les valeurs minimées, attachés aux points $\beta_{n,j}$.

D'après la formule d'interpolation de Lagrange, on a, ayant égard à (8), (4'') et à la symétrie de $h_n(t\,|\,\alpha)$, (*cf.* (4))

$$(15)\qquad \begin{aligned} h_n(t\,|\,\alpha) &= \sum_{j=0}^{n} h_n(\beta_{n,j}\,|\,\alpha)\, f_n(t\,|\,\beta_{n,j}) = \sum_{j=0}^{n} f_n(t\,|\,\beta_{n,j})\, h_n(\alpha\,|\,\beta_{n,j}) = \\ &= \sum_{j=0}^{n} \frac{f_n(t\,|\,\beta_{n,j})\, f_n(\alpha\,|\,\beta_{n,j})}{\varrho_n(\beta_{n,j})}.^{1} \end{aligned}$$

Spécialement, pour $t = \alpha$,

$$(15') \qquad \frac{1}{\varrho_n(\alpha)} = h_n(\alpha\,|\,\alpha) = \sum_{j=0}^{n} \frac{f_n^2(\alpha\,|\,\beta_{n,j})}{\varrho_n(\beta_{n,j})}.$$

Posons notamment $\beta = 0$. En remplaçant les valeurs $\beta_{n,j}$ par les valeurs $\theta_{n,j}$, il vient

$$(16) \qquad h_n(t\,|\,\alpha) = \frac{f_n(t)\, f_n(\alpha)}{\varrho_n(0)} + \sum_{j=1}^{n} \frac{f_n(t\,|\,\theta_{n,j})\, f_n(\alpha\,|\,\theta_{n,j})}{\varrho_n(\theta_{n,j})}$$

et

$$(16') \qquad \frac{1}{\varrho_n(\alpha)} = \frac{f_n^2(\alpha)}{\varrho_n(0)} + \sum_{j=1}^{n} \frac{f_n^2(\alpha\,|\,\theta_{n,j})}{\varrho_n(\theta_{n,j})}.$$

En vue d'une application de la plus haute importance, nous mettrons encore ces identités sous une autre forme. De l'expression (4) pour $h_n(t\,|\,\alpha)$, il découle par un théorème connu de JACOBI

$$(17)\qquad \begin{aligned} h_n(t\,|\,\alpha) &= -\frac{\begin{vmatrix} 0 & \alpha^k \\ t^i & c_{i+k} \end{vmatrix}_0^n}{|\,c_{i+k}\,|_0^n} = -\frac{\begin{vmatrix} 0 & \alpha^k \\ t^i & c_{i+k} \end{vmatrix}_1^n}{|\,c_{i+k}\,|_1^n} + \\ &+ \frac{\begin{vmatrix} 1 & \alpha & \dots & \alpha^n \\ c_1 & c_2 & \dots & c_{n+1} \\ \cdot & \cdot & \cdot & \cdot \\ c_n & c_{n+1} & \dots & c_{2n} \end{vmatrix}}{|\,c_{i+k}\,|_0^n}\,\frac{\begin{vmatrix} 1 & t & \dots & t^n \\ c_1 & c_2 & \dots & c_{n+1} \\ \cdot & \cdot & \cdot & \cdot \\ c_n & c_{n+1} & \dots & c_{2n} \end{vmatrix}}{|\,c_{i+k}\,|_1^n} = \\ &= \frac{f_n(\alpha)\, f_n(t)}{\varrho_n(0)} - \alpha t\, \frac{\begin{vmatrix} 0 & \alpha^k \\ t^i & c_{2+i+k} \end{vmatrix}_0^{n-1}}{|\,c_{i+k}\,|_1^n}. \end{aligned}$$

---

[1] La formule (15) et nos autres formules de structure analogue subsistent évidemment, si l'on y remplace $t^0, t^1, \dots t^n$ et $\alpha^0, \alpha^1, \dots \alpha^n$ par des nombres complexes quelconques $x_0, x_1, \dots x_n$, et $y_0, y_1, \dots y_n$ respective-

En particulier, pour $t=\alpha$, ayant encore égard à la formule (13), il vient

(18) $$\frac{1}{\varrho_n(\alpha)}=\frac{f_n^2(\alpha)}{\varrho_n(0)}+\frac{\alpha^2}{\varrho_n^{(2)}(\alpha)}.$$

La comparaison des formules (16) et (17) fournit

(19) $$\frac{-\begin{vmatrix} 0 & \alpha^k \\ t^i & c_{2+i+k} \end{vmatrix}_0^{n-1}}{|c_{i+k}|_1^n}=\sum_{j=1}^{n}\frac{g_{n,j}(\alpha)\,g_{n,j}(t)}{\varrho_n(\theta_{n,j})}$$

où l'on a posé

$$\frac{f_n(t\,|\,\theta_{n,j})}{t}=g_{n,j}(t) \qquad (j=1,2,\ldots n).$$

En posant $t=\alpha$, il vient

(19′) $$\frac{1}{\varrho_n^{(2)}(\alpha)}=\sum_{j=1}^{n}\frac{g_{n,j}^2(\alpha)}{\varrho_n(\theta_{n,j})},$$

et spécialement pour $\alpha=0$,

(19″) $$\frac{1}{\varrho_n^{(2)}(0)}=\sum_{j=1}^{n}\frac{g_{n,j}^2(0)}{\varrho_n(\theta_{n,j})}.$$

On voit facilement que

$$\frac{g_{n,j}(\alpha)}{g_{n,j}(0)}=\frac{f_n(\alpha)}{1-\dfrac{\alpha}{\theta_{n,j}}}.$$

On a donc en vertu de (19′) et de (19″)

---

ment. Notons encore l'identité équivalente à (15)

$$h_n(t\,|\,\alpha)=\sum_{j=0}^{n}\frac{h_n(t\,|\,\beta_{n,j})\,h_n(\alpha\,|\,\beta_{n,j})}{h_n(\beta_{n,j}\,|\,\beta_{n,j})}.$$

$$\frac{1}{\varrho_n^{(2)}(\alpha)} \leqq \frac{1}{\varrho_n^{(2)}(0)} \operatorname*{Max}_{j=1,2,\ldots n} \left(\frac{f_n(\alpha)}{1 - \frac{\alpha}{\theta_{n,j}}} \right)^2,$$

par conséquent, d'après (18),

$$\text{(20)} \qquad \frac{1}{\varrho_n(\alpha)} \leqq \frac{f_n^2(\alpha)}{\varrho_n(0)} + \frac{\alpha^2}{\varrho_n^{(2)}(0)} \operatorname*{Max}_{j=1,2,\ldots n} \left(\frac{f_n(\alpha)}{1 - \frac{\alpha}{\theta_{n,j}}} \right)^2.$$

Une évaluation analogue donne, si l'on désigne par x et y deux nombres complexes quelconques,

$$\text{(21)} \quad |h_n(x|y)| \leqq \frac{|f_n(x) f_n(y)|}{\varrho_n(0)} + \frac{|xy|}{\varrho_n^{(2)}(0)} \operatorname*{Max}_{j=1,2,\ldots n} \left| \frac{f_n(x) f_n(y)}{\left(1 - \frac{x}{\theta_{n,j}}\right)\left(1 - \frac{y}{\theta_{n,j}}\right)} \right|.$$

Nous reproduisons encore le développement connu que voici:

$$\text{(22)} \qquad h_n(x|y) = \frac{1}{c_0} + \sum_{p=1}^{n} \frac{q_p(x) q_p(y)}{|c_{i+k}|_0^p \, |c_{i+k}|_0^{p-1}}$$

où l'on a posé

$$q_p(x) = \begin{vmatrix} 1 & x & \ldots & x^p \\ c_0 & c_1 & \ldots & c_p \\ \cdot & \cdot & \cdot & \cdot \\ c_{p-1} & c_p & \ldots & c_{2p-1} \end{vmatrix}.$$

La formule s'obtient aisément en évaluant $h_p(x|y) - h_{p-1}(x|y)$ par le théorème de JACOBI.

8. Posons $s_{-k} = \sum_{j=1}^{n} \theta_{n,j}^{-k}$, en désignant, comme ci-dessus, par $\theta_{n,j}$ les zéros de $f_n(t)$. Nous allons déduire des inégalités importantes pour s_{-1} et s_{-2}.

Nous pouvons écrire

$$f_n(t) = 1 - s_{-1} t + \cdots + \omega_n t^n$$

et

$$K(t) = f_n(t) - 1 = -s_{-1} t + \cdots + \omega_n t^n.$$

On a évidemment $K(t) = -1$, pour $t = \theta_{n,1}, \theta_{n,2}, \ldots \theta_{n,n}$. En outre, $K(t)$ commence par $s_{-1}t$ et est de degré n au plus. Par suite, on a d'après la propriété extrémale de $\varrho_n^{(2)}(0)$ (*cf.* p. 7)

$$s_{-1}^2 \varrho_n^{(2)}(0) < K^2(c) = \sum_{j=1}^{n} \varrho_n(\theta_{n,j}) K^2(\theta_{n,j}) =$$

$$= \sum_{j=1}^{n} \varrho_n(\theta_{n,j}) = c_0 - \varrho_n(0).$$

On en tire

$$(23) \qquad s_{-1}^2 < \frac{c_0 - \varrho_n(0)}{\varrho_n^{(2)}(0)}.[1]$$

Afin d'obtenir une inégalité pour s_{-2}, nous formons un polynome $L(t)$ de degré $\leqq n$, sans terme constant et qui, pour $t = \theta_{n,1}, \theta_{n,2}, \ldots \theta_{n,n}$, satisfait aux équations $L(t) = -t^{-1}$. Un calcul simple donne

$$L(t) = \frac{(1 + s_{-1}t) f_n(t) - 1}{t} = -\frac{1}{2}(s_{-1}^2 + s_{-2})t + \cdots + \omega'_n t^n.$$

On conclut

$$\frac{1}{4}(s_{-1}^2 + s_{-2})^2 \varrho_n^{(2)}(0) < L^2(c) = \sum_{j=1}^{n} \varrho_n(\theta_{n,j}) L^2(\theta_{n,j}) =$$

$$= \sum_{j=1}^{n} \varrho_n(\theta_{n,j}) \theta_{n,j}^{-2} < s_{-2} \sum_{j=1}^{n} \varrho_n(\theta_{n,j}) =$$

$$= s_{-2}(c_0 - \varrho_n(0)) \leqq (s_{-1}^2 + s_{-2})(c_0 - \varrho_n(0)),$$

c.-à-d.

$$(24) \qquad s_{-1}^2 + s_{-2} < \frac{4(c_0 - \varrho_n(0))}{\varrho_n^{(2)}(0)},$$

et à plus forte raison

$$(24') \qquad s_{-2} < \frac{4(c_0 - \varrho_n(0))}{\varrho_n^{(2)}(0)}.$$

---

[1] Les mêmes inégalités résultent par l'application du théorème de Jacobi au tableau défini $|c_{i+k}|$.

9. Soit maintenant $g(t)=\left(1-\frac{t}{\lambda_1}\right)\ldots\left(1-\frac{t}{\lambda_n}\right)$ un polynome quelconque dont les zéros satisfont aux inégalités

$$|\lambda_1^{-1}+\cdots+\lambda_n^{-1}|<U,$$

et

$$|\lambda_1|^{-2}+\cdots+|\lambda_n|^{-2}<V^2 \qquad (V>0).$$

En appliquant l'inégalité facile à démontrer, et valable pour tout nombre complexe x,

$$|(1-x)e^x|\leqq e^{\frac{1}{2}|x|^2},$$

on aura pour toute valeur de t,

$$|g(t)|\leqq e^{\frac{1}{2}V^2|t|^2+U|t|}$$

et pour $j=1,2,\ldots n$,

$$\left|\frac{g(t)}{1-\frac{t}{\lambda_j}}\right|\leqq e^{\frac{1}{2}V^2|t|^2+(U+V)|t|}.$$

En appliquant ceci à $f_n(t)$ et ayant égard aux inégalités (23) et (24′), on aura, en posant

$$\sqrt{\frac{c_0-\varrho_n(0)}{\varrho_n^{(2)}(0)}}=\mu_n>0, \tag{25}$$

pour toute valeur de t, et, pour $j=1,2,\ldots n$,

$$|f_n(t)|\leqq e^{2\mu_n^2|t|^2+\mu_n|t|}, \tag{26}$$

$$\left|\frac{f_n(t)}{1-\frac{t}{\theta_{n,j}}}\right|\leqq e^{2\mu_n^2|t|^2+3\mu_n|t|}. \tag{26′}$$

On aura donc, d'après (21), x et y désignant des nombres complexes dont les valeurs absolues sont plus petites que R,

$$|h_n(x\,|\,y)| = |h_n(y\,|\,x)| \leq \frac{e^{4\mu_n^2 R^2 + 2\mu_n R}}{\varrho_n(0)} + \frac{R^2 e^{4\mu_n^2 R^2 + 6\mu_n R}}{\varrho_n^{(2)}(0)} <$$

(27)

$$\leq e^{4\mu_n^2 R^2 + 6\mu_n R}\left(\frac{1}{\varrho_n(0)} + \frac{R^2}{\varrho_n^{(2)}(0)}\right),$$

et, en particulier,

(27′)
$$\frac{1}{\varrho_n(\alpha)} < e^{4\mu_n^2 \alpha^2 + 6\mu_n|\alpha|}\left(\frac{1}{\varrho_n(0)} + \frac{\alpha^2}{\varrho_n^2(0)}\right).$$

10. Nous établirons maintenant une de nos inégalités les plus importantes, en démontrant

(28)
$$\varrho_n^{(4)}(0) > \frac{(\varrho_n^{(2)}(0))^2}{c_0 - \varrho_n(0)}.$$

Considérons à cet effet le polynome

(29)
$$H(t) = \frac{1}{|c_{i+k}|_s^n}\begin{vmatrix} t^2 & t^3 \ldots\ldots & t^n \\ c_5 & c_6 \ldots\ldots & c_{n+3} \\ \ldots & \ldots & \ldots \\ c_{n+2} & c_{n+3} \ldots & c_{2n} \end{vmatrix} = t^2 + \cdots + \nu_n t^n.$$

D'après (13″) et (14), on a $H^2(c) = \varrho_n^{(4)}(0)$. De plus, on voit que

(29′)
$$c^k H(c) = 0, \quad \text{pour } k = 3, 4, \ldots n.$$

Or, en vertu de l'inégalité de CAUCHY, nous avons

$$\varrho_n^{(4)}(0)(c_0 - \varrho_n(0)) = \varrho_n^{(4)}(0)\sum_{j=1}^{n}\varrho_n(\theta_{n,j}) = H^2(c)\sum_{j=1}^{n}\varrho_n(\theta_{n,j}) =$$

$$= \sum_{j=1}^{n}\varrho_n(\theta_{n,j})H^2(\theta_{n,j}) \cdot \sum_{j=1}^{n}\varrho_n(\theta_{n,j}) \geq$$

$$\geq \left(\sum_{j=1}^{n}\varrho_n(\theta_{n,j})H(\theta_{n,j})\right)^2 = [H(c)]^2.$$ [1]

[1] Comme nous l'avons déjà remarqué plus haut, nous entendons par $[H(c)]^2$ le carré de la valeur $H(c)$. La dernière inégalité résulte aussi de l'inégalité générale $[A(c)B(c)]^2 \leq [A^2(c)][B^2(c)]$ analogue à celle de SCHWARZ, valable pour deux polynomes réels quelconques $A(t)$ et $B(t)$ et résultant de l'inégalité $(A(t) + \lambda B(t))^2 \geq 0$, pour toute valeur réelle de λ. Dans ledit mémoire, nous démontrerons (28) de deux autres manières différentes.

Pour aller plus loin, on forme un polynome du second degré $P(t)$ de sorte que le produit $t^{-2} H(t) P(t)$ soit de la forme $K(t) = 1 + b_3 t^3 + \cdots + b_n t^n$. Tous les zéros de $H(t)$ étant réels et $H(t)$ étant positif autour de l'origine, le polynome $K(t)$ aura, en vertu du théorème de DESCARTES, toujours le même signe que $H(t)$. D'après (29'), on a d'autre part

$$H(c) K(c) = H(c).$$

En nous rappelant la définition de $\varrho_n^{(2)}(0)$ et remarquant que le polynome $H(t) K(t) = t^2 + \cdots + \xi_{2n} t^{2n}$ n'est jamais négatif, qu'il commence par t^2 et qu'il est de degré $2n$, au plus, nous obtenons

$$H(c) = H(c) K(c) > \varrho_n^{(2)}(0),$$

ce qui, ayant égard à l'inégalité précédente, achève la démonstration.

Remarquons, en vue d'une application ultérieure, que, dans cette démonstration, nous n'avons utilisé que le fait que la forme quadratique à $n + 1$ variables $\sum_{i,k=0}^{n} c_{i+k} x_i x_k$ était définie positive.

11. Nous rappellerons maintenant brièvement les inégalités célèbres de TCHÉBYCHEF.

Nous appelons distribution d'ordre $2n$ un groupe de $n + 1$ points réels $\lambda_0, \lambda_1, \ldots \lambda_n$ et de masses positives correspondantes $\sigma_0, \sigma_1, \ldots \sigma_n$, les points et les masses étant choisis de sorte que les $2n + 1$ équations

$$c_k = \sum_{j=0}^{n} \sigma_j \lambda_j^k \qquad (k = 0, 1, 2, \ldots 2n) \tag{30}$$

aient lieu. Nous désignerons les points λ_j comme points de concentration de la distribution en question. En admettant p. ex. $\lambda_0 < \lambda_1 < \lambda_2 < \ldots < \lambda_n$ et en posant

$$\varphi_n(t) = \sigma_0 + \sigma_1 + \cdots + \sigma_p, \qquad \text{pour } \lambda_p < t \leqq \lambda_{p+1}, \tag{31}$$

les formules (30) pourront s'écrire

$$c_k = \int_{-\infty}^{+\infty} t^k \, d\varphi_n(t) \qquad (k = 0, 1, 2, \ldots 2n). \tag{30'}$$

Nous dirons que la fonction $\varphi_n(t)$ est attachée à la susdite distribution.

La comparaison de (10′) et (30) montre que le système $(\alpha_{n,j}, \varrho_n(\alpha_{n,j}))$ $(j = 0, 1, \ldots n)$ où $\alpha_{n,0} = \alpha$ est un nombre réel quelconque et les $\alpha_{n,j}$ désignent, comme ci-haut les zéros de $F_n(t \mid \alpha)$, fournit une distribution d'ordre $2n$. Inversement, on voit sans peine que toute distribution d'ordre $2n$ est dudit type.

Soit α un nombre réel quelconque. Nous numérotons, avec M. HAMBURGER, les $\alpha_{n,j}$ par ordre de grandeur croissante

$$
\begin{aligned}
(32)\qquad \alpha_{n,-p} < \alpha_{n,-p+1} < \ldots < \alpha_{n,-1} < \alpha_{n,0} = \\
= \alpha < \alpha_{n,1} < \ldots < \alpha_{n,q-1} < \alpha_{n,q} \qquad (p+q=n).
\end{aligned}
$$

En faisant correspondre à ces valeurs les masses $\varrho_n(\alpha_{n,j})$, formons d'après le schéma (31) la fonction $\varphi_n(t)$ attachée à cette distribution.

Les inégalités de TCHÉBYCHEF consistent en ceci:

Soit $\varphi(t)$ $[\varphi(t) \not\equiv \varphi_n(t), \varphi(-\infty) = 0]$ *une solution du problème des moments relatif à la suite* $c_0, c_1, \ldots$ *ou une fonction attachée à une distribution d'ordre non inférieur à* $2n$ *ou enfin une fonction jamais décroissante quelconque satisfaisant aux équations (30′); on aura*

$$
(33)\qquad \varphi_n(\alpha - 0) < \varphi(\alpha - 0) \leqq \varphi(\alpha + 0) < \varphi_n(\alpha + 0).
$$

Si α *est le plus petit ou le plus grand parmi les valeurs* $\alpha_{n,j}$, *on doit écrire respectivement* $\varphi_n(\alpha - 0) \leqq \varphi(\alpha - 0)$ *ou* $\varphi(\alpha + 0) \leqq \leqq \varphi_n(\alpha + 0)$.

Parmi les nombreuses conséquences de ces inégalités, nous ne citerons que les suivantes:

On a pour toute solution $\varphi(t)$ du problème des moments

$$
(34)\qquad \varphi(\alpha + 0) - \varphi(\alpha - 0) < \varphi_n(\alpha + 0) - \varphi_n(\alpha - 0) = \varrho_n(\alpha)
$$

et]

$$
(34')\qquad \varphi(\alpha + 0) - \varphi(\alpha - 0) \leqq \varrho(\alpha) = \lim_{n \to \infty} \varrho_n(\alpha),
$$

et en désignant par $\varphi^*(t)$ une seconde solution,

$$
(34'')\qquad |\varphi^*(\alpha) - \varphi(\alpha)| \leqq \varrho(\alpha) = \lim_{n \to \infty} \varrho_n(\alpha).
$$

Une autre conséquence immédiate est la suivante:

S'il existe le zéro $\alpha_{n,r}$ ou le zéro $\alpha_{n,-s}$, il existe aussi les zéros $\alpha_{n+1,r}, \alpha_{n+2,r} \ldots$ ou les zéros $\alpha_{n+1,-s}, \alpha_{n+2,-s} \ldots$, et ces zéros se rangent de la façon suivante:

(35) $$\alpha_{n,r} > \alpha_{n+1,r} > \alpha_{n+2,r} > \ldots > \alpha,$$

(35') $$\alpha_{n,-s} < \alpha_{n+1,-s} < \alpha_{n+2,-s} < \ldots < \alpha.$$

Par suite, il existe aussi les valeurs-limites

(36) $$\alpha_r = \lim_{n\to\infty} \alpha_{n,r}$$

(36') $$\alpha_{-s} = \lim_{n\to\infty} \alpha_{n,-s},$$

fait particulièrement important, lorsque le problème est indéterminé.

Le problème des moments.

12. On connaît le théorème important de M. HELLY[1], d'après lequel, *de toute suite de fonctions monotones, bornées dans leur ensemble, définies dans un certain intervalle, on peut tirer une suite partielle, tendant partout, dans cet intervalle, vers une fonction monotone déterminée.*

En appliquant ce théorème à une suite de fonctions $\varphi_n(t)$, attachées à des distributions d'ordres finis et indéfiniment croissants, on voit que de chaque telle suite, on peut tirer une suite partielle $\varphi_1^*(t), \varphi_2^*(t), \ldots$, qui, pour toute valeur de t, tend vers une fonction jamais décroissante $\varphi(t)$. Il est facile de voir que *cette fonction $\varphi(t)$ est une solution du problème des moments.*

Désignons, en effet, par $\varphi_1^*(t), \varphi_2^*(t), \ldots \varphi_p^*(t), \ldots$ les fonctions qui forment ladite suite partielle, par k un indice fixe et par m un nombre entier fixe, tel que $2m > k$. Les fonctions $\varphi_p^*(t)$ étant attachées à des distributions dont les ordres tendent vers l'infini, on a pour p assez grand,

$$c_k = \int_{-\infty}^{+\infty} t^k \, d\varphi_p^*(t)$$

[1] Über lineare Funktionaloperationen, Sitzungsber. K. Ak. d. Wiss. Wien, Math.-natw. Kl. 121, Abt. II a. (1912) p. 265—297, *voir* surtout p. 284—287.

et

$$c_{2m} = \int_{-\infty}^{+\infty} t^{2m}\, d\,\varphi_p^*(t).$$

En désignant par T un nombre positif, on tire des dernières formules, pour $T_1 \geqq T$ et $T_2 \geqq T$,

$$\left| c_k - \int_{-T_1}^{+T_2} t^k\, d\,\varphi_p^*(t) \right| \leqq \frac{c_{2m}}{T^{2m-k}}.$$

D'autre part, en intégrant par parties, on conclut, moyennant un théorème connu de M. LEBESGUE,

$$\lim_{p\to\infty} \int_{-T_1}^{+T_2} t^k\, d\,\varphi_p^*(t) = \int_{-T_1}^{+T_2} t^k\, d\,\varphi(t).$$

On a donc l'inégalité

$$\left| c_k - \int_{-T_1}^{+T_2} t^k\, d\varphi(t) \right| \leqq \frac{c_{2m}}{T^{2m-k}},$$

d'où, en faisant tendre T vers l'infini, il résulte

$$c_k = \int_{-\infty}^{+\infty} t^k\, d\,\varphi(t).$$

Si, en particulier, on applique le principe de choix de M. HELLY aux fonctions attachées aux distributions d'ordres finis qui ont un point de concentration en α, on arrive à une solution $\tilde{\varphi}(t)$, pour laquelle, $\tilde{\varphi}(\alpha+0) - \tilde{\varphi}(\alpha-0) = \varrho(\alpha)$.

Nous dirons avec STIELTJES que le problème des moments est *déterminé*, si toutes les solutions possibles $\varphi(t)$ ($\varphi(-\infty) = 0$) coïncident dans tous leurs points de continuité. Dans le cas contraire, le problème sera dit *indéterminé*.

On voit sans peine que, pourqu'il y ait indétermination, il faut et il suffit qu'il existe deux solutions, qui, dans un certain intervalle (a, b), diffèrent de plus de δ, δ désignant un certain nombre positif. On en conclut immédiatement:

Pour que le problème des moments soit indéterminé, il faut et il suffit qu'il existe un intervalle (a, b) et un nombre positif δ de sorte que $\varrho(\alpha) > \delta$ pour $a \leq \alpha \leq b$.

En effet, dans le cas d'indétermination, ladite propriété de $\varrho(\alpha)$ résulte de la remarque précédente et des inégalités de TCHÉBYCHEF (*cf.* (34'')). Inversement, si $\varrho(\alpha) > \delta$, pour $a \leq \alpha \leq b$, on peut, d'après ce que nous venons de voir, pour chaque valeur α, située entre a et b, construire une solution qui, au point α, présente la discontinuité $\varrho(\alpha) > \delta$. Les points de discontinuité d'une fonction monotone formant au plus un ensemble dénombrable, les solutions attachées, de cette manière, aux différents points α de l'intervalle (a, b), ne pourront être toutes identiques.

13. Ces préliminaires posés, nous sommes à même de démontrer le théorème principal de M. HAMBURGER:

Pour que le problème des moments soit déterminé, il faut et il suffit que l'un au moins des deux nombres non négatifs $\varrho(0)$ et $\varrho^{(2)}(0)$ soit égal à zéro.

Ou le théorème équivalent:

Pour que le problème des moments soit indéterminé, il faut et il suffit que $\varrho(0) > 0$ et $\varrho^{(2)}(0) > 0$.

Chez M. HAMBURGER, c'est la première forme du théorème qui sert de point de départ. A notre avis, c'est la seconde forme qui offre le chemin le plus direct.

Nous démontrons d'abord:

Si $\varrho(0) > 0$ et $\varrho^{(2)}(0) > 0$, on a pour tout point α, $\varrho(\alpha) > 0$. Par conséquent, sous ces conditions, le problème est indéterminé.

Ceci est une conséquence immédiate de l'inégalité (27'). En y passant à la limite, nous obtenons

$$\frac{1}{\varrho(\alpha)} \leq e^{4\mu^2\alpha^2 + 6\mu|\alpha|}\left(\frac{1}{\varrho(0)} + \frac{\alpha^2}{\varrho^{(2)}(0)}\right), \tag{37}$$ [1]

où l'on a posé

$$\mu = \sqrt{\frac{c_0 - \varrho(0)}{\varrho^{(2)}(0)}} > 0.$$

14. Pour établir la réciproque, nous démontrerons d'abord le théorème suivant:

[1] On a aussi $\frac{1}{\rho(\alpha)} = e^{\varepsilon(r) r^2}$ où $r = |\alpha|$ et $\varepsilon(r)$ tend vers zéro avec $\frac{1}{r}$.

Si $\varrho(0) > 0$, *et qu'il existe des distributions d'ordres finis, ces ordres tendant vers l'infini, telles que les distances entre les points de concentration correspondants et l'origine aient une borne inférieure positive, alors on aura* $\varrho^{(2)}(0) > 0$. *Par conséquent, d'après ce qui précède, on aura pour toute valeur* α, $\varrho(\alpha) > 0$, (*et le problème sera indéterminé*).

Soit d un nombre positif tel que les points de concentration desdites distributions à grands indices soient de l'origine à des distances $> d$. Soient $\lambda_0, \lambda_1, \ldots \lambda_m$ les points de concentration d'une telle distribution et $\sigma_0, \sigma_1, \ldots \sigma_m$ les masses correspondantes. Alors, d'une part, d'après notre hypothèse

$$|\lambda_j| > d \qquad (j = 0, 1, 2, \ldots m),$$

d'autre part

$$c_k = \sum_{j=0}^{m} \sigma_j \lambda_j^k \qquad (k = 0, 1, 2, \ldots m).$$

En posant

$$\sum_{j=0}^{m} \sigma_j \lambda_j^{-1} = c_{-1} \quad \text{et} \quad \sum_{j=0}^{m} \sigma_j \lambda_j^{-2} = c_{-2} < c_0 d^{-2},$$

et en désignant par ε un nombre positif quelconque, écrivons la forme quadratique, définie positive

$$\varepsilon x_0^2 + \sum_{j=0}^{m} \sigma_j (\lambda_j^{-1} x_0 + x_1 + \lambda_j x_2 + \cdots + \lambda_j^m x_{m+1})^2 =$$

$$= \varepsilon x_0^2 + \sum_{i,k=0}^{m+1} c_{-2+i+k} x_i x_k = (c_{-2} + \varepsilon) x_0^2 + \cdots$$

Le minimum de cette forme, sous la condition $x_0 = 1$, est égal à ε.

En appliquant — *mutatis mutandis* — à cette forme l'inégalité (28), c.-à-d. en y remplaçant $c_0 - \varrho_n(0)$, $\varrho_n^{(2)}(0)$ et $\varrho_n^{(4)}(0)$ par $c_{-2} (= c_{-2} + \varepsilon - \varepsilon)$, $\varrho_m(0)$ et $\varrho_m^{(2)}(0)$, nous obtenons, grâce à une remarque, faite à la fin du nº 10,

$$\varrho_m^{(2)}(0) > \frac{(\varrho_m(0))^2}{c_{-2}} > \frac{(\varrho_m(0))^2}{c_0 d^{-2}} > \frac{(\varrho(0))^2 d^2}{c_0},$$

par suite aussi

$$\varrho^{(2)}(0) \geqq \frac{(\varrho(0))^2 d^2}{c_0}.$$

L'origine ne jouant aucun rôle particulier pour le problème des moments, nous pouvons généraliser ce résultat, en disant:

S'il existe un nombre réel β pour lequel $\varrho(\beta) > 0$ et qu'il existe des distributions d'ordres finis et indéfiniment croissants, telles que les distances entre les points de concentration et le point β ont une borne inférieure positive, on aura, pour tout point α, $\varrho(\alpha) > 0$.

15. Admettons maintenant que *le problème soit indéterminé.* Nous allons établir que, *sous cette hypothèse, il existe un point β qui remplit les conditions du théorème précédent.*

D'après une remarque précédente, il existe, dans le cas d'indétermination, un intervalle (a, b) et un nombre positif δ, tels que $\varrho(t) > \delta$, pour $a \leqq t \leqq b$. Choisissons, à l'intérieur de cet intervalle, un point ω quelconque. Formons la solution $\varphi_n(t)$ d'ordre $2n$ qui a un point de concentration en ω. L'ensemble des points de concentration de $\varphi_n(t)$ est fourni par les zéros de $F_n(t \mid \omega)$, c.-à-d. par les points

$$\omega_{n,-p}, \omega_{n,-p+1}, \ldots \omega_{n,-1}, \omega_{n,0} = \omega, \omega_{n,1}, \ldots \omega_{n,q-1}, \omega_{n,q}$$
$$(p + q = n).$$

Nous désignons par l_n le nombre des points de concentration $\omega_{n,r}$ $(r \gtreqless 0)$, situés dans l'intervalle (a, b). On aura d'abord, pour un tel point

$$\varrho_n(\omega_{n,r}) > \delta;$$

d'autre part, il est

$$\sum_{j=-p}^{q} \varrho_n(\omega_{n,j}) = c_0.$$

On a par conséquent

$$l_n < \frac{c_0}{\delta}$$

c.-à-d. l_n reste borné. En fixant l'indice $r \gtrless 0$, et en faisant tendre n vers l'infini, le point $\omega_{n,r}$ marchera vers le point ω (*cf.* (35) et (35′)) et tendra par suite vers un point-limite ω_r. Le nombre de ces points-limites étant aussi $< \frac{c_0}{\delta}$, sera fini. En choisissant β différent de ω et de ces points-limites ω_r, β remplira les conditions du théorème précédent. On conclut:

Dans le cas d'indétermination, on a pour un nombre α quelconque, $\varrho(\alpha) > 0$. En particulier, on a $\varrho(0) > 0$.

Pour achever la démonstration du théorème de M. HAMBURGER, il ne reste qu'à montrer:

Dans le cas d'indétermination, on a aussi $\varrho^{(2)}(0) > 0$.

Ceci résulte immédiatement du fait qu'en même temps que le problème des moments relatif à la suite $c_0, c_1, c_2, \ldots$, le problème relatif à la suite plus restreinte $c_2, c_3, c_4 \ldots$ sera indéterminé.

16. Ajoutons encore quelques remarques, concernant surtout le cas d'indétermination.

Posons, en général, pour un nombre complexe α quelconque

$$\varrho_n(\alpha) = \frac{|c_{i+k}|_0^n}{-\begin{vmatrix} 0 & \bar{\alpha}^k \\ \alpha^i & c_{i+k} \end{vmatrix}_0^n} = \frac{1}{h_n(\alpha\,|\,\bar{\alpha})}. \tag{38}$$

Sans insister sur le rôle presqu'évident que ces nombres jouent au point de vue du problème des moments, remarquons qu'il ressort p. ex. de la formule (22) que, dans tous les cas

$$0 < \varrho_n(\alpha) \leqq \varrho_{n-1}(\alpha).$$

Ces inégalités sont d'ailleurs des conséquences immédiates de la signification des nombres $\varrho_n(\alpha)$ à laquelle nous venons de faire allusion.

Nous posons aussi dans ce cas,

$$\lim_{n \to \infty} \varrho_n(\alpha) = \varrho(\alpha).$$

Evidemment, l'inégalité (37) subsiste aussi pour un α imaginaire, si l'on remplace, dans son second membre, α par $|\alpha|$.

17. Admettons maintenant que c'est le cas d'indétermination qui a lieu.

D'après les formules (38) et (22) et en vertu de l'inégalité (27), on obtient

$$\frac{1}{\varrho(\alpha)} = \frac{1}{c_0} + \sum_{p=1}^{\infty} \frac{|q_p(\alpha)|^2}{|c_{i+k}|_0^p \, |c_{i+k}|_0^{p-1}}, \tag{39}$$

la série étant absolument convergente pour toute valeur de α.

On en conclut aisément moyennant l'inégalité

$$|q_p(x)\, q_p(y)| \leq \frac{1}{2}(|q_p(x)|^2 + |q_p(y)|^2),$$

que, x et y désignant des nombres complexes quelconques, la limite

$$\lim_{n\to\infty} h_n(x\,|\,y) = h(x\,|\,y)$$

existe, et que, en vertu de (22), on a

$$h(x\,|\,y) = h(y\,|\,x) = \sum_{p=1}^{\infty} \frac{q_p(x)\, q_p(y)}{|c_{i+k}|_0^p \, |c_{i+k}|_0^{p-1}}, \tag{40}$$

la série étant absolument convergente.

Nous montrerons encore que (40) converge uniformément, lorsque x et y varient, chacun dans un domaine limité. En effet, les fonctions $h_n(x\,|\,y)$ étant, d'après (27), bornées dans leur ensemble, lorsque x et y varient de ladite manière, on conclut par la formule de CAUCHY que ces fonctions sont également continues, x et y restant dans des domaines finis. L'uniformité de la convergence résulte alors d'un théorème connu d'ARZELÀ. (Le théorème de STIELTJES-VITALI conduit évidemment au même résultat.) Par suite $h(x\,|\,y)$ est une fonction entière.

En choisissant notamment $x = y = \alpha$, et en posant

$$\frac{1}{\varrho_n^*(\alpha)} = \frac{-\begin{vmatrix} 0 & \alpha^k \\ \alpha^i & c_{i+k} \end{vmatrix}_0^n}{|c_{i+k}|_0^n} = h_n(\alpha\,|\,\alpha),$$

la fonction

$$\frac{1}{\varrho^*(\alpha)} = \lim \frac{1}{\varrho_n^*(\alpha)} = h(\alpha \mid \alpha) = \frac{1}{c_0} + \sum_{p=1}^{\infty} \frac{q_p^2(\alpha)}{|c_{i+k}|_0^p \, |c_{i+k}|_0^{p-1}}$$

est une fonction entière de α dont le genre ne pourra surpasser *deux*. Une étude approfondie de cette fonction serait de grand intérêt.

La fonction

$$h(t \mid \alpha) = \lim_{n \to \infty} h_n(t \mid \alpha)$$

où α est une constante complexe quelconque, est une fonction entière de t, de genre $\leq$ *deux*.

Si α est une constante réelle, on sait que les zéros de $h_n(t \mid \alpha)$ sont tous réels. On en conclut facilement que, dans ce cas, le genre de la fonction $h(t \mid \alpha)$ est $\leq$ *un*.[1] Il serait intéressant de savoir si, pour α imaginaire, le genre *deux* peut être atteint.

Nous n'insistons pas ici sur les conséquences multiples des faits que nous venons de développer. Nous y reviendrons dans le susdit mémoire.

---

[1] *Cf.* Hamburger, *l. c.*

24.II

Sur le problème des moments

Deuxième Note

Ark. Mat. Fys. *16*, no. 19 (1922), 1–21

1. Soit $c_0, c_1, c_2, \ldots$ une *suite définie positive.*[1] Dans la Note précédente, nous avons donné une démonstration simple du fait que, pour une telle suite, il existe toujours une ou plusieurs distributions de masses positives $d\varphi(t)$ dont les *moments* sont $c_0, c_1, c_2, \ldots$, c'est-à-dire qu'il existe une ou plusieurs fonctions jamais décroissantes $\varphi(t)$, satisfaisant aux équations en nombre infini

$$(1) \qquad c_n = \int_{-\infty}^{+\infty} t^n d\varphi(t), \qquad n = 0, 1, 2, \ldots$$

Dans cette première Note, nous avons surtout insisté sur quelques *inégalités* ayant lieu entre certaines quantités rattachées à la suite (1). Chemin faisant, nous avons aussi retrouvé un critère important dû à M. Hamburger pour que le problème soit *déterminé* ou *indéterminé.*

---

[1] Pour les notations voir les n°s 1—3 de la Note précédente: *Sur le problème des moments. Première Note.* Ce recueil, t. 16, n° 12, p. 1—23. Dans cette deuxième Note, nous ne supposons pas que le lecteur connaisse la première Note à fond. Nous renvoyons dans la suite aux endroits dont nous avons besoin, en citant cette Note par la lettre I.

Nous allons reprendre ici l'étude de cette question de détermination ou d'indétermination. Tandis que dans la première Note, notre but était d'arriver aux résultats principaux en restant dans la voie algébrique et dans le domaine réel, sans rien emprunter à la théorie des fonctions analytiques, la Note présente se rattache directement à une méthode basée sur cette théorie, méthode due à M. MARKOFF et dont M. HAMBURGER a su tirer le meilleur parti. Nous commençons par reproduire une formule due à M. MARKOFF et le résultat qu'en a tiré M. HAMBURGER. Après, nous approfondissons l'étude d'une classe de polynomes introduits par M. HAMBURGER, et en y appliquant la formule de M. MARKOFF, nous arrivons, par une voie facile, à un second critère, également dû à M. HAMBURGER, pour qu'il y ait détermination ou indétermination. Enfin, aux n°s 9—12, nous démontrons d'une manière directe l'équivalence des deux critères et nous y arrivons à des inégalités très précises.

2. On peut évidemment se restreindre à des solutions $\varphi(t)$ du problème des moments qui tendent vers zéro lorsque $t \to -\infty$. Nous désignerons, avec STIELTJES, ces solutions comme *identiques* si elles coïncident en tous leurs points de continuité. Si tel est le cas pour toutes les solutions, le problème sera dit *déterminé*, dans le cas contraire, il sera dit *indéterminé.*

$\varphi(t)$ étant une solution du problème des moments, l'intégrale

$$F(z) = \int_{-\infty}^{+\infty} \frac{d\varphi(t)}{z-t}$$

existe pour tout nombre imaginaire z et définit dans les demi-plans supérieur et inférieur deux fonctions analytiques qui, l'une et l'autre, admettent, en général, l'axe réel comme coupure. Il suffit de se restreindre au demi-plan supérieur, par exemple. La fonction $\varphi(t)$ pouvant être calculée à l'aide de la fonction $F(z)$,[1] on voit que deux solutions du problème des moments qui ne sont pas *identiques* dans le sens que nous avons attribué à ce mot, donneront lieu

[1] *Voir* p. ex. PERRON, *Die Lehre von den Kettenbrüchen*, 1913, p. 372.

à des fonctions $F(z)$ différentes. On en tire le théorème connu que voici.

Pour que le problème des moments relatif à la suite $c_0, c_1, \ldots$ soit déterminé, il faut et il suffit qu'en formant l'intégrale

$$F(iy) = \int_{-\infty}^{+\infty} \frac{d\varphi(t)}{iy - t}$$

avec toutes les solutions possibles $\varphi(t)$ du problème des moments, les fonctions $F(iy)$ coïncident et cela pour toute valeur positive de y.[1]

3. Désignons maintenant par $\varphi_1(t)$ et $\varphi_2(t)$ deux solutions quelconques du problème des moments et déduisons la formule de M. Markoff.

Soit $A(t)$ un polynome quelconque. On aura, en supposant $A(z)$ différent de zéro

$$\frac{1}{z-t} = \frac{A(z)-A(t)}{A(z)(z-t)} + \frac{A(t)}{A(z)(z-t)} = P(z, t) + \frac{A(t)}{z-t}$$

où $P(z, t)$ est un polynome de t. On aura donc

$$\int_{-\infty}^{+\infty} \frac{d\varphi_2(t)}{z-t} - \int_{-\infty}^{+\infty} \frac{d\varphi_1(t)}{z-t} =$$

$$= \frac{1}{A(z)} \left(\int_{-\infty}^{+\infty} \frac{A(t)\, d\varphi_2(t)}{z-t} - \int_{-\infty}^{+\infty} \frac{A(t)\, d\varphi_1(t)}{z-t} \right).$$

C'est la formule de M. Markoff.[2]

---

[1] Il suffit évidemment de supposer explicitement que cela ait lieu pour certaines valeurs de $y > 0$. Nous n'y insistons pas.

[2] Deux démonstrations de la convergence de certaines fractions continues, Acta Math. t. 19 (1895). Nous énonçons ici cette identité sous une forme plus particulière que ne le fait M. Markoff. Si l'on suppose seulement que $\varphi_1(t)$ et $\varphi_2(t)$ satisfont aux équations (1) pour $n = 0, 1, \ldots 2m$, l'identité subsiste à condition que le polynome $A(t)$ soit de degré $\leq 2m + 1$. On arrive ainsi par la méthode que nous allons développer, aux critères de convergence d'une certaine fraction continue associée à la suite des c_j. Nous y reviendrons dans un autre travail.

Posons maintenant $z = x + iy$ et $A(t) = B^2(t)$, en admettant que $B(t)$ ait des coefficients réels. Nous obtenons

$$(2) \qquad \left|\int_{-\infty}^{+\infty} \frac{d\varphi_2(t)}{x+iy-t} - \int_{-\infty}^{+\infty} \frac{d\varphi_1(t)}{x+iy-t}\right| <$$

$$< \frac{1}{|yB^2(x+iy)|}\left(\int_{-\infty}^{+\infty} B^2(t)\,d\varphi_2(t) + \int_{-\infty}^{+\infty} B^2(t)\,d\varphi_1(t)\right) =$$

$$= \frac{2B^2(c)}{|yB^2(x+iy)|}$$

et en particulier

$$(2') \qquad \left|\int_{-\infty}^{+\infty} \frac{d\varphi_2(t)}{iy-t} - \int_{-\infty}^{+\infty} \frac{d\varphi_1(t)}{iy-t}\right| < \frac{2B^2(c)}{|yB^2(iy)|}.$$ [1]

Si donc pour toute valeur positive de y il existe des polynomes $B(t)$ tels que l'on puisse rendre (2') aussi petit qu'on voudra, le premier membre de cette inégalité sera égal à zéro pour deux solutions quelconques du problème des moments,

---

[1] Observons que l'on démontre de la même manière pour un polynome $B(t)$ à coefficients complexes

$$\left|\int_{-\infty}^{+\infty} \frac{d\varphi_2(t)}{x+iy-t} - \int_{-\infty}^{+\infty} \frac{d\varphi_1(t)}{x+iy-t}\right| < \frac{2B(c)\bar{B}(c)}{|yB^2(x+iy)|}.$$

En posant en particulier (*cf.* I nos 4 et 16)

$$B(t) = \frac{\begin{vmatrix} 0 & \bar{\alpha}^k \\ t^i & c_{i+k} \end{vmatrix}_0^n}{\begin{vmatrix} 0 & \bar{\alpha}^k \\ \alpha^i & c_{i+k} \end{vmatrix}_0^n}$$

et en écrivant $x + iy = \alpha$, on obtient

$$\left|\int_{-\infty}^{+\infty} \frac{d\varphi_2(t)}{\alpha-t} - \int_{-\infty}^{+\infty} \frac{d\varphi_1(t)}{\alpha-t}\right| < \frac{2\rho_n(\alpha)}{|y|},$$

inégalité bien remarquable à laquelle nous reviendrons à une autre occasion en y supprimant le facteur 2.

c'est-à-dire que d'après ce que nous venons de dire au numéro précédent, le problème des moments sera déterminé.

4. Voici maintenant le critère de M. Hamburger[1] sous une forme légèrement généralisée:

Pour que le problème des moments relatif à la suite $c_0, c_1, \ldots$ *soit déterminé, il faut et il suffit que l'une au moins des deux valeurs positives*

$$\frac{1}{\varrho_n} = \frac{\begin{vmatrix} c_2 & \ldots & c_{n+1} \\ \cdot & \cdot & \cdot \\ c_{n+1} & \ldots & c_{2n} \end{vmatrix}}{\begin{vmatrix} c_0 & \ldots & c_n \\ \cdot & \cdot & \cdot \\ c_n & \ldots & c_{2n} \end{vmatrix}} = \frac{|c_{i+k}|_1^n}{|c_{i+k}|_0^n} \tag{3}$$

et

$$\Delta_n = \frac{-\begin{vmatrix} 0 & c_{-1} & c_0 & \ldots & c_{n-1} \\ c_{-1} & c_0 & c_1 & \ldots & c_n \\ c_0 & c_1 & c_2 & \ldots & c_{n+1} \\ \cdot & \cdot & \cdot & \cdot & \cdot \\ c_{n-1} & c_n & c_{n+1} & \cdot & c_{2n} \end{vmatrix}}{\begin{vmatrix} c_0 & \ldots & c_n \\ \cdot & \cdot & \cdot \\ c_n & \ldots & c_{2n} \end{vmatrix}} = \frac{-\begin{vmatrix} 0 & c_{k-1} \\ c_{i-1} & c_{i+k} \end{vmatrix}_0^n}{|c_{i+k}|_0^n} \tag{4}$$

où c_{-1} *désigne un nombre réel fixe d'ailleurs quelconque,*[2] *tende vers l'infini avec* n.

5. D'abord, nous reproduirons la démonstration de M. Hamburger[3] relative au cas que la valeur (3) tend vers l'infini avec n. Nous admettrons donc

[1] Math. Ann. t. 81 et 82.
[2] M. Hamburger, pose $c_{-1} = 0$.
[3] Math. Ann. t. 81, p. 295.

$$\frac{1}{\varrho_n} = \frac{|c_{i+k}|_1^n}{|c_{i+k}|_0^n} \to \infty.$$ [1]

Posons maintenant avec M. Hamburger dans l'inégalité (2′)

$$B(t) = \frac{1}{|c_{i+k}|_0^n} \begin{vmatrix} 1 & t & \dots & t^n \\ c_1 & c_2 & \dots & c_{n+1} \\ \dots & \dots & \dots & \dots \\ c_n & c_{n+1} & \dots & c_{2n} \end{vmatrix}$$

On a

$$B(0) = \frac{|c_{i+k}|_1^n}{|c_{i+k}|_0^n} = \frac{1}{\varrho_n}.$$

D'autre part, il est

$$c^k B(c) = 0, \quad \text{pour } k = 1, 2, \dots n.$$

Il en résulte que $B(t)$ a des zéros réels[2] et que

$$B^2(c) = \frac{|c_{i+k}|_1^n}{|c_{i+k}|_0^n} B(c) = \frac{|c_{i+k}|_1^n}{|c_{i+k}|_0^n} = \frac{1}{\varrho_n}.$$

De la réalité des zéros de $B(t)$, on conclut

$$|B(iy)| > B(0) = \frac{1}{\varrho_n},$$

c.-à-d.

$$\frac{2\,B^2(c)}{|yB^2(iy)|} < \frac{2\varrho_n}{|y|} \to 0,$$

et, par conséquent, le problème sera déterminé.

6. Démontrons maintenant par la même méthode que l'hypothèse

---

[1] Pour la signification de ρ_n *cf.* les nos 4—5 de I, où nous désignons cette valeur par $\rho_n(0)$.

[2] *Cf.* p. 8 de la Note présente.

$$(5) \qquad \varDelta_n = \frac{-\begin{vmatrix} 0 & c_{k-1} \\ c_{i-1} & c_{i+k} \end{vmatrix}_0^n}{|c_{i+k}|_0^n} \to \infty$$

donne également lieu à un cas déterminé.[1]

Nous considérons à cet effet les polynomes

$$(6) \qquad q(t) = \frac{1}{|c_{i+k}|_0^n} \begin{vmatrix} 1 & t & \dots & t^{n+1} \\ c_{-1} & c_0 & \dots & c_n \\ c_0 & c_1 & \dots & c_{n+1} \\ \cdot & \cdot & \cdot & \cdot \\ c_{n-1} & c_n & \dots & c_{2n} \end{vmatrix}$$

qui, sous une forme un peu plus particulière,[2] furent introduits par M. HAMBURGER.

On a

$$(7) \qquad c^k q(c) = 0, \quad \text{pour } k = 0, 1, 2, \dots n-1.$$

On en conclut facilement que $q(t)$ a tous ses *zéros réels et distincts*. Il suffit pour cela de montrer que $q(t)$ a du moins n zéros réels impairs. Or, dans le cas contraire, on pourrait former un polynome $l(t)$ de degré $\leq n-1$ qui a toujours le même signe que $q(t)$. On aurait alors d'une part

$$q(t)\, l(t) \geqq 0,$$

c'est-à-dire (*cf.* I n° 3)

$$q(c)\, l(c) > 0,$$

[1] M. HAMBURGER ne donne pas une démonstration *directe* de ce fait. Chez lui, il découle d'un assez grand nombre de théorèmes et de théorèmes auxiliaires qui, tout en formant une construction bien solide, s'entrelacent et s'encombrent à un tel point que la pensée directrice se laisse à peine entrevoir. Notre critique s'adresse surtout aux théorèmes auxiliaires sur ces réduites P_{2n}/Q_{2n} qui, étant un outil puissant et naturel dans les cas traités par M. MARKOFF et par STIELTJES, se comportent d'une façon trop irrégulière dans le cas général où se place M. HAMBURGER, ce qui n'a pas empêché M. HAMBURGER d'arriver à des résultats d'une importance éclatante.

[2] M. HAMBURGER pose $c_{-1} = 0$.

d'autre part, d'après les égalités (7)

$$q(c)\, l(c) = 0.$$

Si donc $q(t)$ est de degré $n+1$, et c'est là le cas général, $q(t)$ aura $n+1$ zéros réels et distincts. Si l'on se trouve dans le cas exceptionnel consistant en ce que $q(t)$ est de degré n, c'est-à-dire

$$|c_{-1+i+k}|_0^n = 0, \tag{8}$$ [1]

$q(t)$ aura n zéros réels et distincts.

Admettons qu'on se trouve dans le cas général et désignons les zéros de $q(t)$ par $\lambda_1, \lambda_2, \ldots \lambda_{n+1}$. Un polynome quelconque $G(t)$ de degré $\leqq 2n$ pourra alors s'écrire

$$G(t) = q(t)\,\alpha(t) + \sum_{j=1}^{n+1} G(\lambda_j)\, q_j(t)$$

où $\alpha(t)$ est de degré $\leqq n-1$ et où l'on a posé

$$q_j(t) = \frac{q(t)}{q'(\lambda_j)\,(t-\lambda_j)}.$$

---

[1] Remarquons que c'est bien là un cas exceptionnel aussi sous le rapport suivant. Si l'égalité (8) a lieu pour une certaine valeur de c_{-1}, elle n'aura lieu pour aucune autre valeur de c_{-1}. En effet, (8) entraîne

$$|c_{1+i+k}|_0^{n-1} \neq 0.$$

Autrement, on arriverait par un théorème bien connu de Jacobi à l'égalité manifestement absurde

$$|c_{i+k}|_0^{n-1} = 0.$$

Ajoutons encore qu'il est assez facile de vérifier que, l'égalité (8) ayant lieu, $q(t)$ peut s'écrire sous la forme réduite

$$q(t) = \frac{1}{|c_{i+k}|_0^{n-1}} \begin{vmatrix} 1 & t & \ldots\ldots & t^n \\ c_{-1} & c_0 & \ldots\ldots & c_{n-1} \\ c_0 & c_1 & \ldots\ldots & c_n \\ \cdot & \cdot & \cdot\;\cdot\;\cdot & \cdot \\ c_{n-2} & c_{n-1} & \ldots & c_{2n-2} \end{vmatrix}.$$

Il vient en tenant compte de (7)

$$G(c)=\sum_{j=1}^{n+1}\sigma_j G(\lambda_j) \tag{9}$$

où

$$\sigma_j = q_j(c).$$

Appliquons en particulier (9) à $q_j^2(t)$. On a $q_j(\lambda_j)=1$ et $q_j(\lambda_r)=0$ pour $r\neq j$. On obtiendra donc

$$q_j^2(c)=q_j(c).$$

Il en résulte (I n° 3) que tous les σ_j sont *positifs*.

Introduisons maintenant dans la formule (9) le polynome

$$R(t)=\frac{q(t)-1}{t}=\frac{1}{|c_{i+k}|_0^n}\begin{vmatrix}0 & 1 & \dots & t^n\\ c_{-1} & c_0 & \dots & c_n\\ \cdot & \cdot & \cdot & \cdot\\ c_{n-1} & c_n & \dots & c_{2n}\end{vmatrix}. \tag{10}$$

Nous avons d'une part, $q(t)$ étant égal à zéro pour $t=\lambda_1,\dots\lambda_{n+1}$,

$$R(c)=-\sum_{j=1}^{n+1}\sigma_j\lambda_j^{-1},$$

et d'autre part

$$R(c)=\frac{1}{|c_{i+k}|_0^n}\begin{vmatrix}0 & c_0\; c_1 & \dots & c_n\\ c_{-1} & c_0\; c_1 & \dots & c_n\\ \cdot & \cdot & \cdot & \cdot\\ c_{n-1} & c_n\; c_{n+1} & \dots & c_{2n}\end{vmatrix}=-c_{-1}.$$

Il vient

$$\sum_{j=1}^{n+1}\sigma_j\lambda_j^{-1}=c_{-1}. \tag{11}$$

Posons maintenant, pour abréger,

$$\sum_{j=1}^{n+1} \sigma_j \lambda_j^{-2} = c_{-2} \tag{12}$$

et calculons cette valeur.

Les formules (9), (11) et (12) mettent en évidence que, pour tout polynome $G(t)$ de degré $\leqq 2n+2$, on a

$$c^{-2}G(c) = \sum_{j=1}^{n+1} \sigma_j \lambda_j^{-2} G(\lambda_j).$$

En appliquant ceci à $q(t)$ et en observant que $q(t) = 0$ pour $t = \lambda_1, \ldots \lambda_{n+1}$, on obtient $c^{-2}q(c) = 0$, ou d'après la formule (6)

$$| c_{-2+i+k} |_0^{n+1} = 0$$

ce qui donne, en tenant compte de la notation (4),

$$c_{-2} = \frac{-\begin{vmatrix} 0 & c_{k-1} \\ c_{i-1} & c_{i+k} \end{vmatrix}_0^n}{| c_{i+k} |_0^n} = \Delta_n. \tag{12'}$$

On a aussi moyennant (9)

$$R^2(c) = \sum_{j=1}^{n+1} \sigma_j \lambda_j^{-2} = c_{-2} = \Delta_n. \tag{12''}$$

On voit sans peine que les formules (11)—(12″) subsistent dans le cas exceptionnel mentionné.

Écrivons

$$\sum_{j=1}^{n+1} \frac{\sigma_j}{\lambda_j - z} = \frac{p(z)}{q(z)} \tag{13}$$

où $p(z)$ sera un polynome de degré n. Les σ_j étant positifs, on voit immédiatement que $p(z)$ a tous ses zéros réels et distincts et, de plus, que ces zéros séparent ceux de $q(z)$. Les formules (6), (11) et (13) donnent

$$q(0) = 1 \tag{14}$$

et

$$(14') \qquad p(0) = \sum_{j=1}^{n+1} \sigma_j \lambda_j^{-1} = c_{-1}$$

et en dérivant (13)

$$(15) \qquad p'(0) - c_{-1} q'(0) = c_{-2}.$$

7. Démontrons maintenant l'inégalité importante qui suit.

y désignant un nombre réel quelconque, il est

$$(16) \qquad |q(iy)| > \frac{y^2 c_{-2}}{|y c_{-1}| + c_0}.$$

On a d'abord d'après la formule (13)

$$(17) \qquad \left|\frac{p(iy)}{q(iy)}\right| < \frac{1}{|y|} \sum_{j=1}^{n+1} \sigma_j = \frac{c_0}{|y|},$$

c'est-à-dire

$$(17') \qquad |q(iy)| > \frac{|y|}{c_0} |p(iy)|.$$

Pour mieux faire ressortir l'idée de notre démonstration, supposons d'abord $c_{-1} = 0$. Nous aurons alors

$$p(0) = 0$$

et

$$p'(0) = c_{-2}.$$

$p(z)$ ayant tous ses zéros réels, on en conclut

$$|p(iy)| > |y p'(0)| = |y| c_{-2}$$

et alors, d'après (17'), on arrive à

$$|q(iy)| > \frac{y^2 c_{-2}}{c_0}.$$

Pour passer au cas général, posons

$$h(z) = p(z) - c_{-1}(q)z.$$

On aura d'après (14) et (14')

$$h(0) = 0$$

et d'après (15)

$$h'(0) = c_{-2}.$$

Les zéros de $p(z)$ et $q(z)$ se séparant mutuellement, $h(z)$ aura tous ses zéros réels. On aura donc comme plus haut

$$|h(iy)| > |y| c_{-2}$$

et à plus forte raison

$$|c_{-1}q(iy)| + |p(iy)| > |y| c_{-2},$$

et alors, d'après (17),

$$|q(iy)| \left(|c_{-1}| + \frac{c_0}{|y|}\right) > |y| c_{-2},$$

ce qui achève la démonstration de (16).[1]

---

[1] Notre résultat peut aussi s'énoncer de la manière suivante:

Soient $\theta_1, \theta_2, \ldots \theta_n$ des nombres réels et $\rho_1, \rho_2, \ldots \rho_n$ des masses positives correspondantes. Posons

$$\sum_1^n \rho_j = c_0, \quad \sum_1^n \rho_j \theta_j = c_1, \quad \sum_1^n \rho_j \theta_j^2 = c_2.$$

Alors, on aura l'inégalité

$$\prod_1^n (1 + \theta_j^2) > \left(\frac{c_2}{c_0 + |c_1|}\right)^2.$$

Voici encore une forme du même résultat:

$a_1, a_2, \ldots a_n$ et $b_1, b_2, \ldots b_n$ désignant des nombres réels quelconques $(b_j \pm 0)$, on aura l'inégalité

$$\prod_1^n \left(1 + \frac{a_j^2}{b_j^2}\right) > \left(\frac{\sum_1^n a_j^2}{\left|\sum_1^n a_j b_j\right| + \sum_1^n b_j^2}\right)^2.$$

8. Posons maintenant dans l'inégalité (2') au lieu de $B(t)$ le polynome $R(t)$ défini par la formule (10). On aura alors d'après la formule (12'') et l'inégalité (16)

$$\left|\int_{-\infty}^{+\infty}\frac{d\varphi_2(t)}{iy-t}-\int_{-\infty}^{+\infty}\frac{d\varphi_1(t)}{iy-t}\right|<\frac{2\Delta_n}{y\left(\frac{y^2\Delta_n}{|yc_{-1}|+c_0}-1\right)^2}.$$

Si Δ_n tend vers l'infini avec n, le second membre tendra en même temps vers zéro, c'est-à-dire, le premier membre sera identiquement égal à zéro, et alors, d'après ce que nous avons dit plus haut, le problème des moments sera déterminé.

Ainsi, nous avons démontré que si l'une au moins des valeurs (3) et (4) tend vers l'infini avec n, on aura un problème déterminé. Ajoutons que les valeurs en question sont toujours croissantes (ou du moins non décroissantes). C'est ce qu'on voit ou de leur signification, chacune étant l'inverse d'un certain minimum, ou encore en évaluant $\varrho_{n-1}-\varrho_n$ ou $\Delta_n-\Delta_{n-1}$ par ledit théorème de JACOBI (*cf.* le n° 11). Il existera donc toujours des limites. Par suite, au lieu de supposer que l'une des valeurs en question tende vers l'infini, on aurait pu supposer qu'elle ne reste pas bornée.

Remarquons encore que la première condition peut, en certaine mesure, être considérée comme cas particulier de la seconde, correspondant au cas $c_{-1}=\infty$. En fixant n, on a en effet

$$\lim_{c_{-1}\to\infty}\frac{\Delta_n}{(c_{-1})^2}=\frac{1}{\varrho_n}.$$

9. Voici maintenant la seconde partie du critère de M. HAMBURGER, c'est-à-dire la réciproque de ce qu'on vient d'établir.

Si, n tendant vers l'infini, la valeur (3) reste bornée, et il existe un nombre réel c_{-1} tel que la valeur (4) reste aussi bornée, le problème des moments relatif à la suite $c_0, c_1, \ldots$ sera indéterminé.

Le lecteur pourra trouver une démonstration complète chez M. HAMBURGER (Math. Ann. t. 81 et 82) ou il pourra consulter les n°s 11—12 de notre première Note. Nous indiquerons la démonstration en quelques mots.

Admettons que $\varrho_n \to \varrho > 0$ et que $\Delta_n \to \Delta$ où Δ désigne un nombre fini. On pourra alors à l'aide du principe de choix de M. HELLY[1] construire deux solutions $\varphi(t)$ et $\psi(t)$ du problème, dont la première présentera la discontinuité ϱ à l'origine, tandis que la seconde satisfera à l'inégalité

$$0 \leqq \psi(\varepsilon) - \psi(-\varepsilon) < \varepsilon^2 \Delta$$

et, par conséquent, sera continue à l'origine. Ces solutions seront donc bien distinctes et le problème sera indéterminé. La première solution $\varphi(t)$ se construit moyennant certaines fonctions $\varphi_1(t), \varphi_2(t), \dots \varphi_n(t), \dots$ attachées à des distributions d'ordre fini qui ont un point de concentration à l'origine et y présentent les masses $\varrho_1, \varrho_2, \dots \varrho_n, \dots$ (distributions traitées à fond dans notre première Note) tandis que la seconde solution se construit moyennant des distributions d'ordre fini $\psi_n(t)$ dont les points de concentration sont les points $\lambda_1, \dots \lambda_{n+1}$ du n° 6. Les fonctions $\psi_n(t)$ satisfont à l'inégalité

$$0 \leqq \psi_n(\varepsilon) - \psi_n(-\varepsilon) < \varepsilon^2 \Delta_n.$$

Nous pourrons laisser aux soins du lecteur la démonstration de la généralisation suivante:

S'il existe deux nombres différents c_{-1} et c'_{-1} tels que les valeurs

$$\frac{1}{|c_{i+k}|_0^n} \begin{vmatrix} 0 & c_{-1} & c_0 & \dots & c_{n-1} \\ c_{-1} & c_0 & \dots & \dots & c_n \\ c_0 & \dots & \dots & \dots & \dots \\ \dots & \dots & \dots & \dots & \dots \\ c_{n-1} & c_n & \dots & \dots & c_{2n} \end{vmatrix} \quad et \quad \frac{1}{|c_{i+k}|_0^n} \begin{vmatrix} 0 & c'_{-1} & c_0 & \dots & c_{n-1} \\ c'_{-1} & c_0 & \dots & \dots & c_n \\ c_0 & \dots & \dots & \dots & \dots \\ \dots & \dots & \dots & \dots & \dots \\ c_{n-1} & c_n & \dots & \dots & c_{2n} \end{vmatrix}$$

restent bornées, le problème sera indéterminé.

10. M. HAMBURGER a énoncé son critère aussi sous la forme suivante dont nous nous sommes occupé dans notre première Note:

[1] *Cf.* I n° 12. En réalité, on n'applique le principe de choix que pour des raisons de commodité. Il est assez facile de démontrer que les fonctions $\varphi_n(t)$ et $\psi_n(t)$ tendent respectivement vers $\varphi(t)$ et $\psi(t)$ lorsque n croît indéfiniment.

La condition nécessaire et suffisante pour que le problème des moments soit indéterminé est que l'une et l'autre des deux valeurs

$$(17'') \qquad \frac{1}{\varrho_n} = \frac{|c_{i+k}|_1^n}{|c_{i+k}|_0^n} \quad et \quad \frac{1}{\varrho_n^{(2)}} = \frac{|c_{i+k}|_2^n}{|c_{i+k}|_1^n}$$

restent bornées lorsque n tend vers l'infini.

En comparant ceci aux résultats qu'on vient d'établir, on trouve:

En désignant par c_{-1} un nombre fini fixe, l'hypothèse que les valeurs $\frac{1}{\varrho_n}$ et Δ_n restent bornées est équivalente à la même hypothèse concernant les valeurs $\frac{1}{\varrho_n}$ et $\frac{1}{\varrho_n^{(2)}}$.

Nous nous proposons de démontrer cela d'une façon directe en établissant les inégalités suivantes entre les nombres ϱ_n, $\varrho_n^{(2)}$ et Δ_n:

$$(18) \qquad \frac{1}{\varrho_n^{(2)}} < \frac{\Delta_n}{\varrho_n^2}$$

et

$$(19) \qquad \Delta_n < \frac{1}{\varrho_n}\left(\frac{2c_0^{\frac{3}{2}}}{\sqrt{\varrho_n^{(2)}}} + |c_{-1}|\right)^2$$

et, en particulier, en posant $\Delta_n = \overset{0}{\Delta}_n$, si $c_{-1} = 0$,

$$(19') \qquad \overset{0}{\Delta}_n < \frac{4c_0^3}{\varrho_n \varrho_n^{(2)}}.$$

L'inégalité (18) est une conséquence immédiate de l'inégalité (28) de notre première Note (*voir* aussi p. 19 de la même Note). Nous en donnons ici une seconde démonstration.

11. Considérons le polynome

$$(20)\qquad S(t)=\frac{1}{|c_{i+k}|_1^n}\begin{vmatrix} 1 & t^2 & \dots & t_{n+1} \\ c_0 & c_2 & \dots & c_{n+1} \\ \cdot & \cdot & \cdot & \cdot \\ c_{n-1} & c_{n+1} & \dots & c_{2n} \end{vmatrix}.$$

On a

$$c^k S(c)=0,\quad \text{pour}\quad k=0,1,\dots n-1.$$

On pourra donc, d'après ce que nous avons établi plus haut (n° 6), trouver $n+1$[1] nombres réels et distincts $\lambda_1,\dots\lambda_{n+1}$, savoir les zéros de $S(t)$, et des nombres positifs correspondants $\sigma_1,\dots\sigma_{n+1}$ tels que pour tout polynome $G(t)$ de degré $\leqq 2n$, on ait

$$G(c)=\sum_{j=1}^{n+1}\sigma_j G(\lambda_j).$$

Posons en particulier

$$G(t)=\frac{1-S(t)}{t^2}.$$

On aura

$$(21)\qquad \sum_{j=1}^{n+1}\sigma_j\lambda_j^{-2}=\frac{-\begin{vmatrix} 0 & c_0 & \dots & c_{n-1} \\ c_0 & c_2 & \dots & c_{n+1} \\ \cdot & \cdot & \cdot & \cdot \\ c_{n-1} & c_{n+1} & \dots & c_{2n} \end{vmatrix}}{|c_{i+k}|_1^n}=D_n.$$

En tenant compte de l'égalité $S(0)=1$ et du fait que $S(t)$ manque de terme linéaire, on a évidemment

$$(22)\qquad \sum_{j=1}^{n+1}\lambda_j^{-2}=\frac{2}{|c_{i+k}|_1^n}\begin{vmatrix} c_0 & c_3 & c_4 & \dots & c_{n+1} \\ c_1 & c_4 & c_5 & \dots & c_{n+2} \\ c_2 & c_5 & c_6 & \dots & c_{n+3} \\ \cdot & \cdot & \cdot & \cdot & \cdot \\ c_{n-1} & c_{n+2} & c_{n+3} & \dots & c_{2n} \end{vmatrix}=2C_n,$$

[1] Dans le cas exceptionnel où $S(t)$ est de degré n, il n'y a presque rien à changer aux considérations qui suivent.

on a donc en particulier

$$\begin{vmatrix} c_0 & c_3 & \ldots & c_{n+1} \\ c_1 & c_4 & \ldots & c_{n+2} \\ \cdot & \cdot & \cdot & \cdot \\ c_{n-1} & c_{n+2} & \ldots & c_{2n} \end{vmatrix} > 0.\text{[1]} \tag{23}$$

Il est manifeste que

$$\sum_{j=1}^{n+1} \sigma_j \lambda_j^{-2} < \sum_{j=1}^{n+1} \sigma_j \cdot \sum_{j=1}^{n+1} \lambda_j^{-2} = c_0 \sum_{j=1}^{n+1} \lambda_j^{-2},$$

c'est-à-dire que, en vertu de (21) et (22),

$$D_n < 2 c_0 C_n. \tag{24}$$

D'autre part, en appliquant ledit théorème de Jacobi au déterminant qui figure au numérateur de D_n et en tenant compte de l'inégalité générale (29) qu'on va démontrer tout à l'heure, on obtient

$$C_n^2 < D_n \frac{|c_{i+k}|_2^n}{|c_{i+k}|_1^n} = \frac{D_n}{\varrho_n^{(2)}}. \tag{25}$$

Observons maintenant que tous nos résultats et en particulier l'inégalité (23) reposent sur la seule hypothèse que la forme quadratique $\sum_{i,k=0}^{n} c_{i+k} x_i x_k$ est définie positive. Le nombre ϱ_n étant le minimum de cette forme sous la condition $x_0 = 1$ (*cf.* I n°s 4—5), la forme $(-\varrho_n + \varepsilon) x_n^2 + \sum_{i,k=0}^{n} c_{i+k} x_i x_k$ sera aussi définie positive pour toute valeur positive de ε. L'inégalité (23) subsiste donc, si l'on y remplace c_0 par $c_0 - \varrho_n + \varepsilon$, et l'on trouve par un passage à la limite qu'elle subsiste, du moins avec le signe $\geqq$, si l'on y remplace c_0 par $c_0 - \varrho_n$.[2]

[1] *Cf.* la Note que nous allons publier prochainement: Sur le signe de certains déterminants asymétriques.

[2] En réalité, le signe d'égalité peut être exclu dans cette formule et dans toutes les autres qui en dérivent. Nous n'y insistons pas.

Mais cela est équivalent à l'inégalité

$$\begin{vmatrix} c_0 & c_3 & \dots & c_{n+1} \\ c_1 & c_4 & \dots & c_{n+2} \\ \cdot & \cdot & \cdot & \cdot \\ c_{n-1} & c_{n+2} & \dots & c_{2n} \end{vmatrix} \geqq \varrho_n \, | c_{i+k} |_2^n$$

ou d'après (17″) et (22), à l'inégalité

$$\frac{\varrho_n}{\varrho_n^{(2)}} < C_n. \tag{26}$$

Dès lors, la combinaison de (24) et (25) donne

$$D_n^2 < 4 c_0^2 \frac{D_n}{\varrho_n^{(2)}},$$

c'est-à-dire

$$D_n < \frac{4 c_0^2}{\varrho_n^{(2)}}. \tag{28}$$

En combinant d'autre part (25) et (26), il vient

$$\left(\frac{\varrho_n}{\varrho_n^{(2)}}\right)^2 < \frac{D_n}{\varrho_n^{(2)}}$$

ou

$$\frac{1}{\varrho_n^{(2)}} < \frac{D_n}{\varrho_n^2}. \tag{28}$$ [1]

---

[1] Les inégalités (27) et (28) mettent en évidence que les valeurs $\frac{1}{\rho_n}$ et D_n restant bornées, il en sera de même pour les valeurs $\frac{1}{\rho_n}$ et $\frac{1}{\rho_n^{(2)}}$ et inversement. On obtient ainsi un nouveau critère pour que le problème soit déterminé.

Ajoutons un mot sur C_n. En éliminant D_n de (24) et (25), on obtient aussi

$$C_n < \frac{2 c_0}{\rho_n^{(2)}}$$

ou

12. Pour en arriver aux inégalités (18), (19) et (19'), rappelons les deux théorèmes connus que voici:

La forme $\sum_{i,k=1}^{n} a_{ik}x_i x_k$ *étant définie positive et* $x_1, x_2, \ldots x_n$ *ainsi que* $y_1, y_2, \ldots y_n$ *désignant des nombres réels quelconques et, en outre, les* x_j *n'étant pas tous nuls, on a*

$$\left|\begin{matrix} 0 & x_k \\ x_i & a_{ik} \end{matrix}\right|_1^n < 0 \qquad \text{pour } n \geqq 1 \tag{29}$$

et

$$\left|\begin{matrix} 0 & 0 & x_k \\ 0 & 0 & y_k \\ x_i & y_i & a_{ik} \end{matrix}\right|_1^n \geqq 0 \qquad \text{pour } n \geqq 2. \tag{30}$$

L'inégalité (29) exprime le fait évident que la forme réciproque d'une forme quadratique définie positive a le même caractère. Voici d'ailleurs une démonstration directe qui donne un complément important à ce que nous venons d'énoncer.

En évaluant par le théorème de Jacobi la différence

$$\frac{\left|\begin{matrix} 0 & x_k \\ x_i & a_{ik} \end{matrix}\right|_1^m}{|a_{ik}|_1^m} - \frac{\left|\begin{matrix} 0 & x_k \\ x_i & a_{ik} \end{matrix}\right|_1^{m-1}}{|a_{ik}|_0^{m-1}},$$

on voit que l'expression

$$\left|\begin{matrix} c_0 & c_3 & \ldots & c_{n+1} \\ c_1 & c_4 & \ldots & c_{n+2} \\ \cdot & \cdot & \cdot & \cdot \\ c_{n-1} & c_{n+2} & \ldots & c_{2n} \end{matrix}\right| < 2c_0 \, |c_{i+k}|_2^n$$

ou enfin

$$\left|\begin{matrix} -c_0 & c_3 & \ldots & c_{n+1} \\ c_1 & c_4 & \ldots & c_{n+2} \\ \cdot & \cdot & \cdot & \cdot \\ c_{n-1} & c_{n+2} & \ldots & c_{2n} \end{matrix}\right| < 0$$

complément intéressant de (23). *Cf.* notre Note précitée, Sur le signe de certains déterminants asymétriques.

$$(31)\qquad \frac{-\begin{vmatrix} 0 & x_k \\ x_i & a_{ik} \end{vmatrix}_1^m}{|a_{ik}|_1^m}$$

n'est pas décroissante. En s'arrangeant maintenant de façon que x_1 soit différent de zéro, il est évident que l'expression (31) est positive pour $m = 1$, et alors l'inégalité générale (29) en résulte.

Pour démontrer (30), remarquons que cette expression est $\geqq 0$ lorsque $n = 2$. Admettons maintenant que cela soit le cas pour tout nombre $< n$. On aura en vertu dudit théorème de JACOBI

$$\begin{vmatrix} 0 & 0 & x_k \\ 0 & 0 & y_k \\ x_i & y_i & a_{ik} \end{vmatrix}_1^n \begin{vmatrix} 0 & x_k \\ x_i & a_{ik} \end{vmatrix}_1^{n-1} = \begin{vmatrix} 0 & x_k \\ x_i & a_{ik} \end{vmatrix}_1^n \begin{vmatrix} 0 & 0 & x_k \\ 0 & 0 & y_k \\ x_i & y_i & a_{ik} \end{vmatrix}_1^{n-1} - A^2,$$

A étant un certain déterminant asymétrique. En tenant compte de (29), l'inégalité (30) résulte par récurrence. On voit aussi qu'on y a le signe $>$, si les systèmes x et y ne sont pas proportionnels.

13. Cela étant, on a les inégalités suivantes entre Δ_n, $\overset{0}{\Delta}_n$ et D_n. En évaluant d'abord $\Delta_n - D_n$ par le théorème de JACOBI, on obtient

$$(32)\qquad D_n \leqq \Delta_n.[1]$$

D'autre part d'après les inégalités (29) et (30), il vient

$$0 < -\begin{vmatrix} 0 & 0 & c_0 & \dots & c_{n-1} \\ 0 & c_0 & c_1 & \dots & c_n \\ c_0 & c_1 & c_2 & \dots & c_{n+1} \\ \dots & & & & \\ c_{n-1} & c_n & c_{n+1} & \dots & c_{2n} \end{vmatrix} = -\begin{vmatrix} 0 & 0 & c_0 & \dots & c_{n-1} \\ 0 & 0 & c_1 & \dots & c_n \\ c_0 & c_1 & c_2 & \dots & c_{n+1} \\ \dots & & & & \\ c_{n-1} & c_n & c_{n+1} & \dots & c_{2n} \end{vmatrix} -$$

$$- c_0 \begin{vmatrix} 0 & c_0 & \dots & c_{n-1} \\ c_0 & c_2 & \dots & c_{n+1} \\ \dots & & & \\ c_{n-1} & c_n & \dots & c_{2n} \end{vmatrix} < -c_0' \begin{vmatrix} 0 & c_0 & \dots & c_{n-1} \\ c_0 & c_2 & \dots & c_{n+1} \\ \dots & & & \\ c_{n-1} & c_{n+1} & \dots & c_{2n} \end{vmatrix},$$

[1] Cette inégalité met en évidence une propriété minimée importante de $S(t)$ sur laquelle nous reviendrons.

c'est-à-dire, d'après les formules (21), (4) et (19')

$$\overset{0}{\Delta}_n < \frac{c_0}{\varrho_n} D_n. \tag{33}$$

Enfin, en développant Δ_n (*voir* (4)) d'après les puissances de c_{-1}, on aura

$$\Delta_n = \overset{0}{\Delta}_n + 2\alpha_n c_{-1} + \frac{c_{-1}^2}{\varrho_n}, \tag{34}$$

α_n étant une certaine valeur indépendante de c_{-1}. La valeur Δ_n étant positive pour toute valeur de c_{-1}, il s'ensuit

$$\alpha_n < \sqrt{\frac{\overset{0}{\Delta}_n}{\varrho_n}}. \tag{35}$$

En combinant maintenant (28) et (32), on arrive à l'inégalité (18). D'autre part (27) et (33) amènent l'inégalité (19'), qui, combinée avec (34) et (35), entraîne l'inégalité (19).

24.III

Sur le problème des moments

Troisième Note[1]

Ark. Mat. Fys. *17*, no. 16 (1923), 1-52

Introduction.

1. Dans la Note présente, nous montrons quel parti on peut tirer de l'idée générale des opérations fonctionnelles linéaires pour le problème des moments.

Soit $c_0, c_1, \ldots$ une suite indéfinie de constantes réelles. A un polynome quelconque

(1) $$f(t) = x_0 + x_1 t + \cdots + x_n t^n,$$

cette suite fait correspondre une valeur déterminée

(2) $$f(c) = x_0 c_0 + x_1 c_1 + \cdots + x_n c_n.$$ [2]

Ces formules définissent pour le champ fonctionnel formé par

[1] Cette Note ne suppose pas la connaissance des deux Notes précédentes (*voir* ce recueil t. 16, n°s 12 (1921), p. 1—23 et 19 (1922), p. 1—21). Nous avons tâché d'en faciliter la lecture même aux dépens de la brièveté, en préférant quelques redites à des renvois trop fréquents. Nous citons lesdites Notes par les lettres *I* et *II*.

[2] Nous empruntons cette notation symbolique à M. E. Stridsberg, Några aritmetiska undersökningar etc., Notes 2 et 3, ce recueil, t. 13, n° 25 (1918), p. 1—70 et t. 15, n° 22 (1921), p. 1—126. C'est également à M. Stridsberg que nous empruntons les dénominations: suites positives, définies et semi-définies positives.

tous les polynomes une *opération fonctionnelle* évidemment *distributive*.

Admettons que la suite en question soit *positive* c.-à-d. telle que l'inégalité $f(t) \geqq 0$ supposée remplie pour toute valeur réelle de t entraîne $f(c) \geqq 0$. Nous dirons que l'opération définie par de telles suites est une *opération distributive et positive.*

On sait que la condition posée est la condition nécessaire et suffisante pour que le *problème des moments*

$$\int_{-\infty}^{+\infty} t^n \, d\varphi(t) = c_n, \qquad n = 0, 1, 2, \ldots, \tag{3}$$

admette une solution $\varphi(t)$ *non décroissante.* Il est évident que la condition est nécessaire et l'on connaît plusieurs démonstrations du fait que, la condition étant remplie, il existe une ou plusieurs solutions non décroissantes. Si, malgré cela, nous nous sommes décidé à présenter ici une nouvelle démonstration du théorème d'existence, basée sur les méthodes connues[1] des opérations fonctionnelles, c'est pour jeter un jour nouveau sur la question de détermination et d'indétermination.

L'idée fondamentale de notre démonstration consiste dans le procédé connu d'étendre l'opération (2) qui n'est définie que pour des polynomes, à des champs fonctionnels plus vastes de façon que le caractère distributif et positif de l'opération soit gardé. En étendant l'opération en particulier au champ fonctionnel obtenu par l'adjonction des fonctions constantes par intervalles, on arrive au théorème d'existence. Ajoutons tout de suite que cette extension ne pourra, en général, s'opérer d'un seul coup, elle ne se fera que de proche en proche.

Jusqu'à ce point, il n'entre aucun élément algébrique dans les raisonnements. La méthode peut donc s'appliquer presque sans modification à des problèmes des moments généralisés où les fonctions $1, t, \ldots, t^n, \ldots$ sont remplacées par d'autres fonctions $\lambda_0(t), \lambda_1(t), \ldots, \lambda_n(t), \ldots$ qui pourront être d'un caractère très général. Nous n'y insistons pas dans le travail actuel.

---

[1] *Voir* en particulier F. Riesz, Démonstration nouvelle d'un théorème concernant les opérations fonctionnelles linéaires, Ann. Éc. Norm. (3) t. 31 (1914), p. 9—14 et E. Helly, Über lineare Funktionaloperationen, Sitzber. K. Ak. d. Wiss. Wien, Math.-natw. Kl. 121, Abt. II a (1912), p. 265—297.

2. Le théorème d'existence établi, il s'agit d'élucider la question d'unicité. Nous retournons pour cela à l'opération fonctionnelle dans son état primaire, c'est-à-dire définie seulement pour le champ des polynomes. Cela étant, nous adjoignons à ce champ une fonction $F(t)$ différente de tout polynome, mais autrement tout à fait arbitraire[1] et nous considérons le champ fonctionnel formé par les combinaisons linéaires à coefficients constants de la fonction $F(t)$ et d'un nombre fini des fonctions $1, t, \ldots, t^n, \ldots$[2] Pour étendre l'opération à ce champ de sorte qu'elle reste distributive et positive, on aura (*voir* le n° 6) à attacher à $F(t)$ une valeur $F(c)$ telle que $\underline{F}(c) \leqq F(c) \leqq \overline{F}(c)$, $\underline{F}(c)$ désignant la limite supérieure des valeurs que l'opération (2) fait correspondre aux polynomes inférieurs à $F(t)$, tandis que $\overline{F}(c)$ signifie la limite inférieure des valeurs correspondant aux polynomes supérieurs à $F(t)$.[3] On a évidemment $-\infty \leqq \underline{F}(c) \leqq \overline{F}(c) \leqq +\infty$. Pour que l'extension ne puisse se faire que d'une manière unique, il faut et il suffit que $\underline{F}(c) = \overline{F}(c)$. Pour que ni $\underline{F}(c)$ ni $\overline{F}(c)$ ne soient infinis, il devra exister deux polynomes $P_1(t)$ et $P_2(t)$ tels que $P_1(t) < F(t) < P_2(t)$.

Un des buts principaux de ce travail est la démonstration du théorème suivant (n° 27):

Soit $F(t)$ une fonction quelconque non polynome, continue pour toute valeur finie de t, dont le module croît moins vite avec $|t|$ qu'une certaine puissance de $|t|$. Pour que le problème des moments attaché à la suite $c_0, c_1, \ldots$ soit déterminé, il faut et il suffit que l'extension de l'opération à cette fonction particulière ne puisse s'opérer que d'une seule manière, c'est-à-dire que $\underline{F}(c) = \overline{F}(c)$. Si l'on ne suppose plus que $F(t)$ est continue, la condition $\underline{F}(c) = \overline{F}(c) =$ une valeur finie est toujours suffisante, mais elle n'est plus nécessaire.

On a un théorème correspondant pour l'intervalle $(0, \infty)$

---

[1] Les fonctions considérées dans ce travail sont, sauf avis contraire, supposées définies pour toute valeur réelle et finie de la variable.

[2] Par combinaisons linéaires formées de fonctions en nombre fini ou infini, nous entendons toujours de telles combinaisons à coefficients constants et à un nombre fini de termes.

[3] Nous employons les mots »inférieur» et »supérieur» toujours au sens large.

(n° 30). La plupart des critères d'unicité connus rentrent comme cas particuliers sous ces théorèmes ou en sont des corollaires. Ainsi p. ex. les critères de STIELTJES et de M. HAMBURGER concernant respectivement la convergence ou la convergence complète de certaines fractions continues. Le critère de M. PERRON (sous une forme bien précisée) est un corollaire immédiat de notre théorème.

La démonstration du théorème général dont nous venons de parler, exige un appareil algébrique considérable. Il s'agit en premier lieu d'un examen approfondi d'une certaine fonction $\varrho(\alpha)$, attachée à la suite $c_0, c_1, \ldots$, dont nous avons abordé l'étude dans notre première Note. Là, tout se tournait autour des propriétés remarquables de cette fonction pour les valeurs réelles de la variable, ici, il s'agira surtout des valeurs imaginaires. Entre autre, il est facile de voir que le rayon de certaines circonférences que M. ROLF NEVANLINNA[1] vient de considérer dans un travail important récemment paru, est, à un facteur simple près, donné par cette fonction (n°s 20—21). Enfin, j'ai réussi à améliorer l'évaluation de l'ordre de grandeur de $\varrho(\alpha)$ en montrant (n° 26) que dans le *cas d'indétermination,* on a

$$\frac{1}{\varrho(\alpha)} < \gamma e^{\varepsilon(r)r}, \quad r = |\alpha|, \quad \lim_{r\to\infty} \varepsilon(r) = 0.$$

Le théorème d'existence.[2]

3. Soit donnée une *suite positive* $c_0, c_1, \ldots$ et une infinité dénombrable de fonctions entièrement arbitraires $h_1(t), h_2(t), \ldots h_n(t) \ldots$ Considérons l'opération distributive et positive définie par les formules (1) et (2) pour toute combinaison linéaire des puissances $1, t, t^2, \ldots$ et montrons qu'on peut l'étendre au

[1] ROLF NEVANLINNA, Asymptotische Entwicklungen beschränkter Funktionen und das STIELTJES'sche Momentenproblem, Ann. Ac. Scient. Fenn. t. XVIII, n° 5 (1922), p. 1—53. Les recherches que j'expose ici, m'ont conduit (en novembre 1921) aux mêmes résultats sur le problème des moments auxquels M. NEVANLINNA est arrivé partant des séries asymptotiques. A ce que j'ai appris depuis, M. NEVANLINNA m'a devancé non seulement dans la publication mais aussi dans la découverte.

[2] J'ai indiqué la démonstration qui suit dans une conférence faite à la Société Mathématique de Stockholm en avril 1918.

champ fonctionnel formé par les combinaisons linéaires des fonctions 1, t, t^2, ... et $h_1(t)$, $h_2(t)$, ... et cela de façon que l'opération reste distributive et positive.

Cette extension, nous la ferons de proche en proche.[1] Considérons d'abord le champ A_1 formé par les combinaisons linéaires desdites puissances de t et de $h_1(t)$. Attachons à $h_1(t)$ une valeur que nous désignerons par $h_1(c)$ en attachant en même temps à toute combinaison linéaire de 1, t, ... et de $h_1(t)$ la combinaison correspondante des valeurs c_0, c_1, ... et de $h_1(c)$. Ainsi on aura défini une opération distributive portant sur le champ fonctionnel A_1.

Pour que cette opération soit positive, on aura évidemment à choisir $h_1(c)$ de façon que $\underline{h}_1(c) \leqq h_1(c) \leqq \bar{h}_1(c)$,[2] où $\underline{h}_1(c)$ signifie la limite supérieure des valeurs que l'opération fait correspondre aux polynomes inférieurs à $h_1(t)$ tandis que $\bar{h}_1(c)$ signifie la limite inférieure des valeurs attachées aux polynomes supérieurs à $h_1(t)$. Ce choix est toujours possible. En effet, l'opération étant distributive et positive dans le champ des polynomes, on aura $\underline{h}_1(c) \leqq \bar{h}_1(c)$. En résumé, *il est toujours possible d'étendre l'opération au champ fonctionnel A_1 en gardant son caractère distributif et positif, dès que ces propriétés ont lieu au champ A_0 formé par les combinaisons linéaires de* 1, t, ... Ajoutons encore que si l'on a $\underline{h}_1(c) = \bar{h}_1(c)$, il n'y a pour $h_1(c)$ qu'une seule valeur admissible, qu'on aura alors à prendre, tandis que, dans le cas $\underline{h}_1(c) < \bar{h}_1(c)$, il y en a une infinité. Dans ce dernier cas, *avant de continuer*, on choisira $h_1(c)$ d'une façon *déterminée*.

Cela étant, on formera le champ A_2 en adjoignant au champ A_1 la fonction $h_2(t)$ et toutes les combinaisons linéaires de cette fonction et des fonctions du champ A_1. En choisissant $h_2(c)$ d'une part $\geqq$ la limite supérieure des valeurs correspondant aux fonctions du champ A_1 inférieures à $h_2(t)$ et

[1] L'extension ne peut, en général, se faire d'un seul coup. En effet, la possibilité d'en faire ainsi, caractérise, pour ainsi dire, le cas déterminé (*cf.* le n° 7). Or, on ne sait d'avance lequel des deux cas se présente.

[2] En effet, l'inégalité $\mu h_1(t) + P(t) \geqq 0$ (où $P(t)$ est un polynome) entraîne $\mu h_1(c) + P(c) \geqq 0$, pour toute valeur de μ, dès qu'il en est ainsi pour $\mu = \pm 1$.

d'autre part $\leqq$ la limite inférieure des valeurs attachées aux fonctions du même champ supérieures à $h_2(t)$, on aura obtenu une opération définie pour tout le champ A_2 et avec les propriétés désirées.

En continuant ainsi, on étend l'opération aux champs $A_1, A_2, \ldots A_n, \ldots$ obtenus par l'adjonction successive des fonctions $h_1(t), h_2(t), \ldots h_n(t), \ldots$

4. Soit maintenant $\xi_1, \xi_2, \ldots \xi_m, \ldots$ un ensemble dénombrable de points dense dans l'intervalle $(-\infty, +\infty)$. Remplaçons dans la construction du numéro précédent les fonctions $h_1(t), h_2(t), \ldots h_m(t), \ldots$ par les fonctions particulières $g_1(t)$, $g_2(t), \ldots g_m(t), \ldots$ définies par les égalités

$$g_m(t) = \begin{matrix} 1 \\ 0 \end{matrix} \quad \text{pour} \quad t \begin{matrix} \leqq \\ > \end{matrix} \xi_m.$$

Cela étant, nous définissons, pour les valeurs $t = \xi_m$, la fonction $\varphi(t)$ par les équations

$$\varphi(\xi_m) = g_m(c).$$

Alors, si l'on a $\xi_\alpha > \xi_\beta$, on aura $\varphi(\xi_\alpha) \geqq \varphi(\xi_\beta)$. En effet, la première inégalité entraîne $g_\alpha(t) \geqq g_\beta(t)$. D'autre part, si nous désignons par γ le plus grand des indices α et β, les fonctions $g_\alpha(t)$ et $g_\beta(t)$ appartiennent toutes les deux au champ A_γ, par suite on aura $g_\alpha(c) \geqq g_\beta(c)$, c'est-à-dire $\varphi(\xi_\alpha) \geqq \varphi(\xi_\beta)$.

La fonction $\varphi(t)$ qui vient d'être définie pour les valeurs $t = \xi_m$, n'est donc pas décroissante. On pourra alors la compléter de sorte qu'elle devienne une fonction non décroissante définie pour toutes les valeurs de t.[1] Nous démontrerons que cette fonction $\varphi(t)$ est une solution du problème des moments.

5. Considérons à cet effet un polynome quelconque $f(t)$ de degré n. Désignons par $2s$ un nombre pair fixe $> n$ et choisissons $-T_1 < 0$ et $T_2 > 0$ de sorte qu'ils appartiennent

[1] Les valeurs $\xi_1, \xi_2, \ldots$ et $\varphi(\xi_1), \varphi(\xi_2), \ldots$ déterminent évidemment $\varphi(t)$ en ses points de continuité. Un autre choix de ces valeurs pourra manifestement conduire à une autre fonction $\varphi(t)$.

à l'ensemble $\xi_1, \xi_2, \ldots$ et que d'autre part T_1 et T_2 soient assez grands pour que, à l'extérieur de l'intervalle $(-T_1, T_2)$, on ait $|f(t)| < \varepsilon t^{2s}$ où ε désigne un nombre positif fixe d'ailleurs arbitrairement petit. Nous divisons[1] l'intervalle $(-T_1, T_2)$ par les points $-T_1 = t_1 < t_2 < \cdots < t_k < t_{k+1} < \cdots t_{p+1} = T_2$, appartenant à l'ensemble $\xi_1, \xi_2, \ldots$ de manière que dans tout intervalle de division (t_k, t_{k+1}) l'oscillation du polynome $f(t)$ soit moindre que ε' où ε' désigne encore un nombre positif arbitrairement petit. Soit τ_k un point de l'intervalle (t_k, t_{k+1}) $(k = 1, 2, \ldots p)$ et formons la fonction $v(t)$ constante par intervalles qui s'annule à l'extérieur de l'intervalle $(-T_1, T_2)$, prend les valeurs $f(\tau_k)$ dans les intervalles $t_k < t \leqq t_{k+1}$ et enfin la valeur zéro pour $t = -T_1$. Alors pour toute valeur de t, on aura

$$v(t) - \varepsilon' - \varepsilon t^{2s} < f(t) < v(t) + \varepsilon' + \varepsilon t^{2s}.$$

La fonction $v(t)$ appartenant à un des champs $A_1, A_2, \ldots$ (et, par conséquent, à tout champ d'indice supérieur) l'opération fonctionnelle se trouve définie pour cette fonction. Il est donc

$$v(c) - \varepsilon' c_0 - \varepsilon c_{2s} \leqq f(c) \leqq v(c) + \varepsilon' c_0 + \varepsilon c_{2s}.$$

D'autre part, les valeurs $t_1, t_2, \ldots t_{p+1}$ étant choisies parmi $\xi_1, \xi_2, \ldots$, on aura $t_1 = \xi_{l_1}, \ldots t_k = \xi_{l_k}, \ldots t_{p+1} = \xi_{l_{p+1}}$ ainsi que

$$v(t) = \sum_{k=1}^{p} f(\tau_k)\,(g_{l_{k+1}}(t) - g_{l_k}(t)).$$

Et alors, puisque

$$g_{l_k}(c) = \varphi(\xi_{l_k}) = \varphi(t_k),$$

il vient

$$v(c) = \sum_{k=1}^{p} f(\tau_k)\,(\varphi(t_{k+1}) - \varphi(t_k)).$$

L'inégalité ci-dessus pourra alors s'écrire

[1] *Cf.* F. RIESZ, *l. c.*

$$\sum_{k=1}^{p} f(\tau_k)(\varphi(t_{k+1})-\varphi(t_k)) - \varepsilon' c_0 - \varepsilon c_{2s} \leqq f(c) \leqq$$

$$\leqq \sum_{k=1}^{p} f(\tau_k)(\varphi(t_{k+1})-\varphi(t_k)) + \varepsilon' c_0 + \varepsilon c_{2s}.$$

En faisant tendre la longueur maximée des intervalles (t_k, t_{k+1}) et avec cela le nombre ε' vers zéro, on aura d'après la définition de l'intégrale de STIELTJES

$$\int_{-T_1}^{T_2} f(t)\,d\varphi(t) - \varepsilon c_{2s} \leqq f(c) \leqq \int_{-T_1}^{T_2} f(t)\,d\varphi(t) + \varepsilon c_{2s}.$$

Enfin, lorsque T_1 et T_2 tendent vers l'infini, on pourra faire tendre ε vers zéro et l'on aura

$$f(c) = \int_{-\infty}^{+\infty} f(t)\,d\varphi(t).$$

Cela étant vrai pour tout polynome et, en particulier, pour les puissances de t, le théorème d'existence se trouve établi. Nous pouvons donc énoncer le théorème suivant:

Pour que le problème des moments

$$\int_{-\infty}^{+\infty} t^n\,d\varphi(t) = c_n, \qquad n = 0, 1, 2, \ldots,$$

admette une solution non décroissante, il faut et il suffit que l'opération définie par les formules (1) *et* (2) *soit une opération positive.*

La question d'unicité. Critères préliminaires.

6. Il est facile de voir que la fonction $\varphi(t)$ qu'on vient de construire, tend vers zéro[1] lorsque t tend vers $-\infty$ (et vers c_0 lorsque t tend vers $+\infty$). Nous dirons avec STIELTJES que

[1] En effet, si ξ_m est négatif, on a $g_m(t) \leqq \xi_m^{-2} t^2$, et alors $\varphi(\xi_m) = g_m(c) \leqq c_2 \xi_m^{-2}$.

le problème des moments est *déterminé*, si toutes les solutions, telles que $\lim\limits_{t \to -\infty} \varphi(t) = 0$, coïncident en tous leurs points de continuité. Dans le cas contraire, le problème sera dit *indéterminé.* Bien entendu, par solution, nous n'entendons que des solutions *non décroissantes.*

Soit $F(t)$ une fonction arbitraire. Nous désignons (comme au n° 3 pour la fonction $h_1(t)$) par $\underline{F}(c)$ la limite supérieure des valeurs que l'opération fait correspondre aux polynomes inférieurs à $F(t)$ et par $\overline{F}(c)$ la limite inférieure des valeurs correspondant aux polynomes supérieurs à $F(t)$. On aura ainsi défini deux opérations portant sur toutes les fonctions $F(t)$ et qu'on pourra appeler respectivement l'*opération inférieure* et l'*opération supérieure.* Ces opérations tout en étant positives, ne seront plus, en général, distributives.[1] On aura toujours $\underline{F}(c) \leqq \overline{F}(c)$. En outre, l'opération inférieure est toujours une opération *concave,* tandis que l'opération supérieure est toujours *convexe.* Cela veut dire que l'égalité $H(t) = F(t) + G(t)$ entraîne les inégalités $\underline{H}(c) \geqq \underline{F}(c) + \underline{G}(c)$ et $\overline{H}(c) \leqq \overline{F}(c) + \overline{G}(c)$.

On pourra désigner encore par $\underline{F}_{(n)}(c)$ et $\overline{F}_{(n)}(c)$ les limites supérieure et inférieure formées d'une façon analogue, mais en se restreignant à des polynomes de degré $\leqq 2n$. On aura évidemment $\underline{F}_{(n)}(c) \leqq \underline{F}_{(n+1)}(c) \leqq \underline{F}(c) \leqq \overline{F}(c) \leqq \overline{F}_{(n+1)}(c) \leqq \overline{F}_{(n)}(c)$,

$$\underline{F}(c) = \lim_{n \to \infty} \underline{F}_{(n)}(c), \quad \overline{F}(c) = \lim_{n \to \infty} \overline{F}_{(n)}(c).$$

S'il n'existe aucun polynome inférieur à $F(t)$ ou aucun polynome supérieur à $F(t)$, on posera respectivement $\underline{F}(c) = -\infty$ ou $\overline{F}(c) = \infty$. Remarque analogue pour $\underline{F}_{(n)}(c)$ et $\overline{F}_{(n)}(c)$.

Nous désignerons le champ des fonctions $F(t)$ pour lesquelles $\underline{F}(c)$ et $\overline{F}(c)$ sont des valeurs finies par D.[2] Une fonction sera

[1] On verra plus loin (n° 7) qu'elles ne seront pas toujours distributives, même si l'on assujettit le champ fonctionnel à des restrictions très grandes. En effet, la distributivité entraîne que ces deux opérations coïncident, et c'est cette coïncidence qui caractérise le cas déterminé. (Pour $G(t) = -F(t)$, on a toujours $\underline{G}(c) = -\overline{F}(c)$ et alors, s'il y a distributivité, $\underline{G}(c) + \underline{F}(c) = 0$, c'est-à-dire $\underline{F}(c) = \overline{F}(c)$).

[2] Remarquons déjà ici que l'attention particulière que nous prêtons dans la suite au champ D se justifie aussi par le fait que ce n'est que pour les fonctions $F(t)$ de ce champ (ou du moins pour ses fonctions conti-

dite *continue,* si elle l'est pour toute valeur finie de la variable. Pour qu'une telle fonction appartienne au champ D, il faut et il suffit évidemment que son module croisse moins vite avec $|t|$ indéfiniment croissant qu'une certaine puissance de $|t|$. Pour ces fonctions, on a le théorème suivant:

$F(t)$ étant une fonction continue appartenant au champ D, l'ensemble des valeurs $\int_{-\infty}^{+\infty} F(t)\,d\varphi(t)$ formées avec toutes les solutions $\varphi(t)$ du problème des moments est identique à celui des valeurs $F(c)$ caractérisées par les inégalités

$$\underline{F}(c) \leqq F(c) \leqq \overline{F}(c).\tag{4}$$ [1]

En effet, $\varphi(t)$ étant une solution du problème des moments, l'intégrale $\int_{-\infty}^{+\infty} F(t)\,d\varphi(t)$ formée avec cette solution définit pour les fonctions continues du champ D une opération distributive et positive. D'après le raisonnement fait au n° 3 pour la fonction $h_1(t)$, la valeur de cette intégrale sera située entre $\underline{F}(c)$ et $\overline{F}(c)$.

Pour démontrer l'inverse, nous modifions notre démonstration du théorème d'existence (*cf.* le n° 4) de la manière suivante:

Soit A_0 le champ formé de tous les polynomes, A_1 le champ formé des combinaisons linéaires d'une certaine fonction $F(t)$ continue et appartenant au champ D et de tous les polynomes, A_2 le champ formé des combinaisons linéaires des fonctions du champ A_1 et de la fonction $g_1(t)$, et ainsi de suite.

---

nues) que l'intégrale $\int_{-\infty}^{+\infty} F(t)\,d\varphi(t)$ a une valeur finie, $\varphi(t)$ étant une solution quelconque d'un problème des moments quelconque. Pour chaque problème particulier, il existe aussi d'autres fonctions qui ont cette propriété et même des fonctions pour lesquelles la valeur de cette intégrale est indépendante du choix de la solution $\varphi(t)$ (*cf.* le n° 31).

[1] La notion de l'intégrale de STIELTJES étant préalablement généralisée, ce théorème subsiste pour de vastes catégories de fonctions discontinues. Tel est le cas, en particulier, pour les fonctions constantes par intervalles et notamment pour les fonctions $g_m(t)$ du texte. Voilà, en germe, l'idée de notre démonstration du théorème d'existence.

En étendant l'opération au champ A_1, on pourra faire correspondre à $F(t)$ une valeur quelconque $F(c)$ satisfaisant aux inégalités (4). Alors en reprenant mot à mot le raisonnement des n°s 4 et 5, on voit que la solution $\varphi(t)$ à laquelle on arrive, satisfait à l'équation $\int_{-\infty}^{+\infty} F(t)\,d\varphi(t) = F(c)$.

On a évidemment en même temps le théorème:

En désignant par $\varphi(t)$ les solutions non décroissantes des équations en nombre $(2n+1)$

$$\int_{-\infty}^{+\infty} t^k\,d\varphi(t) = c_k, \qquad k = 0, 1, \ldots 2n, \tag{5}$$

l'ensemble des valeurs $\int_{-\infty}^{+\infty} F(t)\,d\varphi(t)$ est identique à celui des valeurs $F_{(n)}(c)$ satisfaisant aux inégalités $\underline{F}_{(n)}(c) \leqq F_{(n)}(c) \leqq \overline{F}_{(n)}(c)$.

Soit maintenant $F(t) = X(t) + i\,Y(t)$ une fonction *complexe* de la variable réelle t, $X(t)$ et $Y(t)$ étant des fonctions continues appartenant au champ D. On voit de suite que l'ensemble des valeurs $\int_{-\infty}^{+\infty} F(t)\,d\varphi(t)$ formées avec toutes les solutions du problème des moments donne *un domaine convexe fermé, ce domaine pouvant se réduire à un segment de droite ou à un point.*

En effet, $\varphi_1(t)$ et $\varphi_2(t)$ étant deux solutions, $\varphi(t) = a\varphi_1(t) + b\varphi_2(t)$ en est aussi une, si a et b sont des nombres positifs tels que $a + b = 1$. Donc, si deux points appartiennent au domaine en question, le segment qui les réunit y appartiendra aussi. Le domaine est donc convexe. Le fait qu'il est fermé, résulte immédiatement du principe de choix de M. Helly.[1] On le voit aussi par le théorème suivant qui fournit en même temps un moyen de déterminer ce domaine.

Soient ξ et η deux nombres réels quelconques. Formons la fonction $F_{\xi\eta}(t) = \xi X(t) + \eta\,Y(t)$. L'ensemble des points

[1] *Cf. I,* p. 16.

$$(6)\qquad X + iY = \int_{-\infty}^{+\infty} F(t)\,d\varphi(t) = \int_{-\infty}^{+\infty} (X(t) + iY(t))\,d\varphi(t)$$

coïncide avec ceux du domaine convexe $\varDelta$ caractérisé par les inégalités

$$\underline{F}_{\xi\eta}(c) \leqq \xi X + \eta Y \leqq \overline{F}_{\xi\eta}(c),$$

ces inégalités étant remplies pour toutes les valeurs de ξ et de η.

On démontre ce théorème en reprenant l'ordre d'idées des n°s 3—5 et en adjoignant comme première et seconde fonctions $X(t)$ et $Y(t)$ respectivement.

L'ensemble des valeurs (6) formées avec toutes les solutions non décroissantes des équations (5) coïncide avec un domaine convexe fermé $\varDelta_n$, obtenu d'une manière analogue. Il est évident que $\varDelta_n$ contient $\varDelta_{n+1}$ et que $\varDelta$ est le domaine commun à $\varDelta_0, \varDelta_1, \ldots \varDelta_n, \ldots$

7. De ce qui précède, on déduit facilement le critère préliminaire que voici:

Pour que le problème des moments relatif à la suite $c_0, c_1, \ldots$ *soit déterminé, il faut* et il suffit *que l'on ait* $\underline{F}(c) = \overline{F}(c)$ *pour toute fonction continue du champ D.*

La première partie de ce critère est un corollaire immédiat du premier théorème du numéro précédent. L'inverse se vérifie, si l'on pose $F(t) = x - t$ pour $t \leqq x$ et $F(t) = 0$ pour $t > x$. On trouve alors que la valeur de l'intégrale $\int_{-\infty}^{x} \varphi(t)\,dt$ est la même pour toutes les solutions, et alors ces solutions coïncident en tous leurs points de continuité.

La condition ci-dessus, considérée comme condition suffisante, manque tout à fait d'intérêt. C'est qu'il y entre une fonction arbitraire $F(t)$, c'est-à-dire qu'elle implique une infinité de conditions.[1] Un de nos résultats principaux énoncé

[1] Il est inutile d'insister sur les améliorations banales de cette condition suffisante, consistant p. ex. en ce que, à la totalité des fonctions $F(t)$, on pourrait substituer un système complet etc. D'ailleurs de tels systèmes sont implicitement considérés dans le texte.

déjà dans l'Introduction est précisément qu'il suffit de faire l'hypothèse en question sur une *seule* fonction d'ailleurs arbitraire. En attendant, remplaçons le critère ci-dessus par un critère plus particulier très accessible au calcul.

8. Si $\psi(t)$ est une fonction non décroissante variant entre des limites finies lorsque t va de $-\infty$ à $+\infty$, l'intégrale

$$H(\alpha) = \int_{-\infty}^{+\infty} \frac{d\psi(t)}{\alpha - t}$$

existe pour tout nombre imaginaire (c.-à-d. non réel) α et la fonction $\psi(t+0) + \psi(t-0)$ peut se calculer par la fonction $H(\alpha)$.[1] On obtient alors comme second critère préliminaire pour que le problème des moments soit déterminé la règle suivante, utilisée p. ex. dans les recherches de M. HAMBURGER et dans notre deuxième Note.

Pour que le problème des moments soit déterminé, il faut et il suffit que les fonctions

$$H(\alpha) = \int_{-\infty}^{+\infty} \frac{d\varphi(t)}{\alpha - t}$$

formées avec toutes les solutions $\varphi(t)$ du problème des moments coïncident pour toute valeur imaginaire de α. Il suffit aussi que cette coïncidence ait lieu dans une infinité de points qui ont au moins un point limite imaginaire.

Or on a vu au n° 6 que α étant fixé, l'ensemble des valeurs $H(\alpha)$ donne un domaine convexe. Le critère revient donc à ceci que ces domaines se réduisent à des points pour toute valeur imaginaire de α ou du moins pour un ensemble ayant un point limite imaginaire. Or on verra plus loin (nos 20—21) que ces domaines sont des cercles dont le rayon décroît, quand la valeur absolue de l'ordonnée croît. Il suffira donc que ce rayon soit zéro en un seul point.

[1] *Voir* p. ex. O. PERRON, Die Lehre von den Kettenbrüchen (1913), p. 372.

Pour démontrer ces énoncés, il nous faut passer à des considérations algébriques. Les recherches concernant l'ensemble des valeurs de $H(\alpha)$ pourront, d'après le n° 6, être interprétées comme la recherche des opérations inférieure et supérieure pour les parties imaginaire et réelle de $\frac{1}{\alpha - t}$ et pour les combinaisons linéaires de ces deux fonctions. C'est la voie que nous avons suivie dans nos recherches, ces opérations étant assez faciles à calculer. Ici, nous suivrons une autre marche, en nous contentant d'indiquer (p. 32—33) le calcul auquel nous venons de faire allusion.

Dans nos considérations algébriques, nous nous restreindrons à des suites *définies positives*, c'est-à-dire à des suites, telles que les hypothèses $f(t) \geqq 0$ et $f(t)$ non identiquement nul entraînent $f(c) > 0$. De telles suites n'admettent que des solutions à une *infinité de points de croissance.* Les suites semi-définies positives, c'est-à-dire les suites pour lesquelles il existe au moins un polynome non négatif et non identiquement nul tel que $f(c) = 0$, n'ont, au contraire, que des solutions à un nombre fini de points de croissance. Ces dernières suites sont, d'une part, faciles à traiter directement[1], d'autre part, on voit déjà par les résultats préliminaires qui précèdent qu'elles donnent toujours lieu à des cas déterminés. Nous pourrons donc définitivement les laisser de côté.

Préliminaires algébriques. Les polynomes quasi-orthogonaux. Un problème de minimum. Les quadratures mécaniques. Les réduites généralisées.[2]

9. On sait que tout polynome $f(t)$ non négatif peut se mettre sous la forme $f(t) = p^2(t) + q^2(t)$, $p(t)$ et $q(t)$ étant des polynomes à coefficients réels.[3] On en conclut le théorème suivant: *Pour que la suite* $c_0, c_1, \ldots$ *soit définie positive, il faut et il suffit qu'il en soit de même des formes quadratiques en*

[1] E. STRIDSBERG, *l. c.* Note 2, p. 35 et suiv.; Note 3, p. 17—18, 125—126.

[2] L'exposition simple et la commodité du lecteur justifient la longueur de ce chapitre qui contient peu de matière essentiellement nouvelle.

[3] En effet, on pourra écrire $f(t) = h^2(t)\,[r(t) + i\,s(t)]\,[r(t) - i\,s(t)]$.

nombre infini $\sum_{i,k=0}^{n} c_{i+k} x_i x_k$ $(n = 0, 1, 2, \ldots)$. *Cette dernière hypothèse s'exprime par les inégalités*

$$|c_{i+k}|_0^n > 0 \qquad (n = 0, 1, 2, \ldots).$$

Plus haut, nous avons défini la notation symbolique $f(c)$ où $f(t)$ désignait un polynome de t. En désignant par $g(t)$ un second polynome, nous entendons encore avec M. STRIDSBERG par $f(c)\,g(c)$ ou $f^2(c)$ les valeurs qu'on obtient en remplaçant dans les polynomes $f(t)\,g(t)$ ou $f^2(t)$, développés suivant les puissances de t, les puissances t^j par les nombres correspondants c_j. En posant

$$f(t) = x_0 + x_1 t + \cdots + x_n t^n,$$

on écrira en particulier

$$c^k f(c) = x_0 c_k + x_1 c_{k+1} + \cdots + x_n c_{k+n}.$$

10. Nous dirons qu'un polynome $q(t)$ de degré $(n+1)$[1] non identiquement nul est *quasi-orthogonal* par rapport à la suite $c_0, c_1, \ldots$ s'il remplit les conditions

(7) $$c^k q(c) = 0, \qquad k = 0, 1, \ldots n-1.$$

Si le polynome remplit, en outre, la condition

(8) $$c^n q(c) = 0,$$

nous disons que ce polynome est *orthogonal* par rapport à la suite.

Entre trois polynomes quasi-orthogonaux de même degré, il existe toujours une relation linéaire et homogène à coefficients constants.

[1] Sous certains points de vue, il serait plus commode et surtout plus naturel de désigner le degré des polynomes quasi-orthogonaux considérés dans le texte par n au lieu de $(n+1)$. Nous avons pourtant préféré cette dernière notation, parce que c'est des polynomes quasi-orthogonaux de degré $(n+1)$ qu'on a besoin dans la quadrature mécanique des polynomes de degré $2n$.

Soient en effet $q_1(t)$, $q_2(t)$ et $q_3(t)$ les trois polynomes, leur degré étant $(n+1)$. On pourra évidemment déterminer trois constantes A_1, A_2 et A_3 non toutes nulles en sorte que le polynome $h(t) = A_1 q_1(t) + A_2 q_2(t) + A_3 q_3(t)$ soit de degré $\leqq (n-1)$. Ce polynome remplira les équations

$$c^k h(c) = 0, \qquad k = 0, 1, \ldots n-1.$$

Or, le déterminant $|c_{i+k}|_0^{n-1}$ étant différent de zéro, on en conclut que tous les coefficients de $h(t)$ s'annulent.

En désignant donc par $q_1(t)$ et $q_2(t)$ deux polynomes quasi-orthogonaux de même degré et de rapport non constant, tout polynome quasi-orthogonal du même degré pourra s'écrire sous la forme $\mu_1 q_1(t) + \mu_2 q_2(t)$, μ_1 et μ_2 étant des constantes. Inversement, il est clair qu'une telle combinaison linéaire donne toujours un polynome quasi-orthogonal.

Deux polynomes quasi-orthogonaux de même degré et de rapport non constant ne pourront avoir aucun zéro commun. En effet, le déterminant $|c_{i+k}|_0^{n-1}$ étant différent de zéro, l'équation $q(\alpha) = 0$ et les équations (7) déterminent, à un facteur constant près, les coefficients de $q(t)$.

Un polynome quasi-orthogonal à coefficients réels $q(t)$ a tous ses zéros réels et distincts. Il suffit pour cela de démontrer que $q(t)$ a au moins n zéros réels et impairs. Or, s'il n'en était pas ainsi, on pourrait construire un polynome $r(t)$ de degré $\leqq (n-1)$, tel que $q(t)r(t)$ soit toujours $\geqq 0$. La suite $c_0, c_1, \ldots$ étant définie positive, on devrait avoir $q(c)r(c) > 0$, tandis que (7) donne $q(c)r(c) = 0$.

Un polynome quasi-orthogonal dont les coefficients ne sont pas réels et ne peuvent être rendus réels par division avec un facteur commun convenable, aura, au sens étroit du mot, tous ses zéros situés ou dans le demi-plan supérieur ou dans le demi-plan inférieur. En effet, un tel polynome pourra s'écrire $q(t) = q_1(t) + iq_2(t)$ où $q_1(t)$ et $q_2(t)$ sont quasi-orthogonaux et à coefficients réels. Le quotient de $q_1(t)$ et $q_2(t)$ ne pourra être réel en aucun point imaginaire. (Autrement, il existerait une combinaison réelle de ces polynomes, c'est-à-dire un polynome quasi-orthogonal à coefficients réels, qui s'annule en un point imaginaire, contrairement à ce que nous venons de démontrer.)

Il s'ensuit que la partie imaginaire de ce quotient garde un signe constant dans les demi-plans supérieur et inférieur, ce signe étant opposé dans les deux demi-plans. On en conclut d'abord que $q_1(t) + iq_2(t)$ ne pourra s'annuler que dans un seul de ces demi-plans. En s'approchant des pôles d'un tel quotient, on en conclut aussi que tous les résidus sont de même signe, d'où l'on tire que *les zéros de deux polynomes quasi-orthogonaux à coefficients réels de même degré* (d'ailleurs quelconques) *se séparent.*

Disons encore un mot des propriétés particulières des *polynomes orthogonaux.* Ces polynomes sont évidemment déterminés, à un facteur constant près, par les équations (7) et (8). Il en résulte que tout polynome orthogonal est, à un facteur constant près, égal à un polynome orthogonal déterminé à coefficients réels, p. ex. au polynome donné par la formule (10). Tous les énoncés ci-dessus subsistent si l'on compte le polynome orthogonal de degré n, soit parmi les polynomes quasi-orthogonaux de degré $(n + 1)$, soit parmi ceux de degré n. Dans certains de ces énoncés, on y est même forcé. Nous adopterons donc ce point de vue dans la suite. (Quant aux résultats concernant en particulier la quadrature mécanique, il faudra faire certaines distinctions légères.) Notons dans cet ordre d'idées que *tout polynome quasi-orthogonal de degré* $(n + 1)$ *pourra s'écrire comme combinaison linéaire de deux polynomes orthogonaux de degré* n *et* $(n + 1)$ *respectivement.*

11. Donnons maintenant des formules explicites pour les polynomes quasi-orthogonaux. Posons, $\eta_0, \eta_1, \ldots \eta_{n+1}$ désignant des paramètres,

$$(9) \quad q(t) = \begin{vmatrix} c_0 & c_1 & \ldots c_{n-1} & \eta_0 & 1 \\ c_1 & c_2 & \ldots c_n & \eta_1 & t \\ \cdot & \cdot & \cdot\ \cdot\ \cdot & \cdot & \cdot \\ c_{n+1} & c_{n+2} & \ldots c_{2n} & \eta_{n+1} & t^{n+1} \end{vmatrix} = \left| c_i c_{i+1} \ldots c_{i+n-1} \eta_i t^i \right|_0^{n+1}.$$

Il est clair que $q(t)$ remplit les équations (7), en sorte que c'est un polynome quasi-orthogonal de degré $\leqq (n + 1)$, et inversement, il est facile de voir que tout tel polynome pourra se mettre sous la forme ci-dessus. En effet, on pourra choisir $\eta_0, \eta_1, \dots \eta_{n-1}$ d'une façon arbitraire et déterminer après η_n et η_{n+1} de manière que t^n et t^{n+1} aient les mêmes coefficients dans (9) et dans le polynome donné. Alors la différence des deux polynomes sera d'une part de degré $\leqq (n-1)$, d'autre part elle satisfera aux équations (7) et alors, d'après un raisonnement qu'on a fait plus haut, elle sera identiquement nulle.

En posant dans (9) $\eta_0 = c_n, \dots \eta_i = c_{n+i}, \dots \eta_{n+1} = c_{2n+1}$, on obtiendra un polynome orthogonal de degré $(n+1)$

$$\text{10)} \qquad q(t) = | c_i c_{i+1} \dots c_{i+n} t^i |_0^{n+1}.$$

En posant, au contraire, $\eta_0 = \eta_1 = \dots = \eta_n = 0$, $\eta_{n+1} \neq 0$, on obtiendra un polynome orthogonal de degré n.

Retournons au cas général. On a vu plus haut que dans un polynome quasi-orthogonal, il n'entre en réalité que deux paramètres et que, si l'on fait abstraction d'un facteur constant, il n'y aura qu'un seul paramètre essentiel. Il y a donc lieu de normer le polynome (9) en spécialisant les paramètres. Cela pourra se faire de plusieurs manières.

En posant p. ex. $\eta_0 = c_{-1}, \eta_1 = c_0, \dots \eta_{n+1} = c_n$ où c_{-1} désigne un paramètre arbitraire, on arrivera, à un facteur constant près, aux polynomes considérés dans notre deuxième Note (p. 7 et suiv.)

$$(11) \qquad q(t) = | t^i c_{i-1} c_i \dots c_{i+n-1} |_0^{n+1} : | c_{i+k} |_0^n.$$

On voit sans difficulté que tout polynome quasi-orthogonal qui ne s'annule pas à l'origine pourra, à un facteur constant près, se mettre sous cette forme. Le polynome ci-dessus est normé par la condition $q(0) = 1$.

Tout polynome quasi-orthogonal qui s'annule à l'origine pourra, à un facteur constant près, s'écrire

$$(12) \qquad q(t) = | t^i c_i c_{i+1} \dots c_{i+n-1} |_1^{n+1} : | c_{i+k} |_0^n.$$

Plus généralement, tout polynome quasi-orthogonal qui s'annule en un point donné α, est, toujours à un facteur constant près,

$$(13) \qquad q(t) = | t^i \alpha^i c_i \dots c_{i+n-1} |_0^{n+1} : | c_{i+k} |_0^n.$$

Dans un instant, nous trouverons encore d'autres expressions pour ce polynome.

12. Soit α un nombre réel ou imaginaire. Posons *le problème de minimum* que voici.

Parmi tous les polynomes $h(t)$ de degré $\leqq n$ qui satisfont à la relation $h(\alpha) = 1$, trouver celui qui rend l'expression $h(c)\bar{h}(c)$ minimum. Ici, $\bar{h}(t)$ désigne le polynome imaginaire conjugué de $h(t)$.

En posant $h(t) = x_0 + x_1 t + \cdots + x_n t^n$, on a évidemment $h(c)\bar{h}(c) = \sum_{i,k=0}^{n} c_{i+k} x_i \bar{x}_k$ qu'il s'agit de rendre minimum sous la condition $x_0 + x_1\alpha + \cdots + x_n\alpha^n = 1$. Ou encore, en désignant par $\varphi(t)$ une solution du problème des moments ou une solution des équations (5), il est question de rendre minimum l'intégrale $\int_{-\infty}^{+\infty} |h(t)|^2 d\varphi(t)$ sous la condition $h(\alpha) = 1$.

En désignant par $h(t)$ le polynome qui fournit ce minimum, on devra avoir pour tout polynome $v(t)$ de degré $\leqq n$ qui satisfait à $v(\alpha) = 0$, en désignant encore par $\mu = \varrho e^{i\theta}$ un paramètre,

$$h(c)\bar{h}(c) < (h(c) + \mu v(c))(\bar{h}(c) + \bar{\mu}\bar{v}(c)) =$$
$$= h(c)\bar{h}(c) + \bar{\mu} h(c)\bar{v}(c) + \mu \bar{h}(c) v(c) + |\mu|^2 v(c)\bar{v}(c)$$

ou

$$e^{-i\theta} h(c)\bar{v}(c) + e^{i\theta}\bar{h}(c) v(c) + \varrho v(c)\bar{v}(c) > 0.$$

On voit de suite que cela revient à ce que l'hypothèse $v(\alpha) = 0$ ou, ce qui est la même chose, $\bar{v}(\bar{\alpha}) = 0$, doit entraîner $h(c)\bar{v}(c) = 0$. En écrivant alors $h(t) = x_0 + x_1 t + \cdots + x_n t^n$, on trouve

$$\sum_{k=0}^{n} c_{i+k} x_k = \lambda \bar{\alpha}^i, \qquad i = 0, 1, \ldots n,$$

λ étant indépendant de i. Par ces relations et par $h(\alpha) = 1$, on trouve pour le polynome $h(t)$ qui est la solution de ce problème,

(14) $$h(t) = \frac{\begin{vmatrix} 0 & t^k \\ \bar{\alpha}^i & c_{i+k} \end{vmatrix}_0^n}{\begin{vmatrix} 0 & \alpha^k \\ \bar{\alpha}^i & c_{i+k} \end{vmatrix}_0^n}$$

et pour la valeur $\varrho_n(\alpha)$ de ce minimum,

(15) $$\varrho_n(\alpha) = h(c)\bar{h}(c) = \frac{|c_{i+k}|_0^n}{-\begin{vmatrix} 0 & \alpha^k \\ \bar{\alpha}^i & c_{i+k} \end{vmatrix}_0^n} > 0.$$

On peut interpréter ce résultat aussi de la manière suivante:

(15′) $$\frac{1}{\varrho_n(\alpha)} = \frac{-\begin{vmatrix} 0 & \alpha^k \\ \bar{\alpha}^i & c_{i+k} \end{vmatrix}_0^n}{|c_{i+k}|_0^n}$$

est la plus grande valeur que pourra avoir $|f(\alpha)|^2$, le polynome $f(t)$ de degré $\leqq n$ étant assujetti à la relation

$$\int_{-\infty}^{+\infty} |f(t)|^2 d\varphi(t) = 1,$$

$\varphi(t)$ désignant une solution du problème des moments.

En tenant encore compte de ce que tout polynome non négatif $G(t)$ de degré $\leqq 2n$ peut s'écrire $G(t) = p^2(t) + q^2(t)$, p et q étant de degré $\leqq n$, on a encore le corollaire suivant qui va jouer un rôle fondamental dans nos recherches:

$G(t)$ étant un polynome non négatif de degré $\leqq 2n$, on aura en chaque point α

(16) $$|G(\alpha)| \leqq \frac{G(c)}{\varrho_n(\alpha)}.$$ [1]

[1] Il est facile de voir qu'en tout point imaginaire, il y aura le signe $<$ dans cette formule. La question se pose alors, si la limite trouvée pour $|G(\alpha)| : G(c)$ pourra être notablement abaissée. Or, il est à peu près évident qu'il existe des polynomes non négatifs tels que ce quotient soit

13. Le polynome $h(t)$ peut être facilement exprimé par les polynomes quasi-orthogonaux de degré $(n + 1)$ qui s'annulent en $\bar{\alpha}$. Reprenons en effet le raisonnement par lequel nous avons déterminé ce polynome. Tout polynome $v(t)$ de degré $\leqq n$ qui s'annule en α, peut s'écrire $v(t) = (t - \alpha)\,(\xi_0 + \xi_1 t + \cdots + \xi_{n-1} t^{n-1})$. La relation $h(c)\bar{v}(c) = 0$ étant remplie pour tous ces polynomes, on obtient

$$(c - \bar{\alpha})\, c^k h(c) = 0 \qquad k = 0, 1, \ldots n - 1.$$

Ces relations montrent que $(t - \bar{\alpha})\,h(t)$ est un polynome quasi-orthogonal de degré $(n + 1)$ qui manifestement s'annule en $\bar{\alpha}$. Or les polynomes qui ont ces propriétés, sont déterminés à un facteur constant près. On aura donc en désignant un tel polynome par $q(t)$ et en tenant compte de la relation $h(\alpha) = 1$, pour α imaginaire

$$h(t) = \frac{q(t)\,(\alpha - \bar{\alpha})}{q(\alpha)\,(t - \bar{\alpha})} \tag{17}$$

et pour α réel

$$h(t) = \frac{q(t)}{q'(\alpha)\,(t - \alpha)}. \tag{17'}$$

$> 1 : 4\varrho_n(\alpha)$. Tel est le cas p. ex. ou pour le carré de la partie réelle ou pour celui de la partie imaginaire de $h(t)$. M. H. PRAWITZ qui a bien voulu se charger de calculer la limite exacte, a montré par un calcul très intéressant que cette limite $T_n(\alpha)$ est donnée par la formule

$$T_n(\alpha) = \frac{-\begin{vmatrix} 0 & \alpha^k \\ \bar{\alpha}^i & c_{i+k} \end{vmatrix}_0^n + \left| \begin{vmatrix} 0 & \alpha^k \\ \alpha^i & c_{i+k} \end{vmatrix}_0^n \right|}{2\,\left| c_{i+k} \right|_0^n}.$$

Le polynome $G(t)$ pour lequel la limite sera atteinte, est naturellement le carré d'un certain polynome $g(t)$. Ce dernier polynome s'exprime facilement par le polynome $h(t)$ du texte. La formule ci-dessus montre immédiatement que l'on a

$$\frac{1}{2\,\varrho_n(\alpha)} \leqq T_n(\alpha) < \frac{1}{\varrho_n(\alpha)}.$$

Ajoutons que si $\sum\limits_0^n a_{ik} x_i \bar{x}_k$ est un hermitien défini positif quelconque, le problème de le rendre minimum sous la condition $|x_0\alpha_0 + x_1\alpha_1 + \cdots + x_n\alpha_n| = 1$, les α étant des nombres complexes, conduit à une formule analogue à (15) ou (15'). Le problème restreint à des valeurs réelles de x conduit à une expression analogue à $T_n(\alpha)$.

14. Il est clair que $\varrho_n(\alpha)$ est positif. Des propriétés minima on conclut en outre que $\varrho_n(\alpha)$ n'est pas croissant avec n. (On voit facilement que pour α imaginaire $\varrho_n(\alpha)$ décroît quand n croît. *Cf.* p. ex. l'expression dans la note[2] de la p. 39 et, aussi pour α réel, *I*, p. 4—5.)

Il existera donc la limite

$$\varrho(\alpha) = \lim_{n \to \infty} \varrho_n(\alpha). \tag{18}$$

Nous avons déjà fait allusion au rôle important qui revient aux fonctions $\varrho_n(\alpha)$ et $\varrho(\alpha)$.[1]

Mettons en évidence une propriété remarquable de $\varrho_n(\alpha)$ en démontrant que $\varrho_n(\alpha)$ *est une fonction décroissante de la valeur absolue de l'ordonnée de* α. Cela résulte facilement de la propriété minimum de $\varrho_n(\alpha)$. Posons $\alpha = x + iy$ et admettons que $y \geqq 0$. D'après ce qu'on a vu au n° 10, le polynome $q(t)$ figurant au numéro précédent aura tous ses zéros au demiplan inférieur.[2] Il en sera donc de même de $h(t)$. En posant alors $\alpha' = x + iy'$ avec $y' > y$, on aura $|h(\alpha')| > |h(\alpha)| = 1$. En posant encore $g(t) = h(t) : h(\alpha')$, il vient $g(\alpha') = 1$, et par conséquent $\varrho_n(\alpha') \leqq g(c)\bar{g}(c) = \varrho_n(\alpha) : |h(\alpha')|^2 < \varrho_n(\alpha)$.

15. Soit $q(t)$ un polynome quasi-orthogonal de degré $(n+1)$ et à coefficients réels. Nous savons déjà qu'un tel polynome a tous ses zéros réels et distincts. Désignons ces zéros par $\lambda_1, \lambda_2, \ldots \lambda_{n+1}$ et posons

$$q_j(t) = \frac{q(t)}{(t - \lambda_j)\, q'(\lambda_j)}.$$

---

[1] Le réciproque de l'hermitien $\sum_0^n c_{i+k} x_i \bar{x}_k$ étant

$$-\begin{vmatrix} 0 & x_k \\ \bar{x}_i & c_{i+k} \end{vmatrix}_0^n : |c_{i+k}|_0^n,$$

$\dfrac{1}{\varrho_n(\alpha)}$ est d'après (15') la valeur de ce dernier hermitien pour $x_i = \alpha^i$. Ce fait explique en quelque sorte *a priori* l'importance de $\varrho_n(\alpha)$ et de $\varrho(\alpha)$.

[2] Ce résultat concernant la décroissance de $\varrho_n(\alpha)$ découle immédiatement de la formule explicite donnée dans la note[2] de la p. 39. Pour le raisonnement du texte, il suffit d'ailleurs qu'il n'y ait aucun zéro dans le demi-plan supérieur. Or cela est évident. En effet, le polynome $k(t)$ où un tel zéro serait échangé par le zéro imaginaire conjugué, donnerait $|k(\alpha)| < |h(\alpha)| = 1$ et, en vertu de $|k(t)|^2 = |h(t)|^2$ pour t réel, on aurait toujours $k(c)\bar{k}(c) = \varrho_n(\alpha)$.

Soit maintenant $g(t)$ un polynome quelconque de degré $\leqq 2n$. D'après la formule d'interpolation de LAGRANGE, on aura

$$g(t) = q(t)l(t) + \sum_{j=1}^{n+1} g(\lambda_j) q_j(t)$$

où le degré de $l(t)$ est $\leqq (n-1)$. En vertu des formules (7), il en découle

$$g(c) = \sum_{j=1}^{n+1} \sigma_j g(\lambda_j) \tag{19}$$ [1]

où l'on a posé

$$\sigma_j = q_j(c).$$

En particulier, on a

$$c_k = \sum_{j=1}^{n+1} \sigma_j \lambda_j^k, \qquad k = 0, 1, 2, \ldots 2n. \tag{20}$$

Démontrons que *les nombres σ_j sont positifs.* En posant dans (19) $g(t) = q_j^2(t)$ et en observant que ce polynome non négatif est *nul* pour $t = \lambda_p$ si $p \neq j$ et *un* pour $t = \lambda_j$, on obtient

$$\sigma_j = q_j^2(c) > 0.$$

Remarquons encore que, d'après les formules (15) et (17'), on a aussi

$$\sigma_j = \varrho_n(\lambda_j). \tag{21}$$

Tout polynome quasi-orthogonal ne différant de $q(t)$ que par un facteur constant, fournira évidemment les mêmes λ_j et σ_j, tandis que tout autre polynome quasi-orthogonal aura tous ses zéros différents de $\lambda_1, \lambda_2, \ldots \lambda_{n+1}$.

Si $q(t)$ désigne un polynome orthogonal de degré $(n+1)$ la formule (20) sera valable pour $k = 0, 1, \ldots 2n, 2n+1$, mais

[1] Inversement, on voit facilement que le polynome $q(t)$ de degré $(n+1)$ doit être quasi-orthogonal pour que la relation (19) ait lieu pour tout polynome de degré $\leqq 2n$.

si c'est un polynome orthogonal de degré n, la formule ne sera valable que pour $k = 0, 1, \ldots 2n - 1$.

Retournons au cas général et désignons par $\varphi(t)$ une solution quelconque du problème des moments ou une solution des équations (5). Dès lors, on aura d'après (19) pour tout polynome $g(t)$ de degré $\leqq 2n$,

$$\int_{-\infty}^{+\infty} g(t)\, d\varphi(t) = \sum_{j=1}^{n+1} \sigma_j g(\lambda_j). \tag{22}$$

On voit alors que les zéros λ_j du polynome $q(t)$ et les masses correspondantes σ_j fournissent une espèce de *quadrature mécanique* pour l'intégrale figurant au premier membre. En supposant que $\lambda_1 < \lambda_2 < \cdots < \lambda_{n+1}$ et en posant

$$\begin{aligned} &\varphi^*(t) = \sigma_1 + \sigma_2 + \cdots + \sigma_j \quad \text{pour} \quad \lambda_j < t \leqq \lambda_{j+1}, \\ &(\varphi^*(t) = 0 \quad \text{pour} \quad t \leqq \lambda_1), \qquad (j = 1, 2, \ldots n+1), \end{aligned} \tag{23}$$

$\varphi^*(t)$ sera aussi une solution des équations (5). Ajoutons que c'est la *seule* solution non décroissante qui n'a pas d'autres points de croissance que $\lambda_1, \lambda_2, \ldots \lambda_{n+1}$. En effet, toute autre solution de cette nature donnerait lieu à des équations de la forme (20). Or les σ_j sont entièrement déterminés par ces équations (ils sont d'ailleurs déjà déterminés par les $(n+1)$ premières équations). Nous appellerons les solutions telles que $\varphi^*(t)$ des *solutions du type* (λ_j, σ_j).

16. Soit $q(z)$ un polynome quelconque de degré $(n+1)$ et considérons le produit formel

$$q(z)\left(\frac{c_0}{z} + \frac{c_1}{z^2} + \cdots + \frac{c_{2n}}{z^{2n+1}} + \cdots\right) = p(z) + \frac{q(c)}{z} + \frac{c\,q(c)}{z^2} + \cdots, \tag{24}$$

$p(z)$ signifiant le polynome de degré n qui forme la partie entière de ce produit. On aura, en étendant un peu le symbolisme adopté,

$$p(z) = \frac{q(z) - q(c)}{z - c} \tag{25}$$

ou encore, en désignant par $\varphi(t)$ une solution quelconque des équations (5) (ou une solution des $(n+1)$ premières d'entre elles)

$$(25') \qquad p(z) = \int_{-\infty}^{+\infty} \frac{q(z) - q(t)}{z - t} d\varphi(t).$$

En admettant maintenant que $q(z)$ est quasi-orthogonal, on voit par les relations (7) que le produit (24) ne contiendra aucun terme en $z^{-1}, z^{-2}, \dots z^{-n}$. On en tire de suite le développement convergent pour des valeurs assez grandes de $|z|$,

$$(26) \qquad \frac{p(z)}{q(z)} = \frac{c_0}{z} + \frac{c_1}{z^2} + \cdots + \frac{c_{2n}}{z^{2n+1}} + \frac{c'_{2n+1}}{z^{2n+2}} + \cdots$$

Nous dirons que $p(z)$ est le *numérateur* correspondant à $q(z)$.

Écrivons $q(z)$ sous la forme générale (9) et posons

$$(27) \qquad R_{-1} = 0, \ R_0 = c_0,$$
$$R_1 = c_0 z + c_1, \dots R_n = c_0 z^n + c_1 z^{n-1} + \cdots + c_n;$$

le numérateur correspondant à $q(z)$ sera

$$(28) \qquad p(z) = | c_i c_{i+1} \dots c_{i+n-1} \eta_i R_{i-1} |_0^{n+1}.$$

En désignant par τ un paramètre et par $p_1(z)$ et $p_2(z)$ les numérateurs correspondant à deux polynomes quasi-orthogonaux $q_1(z)$ et $q_2(z)$, le numérateur correspondant à $q_1(z) + \tau q_2(z)$ sera évidemment $p_1(z) + \tau p_2(z)$.

Écrivons encore

$$\frac{p_1(z)}{q_1(z)} - \frac{p_2(z)}{q_2(z)} = \frac{p_1(z) q_2(z) - p_2(z) q_1(z)}{q_1(z) q_2(z)}.$$

En développant le premier membre suivant les puissances négatives de z, ce développement ne contiendra d'après la formule (26) aucun terme en $z^{-1}, z^{-2}, \dots z^{-(2n+1)}$. Le dénominateur du second membre étant de degré $(2n+2)$, son numérateur devra être constant, p. ex.

$$(29) \qquad p_1(z) q_2(z) - p_2(z) q_1(z) = p_1(0) q_2(0) - p_2(0) q_1(0).$$

Ajoutons qu'on voit par une modification légère des raisonnements que les résultats essentiels subsistent sauf un, si $q(z)$ désigne un polynome orthogonal de degré n. Le polynome $p(z)$ sera manifestement de degré $(n-1)$ et l'on aura au lieu de (26)

$$\frac{p(z)}{q(z)} = \frac{c_0}{z} + \frac{c_1}{z^2} + \cdots + \frac{c_{2n-1}}{z^{2n}} + \frac{c'_{2n}}{z^{2n+1}} + \cdots \tag{30}$$

17. On peut arriver aux mêmes fractions $p(z):q(z)$ par la voie suivante:

Étant donnée une suite définie positive $c_0, c_1, \ldots$, il existe toujours[1] une *fraction continue*

$$K(z) = \cfrac{k_1}{z + l_1 - \cfrac{k_2}{z + l_2 - \ddots \cfrac{-k_n}{z + l_n - \ddots}}}$$

avec des k positifs et des l réels, telle que sa réduite d'ordre n, $U_n(z) : V_n(z)$ admette un développement de la forme (30). $U_n(z)$ et $V_n(z)$ sont des polynomes de degré $(n-1)$ et n respectivement. On dit que cette fraction continue est *associée* à la série (considérée au point de vue formel ou asymptotique)

$$\frac{c_0}{z} + \frac{c_1}{z^2} + \cdots + \frac{c_{2n}}{z^{2n+1}} + \cdots \tag{31}$$

De (30) on tire de suite

$$c^k V_n(c) = 0, \quad k = 0, 1, \ldots n-1.$$

Par conséquent V_n est un polynome orthogonal de degré n.

En écrivant $l_{n+1} + \tau$ au lieu de l_{n+1}, on obtient avec M. Hamburger[2] une *réduite généralisée* d'ordre $(n+1)$

[1] *Voir* p. ex. Perron, *l. c.*, p. 393—396.

[2] H. Hamburger, Über eine Erweiterung des Stieltjes'schen Momentenproblems, Math. Ann., I, t. 81, p. 235—319, II, t. 82, p. 120—164, *III*, t. 82, p. 168—187. *Voir* en particulier I, p. 253.

$$(32)\qquad K_{n+1}(z;\tau)=\cfrac{k_1}{z+l_1-\cfrac{\ddots}{\quad\cfrac{-k_{n+1}}{z+l_{n+1}+\tau}}}=\frac{U_{n+1}(z)+\tau U_n(z)}{V_{n+1}(z)+\tau V_n(z)}.$$

Le dénominateur $q(z)=V_{n+1}(z)+\tau V_n(z)$ est un polynome de degré $(n+1)$ qui satisfait aux équations (7). Ce dénominateur est par conséquent un polynome quasi-orthogonal. Inversement, il résulte de la dernière remarque du n° 10 et de la comparaison des formules (26) et (30) que, $q(z)$ étant un polynome quasi-orthogonal et $p(z)$ le numérateur correspondant défini par la formule (24) (ou par (26)), il existe une réduite généralisée, identique à la fraction $p(z):q(z)$. Si $q(z)$ a des coefficients réels, le paramètre τ sera évidemment réel.

18. On a vu plus haut (n° 10) que les polynomes quasi-orthogonaux $q(z)$ à coefficients réels ont tous leurs zéros réels et distincts. En désignant de nouveau ces zéros par $\lambda_1, \lambda_2, \ldots \lambda_{n+1}$, on aura un développement de $p(z):q(z)$ en fractions simples

$$(33)\qquad \frac{p(z)}{q(z)}=\sum_{j=1}^{n+1}\frac{\sigma_j}{z-\lambda_j}.$$

En développant ici suivant les puissances décroissantes de z, la comparaison avec (26) donne

$$c_k=\sum_{j=1}^{n+1}\sigma_j\lambda_j^k,\qquad k=0,1,\ldots 2n.$$

Les σ_j étant déterminés par ces équations, on a retouvé les σ_j du n° 15. On voit en particulier que les σ_j sont *positifs*. Dès lors, la formule (33) montre que $p(z):q(z)$ devra changer de signe entre deux zéros consécutifs de $q(z)$. Il en résulte que *le polynome $p(z)$ de degré n a tous ses zéros réels et distincts et que ces zéros séparent ceux de $q(z)$.*

La formule (33) définit pour la fonction $\dfrac{1}{z-t}$ (où z est considéré comme paramètre) une espèce de quadrature mécanique, analogue à celle donnée par la formule (19). Il y a cependant une différence essentielle. La valeur que (33) fait correspondre à la fonction $\frac{1}{z-t}$ dépend du choix de $q(z)$, tandis que la valeur que (19) fait correspondre à tout polynome $g(t)$ de degré $\leqq 2n$, est indépendante de ce choix.

La formule (32) met en évidence que, z étant un point imaginaire fixe, les réduites généralisées (33) formées avec tous les polynomes quasi-orthogonaux de degré $(n+1)$ et à coefficients réels, décrivent une circonférence. La formule (33) montre d'autre part (les σ_j étant positifs) que cette circonférence sera située au demi-plan opposé à celui de z. Observons encore que, $q_1(z)$ et $q_2(z)$ étant deux polynomes quasi-orthogonaux de degré $(n+1)$, à coefficients réels et de rapport non constant, $p_1(z)$ et $p_2(z)$ étant les numérateurs correspondants et τ étant de nouveau un paramètre réel, la formule

$$(34) \qquad \frac{p_1(z) + \tau p_2(z)}{q_1(z) + \tau q_2(z)}$$

définira évidemment la même circonférence C_n. Il est facile de calculer le centre et le rayon de cette circonférence. D'après des principes connus concernant les transformations linéaires, le centre Ω_n de C_n est donné par la formule

$$\Omega_n = \frac{p_1(z)q_2(\bar{z}) - p_2(z)q_1(\bar{z})}{q_1(z)q_2(\bar{z}) - q_2(z)q_1(\bar{z})},$$

c'est-à-dire que c'est la valeur en z de la réduite généralisée qui a un pôle en $\bar{z}$. Quant au rayon $r_n(z)$, on voit par les mêmes principes

$$(35) \quad r_n(z) = \frac{|p_1(z)q_2(z) - p_2(z)q_1(z)|}{|q_1(z)q_2(\bar{z}) - q_2(z)q_1(\bar{z})|} = \frac{|p_1(0)q_2(0) - p_2(0)q_1(0)|}{|2\Im(q_1(z)q_2(\bar{z}))|}.$$ [1]

On déduit de cette expression presqu'immédiatement que, en posant $z = \xi + i\eta$, on a

[1] Chez M. NEVANLINNA, *l. c.*, p. 23 il y a une erreur légère sur ce point. La valeur qu'il donne pour le rayon est le double de la valeur exacte.

$$(35') \qquad r_n(z) = \frac{\varrho_n(z)}{2|\eta|}.$$

Nous y arriverons en établissant d'abord une formule générale.

19. Les polynomes $q_1(z)$ et $q_2(z)$ ayant la même signification que dans le numéro précédent, on aura

$$q_1(x)q_2(y) - q_2(x)q_1(y) = C(x-y)\begin{vmatrix} 0 & y^k \\ x^i & c_{i+k} \end{vmatrix}_0^n : |c_{i+k}|_0^n,$$

C étant une constante. En effet, en considérant x comme variable et y comme paramètre, le premier membre ainsi que le facteur de C dans le second membre sont tous les deux des polynomes quasi-orthogonaux de x (pour le second membre *cf.* le n° 13) qui sont de même degré et qui s'annulent pour $x = y$. Il en résulte d'après le n° 10 que C est indépendant de x et alors par raisons de symétrie aussi de y. Pour calculer la constante C, divisons le premier membre par $(x-y)$ et désignons le quotient par $F(x, y)$. On aura d'après (25), (29) et la formule qu'on vient d'écrire

$$F(c, y) = p_1(y)q_2(y) - p_2(y)q_1(y) = p_1(0)q_2(0) - p_2(0)q_1(0) =$$
$$= C\begin{vmatrix} 0 & y^k \\ c_i & c_{i+k} \end{vmatrix}_0^n : |c_{i+k}|_0^n = -C.$$

On aura alors la formule définitive

$$-(x-y)\frac{\begin{vmatrix} 0 & y^k \\ x^i & c_{i+k} \end{vmatrix}_0^n}{|c_{i+k}|_0^n} = \frac{q_1(x)q_2(y) - q_2(x)q_1(y)}{p_1(0)q_2(0) - p_2(0)q_1(0)}.$$

En tenant compte de (15), il vient pour $z = \xi + i\eta$ imaginaire,

$$(36) \quad \frac{1}{\varrho_n(z)} = \frac{q_1(z)q_2(\bar{z}) - q_2(z)q_1(\bar{z})}{2i\eta\,(p_1(0)q_2(0) - p_2(0)q_1(0))} = \frac{\Im\,(q_1(z)q_2(\bar{z}))}{\eta\,(p_1(0)q_2(0) - p_2(0)q_1(0))}$$

et pour z réel,

$$(36') \qquad \frac{1}{\varrho_n(z)} = \frac{q'_1(z)\, q_2(z) - q_1(z)\, q'_2(z)}{p_1(0)\, q_2(0) - p_2(0)\, q_1(0)}.$$

La comparaison de (35) et de (36) donne la formule (35').

Sur les intégrales $\int\limits_{-\infty}^{+\infty} \frac{d\varphi(t)}{\alpha - t}$.

20. Nous allons démontrer le théorème suivant[1]:

Soit $\alpha = x + iy$ un nombre imaginaire. Formons avec toutes les solutions non décroissantes $\varphi(t)$ des équations (5) *les valeurs*

$$(37) \qquad H(\alpha) = \int\limits_{-\infty}^{+\infty} \frac{d\varphi(t)}{\alpha - t}.$$

L'ensemble de ces valeurs donne l'intérieur et la périphérie de la circonférence C_n de rayon $\frac{\varrho_n(\alpha)}{2|y|}$ décrite par les réduites généralisées $p(\alpha):q(\alpha)$ à coefficients réels et d'ordre $(n+1)$. A tout point intérieur à cette circonférence, il correspond une infinité de solutions, tandis qu'aux points périphériques, il n'en correspond qu'une seule.

Occupons-nous d'abord des points périphériques. L'affixe d'un tel point pourra s'écrire comme réduite généralisée à coefficients réels $p(\alpha):q(\alpha)$. En calculant cette réduite par la formule de quadrature (33), on aura trouvé une solution particulière $\varphi^*(t)$ du type (λ_j, σ_j) qui, substituée en (37), fournit le point périphérique en question. Démontrons que c'est la seule solution non décroissante fournissant ce point. Soit, en effet, $\varphi(t)$ une telle solution. On aura alors d'après (25')

[1] Cf. Nevanlinna, *l. c.*, p. 14.

(38) $$\int_{-\infty}^{+\infty} \frac{d\varphi(t)}{\alpha - t} = \frac{p(\alpha)}{q(\alpha)} = \frac{1}{q(\alpha)} \int_{-\infty}^{+\infty} \frac{q(\alpha) - q(t)}{\alpha - t}\, d\varphi(t),$$

c'est-à-dire que

$$\int_{-\infty}^{+\infty} \frac{q(t)\, d\varphi(t)}{\alpha - t} = 0.$$

En séparant les parties réelle et imaginaire, on trouve

(38′) $$\int_{-\infty}^{+\infty} \frac{q(t)}{|\alpha - t|^2}\, d\varphi(t) = 0$$

et

(38″) $$\int_{-\infty}^{+\infty} \frac{t q(t)}{|\alpha - t|^2}\, d\varphi(t) = 0.$$

Soit maintenant $L(t)$ un polynome quelconque de degré $\leqq (n+1)$. On pourra écrire

$$L(t) = A + Bt + (\alpha - t)(\bar{\alpha} - t)\, l(t),$$

A et B étant des constantes et $l(t)$ étant de degré $\leqq (n-1)$. Alors $q(t)$ étant d'une part quasi-orthogonal, d'autre part les formules (38′) et (38″) ayant lieu, on trouve

$$\int_{-\infty}^{+\infty} \frac{q(t)\, L(t)}{|\alpha - t|^2}\, d\varphi(t) = 0$$

et, en particulier, pour $L(t) = q(t)$,

(39) $$\int_{-\infty}^{+\infty} \frac{q^2(t)}{|\alpha - t|^2}\, d\varphi(t) = 0.$$

On voit que $\varphi(t)$ ne pourra admettre d'autres points de croissance que les zéros de $q(t)$, mais alors d'après le n° 15, $\varphi(t)$ sera identique à la solution du type (λ_j, σ_j) déterminé par $q(t)$.

Le dernier résultat met aussi en évidence que nulle solution décroissante ne donnera une valeur $H(\alpha)$ qui tombe à l'extérieur de la circonférence C_n. Admettons en effet, par impossible, que $\psi(t)$ soit une telle solution et posons $H(\alpha) = Z$. Tirons par le point Z une ligne droite qui coupe la circonférence en deux points Z_1 et Z_2, Z_1 étant plus rapproché de Z que Z_2. Il existe deux réduites généralisées $p_1(\alpha):q_1(\alpha)$ et $p_2(\alpha):q_2(\alpha)$ égales respectivement à Z_1 et à Z_2. Désignons les solutions du type (λ_j, σ_j) correspondant à ces réduites par $\varphi_1(t)$ et $\varphi_2(t)$ respectivement. On pourra évidemment trouver deux nombres positifs a et b tels que $a + b = 1$ et $Z_1 = aZ + bZ_2$. La fonction $\varphi(t) = a\psi(t) + b\varphi_2(t)$ sera alors une solution non décroissante des équations (5) qui, substituée en (37), donne la valeur Z_1. Mais alors, d'après ce qui précède, $\varphi(t)$ sera identique à $\varphi_1(t)$. Or, cela est impossible puisque parmi les points de croissance de $\varphi(t)$ figurent ceux de $\varphi_2(t)$, c'est-à-dire les zéros de $q_2(t)$, différents des zéros de $q_1(t)$, seuls points de croissance de $\varphi_1(t)$. On voit donc que les solutions non décroissantes et différentes de celles du type (λ_j, σ_j) fournissent des points intérieurs à la circonférence.[1] Inversement, un tel point étant donné, il est clair que, à l'aide des solutions (λ_j, σ_j), on peut construire une infinité de solutions différentes qui fournissent ce point.

---

[1] Voici d'ailleurs, pour le théorème qui précède, une seconde démonstration, à laquelle nous avons déjà fait allusion au n° 8. Posons de nouveau $\alpha = x + iy$. Désignons par ξ et η deux nombres réels quelconques et déterminons deux polynomes quasi-orthogonaux de degré $(n + 1)$, à coefficients réels, $q_1(t)$ et $q_2(t)$, tels que l'on ait pour $t = \alpha$ et $t = \bar{\alpha}$

$$q_1^2(t) = \xi(x - t) - \eta y = - q_2^2(t).$$

On a alors, $\varphi(t)$ désignant une solution non décroissante des équations (5),

$$\xi \Re[H(\alpha)] + \eta \Im[H(\alpha)] = \int_{-\infty}^{+\infty} \frac{\xi(x - t) - \eta y}{|\alpha - t|^2}\, d\varphi(t) =$$

21. En passant de n à $(n+1)$, le nombre des équations (5) aura augmenté. Les solutions du nouveau système formeront donc seulement une partie de celles de l'ancien système. Substituées en (37), elles donneront donc lieu à un cercle C_{n+1} de rayon $\frac{\varrho_{n+1}(\alpha)}{2|y|}$, intérieur à C_n. Le polynome orthogonal de degré $(n+1)$ pouvant être considéré comme un polynome quasi-orthogonal, soit de degré $(n+1)$, soit de degré $(n+2)$, les deux circonférences auront un point commun où elles se touchent. On arrivera enfin à une *circonférence-limite* *C de rayon* $\frac{\varrho(\alpha)}{2|y|}$, intérieure à toutes ces circonférences. Il

---

$$= \int_{-\infty}^{+\infty} \frac{\xi(x-t) - \eta y - q_1^2(t)}{(\alpha - t)(\bar{\alpha} - t)} d\varphi(t) + \int_{-\infty}^{+\infty} \frac{q_1^2(t)\, d\varphi(t)}{|\alpha - t|^2}.$$

Le facteur de $d\varphi(t)$ dans le premier terme du dernier membre est un polynome de t de degré $\leqq 2n$. L'intégrale en question pourra donc se calculer par quadrature mécanique et l'on pourra en particulier choisir comme points de concentration les zéros de $q_1(t)$. En désignant, comme d'habitude, le numérateur correspondant à $q_1(t)$ par $p_1(t)$, on aura

$$\xi \Re[H(\alpha)] + \eta \Im[H(\alpha)] = \xi \Re\left(\frac{p_1(\alpha)}{q_1(\alpha)}\right) + \eta \Im\left(\frac{p_1(\alpha)}{q_1(\alpha)}\right) + \int_{-\infty}^{+\infty} \frac{q_1^2(t)}{|\alpha - t|^2} d\varphi(t) \geqq$$

$$\geqq \xi \Re\left(\frac{p_1(\alpha)}{q_1(\alpha)}\right) + \eta \Im\left(\frac{p_1(\alpha)}{q_1(\alpha)}\right).$$

On trouve d'une façon analogue en changeant ξ et η en $-\xi$ et $-\eta$,

$$\xi \Re[H(\alpha)] + \eta \Im[H(\alpha)] = \xi \Re\left(\frac{p_2(\alpha)}{q_2(\alpha)}\right) + \eta \Im\left(\frac{p_2(\alpha)}{q_2(\alpha)}\right) - \int_{-\infty}^{+\infty} \frac{q_2^2(t)}{|\alpha - t|^2} d\varphi(t) \leqq$$

$$\leqq \xi \Re\left(\frac{p_2(\alpha)}{q_2(\alpha)}\right) + \eta \Im\left(\frac{p_2(\alpha)}{q_2(\alpha)}\right),$$

le signe d'égalité ne pouvant arriver dans l'une ou l'autre des relations ci-dessus que si la solution est du type (λ_j, σ_j), les λ_j étant les zéros de $q_1(t)$ ou de $q_2(t)$. On aura alors respectivement $H(\alpha) = p_1(\alpha) : q_1(\alpha)$ ou $H(\alpha) = p_2(\alpha) : q_2(\alpha)$.

est alors évident que $\varphi(t)$ *étant une solution non décroissante du problème des moments,* c'est-à-dire des équations (3) en nombre infini, *la valeur* (37) *correspondante tombera ou sur la circonférence C ou à son intérieur.* Inversement, en appliquant les principes généraux développés au n° 6, on voit *qu'il existe au moins une solution telle que la valeur* (37) *formée avec cette solution coïncide avec un point arbitraire du domaine circulaire C.*

22. Cela étant, on voit par le n° 8 que la condition nécessaire et suffisante pour que le problème des moments soit déterminé est que le rayon de la circonférence-limite soit zéro et cela pour tout nombre imaginaire. D'après ce qu'on a dit au même numéro, il suffit aussi qu ce rayon soit nul p. ex. sur un segment de droite parallèle à l'axe imaginaire. Or, on vient de voir que pour $\alpha = x + iy$, ce rayon est égal à $\frac{\varrho(\alpha)}{2|y|}$. D'autre part, d'après le n° 14, cette dernière valeur est une fonction non croissante du module de l'ordonnée, elle s'annule donc sur tout un segment (infini), si elle s'annule en un seul point réel ou imaginaire. Cette dernière hypothèse suffit donc pour que le problème des moments soit déterminé. Or, dans ce cas, le rayon en question et par conséquent $\varrho(\alpha)$ devra être zéro en tout point imaginaire. En somme, on a démontré:

Pour que le problème des moments soit déterminé, il faut que l'on ait $\varrho(\alpha) = 0$ *en tout point imaginaire. Il en sera ainsi, dès que cette égalité a lieu en un seul point réel ou imaginaire.*

Ajoutons encore que les résultats de notre première Note (*cf. I*, p. 18) mettent en évidence que, dans le cas déterminé, on a $\varrho(\alpha) = 0$, sauf au plus dans un ensemble dénombrable de valeurs réelles.

Par ce qui précède, on a aussi retrouvé un critère remarquable dû à M. HAMBURGER et établi d'une manière analogue à la nôtre par M. NEVANLINNA[1] et par M. CARLEMAN.[2] M. HAMBURGER[3] dit que la fraction continue considérée au n° 17

---

[1] *L. c.*, p. 21—34.

[2] T. CARLEMAN, Sur le problème des moments, C. R. t. 174 (1922), p. 1680—1682.

[3] *L. c., I*, p. 289.

converge complètement au point α, s'il existe une valeur $K(\alpha)$ telle que, en choisissant n assez grand, $|K(\alpha) - K_n(\alpha; \tau)|$ peut être rendu aussi petit que l'on veut et cela indépendamment du paramètre réel τ. M. HAMBURGER a démontré que la condition nécessaire et suffisante pour que le problème des moments soit déterminé est que la fraction continue converge complètement en tout point imaginaire α. Or lesdites réduites généralisées décrivent les circonférences C_n, la convergence complète revient donc à ce que les rayons de ces circonférences tendent vers zéro avec $\frac{1}{n}\cdot$

23. N'en ayant pas besoin dans la suite, nous pourrons sans inconvénient supprimer ici une partie considérable de nos recherches, en renvoyant au beau Mémoire de M. NEVANLINNA, qui, comme nous l'avons dit plus haut, est arrivé aux mêmes résultats, son point de départ et sa méthode ayant été absolument différents des nôtres. Je vise ici, en particulier, les résultats suivants concernant le cas d'indétermination.

A tout point de la circonférence-limite appartenant à un point imaginaire α, il ne correspond qu'une seule solution du problème des moments telle que la valeur $H(\alpha)$ donnée par la formule (37) tombe en ce point. L'ensemble de ces solutions correspondant à différents points de la circonférence est indépendant du choix de α. Les seuls points de croissance d'une telle solution $\varphi^*(t)$ sont les pôles d'une certaine fonction méromorphe

$$H(t) = \int_{-\infty}^{+\infty} \frac{d\varphi^*(t)}{z-t} = \frac{P(z)}{Q(z)}\cdot$$

$P(z)$ et $Q(z)$ sont des fonctions entières transcendantes, fonctions-limites des numérateurs et dénominateurs de certaines réduites généralisées. En choisissant parmi ces fonctions méromorphes deux fonctions différentes, $P_1(z) : Q_1(z)$ et $P_2(z) : Q_2(z)$, toute autre fonction $P(z) : Q(z)$ pourra s'écrire sous la forme

$$\frac{P(z)}{Q(z)} = \frac{P_1(z) + \tau P_2(z)}{Q_1(z) + \tau Q_2(z)},$$

τ étant un paramètre réel. En désignant par $\varphi(t)$ une solution différente des solutions particulières ci-dessus et en admettant, pour fixer les idées, que z est situé au demi-plan supérieur, on aura

$$H(z) = \int_{-\infty}^{+\infty} \frac{d\varphi(t)}{z - t} = \frac{P_1(z) + \omega(z) P_2(z)}{Q_1(z) + \omega(z) Q_2(z)},$$

$\omega(z)$ étant une fonction holomorphe dans ce demi-plan, dont la partie imaginaire y est de signe constant, ce signe étant le même que celui de la partie imaginaire de $Q_1(z) : Q_2(z)$. Inversement, à chaque fonction $\omega(z)$ qui possède ces propriétés, il correspond une solution du problème des moments satisfaisant à la relation ci-dessus. Les valeurs $H(z)$ correspondant à ces solutions tombent à l'intérieur de la circonférence-limite appartenant aux différents points z. A tout point intérieur à la circonférence appartenant à un certain point z, il correspond une infinité de solutions.

Lesdites fonctions méromorphes interviennent déjà chez M. Hamburger qui s'occupe surtout d'une autre propriét extrémale très remarquable des solutions correspondantes. L'existence de ces fonctions se démontre facilement par les méthodes développées dans notre première Note.

Étude approfondie de la fonction $\varrho(\alpha)$.

24. Le rôle central que cette fonction joue dans toute la théorie et qui est assez mis en évidence par les résultats précédents et par ceux de notre première Note, se manifestera encore davantage dans la démonstration de notre théorème final (n° 27). Cependant, pour y arriver, il nous faudra approfondir encore l'étude de cette fonction.

Nous admettrons qu'on se trouve dans le cas *d'indétermination.* Nous savons par ce qui précède que, dans ce cas, $\frac{1}{\varrho(\alpha)}$ sera partout fini. Nous avons d'ailleurs démontré dans notre première Note (p. 22) que, dans ce cas, $\frac{1}{\varrho(\alpha)}$ était une fonction entière de α et de $\bar{\alpha}$ (résultat dont nous n'aurons pas besoin ici) et qu'on avait l'inégalité (*cf.* *I*, p. 18)

$$\frac{1}{\varrho(\alpha)} < \gamma e^{\gamma' r^2}, \qquad r = |\alpha|,$$

γ et γ' étant deux constantes positives. Nous y avons aussi remarqué qu'il était facile de démontrer que

$$\frac{1}{\varrho(\alpha)} < \gamma e^{\varepsilon(r) r^2}, \qquad \lim_{r \to \infty} \varepsilon(r) = 0.$$

Or, ici, nous nous proposons d'établir une inégalité beaucoup plus précise, savoir

$$\frac{1}{\varrho(\alpha)} < \gamma e^{\varepsilon(r) r}.$$

25. Notre démonstration repose essentiellement sur les relations qui ont lieu entre $\varrho_n(\alpha)$ et les polynomes quasi-orthogonaux de degré $(n+1)$.

Envisageons, pour fixer les idées, les polynomes quasi-orthogonaux particuliers, utilisés dans les travaux de MM. Hamburger et Nevanlinna et dans les nôtres (*cf.* aussi les formules (11) et (12))

$$q_n^*(t) = \begin{vmatrix} t & c_1 & c_2 & \dots c_n \\ t^2 & c_2 & c_3 & \dots c_{n+1} \\ \cdot & \cdot & \cdot & \cdot \\ t^{n+1} & c_{n+1} & c_{n+2} & \dots c_{2n} \end{vmatrix} : |c_{i+k}|_0^n$$

et

$$\dot{q}_n(t) = \begin{vmatrix} 1 & 0 & c_0 & \dots c_{n-1} \\ t & c_0 & c_1 & \dots c_n \\ \cdot & \cdot & \cdot & \cdot \\ t^{n+1} & c_n & c_{n+1} & \dots c_{2n} \end{vmatrix} : |c_{i+k}|_0^n.$$

Désignons par $p_n^*(t)$ et $\dot{p}_n(t)$ les numérateurs correspondants. On aura $q_n^*(0)=0$, $\dot{q}_n(0)=1$ et d'après (28), $p_n^*(0)=1$, $\dot{p}_n(0)=0$, Par conséquent, en vertu de (29)

$$\left|\frac{p_n^*(\alpha)}{q_n^*(\alpha)}-\frac{\dot{p}_n(\alpha)}{\dot{q}_n(\alpha)}\right|=\frac{1}{|q_n^*(\alpha)\,\dot{q}_n(\alpha)|}.$$

Puisque la distance de deux points d'une circonférence n'est pas supérieure au diamètre, on aura (*voir* le n° 20) en posant $\alpha=x+iy$ et en supposant $y>0$,

$$\frac{1}{|q_n^*(\alpha)\,\dot{q}_n(\alpha)|}\leqq\frac{\varrho_n(\alpha)}{y}.\text{[1]}\tag{40}$$

L'inégalité manifeste $\varrho_n(\alpha)<\varrho_0(\alpha)=c_0$ donne alors

$$\frac{1}{|q_n^*(\alpha)\,\dot{q}_n(\alpha)|}<\frac{c_0}{y}.\text{[2]}\tag{41}$$

D'autre part on trouve facilement (*cf.* (15)) par un calcul direct que, en posant $g(t)=\dfrac{q_n^*(t)}{t}$, on aura

$$g^2(c)=\frac{|c_{i+k}|_1^n}{|c_{i+k}|_0^n}=\frac{1}{\varrho_n(0)},$$

et alors, en vertu de (16),

$$|q_n^*(\alpha)|^2=|\alpha|^2|g(\alpha)|^2<\frac{|\alpha|^2g^2(c)}{\varrho_n(\alpha)}=\frac{|\alpha|^2}{\varrho_n(0)\,\varrho_n(\alpha)}.\tag{42}$$

En posant enfin

---

[1] Cette inégalité résulte aussi de la formule explicite (35).

[2] La formule (33) donne immédiatement une inégalité analogue avec $\dfrac{2c_0}{y}$ dans le second membre. Cette inégalité suffirait aussi à nos besoins.

$$R(t) = \frac{q_n(t) - 1}{t},$$

on aura d'après notre deuxième Note (p. 8—9)

$$R^2(c) = \begin{vmatrix} 0 & 0 & c_0 & \dots & c_{n-1} \\ 0 & c_0 & c_1 & \dots & c_n \\ c_0 & c_1 & c_2 & \dots & c_{n+1} \\ \cdot & \cdot & \cdot & \cdot & \cdot \\ c_{n-1} & c_n & c_{n+1} & \dots & c_{2n} \end{vmatrix} : \left| c_{i+k} \right|_0^n = \Delta_n^0.^{1}$$

Or, M. Hamburger a démontré et nous en avons donné une autre démonstration (*II*, p. 13) que, dans le cas indéterminé, Δ_n^0 reste borné[2], c'est-à-dire qu'il existe un nombre positif $\varkappa$

[1] Ceci se vérifie facilement. Écrivons, en effet, $R(t) = R(0) + tL(t)$. On a $R^2(t) = R(t)[R(0) + tL(t)] = [q_n(t) - 1]\, L(t) + R(0)\, R(t)$. En observant que $R(c) = 0$ et que $q_n(t)$ est quasi-orthogonal, on trouve $R^2(c) = -L(c)$ ce qui vérifie la formule en question.

[2] En voici encore une démonstration qui, en principe, se trouve chez M. Nevanlinna (*l. c.*, p. 25). Écrivons avec les notations (27) pour $j > 0$,

$$C_j(z) = \left| R_{i-1} c_i \dots c_{i+j-1} \right|_0^j : \sqrt{\left| c_{i+k} \right|_0^j \left| c_{i+k} \right|_0^{j-1}},$$

$$D_j(z) = \left| z^i c_i \dots c_{i+j-1} \right|_0^j : \sqrt{\left| c_{i+k} \right|_0^j \left| c_{i+k} \right|_0^{j-1}},$$

$$C_0(z) = 0, \quad D_0(z) = \frac{1}{c_0}.$$

$D_j(z)$ sera alors un polynome orthogonal de degré j normé par la condition $D_j^2(c) = 1$, $C_j(z)$ le numérateur correspondant. On aura d'autre part, en évaluant $\Delta_j^0 - \Delta_{j-1}^0$ et $\frac{1}{\varrho_j(z)} - \frac{1}{\varrho_{j-1}(z)}$ par un théorème connu de Jacobi (*cf.* *I*, p. 10),

$$\Delta_n^0 = \sum_{j=0}^{n} C_j^2(0), \qquad \frac{1}{\varrho_n(z)} = \sum_{j=0}^{n} \left| D_j(z) \right|^2.$$

Puisque tous les zéros de $C_j(z)$ sont réels, on a $|C_j(0)| < |C_j(iy)|$. D'autre part, d'après (33), on a $|C_j(iy)| < \frac{c_0}{y} |D_j(iy)|$. On aura donc en définitive

tel que $\Delta_n^0 < \varkappa^2$. On aura donc de nouveau d'après la formule (16)

$$|R(\alpha)|^2 < \frac{\varkappa^2}{\varrho_n(\alpha)}$$

et

$$|q_n(\alpha)| < \frac{\varkappa|\alpha|}{\sqrt{\varrho_n(\alpha)}} + 1 < \frac{\varkappa|\alpha| + \sqrt{c_0}}{\sqrt{\varrho_n(\alpha)}}. \tag{43}$$

Les inégalités (42) et (43) mettent en évidence le fait connu que, dans le cas indéterminé, $q_n^*(\alpha)$ et $\dot{q}_n(\alpha)$ restent bornés[1] pour toute valeur finie de α. En posant maintenant $Q_n(\alpha) = q_n^*(\alpha)\,\dot{q}_n(\alpha)$, on aura, d'après (40)—(43), les relations

$$\frac{y}{c_0} < \frac{y}{\varrho_n(\alpha)} < |Q_n(\alpha)| < \frac{K|\alpha| + L|\alpha|^2}{\varrho_n(\alpha)} < \frac{K|\alpha| + L|\alpha|^2}{\varrho(\alpha)}, \tag{44}$$

K et L désignant des constantes positives.

26. Soit[2] $\eta > c_0$. Nous aurons alors d'après la dernière inégalité

$$\log|Q_n(t+i\eta)| > \log\frac{\eta}{\varrho_n(t+i\eta)} > \log\frac{\eta}{c_0} > 0. \tag{45}$$

D'autre part, tous les zéros de Q_n étant réels, on aura par le calcul des résidus

$$\Delta_n^0 < \frac{c_0^2}{y^2\varrho_n(iy)} < \frac{c_0^2}{y^2\varrho(iy)}$$

ce qui montre que Δ_n^0 reste borné dans le cas indéterminé.

[1] $q_n(t)$ étant un polynome quasi-orthogonal de degré $(n+1)$ et $p_n(t)$ le numérateur correspondant, on aura toujours $q_n(t) = q_n(0)\,\dot{q}_n(t) + p_n(0)\,q_n^*(t)$. En combinant cette formule avec (42) et (43) on obtient une inégalité générale qui pourra être très utile.

[2] Nous arrangeons le calcul de façon que les fonctions à intégrer soient positives. Cela est commode sans être essentiel.

$$(46) \qquad \log |Q_n(i\eta + i)| = \frac{1}{\pi} \int_{-\infty}^{+\infty} \frac{\log |Q_n(t + i\eta)|}{1 + t^2} dt.$$

Or, grâce à l'inégalité (44), le premier membre reste borné, lorsque n tend vers l'infini. Il en sera donc de même du second membre et à plus forte raison de l'intégrale

$$\frac{1}{\pi} \int_{-\infty}^{+\infty} \frac{\log \dfrac{\eta}{\varrho_n(t + i\eta)}}{1 + t^2} dt.$$

Or, la fonction à intégrer étant croissant avec n, il en résulte par un raisonnement connu la convergence de l'intégrale

$$\int_{-\infty}^{+\infty} \frac{\log \dfrac{\eta}{\varrho(t + i\eta)}}{1 + t^2} dt,$$

ou encore la convergence de

$$(47) \qquad \int_{-\infty}^{+\infty} \frac{\log \dfrac{c_0}{\varrho(t + i\eta)}}{1 + t^2} dt,$$

la fonction à intégrer étant positive (*cf.* (44)). Or, $\varrho(\alpha)$ étant une fonction décroissante du module de l'ordonnée, cette intégrale sera convergente sans aucune restriction sur η. On en conclut, en particulier, la convergence (absolue) de l'intégrale

$$(48) \qquad \int_{-\infty}^{+\infty} \frac{\log \dfrac{c_0}{\varrho(t)}}{1 + t^2} dt,$$

la fonction à intégrer étant encore positive. *Ces derniers résultats sont fondamentaux pour ce qui suit.*

On a d'abord pour t réel, en utilisant l'inégalité (44) et en désignant par M et N deux constantes positives assez grandes

$$\text{(49)} \qquad \log |Q_n(t)| < \log \frac{1}{\varrho(t)} + \log (K|t| + Lt^2) <$$

$$< \log \frac{c_0}{\varrho(t)} + \log (M + Nt^2).$$

Ce point établi, on aura pour $y > 0$,

$$\text{(50)} \qquad \log |Q_n(x+iy)| = \frac{y}{\pi} \int_{-\infty}^{+\infty} \frac{\log |Q_n(t)|}{(t-x)^2 + y^2} dt <$$

$$< \frac{y}{\pi} \int_{-\infty}^{+\infty} \frac{\log \frac{c_0}{\varrho(t)} + \log (M + Nt^2)}{(t-x)^2 + y^2} dt.$$

Admettons maintenant que $\alpha = x + iy$ se trouve dans l'angle

$$\text{(51)} \qquad \theta \leqq \arg \alpha \leqq \pi - \theta,$$

θ étant un nombre positif fixe $< \frac{\pi}{2}\cdot$ Il sera évidemment

$$(t-x)^2 + y^2 \geqq \frac{t^2}{\sin^2 \theta}$$

et, pour $|t| \geqq 1$,

$$(t-x)^2 + y^2 \geqq \frac{1+t^2}{2 \sin^2 \theta} = C(1+t^2).$$

Cela étant, on a pour $T \geqq 1$,

$$(52)\qquad \log|Q_n(x+iy)| < \frac{y}{\pi}\int_{-T}^{+T}\frac{\log\frac{c_0}{\varrho(t)}+\log(M+Nt^2)}{y^2}\,dt+$$

$$+\frac{y}{\pi C}\left(\int_{-\infty}^{-T}+\int_{+T}^{+\infty}\right)\frac{\log\frac{c_0}{\varrho(t)}+\log(M+Nt^2)}{1+t^2}\,dt.$$

Puisque l'intégrale (48) est convergente, on pourra choisir T assez grand pour que le dernier terme du second membre soit $<\frac{\varepsilon y}{2}$, ε étant un nombre positif arbitrairement petit. Dès lors, on pourra choisir y assez grand pour que le premier terme du second membre satisfasse à la même inégalité. Alors, on aura à plus forte raison dans l'angle (51) pour $|\alpha|$ assez grand

$$\log|Q_n(\alpha)| < \varepsilon|\alpha|,$$

et à plus forte raison, grâce à l'inégalité (44),

$$\log\frac{1}{\varrho_n(\alpha)} < \varepsilon|\alpha|,$$

ε étant indépendant de n. On en tire d'abord par le fait souvent utilisé que $\varrho_n(\alpha)$ est une fonction décroissante de $|y|$, pour le demi-plan supérieur, et puis, par symétrie, pour le demi-plan inférieur, l'inégalité, valable pour $|\alpha|$ *assez grand dans tout le plan,*

$$\log\frac{1}{\varrho_n(\alpha)} < \frac{\varepsilon|\alpha|}{\cos\theta}.$$

En faisant tendre n vers l'infini, on aura une inégalité analogue pour $\varrho(\alpha)$, inégalité qui pourra s'écrire

$$(53)\qquad \log\frac{1}{\varrho(\alpha)} < \varepsilon(r)\,r,\quad r=|\alpha|,$$

$\varepsilon(r)$ étant une fonction bornée de r tendant vers zéro avec $\frac{1}{r}\cdot$

On en tire l'inégalité fondamentale, valable pour toute valeur de α,

$$\frac{1}{\varrho(\alpha)} < \gamma e^{\varepsilon(r) r}, \tag{54}$$

γ étant une constante positive.

En rapprochant cette inégalité de celles exprimées par (44), on obtiendra des inégalités analogues à (54) pour les fonctions entières transcendantes vers lesquelles $q_n^*(z)$, $\dot{q}_n(z)$, $p_n^*(z)$, $\dot{p}_n(z)$ tendent quand n tend vers l'infini et pour toutes les combinaisons linéaires de ces fonctions entières. Il en résulte, en particulier, que le genre et l'ordre réel de ces fonctions est $\leqq 1$. Cela permet de préciser certains résultats énoncés à la fin de ma première Note.

Un théorème général. Conclusions.

27. Le critère d'unicité du n° 22 admet une interprétation remarquable que nous allons exposer.

Soit $\alpha = x + iy$ un nombre imaginaire fixe. Posons $F(t) = \Im\left(\frac{1}{\alpha - t}\right) = \frac{-y}{(x-t)^2 + y^2}$ et formons les valeurs $\underline{F}(c)$ et $\overline{F}(c)$ que les opérations inférieure et supérieure font correspondre à $F(t)$. La différence de ces valeurs est évidemment égale au diamètre de la circonférence C appartenant au point α, c'est-à-dire égale à $\frac{\varrho(\alpha)}{|y|}$. D'après les résultats dudit numéro, on pourra donc dire que la condition nécessaire et suffisante pour que le problème des moments soit déterminé, est que l'on ait $\underline{F}(c) = \overline{F}(c)$ pour cette fonction particulière $F(t)$. On aura un résultat analogue, si l'on entend par $F(t)$ une quelconque des fonctions $\xi\Re\left(\frac{1}{\alpha - t}\right) + \eta\Im\left(\frac{1}{\alpha - t}\right)$.

Soit d'autre part x un nombre réel fixe, et considérons la fonction discontinue $F(t)$ définie par les égalités $F(t) = 0$ pour $t \neq x$ et $F(t) = 1$ pour $t = x$. On aura manifestement

$\underline{F}(c) = 0$ et $\overline{F}(c) = \varrho(x)$.[1] Un des résultats du n° 22 pourra donc s'exprimer de la façon suivante: Si, pour cette fonction particulière, on a $\underline{F}(c) = \overline{F}(c)$, le problème des moments est déterminé.[2]

Or nous allons montrer que ces résultats ne tiennent nullement au caractère particulier des fonctions considérées. Admettons en effet qu'on se trouve dans le cas *indéterminé*. Nous montrerons que *la relation*

$$\underline{F}(c) = \overline{F}(c) = \textit{nombre fini},$$

évidemment satisfaite par tout polynome, ne pourra avoir lieu pour aucune autre fonction.

En effet, si la relation est remplie, on pourra trouver une suite de polynomes $p_1(t), p_2(t), \ldots p_m(t), \ldots$ tous inférieurs à $F(t)$ et une autre suite de polynomes $P_1(t), P_2(t), \ldots P_n(t) \ldots$ tous supérieurs à $F(t)$, tels que les valeurs $p_1(c), p_2(c), \ldots p_m(c) \ldots$ et les valeurs $P_1(c), P_2(c), \ldots P_n(c) \ldots$ tendent vers la même limite. On aura donc $P_n(c) - p_m(c) = \varepsilon_{mn}$ où $\varepsilon_{mn} \rightarrow 0$, quand m et n tendent vers l'infini. Désignons par α un nombre quelconque et évaluons la valeur $|P_n(\alpha) - p_m(\alpha)|$. Le polynome $P_n(t) - p_m(t)$ étant positif pour les valeurs réelles de t, on aura d'après (16)

$$|P_n(\alpha) - p_m(\alpha)| < \frac{\varepsilon_{mn}}{\varrho(\alpha)}.$$

On en conclut que les polynomes $P_n(\alpha)$ et $p_m(\alpha)$ convergent

---

[1] La fonction discontinue qu'on vient d'écrire, satisfait à la relation $F(t) = \lim\limits_{y \rightarrow 0} \dfrac{y^2}{(x-t)^2 + y^2}$. Ceci met bien en évidence le lien étroit qui existe entre les significations de $\varrho(\alpha)$ pour les valeurs réelles et imaginaires de la variable. *Cf.* aussi *I*, p. 3—4.

[2] Disons aussi quelques mots de la fonction $F(t) = t^{-2}$ qui ne rentre pas dans l'ordre d'idées que nous poursuivons ici. Pour cette fonction, on a naturellement $\overline{F}(c) = +\infty$, tandis que, comme on montre aisément, $\underline{F}(c) = \lim D_n$, D_n étant une valeur définie dans notre deuxième Note. Nous y avons aussi démontré que $D_n \rightarrow \infty$ entraîne que le problème sera déterminé. Cette condition suffisante peut donc encore s'écrire sous la forme $\underline{F}(c) = \overline{F}(c)$. (*Cf.* pour tout cela *II*, p. 16 et les notes des p. 18 et 20.)

vers la même fonction entière et que cette fonction sera en valeur absolue $< |p_1(\alpha)| + \frac{\text{const.}}{\varrho(\alpha)}$. Or, pour t réel, ces polynomes convergent évidemment vers $F(t)$. On aura donc une fonction entière[1] $F(\alpha)$ qui, d'après (54), satisfait partout à l'inégalité

$$|F(\alpha)| < Ce^{\varepsilon(r)r}, \quad |\alpha| = r, \quad \lim_{r \to \infty} \varepsilon(r) = 0$$

et qui, pour α réel, croît moins vite qu'une certaine puissance de $|\alpha|$. Cette fonction sera donc, d'après un théorème connu de MM. PHRAGMÉN et LINDELÖF[2], un polynome.

En rapprochant ce résultat des définitions du n° 6 et du théorème du n° 7, nous voilà arrivé au théorème suivant, cité déjà dans l'Introduction:

Soit $F(t)$ une fonction continue du champ D, différente de tout polynome, mais d'ailleurs entièrement arbitraire. Pour que le problème des moments soit déterminé, il faut et il suffit que l'on ait pour cette fonction $\underline{F}(c) = \overline{F}(c)$. Si l'on ne suppose pas la continuité, la condition est toujours suffisante sans être nécessaire.

Voici une autre forme de ce théorème.

Ou l'égalité $\underline{F}(c) = \overline{F}(c)$ a lieu pour toute fonction continue du champ D ou bien elle n'a lieu pour aucune fonction non polynome de ce champ. Le problème des moments est déterminé dans le premier cas et indéterminé dans le second.

28. Dans les applications, il s'agit surtout de trouver des conditions *suffisantes* aussi simples et aussi générales que possible. En voici une qui est un corollaire immédiat de notre théorème général.

Si $\frac{\sqrt[2n]{c_{2n}}}{n}$ ne tend pas vers l'infini, le problème des moments est déterminé.

---

[1] Remarquons que les critères rappelés au numéro précédent résultent déjà du fait que les fonctions $F(t)$ qu'on y considère, ne sont pas de fonctions entières.

[2] E. PHRAGMÉN et ERNST LINDELÖF, Sur une extension d'un principe classique de l'Analyse etc., Acta Math. t. 31 (1908), p. 381—406 (*voir* p. 387).

Considérons à cet effet la fonction $F(t) = \cos at$, a étant un paramètre positif. On vérifie aisément que les sommes partielles de la série de MACLAURIN

$$1 - \frac{a^2 t^2}{2!} + \frac{a^4 t^4}{4!} - \cdots + (-1)^n \frac{a^{2n} t^{2n}}{(2n)!} + \cdots$$

sont alternativement plus grandes ou plus petites que $\cos at$.[1] En nous rappelant la définition de $\underline{F}(c)$ et $\overline{F}(c)$, il en résulte que $\overline{F}(c) - \underline{F}(c) < \frac{a^{2n} c_{2n}}{(2n)!}$. Mais alors, d'après notre théorème général, on aura dans le cas d'indétermination

$$\frac{c_{2n} a^{2n}}{(2n)!} > \overline{F}(c) - \underline{F}(c) = \gamma > 0.$$

On en tire

$$\sqrt[2n]{\frac{c_{2n}}{(2n)!}} > \frac{\sqrt[2n]{\gamma}}{a}$$

et alors

$$\liminf \sqrt[2n]{\frac{c_{2n}}{(2n)!}} \geqq \frac{1}{a}.$$

Puisque cette inégalité a lieu pour toute valeur de a, la limite inférieure en question sera infinie, ou ce qui revient au même,

$$\lim_{n \to \infty} \frac{\sqrt[2n]{c_{2n}}}{n} = \infty.$$

[1] *Voir* p. ex. T. J. I'A. BROMWICH, An introduction to the theory of infinite series, 1908, p. 146—147. Le fait en question découle immédiatement de la formule suivante que je dois à M. S. WIGERT,

$$\cos x - \left(1 - \frac{x^2}{2!} + \cdots + (-1)^n \frac{x^{2n}}{(2n)!}\right) =$$

$$= \frac{(-1)^{n-1}}{(2n-1)!} x^{2n} \int_0^1 (1 - \cos xv)(1 - v)^{2n-1} dv.$$

Cette relation étant alors toujours valable quand il y a indétermination, le théorème ci-dessus se trouve démontré.

Le critère ci-dessus pourra encore s'énoncer de la manière suivante:

Si la limite inférieure de $\frac{\sqrt[2n]{c_{2n}}}{n}$ est finie, le problème des moments est déterminé.

On voit l'avantage que ce critère a sur celui de M. PERRON[1] où n^2 figure au dénominateur au lieu de n, et sur celui de MM. HAMBURGER[2] et STRIDSBERG[3] où le dénominateur est le même, mais la limite supérieure figure au lieu de la limite inférieure.

Or, le critère ci-dessus est lui-même compris[4] dans le critère récent très remarquable, dû à M. CARLEMAN[5] que voici:

Si la série $\sum \frac{1}{\sqrt[2n]{c_{2n}}}$ *diverge, le problème des moments est déterminé.* Il serait intéressant de voir si ce critère peut aussi se déduire de notre théorème général ou de théorèmes analogues (*cf.* le n° 31).

---

[1] O. PERRON, Erweiterung eines MARKOFFschen Satzes über die Konvergenz gewisser Kettenbrüche, Math. Ann. t. 74 (1913), p. 545—554. M. PERRON y traite en réalité le problème concernant l'intervalle $(0, \infty)$. Pour ce cas, il trouve comme condition suffisante que la limite inférieure de $\frac{\sqrt[n]{c_n}}{n}$ soit finie, tandis que le critère ci-dessus se traduit dans une condition exigeant seulement qu'il en soit ainsi de $\frac{\sqrt[n]{c_n}}{n^2}$.

[2] H. HAMBURGER, Beiträge zur Konvergenztheorie der STIELTJESschen Kettenbrüche, Math. Zschrft. t. 4 (1919), p. 186—222, *cf.* p. 214—215.

[3] E. STRIDSBERG, *l. c.* Note 3, p. 110.

[4] En effet, les valeurs $\sqrt[2n]{\frac{c_{2n}}{c_0}}$ étant toujours croissantes, la convergence de la série de M. CARLEMAN implique que $\frac{\sqrt[2n]{c_{2n}}}{n} \to \infty$.

[5] T. CARLEMAN, Sur les séries asymptotiques, C. R., t. 174 (1922), p. 1527—1530. Ce critère résulte p. ex. de la convergence de l'intégrale (47), si l'on tient encore compte des inégalités évidentes $\varrho(\alpha) \leqq c_{2n} |\alpha|^{-2n}$. On obtient une démonstration plus directe, si l'on combine les inégalités $|Q_n(\alpha)| > \frac{y}{\varrho_n(\alpha)} > \frac{y |\alpha|^{2k}}{c_{2k}}$ $(k \leqq n)$ avec la formule (46).

29. Envisageons encore le théorème général démontré tout à l'heure sous le point de vue des opérations fonctionnelles qui y figurent. Nous nous restreignons, pour fixer les idées, au champ fonctionnel D' formé par les fonctions continues du champ D. On y a, en somme, défini deux opérations, les opérations inférieure et supérieure qui ont les propriétés suivantes. La valeur attribuée à une fonction quelconque par la première n'est jamais supérieure à la valeur attribuée à cette fonction par la seconde, ces valeurs étant les mêmes quand la fonction en question est un polynome. La première opération est toujours concave, tandis que la seconde est toujours convexe (*voir* le n° 6). Ces opérations ne sont distributives, c'est-à-dire linéaires, que si elles coïncident dans tout le champ D'. Cette coïncidence aura lieu, dès qu'il en est ainsi pour une seule fonction non polynome de ce champ.

30. Disons encore quelques mots du problème de Stieltjes, c'est-à-dire du problème des moments concernant l'intervalle $(0, \infty)$. Dans ce cas, on ne s'occupera que des valeurs que les fonctions $F(t)$ prennent dans cet intervalle plus restreint. On définit les opérations inférieure et supérieure en considérant les polynomes qui, dans cet intervalle, sont respectivement inférieurs ou supérieurs à $F(t)$. On a alors un théorème identique à notre théorème général. En choisissant, en particulier, $F(t) = \frac{1}{t+a}$, a désignant un nombre positif, $\underline{F}(c)$ sera égal à la limite vers laquelle convergent, au point a, les réduites d'ordre pair de la fraction continue de Stieltjes, tandis que $\overline{F}(c)$ est la limite des réduites d'ordre impair. On retrouve alors comme cas particulier du théorème général le critère de Stieltjes: *Pour qu'il y ait détermination, il faut et il suffit que ces deux limites soient égales, c'est-à-dire que la fraction continue converge pour une valeur positive a.*

31. Nous finissons par une extension de nos concepts, extension, dont il serait intéressant d'examiner la portée.

Notre définition des opérations supérieure et inférieure est entièrement analogue à celle de DARBOUX pour les intégrales par excès et par défaut. Là, l'opération positive c'est l'intégrale; on la connaît *a priori* pour les fonctions discontinues constantes par intervalles; ici, on connaît l'opération pour les polynomes. Or, dans la théorie de l'intégrale de RIEMANN, l'extension de la notion d'intégrale se fait de la manière suivante: Une fonction est dite intégrable si ses intégrales par excès et par défaut sont finies et s'accordent, et l'on désigne alors leur valeur commune comme intégrale. Or, nous avons vu que, dans le cas déterminé, ce même principe permet d'attribuer à toute fonction continue du champ D une valeur déterminée et conduit ainsi à une extension de l'opération, tandis que, dans le cas indéterminé, ce principe ne pourra conduire à aucune extension, les opérations inférieure et supérieure étant différentes pour toute fonction non polynome. On est ainsi amené à essayer une autre méthode d'extension[1] en utilisant le procédé des *suites monotones*, appliqué avec tant de succès par M. LEBESGUE, M. F. RIESZ et surtout par M. W. H. YOUNG.[2]

Soit $E(t)$ une fonction de la variable réelle t, telle qu'il existe une *suite monotone* (c'est-à-dire croissante ou décroissante) de polynomes $p_n(t)$ qui, pour toute valeur finie de t, tendent vers cette fonction. On obtiendra ainsi deux catégories de fonctions, les polynomes seuls appartenant à toutes les deux. Les valeurs correspondantes $p_n(c)$ tendront vers une limite déterminée, finie ou infinie. Il est facile de montrer, en modifiant un peu un raisonnement dû à M. F. RIESZ[3], que toute autre suite de même caractère donnera la même valeur limite. On conviendra donc de poser $E(c) = \lim p_n(c)$ et l'on se restreindra aux fonctions pour lesquelles $E(c)$ est *fini*. Cela étant, on étendra encore l'opération à toutes les combinaisons linéaires des fonctions $E(t)$ par le principe distributif, ayant

[1] Bien entendu, il n'est pas question ici d'extensions en quelque sorte arbitraires, comme celle qui intervient dans notre démonstration du théorème d'existence.

[2] *Cf.* p. ex. W. H. YOUNG, On the new theory of integration, Roy. Soc. Proc. A, t. 88 (1913), p. 170—178; On integration with respect to a function of bounded variation, Lond. Math. Soc. Proc. (2), t. 13. (1913), p. 109—150.

[3] *L. c.*, p. 11.

soin de démontrer, ce qui se fait sans difficulté, que cette coordination n'amène aucune ambiguïté.

On aura ainsi étendu l'opération à un champ fonctionnel M[1] plus vaste que celui des polynomes[2] en gardant la distributivité. Or il est facile de voir que l'opération ainsi étendue est aussi positive. Pour une fonction arbitraire $F(t)$, on pourra maintenant définir les valeurs $\underline{F}(c)$ et $\overline{F}(c)$ en remplaçant les polynomes par les fonctions du nouveau champ fonctionnel.[3]

On montre sans peine que pour une fonction $K(t)$ du champ M et pour une solution quelconque $\varphi(t)$ du problème des moments, on a $\int_{-\infty}^{+\infty} K(t)d\varphi(t) = K(c)$, tandis que pour toute fonction continue $F(t)$, telle que les valeurs modifiées $\underline{F}(c)$ et $\overline{F}(c)$ sont finies, l'ensemble des valeurs $\int_{-\infty}^{+\infty} F(t)d\varphi(t)$ est identique à l'ensemble des points du segment $(\underline{F}(c), \overline{F}(c))$, toutes les intégrales en question étant absolument convergentes.[4]

Il est facile de voir par le raisonnement du n° 27 que, dans le *cas d'indétermination,* les fonctions $E(t)$ que nous venons de considérer sont des *fonctions entières* et que la relation $\underline{F}(c) = \overline{F}(c) =$ un nombre fini, ne pourra avoir lieu que dans le cas banal, c'est-à-dire quand $F(t)$ appartient elle-même au champ M. On démontre de même que la répétition du procédé qui a conduit aux fonctions $E(t)$, c'est-à-dire l'emploi

---

[1] Il est manifeste que toute fonction de ce champ peut s'écrire comme la différence de deux fonctions appartenant à la même catégorie.

[2] Ce champ est composé des fonctions $F(t) = \Sigma b_n P_n^2(t)$, les $P_n(t)$ étant des polynomes et $\Sigma b_n P_n^2(c)$ étant absolument convergent. Ce champ comprendra donc p. ex. toute fonction entière $\Sigma a_{2n} t^{2n}$ telle que $\Sigma a_{2n} c_{2n}$ converge absolument.

[3] En formant ces valeurs modifiées $\overline{F}(c)$ et $\underline{F}(c)$, il suffit évidemment de considérer parmi les fonctions $E(t) \geqq F(t)$ celles de la première catégorie et parmi les fonctions $E(t) \leqq F(t)$ celles de la seconde. Remarquons que, pour les fonctions continues du champ D, les nouvelles valeurs seront identiques aux anciennes.

[4] Il est manifeste que ces considérations ont très peu d'intérêt si l'on sait d'avance qu'on se trouve dans le cas déterminé, toute extension pouvant alors se faire directement par l'intégrale de STIELTJES.

de suites monotones de fonctions tirées du champ M ne conduit à aucune nouvelle extension. Cela étant, la question suivante se pose. A-t-on obtenu par les fonctions du champ M *toutes* les fonctions $F(t)$ pour lesquelles les intégrales $\int_{-\infty}^{+\infty} F(t)\, d\varphi(t)$ convergent absolument et coïncident, sinon, comment les obtenir?[1]

---

[1] Il y a un rapport intime entre ces questions et celle du développement d'une fonction arbitraire en série suivant les polynomes orthogonaux et normés $D_n(t)$ (*voir* la note[2] de la p. 39). Voici quelques observations sur ces développements ajoutées pendant la correction des épreuves. Soit $d\varphi(t)$ une distribution de masse dont les moments soient $c_0, c_1, \ldots$ et $f(t)$ une fonction telle que l'intégrale $I = \int_{-\infty}^{+\infty} f^2(t)\, d\varphi(t)$ existe. Admettons en outre, pour fixer les idées, que $f(t)$ soit une fonction continue. Alors les intégrales $a_n = \int_{-\infty}^{+\infty} f(t) D_n(t)\, d\varphi(t)$ existent aussi et donnent lieu (d'abord à un point de vue tout à fait formel) à une espèce de série de FOURIER $\Sigma a_n D_n(t)$. On aura toujours l'inégalité de BESSEL $\Sigma a_n^2 \leqq I$. Si $d\varphi(t)$ donne lieu à un *cas indéterminé*, $\Sigma |D_n^2(\alpha)| = 1/\varrho(\alpha)$ sera borné dans tout domaine borné et la série de FOURIER de $f(t)$ convergera, par conséquent, vers une fonction entière $G(t)$ en général différente de la fonction *arbitraire* $f(t)$ dont on est parti, les coefficients de FOURIER de ces fonctions étant pourtant les mêmes. On aura

$$\Sigma a_n^2 = \int_{-\infty}^{+\infty} G^2(t)\, d\varphi(t) \leqq \int_{-\infty}^{+\infty} f^2(t)\, d\varphi(t),$$

le dernier signe d'égalité étant exclu si l'on n'a pas $f(t) = G(t)$ en tous les points de croissance de $\varphi(t)$. C'est-à-dire que, dans le *cas indéterminé*, le théorème de PARSEVAL n'aura lieu que pour des *fonctions entières* très particulières, savoir précisément celles qui peuvent s'écrire sous la forme $\Sigma a_n D_n(t)$, la série Σa_n^2 étant convergente, et pour les fonctions égales à une telle fonction entière en tous les points de croissance de $\varphi(t)$. Or, pendant la correction des épreuves, j'ai réussi à démontrer que, *dans le cas déterminé, le théorème de* PARSEVAL *a lieu pour toute fonction continue* (*ou toute fonction de* BAIRE) *pour laquelle la valeur I est finie.*

25.

Sur la sommation des séries de Fourier

Acta Szeged Sect. Math. *1* (1923), 104–113

1. Soit $f(x)$ une fonction périodique de période 2π intégrable au sens de M. *Lebesgue*. Concernant la série de *Fourier* d'une telle fonction, on connaît le théorème important de M. *Fejér*[1]) que voici

La série de Fourier de $f(x)$ est sommable par le procédé des moyennes arithmétiques en tout point de continuité de $f(x)$ avec la somme $f(x)$, ou, plus généralement, avec la somme $S(x)$, en tout point x où la limite $S(x) = \lim\limits_{t \to 0} {}^1/_2 (f(x+t) + f(x-t))$ existe.

On connaît aussi la généralisation importante qu'a donnée de ce théorème M. *Lebesgue*[2]) :

En posant $\varphi(t) = f(x+t) + f(x-t) - 2f(x)$ et $\Phi(t) = \int\limits_0^t |\varphi(v)|\, dv$, la série de Fourier de $f(x)$ est sommable par le procédé des moyennes arithmétiques avec la somme $f(x)$ en tout point où l'on a $\Phi'(0) = 0$, c'est-à-dire presque partout.

2. On sait que le procédé des moyennes arithmétiques simples (ce sont les moyennes dont il a été question plus haut) fut généralisé de plusieurs manières. Nous nous restreindrons ici à la généralisation donnée pour des indices entiers par *Cesàro* et étendue à des indices non entiers par MM. *Hadamard* et *Knopp*[3]).

---

1) *L. Fejér*, Untersuchungen über *Fourier*sche Reihen, Math. Ann. 58. (1904) p. 51—69.

2) *H. Lebesgue.* Recherches sur la convergence des séries de *Fourier*, Math. Ann. 61 (1905) p. 251—280; Sur les intégrales singulières, Ann. de Toulouse (3) 1 (1909) p. 25—117, particulièrement p. 88—90.

3) Pour tout ce qui concerne ce sujet, *voir L. Bieberbach*, Neuere Untersuchungen über Funktionen von komplexen Variablen, Encykl. d. math Wissensch, II. C 4, p. 379—532, partic. p. 477 et suiv.

$\sum_{n=0}^{\infty} u_n$ étant une série convergente ou divergente quelconque et k un nombre réel quelconque, posons

(1) $$(1-x)^{-(k+1)} = \sum_{n=0}^{\infty} C_n^{(k)} x^n$$

et écrivons l'égalité formelle

(2) $$(1-x)^{-(k+1)} \sum_{n=0}^{\infty} u_n x^n = \sum_{n=0}^{\infty} S_n^{(k)} x^n.$$

D'ici on tire

(3) $$S_n^{(k)} = \sum_{r=0}^{n} C_{n-r}^{(k)} u_r.$$

Les valeurs

(4) $$S_n^{(k)} / C_n^{(k)}$$

sont les moyennes de *Cesàro* d'ordre k de la série Σu_n. Si la limite

(5) $$\lim_{n \to \infty} S_n^{(k)} / C_n^{(k)} = \sigma$$

existe, la série Σu_n est dite sommable par les moyennes d'ordre k avec la somme σ.

On se restreint en général à des valeurs de $k > -1$. Ici, nous ne considérons que les valeurs $k > 0$.

On voit que les moyennes arithmétiques simples, utilisées par M. *Fejér* correspondent à $k = 1$. On sait aussi qu'une série étant sommable par des moyennes d'un certain ordre, il en est de même pour tout ordre supérieur et que les sommes coïncident. Le théorème qui suit, ne donne donc rien de nouveau que pour $0 < k < 1$.

3. En 1909, dans une Note des *Comptes rendus*,[4]) j'ai énoncé sans démonstration une généralisation du théorème de M. *Fejér* qui revient à ceci:

La série de Fourier est sommable par les moyennes d'ordre k avec la somme $S(x)$ en tout point x où la limite

$$S(x) = \lim_{t \to 0} {}^1/_2 (f(x+t) + f(x-t))$$

existe, et cela pour toute valeur positive de k.

---

[4]) *M. Riesz,* Sur les séries de *Dirichlet* et les séries entières, Comptes rendus, 149 (1909) p. 309—312.

Plus tard mais indépendamment de moi, M. *Chapman*[5]) a retrouvé ce théorème et il en publiait aussi une démonstration.

Plus tard encore M. *Hardy* a généralisé le théorème de M. *Lebesgue* dans la même direction que j'ai généralisé le théorème de M *Fejér*, c'est-à-dire qu'il montrait que *dans les conditions posées par M. Lebesgue la série de Fourier est sommable par des moyennes d'ordre k pour toute valeur positive de k.*[6])

4. Au cours des années, on a de plus en plus reconnu l'importance des théorèmes que nous venons de rappeler, importance dont — pourquoi ne pas le dire — je ne me doutais guère lorsque j'ai énoncé le premier de ces théorèmes. Ainsi on a établi des théorèmes plus ou moins analogues portant sur les séries de *Dirichlet*,[7]) séries de *Laplace*[8]) etc. Mais, c'est un autre point sur lequel je veux insister.

Ces théorèmes sur la sommation à indices non entiers conduisent bien souvent à des théorèmes sur la convergence ordinaire ou sur la sommation à indices entiers. Tel est le cas p. ex. pour les séries de Dirichlet où j'ai démontré[9]) que, *la série Σa_n étant sommable par les moyennes d'ordre k, la série $\Sigma a_n n^{-s}$ sera convergente dès que la partie réelle de s est plus grande que k.* Ce théorème est bien utilisable dans la théorie des séries de *Dirichlet*. Mais il y a plus. M. *Hardy* a remarqué que la combinaison de ce théorème avec la généralisation mentionnée qu'il a donné du théorème de M. *Lebesgue*, conduisait immédiatement au théorème très intéressant, dû à M. *Young*, que voici:

$\Sigma a_n \cos nx + b_n \sin nx$ *étant la série de Fourier d'une fonction*

---

[5]) *S. Chapman*, On non-integral orders of summability of series and integrals, Lond. Math. Soc. Proc. (2) 9 (1910—11) p. 369—409.

[6]) *G. H. Hardy*, On the summability of *Fourier*'s series, Lond. Math. Soc. (2) 12 (1913) p. 365—372.

[7]) *Voir G. H. Hardy* and *M. Riesz*, The general theory of *Dirichlet*'s series, Cambridge Tracts of Math. No. 18 (1915).

[8]) *E. Hilb*, Über die *Laplace*sche Reihe, Math. Zschrft, t. 5 (1919) p. 17—25 et t. 8 (1920) p. 79—90. M. *Hilb* travaille, comme M. *Hardy*, avec les moyennes dont j'ai démontré l'équivalence aux moyennes de *Cesàro*. (Une méthode de sommation équivalente à la méthode des moyennes arithmétiques, Comptes rendus 152 (1911) p. 1651—1654). Pour la série de *Laplace*, *voir* encore *F. Lukács*, Über die *Laplace*sche Reihe, Math. Zschrft, t. 14 (1922) p. 250—262.

[9]) *G. H. Hardy* and *M. Riesz* l. c. p. 57—59.

intégrable au sens de Lebesgue, la série $\Sigma \frac{a_n \cos nx + b_n \sin nx}{n^\epsilon}$ $(\epsilon > 0)$ *convergera presque partout.*[10])

5. Voici maintenant un résultat de MM. *Hardy* et *Littlewood* qui, avec sa démonstration, met le mieux en évidence le rôle de la sommation d'ordre non entier :

La fonction $f(x)$ étant bornée au voisinage d'un certain point x, sa série de Fourier est sommable au point x par les moyennes d'ordre k ou pour toute valeur positive de k ou bien pour aucune valeur de k.[11])

Voici l'idée de la démonstration. On a d'abord le théorème suivant, qui se démontre de la même manière et même un peu plus simplement que notre généralisation du théorème de M. *Fejér* :

Si la fonction $f(x)$ est bornée dans un certain intervalle, les moyennes d'ordre k $(k > 0)$ de sa série de Fourier sont uniformément bornées dans tout intervalle intérieur à cet intervalle.

Il en résulte, en particulier, que ces moyennes sont bornées au point x dont il est question dans le théorème de MM. *Hardy* et *Littlewood*, et cela pour une valeur positive quelconque d'ailleurs arbitrairement petite de k.

D'autre part, on a le théorème suivant que j'ai démontré il y a une dizaine d'années, mais dont la démonstration, que j'ai communiquée à MM. *Hardy* et *Littlewood*, n'a pas encore paru :[12])

Admettons qu'une série est sommable par les moyennes d'un certain ordre et que ses moyennes d'ordre k sont bornées. Alors la série est sommable par des moyennes d'ordre quelconque $> k$.[13])

---

[10]) *G. H. Hardy*, dans le travail cité dans la note[3]) de la p. 2. Toutefois, pour améliorer le résultat de M. *Young* en remplaçant n^ϵ par $\log n$, M. *Hardy* devait recourir à un examen plus direct des sommes partielles de la série de Fourier.

[11]) MM. *Hardy* et *Littlewood* donnent l'énoncé de ce théorème dans leur travail, *Abel*'s theorem and its converse, Lond. Math. Soc. Proc. (2) t. 18 (1918) p. 205—235, *voir* p. 235. La démonstration se trouve dans la Note, On the *Fourier* series of a bounded function, ibid, t. 17 (1917) p. XIII—XV. (Records of Proceedings at Meetings). Nous ne reproduisons ici qu'une partie du résultat en question. Je viens d'apprendre que, tout récemment, MM. *Hardy* et *Littlewood* ont poussé plus loin leurs recherches concernant ce sujet.

[12]) La démonstration d'un théorème général embrassant ce théorème paraîtra bien prochainement.

[13]) Pour des ordres entiers de sommation, ce théorème fut donné par MM. *Hardy* et *Littlewood* eux-mêmes.

En choisissant maintenant k arbitrairement petit, le théorème de MM. *Hardy* et *Littlewood* résulte par la combinaison des deux derniers théorèmes.

Ce qu'il y a de plus intéressant dans la démonstration précédente, c'est, à mon avis, qu'il fait ressortir le point suivant: même, si l'on ne s'intéresse qu'aux résultats concernant la sommation à indices entiers,[14]) on ne peut se passer de théorèmes sur la sommation à indices non entiers.

6. Dans ces derniers temps, en écrivant avec M. *Hilb* un rapport sur la théorie des séries trigonométriques pour l'édition allemande de l'Encyclopédie, j'ai eu l'occasion de lire ou de relire les différentes démonstrations qu'on a données de la généralisation du théorème de M. *Fejér* que j'ai cité au début. J'ai constaté que toutes ces démonstrations sont moins simples[15]) que celle que je possède et que je me permets d'exposer ici.

Nous donnons d'abord quelques formules concernant les nombres $C_n^{(k)}$. De (1) on tire

$$C_n^{(k)} = \frac{(k+1)(k+2)\ldots(k+n)}{n!} = \frac{\Gamma(k+n+1)}{\Gamma(k+1)\,n!} \tag{6}$$

d'où il vient

$$\lim_{n\to\infty} \frac{C_n^{(k)}}{n^k} = \frac{1}{\Gamma(k+1)}. \tag{7}$$

En multipliant la relation (1) par $(1-x)$, on obtient

$$C_n^{(k)} - C_{n-1}^{(k)} = C_n^{(k-1)}. \tag{8}$$

D'autre part, en écrivant $(1-x)^{-(k+1)} = (1-x)^{-k}(1-x)^{-1}$, il vient par (2)

$$S_n^{(k)} = \sum_{r=0}^{n} C_{n-r}^{(k-1)} S_r^{(0)} = \sum_{r=0}^{n} C_{n-r}^{(k-1)} (u_0 + u_1 + \ldots + u_r). \tag{9}$$

---

[14]) Bien entendu, ce n'est pas le cas de MM. *Hardy* et *Littlewood*. Disons encore que MM. *Hardy* et *Littlewood*, en combinant leur résultat avec un résultat de M. *Lebesgue* sur la sommabilité de la série de Fourier par les moyennes d'ordre 2, trouvent un nouveau théorème remarquable, concernant les moyennes les plus intéressantes, les moyennes simples.

[15]) Outre les démonstrations citées *voir T. H. Gronwall*, On the summability of *Fourier*'s series. Am. Math. Soc. Bull. t. 20 (1913—1914) p. 139—146. Une généralisation intéressante du théorème en question fut encore donnée par M. *F. Nevanlinna*, Über die Summation der *Fourier*schen Reihen und Integrale, Övers. av Finska Vetensk. Soc. Förhandl. t. LXIV. 1920—1921. Avd. A. No. 3. p. 1—14.

Cela posé et la série de *Fourier* de $f(x)$ étant

$$\frac{a_0}{2}+\sum_{n=1}^{\infty}(a_n\cos nx+b_n\sin nx),$$

on a pour les moyennes d'ordre k de cette série

$$S_n^{(k)}(x)=\frac{C_n^{(k)}a_0}{2}+\sum_{r=1}^{n}C_{n-r}^{(k)}(a_r\cos rx+b_r\sin rx)=$$

$$(10)\qquad =\frac{1}{\pi}\int_{x-\pi}^{x+\pi}f(\alpha)\left\{\frac{C_n^{(k)}}{2}+\sum_{r=1}^{n}C_{n-r}^{(k)}\cos r(\alpha-x)\right\}d\alpha=$$

$$=\frac{1}{\pi}\int_0^{\pi}(f(x+t)+f(x-t))\left\{\frac{C_n^{(k)}}{2}+\sum_{r=1}^{n}C_{n-r}^{(k)}\cos rt\right\}dt=$$

$$=\frac{C_n^{(k)}}{\pi}\int_0^{\pi}(f(x+t)+f(x-t))\,H_n(t)\,dt$$

où l'on a posé

$$(11)\qquad C_n^{(k)}H_n(t)=\frac{C_n^{(k)}}{2}+\sum_{r=1}^{n}C_{n-r}^{(k)}\cos rt.$$

La différence essentielle entre le cas traité par M. *Fejér* et celui qui nous occupe c'est que, là, le noyau était positif, tandis qu'ici, le nombre des changements de signe de $H_n(t)$ croît indéfiniment avec n.

On a évidemment

$$(12)\qquad \frac{2}{\pi}\int_0^{\pi}H_n(t)\,dt=1.$$

Désignons alors par S un nombre qu'on va préciser plus loin, mais qui jusqu'à nouvel ordre reste arbitraire. En posant alors

$$(13)\qquad \varphi(t)=f(x+t)+f(x-t)-2S,$$

les relations (10) et (12) donneront

$$(14)\qquad \frac{S_n^{(k)}(x)}{C_n^{(k)}}-S=\frac{1}{\pi}\int_0^{\pi}\varphi(t)\,H_n(t)\,dt.$$

Cela posé, toute la démonstration du théorème en question se basera sur les deux inégalités suivantes.

On a pour $0 \leqq t \leqq \pi$, $0 < k < 1$[16])

(15) $$|H_n(t)| < An$$

et

(16) $$|H_n(t)| < Bt^{-(k+1)} n^{-k},$$

A et B étant deux nombres ne dépendant que de k.

L'inégalité (15) résulte immédiatement de la relation (11) et du fait que les nombres $C_n^{(k)}$ forment une suite croissante. On trouve en effet

$$|H_n(t)| < {}^1\!/_2 + n < An.$$

Pour établir (16), désignons par $K_n(t)$ le second membre de (11) et écrivons d'après (9)

$$K_n(t) = \sum_{r=0}^{n} C_{n-r}^{(k-1)} \left({}^1\!/_2 + \cos t + \ldots + \cos r t\right) = \sum_{r=0}^{n} C_{n-r}^{(k-1)} \frac{\sin(r+{}^1\!/_2)t}{\sin {}^t\!/_2} =$$

$$= \frac{1}{\sin {}^t\!/_2} \mathfrak{J} \left\{ \sum_{r=0}^{n} C_{n-r}^{(k-1)} e^{(r+{}^1\!/_2) i t} \right\} =$$

$$= \frac{1}{\sin {}^t\!/_2} \mathfrak{J} \left\{ e^{(n+{}^1\!/_2) i t} \sum_{r=0}^{n} C_r^{(k-1)} e^{-r i t} \right\}$$ [17])

Par conséquent

$$\left| K_n(t) \right| \sin\frac{t}{2} \leqq \left| \sum_{r=0}^{n} C_r^{(k-1)} e^{r i t} \right| = \left| (1-e^{i t})^{-k} - \sum_{r=n+1}^{\infty} C_r^{(k-1)} e^{r i t} \right|$$

$$\leqq \left| 1 - e^{i t} \right|^{-k} + \left| \sum_{r=n+1}^{\infty} C_r^{(k-1)} e^{r i t} \right|.$$

Les nombres positifs $C_r^{(k-1)}$ formant une suite décroissante et tendant vers zéro, on peut écrire

$$\left| \sum_{r=n+1}^{\infty} C_r^{(k-1)} e^{r i t} \right| = \left| \sum_{r=n+1}^{\infty} (C_r^{(k-1)} - C_{r+1}^{(k-1)}) (e^{(n+1) i t} + \ldots + e^{r i t}) \right| =$$

$$= \left| \sum_{r=n+1}^{\infty} (C_r^{(k-1)} - C_{r+1}^{(k-1)}) \frac{e^{(r+1) i t} - e^{(n+1) i t}}{e^{i t} - 1} \right| <$$

$$< \frac{2}{|e^{i t} - 1|} \sum_{r=n+1}^{\infty} (C_r^{(k-1)} - C_{r+1}^{(k-1)}) = \frac{2 C_{n+1}^{(k-1)}}{|e^{i t} - 1|} < \frac{B_1 n^{k-1}}{|e^{i t} - 1|}$$ [18]).

[16]) Ces inégalités subsistent encore pour d'autres valeurs de k. Pour $k = 0$ et pour $k = 1$, on les tire de formules explicites bien connues.

[17]) Nous posons comme d'ordinaire $\mathfrak{J}(a + i b) = b$.

[18]) Nous désignons par B_1, B_2, ... des nombres finis ne dépendant que de k.

Par suite

$$\left|K_n(t)\right| < \frac{1}{\sin\frac{t}{2}}\left(\left|1-e^{it}\right|^{-k} + B_1 n^{k-1}\left|1-e^{it}\right|^{-1}\right),$$

et puisque

$$\left|1-e^{it}\right| = 2\sin\frac{t}{2},$$

il vient

$$\left|K_n(t)\right| < 2^{-k}\left(\sin\frac{t}{2}\right)^{-(k+1)} + \frac{B_1}{2} n^{k-1}\left(\sin\frac{t}{2}\right)^{-2}.$$

On a maintenant pour $\frac{1}{n} < t$, $n^{k-1} < t^{1-k}$. Par conséquent, d'après la dernière inégalité

$$\left|K_n(t)\right| < 2^{-k}\left(\sin\frac{t}{2}\right)^{-(k+1)} + \frac{B_1}{2}\left(\sin\frac{t}{2}\right)^{-2} t^{1-k} < B_2 t^{-(k+1)},$$

D'autre part, pour $0 \leqq t \leqq \frac{1}{n}$, on a d'après (15)

$$\left|K_n(t)\right| < A n^{k+1} \leqq A t^{-(k+1)}.$$

Les deux derniéres inégalités et la formule (7) montrent que (16) est valable pour tout l'intervalle $0 \leqq t \leqq \pi$.

7. Ce point établi, la démonstration de la généralisation que nous avons donnée du théorème de M. *Fejér*, s'achève dans la voie ordinaire.

Montrons d'abord que

$$\frac{1}{\pi}\int_0^\pi \left|H_n(t)\right| dt$$

reste borné, lorsque n tend vers l'infini. Écrivons pour cela, en nous appuyant sur (15), (16) et sur (7)

$$(17)\quad \frac{1}{\pi}\int_0^\pi \left|H_n(t)\right| dt = \frac{1}{\pi}\int_0^{\frac{1}{n}} + \frac{1}{\pi}\int_{\frac{1}{n}}^{\pi} < \frac{A}{\pi}\frac{n}{n} + \frac{B n^{-k}}{\pi}\int_{\frac{1}{n}}^{\pi} t^{-(k+1)} dt < B_4.$$[19])

[19]) On voit par ce qui précède que les conditions d'un théorème général bien connu concernant les intégrales singulières et dû à M. *Lebesgue* et à M. *Haar* sont toutes remplies. Au lieu d'appliquer ce théorème, nous exécutons le raisonnement qui y conduit, dans le cas particulier qui nous occupe.

Admettons maintenant avec M. *Fejér* que $\lim_{t \to 0} \frac{1}{2}(f(x+t)+f(x-t)) = S(x)$ existe. En introduisant dans $\varphi(t)$ (*voir* (13)) cette valeur de S, on aura $\lim_{t \to 0} \varphi(t) = 0$. En choisissant alors ε arbitrairement petit et en déterminant δ de façon que $|\varphi(t)| < \varepsilon$ pour $0 \leqq t \leqq \delta$, on aura d'après (14)

$$\left| \frac{S_n^{(k)}(x)}{C_n^{(k)}} - S(x) \right| \leqq \frac{1}{\pi} \int_0^\pi |\varphi(t)|\,|H_n(t)|\,dt \leqq$$

$$\leqq \frac{\varepsilon}{\pi} \int_0^\delta H_n(t)\,|\,dt + \frac{B\,\delta^{-(k+1)}\,n^{-k}}{\pi} \int_\delta^\pi |\varphi(t)|\,dt.$$

La fonction $f(t)$ étant intégrable au sens de Lebesgue, il en sera de même de la fonction $\varphi(t)$ et l'on aura $\int_\delta^\pi |\varphi(t)|\,dt \leqq \int_0^\pi |\varphi(t)|\,dt < C$, C étant une constante. Il vient alors d'après (17)

$$\left| \frac{S_n^{(k)}(x)}{C_n^{(k)}} - S(x) \right| < B_4\,\varepsilon + C\,\frac{B\,\delta^{-(k+1)} n^{-k}}{\pi}.$$

Le dernier terme du second membre tendant vers zéro avec $\frac{1}{n}$, ce second membre devient pour n assez grand $< B_5\,\varepsilon$.

8. Plaçons nous maintenant dans les hypothèses du théorème de M. *Lebesgue* et démontrons la généralisation que M. *Hardy* a donnée de ce théorème. Soit $\varphi(t) = f(x+t) + f(x-t) - 2f(x)$, $\Phi(t) = \int_0^t |\varphi(v)|\,dv$ et $\Phi'(0) = 0$. Le nombre ε étant arbitrairement petit, déterminons δ de façon que $|\Phi(t)| < \varepsilon t$ pour $0 < t < \delta$. Écrivons enfin $S = f(x)$ dans (13). Il n'y a rien à changer à l'évaluation de l'intégrale $\int_\delta^\pi$. Quant à l'intégrale $\int_0^\delta$, écrivons

$$\int_{\frac{1}{n}}^{\delta} |\varphi(t)|\,|H_n(t)|\,dt < B\,n^{-k} \int_{\frac{1}{n}}^{\delta} |\varphi(t)|\; t^{-(k+1)}\,dt =$$

$$= B\,n^{-k}\left[\Phi(t)\,t^{-(k+1)}\right]_{\frac{1}{n}}^{\delta} + B(k+1)\,n^{-k} \int_{\frac{1}{n}}^{\delta} \Phi(t)\,t^{-(k+2)}\,dt <$$

$$< B\varepsilon + \varepsilon B(k+1)n^{-k}\int_{\frac{1}{n}}^{\delta} t^{-(k+1)}\,dt < B_6\,\varepsilon.$$

Enfin

$$\int_0^{\frac{1}{n}} |\varphi(t)|\;|H_n(t)|\,dt < A\,n\int_0^{\frac{1}{n}} |\varphi(t)|\,dt = A\,n\,\Phi\left(\tfrac{1}{n}\right) < A\,\varepsilon.$$

En résumé, on voit que

$$\frac{1}{\pi}\int_0^{\delta} |\varphi(t)|\;|H_n(t)|\,dt < B_7\,\varepsilon,$$

ce qui achève la démonstration.

Les théorèmes précédents et leurs démonstrations subsistent — *mutatis mutandis* — si $\Sigma\,(a_n \cos n\,x + b_n \sin n\,x)$, au lieu d'être la série de *Fourier* d'une fonction intégrable au sens de *Lebesgue*, est la dérivée formelle de la série de *Fourier* d'une fonction à variation bornée.

26.

Sur un théorème de la moyenne et ses applications

Acta Szeged Sect. Math. *1* (1923), 114–126

1. Soit $\varphi(x)$ une fonction définie pour $x \leqq 0$, intégrable au sens de M. *Lebesgue*. Pour fixer les idées, nous supposons encore que $\varphi(x)$ est une fonction bornée dans tout intervalle fini, hypothèse qui pourrait être remplacée par des hypothèses bien plus générales. Nous entendons par intégrale de *Liouville-Riemann* d'ordre α de la fonction $\varphi(x)$ l'expression

$$(1) \qquad I^{(\alpha)}(\varphi(x)) = \Phi_\alpha(x) = \frac{1}{\Gamma(\alpha)} \int_0^x \varphi(t)(x-t)^{\alpha-1}\, dt, \qquad 0 < \alpha.$$ [1]

D'abord, on voit facilement que $\Phi_\alpha(x)$ est une fonction continue (et même que $\Phi_\alpha(x)$ satisfait à une condition de *Lipschitz* d'ordre α). D'autre part, en exprimant l'intégrale d'*Euler* de la première espèce par la fonction Γ, on établit le fait connu que l'opération en question est distributive par rapport à l'indice α, c'est-à-dire qu'on a

$$(2) \qquad I^{(\alpha+\beta)}(\varphi(x)) = I^{(\alpha)}\left[I^{(\beta)}\varphi(x)\right].$$

On a en particulier

$$(3) \qquad \Phi_1(x) = \int_0^x \varphi(t)\, dt,$$

$$(4) \qquad \Phi_{\alpha+1}(x) = \int_0^x \Phi_\alpha(t)\, dt$$

et

$$(5) \qquad \Phi'_{\alpha+1}(x) = \Phi_\alpha(x).$$

---

[1]) La limite inférieure *zéro* dans l'intégrale (1) peut évidemment être remplacée par un nombre fini quelconque et, avec des conditions supplémentaires sur $\varphi(x)$, aussi par $-\infty$.

2. Admettons maintenant que $0 < \alpha < 1$. Posons

$$(6) \qquad g(x,\xi) = \frac{1}{\Gamma(\alpha)} \int_0^\xi \varphi(t)\,(x-t)^{\alpha-1}\,dt\,, \qquad 0 \leqq \xi < x.$$

On a alors *le théorème de la moyenne*

$$(7) \qquad g(x,\xi) = \Phi_\alpha(\tau)$$

où

$$0 \leqq \tau < \xi.$$

Pour démontrer ce théorème, admettons d'abord que la dérivée $\Phi'_\alpha(x)$ existe et que, par exemple, cette dérivée soit une fonction continue de x. Alors, évidemment, à cause de la distributivité de l'opération, $\varphi(x)$ sera l'intégrale d'ordre $(1-\alpha)$ de $\Phi'_\alpha(x)$, c'est-à-dire qu'on aura

$$(8) \qquad \varphi(t) = \frac{1}{\Gamma(1-\alpha)} \int_0^t \Phi'_\alpha(v)\,(t-v)^{-\alpha}\,dv.$$

En portant cela dans (6) et en observant que, sous les conditions admises, on peut intervertir l'ordre des intégrations, il vient

$$(9) \qquad g(x,\xi) = \frac{1}{\Gamma(\alpha)\,\Gamma(1-\alpha)} \int_0^\xi (x-t)^{\alpha-1}\,dt \int_0^t \Phi'_\alpha(v)\,(t-v)^{-\alpha}\,dv =$$

$$= \frac{1}{\Gamma(\alpha)\,\Gamma(1-\alpha)} \int_0^\xi \Phi'_\alpha(v)\,dv \int_v^\xi (x-t)^{\alpha-1}\,(t-v)^{-\alpha}\,dt =$$

$$= \int_0^\xi \Phi'_\alpha(v)\,M(v)\,dv$$

où l'on a posé

$$(10) \qquad M(v) = \frac{1}{\Gamma(\alpha)\,\Gamma(1-\alpha)} \int_v^\xi (x-t)^{\alpha-1}\,(t-v)^{-\alpha}\,dt.$$

Comme on a, indépendamment de v et x,

$$(11) \qquad \frac{1}{\Gamma(\alpha)\,\Gamma(1-\alpha)} \int_v^x (x-t)^{\alpha-1}\,(t-v)^{-\alpha}\,dt = 1,$$

il résulte

$$(12) \qquad M(v) = 1 - \frac{1}{\Gamma(\alpha)\,\Gamma(1-\alpha)} \int_\xi^x (x-t)^{\alpha-1}\,(t-v)^{-\alpha}\,dt\,.$$

L'intégrale qui figure au second membre et dont les limites sont indépendantes de v, est évidemment croissante lorsqe v va de 0 à ξ. Par conséquent, $M(\nu)$ est une fonction décroissante. On pourra alors appliquer le second théorème de la moyenne du calcul intégral et écrire

$$(13)\qquad g(x,\xi)=M(0)\int_0^{\tau_1}\Phi'_\alpha(\nu)\,d\nu=M(0)\,\Phi_\alpha(\tau_1),\qquad 0\leq\tau_1\leq\xi.$$

Dans cette dernière formule, on a déjà utilisé le fait que $\Phi_\alpha(0)=0$. Si l'on observe en outre que Φ_α est une fonction continue et que d'après (10), (11) et (12) on a $0\leq M(0)<1$, on voit qu'il existe toujours un nombre τ $(0\leq\tau<\tau_1)$ tel que $M(0)\,\Phi_\alpha(\tau_1)=\Phi_\alpha(\tau)$ et alors notre théorème se trouve démontré.

3. Dans la démonstration qui précède, les hypothèses concernant Φ'_α nous permettaient d'écrire la formule (8) qui conduisait à la formule (9). Or, en utilisant l'intégrale de *Stieltjes*, on arrive sans aucune hypothèse sur Φ'_α à la formule analogue à (9)

$$(14)\qquad g(x,\xi)=\int_0^{\xi}M(\nu)\,d\,\Phi_\alpha(\nu).$$

Pour éviter le raisonnement assez délicat dont on a besoin pour arriver à (14), nous suivrons ici une autre voie plus laborieuse, qui pourtant a l'avantage de n'exiger que des calculs formels.

L'intégration par parties de (9), faite à un point de vue tout à fait formel, donne, en observant que $\Phi_\alpha(0)=0$ et que $M(\xi)=0$,

$$(15)\qquad g(x,\xi)=-\int_0^{\xi}\Phi_\alpha(\nu)\,M'(\nu)\,d\nu.$$

L'intégrale au second membre est absolument convergente, puisque $M'(\nu)$ ne devient infini, pour $\nu=\xi$, que de l'ordre α. Donnons maintenant une vérification directe de la dernière formule sans aucune hypothèse sur Φ'_α.

---

[2]) La formule (14) se vérifie immédiatement, si l'on démontre d'abord que l'on a, sous la condition que $\varphi(x)$ est bornée, presque partout

$$\varphi(x)=\frac{1}{\Gamma(1-\alpha)}\int_0^{x}(x-t)^{-\alpha}\,d\,\Phi_\alpha(t).$$

Dans un travail plus élaboré, nous mettrons au point le rôle qui revient à l'intégrale de *Stieltjes* dans toutes ces questions.

On a

$$-\int_0^\xi \Phi_\alpha(v)\, M'(v)\, dv = -\frac{1}{\Gamma(\alpha)}\int_0^\xi M'(v)\, dv \int_0^v \varphi(t)\, (v-t)^{\alpha-1}\, dt =$$

(16)
$$= -\frac{1}{\Gamma(\alpha)}\int_0^\xi \varphi(t)\, dt \int_t^\xi M'(v)\, (v-t)^{\alpha-1}\, dv.$$

Or, on obtient par la distributivité (*voir* les formules (2)—(5)) et en intégrant par parties

$$-\int_t^\xi M'(v)\,(v-t)^{\alpha-1}\, dt = \frac{1}{\alpha}\,\frac{d}{dt}\int_t^\xi M'(v)\,(v-t)^{\alpha}\, dv =$$

$$= \frac{d}{dt}\int_t^\xi M(v)\,(v-t)^{\alpha-1}\, dv =$$

$$= -\frac{1}{\Gamma(\alpha)\,\Gamma(1-\alpha)}\,\frac{d}{dt}\left(\int_t^\xi (v-t)^{\alpha-1}\, dv \int_v^\xi (x-u)^{\alpha-1}\,(u-v)^{-\alpha}\, du\right) =$$

$$= -\frac{1}{\Gamma(\alpha)\,\Gamma(1-\alpha)}\,\frac{d}{dt}\left(\int_t^\xi (x-u)^{\alpha-1}\, du \int_t^u (v-t)^{\alpha-1}\,(u-v)^{-\alpha}\, dv\right).$$

En tenant encore compte de (11), cette expression se réduit à

$$(x-t)^{\alpha-1}.$$

Alors, en introduisant cela dans (16), la formule (15) se trouve établie.

En observant maintenant que $M'(v)$ est de signe constant, que $0 < M(v) < 1$ et que $M(\xi) = 0$, (7) se déduit de (15) par l'application du premier théorème de la moyenne du calcul intégral.

4. Nous passons maintenant aux applications de notre théorème. Ce théorème, je l'ai déjà donné sous des conditions plus restrictives dans une Note des *Comptes rendus.* (Une méthode équivalente à la méthode des moyennes arithmétiques, 12 juin 1911). C'est par ce théorème que j'y établis pour des indices non entiers que certaines moyennes que j'ai introduites sont entièrement équivalentes à celles de *Cesàro.* Les dites moyennes (sur lesquelles nous reviendrons au No. 11) sont sous plusieurs rapports plus maniables que celles de *Cesàro,* et ont un grand avantage sur ces dernières notamment dans la théorie des séries de *Dirichlet.*[3]) Or, dans cette théorie, le théorème de la moyenne que

[3]) Ainsi M. *Schnee* en travaillant (Acta Math. t. 35) avec les moyennes de *Cesàro,* obtenait d'abord des résultats très peu satisfaisants, pour arriver dans un supplément ajouté plus tard, en utilisant mon théorème d'équivalence, à des résultats simples.

nous venons d'établir, joue (même sous une forme bien plus particulière) un rôle fondamental pour tout ce qui concerne la sommation à ordres non entiers. (*Voir G. H. Hardy* and *M. Riesz*, The general theory of *Dirichlet*'s series, Cambr. Tracts of Math. No. 18 (1915). *Voir* en particulier le lemme 7 et ses applications). Disons enfin que notre théorème est trés utilisable dans le calcul aux différences à ordres non entiers, en élucidant le lien qui existe entre ces différences et les dérivées (et intégrales) de *Liouville—Riemann*. En remettant cette question à une autre occasion, je passe aux applications concernant certains théorèmes de *convexité*.

5. M. *H. Bohr*[4]) a démontré en se restreignant à *des séries de Dirichlet de la forme* $\Sigma a_n n^{-s}$ *et à des indices entiers de sommation* que *l'abscisse de sommabilité* des moyennes de Cesàro était une *fonction convexe de l'indice*. En combinant le théorème général qu'on va démontrer tout à l'heure avec les formules explicites pour les abscisses de sommabilité qui se trouvent dans le *Tract* cité (p. 45), on arrive à un théorème énoncé dans ce *Tract* (p. 60) qui dit que *le résultat de M. Bohr subsiste pour des séries de Dirichlet de la forme générale* $\Sigma a_n e^{-\lambda_n s}$ *et pour des indices de sommation quelconques, la méthode de sommation étant celle des moyennes typiques.*

Le théorème général qui suit, fut donné pour des indices *entiers* (et sous une forme un peu différente) par MM. *Hardy* et *Littlewood*[5])

I. Soit α *un nombre positif quelconque. Admettons qu'il existe deux fonctions non décroissantes* $V(x)$ *et* $W(x)$, *telles que pour toute valeur positive de* x, *on ait*

$$|\varphi(x)| < V(x) \quad \text{et} \quad |\Phi_\alpha(x)| < W(x). \tag{17}$$

Soit en outre $0 < \alpha' < \alpha$. *Alors, il existe toujours une constante* C, *telle que*

$$|\Phi_{\alpha'}(x)| < C\left(V(x)\right)^{1-\frac{\alpha'}{\alpha}}\left(W(x)\right)^{\frac{\alpha'}{\alpha}}. \tag{18}$$

La démonstration qui suit montre que C est même indépendant de α'.

---

[4]) *H. Bohr*, Bidrag til de Dirichlet'ske Raekkers Theorie. Thèse, Copenhague, 1910, p. 101.

[5]) *G. H. Hardy* and *J. E. Littlewood*, Contributions to the arithmetic theory of series, Lond. Math. Soc. Proc. (2) vol. 11. (1912).

Nous nous restreignons d'abord au cas $0<\alpha<1$, le cas général pouvant facilement être réduit sur ce cas particulier (*voir* le *n°* 10).

On a d'abord

$$\Phi_{\alpha'}(x) = \frac{1}{\Gamma(\alpha')}\int_0^x \varphi(t)\,(x-t)^{\alpha'-1}\,dt.$$

Nous déterminons ξ de façon que

$$(19) \qquad x-\xi = \left(\frac{W(x)}{V(x)}\right)^{\frac{1}{\alpha}}.$$

Si cette équation donne pour ξ une valeur négative, nous posons $\xi = 0$. (Dans ce dernier cas, la démonstration se simplifie, car l'intégrale que nous désignons dans la suite par I_1 n'intervient pas). On aura

$$\Gamma(\alpha')\,\Phi_{\alpha'}(x) = \int_0^x \varphi(t)\,(x-t)^{\alpha'-1}\,dt = \int_0^\xi + \int_\xi^x = I_1 + I_2$$

et

$$(20) \qquad |I_2| = \left|\int_\xi^x \varphi(t)\,(x-t)^{\alpha'-1}\,dt\right| < V(x)\int_\xi^x (x-t)^{\alpha'-1}\,dt =$$

$$= V(x)\,\frac{(x-\xi)^{\alpha'}}{\alpha'} = \frac{1}{\alpha'}\left(V(x)\right)^{1-\frac{\alpha'}{\alpha}}\left(W(x)\right)^{\frac{\alpha'}{\alpha}}.$$

D'autre part

$$I_1 = \int_0^\xi \varphi(t)\,(x-t)^{\alpha'-1}\,dt = \int_0^\xi \varphi(t)\,(x-t)^{\alpha-1}\,(x-t)^{\alpha'-\alpha}\,dt.$$

Puisque $\alpha'-\alpha$ est négatif, on a d'après le second théorème de la moyenne

$$I_1 = (x-\xi)^{\alpha'-\alpha}\int_u^\xi \varphi(t)\,(x-t)^{\alpha-1}\,dt, \quad 0 \leq u \leq \xi.$$

Alors, d'après notre théorème (7), la dernière intégrale est

$$= \Gamma(\alpha)\left(\Phi_\alpha(\tau_2) - \Phi_\alpha(\tau_1)\right), \quad 0<\tau_1<\xi<x,\ 0<\tau_2<\xi<x,$$

c'est-à-dire

$$(21) \qquad |I_1| \leq 2\,\Gamma(\alpha)\,(x-\xi)^{\alpha'-\alpha}\,W(x) =$$

$$= 2\,\Gamma(\alpha)\left(V(x)\right)^{1-\frac{\alpha'}{\alpha}}\left(W(x)\right)^{\frac{\alpha'}{\alpha}}. \quad C.\,Q.\,F.\,D.$$

6. On a encore le théorème

II. Si pour $x \to \infty$, on a

(22) $$\varphi(x) : V(x) \to 0 \quad \text{et} \quad |\Phi_\alpha(x)| < W(x)$$

ou

(23) $$|\varphi(x)| < V(x) \quad \text{et} \quad \Phi_\alpha(x) : W(x) \to 0,$$

on aura aussi pour $0 < \alpha' < \alpha$

(24) $$\Phi_{\alpha'}(x) : \left(V(x)\right)^{1-\frac{\alpha'}{\alpha}} \left(W(x)\right)^{\frac{\alpha'}{\alpha}} \to 0.$$

Remarquons tout de suite que, si sous l'hypothèse (22) la fonction $V(x)$, ou sous l'hypothèse (23) la fonction $W(x)$ tend vers l'infini avec x, le dernier théorème se réduit facilement au théorème précédent. P. ex., sous l'hypothèse (22), on pourra alors trouver une fonction non décroissante $V_1(x)$ telle que $V_1(x) : V(x) \to 0$ et que, en remplaçant dans (22) $V(x)$ par $V_1(x)$, l'hypothèse (22) soit encore remplie. En appliquant alors le théorème *I*, le nouveau théorème découle. Nous pourrions donc dans la démonstration qui suit nous restreindre au cas où $V(x)$ ou $W(x)$ restent bornées respectivement. Or, du moins dans la démonstration de la première partie du théorème, cette hypothèse particulière n'amène aucune simplification. Quant à la seconde partie, nous la démontrons par un lemme que nous voulons établir sous des conditions aussi générales que possible. C'est pour ces raisons que, dans la démonstration suivante, la seule hypothèse faite au sujet de $V(x)$ et $W(x)$ c'est que ces fonctions sont non décroissantes.

7. Aussi dans la démonstration du théorème *II*, nous supposons d'abord $0 < \alpha < 1$.

Admettons d'abord l'hypothèse (22) et déterminons t_0 de façon que l'on ait

$$|\varphi(t)| < \varepsilon V(t), \qquad t > t_0,$$

ε étant un nombre arbitrairement petit. Cela étant, on détermine ξ par l'équation

$$x - \xi = \left(\frac{W(x)}{\varepsilon V(x)}\right)^{\frac{1}{\alpha}},$$

en posant $\xi = t_0$, si cette équation donne pour ξ une valeur $\leqq t_0$.

a) Supposons d'abord $\xi > t_0$.

En posant pour, abréger,

$$(25)\qquad \left(V(x)\right)^{1-\frac{\alpha'}{\alpha}}\left(W(x)\right)^{\frac{\alpha'}{\alpha}}=U(x),$$

on aura par le même raisonnement que plus haut

$$|I_2|=\left|\int_\xi^x \varphi(t)\,(x-t)^{\alpha'-1}\,dt\right|<\frac{\varepsilon^{1-\alpha'/\alpha}}{\alpha'}\,U(x)$$

et

$$|I_1|=\left|\int_0^\xi \varphi(t)\,(x-t)^{\alpha'-1}\,dt\right|<\frac{2\,\varepsilon^{1-\alpha'/\alpha}}{\Gamma(\alpha)}\,U(x).$$

b) Si l'équation donne pour ξ une valeur $\leqq t_0$, nous posons

$$|I_2|=\left|\int_{t_0}^x\right|<\frac{V(x)}{\alpha'}\,(x-t_0)^{\alpha'}<\frac{\varepsilon^{1-\alpha'/\alpha}}{\alpha'}U(x)$$

et

$$|I_1|=\left|\int_0^{t_0}\right|<(x-t_0)^{\alpha'-1}\int_0^{t_0}|\varphi(t)|\,dt\to 0\qquad \text{quand } x\to\infty.$$

Ainsi la démonstration de la première partie du théorème *II* se trouve achevée.

8. Pour arriver à la seconde partie, nous démontrons d'abord le lemme suivant:

Sous l'hypothèse (23), *on a pour* $0<\xi<x$, *uniformément en* ξ,

$$g(x,\xi):\ W(x)\to 0,\ x\to\infty,$$

la fonction $g(x,\xi)$ *étant encore définie par la formule* (6) *et* α *étant* <1.

La démonstration de ce lemme est immédiate, lorsque $W(x)$ tend vers l'infini avec x. En effet, en vertu de (7), on a

$$(26)\qquad g(x,\xi)=\Phi_\alpha(\tau),\qquad 0\leqq\tau\leqq\xi\leqq x.$$

On pourra d'autre part d'après (23) déterminer t_1 de façon que

$$|\Phi_\alpha(t)|<\varepsilon\,W(t)\quad\text{pour}\quad t\geqq t_1.$$

$W(x)$ tendant vers l'infini on pourra choisir x assez grand pour que

$$W(t_1)<\varepsilon\,W(x).$$

Le nombre τ figurant en (26) étant alors $\gtreqless t_1$, on aura toujours

$$|g(x,\xi)|<\varepsilon\,W(x).$$

Lorsque $W(x)$ ne tend pas vers l'infini avec x, la démonstration de ce lemme devient un peu plus délicate.

Nous pourrons poser dans ce cas $W(x) \equiv 1$. Déterminons alors t_2 de façon que

(27) $$|\Phi_\alpha(t)| < \varepsilon \quad \text{pour} \quad t > t_2$$

et désignons le maximum de $|\Phi_\alpha(t)|$ pour $0 \leqq t \leqq t_2$ par K.

Nous aurons par la formule (15) pour $\xi \leqq t_2$

$$|g(x,\xi)| = \left|\int_0^\xi M'(v)\,\Phi_\alpha(v)\,dv\right| \leqq K \left|\int_0^\xi M'(v)\,dv\right| = K\,M(0) =$$

$$= \frac{K}{\Gamma(\alpha)\,\Gamma(1-\alpha)} \int_0^\xi (x-t)^{\alpha-1}\,t^{-\alpha}\,dt <$$

$$< \frac{K}{\Gamma(\alpha)\,\Gamma(1-\alpha)} \int_0^{t_2} (x-t)^{\alpha-1}\,t^{-\alpha}\,dt \rightarrow 0, \quad x \rightarrow \infty.$$

Pour $\xi > t_2$, on a d'autre part

$$|g(x,\xi)| \leqq \left|\int_0^{t_2} M'(v)\,\Phi_\alpha(v)\,dv\right| + \left|\int_{t_2}^\xi M'(v)\,\Phi_\alpha(v)\,dv\right| = \mathcal{J}_1 + \mathcal{J}_2.$$

Il vient

$$|\mathcal{J}_1| < K\left(M(0) - M(t_2)\right) =$$

$$= \frac{K}{\Gamma(\alpha)\,\Gamma(1-\alpha)} \left(\int_0^\xi (x-t)^{\alpha-1}\,t^{-\alpha}\,dt - \int_{t_2}^\xi (x-t)^{\alpha-1}(t-t_2)^{-\alpha}\,dt\right)$$

$$< \frac{K}{\Gamma(\alpha)\,\Gamma(1-\alpha)} \int_0^{t_2} (x-t)^{\alpha-1}\,t^{-\alpha}\,dt \rightarrow 0,\ x \rightarrow \infty.$$

Enfin, d'après (27), on a

$$|\mathcal{J}_2| < \varepsilon\,M(t_2) < \varepsilon$$

ce qui achève la démonstration du lemme.

9. Ce point établi, la seconde partie du théorème *II* résulte immédiatement. On choisit d'abord x assez grand pour que $|g(x,\xi)| < \varepsilon$ pour toute valeur de ξ, telle que $0 \leqq \xi \leqq x$. Aprés, on n'a qu'à déterminer ξ par l'équation

$$x - \xi = \left(\frac{\varepsilon\,W(x)}{V(x)}\right)^{\frac{1}{\alpha}}$$

(en posant $\xi = 0$ si cette équation donne une valeur négative) et appliquer le raisonnement qui a conduit à (18).

On pourra compléter le résultat qu'on vient de démontrer par le théorème suivant :

Si

$$|\varphi(x)| < V(x) \quad et \quad \Phi_\alpha(x) \rightarrow A,$$

$V(x)$ étant toujours une fonction non décroissante et A étant un nombre fini, on aura $\Phi_{\alpha'}(x) : \left(V(x)\right)^{1-\frac{\alpha'}{\alpha}} \rightarrow 0.$

Pour $A = 0$, ce théorème est un cas particulier de la seconde partie du théorème *II*. Pour réduire le cas de A quelconque au cas $A = 0$, posons

$$\bar{\varphi}(t) = \varphi(t) - \frac{A}{\Gamma(1-\alpha)}\, t^{-\alpha}.$$

Alors on aura[6]) [*cf.* la formule (11)]

$$\bar{\Phi}_\alpha(x) = \frac{1}{\Gamma(\alpha)}\int_0^x \bar{\varphi}(t)(x-t)^{\alpha-1}\,dt = \Phi_\alpha(x) - A \rightarrow 0$$

et

$$\bar{\Phi}_{\alpha'}(x) = \frac{1}{\Gamma(\alpha')}\int_0^x \bar{\varphi}(t)(x-t)^{\alpha'-1}\,dt = \Phi_{\alpha'}(x) - \frac{A}{\Gamma(1+\alpha'-\alpha)}\, x^{\alpha'-\alpha}.$$

Par ces dernières formules, on a obtenu la réduction désirée.

10. Reste à se débarrasser dans les théorèmes précédents de la restriction $\alpha < 1$. Soit maintenant α un nombre positif quelconque. Déterminons un nombre entier n de façon que $\frac{\alpha}{n} < 1/2$ et posons $\alpha_k = \frac{k\alpha}{n}$ $(k = 1, 2, \ldots n-1)$. Si l'on établit maintenant les théorèmes précédents pour les valeurs particulières $\alpha' = \alpha_k$, on n'aura qu'à appliquer encore une fois ces théorèmes sous la restriction $\alpha < 1$, pour les obtenir sans aucune restriction.

Occupons-nous d'abord du théorème *I*. Posons

$$\Psi_\beta(x) = \text{Max}\,|\Phi_\beta(t)| \text{ pour } 0 \leq t \leq x.$$

Or, il est clair que les fonctions Ψ ne sont pas décroissantes. On trouve alors d'après le théorème démontré tout à l'heure (en y remplaçant V, $\Phi_{\alpha'}$, W par $\Psi_{\alpha_{k-1}}$, Ψ_{α_k}, $\Psi_{\alpha_{k+1}}$)

$$\Psi_{\alpha_k}(x) < C\sqrt{\Psi_{\alpha_{k-1}}(x)\,\Psi_{\alpha_{k+1}}(x)} \tag{28}$$

[6]) L'infinitude que $\varphi(t)$ présente à l'origine n'améne aucune difficulté.

Cette inégalité est valable pour $k = 1, 2, \ldots n-1$, si l'on pose $\mathfrak{M}_{\alpha_0} = \mathfrak{M}_0 = V$ et $\mathfrak{M}_{\alpha_n} = \mathfrak{M}_\alpha = W$.

Soit maintenant l un quelconque des nombres $1, 2, \ldots n-1$. Élevons ces inégalités correspondant à $k = 1, 2, \ldots l-1, l, l+1, \ldots n-2, n-1$ aux puissances repectives $(n-l)$, $2(n-l)$, $\ldots, (l-1)(n-l)$, $l(n-l)$, $(n-l-1)\,l, \ldots, 2\,l$, l. On trouve par multiplication

$$\mathfrak{M}_{\alpha_l}(x) < C^{l(n-l)} \left(V(x)\right)^{1-\frac{l}{n}} \left(W(x)\right)^{\frac{l}{n}} =$$

$$= C^{l(n-l)} \left(V(x)\right)^{1-\frac{\alpha_l}{\alpha}} W(x)^{\frac{\alpha_l}{\alpha}}$$

et alors, d'après ce qu'on a dit, le théorème *I* résulte sans aucune restriction.

En s'appuyant maintenant sur ce théorème, la première partie de *II* se démontre de proche en proche pour $\alpha' = \alpha_1, \alpha_2, \ldots \alpha_{n-1}$, tandis que la seconde se démontre de proche en proche pour $\alpha' = \alpha_{n-1}, \alpha_{n-2}, \ldots \alpha_1$. Cela étant, ce théorème résulte aussi sans aucune restriction.

11. Quant aux applications des théorèmes qu'on vient de démontrer, en voici une concernant les moyennes auxquelles nous avons fait allusion au n^0 4.

Soit $\sum\limits_0^\infty a_n$ une série quelconque. Posons

$$\sigma_k(\omega) = \sum_{n=0}^{[\omega]} a_n (\omega - n)^k \qquad k \geqq 0.$$

Si la limite

$$\lim_{\omega \to \infty} \omega^{-k} \sigma_k(\omega) = \sigma$$

existe et est finie, nous disons que la série Σa_n est *sommable* d'ordre k avec la somme σ.

Les moyennes $\omega^{-k} \sigma_k(\omega)$ sont équivalentes aux moyennes de *Cesàro* d'ordre k dans le sens qu'une série quelconque est sommable en même temps et avec la même somme par ces deux espèces de moyennes. De plus, si pour une série, les moyennes d'ordre k de *Cesàro* sont bornées, il en sera de même pour nos moyennes d'ordre k et inversement.

Cela posé, on voit que dans le théorème qui suit, on pourra entendre par „moyennes" celles de *Cesàro* ou les nôtres. Ajoutons encore qu'on peut donner une démonstration directe pour les moyennes de *Cesàro,* cette démonstration étant très analogue à celle du texte.

Voici le théorème en question.

Si une série Σa_n est sommable par des moyennes d'un certain ordre et que ses moyennes d'ordre k sont bornées, alors la série est sommable par des moyennes d'ordre quelconque $> k$.[7])

La démonstration est immédiate. La série soit sommable d'ordre l. Il suffit évidemment de considérer des indices $< l$.

En changeant au besoin a_0, on pourra supposer que la somme de la série est zéro. On appliquera alors la seconde partie du théorème *II* en posant $\varphi(x) = \sigma_k(x)$, $\Phi_{k'-k}(x) = \text{const.}\, \sigma_{k'}(x)$, $\Phi_{l-k}(x) = \text{const.}\, \sigma_l(x)$, $V(x) = x^k$, $W(x) = x^l$, on trouve pour $k < k' < l$

$$\frac{\sigma_{k'}(x)}{x^{k'}} \to 0, \quad x \to \infty \qquad C.Q.F.D.$$

12. Voici encore une application. En combinant nos théorèmes avec un théorème sur les séries de *Dirichlet* que j'ai donné il y a bien longtemps et dont la démonstration paraîtra prochainement,[8]) on trouve la généralisation suivante d'un théorème de MM. *Hardy* et *Littlewood.*[9])

Si la série de Dirichlet $\Sigma a_n e^{-\lambda_n s}$ est sommable par les moyennes typiques[10]) *d'ordre r pour $s = 0$ et qu'elle est sommable d'ordre $(r-l)$ $(l < r)$ pour une valeur s dont la partie réelle est β (> 0), la série sera sommable d'ordre $(r-k)$ $(k < l)$ pour toute valeur s dont la partie réelle est $\geqq \frac{k\beta}{l}$. Le théorème subsiste, si dans l'une des hypothèses, au lieu de la sommabilité, on suppose seulement que les moyennes en question restent bornées.*

---

[7]) MM. *Hardy* et *Littlewood* ont donné une application très remarquable de ce théorème dans leur Note : On the *Fourier* series of a bounded function, Lond. Math. Soc. Proc. (2) vol 17 (1917) p. XIII—XV (Records of Proceedings at Meetings).

[8]) Über die Summierbarkeit durch typische Mittel (paraîtra dans ce journal). Au lieu de combiner les résultats ci-dessus avec ceux du travail que je viens de citer, on pourra rester dans la voie indiquée plus haut en utilisant certaines formules de transformation données dans le *Tract* cité.

[9]) *Voir* le théorème 19 du travail cité au *n°* 5.

[10]) Par moyennes typiques on pourra entendre ici, soit celles de première espèce, soit celles de seconde espèce ; *cf.* le *Tract* cité p. 21 et suiv.

On trouvera beaucoup d'applications très intéressantes de notre théorème de la moyenne et des théorèmes *I* et *II* dans un Mémoire étendu de M. *K. Ananda Rau.*[11]) On y trouve, en particulier, un examen approfondi des questions de convexité concernant les abscisses de sommabilité et une généralisation remarquable du théorème de *Schnee-Landau.*[12])

---

[11]) *K. Ananda Rau,* On the convergence and summability of *Dirichlet's* series. Je ne sais pas si ce Mémoire a déjà paru ou non.

[12]) *E. Landau, Handbuch,* p. 853 et suiv.

27.

Sur le problème des moments et le théorème de Parseval correspondant

Acta Szeged Sect. Math. *1* (1923), 209–225

M. E. Phragmén à l'occasion de son soixantième anniversaire le 2 octobre 1923.

Généralités. Le théorème de Parseval pour un intervalle fini.

1. On connaît le rôle que joue le théorème de *Parseval* dans la théorie des séries de *Fourier* et, plus généralement, dans celle des développements suivant des fonctions orthogonales. On sait aussi qu'on peut généraliser la notion d'orthogonalité et le théorème de *Parseval* de différentes manières, p. ex. en se servant de l'intégrale de *Stieltjes* correspondant à une distribution de masses positives.

Pour arriver à cette extension, disons d'abord quelques mots de l'intégrale de *Stieltjes*. Etant données dans l'intervalle $a < x \leq b$ une fonction continue $f(x)$ et une fonction bornée non décroissante (ou, plus généralement, à variation bornée) $\varphi(x)$, on entend par l'intégrale de *Stieltjes*[1]) $\int_a^b f(x)\, d\varphi(x)$ la limite de l'expression

$$\sum_{k=0}^{n-1} f(\xi_k)(\varphi(x_{k+1}) - \varphi(x_k)); \quad x_0 = a, \quad x_n = b, \quad x_k \leq \xi_k \leq x_{k+1},$$

lorsqu'on fait tendre vers zéro le maximum des longueurs $x_{k+1} - x_k$.

Cette définition s'étend aisément à un intervalle infini, p. ex. à l'intervalle $(-\infty, +\infty)$. On posera, par définition,

$$\int_{-\infty}^{+\infty} f(x)\, d\varphi(x) = \lim_{\substack{a \to -\infty \\ b \to +\infty}} \int_a^b f(x)\, d\varphi(x)$$

dès que cette limite a une valeur finie déterminée.

---

[1]) *T.-J. Stieltjes*, Recherches sur les fractions continues, Ann. de Toulouse t. 8 (1894), p. J. 1—122 et t. 9 (1895), p. A. 1—47; *voir* en particulier t. 8, p. J. 71 et *suiv.*

2. Dans el cas où, $\varphi(x)$ étant toujours non décroissante, $f(x)$ cesse d'être continue,[2]) nous adopterons, au lieu de la définition précédente, la généralisation donnée par M. *Lebesgue* qui se rattache à l'ordre d'idées suivant. On considère d'abord l'intégrale de *Stieltjes* comme intégrale curviligne $\int f\,dy$ prise le long d'une courbe continue C, composée des points $y=\varphi(x)$ et des segments verticaux représentant les sauts de $\varphi(x)$; puis, au lieu d'évaluer cette intégrale par le procédé de *Riemann* adopté par *Stieltjes*, on y applique le procédé connu de M. *Lebesgue*.[3]) C'est-à-dire que, en faisant l'inversion de la fonction $y=\varphi(x)$ et cela avec des conventions évidentes pour les valeurs de x et de y qui correspondent respectivement à des segments horizontaux et verticaux, on arrive à une intégrale rectiligne de la forme $\int_\alpha^\beta F(y)\,dy$ où $F(y)=f(x(y))$. Si cette intégrale existe[4]) au sens de M. *Lebesgue*, on dira que *$f(x)$ est intégrable par rapport à $d\varphi(x)$* et on posera, par définition,

$$\int_a^b f(x)\,d\varphi(x)=\int_\alpha^\beta F(y)\,dy.$$

L'extension à un intervalle infini est immédiate.

3. Considérons un cas particulier très simple dont nous aurons besoin dans la suite. Admettons que $\varphi(x)$ est constante par intervalles et ne croît que par un nombre fini de sauts brusques $\varrho_1,\ldots,\varrho_n$ correspondant aux points de discontinuité $x_1,\ldots,x_n$. Dans ce cas, la définition de *Stieltjes* ne s'applique que si $f(x)$ est continue aux points de discontinuité de $\varphi(x)$, tandis que, d'après la définition de M. *Lebesgue*, l'intégrale existe pour toute fonction $f(x)$

---

[2]) C'est le cas où $f(x)$ est continue qui est essentiel pour le but de notre travail, et en se bornant à ce cas, on pourra se contenter de la définition élémentaire que nous venons de donner.

[3]) *H. Lebesgue*, Sur l'intégrale de *Stieltjes* et sur les opérations linéaires, Comptes rendus Paris, t. 150 (1910), p. 86—88. Pour des définitions équivalentes *voir J. Radon*, Theorie und Anwendungen der absolut additiven Mengenfunktionen, Sitzungsb. Akad. Wien, t. 122 IIa (1913), p. 1295–1438, *W. H. Young*, On integration with respect to a function of bounded variation, Proc. London math. Soc. (2) t. 13 (1913). p. 109—150.

[4]) La fonction $F(y)$ sera en général indéterminée pour les valeurs de y qui correspondent aux segments horizontaux de la courbe C, cependant ces valeurs formant toujours un ensemble dénombrable n'exerceront aucune influence sur l'intégrale si l'on la prend au sens de M. *Lebesgue*.

qui est finie aux points x_k. On a dans les deux cas

(1) $$\int_{-\infty}^{+\infty} f(x)\, d\varphi(x) = \sum \varrho_k f(x_k).$$

Dans le cas où les points x_k forment un ensemble dénombrable de points isolés, la formule (1) subsiste dès que la série au second membre converge absolument.[5])

4. Passons maintenant à la question des polynomes orthogonaux, en nous restreignant d'abord au cas d'un intervalle fini (a, b). La fonction $\varphi(x)$ étant non décroissante et bornée dans cet intervalle, nous nous proposons de trouver des polynomes $q_0(x), q_1(x), \ldots$ tels que les relations

(2) $$\int_a^b q_m(x)\, q_n(x)\, d\varphi(x) = \begin{matrix} 0, & m \neq n \\ 1, & m = n \end{matrix}$$

aient lieu. Ces polynomes[6]) se calculent de proche en proche par les formules

(3) $$q_n(x) = C_n \left[x^n - \sum_{j=0}^{n-1} \left(\int_a^b y^n q_j(y)\, d\varphi(y) \right) q_j(x) \right]$$

où le facteur C_n se détermine au signe près par la relation

(2′) $$\int_a^b q_n^2(x)\, d\varphi(x) = 1.$$

En général, ce procédé pourra être continué indéfiniment. Le cas exceptionnel est presqu'évident. En effet, les $q_r(x)$ étant déjà calculés pour $r \leqq n-1$, on pourra déterminer $q_n(x)$ de façon qu'on vient de dire, sauf dans le cas où le polynome $h_n(x) = x^n - \ldots$ figurant au second membre de (3) est tel que

(4) $$\int_a^b h_n^2(x)\, d\varphi(x) = 0.$$

Or si une telle relation a lieu pour un polynome $h(x)$, la fonction $\varphi(x)$ devra être constante dans tout intervalle où $h(x)$ est différent de zéro, c'est-à-dire qu'on se trouve dans le cas particulier très simple dont nous venons de parler. Inversement, dans ce cas particulier, en désignant par n le nombre des sauts, le polynome $h_n(x)$ s'annule[7]) aux points de discontinuité de $\varphi(x)$ et, par con-

---

[5]) D'une façon plus générale, il en est de même lorsque $\varphi(x)$ se réduit à une fonction des sauts; pour cette notion *cf.* p. ex. *C. de la Vallée Poussin*, Intégrales de *Lebesgue* etc. (Collection *Borel*) 1916, p. 89—90.

[6]) Il est bien connu que ces polynomes sont les dénominateurs d'une certaine fraction continue.

[7]) *Voir* la note suivante.

séquent, la relation (4) sera satisfaite. Nous convenons de poser dans ce cas exceptionnel $q_r(x) \equiv 0$ pour $r \geqq n$.

5. Nous passons au théorème de *Parseval* en restant toujours dans le cas d'un intervalle fini. Soit $f(x)$ une fonction intégrable ainsi que f^2 par rapport à $d\varphi(x)$. Posons

$$a_n = \int_a^b f(x)\, q_n(x)\, d\varphi(x).$$

Alors on a le théorème de *Parseval*

$$\int_a^b f^2(x)\, d\varphi(x) = \sum_{n=0}^{\infty} a_n^2.$$

Indiquons brièvement la démonstration de ce théorème. La série $\sum a_n q_n(x)$, considérée d'abord à un point de vue purement formel, est une espèce de série de *Fourier*. La somme d'ordre n, $s_n(x) = a_0 q_0(x) + \ldots + a_n q_n(x)$ est caractérisée par le fait que parmi les polynomes $h(x)$ de degré $\leqq n$ c'est la somme $s_n(x)$ qui donne la meilleure approximation de $f(x)$ au sens des moindres carrés, c'est-à-dire qu'elle rend l'intégrale $\int_a^b (f(x) - h(x))^2\, d\varphi(x)$ minimum.[8]) Cela posé, le théorème de *Parseval* est équivalent au fait que cette meilleure approximation tend vers zéro avec $1/n$. Tout cela se voit par les formules

$$\int_a^b (f(x) - h(x))^2\, d\varphi(x) = \int_a^b (f(x) - s_n(x))^2\, d\varphi(x) + \int_a^b (h(x) - s_n(x))^2\, d\varphi(x) \tag{5}$$

et

$$\int_a^b (f(x) - s_n(x))^2\, d\varphi(x) = \int_a^b f^2(x)\, d\varphi(x) - \sum_{k=0}^{n} a_k^2. \tag{6}$$

Cela étant, la démonstration du théorème de *Parseval* sera achevée dès qu'on aura montré qu'il existe des polynomes $h_n(x)$ rendant l'erreur $\int_a^b (f(x) - h_n(x))^2\, d\varphi(x)$ arbitrairement petite. Or, dans le cas où $f(x)$ est continue, cela est une conséquence immédiate du théorème de *Weierstrass*. Dans le cas

---

[8]) Il s'ensuit immédiatement que, dans le cas particulier considéré au numéro précédent, le nombre des sauts étant désigné par n, $s_{n-1}(x)$ se confond avec le polynome de degré $n-1$ qui en tous les points de discontinuité de $\varphi(x)$ est égal à $f(x)$. Le théorème de *Parseval* pour ce cas particulier en résulte sur-le-champ.

général, on n'aura qu'à combiner ce fait avec un autre que voici, savoir que toute fonction $f(x)$ intégrable avec son carré par rapport à $d\varphi(x)$ peut être approchée par une fonction continue $g(x)$ de sorte que l'intégrale $\int_a^b (f(x)-g(x))^2 d\varphi(x)$ devienne aussi petite qu'on voudra.[9])

6. Si $f(x)$ est une fonction complexe dont les parties réelle et imaginaire sont intégrables avec leurs carrés par rapport à $d\varphi(x)$, nous entendrons par théorème de *Parseval* la formule

$$\int_a^b |f(x)|^2 d\varphi(x) = \sum_{n=0}^{\infty} |a_n|^2.$$

La somme de *Fourier* $s_n(x)$ est alors le polynome qui, parmi tous les polynomes de degré $\leq n$ minime l'intégrale

$$\int_a^b |f(x) - h(x)|^2 d\varphi(x). \tag{7}$$

On a encore l'identité

$$\int_a^b |f(x) - s_n(x)|^2 d\varphi(x) = \int_a^b |f(x)|^2 d\varphi(x) - \sum_{k=0}^{n} |a_k|^2 \tag{7'}$$

et le théorème de *Parseval* est équivalent au fait qu'il existe des polynomes $h_n(x)$ qui rendent l'intégrale (7) arbitrairement petite.

7. En substituant maintenant l'intervalle infini $(-\infty, +\infty)$ à l'intervalle fini (a, b), alors, pour que le problème traité plus haut garde son sens, il faudra évidemment faire l'hypothèse complémentaire que les moments de $d\varphi(x)$

$$c_n = \int_{-\infty}^{+\infty} x^n d\varphi(x) \tag{8}$$

existent. Cela posé, la formation des polynomes orthogonaux se fait comme plus haut.

Quant à la validité du théorème de *Parseval*, il y a une différence essentielle entre les deux cas. Dans le cas d'un intervalle infini, ce théorème ne tient pas sans exception comme dans le cas d'un intervalle fini, mais sa validité dépend de la nature du problème des moments attaché aux constantes (8). En remettant à plus tard les explications nécessaires, nous nous bornons à énoncer notre résultat principal.

---

[9]) Cela se voit par des considérations analogues à celles de M. *F. Riesz* Untersuchungen über Systeme integrierbarer Funktionen, Math. Ann. t. 69 (1910), p. 449–497, *cf.* p. 459.

Pour que le théorème subsiste il faut et il suffit que, ou bien le problème des moments en question soit déterminé ou bien, s'il est indéterminé, que $\varphi(x)$ en soit une solution extrémale.[10])

Pour arriver à ce théorème, nous commençons par réduire le problème général à un cas particulier.

Quelques théorèmes d'approximation. Réduction du problème général au cas particulier où $f(x)=\frac{1}{\alpha-x}$. La meilleure approximation de $\frac{1}{\alpha-x}$.

8. En traitant le cas d'un intervalle fini, nous avons vu que la question de la validité du théorème de *Parseval* était, en somme, une question d'approximation. Pour attaquer le cas d'un intervalle infini, nous commençons par établir quelques théorèmes d'approximation très simples.

I. *Toute fonction $F(x)$ continue dans l'intervalle* $(0,\infty)$, *extrémités y-comprises,*[11]) *peut être approchée uniformément et indéfiniment par des combinaisons linéaires de la fonction constante* 1 *et de fonctions de la forme* $1/(x+a)$ *où les a désignent des nombres positifs.* C'est-à-dire que, ε étant un nombre positif arbitrairement petit, on pourra trouver une combinaison linéaire $S(x)$ formée d'un nombre fini de telles fonctions en sorte que $|F(x)-S(x)|<\varepsilon$, $0\leqq x\leqq\infty$.

La démonstration est immédiate. La transformation $t=1/(x+a)$ fait correspondre à l'intervalle infini $(0,\infty)$ l'intervalle fini $(0, 1/a)$. Donc, d'après le théorème de *Weierstrass,* on pourra trouver des constantes ν de façon que

---

10) Parmi les résultats connus qui sont des cas particuliers du nôtre, nous nous bornons à citer ceux de MM. *Weyl* et *Wigert* qui ont démontré le théorème de *Parseval* pour les distributions $e^{-1/2 x^2}dx$ et $e^{-x}dx\ (x>0)$, les polynomes orthogonaux correspondants étant respectivement les polynomes d'*Hermite* et ceux de *Laguerre*; *voir H. Weyl,* Singuläre Integralgleichungen, Math. Annalen t. 66 (1909), p. 273—324; *S. Wigert,* Contributions à la théorie des polynomes d'*Abel-Laguerre,* Arkiv för Mat. etc. t. 15, nº 25 (1921), p. 1—22. Lesdites distributions donnent lieu à des problèmes des moments déterminés.

11) Nous dirons qu'une fonction $f(x)$ est continue pour une valeur infinie de la variable lorsque la valeur limite correspondante existe et qu'elle est finie.

$$\left| F(x) - \left(v_0 + \frac{v_1}{x+a} + \dots + \frac{v_n}{(x+a)^n} \right) \right| < \varepsilon.$$

On pourra poser $a_j = a + \delta_j$ $(j = 1, \dots, n)$ et choisir les δ_j tous différents les uns des autres et assez petits pour que l'inégalité subsiste si l'on remplace chaque puissance $(x+a)^k$ par le produit $(x+a_1)\dots(x+a_k)$. En développant en fractions simples, on obtiendra une expression de la forme exigée. Ajoutons que si la fonction $F(x)$ s'annule à l'infini, le terme constant pourra être supprimé.[12])

II. *Toute fonction $G(x)$ continue entre $-\infty$ et $+\infty$ qui s'annule pour $x = \pm\infty$, peut être approchée uniformément et indéfiniment par des combinaisons linéaires de fonctions de la forme $1/(x^2+b^2)$ et $x/(x^2+b^2)$.*

Pour la démonstration, observons d'abord que la fonction $G(x)$ pourra être approchée uniformément et indéfiniment par des fonctions qui sont paires au voisinages de $x=0$ et de $x=\pm\infty$. On pourra donc supposer qu'il en est déjà ainsi pour la fonction $G(x)$ elle-même.

Cela posé, les deux fonctions paires

$$G_1(x) = G(x) + G(-x), \quad G_2(x) = \frac{1+x^2}{2x}(G(x) - G(-x))$$

seront des fonctions continues de x^2 qui s'annulent à l'infini et alors elles pourront, d'après le théorème I et la remarque que nous venons d'y ajouter, être approchées uniformément et indéfiniment par des combinaisons linéaires de fonctions la forme $1/(x^2+b^2)$. De plus, en multipliant les expressions approchées de $G_2(x)$ par le facteur $2x/(1+x^2)$ dont le module est ≤ 1, on aura pour $G(x) - G(-x)$ des expressions approchées de la forme

$$\frac{2x}{1+x^2} \sum \frac{\mu_k}{x^2+b_k^2}.$$

On peut évidemment supposer que toutes les valeurs b_k^2 sont différentes de 1. Alors cette expression pourra s'écrire sous la forme

$$\frac{v_0 x}{x^2+1} + \sum \frac{v_k x}{x^2+b_k^2},$$

ce qui achève la démonstration.

[12]) On voit aussi par le raisonnement du texte que l'on peut assujettir les a_j à la restriction d'appartenir à un ensemble donné d'avance n'admettant qu'un seul point $a > 0$ comme point limite. La question concernant les ensembles les plus généraux d'où l'on pourra choisir les a_j de façon que le théorème subsiste, est facile à résoudre.

9. Le troisième théorème d'approximation — c'est précisément le théorème dont nous aurons besoin dans la suite — ne concerne plus l'approximation uniforme; il s'y agit d'une approximation dans le sens des moindres carrés, telle que nous en avons déjà parlé plus haut. Avant d'énoncer ce théorème, il nous sera utile d'introduire quelques notations.

Soit $\varphi(x)$ une fonction non décroissante et bornée, définie dans l'intervalle $-\infty < x < +\infty$. Nous ne considérons que des fonctions $f(x)$ réelles ou complexes qui avec $|f|^2$ sont intégrables par rapport à $d\varphi(x)$ dans l'intervalle en question. La classe de ces fonctions sera désignée par Ω. Les fonctions $f_1, \ldots, f_n$ appartenant à cette classe, toute combinaison linéaire y appartiendra également. Nous appellerons *distance* de f_1 et de f_2 la valeur

$$[f_1, f_2] = \left(\int_{-\infty}^{+\infty} |f_1(x) - f_2(x)|^2 \, d\varphi(x)\right)^{\frac{1}{2}}.$$

D'après une généralisation connue de l'inégalité de *Schwarz*, on obtient immédiatement

$$(9) \qquad [f_1, f_3] \leq [f_1, f_2] + [f_2, f_3].$$

Enfin, étant donné un certain ensemble de fonctions $g(x)$ appartenant à la classe Ω, nous dirons qu'une fonction $f(x)$ de cette classe peut être approchée indéfiniment *en moyenne* (par rapport à $d\varphi(x)$) par les fonctions $g(x)$, si l'on peut choisir $g(x)$ de sorte que la distance de f et de g devienne arbitrairement petite.

Cela étant, voici notre théorème.

III. *Toute fonction $f(x)$ de la classe Ω peut être approchée indéfiniment en moyenne (par rapport à $d\varphi(x)$) par des combinaisons linéaires de fonctions de la forme $1/(\alpha - x)$ où les α sont des nombres imaginaires* (c'est-à-dire non réels).

En effet, toute fonction $f(x)$ de la classe Ω peut être approchée indéfiniment en moyenne par des fonctions continues $g(x)$ de la même classe qui s'annulent à l'infini.[13]) D'après le théorème II, ces dernières fonctions peuvent être approchées uniformément et indéfiniment par des combinaisons linéaires de fonctions de la forme

$$\frac{1}{x^2 + b^2} = \frac{1}{2ib}\left(\frac{1}{x - ib} - \frac{1}{x + ib}\right) \text{ et } \frac{x}{x^2 + b^2} = \frac{1}{2}\left(\frac{1}{x - ib} + \frac{1}{x + ib}\right)$$

---

[13]) On peut même supposer que la fonction $g(x)$ s'annule pour toute valeur suffisamment grande de $|x|$. On le voit en observant que la variation de $\varphi(x)$ est très petite dans des intervalles assez éloignés; *cf.* encore la note[9]).

c'est-à-dire par des combinaisons linéaires de fonctions de la forme $1/(\alpha-x)$. Supposons maintenant que la fonction $f(x)$ de la classe Ω fût approchée en moyenne par la fonction continue $g(x)$ s'annulant à l'infini avec une erreur moindre que ε_1 (c'est-à-dire que $[f, g] < \varepsilon_1$) et que la fonction $g(x)$ fût approchée uniformément, avec une erreur moindre que ε_2, par une combinaison linéaire $S(x)$ de fonctions de la forme $1/(\alpha-x)$, alors, par l'inégalité (9), $S(x)$ approchera en moyenne $f(x)$ avec une erreur moindre que $\varepsilon_1+\varepsilon_2(\int d\varphi(x))^{\frac{1}{2}}$ ce qui achève la démonstration.

10. Revenons maintenant au théorème de *Parseval*. Soit donc $\varphi(x)$ une fonction non décroissante telle que les moments (8) existent. Nous allons démontrer, en nous appuyant sur le théorème III, que *la question de la validité du théorème de Parseval pour toutes les fonctions $f(x)$ de la classe Ω se réduit au cas particulier $f(x)=1/(\alpha-x)$.*

En effet, admettons un instant que la distribution $d\varphi(x)$ soit telle que le théorème de *Parseval* ait lieu pour les fonctions $1/(\alpha-x)$. D'après une remarque faite déjà au n⁰ 5, cela revient à ce que ces fonctions peuvent être approchées indéfiniment en moyenne par des polynomes $h(x)$. Alors il en sera de même de toute combinaison linéaire $S(x)$ de ces fonctions. Ceci résulte immédiatement de l'inégalité analogue à l'inégalité (9)

$$[(\mu_1 f_1+\ldots+\mu_n f_n), (\mu_1 h_1+\ldots+\mu_n h_n)] \leq \tag{10}$$
$$\leq |\mu_1|\,[f_1, h_1]+\ldots+|\mu_n|\,[f_n, h_n].$$

Cela posé, étant donnée une fonction arbitraire $f(x)$ de la classe Ω, on l'approchera d'abord par une combinaison linéaire $S(x)$ des fonctions $1/(\alpha-x)$ avec une erreur $<\varepsilon_1$. Alors, si le théorème de *Parseval* a lieu pour ces dernières fonctions, on pourra approcher $S(x)$ par un polynome $h(x)$ avec une erreur $<\varepsilon_2$. Enfin, en vertu de (9), ce polynome approchera $f(x)$ avec une erreur $<\varepsilon_1+\varepsilon_2$ et alors le théorème de *Parseval* tiendra aussi pour $f(x)$. Tout revient donc à ceci: *Quelles sont les distributions $d\varphi(x)$ telles que le théorème de Parseval a lieu pour les fonctions particulières* $1/(\alpha-x)$?

11. Soit $\sum a_j q_j(x)$ le développement de *Fourier* de $1/(\alpha-x)$. En vue du théorème de Parseval, il s'agit d'évaluer le reste

$$R_n(\alpha)=\int_{-\infty}^{+\infty}\left|\frac{1}{\alpha-x}-s_n(x)\right|^2 d\varphi(x),\; s_n(x)=\sum_{j=0}^{n} a_j q_j(x)$$

(*cf.* la formule (7')). Le polynome $s_n(x)$ est caractérisé par la propriété suivante. Parmi tous les polynomes $h(x)$ de degré $\leqq n$, c'est $s_n(x)$ qui rend l'intégrale

$$\int_{-\infty}^{+\infty}\left|\frac{1}{\alpha-x}-h(x)\right|^2 d\varphi(x)$$

minimum. En posant alors $1-(\alpha-x)\,s_n(x)=l_n(x)$, on voit que le polynome $l_n(x)$ est la solution du problème de minimum suivant.

Parmi tous les polynomes $l(x)$ de degré $\leqq n+1$ tels que $l(\alpha)=1$, trouver celui qui rend l'intégrale

$$\int_{-\infty}^{+\infty}|l(x)|^2\frac{d\varphi(x)}{|\alpha-x|^2}$$

minimum. La valeur de ce minimum sera précisement $R_n(\alpha)$. Le théorème de *Parseval* aura lieu ou n'aura pas lieu suivant que cette valeur minimum tend ou ne tend pas vers zéro. C'est une discussion approfondie du problème des moments correspondant à la distribution $d\varphi(x)$ qui nous permettra de remplacer cette réponse préliminaire par la réponse définitive.

Exposé rapide de quelques points de la théorie du problème des moments.[14])

12. Etant donnée une suite $c_0, c_1, \ldots, c_n, \ldots$, on entend par problème des moments (par rapport à l'intervalle $(-\infty, +\infty)$) les deux questions suivantes.

Les équations

$$\int_{-\infty}^{+\infty} x^n\, d\varphi(x)=c_n \tag{10}$$

admettent-elles au moins une solution non décroissante $\varphi(x)$ (*question d'existence*) et si tel est le cas, cette solution est-elle unique (*question d'unicité*)?

---

[14]) *Cf.* outre le Mémoire cité de *Stieltjes* les travaux suivants: *H. Hamburger,* Über eine Erweiterung des *Stieltjes*schen Momentenproblems, *I, II* et *III,* Math. Ann. t. 81 (1920), p. 235—319, t. 82 (1921), p. 120—164, *ib.* p. 168—187; *E. Stridsberg,* Några aritmetiska undersökningar etc. Notes 2 et 3, Arkiv för Mat. etc. t. 13, n⁰ 25 (1918), p. 1—70 et t. 15, n⁰ 22 (1921), p. 1—126; *M. Riesz,* Sur le problème des moments, Notes *I, II* et *III,* Arkiv för Mat. etc. t. 16, n⁰ˢ 12 (1921), p. 1—23 et 19 (1922), p. 1—21, t. 17, n⁰ 16 (1923), p. 1—52; *R. Nevanlinna,* Asymptotische Entwicklungen beschränkter Funktionen und das *Stieltjes*sche Momentenproblem, Ann. Ac. Scient. Fenn., t. 18, n⁰ 5 (1922), p. 1—53; *T. Carleman,* Sur le problème des moments, Comptes rendus Paris, t. 174 (1922), p. 1680—1682.

Par des raisons qui proviennent de la nature de l'intégrale de *Stieltjes*, on considère comme identiques deux solutions coïncidant, à une constante additive près, en tous leurs points de continuité; en effet, la valeur de l'intégrale de *Stieltjes* (10) ne dépend pas des valeurs qu'on attribue à $\varphi(x)$ en ses points de discontinuité. Si toutes les solutions sont identiques d'après la convention que nous venons de faire, on dira que le problème des moments est *déterminé;* le problème sera dit *indéterminé* dans le cas contraire.

Ici nous n'aurons pas à nous occuper de la question d'existence (question d'ailleurs bien facile) *les moments c_n étant toujours définis par une distribution donnée d'avance.* Il ne reste que la question d'unicité. La réponse à cette question dépend de la solution d'un problème de minimum que voici.

13. Etant donné un nombre complexe arbitraire α, déterminer parmi tous les polynomes $h(x)$ de degré $\leqq n$ tels que $h(\alpha)=1$, celui qui rend l'intégrale

$$\int_{-\infty}^{+\infty} |h(x)|^2 d\varphi(x)$$

minimum.

Le polynome $h(x)$ en question, comme d'ailleurs tout polynome de degré $\leqq n$, pourra être mis sous la forme $h(x)=a_0 q_0(x)+\ldots+a_n q_n(x)$. Soit d'autre part $g(x)=b_0 q_0(x)+ +\ldots+b_n q_n(x)$ un polynome de degré $\leqq n$ s'annulant pour $x=\alpha$, d'ailleurs arbitraire, et soit $\bar{g}(x)$ le polynome imaginaire conjugué. Pour que $h(x)$ fournisse le minimum en question il faut el il suffit, d'après un raisonnement connu,[15]) que l'on ait

$$\int_{-\infty}^{+\infty} h(x)\,\bar{g}(x)\,d\varphi(x)=0.$$

Ceci est équivalent au fait suivant: la relation $g(\alpha)=0$ ou ce qui revient au même, la relation

$$\bar{g}(\bar{\alpha})=\bar{b}_0 q_0(\bar{\alpha})+\ldots+\bar{b}_n q_n(\bar{\alpha})=0$$

entraîne

$$\bar{b}_0 a_0+\ldots+\bar{b}_n a_n=0.$$

Alors $h(x)$ est de la forme $C(q_0(\bar{\alpha})q_0(x)+\ldots+q_n(\bar{\alpha})q_n(x))$ où la constante C se détermine par la condition $h(\alpha)=1$, ce qui donne en définitive

$$h(x)=\sum^{n} q_j(\bar{\alpha})\,q_j(x):\sum_{j=0}^{n}|q_j(\alpha)|^2 \tag{11}$$

---

[15]) *Cf.* p. ex. *M. Riesz, III,* p. 19—21. *l. c.* [14]).

et le minimum en question $\varrho_n(\alpha)$ est donné par la formule

(12) $$\frac{1}{\varrho_n(\alpha)} = \sum_{j=0}^{n} |q_j(\alpha)|^2.$$

Pour n croissant, la valeur positive $\varrho_n(\alpha)$ ira en décroissant vers une valeur limite déterminée $\varrho(\alpha)$ positive ou zéro. Cette valeur est donnée par la formule

(12') $$\frac{1}{\varrho(\alpha)} = \sum_{j=0}^{\infty} |q_j(\alpha)|^2,$$

cette série étant convergente lorsque $\varrho(\alpha) > 0$ et divergente lorsque $\varrho(\alpha) = 0$. La distinction entre ces deux cas est d'une importance décisive pour le problème des moments correspondant à la distribution $d\varphi(x)$.

14. Voici maintenant la solution de la question d'unicité.[16]) *Pour que le problème soit déterminé, il faut que la valeur $\varrho(\alpha)$ définie plus haut s'annule en tout point imaginaire. Il en sera ainsi, dès que $\varrho(\alpha)$ s'annule en un seul point réel ou imaginaire.*

Rappelons aussi que, en vertu de la formule (12'), on pourra encore dire : Pour que le problème des moments soit déterminé, il faut que la série (12') soit divergente en *tout* point imaginaire. Il en sera ainsi dès que la série diverge en un *seul* point réel ou imaginaire. Il s'ensuit que dans le cas indéterminé la série (12') converge pour toute valeur finie de α. Ajoutons que la convergence est uniforme dans tout domaine borné.

15. Admettons maintenant qu'on se trouve dans le cas indéterminé. Soit $\alpha = \xi + i\eta$ un nombre imaginaire fixe. Formons avec toutes les solutions $\psi(x)$ du problème des moments, attaché à la suite $c_0, c_1, \ldots c_n, \ldots$, l'intégrale

(13) $$H(\alpha) = \int_{-\infty}^{+\infty} \frac{d\psi(x)}{\alpha - x}.$$

M. *R. Nevanlinna* a démontré[17]) que l'ensemble des valeurs $H(\alpha)$ donne la périphérie et l'intérieur d'une certaine circonférence C de rayon $\varrho(\alpha)/2|\eta|$. A chaque point intérieur à cette circonférence il correspond une *infinité* de solutions différentes du problème des moments, telles que la valeur (13) correspondante coïncide avec ce point. A un point périphérique il n'en correspond qu'une *seule*. L'ensemble des solutions correspondant à différents points de la circonférence C ne dépend pas du choix de α. Nous appel-

[16]) *Voir* p. 34 du travail cité dans la note précédente.

[17]) *R. Nevanlinna*, l. c. [14]).

lerons ces solutions *solutions extrémales.* Les seuls points de croissance d'une telle solution sont les pôles d'une certaine fonction méromorphe

$$H(z) = \int_{-\infty}^{+\infty} \frac{d\,\psi^*(x)}{z-x} = \frac{P(z)}{Q(z)},$$

$P(z)$ et $Q(z)$ étant des fonctions entières transcendantes à une infinité de zéros réels. En choisissant parmi ces fonctions méromorphes deux fonctions différentes $P_1(z) : Q_1(z)$ et $P_2(z) : Q_2(z)$, toute autre fonction $P(z) : Q(z)$ pourra être mise sous la forme

$$\frac{P(z)}{Q(z)} = \frac{P_1(z) + \tau P_2(z)}{Q_1(z) + \tau Q_2(z)},$$

τ désignant un paramètre réel. En faisant varier ce paramètre de $-\infty$ à $+\infty$, le rapport $P(z) : Q(z)$ décrit la circonférence C correspondant au point z. Quant à la structure d'une solution extrémale, nous avons vu qu'une telle solution ne croît que par des sauts brusques en des points isolés en nombre infini, savoir aux pôles dont nous venons de parler. On peut toujours choisir τ de façon qu'une valeur réelle arbitraire devienne un pôle de la fonction méromorphe correspondant à τ. En désignant un tel pôle par β, le saut en question sera précisément égal à $\varrho(\beta)$. Rappelons enfin une seconde propriété extrémale[18]) très remarquable des solutions considérées. C'est qu'une telle solution présente, en chacun de ses points de discontinuité, un saut plus grand que toute autre solution.

Le théorème de Parseval pour $\frac{1}{\alpha - x}$. Le théorème général.

16. Revenons maintenant au théorème de *Parseval.* Rappelons que nous avons montré au n⁰ 10 que la validité de ce théorème pour des fonctions arbitraires $f(x)$ ne dépend que de la validité du théorème pour les fonctions particulières $1/(\alpha - x)$. De plus, au n⁰ 11, nous avons trouvé que la condition nécessaire et suffisante pour que le théorème ait lieu pour $1/(\alpha - x)$ c'est qu'un certain minimum $R_n(\alpha)$ tende vers zéro avec $1/n$. En rapprochant ce résultat de ceux exposés au n⁰ 14, on pourra dire: Pour que $R_n(\alpha)$ tende vers zéro, c'est-à-dire pour que le théorème de *Parseval* ait lieu pour la fonction $1/(\alpha - x)$ il faut

[18]) *Cf. H. Hamburger, III,* et *M. Riesz, I, l. c.* [14]).

et il suffit que la distribution $\frac{d\varphi(x)}{|\alpha-x|^2}$ donne lieu à un problème des moments *déterminé*. On voit facilement que cela revient à ce que les relations

$$(14)\qquad \int_{-\infty}^{+\infty}\frac{x^n\,d\psi(x)}{|\alpha-x|^2}=\int_{-\infty}^{+\infty}\frac{x^n\,d\varphi(x)}{|\alpha-x|^2},\quad n=0,1,2,\ldots$$

entraînent que la fonction non décroissante $\psi(x)$ est, au point de vue adopté, identique à la fonction donnée $\varphi(x)$.

La condition que nous venons de formuler dépend encore de la nature du problème des moments défini par la distribution $\frac{d\varphi(x)}{|\alpha-x|^2}$, et comme α est arbitraire, il paraît s'agir d'une infinité de problèmes des moments. Or il n'en est rien. On passe immédiatement au problème des moments primitif défini par la distribution $d\varphi(x)$. Pour le voir, remarquons que le système (14) est équivalent au système

$$(15)\qquad \int_{-\infty}^{+\infty}x^n\,d\psi(x)=\int_{-\infty}^{+\infty}x^n\,d\varphi(x),\quad n=0,1,2,\ldots,$$

complété par la relation

$$(16)\qquad \int_{-\infty}^{+\infty}\frac{d\psi(x)}{\alpha-x}=\int_{-\infty}^{+\infty}\frac{d\varphi(x)}{\alpha-x}.$$

En effet, cette dernière relation pourra aussi s'écrire

$$(16')\qquad \int_{-\infty}^{+\infty}(\bar{\alpha}-x)\frac{d\psi(x)}{|\alpha-x|^2}=\int_{-\infty}^{+\infty}(\bar{\alpha}-x)\frac{d\varphi(x)}{|\alpha-x|^2}$$

et alors on voit que les égalités (14) entraînent (16). Inversement, en posant, en (16'), $\alpha=\xi+i\eta$, en séparant les parties réelles et imaginaires et en observant que η est différent de zéro, on obtient immédiatement les relations (14) pour $n=0$ et $n=1$. Afin d'obtenir les relations (14) pour $n>1$, observons d'abord que, pour x réel, $|\alpha-x|^2=(\xi-x)^2+\eta^2=g(x)$ est un polynome du second degré de x. En écrivant $x^n=g(x)\,t_n(x)+\lambda_n+\mu_n x$ où $t_n(x)$ est un polynome, λ_n et μ_n sont des constantes, il vient de (15) et des deux premières relations (14) que nous venons d'obtenir

$$\int_{-\infty}^{+\infty}\frac{x^n\,d\psi(x)}{g(x)}=\int_{-\infty}^{+\infty}\left(t_n(x)+\frac{\lambda_n+\mu_n x}{g(x)}\right)d\psi(x)=$$

$$=\int_{-\infty}^{+\infty}\left(t_n(x)+\frac{\lambda_n+\mu_n x}{g(x)}\right)d\varphi(x)=\int_{-\infty}^{+\infty}\frac{x^n\,d\varphi(x)}{g(x)}.$$

D'ici on conclut: Pour que la distribution $\frac{d\varphi(x)}{|\alpha-x|^2}$ donne lieu à un problème des moments déterminé, il faut et il suffit qu'il n'existe aucune solution $\psi(x)$ du problème original différente de $\varphi(x)$ et telle que la relation (16) ait lieu. Tel sera évidemment le cas, quand le problème original est déterminé. D'autre part, pour que ceci ait lieu lorsque le problème original est indéterminé, il faut et il suffit, d'après un théorème cité de M. *R. Nevanlinna*, que $\varphi(x)$ en soit une solution extrémale. Remarquons que ces conditions sont *indépendantes* de α.

En définitive, on pourra dire:

Pour que le théorème de Parseval ait lieu pour toute fonction $f(x)$ intégrable avec $|f|^2$ par rapport à $d\varphi(x)$, il faut et il suffit que, ou bien le problème des moments défini par la distribution $d\varphi(x)$ soit déterminé ou, si ce problème est indéterminé, que $\varphi(x)$ en soit une solution extrémale.[19])

Une condition équivalente à celle-ci c'est que *le problème des moments correspondant à la distribution* $\frac{d\varphi(x)}{1+x^2}$ *soit déterminé.* Dans cette condition, $1+x^2$ pourra être remplacé par un polynome quadratique positif quelconque.

La série de Fourier dans le cas indéterminé.

17. Nous finissons par quelques remarques supplémentaires concernant le cas indéterminé.

Soit $d\varphi(x)$ une distribution qui donne lieu à un problème des moments indéterminé. Nous venons de voir que si $\varphi(x)$ n'est pas une solution extrémale de ce problème (tel sera p. ex. toujours le cas si $\varphi(x)$ est continue), le théorème de Parseval ne tiendra pas pour toute f, intégrable avec $|f|^2$ par rapport à $d\varphi(x)$ et, en particulier, que ce théorème ne tiendra pour aucune des fonctions $1/(\alpha-x)$. On peut alors se demander s'il existe des fonctions particulières $f(x)$ pour lesquelles le théorème a lieu. Telles fonc-

[19]) Dans son Mémoire classique *l. c.* [1]) *Stieltjes* n'a considéré que des $d\varphi(x)$ réparties sur l'intervalle $(0, \infty)$. Ce cas entre comme cas particulier dans celui du texte et le lecteur familier avec la théorie de *Stieltjes* pourra sans difficulté exprimer notre résultat dans le langage de cette théorie. En particulier, le théorème de *Parseval* tiendra toujours lorsque la distribution $d\varphi(x)$ donne lieu à un problème des moments qui est déterminé au point de vue de *Stieltjes*.

tions sont p. ex. tous les polynomes. Nous allons voir que le théorème subsiste pour une classe de fonctions bien plus générale.

Soit $\sum a_n q_n(x)$ la série de *Fourier* d'une fonction f intégrable avec $|f|^2$ par rapport à une distribution $d\varphi(x)$. Alors la formule (7') donne l'inégalité

$$\sum_{n=0}^{\infty} |a_n|^2 \leq \int_{-\infty}^{+\infty} |f(x)|^2 \, d\varphi(x)$$

ce qui met entre autre en évidence la convergence de la série au premier membre. Admettons que $d\varphi(x)$ donne lieu à un problème indéterminé. Alors l'allure de la série de *Fourier* sera entièrement différente de celle que l'on connaît de la théorie classique. En effet, la série (12') converge uniformément dans tout domaine borné. On en conclut par un raisonnement connu que la série $\sum a_n q_n(z)$ converge absolument pour toute valeur de z, la convergence étant uniforme en tout domaine borné. Par conséquent, cette série représentera dans tout le plan une *fonction entière* $F(z)$.[20]) Evidemment, les valeurs de $F(z)$ sur l'axe réel seront, en général, différentes de celles de la fonction $f(x)$.[21]) La série de *Fourier* $\sum a_n q_n(x)$ converge aussi en moyenne (par rapport à $d\varphi(x)$) vers $F(x)$ dont, par conséquent, elle est aussi la série de *Fourier*. Il s'ensuit que les constantes de *Fourier* de la différence $f(x) - F(x)$ sont tous zéro et que, par suite, cette différence est orthogonale à toutes les puissances x^n $(n = 0, 1, \ldots)$ et en outre à la fonction $F(x)$ et que l'on a

$$(17) \qquad \int_{-\infty}^{+\infty} |f^2| \, d\varphi(x) = \sum_{n=0}^{\infty} |a_n|^2 + \int_{-\infty}^{+\infty} |f-F|^2 \, d\varphi(x).$$

Le théorème de *Parseval* pour $f(x)$ aura lieu ou n'aura pas lieu suivant que l'intégrale au second membre est zéro ou non. En particulier, le théorème tiendra lorsque $f(x) \equiv F(x)$ c'est-à-dire que $f(x)$ est partout représentée par sa série de Fourier. Pour $f(x)$ continue, cette condition est aussi nécessaire dans le cas où $\varphi(x)$ n'est constante dans aucun intervalle. Dans le cas contraire, on pourra laisser de côté l'intérieur des intervalles où $\varphi(x)$ est con-

[20]) D'après un résultat que nous avons donné récemment *l. c.* [15]) p. 37. on aura $|F(z)| = e^{\epsilon(r) r}$ où $r = |z|$ et $\epsilon(r) \to 0$ lorsque $r \to \infty$

[21]) Avant que je fusse arrivé à ces résultats M. *Wigert* y est parvenu par un calcul direct très intéressant dans un cas particulier qu'il a eu la bonté d'examiner; *S. Wigert*, Sur les polynomes orthogonaux et l'approximation des fonctions continues, Arkiv för Mat. etc. t. 17, n⁰ 18 (1923), p. 1—15.

stante et ne prêter attention qu'aux points de croissance qui restent. Dans tous les cas, la formule (17) met en évidence la condition nécessaire et suffisante pour que le théorème de *Parseval* tienne pour $f(x)$, continue ou discontinue; c'est que $f(x)$ soit représentée par sa série de *Fourier* en tous les points de croissance de $\varphi(x)$, sauf peut-être dans un ensemble de mesure nulle par rapport à $d\varphi(x)$, c'est-à-dire dans un ensemble sur lequel la variation totale de $\varphi(x)$ est égale à zéro. Dans le cas où $\varphi(x)$ est une solution extrémale, il s'ensuit, d'après notre théorème général, que *toute* fonction f intégrable avec $|f|^2$ par rapport à $d\varphi(x)$ sera représentée par sa série de *Fourier* en tous les points de croissance de $\varphi(x)$. Rappelons que ces points sont des points isolés et ajoutons que ce résultat se vérifie facilement par un calcul direct.

18. Disons encore un mot d'une question rentrant dans l'ordre d'idées du théorème de *Riesz-Fischer*. Etant données des constantes a_n telles que $\sum |a_n|^2$ est convergente, existe-t-il une fonction $f(x)$ intégrable avec $|f|^2$ par rapport à $d\varphi(x)$ dont ces a_n sont les constantes de *Fourier* suivant les polynomes orthogonaux $q_n(x)$? Alors, que le cas soit déterminé ou indéterminé, la réponse sera affirmative; on le voit par une adaptation facile de l'une ou l'autre des démonstrations connues. Or, il ne sera pas sans intérêt d'observer que, dans le cas indéterminé, cette réponse affirmative est déjà fournie par le raisonnement du numéro précédent. En effet, les a_n étant tels que nous venons de le dire, la fonction entière $F(x)$ donnée par la série $\sum a_n q_n(x)$ aura la propriété désirée.[22])

Győr, le 9 septembre 1923.

---

[22]) Signalons encore une propriété des fonctions telles que $F(x)$, c'est que la valeur de l'intégrale $\int_{-\infty}^{+\infty} F(x)\, d\varphi(x)$ est la même pour toutes les solutions $\varphi(x)$ de notre problème des moments. Il en est de même des fonctions du type $x^n F(x)$, $|F(x)|^2$, $F^2(x)$, $F_1(x) F_2(x)$. D'autre part, à la fin de la troisième de nos Notes citées plus haut, nous sommes arrivé par une voie différente à une classe de fonctions que nous avons désignées par $E(x)$, jouissant de la même propriété, c'est-à-dire telles que leur intégrale par rapport aux différentes solutions était toujours la même. Je profite de l'occasion pour observer que toute telle fonction $E(x)$ pourra s'écrire comme la différence de deux fonctions non négatives de la même classe et que ces dernières fonctions pourront, pour les valeurs réelles de x, être mises sous la forme $|F(x)|^2$, $F(x)$ étant du type considéré dans le texte.

15

28.

Sur l'équivalence de certaines méthodes de sommation

Proc. London Math. Soc. (2) *22* (1924), 412–419

Extrait d'une lettre adressée à M. G. H. Hardy.

. . . J'ai indiqué la démonstration de l'équivalence des moyennes $\omega^{-k} \sum_{r<\omega} u_r(\omega-r)^k$ et des moyennes de Cesàro dans les *Comptes Rendus* ("Une méthode de sommation équivalente à la méthode des moyennes arithmétiques", 12 juin 1911).

Permettez-moi de vous indiquer quelques résultats qui, si insignifiants qu'ils soient, donnent l'explication de quelques difficultés concernant les moyennes dont je viens de parler. Il s'agit en premier lieu d'un théorème d'équivalence assez général, relatif à certaines méthodes de sommation. Je ne sais pas si ce théorème est nouveau; en tout cas, il découle immédiatement d'un théorème connu de M. Toeplitz.

Dans ce que suit, je désignerai la série qu'on veut sommer par Σu_n et je poserai $s_n = u_0+u_1+\ldots+u_n$.

Soit $a_0, a_1, \ldots a_n, \ldots$ une suite indéfinie de nombres positifs, et posons $A_n = a_0+a_1+\ldots+a_n$. Admettons que

$$\text{(1)} \qquad \lim_{n\to\infty} \frac{a_n}{a_0+a_1+\ldots+a_n} = \lim_{n\to\infty} \frac{a_n}{A_n} = 0.$$

Alors on sait que la formule

$$\text{(2)} \qquad T_n^{(a)} = \frac{a_n s_0+a_{n-1}s_1+\ldots+a_0 s_n}{A_n}$$

définit une méthode de sommation dans le sens que chaque fois que $\lim_{n\to\infty} s_n$ existe, il en est de même de $\lim_{n\to\infty} T_n^{(a)}$ et les deux limites coïncident.

Soit $b_0, b_1, \ldots, b_n, \ldots$ une autre suite de nombres positifs. Posons d'une façon analogue $B_n = b_0+b_1+\ldots+b_n$ et admettons encore que b_n et B_n remplissent une relation analogue à (1). On aura encore une méthode de sommation $T_n^{(b)}$. Evidemment, $T_n^{(a)}$ et $T_n^{(b)}$ pourront tendre vers des limites sans qu'il en soit ainsi pour s_n.

Au sujet de telles méthodes de sommation, M. Nörlund* a démontré que *chaque fois que* $\lim T_n^{(a)}$ *et* $\lim T_n^{(b)}$ *existent toutes les deux, ces limites seront égales.*

M. Nörlund considère à cet effet la méthode de sommation $T_n^{(c)}$ où $c_n = a_0b_n+a_1b_{n-1}+\ldots+a_nb_0$. Il est alors facile de démontrer que chaque fois que p. ex. $T_n^{(a)}$ tend vers une limite déterminée, $T_n^{(c)}$ tendra vers la même limite. Par cela, le théorème est établi.

Grâce aux hypothèses ci-dessus, on aura

$$\frac{A_{n+1}}{A_n} \to 1, \qquad \frac{B_{n+1}}{B_n} \to 1. \tag{3}$$

Les séries $\Sigma a_n x^n$, $\Sigma b_n x^n$, $\Sigma A_n x^n$, $\Sigma B_n x^n$ seront donc convergentes pour $|x| < 1$. Posons

$$\frac{\Sigma b_n x^n}{\Sigma a_n x^n} = \frac{\Sigma B_n x^n}{\Sigma A_n x^n} = \Sigma p_n x^n, \qquad \frac{\Sigma a_n x^n}{\Sigma b_n x^n} = \frac{\Sigma A_n x^n}{\Sigma B_n x^n} = \Sigma q_n x^n. \tag{4}$$

Je dirai que les méthodes (*a*) et (*b*) sont *équivalentes* si les limites $\lim T_n^{(a)}$ et $\lim T_n^{(b)}$ existent en même temps. (Grâce au théorème de M. Nörlund, on est dispensé de s'occuper de l'égalité.) Je vais indiquer maintenant la démonstration du théorème suivant.

Pour que les méthodes (*a*) *et* (*b*) *soient équivalentes, il faut et il suffit que les séries* $\sum_0^\infty |p_n|$ *et* $\sum_0^\infty |q_n|$ *soient convergentes.*

La démonstration directe est très facile. Je préfère pourtant m'appuyer sur le théorème bien connu de M. Toeplitz que voici.

Posons $y_n = a_{n0}x_0+a_{n1}x_1+\ldots+a_{nn}x_n \quad (n = 0, 1, 2, \ldots).$

* "Sur une application des fonctions permutables", *Lunds Universitets Årsskrift*, N. F. avd. 2, bd. 16, no. 3 (1919), pp. 1–10, *voir* p. 8.

Pour que l'existence de la limite $\lim x_n$ *entraîne celle de la limite* $\lim y_n$ *et de plus que ces deux limites soient égales, il faut et il suffit que*

$$(1) \quad \lim_{n\to\infty} (a_{n0}+a_{n1}+\ldots+a_{nn}) = 1\,;$$

$$(2) \quad \lim_{n\to\infty} a_{n\nu} = 0 \quad (\nu = 0, 1, 2, \ldots)\,;$$

$$(3) \quad |a_{n0}|+|a_{n1}|+\ldots+|a_{nn}| < \text{const.} \quad (n = 0, 1, 2, \ldots).$$

De ce théorème, on tire immédiatement le lemme suivant qui nous sera très utile.

Pour que toute série sommable par la méthode (*a*) *soit aussi sommable par la méthode* (*b*), *il faut et il suffit que*

$$(5) \qquad \frac{|p_0|A_n+|p_1|A_{n-1}+\ldots+|p_n|A_0}{B_n} < C \quad (n = 0, 1, 2, \ldots),$$

*C étant une constante, et que**

$$(6) \qquad \lim_{n\to\infty} \frac{p_n}{B_n} \to 0.$$

En effet (à un point de vue tout à fait formel) on a

$$\Sigma B_r T_r^{(b)} x^r = \Sigma b_r x^r . \Sigma s_r x^r = \Sigma p_r x^r . \Sigma a_r x^r . \Sigma s_r x^r = \Sigma p_r x^r . \Sigma A_r T_r^{(a)} x^r .$$

Par conséquent

$$T_n^{(b)} = \frac{p_0 A_n T_n^{(a)} + p_1 A_{n-1} T_{n-1}^{(a)} + \ldots p_n A_0 T_0^{(a)}}{B_n}$$

Remarquons d'abord que l'on a évidemment

$$\frac{p_0 A_n + p_1 A_{n-1} + \ldots + p_n A_0}{B_n} = 1,$$

et alors la condition (1) de M. Toeplitz est toujours remplie. D'autre part les conditions (2) et (3) de M. Toeplitz reviennent dans le cas actuel aux conditions que nous venons de poser. Remarquons encore que la condition (5) entraîne que A_n/B_n *reste borné.*

Admettons maintenant que les méthodes (*a*) et (*b*) soient équivalentes. Alors, d'après ce que nous venons de dire, A_n/B_n et B_n/A_n resteront bornés, c'est-à-dire que

$$\frac{A_n}{B_n} < \mu_1, \quad \frac{B_n}{A_n} < \mu_2,$$

* Je ne m'arrête pas sur la question si ces deux conditions sont indépendantes.

μ_1 et μ_2 étant des constantes. On aura de plus d'après (5)

$$\frac{|p_0|A_n+|p_1|A_{n-1}+\ldots+|p_j|A_{n-j}}{B_n}$$
$$=\frac{A_n}{B_n}\left(|p_0|+|p_1|\frac{A_{n-1}}{A_n}+\ldots+p_j\frac{A_{n-j}}{A_n}\right)<C,$$

et par conséquent

$$|p_0|+|p_1|\frac{A_{n-1}}{A_n}+\ldots+|p_j|\frac{A_{n-j}}{A_n}<C\mu_2.$$

Faisons tendre n vers l'infini ; il en vient en vertu de (3)

$$|p_0|+|p_1|+\ldots+|p_j|\leqslant C\mu_2.$$

L'indice j étant quelconque, il en résulte la convergence de la série $\Sigma|p_n|$ et alors, par raisons de symétrie, aussi la convergence de la série $\Sigma|q_n|$.

Admettons inversement que ces deux séries soient convergentes et posons $\Sigma|p_n|=P$, $\Sigma|q_n|=Q$. Alors la condition (6) sera manifestement remplie. De plus

$$A_n=B_0q_n+B_1q_{n-1}+\ldots+B_nq_0\leqslant B_n(|q_0|+|q_1|+\ldots+|q_n|)\leqslant B_nQ,$$

et alors

$$\frac{|p_0|A_n+|p_1|A_{n-1}+\ldots+|p_n|A_0}{B_n}\leqslant\frac{A_n}{B_n}P\leqslant PQ,$$

c'est-à-dire que la sommabilité (*a*) entraîne la sommabilité (*b*). Par les mêmes raisons, la sommabilité (*b*) entraîne la sommabilité (*a*) et alors les méthodes sont équivalentes.

Dans les applications les plus importantes, A_n et B_n seront d'un type de croissance régulier et alors les fonctions représentées par les séries ΣA_nx^n et ΣB_nx^n auront en général pour $|x|\leqslant 1$ une singularité unique au point $x=1$. L'équivalence reviendra donc à ce que les quotients $\Sigma B_nx^n:\Sigma A_nx^n$ et $\Sigma A_nx^n:\Sigma B_nx^n$ aient une singularité assez faible au point $x=1$ et à ce que ni ΣA_nx^n ni ΣB_nx^n n'aient de zéros pour $|x|\leqslant 1$ (ou encore, mais ce sera là un cas très exceptionnel, à ce que ces zéros soient communs).

Avant de passer aux applications de ce théorème d'équivalence, je veux indiquer une application du lemme qui précède.

Je désignerai comme *sommation harmonique* la sommation définie par la formule

$$T_n=\frac{\frac{s_0}{n+1}+\frac{s_1}{n}+\ldots+\frac{s_n}{1}}{\log n}.$$

Démontrons que chaque fois que T_n tend vers une limite déterminée, la série en question sera aussi sommable (C, ϵ), ϵ étant un nombre positif quelconque.

On aura dans ce cas

$$\sum_0^\infty a_n x^n = \sum_0^\infty \frac{x^n}{n+1} = -\frac{\log(1-x)}{x},$$

$$\sum_0^\infty b_n x^n = (1-x)^{-\epsilon},$$

$$A_n = O(\log n), \qquad n^{-\epsilon} B_n \to \frac{1}{\Gamma(\epsilon+1)}.$$

D'autre part

$$\sum_0^\infty p_n x^n = \frac{\sum_0^\infty b_n x^n}{\sum_0^\infty a_n x^n} = \frac{-x(1-x)^{-\epsilon}}{\log(1-x)}.$$

Or on sait que les coefficients du développement de

$$\frac{x}{\log(1-x)}$$

sont

$$O\left(\frac{1}{n \log^2 n}\right)$$

et que leur somme est zéro. On en conclut immédiatement

$$p_n = O\left(\frac{n^{\epsilon-1}}{\log n}\right).$$

[D'ailleurs, le même résultat, on le trouve en calculant p_n par le théorème de Cauchy, le chemin d'intégration étant composé d'une circonférence de rayon $R > 1$ et le segment double $(1, R)$.] Par les évaluations qu'on vient d'écrire, on vérifie facilement que les conditions (5) et (6) sont remplies.

Passons maintenant aux applications du théorème d'équivalence. Je vous ai déjà dit que ce théorème élucide les questions concernant l'équivalence des moyennes $\omega^{-k} \sum\limits_{r<\omega} u_r(\omega-r)^k$ et des moyennes (C, k). Permettez-moi de rappeler quelques faits. D'abord (*Comptes Rendus*, 5 juillet 1909) j'ai défini ces moyennes en ne faisant prendre à ω que des valeurs entières et j'ai dit en passant que ces moyennes étaient équivalentes aux moyennes (C, k). Je me suis bientôt aperçu de ce que cet énoncé était inexact du moins dans le cas général, mais qu'il devenait exact, lorsqu'on faisait varier ω d'une façon continue. J'ai rectifié l'erreur dans la Note des *Comptes Rendus* citée au début de cette lettre.

En quoi consistait alors cette erreur? J'ai constaté (sans me soucier beaucoup de tous ces détails) qu'il était facile de donner une série Σu_n pour laquelle la limite

$$\lim_{n\to\infty} \frac{\sum_{r=0}^{n-1} u_r(n-r)^2}{n^2}$$

existe et qui pourtant n'est pas sommable $(C, 2)$. Il est d'autre part facile de voir qu'une telle série est toujours sommable $(C, 3)$ [mais pas nécessairement $(C, 3-\epsilon)$]. En outre, on pouvait donner une série Σv_n, pour laquelle la limite

$$\lim_{n\to\infty} \frac{\sum_{r=0}^{n-1} v_r(n-r)^3}{n^3}$$

existe et qui n'est sommable (C, k) pour aucune valeur de k.

En voici maintenant l'explication. On a pour $k = 2$

$$\sum_0^\infty A_n x^n = \sum_0^\infty (n+1)^2 x^n = \sum_1^\infty n^2 x^{n-1} = \frac{1+x}{(1-x)^3},$$

et alors la fonction génératrice s'annule pour $x = -1$, c'est-à-dire en un point situé sur la circonférence $|x| = 1$. Pour $k = 3$, on a

$$\sum_0^\infty A_n x^n = \sum_0^\infty (n+1)^3 x^n = \sum_1^\infty n^3 x^{n-1} = \frac{1+4x+x^2}{(1-x)^4},$$

et alors la fonction génératrice s'annule pour $x = -2\pm\sqrt{3}$. Or $x = -2+\sqrt{3}$ est situé à l'intérieur du cercle de rayon *un*. Ces faits expliquent tout ce que j'ai dit au sujet de ces moyennes et expliquent aussi pourquoi les méthodes de sommation qui s'y rattachent, et en particulier la seconde, manquent tout à fait d'intérêt.

Or une modification légère de ces méthodes conduit à deux méthodes équivalentes aux méthodes $(C, 2)$ et $(C, 3)$ respectivement.

On a en effet $$\sum_1^\infty n^2 x^{n-2} = \frac{x^2-3x+4}{(1-x)^3}.$$

Le numérateur de cette expression ne s'annule que pour

$$x = \frac{3\pm i\sqrt{7}}{2};$$

c'est-a-dire qu'il ne s'annule pas pour $|x| \leqslant 1$.

De même $$\sum_2^\infty n^3 x^{n-2} = -\frac{x^3-4x^2+5x-8}{(1-x)^4}$$

et le numérateur ne s'annule que pour les valeurs

$$x = 3{\cdot}2\ldots, \qquad x = 0{\cdot}39\ldots \pm i1{\cdot}5,$$

situées toutes à l'extérieur du cercle de rayon *un*.

On en conclut immédiatement que *les moyennes*

$$\frac{\sum\limits_{r=0}^{n-2} u_r(n-r)^2}{n^2}, \qquad \frac{\sum\limits_{r=0}^{n-2} u_r(n-r)^3}{n^3}$$

sont équivalentes aux moyennes $(C, 2)$ *et* $(C, 3)$ *respectivement.*

Les indices $k = 2$ et $k = 3$ sont les seuls indices $k > 1$ que j'ai examinés. Je me hâte d'ajouter que je ne crois pas que ces résultats aient aucun avantage sur l'ancien théorème avec le paramètre continu ω*. En particulier, pour les séries de Dirichlet $f(s) = \Sigma \mu_n n^{-s}$ ces nouvelles moyennes sont bien plus difficiles à exprimer par la fonction $f(s)$ que celles avec le paramètre continu.

Pour $k = 0$ et $k = 1$, aucune question d'équivalence ne se pose, les moyennes

(7) $$\frac{\sum\limits_{r=0}^{n-1} u_r(n-r)^k}{n^k}$$

étant identiques aux moyennes $(C, 0)$ et $(C, 1)$ respectivement.

Pour $-1 < k < 1$, *je peux démontrer maintenant un fait que j'ai toujours présumé, c'est que les moyennes* (7) *sont équivalentes aux moyennes* (C, k).

Premièrement, k étant tel qu'on vient de dire, la fonction génératrice

$$\sum_0^\infty A_n x^n = \sum_1^\infty n^k x^{n-1}$$

ne s'annule pas pour $|x| \leqslant 1$. Pour $k < 0$, on le voit immédiatement par un théorème connu de M. Kakeya (*The Tôhoku Math. Journ.*, 1912, t. 2,

* Il n'est pas sans intérêt de remarquer que pour toute valeur < 1 du paramètre positif τ, la fonction $\sum\limits_{n=0}^{\infty} (\tau+n)^2 x^n$ aura toujours un zéro à l'intérieur du cercle de rayon *un*, c'est-à-dire que le théorème d'équivalence relatif aux moyennes avec le paramètre continu n'est pas une conséquence du théorème que nous venons de démontrer.

pp. 140–142), les n^k étant positifs et décroissants avec $1/n$. Pour $0 < k < 1$, les coefficients de $(1-x)\sum_1^\infty n^k x^{n-1}$ sont positifs et décroissants et alors le théorème de M. Kakeya s'applique encore.

Cela étant, pour déterminer l'ordre de grandeur des coefficients des séries

$$(1-x)^{k+1} \sum_{n=1}^{\infty} n^k x^{n-1} = \sum_{n=0}^{\infty} p_n x^n, \qquad \frac{1}{(1-x)^{k+1} \sum_{n=1}^{\infty} n^k x^{n-1}} = \sum_{n=0}^{\infty} q_n x^n,$$

on n'aura qu'à étudier les singularités au point 1.

En utilisant alors le chemin d'intégration mentionné plus haut et en intégrant par parties, on montre facilement par une formule qu'on trouve dans le "Calcul des résidus" de M. Lindelöf (p. 139) que l'ordre de grandeur des coefficients p_n et q_n est $O(n^{-k-2})$ et, par conséquent, les séries $\Sigma|p_n|$ et $\Sigma|q_n|$ sont convergentes. Le théorème d'équivalence général qui précède pourra alors immédiatement être appliqué pour $0 < k < 1$ et après des modifications légères aussi pour $-1 < k < 0$ . . .

Stockholm, le 6 février 1923.

29.

Les fonctions conjuguées et les séries de Fourier

C. R. Acad. Sci. Paris *178* (1924), 1464–1467

1. Désignons par $f(x)$ une fonction sommable dans l'intervalle $(0, 2\pi)$ et écrivons l'expression

$$(1) \qquad \bar{f}(x) = \frac{1}{2\pi}\int_0^{2\pi} f(t) \cot \frac{t-x}{2} dt.$$

M. Plessner, complétant certains résultats de M. Fatou, a démontré (¹) que l'intégrale du second membre existe presque partout, si on la prend au sens de la valeur principale de Cauchy.

En désignant par L^p, où $p > 1$, la classe des fonctions $f(x)$ telles que $|f(x)|^p$ soit sommable dans l'intervalle $(0, 2\pi)$, on a les deux théorèmes équivalents que voici :

I. *Si $f(x)$ appartient à la classe L^p, $\bar{f}(x)$ y appartient également, de plus on a l'inégalité*

$$(2) \qquad \int_0^{2\pi} |\bar{f}(x)|^p dx \leqq M_p^p \int_0^{2\pi} |f(x)|^p dx,$$

la constante M_p ne dépendant que de p.

II. *Deux séries trigonométriques conjuguées sont en même temps les séries de Fourier de fonctions de la classe L^p.*

Voici l'idée de la démonstration. $U + iV$ étant une fonction holomorphe pour $|z| < R$ et V s'annulant à l'origine, U et V satisfont à une inégalité de la forme (2), les intégrales étant prises le long d'une circonférence quelconque $|z| = r < R$. Pour des exposants entiers pairs ou non entiers, on le voit en appliquant le théorème de Cauchy à la fonction $(U + iV)^p$, en admettant encore, pour les derniers exposants, que U ne change pas de signe, hypothèse qui ne restreint pas la généralité. Enfin, pour les exposants entiers impairs, on donne une démonstration indirecte, en montrant d'une façon générale que le théorème a lieu en

---

(¹) *Thèse*, Giessen, 1923.

même temps pour deux exposants *complémentaires* p et q, liés par la relation $\frac{1}{p} + \frac{1}{q} = 1$, et que l'on a de plus $M_p = M_q$.

2. Désignons par a_n et b_n les coefficients et par $s_n(x)$ les sommes de la série de Fourier de $f(x)$. On a :

$$\left| s_n(x) - \frac{1}{2}(a_n \cos nx + b_n \sin nx) \right| = \frac{1}{2\pi} \left| \int_0^{2\pi} f(t) \sin n(t-x) \cot \frac{t-x}{2} dt \right|$$

$$\leq \frac{1}{2\pi} \left| \int_0^{2\pi} f(t) \sin nt \cot \frac{t-x}{2} dt \right|$$

$$+ \frac{1}{2\pi} \left| \int_0^{2\pi} f(t) \cos nt \cot \frac{t-x}{2} dt \right|.$$

De la comparaison des inégalités trouvées, il résulte que :

III. *Pour toute fonction de la classe* L^p, *l'intégrale de* $|s_n(x)|^p$ *reste bornée lorsque n tend vers l'infini.*

Il est facile de préciser ce théorème en démontrant que l'intégrale de

$$|f(x) - s_n(x)|^p$$

tend vers zéro.

Désignons par α_n et β_n les constantes de Fourier d'une fonction $g(x)$ appartenant à la classe complémentaire L^q. On tire immédiatement du théorème III que

$$\frac{1}{\pi} \int_0^{2\pi} f(x) g(x)\, dx = \frac{1}{2} a_0 \alpha_0 + \sum_{n=1}^{\infty} (a_n \alpha_n + b_n \beta_n), \tag{3}$$

la série au second membre étant *convergente.*

Le théorème III ainsi complété généralise une proposition démontrée depuis longtemps par M. W. H. Young. Cet auteur considérait les moyennes arithmétiques des sommes de la série de Fourier; il prouvait, en particulier, que la série figurant au second membre de (3), sommée par ces moyennes, est égale au premier membre. De plus, M. W. H. Young et Mme Grace Chisholm Young ont déjà, en 1912 (1), posé la question de savoir si le théorème III du texte et son complément étaient vrais ou non. Ils ont démontré que ce théorème III entraînerait la convergence de la série figurant en (3), ce qui, à leur avis, semblait indiquer que la question devrait être résolue par la négative.

(1) *Quarterly Journal*, t. 44, 1912, p. 58.

3. Soit maintenant $f(x)$ une fonction de la classe L^p par rapport à l'intervalle $(-\infty, +\infty)$ et formons l'intégrale

$$(4) \qquad \bar{f}(x) = \frac{1}{\pi}\int_{-\infty}^{+\infty} \frac{f(y)}{y-x}\,dy.$$

La valeur principale de cette intégrale existe presque partout; on a alors une inégalité analogue à (2), le facteur constant étant le même. On le voit en opérant, avec la fonction de la variable complexe z

$$F(z) = \frac{1}{i\pi}\int_{-\infty}^{+\infty} \frac{f(y)}{y-z}\,dy,$$

de la même manière que l'on a opéré plus haut avec la fonction $U + iV$. On démontre aussi sans difficulté que $|F(z)|^p$ est sommable le long d'une droite arbitraire du plan de z et que son intégrale est plus petite qu'un multiple constant de celle de $|f(x)|^p$ prise le long de l'axe réel, le facteur constant ne dépendant que de p. On a aussi l'identité

$$\bar{f}_\varepsilon(x) = \frac{1}{\pi}\left(\int_{-\infty}^{x-\varepsilon} + \int_{x+\varepsilon}^{+\infty}\right)\frac{f(y)}{y-x}\,dy = \frac{1}{\pi^2}\int_{-\infty}^{+\infty} \bar{f}(y+x)\log\frac{y+\varepsilon}{y-\varepsilon}\,\frac{dy}{y}.$$

Il s'ensuit immédiatement que l'intégrale de $|\bar{f}_\varepsilon(x)|^p$ est inférieure à celle de $|\bar{f}(x)|^p$. De là on conclut que :

IV. *Pour deux fonctions $f(x)$ et $g(x)$ appartenant à des classes complémentaires L^p et L^q, on a*

$$\left|\int\!\!\int_{-\infty}^{+\infty} \frac{f(x)\,g(y)}{x-y}\,dx\,dy\right| \leqq M_p\left(\int_{-\infty}^{+\infty} |f(x)|^p\,dx\right)^{\frac{1}{p}}\left(\int_{-\infty}^{+\infty} |g(x)|^q\,dx\right)^{\frac{1}{q}};$$

l'intégrale double existant toujours si on la prend au sens de la valeur principale de Cauchy, c'est-à-dire comme limite de l'intégrale étendue aux valeurs $|x-y| > \varepsilon > 0$. En particularisant les fonctions $f(x)$ et $g(x)$, on obtient un théorème analogue relatif à une forme bilinéaire bien connue que M. Hilbert a considérée le premier.

30.

Über die Summierbarkeit durch typische Mittel

Acta Szeged Sect. Math. 2 (1924), 18–31

1. Im Jahre 1909 habe ich einen Konvergenzsatz für DIRICHLETsche Reihen ausgesprochen, dessen Beweis ich später in den Acta Mathematica[1]) veröffentlicht habe Dieser und ein verwandter Satz haben vielfache zahlentheoretische Anwendungen gefunden.[2]) Ich habe auch die Verallgemeinerung dieser Sätze auf *typische Mittel* angedeutet. Da nun auch diese Verallgemeinerungen in zahlentheoretischen Fragen wiederholt angewandt worden sind, (namentlich im Teilerproblem)[3]) scheint es mir angebracht, diese Verallgemeinerungen mit ihren ausführlichen Beweisen zu veröffentlichen, um damit endlich auch den von verschiedenen Fachgenossen an mich gerichteten Aufforderungen gerecht zu werden.

Es hat sich zweckmässig erwiesen, den Satz gleich für STIELTJEsche Integrale $\int_0^\infty e^{-tz}\,d\,A(t)$ zu fassen, wo also $A(t)$ eine beliebige Funktion bedeutet, die z. B. in jedem endlichen Intervalle von beschränkter Schwankung ist. Dieser Ausdruck umfasst dann sowohl die DIRICHLETschen Reihen, wie auch die Integrale $\int_0^\infty a(t)\,e^{-tz}\,dt$. Ich beschränke mich in dieser Arbeit auf typische

---

1) Ein Konvergenzsatz für DIRICHLETsche Reihen, Acta Math. 40 (1916) p. 349—361.

2) Für die Anwendungen vgl. man den Encyklopädie-Artikel von H. BOHR und H. CRAMÉR, Neuere Entwicklung der analytischen Zahlentheorie. Die ältere Literatur ist auch in meiner hier angeführten Arbeit zusammengestellt.

3) H. CRAMÉR, Über das Teilerproblem von PILTZ, Arkiv f. Math 16 (1922) No. 21, p. 1—40. Diese Arbeit enthält einen Beweis für ganzzahlige Ordnungszahlen. A. WALFISZ, Über die summatorischen Funktionen einiger DIRICHLETschen Reihen, Diss. Göttingen 1922, p. 1–56. Der Verfasser gibt in dieser Arbeit einen den speziellen Zwecken angepassten Beweis.

Mittel erster Art, deute aber die entsprechenden Ergebnisse für typische Mittel zweiter Art an.

2. Wir nennen ein Integral $\int_0^\infty e^{-tz}\,dA(t)$ durch typische Mittel k-ter Ordnung summierbar,[4]) wenn der Grenzwert

$$(1)\quad \lim_{\omega\to\infty} \omega^{-k} A_k(z,\omega) = \lim_{\omega\to\infty} \omega^{-k} \int_0^\omega e^{-tz}(\omega-t)^k\,dA(t),\quad (k \geq 0)$$

existiert. Die Definition der gleichmässigen Summierbarkeit leuchtet ohne weiteres ein. Wir setzen im folgenden speziell

$$(2)\qquad A_k(0,\omega) = \int_0^\omega (\omega-t)^k\,dA(t) = A_k(\omega).$$

Wir sprechen nun einige Sätze aus, deren Beweise später folgen.

Satz I. *Ist*

$$(3)\qquad A_k(\omega) = o(e^{\omega c}\omega^k)\ ,\ (c > 0,\ k \geq 0),$$

so ist das Integral $\int_0^\infty e^{-tz}\,dA(\cdot)$ *in jedem endlichen Gebiet der Halbebene* $\Re(z) > c$ *durch Mittel k-ter Ordnung gleichmässig summierbar und definiert somit in dieser Halbebene eine regulär-analytische Funktion* $G(z)$. *Ist diese Funktion noch in gewissen Punkten der Geraden* $\Re(z) = c$ *regulär,*[5]) *dann ist das Integral auch in diesen Punkten durch Mittel k-ter Ordnung summierbar und zwar gleichmässig auf jeder Regularitätsstrecke der Geraden. Die Bedingung* (3) *ist auch eine notwendige, damit die Gerade wenigstens einen Punkt enthalte, in dem die Reihe durch Mittel k-ter Ordnung summierbar ist.*[6])

---

[4]) Vgl. G. H. Hardy and M. Riesz, The general theory of Dirichlet's series. Cambridge Tracts, No. 18 (1915). Im folgenden als *Tract* zitiert.

[5]) Wir schreiben, wie üblich, $\Re(a+ib) = a$ und $\Im(a+ib) = b$.

[6]) Die Bedingung (3) ist bei $c \leq 0$ nicht mehr notwendig aber noch hinreichend. Somit besteht der erste Teil des Satzes auch in diesem Falle. Da indessen dieser Fall auf denjenigen des Textes leicht zurückgeführt werden kann, und der Beweis sich bei $c > 0$ bequemer gestaltet, begnügen wir uns hier mit diesem letztgenannten Fall. Die Bezeichnung o wenden wir im bekannten Landauschen Sinne an, d. h. $f(\omega) = o(\varphi(\omega))$ bedeutet, dass $\frac{f(\omega)}{\varphi(\omega)}$ mit ins unendliche wachsendem ω nach Null strebt. Dagegen bedeutet $f(\omega) = O(\varphi(\omega))$ wie gewöhnlich, dass derselbe Quotient endlich bleibt. Wird

Satz II. *Ist* $k=0$, *d. h.*

(3') $$A_0(\omega)=A(\omega)=o(e^{\omega c})$$

und die Funktion $G(z)$ *rechts von einer Strecke der Geraden* $\Re(z)=c$ *beschränkt, dann konvergiert das Integral* $\int_0^\infty e^{-zt}\,d\,A(t)$ *in jedem Punkte, in dem die auf dieser Strecke fast überall existierende Randfunktion entweder eine Lipschitzsche Bedingung oder irgendeine der üblichen Konvergenzbedingungen einer Fourierschen Reihe erfüllt und zwar konvergiert das Integral gleichmässig auf jeder inneren Teilstrecke einer jeden Strecke, wo diese Bedingungen gleichmässig erfüllt sind* [7])

Satz III. *Es sei* $k>0$ *und* (3) *erfüllt. Ist die Funktion* $G(z)$ *rechts von einer Strecke der Geraden* $\Re(z)=c$ *beschränkt, dann ist das Integral auf dieser Strecke fast überall durch Mittel* k*-ter Ordnung summierbar. Dies ist namentlich in jedem Stetigkeitspunkte der Randfunktion der Fall. Die Summierbarkeit ist eine gleichmässige in jedem Intervall, das in einem Stetigkeitsintervall ganz enthalten ist.*

Zuletzt mögen hier einige Andeutungen über typische Mittel zweiter Art Platz finden mit besonderer Rücksicht auf den Umstand, dass diese Mittel bei dem besonders wichtigen Fall der gewöhnlichen Dirichletschen Reihen mit den arithmetischen Mitteln gleichwertig sind.

Wir schreiben jetzt unsere Integrale in der Form

$$\int_1^\infty v^{-z}\,d\cdot C(v).$$

Die typischen Mittel zweiter Art sind dann durch den Ausdruck

$$\omega^{-k}C_k(z,\omega)=\omega^{-k}\int_1^\omega v^{-z}(\omega-v)^k\,d\,C(v)$$

gegeben. Die der Bedingung (3) entsprechende Voraussetzung lautet hier

$$C_k(0,\omega)=C_k(\omega)=o\,(\omega^{k+c}).$$

---

in der Bedingung (3) o durch O ersetzt, dann sind die Mittel k-ter Ordnung auf jeder Regularitätsstrecke der Geraden $\Re(z)=c$ gleichmässig beschränkt. Eine ähnliche Bemerkung gilt auch für die Notwendigkeit der Bedingung und für alle anderen Sätze.

[7]) Zur näheren Erläuterung dieses und des folgenden Satzes vgl. den Abschnitt 6.

Für reguläre Punkte gilt ein ähnlicher Satz wie I. Auch das Analogon von II gilt. In III müssen aber über die Randfunktion ähnliche spezielle Voraussetzungen wie in II getroffen werden. (Vgl. hierzu *Tract* p. 55, Bemerkungen über die Sätze 42 u. 43).

3. Wir leiten jetzt einige Hilfssätze ab, aus denen unsere Sätze leicht folgen.

Hilfssatz I. *Es sei $b(t)$ eine (etwa im Lebesgueschen Sinne) integrierbare Funktion, die der Bedingung*

$$(4) \qquad b(t) = o(t^k)$$

genügt. Hierbei bedeutet k eine beliebige reelle Zahl. Dann konvergiert das Integral

$$(5) \qquad F(z) = \int_0^\infty b(t)\, e^{-tz}\, dt$$

in der Halbebene $\Re(z) > 0$ absolut und stellt daselbst eine regulär-analytische Funktion $F(z)$ dar. Ist diese Funktion $F(z)$ auch noch in gewissen Punkten x der imaginären Achse regulär, so ist für diese Werte x und eine beliebige nicht negative ganze Zahl j

$$(6) \quad H(x, \omega) = e^{-\omega x} \frac{d^j}{d x^j} (F(x)\, e^{\omega x}) - \int_0^\omega b(t)\, e^{-tx} (\omega - t)^j dt = o(\omega^k).$$

Diese Limesgleichung gilt gleichmässig auf jeder Regularitätsstrecke der imaginären Achse.[8])

Der Beweis ist demjenigen ganz analog, den ich für einen ähnlichen Satz über Potenzreihen in einer früheren Arbeit[9]) gegeben habe.

Es sei die Funktion $F(z)$ auf der Strecke $(i\tau_1, i\tau_2)$, wo $\tau_1 < \tau_2$, mit Einschluss der Endpunkte regulär. Wir wählen dann die Zahl $a' < 0$ so nahe an 0, dass die Funktion noch im Rechteck $(i\tau_1, i\tau_2, a' + i\tau_2, a' + i\tau_1)$ dieselbe Eigenschaft besitze. Wir setzen

---

[8]) Wie es aus dem folgenden Beweise hervorgeht, gilt sogar in jedem Halbstreifen, $\Re(z) \geq a$, $\tau_1 \leq \Im(z) < \tau_2$ in dem die Funktion regulär ist, gleichmässig

$$e^{\omega z} H(z, \omega) = o(\omega^k).$$

Herr Szegö gab neulich eine schöne Anwendung für einen Spezialfall eines von mir herrührenden analogen Satzes über Potenzreihen ($k = j = 0$, o durch O ersetzt). Vgl. G. Szegö, Tschebyscheffsche Polynome und nicht fortsetzbare Potenzreihen, Math. Ann. 87 (1922) p. 90—111, spez. p. 95 u. ff.

[9]) Sätze über Potenzreihen, Arkiv f. Mat. etc. Bd. 11, No. 12 (1916) p. 1—16, spez. p. 10 ff.

endlich $z_1' = a' + i\tau_1$, $z_2' = a' + i\tau_2$, $z_1'' = a'' + i\tau_1$, $z_2'' = a'' + i\tau_2$ wo a'' eine beliebige, aber feste positive Zahl bedeutet.

In den folgenden Überlegungen nehmen wir zunächst $k \geq 0$ an Der Fall $k < 0$, den wir übrigens im folgenden nicht benötigen, wird sich nachträglich leicht erledigen.

Wir betrachten die Funktion

(7) $$g_\omega(z) = \omega^{-k} e^{\omega z} (z - i\tau_1)^{j+k+1} (z - i\tau_2)^{j+k+1} H(z, \omega)$$

und zeigen, dass bei hinreichend grossem ω auf der ganzen Begrenzung des Rechtecks $(z_1', z_1'', z_2'', z_2')$

(8) $$|g_\omega(z)| < \varepsilon$$

wird, wie klein auch die positive Zahl ε sein mag.

Wie leicht ersichtlich, ist für $\Re(z) > c$

$$H(z, \omega) = \int_\omega^\infty b(t)\, e^{-tz} (t - \omega)^j\, dt.$$

Mithin nach (4) für genügend grosses ω, wenn noch der Kürze halber $\Re(z) = \sigma$ gesetzt wird.

$$|H(z, \omega)| < \eta \int_\omega^\infty t^k e^{-t\sigma} (t - \omega)^j\, dt,$$

wo η eine beliebig kleine Zahl bedeutet. Oder

(9) $$|H(z, \omega)| < \eta e^{-\omega\sigma} \int_0^\infty (\omega + u)^k e^{-u\sigma} u^j\, du <$$

$$< \eta e^{-\omega\sigma} 2^k \int_0^\infty (\omega^k + u^k) e^{-u\sigma} u^j\, du < \frac{\eta K \omega^k e^{-\omega\sigma}}{\sigma^{j+k+1}}.$$ [10])

Hieraus folgt unmittelbar (vgl. die Ungleichung (11)), dass $|g_\omega(z)|$ auf den Strecken $(i\tau_1, z_1'')$, $(z_1'' z_2'')$, $(z_2'', i\tau_2)$ beliebig klein gemacht werden kann.

Wir müssen nun noch $|g_\omega(z)|$ auf den linken Begrenzungsstücken des Rechtecks abschätzen. Zunächst ist es einleuchtend, dass es eine feste Zahl K_1 gibt derart, dass in jedem Punkte des Rechtecks $(i\tau_1, i\tau_2, z_2', z_1')$

(10) $$\left| e^{-\omega z} \frac{d^j}{dz^j} \left(F(z) e^{\omega z}\right) \right| < K_1 \omega^j.$$

Bestimmen wir andererseits ω_0 so, dass für $t \geq \omega_0$ die Ungleichung $|b(t)| < \delta t^k$ gilt, wo δ eine beliebig klein wählbare positive Zahl

[10]) Wir bezeichnen mit $K, K_1, K_2, \ldots$ Zahlen, bei denen es nur darauf ankommt, dass sie endlich sind.

ist. Dann ergibt sich

$$\left|\int_0^\omega b(t)\,e^{-tz}(\omega-t)^j\,dt\right| < \omega^j \int_0^{\omega_0} |\,b(t)\,|\,e^{-t\sigma}dt + \delta\int_{\omega_0}^{\omega} t^k e^{-t\sigma}(\omega-t)^j\,dt = I_1+I_2.$$

Nun ist im ganzen Rechteck

$$I_1 < K_2\,\omega^j,$$

wo K_2 nur von ω_0 abhängt. Ausserdem ist

$$I_2 < \delta\,e^{-\omega\sigma}\int_{\omega_0}^{\omega} t^k e^{-(\omega-t)\sigma}(\omega-t)^j\,dt < \delta\,\omega^k e^{-\omega\sigma}\int_{\omega_0}^{\omega} e^{-(\omega-t)\sigma}(\omega-t)^j\,dt <$$

$$< \frac{\delta\,\omega^k e^{-\omega\sigma}\,\Gamma(j+1)}{|\,\sigma\,|^{j+1}}.$$

Mithin

$$\left|\int_0^\omega b(t)\,e^{-tz}(\omega-t)^j\,dt\right| < K_2\,\omega^j + \frac{\delta\,\omega^k e^{-\omega\sigma}\,\Gamma(j+1)}{|\,\sigma\,|^{j+1}}.$$

Ferner ist auf der Begrenzung des Rechtecks

$$(11)\qquad |\,z-i\tau_1\,|^{j+k+1}\,|\,z-i\tau_2\,|^{j+k+1} < K_3\,|\,\sigma\,|^{j+k+1}.$$

Aus diesen Abschätzungen erhalten wir

$$|\,g_\omega(z)\,| < |\,\sigma\,|^{j+k+1}\left(K_4\,\omega^{j-k}e^{\omega\sigma} + \frac{K_5\,\delta}{|\,\sigma\,|^{j+1}}\right).$$

Da aber bekanntlich für $v \geq 0$, $l \geq 0$

$$e^{-v}\,v^l \leq \left(\frac{l}{e}\right)^l,$$

so ist der letzte Ausdruck kleiner als

$$K_6\,\omega^{-(2k+1)} + K_5\,\delta\,|\,\sigma\,|^k.$$

Für genügend grosses ω wird dieser Ausdruck kleiner als ein beliebig kleines ε. Damit ist (8) für die ganze Begrenzung des Rechtecks $(z_1', z_1'', z_2'', z_2')$ dargelegt.

Nun ist aber die Funktion $g_\omega(z)$ in diesem Rechteck regulär, daher erreicht ihr absoluter Betrag sein Maximum auf der Begrenzung. Mithin ist für einen beliebigen Punkt x der Strecke $(i\tau_1, i\tau_2)$

$$(12)\qquad \left|e^{\omega x}\frac{d^j}{dx^j}\left(e^{\omega x}F(x)\right) - \int_0^\omega b(t)\,e^{-tx}(\omega-t)^j\,dt\right| <$$

$$< \frac{\varepsilon\,\omega^k}{|\,i\tau_1-x\,|^{j+k+1}\,|\,i\tau_2-x\,|^{j+k+1}}$$

d. h. die Beziehung (6) bewiesen. Besteht nun die Strecke (x_1, x_2) ausschliesslich aus Regularitätsstellen, so ist sie in einer grösseren Strecke von derselben Eigenschaft enthalten. Bezeichnet man die Endpunkte dieser grösseren Strecke mit $i\tau_1$ und $i\tau_2$, so folgt aus (12), dass die Limesgleichung (6) für die Punkte der Strecke (x_1, x_2) gleichmässig besteht.

Im obigen war $k \geq 0$ angenommen. Ist $k < 0$, dann ersetzen wir in (7) den Faktor $(z - i\tau_1)^{j+k+1} (z - i\tau_2)^{j+k+1}$ mit $(z - i\tau_1)^{j-2k+1}$ $(z - i\tau_2)^{j-2k+1}$. Statt der Abschätzung (9) erhält man jetzt

$$|H(z, \omega)| < \eta\, \omega^k e^{-\omega\sigma} \int_0^\infty e^{-u\sigma} u^j\, du = \frac{\eta\, \omega^k e^{-\omega\sigma} \Gamma(j+1)}{\sigma^{j+1}}$$

Daher bleibt die Abschätzung $|g_\omega(z)| < \varepsilon$ für die rechte Hälfte der Begrenzung bestehen.

Für I_2 erhält man jetzt $(\omega_0 > 1)$

$$I_2 < \delta\, e^{-\omega\sigma} \int_{\omega_0}^{\omega} t^k e^{(\omega - t)\sigma} (\omega - t)^j\, dt = \delta\, e^{-\omega\sigma} \left(\int_{\omega_0}^{\frac{\omega}{2}} + \int_{\frac{\omega}{2}}^{\omega} \right) <$$

$$< \delta\, e^{-\omega\sigma} \left(\omega^{j+1} e^{\frac{\omega\sigma}{2}} + \left(\frac{\omega}{2}\right)^k \frac{\Gamma(j+1)}{|\sigma|^{j+1}} \right).$$

Aus dieser Ungleichung und den übrigen auch jetzt bestehenden Ungleichungen ergibt sich die Abschätzung $|g_\omega(z)| < \varepsilon$ auch für die linke Hälfte der Begrenzung. Das übrige folgt wie oben.

4. Im folgenden nehmen wir wieder $k \geq 0$ an.

Aus dem Hilfssatze I folgt unmittelbar der

Hilfssatz II. *Es sei $b(t)$ eine integrierbare Funktion, die der Bedingung (4) genügt, wobei $k \geq 0$. Ist die durch das Integral (5) in der Halbebene $\Re(z) > 0$ dargestellte Funktion $F(z)$ noch in gewissen Punkten x der imaginären Achse regulär, dann ist das Integral (5) in diesen Punkten durch Mittel k-ter Ordnung summierbar, d. h. es ist*

$$(13) \qquad F(x) = \lim_{\omega \to \infty} \omega^{-k} \int_0^\omega (\omega - t)^k\, b(t)\, e^{-tx}\, dt.$$

Die Summierbarkeit ist auf jeder Regularitätsstrecke eine gleichmässige.

Der Beweis lässt sich ebenso führen wie der entsprechende Beweis für Potenzreihen in meiner auf S. 5 angeführten Arbeit.

Ist zunächst k ganz, dann ist der Beweis in der Formel (6) enthalten Man setze in der Tat in dieser Formel $j = k$ und schreibe

$$e^{\omega x} \frac{d^k}{dx^k} (f(x) e^{\omega x}) = \omega^k f(x) + \sum_{p=1}^{k} \binom{k}{p} \omega^{k-p} f^{(p)}(x).$$

Bei nicht ganzzahligen k setzen wir

$$B_k(x, \omega) = \int_0^\omega b(t) e^{-tx} (\omega - t)^k \, dt.$$

Dann ist nach bekannten Sätzen über Integration nicht ganzzahliger Ordnung

$$\frac{\Gamma([k]+1)\,\Gamma(k-[k])}{\Gamma(k+1)} B_k(x, \omega) = \int_0^\omega B_{[k]}(x, t) (\omega - t)^{k-[k]-1} \, dt =$$

$$= \int_0^\omega (B_{[k]}(x, t) - t^{[k]} f(x)) (\omega - t)^{k-[k]-1} dt + \frac{\Gamma([k]+1)\,\Gamma(k-[k])}{\Gamma(k+1)} f(x).$$

Das letzte Integral kann geschrieben werden

$$\int_0^\omega = \int_0^{\omega-1} + \int_{\omega-1}^{\omega}$$

Schreibt man in (6) $j = [k]$, erhält man sofort

$$\int_{\omega-1}^{\omega} = o(\omega^k).$$

Andererseits kommt durch partielle Integration

$$\int_0^{\omega-1} = \frac{B_{[k]+1}(x, \omega-1) - f(x)(\omega-1)^{[k]+1}}{[k]+1} +$$

$$+ \frac{k-[k]-1}{[k]+1} \int_0^{\omega-1} (B_{[k]+1}(x, t) - f(x)\, t^{[k]+1}) (\omega - t)^{k-[k]-2} dt.$$

Beachtet man jetzt, dass $k-[k]-2 < -1$ und daher

$$\int_0^{\omega-1} (\omega-t)^{k-[k]-2}\,dt < \int_1^{\infty} t^{k-[k]-2}\,dt = \frac{1}{1+[k]-k}$$

und setzt in (6) $j=[k]+1$, so folgt ohne Schwierigkeit, dass auch $I_1 = o(\omega^k)$ ist. Damit ist aber (13) dargelegt. Da alle Abschätzungen gleichmässig gelten, gilt auch (13) auf jeder Regularitätsstrecke gleichmässig.

Ähnlich beweist man den

Hilfssatz III. *Wenn* $k > k' > -1$, *dann gilt unter der Bedingung* (4) *auf jeder Regularitätsstrecke der imaginären Achse gleichmässig*

$$\int_0^{\omega} b(t)\, e^{-tx} (\omega-t)^{k'}\,dt = o(\omega^k)$$

und

$$\int_0^{\omega} b(t)\,(e^{-tx}-e^{-\omega x})\,(\omega-t)^{k'-1}\,dt = o(\omega^k).$$ [11]

5. Diese Ergebnisse setzen uns in Stand unsere Hauptsätze zu beweisen. Übrigens ist Hilfssatz II ein Spezialfall von Satz I.

Es sei also die Bedingung (3) erfüllt.

a) Zunächst sei k eine ganze Zahl. Dann kommt (vgl. *Tract* p. 40) nach $k+1$ partiellen Integrationen

$$\omega^{-k} A_k(z,\omega) = \omega^{-k}\int_0^{\omega} e^{-tz}(\omega-t)^k\,d\,A(t) = \omega^{-k} e^{-\omega z} A_k(\omega) +$$

$$+\frac{(-1)^{k+1}\omega^{-k}}{k!}\int_0^{\omega} A_k(t)\left(\frac{d}{dt}\right)^{k+1}\left\{e^{-zt}(\omega-t)^k\right\}dt.$$

Hieraus beweist man wie im *Tract* (l. c.), dass infolge der Bedingung (3) $\omega^{-k} A_k(z,\omega)$ mit ins unendliche wachsendem ω für $\Re(z) > c$ gegen eine regulär-analytische Funktion $G(z) = \frac{z^{k+1}}{k!}\int_0^{\infty} A_k(t)\, e^{-zt}dt$ konvergiert. Es habe nun diese Funktion auf der Geraden $\Re(z) = c$

---

[11]) Vgl. hierzu *Tract* p. 43. Das erste Ergebnis besagt nur für nicht ganze k' etwas neues.

eine Regularitätsstrecke. Es ist leicht zu zeigen, dass auch noch auf dieser Strecke $\lim_{\omega\to\infty} \omega^{-k} A_k(z, \omega) = G(z)$ gleichmässig gilt.

Zunächst ist nach (3) auf der Geraden $\Re(z) = c$ gleichmässig

$$e^{\omega z}\, \omega^{-k} A_k(\omega) = o(1).$$

Damit ist der erste Teil des letzten Ausdruckes abgeschätzt. Der zweite Teil kann, wenn man noch $A_k(t)\, e^{-tc} = b(t)$ setzt, geschrieben werden

$$\frac{\omega^{-k}}{k!} z^{k+1} \int_0^{\omega} A_k(t) e^{-tz}(\omega-t)^k dt + \frac{\omega^{-k}}{k!} \sum_{p=0}^{k} H_p z^{p+1} \int_0^{\omega} A_k(t) e^{-tz}(\omega-t)^p dt =$$

$$= \frac{\omega^{-k}}{k!} z^{k+1} \int_0^{\omega} b(t) e^{-t(z-c)}(\omega-t)^k dt + \frac{\omega^{-k}}{k!} \sum_{p=0}^{k} H_p z^{p+1} \int_0^{\omega} b(t) e^{-t(z-c)}(\omega-t)^p dt,$$

wo die H_p Konstanten sind. Wir schreiben nun für $\Re(z) > c$

$$\frac{k!\, G(z)}{z^{k+1}} = \int_0^{\infty} A_k(t)\, e^{-tz}\, dt = \int_0^{\infty} b(t)\, e^{-t(z-c)}\, dt.$$

Laut Voraussetzung besitzt diese Funktion auf der Geraden $\Re(z) = c$ eine Regularitätsstrecke (z_1, z_2). Daher ist die auf der imaginären Achse gelegene Strecke $(z_1 - c,\ z_2 - c)$ eine Regularitätsstrecke der Funktion $\int_0^{\infty} b(t)\, e^{-tz}\, dt$. Wendet man nun die Hilfssätze II und III an, dann folgt unmittelbar, dass das erste Glied auf der rechten Seite des Ausdruckes (14) auf der Strecke (z_1, z_2) gleichmässig gegen die Funktion $G(z)$ konvergiert und dass die übrigen Glieder, d. h. $\omega^{-k} \sum_{p=0}^{k}$ gleichmässig gegen Null konvergieren.

b) Es sei jetzt k keine ganze Zahl. Es genügt den Beweis nur unter der Voraussetzung $0 < k < 1$ explizit auszuführen. Im allgemeinen Falle spielt $A_{[k]+1}(t)$ dieselbe Rolle wie jetzt $A_1(t)$.

Man ersieht, wie im *Tract* (p. 42—43), dass unter der Voraussetzung (3) $\omega^{-k} A_k(z, \omega)$ für $\Re(z) > c$ mit ins unendliche wachsendem ω gegen die regulär-analytische Funktion

$$(15)\qquad G(z) = \frac{z^{k+1}}{\Gamma(k+1)} \int_0^{\infty} A_k(t)\, e^{-tz}\, dt = z^2 \int_0^{\infty} A_1(t)\, e^{-tz}\, dt$$

konvergiert. Es ergibt sich jetzt (vgl. l. c.)

$$\omega^{-k} A_k(z, \omega) = \omega^{-k} \int_0^{\omega} e^{-tz}(\omega - t)^k \, d\, A(t) =$$

$$= \omega^{-k} \int_0^{\omega} A_1(t) \frac{d^2}{dt^2} \left\{ (e^{-tz} - e^{-\omega z}) \, (\omega - t)^k \right\} dt + A_k(\omega) \, e^{-\omega z} \omega^{-k} = I_1 + I_2.$$

Infolge von (3) ist wieder für $\Re(z) = c$ gleichmässig

$$\lim_{\omega \to \infty} I_1 = \lim_{\omega \to \infty} A_k(\omega) \, e^{-\omega z} \omega^{-k} = 0.$$

Andererseits ist, indem wir $A_1(t) e^{-tc} = b(t)$ setzen,

$$I_2 = \omega^{-k} z^2 \int_0^{\omega} A_1(t) e^{-tz} (\omega - t)^k + 2 k \omega^{-k} z \int_0^{\omega} A_1(t) e^{-tz} (\omega - t)^{k-1} dt +$$

$$+ k(k-1) \omega^{-k} \int_0^{\omega} A_1(t) (e^{-tz} - e^{-\omega z}) \, (\omega - t)^{k-2} dt = J_1 + J_2 + J.$$

Aus (3) und aus (vgl. *Tract* p. 27)

$$A_1(\omega) = \frac{1}{\Gamma(k+1)\,\Gamma(1-k)} \int_0^{\omega} A_k(t) \, (\omega - t)^{-k} dt$$

folgt leicht, dass auch

$$A_1(\omega) = o\,(e^{\omega c} \omega^k)$$

ist. Dann folgt aber aus Hilfssatz II wie oben, dass auf jeder Regularitätsstrecke gleichmässig

$$\lim_{\omega \to \infty} J_1 = G(z)$$

und aus Hilfssatz III, dass ebenfalls gleichmässig

$$\lim_{\omega \to \infty} J_2 = 0 \text{ und } \lim_{\omega \to \infty} J_3 = 0$$

ausfällt. Damit ist nun unser Beweis zu Ende.

Dass die Bedingung (3) auch notwendig ist, damit das Integral an wenigstens einer Stelle der Geraden $\Re(z) = c$ durch Mittel k-ter Ordnung summierbar sei, folgt leicht durch Überlegungen, die im *Tract* durchgeführt sind (p. 40—43) und auf die wir schon oben hingewiesen haben.

6. Wir schreiten jetzt zur Verallgemeinerung des Satzes I, d. h. zum Beweise der Sätze II und III.

Es sei also wieder die Bedingung (3) erfüllt und die durch (15) für $\Re(z) > c$ dargestellte Funktion $G(z)$ sei für

$$\Re(z) > c\ ,\quad \tau_1 - \delta \leq \Im(z) \leq \tau_2 + \delta$$

bechränkt. Dann gibt es bekanntlich auf der Strecke $(c + i(\tau_1 - \delta),\ c + i(\tau_2 - \delta))$ fast überall eine Randfunktion, der sich $G(z)$ nähert, wenn z von rechts und der reellen Achse parallel den Punkten dieser Strecke zustrebt. Wir bezeichnen auch diese Randfunktion mit $G(z)$

Bedeutet jetzt z einen beliebigen Punkt der Halbebene $\Re(z) > c$ und bezeichnet man mit C einen beliebigen in dieser Halbebene verlaufenden rektifizierbaren Kurvenzug, der die Punkte $z_1 = c + i\tau_1$ und $z_2 = c + i\tau_2$ mit einander verbindet und so beschaffen ist, dass die durch die Strecke (z_1, z_2) und C gebildete geschlossene Kurve den Punkt z einschliesst, so ist nach bekannten Sätzen

$$G(z) = \frac{1}{2\pi i}\int_{z_2}^{z_1} \frac{G(\zeta)}{\zeta - z}\, d\zeta + \frac{1}{2\pi i}\int_{C} \frac{G(\zeta)}{\zeta - z}\, d\zeta = G_1(z) + G_2(z) \tag{16}$$

wobei also

$$G_1(z) = \frac{1}{2\pi i}\int_{z_2}^{z_1} \frac{G(\zeta)}{\zeta - z}\, d\zeta \tag{17}$$

und

$$G_2(z) = \frac{1}{2\pi i}\int_{C} \frac{G(\zeta)}{\zeta - z}\, d\zeta. \tag{18}$$

Bedeutet nun ζ einen Punkt der Strecke (z_1, z_2), so ist offenbar

$$\frac{1}{\zeta - z} = -\int_0^{\infty} e^{t(\zeta - z)}\, dt$$

daher

$$G_1(z) = \int_0^{\infty} e^{-tz} dt \int_{z_2}^{z_1} \left(-\frac{1}{2\pi i}\, G(\zeta) e^{t\zeta}\right) d\zeta = \int_0^{\infty} p(t)\, e^{-tz}\, dt$$

Nach einem bekannten Satze von RIEMANN-LEBESGUE ist dann

(19) $$p(t) = o(e^{tc}).$$

Ferner folgt aus (17), dass $G_1(z)$ in der ganzen, durch die Strecke (z_1, z_2) aufgeschnittenen Ebene regulär ist und im unendlichen wenigstens von der ersten Ordnung verschwindet. Wir setzen noch

$$P(t) = \int_0^t p(v)\,dv$$

Dann ist wiederum für $\Re(z) > c$ die Funktion $G_2(z)$ gleich dem durch Mittel k-ter Ordnung summierten Ausdruck $\int_0^\infty e^{-tz}(dA(t) - dP(t))$ d. h

$$G_2(z) = \lim_{\omega \to \infty} \omega^{-k} \int_0^\omega e^{-tz}(\omega - t)^k (dA(t) - dP(t))$$

Setzen wir jetzt $A(t) - P(t) = B(t)$, dann folgt aus (3) und (19)

$$B_k(\omega) = \int_0^\omega (\omega - t)^k \, dB(t) = o(\omega^k e^{\omega c})$$

Andererseits ergibt sich aus (18) (und der Willkürlichkeit von C), dass die Funktion $G_2(z)$ in der z. B. durch die Strecken $(z_2, c + i\infty)$ und $(c - i\infty, z_1)$ aufgeschnittenen Ebene regulär ist. Für das Integral $\int_0^\infty e^{-tz}\,dB(t)$ sind also alle Bedingungen des Satzes I erfüllt und somit ist dieses Integral auf jeder in (z_1, z_2) enthaltenen Strecke durch Mittel k-ter Ordnung gleichmässig summierbar.

Aus dem über $G_1(z)$ gesagten (namentlich aus dem Verhalten dieser Funktion im unendlichen) folgt dass für einen beliebigen Punkt x der Strecke (z_1, z_2) der Ausdruck

$$\int_0^\omega p(t)\,e^{-tx}(\omega - t)^k\,dt \qquad (k \geq 0 \text{ oder auch nur } > -1)$$

durch ein absolut konvergentes Integral dargestellt werden kann, das ähnlich gebaut ist wie das PERRONsche Integral (vgl. *Tract* p. 12) oder dessen Verallgemeinerungen (vgl. *Tract* p. 50). Dieses Integral lässt sich dann durch Methoden behandeln, die aus der Theorie der Potenzreihen oder FOURIERschen Reihen geläufig sind (vgl. z. B. meine Arbeit: Über einen Satz des Herrn Fatou, J. f. Math. Bd. 140 (1911) p. 91—99, spez. p. 98—99).

Da die Sätze II und III in der hier gegebenen Allgemeinheit für die genannten zahlentheoretischen Anwendungen wenigstens einstweilen reichlich ausreichen dürften, gehen wir hier auf weitere Verallgemeinerungen nicht ein.[12])

[12]) Vgl. indessen die Andeutungen, die ich in einer kurzen Mitteilung (Sur les séries de Dirichlet et les séries entières, Comptes Rendus 149 (1909) p. 909) gegeben habe und die zum Teil von Herrn Neder (Zur Theorie der trigonometrischen Reihen, Math. Ann. 84 (1921) p. 117—136) ausgeführt worden sind. Z. B. kann die Beschränktheit der Funktion durch viel allgemeinere Bedingungen ersetzt werden. Es lässt sich auch ein dem sogenannten Riemannschen Lokalisationssatz (vgl. für diesen Ausdruck Neder l. c.) nachgebildeter Satz verhältnismässig leicht herleiten.

32.

Sur les maxima des formes bilinéaires et sur les fonctionnelles linéaires

Acta Math. *49* (1927), 465–497

Introduction.

1. Il y a vingt ans, Jensen[1] a publié dans les *Acta mathematica* un mémoire important sur les fonctions convexes et les inégalités qui s'y rattachent. Dès lors le rôle des fonctions convexes en l'Analyse est devenu de plus en plus manifeste.

Une fonction $f(P)$, définie pour les points d'un domaine convexe, est dite convexe, si pour deux points quelconques P_1 et P_2 du domaine et pour toutes les valeurs positives t_1 et t_2 telles que $t_1 + t_2 = 1$, on a

$$f(t_1 P_1 + t_2 P_2) \leqq t_1 f(P_1) + t_2 f(P_2).$$

La signification géométrique de cette inégalité est évidente.

2. Cela étant, considérons une forme bilinéaire à un nombre fini de variables

$$A(x, y) = \sum_{j=1}^{m} \sum_{k=1}^{n} a_{jk} x_j y_k,$$

à des coefficients quelconques, réels ou non.

Désignons par $M_{\alpha\beta}$ le maximum de $|A(x, y)|$ sous les conditions

$$(1) \qquad \sum_{j=1}^{m} |x_j|^{\frac{1}{\alpha}} \leqq 1; \qquad \sum_{k=1}^{n} |y_k|^{\frac{1}{\beta}} \leqq 1.$$

[1] J. L. W. V. Jensen, Sur les fonctions convexes et les inégalités entre les valeurs moyennes, *Acta math.* t. 30 (1906), p. 175—193.

Dans les cas $\alpha = 0$ ou $\beta = 0$, les conditions respectives (1) seront, par des raisons de continuité, remplacées par

$$(1') \qquad |x_j| \leqq 1 \quad (j = 1, 2, \ldots, m) \quad \text{ou} \quad |y_k| \leqq 1 \quad (k = 1, 2, \ldots, n).$$

3. Ces préliminaires posés, nous sommes à même d'énoncer le théorème que voici.

I. *La fonction* $\log M_{\alpha\beta}$ *est, dans le triangle*

$$(2) \qquad \alpha \leqq 1, \quad \beta \leqq 1, \quad \alpha + \beta \geqq 1,$$

une fonction convexe du point (α, β).[1]

Ce théorème contient en germe tous les résultats du travail présent, en particulier ceux concernant les fonctionnelles linéaires, qui de leur côté embrassent sous un aspect général les théorèmes de MM. W. H. Young, F. Hausdorff[2] et F. Riesz[3] sur l'extension de la formule de Parseval et d'autres problèmes de caractère analogue. Il s'agira notamment des fonctions conjuguées et de certaines suites de facteurs attachées aux séries de Fourier et en particulier de telles qui, dans un certain sens, ne changent pas le caractère de la série. Dans la dernière partie de ce travail, en retournant à notre premier théorème, nous discutons la question de la réalité des valeurs des variables qui fournissent le maximum. Nous y verrons qu'il y a une différence essentielle entre le triangle (2) et le triangle complémentaire $0 \leqq \alpha$, $0 \leqq \beta$, $\alpha + \beta \leqq 1$. Nous finissons par donner des indications, d'ailleurs peu satisfaisantes, sur la validité de notre théorème dans le carré composé des triangles dont nous venons de parler.

Démonstration du Théorème I.

4. Nous définirons les nombres α', a, a' par les relations

$$(3) \qquad \alpha + \alpha' = 1, \quad \alpha a = 1, \quad \alpha' a' = 1,$$

c'est-à-dire que

---

[1] Il en ressort que $M_{\alpha\beta}$ et toutes ses puissances à des exposants positifs sont aussi convexes.

[2] F. Hausdorff, Eine Ausdehnung des Parsevalschen Satzes über Fourierreihen, *Math. Zeitschr.* t. 16 (1923), p. 163—169. On y trouve une liste des travaux relatifs de M. Young.

[3] F. Riesz, Über eine Verallgemeinerung der Parsevalschen Formel, *Math. Zeitschr.* t. 18 (1923), p. 117—124.

$$(3') \qquad \alpha' = 1-\alpha, \quad a = \frac{1}{\alpha}, \qquad a' = \frac{1}{\alpha'} = \frac{1}{1-\alpha} = \frac{a}{a-1}.$$

Notations analogues pour la lettre β et pour d'autres lettres.

On admettra toujours

$$0 \leqq \alpha \leqq 1, \quad 0 \leqq \beta \leqq 1; \quad a \geqq 1, \quad b \geqq 1.$$

La démonstration de notre théorème se fonde entièrement sur l'inégalité connue de Hölder[1]

$$(4) \qquad \left|\sum_{j=1}^{m} \xi_j \eta_j\right| \leqq \left(\sum_{j=1}^{m} |\xi_j|^a\right)^{\alpha} \left(\sum_{j=1}^{m} |\eta_j|^{a'}\right)^{\alpha'},$$

qui résulte immédiatement de ce que x^a est, pour $a > 1$ et pour $x \geqq 0$, une fonction convexe de x. On peut encore préciser cette inégalité en rappelant le fait que *le signe d'égalité intervient seulement dans le cas où le rapport* $|\xi_j|^a : |\eta_j|^{a'}$ *et l'amplitude du produit* $\xi_j \eta_j$ *sont indépendants de* j.

On aura encore besoin du cas particulier suivant de (4):

$$(4') \qquad \sum_{j=1}^{m} |x_j|^{h_1 \sigma_1 + h_2 \sigma_2} \leqq \left(\sum_{j=1}^{m} |x_j|^{h_1}\right)^{\sigma_1} \left(\sum_{j=1}^{m} |x_j|^{h_2}\right)^{\sigma_2},$$

où $0 \leqq \sigma_1$, $0 \leqq \sigma_2$, $\sigma_1 + \sigma_2 = 1$, h_1 et h_2 étant des nombres réels quelconques.

5. En posant

$$(5) \qquad X_k = \sum_{j=1}^{m} a_{jk} x_j \ (k = 1, 2, \ldots, n); \quad Y_j = \sum_{k=1}^{n} a_{jk} y_k \ (j = 1, 2, \ldots, m),$$

on pourra écrire

$$(6) \qquad A(x, y) = \sum_{j=1}^{m} x_j Y_j = \sum_{k=1}^{n} y_k X_k.$$

En écrivant alors les conditions (1) sous la forme

$$(7) \qquad \sum_{j=1}^{m} |x_j|^a \leqq 1; \quad \sum_{k=1}^{n} |y_k|^b \leqq 1,$$

[1] O. Hölder, Ueber einen Mittelwerthssatz, *Götting. Nachr.* 1889, p. 38—47. *Cf.* aussi Jensen, *l. c.* p. 182.

nous allons montrer que *l'expression*

$$(8) \qquad \left(\sum_{k=1}^{n} |X_k|^{b'}\right)^{\beta'},$$

variée sous la première des conditions (7), *et l'expression*

$$(8') \qquad \left(\sum_{j=1}^{m} |Y_j|^{a'}\right)^{\alpha'},$$

variée sous la seconde de ces conditions, admettent aussi $M_{\alpha\beta}$ comme maximum. Ou, en se débarrassant des conditions (7), on pourra dire:

Les rapports

$$(9) \qquad |A(x,y)| : \left(\sum_{j=1}^{m} |x_j|^{a}\right)^{\alpha} \left(\sum_{k=1}^{n} |y_k|^{b}\right)^{\beta},$$

$$(10) \qquad \left(\sum_{k=1}^{n} |X_k|^{b'}\right)^{\beta'} : \left(\sum_{j=1}^{m} |x_j|^{a}\right)^{\alpha},$$

$$(10') \qquad \left(\sum_{j=1}^{m} |Y_j|^{a'}\right)^{\alpha'} : \left(\sum_{k=1}^{n} |y_k|^{b}\right)^{\beta}$$

admettent le même maximum $M_{\alpha\beta}$.

En effet, on a d'abord, en appliquant l'inégalité de Hölder à (6),

$$(11) \qquad |A(x,y)| \leqq \left(\sum |Y_j|^{a'}\right)^{\alpha'} \left(\sum |x_j|^{a}\right)^{\alpha}.$$

C'est-à-dire que le rapport (9) ne pourra dépasser le rapport (10').

D'autre part, d'après la remarque dont on a fait suivre l'inégalité de Hölder, on pourra, pour des y_k arbitraires, déterminer les x_j de façon qu'il y ait égalité entre les deux membres de (11), et alors aussi les rapports (9) et (10') seront égaux. Il y a donc égalité entre les maxima de (9) et de (10') et dès lors, par symétrie, aussi entre ceux de (9) et de (10).

6. Montrons maintenant que *tout système $(x_j; y_k)$ qui rend $|A(x,y)|$ maximum, sous les conditions* (7), *satisfait aux relations*

$$(12) \qquad \begin{cases} |Y_j| = M_{\alpha\beta} |x_j|^{a-1} & (j = 1, 2, \ldots, m); \\ |X_k| = M_{\alpha\beta} |y_k|^{b-1} & (k = 1, 2, \ldots, n). \end{cases}$$

En effet, désignons un instant par μ le maximum de $|A(x,y)|$ sous la première des conditions (7), les y_k étant fixés. L'inégalité (11) et notre remarque ajoutée à l'inégalité de Hölder montrent que pour que ce maximum soit atteint, il faudra choisir les x_j de façon que les rapports

$$|Y_j|^{a'} : |x_j|^a = |Y_j|^{\frac{a}{a-1}} : |x_j|^a$$

et alors les rapports

$$|Y_j| : |x_j|^{a-1} = \lambda$$

soient indépendants de j et qu'il en soit de même des amplitudes des $x_j Y_j$. Il vient

$$\mu = \left|\sum x_j Y_j\right| = \sum |x_j||Y_j| = \lambda \sum |x_j|^a = \lambda,$$

et alors on aura les équations

$$|Y_j| = \mu |x_j|^{a-1}.$$

Or, si l'on considère un système $(x_j; y_k)$ qui rend $|A(x,y)|$ maximum sous les conditions (7), ce système rendra encore $|A(x,y)|$ maximum, si l'on fixe les y_k et ne fait varier que les x_j. Alors le premier système des équations (12) se trouve établi. Le second système en résulte par symétrie.

7. Cela étant, désignons par (α,β), (α_1,β_1), (α_2,β_2) trois points du triangle (2), situés en ligne droite, le point (α,β) séparant les deux autres. Posons encore pour abréger $M_{\alpha\beta} = M$, $M_{\alpha_1\beta_1} = M_1$ et $M_{\alpha_2\beta_2} = M_2$. Notre théorème revient alors à *l'inégalité*

$$M \leqq M_1^{\frac{\alpha_2-\alpha}{\alpha_2-\alpha_1}} M_2^{\frac{\alpha-\alpha_1}{\alpha_2-\alpha_1}} = M_1^{t_1} M_2^{t_2}, \tag{13}$$

où $t_1 + t_2 = 1$.

Or, des relations (12) et de ce qu'on a dit au sujet du maximum de (10) et de (10'), on obtient immédiatement que tout système $(x_j; y_k)$ qui fournit le maximum de $|A(x,y)|$, sous les conditions (7), satisfait aux inégalités

$$M\left(\sum |x_j|^{(a-1)\,a_1'}\right)^{\alpha_1'} = \left(\sum |Y_j|^{a_1'}\right)^{\alpha_1'} \leqq M_1 \left(\sum |y_k|^{b_1}\right)^{\beta_1};$$

$$M\left(\sum |y_k|^{(b-1)\,b_2'}\right)^{\beta_2'} = \left(\sum |X_k|^{b_2'}\right)^{\beta_2'} \leqq M_2 \left(\sum |x_j|^{a_2}\right)^{\alpha_2}.$$

De là

$$(13') \qquad M\left(\sum |x_j|^{(a-1)a_1'}\right)^{\alpha_1' t_1}\left(\sum |y_k|^{(b-1)b_2'}\right)^{\beta_2' t_2} \leqq$$

$$\leqq M_1^{t_1} M_2^{t_2}\left(\sum |x_j|^{a_2}\right)^{\alpha_2 t_2}\left(\sum |y_k|^{b_1}\right)^{\beta_1 t_1}.$$ [1]

Déterminons maintenant λ et μ de façon que[2]

$$a_2 = \lambda(a-1)a_1' + (1-\lambda)a; \quad b_1 = \mu(b-1)b_2' + (1-\mu)b.$$

On trouve par un calcul simple

$$\lambda = \frac{\alpha_1'}{\alpha_2}\frac{\alpha_2-\alpha}{\alpha-\alpha_1}; \qquad \mu = \frac{\beta_2'}{\beta_1}\frac{\beta-\beta_1}{\beta_2-\beta} = \frac{\beta_2'}{\beta_1}\frac{\alpha-\alpha_1}{\alpha_2-\alpha}.$$

Admettons un instant que les points $(\alpha, \beta), (\alpha_1, \beta_1), (\alpha_2, \beta_2)$ soient tels que les valeurs λ et μ, qui sont toujours positives, soient aussi $\leqq 1$. Alors on pourra poser dans $(4')$, $h_1 = (a-1)a_1'$, $h_2 = a$, $\sigma_1 = \lambda$, $\sigma_2 = 1-\lambda$ et, en remplaçant les x_j par les y_k, $h_1 = (b-1)b_2'$, $h_2 = b$, $\sigma_1 = \mu$, $\sigma_2 = 1-\mu$. Il viendra alors, en tenant encore compte des inégalités (7),

$$\sum |x_j|^{a_2} \leqq \left(\sum |x_j|^{(a-1)a_1'}\right)^{\lambda}; \qquad \sum |y_k|^{b_1} \leqq \left(\sum |y_k|^{(b-1)b_2'}\right)^{\mu},$$

et de là, grâce aux relations faciles à vérifier $\lambda\alpha_2 t_2 = \alpha_1' t_1$, $\mu\beta_1 t_1 = \beta_2' t_2$,

$$\left(\sum |x_j|^{a_2}\right)^{\alpha_2 t_2} \leqq \left(\sum |x_j|^{(a-1)a_1'}\right)^{\alpha_1' t_1}; \qquad \left(\sum |y_k|^{b_1}\right)^{\beta_1 t_1} \leqq \left(\sum |y_k|^{(b-1)b_2'}\right)^{\beta_2' t_2}.$$

Par là, l'inégalité $(13')$ se réduit à l'inégalité (13).

8. Tout revient donc à examiner les inégalités

$$\lambda = \frac{\alpha_1'}{\alpha_2}\frac{\alpha_2-\alpha}{\alpha-\alpha_1} \leqq 1; \qquad \mu = \frac{\beta_2'}{\beta_1}\frac{\alpha-\alpha_1}{\alpha_2-\alpha} \leqq 1,$$

ou les inégalités équivalentes

---

[1] Jusqu'ici il aurait suffi de supposer que nos points se trouvent dans le carré $0 \leqq \alpha \leqq 1$, $0 \leqq \beta \leqq 1$.

[2] *Cf.* pour un calcul analogue F. Hausdorff et F. Riesz, *l. c.*

(14) $$\frac{\alpha_1'}{\alpha_2} \leqq \frac{\alpha-\alpha_1}{\alpha_2-\alpha} \leqq \frac{\beta_1}{\beta_2'}.$$

Remarquons d'abord, que si les points (α_1, β_1) et (α_2, β_2) sont intérieurs au triangle (2), on a
$$0 < \alpha_1' = 1 - \alpha_1 = 1 - (\alpha_1 + \beta_1) + \beta_1 < \beta_1,$$
et, d'une manière analogue,
$$0 < \beta_2' < \alpha_2.$$
Il s'ensuit que
$$0 < \frac{\alpha_1'}{\alpha_2} < \frac{\beta_1}{\beta_2'} < +\infty.$$

D'autre part en faisant varier α de α_1 à α_2, le rapport $(\alpha - \alpha_1) : (\alpha_2 - \alpha)$ ira d'une manière continue de 0 à $+\infty$; on pourra donc choisir α entre α_1 et α_2 de sorte que les inégalités (14) soient remplies. C'est-à-dire que le calcul du numéro précédent et l'inégalité (13) sont valables pour les extrémités (α_1, β_1), (α_2, β_2) d'un segment de droite quelconque intérieur au triangle (2) et pour certains points intermédiaires (α, β) du segment.

Cela signifie en langage géométrique que la surface qui, pour (α, β) intérieur au triangle (2), représente la fonction évidemment continue $\log M_{\alpha\beta}$, est telle que aucune de ses cordes n'est située entièrement au-dessous de la surface. Je dis que *la surface est convexe.* En effet, dans le cas contraire, il existerait un segment de droite intérieur au triangle (2) et tel que l'arc de courbe qui y correspond sur la surface, contienne au moins un point situé au-dessus de la corde joignant les extrémités de l'arc en question. En partant de ce point et en cheminant sur l'arc dans les deux directions jusqu'à la première rencontre avec la corde, on aura construit un arc situé entièrement au-dessus de sa corde, ce qui serait en contradiction avec le fait que nous venons d'établir. Notre théorème étant démontré pour l'intérieur du triangle (2), le passage au triangle fermé se fait par continuité.[1]

Seconde forme du théorème I et généralisation.

9. Dans les applications, il sera souvent préférable de considérer, au lieu de la forme bilinéaire $A(x, y)$, les substitutions linéaires, définies par les formu-

[1] Le raisonnement qui précède tient aussi sans passage à la limite pour les points $\alpha + \beta = 1$, $\alpha > 0, \beta > 0$.

les (5), qui y correspondent. On pourra, par des raisons de symétrie, se borner à la substitution (X_k) portant sur les (x_j). On a vu que l'expression (8), variée sous la première des conditions (1), admet le même maximum que $|A(x,y)|$, variée sous ces conditions (1). De là le théorème, où nous avons posé γ au lieu de $\beta' = 1 - \beta$.

II. *En désignant par* $M^*_{\alpha\gamma}$ *le maximum de l'expression*

$$\left(\sum_{k=1}^{n} |X_k|^{\frac{1}{\gamma}}\right)^{\gamma},$$

sous la condition

$$\sum_{j=1}^{m} |x_j|^{\frac{1}{\alpha}} \leqq 1,$$

log $M^*_{\alpha\gamma}$ *sera, dans le triangle*

$$0 \leqq \gamma \leqq \alpha \leqq 1, \tag{15}$$

une fonction convexe du point (α, γ).

10. Voici maintenant une généralisation du théorème I, qui se démontre par un calcul analogue, où le rôle des formes linéaires X_k et Y_j est joué par les formes $\dfrac{X_k}{\sigma_k}$ et $\dfrac{Y_j}{\varrho_j}$.

I'. *Le théorème* I *reste valable lorsqu'on remplace les conditions* (1) *par les conditions*

$$\sum_{j=1}^{m} \varrho_j |x_j|^{\frac{1}{\alpha}} \leqq 1; \quad \sum_{k=1}^{n} \sigma_k |y_k|^{\frac{1}{\beta}} \leqq 1, \tag{16}$$

les ϱ_j *et les* σ_k *désignant des nombres positifs.*

La généralisation correspondante du théorème II s'obtient en appliquant le théorème précédent à la forme bilinéaire $\Sigma \sigma_k y_k X_k$.

II'. *En désignant par* $M^*_{\alpha\gamma}$ *le maximum de l'expression*

$$\left(\sum_{k=1}^{n} \sigma_k |X_k|^{\frac{1}{\gamma}}\right)^{\gamma},$$

sous la condition

$$\sum_{j=1}^{m} \varrho_j |x_j|^{\frac{1}{\alpha}} \leqq 1,$$

log $M^*_{\alpha\gamma}$ *sera, dans le triangle* (15), *une fonction convexe du point* (α, γ).

Une application.

11. Avant d'aller aux applications exigeant des passages à la limite, déduisons du théorème II le théorème suivant dû à M. F. Riesz:[1]

Soit

$$X_k = \sum_{j=1}^{m} a_{jk}\, x_j \qquad (k = 1, 2, \ldots, m),$$

une substitution orthogonale et normée, c'est-à-dire une substitution telle que pour tout système (x_j) *on ait*

$$\sum_{k=1}^{m} |X_k|^2 = \sum_{j=1}^{m} |x_j|^2.$$

Désignons encore par $\mathfrak{M}$ *la plus grande des valeurs* $|a_{jk}|$. *Alors on aura pour tout nombre* a *tel que* $1 \leqq a \leqq 2$,

$$\left(\sum_{k=1}^{m} |X_k|^{\frac{a}{a-1}}\right)^{\frac{a-1}{a}} \leqq \mathfrak{M}^{\frac{2-a}{a}} \left(\sum_{j=1}^{m} |x_j|^a\right)^{\frac{1}{a}}.$$

Posons, comme d'habitude, $\frac{1}{a} = \alpha$ et considérons, dans le plan (α, γ), le segment de droite qui est limité par les points $(1, 0)$ et $(\frac{1}{2}, \frac{1}{2})$ et qui, évidemment, satisfait à l'équation $\gamma = 1 - \alpha$. Ce segment étant situé dans le triangle $0 \leqq \gamma \leqq \alpha \leqq 1$, on pourra y appliquer le théorème II. Or on a évidemment $M^*_{\frac{1}{2}, \frac{1}{2}} = 1$. De plus on a $M^*_{1,0} = \mathfrak{M}$. En effet, (*cf.* le n° 2) $M^*_{1,0}$ sera égal à la plus grande valeur que pourra atteindre l'une ou l'autre des $|X_k|$, variées sous la condition $\Sigma |x_j| = 1$, c'est-à-dire à la plus grande des valeurs $|a_{jk}|$. Il viendra donc pour tout point (α, γ) situé sur ledit segment

$$M^*_{\alpha\gamma} \leqq (M^*_{1,0})^{\frac{\alpha - \frac{1}{2}}{1 - \frac{1}{2}}} \left(M^*_{\frac{1}{2}, \frac{1}{2}}\right)^{\frac{1-\alpha}{1-\frac{1}{2}}} = \mathfrak{M}^{2\alpha - 1} = \mathfrak{M}^{\frac{2-a}{a}}.$$

On voit que l'orthogonalité intervient tres peu et que le théorème subsiste dans tous les cas où l'on a

---

[1] *l. c.* p. 124. Nous reviendrons plus loin sur la généralisation du théorème de Parseval donnée dans le travail cité et dont le théorème du texte est un cas particulier. M. F. Riesz a observé que ce théorème particulier fournit de son côté, par un passage à la limite, la généralisation en question du théorème de Parseval.

$$\sum_{k=1}^{n} |X_k|^2 \leqq \sum_{j=1}^{m} |x_j|^2,$$

c'est-à-dire où la borne de la forme bilinéaire $A(x, y)$, prise dans le sens classique, est $\leqq 1$.

Opérations bilinéaires et transformations linéaires.

12. En quittant maintenant le point de vue algébrique et en laissant aux soins du lecteur l'extension presqu'immédiate de nos résultats aux formes bilinéaires et aux substitutions linéaires à une infinité de variables, nous passons aux applications à l'Analyse fonctionnelle.

Rappelons d'abord quelques faits connus concernant l'intégrale de Stieltjes-Lebesgue et les opérations fonctionnelles.[1] Ce n'est que pour fixer les idées que nous nous bornons à des fonctions d'une seule variable, nos résultats s'étendant immédiatement à des champs plus généraux. Pour ces champs, on se servira avec avantage de l'idée de fonctions d'ensemble.

Envisageons alors une *intégrale de Stieltjes-Lebesgue* de la forme

$$\int_{\varkappa}^{\lambda} f(x)\, d\varphi(x), \tag{17}$$

$\varphi(x)$ étant une *fonction non décroissante* définie dans l'intervalle $\varkappa \leqq x \leqq \lambda$.

Dans les applications particulières de nos résultats, $\varphi(x)$ sera ou bien identique à la variable indépendante x, ou bien $\varphi(x)$ sera constante par intervalles, ne croissant que par des sauts brusques. Dans le premier cas, il s'agira tout simplement de l'intégrale ordinaire de Lebesgue $\int f(x)\, dx$, dans le second, l'intégrale de Stieltjes-Lebesgue se réduira à une somme de la forme $\Sigma \varrho_k f(x_k)$,

---

[1] Pour ces faits et pour les indications bibliographiques *voir* entre autres: J. RADON, Theorie und Anwendungen der absolut additiven Mengenfunktionen, *Sitzungsber. Akad. Wien,* t. 122, Abt. II a, (1913), p. 1295—1438; T. H. HILDEBRANDT, On integrals related to and extensions of the Lebesgue integrals, *Am. M. S. Bull.* t. 24 (1918), p. 113—144, 177—202. C. DE LA VALLÉE POUSSIN, Les fonctions à variation bornée et les questions qui s'y rattachent, *Bull. des Sciences Math.* (2) t. 44 (1920), p. 267—296.

Dans toutes les applications qui suivent, même dans celles concernant des séries, on pourrait s'arranger avec l'intégrale de Lebesgue, mais nous préférons exposer nos résultats dans le langage de l'intégrale plus générale de Stieltjes-Lebesgue, qui permet d'une manière naturelle de considérer séries et intégrales sous le même point de vue.

les x_k étant les points de discontinuité de $\varphi(x)$ et les ϱ_k les sauts correspondants. On voit que, dans ce second cas, ce ne sont que les valeurs particulières prises par la fonction $f(x)$ aux points x_k qui interviennent dans l'intégrale.

Dans le cas général, l'intégrale ci-dessus se définit de la manière suivante. On considère la courbe $y = \varphi(x)$ et on forme la fonction inverse $x = x(y)$, avec des conventions évidentes pour les valeurs de x et de y qui correspondent à des segments horizontaux ou verticaux[1] de la courbe. En posant alors $f(x) = f(x(y)) = h(y)$,[2] l'intégrale ci-dessus se transforme dans une intégrale de la forme $\int_{\tau}^{\omega} h(y)\,dy$. Si cette dernière intégrale existe au sens de M. Lebesgue, on dira que $f(x)$ est intégrable par rapport à $\varphi(x)$ et on posera par définition

$$\int_{\varkappa}^{\lambda} f(x)\,d\varphi(x) = \int_{\tau}^{\omega} h(y)\,dy.$$

13. On dira que la fonction $f(x)$, définie dans le même intervalle que $\varphi(x)$, appartient à la classe L_{φ}^{a} $(a \geqq 1)$, si $\int f(x)\,d\varphi(x)$ existe dans tout intervalle où la variation de $\varphi(x)$ est finie et si l'intégrale

$$\int_{\varkappa}^{\lambda} |f(x)|^{a}\,d\varphi(x)$$

existe.

Cela étant, considérons encore une seconde fonction non décroissante $\psi(x)$, définie dans un certain intervalle $\mu \leqq x \leqq \nu$, et la classe L_{ψ}^{b} $(b \geqq 1)$ qui y correspond.

Dans les développements qui suivent, les fonctions $\varphi(x)$ et $\psi(x)$, une fois définies, *resteront toujours les mêmes.*

14. Envisageons maintenant à la fois les classes L_{φ}^{a} et L_{ψ}^{b} et faisons correspondre à chaque couple de fonctions $f(x)$ et $g(x)$ tirées des classes respectives un nombre $A(f, g)$ de sorte que les hypothèses suivantes soient remplies.

---

[1] A toutes les valeurs de y appartenant à un segment vertical, il correspond une seule valeur de x, l'abscisse de tous les points du segment; à une valeur de y appartenant à un segment horizontal, on fait correspondre une quelconque des abscisses des points du segment.

[2] La fonction $h(y)$ sera, en général, indéterminée pour les valeurs de y qui correspondent à des segments horizontaux; cependant ces valeurs formant un ensemble dénombrable, n'exerceront aucune influence sur l'intégrale, si l'on la prend au sens de Lebesgue.

a) La correspondance est *distributive*, c'est-à-dire que

(18) $$A(f_1+f_2, g_1+g_2) = A(f_1, g_1) + A(f_1, g_2) + A(f_2, g_1) + A(f_2, g_2)$$

et

(19) $$A(\lambda f, \mu g) = \lambda \mu A(f, g)$$

où λ et μ désignent des constantes arbitraires;

b) elle est *bornée*, c'est qu'il existe un nombre M tel que l'on ait toujours

(20) $$|A(f, g)| \leq M\left(\int |f|^a d\varphi\right)^{\frac{1}{a}} \left(\int |g|^b d\psi\right)^{\frac{1}{b}}.$$ [1]

Sous ces hypothèses $A(f, g)$ sera dite une *opération bilinéaire*[2] *portant sur le champ* (L_φ^a, L_ψ^b). Le plus petit nombre M qui suffit pour que l'inégalité (20) soit remplie, sera appelé la *borne* de l'opération bilinéaire.

On dira qu'une opération bilinéaire A porte sur plusieurs champs, si elle est définie pour tous ces champs, et cela d'une façon univoque pour les éléments qui appartiennent à la fois à deux ou à plusieurs de ces champs. La borne de l'opération dépendra alors des exposants a et b, caractérisant les champs, et c'est précisément cette dépendance que nous étudierons dans la suite.

15. Il résulte immédiatement de l'inégalité (20) qu'une opération bilinéaire portant sur un certain champ (L_φ^a, L_ψ^b) est déterminée d'une manière univoque lorsqu'on connaît ses valeurs pour les éléments de deux ensembles de fonctions tirés des deux classes L_φ^a et L_ψ^b, *partout denses* dans les classes respectives, *l'écart* de deux fonctions f_1 et f_2 ou g_1 et g_2 étant toujours mesuré par

$$\left(\int |f_1 - f_2|^a d\varphi\right)^{\frac{1}{a}} \text{ ou } \left(\int |g_1 - g_2|^b d\psi\right)^{\frac{1}{b}}$$

respectivement. Mais il y en a plus. Admettons qu'on ne définisse d'abord une opération que pour les éléments de deux ensembles tels qu'on vient de dire et qui, en outre, contiennent les combinaisons linéaires de leurs éléments. Alors, si les hypothèses a) et b) sont satisfaites pour les éléments desdits ensembles, *l'opération se définit immédiatement par continuité dans tout le champ* (L_φ^a, L_ψ^b), *la borne*

[1] Dans la suite, pour simplifier l'écriture, nous supprimerons, en général, la variable et les limites d'intégration.

[2] Ce n'est que pour des raisons de rédaction que nous traitons les opérations bilinéaires avant d'avoir parlé des opérations linéaires.

étant la même pour tout le champ que celle qui correspond à nos deux ensembles particuliers.

16. Un exemple presqu'évident d'ensembles particuliers tels qu'on vient de dire est fourni par les *fonctions constantes par intervalles,* ne présentant qu'un nombre fini de sauts et s'annulant dans ceux des intervalles en question où la variation respective de $\varphi(x)$ ou de $\psi(x)$ est infinie.[1]

Sur bien d'autres exemples qu'il serait facile de former, le susdit a l'avantage de faire immédiat le passage aux formes bilinéaires à un nombre fini de variables dont nous allons parler tout à l'heure. D'autre part, ces fonctions appartenant à tous les champs (L^a_φ, L^b_ψ), on pourra les utiliser pour étendre l'opération en passant d'un champ à un autre. Tout dépend alors de savoir si *l'opération appliquée à ces fonctions particulières possède une borne finie* par rapport aux exposants variés. S'il en est ainsi, on fera *l'extension* en passant *par continuité* de nos fonctions *au champ complet.*

17. En désignant, comme plus haut pour les formes bilinéaires, la borne correspondant aux exposants a et b par $M_{\alpha\beta}$,[2] où $\alpha = \frac{1}{a}$, $\beta = \frac{1}{b}$, nous sommes à même d'énoncer le théorème suivant, extension des théorèmes I et I'.

III. *Soit A une opération bilinéaire portant sur tous les champs (L^a_φ, L^b_ψ) qui correspondent aux points $\left(\alpha = \frac{1}{a}, \beta = \frac{1}{b}\right)$ d'un segment de droite appartenant au triangle* (2), *alors* $\log M_{\alpha\beta}$ *est une fonction convexe des points* (α, β) *du segment en question.*

La convexité de la fonction $\log M_{\alpha\beta}$ suggère encore le théorème suivant qu'on va démontrer en même temps.

IV. *Chaque fois que l'opération est définie pour les champs correspondant à deux points du triangle* (2), *elle pourra être étendue à tous les champs correspondant aux points du segment joignant les deux points donnés, cette extension étant déterminée d'une manière univoque.*

Admettons d'abord les hypothèses du théorème III. Considérons alors les ensembles de fonctions f et g, constantes par intervalles et telles qu'on vient de dire, les intervalles étant fixés jusqu'à nouvel ordre. En numérotant ces inter-

---

[1] Evidemment, cela ne pourra arriver que dans les intervalles extrêmes.

[2] Dans la définition de $M_{0\beta}$, on aura à remplacer dans le second membre de (20) le facteur correspondant à f par la borne supérieure de $|f|$ au sens de Lebesgue, c'est-à-dire par le plus petit nombre que $|f|$ ne dépasse que dans un ensemble nul au plus. La définition de $M_{\alpha 0}$ se fait d'une manière analogue.

valles et en désignant les valeurs respectives de $f(x)$ et de $g(x)$ par $x_1, x_2, \ldots, x_m$ et $y_1, y_2, \ldots, y_n$, l'opération $A(f, g)$ devient, pour ces fonctions particulières, une forme bilinéaire, tandis que les intégrales $\int |f|^a \, d\varphi$ et $\int |g|^b \, d\psi$ se réduisent à des expressions de la forme $\sum \varrho_j |x_j|^a$ et $\sum \sigma_k |y_k|^b$.

Or il est clair qu'on aura toujours $M'_{\alpha\beta} \leqq M_{\alpha\beta}$, ces lettres désignant les bornes respectives de la forme et de l'opération bilinéaires. D'autre part, (α, β), (α_1, β_1) et (α_2, β_2) étant trois points dudit segment, le premier séparant les deux autres, il viendra par le théorème I' que

$$M'_{\alpha\beta} \leqq (M'_{\alpha_1\beta_1})^{\frac{\alpha_2-\alpha}{\alpha_2-\alpha_1}} (M'_{\alpha_2\beta_2})^{\frac{\alpha-\alpha_1}{\alpha_2-\alpha_1}}$$

et alors, à plus forte raison,

$$M'_{\alpha\beta} \leqq (M_{\alpha_1\beta_1})^{\frac{\alpha_2-\alpha}{\alpha_2-\alpha_1}} (M_{\alpha_2\beta_2})^{\frac{\alpha-\alpha_1}{\alpha_2-\alpha_1}}.$$

Cette inégalité ayant lieu pour chaque division d'intervalles, on pourra, par ce qui précède, y remplacer $M'_{\alpha\beta}$ par $M_{\alpha\beta}$, et alors le théorème III se trouve démontré.

Le théorème IV résulte par les remarques du n° 15 de la même inégalité.

18. Avant de passer aux transformations fonctionnelles, nous aurons à poser quelques conventions évidentes et rappeler quelques faits connus.

On dira qu'un ensemble est un *ensemble nul par rapport à* $\varphi(x)$ ou, plus brièvement, un ensemble nul, si l'on peut l'enfermer dans des systèmes d'intervalles sur lesquels la croissance totale de $\varphi(x)$ est arbitrairement petite. Une fonction sera dite une *fonction nulle par rapport à* $\varphi(x)$ ou, plus brièvement, une fonction nulle, si elle s'annule sauf peut-être dans un ensemble nul. Deux fonctions qui ne diffèrent que par une fonction nulle ayant des intégrales identiques par rapport à $\varphi(x)$ sur des intervalles quelconques, de telles fonctions seront considérées comme identiques.

Quand $\varphi(x) = x$, c'est-à-dire dans le cas de l'intégrale de Lebesgue, les ensembles nuls sont ceux qu'on appelle en général ensembles de mesure nulle, tandis que les fonctions nulles sont celles qui s'annulent, sauf peut-être dans un tel ensemble.

Dans le cas où $\varphi(x)$ est constante par intervalles, les ensembles nuls sont ceux qui ne contiennent aucun point de discontinuité de $\varphi(x)$, tandis que une fonction arbitraire qui s'annule à tous les points de discontinuité sera une fonction nulle.

On posera des conventions analogues en ce qui concerne la fonction $\psi(x)$.

19. On dira que B est une *opération fonctionnelle linéaire* portant sur la classe L_φ^a, si B fait correspondre à toute fonction f de cette classe un nombre déterminé $B(f)$ de sorte que les hypothèses suivantes soient satisfaites:

a) L'opération est *distributive*, c'est-à-dire que pour des constantes λ_1 et λ_2 quelconques, on a

$$B(\lambda_1 f_1 + \lambda_2 f_2) = \lambda_1 B(f_1) + \lambda_2 B(f_2). \tag{21}$$

b) Il existe une constante M telle que

$$|B(f)| \leqq M\left(\int |f|^a d\varphi\right)^{\frac{1}{a}}. \tag{22}$$

On sait[1] *qu'il existe alors une fonction génératrice* $\omega(x)$, *appartenant à la classe complémentaire* $L_\varphi^{a'}$ $\left(\frac{1}{a}+\frac{1}{a'}=1\right)$ *et déterminée à une fonction nulle près telle que l'on ait*

$$B(f) = \int_\varkappa^\lambda f(x)\,\omega(x)\,d\varphi(x) = \int f\omega\,d\varphi.$$

Si M *désigne la borne de* B, *c'est-à-dire le plus petit nombre qui suffit pour que l'inégalité* (22) *soit remplie, on aura*

$$\left(\int |\omega|^{a'} d\varphi\right)^{\frac{1}{a'}} = M.$$

Ces faits importants permettront de traduire nos résultats concernant les opérations bilinéaires en de tels concernant les transformations fonctionnelles.

20. Admettons qu'une transformation $T = T(f)$ fasse correspondre aux fonctions d'une certaine classe L_φ^a ou, plus généralement, à un ensemble de fonctions embrassant plusieurs classes L_φ^a certaines fonctions, définies dans le même intervalle que $\psi(x)$, la correspondance étant *distributive*, c'est-à-dire que pour des constantes arbitraires λ_1 et λ_2 on ait

$$T(\lambda_1 f_1 + \lambda_2 f_2) = \lambda_1 T(f_1) + \lambda_2 T(f_2). \tag{23}$$

[1] *Cf.* p. ex. Radon, *l. c.* p. 1371.

Supposons en outre que les transformées des fonctions de la classe L_φ^a fassent partie d'une certaine classe L_ψ^c ou qu'elles remplissent cette classe et, de plus, qu'il existe un nombre M^* tel qu'on ait toujours

$$\left(\int |T(f)|^c d\psi\right)^{\frac{1}{c}} \leqq M^* \left(\int |f|^a d\varphi\right)^{\frac{1}{a}}. \tag{24}$$

Alors on dira que T est une *transformation fonctionnelle linéaire de L_φ^a en L_ψ^c*. Pour chaque couple d'exposants envisagés a et c, nous désignerons par $M_{\alpha\gamma}^*$, où $\alpha = \frac{1}{a}$ et $\gamma = \frac{1}{c}$, le plus petit des nombres M^* qui conviennent, ce nombre étant appelé *borne* de la transformation (par rapport aux exposants envisagés).[1]

Par les conventions posées plus haut et par la nature des choses, deux fonctions de la classe L_φ^a ou de la classe L_ψ^c qui ne diffèrent que par une fonction nulle par rapport à $\varphi(x)$ ou à $\psi(x)$ respectivement sont considérées comme identiques. Même le cas est admis où ni f ni $T(f)$ ne sont définies qu'à des fonctions nulles près, c'est-à-dire que ces fonctions prennent des valeurs arbitraires, chacune dans un ensemble nul par rapport à $\varphi(x)$ ou à $\psi(x)$ respectivement. Souvent il sera même naturel de ne pas s'occuper des valeurs que f ou $T(f)$ prennent dans un certain ensemble nul. Tel sera le cas p. ex. pour f, si la fonction non décroissante $\varphi(x)$ est constante par intervalles. Alors, on ne s'occupera que des valeurs que f prend aux points de discontinuité de cette fonction.

21. Cela posé, considérons une opération bilinéaire $A(f, g)$ portant sur un certain champ (L_φ^a, L_ψ^b), avec la borne M. En fixant alors la fonction $f(x)$ et en faisant varier $g(x)$, cette opération fera correspondre à toute fonction $g(x)$ de la classe L_ψ^b un nombre bien déterminé $B(g)$. La correspondance sera distributive, et on aura aussi

$$|B(g)| = |A(f, g)| \leqq M' \left(\int |f|^a d\varphi\right)^{\frac{1}{a}} \left(\int |g|^b d\psi\right)^{\frac{1}{b}},$$

$M' \leqq M$ désignant le plus petit nombre (dépendant en général de f) qui convient.

Il s'agira donc d'une opération linéaire pour les fonctions $g(x)$. Alors, par ce qu'on a dit plus haut, il existera une fonction $\omega(x)$, appartenant à la classe

[1] Pour la signification de $M_{0\gamma}^*$ et de $M_{\gamma 0}^*$ *cf.* la note 2 p. 477.

$L_\psi^{b'}$, où $\frac{1}{b}+\frac{1}{b'}=1$, bien déterminée à une fonction nulle près et telle que l'on ait

$$B(g)=\int g\,\omega\,d\psi$$

avec

$$\left(\int|\omega|^{b'}d\psi\right)^{\frac{1}{b'}}=M'\left(\int|f|^{a}d\varphi\right)^{\frac{1}{a}}.$$

Il en découle que le passage de $f(x)$ à $\omega(x)$ est une transformation fonctionnelle linéaire de L_φ^a en $L_\psi^{b'}$, avec la borne M. Un raisonnement analogue pour le cas où l'on aura interverti le rôle des fonctions $f(x)$ et $g(x)$ conduira à une transformation de L_ψ^b en $L_\varphi^{a'}$, avec la même borne.

C'est-à-dire que *toute opération bilinéaire* qui porte sur un champ (L_φ^a, L_ψ^b) *donne lieu à deux transformations linéaires*, transposées l'une à l'autre, l'une menant de L_φ^a en $L_\psi^{b'}$, l'autre de L_ψ^b en $L_\varphi^{a'}$, les bornes étant identiques. Inversement, de *telles transformations donnent toujours lieu à une opération bilinéaire* portant sur ledit champ. Admettons p. ex. que T soit une transformation de L_φ^a en $L_\psi^{b'}$. Il suffit de poser

$$A(f,g)=\int g\,T(f)\,d\psi$$

pour obtenir l'opération bilinéaire cherchée.

22. Le lien intime qu'on vient d'établir entre les opérations bilinéaires et les transformations linéaires permet d'énoncer les théorèmes suivants, équivalents aux théorèmes III et IV.

V. *Soit T une transformation fonctionnelle linéaire de certaines classes L_φ^a en certaines classes correspondantes L_ψ^c, la correspondance entre a et c étant telle que l'on ait toujours $c \geqq a$ et que, de plus, le point (α,γ), où $\alpha=\frac{1}{a}$ et $\gamma=\frac{1}{c}$, décrive un segment de droite. Alors* $\log M_{\alpha\gamma}^*$ *sera une fonction convexe des points du segment.*

VI. *Chaque fois qu'on a une transformation fonctionnelle linéaire de $L_\varphi^{a_1}$ en $L_\psi^{c_1}$ et de $L_\varphi^{a_2}$ en $L_\psi^{c_2}$, avec $c_1 \geqq a_1$ et $c_2 \geqq a_2$, la transformation pourra être étendue à tous les couples d'exposants correspondant aux points (α,γ) du segment joignant les points (α_1,γ_1) et (α_2,γ_2).*

Applications.

23. Soit $f(x)$ une fonction réelle ou non, définie dans l'intervalle $0 \leqq x \leqq 1$, intégrable au sens de Lebesgue et soient

$$C_k = \int_0^1 f(x)\, e^{-2k\pi i x}\, dx \qquad (k = 0, \pm 1, \pm 2, \ldots),$$

ses constantes de Fourier.[1]

Posons

$$J_a = \left(\int_0^1 |f(x)|^a\, dx\right)^{\frac{1}{a}}; \qquad S_a = \left(\sum_{k=-\infty}^{+\infty} |C_k|^a\right)^{\frac{1}{a}}.$$

Soient de plus $a > 1$ et $c > 1$ deux exposants liés par la relation $\frac{1}{a} + \frac{1}{c} = 1$.

On aura les théorèmes suivants.

A. *Si $a \leqq c$, et si la série $\sum |C_k|^a$ est convergente, les C_k sont les constantes de Fourier d'une fonction f, intégrable ainsi que $|f|^c$, et il est $J_c \leqq S_a$.*

B. *Lorsque $a \leqq c$ et que f est intégrable ainsi que $|f|^a$, la série $\sum |C_k|^c$ sera convergente et $S_c \leqq J_a$.*

Ces théorèmes sont dus à MM. W. H. Young et F. Hausdorff.[2] M. Young ayant d'abord établi le cas particulier où c est un entier pair, M. Hausdorff a démontré les théorèmes pour le cas général.

M. F. Riesz[3] a montré que *les théorèmes ci-dessus subsistent lorsqu'on introduit au lieu des fonctions $e^{2k\pi i x}$, un système quelconque de fonctions orthogonales et normées dont le module ne dépasse pas l'unité, définies d'ailleurs sur un intervalle quelconque.* Plus généralement, *si les fonctions du système ne dépassent pas en module le nombre $\mathfrak{M}$, les inégalités ci-dessus seront à remplacer par*

25) $$J_c \leqq \mathfrak{M}^{\frac{2-a}{a}} S_a \quad \text{et} \quad S_c \leqq \mathfrak{M}^{\frac{2-a}{a}} J_a$$

respectivement.

[1] Ce n'est que pour simplifier les énoncés que nous considérons ici les fonctions $e^{2k\pi i x}$ au lieu des fonctions e^{kix}.

[2] *l. c.* note 2, p. 466.

[3] *l. c.* note 3, p. 466.

24. On pourrait déduire ces théorèmes comme cas particuliers de deux autres théorèmes résultant, par passage à la limite, du théorème II et que le lecteur pourra formuler sans peine.

Il s'y agit de la convexité pour $0 \leqq \gamma \leqq \alpha \leqq 1$, du logarithme de la borne supérieure du rapport

$$\left(\int_A^B \left|\sum C_j \tau_j(x)\right|^{\frac{1}{\gamma}} dx\right)^{\gamma} : \left(\sum |C_j|^{\frac{1}{\alpha}}\right)^{\alpha}$$

et de celle du rapport

$$\left(\sum \left|\int_A^B f(x)\, \tau_j(x)\, dx\right|^{\frac{1}{\gamma}}\right)^{\gamma} : \left(\int_A^B |f(x)|^{\frac{1}{\alpha}} dx\right)^{\alpha},$$

les C_j étant variés dans le premier cas et $f(x)$ dans le second. Les τ_j sont des fonctions données arbitraires, orthogonales ou non.

Nous préférons montrer comment les théorèmes précités rentrent dans l'ordre d'idées de nos résultats concernant les transformations fonctionnelles.

25. Nous établirons à titre d'exemple la seconde des inégalités (25).

Considérons un système de fonctions orthogonales et normées $\omega_k(x)$ définies dans l'intervalle $A \leqq x \leqq B$ et telles que l'on ait $|\omega_k(x)| \leqq \mathfrak{M}$. Soit de plus $\varphi(x) = x$ pour $A \leqq x \leqq B$ et soit $\psi(x)$, dans l'intervalle $(0, \infty)$[1], égale au plus grand nombre entier compris dans x. La transformée $T(f)$ de f sera, par définition, égale à

$$C_k = \int_A^B f(t)\, \overline{\omega_k(t)}\, dt,^{2}$$

pour les valeurs entières k de x, tandis qu'elle sera arbitraire pour x non entier.

On aura

[1] Dans le cas du théorème de Young-Hausdorff, on définira $\psi(x)$ d'une manière analogue dans l'intervalle $(-\infty, +\infty)$. Ajoutons qu'on pourrait éviter les intégrales de Stieltjes-Lebesgue et se contenter d'intégrales ordinaires par le procédé un peu artificiel de poser, dans le cas du texte p. ex., $\psi(x) = x$, pour $0 \leqq x \leqq \infty$ et $T(f) = C_k$ pour $k \leqq x < k + 1$.

[2] $\overline{\omega_k(t)}$ désigne la valeur conjuguée de $\omega_k(t)$.

$$\int |T(f)|^{\frac{1}{\gamma}} d\psi = \sum_{k=0}^{\infty} |C_k|^{\frac{1}{\gamma}}; \tag{26}$$

de plus $M^*_{1,0} \leqq \mathfrak{M}$ et $M^*_{\frac{1}{2},\frac{1}{2}} \leqq 1$. En effet, la première inégalité résulte de ce que $M^*_{1,0}$ signifie la borne supérieure des $|C_k|$, f étant variée sous la condition $\int_A^B |f(x)|\,dx \leqq 1$,[1] tandis que la seconde n'est que l'inégalité bien connue de Bessel.

Le segment limité par les points $(1,0)$ et $(\frac{1}{2},\frac{1}{2})$ ayant pour équation $\gamma = 1-\alpha$, où $\frac{1}{2} < \alpha = \frac{1}{a} < 1$, il résulte du théorème VI que, chaque fois que f et $|f|^a$ sont intégrables, la série $\sum |C_k|^c$, où $c = \frac{1}{\gamma} = \frac{1}{1-\alpha}$, sera convergente. L'inégalité à démontrer résulte alors de la convexité de $\log M^*_{\alpha,1-\alpha}$, assurée par le théorème V. On a en effet, tout comme au n° 11,

$$M^*_{\alpha,1-\alpha} \leqq (M^*_{1,0})^{\frac{\alpha-\frac{1}{2}}{1-\frac{1}{2}}} (M^*_{\frac{1}{2},\frac{1}{2}})^{\frac{1-\alpha}{1-\frac{1}{2}}} \leqq \mathfrak{M}^{2\alpha-1} = \mathfrak{M}^{\frac{2-a}{a}}.$$

La première des inégalités (25) résulte par un raisonnement analogue ou encore par le fait général établi plus haut savoir que deux transformations transposées admettent la même borne, égale à celle de l'opération bilinéaire correspondante.

26. L'analogue des théorèmes de Young-Hausdorff pour l'intégrale de Fourier fut démontré par M. Titchmarsh.[2] Les raisonnements qui précèdent s'appliquent sans difficulté à ce cas-ci[3] et à d'autres intégrales analogues; nous avons l'intention d'y revenir à une autre occasion.

27. Passons maintenant à une autre application, concernant quelques théorèmes que j'ai établis ailleurs[4] au sujet des fonctions conjuguées.[5]

---

[1] *Cf.* la note 2 p. 477 et la note 1 p. 480.

[2] E. C. Titchmarsh, A contribution to the theory of Fourier transforms, *Proc. London Math. Soc.* (2), t. 23 (1924), p. 279—287.

[3] Au lieu de l'inégalité de Bessel on s'appuie sur un théorème connu de M. Plancherel.

[4] M. Riesz, 1) Les fonctions conjuguées et les séries de Fourier, *Comptes rendus*, t. 178 (28 avril 1924) p. 1464; 2) Sur les fonctions conjuguées, va paraître dans la *Math. Zeitschr.*

[5] *Cf.* aussi E. C. Titchmarsh, Reciprocal formulæ involving series and integrals, *Math. Zeitschr.* t. 25 (1926) p. 321—347.

Désignons par $f(x)$ une fonction intégrable au sens de M. Lebesgue dans l'intervalle $(0, 2\pi)$ et écrivons l'expression

$$\bar{f}(x) = \frac{1}{2\pi}\int_0^{2\pi} f(t) \cot \frac{t-x}{2}\, dt. \tag{27}$$

Complétant certains résultats de M. FATOU, M. PLESSNER a démontré que l'intégrale du second membre existe presque partout, si l'on la prend au sens de la valeur principale de CAUCHY.[1]

En désignant par L^a, où $a > 1$, la classe des fonctions $f(x)$ intégrables avec $|f|^a$, on a les théorèmes que voici.

A. *Si la fonction périodique $f(x)$, de période 2π, appartient à la classe L^a, la fonction $\bar{f}(x)$ y appartient également, et l'on a*

$$\left(\int_0^{2\pi} |\bar{f}(x)|^a\, dx\right)^{\frac{1}{a}} \leqq M_a \left(\int_0^{2\pi} |f(x)|^a\, dx\right)^{\frac{1}{a}}. \tag{28}$$

B. *Soit $f(x)$ une fonction de la classe L^a par rapport à l'intervalle $(-\infty, +\infty)$ et formons l'intégrale*

$$\bar{f}(x) = \frac{1}{\pi}\int_{-\infty}^{+\infty} \frac{f(t)}{t-x}\, dt \tag{29}$$

dont la valeur principale existe presque partout. On a alors une inégalité analogue à (28), *le facteur constant étant le même.*

C. *Pour deux fonctions $f(x)$ et $g(x)$ appartenant à des classes complémentaires L^a et $L^{a'}$, où $\frac{1}{a} + \frac{1}{a'} = 1$, on a, sous les hypothèses du premier théorème,*

$$\frac{1}{2\pi}\left|\int_0^{2\pi}\int_0^{2\pi} f(x)\, g(y) \cot \frac{x-y}{2}\, dx\, dy\right| \leqq M_a \left(\int_0^{2\pi} |f(x)|^a\, dx\right)^{\frac{1}{a}} \left(\int_0^{2\pi} |g(x)|^{a'}\, dx\right)^{\frac{1}{a'}} \tag{30}$$

et sous celles du second

---

[1] A. PLESSNER, Zur Theorie der konjugierten trigonometrischen Reihen, *Inauguraldissertation*, Giessen, 1923.

$$(31)\qquad \frac{1}{\pi}\left|\int_{-\infty}^{+\infty}\int_{-\infty}^{+\infty}\frac{f(x)\,g(y)}{x-y}\,dx\,dy\right| \leqq M_a\left(\int_{-\infty}^{+\infty}|f(x)|^a\,dx\right)^{\frac{1}{a}}\left(\int_{-\infty}^{+\infty}|g(x)|^{a'}\,dx\right)^{\frac{1}{a'}},$$

les intégrales doubles existant toujours, si l'on les prend au sens de la valeur principale de Cauchy, *c'est-à-dire comme limite des intégrales étendues aux valeurs* $|x-y| \geqq \varepsilon > 0$.

Bien entendu, nous avons désigné par M_a la plus petite constante convenable.

28. Les intégrales doubles figurant en (30) et (31) définissent évidemment des opérations bilinéaires pour les classes complémentaires L^a et $L^{a'}$. La valeur de ces intégrales restant la même, au signe près, si l'on intervertit le rôle des fonctions f et g, l'opération est symétrique gauche. Il s'ensuit que $M_{a'} = M_a$. On a de plus $M_2 = 1$ et l'on démontre facilement que $M_a > 1$ pour toute autre valeur de a et que M_a devient infini lorsque a tend vers 1 ou vers l'infini. D'autre part, il résulte de nos théorèmes III ou V que $\log M_a$ est une fonction convexe de $\alpha = \frac{1}{a}$. En résumé, $\log M_a$ et alors M_a sont des fonctions convexes de $\alpha = \frac{1}{a}$ pour $0 \leqq \alpha \leqq 1$, symétriques par rapport à $\alpha = \frac{1}{2}$, et vont en croissant vers l'infini quand α va de $\frac{1}{2}$ à 0 ou à 1.

Quant à l'ordre de croissance de M_a, il ressort des recherches de M. Titchmarsh qu'il existe deux constantes H et K telles que pour $a \geqq 2$,

$$H\,a \leqq M_a \leqq K\,a^{\frac{4}{3}}.\ [1]$$

Or, dans le second de mes travaux cités, j'ai démontré que pour a entier et pair, on a $M_a < a : \log 2$. Par la croissance de M_a pour $a > 2$, il en résulte l'existence d'une constante K telle que l'on a pour tout exposant $a \geqq 2$, $M_a \leqq K\,a$. La combinaison de notre résultat avec celui de M. Titchmarsh donne en définitive que pour $a \geqq 2$,

$$H\,a \leqq M_a \leqq K\,a.$$

29. Voyons maintenant quel parti on peut tirer des résultats généraux du travail présent pour la démonstration des théorèmes du n° 27. Nous nous contentons de quelques indications sommaires. Considérons p. ex. le théorème A. Pour $a = 2$, la démonstration de ce théorème est immédiate. On l'obtient p. ex.

---

[1] *l. c.* p. 324.

par le théorème de Parseval en observant que les séries de Fourier de $f(x)$ et de $\bar{f}(x)$ sont conjuguées.

La démonstration n'étant immédiate que pour l'exposant 2, on ne pourra imiter le procédé suivi dans les applications précédentes, où l'on disposait de deux exposants distingués. Voici tout de même comment on peut ramener le cas d'un exposant arbitraire à celui, beaucoup plus facile, des entiers pairs.

Admettons que le théorème A soit démontré pour l'entier pair $a = 2n$. Alors, par la symétrie gauche, le théorème se trouvera en même temps démontré pour $a' = 2n:(2n-1)$ et enfin, par le théorème VI, pour tout exposant situé entre a et a'.

30. Comme dernière application considérons le problème suivant.

Soit

$$f(x) \sim \sum_{k=-\infty}^{+\infty} C_k e^{kix}$$

une série de Fourier, c'est-à-dire que les C_k, les constantes de Fourier de $f(x)$, se déterminent par les relations

$$C_k = \frac{1}{2\pi} \int_0^{2\pi} f(x) e^{-kix} dx.$$

Considérons une suite de nombres λ_k telle que, chaque fois que les C_k sont les constantes de Fourier d'une fonction de la classe L^a, les produits $\lambda_k C_k$ soient les constantes de Fourier d'une fonction de la classe L^c. Cette fonction sera désignée par $T(f)$ et appelée la *transformée* de $f(x)$, tandis que la suite sera dite une *suite de facteurs du type* (a, c).

Il est très facile d'établir, par un raisonnement connu remontant à M. Lebesgue et à M. Haar, qu'il existe un nombre M, attaché à la suite, tel qu'on ait

$$(32) \qquad \left(\int_0^{2\pi} |T(f)|^c dx\right)^{\frac{1}{c}} \leqq M \left(\int_0^{2\pi} |f(x)|^a dx\right)^{\frac{1}{a}}.$$

Ici encore, c'est évidemment le plus petit nombre M convenable qui intéresse et qui est la *borne* de la transformation.

31. Les suites du type (2, 2) ne sont que les suites bornées; cela ressort immédiatement du théorème de Riesz-Fischer.

Les suites du type (1, 1) sont caractérisées par le fait que la série

$$\sum_{k=-\infty}^{+\infty}{}' \lambda_k \frac{e^{kix}}{ki}, \tag{33}$$

où l'accent indique la suppression de l'indice zéro, est la série de Fourier d'une *fonction à variation bornée* $G(x)$.

Ces suites (λ_k) sont aussi identiques à celles qui transforment les fonctions bornées en des fonctions bornées,[1] c'est-à-dire que ces suites sont en même temps celles du type (∞, ∞), les fonctions bornées correspondant d'une manière connue au cas limite $a = \infty$.

Ajoutons que, toutes les fois que la fonction $G(x)$ donnée par la série (33) est à variation bornée, la transformation peut être mise sous la forme d'une intégrale de Stieltjes-Lebesgue

$$T[f(x)] = \frac{1}{2\pi}\int_0^{2\pi} f(x-t)\, d[G(t) + \lambda_0 t]. \tag{34}$$

On sait assez peu des suites du type (a, a) pour d'autres valeurs que $a = 1, 2, \infty$. On démontre sans peine — nous y reviendrons tout à l'heure — que, tout comme les suites des types (1, 1) et (∞, ∞) sont identiques, il en est de même des types (a, a) et (a', a'), où, comme toujours, $\frac{1}{a} + \frac{1}{a'} = 1$. De plus, il est très facile de montrer que toute suite du type (a, a) est bornée, c'est-à-dire qu'elle est en même temps du type (2, 2). Pour le voir, il suffit d'appliquer l'inégalité (32) aux fonctions e^{kix}. On démontre encore assez facilement que toute suite du type (1, 1) ou (∞, ∞) est en même temps du type (a, a) pour a quelconque. Ce fait est un cas particulier du théorème qui suit; d'ailleurs il résulte aussi d'une manière directe en appliquant l'inégalité de Hölder à la formule (34).

32. Les faits qu'on vient d'indiquer suggèrent le théorème suivant.

[1] Les résultats concernant les suites des types (1, 1) ou (∞, ∞) que nous avons indiqués sont dus à MM. W. H. Young et S. Szidon; W. H. Young, On Fourier series and functions of bounded variation, *Lond. Roy. Soc. Proc.* t. 88 (1913), p. 561—568, On a condition that a trigonometrical series should have a certain form, *ibid.* p. 569—574; S. Szidon, Reihentheoretische Sätze und ihre Anwendungen in der Theorie der Fourierschen Reihen, *Math. Zeitschr.* t. 10 (1921), p. 121—127. M. Fekete a montré (Über Faktorenfolgen welche die »Klasse« einer Fourierschen Reihe unverändert lassen, *Szeged Acta Univ. Franc.-Jos.* t. 1 (1923), p. 148—166) que lesdites suites sont aussi les suites de facteurs qui transforment bien d'autres classes de fonctions (fonctions continues, fonctions intégrables au sens de Riemann etc.) en elles-mêmes.

Toute suite du type (a_0, a_0) est en même temps une suite de type (a, a) pour tout nombre a situé entre a_0 et a'_0, les bornes correspondant aux exposants intermédiaires ne dépassant pas la borne commune correspondant aux extrémités.

Ce théorème sera une conséquence immédiate des théorèmes VI et V, dès qu'on aura démontré que les suites des types (a, a) et (a', a') sont identiques et que les bornes correspondant à une suite particulière sont les mêmes, cette suite étant considérée comme une suite du premier ou du second type.

Or ceci ressort du théorème suivant qui est de caractère symétrique et qui, de son côté, découle assez facilement de ce que j'ai démontré ailleurs au sujet du théorème de Parseval.[1]

Pour que la suite (λ_k) soit une suite du type (a, a) ou (a', a'), il faut et il suffit que la série $\sum\limits_{k=-\infty}^{+\infty} \lambda_k C_k D_k$ soit convergente, les C_k et les D_k désignant les constantes de Fourier de fonctions arbitraires $f(x)$ et $g(x)$ des classes respectives L^a et $L^{a'}$. Le plus petit nombre M tel qu'on ait

$$\left|\sum_{k=-\infty}^{+\infty} \lambda_k C_k D_k\right| \leqq \frac{M}{2\pi} \left(\int_0^{2\pi} |f(x)|^a dx\right)^{\frac{1}{a}} \left(\int_0^{2\pi} |g(x)|^{a'} dx\right)^{\frac{1}{a'}}, \tag{35}$$

est la borne commune, la suite étant envisagée comme suite des types (a, a) ou (a', a').

On pourrait aussi s'appuyer sur le fait plus élémentaire qui suit et qui ne fait intervenir que des sommes finies.

Pour que la suite (λ_k) soit du type (a, a) ou du type (a', a'), avec la borne M, il faut et il suffit que (35) ait lieu pour toutes les fonctions $f(x)$ et $g(x)$ qui sont des combinaisons linéaires à un nombre fini de termes des fontions e^{kix}, c'est-à-dire des polynomes trigonométriques, M étant le plus petit nombre qui y convient.

La nécessité de la condition est évidente. D'autre part, on voit facilement que la condition est suffisante; on n'aura qu'à observer que lesdits polynomes sont partout denses dans les classes respectives au sens du n° 15.

Voici encore une généralisation qui se démontre par un raisonnement analogue.

---

[1] *Comptes rendus*, l. c., *Math. Zeitschr.* l. c.

Considérons des suites du type (a, c), avec $c \geqq a$.[1] Posons, comme plus haut, $\alpha = \frac{1}{a}$, $\gamma = \frac{1}{c}$.

Cela posé, *toute suite du type* (a_0, c_0) *est aussi du type* (a, c) *pour tout point* (α, γ) *appartenant au segment de droite limité par les points* (α_0, γ_0) *et* (α_0', γ_0'), *les bornes correspondant aux points intermédiaires ne dépassant pas la borne commune qui correspond aux extrémités.*

33. Considérons enfin la suite de facteurs particulière $\lambda_0 = 0$, $\lambda_k = \pm i$ pour $k \lessgtr 0$ qui transforme toute série trigonométrique

$$\sum_{k=-\infty}^{+\infty} C_k e^{kix}$$

dans sa conjuguée

$$\sum_{k=-\infty}^{-1} i\, C_k e^{kix} - \sum_{k=1}^{+\infty} i\, C_k e^{kix}.$$

En particulier, la première série étant la série de Fourier d'une fonction $f(x)$ de la classe L^a, la seconde sera celle de la fontion $\bar{f}(x)$, donnée par la formule (27) qui, par le théorème A du n° 27, appartient à la même classe. Le problème des fonctions conjuguées rentre donc dans l'ordre d'idées des suites de facteurs. D'autre part, on voit que ladite suite particulière, et alors celle où l'on aura remplacé $\pm i$ par ± 1, est, pour tout a fini > 1, une *suite de facteurs du type* (a, a) *avec la borne* M_a. Toute translation d'indices fournissant une suite analogue, on en déduit par soustraction les suites $\ldots, 0, 0, \frac{1}{2}, 1, 1, \ldots, 1, 1, \frac{1}{2}, 0, 0, \ldots$, les bornes de ces suites, considérées comme suites de facteurs du type (a, a), étant encore $\leqq M_a$. Il s'ensuit que les bornes des suites $\ldots, 0, 0, 1, 1, \ldots, 1, 1, 0, 0, \ldots$ sont $\leqq M_a + 1$. Par ces dernières suites, on passe de la fonction $f(x)$ aux *sommes partielles* de sa série de Fourier et l'on trouve en particulier

$$\int_0^{2\pi} |s_{m,n}(x)|^a dx = \int_0^{2\pi} \left| \sum_{k=m}^{n} C_k e^{kix} \right|^a dx \leqq (M_a + 1)^a \int_0^{2\pi} |f(x)|^a dx,$$

résultat que j'ai déjà donné ailleurs sous une forme moins précise.

[1] Pour de telles suites *cf.* G. H. HARDY and J. E. LITTLEWOOD, Some properties of fractional integrals, *Proc. London Math. Soc.* t. 23 (1924), *Records*, p. XXXVII—XLI.

Voici une autre conséqence des remarques qui précèdent qu'on obtient en appliquant la transformation d'Abel; c'est que *toute suite monotone bornée* ou, plus généralement, *toute suite* (λ_k) *à variation bornée*, c'est-à-dire telle que la série $V=\Sigma|\lambda_{k+1}-\lambda_k|$ converge, *est* encore pour tout a fini >1 *une suite du type* (a, a).

Questions de réalité.

34. Retournons maintenant aux formes bilinéaires et substitutions linéaires à un nombre fini de variables et supposons que *les coefficients* a_{jk} qui y interviennent *soient réels*. Cela étant, nous allons établir que, tant qu'on se trouve *dans le triangle*

$$\alpha\leqq 1,\ \beta\leqq 1,\ \alpha+\beta>1, \tag{36}$$

les systèmes (x_j) *et* (y_k) qui rendent $|A(x, y)|$ maximum, sous les conditions (1), *sont réels*, chacun à un facteur arbitraire $e^{i\theta}$ près.

Il sera un peu plus commode d'établir le fait équivalent, savoir que les systèmes (x_j) qui rendent

$$\left(\sum_{k=1}^{n}|X_k|^c\right)^{\frac{1}{c}}:\left(\sum_{j=1}^{m}|x_j|^a\right)^{\frac{1}{a}} \tag{37}$$

maximum sont, pour $c>a\geqq 1$, réels à un facteur arbitraire près.

Or ceci résulte immédiatement du lemme suivant.

Soient x_j', x_j'', X_k', X_k'' $(j=1, 2, \ldots, m;\ k=1, 2, \ldots, n)$ des nombres réels et soit $\zeta=\xi+i\eta$ une variable complexe. Admettons en outre que les deux systèmes (x_j') et (x_j'') ne soit pas proportionnels l'un à l'autre. Soit de plus $c>a\geqq 1$. Alors en posant

$$\varphi(\zeta)=\left(\sum_{k=1}^{n}|X_k'+\zeta X_k''|^c\right)^{\frac{1}{c}}:\left(\sum_{j=1}^{m}|x_j'+\zeta x_j''|^a\right)^{\frac{1}{a}},$$

$\varphi(\zeta)$ atteint son maximum sur l'axe réel, l'infini y-compris.

Admettons en effet le contraire, c'est-à-dire que $\varphi(\zeta)$ atteigne son maximum pour une valeur finie $\zeta=\xi+i\eta$ et telle que $\eta\neq 0$. Alors on aura

$$\frac{\partial \varphi}{\partial \xi} = 0, \quad \frac{\partial \varphi}{\partial \eta} = 0, \quad \text{ou} \quad \left(\frac{\partial}{\partial \xi} - i\frac{\partial}{\partial \eta}\right)\varphi = 0.$$

Or on en déduit facilement en utilisant l'inégalité de Hölder et l'hypothèse $c > a$, que l'on a au même point

$$\frac{\partial^2 \varphi}{\partial \xi^2} + \frac{\partial^2 \varphi}{\partial \eta^2} = \left(\frac{\partial}{\partial \xi} + i\frac{\partial}{\partial \eta}\right)\left(\frac{\partial}{\partial \xi} - i\frac{\partial}{\partial \eta}\right)\varphi > 0.$$

Par conséquent, il ne pourra s'agir de maximum dans le point en question.

Le lemme établi, notre énoncé en ressort, en décomposant les variables $x_j = x_j' + i\,x_j''$ et les formes $X_k = X_k' + i\,X_k''$ en parties réelle et imaginaire et en supposant par impossible $\zeta = i$.

35. Quant aux points du triangle situés sur le côté $\alpha + \beta = 1$, il résulte par continuité qu'il existe toujours des systèmes réels des variables fournissant le maximum. Mais il peut arriver que la même valeur maximée sera aussi atteinte pour des systèmes qui ne sont pas proportionnels à des systèmes réels. Il suffit de montrer que, pour $c = a$, le fait analogue pourra arriver pour l'expression (37). Or il en sera ainsi dans l'exemple évident

$$X_j = x_j \qquad (j = 1, 2, \ldots, m).$$

Pour $c = a$, le rapport étant constant, il atteint son maximum pour n'importe quel système, *réel ou non.*

Il sera utile de considérer le même exemple pour $c > a$ et $c < a$.

Dans le premier cas, les systèmes fournissant le maximum sont ceux où toutes les variables s'annulent, sauf une d'entre elles. De tels systèmes sont réels à un facteur arbitraire près, ce qui est bien d'accord avec le résultat établi au numéro précédent.

Pour $c < a$, l'inégalité de Hölder ou l'inégalité (4') donnent

$$(38) \qquad \left(\sum_{j=1}^{m} |x_j|^c\right)^{\frac{1}{c}} \leqq m^{\frac{1}{c}-\frac{1}{a}} \left(\sum_{j=1}^{m} |x_j|^a\right)^{\frac{1}{a}},$$

le signe d'égalité n'ayant lieu que quand tous les $|x_j|$ *sont égaux.* Tout tel système, réel ou non, fournira alors, dans l'exemple particulier qu'on considère, le maximum de (37).

En retournant à la forme bilinéaire, on voit que tant que le point (α, β) se trouve *dans le triangle*

$$(39) \qquad 0 \leqq \alpha, \ 0 \leqq \beta, \ \alpha + \beta \leqq 1,$$

les systèmes qui fournissent le maximum pourront être réels ou non.

36. En supprimant maintenant l'hypothénuse du triangle en question, on aura le triangle

$$(40) \qquad 0 \leqq \alpha, \ 0 \leqq \beta, \ \alpha + \beta < 1.$$

Nous allons montrer que pour chaque point de ce dernier triangle, on pourra construire des exemples où *le maximum n'est fourni que par des systèmes non réels.* En comparant ce fait avec le résultat établi au n° 34, on voit qu'il y a une différence essentielle entre les deux triangles (36) et (40), le côté commun possédant dans un certain sens le caractère de tous les deux.

Pour construire les exemples en question, il suffira encore de considérer l'expression (37) pour $1 \leqq c < a$.

En posant[1] $m = n = 3$ et en modifiant un peu l'exemple précédent, considérons la substitution

$$(41) \qquad X_j = x_j - \varepsilon (x_1 + x_2 + x_3) \qquad (j = 1, 2, 3),$$

ε étant un nombre positif très petit. Montrons que, étant donné un couple de valeurs (a, c) telles que $1 \leqq c < a$, on pourra choisir ε assez petit pour que le maximum de

$$(42) \qquad (|X_1|^c + |X_2|^c + |X_3|^c)^{\frac{1}{c}} : (|x_1|^a + |x_2|^a + |x_3|^a)^{\frac{1}{a}}$$

ne soit atteint pour aucun système réel.

Remarquons d'abord que le rapport (42) a évidemment la valeur $3^{\frac{1}{c} - \frac{1}{a}}$ pour tout système tel que $|x_1| = |x_2| = |x_3|$ et que $x_1 + x_2 + x_3 = 0$, p. ex. pour le système $x_1 = 1$, $x_2 = e^{\frac{2\pi i}{3}}$, $x_3 = e^{-\frac{2\pi i}{3}}$; chaque système qui remplit les conditions précédentes étant d'ailleurs, à l'ordre près, proportionnel au système particulier indiqué.

Pour voir que le maximum n'est atteint pour aucun système réel, il suffit de montrer que les valeurs que le rapport (42) admet pour les valeurs réelles des variables, sont inférieures à $3^{\frac{1}{c} - \frac{1}{a}}$.

---

[1] Le raisonnement qui suit tient pour m impair quelconque.

On peut admettre, sans restreindre la généralité, que $|x_1|^a + |x_2|^a + |x_3|^a = 3$. En fixant alors un voisinage très étroit des systèmes $(1, 1, 1)$ et $(1, 1, -1)$, il est clair, d'après la remarque ajoutée à l'inégalité (38), qu'on pourra déterminer ε assez petit pour que tout système qui à un facteur commun ou à l'ordre près, ne se trouve dans ledit voisinage fournisse pour le rapport (42) une valeur inférieure à $3^{\frac{1}{c}-\frac{1}{a}}$. Quant aux systèmes qui se trouvent auxdits voisinages, on aura, suivant le cas,

$$|X_1|^c + |X_2|^c + |X_3|^c = |x_1|^c + |x_2|^c + |x_3|^c - c\varepsilon(|x_1|^{c-1} + |x_2|^{c-1} \pm |x_3|^{c-1})(x_1 + x_2 + x_3) + O(\varepsilon^2).$$

On a d'abord, d'après (38), $|x_1|^c + |x_2|^c + |x_3|^c \leqq 3$. D'autre part, la valeur approchée du terme négatif sera, suivant le cas, $-9c\varepsilon$ ou $-c\varepsilon$.

On aura donc dans tous les cas, les voisinages en question étant assez petits,

$$(|X_1|^c + |X_2|^c + |X_3|^c)^{\frac{1}{c}} < 3^{\frac{1}{c}} - \frac{\varepsilon}{2}$$

c'est-à-dire que le rapport (42) est inférieur à $3^{\frac{1}{c}-\frac{1}{a}}$.

Remarques additionnelles concernant la convexité.

37. Il reste à élucider la question de savoir si nos résultats concernant la convexité de $\log M_{\alpha\beta}$ subsistent dans le triangle (39), ou même dans tout le carré $0 \leqq \alpha \leqq 1$, $0 \leqq \beta \leqq 1$. Malheureusement, c'est bien peu de choses que je pourrai dire sur ce sujet. J'espère pouvoir y revenir bientôt.

Tout d'abord, on voit très simplement que, dans le cas particulier où *tous les a_{jk} sont positifs ou nuls, il y a convexité dans tout le carré et même dans tout le plan.*

En effet, soient t_1 et t_2 des nombres positifs tels que $t_1 + t_2 = 1$. Posons $\alpha = t_1\alpha_1 + t_2\alpha_2$, $\beta = t_1\beta_1 + t_2\beta_2$ et désignons par (x_j) et (y_k) des systèmes positifs satisfaisant aux inégalités

$$\sum_{j=1}^{m} x_j^{\frac{1}{\alpha}} \leqq 1; \quad \sum_{k=1}^{m} y_k^{\frac{1}{\beta}} \leqq 1. \tag{43}$$

Il vient alors par l'inégalité de Hölder

$$\sum\sum a_{jk} x_j y_k = \sum\sum a_{jk}^{t_1+t_2} x_j^{\frac{1}{\alpha}(t_1\alpha_1+t_2\alpha_2)} y_k^{\frac{1}{\beta}(t_1\beta_1+t_2\beta_2)} \leqq$$

$$\leqq \left(\sum\sum a_{jk} x_j^{\frac{\alpha_1}{\alpha}} y_k^{\frac{\beta_1}{\beta}}\right)^{t_1} \left(\sum\sum a_{jk} x_j^{\frac{\alpha_2}{\alpha}} y_k^{\frac{\beta_2}{\beta}}\right)^{t_2}.$$

En observant maintenant que les systèmes

$$x_j' = x_j^{\frac{\alpha_1}{\alpha}}, \quad y_k' = y_k^{\frac{\beta_1}{\beta}}; \quad x_j'' = x_j^{\frac{\alpha_2}{\alpha}}, \quad y_k'' = y_k^{\frac{\beta_2}{\beta}}$$

satisfont à des inégalités analogues à (43), les exposants respectifs étant $\frac{1}{\alpha_1}, \frac{1}{\beta_1}, \frac{1}{\alpha_2}, \frac{1}{\beta_2}$, notre remarque se trouve établie.

38. Le fait presqu'évident qui précède admet des applications assez intéressantes et remplace dans bien des cas le théorème I beaucoup plus caché. Considérons à titre d'exemple l'application suivante.

Soient, pour la forme bilinéaire $A(x,y) = \sum_{j=1}^{m}\sum_{k=1}^{n} a_{jk} x_j y_k$, $\sum_{k=1}^{n} |a_{jk}| \leqq G$ $(j = 1, 2, \ldots m)$, $\sum_{j=1}^{m} |a_{jk}| \leqq H$ $(k = 1, 2. \ldots n)$.

On aura alors $M_{1,0} \leqq G$, $M_{0,1} \leqq H$ et *par conséquent* $M_{\alpha,\,1-\alpha} \leqq G^\alpha H^{1-\alpha}$. En particulier $M_{\frac{1}{2},\frac{1}{2}} \leqq \sqrt{GH}$, inégalité due à Frobenius et à M. I. Schur.[1]

L'extension aux opérations bilinéaires et, en particulier, aux intégrales doubles de la forme $\int\int K(x,y) f(x) g(y)\, dx\, dy$ est immédiate.

39. Le fait établi au n° 37 et ses généralisations évidentes embrassent à peu près tout[2] ce que je sais dire d'affirmatif sur la convexité dans des domaines plus étendus que le triangle (2). Voici maintenant ce qui est de caractère négatif.

Considérons la forme bilinéaire réelle $A(x,y) = X_1 y_1 + X_2 y_2$ où $X_1 = x_1 + x_2$ et $X_2 = x_1 - x_2$. Etudions le maximum de cette forme en supposant les variables de rester réelles et de varier sous les conditions

---

[1] I. Schur, Bemerkungen zur Theorie der beschränkten Bilinearformen mit unendlich vielen Veränderlichen, *Journ. f. Math.* t. 140 (1911), p. 1—28, *voir,* en particulier, p. 6.

[2] On pourrait encore y ajouter le fait presqu'évident que, pour des a_{jk} quelconques, il y a convexité dans tout le plan sur les droites $\alpha =$ const. et $\beta =$ const.

$$(44)\qquad |x_1|^{\frac{1}{\alpha}} + |x_2|^{\frac{1}{\alpha}} \leqq 1; \quad |y_1|^{\frac{1}{\beta}} + |y_2|^{\frac{1}{\beta}} \leqq 1.$$

Cela revient à considérer le maximum de

$$(45)\qquad \left(|X_1|^{\frac{1}{1-\beta}} + |X_2|^{\frac{1}{1-\beta}}\right)^{1-\beta}$$

sous la première des conditions (44).

On voit facilement que le logarithme de ce maximum n'est pas convexe sur le segment limité par les points $(0, 0)$ et $(\frac{1}{2}, \frac{1}{2})$. En effet, on a $M_{0,0} = M^*_{0,1} = 2$, $M_{\frac{1}{2},\frac{1}{2}} = M^*_{\frac{1}{2},\frac{1}{2}} = 2^{\frac{1}{2}}$. D'autre part, en calculant le rapport

$$(46)\qquad \left(|X_1|^{\frac{1}{1-\alpha}} + |X_2|^{\frac{1}{1-\alpha}}\right)^{1-\alpha} : \left(|x_1|^{\frac{1}{\alpha}} + |x_2|^{\frac{1}{\alpha}}\right)^{\alpha}$$

pour un système arbitraire différent des systèmes $x_1 = \pm x_2$ et des systèmes où l'une des variables s'annule, on trouve une valeur plus grande que

$$2^{1-\alpha} = (M_{0,0})^{\frac{\frac{1}{2}-\alpha}{\frac{1}{2}}} \left(M_{\frac{1}{2},\frac{1}{2}}\right)^{\frac{\alpha}{\frac{1}{2}}}.$$

Cela se voit aussi sans calcul, en observant que pour $\alpha = 0$ et pour $\alpha = \frac{1}{2}$ — et seulement pour ces valeurs-là — le rapport (46) ne change pas avec les variables, et que le *logarithme du minimum du rapport est concave*, ce dernier fait ressortant du théorème I appliqué à la forme réciproque. Le minimum et le maximum coïncidant pour les extrémités du segment, le logarithme du maximum ne pourra être convexe.

Par conséquent, *si l'on se restreint, pour les formes réelles, aux valeurs réelles des variables, nos théorèmes ne tiennent plus dans le triangle* (39).

40. Dans l'exemple qui précède, le vrai maximum est fourni par des systèmes complexes tels que $x_1 : x_2 = 1 : \pm i$. La valeur du logarithme du maximum est donnée par la fonction $\log 2 . (\frac{3}{2} - 2\alpha)$; c'est-à-dire que ce logarithme est une fonction linéaire et alors une fonction convexe dans le sens que nous avons attribué à ce mot. On voit aussi sans peine que, dans cet exemple, le logarithme du vrai maximum est convexe dans tout le carré $0 \leqq \alpha \leqq 1$, $0 \leqq \beta \leqq 1$.

Nos considérations font voir que si, dans le cas général, la question de la validité de nos théorèmes sera à résoudre par l'affirmative pour tout le carré — ce qui d'ailleurs paraît peu probable — les relations (12), desquelles nous avons déduit nos théorèmes pour le triangle (2), ne suffiront certainement pas pour fournir ce résultat. En effet, ces relations tiennent évidemment aussi pour les systèmes qui fournissent le maximum d'une forme bilinéaire réelle, quand on se restreint aux valeurs réelles des variables.

33.

Sur les fonctions conjuguées

Math. Zeitschrift *27* (1927), 218–244

Introduction.

Dans ce travail[1]), nous commençons par établir une inégalité qui a lieu entre les intégrales de $|U|^p$ et de $|V|^p$, où U et V sont respectivement la partie réelle et imaginaire d'une fonction holomorphe, les intégrales étant prises le long de certaines circonférences concentriques. En désignant alors par L^p, où $p > 1$, la classe des fonctions sommables $f(x)$ telles que $|f(x)|^p$ soit aussi sommable, nous démontrons, entre autre, que deux séries trigonométriques conjuguées sont en même temps les séries de Fourier de fonctions de la classe L^p, $f(x)$ et $\bar{f}(x)$, la fonction $\bar{f}(x)$ étant, à une constante additive près, donnée par la formule

$$\bar{f}(x) = \frac{1}{2\pi}\int_0^{2\pi} f(t) \cot \frac{x-t}{2}\, dt:$$

l'intégrale est prise au sens de la valeur principale de Cauchy.

---

[1]) Ich habe diese Arbeit vor etwa zwei Jahren in der vorliegenden Form niedergeschrieben, nachdem ich etwas früher eine kurze Mitteilung über den Gegenstand (Sur les fonctions conjuguées et les séries de Fourier, Comptes rendus 178, p. 1464, 28 avril 1924) veröffentlicht hatte. Ich hatte von Anfang an die Absicht, die Arbeit in erweiterter und umgearbeiteter Form, ins Deutsche übertragen, in der *Mathematischen Zeitschrift* zu veröffentlichen. Da nun sowohl von mir wie auch von verschiedenen anderen Verfassern auf Grund der gedruckten oder persönlichen Mitteilungen auf die Arbeit mehrfach Bezug genommen wurde (vgl. u. a. E. Hilb-M. Riesz, Encykl. d. math. Wissensch. II C 10, S. 1211 u. 1213, E. Hobson, Theory of Funct. of a Real Variable II (1926), S. 610–614, 698; E. C. Titchmarsh, Reciprocal formulae involving series and integrals. Math. Zeitschr. 25 (1926), S. 321–347), will ich die Veröffentlichung nicht länger hinauszögern, und gebe die Arbeit in unveränderter Form wieder.

Stockholm, den 15. Juli 1926. M. R.

En désignant par $s_n(x)$ les sommes de la série de Fourier de $f(x)$, on tire de l'inégalité qui a lieu entre l'intégrale de $|f(x)|^p$ et celle de $|\bar{f}(x)|^p$ le fait que, pour toute fonction $f(x)$ de la classe L^p, l'intégrale de $|s_n(x)|^p$ reste bornée lorsque n tend vers l'infini. De ce théorème, il résulte immédiatement le théorème de Parseval pour le produit de deux fonctions appartenant à des classes complémentaires L^p et L^q, où $1/p+1/q=1$, théorème bien connu dans le cas particulier $p=q=2$.

Il y a pourtant une différence essentielle entre ce cas particulier et le cas général. Tandis que la série intervenant dans le théorème général est toujours convergente, ce n'est que dans le cas particulier qu'il y a nécessairement convergence absolue.

Dans la seconde partie, nous établissons des théorèmes analogues pour l'intervalle $(-\infty, +\infty)$. En particulier, nous démontrons que, pour deux fonctions $f(x)$ et $g(x)$ appartenant à des classes complémentaires, l'intégrale double

$$\int\int_{-\infty}^{+\infty} \frac{f(x)\, g(y)}{x-y}\, dx\, dy$$

existe toujours, si on la prend au sens de la valeur principale de Cauchy et que cette intégrale satisfait à une certaine inégalité. En particularisant les fonctions $f(x)$ et $g(x)$, on trouve que la forme bilinéaire

$$\sum_{i,k=-\infty}^{+\infty}{}' \frac{x_i y_k}{i-k}, \qquad i \neq k,$$

reste bornée sous les conditions

$$\sum |x_i|^p \leqq 1 \qquad \sum |y_k|^q \leqq 1, \qquad 1/p+1/q=1.$$

Comme on sait, c'est M. Hilbert qui a considéré le premier cette forme bilinéaire bien remarquable, en démontrant qu'elle était bornée sous les conditions $\sum |x_i|^2 \leqq 1$, $\sum |y_k|^2 \leqq 1$ qui correspondent au cas $p=q=2$.

Dans le même chapitre nous faisons quelques remarques sur certaines substitutions particulières.

On étudie aussi, $f(x)$ étant une fonction de la classe L^p, la fonction de la variable complexe z,

$$F(z) = \frac{1}{i\pi} \int_{-\infty}^{+\infty} \frac{f(y)}{y-z}\, dy,$$

dont la partie réelle prend les valeurs $\pm f(x)$ presque partout sur les deux bords de l'axe réel.

Les résultats et les méthodes du présent travail admettent bien des applications, entre autre dans la théorie des équations intégrales singulières.

Tel est le cas, en particulier, pour deux formules de *réciprocité* que nous démontrons sous des conditions beaucoup plus générales qu'on ne l'a fait jusqu'ici. Nous reviendrons à ces applications dans un travail prochain.

I. Le cas d'un intervalle fini.

1. Nous dirons que $f(x)$ appartient à la classe L^p, où $p > 1$, par rapport à l'intervalle (a, b), si $f(x)$ et $|f(x)|^p$ sont sommables[2]) dans cet intervalle. On désignera dans la suite par p et q deux nombres positifs > 1 tels que

$$\frac{1}{p} + \frac{1}{q} = 1. \tag{1}$$

On appellera respectivement les exposants p et q et les classes L^p et L^q exposants et classes *complémentaires.*

Désignons maintenant par $f(x)$ et $g(x)$ des fonctions appartenant à de telles classes. On connaît l'inégalité de Hölder-Schwarz

$$\left|\int_a^b f(x)\,g(x)\,dx\right| \leqq \left(\int_a^b |f(x)|^p\,dx\right)^{\frac{1}{p}} \left(\int_a^b |g(x)|^q\,dx\right)^{\frac{1}{q}}, \tag{2}$$

et celle de Minkowski[3]), valable pour deux fonctions $f_1(x)$ et $f_2(x)$ de la classe L^p,

$$\left(\int_a^b |f_1(x) + f_2(x)|^p\,dx\right)^{\frac{1}{p}} \leqq \left(\int_q^v |f_1(x)|^p\,dx\right)^{\frac{1}{p}} + \left(\int_a^b |f_2(x)|^p\,dx\right)^{\frac{1}{p}}. \tag{3}$$

Bien entendu, il s'ensuit de nos hypothèses que toutes les intégrales écrites existent.

2. Moyennant ces inégalités, on établit facilement le théorème suivant:

I. *Soit* $F(z) = F(re^{ix}) = U(x, r) + iV(x, r)$ *une fonction holomorphe pour* $|z| < R$, *dont la partie imaginaire s'annule à l'origine. On aura alors pour* $|z| = r < R$ *une inégalité de la forme*

$$\int_0^{2\pi} |V(x, r)|^p\,dx \leqq M_p^p \int_0^{2\pi} |U(x, r)|^p\,dx, \tag{4}$$

la constante M_p *ne dépendant que de* p.

Ce théorème est bien connu pour $p = 2$ et s'obtient dans ce cas particulier comme un corollaire immédiat du théorème de Parseval. Or nous allons voir par la démonstration générale qui suit que c'est le théorème de Cauchy qui est la vraie source du théorème. Dans cette

---

[2]) La sommabilité de $|f|^p$ entraîne celle de $|f|$, mais non pas celle de f.

[3]) Voir pour ces inégalités F. Riesz, Untersuchungen über Systeme integrierbarer Funktionen, Math. Annalen **69** (1910), p. 456.

démonstration, nous distinguerons trois cas suivant que p est un entier pair, un nombre non entier, ou un entier impair.

3. Posons $F(0) = u$. Soit d'abord p un entier pair. On aura alors d'après le théorème de Cauchy

$$(5) \qquad \frac{1}{2i\pi}\int\limits_{|z|=r}\frac{(F(z))^p}{z}\,dz = \frac{1}{2\pi}\int\limits_0^{2\pi}(U(x,r)+iV(x,r))^p\,dx = u^p.$$

En passant à la partie réelle, on trouve

$$\int V^p < \int\left(\binom{p}{2}V^{p-2}U^2 + \binom{p}{6}V^{p-6}U^6 + \dots\right) + (-1)^{\frac{p}{2}}\,2\pi u^p.$$

En observant d'abord que

$$2\pi u^p = (2\pi)^{1-p}\left(\int U\right)^p < \int U^p,$$

en posant

$$(6) \qquad \int V^p = X^p \int U^p$$

et en remplaçant, dans (2), p et q par $p/(p-2k)$ et $p/2k$ respectivement, on obtient

$$(7) \qquad X^p < \binom{p}{2}X^{p-2} + \binom{p}{6}X^{p-6} + \dots + 1.$$

X est donc plus petit que la racine positive de l'équation

$$X^p - \binom{p}{2}X^{p-2} - \binom{p}{6}X^{p-6} - \dots - 1 = 0.$$

Il en résulte, en particulier, que les intégrales figurant en (4) restent bornées en même temps, et que la borne supérieure M_p des valeurs X définies par la relation (6) est aussi limitée par ladite racine.

4. Désignons maintenant par p un nombre > 1 d'ailleurs tout à fait arbitraire. On réduit facilement le théorème à démontrer au cas où les conditions originales sont remplacées par les conditions suivantes: $U + iV$ est holomorphe pour $|z| < r$ et continue pour $|z| \leq r$ et on a $U \geqq 0$ pour $|z| = r$.

En effet, il vient par une formule connue,

$$(8) \qquad F(z) = \frac{1}{2\pi}\int\limits_0^{2\pi} U(x,r)\frac{re^{ix}+z}{re^{ix}-z}\,dx, \qquad |z| < r.$$

En posant alors $U(x,r) = U_1(x,r) - U_2(x,r)$, où $U_1(x,r) = 0$ pour toute valeur de x où $U(x,r) \leq 0$, et $U_2(x,r) = 0$ pour toute valeur où $U(x,r) \geqq 0$, on obtiendra deux fonctions non négatives, à des dérivées simplement discontinues. En introduisant ces fonctions en (8), on obtient

$F(z) = F_1(z) - F_2(z)$. En intégrant par parties, on voit que les fonctions $F_1(z)$ et $F_2(z)$ tendent uniformément vers des fonctions limites de la forme $U_1(x, r) + i V_1(x, r)$ et $U_2(x, r) + i V_2(x, r)$ lorsque z tend vers la circonférence du rayon r.

En admettant maintenant que le théorème soit vrai sous les conditions modifiées, la constante M_p étant remplacée par une certaine constante M'_p, on aura d'après (3)

$$\left(\int_0^{2\pi} |V(x,r)|^p dx\right)^{\frac{1}{p}} \leqq \left(\int_0^{2\pi} |V_1(x,r)|^p dx\right)^{\frac{1}{p}} + \left(\int_0^{2\pi} |V_2(x,r)|^p dx\right)^{\frac{1}{p}}$$

$$\leqq M'_p \left(\int_0^{2\pi} |U_1(x,r)|^p dx\right)^{\frac{1}{p}} + M_p \left(\int_0^{2\pi} |U_2(x,r)|^p dx\right)^{\frac{1}{p}}$$

$$\leqq 2 M'_p \left(\int_0^{2\pi} |U(x,r)|^p dx\right)^{\frac{1}{p}}.$$

Par conséquent, on aura $M_p \leqq 2 M'_p$, et alors on a montré que le problème original se réduit au problème modifié.

5. Cela posé, admettons donc que les conditions modifiées aient lieu. Alors, U étant $\geqq 0$ sur la circonférence $|z| = r$, la fonction $U + iV$ ne s'annule pas à l'intérieur de cette circonférence. Cela étant, considérons celle des branches de $(U + iV)^p$ qui est réelle et positive à l'origine. On pourra écrire

$$(U + iV)^p = (iV)^p + \sigma = e^{\pm \frac{ip\pi}{2}} |V|^p + \sigma$$

suivant le signe de V, et où l'on trouve par un calcul simple (évaluation d'une intégrale rectiligne)

$$|\sigma| < 2^{\frac{p-1}{2}} p(U|V|^{p-1} + U^p).$$

En passant à la partie réelle et en définissant encore X par l'équation (6), où l'on remplace V par $|V|$, on obtient immédiatement une inégalité de la forme

$$\left|\cos\frac{p\pi}{2}\right| X^p < C_1 X^{p-1} + C_2,$$

les constantes C_1 et C_2 ne dépendant que de p. La valeur X est donc plus petite que la racine positive de l'équation

$$\left|\cos\frac{p\pi}{2}\right| X^p - C_1 X^{p-1} - C_2 = 0.$$

6. Par le raisonnement général qui précède, le théorème est établi pour tout exposant qui n'est pas un entier impair. Par contre, pour ces derniers exposants, la démonstration tombe en défaut, et il faut recourir

à une démonstration indirecte. A cet effet, nous établirons que le théorème a lieu *en même temps* pour des exposants complémentaires p et q et que l'on a de plus

$$M_p = M_q.$$

La démonstration de ce fait repose sur le théorème élémentaire bien connu[4]) qui suit.

Si la fonction continue $g(x)$ remplit l'inégalité

$$(9) \qquad \left|\int_a^b f(x)\, g(x)\, dx\right| \leqq A\left(\int_a^b |f(x)|^p dx\right)^{\frac{1}{p}},$$

pour toute fonction continue $f(x)$, alors on a

$$\int_a^b |g(x)|^q dx \leqq A^q.$$

Voici en quelques mots la démonstration de ce théorème. On pourra admettre, sans restreindre la généralité, que $g(x) \geqq 0$. En posant alors $f = g^{\frac{q}{p}}$, on trouve, en tenant compte de (1) et de (2),

$$\int_a^b (g(x))^{1+\frac{q}{p}} dx = \int_a^b (g(x))^q dx \leqq A\left(\int_a^b (g(x))^q dx\right)^{\frac{1}{p}}.$$

De là on tire

$$\int_a^b (g(x))^q dx \leqq A^q.$$

Il est manifeste que ce théorème subsiste encore, si l'on assujettit $f(x)$ à appartenir à certaines classes particulières de fonctions continues, p. ex., s'il est question de l'intervalle $(0, 2\pi)$, à celles des sommes trigonométriques d'ordre fini quelconque.

Soit maintenant $F(z) = U + iV$ une fonction holomorphe pour $|z| < R$ telle que V s'annule à l'origine, et désignons par $F_1(z) = U_1 + iV_1$ un polynome arbitraire de z dont la partie imaginaire est également nulle à l'origine. Alors il en sera de même de la partie imaginaire du produit $F(z)\, F_1(z)$ et l'on pourra écrire, l'intégrale étant prise le long d'une circonférence quelconque $|z| = r < R$,

$$\int_0^{2\pi} (U_1(x, r)\, V(x, r) + U(x, r)\, V_1(x, r))\, dx = 0.$$

On en tire, le théorème étant supposé vrai pour l'exposant p,

$$\left|\int U_1 V\right| = \left|\int U V_1\right| \leqq \left(\int |U|^q\right)^{\frac{1}{q}} \left(\int |V_1|^p\right)^{\frac{1}{p}} \leqq M_p \left(\int |U|^q\right)^{\frac{1}{q}} \left(\int |U_1|^p\right)^{\frac{1}{p}},$$

[4]) Pour une forme approfondie de ce théorème, voir F. Riesz, loc. cit. p. 457.

et alors, par le théorème élémentaire que nous venons de citer,

$$\int |V|^q \leqq M_p^q \int |U|^q.$$

Cette inégalité montre que l'on a toujours $M_q \leqq M_p$ et alors, par symétrie, $M_q \leqq M_p$, ce qui achève la démonstration.

7. Cela étant, on voit que M_p considéré comme fonction de $1/p$ est symétrique par rapport à $1/2$. On a $M_2 = 1$, et il est facile de montrer que $M_p > 1$ pour toute autre valeur de p. Il est aussi facile de voir que M_p tend vers ∞ lorsque $p \to \infty$, ou $p \to 1$. En effet, s'il n'en était pas ainsi, le théorème ci-dessus tiendrait aussi pour ces exposants limites ce qui est en contradiction avec des faits connus. Il serait d'un certain intérêt d'étudier de plus près les propriétés de croissance, de convexité etc. de M_p [4']). Nous nous restreignons ici à donner une borne supérieure pour M_p dans le cas où p est un entier pair, en ajoutant que le raisonnement des n°s 5 et 6 permet de démontrer que pour tout $p \geqq 2$, on a $M_p < Cp$, la constante C étant indépendante de p.

Admettons donc que p soit un nombre entier et pair. De l'inégalité (7), il vient, à plus forte raison,

$$2X^p < (X+1)^p,$$

et alors

$$X < \frac{1}{2^{\frac{1}{p}} - 1} < \frac{p}{\log 2}.$$

8. Soit $f(x)$ une fonction de la classe L^p par rapport à l'intervalle $(0, 2\pi)$. Formons la fonction qui est holomorphe pour $|z| < 1$ et dont la partie imaginaire s'annule à l'origine

$$F(z) = F(re^{ix}) = \frac{1}{2\pi}\int_0^{2\pi} f(t)\frac{e^{it}+z}{e^{it}-z}\,dt. \tag{9}$$

Des recherches de M. Fatou [5]) et de celles de M. Plessner [6]), il résulte que, si l'on fait tendre z vers les points de la circonférence $|z| = 1$ le long des rayons de cette circonférence, la partie réelle de $F(z)$ tend presque partout vers la fonction $f(x)$, tandis que la partie imagi-

---

[4']) [Note ajoutée pendant la correction des épreuves.] Pour une telle étude, voir M. Riesz, Sur les maxima des formes bilinéaires et sur les fonctionnelles linéaires, Acta Math. **49** (1927), p. 465–497; p. 486.

[5]) P. Fatou, Séries trigonométriques et séries de Taylor, Acta math. **30** (1906), p. 335–400.

[6]) A. Plessner, Zur Theorie der konjugierten trigonometrischen Reihen, Thèse, Giessen, 1923, 36 p.

naire $V(x, r)$ tend presque partout [7]) vers la valeur principale, *existant presque partout*, de l'intégrale

$$(10) \qquad \bar{f}(x) = \frac{1}{2\pi} \int_0^{2\pi} f(y) \cot \frac{x-y}{2}\, dy.$$

Cela étant, notre théorème I conduit facilement au théorème suivant:

II. *Si* $f(x)$ *appartient à la classe* L^p, $\bar{f}(x)$ *y appartient également; de plus on a l'inégalité*

$$(11) \qquad \int_0^{2\pi} |\bar{f}(x)|^p\, dx \leqq M_p^p \int_0^{2\pi} |f(x)|^p\, dx.$$

On a, en effet, pour la partie réelle $U(x, r)$ de $F(z)$, si l'on définit encore $f(x)$ à l'extérieur de l'intervalle $(0, 2\pi)$ par la propriété d'être périodique et de période 2π,

$$(12) \qquad U(x, r) = \frac{1}{2\pi} \int_0^{2\pi} f(x+y) \frac{1-r^2}{1+r^2-2r\cos y}\, dy.$$

Or, $h(x)$ désignant une fonction quelconque de la classe L^q, on conclut par l'inégalité de Hölder-Schwarz d'une façon générale:

$$(13) \qquad \left|\int_0^{2\pi} f(x+y) h(y)\, dy\right|^p \leqq \left(\int_0^{2\pi} |f(x+y)|^p |h(y)|\, dy\right) \left(\int_0^{2\pi} |h(y)|\, dy\right)^{\frac{p}{q}}$$

et alors, en tenant compte de la périodicité de $f(x)$,

$$(14) \qquad \int_0^{2\pi} \left|\int_0^{2\pi} f(x+y) h(y)\, dy\right|^p dx \leqq \left(\int_0^{2\pi} |f(x)|^p\, dx\right) \left(\int_0^{2\pi} |h(x)|\, dx\right)^p.$$

En utilisant alors la relation

$$(15) \qquad \frac{1}{2\pi} \int_0^{2\pi} \frac{1-r^2}{1+r^2-2r\cos y}\, dy = 1,$$

on trouve [8]), en particulier,

$$(16) \qquad |U(x,r)|^p \leqq \frac{1}{2\pi} \int_0^{2\pi} |f(x+y)|^p \frac{1-r^2}{1+r^2-2r\cos y}\, dy$$

et

$$(17) \qquad \int_0^{2\pi} |U(x,r)|^p\, dx \leqq \int_0^{2\pi} |f(x)|^p\, dx.$$

---

[7]) En réalité, $V(x, r)$ tend vers $\bar{f}(x)$ en tout point où cette valeur principale existe. Cf. Hilb-Riesz, Encykl. der math. Wissensch. II C 10, p. 1208, note 79a.

[8]) Cette inégalité est due à W. Gross, Wiener Ber. **124** (1915), p. 1017–1037.

Cela étant, on trouve par le théorème I

$$(18) \qquad \int_0^{2\pi} |V(x, r)|^p \, dx \leqq M_p^p \int_0^{2\pi} |f(x)|^p \, dx .$$

Alors, en appliquant un théorème connu de M. Fatou[9]), on voit que $\bar{f}(x)$ appartient également à la classe L^p et que l'inégalité (11) est remplie.

9. Un corollaire immédiat du théorème qui précède est le théorème suivant:

III. *Deux séries trigonométriques conjuguées sont en même temps les séries de Fourier de fonctions de la classe L^p.*

On entend par séries trigonométriques conjugées deux séries de la forme

$$S = \frac{1}{2} a_0 + \sum_{n=1}^{\infty} (a_n \cos nx + b_n \sin nx)$$

et

$$\bar{S} = \frac{1}{2} b_0 + \sum_{n=1}^{\infty} (b_n \cos nx - a_n \sin nx)$$

considérées à un point de vue tout à fait formel et satisfaisant à la relation encore formelle

$$S - i\bar{S} = \frac{a_0 - i b_0}{2} + \sum_{n=1}^{\infty} (a_n - i b_n) e^{nix} .$$

Admettons donc que S soit la série de Fourier d'une fonction $f(x)$ de la classe L^p. Introduisons alors, pour $r < 1$, les fonctions harmoniques, conjuguées

$$U(x, r) = \frac{a_0}{2} + \sum_{n=1}^{\infty} (a_n \cos nx + b_n \sin nx) r^n$$

et

$$V(x, r) = -\left(\frac{b_0}{2} + \sum_{n=1}^{\infty} (b_n \cos nx - a_n \sin nx) r^n \right) .$$

On peut, sans restreindre la généralité, supposer que $b_0 = 0$.

Les constantes de Fourier de $V(x, r)$ considérée comme fonctions de x sont évidemment $-b_n r^n$ et $a_n r^n$. D'autre part, on a entre $U(x, r)$ et $f(x)$ la relation (12) et l'inégalité (17) qui, de son côté, donne pour $V(x, r)$ l'inégalité (18). En désignant maintenant la fonction limite de

9) Voir Fatou, loc. cit. p. 375–376. Le théorème en question dit que, si $f_n(x)$ tend presque partout vers $f(x)$ et que l'intégrale de $|f_n(x)|$ reste $\leqq A$, il en sera de même de l'intégrale de $|f(x)|$.

$V(x, r)$ par $\bar{f}(x)$, cette fonction appartiendra à la classe L^p et on voit par un théorème connu[10]) sur le passage à la limite sous le signe d'intégration que cette fonction admet comme constantes de Fourier les nombres $-b_n$ et a_n respectivement.

Le raisonnement montre aussi qu'il y a (au signe et à une constante additive près) une réciprocité complète entre les deux fonctions $f(x)$ et $\bar{f}(x)$ de la classe L^p. La formule (10) admet *la réciproque*:

$$f(x) = \frac{1}{2\pi}\int_0^{2\pi} \bar{f}(y) \cot \frac{y-x}{2}\, dy + \frac{1}{2\pi}\int_0^{2\pi} f(y)\, dy. \tag{19}$$

10. Soient $f(x)$ et $g(x)$ deux fonctions appartenant par rapport à l'intervalle $(0, 2\pi)$ aux classes complémentaires L^p et L^q. Nous démontrons le théorème suivant:

IV. *On a*

$$\frac{1}{2\pi}\left|\int_0^{2\pi}\int_0^{2\pi} f(x)\, g(y) \cot \frac{x-y}{2}\, dx\, dy\right| \tag{20}$$

$$\leqq M_p \left(\int_0^{2\pi} |f(x)|^p\, dx\right)^{\frac{1}{p}} \left(\int_0^{2\pi} |g(x)|^q\, dx\right)^{\frac{1}{q}}$$

l'intégrale double existant toujours, si on la prend au sens de la valeur principale de Cauchy, c'est-à-dire comme limite de l'intégrale étendue aux valeurs $|x-y| > \varepsilon > 0$.

11. Pour arriver à ce théorème, écrivons $\bar{f}(x)$ sous la forme

$$\bar{f}(x) = -\frac{1}{2\pi}\int_0^{2\pi} f(x+y) \cot \frac{y}{2}\, dy$$

et formons la fonction

$$\bar{f}_\varepsilon(x) = -\frac{1}{2\pi}\int_\varepsilon^{2\pi-\varepsilon} f(x+y) \cot \frac{y}{2}\, dy, \qquad 0 < \varepsilon < \pi. \tag{21}$$

Alors $f(x)$ est une fonction continue x. Démontrons l'inégalité[11])

$$\int_0^{2\pi} |\bar{f}_\varepsilon(x)|^p\, dx \leqq \left(1 - \frac{\varepsilon}{\pi}\right)^p \int_0^{2\pi} |\bar{f}(x)|^p\, dx. \tag{22}$$

---

[10]) Ce théorème, dû à M. F. Riesz, dit que si les fonctions $f_n(x)$ tendent presque partout vers une fonction $f(x)$ et si l'intégrale de $|f_n(x)|^p$ reste bornée, on a pour toute fonction $g(x)$ de la classe L^q

$$\lim_{n\to\infty} \int f_n(x)\, g(x)\, dx = \int f(x)\, g(x)\, dx.$$

[11]) Cette relation tient si l'on remplace p par un nombre k quelconque $\geqq 1$, le second membre pouvant devenir infini pour $k > p$.

Nous exprimerons $\bar{f}_\varepsilon(x)$ par $\bar{f}(x)$. Considérons à cet effet la fonction $g(y)$ égale à $\cot\frac{y}{2}$ pour $\varepsilon < y < 2\pi - \varepsilon$ et nulle ailleurs. Désignons par $\bar{g}(y)$ une quelconque des fonctions conjuguées de $g(y)$, ces fonctions n'étant déterminées qu'à une constante additive près. Or la partie imaginaire de la fonction holomorphe correspondant à $f(y) + i\bar{f}(y)$ est nulle à l'origine. La fonction $g(y)$ étant impaire, la partie réelle de la fonction holomorphe correspondant à $g(y) + i\bar{g}(y)$ s'annulera également à l'origine. Par conséquent, il en sera encore de même de la partie réelle du produit des deux fonctions holomorphes. On obtient donc par le théorème de Cauchy[12])

$$(23)\qquad \int_0^{2\pi} (f(x+y)g(y) - \bar{f}(x+y)\bar{g}(y))\,dy = 0,$$

ce qui donne en définitive

$$(24)\qquad \bar{f}_\varepsilon(x) = -\frac{1}{2\pi}\int_0^{2\pi} \bar{f}(x+y)\bar{g}(y)\,dy.$$

Reste à calculer les fonctions $\bar{g}(y)$. Une telle fonction sera fournie par les valeurs que prend sur la circonférence de rayon *un* la partie imaginaire de la fonction holomorphe

$$G(z) = \frac{1}{2\pi}\int_\varepsilon^{2\pi-\varepsilon} \frac{e^{iy}+z}{e^{iy}-z}\cot\frac{y}{2}\,dy.$$

Observons tout de suite que l'on a $G(0) = 0$. De plus, on trouve par un calcul facile

$$G(z) = \frac{1}{\pi}\left(i(\pi-\varepsilon) - i(\pi-\varepsilon)\frac{z+1}{z-1} + \frac{z+1}{z-1}\log\left(\frac{e^{i(2\pi-\varepsilon)}-z}{e^{i\varepsilon}-z}\right)\right),$$

le logarithme qui intervient étant déterminé par la condition $G(0) = 0$. La partie imaginaire de cette fonction, sur la circonférence de rayon *un*, sera, en posant $z = e^{iy}$,

$$1 - \frac{\varepsilon}{\pi} - \frac{1}{\pi}\cot\frac{y}{2}\log\left|\frac{e^{iy}-e^{-i\varepsilon}}{e^{iy}-e^{i\varepsilon}}\right|.$$

La relation $G(0) = 0$ donne

$$(25)\qquad \frac{1}{2\pi^2}\int_0^{2\pi}\cot\frac{y}{2}\log\left|\frac{e^{iy}-e^{-i\varepsilon}}{e^{iy}-e^{i\varepsilon}}\right|dy = 1 - \frac{\varepsilon}{\pi}.$$

---

[12]) Pour démontrer la légitimité de cette formule, on n'a qu'à recourir au théorème sur le passage à la limite qu'on vient de citer au numéro précédent.

D'autre part, en vertu de (24), on peut écrire

$$\bar{f}_\varepsilon(x) = \frac{1}{2\pi^2}\int_0^{2\pi} \bar{f}(x+y)\cot\frac{y}{2}\log\left|\frac{e^{iy}-e^{-i\varepsilon}}{e^{iy}-e^{i\varepsilon}}\right| dy. \tag{26}$$

En désignant le facteur par lequel $\bar{f}(x+y)$ est multiplié par $h(y)$, on observe d'abord que $h(y) > 0$ et que l'on a d'après (25)

$$\int_0^{2\pi} h(y)\,dy = 1 - \frac{\varepsilon}{\pi}. \tag{27}$$

On en tire par l'inégalité (14), en tenant encore compte de (27),

$$\int_0^{2\pi} |\bar{f}_\varepsilon(x)|^p dx \leqq \left(1 - \frac{\varepsilon}{\pi}\right)^p \int_0^{2\pi} |\bar{f}(x)|^p dx. \tag{28}$$

12. Cela étant, la démonstration du théorème IV s'achève facilement. En effet, on a d'une part pour presque toutes les valeurs de x,

$$\lim_{\varepsilon\to 0} \bar{f}_\varepsilon(x) = \bar{f}(x).$$

D'autre part, d'après (11) et (28),

$$\int_0^{2\pi} |f_\varepsilon(x)|^p dx \leqq M_p^p \left(1 - \frac{\varepsilon}{\pi}\right)^p \int_0^{2\pi} |f(x)|^p dx. \tag{29}$$

On pourra donc appliquer le théorème sur le passage à la limite cité plus haut et écrire

$$\lim_{\varepsilon\to 0} \int_0^{2\pi} \bar{f}_\varepsilon(y)\, g(y)\,dy = \int_0^{2\pi} \bar{f}(y) g(y)\,dy.$$

Par conséquent, l'intégrale double existe toujours au sens de la valeur principale de Cauchy. Il suffit maintenant d'appliquer encore une fois l'inégalité de Hölder-Schwarz pour arriver à l'inégalité (20).

Ajoutons que si l'on n'étend, en (20), l'intégrale double qu'aux valeurs $|x-y| > \varepsilon$, l'inégalité subsiste, même si l'on multiplie le second membre par $1 - \frac{\varepsilon}{\pi}$.

Quelques applications à la série de Fourier.

13. Soit toujours $f(x)$ une fonction de la classe L^p par rapport à l'intervalle $(0, 2\pi)$ et désignons par a_n, b_n les coefficients et par $s_n(x)$ les sommes de la série de Fourier de $f(x)$. On aura alors:

$$\left| s_n(x) - \frac{1}{2}(a_n \cos nx + b_n \sin nx) \right| = \frac{1}{2\pi} \left| \int_0^{2\pi} f(t) \sin n(t-x) \cot \frac{t-x}{2}\, dt \right|$$

$$= \frac{1}{2\pi} \left| \cos nx \int_0^{2\pi} f(t) \sin nt \cot \frac{t-x}{2}\, dt - \sin nx \int_0^{2\pi} f(t) \cos nt \cot \frac{t-x}{2}\, dt \right|$$

$$\leqq \frac{1}{2\pi} \left| \int_0^{2\pi} f(t) \sin nt \cot \frac{t-x}{2}\, dt \right| + \frac{1}{2\pi} \left| \int_0^{2\pi} f(t) \cos nt \cot \frac{t-x}{2}\, dt \right|.$$

Puisque $|\sin nt| \leqq 1$ et $|\cos nt| \leqq 1$, on n'a qu'à appliquer les inégalités (11) et (3) pour conclure que

$$\int_0^{2\pi} |s_n(x)|^p dx \leqq C M_p^p \int_0^{2\pi} |f(x)|^p dx, \tag{30}$$

C étant une constante numérique, c'est-à-dire, indépendante de p. En particulier, on a démontré:

V. *Pour toute fonction $f(x)$ de la classe L^p, l'intégrale de $|s_n(x)|^p$ reste bornée lorsque n tend vers l'infini.*

14. Il est facile de préciser ce théorème en montrant que

$$\lim_{n \to \infty} \int_0^{2\pi} |f(x) - s_n(x)|^p dx = 0. \tag{31}$$

Pour démontrer cette relation, observons d'abord qu'on peut approcher $f(x)$ par des fonctions[13]) $h(x)$ dont la série de Fourier converge uniformément, de façon que l'on ait

$$\int_0^{2\pi} |f(x) - h(x)|^p dx < \varepsilon. \tag{32}$$

En désignant alors par $t_n(x)$ les sommes de la série de Fourier de $h(x)$, il suffit d'écrire

$$f - s_n = (f-h) + (h - t_n) + (t_n - s_n).$$

L'intégrale de $|t_n - s_n|^p$ est alors d'après (30) moindre qu'un multiple constant de (32). Cela étant, on n'aura qu'à utiliser encore une fois l'inégalité (3) pour arriver à la relation (31).

15. Le théorème V admet comme corollaire un théorème concernant la formule de Parseval pour le produit de deux fonctions.

---

[13]) Telles fonctions sont d'après H. W. Young (Proc. London Math. Soc. (2) **11** (1911), p. 89 et suiv.) les moyennes arithmétiques de la série de Fourier de $f(x)$. Pour d'autres fonctions, pour lesquelles la démonstration est plus simple, du moins en principe, voir F. Riesz, loc. cit. § 4.

VI. *Soient $f(x)$ et $g(x)$ deux fonctions appartenant à des classes complementaires, L^p et L^q. Désignons par a_n, b_n et α_n, β_n leurs constantes de Fourier. Alors on a*

$$(33)\qquad \frac{1}{\pi}\int_0^{2\pi} f(x)g(x)\,dx = \frac{a_0\alpha_0}{2} + \sum_{n=1}^{\infty}(a_n\alpha_n + b_n\beta_n),$$

la série au second membre étant convergente.

En désignant toujours par $s_n(x)$ les sommes de la série de Fourier de $f(x)$, ce théorème revient à la relation

$$(34)\qquad \int_0^{2\pi} f(x)g(x)\,dx = \lim_{n\to\infty}\int_0^{2\pi} s_n(x)g(x)\,dx.$$

Or on sait que, même pour toute fonction sommable, on a uniformément

$$\lim_{n\to\infty}\int_0^{2\pi} s_n(x)\,dx = \int_0^{2\pi} f(x)\,dx.$$

Alors, grâce au théorème V, la relation (33) s'ensuit immédiatement par un théorème connu de M. F. Riesz[14]).

[14]) F. Riesz, loc. cit. p. 464 et suiv. Voici ce théorème. Admettons que $g(x)$ appartienne à la classe L^q, que l'on ait

$$\lim\int_a^x f_n(x)\,dx = \int_a^x f(x)\,dx$$

et qu'il existe une constante M telle que

$$\int_a^b |f_n(x)|^p\,dx \leqq M.$$

Alors on a

$$\lim\int_a^b f_n(x)\,g(x)\,dx = \int_a^b f(x)g(x)\,dx.$$

La démonstration de M. F. Riesz donne aussi que, f et les f_n étant supposés périodiques, de période $(b-a)$, on aura uniformément

$$\lim\int_a^b f_n(x+t)\,g(x)\,dx = \int_a^b f(x+t)\,g(x)\,dx.$$

Cette remarque suffit pour démontrer un fait énoncé dans l'Encykl. d. math. Wissensch. II C 10, p. 1211, cf. note 95.

[Remarque ajoutée pendant la correction des épreuves.] On peut arriver au théorème VI par une voie plus courte. En effet, en appliquant l'inégalité de Hölder-Schwarz et la relation (31), on voit immédiatement que l'intégrale

$$\int_0^{2\pi}(f(x)-s_n(x))\,g(x)\,dx$$

tend vers zéro.

Le théorème VI est connu depuis longtemps pour $p = q = 2$. Dans ce cas, on le déduit immédiatement de la relation plus particulière

$$\frac{1}{\pi}\int_0^{2\pi} f^2(x)\,dx = \frac{a_0^2}{2} + \sum_{n=1}^{\infty}(a_n^2 + b_n^2).$$

En effet, il suffit d'écrire cette relation pour $f(x) + g(x)$ et $f(x) - g(x)$ et soustraire l'une des relations de l'autre. Evidemment, dans ce cas, la série figurant au second membre est absolument convergente. Or, c'est le seul cas où il y a nécessairement convergence absolue. Ou plus précisément, pour tout autre couple de classes complémentaires L^p et L^q, on peut donner des fonctions $f(x)$ et $g(x)$ appartenant à ces classes et telles que la série en (33) ne converge pas absolument. Cette remarque explique pourquoi le cas général est plus difficile que le cas particulier.

Notre remarque résulte immédiatement du fait suivant, facile à vérifier. Le nombre k étant $< \frac{1}{2}$, il existe des fonctions continues $f(x)$, avec les constantes de Fourier a_n et b_n, telles que les séries

$$\sum \frac{|a_n|}{n^k} \quad \text{et} \quad \sum \frac{|b_n|}{n^k}$$

soient divergentes.

En effet, par un résultat très connu de M. Serge Bernstein, pour tout nombre $c < 1/2$, on peut former des fonctions périodiques qui satisfont à une condition de Lipschitz d'ordre c et telles que les séries formées avec les modules de leurs constantes de Fourier A_n et B_n soient divergentes. En choisissant alors $c > k$, les séries de Fourier $\sum n^k A_n \cos nx = \sum a_n \cos nx$ et $\sum n^k B_n \sin nx = \sum b_n \sin nx$ représenteront des fonctions continues avec la propriété désirée. Cela étant, pour achever la vérification de notre remarque, il suffit d'observer que toute fonction continue appartient à toutes les classes L^p, tandis que les fonctions $\sum n^{-k} \cos nx$ et $\sum n^{-k} \sin nx$ appartiennent à toutes les classes L^q telles que $q < \frac{1}{1-k}$. On a évidemment $\frac{1}{1-k} < 2$.

Observons, en finissant ce numéro, que les constantes de Fourier de $\bar{f}(x)$ étant $-b_n$ et a_n, on a encore la formule remarquable, l'intégrale double étant prise au sens de la valeur principale de Cauchy

$$(35) \qquad \frac{1}{2\pi^2}\int_0^{2\pi}\int_0^{2\pi} f(x)\,g(y)\cot\frac{x-y}{2}\,dx\,dy = \sum_{n=1}^{\infty}(a_n\beta_n - b_n\alpha_n).$$

16. Voici encore quelques indications historiques.

Le théorème V et son complément que nous venons d'établir pour les sommes de la série de Fourier, M. W. H. Young les a donnés, il y a bien longtemps, pour les moyennes arithmétiques de ces sommes, en démontrant aussi que la série figurant au second membre de (33), sommée par ces moyennes, est égale au premier membre. Mais il y en a plus. M. W. H. Young et Mme Grace Chisholm Young ont déjà, en 1912[15]), posé la question de savoir si le théorème V du texte et son complément étaient vrais ou non. Ils ont même démontré[16]), que ce théorème entraînerait la convergence de la série figurant en (33), ce qui, à leur avis, semblait indiquer que la question devrait être résolue par la négative.

Dans le travail important auquel nous venons de faire allusion, M. Young et Mme Young rassemblent les différentes démonstrations qu'on a données jusqu'alors pour le théorème de Riesz-Fischer et ils donnent le théorème suivant qui comprend ledit théorème comme cas particulier.

Désignons par $s_n(x)$ les sommes de la série trigonométrique

$$\frac{a_0}{2}+\sum_{n=1}^{\infty}(a_n \cos nx + b_n \sin nx)$$

et par

$$S_n(x)=\frac{s_0(x)+s_1(x)+\ldots+s_{n-1}(x)}{n}$$

les moyennes arithmétiques de ces sommes. Pour que cette série soit la série de Fourier d'une fonction de la classe L^p, il faut et il suffit que l'intégrale

$$\int_0^{2\pi} |S_n(x)|^p \, dx$$

reste bornée, lorsque n tend vers l'infini.

M. Young et Mme Young observent aussi que la dernière condition sera toujours remplie, si

$$\int_0^{2\pi} |s_n(x)|^p \, dx$$

reste bornée. Cela se voit[17]) p. ex. par l'inégalité facile à vérifier

$$\int_0^{2\pi} |S_n(x)|^p \, dx \leqq \int_0^{2\pi} |s_{n-1}(x)|^p \, dx.$$

C'est-à-dire que la dernière condition est aussi une condition suffisante. Par le théorème V, on voit que cette condition est aussi nécessaire, c'est-à-dire que, dans le théorème ci-dessus, $S_n(x)$ *peut être remplacé par* $s_n(x)$.

[15]) W. H. Young et G. C. Young, On the theorem of Riesz-Fischer, Quarterly Journal **44** (1912–13), voir en particulier p. 58.

[16]) Leur démonstration fait appel à un théorème plus difficile que ne fait la nôtre.

[17]) Cette démonstration est un peu différente de celle des auteurs cités.

Pour $p=2$, on retombe évidemment sur le théorème de Riesz-Fischer. On a en effet

$$\frac{1}{\pi}\int_0^{2\pi} s_n^2(x)\,dx = \frac{a_0^2}{2} + \sum_{k=1}^{n}(a_k^2+b_k^2).$$

Les problèmes correspondant à l'intervalle $(-\infty, +\infty)$.

17. Soit maintenant $f(x)$ une fonction de la classe L^p par rapport à l'intervalle $(-\infty, +\infty)$. On démontre facilement le théorème suivant:

VII. *La valeur principale de l'intégrale*

$$\bar{f}(x) = \frac{1}{\pi}\int_{-\infty}^{+\infty} \frac{f(y)}{x-y}\,dy \tag{36}$$

existe presque partout et l'on a

$$\int_{-\infty}^{+\infty} |\bar{f}(x)|^p\,dx \leqq M_p^p \int_{-\infty}^{+\infty} |f(x)|^p\,dx; \tag{37}$$

la constante M_p étant identique à celle de l'inégalité (11).

Nous attaquons ce problème par deux méthodes différentes. La première de ces méthodes, qui nous fournira une démonstration complète, se rattache à nos résultats sur l'intervalle fini, tandis que la seconde se rattache à l'étude d'une certaine fonction holomorphe. Cette dernière méthode fournit aussi tout le résultat en question, sauf l'identité des constantes. Elle nous conduira à la réciproque de (36).

18. Soit donc $f(x)$ une fonction de la classe L^p par rapport à l'intervalle $(-\infty, +\infty)$. Posons

$$\bar{f}_n(x) = \frac{1}{2n\pi}\int_{-n\pi}^{+n\pi} f(y)\cot\frac{x-y}{2n}\,dy. \tag{38}$$

Il est clair que $\bar{f}(x)$ et $\bar{f}_n(x)$ existent presque partout, et l'on démontre très facilement que

$$\lim_{n\to\infty} \bar{f}_n(x) = \bar{f}(x).$$

Nous ne nous y arrêtons pas.

La formule (38) peut aussi s'écrire

$$\bar{f}_n(nx) = \int_{-\pi}^{+\pi} f(ny)\cot\frac{x-y}{2}\,dy.$$

En vertu du théorème II, on a maintenant

$$\int_{-n\pi}^{+n\pi} |\bar{f}_n(x)|^p\, dx = n \int_{-\pi}^{+\pi} |\bar{f}_n(nx)|^p\, dx \leqq n M_p^p \int_{-\pi}^{+\pi} |f(nx)|^p\, dx$$

$$= M_p^p \int_{-n\pi}^{+n\pi} |f(x)|^p\, dx \leqq M_p^p \int_{-\infty}^{+\infty} |f(x)|^p\, dx.$$

Alors m étant un nombre quelconque $\leqq n$, on a, à plus forte raison,

$$\int_{-m\pi}^{+m\pi} |\bar{f}_n(x)|^p\, dx \leqq M_p^p \int_{-\infty}^{+\infty} |f(x)|^p\, dx.$$

En gardant m fixe et en faisant tendre n vers l'infini, on voit, par le théorème cité de M. Fatou, que la même inégalité tient si l'on y remplace $\bar{f}_n(x)$ par $\bar{f}(x)$. Enfin, en faisant tendre m vers l'infini, l'inégalité (37) résulte.

19. Il nous reste à démontrer que la constante M_p dans l'inégalité (37) ne pourra être abaissée. Admettons, en effet, que cette inégalité tienne avec une certaine constante N_p, où N_p est la moindre valeur admissible de la constante en question. Nous savons déjà que l'on a $N_p \leqq M_p$. Pour conclure l'égalité des deux constantes, il suffit de montrer que $M_p \leqq N_p$. Cela se fait de la manière suivante.

Entendons par $f(x)$ et $g(x)$ deux fonctions appartenant respectivement aux classes L^p et L^q par rapport à l'intervalle $(0, 2\pi)$. A l'extérieur de cet intervalle on les définit par les relations $f(x+2\pi) = f(x)$ et $g(x+2\pi) = g(x)$.

Démontrons que l'on a

$$(39)\quad \lim_{n\to\infty} \frac{1}{2n\pi} \int_{-2n\pi}^{2n\pi} \int_{-2n\pi}^{2n\pi} \frac{f(x)g(y)}{x-y}\, dx\, dy = \frac{1}{2\pi} \int_0^{2\pi}\int_0^{2\pi} f(x)g(y) \cot\frac{x-y}{2}\, dx\, dy.$$

En effet, avant le passage à la limite, le premier membre pourra s'écrire

$$\frac{1}{2n\pi} \int_0^{2\pi}\int_0^{2\pi} f(x)g(y) \left(\sum_{j=-n}^{n-1} \sum_{k=-n}^{n-1} \frac{1}{x-y+2\pi(j-k)} \right) dx\, dy.$$

Or on a

$$\frac{1}{2n} \sum_{j=-n}^{n-1} \sum_{k=-n}^{n-1} = \frac{1}{x-y} + \frac{1}{2n} \sum_{s=1}^{2n-1} (2n-s) \left(\frac{1}{x-y+2\pi s} + \frac{1}{x-y-2\pi s} \right)$$

$$= \frac{1}{x-y} + \sum_{s=1}^{2n-1} \left(\frac{1}{x-y+2\pi s} + \frac{1}{x-y-2\pi s} \right) - \frac{1}{2n} \left(\frac{1}{x-y+2\pi} + \frac{1}{x-y-2\pi} \right) + \sigma$$

avec

$$\sigma = \frac{1}{2n}\sum_{s=2}^{2n-1} \frac{2s(x-y)}{4\pi^2 s^2 - (x-y)^2}.$$

Or dans cette formule, on a $|x-y| \leqq 2\pi$. Par conséquent, il est

$$|\sigma| \leqq \frac{1}{2n}\sum_{s=2}^{2n-1} \frac{4s}{4\pi^2(s^2-1)} = O\left(\frac{\log n}{n}\right) \to 0.$$

En outre, puisque les valeurs principales des intégrales doubles qui suivent existent, il vient

$$\frac{1}{2n}\left|\int_0^{2\pi}\int_0^{2\pi} \frac{f(x)g(y)}{x-y\pm 2\pi}\,dx\,dy\right| \to 0.$$

Enfin puisqu'on a uniformément pour

$$0 \leqq x \leqq 2\pi \quad \text{et} \quad 0 \leqq y \leqq 2\pi,$$

$$\lim_{n\to\infty}\left(\frac{1}{x-y} + \sum_{s=1}^{2n-1}\left(\frac{1}{x-y+2\pi s} + \frac{1}{x-y-2\pi s}\right)\right) = \frac{1}{2}\cot\frac{x-y}{2},$$

la relation (39) s'ensuit.

Cela étant, on obtient par la périodicité de $f(x)$ et de $g(x)$

$$\frac{1}{2\pi}\left|\int_0^{2\pi}\int_0^{2\pi} f(x)g(y)\cot\frac{x-y}{2}\,dx\,dy\right| = \lim_{n\to\infty}\frac{1}{2n\pi}\left|\int_{-2n\pi}^{2n\pi}\int_{-2n\pi}^{2n\pi}\frac{f(x)g(y)}{x-y}\,dx\,dy\right|$$

$$\leqq \frac{N_p}{2n}\left(\int_{-2n\pi}^{2n\pi}|f(x)|^p\,dx\right)^{\frac{1}{p}}\left(\int_{-2n\pi}^{2n\pi}|g(x)|^q\,dx\right)^{\frac{1}{q}}$$

$$= N_p\left(\int_0^{2\pi}|f(x)|^p\,dx\right)^{\frac{1}{p}}\left(\int_0^{2\pi}|g(x)|^q\,dx\right)^{\frac{1}{q}}.$$

Cela étant vrai pour toute fonction $g(x)$ de la classe L^q, on conclut comme au n° 6, que la fonction $\bar{f}(x)$ définie par la formule (10) satisfait à la relation

$$\int_0^{2\pi}|\bar{f}(x)|^p\,dx \leqq N_p^p \int_0^{2\pi}|f(x)|^p\,dx,$$

c'est-à-dire que l'on a $M_p \leqq N_p$, ce qui achève la démonstration.

20. Nous passons à la seconde démonstration qui se fonde sur la considération de la fonction de la variable complexe z

$$(40) \qquad F(z) = \frac{1}{i\pi}\int_{-\infty}^{+\infty}\frac{f(y)}{y-z}\,dy.$$

Cette expression représente deux fonctions holomorphes respectivement dans les demi-plans supérieur et inférieur, l'axe réel formant, en général, une coupure. Il suffira de considérer le demi-plan supérieur.

On pourra supposer, sans restreindre la généralité, que $f(x)$ ne prend pas de valeurs négatives. Il sera aussi commode de supposer d'abord que la fonction s'évanouit à l'extérieur d'un certain intervalle $(-A, +A)$ et, après, faire tendre A vers l'infini.

Admettons donc d'abord que $f(x)$ soit $\geqq 0$ dans l'intervalle $(-A, +A)$ et nulle ailleurs et que $f(x)$ appartienne dans cet intervalle à la classe L^p. Posons $z = \xi + i\eta$. La partie réelle de $F(z) = U(\xi, \eta) + iV(\xi, \eta)$ s'écrit

$$U(\xi, \eta) = \frac{\eta}{\pi}\int_{-\infty}^{+\infty} \frac{f(y)}{(\xi - y)^2 + \eta^2}\, dy. \tag{41}$$

Par conséquent, on a $U(\xi, \eta) > 0$ dans le demi-plan supérieur. Cela étant, $(F(z))^p$ sera holomorphe dans ce demi-plan. Désignons alors par c un nombre positif. On vérifie facilement que l'on a

$$\int_{-\infty+ic}^{+\infty+ic} (F(z))^p\, dz = 0, \tag{42}$$

la ligne d'intégration étant la droite dont on a indiqué les extrémités.

Pour le voir, il suffit d'appliquer le théorème de Cauchy à une figure située dans le demi-plan supérieur et composée d'une demi-circonférence de centre ic et de rayon très grand, et du segment que cette demi-circonférence découpe de la droite dont on vient de parler. La fonction $f(x)$ s'annulant à l'extérieur de l'intervalle $(-A, +A)$, $F(z)$ sera de l'ordre de grandeur de $\frac{1}{|z|}$, pour $|z|$ assez grand. En faisant alors tendre le rayon de la demi-circonférence vers l'infini, on obtient la relation (42).

Cela étant, un raisonnement tout à fait analogue à celui qu'on a fait au sujet du théorème I, établit l'existence d'une constante — qui, d'après les résultats des numéros précédents, pourra être désignée par M_p — telle que, pour toute valeur de η, on ait

$$\int_{-\infty}^{+\infty} |V(\xi, \eta)|^p\, d\xi \leqq M_p^p \int_{-\infty}^{+\infty} |U(\xi, \eta)|^p\, d\xi. \tag{43}$$

Enfin puisqu'on a

$$\frac{\eta}{\pi}\int_{-\infty}^{+\infty} \frac{d\xi}{(\xi - y)^2 + \eta^2} = 1, \tag{44}$$

on tire de (14), où l'on substitue l'intervalle $(-\infty, +\infty)$ à l'intervalle $(0, 2\pi)$ et de (44) que

$$(45) \qquad \int_{-\infty}^{+\infty} |U(\xi, \eta)|^p d\xi \leqq \int_{-\infty}^{+\infty} |f(x)|^p dx,$$

et alors

$$(46) \qquad \int_{-\infty}^{+\infty} |V(\xi, \eta)|^p d\xi \leqq M_p^p \int_{-\infty}^{+\infty} |f(x)|^p dx.$$

On démontre, comme M. Fatou et M. Plessner l'ont fait au cas d'un intervalle fini, que la valeur principale de (36) existe presque partout et que $V(x, \eta)$ tend presque partout vers $\bar{f}(x)$ lorsque η tend vers zéro. Enfin l'inégalité (37) résulte aussi par la théorème de M. Fatou cité à la fin du n° 8. Ce même théorème nous permet de nous débarrasser de la restriction posée au sujet de $f(x)$, savoir que $f(x)$ s'annulait à l'extérieur d'un certain intervalle. Dans le cas général, on remplacera $f(x)$ par une autre fonction égale à $f(x)$ pour $-A \leqq x \leqq +A$ et nulle ailleurs. Le théorème étant démontré pour la fonction modifiée, il suffira de faire tendre A vers l'infini et d'appliquer encore une fois le théorème de M. Fatou. Par le même théorème, on voit que les inégalités (45) et (46) tiennent aussi dans le cas général.

Remarquons encore que la relation (42) subsiste également après la suppression de ladite restriction. Pour le voir, il suffit d'écrire $f(x) = f_1(x) + f_2(x)$, où $f_1(x) = 0$, à l'extérieur de l'intervalle $(-A, +A)$ et $f_2(x) = 0$, à son intérieur, A étant assez grand pour que

$$\int_{-\infty}^{+\infty} |f_2(x)|^p dx$$

soit très petite. A ces fonctions, il correspondra deux fonctions holomorphes $F_1(z)$ et $F_2(z)$, et l'on aura $F(z) = F_1(z) + F_2(z)$. Alors la relation (42) tient pour $F_1(z)$. D'autre part, en vertu de (45), de (46) et de (3), l'intégrale de $|F_2(z)|^p$, prise le long de la droite $(-\infty + ic, +\infty + ic)$, sera aussi très petite, et alors il suffira d'appliquer encore une fois l'inégalité (3) pour conclure la relation (42).

Dans tout ce qui suit, on pourra toujours, si cela est plus commode, faire les démonstrations d'abord dans le cas partieulier et passer, par le dernier artifice, au cas général. Nous ne nous arrêtons plus sur les difficultés de ce caractère-ci.

Pour arriver à la réciproque de la formule (36), observons qu'on pourra exprimer $F(z)$ aussi par la formule de Cauchy, où l'on désigne par $F(x)$ la fonction-limite $f(x)+i\bar{f}(x)$ vers laquelle $F(z)$ tend presque partout lorsque $z=x+i\eta$ tend vers x. On a

$$(47)\qquad F(z)=\frac{1}{2i\pi}\int_{-\infty}^{+\infty}\frac{F(y)}{y-z}\,dy=\frac{1}{2i\pi}\int_{-\infty}^{+\infty}\frac{f(y)+i\bar{f}(y)}{y-z}\,dy.$$

On le voit, en démontrant la formule d'abord au cas où l'intégrale est prise le long d'une droite $(-\infty+ic,\ +\infty+ic)$, c désignant un nombre positif assez petit. Après, on fait tendre c vers zéro.

Dès lors, la comparaison de (40) et de (47) donne

$$(48)\qquad F(z)=\frac{1}{\pi}\int_{-\infty}^{+\infty}\frac{\bar{f}(y)}{y-z}\,dy.$$

Par cela, la réciprocité entre $f(x)$ et $\bar{f}(x)$ est établie et on peut écrire la réciproque de la formule (36)

$$(49)\qquad f(x)=\frac{1}{\pi}\int_{-\infty}^{+\infty}\frac{\bar{f}(y)}{y-x}\,dy.$$

Évidemment, cette formule peut aussi se déduire, par un passage à la limite, de la formule (19).

21. Nous établirons maintenant l'inégalité[18]) suivante dont nous donnerons quelques applications assez remarquables dans un autre travail. On a

$$(50)\qquad \int_D |F(z)|^p\,|dz|\leqq(1+M_p)^p\int_{-\infty}^{+\infty}|f(x)|^p\,dx,$$

où le chemin d'intégration D désigne une droite arbitraire du plan de la variable complexe z. Cet énoncé s'ensuit d'une combinaison facile des inégalités (37) et (3) et de l'inégalité

$$(51)\qquad \int_D |F(z)|^p\,|dz|\leqq\int_{-\infty}^{+\infty}|f(x)+i\bar{f}(x)|^p\,dx.$$

Cette dernière relation, on peut la ramener au théorème suivant de MM. L. Fejér et F. Riesz[19]).

Si $F(z)$ est une fonction holomorphe pour $|z|<R$ et

$$\int_{|z|=r}|f(z)|^p\,|dz|\leqq M\qquad (p\geqq 0),$$

[18]) La constante figurant dans cette inégalité peut être abaissée par des évaluations plus précises que celles du texte.

[19]) L. Fejér und F. Riesz, Über einige funktionentheoretische Ungleichungen, Math. Zeitschr. **11** (1921), p. 305–314.

pour tout $r < R$, alors $|\bar{f}(z)|^p$ est sommable le long d'un diamètre quelconque de la circonférence de rayon R et son intégrale est $\leqq \frac{M}{2}$.

Indiquons maintenant la démonstration de (51). Soit d'abord D' une demi-droite issue d'un point de l'axe réel et perpendiculaire à cet axe. Par une substitution linéaire, on transforme le demi-plan supérieur dans un cercle. On pourra choisir cette substitution de façon que la demi-droite donnée devienne un diamètre du cercle. Alors une adaption facile du théorème cité montre que le double de l'intégrale de $|F|^p$ prise le long d'une telle demi-droite est inférieur au second membre de (51). Dès lors, par des représentations conformes répétées, on obtient l'inégalité (51) pour toute droite qui forme avec l'axe réel un angle de la forme $\frac{m\pi}{2^n}$, m et n étant entiers, et alors, par un passage à la limite, pour une droite quelconque.

On peut facilement préciser l'inégalité (51), en démontrant que, dans tout faisceau de demi-droites issues d'un point de l'axe réel et situées dans l'un ou l'autre des demi-plans supérieur ou inférieur, le logarithme de l'intégrale de $|F|^p$, suivant ces demi-droites est une fonction convexe de l'angle que ces demi-droites forment avec l'axe réel positif.

22. Voici maintenant l'analogue du théorème IV.

IV'. *Pour deux fonctions $f(x)$ et $g(x)$ appartenant à des classes complémentaires L^p et L^q, on a* .

$$(52) \qquad \left|\int_{-\infty}^{+\infty}\int_{-\infty}^{+\infty} \frac{f(x)\,g(y)}{x-y}\,dx\,dy\right| \leqq M_p\left(\int_{-\infty}^{+\infty}|f(x)|^p\,dx\right)^{\frac{1}{p}}\left(\int_{-\infty}^{+\infty}|g(x)|^q\,dx\right)^{\frac{1}{q}};$$

l'intégrale double existant toujours si on la prend au sens de la valeur principale de Cauchy, c'est-à-dire comme limite de l'intégrale étendue aux valeurs $|x-y| > \varepsilon > 0$.

La démonstration est toute pareille à celle qu'on a donnée pour le théorème IV. On forme ici la fonction

$$(53) \qquad \bar{f}_\varepsilon(x) = \frac{1}{\pi}\left(\int_{-\infty}^{x-\varepsilon} + \int_{x+\varepsilon}^{\infty}\right)\frac{f(y)}{x-y}\,dy,$$

et l'on démontre que

$$(54) \qquad \bar{f}_\varepsilon(x) = \frac{1}{\pi^2}\int_{-\infty}^{+\infty}\bar{f}(x+y)\log\left|\frac{y+\varepsilon}{y-\varepsilon}\right|\frac{dy}{y}.$$

La fonction

$$(55) \qquad \frac{1}{\pi y}\log\left|\frac{y+\varepsilon}{y-\varepsilon}\right|$$

est la partie imaginaire, sur l'axe réel, de la fonction, holomorphe dans le demi-plan supérieur,

$$G(z) = \frac{i}{\pi z} \log \frac{\varepsilon + z}{\varepsilon - z},$$

le dernier logarithme étant déterminé par la condition de s'annuler à l'origine. La partie réelle de $G(z)$ sur l'axe réel est $-\frac{1}{y}$ pour $|y| > \varepsilon$ et zéro pour $|y| < \varepsilon$.

En faisant tendre z vers l'infini le long de l'axe imaginaire, le module de $zG(z)$ tend vers 1, ce qui donne immédiatement (voir la formule (48))

$$(56) \qquad \frac{1}{\pi^2} \int_{-\infty}^{+\infty} \log \left| \frac{y+\varepsilon}{y-\varepsilon} \right| \frac{dy}{y} = 1.$$

En observant encore que la fonction (55) ne change pas de signe, et en appliquant — *mutatis mutandis* — l'inégalité (14), on voit que l'intégrale de $|\bar{f}_\varepsilon(x)|^p$ est inférieure à celle de $|\bar{f}(x)|^p$. Dès lors, la démonstration s'achève comme plus haut. Remarquons encore que l'inégalité (52) subsiste, si, dans l'intégrale double, on n'étend l'intégration qu'aux valeurs $|x - y| > \varepsilon$.

23. Le dernier résultat conduit immédiatement à un théorème analogue concernant certaines formes bilinéaires. Nous nous contentons, ici, de l'indiquer en quelques mots, ayant l'intention d'y revenir à une autre occasion.

Posons $f(x) = x_k$, $y(x) = g_k$ pour $k < x \leqq k+1$ (pour tout nombre entier k) les valeurs x_k, y_k étant telles que l'on ait

$$(57) \qquad \sum_{k=-\infty}^{+\infty} |x_k|^p \leqq 1, \qquad \sum_{k=-\infty}^{+\infty} |y_k|^q \leqq 1, \qquad \frac{1}{p} + \frac{1}{q} = 1.$$

L'intégrale double étendue aux valeurs $|x - y| > 1$ deviendra une forme bilinéaire

$$\sum_{j,k=-\infty}^{+\infty} a_{jk} x_j y_k.$$

Or on a $a_{jj} = 0$, et une évaluation toute simple montre que l'on a, pour $j \neq k$,

$$(58) \qquad \left| a_{jk} - \frac{1}{j-k} \right| < \frac{C}{(j-k)^2}.$$

C étant une constante. On en tire que les formes bilinéaires

$$(59) \qquad \sum_{j,k=-\infty}^{+\infty} \frac{x_j y_k}{j-k} \qquad (j \neq k),$$

et

$$\frac{\sin \alpha \pi}{\pi} \sum_{j,\, k=-\infty}^{+\infty} \frac{x_j y_k}{\alpha + j - k} \tag{60}$$

sont bornées sous les conditions (57).

Il en résulte, par un théorème connu de M. Landau, que, en définissant les substitutions H et K_α par les formules respectives

$$x_j' = \sum_{k=-\infty}^{+\infty} \frac{x_k}{j-k} \qquad (j \neq k), \tag{61}$$

et

$$x_j' = \frac{\sin \pi \alpha}{\pi} \sum_{k=-\infty}^{+\infty} \frac{x_k}{\alpha + j - k}, \tag{62}$$

les séries $\sum |x_j'|^p$ sont convergentes.

24. J'ajoute quelques remarques sur les dernières substitutions, en réservant les détails pour un petit travail que je suis en train de rédiger. Nous n'envisageons que le cas $p = 2$, c'est-à-dire que nous supposons que la série $\sum x_k^2$ est convergente.

Remarquons d'abord que la substitution K_α devient, pour des valeurs entières de α,

$$x_j' = (-1)^\alpha x_{j+\alpha}. \tag{63}$$

Les substitutions K_α forment un *groupe* et l'on a $K_\alpha K_\beta = K_{\alpha+\beta}$. Par conséquent $K_{-\alpha}$ est la *réciproque* de K_α. En outre les substitutions K_α sont *orthogonales*.

Tout cela se vérifie facilement par un calcul direct. Mais il est plus intéressant de rattacher ces remarques à la théorie des formes de M. Toeplitz.

Posons, en effet,

$$a_j = \frac{1}{2\pi} \int_0^{2\pi} a(t)\, e^{-jit}\, dt \tag{64}$$

et

$$x_j = \frac{1}{2\pi} \int_0^{2\pi} x(t)\, e^{-jit}\, dt,$$

$a(t)$ et $x(t)$ étant des fonctions à carré sommable. Il vient par le théorème de Parseval

$$x_j' = \sum_{k=-\infty}^{+\infty} a_{j-k} x_k = \frac{1}{2\pi} \int_0^{2\pi} a(t)\, x(t)\, e^{-jit}\, dt. \tag{65}$$

Puisque, encore par le théorème de Parseval, on a

$$\sum_{j=-\infty}^{+\infty} |x_j|^2 = \frac{1}{2\pi}\int_0^{2\pi} |x(t)|^2 dt$$

et, en admettant que la fonction $a(t)$ soit bornée,

$$\sum_{j=-\infty}^{+\infty} |x_j'|^2 = \frac{1}{2\pi}\int_0^{2\pi} |x(t)\,a(t)|^2 dt,$$

on voit que la condition nécessaire et suffisante pour que (65) donne une substitution orthogonale, c'est que l'on ait presque partout $|a(t)| = 1$.

D'autre part on sait que, $b(t)$ étant une seconde fonction bornée, la substitution obtenue par la composition des substitutions a_{j-k} et b_{j-k} correspondra au produit $a(t)\,b(t)$. Cela étant, on voit que si $a(t)$ dépend d'un paramètre α, c'est-à-dire que $a(t) = a_\alpha(t)$, alors, pour que (65) donne un groupe avec la propriété $K_\alpha K_\beta = K_{\alpha+\beta}$, il faut et il suffit que $a_\alpha(t)$ soit de la forme $a_\alpha(t) = e^{\alpha\varphi(t)}$.

La substitution (62) correspond à $a_\alpha(t) = e^{i\alpha(\pi-t)}$. La substitution H correspond, de son côté, à la fonction

$$\lim_{\alpha\to 0} \frac{e^{i\alpha(\pi-t)} - 1}{\alpha} = i(\pi - t).$$

On retrouve ainsi pour cette dernière substitution, qui, évidemment, n'est pas orthogonale, la borne π.

En posant $a_\alpha(t) = e^{i\alpha\sin t}$ on arrive à des substitutions, dont les coefficients sont des fonctions de Bessel.

Nous finissons ce numéro en donnant un résultat analogue pour un nombre fini des variables.

Les substitutions

$$(66) \qquad x_j' = \frac{\sin\pi\alpha}{\pi n} \sum_{k=0}^{n-1} \frac{x_k}{\sin(\alpha + j - k)\frac{\pi}{n}} \qquad (j = 0, \ldots, n-1)$$

sont aussi *orthogonales*, elles forment également un *groupe* et satisfont à la relation $K_\alpha K_\beta = K_{\alpha+\beta}$.

Si l'on pose encore $s = bn + k$, avec $k = 0, 1, \ldots, n-1$, et l'on définit x_s par $x_s = (-1)^b x_k$, on voit qu'à toute valeur entière de α, il correspond la substitution

$$x_k' = (-1)^a x_{k+a}.$$

On obtient des généralisations intéressantes en considérant, dans un ordre d'idées général, les substitutions de la forme

$$x_j' = \sum_{k=0}^{n-1} a_{j-k} x_k.$$

25. On traduit facilement nos résultats au sujet de la série de Fourier en des résultats concernant l'intégrale de Fourier.

Nous nous contentons d'écrire la formule, valable pour tout couple de fonctions $f(x)$ et $g(x)$ appartenant respectivement aux classes L^p et L^q,

$$\lim_{\omega \to \infty} \frac{1}{\pi} \int_{-\infty}^{+\infty} \int_{-\infty}^{+\infty} f(x)\, g(y) \frac{\sin \omega (x-y)}{x-y}\, dx\, dy = \int_{-\infty}^{+\infty} f(x) g(x)\, dx.$$

Stockholm, le 30 mai 1924.

(Eingegangen am 23. Juli 1926.)

34.

Sur certaines inégalités dans la théorie des fonctions avec quelques remarques sur les géometries non-euclidiennes

Proc. Roy. Physiogr. Soc. Lund *1* (1931), 1–21

Introduction.

Dans le présent travail, nous considérons d'abord une expression dépendant de deux points et d'une circonférence et invariante par rapport aux substitutions linéaires de la variable complexe. Cette expression définit donc une espèce d'écart non-euclidien. Cet écart se trouve d'ailleurs dans un rapport très intime avec l'écart non euclidien ordinaire et diminue comme celui-ci quand on applique à la variable certaines transformations non linéaires. La considération de ladite expression conduit ainsi à une inégalité qui renferme l'inégalité appelée lemme de Schwarz et qui, d'autre part, fournit facilement les théorèmes de Julia et de Carathéodory (n°s 1—5). Dans les n°s 6—9 nous approfondissons un peu l'étude de ces deux théorèmes et de leur rapport avec un théorème de MM. F. Riesz et Herglotz sur la représentation des fonctions dont la partie réelle est positive. Nous y donnons aussi la solution d'un problème qui nous fut posé par M. Myrberg.

La dernière partie du travail (n°s 10—14), qui est indépendante des parties précédentes, se rattache à l'idée de groupes de transformations dont l'importance pour la géométrie fut mise en évidence par les recherches de Cayley, Klein, Lie, Helmholtz, Poincaré et d'autres. On montre d'abord qu'il est d'un certain intérêt de considérer les géométries non-euclidiennes, invariantes par rapport à des transformations conformes, où l'infini (ou l'absolu) est représenté, non par une circonférence, mais par la frontière d'un domaine simplement connexe quelconque. Il est surtout utile de comparer les géométries relatives à des domaines dont l'un renferme l'autre. On obtient ainsi, entre autres, une interprétation instructive du principe de Lindelöf.

Une fois la géométrie non-euclidienne par rapport à un cercle développée, on peut arriver auxdites géométries généralisées par une extension évidente, la géométrie du domaine étant définie comme identique à celle de son image conforme sur le cercle. Il nous paraît pourtant préférable de suivre une voie différente où, au lieu de supposer la géométrie du cercle toute prête, on construit cette géométrie au moyen du principe même par lequel on vient de la généraliser. Il est en effet clair que cette géométrie, aussi bien que celle concernant un domaine simplement connexe quelconque, est entièrement définie par son invariance par rapport aux transformations conformes, les éléments géométriques n'étant définis que dans le voisinage infinitésimal d'un seul point d'un domaine simplement connexe donné, p. ex. au centre d'un cercle de rayon *un*.

Le groupe des transformations conformes étant remplacé par celui des transformations projectives, on arrive facilement par le même procédé à la géométrie Cayleyenne.

Un invariant.

1. Par *domaine circulaire* nous entendons l'intérieur ou l'extérieur d'un cercle ou encore un demi-plan.

C étant un domaine circulaire délimité par un cercle fini, désignons par R le rayon du cercle et par r la distance d'un point z quelconque du plan complexe au centre du cercle. Posons

$$\delta_z^C = \frac{|R^2 - r^2|}{R}. \tag{1}$$

C étant un demi-plan, on pose

$$\delta_z^C = 2d, \tag{2}$$

d désignant la distance du point z à la droite délimitant le demi-plan.

On exécute la transformation linéaire

$$w = \frac{\alpha z + \beta}{\gamma z + \delta}, \tag{3}$$

qui fait correspondre à C un domaine circulaire Γ et on définit d'une manière analogue l'élément δ_w^Γ. Cela étant, on a

(4) $$D^C_{z_1, z_2} = \frac{|z_1 - z_2|^2}{\delta^C_{z_1}\delta^C_{z_2}} = D^\Gamma_{w_1, w_2} = \frac{|w_1 - w_2|^2}{\delta^\Gamma_{w_1}\delta^\Gamma_{w_2}},$$

w_1 et w_2 étant respectivement les points que la transformation fait correspondre aux points z_1 et z_2.

La démonstration est immédiate. Désignons par z_1' et z_2' les images respectives de z_1 et de z_2 par rapport au cercle C. On trouve facilement pour le rapport anharmonique (z_1, z_2', z_2, z_1')

$$|(z_1, z_2', z_2, z_1')| = \left| \frac{z_2 - z_1}{z_2 - z_2'} : \frac{z_1' - z_1}{z_1' - z_2'} \right| = D^C_{z_1, z_2}.$$

Ce rapport anharmonique étant invariant relativement à toute substitution de la forme (3), l'égalité (4) s'ensuit. Du reste, on arrive facilement à la même égalité en écrivant l'équation du cercle C sous la forme

$$az\bar{z} + \bar{b}z + b\bar{z} + c = 0.$$

Une inégalité.

2. Désignons maintenant par $w = w(z)$ une fonction qui est méromorphe à l'intérieur du domaine circulaire C et qui n'y prend que des valeurs w situées à l'intérieur du domaine circulaire Γ. (Bien entendu, si $w(z)$ admet des pôles à l'intérieur de C, Γ ne pourra être que l'extérieur d'un certain cercle.) Alors on a, en désignant par z_1 et z_2 deux points du domaine C et par $w_1 = w(z_1)$ et $w_2 = w(z_2)$ les valeurs correspondantes,

(5) $$D^\Gamma_{w_1, w_2} \leqq D^C_{z_1, z_2},$$

le signe d'égalité n'ayant lieu que dans le cas où $w(z)$ se réduit à une fonction de la forme (3) *qui donne la représentation conforme du domaine C sur le domaine Γ.*[1]

[1] Pendant la rédaction de ce travail, j'ai observé que l'inégalité (5) se trouve déjà sous une forme un peu moins générale chez J. L. W. V. JENSEN, Undersøgelser over en Klasse fundamentale Uligheder i de analytiske Funktioners Theori. I, Danske Vidensk. Selsk. Skrifter, Naturv. og Math. Afd., 8 Række II. 3 (1916), p. 220. JENSEN, dont le point de vue est tout différent du nôtre, ne donne aucune application de l'inégalité en question.

On peut facilement réduire cette inégalité au lemme de SCHWARZ. On peut en effet trouver une substitution linéaire ζ qui transforme le domaine C dans le cercle $|z| < 1$, désigné par U et, en outre, fait correspondre l'origine au point z_1, le point correspondant à z_2 étant désigné par ζ_2. Il vient alors par la formule (4)

$$D^C_{z_1, z_2} = D^U_{0, \zeta_2} = \frac{|\zeta_2|^2}{1 - |\zeta_2|^2}.$$

En déterminant une substitution linéaire analogue ω pour le domaine Γ qui transforme ce domaine dans l'intérieur du cercle U, les points correspondant à w_1 et à w_2 étant les points 0 et ω_2 respectivement, l'inégalité (5) se réduit à l'inégalité

$$\frac{|\omega_2|^2}{1 - |\omega_2|^2} \leqq \frac{|\zeta_2|^2}{1 - |\zeta_2|^2}$$

qui est équivalente au lemme de SCHWARZ

$$|\omega_2| \leqq |\zeta_2|.$$

Le signe d'égalité n'ayant lieu dans ce dernier lemme que si l'on a $\omega = e^{i\theta}\zeta$, le signe d'égalité ne pourra avoir lieu dans (5) que dans le cas qu'on a précisé.

Sur la relation entre $D^C_{z_1, z_2}$ et l'écart non-euclidien ordinaire.[1]

3. Considérons d'abord la fonction w définie par la formule (3). En passant à la limite, la formule (4) donne

$$\frac{|dw|}{\delta^\Gamma_w} = \frac{|dz|}{\delta^C_z}. \tag{6}$$

Cette relation exprime le *caractère invariant* de l'élément figurant au second membre de (6) par rapport à des substitutions linéaires, propriété bien connue (en général on admet que les domaines C et Γ coincident) et qui fait qu'on définit $\frac{dz}{\delta^C_z}$ comme *l'élément*

[1] Pour les applications on n'a pas besoin des remarques qui se trouvent sous ce numéro.

linéaire de la géométrie non-euclidienne relative au domaine circulaire C.

Pour toute fonction $w(z)$ méromorphe dans le domaine C qui fait correspondre aux points de ce domaine des points intérieurs au domaine Γ, il vient de la formule (5) par un passage à la limite

$$\frac{|dw|}{\delta_w^\Gamma} \leqq \frac{|dz|}{\delta_z^C}. \quad (7)$$ [1]

On voit aussi que le signe d'égalité ne pourra arriver que dans le cas particulier précisé au numéro précédent. En effet, en reprenant les notations de ce numéro, on voit que

$$\left|\frac{d\omega}{d\zeta}\right|_{\zeta=0} = \left|\frac{\omega}{\zeta}\right|_{\zeta=0}$$

et cette dernière valeur est toujours < 1 sauf dans le cas où $\omega = e^{i\theta}\zeta$.

Dans la géométrie non-euclidienne, l'élément linéaire étant défini par $\frac{dz}{\delta_z^C}$, les *géodésiques*, c'est-à-dire les lignes les plus courtes, sont les arcs de cercles orthogonaux au cercle C. [2] L'écart non-euclidien de deux points z_1 et z_2 se calcule facilement comme

$$[z_1, z_2]^C = \log\left(\varkappa + \sqrt{1+\varkappa^2}\right),$$

où l'on a posé $\varkappa^2 = D^C_{z_1, z_2}$.

On a donc

$$D^C_{z_1, z_2} = \sin^2 \text{hyp}\,[z_1, z_2]^C. \text{ [3]}$$

Par conséquent *les deux écarts sont des fonctions croissantes l'un de l'autre* et ils deviennent zéro et infinis en même temps,

---

[1] M. Rolf Nevanlinna, en adoptant toutefois un point de vue un peu moins général ($C \equiv \Gamma$), a déjà signalé comme une des propriétés fondamentales des transformations conformes de faire diminuer l'élément linéaire non-euclidien: Über beschränkte analytische Funktionen, Commentationes in honorem Ernesti Leonardi Lindelöf, Helsinki 1929, p. 26.

[2] *Voir* pour ces questions p. ex. le n° 10 du présent travail.

[3] On connaît le rôle important que joue le sinus hyperbolique de l'écart de deux points dans la trigonométrie de Bolyai-Lobatchefskij; *cf.* p. ex. F. Enriques, Fragen der Elementargeometrie I, 1911, p. 336, § 39.

à savoir zéro lorsque les points coïncident à l'intérieur du domaine et infinis lorsque au moins l'un des points se trouve à la frontière du domaine.

Le théorème de Julia.

4. Le théorème de JULIA [1] consiste en ceci:

Admettons que la fonction $w = w(z)$, *qui est méromorphe à l'intérieur du domaine circulaire C et n'y prend que des valeurs intérieures au domaine circulaire, est encore holomorphe au point fini* z_0 *de la frontière de C et qu'elle fait correspondre à ce point le point fini* w_0 *de la frontière de* Γ. *Alors on a, en posant* $|w'(z_0)| = \gamma$,

$$\frac{|w - w_0|^2}{\delta_w^\Gamma} \leqq \gamma \frac{|z - z_0|^2}{\delta_z^C}. \tag{8}$$

Ce théorème est une conséquence immédiate de l'inégalité (5). Remplaçons en effet dans cette inégalité z_2 par z, w_2 par w et faisons tendre z_1 vers z_0 le long de la normale en z_0. Le point w_1 tendra en même temps vers w_0. D'après (5) on aura

$$\frac{|w - w_1|^2}{\delta_w^\Gamma} \leqq \frac{\delta_{w_1}^\Gamma}{\delta_{z_1}^C} \frac{|z - z_1|^2}{\delta_z^C} = \frac{\delta_{w_1}^\Gamma}{|w_1 - w_0|} \frac{|z_1 - z_0|}{\delta_{z_1}^C} \left| \frac{w_1 - w_0}{z_1 - z_0} \right| \frac{|z - z_1|^2}{\delta_z^C}. \tag{9}$$

Le premier facteur du dernier produit est d'après les formules (1) et (2) à la limite $\leqq 2$, le deuxième facteur devient à la limite $\frac{1}{2}$, grâce à la façon particulière dont z_1 tend vers z_0. Enfin le troisième facteur tend vers $|w'(z_0)| = \gamma$, ce qui achève la démonstration.

En réalité on n'a pas utilisé l'hypothèse que $w(z)$ était holomorphe en z_0. Il suffit de supposer qu'il existe sur la normale en z_0 une suite de points z tels que $\left| \frac{w - w_0}{z - z_0} \right|$ reste borné. En désignant par γ la limite supérieure de ces valeurs, l'inégalité (8) subsiste. Ce n'est pourtant que par le théorème de CARATHÉODORY que la question fut mise en pleine lumière.

[1] G. JULIA, Extension d'un lemme de SCHWARZ, Acta math. 42 (1920).

Le théorème de Carathéodory.

5. Nous démontrons d'abord le théorème suivant de MM. LANDAU et VALIRON [1]:

Soit $F(z)$ *holomorphe et* $\Re(F(z)) \geqq 0$ *pour* $x = \Re(z) \geqq 0$.[2] *Il existe alors une constante* $c \geqq 0$, *telle que en écrivant*

$$G(z) = F(z) - cz,$$

on a $\Re(G(z)) \geqq 0$ *et que, dans tout angle fixe* $|\arg z| \leqq \theta < \frac{\pi}{2}$, *on a uniformément*

$$\lim_{|z| \to \infty} \frac{G(z)}{z} = 0,$$

c'est-à-dire que

$$\lim_{|z| \to \infty} \frac{F(z)}{z} = c. \tag{10}$$

Nous déterminons c avec MM. LANDAU et VALIRON comme la borne inférieure de $\frac{\Re(F(z))}{x}$ pour $x > 0$. Alors on a évidemment $\Re(G(z)) \geqq 0$. La fonction $\Re(G(z))$ étant harmonique, le signe d'égalité sera manifestement exclu en chaque point sauf dans le cas où $G(z) \equiv 0$, c'est-à-dire que $F(z) = cz$. Nous posons dans le cas général $G(z) = w = \xi + i\eta$ et nous choisissons le point particulier $z' = x' + iy'$ de façon que, en posant $G(z') = w' = \xi' + i\eta'$, on ait $\frac{\xi'}{x'} < \varepsilon$. Il vient alors par (2) et (5) quand z tend vers l'infini dans l'angle précisé tout à l'heure

$$\varepsilon > \frac{\xi'}{x'} > \frac{|w - w'|^2}{|z - z'|^2} \frac{x}{\xi} = \frac{|w - w'|}{\xi} \frac{|w - w'|}{|z - z'|} \frac{x}{|z - z'|} >$$

$$> \frac{|w - w'|}{\xi} \frac{|w - w'|}{|z - z'|} (1 - \varepsilon) \cos\theta.$$

---

[1] $\Re(a)$ désigne la partie réelle de a.

[2] E. LANDAU and G. VALIRON, A deduction from SCHWARZ's lemma, Journ. of the Lond. Math. Soc. 4 (1929).

Si $|w| > 2|w'|$, le premier facteur sera $> \frac{1}{2}$. On aura donc dans ce cas

$$\left|\frac{w}{z}\right| < 2\left|\frac{w-w'}{z-z'}\right| < \frac{4\varepsilon}{(1-\varepsilon)\cos\theta}.$$

Si $|w| \leqq 2|w'|$, le premier membre de la dernière relation sera encore très petit quand $|z|$ est assez grand.

Démontrons encore *qu'on a uniformément dans l'angle considéré*

$$\lim_{|z|\to\infty} F'(z) = c,$$

c'est-à-dire que

$$\lim_{|z|\to\infty} G'(z) = 0.$$

Considérons à cet effet l'angle $|\arg z| \leqq \theta' > \theta$. Autour de chaque point z_0 de l'angle $|\arg z| \leqq \theta$ comme centre on peut décrire un cercle de rayon $|z_0|\sin(\theta'-\theta) = k|z_0|$ qui se trouve encore tout entier dans l'angle augmenté. Le théorème précédent étant vrai dans le dernier angle, on aura, pour $|z_0|$ assez grand, sur la frontière du cercle $|G(z)| < \varepsilon|z_0|$. Il en résulte par une inégalité classique

$$|G'(z_0)| < \frac{\varepsilon|z_0|}{k|z_0|} = \frac{\varepsilon}{k}.$$

Par des substitutions linéaires on déduit de ce qui précède immédiatement le théorème suivant dû à M. Carathéodory[1]:

Soient C et Γ des domaines circulaires, $w = w(z)$ une fonction méromorphe dans C qui n'y admet que des valeurs comprises dans Γ, z_0 et w_0 des points des frontières respectives de C et de Γ, φ_0 et ψ_0 les arguments des normales extérieures respectives en z_0 et w_0. Cela étant, si z tend vers z_0 en restant dans un triangle compris dans C qui a un de ses sommets en z_0, la valeur limite

$$e^{-i(\psi_0-\varphi_0)} \lim_{z\to z_0} \frac{w-w_0}{z-z_0} = \gamma \tag{11}$$

existera d'une manière uniforme et γ sera un nombre positif ou

[1] C. Carathéodory, Über die Winkelderivierten von beschränkten analytischen Funktionen, Sitzungsber. d. preuss. Akad. d. Wiss. 32 (1929).

l'infini. Dans le cas où γ est fini,[1] *on aura uniformément, z tendant vers z_0 dans le même domaine,*

$$e^{-i(\psi_0 - \varphi_0)} \lim_{z \to z_0} w'(z) = \gamma .$$

Dans ce cas on a aussi pour toute valeur de z dans C,

$$\frac{|w - w_0|^2}{\delta_w^\Gamma} \leqq \gamma \frac{|z - z_0|^2}{\delta_z^C}, \tag{12}$$ [2]

le signe d'égalité n'ayant lieu pour aucune valeur de z sauf dans le cas où w est une substitution linéaire qui représente le domaine C sur le domaine Γ, faisant correspondre le point w_0 au point z_0.

6. On doit à MM. F. Riesz[3] et Herglotz[4] le théorème suivant:

Soit $w(z) = u + iv$ $(w(0) = \alpha + i\beta)$ une fonction holomorphe pour $|z| < 1$ dont la partie réelle est positive pour les mêmes valeurs de z. Il existe alors une fonction bornée et non décroissante $\mu(\theta)$ telle qu'on ait pour $|z| < 1$

$$w(z) = \frac{1}{2\pi} \int_{-\pi}^{+\pi} \frac{e^{i\theta} + z}{e^{i\theta} - z} \, d\mu(\theta) + i\beta . \tag{13}$$

M. Rolf Nevanlinna[5] a montré que cette représentation

---

[1] Le cas $\gamma = \infty$ correspond au cas $c = 0$ dans le théorème de MM. Landau et Valiron. Remarquons que là, évidemment, il ne fallait pas faire exception pour le cas $c = 0$.

[2] L'énoncé du texte suppose que z_0 et w_0 sont finis. Quand l'un d'eux ou tous les deux sont infinis, on remplacera suivant le cas $z - z_0$ par $\frac{1}{z}$, $w - w_0$ par $\frac{1}{w}$, $\frac{|z - z_0|^2}{\delta_z^C}$ par $\frac{1}{\delta_z^C}$ et $\frac{|w - w_0|^2}{\delta_w^\Gamma}$ par $\frac{1}{\delta_w^\Gamma}$. (*Cf.* pour cela le théorème de Carathéodory sous les formes traitées respectivement par MM. Landau et Valiron et par M. Nevanlinna (n° 6).)

[3] F. Riesz, Sur certains systèmes singuliers d'équations intégrales, Annales de l'Ecole Norm. Sup. (3) 28 (1911).

[4] G. Herglotz, Über Potenzreihen mit positivem reellem Teil im Einheitskreis, Leipz. Ber. 63 (1911).

[5] Rolf Nevanlinna, Remarques sur le lemme de Schwarz, Comptes rendus 188 (15 avril 1929). Dans le même travail, M. Nevanlinna donne

conduit facilement à un théorème équivalent à celui de M. Carathéodory. Voici le raisonnement de M. Nevanlinna:

Démontrons le théorème suivant qui par la considération de la fonction $\frac{1}{w}$ se transforme dans le théorème de M. Carathéodory (avec $w_0 = 0$, C étant le cercle $|z| < 1$ et Γ le demi-plan $\Re(w) > 0$).

Le produit $e^{-i\varphi}(e^{i\varphi} - z)w$ *tend uniformément vers une valeur finie positive ou nulle lorsque le point* z *tend vers* $z = e^{i\varphi}$ *en restant dans le domaine fixe*

$$(14) \qquad |e^{i\varphi} - z| \leqq k(1 - |z|),$$

k étant un nombre positif arbitraire > 1.

En effet on peut écrire

$$(15) \qquad w(z) = c\frac{e^{i\varphi} + z}{e^{i\varphi} - z} + \frac{1}{2\pi}\int_{-\pi}^{+\pi} \frac{e^{i\theta} + z}{e^{i\theta} - z}\, d\nu(\theta) + i\beta,$$ [1]

où $2\pi c = \mu(\varphi + 0) - \mu(\varphi - 0) \geqq 0$ désigne le saut au point φ et $\nu(\theta) = \mu(\theta)$ pour $-\pi \leqq \theta < \varphi$ et $\mu(\theta) - 2\pi c$ pour $\varphi < \theta \leqq +\pi$. En divisant alors la dernière intégrale en trois parties correspondant aux intervalles $(-\pi, \varphi - \theta_0)$, $(\varphi - \theta_0, \varphi + \theta_0)$, $(\varphi + \theta_0, +\pi)$, on voit que la première et la troisième intégrale restent bornées lorsque $z \to e^{i\varphi}$. Pour la deuxième on aura

---

une généralisation du théorème de M. Carathéodory qu'il tire d'un théorème de MM. Ostrowski et Plessner, généralisation du théorème qui se trouve au début de ce numéro.

[1] Pour les fonctions w qui sont holomorphes et de partie réelle positive pour les valeurs de z dont la partie réelle est positive, on déduit facilement la représentation suivante

$$w(z) = cz + \frac{1}{\pi}\int_{-\infty}^{+\infty}\left(\frac{1}{z - it} - \frac{it}{1 + t^2}\right) d\mu(t) + i\beta,$$

c étant une constante $\geqq 0$ et $\mu(t)$ une fonction non décroissante telle que l'intégrale

$$\int_{-\infty}^{+\infty} \frac{d\mu(t)}{t^2 + 1}$$

est convergente. Cette formule fait bien ressortir le terme cz du théorème de MM. Landau et Valiron et, par ce terme, le théorème de Julia.

$$|e^{i\varphi}-z|\left|\int\limits_{\varphi-\theta_0}^{\varphi+\theta_0}\right| < 2\,\frac{|e^{i\varphi}-z|}{1-|z|}\int\limits_{\varphi-\theta_0}^{\varphi+\theta_0} d\nu(\theta) \leqq 2k\lambda_0,$$

λ_0 étant la variation de $\nu(\theta)$ dans l'intervalle $(\varphi-\theta_0,\ \varphi+\theta_0)$. Or $\nu(\theta)$ étant continue pour $\theta=\varphi$, λ_0 s'annule avec θ_0. On a donc uniformément $e^{-i\varphi}(e^{i\varphi}-z)\,w(z)\to 2c$. Le fait que la partie réelle de l'intégrale figurant dans (15) est positive exprime le théorème de Julia.

7. La démonstration de M. Nevanlinna met en évidence les valeurs de φ pour lesquelles

$$\lim_{z\to e^{i\varphi}} (e^{i\varphi}-z)w$$

est différent de zéro. Ces valeurs forment manifestement un ensemble dénombrable. En passant à la fonction $\frac{1}{w}$ qu'on désignera encore par w, ce seront les valeurs pour lesquelles

$$\frac{w(z)}{e^{i\varphi}-z}$$

reste borné. Or ce n'est pas là la seule question dont il s'agit. Dans les hypothèses où l'on se trouve (la partie réelle de w est positive) il s'agit de savoir pour quelles valeurs de φ il existe une valeur réelle de A de sorte que

$$\frac{w(z)-iA}{e^{i\varphi}-z}$$

reste borné. Nous allons voir qu'on peut caractériser ces valeurs de φ par les propriétés d'une seule fonction décroissante, celle qui figure dans la représentation (13) de $w(z)$. (La méthode précédente serait à appliquer, pour toute valeur particulière de A, à la représentation (13) de $(w(z)-iA)^{-1}$.)

Soit donc $w(z)$ la fonction caractérisée au début du n° 6 et représentée par la formule (13). On a alors le théorème suivant:

L'intégrale

$$\gamma = \frac{1}{4\pi}\int_{-\pi}^{+\pi} \frac{d\mu(\theta)}{\sin^2 \frac{\theta - \varphi}{2}} \tag{16}$$ [1]

est évidemment un nombre positif fini ou infini. Dans le cas où γ *est fini, on aura uniformément, lorsque* z *tend vers* $e^{i\varphi}$ *dans le domaine* (14),

$$e^{i\varphi}\frac{w(z) - iA}{e^{i\varphi} - z} \to \gamma, \tag{17}$$

si l'on pose

$$A = -\frac{1}{2\pi}\int_{-\pi}^{+\pi} \cot\frac{\theta - \varphi}{2}\, d\mu(\theta) + \beta, \tag{18}$$

la convergence de la dernière intégrale étant une conséquence de la convergence de l'intégrale de la formule (16).

Cela étant, il est clair que pour toute autre valeur de A *le rapport* (17) *tendra uniformément vers l'infini. Il en sera de même pour toute valeur réelle de* A *lorsque* γ *est infini.*

Dans la démonstration on pourra admettre que $\varphi = 0$. En supposant alors que γ est fini et en définissant A par la formule (18), en y posant encore $\varphi = 0$, on aura par des calculs simples

$$w(z) - iA = \frac{1}{2\pi}\int_{-\pi}^{+\pi}\left[\frac{e^{i\theta} + z}{e^{i\theta} - z} + i\cot\frac{\theta}{2}\right] d\mu(\theta) =$$

$$= \frac{1 - z}{2\pi}\int_{-\pi}^{+\pi} \frac{ie^{\frac{i\theta}{2}}}{\sin\frac{\theta}{2}\left(e^{i\theta} - z\right)}\, d\mu(\theta).$$

[1] Pour les fonctions considérées dans la note précédente, l'intégrale

$$\frac{1}{\pi}\int_{-\infty}^{+\infty} \frac{d\mu(t)}{t^2}$$

jouera un rôle analogue à celui de l'intégrale (16).

Or la dernière intégrale tend uniformément vers

$$\int_{-\pi}^{+\pi} \frac{d\mu(\theta)}{2\sin^2\frac{\theta}{2}}$$

lorsque z tend vers 1 dans le domaine (14) correspondant à $e^{i\varphi}=1$, ce qui démontre la première partie du théorème. Pour aller plus loin, écrivons $z=re^{i\psi}$. Il vient

$$\frac{e^{i\theta}+z}{e^{i\theta}-z}=\frac{(e^{i\theta}+z)(e^{-i\theta}-\bar{z})}{|e^{i\theta}-z|^2}=\frac{(1-r^2)-2ir\sin(\theta-\psi)}{|e^{i\theta}-z|^2}.$$

Par conséquent d'après (13)

$$\frac{|w(re^{i\psi})-iA|}{1-r}>\frac{1}{2\pi(1-r)}\int_{-\pi}^{+\pi}\frac{1-r^2}{|e^{i\theta}-z|^2}d\mu(\theta)=\frac{1+r}{2\pi}\int_{-\pi}^{+\pi}\frac{d\mu(\theta)}{|e^{i\theta}-z|^2}.$$

Or si la valeur γ définie par (16) est infinie pour $\varphi=0$, la dernière intégrale tend vers l'infini lorsque $r\to 1$ et $\psi\to 0$ d'une manière quelconque. En effet

$$\int_{-\pi}^{+\pi}>\int_{-\pi}^{-\varepsilon}+\int_{\varepsilon}^{+\pi}\to\left(\int_{-\pi}^{-\varepsilon}+\int_{\varepsilon}^{+\pi}\right)\frac{d\mu(\theta)}{4\sin^2\frac{\theta}{2}}.$$

On aura donc aussi uniformément,

$$\left|\frac{w(z)-iA}{1-z}\right|\to\infty,$$

si z tend vers 1 dans le domaine (14), où l'on pose encore $e^{i\varphi}=1$.

8. Voici maintenant quelques applications des derniers résultats. Dans un travail concernant les groupes fuchsiens, M. Myrberg[1] s'est occupé de fonctions $f(z)$ holomorphes et telles que $|f(z)|<1$ pour $|z|<1$ et de plus telles que $|f(z)|\to 1$ *presque*

[1] P. J. Myrberg, Ein Satz über die Fuchsschen Gruppen und seine Anwendung in der Funktionentheorie, Commentationes in honorem Ernesti Leonardi Lindelöf, Helsinki 1929; *voir* en particulier p. 28.

partout lorsque z tend vers $e^{i\varphi}$ en suivant le rayon qui aboutit en ce point. Il est clair que ces fonctions se réduisent immédiatement aux fonctions $w(z) = u + iv$ telles que pour $|z| < 1$ on ait $u > 0$ et $u \to 0$ presque partout lorsque z tend vers $e^{i\varphi}$ suivant ledit rayon. Il est très facile de donner la condition nécessaire et suffisante pour qu'il en soit ainsi. En effet, la fonction $\mu(\theta)$ étant monotone, elle possède d'après un théorème connu de M. Lebesgue presque partout une dérivée $\mu'(\theta)$, et l'on sait, d'après un théorème de Fatou, que $u(re^{i\theta}) \to \mu'(\theta)$ presque partout lorsque $r \to 1$. Les fonctions satisfaisant à la condition de M. Myrberg sont donc caractérisées par le fait que la relation $\mu'(\theta) = 0$ doit avoir lieu presque partout. Or de telles fonctions $\mu(\theta)$ peuvent s'écrire sous la forme

$$\mu(\theta) = \alpha(\theta) + \beta(\theta)$$

où $\alpha(\theta)$ est une fonction discontinue non décroissante, dite *fonction des sauts,* et $\beta(\theta)$ une fonction continue non décroissante dont la dérivée est nulle presque partout, dite *fonction singulière.*[1]

M. Myrberg nous a posé la question si l'on pouvait affirmer que, pour les fonctions $w(z)$ considérées, *la valeur limite de* M. Carathéodory *était finie presque partout ou non.* Or, en s'appuyant sur le critère donné plus haut, il est très facile de construire des fonctions satisfaisant à la condition de M. Myrberg et telles que *la valeur limite de* M. Carathéodory *devient infinie pour toute valeur de φ et toute valeur réelle de A.* A cet effet il est nécessaire et suffisant que l'intégrale (16) devienne infinie pour toute valeur de φ. Or il est facile de construire une fonction de sauts $\mu(\theta)$ qui ait la propriété désirée. En désignant les valeurs de θ où les sauts ont lieu par θ_k et les sauts correspondants par ρ_k, l'intégrale ci-dessus prendra la forme

$$h(\varphi) = \frac{1}{4\pi} \sum \frac{\rho_k}{\sin^2 \dfrac{\theta_k - \varphi}{2}}. \tag{19}$$

Les ρ_k ne sont assujettis qu'aux conditions d'être positifs et tels que $\sum \rho_k$ soit convergent. Or nous allons montrer que, si $\sum \sqrt{\rho_k}$ est divergent, on peut toujours choisir les valeurs θ_k de façon

[1] *Cf.* p. ex. C. de la Vallée Poussin, Les fonctions à variation bornée et les questions qui s'y rattachent, Bull. des Sciences Math., 44 (1920).

que l'expression (19) soit infinie pour *toute* valeur de φ. Posons en effet

$$\theta_k \equiv \sum_{j=0}^{k} \sqrt{\rho_j} \pmod{2\pi}.$$

(Si certaines des valeurs θ_k coincident, le saut correspondant sera simplement la somme des ρ_k correspondants). La série formée des racines carrées étant divergente en même temps que les termes de la série tendent vers zéro (puisque $\Sigma\, \rho_k$ est convergent), toute valeur de φ sera intérieure à une *infinité* d'arcs (θ_{k-1}, θ_k). Les termes de (19) qui correspondent à de tels θ_k seront à un facteur constant près

$$> \frac{\rho_k}{\sin^2 \frac{\sqrt{\rho_k}}{2}} > 4.$$

Puisqu'il y a une infinité de tels termes, la série sera divergente.

Un complément intéressant à ce résultat est le théorème suivant.

La distribution des θ_k étant quelconque, chaque fois que la série $\Sigma \sqrt{\rho_k}$ est convergente, la série (19) *sera convergente presque partout.*

Posons à cet effet

$$\Sigma \sqrt{\rho_k} = \alpha,$$

désignons par ε un nombre positif très petit et excluons autour de chaque point θ_k un arc de longueur $2\varepsilon \sqrt{\rho_k}$ symétrique par rapport à θ_k. L'ensemble E qui reste sera de mesure $\geqq 2\pi - 2\varepsilon\alpha$. On aura, en désignant encore par C_k l'arc qui reste après l'exclusion relative à θ_k,

$$4\pi \int_E h(\varphi)\, d\varphi = \sum \rho_k \int_E \frac{d\varphi}{\sin^2 \frac{\theta_k - \varphi}{2}} \leqq \sum \rho_k \int_{C_k} \frac{d\varphi}{\sin^2 \frac{\theta_k - \varphi}{2}} =$$

$$= 4 \sum \rho_k \cot \frac{\varepsilon \sqrt{\rho_k}}{2} < \frac{8}{\varepsilon} \sum \sqrt{\rho_k} = \frac{8\alpha}{\varepsilon},$$

c'est-à-dire que l'intégrale est finie. Par conséquent $h(\varphi)$ est fini presque partout dans E et alors, ε étant arbitrairement petit, presque partout dans $(-\pi, +\pi)$.

2

En remplaçant les sauts par des fonctions singulières qui sont constantes sauf dans de très petits intervalles autour des θ_k, on voit facilement que les deux possibilités qu'on vient de considérer pourront aussi se présenter pour ces fonctions-là. Ou d'une façon plus précise:

Il existe des fonctions singulières bornées non décroissantes $\mu(\theta)$ telles que l'intégrale (16) est infinie pour toute valeur de φ.

Il existe des fonctions singulières bornées non décroissantes $\mu(\theta)$, qui ne sont constantes dans aucun intervalle,[1] telles que l'intégrale (16) est finie pour presque toutes les valeurs de φ.

9. Il est aussi facile de construire une fonction $w(z) = u + iv$ qui a les propriétés suivantes, analogues aux propriétés d'une fonction que M. Carathéodory considère à la fin de son travail.

$w(z)$ est holomorphe et $u > 0$ pour $|z| < 1$, $\frac{w(z)}{1-z}$ tend vers un nombre positif fini si z tend vers 1 dans un domaine arbitraire de la forme (14) (avec $e^{i\varphi} = 1$), et cela uniformément pour toute valeur fixe de k, quoique $|w(z)|$ prenne des valeurs arbitrairement grandes dans un voisinage arbitraire de $z = 1$.

Il suffit pour cela de choisir une fonction de sauts construite avec des valeurs θ_k tendant vers l'origine, les sauts ρ_k appartenant aux deux valeurs $\pm\theta_k$ étant égaux entre eux et choisis de façon que $\sum \frac{\rho_k}{\theta_k^2}$ soit convergent.

Il est d'ailleurs facile de former une fonction positive paire $g(\theta)$, définie dans $(-\pi, +\pi)$, continue en tout point sauf à l'origine, telle que $\limsup_{\theta \to 0} g(\theta) = \infty$, en même temps que l'intégrale

$$\int_{-\pi}^{+\pi} \frac{g(\theta)}{\theta^2}\, d\theta$$

soit convergente. En posant alors

$$\mu(\theta) = \int_{-\pi}^{\theta} g(\theta)\, d\theta,$$

[1] Evidemment, les fonctions figurant dans l'énoncé qui précède ne sont constantes non plus dans aucun intervalle.

la formule (13) définit une fonction $w(z)$ qui a les propriétés désirées.

Quelques remarques sur les géométries non-euclidiennes.

10. Soit donné un domaine simplement connexe W et supposons d'abord que ce domaine ne se recouvre pas lui-même. Effectuons une transformation conforme $z = z(w)$ du domaine sur le cercle $|z| < 1$ de façon qu'un certain point w_0 tombe au point $z = 0$. La fonction $z(w)$ sera déterminée à un facteur $e^{i\theta}$ près. A l'élément linéaire euclidien ds au point particulier w_0, il correspondra par la représentation conforme un élément linéaire $d\sigma$. Cela posé, nous définirons l'élément linéaire non euclidien $ds_{w_0}^W$ relatif au domaine W par

(20) $$ds_{w_0}^W = d\sigma = |z'(w_0)|\, ds.$$ [1]

Pour l'élément d'aire on procède d'une façon analogue. Enfin, puisqu'il s'agira d'une invariance par rapport à des transformations conformes, l'angle non-euclidien est défini comme identique à l'angle euclidien. La longueur non-euclidienne d'une courbe L sera mesurée par $\int_L ds^W$ et l'on posera une définition analogue pour le calcul des aires. Les lignes géodésiques ou droites non euclidiennes (dont on démontrera l'existence tout à l'heure [2]) sont des courbes telles que l'intégrale ci-dessus prise entre deux points quelconques d'une telle courbe est moindre que la même intégrale prise le long de toute autre courbe reliant les mêmes points. On verra qu'il y a toujours une et une seule ligne droite qui relie deux points donnés et qu'il y a toujours par un point non situé sur la droite deux parallèles à une ligne droite donnée, c'est-à-dire qu'il y a deux droites passant par le point qui, dans un sens

[1] On pourrait, en adoptant le point de vue habituelle de la géométrie non-euclidienne, poser $\frac{d\sigma}{\varkappa}$ au lieu de $d\sigma$, $\varkappa$ désignant une constante absolue, mais cela n'aurait aucun intérêt pour ce qui suit.

[2] Comme nous l'avons dit dans l'Introduction, nous ne nous appuyons pas dans ces considérations sur les faits connus concernant la géométrie non-euclidienne du cercle.

facile à préciser, rencontrent la droite donnée à l'infini, savoir à la frontière du domaine W.

On aura ainsi construit une géométrie non-euclidienne hyperbolique relative à un domaine simplement connexe quelconque. Cette géométrie est, par sa définition même, invariante par rapport à des transformations conformes quelconques.

Il reste à établir l'existence et les propriétés énoncées des droites Par ce qu'on vient de dire, il suffira de le faire pour le cercle $|z|<1$, et on pourra le faire presque sans calcul.

Démontrons en effet d'abord que tout segment d'un rayon quelconque du cercle a la propriété géodésique. En effet il est clair par symétrie que, en posant $|z|=r$, l'élément linéaire non-euclidien aura la forme $f(r)\,ds$, où $f(r)$ est une fonction positive et ds désigne l'élément linéaire euclidien. Cela étant, désignons par z_1 et z_2 deux points d'un même rayon et supposons p. ex. que $|z_1|<|z_2|<1$. Alors il est évident que l'intégrale $\int_{z_1}^{z_2} f(r)\,ds$ prise le long du rayon est moindre que la même intégrale prise le long de toute autre courbe reliant les deux points.

Or on se rappelle que toute substitution linéaire change les cercles en des cercles, conserve les angles, et qu'on peut, par une substitution linéaire qui change le cercle $|z|<1$ en lui-même, transporter deux points intérieurs à ce cercle en deux points tels qu'on vient de considérer. Il suffira en effet de transporter l'un des points au centre du cercle. On retrouve alors le fait connu que les droites de la géométrie dudit cercle sont les arcs de cercles coupant $|z|=1$ sous un angle droit. Ces arcs de cercles possèdent évidemment les propriétés énoncées au sujet des droites non-euclidiennes.

11. Soient maintenant W_1 et W_2 deux domaines simplement connexes non identiques, le second renfermant le premier, et w_0 un point intérieur à tous les deux. En désignant par $z_1(w)$ et $z_2(w)$ les fonctions qui donnent les représentations conformes respectives des deux domaines sur le cercle $|z|<1$ en transportant w_0 à l'origine, on trouve

$$|z_1'(w_0)|>|z_2'(w_0)|, \tag{21}$$

c'est-à-dire que

(22) $$ds^{W_1} > ds^{W_2}.$$

Cela veut dire que l'élément linéaire non-euclidien par rapport au domaine W_1 est plus grand que l'élément correspondant au domaine W_2 qui renferme W_1.

L'inégalité (21) est en effet une conséquence immédiate du lemme de SCHWARZ, si l'on considère z_2 comme fonction de z_1.

Cela étant, soient w' et w'' deux points de W_1 et L la ligne droite par rapport à W_1 qui les relie. L'écart non-euclidien des deux points par rapport à W_1 est donné par l'intégrale

(23) $$[w', w'']^{W_1} = \int_L ds^{W_1} > \int_L ds^{W_2} \geqq [w', w'']^{W_2}.$$

C'est-à-dire que *l'écart non-euclidien de deux points situés dans W_1 formé par rapport à ce domaine est plus grand que l'écart formé par rapport au domaine W_2 qui renferme W_1*.[1]

Ce résultat n'est au fond que le principe de LINDELÖF.[2]

12. Envisageons maintenant de plus près l'élément linéaire ds^W. On a déjà utilisé l'expression

(20) $$ds^W_{w_0} = |z'(w_0)|\, ds.$$

La fonction

$$\frac{z(w)}{|z'(w_0)|}$$

dont la dérivée au point w_0 a évidemment la valeur absolue 1, donne la représentation conforme du domaine W sur un cercle de rayon

(24) $$R^W_{w_0} = \frac{1}{|z'(w_0)|}.$$

---

[1] En considérant l'angle entre deux droites non-euclidiennes et en admettant que les frontières de W_1 et de W_2 ont un arc ou un élément d'arc en commun, on peut faire rentrer le théorème de LÖWNER dans l'ordre d'idées du texte; nous n'y insistons pas. (*Cf.* le Mémoire cité de M. NEVANLINNA, p. 25).

[2] E. LINDELÖF, Mémoire sur certaines inégalités dans la théorie des fonctions, etc., Acta Soc. Fenn. 35 (1908). *Voir* aussi L. BIEBERBACH, Lehrbuch der Funktionentheorie II, 1931, p. 106. En énonçant son principe, M. LINDELÖF admet des domaines plus généraux que ceux du texte. (*Cf.* le n° 13 du présent travail.)

On pourra donc écrire

$$ds_{w_0}^{W} = \frac{ds}{R_{w_0}^{W}}. \tag{25}$$

Si le domaine W est un cercle C de rayon R, l'écart euclidien de w_0 au centre étant r, on trouve en conformité avec la forme habituelle de l'élément linéaire non-euclidien

$$R_{w_0}^{C} = \frac{R^2 - r^2}{R}, \tag{26}$$

expression qui, avec la notation (1), pourrait être désignée par $\delta_{w_0}^{C}$. Quand w_0 se trouve au centre du cercle, cette expression se réduit à R.

Quand le domaine W est un demi-plan, l'expression (24) deviendra $2d$, où d est la distance du point à la droite-frontière du demi-plan. On retrouve donc aussi dans ce cas-là l'expression $\delta_{w_0}^{C}$ de la formule (2).

Cela étant, en désignant respectivement par $\overline{R}_{w_0}^{W}$ et $\underline{R}_{w_0}^{W}$ la distance euclidienne maximum et minimum entre w_0 et la frontière de W, il vient par les dernières remarques et par (21) et (24)

$$\overline{R}_{w_0}^{W} \geqq R_{w_0}^{W} \geqq \underline{R}_{w_0}^{W},$$

le signe d'égalité n'ayant lieu que dans des cas évidents.

Pour les domaines *convexes* il vient encore

$$R_{w_0}^{W} \leqq 2\, \underline{R}_{w_0}^{W},$$

le signe d'égalité n'ayant lieu que lorsque W est un demi-plan.

13. On a supposé jusqu'ici que les domaines considérés ne se recouvraient pas eux-mêmes. Il est à peu près évident, et nous y avons déjà fait allusion, que cette restriction n'est nullement nécessaire. La seule hypothèse essentielle c'était que les domaines admettent chacun une représentation conforme sur un domaine circulaire C. Ainsi p. ex. $w(z)$ étant une fonction méromorphe dans un tel domaine C, cette fonction donne la représentation conforme du domaine C sur un domaine simplement connexe W qui pourra se recouvrir plusieurs ou même une infinité de fois. En désignant alors pour plus de netteté les éléments linéaires relatifs à C et à W par $|dz|^C$ et $|dw|^W$ respectivement, on aura par définition

$$|dw|^W = |dz|^C. \tag{29}$$

Admettons maintenant que W soit compris dans un autre domaine W' du même caractère que W. On montre alors comme plus haut que

$$|dw|^{W'} < |dw|^W,$$

c'est-à-dire que

$$|dw|^{W'} < |dz|^C.$$

Si, en particulier, W' est un domaine circulaire Γ, cela se confond avec l'inégalité (6).

14. Parmi d'autres groupes de transformations que celui des transformations conformes signalons en particulier le groupe des transformations projectives. Dans un espace à un nombre de dimensions quelconque, on définira au centre d'un cercle ou d'une sphère fixés les éléments d'arc, de surface, de volume etc. par les éléments euclidiens respectifs. On en fera de même pour les angles. Pour d'autres points du cercle ou de la sphère, ou pour les points de domaines limités par des coniques ou des quadriques quelconques, on définira les éléments respectifs au moyen de leur invariance par rapport aux transformations projectives. On retrouve ainsi la géométrie hyperbolique Cayleyenne.

Des considérations analogues s'appliquent aussi aux géométries elliptiques, conformes ou Cayleyennes.

35.

Sur les ensembles compacts de fonctions sommables*

Acta Szeged Sect. Math. *6* (1933), 136–142

1. Soit R un espace euclidien à un nombre fini de dimensions. En désignant par x un point quelconque de cet espace, on entend par dx l'élément de volume correspondant. Par $x+h$ on désigne l'effet du déplacement h au point x. La valeur absolue $|h|$ du déplacement sera mesurée au sens euclidien.

Nous désignons, pour $p \geqq 1$, par L^p la classe des fonctions $f(x)$ définies et mesurables dans l'espace R et telles que l'intégrale de $|f(x)|^p$, étendue à tout l'espace, existe. Nous posons pour une telle fonction

$$(1) \qquad \|f\| = \|f(x)\| = \left(\int_R |f(x)|^p \, dx\right)^{\frac{1}{p}}$$

et encore, en désignant par E un sous-ensemble mesurable quelconque de l'espace R,

$$(2) \qquad \|f\|_E = \|f(x)\|_E = \left(\int_E |f(x)|^p \, dx\right)^{\frac{1}{p}}$$

Nous aurons souvent à nous servir de l'inégalité de MINKOWSKI

$$(3) \qquad \|f+g\| \leqq \|f\| + \|g\|,$$

qui met en évidence que (1) définit un espace métrique au sens de M. FRÉCHET, la dénomination étant celle de M. HAUSDORFF.

Si $f(x)$ satisfait à l'inégalité

$$(4) \qquad \|f\| \leqq M,$$

nous dirons que M est une *borne* de $f(x)$.

*) Conférence faite au Séminaire mathématique de l'Université de Lund le 21 février 1933.

Cela étant, nous pouvons énoncer le théorème suivant dû à M. LEBESGUE.

Pour toute fonction $f(x)$ de la classe L^p, on a

$$(5)\quad \lim_{|h|\to 0} \|f(x+h)-f(x)\| = \lim_{|h|\to 0} \left(\int_R |f(x+h)-f(x)|^p\,dx\right)^{\frac{1}{p}} = 0.$$

M. LEBESGUE n'a donné ce théorème que pour $p=1$, mais la démonstration s'étend immédiatement au cas général. Pour préciser le sens de (5), nous l'écrirons encore sous la forme

$$(5')\qquad \|f(x+h)-f(x)\| \leqq \varepsilon, \qquad \text{pour } |h| \leqq \delta.$$

Ici ε et δ sont des nombres positifs, le premier arbitraire, le second suffisamment petit.

En désignant encore par E_A l'ensemble des points de R dont a distance à un point fixe de R est $>A$, on a évidemment

$$(6)\qquad \lim_{A\to\infty} \|f\|_{E_A} = 0,$$

ou

$$(6')\qquad \|f\|_{E_A} \leqq \varepsilon, \qquad \text{pour } A \text{ assez grand.}$$

Nous dirons avec M. FRÉCHET qu'un ensemble de fonctions $\{f(x)\}$ appartenant à la classe L^p est *compact* au sens L^p, si, de tout sous-ensemble infini, on peut tirer une suite qui tend vers une fonction limite au sens L^p. Cette définition implique que tout ensemble de fonctions qui ne contient qu'un nombre fini d'éléments, est compact.

Le but principal de cette Note est de démontrer le théorème suivant.

Pour qu'un ensemble $\{f(x)\}$ de fonctions appartenant à la classe L^p soit compact au sens L^p, il faut et il suffit qu'il existe une borne commune M telle que (4) *tienne, et que les relations* (5) *et* (6) *aient lieu uniformément pour toutes les fonctions de l'ensemble $\{f(x)\}$.*

On pourra aussi exprimer ces conditions en disant que *les fonctions $f(x)$ ont à être également bornées et également continues dans l'espace R et encore également infiniment petites dans la partie infiniment éloignée de cet espace, tout cela étant entendu au sens L^p.*

En admettant un instant que nos fonctions s'annulent à l'extérieur d'un certain ensemble borné et fermé, la condition (6) disparaît et l'on retrouve, pour $p\to\infty$, comme cas limite les conditions d'ARZELÀ pour qu'un système de fonctions continues dans un tel ensemble y soit compact dans le sens de la convergence uniforme, à savoir que les fonctions soient également bornées et également continues au sens ordinaire.

2. *Pour* $p > 1$, on possède déjà un critère pour la compacticité au sens L^p dû à M. KOLMOGOROFF[1]) et à M. TAMARKIN.[2]) M. KOLMOGOROFF s'est restreint à des fonctions qui s'annulent à l'extérieur d'un ensemble borné. Il trouve comme conditions nécessaires et suffisantes que les fonctions $\{f(x)\}$ soient également bornées et que ces fonctions puissent être approchées uniformément par certaines valeurs moyennes qu'on retrouvera plus loin au n° **4** (formule (8)). Là on retrouve aussi la condition de M. KOLMOGOROFF sous forme explicite (inégalité (9)). M. TAMARKIN a étendu le résultat de M. KOLMOGOROFF en supprimant la condition que les fonctions $\{f(x)\}$ s'annulent à l'extérieur d'un ensemble borné. Il ajouta en échange aux conditions nécessaires et suffisantes de M. KOLMOGOROFF celle que nous avons exprimée en disant que les fonctions $\{f(x)\}$ ont à être également infiniment petites dans la partie infiniment éloignée de l'espace R.

En résumé, *pour* $p > 1$, deux de nos conditions se trouvent déjà chez les auteurs cités, tandis que celle exigeant que nos fonctions soient également continues au sens L^p, remplace la condition la plus marquée de M. KOLMOGOROFF se rattachant aux valeurs moyennes (8). Notre condition permet de résoudre la question aussi dans le cas $p = 1$.

3. Avant d'aborder la démonstration de notre théorème, nous donnons encore une condition alternative qui découle immédiatement d'un théorème très général et très important de M. HAUSDORFF sur les espaces métriques. M. HAUSDORFF démontre[3]) qu'un ensemble d'éléments d'un espace métrique (complet) est compact en même temps qu'il est totalement borné. Appliqué au cas particulier qui nous occupe, le théorème de M. HAUSDORFF revient à ceci:

---

[1]) A. KOLMOGOROFF, Über die Kompaktheit der Funktionenmengen bei der Konvergenz im Mittel, *Nachrichten Ges. Wiss. Göttingen,* 1931, p. 60—63.

[2]) J. D. TAMARKIN, On the compactness of the space L_p, *Bulletin American Math. Soc.,* **38** (1932), p. 79—84.

[3]) F. HAUSDORFF, *(Grundzüge der) Mengenlehre,* 1ère éd. 1914, p. 311 et *suiv.,* 2ième éd. 1927, p. 107 et *suiv.*

Pour que l'ensemble $\{f(x)\}$ *de la classe* L^p *soit compact au sens* L^p *il faut et il suffit que, pour tout nombre positif* ε, *il existe un nombre f i n i de fonctions* $g_1(x), g_2(x), \ldots, g_m(x)$ *appartenant elles-mêmes à la classe* L^p *de sorte que, pour chaque fonction* $f(x)$, *l'une au moins des inégalités*

$$(7) \qquad \|f(x)-g_k(x)\| \leqq \varepsilon \qquad (k=1, 2, \ldots, m)$$

soit remplie.

Il est pour ainsi dire évident que les fonctions $g_k(x)$ pourront être spécialisées de beaucoup de manières. Ainsi p. ex. on pourra les choisir parmi les fonctions $\{f(x)\}$, ou encore on pourra prescrire qu'elles soient continues ou qu'elles n'admettent qu'un nombre fini de valeurs etc.

Le critère de M. Hausdorff admet bien des applications directes. Ainsi p. ex., on voit sur le champ que tout système de fonctions également bornées et à variation également bornée sur un segment fini, y est compact L^p pour toute valeur finie de p.

4. Montrons maintenant comment on peut déduire le théorème énoncé au n° **1** du théorème de M. Hausdorff. Il s'ensuit immédiatement de ce dernier théorème et des remarques faites au n° **1** que nos conditions sont nécessaires. En effet, ces conditions étant remplies pour chacune des fonctions $g_1(x), g_2(x), \ldots, g_m(x)$, on n'aura qu'à prendre les bornes les moins favorables et à appliquer l'inégalité (3).

Il reste à démontrer que nos conditions sont suffisantes. Le nombre de dimensions de notre espace étant n, désignons avec M. Kolmogoroff la sphère à n dimensions de centre x et de rayon δ par $S(x, \delta)$ et formons avec lui la valeur moyenne

$$(8) \qquad f_\delta(x) = \frac{1}{V(\delta)} \int\limits_{S(x,\delta)} f(y)\, dy,$$

$V(\delta)$ désignant le volume de la sphère.

Nous allons démontrer que si les conditions (5′) et (6′) sont remplies par la fonction $f(x)$, on a aussi

$$(9) \qquad \|f(x)-f_\delta(x)\| \leqq \varepsilon$$

et

$$(10) \qquad \|f_\delta(x)\|_{E_A} \leqq 2\varepsilon.$$

On a encore, pour *tout point* x, les deux inégalités, la dernière étant valable pour h quelconque,

$$|f_\delta(x)| \leqq (V(\delta))^{-\frac{1}{p}} \|f\| \tag{11}$$

et

$$|f_\delta(x+h) - f_\delta(x)| \leqq (V(\delta))^{-\frac{1}{p}} \|f(x+h) - f(x)\|. \tag{12}$$

Ces deux dernières relations disent que nos conditions (4) et (5) étant également remplies par les fonctions $\{f(x)\}$, les fonctions $\{f_\delta(x)\}$ seront *également bornées et également continues au sens ordinaire, pour toute valeur fixe de* δ. Les mêmes conclusions découlent, pour $p > 1$, aussi des hypothèses de M. Kolmogoroff et jouent un rôle important dans sa démonstration.

Admettons un instant que les inégalités (9)—(12) soient déjà établies. Alors il est facile de trouver des fonctions $g_k(x)$ qui pourront figurer au théorème de M. Hausdorff et de réduire ainsi notre théorème à celui-là.

En effet, on peut d'abord, en vertu de (9), choisir δ de façon que les valeurs $\|f(x) - f_\delta(x)\|$ soient *également*[4]) très petites. Ensuite, grâce à (10), on peut déterminer un domaine rectangulaire D assez grand pour que les fonctions $\{f'_\delta(x)\}$ identiques aux fonctions $\{f_\delta(x)\}$ à l'intérieur de D et s'annulant à son extérieur, rendent les valeurs $\|f_\delta(x) - f'_\delta(x)\|$ très petites. Puisque les fonctions $\{f'_\delta(x)\}$ sont également bornées, on pourra trouver des valeurs $\alpha_1, \alpha_2, \ldots, \alpha_r$, en nombre fini, de manière que chaque valeur particulière $f'_\delta(x)$ soit très voisine de l'une au moins de ces valeurs α_j. D'autre part, les fonctions $\{f'_\delta(x)\}$ étant aussi également continues, on peut diviser D en un nombre fini s de domaines rectangulaires partiels de sorte que l'oscillation d'une fonction quelconque $f'_\delta(x)$ dans un tel domaine soit très petite. Toutes les valeurs qu'une certaine fonction $f'_\delta(x)$ admet dans un tel domaine partiel seront alors très voisines de l'une au moins des valeurs α_j que nous venons d'introduire. Cela étant, nous définissons un système de r^s fonctions $g_k(x)$ ($k = 1, 2, \ldots, r^s$) s'annulant toutes à l'extérieur de D et admettant dans chaque domaine partiel une valeur constante, égale à l'une des valeurs α_j. Il ressort

[4]) Dans la suite, nous ne répétons pas ce mot à chaque occasion.

de la construction ci-dessus que toute fonction $f'_\delta(x)$ sera très voisine — au sens ordinaire — de l'une au moins des fonctions $g_k(x)$. Les deux fonctions étant identiques à l'extérieur de D, une approximation assez bonne, au sens ordinaire, rendra la valeur $\|f'_\delta(x) - g_k(x)\|$ aussi petite qu'on voudra. Les valeurs $\|f(x) - f_\delta(x)\|$, $\|f_\delta(x) - f'_\delta(x)\|$ et $\|f'_\delta(x) - g_k(x)\|$ pouvant être rendues arbitrairement petites, il en sera de même, grâce à l'inégalité fondamentale (3), de la valeur $\|f(x) - g_k(x)\|$.

5. Tout ce qui reste, c'est à démontrer les inégalités (9)—(12). Nous partons, pour y arriver, d'une inégalité bien connue, qui est une conséquence immédiate de l'inégalité de Hölder. On a pour toute fonction $l(x)$ de la classe L^p

$$\left|\frac{1}{V(\delta)}\int\limits_S l(y)\,dy\right| \leqq \left(\frac{1}{V(\delta)}\int\limits_S |l(y)|^p\,dy\right)^{\frac{1}{p}}, \tag{14}$$

où l'on a posè, pour abréger, S au lieu de $S(x, \delta)$.

Posons ici d'abord $l(y) = f(y)$. Il vient d'après (8)

$$|f_\delta(x)| \leqq (V(\delta))^{-\frac{1}{p}} \|f\|_S \leqq (V(\delta))^{-\frac{1}{p}} \|f\|,$$

c'est-à-dire l'inégalité (11).

En posant maintenant dans (14) $l(y) = f(y+h) - f(y)$, nous obtenons, par un calcul analogue,

$$|f_\delta(x+h) - f_\delta(x)| \leqq (V(\delta))^{-\frac{1}{p}} \|f(x+h) - f(x)\|,$$

c'est-à-dire l'inégalité (12).

En posant encore $l(y) = f(x) - f(y)$, il vient d'abord

$$|f(x) - f_\delta(x)| \leqq \left(\frac{1}{V(\delta)}\int\limits_S |f(x) - f(y)|^p\,dy\right)^{\frac{1}{p}}$$

ou

$$|f(x) - f_\delta(x)|^p \leqq \frac{1}{V(\delta)}\int\limits_{S(x,\,\delta)} |f(x) - f(y)|^p\,dy =$$

$$= \frac{1}{V(\delta)}\int\limits_{S(0,\,\delta)} |f(x) - f(x+h)|^p\,dh.$$

En intégrant cela par rapport à x, l'intégrale étant étendue à l'espace entier R, il résulte

$$\int_R |f(x)-f_\delta(x)|^p\,dx \leqq \frac{1}{V(\delta)} \int_{S(0,\,\delta)} dh \int_R |f(x+h)-f(x)|^p\,dx.$$

L'intégrale étendue à l'espace R qui figure au second membre étant $<\varepsilon^p$ d'après (5'), on en tire immédiatement l'inégalité (9).

Enfin l'inégalité (10) résulte immédiatement par (3) de la combinaison de (9) et de (6').

(Reçu le 12 juin 1933)

36.

Zum Eindeutigkeitssatz der fastperiodischen Funktionen

Proc. Roy. Physiogr. Soc. Lund *3*, no. 10 (1933), 1–9

Einleitung.

Im folgenden wird das Wort *»fastperiodisch» im ursprünglichen Bohrschen Sinne gebraucht.* Bekanntlich besteht die Hauptschwierigkeit der Bohrschen Theorie in dem Beweise des folgenden Satzes. *(Eindeutigkeitssatz.)*[1]

Verschwindet die fastperiodische Funktion $f(x)$ nicht identisch, so gibt es wenigstens einen reellen Exponenten λ derart, dass

$$M\left\{f(t)e^{-i\lambda t}\right\} = \lim_{T\to\infty} \int_0^T f(t)e^{-i\lambda t}dt \neq 0. \tag{1}$$

Aus diesem Satze folgt dann leicht der Parsevalsche Satz *(Vollständigkeitssatz)* der Bohrschen Theorie. Dieser letztere besagt u. a., dass jeder fastperiodischen Funktion, in einer hier nicht näher zu beschriebenden Weise, eine ganz bestimmte »Fourierreihe» entspricht

$$f(x) \sim \sum a_n e^{i\lambda_n x}. \tag{2}$$

[1] Für die Literatur, die Bezeichnungen und die einfachsten Eigenschaften der fastperiodischen Funktionen, insbesondere die gleichmässige Existenz der Bohrschen Mittelbildungen, welche die Zulässigkeit gewisser hier zu verwendender Grenzübergänge sicherstellt, verweise ich auf das glänzende Buch: Harald Bohr, Fastperiodische Funktionen, Berlin 1932. Im übrigen wird die Kenntnis der in der Einleitung besprochenen Arbeiten nicht vorausgesetzt.

Für den Eindeutigkeitssatz wurden im Laufe der Zeit verschiedene Beweise erbracht. Der ursprüngliche Bohrsche Beweis wurde von de la Vallée Poussin bedeutend vereinfacht. (Wie ich Herrn Bohr brieflich mitgeteilt habe, lässt sich der letzte Teil dieses Beweises noch wesentlich verkürzen.) Einen sehr schönen Beweis, dessen Grundgedanke auf Wiener zurückgeht, gibt Bochner in seinem neuerschienenen Buch über das Fouriersche Integral. Uns interessieren hier in erster Linie zwei Beweise, die an die Theorie der Integralgleichungen anknüpfen, diejenigen von WEYL [1] und HAMMERSTEIN.[2]

Wir nehmen einen Augenblick an, dass die Existenz der in (2) angedeuteten Entwicklung schon in einem gewissen Sinne dargetan sei. Es seien etwa $a_1, a_2, \ldots, a_r$ die Koeffizienten mit dem grössten absoluten Betrage $= |a_1|$. WEYL bildet dann durch ein wohlbekanntes Iterationsverfahren die Funktionen, denen die Reihen $\sum |a_n|^2 e^{i\lambda_n x}$, $\sum |a_n|^4 e^{i\lambda_n x}$, $\sum |a_n|^6 e^{i\lambda_n x}$ u. s. w. entsprechen. Durch eine vortreffliche Handhabung des Bernoulli-Schwarz-Schmidtschen Verfahrens trennt er dann die Funktion

$$e^{i\lambda_1 x} + e^{i\lambda_2 x} + \ldots + e^{i\lambda_r x}$$

ab. Von dieser Funktion weiss man natürlich nicht, dass sie diese einfache Form besitzt; man weiss nur, dass sie fastperiodisch ist und sich bei der beschriebenen Iteration reproduziert. Hierauf fussend, sondert dann Weyl, mit Hilfsmitteln, die der Theorie der Integralgleichungen und derjenigen der unitären Matrizen entspringen, die angeschriebenen reinen Schwingungen ab.

HAMMERSTEIN geht von einem Variationsansatz aus, der zu einer Integralgleichung führt und den wir weiter unten genau angeben. In der Ausdrucksweise der (natürlich noch zu beweisenden) Reihenentwicklung läuft der Ansatz auf die Bestimmung einer Funktion $g(x) \sim \sum b_n e^{i\lambda_n x}$ hinaus, für welche der Ausdruck

$$\frac{\sum |a_n|^2 \, |b_n|^2}{\sum |b_n|^2} \tag{3}$$

[1] H. WEYL, Integralgleichungen und fastperiodische Funktionen, Math. Ann. 97 (1926) S. 338—356.

[2] A. HAMMERSTEIN, Über die Vollständigkeitsrelation in der Theorie der fastperiodischen Funktionen, S.-B. preuss. Ak. Wiss., Phys.-Math. Kl. 1928 S. 17—20.

zu einem Maximum wird. Man erhält zunächst eine Funktion der Gestalt

$$b_1 e^{i\lambda_1 x} + b_2 e^{i\lambda_2 x} + \ldots + b_r e^{i\lambda_r x},$$

wo $b_1, b_2, \ldots, b_r$ ganz beliebig sind. Alle diese Funktionen liefern nämlich das gesuchte Maximum $|a_1|^2$. Durch Betrachtung des Additionstheorems solcher Funktionen gewinnt dann Hammerstein Anschluss an den Weylschen Apparat der unitären Matrizen.

Bei beiden Verfahren wird die Hauptschwierigkeit durch den Umstand verursacht, dass es in der Reihe mehrere Glieder geben kann, deren Koeffizienten den maximalen absoluten Betrag haben. Es liegt in der Natur der Sache, dass weder das Weylsche Iterationsverfahren, noch der Hammersteinsche Variationsansatz im Stande sind, diese Glieder von einander zu trennen.

Wir werden unten zeigen, dass der Hammersteinsche Ansatz so abgeändert werden kann, dass dabei nur ein einziges Exponentialglied $b_k e^{i\lambda_k x}$ $(k = 1, 2, \ldots, r)$ herauskommt. Unser Ansatz lautet — hier zunächst in Reihensprache — folgendermassen:

Es sind *zwei* Funktionen $g(x) \sim \sum b_n e^{i\lambda_n x}$ und $h(x) \sim \sum c_n e^{i\lambda_n x}$ so zu bestimmen, dass der Ausdruck

$$\frac{\sum |a_n|^2 |b_n|^2 |c_n|^2}{\sum |b_n|^2 \cdot \sum |c_n|^2} \tag{4}$$

zu einem Maximum wird. Dieses Maximum $= |a_1|^2$ tritt dann und nur dann ein, wenn alle b_n und c_n verschwinden ausser einem einzigen Paare b_k, c_k $(k = 1, 2, \ldots, r)$.

Bezeichnungen und Hilfsbetrachtungen.

1. Es sei $f(x)$ eine fastperiodische Funktion der reellen Veränderlichen x, die eventuell noch von gewissen Parametern abhängen mag. Wir setzen dann mit Bohr

$$M_x\{f(x)\} = \lim_{T \to \infty} \int_0^T f(x)\,dx. \tag{5}$$

Es seien ferner $f(x)$, $g(x)$, $h(x)$, $k(x)$, u. s. w. fastperiodische Funktionen in einer endlichen Anzahl. Dann bilden wir die gefalteten Funktionen

(6) $$fg(x) = M_r\left\{f(x-r)\,g(r)\right\}$$

bzw.

(7) $$fghk\ldots(x) = M_r M_s M_t \ldots \left\{f(x-r-s-t-\ldots)g(r)h(s)k(t)\ldots\right\}.$$

Die Bildung ist assoziativ und kommutativ. Z. B. um die erste Behauptung zu illustrieren, ist

$$fgh(x) = f\,.\,gh(x) = fg\,.\,h(x).$$

Wir setzen noch

$$f^*(x) = \overline{f(-x)},$$

wo, wie gebräuchlich, $\bar{a}$ den zu a konjugiert-komplexen Wert bedeutet.

Wir bilden ferner

(8) $$F(x) = ff^*(x) = M_r\left\{f(x-r)f^*(r)\right\} = M_r\left\{f(x+r)\overline{f(r)}\right\}.$$

$F(x)$ (oder eigentlich richtiger $F(x-y)$) ist dann von *Hermiteschem* Charakter, d. h. $F^*(x) = F(x)$. Die Funktionen $G(x)$, $H(x)$, $K(x)$ u. s. w. werden in entsprechender Weise aus den Funktionen $g(x)$, $h(x)$, $k(x)$ u. s. w. gebildet. Wir schreiben hier noch die bald zu gebrauchende Ungleichung

$$K(0) = M_r\left\{k(r)\overline{k(r)}\right\} = M_r\left\{|k(r)|^2\right\} \geqq 0$$

an.

Wie Bohr beweist (a. a. O. § 72), ist $K(0) = 0$ *dann und nur dann, wenn $k(x)$ identisch verschwindet.* Aus diesen Tatsachen fliesst sofort eine wichtige Ungleichung.

Es ist

(9) $$|fg^*(0)|^2 \leqq F(0)\,G(0),$$

wobei das Gleichheitszeichen dann und nur dann stattfindet, wenn

(10) $$f(x) = \mu g(x)$$

ist, mit einem konstanten μ. (Schwarzsche Ungleichung.)

Um den Beweis zu führen, setzen wir $k(x) = f(x) - \mu g(x)$. Dann ergibt sich

$$0 \leqq K(0) = M_r\left\{\left(f(r) - \mu g(r)\right)\left(\overline{f(r)} - \bar{\mu}\,\overline{g(r)}\right)\right\} =$$
$$= F(0) - \bar{\mu} fg^*(0) - \mu f^* g(0) + |\mu|^2 G(0).$$

Hier sind die mittleren Glieder zu einander konjugiert. Aus der Tatsache, dass der Wert dieser Hermiteschen Form immer $\geqq 0$

ausfällt, folgt der erste Teil unserer Behauptung. Der zweite Teil folgt aus der Tatsache, dass die Form, nach dem obigen, für irgendeinen Wert von μ dann und nur dann verschwindet, wenn der Integrand für denselben Wert von μ identisch verschwindet.

Setzt man noch einen Augenblick $f'(r) = f(r - x)$, so folgt aus (6) und (8) $F(x) = f'f^*(0)$, daher, nach (9), $|F(x)|^2 \leqq F'(0)F(0)$, was, mit Rücksicht auf $F'(0) = F(0)$,

$$|F(x)| \leqq F(0) \tag{11}$$

ergibt.

Der Variationsansatz.

2. Es sei nun $f(x)$ eine gegebene, nicht identisch verschwindende fastperiodische Funktion, und die ebenfalls nicht identisch verschwindenden, fastperiodischen Funktionen $g(x)$ und $h(x)$ seien so zu bestimmen, dass das Verhältnis[1]

$$A(g, h) = \frac{FGH(0)}{G(0)\,H(0)} \tag{12}$$

seinen grösstmöglichen Wert annimmt. Es sei schon bewiesen (das wird erst im letzten Teil nachgeholt), dass es mindestens ein solches Funktionenpaar gibt. Wir beweisen dann, dass $g(x)$ und $h(x)$ beide von der Form $be^{i\lambda x}$ (mit demselben λ) sind. Dieses geschieht so. Es sei

$$A(g, h) = \frac{FGH(0)}{G(0)\,H(0)} = M, \tag{13}$$

wo also M den Maximalwert bedeutet. Wir setzen dann[2]

$$h'(x) = \frac{fgh(x)}{\sqrt{G(0)\,H(0)}}. \tag{14}$$

(Der Nenner, der den Wert (15) nicht beeinflusst, dient nur zu einer gewissen Normierung, die erst später beim Existenzbeweis benötigt wird.) Es ergibt sich

---

[1] Der Hammersteinsche Ansatz bezieht sich (in anderen Bezeichnungen) auf das Verhältnis

$$\frac{FG(0)}{G(0)}.$$

[2] Vgl. hierzu Hammerstein, a. a. O.

$$(15)\qquad A(g,\ h') = \frac{FGFGH(0)}{G(0)\,FGH(0)}.$$

Ersetzt man nun in (9) und (10) f durch FGh und g durch h, so folgt unmittelbar

$$(15')\qquad FGFGH(0)\,.\,H(0) \geqq (FGH(0))^2,$$

wo das Gleichheitszeichen dann und nur dann besteht, wenn

$$(16)\qquad FGh(x) = \mu h(x)$$

ist, wo μ eine Konstante bedeutet.

Die Ungleichung (15′) ist aber nach (12) und (15) gleichbedeutend mit der Beziehung

$$(17)\qquad A(g,\ h') \geqq A(g,\ h)$$

Wegen der Maximaleigenschaft des Paares $(g,\ h)$ muss aber hier und somit auch in (15′) das Gleichheitszeichen stattfinden, mithin besteht auch die Gleichung (16). Indem man nun diese Gleichung mit $\overline{h(x)}$ multipliziert, ergibt sich durch Mittelbildung $FGH(0) = \mu H(0)$, woraus, mit Rücksicht auf (13), $\mu = M\,G(0)$ folgt. Die Gleichung (16) lautet also in ihrer endgültigen Form

$$(18)\qquad FGh(x) = M\,G(0)\,.\,h(x).$$

Besteht umgekehrt diese Gleichung, wo M den obengenannten Maximalwert bedeutet, so schliesst man durch das soeben angewandte Verfahren, dass auch $FGH(0) = M\,G(0)H(0)$, oder nach (13), auch $A(g,\ h) = M$ besteht, d. h. das Funktionenpaar $(g,\ h)$ besitzt die erstrebte Maximaleigenschaft. Zusammenfassend können wir sagen:

Das Bestehen der Gleichung (18) *ist eine notwendige und hinreichende Bedingung für die Maximaleigenschaft des Funktionenpaares* $(g,\ h)$.

Aus (18) folgt noch durch Faltung mit $h^*(x)$

$$FGH(x) = M\,G(0)\,.\,H(x)$$

und dann aus Symmetriegründen

$$FGH(x) = M\,H(0)\,.\,G(x),$$

d.h.

$$(19)\qquad H(x) = \varrho G(x)$$

wo ϱ eine Konstante bedeutet. *Diese Gleichung stellt also eine notwendige Bedingung für die Maximaleigenschaft dar.*[1]

Wir setzen jetzt

$$k(x) = h(x) + \alpha h(x + t),\,^{2}$$

wo α einen komplexen und t einen reellen Parameter bedeuten. Selbstverständlich genügt auch diese Funktion der Gleichung (18). Mithin liefert auch das Funktionenpaar (g, k) das fragliche Maximum. Daher muss nach (19) auch eine Gleichung der Form $K(x) = \varrho' G(x)$ und somit auch eine Gleichung

$$K(x) = \omega H(x)$$

bestehen, wo ω wiederum eine Konstante bedeutet. In ausführlicher Form lautet diese Gleichung

$$H(x) + \alpha H(x + t) + \bar{\alpha} H(x - t) + |\alpha|^2 H(x) = \omega H(x).$$

Daher ist auch (für ein beliebiges konstantes t) das Verhältnis $H(x + t) : H(x)$ konstant. Hieraus und aus der Fastperiodizität von $H(x)$ folgt sofort

$$H(x) = Be^{i\lambda x}. \tag{20}$$

Damit sind wir schon eigentlich am Ziele. Denn, da $FGH(0) > 0$ ist und $FGH(0) = FHG(0)$ nach (6) und (11) immer $\leqq FH(0) \,.\, G(0)$ ausfällt, ist auch $FH(0) > 0$, d. h. nach (6) und (20)

$$M_r\left\{F(r)e^{-i\lambda r}\right\} = A \neq 0,$$

und daher auch

$$M_r\left\{f(r)e^{-i\lambda r}\right\} = a \neq 0,$$

da $A = |a|^2$ (vgl. Bohr, a. a. O. § 66). Wir beweisen aber noch, der Vollständigkeit halber, die am Anfang behauptete, nunmehr

---

[1] Hieraus folgt natürlich, dass wir statt des Ausdruckes (12) gleich den formal einfacheren Ausdruck

$$\frac{FGG(0)}{(G(0))^2}$$

hätten betrachten können. Der entsprechende Variationsansatz führt nun tatsächlich zum erstrebten Ergebnis, aber auf einem etwas mühsamen und langweiligen Weg.

[2] Für jedes t kann α natürlich so gewählt werden, dass $k(x)$ nicht identisch verschwindet.

fast evidente, Tatsache, dass $h(x)$ (und damit natürlich auch $g(x)$) von der Form $be^{i\lambda x}$ ist. Zufolge (20) ist nämlich

$$M_r\left\{|h(x+r)-e^{i\lambda x}h(r)|^2\right\}=2H(0)-e^{-i\lambda x}H(x)-e^{i\lambda x}H(-x)=0,$$

mithin

$$h(x+r)-e^{i\lambda x}h(r)\equiv 0,$$

also

$$h(x)=h(0)e^{i\lambda x}.$$

Beweis für die Existenz des Maximums.

3. $f(x)$ sei die gegebene fastperiodische Funktion, $g(x)$ und $h(x)$ beliebige fastperiodische Funktionen. Keine der genannten Funktionen soll identisch verschwinden. Zu beweisen ist, dass es mindestens ein Paar von Funktionen $g(x)$ und $h(x)$ gibt, die dem Ausdrucke (12) sein Maximum erteilen. Wir bezeichnen die obere Grenze der Werte, deren (12) fähig ist, mit M und zeigen zunächst, dass M eine endliche positive Zahl ist. Die Endlichkeit ergibt sich aus den Beziehungen (7) und (11), wonach

$$\frac{FGH(0)}{G(0)H(0)}\leqq\frac{F(0)G(0)H(0)}{G(0)H(0)}=F(0).$$

Setzt man dann noch z. B. $g=f$, $h=F$, so geht der Zähler von (12) in $FFFF(0)$ über. Da nun $f(x)$ nicht identisch verschwindet, verschwinden, nach einem oben angeführten Satze von Bohr, $F(x)$ und daher auch $FF(x)$ nicht identisch, woraus sich dann, durch eine nochmalige Anwendung des Satzes, $FFFF(0)>0$ ergibt. Daher ist auch die gesuchte obere Grenze $M>0$.

Wir können uns nunmehr auf solche Funktionen g und h beschränken, für welche $A(g,h)$ grösser als etwa $\frac{M}{2}$ ausfällt. Es sei also g und h ein solches Funktionenpaar. Wir ersetzen dann h durch die in der Formel (14) definierten Funktion h'. Dann wird einerseits nach (17)

$$A(g,h')\geqq A(g,h), \tag{17}$$

andererseits

$$H'(0)=A(g,h)>\frac{M}{2}. \tag{21}$$

Ersetzen wir noch g durch

$$g'(x) = \frac{fgh'(x)}{\sqrt{G(0)H'(0)}},$$

dann wird wieder (vgl. (17))

$$A(g', h') \geqq A(g, h') \geqq A(g, h) \tag{22}$$

und

$$G'(0) = A(g, h') > \frac{M}{2}. \tag{23}$$

Aus (22) geht hervor, dass wir uns bei der Aufsuchung der oberen Grenze auf Funktionen vom Typus $g'(x)$ und $h'(x)$ beschränken dürfen. Diese Funktionen genügen den für unsere Zwecke wichtigen Ungleichungen (21) und (23), haben aber ausserdem noch folgende Eigenschaften: Sie sind gleichgradig beschränkt, gleichgradig stetig und gleichgradig fastperiodisch. Es genügt, wenn wir das für die Funktion $h'(x)$ darlegen. Diese Funktion wird nun, nach (14) und (6), aus $f(x-r)$ durch eine Mittelbildung erhalten. Das »Gewicht» ist dabei

$$\frac{M_r\{|gh(r)|\}}{\sqrt{G(0)H(0)}} \leqq \frac{\sqrt{M_r\{|gh(r)|^2\}}}{\sqrt{G(0)H(0)}} = \sqrt{\frac{GH(0)}{G(0)H(0)}} \leqq \sqrt{\frac{G(0)H(0)}{G(0)H(0)}} = 1,$$

wo die erste Ungleichung aus der Schwarzschen unmittelbar folgt. Damit ist alles behauptete bewiesen. Aus den angegebenen Eigenschaften folgt durch ein bekanntes Auswahlsverfahren (vgl. etwa Hammerstein, a. a. O.), dass es unter den Funktionenpaaren (g', h') eine Teilfolge gibt, für die $A(g', h') \to M$, während g' und h' gleichmässig gegen gewisse Grenzfunktionen konvergieren, die wir dann wieder mit g und h bezeichnen. Aus (21) und (23) folgt, dass diese Grenzfunktionen nicht identisch verschwinden. Daher ist für diese Funktionen der Ausdruck (12) nicht sinnlos und es wird $A(g, h) = M =$ Maximum.

37.

Eine Bemerkung über den Eindeutigkeitssatz der Theorie der fastperiodischen Funktionen

Mat. Tidskr. B, *1* (1934), 11–13

(Auszug aus einem Brief an H. Bohr).

Einleitende Bemerkungen von H. Bohr.

Für den im Titel genannten Eindeutigkeitssatz, welcher besagt, *dass eine fastperiodische Funktion $f(x)$, für welche der Mittelwert*

$$a(\lambda) = M\{f(t)e^{-i\lambda t}\} = \lim_{T\to\infty}\frac{1}{T}\int_0^T f(t)\,e^{-i\lambda t}\,dt$$

gleich 0 ist für alle λ, identisch verschwinden muss, gibt es mehrere verschiedene Beweise, die zum Teil mit gänzlich verschiedenen Mitteln operieren. Der Beweis, welcher sich am meisten eingebürgert zu haben scheint, und auch in den verschiedenen Monographien über fastperiodische Funktionen*) dargestellt wird, rührt von *de la Vallée Poussin* her und kann als weitgehende Vereinfachung — teilweise unter Heranziehung *Weyl*'scher Hilfsmittel — des ursprünglich gegebenen Beweises angesehen werden. Professor Riesz hat den letzten Teil des Beweises von de la Vallée Poussin wesentlich vereinfacht und mir diese Vereinfachung brieflich mitgeteilt. Da die Beweisvariante von Professor Riesz allgemeineres Interesse beanspruchen kann, wird sein Brief, mit seiner freundlichen Zustimmung, in der vorliegenden Note abgedruckt.

In seinem Brief knüpft Professor Riesz direkt an die Darstellung in F. F. an, und es mag daher dem Leser bequem sein, wenn ich zuerst kurz an die Terminologie des Buches und einige in Betracht kommende Sätze erinnere.

Es bezeichne

$$f(x) \sim \sum A_n\, e^{i\Lambda_n x}$$

eine fastperiodische Funktion und Γ die obere Grenze von $|f(x)|$; ferner sei

$$g(x) = M\{f(x+t)\overline{f(t)}\} \sim \sum |A_n|^2 e^{i\Lambda_n x}$$

die „gefaltete" Funktion. Damit $f(x)$ identisch 0 sei (also $\Gamma = 0$),

---

*) *A. S. Besicovitch:* Almost periodic functions (Cambridge 1932), *H. Bohr:* Fastperiodische Funktionen (Ergebn. d. Math. Bd. 1, H. 5, Berlin 1932) [Im Folgenden mit F. F. zitiert]. *J. Favard:* Leçons sur les fonctions presque-périodiques (Paris 1933).

ist notwendig und hinreichend, dass $g(0) = M\{|f(t)|^2\} = 0$ ist. Der auch von Riesz verwendete „Haupthilfssatz" des de la Vallée Poussin'schen Beweises besagt, dass falls die Gleichung $a(\lambda) = M\{f(t)e^{-i\lambda t}\} = 0$ für alle λ besteht, sie dann auch gleichmässig in λ gilt, also

$$\left| \frac{1}{T} \int_0^T f(t)\, e^{-i\lambda t}\, dt \right| < \varepsilon \qquad \text{für } T > T_0 \qquad \text{und alle } \lambda.$$

Mit

$$F_T(x) \sim \sum_{-\infty}^{\infty} a_n e^{i\frac{2\pi}{T}nx}$$

sei die reinperiodische Funktion mit der Periode T bezeichnet, welche im Periodenintervall $0 < x < T$ gleich $f(x)$ ist. Mit $g_T(x)$ bezw. $G_T(x)$ werden die beiden gefalteten Funktionen

$$g_T(x) = \frac{1}{T} \int_0^T f(x+t)\ \overline{f(t)}\, dt$$

und

$$G_T(x) = \frac{1}{T} \int_0^T F_T(x+t)\ \overline{F_T(t)}\, dt$$

bezeichnet, von denen $G_T(x)$ reinperiodisch ist und nach bekannten Sätzen durch die absolut konvergente Reihe

$$\sum_{-\infty}^{\infty} |a_n|^2 e^{i\frac{2\pi}{T}nx}$$

dargestellt wird. In F. F. wird bewiesen, dass $g_T(x)$ für $T \to \infty$ gleichmässig in $-\infty < x < \infty$ gegen $g(x)$ konvergiert, und ferner, dass für gewisse hinreichend grosse T (nämlich „feine" Verschiebungszahlen) die Ungleichung

$$|g_T(x) - G_T(x)| \leqq \varepsilon \Gamma$$

für jedes x in $0 < x < T$ besteht.

An diese letzte Ungleichung knüpft nun der Riesz'sche Brief an und führt den Beweis des Eindeutigkeitssatzes in einfacherer und viel kürzerer Weise zu Ende, als es in F. F. nach de la Vallée Poussin geschieht. Die Ausführungen des Briefes sind nur an Hand von F. F. in Einzelheiten zu verstehen, und zwar ist der Brief so abgefasst, dass er die letzte Hälfte des § 83 (von der Mitte der Seite 67 an) durch eine geänderte Schlussweise ersetzt.

Auszug aus dem Briefe von Professor Riesz.

. . . . . .

$$| g_T(x) - G_T(x) | \leqq \varepsilon \Gamma. \qquad (0 < x < T).$$

Hieraus folgt für hinreichend grosses T

$$| g(x) - G_T(x) | < 2\varepsilon \Gamma. \qquad (0 < x < T).$$

Andererseits folgt aus § 66 (7), dass alle Fourierkoeffizienten von $g(x)$ und daher auch diejenigen von $\overline{g(x)}$ gleich Null sind. Mithin ist nach dem Haupthilfssatz von § 81

$$\frac{1}{T}\left| \int_0^T \overline{g(x)}\, e^{i\lambda x}\, dx \right| < \varepsilon$$

für alle λ und für genügend grosses T. Nun ist aber

$$G_T(x) = \sum | a_n |^2 e^{i\frac{2\pi}{T} nx},$$

wo $\sum | a_n |^2 \leqq \Gamma^2$. Daher ist

$$\frac{1}{T}\left| \int_0^T \overline{g(x)}\, G_T(x)\, dx \right| < \varepsilon \Gamma^2.$$

Mithin, da $|\overline{g(x)}| \leqq \Gamma^2$,

$$\frac{1}{T}\int_0^T | g^2(x) |\, dx = \frac{1}{T}\int_0^T \overline{g(x)} (g(x) - G_T(x))\, dx + \frac{1}{T}\int_0^T \overline{g(x)}\, G_T(x)\, dx$$
$$< \varepsilon(\Gamma^2 \cdot 2\Gamma + \Gamma^2),$$

also $M\{| g^2(x) |\} = 0$. Daher ist nach § 72 $g(x) \equiv 0$, also insbesondere $g(0) = 0$ und mithin nochmals nach § 72 $f(x) \equiv 0$.

38.

Intégrale de Riemann-Liouville et solution invariantive du problème de Cauchy pour l'équation des ondes

C. R. Congrès. Int. des Math. Oslo (1937) II, 44–45

On définit dans l'espace à m dimensions l'intégrale

$$I^{\alpha} u(P) = \frac{1}{H_m(\alpha)} \int_{D_S^P} u(Q) r_{PQ}^{\alpha-m} dQ.$$

Ici $r = r_{PQ}$ désigne la distance hyperbolique $\sqrt{(x_1-\xi_1)^2-(x_2-\xi_2)^2-\cdots}$ des points $P=(x_1, x_2, \cdots)$ et $Q=(\xi_1, \xi_2, \cdots)$, D_S^P le domaine que la surface (à $m-1$ dimensions) S «orientée dans l'espace» découpe du cône lumière rétrograde au sommet P et

$$H_m(\alpha) = \pi^{\frac{m-2}{2}} 2^{\alpha-1} \Gamma\left(\frac{\alpha}{2}\right) \Gamma\left(\frac{\alpha+2-m}{2}\right).$$

L'intégrale converge pour $\alpha > m-2$ et satisfait aux relations $I^{\alpha} I^{\beta} = I^{\alpha+\beta}$ et $\Delta I^{\alpha+2} = I^{\alpha}$, où $\Delta = \partial^2/\partial x_1^2 - \partial^2/\partial x_2^2 - \cdots$. La définition s'étend par prolongement analytique par rapport à α aux indices $\alpha \leqq m-2$ sous des conditions de régularité convenables. On a la formule de Green

$$I^{\alpha} u(P) = I^{\alpha+2} \Delta u(P) - \frac{1}{H_m(\alpha+2)} \int_{S^P} \left(\frac{\partial u(Q)}{\partial n} r_{PQ}^{\alpha+2-m} - u(Q) \frac{\partial r_{PQ}^{\alpha+2-m}}{\partial n} \right) dS.$$

Ici S^P désigne la portion de surface découpée de S par le cône et n la conormale extérieure. Par prolongement analytique la formule reste valide pour $\alpha=0$. Le premier membre devient alors $u(P)$ et le second membre fournit la solution du problème de Cauchy, qui se présente ainsi sous la même forme pour m impair et pair. Pour m impair, on retrouve la solution donnée par M. Hadamard (aussi pour des coefficients non constants) à l'aide de la partie finie des intégrales. Pour m pair, l'évanouissement des facteurs devant les intégrales fait surgir le principe de Huygens.

Pour $m=4$, on déduit de la formule générale pour l'équation homogène $\Delta u=0$, à laquelle nous nous bornons ici, la solution suivante

$$u(P)=\frac{1}{\pi}\int\limits_{s}\left\{\frac{1}{2}\left(\frac{1}{R_1}+\frac{1}{R_1}\right)\cdot u-\left(\frac{\partial u}{\partial R}\right)_{\nu}\right\}ds.$$

Ici s est le bord de S^P, l'élément ds est mesuré dans le sens *lorentzien*, $R=R_{PQ}=r_{PQ}^2$. En chaque point de s il n'intervient qu'une seule dérivée, elle est prise par rapport à R suivant la génératrice ν de la nappe intérieure d'une certaine surface, enveloppée par les cônes rétrogrades qui ont leurs sommets sur s. La congruence des ν admet sur chacune des ν deux foyers Q_1 et Q_2. Dans la formule ci-dessus on a posé $R_{PQ_1}=R_1$ et $R_{PQ_2}=R_2$.

39.

Modules réciproques

C. R. Congrès Int. des Math. Oslo (1937) II, 36–37

Soit M un module constitué de certains points (vecteurs) x de l'espace à n dimensions. L'ensemble des points (vecteurs) v tels que le produit scalaire (x, v) soit entier pour chaque x, forme un *module fermé*, c.-à-d. un module qui contient tous ses points limites. En désignant ce module par M^{-1}, on a la relation fondamentale

$$(1) \qquad (M^{-1})^{-1} = [M],$$

où $[M]$ désigne la fermeture de M. Dans le cas où M est fermé lui-même, cette relation se réduit à une vraie relation de réciprocité, savoir à $(M^{-1})^{-1} = M$. En retournant au cas général et en désignant par N un second module, formons au sens de l'addition vectorielle le module $M+N$. On a alors

$$(2) \qquad (M+N)^{-1} = M^{-1}N^{-1},$$

le second membre signifiant d'après une notation usuelle l'ensemble des points communs à M^{-1} et à N^{-1}.

Les théorèmes de Kronecker sur les approximations diophantiques sont des conséquences évidentes des relations (1) *et* (2).

Soit maintenant M un module discret (c.-à-d. un grillage de points) du rang n. Il en sera alors de même du module réciproque M^{-1}. Par $|b|$ nous entendons la longueur d'un vecteur b quelconque, mesurée dans le sens de Minkowski par un certain corps convexe C, et par $|u|^*$ la longueur d'un vecteur u quelconque, mesurée par le corps dual (réciproque) C^*. On a alors le théorème suivant.

Il existe une base $b_1, b_2, \cdots b_n$ de M et une base $u_1, u_2, \cdots u_n$ de M^{-1} telles que

$$1 \leqq |b_k|\,|u_k|^* \leqq a_n, \quad k = 1, 2, \cdots n,$$

le nombre a_n ne dépendant que de n, c.-à-d. que ce nombre est indépendant du module M et du corps C. On peut aussi admettre que les bases soient réciproques, c.-à-d. que $(b_k, u_l) = \delta_{kl}$, δ_{kl} étant le symbole de Kronecker.

40.

Potentiels de divers ordres et leurs fonctions de Green

C. R. Congrès Int. des Math. Oslo (1937) II, 62–63

L'autre jour je vous ai indiqué une intégration du type hyperbolique. Dans le même ordre d'idées on peut aussi considérer une intégration du type elliptique. Ω étant l'espace à n dimensions et r_{PQ} désignant la distance euclidienne des points P et Q, nous posons, pour $\alpha > 0$,

$$I^\alpha f(P) = \frac{1}{H_m(\alpha)} \int_\Omega f(Q) r_{PQ}^{\alpha-m}\, dQ; \quad H_m(\alpha) = \frac{\pi^{\frac{m}{2}} 2^\alpha \Gamma\left(\frac{\alpha}{2}\right)}{\Gamma\left(\frac{m-\alpha}{2}\right)}.$$

On a $I^\alpha I^\beta = I^{\alpha+\beta} (\alpha+\beta < m)$ et $\Delta I^{\alpha+2} = -I^\alpha$, Δ désignant l'opérateur de Laplace. C'est la valeur $\alpha = 2$ qui donne le potentiel newtonien, sauf pour $m=2$, où un passage à la limite facile conduit au potentiel logarithmique. En remplaçant $f(Q)dQ$ par une différentielle de Stieltjes, on peut aussi considérer des potentiels de la forme

$$v(P) = \int_F r_{PQ}^{\alpha-m}\, d\mu(Q),$$

F étant un ensemble fermé. Dans sa Thèse, M. Frostman a étudié la capacité des ensembles par rapport à ces potentiels et, sous la condition indispensable $\alpha \leqq 2$, il a démontré l'existence du potentiel d'équilibre pour tout ensemble fermé de capacité positive. De ce résultat de M. Frostman, on déduit par la transformation de Kelvin l'existence d'une distribution unique μ_M telle que la fonction de Green

$$G_M(P) = r_{MP}^{\alpha-m} - \int_F r_{PQ}^{\alpha-m}\, d\mu_M(Q)$$

s'annule en tout point de F, sauf peut-être dans un ensemble de capacité nulle. Une différence essentielle entre le cas newtonien et le cas $\alpha < 2$, c'est que, dans le dernier cas, la masse se répartit — dans un sens facile à préciser — sur l'ensemble F tout entier et non seulement sur sa frontière. Un probléme central est de donner les conditions pour qu'une fonction donnée sur un ensemble F puisse s'écrire comme potentiel de masses portées par cet ensemble. Nous arrivons dans cet ordre d'idées à une extension naturelle des fonctions surharmoniques (sousharmoniques).

Un résumé assez complet de ces recherches va paraître dans les *Acta Szeged.*

41.

Volumes mixtes et facteurs invariants dans la théorie des modules

C. R. Congrès Int. des Math. Oslo (1937) II, 16

Il y a une correspondance intime entre la théorie des corps convexes et celle des modules. Cela tient au fait que ce sont les mêmes opérations $P=\lambda_1 P_1+\lambda_2 P_2$ qui, sous des conditions différentes, conservent d'une part les corps convexes ($\lambda_1 \geqq 0$, $\lambda_2 \geqq 0$, $\lambda_1+\lambda_2=1$), d'autre part les modules (λ_1 et λ_2 entiers). En nous bornant ici à des modules M de points entiers dans l'espace à n dimensions, nous entendons par le « volume » de M le *module de nombres* formé des multiples du volume du parallélépipède élémentaire. Ce module est identique à l'ensemble des volumes des parallélépipèdes construits sur les vecteurs de M.

$M_1, M_2, \cdots M_s$ étant encore des modules de points entiers et $\lambda_1, \lambda_2, \cdots \lambda_s$ des nombres entiers, le « volume » du module $\lambda_1 M_1+\lambda_2 M_2+\cdots+\lambda_s M_s$ est un polynome homogène du degré n dans les λ, dont les coefficients, les « *volumes mixtes* », sont des *modules de nombres*. Le coefficient de $\Pi \lambda_j^{k_j}$ s'obtient comme l'ensemble des volumes des parallélépipèdes construits sur des vecteurs dont exactement k_j appartiennent à M_j $(j=1, 2, \cdots s)$.

Il y a un grand rapport entre les volumes mixtes et les facteurs invariariants des modules. Au lieu d'y insister, nous montrons ici, comment on peut caractériser chacun des facteurs invariants $e_1, e_2, \cdots e_k, \cdots$ d'un module M indépendamment des autres. En désignant par (a) l'ensemble des multiples d'un nombre arbitraire a, on exprime le fait que a est diviseur de b par l'inégalité $(b) \subseteqq (a)$, entendue au sens de la théorie des ensembles. On sait d'autre part que le module (e_1) est identique à l'ensemble des produits scalaires (x, v), quand x parcourt tous les points de M et v tous les points entiers de l'espace. (e_1) est donc une espèce de maximum, embrassant tous les modules qu'on peut former des produits scalaires (x, v). Le module (e_k) peut s'obtenir de la manière suivante. Donnons nous arbitrairement $k-1$ points entiers et restreignons x à ceux des points de M dont le produit scalaire avec chacun des points donnés est zéro. L'ensemble des valeurs (x, v) ainsi obtenues forme un module (e'_k). Parmi ces modules il y en a *un* (e_k) qui est *minimum*: $(e_k) \subseteqq (e'_k)$. Le module (e_k) s'obtient donc comme une espèce de *minimum-maximorum*. On sait quel parti heureux M. Courant a su tirer de la caractérisation analogue des valeurs singulières des équations intégrales. Il en est de même ici. P. ex. les conditions, dues à Frobenius, pour qu'une forme bilinéaire contienne une autre, résultent en quelques mots de cette interprétation des facteurs invariants.

42.

Intégrales de Riemann-Liouville et potentiels

Acta Szeged Sect. Math. 9 (1937–38), 1–42

Introduction.

La définition de l'intégrale de RIEMANN—LIOUVILLE peut s'étendre de beaucoup de manières à l'espace à plusieurs dimensions. On peut en particulier considérer certains procédés d'intégration de caractère elliptique, hyperbolique et parabolique respectivement. Dans tous ces procédés l'intégrale d'ordre deux joue un rôle particulier, elle constitue l'inverse des opérations qui figurent respectivement dans l'équation de LAPLACE, celle des ondes et celle de la chaleur. Nous nous sommes occupé en particulier des deux premiers procédés et nous avons l'intention de rassembler nos recherches dans un mémoire élaboré. En attendant, nous donnons dans le présent travail un résumé assez détaillé de nos résultats concernant l'intégration elliptique et les potentiels[1]) qui y correspondent. En ce qui concerne l'intégration hyperbolique, une conférence faite récemment[2]) à Paris à la Réunion de la *Société Mathématique de France* sur ce procédé et son application au probléme de CAUCHY pour l'équation des ondes va paraître dans le *Bulletin* de la Société.

Les recherches indiquées dans le Chapitre I (et la plus grande partie de celles concernant l'intégration hyperbolique) datent de mars—avril 1933, tandis que celles concernant les masses et la fonction de GREEN et leurs applications ne furent commen-

[1]) Ces potentiels sont considérés depuis longtemps en Analyse et en Physique mathématique.

[2]) Le 9 juillet 1937.

cées qu'en novembre 1935 et terminées en mai 1936.[3]) Nos recherches embrassent aussi le cas newtonien. La fonction de GREEN est toujours considérée comme potentiel défini dans l'espace entier. Dans les chapitres VIII et IX il est aussi question des rapports entre le problème de DIRICHLET (généralisé) de la théorie classique et le potentiel newtonien.

Nous venons de dire que nous donnons ici un résumé assez détaillé. Bien entendu, nous avons omis beaucoup de démonstrations, les unes parce qu'elles sont très simples, les autres parce qu'elles sont trop longues. En ce qui concerne les dernières, nous avons tâché d'en faire ressortir les points essentiels.

Chapitre I. — Les intégrales de Riemann—Liouville.

1. L'intégrale de R.—L.

$$(1) \qquad I^\alpha f(x) = \frac{1}{\Gamma(\alpha)} \int_a^x f(t)\,(x-t)^{\alpha-1} dt; \qquad \alpha > 0,$$

satisfait aux relations fondamentales

$$(2) \qquad I^\alpha\,(I^\beta f(x)) = I^{\alpha+\beta} f(x)$$

et

$$(3) \qquad \frac{d}{dx} I^{\alpha+1} f(x) = I^\alpha f(x).$$

En tout point de continuité de $f(x)$ l'intégrale peut se définir pour l'indice $\alpha = 0$ par un passage à la limite; on obtient alors $I^0 f(x) = f(x)$. Dans des conditions de dérivabilité appropriées on peut définir cette intégrale, comme l'a montré M. HADAMARD, aussi pour des indices négatifs non entiers en retranchant de l'intégrale divergente certaines parties infinies d'ordre fractionnaire. On arrive ainsi à la notion importante de la partie finie des intégrales. On peut atteindre le même but, et cela aussi pour l'indice zéro et les indices entiers négatifs par une méthode qui nous semble très naturelle, celle du prolongement analytique par rapport à l'indice α. Ce prolongement peut s'effectuer p. ex. au moyen de la formule obtenue par un certain nombre d'intégrations par parties

---

[3]) Deux aperçus succincts de ces procédés d'intégration et de leurs applications viennent de paraître dans les *Comptes rendus du Congrès International des Mathématiciens, Oslo,* 1936, t. II.

(4) $$I^{\alpha} f(x) = \sum_{k=0}^{n-1} \frac{f^{(k)}(a)}{\Gamma(\alpha+k+1)} (x-a)^k + I^{\alpha+n} f^{(n)}(x).$$

La dernière expression s'écrit

$$\frac{1}{\Gamma(\alpha+n)} \int_a^x f^{(n)}(t)\,(x-t)^{\alpha+n-1},$$

l'intégrale étant convergente pour $\alpha > -n$. Le prolongement analytique se trouve donc effectué pour tous ces indices. On trouve en particulier $I^{-k} f(x) = f^{(k)}(x)$. La présence du facteur $1/\Gamma(\alpha)$ devant l'intégrale est essentielle pour la validité de la dernière relation.

2. La généralisation de l'intégrale de R.—L. dont il sera question ici se rapporte à un espace Ω_m ou plus brièvement Ω à un nombre quelconque $m \geqq 1$ de dimensions. La distance (euclidienne) de deux points P et Q sera désignée par $r_{PQ} = r_{QP}$. Nous posons pour α positif

(5) $$I^{\alpha} f(P) = \frac{1}{H_m(\alpha)} \int_{\Omega} f(Q)\, r_{PQ}^{\alpha-m}\, dQ,$$

avec

(6) $$H_m(\alpha) = \frac{\pi^{\frac{m}{2}}\, 2^{\alpha}\, \Gamma\left(\frac{\alpha}{2}\right)}{\Gamma\left(\frac{m-\alpha}{2}\right)},$$

l'intégration s'étendant à l'espace entier et dQ désignant l'élément de volume de cet espace. Quant à la fonction f, nous supposons seulement que toutes les intégrales où elle intervient sont absolument convergentes. On a alors les formules fondamentales, où Δ désigne l'opérateur de LAPLACE,

(7) $$I^{\alpha}(I^{\beta} f(P)) = I^{\alpha+\beta} f(P); \quad \alpha > 0,\ \beta > 0,\ \alpha+\beta < m$$

et

(8) $$\Delta I^{\alpha+2} f(P) = -I^{\alpha} f(P),$$

qui d'ailleurs déterminent entièrement le facteur $H_m(\alpha)$.

En tout point de continuité de f, on obtient par un passage à la limite $I^0 f(P) = f(P)$. Dans des hypothèses de dérivabilité convenables, on peut par prolongement analytique étendre la définition de l'intégrale à des valeurs négatives de α. On appliquera la formule (8), ou bien on remplacera la formule (4) par une autre qu'on obtient par une application (itérée) de la formule de Green. On trouve en particulier

$$I^{-2k}f(P) = (-1)^k \Delta^k f(P).$$

Observons encore qu'il peut avoir un certain intérêt d'étendre la formule (7) aussi au cas où $\alpha+\beta \geqq m$. On le fait encore par prolongement analytique. A la fin du présent travail on trouve un exemple de l'utilité de cette extension.

3. La formule (8) met en évidence le fait que l'opérateur I^2 est dans un certain sens l'inverse de l'opérateur de Laplace (au signe près) et on comprend alors qu'il mérite une attention particulière. On trouve pour $m \neq 2$

$$I^2 f(P) = \frac{\Gamma\left(\frac{m-2}{2}\right)}{4\pi^{\frac{m}{2}}} \int_{\Omega} f(Q) r_{PQ}^{2-m} dQ$$

qu'on peut désigner sous le nom de *potentiel newtonien* correspondant à la couche de densité spatiale $f(Q)$, bien que le facteur devant l'intégrale n'intervienne pas dans la forme usuelle de ce potentiel.

L'interprétation qui précède échoue pour $m=2$. L'intégrale est alors indépendante de P et le facteur devant l'intégrale devient infini. Cependant, dans l'hypothèse particulière que l'intégrale de $f(Q)$ étendue à l'espace entier, c'est-à-dire la masse totale, s'annule, on peut poser

$$I^2 f(P) = \lim_{\varepsilon \to 0} I^{2-\varepsilon} f(P) = \lim_{\varepsilon \to 0} \frac{\Gamma\left(\frac{\varepsilon}{2}\right)}{\pi 2^{2-\varepsilon}\Gamma\left(1-\frac{\varepsilon}{2}\right)} \int_{\Omega_2} f(Q)\, r_{PQ}^{-\varepsilon}\, dQ =$$

$$= \lim_{\varepsilon \to 0} \frac{\Gamma\left(\frac{\varepsilon}{2}\right)}{\pi 2^{2-\varepsilon}\Gamma\left(1-\frac{\varepsilon}{2}\right)} \int_{\Omega_2} f(Q)\,(r_{PQ}^{-\varepsilon} - 1)\, dQ =$$

$$= \frac{1}{2\pi} \int_{\Omega_2} f(Q) \log \frac{1}{r_{PQ}}\, dQ,$$

c'est le potentiel logarithmique.

Dans le cas où la masse totale est différente de zéro, on arrive encore au potentiel logarithmique par un passage à la limite modifié. En effet, on peut poser

$$I^2 f(P) = \lim_{\varepsilon \to 0} \frac{1}{2} (I^{2-\varepsilon} f(P) + I^{2+\varepsilon} f(P)) = \frac{1}{2\pi} \int_{\Omega_2} f(Q) \log \frac{1}{r_{PQ}} dQ.$$

Les choses se passent d'une manière analogue pour une dimension m quelconque, bien entendu non plus pour $\alpha = 2$ mais pour $\alpha = m$. Le potentiel $I^m f(P)$ pourra toujours être interprété comme un potentiel logarithmique, la seule différence étant que l'indice $\alpha = m$ n'est un indice particulièrement important que dans le cas $m = 2$ où les indices $\alpha = m$ et $\alpha = 2$ deviennent identiques.

4. On peut aussi remplacer la fonction f par une différentielle de STIELTJES et poser[4])

$$(9) \qquad U^\alpha(P) = \frac{1}{H_m(\alpha)} \int_\Omega r_{PQ}^{\alpha - m} d\mu(Q); \qquad \alpha > 0,$$

μ étant une fonction additive d'ensemble. On aura alors

$$(10) \qquad I^\alpha U^\beta(P) = U^{\alpha+\beta}(P); \qquad \alpha > 0,\ \beta > 0,\ \alpha + \beta < m$$

et

$$(11) \qquad \Delta U^{\alpha+2}(P) = - U^\alpha(P).$$

En formant dans le même ordre d'idées la fonction

$$V^\beta(\beta) = \frac{1}{H_m(\beta)} \int_\Omega r_{PQ}^{\beta - m} d\nu(Q),$$

on a la formule de composition ($\alpha > 0$, $\beta > 0$, $\alpha + \beta < m$)

$$(12) \quad \int_\Omega U^\alpha(T)\, V^\beta(T)\, dT = \frac{1}{H_m(\alpha+\beta)} \iint_\Omega r_{PQ}^{\alpha+\beta-m} d\mu(P)\, d\nu(Q).$$

En spécialisant les fonctions et les indices, on en tire pour $0 < \alpha < m$

$$(13) \qquad \iint_\Omega r_{PQ}^{\alpha - m} d\mu(P)\, d\mu(Q) = H_m(\alpha) \int_\Omega \left[U^{\frac{\alpha}{2}}(T)\right]^2 dT,$$

formule qui joue un rôle décisif dans les recherches de M. FROSTMAN sur le potentiel d'équilibre.

Le second membre de la formule ci-dessus étant non négatif, on voit que l'intégrale double, qui n'est autre que l'intégrale

---

4) Jusqu'à nouvel ordre nous supposons que toutes les intégrales de STIELTJES figurant dans le texte soient absolument convergentes. Pour une hypothèse plus générale *voir* le chapitre III.

d'énergie, est $\geqq 0$ pour $0 < \alpha < m$.[5]) On peut facilement préciser cette inégalité en démontrant que le signe d'égalité ne peut avoir lieu que dans le cas évident où la fonction d'ensemble μ s'annule identiquement. En effet on voit par la formule (13) que l'égalité entraîne que la fonction $U^{\frac{\alpha}{2}}(T)$ s'annule presque partout, et alors le fait énoncé découle du théorème d'unicité démontré plus loin (nº **10**).

Dans le cas où la masse totale s'annule, la formule (13) devient à la limite

$$\iint\limits_{\Omega} \log \frac{1}{r_{PQ}} \, d\mu(P) \, d\mu(Q) = 2^{m-1} \pi^{\frac{m}{2}} \Gamma\left(\frac{m}{2}\right) \int\limits_{\Omega} \left[U^{\frac{m}{2}}(T)\right]^2 dT,$$

identité qui joue le même rôle dans la théorie du potentiel logarithmique que l'identité (13) dans le cas général.

Chapitre II. — Capacité et potentiel d'équilibre.

5. En supprimant le facteur $1/H_m(\alpha)$ qui figure dans les considérations précédentes, nous allons considérer des potentiels de la forme $\int r_{PQ}^{\alpha-m} \, d\mu(Q)$. Quant à l'exposant α, nous admettons ici et sauf avis contraire dans la suite qu'il satisfait aux inégalités $0 < \alpha \leqq 2$,[6]) l'exposant $\alpha = 2$ donnant le cas le plus important, le potentiel newtonien. Pour $m = 2$ (le cas du plan) nous excluons cet exposant, le potentiel newtonien étant alors à remplacer (*voir* le nº **3**) par le potentiel logarithmique qui n'interviendra plus dans la suite. Pour $m = 1$, nous nous restreignons en général à ceux des exposants α qui satisfont aux inégalités $0 < \alpha < 1$ (*voir* pourtant le nº **34**). D'ailleurs en ce qui concerne les dimensions $m = 1$ et 2, le lecteur trouvera souvent dans cet exposé des lacunes faciles à combler.

Ce chapitre traite de la *capacité* des ensembles et de leur *potentiel d'équilibre*. En nous contentant de rendre compte de l'état actuel de ces problèmes, nous aurons avant tout à parler du pro-

---

[5]) Récemment (O. Frostman, La méthode de variation de Gauss et les fonctions sousharmoniques, *ces Acta*, **8** (1937), p. 149—159) M. Frostman a montré que l'intégrale d'énergie est encore $\geqq 0$, si l'on remplace le noyau du texte par la fonction de Green.

[6]) Pour la dernière limitation *voir* p. 40 de la Thèse de M. Frostman citée ci-après.

grès marqué par un travail fondamental de M. DE LA VALLÉE POUSSIN[7]) et par la Thèse de M. FROSTMAN.[8])

6. Voici comment on peut définir la capacité d'un ensemble mesurable au sens de BOREL[9]).

Pour toutes les répartitions positives de la masse unité sur l'ensemble E, l'intégrale d'énergie

$$\iint_E r_{PQ}^{\alpha-m}\, d\mu(P)\, d\mu(Q)$$

a une borne inférieure W_E, finie ou infinie. D'autre part le potentiel correspondant à une telle répartition

$$\int_E r_{PQ}^{\alpha-m}\, d\mu(Q)$$

a une borne supérieure, finie ou infinie. Désignons par V_E la borne inférieure, finie ou non, de ces bornes supérieures. On a alors $V_E = W_E$ et la capacité $C(E)$[10]) *de l'ensemble E peut se définir par $C(E) = W_E^{-1} = V_E^{-1}$.*

Voici deux propriétés de la capacité, dont la première est évidente et la seconde facile à démontrer. Si l'ensemble E_1 est compris dans l'ensemble E_2, on a toujours $C(E_1) \leqq C(E_2)$. D'autre part, pour un nombre fini ou une infinité dénombrable d'ensembles E_n on a toujours $C(\Sigma E_n) \leqq \Sigma C(E_n)$.

7. Il faut prêter une attention particulière aux *ensembles de capacité nulle.* D'après ce qu'on vient de dire, un tel ensemble est caractérisé par le fait qu'il ne peut porter aucune masse positive dont le potentiel reste borné partout. Pour un ensemble fermé de capacité nulle on peut même démontrer que le potentiel de toute répartition positive portée par un tel ensemble devient infini en un point au moins de l'ensemble.[11]) On tire des propriétés précéden-

---

[7]) C. DE LA VALLÉE POUSSIN, Extension de la méthode du balayage de Poincaré et problème de Dirichlet, *Annales de l'Institut H. Poincaré,* **2** (1932), p. 169—232.

[8]) O. FROSTMAN, Potentiel d'équilibre et capacité des ensembles avec quelques applications à la théorie des fonctions, *Meddel. Lunds Univ. Mat. Sem.,* **3** (1935), p. 1—118, citée dans la suite comme „*Thèse*".

[9]) *Thèse,* p. 48.

[10]) Notre façon de normer la capacité est différente de celle de M. FROSTMAN. La valeur $C(E)$ du texte s'écrit dans la notation de M. FROSTMAN $[C(E)]^{m-\alpha}$.

[11]) *Thèse,* p. 83.

tes que la masse partielle portée par un ensemble de capacité nulle s'évanouit si le potentiel dû à la masse totale est fini partout ou si l'intégrale d'énergie est finie. Ce fait remarquable permet de *négliger les ensembles de capacité nulle dans toute intégration suivant des répartitions produisant des potentiels finis.* Il s'agissait ici de potentiels dus à des masses positives. Pour un complément important concernant les potentiels dus à des masses de signe variable *voir* le n° **13.**

Notons encore que la somme d'un nombre fini ou d'une infinité dénombrable d'ensembles de capacité nulle est de capacité nulle et qu'un ensemble de capacité nulle est toujours en même temps un ensemble de mesure nulle au sens de LEBESGUE.

8. Passons maintenant aux ensembles de capacité positive et en particulier aux ensembles fermés de cette espèce. Au sujet de tels ensembles M. FROSTMAN donne le théorème fondamental[12]) que voici.

Soit F un ensemble fermé et borné quelconque de capacité $C(F) > 0$; il existe alors une répartition unique de masse positive sur F dont le potentiel est constant et égal à un sur F sauf au plus dans un ensemble de capacité nulle ne contenant aucun point intérieur de F. Le potentiel de cette répartition est partout ≤ 1, sa masse totale et l'intégrale d'énergie correspondante ont la valeur commune $C(F)$. Toute autre répartition de la masse $C(F)$ sur F engendre un potentiel dont la borne supérieure est > 1 et une intégrale d'énergie $> C(F)$.

C'est précisément comme solution des problèmes de minimum auxquels lesdites propriétés donnent lieu que M. FROSTMAN, en rétablissant en toute rigueur un principe de variation dû à GAUSS, obtient la répartition en question. Le potentiel engendré s'appelle *le potentiel d'équilibre* attaché à l'ensemble F. Arrêtons-nous un instant sur les différences entre le cas général $0 < \alpha < 2$ et le cas classique, celui du potentiel newtonien correspondant à $\alpha = 2$, en notant quelques faits importants qui pour ainsi dire sont caractéristiques pour ce dernier cas. Ni la capacité ni le potentiel d'équilibre newtoniens ne sont changés si l'on ajoute ou retranche un ensemble de points faisant partie d'un domaine complètement limité au dehors par F ou un sous-ensemble de F.[13]) Cela pro-

[12]) *Thèse,* p. 56.

[13]) *Thèse,* p. 39 et 55.

vient de la propriété très remarquable, particulière au cas classique, que les masses produisant le potentiel d'équilibre se trouvent toujours sur la frontière (externe). Le potentiel est $=1$ en tout point extérieur à F n'appartenant pas au domaine connexe infini limité par F. Dans ce dernier domaine le potentiel est < 1. Pour les autres exposants le potentiel d'équilibre est < 1 en tout point extérieur à F. Cela découle immédiatement du fait que le laplacien est positif en tous ces points.

Chapitre III. — Quelques propriétés générales du potentiel.

9. Dans sa Thèse M. FROSTMAN démontre un théorème important disant que *le potentiel d'une distribution positive ne devient infini que sauf peut-être dans un ensemble de capacité nulle*[14]).

Remarquons que M. FROSTMAN admet toujours que les masses sont situées dans un domaine borné, hypothèse commode et très naturelle à son point de vue, puisque c'est le potentiel d'équilibre des ensembles bornés qui l'intéresse en premier lieu. *Il nous sera utile de nous placer dans la suite dans des hypothèses un peu plus générales.* Nous admettrons des potentiels dus à des masses qui peuvent s'étendre à l'infini. La masse portée par un ensemble borné sera toujours finie, tandis que la masse totale peut bien devenir infinie. Il faut alors faire une hypothèse complémentaire. En supposant d'abord les masses positives nous admettrons, pour un point fixé P, que le potentiel devient fini si l'on écarte les masses qui se trouvent au voisinage de P. Cette hypothèse concerne en réalité l'influence des masses très éloignées et elle est évidemment indépendante du point P. Il est manifeste que le théorème de M. FROSTMAN subsiste encore dans ce cas et que notre hypothèse est équivalente à celle exigeant que le potentiel soit fini en un point au moins. Quant aux potentiels dus à des masses ν de signe variable, on admettra que ces masses peuvent s'écrire sous la forme $\nu = \nu_1 - \nu_2$, ν_1 et ν_2 étant des masses positives satisfaisant aux conditions ci-dessus. La décomposition la plus avantageuse sera donnée par la formule $\nu = \nu^+ - \nu^-$ où ν^+ est la variation positive et ν^- la variation négative de ν, $\nu^+ + \nu^-$ étant la variation totale. On a alors $d\nu = d\nu^+ - d\nu^-$ et $|d\nu| = d\nu^+ + d\nu^-$. Un tel potentiel sera *bien déterminé* excepté

[14]) *Thèse*, p. 81.'

aux points où les potentiels dus à ν^+ et à ν^- deviennent infinis tous les deux. Grâce au théorème de M. FROSTMAN on voit que ces *points d'indétermination* forment au plus un ensemble de capacité nulle.

10. Nous donnerons ici deux théorèmes d'unicité. Dans le premier nous ne ferons que la restriction $0 < \alpha < m$, tandis que dans le second il s'agira de restrictions de caractère tout à fait différent. La démonstration du premier théorème n'est pas trop simple, tandis que celle du second se fait en quelques lignes.

Tout potentiel est entièrement déterminé par les masses qui l'engendrent. La réciproque est vraie, ce qui revient à dire qu'à un potentiel identiquement nul il correspond une masse identiquement nulle.

$$v(P) = \int_\Omega r_{PQ}^{\alpha-m}\, d\nu(Q)$$

étant un potentiel qui s'annule en tout point P à l'exception au plus d'un ensemble de mesure nulle, la masse ν est identiquement nulle.

Nous indiquons brièvement la démonstration de ce théorème. Gràce à la formule (10) du n° **4** le cas $\alpha > 2$ se réduit facilement au cas $0 < \alpha \leq 2$. Cela étant, soit $K(P)$ une fonction qui avec ses dérivées des deux premiers ordres est continue dans tout l'espace et s'annule identiquement dans la partie éloignée de l'espace. On aura alors d'après la formule qu'on vient de citer

$$K(P) = -I^2 \Delta K(P) = I^\alpha (I^{2-\alpha} h(P)) = I^\alpha k(P),$$

avec $h(P) = -\Delta K(P)$ et $k(P) = I^{2-\alpha} h(P)$. En multipliant le potentiel $v(P)$, qui s'annule presque partout, par $k(P)$, en intégrant et en admettant ici le fait que l'interversion de l'ordre des intégrations est légitime[15]), il vient

$$0 = \int_\Omega k(P)\, v(P)\, dP = \int_\Omega k(P)\, dP \int_\Omega r_{PQ}^{\alpha-m}\, d\nu(Q) =$$

$$= \int_\Omega d\nu(Q) \int_\Omega k(P) r_{PQ}^{\alpha-m}\, dP = H_m(\alpha) \int_\Omega I^\alpha k(Q)\, d\nu(Q) =$$

$$= H_m(\alpha) \int_\Omega K(Q)\, d\nu(Q).$$

[15]) La démonstration de ce fait est assez délicate dans le cas où les masses s'étendent à l'infini.

On en tire facilement par des méthodes d'approximation connues que la dernière intégrale s'annule encore sous des conditions moins restrictives concernant la fonction K; entre autres elle s'annule lorsque K désigne la fonction caractéristique d'un ensemble fermé et borné d'ailleurs quelconque. Il en résulte immédiatement que la fonction d'ensemble ν s'annule identiquement.

11. Nous passons maintenant au second théorème d'unicité dans lequel l'exposant newtonien et certains autres exposants moins intéressants font exception.

Soient F un ensemble fermé et α un exposant $\neq 2+2k$ et $\neq m+2k$ avec $k=0, 1, 2, \ldots$, et admettons que le potentiel

$$v(P) = \int_F r_{PQ}^{\alpha-m}\, d\nu(Q)$$

s'annule identiquement dans un domaine partiel D de $\Omega - F$. Alors la masse ν est identiquement nulle.

Sans restreindre la généralité on peut admettre que le domaine D comprend tout l'extérieur d'une sphère de rayon très grand, le cas général pouvant facilement se réduire à ce cas-là par une transformation de KELVIN (n° **14**). L'ensemble F sera alors évidemment un ensemble borné. Formons maintenant le produit, identiquement nul dans D, $r_{PQ'}^{\alpha-m}\, v(P)$ où Q' désigne un point fixe de F. En faisant tendre P vers l'infini, on trouve que la masse totale $\int_F d\nu(Q) = 0$. En formant d'autre part les laplaciens successifs et en observant que $\Delta r^{\alpha-m} = (\alpha-2)(\alpha-m)\, r^{\alpha-2-m}$, où le facteur $(\alpha-2)(\alpha-m) \neq 0$, on voit que les potentiels qu'on obtient en remplaçant α par $\alpha-2, \alpha-4, \ldots, \alpha-2k, \ldots$ s'annulent aussi dans le domaine D. Cela étant, désignons par $x_1, x_2, \ldots, x_m$ et $\xi_1, \xi_2, \ldots, \xi_m$ les coordonnées respectives des points P et Q et observons que p. ex. $\dfrac{\partial v(P)}{\partial x_1}$ s'annule aussi identiquement dans D, ce qui peut s'écrire

$$0 = \int_F (\xi_1 - x_1)\, r_{PQ}^{\alpha-2-m}\, d\nu(Q) = \int_F \xi_1 r_{PQ}^{\alpha-2-m}\, d\nu(Q).$$

Il en résulte comme plus haut $\int_F \xi_1 d\nu(Q) = 0$. On trouve de la même façon de proche en proche que tous les moments de la distribution ν s'annulent, ce qui — F étant un ensemble borné — entraîne le fait que ν est identiquement nul.

Le théorème est évidemment en défaut dans le cas classique. P. ex. le potentiel dû à la masse $+1$ placée au centre d'une sphère et à la masse -1 distribuée uniformément sur la surface de la sphère est identiquement nul à l'extérieur de la sphère. La raison en est que, de l'évanouissement de ce potentiel, on ne peut pas conclure à celui des potentiels successifs donnés ci-dessus, puisque $\Delta r^{2-m} = 0$.

12. Nous donnons maintenant un théorème de la moyenne qui admet une application intéressante au potentiel d'équilibre et par là, comme on verra plus loin, à la fonction de GREEN.

Soit $v(P)$ le potentiel d'une distribution positive. On a en tout point P

$$v(P) = \lim \frac{1}{s} \int_s v(T)\, dT,$$

s étant une sphère, décrite autour de P, de volume s et de rayon tendant vers zéro[16]). Le théorème subsiste évidemment pour des masses de signe variable en tout point P où le potentiel est bien déterminé.

Pour la démonstration on peut recourir à un procédé utilisé par M. FROSTMAN dans une évaluation analogue[17]).

De ce théorème on conclut facilement que:

Pour que le potentiel d'équilibre u relatif à un ensemble fermé F admette la valeur 1 *en un point frontière P de l'ensemble, il faut que l'on ait*

$$\lim_{T \to P} u(T) = 1$$

et il suffit que cette égalité ait lieu pour les points T extérieurs à F.

13. Le théorème suivant dont nous aurons besoin plus loin et qui est relatif à des potentiels bornés dus à des masses de signe quelconque complète sous certains rapports un théorème de M. FROSTMAN sur les potentiels finis dus à des masses positives (nº **7**).

Si un potentiel est borné supérieurement (inférieurement) sauf

---

16) Ce théorème est bien connu dans le cas classique où il est une conséquence immédiate de la surharmonicité et de la semicontinuité inférieure. Nous donnerons aux nºs **21** et **32** d'autres moyennes de structure plus compliquée qui sont mieux appropriées au cas non-newtonien que la moyenne ci-dessus.

17) *Thèse,* p. 37.

au plus dans un ensemble de capacité nulle, la variation positive (négative) de la masse s'annule sur tout ensemble de capacité nulle. Si donc le potentiel est borné dans les deux sens (sauf au plus etc.) la variation totale de la masse s'annule sur tout ensemble de capacité nulle.

Nous indiquons brièvement la démonstration. Admettons que le potentiel v dû à la masse ν soit borné supérieurement par la constante L sauf au plus dans un certain ensemble de capacité nulle. Soit alors E un ensemble fermé et borné quelconque de capacité nulle et désignons par E_n des ensembles fermés qui comprennent E et tendent en se rétrécissant vers E et qui admettent des potentiels d'équilibre u_n au sens strict. Les masses correspondantes étant désignées par μ_n on aura

$$\int_\Omega u_n d\nu = \int_\Omega v d\mu_n \leqq L \int_\Omega d\mu_n = L\mu_n(E_n).$$

On a $u_n = 1$ sur E_n et par conséquent sur E. D'autre part, puisque la capacité de E est nulle, $\mu_n(E_n)$ tend vers zéro et il en est de même de u_n en tout point extérieur à E. On trouve donc par un passage à la limite (légitime puisque les u_n sont bornés dans leur ensemble) $\nu(E) \leqq 0$, ce qui implique le théorème.

14. Nous terminons ce chapitre par quelques remarques sur la *transformation de* Kelvin. Soient F un ensemble fermé et M un point, extérieur à l'ensemble jusqu'à nouvel ordre. Effectuons une *inversion* de centre M, les points correspondants P et P' situés sur le même rayon issu de M étant liés par la relation

(1) $$r_{MP} \cdot r_{MP'} = 1.$$

Il vient

(2) $$r_{P'Q'} = \frac{r_{PQ}}{r_{MP} \cdot r_{MQ}}.$$

L'ensemble F sera transformé en un ensemble F' fermé et borné. Si deux distributions μ et μ', la première sur F, la seconde sur F', sont liées par la relation

(3) $$d\mu'(Q') = r_{MQ}^{\alpha - m} d\mu(Q),$$

on a évidemment

(4) $$\int_{F'} r_{P'Q'}^{\alpha - m} d\mu'(Q') = r_{MP}^{m - \alpha} \int_F r_{PQ}^{\alpha - m} d\mu(Q).$$

Il ressort immédiatement de cette formule que l'inversion (1) transforme des ensembles de capacité positive ou nulle en des ensembles présentant respectivement le même caractère.

Si le point M appartient à l'ensemble F, il n'y a rien à changer à la formule pourvu qu'aucune masse ne soit concentrée en M.[18]) Dans le cas contraire, la masse concentrée en M étant c, la formule (4) est à remplacer par cette autre

$$(4') \qquad \int_{F'} r_{P'Q'}^{\alpha-m} d\mu'(Q') + c = r_{MP}^{m-\alpha} \int_{F} r_{PQ}^{\alpha-m} d\mu(Q).$$

Le terme c est à considérer comme une contribution venant du point à l'infini que l'inversion fait correspondre au point M.

Dans un ordre d'idées général, à toute fonction donnée $f(P)$ on peut faire correspondre une fonction $f'(P')$ définie par la formule

$$(5) \qquad f'(P') = r_{MP}^{m-\alpha} f(P),$$

les points P et P' étant toujours liés par l'inversion (1). On voit par ce qui précède que la classe des fonctions $v(P)+c$, où $v(P)$ est un potentiel et c une constante, est invariante par rapport à cette transformation. Les fonctions de cette classe interviendront en particulier dans les derniers chapitres du présent travail.

Chapitre IV. — Les masses de Green et la fonction de Green.

15. Soient F un ensemble fermé, non nécessairement borné, de capacité positive, M un point extérieur à F, et désignons par μ' la distribution engendrant le potentiel d'équilibre sur l'ensemble fermé et borné F' que la transformation (1) du numéro précédent fait correspondre à F. En définissant alors la distribution de masse positive μ_M relative à F par la formule (3) du même numéro, il vient

$$(1) \qquad \int_{F} r_{PQ}^{\alpha-m} d\mu_M(Q) = r_{MP}^{\alpha-m},$$

égalité valable pour tout point P de F excepté peut-être pour un ensemble de capacité nulle. Le calcul étant réversible, on conclut que la distribution μ_M est entièrement déterminée par cette propriété-là. Le potentiel d'équilibre étant toujours $\leqq 1$, on a encore l'inégalité fondamentale

[18]) Evidemment l'ensemble F' cesse d'être borné.

$$h_M(P) = \int_F r_{PQ}^{\alpha-m} d\mu_M(Q) \leqq r_{MP}^{\alpha-m}, \tag{2}$$

P étant un point tout à fait arbitraire de l'espace. Nous appellerons les masses μ_M *masses de Green relatives à l'ensemble F et au pôle M.* En désignant maintenant par M et P deux points arbitraires extérieurs à F, on a la formule de symétrie $h_M(P) = h_P(M)$. On a en effet

$$h_M(P) = \int_F r_{PQ}^{\alpha-m} d\mu_M(Q) = \int_F d\mu_M(Q) \int_F r_{QT}^{\alpha-m} d\mu_P(T) =$$
$$= \iint_F r_{QT}^{\alpha-m} d\mu_M(Q)\, d\mu_P(T),$$

formule symétrique en M et en P. Observons que dans les intégrations on a négligé des ensembles de capacité nulle, ce qui est légitime d'après une remarque antérieure.

La fonction de Green se définit par la formule

$$G_M(P) = r_{MP}^{\alpha-m} - h_M(P), \tag{3}$$

M étant toujours extérieur à F. Elle est évidemment $\geqq 0$ pour tout point P, l'égalité ayant lieu en tout point de F sauf au plus dans un ensemble de capacité nulle. On a aussi la relation de symétrie $G_M(P) = G_P(M)$ (M et P extérieurs à F).

Les propriétés énumérées du potentiel d'équilibre se traduisent immédiatement en des propriétés correspondantes de la fonction de Green. Au lieu d'insister ici sur les détails (nous revenons sur quelques points dans la suite) nous nous contentons de formuler le critère suivant qui se déduit immédiatement du critère analogue donné au n° **12** pour le potentiel d'équilibre.

La fonction de Green s'annule en tout point intérieur de F. Pour qu'elle s'annule en un point frontière P, il faut que l'on ait

$$\lim_{T \to P} G_M(T) = 0 \tag{4}$$

et il suffit que cette égalité ait lieu pour les points T extérieurs à F.

Observons le fait capital que les masses de Green s'évanouissent sur tout sous-ensemble de capacité nulle. Cela est une conséquence immédiate du fait qu'elles engendrent un potentiel borné (*cf.* le n° **7**).

Chapitre V. — L'intégrale de Poisson et l'inégalité de Harnack.

16. Pour déterminer la fonction de GREEN relative à l'extérieur ou à l'intérieur d'une sphère nous aurons d'abord à résoudre le problème d'équilibre pour un volume sphérique. La solution due à MM. PÓLYA et SZEGŐ[19]) pour les dimensions $m = 1, 2, 3$ s'étend facilement à une dimension quelconque m.

Soient S^* l'intérieur d'une sphère S de rayon R, r_Q la distance du point Q au centre de la sphère, dQ l'élément de volume, α un exposant satisfaisant aux inégalités $0 < \alpha < 2$, et posons

$$(1) \qquad u(P) = \pi^{-\left(\frac{m}{2}+1\right)} \Gamma\left(\frac{m}{2}\right) \sin\frac{\pi\alpha}{2} \int\limits_{S^*} (R^2 - r_Q^2)^{-\frac{\alpha}{2}} r_{PQ}^{\alpha-m} dQ.$$

On a alors $u(P) = 1$ pour tout point P du domaine S^*, surface comprise.

Nous omettons ici le calcul qui est assez laborieux. Un point essentiel y est l'invariance de l'expression

$$D^S_{P_1 P_2} = \frac{r^2_{P_1 P_2}}{\delta^S_{P_1} \cdot \delta^S_{P_2}}$$

par rapport à toute inversion.[20]) Ici on a posé

$$\delta^S_P = \frac{|R^2 - r_P^2|}{R}.$$

Bien entendu, l'inversion transforme la sphère en même temps que les points. Rien n'est changé si le produit des rayons vecteurs réciproques est $\pm k^2$ au lieu de 1. Dans l'application actuelle on choisit comme centre de l'inversion le point P et on transforme la surface de la sphère S en elle-même. Le fait que $r_{PQ}^{-m} dQ$ est invariant, si l'on remplace Q par Q', facilite encore le calcul.

Du potentiel d'équilibre on déduit comme plus haut la fonction de GREEN. Soient S une sphère, M un point quelconque qui n'est pas situé sur S, et désignons par S^* celui des deux volumes sphériques limités par S auquel M est *extérieur*. On pose

---

[19]) G. PÓLYA und G. SZEGŐ, Über den transfiniten Durchmesser (Kapazitätskonstante) von ebenen und räumlichen Punktmengen, *J. f. Math.*, **165** (1931). p. 4—49.

[20]) *Cf.* M. RIESZ, Sur certaines inégalités dans la théorie des fonctions etc., *Fysiogr. Sällsk. Lund förh.*, **1** (1931), n° 4, p. 18—38.

(2) $$h_M(P) = \int_{S^*} r_{PQ}^{\alpha-m} \lambda_M(Q)\, dQ,$$

la masse de GREEN $\lambda_M(Q)\,dQ$ étant donnée par la formule

(3) $$\lambda_M(Q) = \pi^{-\left(\frac{m}{2}+1\right)} \Gamma\left(\frac{m}{2}\right) \sin\frac{\pi\alpha}{2} |R^2 - r_M^2|^{\frac{\alpha}{2}} |R^2 - r_Q^2|^{-\frac{\alpha}{2}} r_{MQ}^{-m},$$

et on obtient pour tout point P de S^* (surface comprise) $h_M(P) = r_{MP}^{\alpha-m}$.

La structure du noyau (3) est très analogue à celle du noyau classique de POISSON. On peut obtenir celui-ci en faisant tendre α vers 2. En effet, grâce au facteur sinus, qui tend vers zéro, il en sera de même de $\lambda_M(Q)$ en tout point Q intérieur à S^*. Quant aux points frontières Q, la contribution d'une lamelle mince adjacente à un élément dS de la surface autour de Q tendra vers $(\omega_m R)^{-1} |R^2 - r_M^2|\, r_{MQ}^{-m} dS$ où ω_m est la surface totale de la sphère unité, et on retombe sur le noyau classique.

La fonction de GREEN relative au volume S^* et au pôle M s'écrit évidemment

$$G_M(P) = r_{MP}^{\alpha-m} - h_M(P).$$

17. Cela posé, soit $v(P)$ un potentiel engendré par des masses quelconques situées ou à l'extérieur ou à l'intérieur d'une certaine sphère S, et désignons par S^* celui de ces deux domaines qui renferme les masses, la surface S comprise. Nous nous proposons d'exprimer les valeurs admises par le potentiel dans le domaine complémentaire $\Omega - S^*$ par celles admises dans le domaine S^*. On arrivera ainsi à une formule qui peut être considérée comme une généralisation de *l'intégrale de Poisson*. En effet, M étant un point quelconque du domaine $\Omega - S^*$, on a

(4) $$v(M) = \int_{S^*} v(Q)\, \lambda_M(Q) d(Q).$$

Cette formule, qui s'établit facilement par l'interversion de deux intégrations, forme un cas très particulier des formules (1) et (2) du n° **19** qu'on rencontrera plus loin.

Envisageons en particulier le cas où le potentiel $v(P)$ est non négatif sur S^*. Notre formule met en évidence qu'il en sera de même en $\Omega - S^*$. Mais il y en a plus. Soient M_0 et M deux points arbitraires de ce dernier domaine. On trouve alors immédiatement *l'inégalité de* HARNACK

$$(5) \qquad v(M) \leqq \left| \frac{R^2 - r_M^2}{R^2 - r_{M_0}^2} \right|^{\frac{\alpha}{2}} \left| \frac{R - r_M}{R + r_{M_0}} \right|^{-m} v(M_0).$$

Appliquée à une suite de potentiels $v_n(P)$ non négatifs avec des masses situées sur S^*, notre inégalité met en évidence les faits suivants. Lorsque $v_n(M)$ tend vers zéro en un seul point M de $\Omega - S^*$, il en sera de même en tout point de ce domaine et cela uniformément dans tout domaine qui avec sa frontière est intérieur au sens strict à $\Omega - S^*$. Fait analogue lorsque la série Σv_n converge en un seul point M de $\Omega - S^*$. Ces faits s'étendent de proche en proche à tout domaine connexe qui ne renferme pas de masses.

Dans la théorie classique, on parle de fonctions harmoniques positives dans un domaine et non pas de potentiels. Cependant il est bien facile de faire rentrer le cas classique sous le point de vue adopté ici. Soit en effet D un domaine simplement connexe et u une fonction harmonique dans D. En désignant par D' un domaine intérieur au sens strict à D et de frontière assez régulière, u pourra à l'intérieur de D' s'écrire comme potentiel d'une simple couche étalée sur la frontière de D'. Pour l'extérieur de D' ce potentiel est identique à la solution du problème de Dirichlet extérieur, posé pour la frontière de D' avec des valeurs frontières coïncidant avec celles du u. Alors si u est positif, il en sera de même du potentiel dans l'espace entier.

Chapitre VI. — Etude approfondie de la fonction de Green.

18. On suppose dans ce chapitre et dans toute la suite que *F désigne un ensemble fermé de capacité positive.* Quant à l'exposant α on suppose ici $0 < \alpha \leqq 2$, le signe d'égalité inclus.

Nous donnerons d'abord une généralisation d'un théorème important de M. Bouligand.

Soient Q' un point frontière de l'ensemble F et D un domaine connexe dont la frontière est formée par F ou un sous-ensemble de F. Si la fonction de Green s'annule en Q' pour une seule position du pôle M dans D, elle s'annulera en Q' pour toute autre position du pôle dans D.

En tenant compte du critère donné à la fin du chap. IV, le théorème ci-dessus se réduit au théorème suivant.

$P_1, P_2, \ldots, P_n, \ldots$ étant des points extérieurs à F tendant vers Q', si la relation

$$\lim_{P_n \to Q'} G_M(P_n) = 0$$

a lieu pour une position de M dans D, elle a lieu pour toute autre position de ce point dans D.

Observons d'abord que l'on a $G_M(P_n) = G_{P_n}(M)$. Le potentiel $G_{P_n}(M)$, dû à des masses situées sur F et en P_n, étant partout $\geqq 0$, la première application que nous avons donnée de l'inégalité de HARNACK met en évidence que si ce potentiel tend vers zéro avec $n \to \infty$ pour *un* point M de D, il tend vers zéro en *tout* point de D.

Cela étant, les points exceptionnels de la frontière, c'est-à-dire les points Q'' pour lesquels on a $G_M(Q'') > 0$, sont les mêmes pour tout point M du domaine D et, grâce aux résultats sur les points exceptionnels de la fonction de Green à pôle fixe, ils forment un ensemble F''_D de capacité nulle ne dépendant que du domaine adjacent D. En remarquant qu'il existe au plus un ensemble dénombrable de domaines connexes D limités par F, on conclut que l'ensemble F'' somme des F''_D est encore un ensemble de capacité nulle, évidemment mesurable au sens de BOREL. Les points Q'' de l'ensemble F'' sont appelés *points irréguliers de F*. Ils sont caractérisés par le fait qu'il existe au moins un point M extérieur à F, de sorte que $G_M(Q'') > 0$. Nous introduisons aussi les ensembles $F'_D = F - F''_D$ et leur partie commune $F' = F - F''$. Les points Q' de F'_D sont caractérisés par la propriété: $G_M(Q') = 0$ pour tout point M appartenant à D et ceux de F' par la propriété que cette égalité a lieu pour tout point M extérieur à F. Nous appelons les points de F' *points réguliers de F* et ceux de F'_D *points réguliers de F par rapport au domaine D*. Ce sont ces derniers points que d'ordinaire on désigne comme points réguliers quand il s'agit d'un problème de DIRICHLET posé pour le domaine D.

Notons encore les faits suivants qui ont lieu *dans le cas newtonien* et qui résultent immédiatement des propriétés correspondantes du potentiel d'équilibre. M étant un point de D, la masse de GREEN μ_M se trouve sur la frontière F_D de D et l'ensemble F_D renferme l'ensemble irrégulier F''_D. Les masses de GREEN μ_M relatives aux ensembles F_D, F et $\Omega - D$ sont identiques. Quelques conséquences importantes de ces faits seront rassemblées au n° **24**.

19. Nous sommes maintenant en mesure de résoudre le problème suivant que nous formulons d'abord d'une manière un peu vague. Soit $v(P)$ un potentiel engendré par des masses (inconnues) réparties sur l'ensemble F. Les valeurs de v étant connues sur F, calculer v en un point extérieur à F. Nous donnerons la solution de ce problème sous certaines conditions restrictives très naturelles en déduisant d'abord une formule tout à fait générale.

Le potentiel v étant engendré par des masses ν situées sur F, on a en tout point M extérieur à F

$$(1) \qquad v(M) = \int_F v(Q)\, d\mu_M(Q) + \int_{F''} G_M(Q)\, d\nu(Q).$$

Cette formule résulte immédiatement des formules (1) et (3) du n° **15** et du fait que la fonction de Green $G_M(Q)$ s'annule en tout point régulier. Il est d'ailleurs clair que la dernière intégrale peut être prise sur F''_D au lieu de F''. Si les masses ν ne sont pas de signe constant, $v(Q)$ peut devenir indéterminé dans un sous-ensemble de F de capacité nulle, ce qui n'exerce aucune influence sur la validité de notre formule, puisque les masses de Green s'évanouissent sur un tel sous-ensemble (n° **15**).

La formule (1) se simplifie beaucoup, si les masses ν s'annulent sur l'ensemble F'' (ou F''_D). Il vient alors

$$(2) \qquad v(M) = \int_F v(Q)\, d\mu_M(Q).$$

Signalons quelques cas importants où cette formule simplifiée a lieu. D'abord il en est ainsi si l'ensemble F'' est vide, c'est-à-dire que tous les points de F sont réguliers. P. ex. cela arrive toujours, d'après les résultats de M. Frostman, si tout point de l'ensemble F satisfait à la condition de Poincaré, ce qui consiste en ce que chaque point de F peut être considéré comme le sommet d'un cône de révolution dont la pointe fait partie de F. Les volumes sphériques traités au chapitre précédent fournissent les exemples les plus simples. La formule (2) est encore valide si 1) $v(P)$ est un potentiel borné partout sauf au plus dans un ensemble de capacité nulle, 2) $v(P)$ est un potentiel partout fini, dû à des masses positives ou la différence de tels potentiels. En effet, dans ces circonstances les masses ν s'annulent sur l'ensemble F'' qui est de capacité nulle (n^os^ **13** et **7**).

Voici deux applications de ces résultats; on y suppose toujours que les masses sont situées sur F.

La formule simplifiée (2) *ayant lieu pour une raison ou une autre et le potentiel v étant* $\geqq 0$ *sur F en tout point où il est bien déterminé, il sera* $\geqq 0$ *en tout point extérieur à F.*

La masse totale de GREEN $\mu_M(F)$ est évidemment toujours $\leqq 1$.[21]) On peut préciser ce résultat dans le cas où F admet un potentiel d'équilibre u, ce qui, comme on sait, a toujours lieu si F est borné. On a alors $\mu_M(F)=u(M)$. Ces préliminaires posés, on comprendra l'intérêt de l'énoncé suivant qui se déduit facilement de la formule (1).

En désignant par l et L les bornes inférieure et supérieure sur F d'un potentiel v bien déterminé en tout point de F, on aura en tout point M extérieur à F

$$l\mu_M(F) \leqq v(M) \leqq L\mu_M(F).$$

Chapitre VII. — Le balayage de masses.

20. Soit ν une distribution quelconque de masses, situées en partie sur l'ensemble F, en partie sur l'ensemble complémentaire $\Omega - F$. Le problème de balayage consiste à remplacer ces dernières masses par des masses réparties sur F, de sorte que le potentiel reste conservé sur F (pour autant que cela est possible). Dans le cas où la masse extérieure à F est concentrée en un seul point M et que c'est la masse unité, ce sont les masses de GREEN μ_M qui fournissent la solution de ce problème. Le cas général se réduit facilement à ce cas particulier.

Soit e un sous-ensemble de F, mesurable au sens de BOREL, d'ailleurs arbitraire. La formule

$$\nu^*(e) = \int_{\Omega - F} \mu_M(e)\, d\nu(M) \tag{1}$$

définit une distribution de masses sur F dont on va préciser les propriétés tout à l'heure. P étant un point arbitraire de l'espace, il vient par la dernière formule et les formules (1) et (3) du n° **15**

$$\int_{\Omega - F} r_{PM}^{\alpha - m}\, d\nu(M) = \int_F r_{PO}^{\alpha - m}\, d\nu^*(O) + \int_{\Omega - F} G_M(P)\, d\nu(M).$$

En posant maintenant, sur l'ensemble F, $\bar{\nu} = \nu + \nu^*$, nous aurons

[21]) Pour le voir, on peut p. ex. faire tendre P vers l'infini dans l'inégalité (2) du n° **15**.

$$v(P)=\int_{\Omega} r_{PM}^{\alpha-m}\,d\nu(M)=\bar{v}(P)+\int_{\Omega-F} G_M(P)\,d\nu(M),$$

où

$$\bar{v}(P)=\int_{F} r_{PQ}^{\alpha-m}\,d\bar{\nu}(Q).$$

On a donc $\bar{v}(Q')=v(Q')$ *en tout point régulier* Q' *de* F.

Évidemment, cette égalité n'a pas lieu en général dans l'ensemble irrégulier F'' et dans l'ensemble ouvert $\Omega-F$. On peut préciser cet énoncé dans le cas important où la masse ν est de signe constant dans $\Omega-F$, p. ex. de signe *positif. Alors par le balayage le potentiel n'est jamais augmenté.* Il est diminué en tout point de $\Omega-F$ dans le cas non-newtonien, tandis que dans le cas newtonien il n'est diminué que dans ceux des domaines connexes adjacents qui, avant le balayage, contenaient de la masse.

La masse $\mu_M(e)$ s'annulant sur tout ensemble e de capacité nulle, on voit par la formule de définition (1) que la masse $\nu^*(e)$ résultant du balayage s'annule aussi sur tout tel ensemble. Par conséquent, elle s'annule sur l'ensemble des points irréguliers F'', et alors la distribution $\bar{\nu}=\nu+\nu^*$ est identique à ν sur cet ensemble. En tenant compte de ceci et de ce que $\bar{v}(Q)=v(Q)$ sauf au plus dans l'ensemble F'', on obtient à l'aide de la formule (1) du n° **19** l'expression suivante pour le potentiel $\bar{v}(M)$ en un point M extérieur à F

$$\bar{v}(M)=\int_{F} v(Q)\,d\mu_M(Q)+\int_{F''} G_M(Q)\,d\nu(Q). \tag{2}$$

Si en particulier l'ensemble irrégulier F'' n'existe pas ou si la variation totale de ν sur cet ensemble s'évanouit, l'expression devient

$$\bar{v}(M)=\int_{F} v(Q)\,d\mu_M(Q). \tag{3}$$

On pourra juger tout à l'heure de l'utilité des formules obtenues.

21. Un cas particulier intéressant, qui nous donnera aussi une formule de la moyenne, est celui où l'on balaye les masses sur l'extérieur S^* d'une sphère S de rayon R et de centre M. On trouve par spécialisation de la fonction $\lambda_M(Q)$ (formule (3) du n° **16**) que le nouveau potentiel aura en M la valeur

(4) $$v_{S^*}(M) = \int_{S^*} v(Q)\varkappa_M(Q)\,dQ,$$

avec

(5) $$\varkappa_M(Q) = \pi^{-\left(\frac{m}{2}+1\right)}\Gamma\left(\frac{m}{2}\right)\sin\frac{\pi\alpha}{2}\,R^\alpha(r_Q^2 - R^2)^{-\frac{\alpha}{2}}r_Q^{-m},$$

où l'on a posé $r_{MQ} = r_Q$. De nos résultats généraux sur le balayage on conclut immédiatement que si les masses engendrant v sont *positives*, on a

(6) $$v_{S^*}(M) \leqq v_M.$$

On en déduit facilement

(7) $$v(M) = \lim_{R\to 0} v_{S^*}(M),$$

égalité valable pour des potentiels dus à des masses de signe quelconque en tout point de détermination M. Il suffit de le montrer pour des masses positives. Or dans ce cas on le déduit de la semicontinuité inférieure, de l'identité fournie par un calcul direct

$$\int_{S^*} \varkappa_M(Q)\,dQ = 1$$

et du fait que la même intégrale étendue à la partie commune de S^* et de la sphère $r_Q \leqq \delta$ (δ un nombre positif arbitraire) tend vers 1 lorsque R tend vers 0. Ajoutons encore qu'il est à peu près évident que $v_{S^*}(M)$ croît lorsque le rayon R de la sphère décroît (*cf.* d'ailleurs les remarques du n° **24** sur les balayages successifs).

Cette valeur moyenne joue le même rôle pour les potentiels non-newtoniens que joue la valeur moyenne de Gauss

(8) $$\frac{1}{S}\int_S v(Q)\,dS$$

pour le potentiel newtonien. La dernière intégrale est étendue à la surface de la sphère S de centre M, et nous avons désigné en même temps par S la surface totale de la sphère. Pour $\alpha = 2$ la valeur moyenne (4) devient à la limite identique à celle de Gauss (*cf.* le n° **16**).

22. Nous aurons besoin dans le dernier chapitre d'une extension d'un théorème de M. Evans[22]) concernant la fonction

[22]) C. G. Evans, Potentials of Positive Mass I, *Transactions American Math. Soc.,* **37** (1935), p. 226—253, spéc. p. 231.

limite d'une suite croissante de potentiels v_n dus à des masses positives. M. EVANS suppose que les masses sont portées par un ensemble fermé et *borné* F et trouve sous cette hypothèse que les potentiels v_n tendent vers un potentiel v dû à des masses positives portées par F, sauf dans le cas où ils tendent identiquement vers ∞. Nous aurons à nous placer dans des conditions un peu plus générales en admettant que les masses peuvent s'étendre à l'espace entier. On a le théorème suivant, valable pour tout exposant $0 < \alpha \leqq 2$.

Une suite croissante de potentiels $v_n(P)$ dus à des masses positives qui ne tend pas identiquement vers l'infini, tend vers une fonction de la forme $v(P)+c$ où $v(P)$ est encore un potentiel dû à des masses positives et c une constante $\geqq 0$.

La démonstration dans le cas non-newtonien est tout analogue à celle ayant rapport au cas classique, on n'aura qu'à remplacer la valeur moyenne de GAUSS par celle donnée par la formule (4). Quant au cas classique, le lecteur n'aura aucune difficulté à compléter la démonstration de M. EVANS en ce qui concerne la légère extension donnée ci-dessus.

23. Comme application de nos résultats sur le balayage nous examinerons la distribution de la masse de GREEN sur F lorsque M est un point très voisin d'un point frontière régulier Q' de F, en montrant que presque toute la masse se trouve au voisinage de Q'. La démonstration se base sur l'existence de potentiels auxiliaires convenables dus à des masses positives réparties sur F. La méthode du balayage nous fournira de tels potentiels.

Admettons un instant qu'on puisse former un potentiel $v(P)$ dû à des masses positives réparties sur F qui ait les propriétés suivantes. 1) $v(P)=1$ au point frontière régulier Q', 2) $v(P)$ est partout $\leqq 1$, 3) $v(P) < \varepsilon$ à l'extérieur d'une sphère S de centre Q' et de rayon R, ce rayon et le nombre ε étant très petits. Cela posé, on a par la formule (2) du n° **19**, en désignant par F_S le sous-ensemble de F contenu dans S,

$$v(M)=\int_F v(Q)\,d\mu_M(Q) \leqq \int_{F_S} d\mu_M(Q)+\varepsilon \int_{F-F_S} d\mu_M(Q) =$$

$$=\mu_M(F_S)+\varepsilon\mu_M(F-F_S) \leqq \mu_M(F_S)+\varepsilon,$$

où l'on a tenu compte des inégalités $\mu_M(F-F_S) \leqq \mu_M(F) \leqq 1$. Le point M étant voisin de Q', on aura à cause de la semicontinuité inférieure, $v(M) > v(Q') - \delta = 1 - \delta$, où δ est arbitrairement petit, et alors

$$1 \geqq \mu_M(F_S) > 1 - (\varepsilon + \delta)$$

et par suite

$$\mu_M(F - F_S) < \varepsilon + \delta.$$

Il reste à construire le potentiel auxiliaire $v(P)$. Soit s une sphère de centre Q' et de rayon $r < R$. Le potentiel d'équilibre $u(P)$ du volume sphérique s est très petit à l'extérieur de la sphère S si le rapport r/R est assez petit. En outre on a $u(Q') = 1$ et $u(P) \leqq 1$. Le balayage des masses de u sur F fournit maintenant un potentiel $v(P)$ qui remplit toutes les conditions posées. En résumé :

Si Q' est un point frontière régulier de l'ensemble F, la masse de Green μ_M qui se trouve dans un voisinage arbitrairement petit de Q' tend vers 1, *et la masse extérieure à ce voisinage tend vers* 0, *lorsque le point M tend d'une manière quelconque vers Q'. La réciproque est vraie.*

Le lecteur n'aura aucune difficulté à vérifier ce dernier énoncé en observant que ladite concentration de la masse de GREEN entraîne $\lim\limits_{P \to Q'} G_M(P) = 0$, pour toute position du pôle M, ce qui d'après le critère du n° **15** suffit pour assurer la régularité de Q'.

On peut préciser le théorème ci-dessus par l'énoncé suivant que nous donnons ici sans démonstration.

$V(P)$ étant un potentiel quelconque, on a avec les notations utilisées plus haut

$$\lim_{M \to Q'} \int_{F - F_S} V(Q)\, d\mu_M(Q) = 0.$$

24. Nous rassemblons ici quelques *remarques sur le cas newtonien* qui sont des conséquences faciles des faits indiqués à la fin du n° **18.**

M étant toujours un point d'un certain domaine connexe D adjacent à F, la formule (1) du n° **19** a lieu même dans le cas où les masses sont réparties sur l'ensemble (fermé) $\Omega - D$. Il en est de même de la formule (2) de ce numéro sous les conditions qu'on y trouve précisées.

Quant au balayage des masses, il ressort du calcul donné au n° **20** que si l'on ne balaye sur F que celles des masses qui se trouvent dans le domaine D, le potentiel est conservé en tout point de F'_D, c'est-à-dire en tout point régulier par rapport au domaine D[23]) (*cf.* le n° **18**).

Cela posé, il est facile de préciser les résultats du numéro précédent dans le cas newtonien. En effet, *ces résultats subsistent pour tout point Q' de F'_D.*[24]) Pour former le potentiel auxiliaire $v(P)$, on ne balaye sur F que celles des masses du potentiel d'équilibre de la sphère s qui se trouvent dans le domaine D. Le nouveau potentiel $v(P)$ a les mêmes propriétés essentielles que l'ancien sauf que ces masses sont réparties sur $\Omega - D$ au lieu de F. Alors, d'après ce qu'on vient de dire, on peut encore appliquer la formule (2) du n° **19**, et la démonstration s'achève comme plus haut.

25. Nous allons maintenant étudier de plus près la masse $\mu_M(e)$ portée par un sous-ensemble arbitraire e de F mesurable au sens de Borel. Soient $\overline{\overline{F}}$ un second ensemble fermé comprenant F, M un point extérieur à $\overline{\overline{F}}$. Il est presque évident qu'on peut trouver les masses de Green μ_M par rapport à F en formant d'abord les masses de Green $\overline{\overline{\mu}}_M$ par rapport à $\overline{\overline{F}}$ et en balayant ensuite sur F la partie de la masse portée par $\overline{\overline{F}} - F$. En effet le procédé en question conduit à des masses positives distribuées sur F dont le potentiel est égal à $r_{MQ}^{\alpha - m}$ en tout point Q de F à l'exception au plus d'un ensemble de capacité nulle. Or, comme on a vu plus haut, il n'y a que la distribution μ_M qui engendre un tel potentiel. Ayant recours à la formule (1), ce résultat peut s'écrire

$$\mu_M(e) = \overline{\overline{\mu}}_M(e) + \int\limits_{\overline{\overline{F}} - F} \mu_Q(e)\, d\overline{\overline{\mu}}_M(Q). \tag{9}$$

Cela posé, soit M un point extérieur à F et décrivons une sphère s contenant le point M à son intérieur et assez petite pour qu'elle ne renferme aucun point de F. Comme ensemble $\overline{\overline{F}}$ on pourra alors prendre l'extérieur de cette sphère, frontière comprise. Les formules (2) et (3) du n° **16** mettent alors en évidence que $d\overline{\overline{\mu}}_M(Q)$ peut s'écrire $\lambda_M(Q)\,dQ$, où $\lambda_M(Q)$ est une fonction analytique et régulière de M. Par conséquent, il en est de même de $\mu_M(e)$. On voit aussi que dans le cas non newtonien tout sous-ensemble de F mesurable au sens de Borel et de mesure posi-

[23]) Cela a lieu aussi dans le cas général $0 < \alpha \leqq 2$.

[24]) *Cf.* C. de la Vallée Poussin, *loc. cit.*, p. 183 et 203.

tive porte une masse positive. Dans le cas newtonien la masse $\bar{\bar{\mu}}_M$ est toute entière située sur la surface de s et alors $\bar{\bar{\mu}}_M(e)$ s'évanouit. Par dQ on entend alors l'élément d'aire de la sphère s et $\lambda_M(Q)$ devient le noyau classique de POISSON, manifestement harmonique en M. Il en est donc de même de la masse $\mu_M(e)$.[25])

Soit maintenant de nouveau α un exposant quelconque, $0 < \alpha \leq 2$. Alors, M_0 et M étant deux points fixes, d'ailleurs arbitraires, d'un domaine connexe adjacent à F, il existe un nombre positif K, *indépendant du sous-ensemble e de F*, tel que l'on ait

(10) $$\mu_M(e) \leq K\mu_{M_0}(e).$$

Pour le voir, admettons d'abord qu'il existe une sphère s entourant les points M_0 et M et laissant F à son extérieur. Alors, avec les notations de tout à l'heure, l'inégalité en question pour les masses $\bar{\bar{\mu}}_{M_0}$ et $\bar{\bar{\mu}}_M$ s'établit par le même calcul qui donnait l'inégalité de HARNACK (n° **17**). Le passage aux masses μ_{M_0} et μ_M se fait alors par la formule (9). Pour M_0 et M arbitraires, on opère de proche en proche en intercalant un nombre fini de points auxiliaires.

Le résultat ci-dessus implique qu'il existe une fonction positive et bornée $\gamma_M(Q)$ telle que

$$\mu_M(e) = \int_e \gamma_M(Q)\, d\mu_{M_0}(Q).$$

La fonction $\gamma_M(Q)$ joue dans le cas général ($0 < \alpha \leq 2$, F un ensemble fermé de capacité positive quelconque) le rôle du noyau de POISSON, l'intégrale de POISSON étant alors à former par rapport à la distribution de masses μ_{M_0}. P. ex. la formule (2) du n° **19** s'écrit pour tout point M appartenant au même domaine connexe adjacent que M_0

$$v(M) = \int_F v(Q)\, \gamma_M(Q)\, d\mu_{M_0}(Q).$$

Chapitre VIII. — Prolongement d'une fonction donnée sur un ensemble fermé.

26. Soit $f(Q)$ une fonction donnée sur l'ensemble F, et définissons pour les points M extérieurs à F un *prolongement* $\bar{f}(M)$ de f par la formule

---

25) *Cf.* C. DE LA VALLÉE POUSSIN, *loc. cit.*, p. 182 et 203.

$$\bar{f}(M) = \int_{F} f(Q)\, d\mu_M(Q). \tag{1}$$

Pour que l'intégrale ait un sens, nous admettrons que la fonction donnée $f(Q)$ est mesurable au sens de BOREL et que $|f(Q)|$ est majoré par une fonction de la forme $v(Q)+c$, où v est un potentiel dû à des masses positives et c une constante positive. Dans le cas déjà considéré où $f(Q)$ est un potentiel engendré par des masses situées sur F et s'évanouissant sur la partie irrégulière F'' de F, $\bar{f}(M)$ sera, d'après la formule (2) du n° **19,** identique à ce potentiel à l'extérieur de F, $\bar{f}(M) = v(M)$. Plus généralement, si $f(Q)$ est le potentiel de masses situées n'importe où, sauf sur F'', $\bar{f}(M)$ ne sera, d'après la formule (3) du n° **20,** que le potentiel résultant du balayage de ces masses sur F, $\bar{f}(M) = \bar{v}(M)$. Enfin dans le cas où $f(Q) = 1$ dans un certain sous-ensemble e de F et $= 0$ dans $F-e$, on a $\bar{f}(M) = \mu_M(e)$.

Pour ce qui suit, il sera commode de définir $\bar{f}$ aussi pour les points de F en y posant $\bar{f}(Q) = f(Q)$. Cela posé, on voit facilement que $\bar{f}$ est continu en tout point extérieur à F et en tout point régulier de F qui est en même temps point de continuité de f (n° **23**). Pour les points extérieurs on peut même dire davantage, savoir que $\bar{f}$ y est analytique et régulière (n° **25**).

Soit de nouveau $\bar{\bar{F}}$ un ensemble fermé comprenant notre ensemble F. La fonction $\bar{f}$ étant définie par ce qui précède sur tout l'ensemble $\bar{\bar{F}}$, on pourra en former le prolongement pour l'extérieur de $\bar{\bar{F}}$ par une formule analogue à (1) portant sur $\bar{\bar{F}}$. *Ce nouveau prolongement est identique au prolongement original dans tout l'extérieur de* $\bar{\bar{F}}$. En effet, cela découle immédiatement de la formule de composition pour les masses de GREEN correspondantes donnée au numéro précédent.

Nous indiquerons maintenant la démonstration d'un théorème d'unicité où il faut de nouveau exclure l'exposant newtonien $\alpha = 2$.

Si le prolongement d'une fonction f donnée sur un ensemble fermé F s'annule identiquement dans un domaine D de $\Omega - F$, ce prolongement s'annule identiquement dans tout $\Omega - F$ et la fonction donnée f est zéro presque partout[26]).

[26]) On sait que ce théorème est faux dans le cas newtonien. Le meilleur théorème de ce genre qui puisse être vrai c'est que, sous la condition posée, f s'annule presque partout par rapport à la masse de GREEN. Je n'ai réussi à démontrer ce théorème que dans le cas bien facile où f est continu (presque partout).

Ce théorème est déjà démontré au n° **11** dans le cas particulier où f est le potentiel d'une distribution de masses sur F. Pour démontrer le théorème donné ici, on peut, sans restreindre la généralité[27]), supposer que le domaine D embrasse l'extérieur d'une sphère de rayon très grand. L'ensemble F est alors borné et intérieur à une certaine sphère S de rayon R. On pourra, pour arriver au prolongement dans D, effectuer le prolongement dans S et le poursuivre ensuite dans l'extérieur de S. Par cela on aura réduit le théorème au cas où f est donné de prime abord à l'intérieur S^* d'une sphère S. On aura alors dans les notations des n^{os} **16** et **17** pour tout point M extérieur à S

$$\bar{f}(M) = \int_{S^*} f(Q)\, \lambda_M(Q)\, dQ,$$

cette intégrale s'annulant identiquement dans D. D'où l'on voit par le raisonnement du n° **11** que tous les moments de $f(Q)\,(R^2 - r_Q^2)^{-\frac{\alpha}{2}}$ s'annulent.

27. Terminons ce chapitre par quelques indications sur le cas newtonien. Pour nous approcher de l'ordre d'idées habituel, nous admettons que l'ensemble fermé F ne contienne aucun point intérieur et que la fonction f donnée sur F soit continue. Soit alors D un domaine dont la frontière est constituée par F. Le prolongement de f est harmonique dans D (n° **25**), il est borné et il admet en vertu de la propriété de la masse de Green démontrée au n° **23**, comme valeurs limites les valeurs données f en tout point régulier de F (et selon le n° **24** même en tout point de F'_D)[28]). Alors il est évident d'après le théorème d'unicité de M. Vasilesco[29]) que notre prolongement est identique à la solution

---

[27]) En effet la correspondance donnée par la formule (1) est invariante par rapport à la transformation de Kelvin.

[28]) *Cf.* C. de la Vallée Poussin, *loc. cit.*, p. 184 et 205.

[29]) Le théorème de M. Vasilesco dit qu'une fonction qui est harmonique et bornée dans D et qui admet la valeur limite zéro en tout point frontière de D sauf au plus dans un ensemble de capacité nulle, s'annule identiquement dans D; *voir* F. Vasilesco, Sur la méthode du balayage de Poincaré, son extension par M. de la Vallée Poussin et le problème de Dirichlet généralisé, *Journal de Math.*, 14 (1935), p. 209—227 et (aussi pour une généralisation) O. Frostman, La méthode de variation de Gauss et les fonctions sousharmoniques, *ces Acta*, 8 (1937), p. 149—159, spéc. p. 158. Ajoutons que le théorème de M. Vasilesco subsiste, si l'on remplace l'en-

dans le sens de M. Wiener du problème de Dirichlet pour le domaine D avec les valeurs limites f.

Chapitre IX. — Les suites bornées de potentiels et leurs applications au problème du prolongement et au problème de Dirichlet.

28. On a vu plus haut combien il est simple de caractériser, *sans recourir aux masses de* Green, le prolongement d'une fonction f donnée sur un ensemble fermé F dans le cas où f coïncide sur F avec un potentiel dont les masses sont situées sur F. En effet, sous des conditions très générales, il y aura alors coïncidence dans l'espace entier. Dans le chapitre suivant nous donnerons certaines conditions sous lesquelles une fonction peut s'écrire comme potentiel. On sait par la théorie classique que la continuité de la fonction ne suffit pas pour qu'une telle représentation soit possible même dans le cas où F est un ensemble très régulier, p. ex. la surface d'une sphère. Nous allons montrer ici que par les suites bornées de potentiels on arrive dans des cas très généraux au prolongement des fonctions et alors, en particulier dans le cas newtonien, à la solution du problème de Dirichlet généralisé.

Soient $v_1, v_2, \ldots, v_n, \ldots$ des potentiels tels que $|v_n| \leqq c$, c étant une constante positive. Nous allons considérer les fonctions limites de telles suites bornées. Avant d'aborder le problème indiqué plus haut remarquons en passant que les potentiels de double couche de la théorie classique peuvent être écrits comme fonctions limites de cette espèce. A titre d'exemple considérons dans l'espace à trois dimensions la double couche de la densité $1/4\pi$ étalée sur la surface de la sphère unité. On sait que le potentiel correspondant est égal à 1, $^1/_2$, 0 à l'intérieur, sur la surface et à l'extérieur de la sphère. Étalons maintenant uniformément les masses respectives $1/2\varepsilon$ et $-1/2\varepsilon$ sur les sphères concentriques de rayons $1/(1+\varepsilon)$ et $1/(1-\varepsilon)$. Le potentiel de simple couche correspondant v_ε a la valeur 1 à l'intérieur de la première sphère et la valeur 0 à l'extérieur de la seconde. Entre les deux sphères on a

semble exceptionnel de capacité nulle par un ensemble qui est de mesure nulle par rapport à la masse de Green relative à la frontière de D et à un point arbitraire de D.

$$0 < v_\varepsilon = \frac{1}{2\varepsilon}\left[\frac{1}{r} - (1-\varepsilon)\right] < \frac{1}{2\varepsilon}\,[1+\varepsilon-(1-\varepsilon)] = 1$$

et en particulier sur la sphère de rayon 1, $v_\varepsilon = \frac{1}{2}$. La suite v_ε, majorée par la constante 1, admet donc bien comme fonction limite le potentiel de double couche en question.

Retournons maintenant à notre problème original et montrons que le prolongement de toute fonction *continue* f peut s'obtenir comme fonction limite d'une suite bornée de potentiels. Nous supposons encore pour simplifier que l'ensemble F soit borné. Soit alors f une fonction continue donnée sur cet ensemble. Nous construisons d'abord une suite bornée de potentiels, avec leurs masses sur F, qui tendent vers f en tout point régulier de F. Complétons à cet effet la fonction f de façon à obtenir une fonction continue définie dans l'espace entier qui s'annule dans la partie éloignée de l'espace. On peut évidemment trouver des fonctions f_n continues avec leurs dérivées des deux premiers ordres, s'annulant aussi identiquement à l'infini et tendant (uniformément) vers la fonction f dans l'espace entier et en particulier sur l'ensemble F. Ces fonctions f_n peuvent s'écrire (n° **10**)

$$f_n(P) = I^\alpha k_n(P),$$

la signification de $k_n(P)$ étant évidente. On aura donc en particulier pour les points P de F

$$f_n(P) = \int\limits_{\Omega} r_{PQ}^{\alpha-m} \frac{k_n(Q)}{H_m(\alpha)}\, dQ.$$

En balayant les masses sur F, on aura de nouveaux potentiels coïncidant avec les premiers en tout point régulier de F. Je dis que la nouvelle suite est encore bornée. En effet, les masses de f_n ayant une dérivée spatiale continue, ces masses s'évanouissent sur tout ensemble de capacité nulle et en particulier sur la partie irrégulière de F. Par conséquent le potentiel résultant du balayage sera donné par la formule (3) du n° **20** (en y remplaçant $v(Q)$ par $f_n(Q)$), ce qui met en évidence que toute borne de la première suite est en même temps une borne de la seconde.

Il est d'un certain intérêt de s'affranchir dans le résultat qui précède de l'hypothèse de continuité. Il suffit p. ex. que la fonction f puisse, sur l'ensemble F, être approchée par une suite bornée de fonctions f_n de la nature indiquée ou, ce qui revient au même, par une suite bornée de fonctions continues d'ailleurs quelconques[30]).

Nous allons montrer que les potentiels qu'on vient de construire admettent une fonction limite aussi en dehors de l'ensemble F et que cette fonction limite est précisément le prolongement de la fonction f. En effet, on peut énoncer à ce sujet le théorème suivant.

Soit $v_1, v_2, \ldots, v_n, \ldots$ *une suite bornée de potentiels dont les masses sont situées sur l'ensemble F. Si ces potentiels tendent vers une fonction f, sur tout l'ensemble F, sauf au plus sur un sous-ensemble de capacité nulle, ils tendent en dehors de F vers le prolongement* $\bar{f}_F$ *de f.*

En effet, puisque les v_n sont bornés, leurs masses s'annulent d'après le n° **13** sur la partie irrégulière de F et on aura

$$v_n(M) = \int_F v_n(Q)\, d\mu_M(Q),$$

tandis que le prolongement de f est donné par la formule analogue

$$\bar{f}_F(M) = \int_F f(Q)\, d\mu_M(Q).$$

Puisque $f(Q) = \lim v_n(Q)$ sauf au plus dans un ensemble de capacité nulle, où la masse de GREEN s'annule, on n'aura qu'à intervertir l'ordre de l'intégration et du passage à la limite pour arriver à notre énoncé. Cette interversion est légitime, puisque les v_n sont bornés[31]) dans leur ensemble.

Voici un exemple pour illustrer ce résultat. Prenons comme ensemble F l'intérieur et la surface de la sphère unité et comme fonction f la fonction $=1$ à l'intérieur et $=0$ sur la surface. Cette fonction appartient à la classe de fonctions considérée tout

---

[30]) C'est-à-dire qu'il suffit que la fonction f soit bornée et qu'elle appartienne à la première classe de BAIRE. Tout à l'heure on trouvera une généralisation qui va beaucoup plus loin.

[31]) On peut outre les suites bornées de potentiels introduire la classe plus générale des *suites majorées de potentiels,* c'est-à-dire des suites v_n telles que $|v_n| \leqq V + C$, V étant un potentiel fixe dû à des masses positives et C une constante positive. L'intégrale $\int (V+C)\, d\mu_M$ étant toujours convergente, l'interversion effectuée dans le texte est, d'après un théorème bien connu de M. LEBESGUE, toujours légitime pour les suites majorées convergentes.

à l'heure. Les suites bornées de potentiels avec leurs masses à l'intérieur et sur la surface de la sphère qui y tendent vers la fonction donnée tendent à l'extérieur vers le prolongement de cette fonction. Or ce prolongement est de caractère tout différent dans le cas newtonien et dans le cas non-newtonien. Dans le premier cas, ce prolongement donné par l'intégrale classique de POISSON pour l'extérieur de la sphère est identiquement zéro. Pour un exposant α non-newtonien, on obtient comme prolongement le potentiel d'équilibre d'ordre α de la sphère.

Indiquons encore succinctement la généralisation suivante, facile à démontrer. Toute fonction bornée f appartenant à l'une quelconque des classes de BAIRE, donnée sur l'ensemble F, peut être approchée par une suite bornée de potentiels dus à des masses portées par F, cela excepté au plus sur un sous-ensemble qui est de mesure nulle par rapport aux masses de GREEN. Toute suite pareille tend en dehors de F vers le prolongement de f.

Le seconde partie de notre énoncé est immédiate. Pour établir la première, observons d'abord que, d'après l'inégalité (10) du n° **25**, les ensembles de mesure nulle par rapport aux masses de GREEN μ_M sont les mêmes pour autant que le pôle M varie dans le même domaine connexe adjacent. Cela étant, on prend un point M_k dans chacun des domaines adjacents D_k et on forme la fonction additive d'ensemble $\mu = \Sigma a_k \mu_{M_k}$ avec des $a_k > 0$ et tels que la série Σa_k converge. La fonction f étant mesurable par rapport à la fonction d'ensemble μ, on sait former, tout comme s'il s'agissait de la mesure ordinaire, une suite bornée de fonctions continues tendant vers f sauf au plus sur un sous-ensemble de mesure nulle par rapport à μ et à plus forte raison par rapport aux masses de GREEN μ_{M_k}. Enfin on remplace les fonctions continues par des fonction f_n de la nature indiquée plus haut.

29. Nos résultats permettent aussi de donner une nouvelle solution du problème de DIRICHLET généralisé. Soit F un ensemble fermé et borné et f une fonction continue (ou une fonction appartenant à la classe considérée) définie sur cet ensemble. Cette fonction peut, sauf au plus sur un sous-ensemble de capacité nulle (la partie irrégulière de F), s'écrire comme fonction limite d'une suite bornée de potentiels newtoniens dont les masses sont situées sur F. D'autre part, toute suite bornée de potentiels newtoniens dus à des masses situées sur F qui tend vers la fonction

f en tout point de F, sauf au plus dans un sous-ensemble de capacité nulle, tend en dehors de F vers une fonction bien déterminée f_F. Cette fonction est harmonique en tout domaine connexe adjacent à F et forme pour un tel domaine la solution du problème de DIRICHLET, aux valeurs limites f, dans le sens de M. WIENER.

Chapitre X. — Le problème de Green et les fonctions surharmoniques.

30. Dans la théorie classique du potentiel, on désigne par problème de GREEN[32]) — appelé aussi problème de GAUSS — le problème suivant, auquel nous avons déjà fait allusion tout à l'heure. Étant donné sur un certain nombre de domaines et de surfaces, désignés dans leur ensemble par A, une certaine fonction f, écrire cette fonction comme potentiel de masses portées par A. Ce problème est résoluble et se réduit à un nombre fini de problèmes de DIRICHLET, si les surfaces données et celles formant les frontières des domaines donnés sont assez régulières et s'il en est de même de la fonction f, p. ex. si cette fonction admet un assez grand nombre de dérivées.

Dans le problème général on entendra par A un ensemble fermé arbitraire de capacité positive et on remplacera le potentiel newtonien par un potentiel d'ordre α quelconque, $0<\alpha\leqq 2$. Ce problème, nous l'avons envisagé dans le chapitre précédent pour des fonctions de caractère très régulier (les fonctions f_n). La méthode utilisée consiste à représenter la fonction par un potentiel avec des masses situées n'importe où et à balayer ensuite ces masses sur l'ensemble donné. Cette méthode peut encore s'appliquer à des cas plus généraux; nous n'insistons pas ici. Une autre méthode sera élucidée à la fin de ce chapitre par un problème particulier très simple. La plus grande partie du présent chapitre est consacrée au problème où les masses cherchées sont *positives.* Nous commençons par le cas où l'ensemble fermé A est identique à l'espace entier, et nous allons donner les conditions nécessaires et suffisantes pour qu'une fonction f définie en tout point de l'espace puisse s'écrire comme *potentiel de masses positives.* Dans le

---

[32]) W. THOMSON and P. G. TAIT, *Treatise on Natural Philosophy,* II, Cambridge, 1883, p. 52.

n° **33** nous donnons la solution du problème correspondant pour un ensemble fermé quelconque.

31. Soit donc f une fonction, donnée dans l'espace entier, qu'il s'agit d'écrire comme potentiel de masses positives. *Une condition nécessaire, toujours admise dans la suite, est évidemment que la fonction f soit semicontinue inférieurement.* Pour obtenir d'autres conditions, on n'aura qu'à recourir à nos résultats sur le balayage de masses positives (n° **20**). Soient en effet F un ensemble fermé quelconque dont tous les points sont réguliers[33]) et M un point extérieur à cet ensemble. Alors, si f est un potentiel dû à des masses positives, le prolongement

$$\bar{f}_F(M) = \int_F f(Q)\, d\mu_M(Q) \tag{1}$$

devra satisfaire à l'inégalité

$$\bar{f}_F(M) \leqq f(M), \tag{2}$$

ce qui donne une nouvelle condition nécessaire. Remarquons que, grâce aux propriétés de la masse totale de Green, cette inégalité subsiste si f est une constante positive et alors aussi lorsque f est la somme d'un potentiel dû à des masses positives et d'une constante positive. Inversement on peut montrer que *si l'inégalité* (2) *a lieu pour tout ensemble F de l'espèce considérée et si la fonction f est semicontinue inférieurement, cette fonction peut, à une constante positive additive près, se mettre sous la forme d'un potentiel de masses positives.*

Les conditions données ici rappellent la définition des fonctions surharmoniques, due à M. F. Riesz[34]). En effet, on a vu plus haut que dans le cas classique le prolongement de f n'est que la solution du problème de Dirichlet (n° **27**).

Comme dans le cas classique, il est avantageux de remplacer notre condition (2) ayant rapport à des ensembles F généraux par une autre se rapportant à des ensembles très particuliers. On peut en effet démontrer (non sans quelque difficulté) que, la semicontinuité étant toujours admise, la condition (2) est entièrement équivalente à celles-ci

[33]) Il est commode pour simplifier les démonstrations de se restreindre à de tels ensembles. Nos énoncés restent valides, et même à plus forte raison, si l'on omet cette hypothèse restrictive.

[34]) F. Riesz, Sur les fonctions subharmoniques et leur rapport à la théorie du potentiel, *Acta Math.*, **48** (1926), p. 329—343 et **54** (1930), p. 321—360.

(2′) $$f \geqq 0$$

et la condition (2) posée pour l'extérieur (surface comprise) S^* d'une sphère S quelconque de centre M et de rayon R très petit, c'est-à-dire que (*cf.* le n° **21**) la valeur moyenne

(2″) $$\bar{f}_{S^*}(M) = \int_{S^*} f(Q)\,\varkappa_M(Q)\,dQ \leqq f(M),$$

inégalité dont le premier membre est à remplacer dans le cas classique par la valeur moyenne de GAUSS.

Nous appellerons les fonctions remplissant l'un ou l'autre de nos systèmes équivalents de conditions *fonctions surharmoniques d'ordre* α[35]), en supprimant le dernier attribut quand aucune confusion n'est à craindre. Parmi les propriétés presque évidentes de ces fonctions, nous nous contentons de signaler la suivante. *La fonction f étant surharmonique, il en est de même de la fonction définie par la valeur moyenne* (2″), S^* désignant l'extérieur d'une sphère variable, de rayon constant R, décrite autour du point variable M. Cette fonction qui ne dépend que du point M et du rayon R sera dans la suite désignée par $f_{(R)}(M)$. On montre comme pour les potentiels de masses positives (n° **21**) que $f_{(R)}(M)$ *tend en croissant vers* $f(M)$ *quand le rayon* R *tend en décroissant vers zéro.*

Ajoutons qu'on démontre à peu près comme dans le cas classique que f étant surharmonique dans notre sens, il en sera de même de $\bar{f}_F$.

Nous avons vu que $f_{(R)}(M)$ tend en croissant vers $f(M)$ lorsque le rayon R tend vers zéro. De la même manière on voit que $f_{(R)}(M)$ tend en décroissant vers une valeur $C \geqq 0$ quand R tend vers l'infini. On montre facilement que la valeur limite C est indépendante de la position du point M et que c'est précisément la constante additive entrant dans la représentation de notre fonction f.

On peut montrer que la fonction $f-C$ est encore positive et alors manifestement surharmonique. En désignant donc $f-C$ de nouveau par f, on aura obtenu une fonction dont la valeur moyenne tend vers zéro lorsque M tend vers l'infini. Or cette propriété appartient aussi à tout potentiel. On voit alors que le théorème

---

[35]) En réalité, on devrait parler de fonctions surharmoniques d'ordre α *dans l'espace entier.*

à démontrer est équivalent au suivant: Pour qu'une fonction puisse s'écrire comme potentiel de masses positives, il faut et il suffit qu'elle soit surharmonique et que sa valeur moyenne tende vers zéro quand le rayon R tend vers l'infini.

32. Nous allons maintenant indiquer la démonstration du fait que nos conditions sont suffisantes. Il s'agit d'abord d'améliorer la fonction donnée sous certains rapports, en sorte qu'on obtienne une fonction que l'on puisse facilement écrire sous la forme d'un potentiel d'ordre α. Une première opération, consistant dans la formation du prolongement de la fonction donnée par rapport à *l'intérieur* d'une sphère très grande, donne une fonction surharmonique d'ordre α qui à *l'extérieur* de cette sphère est analytique et régulière et s'annule avec ses dérivées à l'infini d'une manière convenable. A l'intérieur de la sphère, cette fonction est identique à la fonction donnée et on n'y peut donc rien dire de la dérivabilité de la fonction. Or, une seconde opération qui n'est qu'une espèce d'amélioration de l'opération (2''), appliquée à la nouvelle fonction, conduit à une fonction surharmonique d'ordre α qui dans l'espace entier est continue avec ses dérivées des deux premiers ordres et qui encore se comporte à l'infini d'une manière convenable. Cette fonction pourra donc s'écrire d'abord comme potentiel newtonien et ensuite comme potentiel d'ordre α. Grâce à la surharmonicité d'ordre α, ce dernier potentiel sera dû à des masses positives. Un passage à la limite itéré effectué sur des potentiels croissants dus à des masses positives conduit enfin à la fonction donnée qui d'après un théorème donné au n° **22** pourra elle-même, à une constante positive additive près, s'écrire comme potentiel de masses positives. On voit que la marche est bien analogue à celle suivie par M. F. Riesz dans l'une de ses démonstrations concernant le cas classique. Ne cachons pourtant pas qu'on a à surmonter des difficultés considérables.

En ce qui concerne les détails, nous devons nous contenter ici de quelques indications. La première opération signalée plus haut s'effectue par la formule (*cf.* le n° **16**)

$$\bar{f}_{S'}(M) = \int_{S'} f(Q)\, \lambda_M(Q)\, dQ, \tag{3}$$

où S' désigne *l'intérieur* d'une sphère décrite autour de l'origine O, dont on va faire tendre le rayon vers l'infini, et M signifie un point arbitraire à l'extérieur de cette sphère. Il est manifeste que dans le cas $\alpha = 2$ le second membre se réduit à l'intégrale classique de Poisson formée pour l'extérieur de la sphère. On voit que la fonction $\bar{f}_{S'}$ est analytique et régulière à l'extérieur de la sphère, qu'elle s'annule à l'infini comme $r_{OM}^{\alpha - m}$ et que ses dérivées d'ordre k s'y annulent comme $r_{OM}^{\alpha - k - m}$. A l'intérieur et sur la surface on pose évidemment $\bar{f}_{S'}(M) = f(M)$. Nous appelons la fonction ainsi obtenue, qui est manifestement encore surharmonique (d'ordre α) une fonction *tronquée*. Ajoutons que *ces fonctions tendent en croissant vers la fonction donnée f quand le rayon de la sphère S' tend vers l'infini.*

La seconde opération, qui finalement sera appliquée à ces fonctions tronquées, consiste dans le procédé suivant que, pour simplifier l'écriture, on explique pour la fonction f. On forme, γ étant une constante positive arbitraire, une nouvelle moyenne $F_{(R)}(M)$ de la fonction $f_{(R)}(M)$ (définie par la formule (2″)) en posant

$$F_{(R)}(M) = \gamma \int_1^{\infty} \varrho^{-\gamma - 1} (\varrho^2 - 1)^{\frac{\gamma}{2} - 1} f_{(R\varrho)}(M)\, d\varrho. \tag{4}$$

Il est bien clair que la fonction $F_{(R)}(M)$ est encore *surharmonique et qu'elle tend en croissant vers $f(M)$ lorsque R tend vers zéro,* puisqu'il en est ainsi des fonctions $f_{(R\varrho)}(M)$. En désignant d'autre part par $S_{(R)}^*$ l'extérieur de la sphère de rayon R décrite autour de M, on trouve par un calcul simple la formule explicite

$$F_{(R)}(M) = \int_{S_{(R)}^*} f(Q)\, \sigma_M(Q)\, dQ, \tag{5}$$

avec

$$\sigma_M(Q) = C_{\alpha\gamma m} R^{\alpha} (r_{MQ}^2 - R^2)^{\frac{\gamma - \alpha}{2}} r_{MQ}^{-\gamma - m}. \tag{6}$$

La valeur de la constante $C_{\alpha\gamma m}$, qui d'ailleurs joue un rôle peu important dans la suite, se calcule facilement si l'on observe que toute fonction constante se reproduit par nos opérations, ce qui implique que l'intégrale de $\sigma_M(Q)$ étendue à $S_{(R)}^*$ est $= 1$. On en tire

$$C_{\alpha\gamma m} = \frac{\pi^{-\frac{m}{2}} \Gamma\left(\frac{m}{2}\right) \Gamma\left(1 + \frac{\gamma}{2}\right)}{\Gamma\left(\frac{\alpha}{2}\right) \Gamma\left(1 + \frac{\gamma - \alpha}{2}\right)}. \tag{7}$$

Remarquons en passant que d'après les dernières formules, $F_{(R)}(M)$ se réduit à $f_{(R)}(M)$ pour $\gamma = 0$ et que, d'autre part, pour $\gamma > 0$, la fonction $F_{(R)}(M)$ est bien définie aussi dans le cas classique et qu'elle peut alors aussi s'obtenir en remplaçant dans la formule (4) la fonction $f_{(R)}(M)$ par la valeur moyenne de GAUSS.

Le point M étant fixé, le noyau $\sigma_M(Q)$ s'annule à l'infini, indépendamment de γ, comme $r_{MQ}^{-\alpha-m}$ et ses dérivées d'ordre k s'y annulent comme $r_{MQ}^{-\alpha-k-m}$. Pour γ *assez grand*, $\sigma_M(Q)$ et ses dérivées jusqu'à un certain ordre s'annulent encore sur la sphère $r_{MQ} = R$. Le noyau étant symétrique en M et en Q, on en conclut que $F_{(R)}(M)$ *possède des dérivées d'ordre arbitrairement donné* et que toutes les dérivations peuvent être exécutées sur $\sigma_M(Q)$ sous le signe $\int$, la variation avec M du domaine $S^*_{(R)}$ n'exerçant aucune influence sur l'expression de ces dérivées.

Nous posons maintenant les *conditions supplémentaires* que, pour M éloigné, la fonction donnée $f(M)$ possède elle-même des dérivées des deux premiers ordres et que la fonction, ses dérivées premières et son laplacien aient les ordres respectifs $O(r^{\alpha-m})$, $O(r^{\alpha-1-m})$ et $O(r^{\alpha-2-m})$, où r désigne la distance de M à l'origine. On démontre alors par un calcul assez laborieux, que nous devons omettre ici, que *la fonction* $F_{(R)}(M)$, *ses dérivées premières et son laplacien admettent les mêmes limitations.*

Ce point acquis, il est immédiat que si la fonction $f(M)$ surharmonique (d'ordre α) satisfait à nos conditions supplémentaires, la fonction $F_{(R)}(M)$ pourra, à un facteur constant près, s'écrire comme le potentiel newtonien absolument convergent de son laplacien. Donc, par le schéma donné aux n^{os} **10** et **28**, $F_{(R)}(M)$ pourra encore s'écrire comme potentiel d'ordre α absolument convergent d'une masse due à une fonction de densité continue. Grâce à la surharmonicité d'ordre α de $F_{(R)}(M)$ il est presque évident que cette fonction de densité est partout $\geqq 0$. En effet, si elle était négative en un point, elle le serait encore dans une petite sphère entourant ce point. En désignant par S^* l'extérieur de cette sphère, le prolongement de la fonction surharmonique $F_{(R)}$ par rapport à S^* aura à diminuer la fonction. Or ce prolongement, identique au potentiel obtenu par le balayage sur S^* des masses négatives intérieures à la petite sphère, augmenterait au sens strict la fonction $F_{(R)}$ au lieu de la diminuer. On arrive donc à une contradiction. En résumé, si la fonction surharmonique $f(M)$

satisfait aux conditions supplémentaires, $F_{(R)}(M)$ peut s'écrire comme potentiel de masses positives. En faisant tendre R vers zéro, $F_{(R)}$ tend en croissant vers f, par conséquent f peut encore s'écrire comme potentiel de masses positives[36]).

Soit maintenant f une fonction surharmonique générale. Alors la fonction tronquée $\overline{f}_{S'}$ satisfait aux conditions supplémentaires, elle peut par conséquent s'écrire comme potentiel de masses positives. Ces fonctions tronquées tendent en croissant vers f lorsque le rayon de la sphère S' tend vers l'infini. Toute fonction surharmonique peut donc, à une constante positive additive près, s'écrire comme potentiel de masses positives.

33. Pour un *ensemble fermé* général A, il suffit d'indiquer la solution du problème suivant. Trouver les conditions dans lesquelles une fonction f donnée sur A peut, sur la partie régulière de A, à une constante positive additive près, s'écrire comme potentiel de masses positives portées par A. Il est clair que la positivité et la semicontinuité inférieure sont encore des conditions nécessaires. On peut alors former le prolongement de la fonction par rapport à l'ensemble A. Ce prolongement définit une fonction $\overline{f}_A$ en tout point extérieur à l'ensemble A. Nous posons aux points réguliers de A, $\overline{f}_A = f$, tandis que, en tout point irrégulier, nous définissons $\overline{f}_A$ comme la limite inférieure des valeurs de $\overline{f}_A$ admises au voisinage de ce point. Alors la condition nécessaire et suffisante pour que f puisse, sur la partie régulière de A et à une constante positive additive près, s'écrire comme potentiel de masses positives portées par A, est que la fonction $\overline{f}_A$ ainsi définie soit une fonction surharmonique.

34. La dernière méthode est évidemment applicable aussi dans le cas où l'on n'exige pas de masses positives. Cependant il paraît alors très difficile de donner des conditions effectives qui soient en même temps nécessaires et suffisantes. Au lieu d'insister sur cette question, nous donnons ici une solution explicite de caractère purement formel de notre problème dans le cas particulier où la dimension m de l'espace est $= 1$ et l'ensemble A se réduit à un intervalle. Ce problème et sa première solution

---

[36]) La constante additive s'annule puisque d'après la première condition supplémentaire, $f(M)$ tend vers zéro quand M tend vers l'infini.

sont dus à M. CARLEMAN[37]). Une solution différente fut donnée par M. ZEILON[38]).

On peut admettre que l'intervalle en question est limité par les points -1 et $+1$. Il faut donc résoudre l'équation

$$(8)\qquad \int_{-1}^{+1} \varphi(y)\,|x-y|^{\alpha-1}dy = f(x),$$

où $f(x)$ est donné dans l'intervalle $(-1, +1)$. On forme le prolongement de $f(x)$ pour les points extérieurs à cet intervalle

$$(9)\qquad \bar{f}(x) = \int_{-1}^{+1} f(y)\,\lambda_x(y)\,dy,$$

avec (*voir* la formule (3) du n° **16**)

$$(10)\qquad \lambda_x(y) = \pi^{-1}\sin\frac{\pi\alpha}{2}\,(x^2-1)^{\frac{\alpha}{2}}(1-y^2)^{-\frac{\alpha}{2}}|x-y|^{-1}.$$

En remplaçant pour $|x|>1$ la notation $\bar{f}(x)$ par $f(x)$, la formule (8) sera valable pour toute valeur de x. Or cette formule peut aussi s'écrire (formules (5) et (6) du n° **2**)

$$I^\alpha\varphi(x) = \frac{f(x)}{H_1(\alpha)},$$

avec

$$H_1(\alpha) = \frac{\pi^{\frac{1}{2}}2^\alpha\,\Gamma\left(\frac{\alpha}{2}\right)}{\Gamma\left(\frac{1-\alpha}{2}\right)}.$$

Il vient par la formule (7) du même numéro

$$-\frac{1}{2}\int_{-1}^{+1}\varphi(y)\,|x-y|\,dy = I^2\varphi(x) = I^{2-\alpha}(I^\alpha\varphi(x)) =$$

$$= \frac{1}{H_1(2-\alpha)\,H_1(\alpha)}\int_{-\infty}^{+\infty}|x-y|^{1-\alpha}f(y)\,dy = g(x),$$

et alors

$$\varphi(x) = -g''(x).$$

Il y a pourtant ici une difficulté qu'on ne peut pas passer

[37]) T. CARLEMAN, Über die Abelsche Integralgleichung mit konstanten Integrationsgrenzen, *Math. Zeitschrift*, **15** (1922), p. 111—120.

[38]) N. ZEILON, Sur quelques points de la théorie de l'équation intégrale d'Abel, *Arkiv för Mat.*, **18** (1924), n° **5**, p. 1—19.

sous silence, c'est que la dernière intégrale diverge en général[39]. En effet, la fonction $f(x)\ (=\bar{f}(x))$ est dans le cas général du même ordre de grandeur que $|x|^{\alpha-1}$. On peut remédier à cet inconvénient en définissant l'intégrale $I^{2-\alpha}f(x)$ au moyen du prolongement analytique par rapport à l'exposant. Cela peut s'effectuer de beaucoup de manières, en voici une. La fonction $f(x)\ (=\bar{f}(x))$ définie par la formule (9) peut s'écrire pour $|x|>1$

$$f(x)=A|x|^{\alpha-1}+Bx|x|^{\alpha-3}+O(|x|^{\alpha-3}),$$

A et B étant à des facteurs numériques près les moments d'ordres zéro et un de $f(x)(1-x^2)^{-\frac{\alpha}{2}}$ par rapport à l'intervalle $(-1,+1)$. Soit alors $\varkappa(x)=a+bx$ et posons

$$k(x)=\int_{-1}^{+1}\varkappa(y)|x-y|^{\alpha-1}\,dy.$$

On peut déterminer les constantes a et b de sorte que les moments d'ordres zéro et un de $k(x)(1-x^2)^{-\frac{\alpha}{2}}$ soient identiques à ceux de $f(x)(1-x^2)^{-\frac{\alpha}{2}}$. En posant alors pour $|x|>1$, avec une notation évidente, $k(x)=\bar{k}(x)$, on aura la formule définitive

$$\frac{1}{2}\int_{-1}^{+1}\varphi(y)|x-y|\,dy=\frac{-1}{H_1(\alpha)\,H_1(2-\alpha)}\int_{-\infty}^{+\infty}(f(y)-k(y))|x-y|^{1-\alpha}\,dy+$$

$$+\frac{1}{2}\int_{-1}^{+1}\varkappa(y)|x-y|\,dy.$$

Notre méthode s'applique pour $0<\alpha<2$, excepté pour $\alpha=1$, où le problème n'admet pas de solution sauf dans le cas banal où la fonction $f(x)$ est constante. Pour arriver à un problème naturel correspondant à $\alpha=1$, il faut le poser pour le potentiel logarithmique (*cf.* le n° **3**), problème qui est facile à résoudre.

(Reçu le 26 août 1937)

[39]) Cette difficulté dérive du fait que la formule (7) du n° **2** n'est valable en général que pour $\alpha+\beta<m$. Or on l'applique ici pour $m=1$ et $\alpha+\beta=2$. On voit que la difficulté n'intervient que pour $m=1$ et $m=2$ et que, en résolvant le même problème pour un volume sphérique, on aura convergence dès que la dimension de l'espace est $\geqq 3$.

42. Rect

Rectification au travail "Intégrales de Riemann-Liouville et potentiels"[1]

Acta Szeged Sect. Math. *9* (1937–38), 116–118

M. BRELOT a bien voulu me signaler que dans la définition des masses de GREEN μ_M relatives à un ensemble fermé F de capacité positive et à un pôle M, il ne suffit pas de demander que le potentiel $r_{MP}^{\alpha-m}$ soit conservé en tout point P de F sauf peut-être dans un ensemble de capacité nulle, mais qu'il faut encore ajouter quelque condition restrictive pour assurer l'unicité de la distribution de GREEN. L'erreur vient d'un oubli fâcheux quoique banal qui m'est arrivé en énonçant le théorème d'équilibre de M. FROSTMAN (p. 8). Voici l'énoncé correct de ce théorème.

Soit F un ensemble fermé et borné de capacité positive; il existe alors une répartition unique de masse positive engendrant un potentiel qui est $=1$ sur F, sauf au plus dans un ensemble de capacité nulle, et $\leqq 1$ dans l'espace entier[2]).

J'obtiens les masses de GREEN de celles qui donnent le potentiel d'équilibre par une transformation de KELVIN (p. 14—15). Pour rendre le texte exact au point de vue de l'unicité on n'a en réalité qu'à intervertir l'ordre des deux dernières phrases de la page 14 après avoir remplacé les derniers mots de la première

[1]) Ces *Acta*, **9** (1938), p. 1—42.

[2]) La dernière inégalité figure dans mon travail parmi les propriétés du potentiel d'équilibre, mais elle n'intervient pas dans la définition. Or sans une telle restriction, la répartition consistant p. ex. en la masse unité concentrée au centre d'une sphère de rayon *un* pourrait aussi être considérée comme une répartition d'équilibre pour l'ensemble fermé F composé de la surface de la sphère et de son centre. En effet, son potentiel est $=1$ sur tout l'ensemble F sauf au centre, c'est-à-dire sauf dans un ensemble de capacité nulle.

phrase "cette propriété-là" par les mots "les propriétés (1) et (2)". Pour la commodité du lecteur nous donnons ici la définition corrigée complète:

15. Soient F un ensemble fermé, non nécessairement borné, de capacité positive, M un point extérieur à F, et désignons par μ' la distribution engendrant le potentiel d'équilibre sur l'ensemble fermé et borné F' que la transformation (1) du numéro précédent fait correspondre à F. En définissant alors la distribution de masse positive μ_M relative à F par la formule (3) du même numéro, il vient

$$\int_F r_{PQ}^{\alpha-m}\, d\mu_M(Q) = r_{MP}^{\alpha-m}, \tag{1}$$

égalité valable pour tout point P de F excepté peut-être pour un ensemble de capacité nulle. Le potentiel d'équilibre étant toujours $\leqq 1$, on a encore l'inégalité fondamentale

$$h_M(P) = \int_F r_{PQ}^{\alpha-m}\, d\mu_M(Q) \leqq r_{MP}^{\alpha-m}, \tag{2}$$

P étant un point tout à fait arbitraire de l'espace. Le calcul étant réversible, on conclut que la distribution μ_M est entièrement déterminée par les propriétés (1) et (2). Nous appellerons les masses μ_M *masses de Green relatives à l'ensemble F et au pôle M.*

Le lecteur voudra bien tenir compte de cette définition précisée des masses de GREEN dans toute question d'unicité[3]), et cela aussi en ce qui concerne le travail de M. FROSTMAN "Sur le balayage des masses" publié dans ce recueil immédiatement après le mien.

Dans mon travail je ne m'occupe pas de questions d'unicité concernant le balayage d'une distribution générale que je définis par des formules explicites au moyen des masses de GREEN. Dans un travail récent[4]) M. BRELOT donne sur ce sujet quelques théorèmes très précis. Je profite de l'occasion pour montrer comment les résultats de M. BRELOT ressortent de ceux donnés dans mon travail. M. BRELOT montre par exemple qu'en balayant sur l'ensemble fermé F de capacité positive des masses $\geqq 0$, il y a unicité si le potentiel reste conservé sur F sauf au plus sur un ensemble de capacité nulle et si en outre l'une au moins des deux conditions suivantes est remplie:

a) Le potentiel n'est pas augmenté aux points extérieurs à F.

---

[3]) Cette question intervient en particulier au n° 25 où il faut observer que les deux potentiels comparés sont partout $\leqq r_{MQ}^{\alpha-m}$.

[4]) M. BRELOT, Fonctions sous-harmoniques et balayage, *Bulletin de l'Académie Royale de Belgique,* **24** (1938), p. 301—312 et p. 421—436.

b) La répartition obtenue sur F ne comporte pas de masses sur les ensembles de capacité nulle.

Démontrons d'abord l'unicité dans la condition *b)*. On voit immédiatement que le potentiel obtenu par le balayage peut à l'extérieur de F être représenté par la formule (2) du n° **19** de mon travail. Alors le potentiel est déterminé sauf au plus dans un ensemble de capacité nulle, et dès lors de mesure nulle; les masses sont par conséquent entièrement déterminées par le théorème d'unicité du n° **10.**

Par le théorème du n° **13,** appliqué à la différence entre le potentiel original et celui obtenu par le balayage, la condition *a)* se réduit facilement à la condition *b).*

(Reçu le 14 novembre 1938)

43.

L'intégrale de Riemann-Liouville et le problème de Cauchy pour l'équation des ondes

Soc. Math. de France. Conf. à la Réunion Int. des Math. Paris, Juillet 1937, 153–170

C'est la notion de la *partie finie* de certaines intégrales due à M. Hadamard qui forme le point de départ de ces recherches. On connaît les applications brillantes à la théorie des équations aux dérivées partielles que M. Hadamard a données de la notion qu'il a créée [1]. Chez lui il s'agit surtout d'équations à coefficients variables, tandis que la méthode que nous allons développer ne s'applique qu'à des équations à coefficients constants [2]. En revanche, elle permet de donner pour ces équations une solution du problème de Cauchy qui est la même pour les dimensions impaires et paires.

Considérons d'abord, comme M. Hadamard, l'intégrale de Riemann-Liouville

$$I^{\alpha} f(x) = \frac{1}{\Gamma(\alpha)} \int_a^x f(t)(x-t)^{\alpha-1}\, dt \quad [3].$$

Elle converge pour $\alpha > 0$ et satisfait aux relations fondamentales

$$(1) \qquad I^{\alpha}(I^{\beta}) = I^{\alpha+\beta}, \qquad \frac{d}{dx}(I^{\alpha+1}) = I^{\alpha}.$$

---

(1) *Cf.* (aussi pour la littérature) J. Hadamard, *Le problème de Cauchy et les équations aux dérivées partielles linéaires hyperboliques*. Paris, Hermann, 1932.

(2) *Voir* pourtant pour l'extension de la méthode la Note additionnelle.

(3) Le facteur $\frac{1}{\Gamma(\alpha)}$ ne figure pas chez M. Hadamard.

Dans des conditions de dérivabilité convenables, il s'agit maintenant d'étendre la définition de cette intégrale à des valeurs de α pour lesquelles elle cesse de converger. M. Hadamard arrive à cette extension pour des indices négatifs non entiers en retranchant de l'intégrale divergente certaines parties infinies d'ordre fractionnaire. Ce qui reste c'est alors la partie finie de l'intégrale. On peut atteindre le même but, et cela aussi pour l'indice zéro et les indices entiers négatifs, par un procédé qui me semble très naturel, celui du prolongement analytique par rapport à l'indice α. En supposant l'existence d'un assez grand nombre de dérivées, on obtient au moyen des intégrations par parties

$$I^\alpha f(x) = \sum_{k=0}^{n-1} \frac{f^{(k)}(a)}{\Gamma(\alpha+k+1)}(x-a)^{\alpha+k} + I^{\alpha+n} f^{(n)}(x).$$

La dernière expression s'écrit

$$\frac{1}{\Gamma(\alpha+n)}\int_a^x f^{(n)}(t)(x-t)^{\alpha+n-1}\,dt,$$

l'intégrale étant convergente pour $\alpha > -n$. Le prolongement analytique se trouve donc effectué pour tous ces indices. Bien entendu, le prolongement en question peut être obtenu par bien d'autres procédés. Signalons en particulier la formule suivante qui est très utile dans les applications et par laquelle nous nous approchons beaucoup de l'ordre d'idées de M. Hadamard.

$$I^\alpha f(x) = \frac{1}{\Gamma(\alpha)}\int_a^x [f(t)-P(t)](x-t)^{\alpha-1}\,dt$$
$$+\frac{1}{\Gamma(\alpha)}\sum_{k=0}^{n-1}\frac{(-1)^k f^{(k)}(x)(x-a)^{k+\alpha}}{(k+\alpha)k!},$$

avec

$$P(t) = \sum_{k=0}^{n-1}\frac{f^{(k)}(x)}{k!}(t-x)^k \quad (^1).$$

(1) Nos procédés s'étendent facilement à des intégrales qui possèdent plusieurs singularités, par exemple à des intégrales de la forme

$$I^{\alpha,\beta} f = \frac{\Gamma(\alpha+\beta)}{\Gamma(\alpha)\Gamma(\beta)}\int_y^x f(t)(t-y)^{\alpha-1}(x-t)^{\beta-1}\,dt.$$

L'intégrale prolongée satisfait, sauf dans des cas exceptionnels faciles à préciser, aux relations fondamentales (1) dont la seconde pourra aussi servir à effectuer le prolongement en question.

Pour les indices $\alpha = 0$ et entiers négatifs $\alpha = -n$ on trouve, comme il fallait s'y attendre,

$$I^0 f(x) = f(x), \qquad I^{-n} f(x) = f^{(n)}(x).$$

Arrêtons-nous un instant sur les dernières relations. On voit que, pour tout autre indice, $I^\alpha f(x)$ est une fonctionnelle dépendant de toutes les valeurs que $f(t)$ admet entre a et x, tandis que, pour $\alpha = 0$ ou entier négatif, $I^\alpha f(x)$ a une valeur de caractère local; ce ne sont que les valeurs admises au voisinage infinitésimal de x qui interviennent. *Voilà en germe le principe de Huygens* [1]. C'est évidemment au facteur $\frac{1}{\Gamma(\alpha)}$, s'annulant pour les indices en question, que ces indices doivent leur caractère exceptionnel. D'ailleurs, même sans effectuer le prolongement analytique d'une façon explicite, on voit nettement que le prolongement de

$$\frac{1}{\Gamma(\alpha)} \int_a^{x-\delta} f(t)(x-t)^{\alpha-1}\, dt \qquad (\delta > 0),$$

est zéro pour $\alpha = 0$ ou entier négatif, parce que l'intégrale ne cesse jamais de converger. Les valeurs de $f(t)$ qui interviennent dans cette intégrale ne donneront donc aucune contribution à la valeur finale.

J'ai insisté sur ces questions élémentaires bien familières à vous tous, parce qu'elles nous permettront de bien saisir la différence qu'on rencontre dans la solution du problème de Cauchy de l'équation des ondes, posé pour un nombre pair ou pour un nombre impair de variables.

Considérons dans l'espace à m dimensions les points P et Q aux coordonnées respectives $x_1, x_2, \ldots, x_m$ et $\xi_1, \xi_2, \ldots, \xi_m$. Nous intro-

---

(1) Nous entendons ici et dans tout ce qui suit par principe de Huygens la *mineure* du principe dans la terminologie de M. Hadamard (*loc. cit.*, p. 75). D'après ce principe, une perturbation lumineuse étant localisée à l'instant $t = 0$ au voisinage immédiat du point O, son effet sera localisé, pour $t = t'$, au voisinage immédiat de la surface d'une sphère de centre O et de rayon ct', c désignant la vitesse de la lumière.

duisons la distance lorentzienne de ces points

$$r_{PQ} = \sqrt{(x_1 - \xi_1)^2 - (x_2 - \xi_2)^2 - \ldots - (x_m - \xi_m)^2},$$

en supposant que l'expression qui figure sous le signe racine carrée soit ≥ 0. En considérant le point P comme fixe et le point Q comme variable, $r_{PQ} = 0$ définit la nappe du cône-lumière au sommet P, $r_{PQ}^2 > 0$ définit son intérieur, $x_1 - \xi_1 > 0$ le cône rétrograde et $x_1 - \xi_1 < 0$ le cône direct. C'est le cône rétrograde que nous allons considérer en général, en le désignant par D^P. Soit encore S une surface (c'est-à-dire une variété à $m - 1$ dimensions) que les génératrices des cônes rétrogrades appartenant aux points P considérés ne coupent qu'en un seul point. On admet aussi que la surface soit assez régulière pour que les dérivations qui interviendront plus tard puissent être effectuées. Le domaine limité par la nappe rétrograde et la surface S sera désigné par D_S^P.

Cela étant, nous posons

$$(2) \qquad I^\alpha f(P) = \frac{1}{H_m(\alpha)} \int_{D_S^P} f(Q)\, r_{PQ}^{\alpha - m}\, dQ,$$

avec

$$(3) \qquad H_m(\alpha) = \pi^{\frac{m-2}{2}}\, 2^{\alpha-1}\, \Gamma\left(\frac{\alpha}{2}\right) \Gamma\left(\frac{\alpha + 2 - m}{2}\right).$$

Cette intégrale converge pour $\alpha > m - 2$ et l'on vérifie facilement qu'elle satisfait aux relations fondamentales

$$(4) \qquad I^\alpha I^\beta = I^{\alpha+\beta}, \qquad \Delta I^{\alpha+2} = I^\alpha,$$

où l'on a désigné par Δ l'opérateur des ondes :

$$\Delta = \frac{\partial^2}{\partial x_1^2} - \frac{\partial^2}{\partial x_2^2} - \ldots - \frac{\partial^2}{\partial x_m^2}.$$

Ajoutons que ces propriétés de notre intégrale déterminent entièrement le facteur $H_m(\alpha)$.

Pour les indices $\alpha \leq m - 2$ l'intégrale (2) se définit au moyen de prolongement analytique par rapport à α, bien entendu dans des conditions de régularité convenables portant sur la fonction f et sur la surface S. *On trouve en particulier* $I^0 f(P) = f(P)$. Remarquons que ce fait, qui est d'une importance capitale pour la suite, n'est nullement évident. En effet, I^0 est défini ici comme prolongement analytique de I^α qui ne converge que pour $\alpha > m - 2$, et le prolongement en question n'existe que dans certaines conditions de dérivabilité (1). Des remarques analogues s'appliquent aux relations $I^{-2k} = \Delta^k$.

Il est clair par ce qui précède que notre procédé d'intégration tire son origine de l'opérateur différentiel du second ordre Δ. On a vu en effet que $\Delta I^2 = 1$, c'est-à-dire que I^2 est dans un certain sens l'inverse de l'opérateur des ondes. Or on peut construire un procédé qui, dans le même sens, forme l'inverse de l'opérateur différentiel du premier ordre *gradient* ou *nabla*, cet opérateur étant pris au sens lorentzien. Il est manifeste que ce procédé devra être de caractère vectoriel (2).

Soient $X_1, X_2, \ldots, X_m$ les composantes d'un vecteur arbitraire X et définissons *les nombres de Clifford* comme symboles de calcul ou comme vecteurs unités par la relation

$$X^2 = (X_1 e_1 + X_2 e_2 + \ldots + X_m e_m)^2 = X_1^2 - X_2^2 - \ldots - X_m^2.$$

Cette relation met en évidence les règles de calcul

$$e_1^2 = -e_2^2 = \ldots = -e_m^2 = 1$$

et

$$e_j e_k + e_k e_j = 0 \qquad (j \neq k).$$

L'opérateur $\nabla = \sum_{k=1}^{m} e_k \frac{\partial}{\partial x_k}$ a manifestement la propriété $\nabla^2 = \Delta$ (3).

Nous posons

$$J^\alpha f(P) = e^{\frac{i\pi\alpha}{2}} \left[\cos \frac{\pi\alpha}{2} I^\alpha f(P) - i \sin \frac{\pi\alpha}{2} \nabla [I^{\alpha+1} f(P)] \right],$$

l'opération ∇ pouvant être exécutée sous le signe $\int$. On a alors

$$J^\alpha J^\beta = J^{\alpha+\beta}, \qquad \nabla J^{\alpha+1} = J^\alpha, \qquad J^{2k} = I^{2k},$$

k étant un nombre entier. En particulier on a $J^0 f(P) = f(P)$ ou plus

(1) Pour m impair ou pair il suffit de supposer l'existence des dérivées d'ordre $\leq \frac{m+1}{2}$ ou $\leq \frac{m}{2}$ respectivement.

(2) L'intégration vectorielle que j'esquisse ici ne figurait pas dans ma conférence. Je l'ai indiquée succinctement après la conférence en réponse à une question de M. Fréchet qui présidait à la séance.

(3) Depuis les travaux de M. Dirac, ces choses sont certainement bien familières à la plupart des lecteurs.

brievement $J^0 = 1$. Signalons encore l'opérateur remarquable

$$\mathrm{O}f(\mathrm{P}) = \nabla[\mathrm{I}^1 f(\mathrm{P})]$$

qui satisfait à la relation $\mathrm{O}^2 = 1$.

En remarquant que la plupart des considérations qui suivent s'appliquent — *mutatis mutandis* — aussi à nos intégrales vectorielles, retournons aux intégrales scalaires.

Définissons avec M. d'Adhémar *la conormale n* à la surface S au point Q (*la transversale* suivant la terminologie de M. Hadamard) par la relation

$$dx_1\,\delta x_1 - dx_2\,\delta x_2 - \ldots - dx_m\,\delta x_m = 0,$$

où d et δ désignent respectivement un déplacement arbitraire sur la surface et un déplacement sur la conormale. Cela étant, nous choisissons celle des deux directions admissibles QT qui rend le produit scalaire lorentzien des vecteurs QP et QT positif. Il vient alors par la formule de Green

$$\begin{aligned} \mathrm{I}^\alpha u(\mathrm{P}) = {} & \frac{1}{H_m(\alpha+2)} \int_{\mathrm{D}_\mathrm{S}^\mathrm{P}} \Delta u(\mathrm{Q})\, r_{\mathrm{PQ}}^{\alpha+2-m}\, d\mathrm{Q} \\ & + \frac{1}{H_m(\alpha+2)} \int_{\mathrm{S}^\mathrm{P}} \left[\frac{\partial u(\mathrm{Q})}{\partial n}\, r_{\mathrm{PQ}}^{\alpha+2-m} - u(\mathrm{Q})\, \frac{\partial r_{\mathrm{PQ}}^{\alpha+2-m}}{\partial n} \right] d\mathrm{S}, \end{aligned}$$

tous les éléments géométriques étant mesurés, ici et dans la suite, dans le sens lorentzien. (Il faut prendre quelques précautions aux points de S où le plan tangent est en même temps tangent à un cône-lumière. D'ailleurs ce cas sera exclu plus loin quand il s'agira de l'application de la formule au problème de Cauchy.)

La formule ci-dessus qui contient une intégrale spatiale, une simple couche et une double couche peut se mettre sous une forme plus condensée. En effet, en posant d'une manière générale

$$(5) \quad \begin{aligned} \mathrm{I}^{*\alpha} f, g, h(\mathrm{P}) = {} & \frac{1}{H_m(\alpha)} \int_{\mathrm{D}_\mathrm{S}^\mathrm{P}} f(\mathrm{Q})\, r_{\mathrm{PQ}}^{\alpha-m}\, d\mathrm{Q} \\ & + \frac{1}{H_m(\alpha)} \int_{\mathrm{S}^\mathrm{P}} \left[g(\mathrm{Q})\, r_{\mathrm{PQ}}^{\alpha-m} - h(\mathrm{Q})\, \frac{\partial r_{\mathrm{PQ}}^{\alpha-m}}{\partial n} \right] d\mathrm{S}, \end{aligned}$$

on peut écrire

$$(6) \qquad \mathrm{I}^\alpha u(\mathrm{P}) = \mathrm{I}^{*\alpha+2} \Delta u, \frac{\partial u}{\partial n}, u(\mathrm{P}).$$

Avant d'aller plus loin, signalons quelques propriétés importantes du symbole I^*. On a

$$(7) \qquad I^{\alpha} I^{*\beta} = I^{*\alpha+\beta}, \qquad \Delta I^{*\alpha+2} = I^{*\alpha},$$

et en outre

$$(8) \qquad I^{*0} f,\, g,\, h(P) = f(P).$$

La formule (6) conduit facilement à la solution du problème de Cauchy pour l'équation des ondes. En effet il s'ensuit pour $\alpha = 0$, grâce à la relation $I^0 u = u$,

$$(9) \qquad u(P) = I^{*2} \Delta u,\, \frac{\partial u}{\partial n},\, u(P),$$

c'est-à-dire que la valeur de u au point P peut se calculer par une intégrale spatiale portant sur $\Delta(u)$ et étendue à un certain volume D_S^P, par une simple couche et une double couche portant respectivement sur $\frac{\partial u}{\partial n}$ et u et étendues à une certaine portion S^P de la surface S. Malheureusement, d'après ce que nous avons dit plus haut, ces intégrales sont en général divergentes. En effet elles ne convergent toutes que si l'on a $m \leqq 2$ (1), l'une d'elles, l'intégrale de double couche, diverge déjà pour l'équation de M. Volterra ($m = 3$) et elles divergent toutes à partir de $m = 4$, correspondant à l'équation des ondes relative à l'espace ordinaire. Mais on pourra toujours définir ces intégrales par prolongement analytique, ce qui veut dire que $I^{*\alpha} \Delta(u)$, $\frac{\partial u}{\partial n}$, u étant convergent pour α assez grand ($\alpha > m$), on peut, dans les conditions de dérivabilité admises, obtenir I^{*2} par prolongement analytique.

La formule fait nettement ressortir la différence qui a lieu entre les cas de m impair et de m pair. Pour m impair, où l'on retrouve la formule de M. Hadamard, le facteur $\frac{1}{H_m(2)}$ ne s'évanouit pas, la solution sera donc fournie par une intégrale spatiale étendue au

(1) Néanmoins, même pour $m = 2$, la formule (9) donne une solution erronée, indépendante des valeurs de u sur S, si on l'applique au sens strict, c'est-à-dire en *substituant* $\alpha = 0$ dans le second membre de la formule (6). Par contre, la formule obtenue de (6) par prolongement analytique donne la solution correcte bien connue, où il entre les valeurs de u aux points d'intersection de S et des deux rayons de lumière issus de P.

volume entier D_S^P et des intégrales étendues à la portion de surface entière S^P. Au contraire dans le cas de m pair (> 2), $\frac{1}{H_m(2)}$ s'annule à cause du facteur $\frac{1}{\Gamma\left(\frac{2+2-m}{2}\right)}$, et la formule finale ne sera pas influencée par les valeurs que u et ses dérivées prennent à l'intérieur du cône. L'intégrale spatiale se réduira à une intégrale étendue à la nappe du cône et les intégrales de surface à des intégrales étendues à l'intersection du cône avec la surface S. *Voilà le principe de Huygens : ce ne sont que les points d'univers situés sur la nappe rétrograde du cône au sommet* P *qui agissent sur ce point* (¹). Le principe dérive donc du fait que l'opérateur I^{*2} a pour m pair (> 2) un caractère quasi local, il n'y intervient que des points qui sont à distance zéro du point P (²).

En réalité notre formule n'était jusqu'ici qu'une formule de représentation du même caractère que la formule classique de Cauchy l'est pour les fonctions monogènes. Mais elle donne aussi la solution d'un problème aux limites, notamment celle du problème de Cauchy (³), si l'on admet en plus que la surface S ait, suivant la terminologie de M. Hadamard, une orientation d'espace. Cela veut dire que l'on a pour tout déplacement sur la surface $dx_1^2 - dx_2^2 - \ldots - dx_m^2 < 0$. En effet dans ce cas (et dans ce cas seulement) les domaines d'intégration deviennent infinitésimaux lorsque P s'approche indéfiniment de la surface (⁴).

Je finis cet exposé déjà trop long en donnant *une solution explicite invariante par rapport aux transformations de Lorentz* pour le cas de l'espace ordinaire, c'est-à-dire pour $m = 4$. En écrivant l'équation sous la forme $\Delta(u) = f$, le second membre donne lieu à une intégrale étendue à la nappe du cône qui n'est que le potentiel retardé bien connu. Puisque je ne saurais rien dire de nouveau sur cette partie de la solution, je me restreins ici à l'équation homogène.

(¹) *Cf.* Hadamard, *loc. cit.*, p. 239 et 324.

(²) Il en est de même, pour m pair, des opérateurs I^{*2k}, k entier positif, $2k \leqq m - 2$, et pour m impair des opérateurs I^{*2k+1}, k entier, $2k+1 \leqq m - 2$. I^{*-2k} a pour k entier $\geqq 0$ un caractère strictement local indépendamment de la parité de m. En effet on a (*voir* plus haut) $I^{*0} = u$ (P) et $I^{*-2k} = \Delta^k u(\mathrm{P})$.

(³) C'est le problème de résoudre l'équation $\Delta u = f$, la fonction f étant connue dans l'espace et les valeurs de u et de $\frac{\partial u}{\partial n}$ étant données sur la surface S.

(⁴) *Cf.* Hadamard, *loc. cit.*, p. 249.

Posons d'une manière générale $R = R_{PQ} = r^2_{PQ}$ et désignons par s^P la variété à deux dimensions formée par l'intersection du cône au sommet P et de la surface S. En remettant à tout à l'heure l'explication des autres notations, nous écrivons ici la solution suivante du problème de Cauchy posé pour l'équation $\Delta(u) = 0$, solution qu'on peut déduire de la solution générale (9).

$$(10) \qquad u(P) = \frac{1}{\pi}\int_{s^P}\left[\frac{1}{2}\left(\frac{1}{R_1}+\frac{1}{R_2}\right)u - \left(\frac{\partial u}{\partial R}\right)_{(\nu)}\right]ds.$$

Vous voyez qu'il n'intervient dans la formule qu'une seule dérivée suivant une certaine direction ν qu'il faudra maintenant définir. Cela est très facile dans le cas classique considéré par Poisson où la surface S est le plan $x_1 = 0$. Alors ν n'est que l'image de la génératrice QP par rapport à ce plan. Dans le cas général la chose est un peu plus compliquée.

Par un élément infinitésimal à deux dimensions du bord s^P contenant le point Q il passe ∞^1 plans à trois dimensions. Formons l'image de la génératrice QP, indéfiniment prolongée au delà de P, par rapport à l'un quelconque de ces plans, par exemple le plan tangent à S en Q, l'image étant prise dans la direction transversale au plan. Cette image, *indépendante du plan choisi*, donne la demi-droite ν appartenant au point Q du bord. Les lignes ν, qui évidemment sont des rayons-lumière, engendrent une surface caractéristique Σ de l'équation des ondes, qui dans un certain sens peut être considérée comme l'image du cône au sommet P par rapport à la surface S. La surface Σ peut aussi être obtenue de la manière suivante. On fait décrire au point d'intégration Q le bord s^P et l'on considère les cônes rétrogrades qui ont leur sommet au point variable Q. Ces cônes enveloppent une surface à deux nappes. La nappe extérieure fait évidemment partie du cône original au sommet P, tandis que la nappe intérieure est identique à la surface Σ.

La dérivée figurant dans notre formule est prise suivant la direction ν. La longueur lorentzienne des segments d'une telle droite étant nulle, on introduit R comme paramètre et forme la dérivée de u par rapport à R suivant ladite direction. Observons que $R = R_{PT}$ varie d'une manière linéaire sur une telle droite. En effet, T désignant un point quelconque de la droite ν appartenant à Q, R_{PT} est égal au double du produit scalaire lorentzien des vecteurs PQ et QT.

Il faut encore définir les valeurs R_1 et R_2 figurant dans notre formule. Remarquons à cet effet que les droites ν, qui sont en nombre ∞^2, forment évidemment une congruence de droites, qui est d'ailleurs d'un caractère bien particulier. En nous trouvant dans quatre dimensions, ce n'est que grâce à cette dernière circonstance que, pour chaque droite fixe ν appartenant à un certain point Q, il existe en général deux directions de déplacements infinitésimaux, $\delta_1 Q$ et $\delta_2 Q$ telles que les droites correspondantes coupent la droite ν. Cela veut dire que la congruence admet en général *deux foyers* Q_1 et Q_2 sur chacune de ses droites. Dans la formule ci-dessus on a posé pour abréger,

$$R_{PQ_1} = R_1, \qquad R_{PQ_2} = R_2.$$

Le lieu géométrique des foyers est la surface focale, *la caustique*, de la congruence, qui dans un certain sens peut être interprétée comme l'image du point P par rapport à la surface S. Dans le cas classique de Poisson elle se réduit à l'image *distincte* de ce point par rapport au plan $x_1 = 0$, et notre solution se réduit immédiatement à celle de Poisson.

En résumé, le principe de Huygens dit que la valeur de u au point P est déterminée par les valeurs que u et ses dérivées premières admettent en des points Q appartenant au passé-lumière du point P. Notre formule y apporte un complément qui doit présenter un certain intérêt, c'est que la seule dérivée qui y entre est prise suivant un certain rayon de lumière, c'est-à-dire que seul le passé-lumière infinitésimal de la fonction u au point Q y intervient. L'intervention de la surface (congruence) Σ et de sa caustique dans notre solution jette un jour nouveau sur les rapports entre l'optique physique et l'optique géométrique.

NOTE ADDITIONNELLE.

Tout récemment j'ai réussi à étendre la méthode d'intégration exposée dans la conférence qui précède aux équations linéaires du type hyperbolique normal *à coefficients variables*. Je me restreins ici à indiquer le procédé pour le cas où le premier membre de l'équation en question est fourni par le paramètre différentiel du second ordre de Beltrami attaché à une certaine forme quadratique. L'extension au cas général est immédiate.

Soit donc

$$ds^2 = g_{ik}\, dx^i\, dx^k \quad (^1)$$

la forme métrique attachée à un espace de Riemann à m dimensions. Nous dirons avec M. Hadamard qu'on se trouve dans le cas hyperbolique normal, si cette forme, transformée en somme de carrés, admet un carré positif et $m-1$ carrés négatifs. En désignant par g la valeur absolue du déterminant $|g_{ik}| \neq 0$ et par g^{ik} les coefficients de la forme réciproque, le paramètre différentiel correspondant $\Delta_2 u = \Delta u$ s'écrit

$$\Delta u = \frac{1}{\sqrt{g}} \frac{\partial}{\partial x_i}\left(\sqrt{g} g^{ik} \frac{\partial u}{\partial x_k}\right).$$

La difficulté nouvelle qui se présente ici c'est la formation de la solution élémentaire de l'équation, problème que M. Hadamard a résolu d'une manière ingénieuse. Or le dualisme fatal entre les espaces à dimensions impaires et paires se manifeste dans le cas actuel d'une manière encore plus fâcheuse que dans le cas de l'équation des ondes. Les solutions élémentaires obtenues sont de caractère tout à fait différent, la méthode de la partie finie n'est applicable que pour les dimensions impaires, et pour les dimensions paires il faut recourir à la méthode de la descente. Nous allons montrer que l'introduction d'un paramètre convenable permet encore de donner la solution du problème de Cauchy sous une forme qui est indépendante de la parité de la dimension de l'espace. Chemin faisant nous allons développer un procédé d'intégration fractionnaire tout analogue à celui que nous avons donné dans notre conférence pour le cas de l'équation des ondes.

Nous supposons que dans la région considérée de l'espace les coefficients g_{ik} sont assez réguliers pour que le problème des géodésiques soit résoluble. Nous admettons aussi que deux points quelconques intérieurs à la région peuvent être joints par une géodésique d'une manière univoque. Les géodésiques de longueur nulle issues d'un point P donnent le conoïde caractéristique ayant pour sommet le point P. Ce conoïde divise l'espace (voisin de P) en trois régions dont l'une est extérieure au conoïde et les deux autres intérieures. Certaines conventions faites, on pourra, tout comme dans le cas de l'équation des ondes, sans ambiguïté parler de la nappe rétrograde et de la nappe directe de ce conoïde.

Soit P un point fixe et Q un point variable intérieur au conoïde qui a son sommet en P. Désignons par $s_{PQ} = s_{QP} = s$ la distance géodésique

(1) Nous appliquons dans tout ce qui suit l'écriture tensorielle.

entre P et Q. Nous nous proposons de construire une fonction $V^\alpha(P, Q)$ dépendant d'un paramètre α qui satisfait à l'équation

$$(1)\qquad \Delta_Q V^{\alpha+2}(P, Q) = V^\alpha(P, Q),$$

et qui devient singulier sur le conoïde comme $s^{\alpha-m}$. A l'exemple de M. Hadamard nous admettons d'abord que les coefficients g_{ik} sont des fonctions holomorphes des variables x^i. Il en sera alors de même de s sauf sur le conoïde et de s^2 partout. Nous posons

$$(2)\qquad V^\alpha = \sum_{k=0}^{\infty} \frac{V_k s^{\alpha+2k-m}}{K_m(\alpha) L_m(\alpha+2k)},$$

les coefficients V_k ne dépendant que de P et de Q, tandis que $K_m(\alpha)$ et $L_m(\alpha)$ ne dépendent que de α (et de la dimension m).

Introduisons maintenant un système de coordonnées normales, admettant le point P comme origine, en gardant la notation g pour la valeur absolue du déterminant des coefficients g_{ik} correspondants. On a alors les formules suivantes faciles à vérifier [1]

$$\Delta_Q s^{\alpha-m} = (\alpha - m) s^{\alpha-m-2}\left(\alpha - 2 + s\frac{\partial \log \sqrt{g}}{\partial s}\right),$$

et plus généralement, U étant une fonction arbitraire,

$$\Delta_Q(s^{\alpha-m} U) = (\alpha - m) s^{\alpha-m-2}\left[\left(\alpha - 2 + s\frac{\partial \log \sqrt{g}}{\partial s}\right) U + 2s\frac{\partial U}{\partial s}\right] + s^{\alpha-m}\Delta_Q U,$$

la notation $\frac{\partial}{\partial s}$ désignant une dérivation suivant la géodésique qui va de P à Q. Cela posé, il vient

$$\Delta V^{\alpha+2} = \sum_{k=0}^{\infty} s^{\alpha+2k-m}\left\{\frac{\alpha+2+2k-m}{K_m(\alpha+2) L_m(\alpha+2+2k)} \times \left[\left(\alpha+2k+s\frac{\partial \log\sqrt{g}}{\partial s}\right) V_k + 2s\frac{\partial V_k}{\partial s}\right] + \frac{\Delta V_{k-1}}{K_m(\alpha+2) L_m(\alpha+2k)}\right\}.$$

---

(1) *Cf.* HADAMARD, *loc. cit.*, p. 117 et W. FELLER, *Über die Lösungen der partiellen Differentialgleichungen zweiter Ordnung vom elliptischen Typus* (*Math. Annalen*, t. 102, 1930, p. 633-649, *voir* p. 639).

On choisit maintenant $L_m(\alpha)$ de sorte qu'on ait

$$\frac{\alpha + 2 - m}{L_m(\alpha + 2)} = \frac{1}{L_m(\alpha)},$$

et l'on détermine V_k par l'équation différentielle récurrente

$$2s\frac{\partial V_k}{\partial s} + \left(2k + s\frac{\partial \log\sqrt{g}}{\partial s}\right)V_k + \Delta V_{k-1} = 0;$$

$$V_{-1} \equiv 0, \qquad V_0(P, P) = 1.$$

Comme chez M. Hadamard, la solution est uniquement déterminée par le postulat que les $V_k(P, Q)$ soient des fonctions régulières de Q. En effet pour $s = 0$, correspondant à $Q = P$, l'équation ci-dessus n'admet qu'une seule solution régulière. Notons que l'on trouve en particulier

$$V_0(P, Q) = \left(\frac{g(Q)}{g(P)}\right)^{-\frac{1}{4}}.$$

Les formules de récurrence posées étant satisfaites, il vient

$$\Delta V^{\alpha+2} = \alpha \sum_{k=0}^{\infty} \frac{V_k s^{\alpha+2k-m}}{K_m(\alpha+2)\,L_m(\alpha+2k)}.$$

En choisissant encore $K_m(\alpha)$ en sorte que

$$\frac{\alpha}{K_m(\alpha+2)} = \frac{1}{K_m(\alpha)},$$

l'équation (1) se trouvera résolue.

$K_m(\alpha)$ et $L_m(\alpha)$ auront les propriétés désirées si l'on pose

$$K_m(\alpha) = K\,2^{\frac{\alpha}{2}}\Gamma\left(\frac{\alpha}{2}\right)$$

et

$$L_m(\alpha) = L\,2^{\frac{\alpha}{2}}\Gamma\left(\frac{\alpha + 2 - m}{2}\right),$$

K et L étant indépendants de α. Nous écrivons encore

$$H_m(\alpha, k) = K_m(\alpha)\,L_m(\alpha + 2k),$$

et nous déterminons le produit KL en sorte qu'on ait

$$(3)\qquad H_m(\alpha, 0) = H_m(\alpha) = \pi^{\frac{m-2}{2}} 2^{\alpha-1} \Gamma\left(\frac{\alpha}{2}\right) \Gamma\left(\frac{\alpha+2-m}{2}\right),$$

la fonction $H_m(\alpha)$ étant le facteur adapté à l'équation des ondes qui figure dans notre conférence. Il vient en définitive

$$\begin{aligned} H_m(\alpha, k) &= K_m(\alpha) L_m(\alpha+2k) \\ &= \pi^{\frac{m-2}{2}} 2^{\alpha+k-1} \Gamma\left(\frac{\alpha}{2}\right) \Gamma\left(\frac{\alpha+2k+2-m}{2}\right). \end{aligned}$$

Signalons encore la formule de récurrence

$$(4)\qquad H_m(\alpha, k) = (\alpha+2k-m) H_m(\alpha, k-1),$$

et observons que cette formule associée à la formule (3) détermine entièrement la fonction $H_m(\alpha, k)$ (¹).

La solution élémentaire

$$s^{2-m} U = \sum_{k=0}^{\infty} U_k s^{2-m+2k}$$

que M. Hadamard forme dans le cas impair est à un facteur constant près identique à notre fonction V^2. Il est instructif de comparer les coefficients U_k à nos coefficients V_k. En normant U par la condition $U(P, P) = 1$, c'est-à-dire $U_0(P, P) = 1$, on a

$$(5)\qquad U_k = \frac{\Gamma\left(2-\frac{m}{2}\right)}{2^k \Gamma\left(2-\frac{m}{2}+k\right)} V_k,$$

(¹) Dans le cas elliptique la fonction correspondante $H_m(\alpha, k)$ se détermine par la formule de récurrence (4) combinée avec la relation

$$H_m(\alpha, 0) = \frac{\pi^{\frac{m}{2}} 2^{\alpha} \Gamma\left(\frac{\alpha}{2}\right)}{\Gamma\left(\frac{m-\alpha}{2}\right)},$$

cette dernière fonction étant le facteur adapté à l'opérateur de Laplace (*cf.* Marcel Riesz, *Intégrales de Riemann-Liouville et potentiels*, *Acta de Szeged*, t. 9, 1938, p. 1-42).

ce qui met en évidence les difficultés qui surgiraient dans le cas où m est un nombre pair, si l'on se bornait à considérer la solution élémentaire V^2 sans introduire un paramètre auxiliaire. En effet $H_m(2, k)$ devient infini pour $2k \leqq m - 4$, ce qui entraîne que V^2 reste fini sur le conoïde. M. Hadamard donne de son côté dans le cas pair une solution élémentaire de la forme

$$s^{2-m} \sum_{k=0}^{\frac{m}{2}-2} U_k s^{2k} + w - 2\mathfrak{U} \log s,$$

où w n'est déterminé qu'à une solution régulière près de l'équation $\Delta u = 0$. Il ne manque pas d'intérêt d'observer que cette solution élémentaire est à un facteur constant et à une fonction régulière additive près identique à

$$\frac{\partial V^\alpha}{\partial \alpha} \quad (\alpha = 2).$$

Soient maintenant P et Q deux points de l'espace de Riemann tels que le conoïde rétrograde appartenant à l'un d'eux et le conoïde direct appartenant à l'autre délimitent un certain domaine B_Q^P. Dès lors on a pour $\alpha > m - 2$ et $\beta > m - 2$

$$(6) \qquad \int_{B_Q^P} V^\alpha(P, T)\, V^\beta(Q, T)\, dT = V^{\alpha+\beta}(P, Q);$$

où dT signifie l'élément de volume de l'espace de Riemann. Pour la démonstration, qui est facile, on met en jeu la propriété (1), la formule de Green et le principe du prolongement analytique par rapport à l'un des indices. On obtient par les mêmes moyens *la propriété d'échange* $V^\alpha(P, Q) = V^\alpha(Q, P)$. De là on tire immédiatement la propriété d'échange des coefficients V_k, démontrée par M. Hadamard au moyen de la méthode de la descente.

Il faut encore dire un mot sur la convergence de nos séries. Or, grâce aux relations (5), on voit immédiatement du résultat correspondant de M. Hadamard que nos séries convergent, dès que la distance géodésique des points P et Q est assez petite.

Soit maintenant S une surface (à $m - 1$ dimensions) dans notre espace de Riemann, et admettons que les conoïdes rétrogrades, par exemple, ayant comme sommet un point arbitraire P situé d'un certain côté de S et assez voisin de S, délimitent avec S certains domaines D_S^P. Nous admettons aussi jusqu'à nouvel ordre que nos séries convergent dans ces domaines. Cela étant, on pose

$$(7) \qquad I^\alpha f(P) = \int_{D_S^P} f(Q)\, V^\alpha(P, Q)\, dQ,$$

Cette intégrale ne converge en général que pour $\alpha > m - 2$. On a, en supposant aussi $\beta > m - 2$,

$$(8) \qquad I^\alpha I^\beta = I^{\alpha+\beta}$$

et

$$(9) \qquad \Delta I^{\alpha+2} = I^\alpha.$$

Or, dans des conditions de dérivabilité convenables portant sur la fonction f et la surface S, on peut étendre la définition (7) et les relations (8) et (9) à tous les indices positifs, par exemple. On peut aussi définir l'intégrale

$$(10) \quad I^{*\alpha} f, g, h(P) = \int_{D_S^P} f(Q)\, V^\alpha(P, Q)\, dQ + \int_{S^P} \left[g(Q)\, V^\alpha(P, Q) - h(Q) \frac{\partial V^\alpha(P, Q)}{\partial n} \right] dS,$$

tous les éléments géométriques étant mesurés au sens riemannien, S^P étant la portion de surface découpée de S par le conoïde et n étant la transversale (conormale) à la surface S au point Q. Cette direction est déterminée au sens près par l'égalité $g_{ik}\, dx^i\, \delta x^k = 0$, qui doit avoir lieu pour tout déplacement d sur la surface S au point Q et pour le déplacement transversal δ. Le sens est donné par l'inégalité $g_{ik}\, dx^i\, \delta x^k > 0$, où d est un déplacement sur la géodésique allant de Q à P. L'intégrale (10) donne lieu aux relations

$$I^\alpha I^{*\beta} = I^{*\alpha+\beta}$$

et

$$\Delta I^{*\alpha+2} = I^{*\alpha}.$$

On a encore les formules capitales

$$I^0 f(P) = I^{*0} f, g, h(P) = f(P)$$

qui, par l'introduction de coordonnées normales, se réduisent facilement aux formules correspondantes données dans notre conférence.

Ces points acquis, la formule de Green donne

$$I^{\alpha} u(P) = I^{*\alpha+2}\left(\Delta u, \frac{\partial u}{\partial n}, u(P)\right),$$

et en particulier

$$u(P) = I^{*2}\left(\Delta u, \frac{\partial u}{\partial n}, u(P)\right). \tag{11}$$

Dans le cas où S a une orientation d'espace, c'est-à-dire que $g_{ik} dx^i dx^k < 0$ pour tout déplacement sur la surface, cette formule de représentation devient une formule de solution pour le problème de Cauchy.

Il s'agit maintenant d'étendre la validité de nos résultats sous deux rapports. L'application de la série (2) donnant V^{α} exigeait en effet d'une part que les coefficients g_{ik} fussent holomorphes et d'autre part que la distance géodésique des points P et Q fût assez petite. Au lieu de supposer que les coefficients sont holomorphes, on n'admettra que l'existence des dérivées jusqu'à un certain ordre. Dans ce cas on ne pourra former la série donnant V^{α}, mais on pourra toujours former un certain nombre des quantités V_k. Dès lors, on pourra, tout comme le fait M. Hadamard, appliquer la méthode de la *parametrix*. Or par cette méthode nous n'obtenons que V^2 (¹) qui ne suffit pas pour donner à la formule (11) la validité générale à laquelle nous tenons. On peut remédier à cet inconvénient de la manière suivante.

Admettons de nouveau un instant que les coefficients g_{ik} soient holomorphes et formons une fonction W^{α} qui joue le même rôle par rapport à l'opérateur $\Delta u + \lambda^2 u$, le nombre λ désignant un nouveau paramètre, que V^{α} joue par rapport à Δu. Posons

$$\begin{aligned} W^{\alpha} &= \sum_{h=0}^{\infty} \binom{-\frac{\alpha}{2}}{h} \lambda^{2h} V^{\alpha+2h} \\ &= \frac{1}{\pi^{\frac{m-2}{2}} 2^{\frac{\alpha+m-2}{2}} \Gamma\left(\frac{\alpha}{2}\right)} \sum_{k=0}^{\infty} V_k \cdot \left(\frac{s}{\lambda}\right)^{\frac{\alpha-m}{2}+k} J_{\frac{\alpha-m}{2}+k}(\lambda s), \end{aligned}$$

où les J sont des fonctions de Bessel. Il est manifeste que W^{α} devient

(¹) Cela tient au fait qu'on connaît $\Delta V^2 (\Delta V^2 = 0)$, mais on ne connaît pas ΔV^{α} d'une manière explicite.

singulier sur le conoïde de la même manière que V^α et l'on vérifie facilement la relation

$$\Delta W^{\alpha+2} + \lambda^2 W^{\alpha+2} = W^\alpha.$$

Dans le cas où les g_{ik} ne sont pas holomorphes mais suffisamment dérivables, on pourra encore par la méthode de la *parametrix* déterminer W^2 et alors V^α s'obtient facilement au moyen d'une intégration complexe par rapport au paramètre λ.

Lund, Institut Mathématique
de l'Université, le 26 juin 1938.

45.

Sur certaines notions fondamentales en théorie quantique relativiste

10e Congr. des Math. Scand. Copenhagen 1946, 123–148

Introduction.

On sait depuis les travaux de FOURIER et de CAUCHY que relations algébriques et équations aux dérivées partielles à coefficients constants peuvent, dans une certaine mesure, se traduire les unes dans les autres par l'intermédiaire de l'intégrale de FOURIER. De telles correspondances se rencontrent fréquemment en théorie quantique, et nous en donnerons quelques exemples dans la suite.

L'espace-temps.

Si dans l'*espace de Lorentz* à quatre dimensions (espace-temps ou univers), la vitesse de la lumière est égalée à 1, le carré scalaire (x,x) du rayon vecteur x mené de l'origine au point $x = (x^0, x^1, x^2, x^3)$ est mesuré par la forme métrique

$$(1) \qquad (x, x) = x^{0^2} - x^{1^2} - x^{2^2} - x^{3^2} = \sum_{j,\, k=0}^{3} g_{jk} x^j x^k ;$$

$$g_{00} = 1,\ g_{11} = g_{22} = g_{33} = -1;\quad g_{jk} = 0,\ j \neq k .$$

En adoptant la convention due à EINSTEIN, à savoir que, sauf indication contraire, une sommation est à effectuer pour tout couple d'indices identiques (l'un supérieur, l'autre inférieur), on peut aussi écrire

$$(1^{bis}) \qquad (x, x) = g_{jk} x^j x^k .$$

La coordonnée x^0 correspond au temps, tandis que x^1, x^2, x^3 sont les coordonnées d'espace. Par une *transformation de Lorentz* on entend toute transformation linéaire des coordonnées qui conserve la forme métrique. U étant un vecteur aux composantes U^j $(j = 0, 1, 2, 3)$, on définit le carré scalaire (U, U) par la formule, analogue à (1), $(U, U) = g_{jk}U^jU^k$. Les coordonnées étant assujetties à une transformation de LORENTZ, la même transformation sera à appliquer aux composantes U^j du vecteur U, ce qui implique qu'une telle transformation conserve aussi le carré scalaire d'un vecteur arbitraire. Les composantes U^j sont appelées les composantes *contrevariantes* du vecteur U. On leur associe les composantes *covariantes*, définies par les formules $U_j = g_{jk}U^k$. Pour garder la symétrie, on pose en même temps $U^j = g^{jk}U_k$, où, dans le cas particulier actuel, $g^{jk} = g_{jk}$. En résumé, le carré scalaire d'un vecteur U peut s'écrire

$$(U, U) = g_{jk}U^jU^k = g^{jk}U_jU_k = U^jU_j\,. \tag{2}$$

Un vecteur, à composantes réelles, est dit 1) vecteur de *temps*, 2) vecteur de *lumière* ou vecteur *isotrope*, 3) vecteur d'*espace*, suivant que son carré scalaire est 1) positif, 2) nul, 3) négatif. Un vecteur au carré scalaire ± 1 est dit *unitaire*. Un vecteur de temps U est dit *positif* ou *négatif* suivant que sa composante de temps U^0 est positive ou négative. Les définitions posées sont invariantes par rapport aux transformations de LORENTZ, la dernière par rapport à celles seulement qui n'échangent pas entre eux les deux demi-cônes de lumière.

On définit le produit scalaire (U, V) de deux vecteurs U et V par la formule

$$(U, V) = (V, U) = g_{jk}U^jV^k = g^{jk}U_jV_k = U^jV_j = U_jV^j. \tag{3}$$

Il est clair que ce produit est encore invariant par rapport à toute transformation de LORENTZ. La transformation relative aux composantes covariantes devra être contregrédiente à celle opérée sur les composantes contrevariantes. Si, pour une raison ou une autre, une expression de la forme $U^\mu V_\mu$ est invariante et que l'on sait que les U^μ sont les composantes contrevariantes d'un certain vecteur U, on peut en conclure que les V_μ sont de leur côté les composantes covariantes d'un certain vecteur V. Appliquant ceci en particulier à la différentielle totale $df = \sum \frac{\partial f}{\partial x^k} dx^k$ d'une fonction f, on voit que les quantités $\partial f/\partial x^k$ sont les composantes covariantes[1]) $\text{grad}_k f$ du gradient de f, dont les composantes contrevariantes sont, par conséquent,

---

[1]) L'indice k figurant dans l'opération $\partial/\partial x^k$ est à considérer comme un indice inférieur (covariant).

$$\text{grad}^j f = g^{jk} \frac{\partial f}{\partial x^k}. \tag{4}$$

Deux vecteurs dont le produit scalaire est nul sont dits *orthogonaux*.

L'espace impulsion-énergie.

Ces préliminaires posés, considérons un point matériel $x = (x^0, x^1, x^2, x^3)$, à la masse de repos m, qui décrit une ligne d'univers. Le déplacement infinitésimal dx sera d'après les conceptions relativistes (toute vitesse mécanique est inférieure à la vitesse de la lumière) un vecteur de temps. Le carré de l'élément d'arc ds (élément du temps propre du corpuscule) est donné par la relation $ds^2 = (dx, dx) > 0$. Sans préciser le sens positif de s, on forme le vecteur quadrivitesse $v = dx/ds$ et le vecteur impulsion-énergie $p = m \cdot dx/ds$, qui sont des vecteurs de temps tous les deux, admettant les carrés scalaires $(v, v) = 1$, $(p, p) = m^2$ respectivement.

Cela étant, nous associons à l'espace x (espace-temps) l'espace p (espace impulsion-énergie). Ils n'interviendront dans nos considérations, de caractère relativiste, que les points p qui remplissent la condition $(p, p) > 0$ (ou, à la limite, la condition $(p, p) = 0$). Géométriquement parlant, nous nous restreignons à l'intérieur (ou à la frontière) du cône de lumière $(p, p) = 0$.

A l'hypothèse, admise dans la suite, à savoir que la masse de repos admet une valeur déterminée m, il correspond la relation algébrique

$$(p, p) = p_k p^k = m^2 . \tag{5}$$

Cette relation signifie en langage géométrique que le point p se trouve sur l'hyperboloïde à deux nappes H_m, dont l'équation est donnée par (5). Bien entendu, cet hyperboloïde est une variété à trois dimensions, dont les deux nappes correspondent respectivement aux valeurs positives et négatives de l'énergie p_0.

L'intégrale de Fourier et l'équation relativiste de Schrödinger.

Cela posé, et égalant $h/2\pi$ (h est la constante de Planck) à 1, nous formons l'intégrale

$$\psi(x) = \frac{1}{(\sqrt{2\pi})^3} \int_{H_m} \varphi(p) e^{i(p,x)} d\sigma , \tag{6}$$

où cette intégrale est étendue à l'hyperboloïde (5) et $d\sigma$ est un élément d'intégration, attaché à cette surface, sur lequel on reviendra tout à l'heure. Quant à la fonction $\varphi(p)$, il suffit de l'avoir définie le long de H_m. Notre intégrale fournit la traduction suivante de la relation (5). En introduisant l'opérateur des ondes

$$\square = \frac{\partial^2}{\partial x^{0^2}} - \frac{\partial^2}{\partial x^{1^2}} - \frac{\partial^2}{\partial x^{2^2}} - \frac{\partial^2}{\partial x^{3^2}} = g^{jk} \frac{\partial^2}{\partial x^j \partial x^k}, \tag{7}$$

il vient dans des conditions très générales

$$\square\, \psi(x) = \frac{-1}{(\sqrt{2\pi})^3} \int g^{jk} p_j p_k \varphi(p) e^{i(p,x)} d\sigma \,.$$

Puisque le point p est restreint à la surface (5) on a $g^{jk} p_j p_k = (p, p) = m^2$, ce qui donne

$$(\square + m^2)\psi(x) = 0\,, \tag{8}$$

c'est-à-dire que la fonction $\psi(x)$ satisfait à l'équation relativiste de SCHRÖDINGER, posée pour un corpuscule libre (absence de champ électromagnétique) à la masse de repos m.[1])

Inversement, on peut prétendre que toute solution assez régulière de l'équation (8) peut être représentée directement ou par des passages à la limite (embrassant les intégrales de FOURIER-STIELTJES) par des intégrales de FOURIER. Nous n'insistons pas ici sur cette question, vu que les intégrales de FOURIER que nous allons considérer appartiendront à une classe assez particulière.

Nous aurons dans la suite souvent à faire à des intégrales étendues à certaines surfaces S et nous aurons besoin de l'élément de surface lorentzien, élément invariant par rapport aux transformations de LORENTZ.

Pour arriver à un tel élément, plaçons nous d'abord dans le cas plus général que voici. Considérons dans un espace à n dimensions une métrique quadratique arbitraire, $(U, U) = g_{ik} U^i U^k$, $(g_{ik} = g_{ki})$,

---

[1]) La remarque suivante n'est peut-être pas dépourvue d'intérêt, quoiqu'il ne s'agisse que d'une transcription géométrique d'un fait bien connu par ailleurs. Considérons dans l'espace p à quatre dimensions, le paraboloïde

$$p_0 = \frac{1}{2m} (p_1^2 + p_2^2 + p_3^2)\,.$$

La fonction $\psi(x)$ (x est un point de l'espace-temps) fournie par une intégrale analogue à (6), mais cette fois-ci étendue à la surface du paraboloïde susdit, satisfait à l'équation non relativiste de SCHRÖDINGER.

qui nous fournit en même temps le produit scalaire $(U, V) = g_{ik}U^iV^k$. Dès lors, $U, V, W, \ldots$ étant un certain nombre, soit l, de vecteurs, finis ou infinitésimaux, nous définissons le « volume » v du parallélépipède à l dimensions construit sur ces vecteurs en égalant v^2 à la valeur absolue du déterminant

$$\begin{vmatrix} (U, U) & (U, V) \ldots \\ (V, U) & (V, V) \ldots \\ \ldots\ldots & \ldots\ldots \end{vmatrix}.$$

Ce volume possède toutes les propriétés essentielles des volumes euclidiens. Ainsi p. ex. lorsque les vecteurs donnés sont mutuellement orthogonaux, le « volume » v devient égal au produit des « longueurs » de ces vecteurs.

Dans le cas actuel d'un élément de surface lorentzien, on aura $n = 4$, $l = 3$ et les vecteurs seront les tangentes infinitésimales de trois familles de courbes tracées sur la surface S.

Ce point élucidé, retournons à nos intégrales de Fourier. Dès lors, en désignant par dH_m l'élément de surface lorentzien de H_m, nous définissons l'élément d'intégration $d\sigma$ figurant dans (6), par la relation $d\sigma = dH_m/m$. Je dis que

$$d\sigma = \frac{dH_m}{m} = \frac{dp_1 dp_2 dp_3}{|p_0|}, \tag{9}$$

en attribuant à toutes les différentielles des valeurs positives.

Pour éviter des calculs explicites, nous pouvons nous laisser guider par les formules euclidiennes correspondantes. Considérons dans l'espace euclidien à n dimensions la sphère $\sum_1^n y_k^2 = r^2$ et désignons par dS son élément de surface. On aura évidemment

$$|dy_1 \ldots dy_{k-1} dy_{k+1} \ldots dy_n| = |dS. \cos(y_k, r)| = \frac{|y_k|}{r} dS$$

et par là

$$\frac{dS}{r} = \frac{|dy_1 \ldots dy_{k-1} dy_{k+1} \ldots dy_n|}{|y_k|}.$$

Ce point acquis, on voit que le dernier membre de (9) est un invariant puisqu'il en est ainsi du second membre. D'autre part celui-ci n'a pas de sens pour $m = 0$, valeur correspondant au cône de lumière, puisque le numérateur et le dénominateur s'annulent tous les deux, tandis que le dernier membre se réduit à l'expression bien connue $dp_1 dp_2 dp_3 : (p_1^2+p_2^2+p_3^2)^{\frac{1}{2}}$. Ces préliminaires posés, nous nous bornons dans la suite au cas où l'intégrale (6) n'est étendue qu'à celle des nappes de H_m pour laquelle l'énergie p_0 est positive, nappe qui sera désignée par H_m^+. Notre intégrale de FOURIER s'écrit en définitive

$$\psi(x) = \frac{1}{(\sqrt{2\pi})^3} \int_{H_m^+} \varphi(p) e^{i(p,x)} d\sigma \,, \tag{10}$$

où $d\sigma$ est défini par (9). En ce qui concerne $\varphi(p)$ nous supposerons dès maintenant que l'intégrale

$$A = \int_{H_m^+} |\varphi(p)|^2 d\sigma \tag{11}$$

admet une valeur finie. Rien n'empêche d'ailleurs de faire des hypothèses supplémentaires sur l'allure de $\varphi(p)$ quand p tend vers l'infini (en restant toujours sur H_m^+) pour assurer p. ex. que l'opération $\Box\psi$ puisse être exécutée sans difficulté.

Les distributions de probabilité se correspondant dans les deux espaces. Une identité.

Il est maintenant assez naturel d'interpréter $A^{-1} |\varphi(p)|^2 d\sigma$ comme la probabilité du fait que le point représentant le vecteur impulsion-énergie d'un corpuscule libre, à la masse de repos donnée m, appartient à l'élément $d\sigma$ autour du point p. (Pour être rigoureux, il faudrait parler de l'élément de surface dH_m^+ correspondant à $d\sigma$).

La question se pose de savoir quelle notion de probabilité correspond dans l'espace des x à l'élément de probabilité relatif à l'espace des p que nous venons de définir. Il est facile de montrer que si la quantité A, donnée par la formule (11), est finie, il en sera des même de l'intégrale triple

$$\iiint_{-\infty}^{+\infty} |\psi|^2 dx_1 dx_2 dx_3$$

pour toute valeur fixée du temps x^0. Or cette intégrale converge même trop bien, elle converge dès qu'il en est ainsi de l'intégrale qu'on obtient de (11) en y remplaçant $d\sigma$ par $d\sigma/p_0$. En outre, en variant x^0, la valeur de l'intégrale ne sera pas conservée. Cet état de choses, bien connu en principe dès les débuts de la théorie quantique relativiste, fait pressentir qu'en cherchant une intégrale formée dans l'espace x qui ait une valeur identique à (11), on devra recourir à une expression pour laquelle il existe une loi de conservation. Une telle expression est fournie par la composante de temps du quadrivecteur considéré par GORDON (1926) et par KLEIN (1927).

Dans sa *Bakerian Lecture* bien connue[1]), M. DIRAC a (dans des termes un peu vagues) constaté que les probabilités fournies d'une part par l'expression $|\varphi|^2\, d\sigma$ et de l'autre par le dit quadrivecteur doivent se correspondre. Nous déduirons tout à l'heure une identité par laquelle cette correspondance sera entièrement précisée.

La fonction $\psi(x)$ étant une solution de l'équation (8), il en sera de même de la fonction imaginaire conjuguée $\bar{\psi}(x)$. Formons les gradients de ces fonctions par rapport aux *quatre* variables x^μ. Le gradient de $\psi(x)$ aura p. ex. les composantes covariantes $\frac{\partial \psi}{\partial x^\mu}$. Posons dès lors le quadrivecteur réel

$$(12) \qquad G = \frac{1}{2i}(\bar{\psi}\,\mathrm{grad}\,\psi - \psi\,\mathrm{grad}\,\bar{\psi}).$$

En raison de l'équation (8) remplie par les fonctions ψ et $\bar{\psi}$, la divergence (c'est-à-dire quadridivergence) de ce vecteur sera nulle. Les composantes contrevariantes G^k de G sont en effet

$$G^k = \frac{1}{2i}(\bar{\psi}\,\mathrm{grad}^k\,\psi - \psi\,\mathrm{grad}^k\,\bar{\psi}),$$

par conséquent

$$\mathrm{div}\,G = \frac{\partial G^k}{\partial x^k} = \frac{1}{2i}\left(\frac{\partial \bar{\psi}}{\partial x^k}\,\mathrm{grad}^k\,\psi + \bar{\psi}\frac{\partial\,\mathrm{grad}^k \psi}{\partial x^k} - \frac{\partial \psi}{\partial x^k}\,\mathrm{grad}^k\,\bar{\psi} - \psi\frac{\partial\,\mathrm{grad}^k \bar{\psi}}{\partial x^k}\right).$$

Les quatre termes se détruisent deux à deux. En effet, eu égard à ce que $\partial\psi/\partial x^k = \mathrm{grad}_k\,\psi$, $\partial\bar{\psi}/\partial x^k = \mathrm{grad}_k\,\bar{\psi}$, on aura d'après (3) deux termes $\pm$ $(\mathrm{grad}\,\psi,\ \mathrm{grad}\,\bar{\psi})$. D'autre part, en raison des formules (7) et (8) on obtiendra deux termes $\pm m^2\psi\bar{\psi}$.

Considérons dès lors une surface Σ orientée dans l'espace[2]) et s'étendant à l'infini. Menons en chaque point la normale unitaire n dirigée vers les x^0 croissants et formons le produit scalaire $(G, n) = G_k n^k$, qui pourra être interprété comme la projection de G sur n ou la composante de G suivant n. Ce produit est évidemment

[1]) Proc. Roy. Soc. A. 180; 1—40; voir en partic. p. 6—7.

[2]) Une surface est dite orientée dans l'espace, si un cône de lumière arbitraire qui a son sommet sur cette surface n'a que ce point en commun avec elle.

invariant par rapport à toute transformation de LORENTZ. Nous posons encore pour abréger $(G, n) = G_{<n>}$. Dans les conditions habituelles concernant l'allure de $\psi(x)$ à l'infini, la relation $\operatorname{div} G=0$ se traduit en vertu du théorème de GAUSS dans le fait que l'intégrale $\int_{\Sigma} G_{<n>} d\Sigma$ est indépendante de la surface Σ. C'est la loi de conservation générale relative au vecteur G. En particulier, si la surface Σ se réduit à un plan $x^0 = \text{const.}$, $G_{<n>}$ devient G^0, et on trouve

$$(13) \qquad \int_{\Sigma} G_{<n>} d\Sigma = \iiint_{-\infty}^{+\infty} G^0(x^0, x^1, x^2, x^3)\, dx^1 dx^2 dx^3 .$$

Le fait que cette dernière intégrale est indépendante de x^0 représente la loi de conservation sous sa forme usuelle.

Nous allons maintenant établir l'identité, valable pour $m \geqq 0$,

$$(14) \qquad \int_{\Sigma} G_{<n>} d\Sigma = \int_{H_m^+} |\varphi|^2 d\sigma ,$$

qui est un corollaire immédiat du théorème de PARSEVAL-PLANCHEREL. Voici le théorème en question.

Dans l'espace à n dimensions nous posons $\sum_0^n x_k y_k = (x, y)$, $dx_1 dx_2 \ldots dx_n = dx$, $dy_1 dy_2 \ldots dy_n = dy$. Soient en outre les fonctions $f(x)$ et $g(x)$ mesurables au sens de LEBESGUE et telles que les intégrales $\int |f|^2 dx$ et $\int |g|^2 dx$, étendues à l'espace entier, soient finies. Dès lors les transformées de FOURIER

$$T(f(x)) = \frac{1}{(\sqrt{2\pi})^n} \int f(y) e^{i(x,y)} dy \quad \text{et} \quad T(g(x)) = \frac{1}{(\sqrt{2\pi})^n} \int g(y) e^{i(x,y)} dy$$

existent presque partout (dans le sens de la convergence en moyenne) et l'on a

$$\int T(f(x))\, \overline{T(g(x))}\, dx = \int f(y) \overline{g(y)}\, dy ,$$

les dernières intégrales étant absolument convergentes. Bien entendu, toutes les intégrales sont étendues à l'espace entier.

Appliquons ce théorème au cas actuel. On a

$$(15) \qquad \begin{cases} \psi(x) = \dfrac{1}{(\sqrt{2\pi})^3} \displaystyle\iiint_{-\infty}^{+\infty} \varphi(p)\, e^{i(p,x)} \frac{dp_1 dp_2 dp_3}{p_0}, \\ \qquad\qquad p_0^2 = m^2 + p_1^2 + p_2^2 + p_3^2, \; p_0 > 0 . \\ \dfrac{\partial}{\partial x^0} \psi(x) = \dfrac{i}{(\sqrt{2\pi})^3} \displaystyle\iiint_{-\infty}^{+\infty} \varphi(p)\, e^{i(p,x)} dp_1 dp_2 dp_3 . \end{cases}$$

On en tire

$$\iiint_{-\infty}^{+\infty} \overline{\psi(x)} \frac{\partial}{\partial x^0} \psi(x)\, dx^1 dx^2 dx^3 =$$

$$\iiint_{-\infty}^{+\infty} \frac{\overline{\varphi(p)}\, e^{-ip_0 x^0}}{p_0} \cdot i\varphi(p)\, e^{ip_0 x^0} dp_1 dp_2 dp_3 = i \int_{H_m^+} |\varphi(p)|^2 d\sigma \,.$$

Par l'intermédiaire de la relation (13) l'identité (14) s'ensuit de l'identité qu'on vient d'obtenir et de son imaginaire conjuguée.

Bien entendu, le facteur p_0 ne figurant pas au dénominateur dans la seconde formule (15), on devra, pour assurer la légitimité de cette déduction, supposer que l'intégrale (11) reste convergente si l'on y remplace l'élément d'intégration $d\sigma$ par $p_0 d\sigma = dp_1 dp_2 dp_3$.

Comme élément de probabilité, on a toujours reproché à l'expression $G^0 dx^1 dx^2 dx^3$ que G^0 n'était pas nécessairement positif. Par notre résultat on voit du moins que la probabilité totale fournie par G^0 peut s'exprimer par une intégrale prise dans l'espace p où l'élément d'intégration $|\varphi|^2 d\sigma$ est toujours $\geqq 0$.

En résumé, à la probabilité scalaire $|\varphi|^2$ relative à l'espace p il correspond dans l'espace x une probabilité fournie par une composante d'un vecteur, à la probabilité positive une probabilité qui peut admettre des valeurs négatives. L'état des choses sera beaucoup plus satisfaisant, si l'on passe de l'équation de Schrödinger à celle de Dirac.

Les nombres de Clifford.

Rappelons d'abord la définition et quelques propriétés des nombres de Clifford. Chaque vecteur U peut évidemment se mettre sous la forme $U = U^k e_k$, e_k étant le vecteur (de base) dont la $k^{\text{ième}}$ composante contrevariante est 1, tandis que les autres sont nulles. Cherchons maintenant les conditions pour qu'on puisse écrire

(16) $$(U, U) = g_{jk} U^j U^k = (U^k e_k)^2 = U^2 \,.$$

Ces conditions sont évidemment

(17) $$e_j e_k + e_k e_j = 2 g_{jk} \,.$$

On tire de l'une quelconque des relations (16) et (17) pour deux vecteurs quelconques

(18) $$UV + VU = 2(U, V)\,,$$

ce qui met en évidence que deux vecteurs orthogonaux arbitraires

$\big((U, V) = 0\big)$ anticommutent. Les considérations précédentes s'appliquent à une forme métrique et à une dimension quelconques. Nous ne nous intéressons ici qu'au cas où la dimension est quatre et que les g_{jk} ont les valeurs données dans (1), c'est-à-dire que les règles de calcul (17) sont précisément adaptées à l'espace de LORENTZ. D'une manière détaillée, nous aurons donc

$$(19) \qquad e_0^2 = 1, \quad e_1^2 = e_2^2 = e_3^2 = -1, \quad e_j e_k + e_k e_j = 0, j \neq k.$$ [1]

En posant encore $e^j = g^{jk} e_k$, on voit de suite que $e^j e^k + e^k e^j = 2g^{jk}$. Observons de plus que, dans le cas actuel, on a $e^0 = e_0$, $e^k = -e_k$ pour $k = 1, 2, 3$.

Avant d'aller plus loin, notons ici le fait suivant qui sera d'une importance capitale dans la suite. On a

$$(20) \qquad e_j^{-1} = e_0 e_j e_0 = e_0 e_j e_0^{-1}, \quad j = 0, 1, 2, 3\,.$$

En effet, ceci équivaut à $e_0 e_j \cdot e_0 e_j = 1$, ce qui n'est qu'un corollaire immédiat des relations (19). La relation ci-dessus est caractéristique de la signature lorentzienne: un seul des carrés e_j^2 est positif.

Les quantités e_j engendrent une algèbre, ou un système hypercomplexe, ou un anneau, d'ordre 16, à multiplication associative. Une base est fournie par les 16 monomes figurant dans le développement du produit $(1+e_0)$ $(1+e_1)$ $(1+e_2)$ $(1+e_3)$. Les monomes en question sont antisymétriques en ce sens que l'intervertissement de deux facteurs quelconques dans $e_j e_k e_l$, par exemple altère le signe: $e_k e_j e_l = -e_j e_k e_l = e_j e_l e_k$ (j, k, l sont supposés inégaux). Nous désignerons ces 16 monomes par $e_A, e_B, \ldots$, l'ordre des facteurs dans chacun étant indifférent, mais déterminé une fois pour toutes. Un nombre de CLIFFORD c sera alors de la forme $c = \Sigma c_A e_A$, les c_A étant des nombres complexes ordinaires, qui commutent avec les éléments $e_A : c_A e_A = e_A c_A$. On pose $c = 0$, si tous les coefficients c_A sont nuls et dans ce cas-là seulement.

Pour éviter toute confusion, nous soulignons ici que nous considérons les symboles primaires e_j et leurs produits e_A comme des quantités *réelles*. Un nombre c de CLIFFORD sera considéré comme une quantité réelle si tous les coefficients c_A sont des nombres réels ordinaires. Nous ne serons donc pas ici gênés par les inconvénients qu'on rencontre quand il s'agit de reconnaître le caractère réel de certaines quantités cliffordiennes relatives à l'espace de LORENTZ

---

[1]) Dans l'espace euclidien on a $e_j^2 = 1$ et $e^j = e_j$ pour toute valeur de j.

et représentées par des matrices inconvénients, auxquels on remédie en passant des matrices γ_k aux matrices α_k. (*Cf.* PAULI, Handb. d. Phys. 24_1 (1934) p. 221 et Rev. of Modern Phys. 13 (1941) p. 221).

L'élément de base scalaire 1, qui commute avec tout nombre de CLIFFORD sera alternativement désigné par e_S. Son coefficient c_S donne la *partie scalaire* du nombre c. Nous notons les propriétés suivantes de cette partie scalaire qui jouent un grand rôle dans ce qui suit.

1) c_S est invariant par rapport à toute transformation de LORENTZ.
2) c' et c'' étant deux nombres de CLIFFORD, on a $(c'c'')_S = (c''c')_S$. (Il en résulte que dans les produits comprenant plusieurs facteurs les permutations circulaires ne changent pas non plus la partie scalaire).

La première propriété dépend du fait que chaque partie homogène forme un tenseur antisymétrique (en y rangeant par extension aussi les scalaires et les vecteurs) et se transforme, d'après les lois de transformation de ces tenseurs, en une forme homogène du même degré.

Pour démontrer la seconde propriété, observons que $(c'c'')_S = \sum_A c'_A c''_A e_A^2$ où $e_A^2 = \pm 1$.

Norme d'un nombre de Clifford. Un tenseur du second ordre.

Nous définirons comme norme $||c||$ d'un nombre de CLIFFORD la quantité $||c|| = \sum_A |c_A|^2$ qui évidemment est *positive*, si $c \neq 0$. Nous allons voir que cette valeur, qui est un scalaire (invariant) dans l'espace euclidien, est, dans l'espace de LORENTZ, la composante T_{00} d'un tenseur du second ordre. Pour arriver à ce résultat, il nous faudra considérer certains automorphismes de l'anneau cliffordien.

1) *Renversement.* Cet automorphisme consiste dans le renversement dans chaque produit e_A de l'ordre des facteurs e_j. P. ex. $e_1e_3 \to e_3e_1$, $e_2e_0e_3 \to e_3e_0e_2$. Cette opération sera désignée par $e_A \to \breve{e}_A$, $c = \sum c_A e_A \to \sum c_A \breve{e}_A = \breve{c}$.

2) *Conjugaison.* Chaque coefficient c_A est remplacé par son imaginaire conjugué $\bar{c}_A$, c'est-à-dire que $c = \sum c_A e_A \to \sum \bar{c}_A e_A = \bar{c}$.

3) *Renversement conjugué.* Les deux opérations ci-dessus sont commutatives. Leur produit consiste en $c \to c^\dagger = (\bar{\breve{c}}) = (\breve{\bar{c}}) = \sum \bar{c}_A \breve{e}_A$.

Remarquons que toutes les opérations définies plus haut sont *indépendantes du système de coordonnées* (invariantes par les transformations qui conservent la forme métrique). Il suffit de le montrer pour le dernier. Celui-ci peut être caractérisé de la manière suivante, où n'interviennent pas les symboles primaires e_j.

L'opération $c \to c^\dagger$ est un automorphisme renversé ($c' \to c'^\dagger$, $c'' \to c''^\dagger$ implique $c'c'' \to c''^\dagger c'^\dagger$) de l'anneau de Clifford, qui conserve les scalaires et les vecteurs réels[1]) et change le scalaire $i = \sqrt{-1}$ en $-i$.

Il résulte immédiatement de ce fait que la norme $||c||$ est scalaire dans un anneau de Clifford relatif à un espace euclidien. En effet dans un tel espace on a, en posant à titre d'exemple $e_A = e_j e_k$: $e_A \bar{e}_A = e_j e_k e_k e_j = e_j e_j = 1$, c'est-à-dire que $||c|| = \Sigma |c_A|^2 = (c\, c^\dagger)_S$. Il n'en est pas ainsi dans l'espace de Lorentz où $e_A \bar{e}_A = \pm 1$. Pour reconnaître le caractère tensoriel de la norme dans un tel espace, il faudra encore définir un quatrième automorphisme qui cependant dépendra du choix du système de coordonnées ou, plus précisément, du choix de la direction de l'axe des x^0, sans dépendre du sens positif de cet axe.

4) *Adjonction hermitique.* $c = \Sigma c_A e_A \to c^* = \Sigma \bar{c}_A e_A^{-1}$.[2]) Cette fois-ci on a bien

$$(21) \qquad ||c|| = (cc^*)_S = (c^* c)_S \,.$$

Il est facile de montrer que $c^* = e_0 c^\dagger e_0 = e_0 c^\dagger e_0^{-1}$ et par là

$$(22) \qquad ||c|| = (e_0 c^\dagger e_0 c)_S \,.$$

En effet, la relation revient à $e_A^{-1} = e_0 \bar{e}_A e_0^{-1}$. Or en vertu de (20) on a p. ex. pour $e_A = e_j e_k$, $e_A^{-1} = e_k^{-1} e_j^{-1} = e_0 e_k e_0^{-1}$, $e_0 e_j e_0^{-1} = e_0 e_k e_j e_0 = e_0 \bar{e}_A e_0^{-1}$.

En formant le tenseur du second ordre

$$(23) \qquad T_{jk} = (e_j c^\dagger e_k c)_S \,,$$

on a évidemment $||c|| = T_{00}$.

Le tenseur T_{jk} est *réel*, mais il n'est pas symétrique en général. Le premier point s'établit de la manière suivante. On a

$$T_{jk} = \sum_{A,B} \bar{c}_A c_B e_j \bar{e}_A e_k e_B \,,$$

---

[1]) Bien entendu, les produits de tels vecteurs ne sont pas conservés en général.

[2]) On vient de voir que dans un anneau cliffordien relatif a un espace euclidien on a $c^* = c^\dagger$.

la sommation étendue aux couples d'indices A et B pour lesquels le produit $e_j\bar{e}_A e_k e_B$ est scalaire. Or tel étant le cas, il en sera de même du produit $e_j\bar{e}_B e_k e_A$ qui figure avec le coefficient $\bar{c}_B c_A$. Si l'on peut démontrer que ces deux produits sont égaux, leur valeur commune figurera avec le coefficient $\bar{c}_A c_B+\bar{c}_B c_A$ qui est réel. Or un produit qui est scalaire n'est altéré, ni par un renversement, ni par une permutation circulaire des facteurs. On a donc $e_j\bar{e}_A e_k e_B = \bar{e}_B e_k e_A e_j = e_j\bar{e}_B e_k e_A$. *C. Q. F. D.*

En ce qui concerne le manque de symétrie dans le cas général, nous en verrons des exemples plus loin. Ici pourtant, pour mettre en évidence tout l'intérêt que mérite le tenseur T_{jk}, nous montrerons que le tenseur impulsion-énergie de MAXWELL-MINKOWSKI qui figure dans la théorie électromagnétique est un cas particulier — d'ailleurs symétrique — du tenseur ci-dessus.

Le tenseur antisymétrique aux composantes deux fois contrevariantes F^{jk}, qui correspond à un champ électromagnétique peut être représenté par le nombre de CLIFFORD $F = \sum_{p<q} F^{pq} e_p e_q$. On sait que la densité d'énergie, exprimée dans les notations de MAXWELL par $\frac{1}{2}(|E|^2+|H|^2)$ et dans celles de MINKOWSKI par $\frac{1}{2}\sum_{p<q} F^{pq^2}$, n'est pas un invariant, mais qu'elle est la composante S_{00} du dit tenseur impulsion-énergie S_{jk}. D'autre part, en observant que le champ et, par là, les F^{jk} sont toujours supposés réels, on a $F^{pq^2} = |F^{pq}|^2$ et alors, en vertu de (21),

$$2S_{00} = \sum F^{pq^2} = ||F|| = (e_0 F^\dagger e_0 F)_S = T_{00}\,,$$

où l'on a posé d'une manière générale $T_{jk} = (e_j F^\dagger e_k F)_S$. Nous montrons que les tenseurs T_{jk} et S_{jk} sont identiques au facteur $\frac{1}{2}$ près. Or on sait que le dernier tenseur est symétrique. Montrons qu'il en est de même de T_{jk}. On a, en effet, à cause de $e_q e_p = -e_p e_q$, $F^\dagger = -F$ et, par une permutation circulaire des facteurs,

$$T_{jk} = (e_j F^\dagger e_k F)_S = -(e_j F e_k F)_S = -(e_k F e_j F)_S = (e_k F^\dagger e_j F)_S = T_{kj}\,.$$

Cela étant, on voit que le tenseur $W_{jk} = S_{jk}-\frac{1}{2}T_{jk}$ est symétrique et de plus que $W_{00} = 0$ dans tout système de coordonnées lorentziennes. Il en résulte par un calcul facile que $W_{jk} = 0$ pour tout couple d'indices j, k.

Revenons au cas général. Nous avons vu que $(e_0 c^\dagger e_0 c)_S > 0$, pour tout nombre de CLIFFORD $c \neq 0$, puisque la quantité ci-dessus est

$= ||c||$. Ce fait admet la généralisation suivante dont nous aurons besoin dans la suite.

U et V étant deux vecteurs de temps positifs (ou négatifs) *et c un nombre de Clifford* $\neq 0$, *on a*

$$(Uc^\dagger Vc)_S > 0 \,. \tag{24}$$

Sans restreindre la généralité, on peut supposer que U et V sont unitaires. Cela posé, l'inégalité (24) est exacte pour $V = U$, comme on le voit en introduisant un système de coordonnées lorentziennes x'^k où le vecteur de base e_0' coïncide avec U[1]). Dans le cas général, nous nous appuyerons sur le lemme suivant que nous allons démontrer tout à l'heure.

U et V étant deux vecteurs de temps positifs et unitaires, il existe un vecteur de temps positif et unitaire tel que les deux relations équivalentes

$$V = KUK \text{ et } U = KVK \tag{25}$$

aient lieu.

Ce lemme admis, on aura

$$(Uc^\dagger Vc)_S = (Uc^\dagger KUKc)_S = \big(U(Kc)^\dagger U(Kc)\big)_S \,,$$

par quoi le cas général se réduit au cas particulier.

Il reste à démontrer le lemme. Observons d'abord que $(U, U) = (V, V)$ peut aussi s'écrire $(U+V, U-V) = 0$, ce qui amène d'après la formule (3) que ces derniers vecteurs anticommutent. En posant maintenant

$$U = \frac{U+V}{2} + \frac{U-V}{2}, \quad V = \frac{U+V}{2} - \frac{U-V}{2}, \quad K = \mu(U+V) \,,$$

μ étant déterminé de façon que $K^2 = (K, K) = 1$, on obtient

$$KUK = K \cdot \frac{U+V}{2} K + K \cdot \frac{U-V}{2} K =$$

$$K^2 \cdot \frac{U+V}{2} - K^2 \cdot \frac{U-V}{2} = \frac{U+V}{2} - \frac{U-V}{2} = V.$$

Notre lemme est intimement lié au théorème connu[2]) que voici.

---

[1]) On aura, en se servant du nouveau système de coordonnées,

$$c = \sum_A c_A e_A = \sum_A c_A' e_A' \text{ et } (e_0' c^\dagger e_0' c)_S = \sum_A |c_A'|^2 .$$

[2]) Ce théorème embrasse les relations (20).

L étant un vecteur non isotrope et U un vecteur quelconque, $V = LUL^{-1}$ est un vecteur, à savoir le symétrique de U par rapport à la droite qui porte L.[1])

Décomposons U en deux vecteurs U' et U'', le premier parallèle, le second orthogonal à L. Dès lors $LUL^{-1} = L(U'+U'')L^{-1} = U'LL^{-1} - U''LL^{-1} = U' - U''$. — D'ailleurs on a aussi (cf. formule (3))

$$(25^{bis}) \quad LUL^{-1} = (LU+UL)L^{-1} - UL \cdot L^{-1} = 2(L, U)L^{-1} - U\,.$$

Diviseurs de zéro.

Un trait essentiel de l'anneau de CLIFFORD est de contenir des diviseurs de zéro, c'est-à-dire des quantités f et g telles que $fg = 0$, tandis que $f \neq 0$ et $g \neq 0$. Si f est scalaire, $fg = 0$, $g \neq 0$, on a nécessairement $f = 0$. En effet, dans le cas contraire f^{-1} existerait et on aurait $g = f^{-1}. fg = f^{-1}. 0 = 0$. Ceci peut aussi s'exprimer en disant qu'*il n'existe pas de diviseurs scalaires de zéro.* Les diviseurs de zéro qui nous intéressent en premier lieu sont de la forme $U - u$, U étant un vecteur $\neq 0$ et u un scalaire. On a à ce sujet le théorème suivant.

U ($\neq 0$) étant un vecteur de temps ou d'espace, il existe deux quantités scalaires $u = \pm \sqrt{(U, U)} \neq 0$ de sorte que $U - u$ est un diviseur de zéro. Tout vecteur isotrope est lui-même un diviseur de zéro.

Le dernier énoncé est évident, puisque U étant isotrope, on a $U^2 = (U, U) = 0$. Pour démontrer le premier énoncé, désignons l'une quelconque des racines carrées $\sqrt{(U, U)}$ par $|U|$. On a alors $(U - |U|)(U + |U|) = U^2 - |U|^2 = U^2 - (U, U) = 0$, ce qui veut dire que $U - |U|$ et $U + |U|$ sont des diviseurs de zéro, fait qui va jouer un rôle décisif dans la suite. Inversement $U - u$ ne peut être diviseur de zéro que pour $u = \pm |U|$. Admettons en effet que $(U - u)c = 0$, tandis que $c \neq 0$. La multiplication de gauche par $U + u$ nous fournit $(U^2 - u^2)\, c = \big((U, U) - u^2\big)\, c = 0$. Le premier facteur étant scalaire, il devra s'annuler. Ce que nous avons dit sur $U - u$ comme diviseur de gauche subsiste évidemment pour $U - u$ comme diviseur de droite. Nous démontrons maintenant le théorème suivant.

Pour que la relation

$$(26) \qquad (U - u)c = 0, \quad u = \pm |U|\,,$$

---

[1]) $L^{-1} = L/(L, L)$.

ait lieu pour un certain nombre de Clifford c, il faut et il suffit qu'il existe un nombre de Clifford g tel que

$$c = (U+u)g\,. \tag{27}$$

En effet, il est clair que la condition (27) est suffisante. D'autre part, tout nombre c peut s'écrire

$$c = \tfrac{1}{2}\left[\left(1-\frac{U}{u}\right)+\left(1+\frac{U}{u}\right)\right]c\,.$$

Il en vient lorsque (26) est rempli, $c = \frac{1}{2}(1+U/u)c$, c'est-à-dire que (27) est aussi rempli, si l'on pose $g = c/2u$.

Le théorème ci-dessus reste valable pour les vecteurs isotropes $U^2 = (U, U) = 0$ ou $u = 0$, quoique la démonstration ci-dessus tombe en défaut. Admettons donc que $Uc = 0$ et choisissons un vecteur V tel que $(U, V) = \frac{1}{2}$ ou (voir formule (18)) $UV+VU = 1$. Dès lors $c = (UV+VU)c = U\cdot Vc = U\cdot g$.

Terminons ces considérations par la remarque, évidente d'après ce qui précède, que la condition nécessaire et suffisante pour que la relation $c\,(U\mp|U|) = 0$ ait lieu est que $c = g(U\pm|U|)$.

Dans nos applications U sera un vecteur de temps. Nous y entendrons par $|U|$ la racine carrée positive de (U, U).

Équation de Dirac. Intégrale de Fourier.

A l'aide des symboles introduits on peut maintenant écrire

$$\square + m^2 = g^{jk}\frac{\partial^2}{\partial x^j \partial x^k} + m^2 = \left(\frac{1}{i}e^j\frac{\partial}{\partial x^j} - m\right)\left(-\frac{1}{i}e^k\frac{\partial}{\partial x^k} - m\right). \tag{28}$$

En posant

$$\nabla = e^k\frac{\partial}{\partial x^k}, \tag{29}$$

l'équation

$$\left(\frac{1}{i}\nabla - m\right)\psi(x) = 0\,, \tag{30}$$

sera appelée l'équation de Dirac pour un corpuscule libre — absence de champ — dont la masse de repos est m. Chez Dirac les quantités e^k sont représentées par des matrices γ^k à quatre lignes et à quatre colonnes, tandis que $\psi(x)$ est une colonne à quatre éléments où si l'on veut un *spineur*. Il n'entre donc dans ψ que quatre composantes complexes, ce qui certainement doit être considéré comme un avantage considérable. Chez nous $\psi(x)$ désigne, jusqu'à nouvel ordre, un nombre cliffordien général (dépendant évidemment de x) ce qui veut dire qu'il s'agit d'une quantité à seize composantes complexes. Après avoir développé la théorie dans le cas général, nous spécialiserons les $\psi(x)$, en sorte que le nombre des composantes se réduit à quatre. Par la formule de décomposition (28), il est clair que $\psi(x)$ satisfait à l'équation de SCHRÖDINGER $(\Box+m^2)\psi(x)=0$. L'opérateur $\Box+m^2$ étant scalaire, toute composante ψ_A de $\psi=\sum_A \psi_A e_A$ devra satisfaire à la même équation.

Posons d'une manière formelle

$$\psi(x) = \frac{1}{(\sqrt{2\pi})^3}\int \varphi(p) e^{i(p,x)} d\tau \,, \tag{31}$$

ni le domaine ni l'élément d'intégration $d\tau$ n'étant précisés pour le moment. Par l'analyse, qui va suivre $d\tau$ deviendra identique à l'élément $d\sigma$, mais *a priori* nous n'excluons pas même le cas où $d\tau$ est un élément à quatre dimensions. Quant à $\varphi(p)$, c'est un nombre de CLIFFORD dépendant (des composantes) du vecteur $p = p_k e^k$ et défini dans le domaine d'intégration qui reste à déterminer. Dans les calculs qui suivent, p est lui-même considéré comme un nombre (vecteur) cliffordien.

En admettant qu'il est permis de différentier sous le signe $\int$, il vient

$$\left(\frac{1}{i}\nabla - m\right)\psi(x) = \left(\frac{1}{i}e^k\frac{\partial}{\partial x^k} - m\right)\psi(x) = \frac{1}{(\sqrt{2\pi})^3}\int (e^k p_k - m)\varphi(p) e^{i(p,x)} d\tau$$
$$= \frac{1}{(\sqrt{2\pi})^3}\int (p-m)\varphi(p) e^{i(p,x)} d\tau \,.$$

Pour que le premier membre soit identiquement 0, on devra avoir

$$(p-m)\varphi(p) d\tau \equiv 0 \,. \tag{32}$$

Ici p est un vecteur, m un scalaire fixe, $\varphi(p)$ un nombre de CLIFFORD général (dépendant de p), $d\tau$ un élément d'intégration, qui n'est pas encore spécifié. D'autre part d'après un théorème démontré plus haut $p-m$ n'est diviseur de zéro que dans le cas où $m = \pm|p| = \pm\sqrt{(p,p)}$ ou en d'autres mots que $(p,p) = m^2$, c'est-à-dire que le point p représentant le vecteur impulsion-énergie est situé sur l'hyperboloïde (5) que nous avons désigné par H_m. La quantité $\varphi(p)d\tau$ devra donc s'annuler en tout point de l'espace p

qui n'est pas situé sur H_m. Cela implique que l'intégration ne s'étend qu'à H_m et que $d\tau$ sera à remplacer par un élément d'intégration relatif à cette surface, soit l'élément invariant $d\sigma$, utilisé plus haut (formules (9) et (6)). L'intégrale a donc exactement la même forme que (6), sauf que cette fois-ci φ est une quantité hypercomplexe. Cela suffit pour que $\psi = \Sigma \psi_A e_A$ et toutes ses composantes ψ_A satisfassent à l'équation $(\Box + m^2)\psi = 0$ ou $(\Box + m^2)\psi_A = 0$. Pour que $\psi(x)$ satisfasse à l'équation de Dirac, la condition nécessaire et suffisante est qu'on ait en plus $(p-m)\varphi(p)=0$ sur H_m. En analysant plus haut les diviseurs de zéro, nous avons montré que cette égalité aura lieu lorsque $\varphi(p) = (p+m)\,\chi(p)$ et dans ce cas seulement, $\chi(p)$ étant une fonction arbitraire — définie sur H_m — qui, bien entendu, satisfait à des conditions de régularité convenables. En résumé:

La solution générale de l'équation de Dirac (30) est fournie par l'intégrale de Fourier

(33) $$\psi(x) = \frac{1}{(\sqrt{2\pi})^3}\int_{H_m} \varphi(p) e^{i(p,x)} d\sigma$$

où

(34) $$(p-m)\varphi(p) = 0\,.$$

Le tenseur T_{jk} formé avec une solution de l'équation de Dirac.

En laissant de côté un instant l'intégrale de Fourier, admettons que $\psi(x)$ soit une quantité cliffordienne (assez régulière comme fonction de x) qui satisfait à l'équation de Dirac

(35) $$\frac{1}{i} e^k \frac{\partial \psi}{\partial x^k} - m\psi = 0\,.$$

Formons, d'après les prescriptions données plus haut, la quantité $\psi^\dagger(x)$ qu'on obtient de ψ par un renversement conjugué. La même opération, appliquée au premier membre, qui est nul, nous donne

(36) $$-\frac{1}{i}\frac{\partial \psi^\dagger}{\partial x_k} e^k - m\psi^\dagger = 0 \quad \text{ou} \quad \frac{1}{i}\frac{\partial \psi^\dagger}{\partial x^k} e^k + m\psi^\dagger = 0\,.$$

Cela étant, formons avec ψ et $\psi^\dagger$ le tenseur T_{jk}, ou plutôt le tenseur mixte correspondant,

(37) $$T_j^k = (e_j \psi^\dagger e^k \psi)_S\,.$$

On trouve par un procédé connu que la divergence tensorielle $\partial T_j^k / \partial x^k = 0$. On a, en effet, eu égard aux équations (35) et (36),

$$(38)\quad \frac{\partial T_j^k}{\partial x^k} = \left(e_j \frac{\partial \psi^\dagger}{\partial x^k} e^k \psi + e_j \psi^\dagger e^k \frac{\partial \psi}{\partial x^k}\right)_S = i(-m\psi^\dagger\psi + m\psi^\dagger\psi)_S = 0\,.$$

Dès lors Σ étant une surface orientée dans l'espace, formons les composantes normales

$$(39)\quad T_{j<n>} = T_{jk}n^k = (e_j\psi^\dagger n^k e_k \psi)_S = (e_j \psi^\dagger n \psi)_S,\ (j = 0, 1, 2, 3)\,.$$

Il s'ensuit de (38) par le théorème de GAUSS une loi de conservation, à savoir que l'intégrale

$$(40)\quad \int_\Sigma T_{j<n>} d\Sigma$$

est indépendante de Σ. En particulier, si la surface Σ se réduit à un plan $x^0 = \text{const.}$, on obtient la loi de conservation particulière qui dit que l'intégrale

$$(41)\quad \iiint_{-\infty}^{+\infty} T_{j0}(x^0, x^1, x^2, x^3)\, dx^1 dx^2 dx^3$$

est indépendante de x^0.

Notons encore que l'application de la relation (24) au dernier membre de (39), avec $j = 0$, met en évidence le fait important que *la composante normale* $T_{0<n>}$ *est toujours positive.*

Remarque. La loi de conservation reste valable (et sa démonstration reste à peu près la même) si l'on substitue à l'équation (35) l'équation plus générale correspondant à la présence d'un champ électromagnétique. Il en est encore ainsi du caractère positif de $T_{0<n>}$, qui est entièrement d'origine algébrique.

Une identité.

Nous retournons maintenant à l'intégrale de FOURIER. En supposant comme plus haut que l'intégrale n'est étendue qu'à une seule des nappes de H_m, à H_m^+ par exemple, nous établirons des identités entre les intégrales (40) et des intégrales de structure analogue étendues à H_m^+. Posons donc en analogie avec la formule (10)

$$(42)\quad \psi(x) = \frac{1}{(\sqrt{2\pi})^3} \int_{H_m^+} \varphi(p)\, e^{i(p,\,x)} d\sigma\,,$$

la quantité cliffordienne $\varphi(p)$ satisfaisant à la relation (34), ce qui entraîne que $\psi(x)$ satisfait à l'équation de DIRAC (30). Formons avec $\varphi(p)$ le tenseur analogue à (32),

$$(43)\quad M_{jk} = (e_j \varphi^\dagger e_k \varphi)_S\,,$$

désignons par n la normale unitaire $n^k e_k$ de H_m^+, dirigée vers l'intérieur de H_m^+ et posons $M_{j<n>} = M_{jk}n^k$. Je dis qu'on a les identités

$$(44) \qquad \int_\Sigma T_{j<n>} d\Sigma = \int_{H_m^+} M_{j<n>} \frac{d\sigma}{m} \quad (j = 0, 1, 2, 3).^{1)}$$

Quant au premier membre, il suffit de le considérer sous la forme particulière (41). Notons d'autre part que la relation (44) écrite pour $\varphi = \Sigma\varphi_A e_A$ et $\psi = \Sigma\psi_A e_A$ implique une relation analogue entre chaque couple de composantes φ_A et ψ_A. En observant enfin que $M_{jk} = \Sigma \pm \varphi_A \bar{\varphi}_B$ et $T_{jk} = \Sigma \pm \psi_A \psi_B$ renferment les mêmes couples d'indices et les mêmes distributions de signes, il vient par le théorème de Parseval-Plancherel, en nous rappelant l'expression (9) de $d\sigma$,

$$(46) \qquad \iiint_{-\infty}^{+\infty} T_{j0} dx^1 dx^2 dx^3 = \iiint_{-\infty}^{+\infty} M_{j0} \frac{dp_1 dp_2 dp_3}{p_0^2}.$$

Il n'y a rien à objecter au premier membre de cette relation. Nous connaissons déjà sa signification invariantive. Il n'en pas de même du second membre; ni M_{j0}, ni $p_0^{-2} dp_1 dp_2 dp_3$, ni leur produit n'ont l'air invariantif. Cela tient pourtant au fait que nous n'avons pas utilisé jusqu' ici la relation fondamentale (34). Cette relation nous donne

$$\varphi = \frac{p}{m}\varphi \text{ et } \varphi^\dagger = \varphi^\dagger \frac{p}{m}.$$

Par là

$$\varphi^\dagger e_0 \varphi = \varphi^\dagger \frac{p}{m} e_0 \frac{p}{m} \varphi,$$

ou en formant la moyenne arithmétique des deux membres

$$(47) \qquad \varphi^\dagger e_0 \varphi = \varphi^\dagger \cdot \tfrac{1}{2}\left(e_0 + \frac{p}{m} e_0 \frac{p}{m}\right)\varphi.$$

Puisque $\left(\frac{p}{m}\right)^2 = \frac{(p, p)}{m^2} = 1$, on trouve, en utilisant (18),

$$\tfrac{1}{2}\left(e_0 + \frac{p}{m} e_0 \frac{p}{m}\right) = \frac{p}{m} \cdot \tfrac{1}{2}\left(\frac{p}{m} e_0 + e_0 \frac{p}{m}\right) = \frac{p}{m} \cdot \left(\frac{p}{m}, e_0\right) = \frac{p}{m}\frac{p_0}{m}.^{2)}$$

---

1) Nous reviendrons tout à l'heure sur la normale de H_m^+. Pour le moment, nous nous contentons d'observer que H_m^+ étant orienté dans l'espace, la normale en question sera un vecteur de temps positif. La composante normale $M_{0<n>}$ est donc positive pour les mêmes raisons que $T_{0<n>}$.

2) On aurait aussi pu appliquer la formule (25bis).

La normale lorentzienne unitaire $n = n^k e_k$ de H_m^+ dirigée vers son intérieur est évidemment identique au rayon vecteur p, rendu unitaire, c'est-à-dire à p/m. Cela étant, le dernier membre de la relation ci-dessus peut aussi s'écrire $n \cdot p_0/m$. En portant cela dans (47), il vient

$$\varphi^\dagger e_0 \varphi = \frac{p_0}{m} \varphi^\dagger n \varphi .$$

Ce point acquis, il s'ensuit

$$M_{j0} = (e_j \varphi^\dagger e_0 \varphi)_S = \frac{p_0}{m} (e_j \varphi^\dagger n \varphi)_S = \frac{p_0}{m} (e_j \varphi^\dagger n^k e_k \varphi)_S$$

$$= \frac{p_0}{m} (e_j \varphi^\dagger e_k \varphi)_S \cdot n^k = \frac{p_0}{m} M_{jk} n^k = \frac{p_0}{m} M_{j<n>} .$$

Si l'on substitue dans (46) à M_{j0} la valeur $\frac{p_0}{m} M_{j<n>}$ que nous venons d'obtenir et que l'on tient compte de (9), l'identité invariantive (44) se trouvera démontrée.

Une identité pour le vecteur-courant de Dirac.

Avant de confronter notre tenseur T_{jk} au vecteur-courant s_k de Dirac, auquel il ressemble sous plusieurs rapports, nous donnerons pour ce dernier une identité analogue à (44). Ce faisant, nous nous servirons du langage matriciel accoutumé. Considérons donc, sans les écrire explicitement, les quatre matrices hermitiques d'ordre quatre de Dirac, désignées habituellement par γ_k et satisfaisant aux relations

$$(48) \qquad \gamma_l \gamma_m + \gamma_l \gamma_m = 2\delta_{lm}, \quad \delta_{lm} = \begin{cases} 1, & l = m, \\ 0, & l \neq m. \end{cases} \quad l, m = 1, 2, 3, 4.$$

En posant $\breve{\gamma}_0 = \gamma_4$, $\breve{\gamma}_j = i\gamma_j$, $j = 1, 2, 3$, les matrices $\breve{\gamma}_k$, ($k = 0, 1, 2, 3$), donnent une représentation de nos symboles e_k, caracté risés par les formules (19). Désignons par ψ une colonne à quatre éléments ψ_r, par ψ^* une ligne dont les éléments sont les imaginaires conjugués ψ_r^* des ψ_r. Formons encore la nouvelle ligne $\psi^\dagger = \psi^* \gamma_0$ et le vecteur s, appelé vecteur-courant, aux composantes covariantes

$$(49) \qquad s_k = \psi^\dagger \breve{\gamma}_k \psi = \psi^* \breve{\gamma}_0 \breve{\gamma}_k \psi, \quad k = 0, 1, 2, 3 .$$

Pour $k = 0$, il vient par la dernière formule $s_0 = \psi^* \psi = \Sigma \psi_r^* \psi_r = \Sigma |\psi_r|^2 > 0$. Si ψ satisfait à l'équation de Dirac,

(50) $$\left(\frac{1}{i}\breve{\gamma}^k\frac{\partial}{\partial x^k}-m\right)\psi=0\,,$$

la quadridivergence $\partial s^k/\partial x^k$ du vecteur s sera nulle.

Tout cela étant admis, reprenons les surfaces Σ d'orientation d'espace considérées plus haut, formons la composante $s_{<n>} = s_k n^k$; on voit comme plus haut (pour $T_{0<n>}$) que $s_{<n>} > 0$. Par le théorème de Gauss, il vient encore, pour toute solution assez régulière de (35), la loi de conservation, consistant en ce que l'intégrale

(51) $$\int_\Sigma s_{<n>}\,d\Sigma$$

est indépendante de Σ.

Quant à l'espace impulsion-énergie, le vecteur p y sera représenté par la matrice $p = p_k\breve{\gamma}^k$. Nous nous restreignons toujours aux points p situés sur la nappe H_m^+, $(p, p) = m^2$, $p_0 > 0$. Si l'on forme une colonne $\varphi = \varphi(p)$, satisfaisant à la relation $(p-m)\varphi(p) = 0$, la colonne $\psi(x)$ fournie par l'intégrale de Fourier

(52) $$\psi(x) = \frac{1}{(\sqrt{2\pi})^3}\int_{H_m^+}\varphi(p)e^{i(p,x)}d\sigma$$

satisfera évidemment à l'équation de Dirac (35).

Tel étant le cas, formons comme plus haut les lignes correspondantes φ^* et $\varphi^\dagger$ et, enfin, le vecteur v aux composantes

(53) $$v_k = \varphi^\dagger\breve{\gamma}_k\varphi = \varphi^*\breve{\gamma}_0\breve{\gamma}_k\varphi\,.$$

En désignant toujours par $n = p/m$ la normale unitaire de H_m^+ (dirigée vers l'intérieur de H_m^+) et par $v_{<n>} = v_k n^k$ la composante normale de v, on obtient, comme plus haut, l'identité

(54) $$\int_\Sigma s_{<n>}\,d\Sigma = \int_{H_m^+} v_{<n>}\frac{d\sigma}{m}.$$

Nombres primitifs et idéaux simples. Colonnes, lignes, spineurs.

Un nombre cliffordien $\pi \neq 0$ est dit *primitif*, si, pour un nombre cliffordien c *arbitraire*, on a une relation de la forme $\pi c\pi = t\pi$, où t est un scalaire (nombre complexe ordinaire) dépendant évidemment de c. Un tel nombre primitif est p. ex. l'idempotent

$$\varepsilon = \frac{1+e_0}{2}\,\frac{1+ie_1e_2}{2}\,,\quad \varepsilon^2 = \varepsilon\,.$$

En effet, pour les nombres $e_A = 1$, e_0, ie_1e_2, $ie_0e_1e_2$, qui, au facteur i près sont des éléments de base de l'anneau cliffordien, on a $\varepsilon e_A \varepsilon = \varepsilon$, tandis que, pour tout autre élément de base e_B, on a $\varepsilon e_B \varepsilon = 0$. Cela tient aux faits suivants: 1) Les e_A et les idempotents $\varepsilon_1 = \frac{1+e_0}{2}$, $\varepsilon_1' = \frac{1-e_0}{2}$, $\varepsilon_2 = \frac{1+ie_1e_2}{2}$, $\varepsilon_2' = \frac{1-ie_1e_2}{2}$ commutent tous l'un avec l'autre. 2) $e_0\varepsilon_1 = \varepsilon_1$, $ie_1e_2\varepsilon_2 = \varepsilon_2$. 3) $\varepsilon_1\varepsilon_1' = \varepsilon_1'\varepsilon_1 = \varepsilon_2\varepsilon_2' = \varepsilon_2'\varepsilon_2 = 0$. 4) Tout e_B remplit, pour l'un au moins des indices $j = 1, 2$, une relation de la forme $\varepsilon_B\varepsilon_j = \varepsilon_j' e_B$.

Le nombre cliffordien a étant arbitraire et π étant primitif, les nombres $a\pi$ et πa seront primitifs. En effet, $a\pi \cdot c \cdot a\pi = a \cdot \pi(ca)\pi = a \cdot t\pi = t \cdot a\pi$.

On entend par un *idéal gauche* I un ensemble de nombres cliffordiens tels que $z_1 \in I$, $z_2 \in I$ entraînent $(z_1 \pm z_2) \in I$ et que $z \in I$ entraîne $az \in I$, a étant un nombre cliffordien arbitraire. Définition analogue pour un *idéal droit*. Un idéal gauche (droit) est dit *simple*, s'il ne renferme aucun idéal gauche (droit) sauf lui-même et l'idéal nul, consistant du seul élément 0. Au sujet de tels idéaux, on a les théorèmes suivants:

I. *Tout nombre primitif* π *engendre un idéal gauche simple* $I = \{a\pi\}$, *où* a *parcourt l'anneau cliffordien entier.*

II. *Tout élément d'un idéal gauche simple est primitif et l'idéal est engendré par l'un quelconque de ses éléments* $\neq 0$. (*Théorèmes analogues pour les idéaux droits*).

Le fait que les nombres $I = \{a\pi\}$ forment un idéal gauche est évident pour tout nombre fixe π, primitif ou non. Le reste de notre énoncé exige des considérations plus délicates. Je reviendrai sur ces questions à une autre occasion, en les approfondissant sous plusieurs rapports. Ici, pour arriver rapidement au but, je me contente de signaler que nos théorèmes découlent presqu'immédiatement du fait connu que, dans le champ complexe, l'anneau cliffordien est isomorphe a une algèbre matricielle totale d'ordre 4, c'est-à-dire à l'ensemble des matrices du dit ordre, admettant des éléments complexes arbitraires.

Aux éléments primitifs π de l'anneau il correspond les matrices P de rang 1, que nous désignerons comme matrices primitives. A étant une matrice arbitraire, les matrices AP et PA seront primitives. Toute matrice primitive peut s'écrire comme produit: colonne fois ligne, d'où il découle $PAP = tP$, t étant un scalaire.

Toute matrice primitive P engendre un idéal gauche simple $I = \{AP\}$, où A parcourt toutes les matrices d'ordre 4, et inversement, un idéal gauche simple ne renferme que de matrices primitives et il est engendré par l'un quelconque de ses éléments $\neq 0$.

On peut aller un peu plus loin. A étant une matrice *arbitraire*, P une matrice *primitive*, c'est-à-dire le produit d'une colonne *fixe* et d'une ligne *fixe*, les matrices AP qui constituent l'idéal I seront *le produit d'une colonne arbitraire et d'une ligne fixe*. Par là, on voit qu'un idéal gauche[1]) simple admet quatre éléments de base, fournis p. ex. par les colonnes particulières $\begin{matrix}1\\0\\0\\0\end{matrix}$ etc.

Les idéaux que nous venons de considérer font partie de l'anneau de CLIFFORD. La fonction ψ figurant dans l'équation de DIRAC, qui de prime abord avait 16 composantes, n'en aura que 4, si l'on la restreint à appartenir à un idéal gauche simple. D'autre part, nous venons de voir que l'anneau cliffordien étant représenté par une algèbre matricielle, un idéal gauche sera représenté par les produits d'une colonne arbitraire et d'une ligne fixe. Or cette dernière variété est évidemment isomorphe à la variété des colonnes arbitraires (spineurs) que l'on rencontre dans les exposés habituels de la théorie de DIRAC, représentation qui est bien à préférer au point de vue technique. Pourtant, par cette simplification très naturelle, l'idéal sort du domaine cliffordien, d'où proviennent les complications connues relatives à la transformation des spineurs. Une autre circonstance qui, justement dans le cas d'une représentation matricielle, est en faveur des idéaux gauches et droits au lieu de colonnes et de lignes est qu'un nombre de CLIFFORD, représenté par une matrice, ne pourra figurer que comme facteur à gauche quand l'autre facteur est une colonne et comme facteur à droite quand l'autre facteur est une ligne. Bien entendu, ce ne sont pas là des points trop sérieux.

Retournons à l'intégrale de FOURIER (42). Lorsque la fonction $\varphi(p)$ appartient à un idéal simple déterminé, il en sera de même de la fonction $\psi(x)$, solution de l'équation de DIRAC. Il est très facile de voir que dans ce cas-là les tenseurs M_{jk} et T_{jk} seront de rang *un*, ce qui veut dire qu'on aura $M_{jk} = a_j b_k$ et $T_{jk} = c_j d_k$. En effet, ε et ε'

[1]) Les faits évidents que nous énonçons ici pour les idéaux gauches se traduisent immédiatement aux idéaux droits.

étant deux nombres primitifs, on voit par la représentation matricielle que, pour un nombre cliffordien arbitraire c, on a $\varepsilon' c \varepsilon = t\varepsilon''$ où t est un scalaire et ε'' un nombre primitif ne dépendant que de ε et de ε'. D'autre part, e_j et e_l étant deux vecteurs de base, on pourra toujours choisir les scalaires v_j et v_l, dont l'un au moins $\neq 0$, de façon que la partie scalaire de $(v_j e_j + v_l e_l)\varepsilon''$ soit nulle.

D'un autre côté le nombre primitif ψ étant dans une représentation matricielle écrit sous la forme μ fois ν, où μ est une colonne et ν une ligne, on formera, comme plus haut, au moyen de μ et de la ligne μ^* dont les éléments sont les imaginaires conjugués de ceux de μ le vecteur-courant s_k de DIRAC. Dès lors, on aura pour $j = 0, 1, 2, 3$, $T_{jk} = t_j s_k$.

Les matrices γ_k de DIRAC sont intimement liées aux idempotents

$$\varepsilon = \frac{1 \pm e_0}{2} \, \frac{1 \pm i e_1 e_2}{2}, \tag{55}$$

qui satisfont aux relations $\varepsilon = \varepsilon^\dagger = \varepsilon^*$. On a donc, si ψ appartient à l'idéal gauche $\{a\varepsilon\}$,

$$T_{jk} = (e_j (a\varepsilon)^\dagger e_k a\varepsilon)_S = (e_j \varepsilon^\dagger a^\dagger e_k a \varepsilon)_S = (e_j \varepsilon a^\dagger e_k a \varepsilon)_S,$$

ou, puisque ε est un nombre primitif,

$$T_{jk} = (e_j s_k \varepsilon)_S,$$

s_k étant un scalaire. Or, pour $j = 1, 2, 3$, la partie scalaire de $e_j \varepsilon$ est nulle. On a par conséquent

$$T_{jk} = 0, \quad j = 1, 2, 3 .$$

Pour $j = 0$, on a, suivant le signe de e_0 dans (55), $e_0 \varepsilon = \pm \varepsilon$ et

$$T_{0k} = \pm s_k \varepsilon_S = \pm \tfrac{1}{4} s_k .$$

Il faut encore ajouter quelques remarques concernant les lois de transformation des quantités T_{jk} (ou M_{jk}). En passant à un nouveau système de coordonnées, les quantités ψ et $\psi^\dagger$ sont invariantes. En effet, ψ_A et e_A se transforment par des transformations contregrédientes en ψ'_A et e'_A de façon que $\psi = \Sigma \psi_A e_A = \Sigma \psi'_A e'_A$ reste invariant. Par contre e_j et e_k se transforment en e'_j et e'_k, de sorte que $T_{jk} = (e_j \psi^\dagger e_k \psi)_S$ se transforme en $T'_{jk} = (e'_j \psi^\dagger e'_k \psi)_S$, ce qui est bien en accord avec l'écriture tensorielle.

Quant à la signification physique du tenseur T_{jk}, il serait prématuré d'émettre une opinion quelconque.

Dans le cas particulier considéré ci-dessus, chaque ligne T_{jk} sera proportionnelle au vecteur-courant s_k de Dirac, les facteurs de proportionnalité étant constants dans tout système de coordonnées et se transformant comme les composantes d'un vecteur.

J'indiquerai ici succinctement comment on peut arriver par le mécanisme développé ci-dessus à les quantités qui se transforment exactement comme les composantes du vecteur de Dirac. On n'a qu'à poser $s_k = (b\psi^\dagger e_k \psi)_S$, où b est un vecteur de temps positif fixe. Par une transformation de Lorentz, $e_k \to o_k^r e_r$, il vient $s_k \to o_k^r s_r$. On peut montrer au moyen d'un théorème donné plus haut (p. 137) que, dans le cas où $o_0^0 > 0$, il existe un nombre de Clifford R, déterminé au signe près, tel que $R^{-1} = R^\dagger$ et que $o_k^r e_r = R^{-1} e_k R$.

Dès lors, la transformation de Lorentz relative à s_k peut aussi s'exprimer par la formule $s_k \to (b\psi^\dagger R^\dagger c_k R\psi)_S = (b\psi'^\dagger c_k \psi')_S$, où $\psi' = R\psi$. Dans la dernière transformation on a laissé e_k invariant et transformé ψ conformément à la théorie de Dirac.

47.

L'intégrale de Riemann-Liouville et le problème de Cauchy

Acta Math. *81* (1949), 1–223

Avant-Propos.

... nonumque prematur in annum.
Horace.

Les recherches que je présente ici datent d'il y a longtemps. Celles dont se composent les cinq premiers Chapitres sont, à quelques exceptions près, des années 1933—1936. J'en ai donné un compte rendu très sommaire au Congrès international des mathématiciens qui a eu lieu à Oslo en 1936, et un résumé assez complet dans une Conférence à la Réunion internationale des mathématiciens, tenue à Paris en Juillet 1937 sous les auspices de la Société mathématique de France. Cette conférence, publiée par les soins de la Société (Paris 1939), servira, légèrement remaniée, d'Introduction au présent ouvrage. Au moment où j'ai donné le résumé en question, mes études visaient exclusivement les équations à coefficients constants, et ce n'est que pendant l'été de 1938 que j'ai réussi à étendre la méthode aux équations à coefficients variables de signature lorentzienne, dont la théorie, due à M. Hadamard, constitue un des plus beaux titres de gloire de ce grand savant. J'ai publié une esquisse de ces dernières études dans une *Note additionnelle* imprimée avec la susdite conférence; ces études constituent ici le dernier Chapitre.

Un ami dévoué, M. K.-G. Hagstroem, a été pour moi un collaborateur inappréciable dans la poursuite de ces recherches. Il en a pris les premières notes, qui m'ont été par la suite d'un grand secours, et c'est encore lui qui m'a aidé dans la rédaction finale. Je tiens aussi à remercier mes amis MM. O. Frostman et W. Feller qui, chacun de son côté, m'ont prêté un précieux concours.

Introduction.[1]

C'est la notion de la *partie finie* de certaines intégrales due à M. Hadamard qui forme le point de départ des recherches dont j'aurai l'honneur de vous entretenir dans cette conférence. On connaît les applications brillantes à la théorie des équations aux dérivées partielles que M. Hadamard a données de la notion qu'il a créée[2]. Chez lui il s'agit surtout d'équations à coefficients variables, tandis que la méthode que nous allons développer ne s'applique qu'à des équations à coefficients constants[3]. En revanche, elle permet de donner pour ces équations une solution du problème de Cauchy qui est la même pour les dimensions impaires et paires.

Considérons d'abord l'intégrale de Riemann-Liouville

$$I^\alpha f(x) = \frac{1}{\Gamma(\alpha)} \int_a^x f(t)(x-t)^{\alpha-1}\,dt.\text{[4]}$$

Elle converge pour $\alpha > 0$ et satisfait aux relations fondamentales

$$(1) \qquad I^\alpha(I^\beta) = I^{\alpha+\beta}, \qquad \frac{d}{dx}(I^{\alpha+1}) = I^\alpha.$$

Dans des conditions de dérivabilité convenables, il s'agit maintenant d'étendre la définition de cette intégrale à des valeurs de α pour lesquelles elle cesse de converger. M. Hadamard arrive à cette extension pour des indices négatifs non entiers en retranchant de l'intégrale divergente certaines parties infinies d'ordre fractionnaire. Ce qui reste c'est alors la partie finie de l'intégrale. On peut atteindre le même but, et cela aussi pour l'indice zéro et les indices entiers négatifs, par un procédé qui me semble très naturel, celui du prolongement

---

(1) M. Riesz 2; les chiffres après les noms d'auteurs se rapportent à la Bibliographie à la fin de cet ouvrage.

(2) *Cf.* Hadamard 1.

(3) L'extension de la méthode à des coefficients variables, qui constitue le dernier Chapitre, est postérieure à la Conférence reproduite ici.

(4) C'est à partir de la même intégrale, mais sans le facteur $\Gamma(\alpha)$ au dénominateur, que M. Hadamard définit la »partie finie» pour les valeurs négatives fractionnaires de α. L'absence dudit facteur est le seul obstacle à l'extension de cette notion à la valeur zéro et aux valeurs négatives entières de α.

analytique par rapport à l'indice α. En supposant l'existence d'un assez grand nombre de dérivées, on obtient au moyen des intégrations par parties

$$I^\alpha f(x) = \sum_{k=0}^{n-1} \frac{f^{(k)}(a)}{\Gamma(\alpha+k+1)}(x-a)^{\alpha+k} + I^{\alpha+n} f^{(n)}(x).$$

La dernière expression s'écrit

$$\frac{1}{\Gamma(\alpha+n)}\int_a^x f^{(n)}(t)(x-t)^{\alpha+n-1}\,dt,$$

l'intégrale étant convergente pour $\alpha > -n$. Le prolongement analytique se trouve donc effectué pour tous ces indices. Bien entendu, le prolongement en question peut être obtenu par beaucoup d'autres procédés. L'intégrale prolongée satisfait, sauf dans des cas exceptionnels faciles à préciser, aux relations fondamentales (1) dont la seconde pourra aussi servir à effectuer le prolongement en question.

Pour les indices $\alpha = 0$ et entiers négatifs $\alpha = -n$ on trouve, comme il fallait s'y attendre,

$$I^0 f(x) = f(x), \qquad I^{-n} f(x) = f^{(n)}(x).$$

Arrêtons-nous un instant sur les dernières relations. On voit que, pour tout autre indice, $I^\alpha f(x)$ est une fonctionnelle dépendant de toutes les valeurs que $f(t)$ admet entre a et x, tandis que, pour $\alpha = 0$ ou entier négatif, $I^\alpha f(x)$ a une valeur de caractère local; ce ne sont que les valeurs admises au voisinage infinitésimal de x qui interviennent. *Voilà en germe le principe de Huygens*[1]. C'est évidemment au facteur $\frac{1}{\Gamma(\alpha)}$, s'annulant pour les indices en question, que ces indices doivent leur caractère exceptionnel. D'ailleurs, même sans effectuer le prolongement analytique d'une façon explicite, on voit nettement que le prolongement de

$$\frac{1}{\Gamma(\alpha)}\int_a^{x-\delta} f(t)(x-t)^{\alpha-1}\,dt \qquad (\delta > 0)$$

est zéro pour $\alpha = 0$ ou entier négatif, parce que l'intégrale ne cesse jamais de converger. Les valeurs de $f(t)$ qui interviennent dans cette intégrale ne donneront donc aucune contribution à la valeur finale.

(1) Nous entendons ici et dans tout ce qui suit par principe de Huygens la *mineure* du principe dans la terminologie de M. Hadamard (*loc. cit.*, p. 75). D'après ce principe, une perturbation lumineuse étant localisée à l'instant $t = 0$ au voisinage immédiat du point O, son effet sera localisé, pour $t = t'$, au voisinage immédiat de la surface d'une sphère de centre O et de rayon ct', c désignant la vitesse de la lumière.

J'ai insisté sur ces questions élémentaires bien familières à vous tous, parce qu'elles nous permettront de bien saisir la différence qu'on rencontre dans la solution du problème de Cauchy de l'équation des ondes, posé pour un nombre pair ou pour un nombre impair de variables.

Considérons dans l'espace à m dimensions les points P et Q aux coordonnées respectives $x_1, x_2, \ldots, x_m$ et $\xi_1, \xi_2, \ldots, \xi_m$. Nous introduisons la distance lorentzienne de ces points

$$r_{PQ} = \sqrt{(x_1 - \xi_1)^2 - (x_2 - \xi_2)^2 - \cdots - (x_m - \xi_m)^2},$$

en supposant que l'expression qui figure sous le signe racine carrée soit $\geqq 0$. En considérant le point P comme fixe et le point Q comme variable, $r_{PQ}^2 = 0$ définit la nappe du cône de lumière au sommet P, $r_{PQ}^2 > 0$ définit son intérieur, $x_1 - \xi_1 > 0$ le cône rétrograde et $x_1 - \xi_1 < 0$ le cône direct. C'est le cône rétrograde que nous allons considérer en général, en le désignant par D^P. Soit encore S une surface (c'est-à-dire une variété à $m-1$ dimensions) qui, suivant la terminologie de M. Hadamard, ait une *orientation d'espace*. Cela veut dire qu'on a pour tout déplacement infinitésimal sur la surface $dx_1^2 - dx_2^2 - \cdots - dx_m^2 < 0$. On admet aussi que la surface soit assez régulière pour que les dérivations qui interviendront plus tard puissent être effectuées. Le domaine limité par la nappe rétrograde et la surface S sera désigné par D_S^P.

Cela étant, nous posons

$$(2) \qquad I^\alpha f(P) = \frac{1}{H_m(\alpha)} \int\limits_{D_S^P} f(Q)\, r_{PQ}^{\alpha-m}\, dQ,$$

avec

$$(3) \qquad H_m(\alpha) = \pi^{\frac{m-2}{2}}\, 2^{\alpha-1}\, \Gamma\left(\frac{\alpha}{2}\right) \Gamma\left(\frac{\alpha+2-m}{2}\right).$$

Cette intégrale converge pour $\alpha > m-2$ et l'on vérifie facilement qu'elle satisfait aux relations fondamentales

$$(4) \qquad I^\alpha I^\beta = I^{\alpha+\beta}, \qquad \Delta I^{\alpha+2} = I^\alpha,$$

où l'on a désigné par Δ l'opérateur des ondes

$$\Delta = \frac{\partial^2}{\partial x_1^2} - \frac{\partial^2}{\partial x_2^2} - \cdots - \frac{\partial^2}{\partial x_m^2}.$$

Pour les indices $\alpha \leqq m-2$ l'intégrale (2) se définit au moyen de prolongement analytique par rapport à α, bien entendu dans des conditions de régularité convenables portant sur la fonction f et sur la surface S. *On trouve en particulier* $I^0 f(P) = f(P)$. Remarquons que ce fait, qui est d'une importance capitale pour la suite, n'est nullement évident. En effet, I^0 est défini ici comme prolongement analytique de I^α qui ne converge que pour $\alpha > m-2$, et le prolongement en question n'existe que dans certaines conditions de dérivabilité[1]. Des remarques analogues s'appliquent aux relations $I^{-2k} = \Delta^k$.

Il ressort de ce qui précède que notre procédé d'intégration tire son origine de l'opérateur différentiel du second ordre Δ. On a vu en effet que $\Delta I^2 = 1$, c'est-à-dire que I^2 est dans un certain sens l'inverse de l'opérateur des ondes. Or on peut construire un procédé qui, dans le même sens, forme l'inverse de l'opérateur différentiel du premier ordre *gradient* ou *nabla*, cet opérateur étant pris au sens lorentzien. Il est manifeste que ce procédé devra être de caractère vectoriel[2].

Soient $X_1, X_2, \ldots, X_m$ les composantes d'un vecteur arbitraire X et définissons *les nombres de Clifford* comme symboles de calcul ou comme vecteurs unitaires par la relation

$$X^2 = (X_1 e_1 + X_2 e_2 + \cdots + X_m e_m)^2 = X^2 - X_2^2 - \cdots - X_m^2.$$

Cette relation met en évidence les règles de calcul

$$e_1^2 = -e_2^2 = \cdots = -e_m^2 = 1$$

et

$$e_j e_k + e_k e_j = 0 \qquad (j \neq k).$$

L'opérateur $\nabla = e_1 \frac{\partial}{\partial x_1} - e_2 \frac{\partial}{\partial x_2} - \cdots - e_m \frac{\partial}{\partial x_m}$ a manifestement la propriété $\nabla^2 = \Delta$[3]. Nous posons

$$J^\alpha f(P) = e^{\frac{i\pi\alpha}{2}} \left[\cos\frac{\pi\alpha}{2}\, I^\alpha f(P) - i \sin\frac{\pi\alpha}{2}\, \nabla\, [I^{\alpha+1} f(P)] \right],$$

l'opération ∇ pouvant être exécutée sous le signe $\int$. On a alors

$$J^\alpha J^\beta = J^{\alpha+\beta}, \qquad \nabla J^{\alpha+1} = J^\alpha, \qquad J^{2k} = I^{2k},$$

(1) Pour m impair ou pair il suffit de supposer l'existence des dérivées d'ordre $\leqq \frac{m+1}{2}$ ou $\leqq \frac{m}{2}$ respectivement.

(2) L'intégration vectorielle que j'esquisse ici ne figurait pas dans ma conférence. Je l'ai indiquée succinctement après la conférence en réponse à une question de M. Fréchet qui présidait à la séance.

(3) Depuis les travaux de M. Dirac, ces choses sont certainement bien familières à la plupart des lecteurs.

k étant un nombre entier. En particulier on a $J^0 f(P) = f(P)$ ou plus brièvement $J^0 = 1$. Signalons encore l'opérateur singulier remarquable

$$Of(P) = \nabla [I^1 f(P)]$$

qui satisfait à la relation $O^2 = 1$.

En remarquant que la plupart des considérations qui suivent s'appliquent — *mutatis mutandis* — aussi à nos intégrales vectorielles, retournons aux intégrales scalaires.

Définissons avec M. d'Adhémar *la conormale* n à la surface S au point Q (*la transversale* suivant la terminologie de M. Hadamard) par la relation

$$dx_1\,\delta x_1 - dx_2\,\delta x_2 - \cdots - dx_m\,\delta x_m = 0,$$

où d et δ désignent respectivement un déplacement arbitraire sur la surface et un déplacement sur la conormale. Cela étant, nous choisissons celle des deux directions admissibles QT qui rend le produit scalaire lorentzien des vecteurs QP et QT positif. Il vient alors par la formule de Green

$$I^\alpha u(P) = \frac{1}{H_m(\alpha+2)} \int\limits_{D_S^P} \Delta u(Q)\, r_{PQ}^{\alpha+2-m}\, dQ + \frac{1}{H_m(\alpha+2)} \int\limits_{S^P} \left[\frac{du(Q)}{dn} r_{PQ}^{\alpha+2-m} - u(Q) \frac{d r_{PQ}^{\alpha+2-m}}{dn} \right] dS,$$

tous les éléments géométriques étant mesurés, ici et dans la suite, dans le sens lorentzien.

La formule ci-dessus qui contient une intégrale spatiale, une simple couche et une double couche peut se mettre sous une forme plus condensée. En effet, en posant d'une manière générale

$$(5)\qquad I_*^\alpha \overline{f, g, h}(P) = \frac{1}{H_m(\alpha)} \int\limits_{D_S^P} f(Q)\, r_{PQ}^{\alpha-m}\, dQ + \frac{1}{H_m(\alpha)} \int\limits_{S^P} \left[g(Q)\, r_{PQ}^{\alpha-m} - h(Q) \frac{d r_{PQ}^{\alpha-m}}{dn} \right] dS,$$

on peut écrire

$$(6)\qquad I^\alpha u(P) = I_*^{\alpha+2}\, \overline{\Delta u, \frac{du}{dn}, u}(P).$$

Avant d'aller plus loin, signalons quelques propriétés importantes du symbole I_*. On a

$$(7)\qquad I^\alpha I_*^\beta = I_*^{\alpha+\beta}, \qquad \Delta I_*^{\alpha+2} = I_*^\alpha,$$

et en outre

$$I_*^0 \overline{f, g, h}(P) = f(P). \tag{8}$$

La formule (6) conduit facilement à la solution du problème de Cauchy pour l'équation des ondes. En effet il s'ensuit pour $\alpha = 0$, grâce à la relation $I^0 u = u$,

$$u(P) = I_*^2 \overline{\Delta u, \frac{du}{dn}, u}(P), \tag{9}$$

c'est-à-dire que la valeur de u au point P peut se calculer par une intégrale spatiale portant sur $\Delta(u)$ et étendue à un certain volume D_S^P, par une simple couche et une double couche portant respectivement sur $\frac{du}{dn}$ et u et étendues à une certaine portion S^P de la surface S. Malheureusement, d'après ce que nous avons dit plus haut, ces intégrales sont en général divergentes. En effet, elles ne convergent toutes que si l'on a $m \leqq 2$ [1], l'une d'elles, l'intégrale de double couche, diverge déjà pour l'équation de M. Volterra ($m = 3$) et elles divergent toutes à partir de $m = 4$, correspondant à l'équation des ondes relative à l'espace ordinaire. Mais on pourra toujours définir ces intégrales par prolongement analytique, ce qui veut dire que, $I_*^\alpha \overline{\Delta(u), \frac{du}{dn}, u}$ étant convergent pour α assez grand ($\alpha > m$), on peut, dans les conditions de dérivabilité admises, obtenir I_*^2 par prolongement analytique.

La formule fait nettement ressortir la différence qu'il y a entre les cas de m impair et de m pair. Pour m impair, où l'on retrouve la formule de M. Hadamard, le facteur $\frac{1}{H_m(2)}$ ne s'évanouit pas, la solution sera donc fournie par une intégrale spatiale étendue au volume entier D_S^P et des intégrales étendues à la portion de surface entière S^P. Au contraire dans le cas de m pair (> 2), $\frac{1}{H_m(2)}$ s'annule à cause du facteur $\frac{1}{\Gamma\left(\frac{2+2-m}{2}\right)}$, et la formule finale ne sera pas influencée par les valeurs que u et ses dérivées prennent à l'intérieur du cône. L'intégrale spatiale se réduira à une intégrale étendue à la nappe du cône et

(1) Néanmoins, même pour $m = 2$, la formule (9) donne une solution erronée, indépendante des valeurs de u sur S, si l'on l'applique au sens strict, c'est-à-dire en *substituant* $\alpha = 0$ dans le second membre de la formule (6). Par contre, la formule obtenue de (6) par prolongement analytique donne la solution correcte bien connue, où il entre les valeurs de u aux points d'intersection de S et des deux rayons de lumière issus de P.

les intégrales de surface à des intégrales étendues à l'intersection du cône avec la surface S. *Voilà le principe de Huygens: seuls les points d'univers situés sur la nappe rétrograde du cône au sommet P agissent sur ce point.* Le principe dérive donc du fait que l'opérateur I_*^2 a pour m pair (>2) un caractère quasi local, il n'y intervient que des points qui sont à distance zéro du point P.

En réalité notre formule n'était jusqu'ici qu'une formule de représentation du même caractère que la formule classique de Cauchy l'est pour les fonctions monogènes. Mais elle donne aussi la solution d'un problème aux limites, notamment celle du problème de Cauchy(¹), si l'on admet, comme plus haut, que la surface S ait une orientation d'espace. En effet dans ce cas (et dans ce cas seulement) les domaines d'intégration deviennent infinitésimaux lorsque P s'approche indéfiniment de la surface.

Je termine cet exposé déjà trop long en donnant *une solution explicite* d'aspect profondément géométrique pour le cas de l'espace ordinaire, c'est-à-dire pour $m = 4$. En écrivant l'équation sous la forme $\Delta(u) = f$, le second membre donne lieu à une intégrale étendue à la nappe du cône qui n'est que le potentiel retardé bien connu. Puisque je ne saurais rien dire de nouveau sur cette partie de la solution, je me restreins ici à l'équation homogène.

Posons d'une manière générale $R = R_{PQ} = r_{PQ}^2$ et désignons par s^P la variété à deux dimensions formée par l'intersection du cône au sommet P et de la surface S. En remettant à tout à l'heure l'explication des autres notations, nous écrivons ici la solution suivante du problème de Cauchy posé pour l'équation $\Delta(u) = 0$, solution qu'on peut déduire de la solution générale (9),

$$u(P) = \frac{1}{\pi} \int_{s^P} \left[\frac{1}{2} \left(\frac{1}{R_1} + \frac{1}{R_2} \right) u - \frac{d_\nu u}{d_\nu R} \right] ds. \tag{10}$$

Vous voyez qu'il n'intervient dans la formule qu'une seule dérivée suivant une certaine direction ν qu'il faudra maintenant définir. Cela est très facile dans le cas classique considéré par Poisson où la surface S est le plan $x_1 = 0$. Alors ν n'est que l'image de la génératrice QP par rapport à ce plan. Dans le cas général la chose est un peu plus compliquée.

Par un élément infinitésimal à deux dimensions du bord s^P contenant le point Q il passe ∞^1 plans à trois dimensions. Formons l'image de la génératrice QP,

(¹) C'est le problème de résoudre l'équation $\Delta u = f$, la fonction f étant connue dans l'espace et les valeurs de u et de $\frac{du}{dn}$ étant données sur la surface S.

indéfiniment prolongée au delà de P, par rapport à l'un quelconque de ces plans, par exemple le plan tangent à S en Q, l'image étant prise dans la direction transversale au plan. Cette image, *indépendante du plan choisi*, donne la droite ν appartenant au point Q du bord. Les lignes ν, qui évidemment sont des rayons de lumière, engendrent une surface caractéristique Σ de l'équation des ondes, qui dans un certain sens peut être considérée comme l'image du cône au sommet P par rapport à la surface S. La surface Σ peut aussi être obtenue de la manière suivante. On fait décrire au point d'intégration Q le bord s^P et l'on considère les cônes rétrogrades qui ont leur sommet au point variable Q. Ces cônes enveloppent une surface à deux nappes. La nappe extérieure fait évidemment partie du cône original au sommet P, tandis que la nappe intérieure fait partie de la surface réglée Σ et la détermine entièrement.

La dérivée figurant dans notre formule est prise suivant la direction ν. La longueur lorentzienne des segments d'une telle droite étant nulle, on introduit R comme paramètre et forme la dérivée de u par rapport à R suivant ladite direction. Observons que $R = R_{PT}$ varie d'une manière linéaire sur une telle droite. En effet, T désignant un point quelconque de la droite ν appartenant à Q, R_{PT} est égal au double du produit scalaire lorentzien des vecteurs PQ et QT.

Il faut encore définir les valeurs R_1 et R_2 figurant dans notre formule. Remarquons à cet effet que les droites ν, qui sont en nombre ∞^2, forment évidemment une congruence de droites, qui est d'ailleurs d'un caractère très particulier. En nous trouvant dans quatre dimensions, ce n'est que grâce à cette dernière circonstance que, pour chaque droite fixe ν appartenant à un certain point Q, il existe en général deux directions de déplacements infinitésimaux $\delta_1 Q$ et $\delta_2 Q$ telles que les droites correspondantes coupent la droite ν. Cela veut dire que la congruence admet en général *deux foyers* Q_1 et Q_2 sur chacune de ses droites. Dans la formule ci-dessus on a posé, pour abréger,

$$R_{PQ_1} = R_1, \qquad R_{PQ_2} = R_2.$$

Le lieu géométrique des foyers est la surface focale, *la caustique*, de la congruence, qui dans un certain sens peut être interprétée comme l'image du point P par rapport à la surface S. Dans le cas classique de Poisson, elle se réduit à l'image *distincte* de ce point par rapport au plan $x_1 = 0$, et notre solution se réduit immédiatement à celle de Poisson.

En résumé, le principe de Huygens dit que la valeur de u au point P est déterminée par les valeurs que u et ses dérivées premières admettent en des points Q appartenant au passé-lumière du point P. Notre formule y apporte un complément qui doit présenter un certain intérêt, c'est que les seules dérivées qui y entrent sont prises suivant certains rayons de lumière, c'est-à-dire que seul le passé-lumière infinitésimal de la fonction u aux points Q y intervient. L'intervention de la surface (congruence) Σ et de sa caustique dans notre solution jette un jour nouveau sur les rapports entre l'optique physique et l'optique géométrique. Bien entendu, il s'agit ici d'une optique géométrique à quatre dimensions.

CHAPITRE I.

Intégrale de Riemann-Liouville.

1. Définition et formule de composition. — Posons

$$I^\alpha f(x) = \frac{1}{\Gamma(\alpha)} \int_a^x f(t)\,(x-t)^{\alpha-1}\,dt \qquad (x > a), \tag{1}$$

a étant un nombre fini ou égal à $-\infty$. Pour fixer les idées, nous supposons que $f(t)$ soit une fonction continue dans l'intervalle d'intégration et de plus que, dans le cas où $a = -\infty$, $f(t)$ tende vers zéro de façon que les contributions provenant du voisinage de $-\infty$ ne donnent lieu à aucune complication. Cela posé, l'intégrale existe pour les valeurs positives de α (et pour les valeurs complexes de α dont la partie réelle $\Re(\alpha)$ est positive). D'autre part, il est clair que cette intégrale diverge en général pour $\alpha = 0$ et pour toutes les valeurs négatives de α.

Dans le cas où les valeurs α et β sont positives (ou ont leurs parties réelles positives), on a la formule de composition

$$I^\alpha (I^\beta f(x)) = I^{\alpha+\beta} f(x). \tag{2}$$

En effet, en intervertissant l'ordre des intégrations, on obtient par une formule connue de Dirichlet[1]

$$\begin{aligned} I^\alpha (I^\beta f(x)) &= \frac{1}{\Gamma(\alpha)\,\Gamma(\beta)} \int_a^x (x-t)^{\alpha-1}\,dt \int_a^t f(v)\,(t-v)^{\beta-1}\,dv \\ &= \frac{1}{\Gamma(\alpha)\,\Gamma(\beta)} \int_a^x f(v)\,dv \int_v^x (x-t)^{\alpha-1}(t-v)^{\beta-1}\,dt. \end{aligned} \tag{3}$$

---

[1] *Cf.* Goursat *Cours d'Analyse* I (1917), p. 309.

La dernière intégrale est évidemment égale à

$$(x-v)^{\alpha+\beta-1}\cdot\int_0^1(1-u)^{\alpha-1}\cdot u^{\beta-1}\cdot du=(x-v)^{\alpha+\beta-1}\cdot\frac{\Gamma(\alpha)\Gamma(\beta)}{\Gamma(\alpha+\beta)},$$

ce qui donne

$$I^\alpha(I^\beta f(x))=\frac{1}{\Gamma(\alpha+\beta)}\int_a^x f(v)(x-v)^{\alpha+\beta-1}\,dv=I^{\alpha+\beta}f(x).$$

Pour $\alpha=1$, l'intégrale I^α se réduit à l'intégrale ordinaire $\int_a^x f(t)\,dt$. D'autre part, on sait que l'intégrale itérée d'ordre n

$$\int_a^x\int_a^{x_1}\cdots\int_a^{x_{n-1}}f(x_n)\,dx_n\,dx_{n-1}\cdots dx_1$$

se transforme facilement (p. ex. par la susdite formule de Dirichlet) dans l'intégrale simple

$$\frac{1}{(n-1)!}\int_a^x f(t)(x-t)^{n-1}\,dt=\frac{1}{\Gamma(n)}\int_a^x f(t)(x-t)^{n-1}\,dt.$$

C'est cette propriété-là qui chez plusieurs auteurs, notamment chez Riemann, fut le point de départ pour la définition d'une intégration généralisée d'ordre fractionnaire. La formule (2) met en évidence qu'il s'agit là d'une généralisation heureuse.

Nous constatons en même temps que l'intégrale $I^\alpha f(x)$ est pour $\Re(\alpha)>0$ une *fonction holomorphe de* α. Bien entendu, on suppose toujours que les conditions respectives posées au début de ce numéro pour la fonction f sont remplies et que la valeur x est fixée.

2. Prolongement analytique. — Admettons maintenant en outre que les dérivées $f^{(k)}(t)$ existent pour $k\leqq p$ et qu'elles remplissent des conditions analogues à celles posées pour $f(t)$. On aura alors après un certain nombre d'intégrations par parties successives, pour $n\leqq p$,

$$(4)\quad I^\alpha f(x)=\sum_{k=0}^{n-1}\frac{f^{(k)}(a)}{\Gamma(\alpha+k+1)}(x-a)^{\alpha+k}+\frac{1}{\Gamma(\alpha+n)}\int_a^x f^{(n)}(t)(x-t)^{\alpha+n-1}\,dt=$$

$$=\sum_{k=0}^{n-1}\frac{f^{(k)}(a)}{\Gamma(\alpha+k+1)}(x-a)^{\alpha+k}+I^{\alpha+n}f^{(n)}(x).$$

Dans le cas où $a = -\infty$ cette formule se réduit à la formule plus simple

$$(5) \quad I^\alpha f(x) = \frac{1}{\Gamma(\alpha)} \int_{-\infty}^{x} f(t)(x-t)^{\alpha-1}\,dt = \frac{1}{\Gamma(\alpha+n)} \int_{-\infty}^{x} f^{(n)}(t)(x-t)^{\alpha+n-1}\,dt = I^{\alpha+n} f^{(n)}(x).$$

Par les formules (4) et (5) on a montré que, dans les conditions posées, l'intégrale $I^\alpha f(x)$, considérée comme fonction analytique de α, peut se prolonger au delà du domaine de convergence de l'intégrale originale, le prolongement s'étendant en réalité à toutes les valeurs de α dont la partie réelle est $> -p$. Pour toutes ces valeurs, *on entendra par $I^\alpha f(x)$ le prolongement analytique dont on vient de démontrer l'existence.* [1]

Le même procédé s'étend facilement à des intégrales qui possèdent plusieurs singularités, p. ex. aux intégrales de la forme

$$\frac{\Gamma(\alpha)\,\Gamma(\beta)}{\Gamma(\alpha+\beta)} \int_{y}^{x} f(t)(t-y)^{\alpha-1}(x-t)^{\beta-1}\,dt.$$

On considère cette intégrale comme fonction analytique des exposants α, β et on en fait le prolongement analytique par une modification légère du procédé appliqué plus haut. Sans nous arrêter sur les détails, nous retournons aux intégrales I^α.

On démontre sans peine par les principes de la théorie des fonctions analytiques que la formule fondamentale (2) reste valide pourvu que β et $\alpha + \beta$ (ou leurs parties réelles) soient $> -p$. Il faut toutefois faire exception, au moins pour l'instant, pour les valeurs de α et de β telles que α ou β ou $\alpha + \beta$ sont des entiers négatifs ou 0. (*Cf.* le n° **4**.) C'est exactement la signification qu'on aura à attribuer à l'opération I^α quant α est égal à 0 ou à un entier négatif que nous allons examiner à présent.

3. L'intégrale I^α pour $\alpha = 0$ et $\alpha =$ entier négatif. — Dans le cas le plus simple, $\alpha = 0$, on sera conduit à poser

$$(6) \qquad I^0 f(x) = f(x).$$

Nous allons en effet montrer que sous la seule condition que $f(x)$ soit continu

$$\lim_{\alpha \to +0} I^\alpha f(x) = f(x).$$

[1] Dans tout ce travail *les intégrales divergentes* dépendant d'un paramètre α seront interprétées comme le *prolongement analytique* de l'intégrale correspondante, convergeant pour certaines valeurs de α, autant que ce prolongement existe et qu'il est défini d'une manière univoque.

Il n'est donc pas nécessaire de recourir au prolongement analytique, même si un prolongement existe grâce à des conditions supplémentaires concernant les dérivées de $f(x)$. Si tel est le cas, le prolongement par passage à la limite et le prolongement analytique conduisent évidemment à des résultats identiques.

Supposons d'abord dans la démonstration que a soit fini. On a

$$I^\alpha f(x) = \frac{1}{\Gamma(\alpha)} \int_a^x [f(t) - f(x)] (x-t)^{\alpha-1} dt + \frac{f(x)}{\Gamma(\alpha)} \int_a^x (x-t)^{\alpha-1} dt. \tag{7}$$

Le dernier terme qui peut encore s'écrire $\dfrac{f(x)(x-a)^\alpha}{\Gamma(\alpha+1)}$ devient $f(x)$ pour $\alpha \to 0$. Le premier terme tend vers 0 avec α. On le voit en divisant l'intervalle (a, x) en deux parties $(a, x-\delta)$ et $(x-\delta, x)$. Si δ est assez petit, le terme correspondant au second intervalle sera très petit puisqu'il en est de même de $f(t) - f(x)$. Le nombre δ étant fixé, le terme correspondant au premier intervalle tend vers 0 avec α grâce au facteur $\dfrac{1}{\Gamma(\alpha)}$. C'est encore grâce à ce facteur que le cas d'un intervalle infini se réduit immédiatement au cas précédent. En effet, si a est égal à $-\infty$, la partie correspondant à l'intervalle $(-\infty, b)$, où b est un nombre fixe $< x$, tendra encore vers 0.

Notons que pour $\alpha = 0$ la formule (4) se réduit à la formule classique de Cauchy

$$f(x) = \sum_{k=0}^{n-1} \frac{f^{(k)}(a)}{k!}(x-a)^k + \frac{1}{(n-1)!} \int_a^x f^{(n)}(t)(x-t)^{n-1} dt. \tag{8}$$

Passons maintenant à la signification de $I^{-n} f(x)$, n étant un entier positif. En faisant sur $f^{(n)}(t)$ les mêmes hypothèses qu'on a fait plus haut sur $f(t)$, on voit de suite par la formule (4) qu'on a à poser

$$I^{-n} f(x) = \lim_{\alpha \to -n+0} I^\alpha f(x) = f^{(n)}(x). \tag{9}$$

Nous avons fait ressortir dans l'Introduction (p. 3) que le caractère exceptionnel de l'indice 0 et des indices entiers négatifs est dû au facteur $\dfrac{1}{\Gamma(\alpha)}$. Nous y avons aussi dit que c'est d'un facteur analogue intervenant dans la solution du problème de Cauchy posé pour l'équation des ondes que le principe de Huygens tire son origine. Nous reviendrons plus loin sur ce point (n° **55**).

4. Différentiation sous le signe $\int$. — Notre point de départ était de considérer l'opération I^α comme une intégration généralisée. Il est temps d'éclaircir les rapports presque évidents qui existent entre cette opération et la dérivation, rapports indiqués d'ailleurs par les derniers résultats.

En admettant d'abord que β (ou sa partie réelle) est positif, on a par les règles usuelles de la dérivation des intégrales

$$\frac{d}{dx} I^{\beta+1} f(x) = \frac{1}{\Gamma(\beta+1)} \frac{d}{dx} \int_a^x f(t)(x-t)^\beta \, dt = \frac{1}{\Gamma(\beta)} \int_a^x f(t)(x-t)^{\beta-1} \, dt = I^\beta f(x).$$

Par la formule de prolongement (4) on voit que cette relation subsiste aussi pour les valeurs $\beta > -p$ et qu'il en est de même de la relation plus générale

$$\frac{d^k}{dx^k} I^{\beta+k} = I^\beta. \tag{10}$$

L'extension à des valeurs complexes de β est évidente.

Par la relation $I^{-k} f(x) = f^{(k)}(x)$ la relation (10) pourra encore s'écrire

$$I^{-k}(I^\beta) = I^{-k+\beta}. \tag{11}$$

En rapprochant ceci de nos résultats antérieurs on trouve en définitive: La formule fondamentale (2) est valable pour toutes les valeurs de α et de β telles que $\beta > -p$ et $\alpha + \beta > -p$ s'il est question d'un intervalle infini. Dans le cas d'un intervalle fini la formule sera encore valide pour toutes les dites valeurs de α et de β sauf pour les β qui sont des entiers négatifs. (Dans le cas évident où α aussi est un entier négatif ou 0 la formule restera encore valide.)

5. Prolongement par développement en série. — La formule (4) qui est très commode quand il s'agit de définir le prolongement analytique de I^α ne l'est pas toujours dans les applications lorsqu'il s'agit d'exécuter les calculs. Il sera donc utile d'observer que, d'après les derniers résultats, le prolongement analytique pourra aussi être exécuté par la formule (10), procédé très avantageux dans certains cas.[1] Nous donnons ici encore une formule pour le calcul effectif du prolongement par laquelle nous nous approchons beaucoup de l'ordre d'idées de M. Hadamard.[2] Posons en effet

---

(1) *Cf.* les n^{os} **52** et **53**.

(2) *Voir l. c.* p. 190.

$$P(t) = \sum_{k=0}^{n-1} \frac{f^{(k)}(x)}{k!}(t-x)^k$$

et en supposant d'abord α positif

$$I^\alpha f(x) = \frac{1}{\Gamma(\alpha)}\int_a^x f(t)(x-t)^{\alpha-1}\,dt$$

$$= \frac{1}{\Gamma(\alpha)}\int_a^x (f(t)-P(t))(x-t)^{\alpha-1}\,dt + \frac{1}{\Gamma(\alpha)}\int_a^x P(t)(x-t)^{\alpha-1}\,dt$$

ou

$$(12) \quad I^\alpha f(x) = \frac{1}{\Gamma(\alpha)}\int_a^x (f(t)-P(t))(x-t)^{\alpha-1}\,dt + \frac{1}{\Gamma(\alpha)}\sum_{k=0}^{n-1}\frac{(-1)^k f^{(k)}(x)(x-a)^{k+\alpha}}{(k+\alpha)\cdot k!}.$$

Si la dérivée $f^{(n)}(t)$ existe et est continue, l'intégrale converge évidemment pour $\alpha > -n$ et la formule finale définit le prolongement analytique pour toutes ces valeurs. $I^\alpha f(x)$ est une fonction holomorphe de α, puisque $\Gamma(\alpha)(k+\alpha) \neq 0$ partout. En particulier, pour $\alpha = -k$, on a $\Gamma(\alpha)(k+\alpha)\cdot k! = (-1)^k$, d'où l'on tire encore une fois

$$(12^{bis}) \qquad I^0 f(x) = f(x), \qquad I^{-k} f(x) = f^{(k)}(x).$$

Ajoutons encore une remarque. C'est que, pour la validité du dernier énoncé, il n'est pas nécessaire que les dérivées $f^{(k)}(t)$ existent au sens strict. Il suffit p. ex. qu'il existe un polynome $P(t)$ tel que $(f(t)-P(t))(x-t)^{-n}$ reste borné lorsque t tend vers x.

Il y a lieu de remarquer ici que la formule (7) n'est qu'un cas particulier de la formule (12).

Pour les applications il est souvent commode de placer la singularité à l'origine, c'est-à-dire de considérer des intégrales de la forme

$$J^\alpha f(0) = \frac{1}{\Gamma(\alpha)}\int_0^b f(t)\,t^{\alpha-1}\,dt.$$

En posant alors

$$P(t) = \sum_{k=0}^{n-1}\frac{f^{(k)}(0)}{k!}t^k,$$

on a la formule

$$J^\alpha f(0) = \frac{1}{\Gamma(\alpha)} \int_0^b (f(t) - P(t))\, t^{\alpha-1}\, dt + \frac{1}{\Gamma(\alpha)} \sum_{k=0}^{n-1} \frac{f^{(k)}(0)\, b^{k+\alpha}}{(k+\alpha)\cdot k!} \tag{13}$$

et, en particulier,

$$J^0 f(0) = f(0), \qquad J^{-k} f(0) = (-1)^k f^{(k)}(0). \tag{13bis}$$

CHAPITRE II.

Intégrale de Riemann-Liouville dans l'espace euclidien. Potentiels.

6. Définition et propriétés de l'intégrale I^α. Remarque. — Les généralisations de l'intégrale de R.-L.(1) dont nous allons nous occuper dans la suite se rapportent à un espace Ω_m (ou plus brièvement Ω) à un nombre quelconque m ($\geqq 1$) de dimensions, pourvu d'une métrique soit *euclidienne*, soit *lorentzienne*. Nous désignons les procédés d'intégration respectifs par *intégration elliptique* et *intégration hyperbolique*. C'est l'intégration hyperbolique qui forme la généralisation la plus directe de l'intégrale de R.-L. qu'on vient d'étudier. C'est encore uniquement l'intégration hyperbolique qui intervient dans l'application à l'équation des ondes qui est le but principal de ce travail. Cependant, ce procédé présentant sous certains rapports des difficultés considérables qui n'interviennent pas dans le procédé elliptique, il nous paraît utile de donner ici un aperçu sur le dernier procédé qui, tout en facilitant la lecture des chapitres suivants, n'y est nullement indispensable. Nous nous dispensons pourtant de faire précéder l'exposé de l'intégration dans plusieurs dimensions par un exposé concernant le cas linéaire ($m = 1$)(2), les difficultés dépendant très peu du nombre de dimensions, dans le cas actuel.

Dans ce chapitre on désigne la distance *euclidienne* de deux points P et Q par $r_{PQ} = r_{QP} = r$. Introduisons de la manière suivante un *potentiel généralisé* d'ordre α. Nous posons, pour α positif et pour des fonctions $f(P)$ dont on va préciser la nature plus loin,

$$I^\alpha f(P) = \frac{1}{H_m(\alpha)} \int_\Omega f(Q)\, r_{PQ}^{\alpha-m}\, dQ, \tag{1}$$

(1) R.-L. = Riemann-Liouville.

(2) Pour un problème de M. Carleman relatif à ce cas, *cf.* M. Riesz 8.

$H_m(\alpha)$ ne dépendant que de α et de m et l'intégration s'étendant à l'espace entier Ω; dQ désigne l'élément de volume de cet espace. Nous allons démontrer que $H_m(\alpha)$, qui joue ici le rôle revenant tout à l'heure à $\Gamma(\alpha)$, peut être déterminé de manière que les formules fondamentales

$$I^\alpha(I^\beta f(P)) = I^{\alpha+\beta} f(P) \tag{2}$$

et

$$\Delta I^{\alpha+2} f(P) = - I^\alpha f(P) \tag{3}$$

soient satisfaites, Δ désignant l'opérateur de Laplace

$$\Delta = \sum_{k=1}^{m} \frac{\partial^2}{\partial x_k^2},$$

où $x_1, x_2, \ldots, x_m$ sont les coordonnées cartésiennes du point P.(¹)

Montrons d'abord qu'on peut déterminer $H_m(\alpha)$ de façon que les relations (2) et (3) ci-dessus aient lieu pour la fonction particulière $f(P) = e^{ix_1}$. On trouve par des transformations évidentes, $\xi_1, \xi_2, \ldots, \xi_m$ désignant les coordonnées du point Q,

$$I^\alpha e^{ix_1} = \frac{1}{H_m(\alpha)} \int_\Omega e^{i\xi_1} \left(\sum (\xi_k - x_k)^2\right)^{\frac{\alpha-m}{2}} d\xi_1\, d\xi_2 \ldots d\xi_m$$

$$= \frac{1}{H_m(\alpha)} e^{ix_1} \int_\Omega e^{i\eta_1} \left(\sum \eta_k^2\right)^{\frac{\alpha-m}{2}} d\eta_1\, d\eta_2 \ldots d\eta_m.$$

Alors, si l'on pose $H_m(\alpha)$ égal à la dernière intégrale, il viendra

$$I^\alpha e^{ix_1} = e^{ix_1}. \tag{4}$$

Par ce choix de $H_m(\alpha)$ les relations (2) et (3) seront évidemment remplies pour la fonction particulière qu'on est en train de considérer.

Ajoutons qu'avec le même choix de $H_m(\alpha)$, il résulte de la formule précédente par une transformation orthogonale des variables

$$I^\alpha e^{i(a_1x_1+a_2x_2+\cdots+a_mx_m)} = (\Sigma a_k^2)^{-\frac{\alpha}{2}} \cdot e^{i(a_1x_1+a_2x_2+\cdots+a_mx_m)}. \text{ (²)}$$

(¹) Il est presque évident que les conditions (2) et (3) et une condition supplémentaire, exigeant p. ex. continuité en général, déterminent $H_m(\alpha)$ d'une manière unique, à un facteur $e^{in\alpha\pi}$ près, n étant un nombre entier.

(²) On pourrait par cette formule rattacher notre procédé d'intégration à la théorie de l'intégrale de Fourier. Mais, comme cette méthode présente de graves inconvénients et ne s'adapte pas au cas de coefficients variables, nous suivons dans le texte une voie différente.

Calculons maintenant l'intégrale à laquelle on a égalé $H_m(\alpha)$. La fonction à intégrer étant constante sur la sphère $\eta_2^2 + \eta_3^2 + \cdots + \eta_m^2 = \varrho^2$, l'intégrale se réduit immédiatement à

$$H_m(\alpha) = \frac{2 \cdot \pi^{\frac{m-1}{2}}}{\Gamma\left(\frac{m-1}{2}\right)} \cdot \int_0^{\infty} \int_{-\infty}^{+\infty} e^{i\eta_1} (\varrho^2 + \eta_1^2)^{\frac{\alpha-m}{2}} \varrho^{m-2} \, d\eta_1 \, d\varrho,$$

le facteur devant les signes $\int$ exprimant la surface totale de ladite sphère pour $\varrho = 1$.[1] Par la substitution $\eta_1 = r \cos\theta$, $\varrho = r \sin\theta$, l'intégrale devient

$$\int_0^{\pi} \sin^{m-2}\theta \, d\theta \int_0^{\infty} e^{ir\cos\theta} r^{\alpha-1} \, dr .$$

Le théorème de Cauchy concernant l'intégrale d'une fonction holomorphe le long d'un contour fermé donne pour la dernière intégrale (simple) la valeur

$$e^{\pm \frac{i\pi\alpha}{2}} |\cos\theta|^{-\alpha} \, \Gamma(\alpha),$$

le signe à choisir étant celui de $\cos\theta$. On a en outre d'abord, pour assurer la convergence, pour $\alpha < 1$ et après, par prolongement analytique,

$$\int_0^{\frac{\pi}{2}} \sin^{m-2}\theta \cos^{-\alpha}\theta \, d\theta = \frac{\Gamma\left(\frac{m-1}{2}\right) \Gamma\left(\frac{1-\alpha}{2}\right)}{2\,\Gamma\left(\frac{m-\alpha}{2}\right)} .$$

De plus

(5) $$\Gamma(\alpha) = 2^{\alpha-1} \pi^{-\frac{1}{2}} \Gamma\left(\frac{\alpha}{2}\right) \Gamma\left(\frac{1+\alpha}{2}\right)$$

et

(6) $$\Gamma(x)\,\Gamma(1-x) = \frac{\pi}{\sin \pi x},$$

et par là

$$\Gamma\left(\frac{1-\alpha}{2}\right) \Gamma\left(\frac{1+\alpha}{2}\right) = \frac{\pi}{\cos \frac{\pi\alpha}{2}} .$$

([1]) *Voir* la *Remarque* à la fin de ce numéro.

La combinaison des formules ci-dessus donne enfin [1]

(7) $$H_m(\alpha) = \frac{\pi^{\frac{m}{2}} \cdot 2^\alpha \cdot \Gamma\left(\frac{\alpha}{2}\right)}{\Gamma\left(\frac{m-\alpha}{2}\right)}.$$

Remarque. — Désignons par v_n le volume et par ω_n la surface totale de la sphère unitaire relative à l'espace euclidien à n dimensions

$$x_1^2 + x_2^2 + \cdots + x_n^2 = 1.$$

La sphère découpe du plan $x_n = t\ (|t| \leqq 1)$ une sphère à $n-1$ dimensions au rayon $\varrho = (1-t^2)^{\frac{1}{2}}$. Le volume (à $n-1$ dimensions) de cette sphère sera $= \varrho^{n-1} \cdot v_{n-1} = (1-t^2)^{\frac{n-1}{2}} \cdot v_{n-1}$. On a donc la formule de récurrence

$$v_n = v_{n-1} \int_{-1}^{+1} (1-t^2)^{\frac{n-1}{2}} dt = 2 v_{n-1} \int_0^1 (1-t^2)^{\frac{n-1}{2}} dt$$

$$= v_{n-1} \cdot \frac{\Gamma\left(\frac{n+1}{2}\right)\Gamma\left(\frac{1}{2}\right)}{\Gamma\left(\frac{n+2}{2}\right)} = v_{n-1} \cdot \pi^{\frac{1}{2}} \cdot \frac{\Gamma\left(\frac{n+1}{2}\right)}{\Gamma\left(\frac{n+2}{2}\right)}.$$

Puisque $v_1 = 2$, il en résulte

$$v_n = \frac{\pi^{\frac{n}{2}}}{\Gamma\left(\frac{n+2}{2}\right)} = \frac{2\pi^{\frac{n}{2}}}{n\,\Gamma\left(\frac{n}{2}\right)}.$$

Par la formule évidente

$$\omega_n = n\, v_n$$

il vient enfin

(8) $$\omega_n = \frac{2\pi^{\frac{n}{2}}}{\Gamma\left(\frac{n}{2}\right)}.$$

7. Formule de composition. — Retournons maintenant à l'opération $I^\alpha f(P)$, $f(P)$ étant une fonction arbitraire assujettie à certaines conditions que nous

---

[1] Nous profitons de l'occasion pour donner ici la déduction de l'expression de $H_m(\alpha)$ et celle de la formule de composition qui suit, déductions que nous regrettons d'avoir omises dans notre travail **3**.

allons préciser à l'instant. Pour le moment nous supposons seulement que toutes les intégrales qui interviennent dans le calcul suivant sont absolument convergentes.

$$I^\alpha(I^\beta f(P)) = \frac{1}{H_m(\alpha)}\int_\Omega r_{PQ}^{\alpha-m}\,dQ \cdot \frac{1}{H_m(\beta)}\int_\Omega f(S)\, r_{QS}^{\beta-m}\,dS \tag{9}$$

$$= \frac{1}{H_m(\alpha)H_m(\beta)}\int_\Omega f(S)\,dS \int_\Omega r_{PQ}^{\alpha-m} r_{QS}^{\beta-m}\,dQ.$$

Par une translation la dernière intégrale se réduit d'abord à une intégrale de la même forme où P est remplacé par l'origine O et S par un point dont la distance à l'origine est r_{PS}. Enfin, par une homothétie on arrive à

$$r_{PS}^{\alpha+\beta-m}\int_\Omega r_{OT}^{\alpha-m} r_{1T}^{\beta-m}\,dT,$$

où r_{OT} et r_{1T} désignent les distances respectives du point T à l'origine et à un certain point à distance unité de l'origine. Évidemment la dernière intégrale

$$\mathrm{B}_m(\alpha,\beta) = \int_\Omega r_{OT}^{\alpha-m} r_{1T}^{\beta-m}\,dT \tag{10}$$

ne dépend que des nombres α, β et m. Ajoutons que cette intégrale est convergente quand les inégalités $\alpha > 0$, $\beta > 0$, $\alpha+\beta < m$ sont remplies.(1)

Admettons pour l'instant qu'il soit déjà démontré que

$$\mathrm{B}_m(\alpha,\beta) = \frac{H_m(\alpha)\,H_m(\beta)}{H_m(\alpha+\beta)}. \tag{10^{bis}}$$

Alors, moyennant les simplifications qu'on y a déjà apportées, la formule (9) se réduit à la formule fondamentale

$$I^\alpha I^\beta = I^{\alpha+\beta}\,(\alpha > 0,\ \beta > 0,\ \alpha+\beta < m).\text{(2)} \tag{11}$$

Il reste à vérifier la formule (10bis). L'intégrale qu'on veut calculer est manifestement une sorte de généralisation de l'intégrale d'Euler de première espèce. On démontrera d'abord l'identité (10bis) dans les conditions restrictives $\alpha > 0$, $\beta > 0$, $\alpha+\beta < 1$ et ensuite on la rendra valide dans le cas général par prolongement analytique. La démonstration se fait par l'intermédiaire de la fonction déjà considérée e^{ix_1}. Observons tout de suite que cette fonction ne remplit nullement les

(1) Nous n'insistons ni ici ni dans la suite sur le fait évident, rappelé souvent au chapitre précédent, que toute condition concernant les exposants peut être remplacée pas une condition identique concernant leurs parties réelles.

(2) Pour une généralisation remarquable où, entre autres choses, l'espace euclidien est remplacé par un groupe localement compact, *voir* H. Cartan 1 et 2.

conditions auxquelles nous avons assujetti nos fonctions générales $f(P)$. Néanmoins, il résulte par une vérification facile que la formule (9) est valable pour cette fonction. Cela donne

$$I^\alpha I^\beta e^{i x_1} = B_m(\alpha, \beta) \cdot \frac{H_m(\alpha + \beta)}{H_m(\alpha) \cdot H_m(\beta)} I^{\alpha+\beta} e^{i x_1}.$$

La comparaison avec la formule (4) donne de son côté que le facteur devant $I^{\alpha+\beta} e^{i x_1}$ se réduit à 1, ce qui vérifie l'identité (10bis).[1]

8. L'opérateur $I^2 = -\Delta^{-1}$. Potentiel newtonien. — Disons un mot sur le second de nos *desiderata* exprimé par la formule (3). Cette formule se vérifie pour $\alpha > 0$ moyennant la relation

$$(12) \qquad \frac{1}{H_m(\alpha + 2)} \Delta r^{\alpha+2-m} = \frac{\Gamma\left(\frac{m-\alpha-2}{2}\right)}{\pi^{\frac{m}{2}} \cdot 2^{\alpha+2} \cdot \Gamma\left(\frac{\alpha+2}{2}\right)} \cdot \alpha(\alpha + 2 - m) \cdot r^{\alpha-m}$$

$$= - \frac{\Gamma\left(\frac{m-\alpha}{2}\right)}{\pi^{\frac{m}{2}} \cdot 2^{\alpha} \cdot \Gamma\left(\frac{\alpha}{2}\right)} \cdot r^{\alpha-m} = - \frac{1}{H_m(\alpha)} \cdot r^{\alpha-m}.$$

Dans cette formule, Δ signifie soit Δ_P, soit Δ_Q.

La formule (3) met en évidence que l'opérateur I^2 est (au signe près) l'inverse de l'opérateur Δ,[2] et on comprend alors qu'il mérite une attention particulière. On trouve pour $m \neq 2$

$$(13) \qquad I^2 f(P) = \frac{\Gamma\left(\frac{m-2}{2}\right)}{4 \cdot \pi^{\frac{m}{2}}} \int_\Omega f(Q)\, r_{PQ}^{2-m}\, dQ$$

---

[1] On peut aussi vérifier cette identité au moyen de la formule

$$r^{-\lambda} = \frac{2}{\Gamma\left(\frac{\lambda}{2}\right)} \int_0^\infty e^{-r^2 t^2} t^{\lambda-1}\, dt \quad (\lambda > 0).$$

[2] Le postulat $\Delta I^{\alpha+2} = + I^\alpha$ pourrait être rempli si l'on remplaçait $H_m(\alpha)$ par $e^{\frac{i\pi\alpha}{2}} \cdot H_m(\alpha)$. Cette valeur est sous un certain rapport plus adéquate que celle adoptée dans le texte puisqu'elle donnerait $I^\alpha e^{i x_1} = e^{-\frac{i\pi\alpha}{2}} e^{i x_1} = i^{-\alpha} \cdot e^{i x_1}$. Notre choix se justifie par le fait qu'il rend $H_m(\alpha)$ réel pour α réel.

que nous désignerons comme *le potentiel newtonien* correspondant à la couche de la densité spatiale $f(Q)$, bien que le facteur devant l'intégrale manque dans la forme usuelle de ce potentiel.

9. L'opérateur I^m. Potentiel logarithmique. — L'opérateur I^2 défini par la formule (13) perd tout intérêt dans le cas $m = 2$. L'intégrale sera indépendante de P et le facteur devant l'intégrale deviendra infini. Cependant, dans la condition particulière que l'intégrale de $f(Q)$ étendue à l'espace entier, c'est à dire la masse totale, s'annule, on pourra poser la définition nouvelle

$$(14) \qquad I^2 f(P) = \lim_{\varepsilon \to 0} I^{2-\varepsilon} f(P) = \lim_{\varepsilon \to 0} \frac{\Gamma\left(\frac{\varepsilon}{2}\right)}{\pi \cdot 2^{2-\varepsilon} \cdot \Gamma\left(1 - \frac{\varepsilon}{2}\right)} \int_{\Omega_2} f(Q) r_{PQ}^{-\varepsilon} d Q$$

$$= \lim_{\varepsilon \to 0} \frac{\Gamma\left(\frac{\varepsilon}{2}\right)}{\pi \cdot 2^{2-\varepsilon} \cdot \Gamma\left(1 - \frac{\varepsilon}{2}\right)} \int_{\Omega_2} f(Q) (r_{PQ}^{-\varepsilon} - 1) d Q = \frac{1}{2\pi} \int_{\Omega_2} f(Q) \log \frac{1}{r_{PQ}} d Q,$$

c'est à dire qu'on obtient le potentiel logarithmique.

Dans le cas général, où la masse totale M est différente de zéro, on pourra toujours écrire au voisinage de la valeur $\alpha = 2$

$$I^\alpha f(P) = \frac{A_{-1}}{\alpha - 2} + A_0 + \sum_{k=1}^{\infty} A_k (\alpha - 2)^k,$$

où $A_{-1} = -M/\pi$ et A_0 est (à une constante additive près) le potentiel logarithmique

$$(15) \qquad \frac{1}{2\pi} \int_{\Omega_2} f(Q) \log \frac{1}{r_{PQ}} d Q = \lim_{\varepsilon \to 0} \frac{1}{2} (I^{2-\varepsilon} f(P) + I^{2+\varepsilon} f(P)).$$

Retournons maintenant au cas général d'une dimension m quelconque, et remarquons que les choses se passent d'une manière analogue, bien entendu, non plus pour $\alpha = 2$, mais pour $\alpha = m$. Le potentiel $I^m f(P)$ pourra toujours être interprété comme un potentiel logarithmique, la seule différence étant que l'indice $\alpha = m$ n'est un indice particulièrement important que dans le cas $m = 2$, où les indices $\alpha = 2$ et $\alpha = m$ deviennent identiques. Une dernière remarque dans cet ordre d'idées concerne les indices $\alpha = m + 2k$, k étant un entier positif. Dans ce cas-là on tombe par un procédé analogue sur des noyaux de la forme $r^{2k} \cdot \log \frac{1}{r} \cdot$

10. Formule de Green pour l'espace entier. Prolongement analytique. — Les considérations qui précèdent s'étendent immédiatement au cas où l'on remplace les couches définies moyennant des fonctions de point représentant la densité spatiale par des couches définies moyennant des fonctions additives d'ensemble, généralisation embrassant entre autres choses les simples couches. (Avec quelques précautions on peut même étendre les résultats à des intégrales définies par des doubles couches.) Ici nous abordons un autre problème, celui des indices zéro et négatifs, où, au lieu de généraliser les couches en question, il faut les assujettir à de nouvelles restrictions.

En effet, il est clair que tout comme dans l'espace de dimension 1, on pourra dans des conditions supplémentaires étendre le procédé à ces derniers indices. L'extension se fera encore par prolongement analytique; pour l'indice zéro on peut y arriver par un passage à la limite. En supposant que $f(Q)$ soit continu au point P, on trouve en copiant les raisonnements qui s'attachaient à la formule (7) du n° **3**

$$I^0 f(P) = \lim_{\alpha \to +0} I^\alpha f(P) = f(P), \tag{16}$$

formule qui, rapprochée de la formule $\Delta I^{\alpha+2} = - I^\alpha$, exprime d'une façon un peu rude la formule de Poisson $(\lim_{\varepsilon \to +0} \Delta I^{2+\varepsilon} f(P) = - \lim_{\varepsilon \to +0} I^\varepsilon f(P) = - f(P))$.

Il reste les indices négatifs. Supposons d'abord que la fonction $f(P)$ admette des dérivées continues de tout ordre $\leqq 2p$ et de plus que $f(P)$ et ses dérivées se comportent à l'infini de façon que les intégrales qui interviendront dans la suite soient absolument convergentes et que les intégrations par parties qu'on exécutera soient légitimes. En appliquant p fois la formule de Green

$$\int_\Omega U(Q)\, \Delta V(Q)\, dQ = \int_\Omega V(Q)\, \Delta U(Q)\, dQ,$$

on trouve moyennant les formules (1) et (12), d'abord pour α positif,

$$I^\alpha f(P) = (-1)^p I^{\alpha+2p} \Delta^p f(P). \tag{17}$$

Par cette formule on a le prolongement analytique de $I^\alpha f(P)$ pour toute valeur de $\alpha > -2p$.

Si l'on admet en particulier que $\Delta^p f(Q)$ soit continu pour $Q = P$ (hypothèse qui n'est pas nécessaire pour la validité de la dernière formule) on trouve comme plus haut en se servant du prolongement par passage à la limite

$$(18)\qquad I^{-2p}f(P)=\lim_{\varepsilon\to+0} I^{-2p+\varepsilon}f(P)=(-1)^p\,\Delta^p f(P).$$

Une formule analogue vaudra à plus forte raison pour tout nombre entier positif $k<p$. On voit aussi sans peine que la formule fondamentale (11) s'étend au domaine $\alpha>-2p$, $\beta>-2p$, $\alpha+\beta>-2p$.

11. Formule de Green pour un domaine borné. Prolongement analytique. — Nos hypothèses concernant l'allure de la fonction $f(P)$ et de ses dérivées, en particulier nos hypothèses de continuité, furent posées pour l'espace entier. Les choses deviennent plus compliquées, si l'on admet certaines surfaces[1] de discontinuité. Pour ne pas nous perdre dans des généralités inutiles, nous nous restreindrons au cas où $f(P)$ est identiquement zéro à l'extérieur d'une surface fermée S, suffisamment régulière, tandis que la fonction $f(P)$ et les dérivées qui interviennent sont continues dans le domaine fermé limité par cette surface, sans s'annuler sur la surface, en général.

En supposant d'abord α positif, on trouve par la formule de Green

$$(19)\qquad I^\alpha f(P)=-I^{\alpha+2}\Delta f(P)+\frac{1}{H_m(\alpha+2)}\int_S\left\{f(Q)\frac{d\,r_{PQ}^{\alpha+2-m}}{dn}-\frac{df(Q)}{dn}r_{PQ}^{\alpha+2-m}\right\}dS$$

ou d'une façon générale

$$(20)\qquad I^\alpha f(P)=(-1)^p I^{\alpha+2p}\Delta^p f(P)$$
$$+\sum_{k=1}^{p}\frac{(-1)^{k-1}}{H_m(\alpha+2k)}\int_S\left\{\Delta^{k-1}f(Q)\frac{d\,r_{PQ}^{\alpha+2k-m}}{dn}-\frac{d\,\Delta^{k-1}f(Q)}{dn}r_{PQ}^{\alpha+2k-m}\right\}dS,$$

n désignant toujours la normale au point Q de S dirigée vers l'intérieur de cette surface et dS l'élément de surface autour du point Q.

Par la dernière formule on aura aussi dans le cas actuel obtenu le prolongement analytique de $I^\alpha f(P)$ pour $\alpha>-2p$. Pourtant il y a une différence essentielle entre le cas plus simple traité tout à l'heure et le cas actuel. *Là* on avait le prolongement pour tout point fini P, *ici* il faut faire exception pour les points P situés sur la surface. Pour les autres points on a comme précédemment

$$(21)\qquad I^{-2k}f(P)=(-1)^k\,\Delta^k f(P)$$

---

[1] Dans ce qui suit on entendra toujours par une *surface* une multiplicité à $m-1$ dimensions. L'expression *hypersurface* ne sera donc pas usitée.

et plus généralement

(22) $$I^{\alpha-2k}f(P)=(-1)^k\Delta^k I^\alpha f(P),$$ [1]

cette formule pouvant aussi servir à effectuer le prolongement analytique.

Les dernières formules donnent évidemment les analogues de la formule (9) du n° **3** et de la formule (10) du n° **4**. La formule fondamentale (11) est encore valide si β n'est pas un nombre entier pair négatif.

12. Analogie avec le cas linéaire. — La surface S joue ici à peu près le même rôle que le point initial a jouait dans le cas de l'intégrale de R.-L. ordinaire. Aux opérateurs $\frac{d^{2k}}{da^{2k}}$ et $\frac{d^{2k+1}}{da^{2k+1}}$ correspondent (abstraction faite de certains facteurs constants, dont le rôle n'est d'ailleurs nullement négligeable) les opérateurs Δ^k et $\frac{d}{dn}\Delta^k$ respectivement. En observant que $I^0 f(P)$ est égal à $f(P)$, la formule (19) devient pour $\alpha=0$ identique à la formule de Green

(23) $$f(P)=\frac{\Gamma\left(\frac{m-2}{2}\right)}{4\cdot\pi^{\frac{m}{2}}}\int_S\left\{f(Q)\frac{d\,r_{PQ}^{2-m}}{dn}-\frac{df(Q)}{dn}r_{PQ}^{2-m}\right\}dS$$
$$-\frac{\Gamma\left(\frac{m-2}{2}\right)}{4\cdot\pi^{\frac{m}{2}}}\int_D\Delta f(Q)\,r_{PQ}^{2-m}\,dQ,$$

(D étant le domaine limité par S) qui de son côté correspond évidemment à la formule

(24) $$f(x)=f(a)+f'(a)(x-a)+\int_a^x f''(t)(x-t)\,dt.$$

A la formule générale de Cauchy (formule (8) du n° **3**) correspond la formule (20) si l'on y pose $\alpha=0$.

13. Renvoi aux problèmes classiques. — On va voir que dans le cas hyperbolique, quand il s'agira p. ex. de résoudre le problème de Cauchy, la formule de

---

[1] Cette formule renferme comme cas particulier la formule de Poisson (qui, bien entendu, n'est démontrée ici que dans des conditions assez restrictives) sous sa forme précise

$$\Delta\, I^2 f(P)\ (=-I^0 f(P))=-f(P).$$

Green est un outil tout-puissant, toute la difficulté, d'ailleurs assez sérieuse, n'étant que de lui donner un sens. Il n'en est rien dans le cas elliptique, où, comme on le sait, la formule de Green n'est qu'un premier pas vers la solution des problèmes difficiles, tels le problème de Dirichlet et les problèmes connexes. Aussi faut-il, si l'on veut approfondir dans ce cas l'étude de nos potentiels, recourir à d'autres instruments, tels la fonction de Green et les masses de Green. Nous renvoyons pour ces questions aux travaux de l'auteur et de M. Frostman.[1]

CHAPITRE III.

Intégrale de Riemann-Liouville dans l'espace lorentzien.

I. Définition de l'intégrale.

14. La forme métrique lorentzienne et l'opérateur des ondes. Remarque. — Nous passons maintenant à la définition de l'intégrale de R.-L. dans l'espace lorentzien à m dimensions. Considérons dans cet espace les points P et Q aux coordonnées respectives $x_1, x_2, \ldots, x_m$ et $\xi_1, \xi_2, \ldots, \xi_m$. Le carré de la distance lorentzienne de ces points — qu'on va d'ailleurs légèrement généraliser dans le cours de ce numéro — est défini par la formule

$$r_{PQ}^2 = r_{QP}^2 = r^2 = (x_1 - \xi_1)^2 - (x_2 - \xi_2)^2 - \cdots - (x_m - \xi_m)^2. \tag{1}$$

L'application du même symbole r_{PQ} dans deux espaces portant des métriques différentes ne devra amener aucune confusion chez le lecteur, d'autant moins que dans la suite il ne sera guère question de la métrique euclidienne. D'ailleurs, on peut, si l'on veut, considérer l'une et l'autre définition comme des cas particuliers de la définition générale

$$r_{PQ}^2 = \sum_{i,k} g_{ik}(x_i - \xi_i)(x_k - \xi_k).$$

Dans le même ordre d'idées on gardera les symboles utilisés dans le cas euclidien pour désigner les différents éléments métriques correspondants et on désignera l'opérateur des ondes par le même symbole Δ que l'opérateur de Laplace, c'est-à-dire que l'on pose ici

$$\Delta = \frac{\partial^2}{\partial x_1^2} - \frac{\partial^2}{\partial x_2^2} - \cdots - \frac{\partial^2}{\partial x_m^2}. \tag{2}$$

En considérant le point P comme fixe et le point Q comme variable,

[1] M. Riesz **3**, O. Frostman **1—3**.

$r_{PQ}^2 = 0$ définit la surface du cône caractéristique ou *cône de lumière* au sommet P, $r_{PQ}^2 < 0$ définit son intérieur, $x_1 - \xi_1 > 0$ le cône *rétrograde*, et $x_1 - \xi_1 < 0$ le cône *direct*.

C'est le *cône rétrograde* que nous allons considérer en général, en le désignant par D^P.

Le *carré scalaire* d'un vecteur $\mathbf{X}$ aux composantes X_k est défini par la forme métrique

$$(\mathbf{X}, \mathbf{X}) = X_1^2 - X_2^2 - \cdots - X_m^2. \tag{3}$$

En nous inspirant de la terminologie relativiste, nous appelons un vecteur *vecteur de temps*, *vecteur de lumière (vecteur isotrope)*, *vecteur d'espace*[1] suivant que son carré scalaire est positif, nul, négatif respectivement. On comprend par là ce qu'on entendra par droite ou direction de temps, de lumière, d'espace (droite ou direction isotrope). Un vecteur *de temps* est dit *unitaire* si son carré scalaire est égal à 1, tandis qu'un vecteur *unitaire d'espace* a son carré scalaire égal à -1.

Le *produit scalaire* de deux vecteurs $\mathbf{X}$ et $\mathbf{Y}$ aux composantes respectives X_k et Y_k est défini par la formule

$$(\mathbf{X}, \mathbf{Y}) = X_1 Y_1 - X_2 Y_2 - \cdots - X_m Y_m. \tag{4}$$

Ces vecteurs sont dits *orthogonaux au sens lorentzien* si leur produit scalaire $(\mathbf{X}, \mathbf{Y}) = 0$. En rapprochant cette définition de celles données plus haut, on voit que les vecteurs de lumière, et ces vecteurs seulement, sont orthogonaux à eux-mêmes.

Par une *transformation de Lorentz* on entend une transformation linéaire et homogène des coordonnées x_k qui transforme la forme métrique

$$x_1^2 - x_2^2 - \cdots - x_m^2 \tag{5}$$

en elle-même. Les composantes d'un vecteur sont à transformer de la même manière que les coordonnées, c'est-à-dire que le carré scalaire $(\mathbf{X}, \mathbf{X})$ et le produit scalaire $(\mathbf{X}, \mathbf{Y}) = \frac{1}{4}\{(\mathbf{X} + \mathbf{Y}, \mathbf{X} + \mathbf{Y}) - (\mathbf{X} - \mathbf{Y}, \mathbf{X} - \mathbf{Y})\}$ sont aussi invariants par rapport à toute transformation de Lorentz. Ces transformations forment évidemment un groupe: *le groupe de Lorentz*. Il y a donc invariance par rapport à ce groupe. On voit d'autre part facilement que les Y_k étant transformés d'une manière inconnue, mais telle que le produit scalaire $(\mathbf{X}, \mathbf{Y})$, formé avec des X_k arbitraires, reste in-

[1] Ce qu'on appelle ici *espace lorentzien* à m dimensions n'est que l'*espace-temps* ou l'*univers* des physiciens, à $m - 1$ dimensions d'espace et *une* dimension de temps.

variant lorsqu'on applique aux X_k une certaine transformation de Lorentz, la transformation des Y_k doit être identique à celle des X_k. En effet cette dernière règle conduit à l'invariance postulée. Inversement, le postulat d'invariance mène pour les coefficients de la transformation inconnue à un système d'équations linéaires à solution unique.

La nomenclature introduite plus haut (vecteur de temps, d'espace etc.) est manifestement indépendante du choix du système de coordonnées lorentziennes. Ajoutons que, par une transformation de Lorentz convenable, toute droite de temps peut devenir l'axe des x_1. De même, toute droite d'espace peut devenir un axe des x_i ($i = 2, 3, \ldots, m$). Tout plan d'espace (*voir* plus loin, n° **19**) peut devenir un plan $x_2 x_3 \ldots x_m$. La métrique lorentzienne dans un tel plan est — au signe du carré scalaire près — identique à la métrique euclidienne relative aux coordonnées $x_2, x_3, \ldots, x_m$. Cela explique le rôle que jouent les sphères euclidiennes dans certains de nos calculs (*voir* p. ex. les n^{os} **16**, **31** et *suiv.*).

En posant

$$\varepsilon_1 = 1, \quad \varepsilon_2 = \cdots = \varepsilon_m = -1, \tag{6}$$

les expressions (3) et (4) s'écrivent respectivement

$$(\mathbf{X}, \mathbf{X}) = \sum_{k=1}^{m} \varepsilon_k X_k^2, \tag{7}$$

$$(\mathbf{X}, \mathbf{Y}) = \sum_{k=1}^{m} \varepsilon_k X_k Y_k. \tag{8}$$

Dans la suite il nous sera utile de nous servir de certaines notions appartenant au calcul tensoriel. Dans ce calcul on désigne les composantes ordinaires X_k (qui se transforment de la même manière que les coordonnées, ou, quand il s'agit de coordonnées curvilignes, de la même manière que les différentielles des coordonnées) comme composantes *contravariantes*, et on les distingue d'habitude par des indices supérieurs (X^k). Cette dernière notation sera adoptée plus loin dans ce travail, notamment dans le dernier Chapitre; ici il nous paraît plus commode de garder la notation X_k.

Chaque forme métrique fait correspondre aux composantes contravariantes d'un vecteur des composantes *covariantes* et inversement. En renvoyant pour le mécanisme général audit chapitre nous nous restreignons ici aux formes métriques de l'aspect (7). Pour mieux élucider les choses, nous admettons cependant que les ε_k figurant dans (7) aient des valeurs constantes arbitraires, telles que

(9) $$\varepsilon_1 > 0, \ \varepsilon_i < 0 \ (i = 2, 3, \ldots, m) \quad \text{et} \quad |\varepsilon_1 \varepsilon_2 \ldots \varepsilon_m| = 1. \text{(}^1\text{)}$$

Le groupe de transformations admis sera donc un groupe de Lorentz légèrement généralisé, à savoir le groupe de transformations qui laissent invariante la forme métrique généralisée.

Nous posons, en désignant les composantes covariantes de **X** par X_k^{cov},

(10) $$X_k^{\text{cov}} = \varepsilon_k X_k \quad \text{et} \quad X_k = \varepsilon_k^{-1} X_k^{\text{cov}}.$$

Dans le cas particulier lorentzien (6) on a évidemment $\varepsilon_k^{-1} = \varepsilon_k$, ce qui fait que les calculs deviennent moins clairs si l'on se restreint au cas particulier.

Dans les notations posées on peut écrire le carré ou le produit scalaires sous les formes simples

(11) $$(\mathbf{X}, \mathbf{X}) = \Sigma X_k X_k^{\text{cov}}, \qquad (\mathbf{X}, \mathbf{Y}) = \Sigma X_k Y_k^{\text{cov}} = \Sigma X_k^{\text{cov}} Y_k.$$

Manifestement, toute transformation des X_k appartenant au groupe (généralisé) induit moyennant (10) une transformation correspondante des X_k^{cov}:

(12) $$X_i = \sum_k a_{ik} X'_k$$

donne

(12*bis*) $$X_i^{\text{cov}} = \sum_k \varepsilon_i \varepsilon_k^{-1} a_{ik} X_k'^{\text{cov}}. \text{(}^2\text{)}$$

Ces transformations (contragrédientes), qui en général sont différentes l'une de l'autre, laissent invariants le carré et le produit scalaires, le dernier, bien entendu, si l'on applique l'une des transformations aux composantes de **X** et l'autre aux composantes de **Y**. Inversement, en utilisant une remarque faite tout à l'heure on voit facilement que si les H_k admettent une loi de transformation telle que $\Sigma X_k \mathrm{H}_k$ formé avec des X_k arbitraires reste invariant lorsqu'on applique

---

(1) La dernière de ces conditions n'est posée que pour des raisons de commodité (*voir* la note (2) à la page 31 et le n° **21***bis*).

(2) Pour que la forme métrique (7) soit invariante par rapport à cette transformation il faut et il suffit que l'on ait $\sum_i \varepsilon_i a_{ik} a_{il} = \begin{cases} \varepsilon_k, & k = l, \\ 0, & k \neq l. \end{cases}$ En passant aux déterminants correspondants, on en tire $|a_{ik}|^2 \Pi \varepsilon_i = \Pi \varepsilon_i$, ou

$$|a_{ik}| = \pm 1.$$

aux X_k les transformations du groupe de Lorentz généralisé, les H_k sont les composantes covariantes d'un vecteur $\mathbf{Y}$, $\mathrm{H}_k = Y_k^{\mathrm{cov}}$.

Parmi les exemples les plus simples de composantes contravariantes d'un vecteur citons les composantes dx_k d'un déplacement infinitésimal $d\mathbf{x}$. D'autre part il résulte de ce que nous venons de dire que les quantités $\partial\varphi/\partial x_k$ sont les composantes covariantes d'un vecteur, à savoir du vecteur $\mathbf{grad}\,\varphi$. En effet, la différentielle totale

$$d\varphi = \sum \frac{\partial \varphi}{\partial x_k} dx_k$$

est manifestement invariante par rapport à toute transformation des coordonnées. Les composantes contravariantes du gradient de φ sont $\varepsilon_k^{-1}\partial\varphi/\partial x_k$:

$$\mathrm{grad}_k^{\mathrm{cov}}\,\varphi = \frac{\partial \varphi}{\partial x_k} \quad \text{et} \quad \mathrm{grad}_k\,\varphi = \varepsilon_k^{-1}\frac{\partial \varphi}{\partial x_k}.$$

On peut aussi exprimer cet état de choses en disant que les opérateurs $\partial/\partial x_k$ sont les composantes covariantes d'un vecteur symbolique, fait dont on verra une conséquence importante tout à l'heure.

Le carré et le produit scalaires exprimés par les formules (7), (8) et (11) peuvent évidemment aussi s'écrire

$$(13) \qquad (\mathbf{X}, \mathbf{X}) = \sum_k \varepsilon_k^{-1}(X_k^{\mathrm{cov}})^2 \quad \text{et} \quad (\mathbf{X}, \mathbf{Y}) = \sum \varepsilon_k^{-1} X_k^{\mathrm{cov}} Y_k^{\mathrm{cov}}.$$

En appliquant cela aux vecteurs aux composantes covariantes respectives $\partial\varphi/\partial x_k$ et $\partial/\partial x_k$, on voit que les expressions

$$(14) \qquad (\mathbf{grad}\,\varphi, \mathbf{grad}\,\varphi) = \Sigma\,\varepsilon_k^{-1}\left(\frac{\partial \varphi}{\partial x_k}\right)^2$$

et

$$(15) \qquad \Delta\varphi = \left(\Sigma\,\varepsilon_k^{-1}\frac{\partial^2}{\partial x_k^2}\right)\varphi = \Sigma\,\varepsilon_k^{-1}\frac{\partial^2 \varphi}{\partial x_k^2}$$

sont invariantes par rapport aux transformations du groupe de Lorentz généralisé. Il en résulte en particulier le fait qui est pour nous d'une importance capitale que *l'opérateur des ondes* (2) *est invariant par rapport aux transformations du groupe de Lorentz.*

Remarque. — Toutes nos formules restent invariantes, même si l'on passe d'une forme métrique $\Sigma\,\varepsilon_k x_k^2$ à une autre d'aspect analogue, dont les coefficients

ε'_k satisfont aux relations correspondant à (9). Un tel passage peut, en particulier, être réalisé par la substitution évidente $x_k = (\varepsilon'_k/\varepsilon_k)^{\frac{1}{2}} \cdot x'_k$. Dans le dernier Chapitre, où nous appliquons l'écriture tensorielle, on trouve nos calculs sous une forme qui est invariante par rapport à toute transformation ponctuelle des variables indépendantes.

15. Définition et premières propriétés de l'intégrale I^α.(¹) — $I^\alpha f(P)$ sera ici défini par une intégrale étendue à l'intérieur du *cône rétrograde* D^P (*cf.* p. 27 et fig. 2, p. 38). Il n'est donc pas nécessaire et pas même naturel de faire des hypothèses sur l'existence et l'allure de nos fonctions $f(P)$ dans l'espace entier, comme on a fait dans le cas euclidien. On supposera ces fonctions définies dans un domaine tel que, lorsque ce domaine renferme un certain point P, il renferme aussi le cône D^P. Pour assurer la convergence absolue des intégrales écrites, on supposera entre autres choses que $f(Q)$ tend vers zéro assez vite dans la partie infinie des cônes qui interviennent.

Cela étant, on pose

$$I^\alpha f(P) = \frac{1}{H_m(\alpha)} \int_{D^P} f(Q)\, r_{PQ}^{\alpha - m}\, d\,Q. \tag{16}$$ (²)

Bien entendu, le facteur $H_m(\alpha)$, qu'on va préciser tout à l'heure, est différent de celui qui figure dans le cas elliptique.

Nous avons d'abord à parler d'une difficulté — bien connue en principe depuis les travaux de Volterra et de M. Hadamard — difficulté que nous rencontrons ici pour la première fois. Supposons, pour simplifier, que P tombe à l'origine O et posons $\varrho^2 = \sum_{2}^{m} \xi_k^2$. Alors si ξ_1 reste constant et ϱ tend vers $|\xi_1|$, Q tend vers la frontière et en même temps r_{OQ} tend vers zéro de la même manière que $(|\xi_1| - \varrho)^{\frac{1}{2}}$. Par là on voit que pour assurer la convergence de nos intégrales nous aurons à supposer que l'exposant $\alpha - m > -2$, c'est-à-dire que $\alpha > m - 2$ $(m \geqq 2)$.

(¹) Les n^{os} **15—17** sont, pour simplifier certains calculs, rédigés en admettant les valeurs originales (6) du n° **14** des ε_k dans la forme lorentzienne. On voit immédiatement que les résultats essentiels restent valables pour la forme généralisée. En particulier, la valeur de $H_m(\alpha)$ reste inaltérée.

(²) Comme au chapitre précédent, on entend ici par dQ le produit $d\xi_1 d\xi_2 \ldots d\xi_m$, qui, d'après le n° **21**^{bis}, convient comme élément de volume à m dimensions dans l'espace lorentzien, même dans le cas où les ε_k admettent leurs valeurs générales, grâce à la dernière des conditions (9) du n° **14**. L'élément de volume est même invariant par les changements de forme métrique donnés dans la Remarque qui se trouve à la fin du n° **14**. *Cf.* aussi la note (¹) à la page 29.

Cette hypothèse étant admise, du moins jusqu'à nouvel ordre, nous allons montrer qu'on peut choisir $H_m(\alpha)$ de manière que les relations

$$I^\alpha(I^\beta f(P)) = I^{\alpha+\beta} f(P) \tag{17}$$

et

$$\Delta I^{\alpha+2} f(P) = I^\alpha f(P), \tag{18}$$

analogues à celles rencontrées plus haut dans le cas elliptique, aient lieu.(¹) L'opérateur Δ est défini par la formule (15). D'ailleurs, en ce qui concerne la seconde formule, elle ne sera démontrée que sous certaines conditions supplémentaires.

16. Calcul de $H_m(\alpha)$. — Le calcul de $H_m(\alpha)$ sera tout analogue au calcul fait dans le chapitre précédent et même plus simple en principe puisque la fonction particulière $f(P) = f(x_1, x_2, \ldots, x_m) = e^{x_1}$, moyennant laquelle le calcul sera exécuté, appartient ici à la classe de fonctions admise. Nous avons

$$I^\alpha e^{x_1} = \frac{1}{H_m(\alpha)} \int_{D^P} e^{\xi_1} r_{PQ}^{\alpha-m}\, dQ = \frac{e^{x_1}}{H_m(\alpha)} \int_{D^O} e^{\xi_1} r_{OQ}^{\alpha-m}\, dQ,$$

O signifiant l'origine. Si donc on pose $H_m(\alpha)$ égal à la dernière intégrale, il viendra

$$I^\alpha e^{x_1} = e^{x_1}. \tag{19}$$

Il reste à calculer $H_m(\alpha)$. En posant

$$\xi_1 = -r\cosh\theta \quad \text{et} \quad \varrho = (\xi_2^2 + \cdots + \xi_m^2)^{\frac{1}{2}} = r\sinh\theta,$$

l'intégrale à laquelle on a égalé $H_m(\alpha)$ se réduit à (*cf.* le n° **6**)

$$H_m(\alpha) = \frac{2 \cdot \pi^{\frac{m-1}{2}}}{\Gamma\left(\frac{m-1}{2}\right)} \int_0^\infty \sinh^{m-2}\theta \cdot d\theta \int_0^\infty e^{-r\cosh\theta} r^{\alpha-1}\, dr.$$

Moyennant la formule, facile à vérifier,

(¹) Comme au cas elliptique, le facteur $H_m(\alpha)$ est essentiellement déterminé par ces deux conditions. *Voir* la note (¹) à la p. 17 et la note (¹) à la p. 46.

$$\int_0^\infty \sinh^{p-1}\theta \cdot \cosh^{1-p-q}\theta \cdot d\theta = \frac{1}{2} \cdot \frac{\Gamma\left(\frac{p}{2}\right)\Gamma\left(\frac{q}{2}\right)}{\Gamma\left(\frac{p+q}{2}\right)}$$

et la formule (5) du n° 6, il en vient

$$H_m(\alpha) = \pi^{\frac{m-2}{2}} \cdot 2^{\alpha-1} \cdot \Gamma\left(\frac{\alpha}{2}\right)\Gamma\left(\frac{\alpha+2-m}{2}\right). \tag{20}$$

Notons que de la formule (19) il résulte immédiatement

$$I^\alpha e^{a_1 x_1 - a_2 x_2 - \cdots - a_m x_m} = (a_1^2 - a_2^2 - \cdots - a_m^2)^{-\frac{\alpha}{2}} \cdot e^{a_1 x_1 - a_2 x_2 - \cdots - a_m x_m}$$

chaque fois que l'expression entre parenthèses et a_1 sont positifs, ce qui montre que la formule (17) subsiste pour ces fonctions. On pourrait en tirer la validité de cette formule pour la classe générale de nos fonctions $f(P)$ en démontrant que les fonctions envisagées forment, dans un sens facile à préciser, un système complet dans cette classe. Cependant, nous suivrons ici une voie différente.

17. Formule de composition. — La formule fondamentale (17) étant valable pour la fonction e^{x_1}, elle le sera encore pour une fonction quelconque $f(P)$ remplissant nos conditions générales. En effet, il vient par un calcul calqué sur celui opéré dans la formule (3) du n° 1

$$\begin{aligned} I^\alpha(I^\beta f(P)) &= \frac{1}{H_m(\alpha)} \int_{D^P} r_{PQ}^{\alpha-m}\, dQ \cdot \frac{1}{H_m(\beta)} \int_{D^Q} f(\theta)\, r_{Q\theta}^{\beta-m}\, d\theta \\ &= \frac{1}{H_m(\alpha)\, H_m(\beta)} \int_{D^P} f(\theta)\, d\theta \int_{D^{P\theta}} r_{PQ}^{\alpha-m}\, r_{Q\theta}^{\beta-m}\, dQ, \end{aligned} \tag{21}$$

$D^{P\theta}$ étant le domaine commun au cône rétrograde au sommet P et au cône direct au sommet θ. Le fait est qu'un point θ étant situé dans le cône rétrograde au sommet Q, ce dernier point sera situé dans le cône direct au sommet θ et *vice versa* (fig. 1). La dernière intégrale se transforme par une translation et une homothétie en l'expression

$$r_{P\theta}^{\alpha+\beta-m} \int_{D^{O1}} r_{1T}^{\alpha-m}\, r_{TO}^{\beta-m}\, dT = r_{P\theta}^{\alpha+\beta-m}\, \mathrm{B}_m(\alpha, \beta),$$

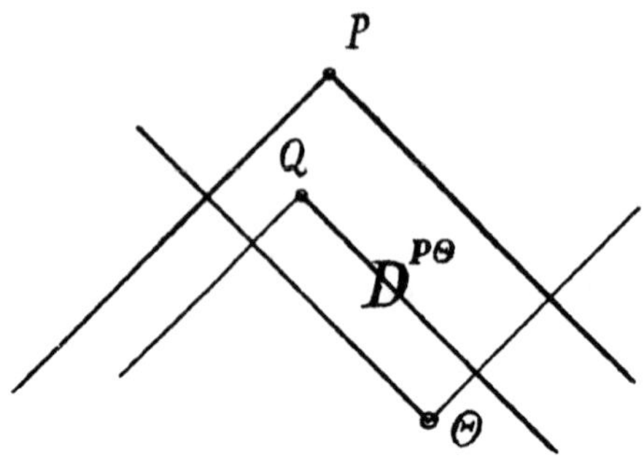

Fig. 1. Figure schématique se référant au cas de deux dimensions.

O désignant l'origine, 1 un certain point intérieur au cône direct appartenant à l'origine et ayant la distance lorentzienne 1 de l'origine. En appliquant ces formules à la fonction particulière $f(P) = e^{x_1}$, on trouve une seconde généralisation de l'intégrale d'Euler de première espèce

$$(22) \qquad \mathrm{B}_m(\alpha, \beta) = \frac{H_m(\alpha)\, H_m(\beta)}{H_m(\alpha + \beta)} \cdot \text{(1)}$$

En raison de cette relation, la formule fondamentale (17) se trouve établie.

II. Formule de Green.

18. Théorème de Gauss. — Dans la suite, nous ferons un grand usage de la formule de Green relative à l'opérateur des ondes. Elle se déduit, comme toute autre formule analogue, du théorème de Gauss, qui donne la transformation d'une certaine intégrale de volume dans une intégrale de surface.

Considérons dans un espace (affine) à m dimensions un domaine D limité par une ou plusieurs surfaces fermées T (à $m - 1$ dimensions) que nous supposons assez régulières. On a alors

$$(23) \qquad \int_D \frac{\partial P_k}{\partial \xi_k} d\xi_1 \, d\xi_2 \ldots d\xi_m = \int_T P_k \cdot (\pm \,| d\xi_1 \ldots d\xi_{k-1} \, d\xi_{k+1} \ldots d\xi_m |),$$

(1) La fonction hyperbolique $\mathrm{B}_m(\alpha, \beta)$ est évidemment bien plus rapprochée de la fonction eulérienne $\mathrm{B}(\alpha, \beta)$ que la fonction elliptique correspondante. La formule explicite montre que $\mathrm{B}_m(\alpha, \beta)$ est indépendant du point désigné par 1 pourvu que la distance lorentzienne de ce point à l'origine soit $= 1$. Une vérification directe de ce fait s'ensuit facilement par une transformation de Lorentz.

le signe étant *positif* aux points de T où les droites parallèles à l'axe des ξ_k, suivant lesquelles on exécute l'intégration par rapport à ξ_k, sortent du domaine D et *négatif* aux points où ces droites entrent dans le domaine.

La frontière T étant rapportée à un système (ou plusieurs systèmes) de paramètres $\lambda_1, \lambda_2, \ldots, \lambda_{m-1}$, on a l'identité connue

$$|d\xi_1 \ldots d\xi_{k-1}\, d\xi_{k+1} \ldots d\xi_m| = |J_k| \cdot d\lambda_1\, d\lambda_2 \ldots d\lambda_{m-1}, \tag{24}$$

où l'on attribue toujours au produit des $d\lambda_i$ une valeur positive et où les J_k sont les jacobiens avec des signes alternés

$$\begin{aligned} J_1 &= \frac{d(\xi_2, \xi_3, \ldots, \xi_m)}{d(\lambda_1, \lambda_2, \ldots, \lambda_{m-1})}, \\ J_2 &= -\frac{d(\xi_1, \xi_3, \ldots, \xi_m)}{d(\lambda_1, \lambda_2, \ldots, \lambda_{m-1})}, \\ &\ldots \quad \ldots \quad \ldots \quad \ldots \\ J_k &= (-1)^{k-1} \frac{d(\xi_1, \xi_2, \ldots, \xi_{k-1}, \xi_{k+1}, \ldots, \xi_m)}{d(\lambda_1, \lambda_2, \ldots \quad \ldots \quad \ldots \lambda_{m-1})}. \\ &\ldots \quad \ldots \quad \ldots \quad \ldots \end{aligned} \tag{25}$$

Tout réarrangement des λ_i conduisant à une permutation paire conserve les J_k, tandis que tout réarrangement conduisant à une permutation impaire change les signes de tous. Nous allons montrer tout à l'heure que l'ordre des λ_i peut être choisi de manière qu'en ajoutant les relations (23), le résultat s'exprime par la formule simple, due en principe à Gauss,[1]

$$\int_D \sum_1^m \frac{\partial P_k}{\partial \xi_k}\, d\xi_1\, d\xi_2 \ldots d\xi_m = \int_T \sum_1^m P_k J_k\, d\lambda_1\, d\lambda_2 \ldots d\lambda_{m-1}. \tag{26}$$

Avant d'aller à la démonstration, observons que l'expression figurant dans l'intégrale de surface peut aussi se mettre sous la forme d'un déterminant

$$\sum_1^m P_k J_k = \begin{vmatrix} P_1 & \ldots & P_k & \ldots & P_m \\ \cdot & \ldots & \cdot & \ldots & \cdot \\ \dfrac{\partial \xi_1}{\partial \lambda_i} & \ldots & \dfrac{\partial \xi_k}{\partial \lambda_i} & \ldots & \dfrac{\partial \xi_m}{\partial \lambda_i} \\ \cdot & \ldots & \cdot & \ldots & \cdot \end{vmatrix} \quad (i = 1, 2, \ldots, m-1). \tag{27}$$

---

(1) *Voir* aussi POINCARÉ, Acta mathematica 9 (1887) et GOURSAT, Journal de mathématiques (6) 4 (1908).

Évidemment, cette identité subsiste si l'on y remplace les P_k par m quantités quelconques.

Notons encore que, pour un déplacement (infinitésimal) $(d\xi_k)$ tangent à la surface, on a

$$\sum_1^m J_k\, d\xi_k = 0. \tag{28}$$

En effet, il en est ainsi pour les $m - 1$ déplacements linéairement indépendants $\left(\frac{\partial \xi_k}{\partial \lambda_i} d\lambda_i\right)$, puisque dans le déterminant correspondant deux colonnes deviennent proportionnelles l'une à l'autre.

Cela étant, un point de la frontière T est dit point régulier par rapport à la représentation paramétrique (λ_i), si les J_k ne s'annulent pas simultanément en ce point. En un tel point la frontière T admet un plan tangent (à $m - 1$ dimensions) déterminé et on aura $\Sigma J_k \delta\xi_k \neq 0$ pour tout déplacement infinitésimal $(\delta\xi_k)$ qui n'est pas situé dans ce plan tangent. Le signe de l'expression $\Sigma J_k \delta\xi_k$ sera différent suivant que le déplacement $(\delta\xi_k)$ est situé d'un côté ou de l'autre du plan tangent, c'est-à-dire suivant que $(\delta\xi_k)$ mène de la frontière à l'extérieur ou à l'intérieur du domaine. Cela posé, voici la règle définitive pour l'ordre des λ_i.

Les λ_i sont à numéroter de manière que l'expression $\Sigma J_k \delta\xi_k$, et dès lors le déterminant qui lui est égal, soient *positifs* dans les points réguliers de la frontière T pour tout[1] déplacement infinitésimal $(\delta\xi_k)$ qui d'un tel point mène à l'*extérieur* du domaine D.

Cette règle de signe se vérifie de la manière suivante. Dans la condition posée, on a $J_k > 0$ en tout point régulier qui est point de sortie d'une droite parallèle à l'axe des ξ_k et $J_k < 0$ en tout point régulier qui est point d'entrée d'une telle droite. En effet, un déplacement dont toutes les composantes sont nulles, sauf $\delta\xi_k$, mène à l'extérieur de D, si l'on a soin de prendre $\delta\xi_k$ positif ou négatif suivant qu'il s'agit d'un point de sortie ou d'entrée respectivement. La condition (de numérotage) étant remplie, on aura donc dans les deux cas $J_k \delta\xi_k > 0$, d'où notre énoncé s'ensuit. Dès lors, la formule (24) peut s'écrire sous la forme précisée

$$\pm\, | d\xi_1 \ldots d\xi_{k-1}\, d\xi_{k+1} \ldots d\xi_m | = J_k \cdot d\lambda_1\, d\lambda_2 \ldots d\lambda_{m-1}, \tag{29}$$

---

[1] D'après ce qu'on vient de voir, il *suffit* pour cela qu'il en soit ainsi pour un *seul* déplacement menant à l'extérieur.

le signe positif se référant encore aux points de sortie et le signe négatif aux points d'entrée. La dernière formule rend immédiat le passage de (23) à (26).

Par la loi de multiplication des déterminants fonctionnels il est clair que le second membre de la formule (26) est invariant par rapport à un changement de paramètres, si l'on tient aussi compte de la règle de signe. Rien n'empêche en outre que différentes portions de la frontière T soient repérées par différents systèmes de paramètres. Ces portions (en nombre fini) peuvent empiéter les unes sur les autres ou peuvent être séparées par des variétés à $m-2$ dimensions, qui sont supposées assez régulières. Dans un ordre d'idées plus général, on peut admettre un nombre fini de variétés exceptionnelles à différentes dimensions p, $0 \leqq p \leqq m-2$. Tel sera forcément le cas si la frontière elle-même admet des points exceptionnels, sans plan tangent déterminé, points anguleux isolés ou arêtes à différentes dimensions etc. La dernière circonstance se présente p. ex. chez les domaines extrêmement simples, limités par des plans parallèles aux plans de coordonnées. Nos domaines d'intégrations usuels, les domaines D_S^P, qu'on rencontrera dans le numéro suivant, présentent deux espèces de singularités, un point anguleux isolé, à savoir le sommet du cône D^P, et une arête à $m-2$ dimensions, l'intersection de la nappe C^P avec une surface S.

19. Domaine d'intégration. — Dans les applications qui vont suivre, le domaine général D sera remplacé par un domaine de caractère très particulier, savoir le domaine borné D_S^P (fig. 2) limité par la nappe C^P d'un cône rétrograde D^P et une certaine surface (ouverte) S, qui devra être assez régulière pour que les dérivations qui interviendront puissent être exécutées. Sauf avis contraire, nous admettrons aussi que S ait, suivant la terminologie de M. Hadamard (¹), une *orientation d'espace*. Cela veut dire que $\Sigma \varepsilon_k d\xi_k^2 < 0$ pour tout déplacement infinitésimal $(d\xi_k)$ le long de la surface. L'exemple le plus simple est fourni par les plans $\xi_1 = \text{const.}$ Géométriquement, une telle surface peut aussi être caractérisée par la propriété que son plan tangent en un point arbitraire Q n'a aucun point en commun avec le cône de lumière qui a son sommet en Q (excepté, bien entendu, le point Q lui-même). La *normale* (²) (lorentzienne) d'une surface à orientation d'espace est toujours une droite de temps (*cf.* n° **14**); propriété qui d'ailleurs pourrait également servir de définition. En effet, s'il n'en était pas ainsi, la

(¹) HADAMARD **1**, p. 53.

(²) Cette expression remplace, dans notre terminologie, celle de la *conormale* chez M. D'ADHÉMAR et de la *transversale* chez M. HADAMARD **1**, p. 87.

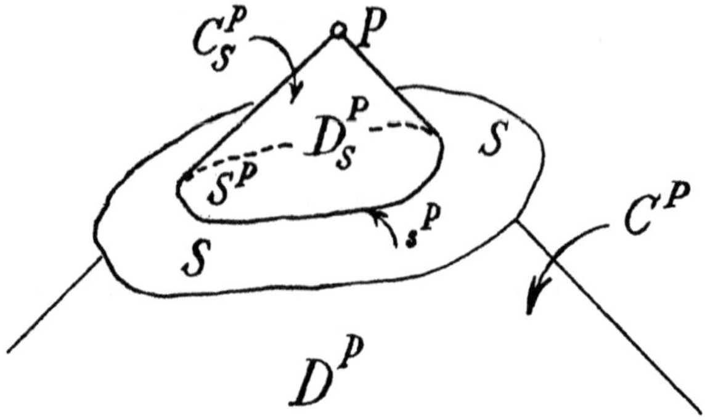

Fig. 2.

forme lorentzienne pourrait être transformée dans une forme qui ne contient que des carrés négatifs. On n'aurait qu'à choisir comme nouvelles directions d'axes $m-1$ directions orthogonales dans le plan tangent et la direction de la normale. Nous désignons par S^P la portion de la surface S qui est intérieure au cône D^P et par C_S^P la portion bornée de la nappe C^P découpée par la surface S. Notons encore le fait évident que toute droite de temps ou de lumière ne coupe une surface telle que S qu'en un seul point.

20. Formule de Green préliminaire. — Dans les applications auxquelles nous venons de faire allusion, les fonctions P_k s'évanouissent sur la portion de nappe C_S^P; la formule (26) se réduira donc à

$$(30) \qquad \int\limits_{D_S^P} \sum_1^m \frac{\partial P_k}{\partial \xi_k} \cdot d\xi_1 d\xi_2 \ldots d\xi_m = \int\limits_{S^P} \sum_1^m P_k J_k \cdot d\lambda_1 \ldots d\lambda_{m-1}.$$

Reprenons maintenant l'ordre d'idées du nº **14**. Posons (formule (15))

$$(31) \qquad \Delta_Q = \Delta = \sum_1^m \varepsilon_k^{-1} \frac{\partial^2}{\partial \xi_k^2}$$

et transformons l'intégrale de volume

$$\int\limits_{D_S^P} (u\,\Delta v - v\,\Delta u)\,d\,Q\ (^1)$$

dans une intégrale de surface. Il suffit de montrer que la fonction à intégrer peut se mettre sous la forme d'une divergence. Or on a

(1) *Voir* la note (2) à la page 31.

$$u \frac{\partial^2 v}{\partial \xi_k^2} - v \frac{\partial^2 u}{\partial \xi_k^2} = \frac{\partial}{\partial \xi_k} \left(u \frac{\partial v}{\partial \xi_k} - v \frac{\partial u}{\partial \xi_k} \right)$$

et

$$u \Delta v - v \Delta u = \sum_1^m \frac{\partial}{\partial \xi_k} \left\{ \varepsilon_k^{-1} \left(u \frac{\partial v}{\partial \xi_k} - v \frac{\partial u}{\partial \xi_k} \right) \right\}.$$

Dans les applications que nous avons en vue, v s'annule avec ses dérivées premières sur la nappe C^P; il vient donc, d'après la formule (30)

$$(32) \qquad \int_{D_S^P} (u \Delta v - v \Delta u) \, d Q = \int_{S^P} \sum_1^n \left(u \frac{\partial v}{\partial \xi_k} - v \frac{\partial u}{\partial \xi_k} \right) \varepsilon_k^{-1} J_k \, d \lambda_1 \, d \lambda_2 \ldots d \lambda_{m-1}.$$

21. Notions géométriques relatives à l'espace lorentzien. Remarque. — Au moyen de certains éléments géométriques on peut simplifier considérablement cette formule. Menons, en un point Q de la portion de surface S^P la normale, qui, comme on a vu plus haut (n° **19**), est une droite de temps. Soit $\mathbf{n}$ un vecteur quelconque porté par cette normale. Nous disons que les composantes covariantes n_k^{cov} de $\mathbf{n}$ sont proportionnelles aux quantités J_k. En effet, ces composantes sont, à un facteur de proportionnalité près, déterminées par l'équation

$$(\mathbf{n}, d \mathbf{x}) = \sum_1^m n_k^{\text{cov}} \, d x_k = 0,$$

valable pour tout déplacement $d \mathbf{x}$ le long de la surface, équation qui, en raison de la formule (28), est remplie par les quantités J_k. On a donc $n_k^{\text{cov}} = \nu \cdot J_k$. Des deux directions admissibles nous choisissons celle de la normale *intérieure*: $(\mathbf{n}, \delta \boldsymbol{\xi}) = \Sigma \, n_k \, \delta \xi_k > 0$ pour tout déplacement infinitésimal $\delta \boldsymbol{\xi}$ qui du point Q mène à l'intérieur de D_S^P. Comme on a (*voir* plus haut) $\Sigma \, J_k \, \delta \xi_k > 0$ pour tout déplacement infinitésimal qui mène à l'extérieur, le facteur de proportionnalité en question sera négatif. Si l'on exige encore que la normale soit unitaire, c'est-à-dire que $(\mathbf{n}, \mathbf{n}) = \Sigma \, \varepsilon_k^{-1} (n_k^{\text{cov}})^2 = 1$, il vient en définitive

$$(33) \qquad n_k^{\text{cov}} = - J_k \cdot J^{-1}$$

où

$$(34) \qquad J = \left(\sum_1^m \varepsilon_k^{-1} J_k^2 \right)^{\frac{1}{2}}$$

et pour les composantes contravariantes (formule (10) du n° **14**)

$$(35)\qquad n_k = -\varepsilon_k^{-1} J_k \cdot J^{-1}.\text{(}^1\text{)}$$

Remarquons que $\mathbf{n}$ étant un vecteur de temps, l'expression dont on a formé la racine carrée est certainement positive.

En restant dans le même ordre d'idées, ajoutons que si la surface S est donnée par l'équation $S(\xi_1, \xi_2, \ldots, \xi_m) = 0$, avec $\partial S/\partial \xi_1 > 0$[2], les composantes covariantes de la normale intérieure seront évidemment données par les formules

$$(36)\qquad n_k^{\text{cov}} = \frac{\partial S}{\partial \xi_k} \cdot N^{-1},$$

où

$$(37)\qquad N = \left\{ \sum_1^m \varepsilon_k^{-1} \left(\frac{\partial S}{\partial \xi_k}\right)^2 \right\}^{\frac{1}{2}} = (\textbf{grad}\ S,\ \textbf{grad}\ S)^{\frac{1}{2}},$$

et les composantes contravariantes par les formules

$$(38)\qquad n_k = \varepsilon_k^{-1} \frac{\partial S}{\partial \xi_k} \cdot N^{-1}.$$

Nous retournons maintenant à la formule (32). Nous posons par définition

$$(39)\qquad J \cdot d\lambda_1\, d\lambda_2 \ldots d\lambda_{m-1} = dS$$

= *élément de surface lorentzien*[3] autour du point $\xi_1, \xi_2, \ldots, \xi_m$.

F désignant une fonction quelconque, nous introduisons la notation

$$(40)\qquad \sum_1^m \frac{\partial F}{\partial \xi_k} n_k = \frac{dF}{dn}$$

= *dérivée de F dans la direction et le sens de la normale intérieure.*[4]

---

(1) La valeur de la composante n_1 et le coefficient ε_1^{-1} étant positifs tous les deux, on a forcément $J_1 < 0$. Ceci peut, pour la surface S, remplacer la règle de signe assez compliquée qu'on a donnée pour les J_k au n° **18**.

(2) Les $\partial S/\partial \xi_k$ étant les composantes covariantes d'un vecteur de temps, $\partial S/\partial \xi_1$ ne peut s'annuler et dès lors, variant d'une manière continue, ne peut changer de signe. On pourra donc s'arranger de manière que $\partial S/\partial \xi_1$ soit positif.

(3) *Voir* le n° **21**^{bis} et, en particulier, les formules (53) et (54).

(4) *Voir* la *Remarque* qui suit.

Ces notations posées et en raison de la formule (35), on peut écrire

$$\sum \frac{\partial F}{\partial \xi_k} \cdot \varepsilon_k^{-1} J_k \cdot d\lambda_1 d\lambda_2 \ldots d\lambda_{m-1} = \sum \frac{\partial F}{\partial \xi_k} \cdot (\varepsilon_k J)^{-1} J_k \cdot J\, d\lambda_1 d\lambda_2 \ldots d\lambda_{m-1}$$

$$= \sum \frac{\partial F}{\partial \xi_k} n_k \cdot dS = \frac{dF}{dn} dS$$

et dès lors la formule (32) du n° **20** se réduit à

$$\int_{D_S^P} (u \Delta v - v \Delta u)\, dQ = - \int_{S^P} \left(u \frac{dv}{dn} - v \frac{du}{dn} \right) dS. \tag{41}$$

Remarque. — D'une manière analogue à ce qu'on vient de faire pour le vecteur unitaire **n**, on peut définir la «dérivée» de la fonction $F(\xi_1, \xi_2, \ldots, \xi_m)$ dans toute direction non isotrope et de sens déterminé. La direction et le sens étant représentés par le vecteur unitaire **l**, aux composantes contravariantes l_k, nous posons

$$\frac{dF}{dl} = \sum_1^m \frac{\partial F}{\partial \xi_k} l_k = (\mathbf{grad}\, F, \mathbf{l}). \tag{42}$$

Nous allons démontrer que cette expression, introduite d'une manière tout à fait formelle, peut, tout comme dans l'espace euclidien, s'interpréter comme une dérivée dans le sens propre du mot.

Que le vecteur **l** soit unitaire ou non ou même isotrope, on a toujours

$$\lim_{\varepsilon \to 0} \frac{F(\xi_1 + \varepsilon l_1, \ldots, \xi_m + \varepsilon l_m) - F(\xi_1, \ldots, \xi_m)}{\varepsilon} = \sum \frac{\partial F}{\partial \xi_k} l_k. \tag{43}$$

En introduisant le vecteur $\boldsymbol{\xi}$ aux composantes ξ_k, ceci peut aussi s'écrire

$$\lim_{\varepsilon \to 0} \frac{F(\boldsymbol{\xi} + \varepsilon \mathbf{l}) - F(\boldsymbol{\xi})}{\varepsilon} = \sum \frac{\partial F}{\partial \xi_k} l_k. \tag{44}$$

Le vecteur $|\varepsilon| \cdot \mathbf{l}$ est un vecteur infinitésimal $d\mathbf{l}$ qui a la direction et le sens de **l**.[1] Nous posons $|d\mathbf{l}| = |(d\mathbf{l}, d\mathbf{l})|^{\frac{1}{2}}$. Si **l** est *unitaire* il vient $|d\mathbf{l}| = |\varepsilon|$. La relation ci-dessus peut donc aussi s'écrire

$$\lim_{d\mathbf{l} \to 0} \frac{F(\boldsymbol{\xi} \pm d\mathbf{l}) - F(\boldsymbol{\xi})}{\pm |d\mathbf{l}|} = \sum \frac{\partial F}{\partial \xi_k} l_k. \tag{45}$$

---

[1] Ces propriétés sont caractéristiques et $d\mathbf{l}$ pourrait aussi être introduit sans l'intermédiaire de la quantité auxiliaire ε comme un vecteur ayant la direction et le sens de **l** et tendant vers zéro.

Il est donc naturel de considérer le second membre de (42) comme la dérivée de F dans la direction et le sens de $\mathbf{l}$.

Notre formule de dérivation (42) donne en particulier $d\xi_k/dl = l_k$; on a donc aussi, comme il fallait s'y attendre,

$$\frac{dF}{dl} = \sum_1^m \frac{\partial F}{\partial \xi_k} \cdot \frac{d\xi_k}{dl}. \tag{46}$$

Considérons à titre d'exemple la fonction

$$F = R_{PQ} = r_{PQ}^2 = \sum_1^m \varepsilon_k (\xi_k - x_k)^2.$$

Le gradient de $R_{PQ} = R$ par rapport au point Q a les composantes covariantes

$$\frac{\partial R}{\partial \xi_k} = 2\,\varepsilon_k (\xi_k - x_k). \tag{47}$$

On aura donc

$$\frac{1}{2}\frac{dR}{dl} = \sum_1^m \varepsilon_k (\xi_k - x_k)\, l_k = (\mathbf{r}_{PQ}, \mathbf{l}) \tag{48}$$

qui n'est que le produit scalaire du vecteur $\mathbf{r}_{PQ}$ allant de P à Q et du vecteur $\mathbf{l}$.

21bis. Mesure des éléments d'une variété de dimension arbitraire. — Soient $\mathbf{X}, \mathbf{Y}, \ldots p$ vecteurs, où $p \leqq m$. Nous définissons le volume V du parallélépipède construit sur ces vecteurs par la formule

$$V = |G|^{\frac{1}{2}} \text{ où } G = \begin{vmatrix} (\mathbf{X},\mathbf{X}) & (\mathbf{X},\mathbf{Y}) & \ldots \\ (\mathbf{Y},\mathbf{X}) & (\mathbf{Y},\mathbf{Y}) & \ldots \\ \cdot & \cdot & \cdot\,\cdot \end{vmatrix}. \tag{49}$$ [1]

Cette définition se justifie par les faits suivants. 1°. L'expression de V est indépendante de l'ordre des vecteurs; 2°. elle est invariante par toute transformation de coordonnées (il en étant ainsi des produits scalaires); 3°. elle est encore invariante si, en gardant $\mathbf{X}$, on remplace $\mathbf{Y}$ par $\mathbf{Y} + \lambda\mathbf{X}$, $\mathbf{Z}$ par $\mathbf{Z} + \mu\mathbf{X} + \nu\mathbf{Y}, \ldots$; 4°. elle se multiplie par $|\lambda\mu\ldots|$ si l'on multiplie $\mathbf{X}, \mathbf{Y}, \ldots$ par $\lambda, \mu, \ldots$

[1] Nous faisons ici abstraction du fait que, dans le cas particulier $p = m$, où le volume peut aussi se définir par le déterminant formé des composantes des vecteurs considérés (cas spécial de la formule (51)), on peut sans ambiguïté attribuer au volume le signe + ou —, ce signe dépendant de l'ordre des vecteurs.

respectivement; 5°. enfin elle se réduit au produit des «longueurs» des arêtes, $|(\mathbf{X}, \mathbf{X})|^{\frac{1}{2}}$, $|(\mathbf{Y}, \mathbf{Y})|^{\frac{1}{2}}, \ldots$, dans le cas où les vecteurs donnés sont orthogonaux l'un à l'autre $((\mathbf{X}, \mathbf{Y}) = \cdots = 0)$. Ajoutons que si l'on peut répartir les vecteurs donnés en deux systèmes *totalement orthogonaux* $\mathbf{X}', \mathbf{Y}', \ldots$ et $\mathbf{X}'', \mathbf{Y}'', \ldots$ de p' et p'' vecteurs respectivement $(p' + p'' = p)$, c'est-à-dire tels que chaque vecteur du premier système soit orthogonal à chaque vecteur du second [1], on aura, dans des notations évidentes, $G = G' G''$ et alors

$$V = V' \cdot V''. \tag{50}$$

En effet, $(\mathbf{X}', \mathbf{X}'') = (\mathbf{X}', \mathbf{Y}'') = \cdots = 0$.

Il nous faut encore exprimer ces volumes par des déterminants formés des composantes de nos vecteurs. Or, on sait, d'après le théorème de Cauchy-Binet [2], que le déterminant $|c_{ik}|$ aux éléments

$$c_{ik} = a_{i1} b_{k1} + a_{i2} b_{k2} + \cdots + a_{im} b_{km} \qquad (i, k = 1, 2, \ldots, p)$$

est égal à

$$\sum \begin{vmatrix} a_{1t}\, a_{1u} \ldots \\ \cdot \quad \cdot \quad \ldots \\ a_{pt}\, a_{pu} \ldots \end{vmatrix} \cdot \begin{vmatrix} b_{1t}\, b_{1u} \ldots \\ \cdot \quad \cdot \quad \ldots \\ b_{pt}\, b_{pu} \ldots \end{vmatrix},$$

la sommation étant étendue aux $\binom{m}{p}$ combinaisons de p indices, $t, u, \ldots$, tirés de $1, 2, \ldots, m$. Ce point admis, on aura

$$G = \sum \varepsilon_t \varepsilon_u \ldots \begin{vmatrix} X_t\, X_u \ldots \\ Y_t\, Y_u \ldots \\ \cdot \quad \cdot \quad \ldots \end{vmatrix}^2, \tag{51}$$

les éléments des derniers déterminants étant les composantes des vecteurs $\mathbf{X}, \mathbf{Y}, \ldots$ En particulier, pour $p = m - 1$, il vient, en raison de la relation

$$\varepsilon_1 \ldots \varepsilon_{k-1} \varepsilon_{k+1} \ldots \varepsilon_m = (-1)^{m-1} \varepsilon_k^{-1}$$

(*cf.* formule (9) du n° **14**),

---

[1] Dans le cas de $m = 3$, p. ex., les droites parallèles à l'axe des coordonnées ξ_1 et les plans parallèles au plan $\xi_2 \xi_3$ sont totalement orthogonaux, mais il n'en est pas ainsi des plans (orthogonaux) parallèles au plan $\xi_1 \xi_2$, d'une part, et au plan $\xi_2 \xi_3$, d'autre part. D'une manière générale, en 3 dimensions il n'existe pas de plans totalement orthogonaux. Par contre, en 4 dimensions, les plans parallèles aux plans $\xi_1 \xi_2$ et $\xi_3 \xi_4$, respectivement, sont totalement orthogonaux.

[2] *Cf.* p. ex. R. Baltzer, *Theorie und Anwendung der Determinanten* (1881), p. 48.

$$G = (-1)^{m-1} \sum \varepsilon_w^{-1} \begin{vmatrix} X_t X_u \ldots \\ Y_t Y_u \ldots \\ \cdot \quad \cdot \quad \cdot\cdot\cdot \end{vmatrix}^2, \tag{52}$$

w étant celui des indices 1, 2, ..., m, qui manque dans la combinaison $t, u, \ldots$

Considérons maintenant une variété (courbée) S à p dimensions, dont les points Q (ξ_k) sont rapportés aux paramètres $\lambda_1, \lambda_2, \ldots, \lambda_p$. En désignant encore le vecteur aux composantes ξ_k par $\boldsymbol{\xi}$, l'élément de volume dS de cette variété sera défini par

$$d S = |\gamma|^{\frac{1}{2}} \cdot d\lambda_1 d\lambda_2 \ldots d\lambda_p, \tag{53}$$

où

$$\gamma = |\gamma_{ik}| = \left|\left(\frac{\partial \boldsymbol{\xi}}{\partial \lambda_i}, \frac{\partial \boldsymbol{\xi}}{\partial \lambda_k}\right)\right| (i, k = 1, 2, \ldots, p) = \sum \varepsilon_t \varepsilon_u \ldots \begin{vmatrix} \frac{\partial \xi_t}{\partial \lambda_1} \frac{\partial \xi_u}{\partial \lambda_1} \ldots \\ \frac{\partial \xi_t}{\partial \lambda_2} \frac{\partial \xi_u}{\partial \lambda_2} \ldots \\ \cdot \quad \cdot \quad \cdot\cdot\cdot \end{vmatrix}^2. \tag{54}$$

La première forme de γ, exprimée par des produits scalaires, met en évidence que $dS = |\gamma|^{\frac{1}{2}} \cdot d\lambda_1 \ldots d\lambda_p$ est invariant par rapport aux transformations des coordonnées ξ_k. On voit par la seconde forme de γ qu'il y a aussi invariance par rapport à toute transformation de paramètres, grâce aux propriétés de transformations des jacobiens d'une part, et de l'élément $d\lambda_1 \ldots d\lambda_p$ de l'autre.

Il arrive souvent qu'on peut répartir les indices $k = 1, 2, \ldots, m$ en deux systèmes (k') et (k'') tels que la variété où seuls les $\lambda_{k'}$ et celle où seuls les $\lambda_{k''}$ varient soient *totalement orthogonales* l'une à l'autre, c'est-à-dire telles que deux déplacements quelconques, en un point, de la forme

$$\sum \frac{\partial \boldsymbol{\xi}}{\partial \lambda_{k'}} d\lambda_{k'} \quad \text{et} \quad \sum \frac{\partial \boldsymbol{\xi}}{\partial \lambda_{k''}} d\lambda_{k''},$$

soient orthogonaux, ce qui revient à $\gamma_{k'k''} = 0$ pour tout couple k' et k''. Dans ces conditions, on aura, en vertu de la formule (50) et dans une notation évidente,

$$d S = d S' \cdot d S''. \tag{55}$$

Pour le cas $p = m - 1$ on a une formule analogue à (52), ce qui justifie notre définition de dS au n° **21** par la formule (39). L'élément de volume à m

dimensions devient d'après (51), si l'on introduit comme paramètres les ξ_k eux-mêmes, $d\,Q = |\varepsilon_1 \ldots \varepsilon_m|^{\frac{1}{2}} \cdot d\,\xi_1 \ldots d\,\xi_m$, ce qui se réduit à $d\,Q = d\,\xi_1 \ldots d\,\xi_m$ par la dernière des conditions (9) du n° **14**.

L'élément d'arc $d\,\sigma$ est dans notre variété donné par

$$(56)\qquad d\sigma^2 = |(d\,\boldsymbol{\xi}, d\,\boldsymbol{\xi})| = \left|\left(\sum \frac{\partial \boldsymbol{\xi}}{\partial \lambda_k} d\,\lambda_k, \sum \frac{\partial \boldsymbol{\xi}}{\partial \lambda_k} d\,\lambda_k\right)\right| = \left|\sum_{i,k=1}^{p} \gamma_{ik}\, d\,\lambda_i\, d\,\lambda_k\right|.$$

Par conséquent, ayant obtenu par un calcul quelconque l'expression de $d\sigma^2$ en (valeur absolue d'une) forme quadratique des $d\lambda_k$, on connaîtra par là automatiquement les γ_{ik} et dès lors l'élément de volume $d\,S = |\gamma|^{\frac{1}{2}} \cdot d\,\lambda_1 \ldots d\,\lambda_p$.

22. Formule de Green pour I^α. — Nous arriverons à la formule de Green relative à nos intégrales I^α en posant dans la formule (41) du n° **21**

$$(57)\qquad v = \frac{r_{PQ}^{\alpha+2-m}}{H_m(\alpha+2)},$$

$P = (x_k)$ et $Q = (\xi_k)$ étant deux points pour lesquels $r_{PQ}^2 > 0$. Nous disons qu'en définissant, comme au n° **14**, r_{PQ}^2 et Δ par les formules

$$r_{PQ}^2 = r^2 = \sum_1^m \varepsilon_k (\xi_k - x_k)^2 \quad \text{et} \quad \Delta_Q = \Delta = \sum_1^m \varepsilon_k^{-1} \frac{\partial^2}{\partial \xi_k^2},$$

on a, *indépendamment* des ε_k,

$$\Delta\, r^{\alpha+2-m} = \alpha(\alpha + 2 - m) \cdot r^{\alpha-m}.$$

En effet,

$$(57^{bis})\qquad \frac{\partial}{\partial \xi_k} r^{\alpha+2-m} = (\alpha + 2 - m)\, \varepsilon_k\, r^{\alpha-m} (\xi_k - x_k)$$

et

$$\varepsilon_k^{-1} \frac{\partial^2}{\partial \xi_k^2} r^{\alpha+2-m} = (\alpha + 2 - m)\, \varepsilon_k^{-1} \varepsilon_k \frac{\partial}{\partial \xi_k} \{ r^{\alpha-m} (\xi_k - x_k) \}$$
$$= (\alpha + 2 - m) \{ r^{\alpha-m} + (\alpha - m)\, r^{\alpha-m-2} \varepsilon_k (\xi_k - x_k)^2 \},$$

d'où

$$\Delta\, r^{\alpha+2-m} = (\alpha + 2 - m) \{ m\, r^{\alpha-m} + (\alpha - m)\, r^{\alpha-m-2} \cdot r^2 \} = \alpha(\alpha + 2 - m)\, r^{\alpha-m}.$$

Si donc $H_m(\alpha)$ satisfait à la relation

(58) $$H_m(\alpha+2) = \alpha(\alpha+2-m)\,H_m(\alpha)\,(^1),$$

il vient

(59) $$\Delta \frac{r^{\alpha+2-m}}{H_m(\alpha+2)} = \frac{r^{\alpha-m}}{H_m(\alpha)}\cdot(^2)$$

Tel est donc le cas pour la fonction $H_m(\alpha)$ du n° **16** (formule (20)) relative au cas lorentzien que nous avons à envisager ici.

Avant d'aller plus loin, observons que dans (59) Δ peut désigner soit Δ_P, soit Δ_Q. Pour $\alpha > m$ la fonction (57) s'annule avec ses dérivées du premier ordre sur la nappe C^P du cône D^P; elle pourra donc bien jouer le rôle de la fonction v dans la formule (41). Cela étant, on trouve par un réarrangement évident

(60) $$\frac{1}{H_m(\alpha)}\int\limits_{D_S^P} u(Q)\,r_{PQ}^{\alpha-m}\,dQ = \frac{1}{H_m(\alpha+2)}\int\limits_{D_S^P} \Delta u(Q)\,r_{PQ}^{\alpha+2-m}\,dQ$$
$$+\frac{1}{H_m(\alpha+2)}\int\limits_{S^P}\left(\frac{du(Q)}{dn}\,r_{PQ}^{\alpha+2-m} - u(Q)\,\frac{d\,r_{PQ}^{\alpha+2-m}}{dn}\right)dS, \qquad (\alpha > m).$$

23. L'intégrale I_*^α. — Dans les intégrales ci-dessus il ne figure que les valeurs que u et ses dérivées du premier et du second ordre admettent à l'intérieur et sur la frontière des domaines D_S^P auxquels la formule (60) sera ap-

(¹) L'équation aux différences (58) est remplie soit par la fonction $H_m(\alpha)$ du n° **16** que nous sommes en train de considérer et que nous désignons ici par $H_m^{\text{hyp.}}(\alpha)$, soit par la fonction, écrite dans une notation correspondante,

$$\overline{H}_m^{\text{ell.}}(\alpha) = e^{\frac{i\pi\alpha}{2}}\cdot H_m^{\text{ell.}}(\alpha) = e^{\frac{i\pi\alpha}{2}}\cdot\pi^{\frac{m}{2}}\cdot 2^\alpha\cdot\Gamma\left(\frac{\alpha}{2}\right):\Gamma\left(\frac{m-\alpha}{2}\right),$$

qui est définie par la formule (7) du n° **6** et dans la note (¹) p. 21 (*cf.* aussi la formule (12) du n° **8**). Le rapport de ces deux fonctions devra donc être une fonction périodique à la période 2. On trouve en effet par un calcul simple

$$\overline{H}_m^{\text{ell.}}(\alpha) : H_m^{\text{hyp.}}(\alpha) = 2\cdot e^{\frac{i\alpha\pi}{2}}\,\sin\frac{(m-\alpha)\pi}{2}.$$

(²) Moyennant la notation $r_{PQ}^{\alpha-m} / H_m(\alpha) = V^\alpha(P, Q)$, cette formule s'écrit $\Delta\, V^{\alpha+2} = V^\alpha$. Bien d'autres formules peuvent à l'aide de cette notation se mettre sous des formes plus condensées; *cf.* le dernier Chapitre.

pliquée. Ces domaines sont tous situés d'un certain côté de la surface S, celui qui est tourné vers les ξ_1 indéfiniment croissants. Nous pourrons donc sans aucun inconvénient admettre que u s'annule identiquement du côté opposé de la surface. Cela posé, les intégrales étendues aux domaines D_S^P seront identiques aux intégrales correspondantes étendues aux cônes D^P et pourront être désignées par $I^\alpha u(P)$ et $I^{\alpha+2} \Delta u(P)$ respectivement.

Remarquons en passant que la relation $I^\alpha I^\beta = I^{\alpha+\beta}$ subsiste si l'on pose *a priori*

$$I^\alpha f(P) = \frac{1}{H_m(\alpha)} \int\limits_{D_S^P} f(Q)\, r_{PQ}^{\alpha-m}\, dQ,$$

la surface S *ayant une orientation d'espace*.[1] En effet, lorsque $f(Q) = 0$ «au-delà de» S, il en est de même de $I^\beta f(Q)$, dans la condition qu'on vient de préciser. Par cela on a assurément

$$I^\alpha_{D_S^P}\left(I^\beta_{D_S^P}\right) = I^\alpha_{D^P}\left(I^\beta_{D^P}\right) = I^{\alpha+\beta}_{D^P} = I^{\alpha+\beta}_{D_S^P}.$$

La formule (60) qui contient des intégrales spatiales, une intégrale de simple couche et une intégrale de double couche, peut se mettre sous une forme plus condensée. En effet, en posant d'une manière générale

$$(61)\quad I_*^\alpha \overline{f, g, h}(P) = \frac{1}{H_m(\alpha)} \int\limits_{D_S^P} f(Q)\, r_{PQ}^{\alpha-m}\, dQ + \frac{1}{H_m(\alpha)} \int\limits_{S^P} \left(g(Q)\, r_{PQ}^{\alpha-m} - h(Q) \frac{d\, r_{PQ}^{\alpha-m}}{dn}\right) dS,$$

on peut écrire

$$(61^{bis})\qquad I^\alpha u(P) = I_*^{\alpha+2} \overline{\Delta u, \frac{du}{dn}, u(P)}.$$

24. Propriétés de I_*^α. — Avant d'aller plus loin, notons encore les propriétés suivantes du symbole I_*

$$(62)\qquad I^\alpha I_*^\beta = I_*^{\alpha+\beta},$$

$$(63)\qquad \Delta I_*^{\alpha+2} = I_*^\alpha$$

---

[1] Dans l'édition originale de notre Conférence, servant d'Introduction à cet ouvrage, il y a une erreur sur ce point en ce qu'on y énonce ce fait sous des conditions sensiblement moins restrictives. L'erreur est éliminée dans la présente édition de cette Conférence.

et

(64) $$I_*^0 \overline{f, g, h}(P) = f(P).$$

La première propriété se démontre pour $\alpha > m-2$ et $\beta > m-2$ de la même manière que la propriété correspondante du symbole I^α (n^os **16**, **17**). L'extension à d'autres valeurs de α et de β se fait par prolongement analytique. La démonstration de la deuxième propriété, facile pour d'assez grandes valeurs de $\alpha\,(\alpha > m)$, sera ajournée jusqu'au moment où nous aurons donné des méthodes pour effectuer les prolongements en question (n° **42**). Quant à la formule (64), d'importance capitale, sur laquelle repose notre méthode de solution(¹) du problème de Cauchy (n° **44**), il n'a aucun sens pour le moment. En effet, les deux premières intégrales figurant dans I_*^α ne convergent que pour $\alpha > m-2$ (n° **15**) et la dernière ne converge que pour $\alpha > m$ (*voir* le numéro suivant en observant que là il est question de la convergence de $I_*^{\alpha+2}$). I_*^0 n'a donc aucun sens avant qu'on ait montré que dans des conditions convenables le prolongement analytique par rapport à α peut être effectué jusqu'à $\alpha = 0$ (n° **37**).

Dans le cas elliptique, la formule de Green et ses itérées livraient un instrument utile pour effectuer le prolongement analytique des intégrales divergeant pour certaines valeurs de α. Or nous allons voir à l'instant (n° **25**) que dans le cas actuel la formule de Green ne se prête guère à cette tâche. Dans ces circonstances nous nous dispensons d'écrire la formule itérée de Green analogue à la formule (20) du n° **11**. C'est dans les n^os **27**—**37** que nous indiquerons une méthode efficace qui nous donne le prolongement en question. Remarquons néanmoins que le principe de prolongement nous permet sur le champ d'étendre la validité de la formule de Green elle-même, établie pour $\alpha > m$, à toute valeur de $\alpha > m-2$. En effet, pour les raisons indiquées tout à l'heure, les intégrales figurant dans (60) convergent pour $\alpha > m-2$ et sont évidemment des fonctions holomorphes de α.

25. Convergence des intégrales. — Examinons maintenant de plus près la formule de Green (60) du n° **22** sous le point de vue de la convergence (divergence) des intégrales qui y interviennent. Nous savons déjà (n° **15**) que l'intégrale figurant au premier membre diverge en général pour $\alpha \leqq m-2$ et que d'autre part les deux premières intégrales dans le second membre ne divergent que pour $\alpha \leqq m-4$. Autant qu'on ne considère que ces deux intégrales, on a donc, par la transformation de Green, amélioré la convergence de l'intégrale

(¹) En réalité, dans cette solution on n'utilise que la relation un peu moins générale $I^0 f(P) = f(P)$.

figurant au premier membre. Il en est tout autrement de la dernière intégrale du caractère d'un potentiel de double couche. En effet, on voit par la formule (40) du n° 21 que le noyau de double couche s'exprime par les dérivées du premier ordre de $r^{\alpha+2-m}$, fonctions qui, d'après la formule (57bis) du n° **22**, présentent au voisinage de l'intersection de la nappe de D^P et de S la singularité $r^{\alpha-m}$, c'est-à-dire que l'intégrale examinée ne converge que pour $\alpha > m - 2$. Ceci met en évidence que cette intégrale diverge en général pour les mêmes valeurs de α que celle du premier membre, ce qui montre qu'en fin de compte on n'a rien gagné en ce qui concerne la convergence.

26. L'intégrale I^α étendue au cône rétrograde entier. (1) — Quant à l'application de la formule de Green au prolongement analytique de $I^\alpha f(P)$, on se trouve dans des circonstances bien plus favorables dans le cas particulier où le domaine d'intégration D^P_S se confond avec le cône entier D^P. Dans des conditions (suffisantes) relatives à f et à ses dérivées, qu'il serait facile de spécifier, il vient en effet, par un passage à la limite de la formule (60) du n° 22,

$$(65)\qquad \frac{1}{H_m(\alpha)}\int\limits_{D^P} f(Q)\, r_{PQ}^{\alpha-m}\, dQ = \frac{1}{H_m(\alpha+2)}\int\limits_{D^P} \Delta f(Q)\, r_{PQ}^{\alpha+2-m}\, dQ$$

ou

$$I^\alpha f(P) = I^{\alpha+2}\,\Delta f(P).$$

On en obtient par itération

$$(66)\qquad \frac{1}{H_m(\alpha)}\int\limits_{D^P} f(Q)\, r_{PQ}^{\alpha-m}\, dQ = \frac{1}{H_m(\alpha+2p)}\int\limits_{D^P} \Delta^p f(Q)\, r_{PQ}^{\alpha+2p-m}\, dQ$$

ou

$$(67)\qquad I^\alpha f(P) = I^{\alpha+2p}\,\Delta^p f(P).$$

Dans les conditions auxquelles on a fait allusion, le second membre de (66) converge, pour toute valeur de $\alpha > m - 2 - 2p$ (n° **15**). Ainsi dans le cas actuel, le prolongement analytique de $I^\alpha f(P)$ se trouve effectué pour toutes ces valeurs.

III. Prolongement analytique. (2)

27. Un système de coordonnées. — Dans la suite, je donne une démonstration explicite de la possibilité du prolongement analytique dans le cas général,

(1) *Cf.* aussi le n° **37**.

(2) Pour obtenir quelques simplifications dans les calculs qui vont suivre, nous admettons dans les n^{os} **27—38** (=Chap. III_{III}) que les ε_k ont leurs valeurs originales, $\varepsilon_1 = 1$, $\varepsilon_2 = \cdots = \varepsilon_m = -1$. Le cas général se réduit facilement au cas particulier. (*Cf.* la *Remarque* à la fin du n° **14**.)

la première que je publie. Les deux démonstrations de ce fait, publiées jusqu'ici, sont dues à mon ancien élève, M. Fremberg[1]. Tout en suivant ici une voie assez différente de la sienne, je profite largement de ses méthodes ainsi que des simplifications que M. Gårding y a apportées et a bien voulu me communiquer.

Pour effectuer le prolongement analytique de l'intégrale $I_{\bullet}^{\alpha}\overline{f, g, h}(P)$, qui jusqu'ici est exprimée dans des coordonnées lorentziennes ξ_k, nous nous servirons de nouvelles coordonnées auxquelles nous arriverons de la manière suivante.

Considérons d'abord comme surface auxiliaire le *demi-hyperboloïde* H, défini par $(\mathbf{z}, \mathbf{z}) = \mathbf{z}^2 = 1$, $z_1 < 0$.[2] On peut évidemment paramétriser H en posant, avec

$$0 \leqq t \leqq 1 \quad \text{et} \quad \sum_2^m \varphi_i^2 = 1\,(^3),$$

$$(68) \qquad z_1 = -\tfrac{1}{2}(t^{-1} + t), \; z_i = -\tfrac{1}{2}(t^{-1} - t)\varphi_i, \quad (i = 2, 3, \ldots, m).$$

On a manifestement

$$(69) \qquad t \cdot z_k = O(1)\,(^4), \quad (k = 1, 2, \ldots, m).$$

Notons qu'à $t = 0$ il correspond les points de H situés à l'infini. En posant encore $t^2 = T$, il vient

$$\frac{d}{dT}\{t(t^{-1} \pm t)\} = \frac{d}{dT}(1 \pm T) = \pm 1,$$

et par conséquent

$$\frac{\partial}{\partial T}(t \cdot z_1) = -1, \; \frac{\partial}{\partial T}(t \cdot z_i) = \varphi_i,$$

c'est-à-dire que

$$(70) \qquad \frac{\partial}{\partial T}(t \cdot z_k) = c_k = \text{const.}\,(^5) \quad (k = 1, 2, \ldots, m).$$

Dans la suite, une *fonction* d'une ou de plusieurs variables, définie dans un certain domaine, sera dite « *suffisamment dérivable* »[6] si toutes ses dérivées,

---

[1] FREMBERG **1** et **2**.

[2] La notation $\mathbf{x}^2$ remplacera dans ce qui suit la notation $(\mathbf{x}, \mathbf{x})$ pour le carré scalaire d'un vecteur $\mathbf{x}$.

[3] Les φ_i peuvent être exprimés par $m - 2$ paramètres indépendants, qu'il est inutile de préciser, en observant que dans les différentiations par rapport à $T = t^2$ qui vont suivre, les φ_i auront à rester constants.

[4] La notation, dite de Landau, $A = O(B)$ signifie que $|A| \leqq$ (constante finie) $\cdot |B|$.

[5] Cela veut dire que les c_k sont indépendants de T.

[6] De temps en temps, on va préciser cette condition et ses conséquences.

dont on pourra avoir besoin dans le calcul, existent et sont des fonctions continues quand le domaine donné est borné et fermé et qu'elles tendent même assez vite vers zéro lorsque le domaine s'étend à l'infini. Une *surface* donnée par une équation de la forme $S = 0$ est dite « *suffisamment dérivable* » s'il en est de même de la fonction S. (¹) Une *fonction* $I(\alpha)$ *holomorphe* pour toute valeur assez grande de α et dépendant de certaines fonctions $f, g, h, S, \ldots$ (des coordonnées) sera dite « *arbitrairement prolongeable* » si elle peut être prolongée vers le gauche, en restant holomorphe, jusqu'à une valeur arbitraire de α qui ne dépend que du degré de dérivabilité des fonctions $f, g, h, S, \ldots$.

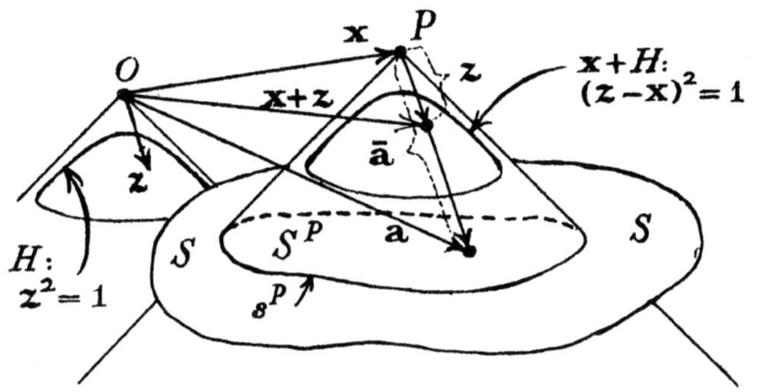

Fig. 3.

Soit maintenant S une surface d'espace(²) suffisamment dérivable donnée par l'équation $S(a_1, a_2, \ldots, a_m) = S(\mathbf{a}) = 0$, les a_k étant les coordonnées des points de S. Soit, d'autre part, $P = \mathbf{x}$ un point aux coordonnées x_k tel que le cône rétrograde au sommet P délimite avec S un domaine borné D_S^P. Nous paramétrisons d'abord la portion de surface S^P, découpée de S par le cône, et ensuite le domaine D_S^P.

Considérons à cet effet le demi-hyperboloïde $\mathbf{x} + H$ au centre $\mathbf{x} (= P)$ provenant de H par une translation, représentée par le vecteur $\mathbf{x}$ (fig. 3 et 4). Nous *projetons* du point $\mathbf{x}$ comme centre les points $\mathbf{a}$ de S^P dans les points $\mathbf{x} + \mathbf{z}$ de $\mathbf{x} + H$, ce qui établit une correspondance biunivoque entre S^P et $\mathbf{x} + H$. Par là, il correspond à tout point $\mathbf{a}$ de S^P un point déterminé $\mathbf{z}$ de H et (autant que $\mathbf{x}$ reste

(¹) Pour les fonctions $g(\mathbf{a})$ et $h(\mathbf{a})$, définissant la simple et la double couches respectivement, il ne sera question que de différentiations exécutées de long de la surface S. Chaque fois qu'il s'agit de préciser la notion qu'on vient d'introduire (nombre de dérivées etc.), on pourra supposer que la surface est exprimée par $m - 1$ paramètres.

(²) On verra plus loin que la plupart des résultats qui suivent restent valides pour des surfaces de temps et des surfaces caractéristiques (nos **61—64**).

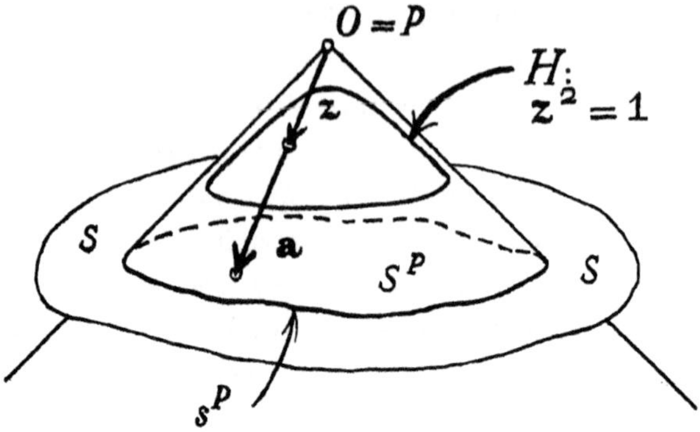

Fig. 4.

fixe) les points **a** peuvent être considérés comme dépendant des paramètres T et φ_i(¹) attachés à **z**. Ayant égard aussi à la variabilité de **x**, chaque coordonnée a_k peut être regardée comme une fonction des variables T, φ_i ($i = 2, 3, \ldots, m$) et x_j ($j = 1, 2, \ldots, m$). Observons que c'est aux points **b** situés sur l'intersection s^P du cône et de la surface S que correspondent les valeurs $t = 0$ ou $T = 0$, c'est-à-dire les points de H situés à l'infini.

28. Dérivabilité par rapport à T. — Notre premier but est de démontrer que *toutes les dérivées partielles des* a_k *par rapport à* T *et aux* x_j, qui interviennent dans nos calculs, *sont continues*, si S est suffisamment dérivable.

Désignons un instant la distance lorentzienne du point **x** et d'un point **y** quelconque par r_y c'est-à-dire que $r_y^2 = (\mathbf{y} - \mathbf{x})^2$. D'autre part, entre un point **a** et sa projection $\mathbf{x} + \mathbf{z}$ on a la relation $\mathbf{a} - \mathbf{x} = \lambda \cdot ((\mathbf{x} + \mathbf{z}) - \mathbf{x}) = \lambda \cdot \mathbf{z}$ ($\lambda > 0$). De là on tire $r_a^2 = \lambda^2 \mathbf{z}^2 = \lambda^2$, ou $r_a = \lambda$, ce qui donne $\mathbf{a} - \mathbf{x} = r_a \cdot \mathbf{z}$ ou plus brièvement

(71) $$\mathbf{a} - \mathbf{x} = r\,\mathbf{z}.$$

Cela étant, pour exprimer deux relations simultanées de la forme $|B| = O(|A|)$ et $|A| = O(|B|)$, nous introduisons la notation abrégée $A \approx B$. Dès lors, il vient d'abord de $0 \leqq t \leqq 1$, $\frac{1}{2} t^{-1} \leqq \frac{1}{2}(t^{-1} + t) \leqq t^{-1}$, et par suite $\frac{1}{2}(t^{-1} + t) \approx t^{-1}$. On en tire, grâce à la relation $|a_1 - x_1| \approx 1$, $r \cdot t^{-1} \approx \frac{r}{2}(t^{-1} + t) = r \cdot |z_1| = |a_1 - x_1| \approx 1$, c'est-à-dire que

(72) $$r \approx t \quad \text{ou} \quad v = \frac{r}{t} \approx 1.$$

(¹) En réalité, les φ_i doivent être remplacés par les $m - 2$ paramètres indépendants auxquels on vient de faire allusion.

De la relation $|a_1 - x_1| = \frac{r}{2}(t^{-1} + t) = \frac{v}{2}(1 + t^2)$ on tire encore pour *les points* **b** *de l'intersection* s^P du cône avec la surface S, vu que t s'annule pour ces points,

$$v = 2\,|b_1 - x_1|. \tag{73}$$

En posant incidemment $a_k - x_k = \bar{a}_k$, nous démontrons pour commencer que — la surface S étant suffisamment dérivable — les dérivées

$$\frac{\partial a_k}{\partial T} = \frac{\partial \bar{a}_k}{\partial T} \tag{74}$$

existent et sont continues. Des relations $\bar{a}_k = r \cdot z_k$, on tire $\bar{a}_k = \frac{r}{t} \cdot t z_k = v \cdot t z_k$ ou $t z_k = v^{-1} \cdot \bar{a}_k$. Il vient alors de (70), si l'on admet un instant que $\partial \bar{a}_k / \partial T$ et $\partial v / \partial T$ existent,

$$\frac{\partial}{\partial T}(v^{-1} \cdot \bar{a}_k) = \frac{\partial \bar{a}_k}{\partial T} \cdot v^{-1} - \bar{a}_k \cdot v^{-2} \frac{\partial v}{\partial T} = c_k$$

et par suite

$$v \cdot \frac{\partial \bar{a}_k}{\partial T} - \bar{a}_k \cdot \frac{\partial v}{\partial T} = c_k \cdot v^2. \tag{75}$$

Ayant égard à $v \approx 1$ (*cf.* (72)) et à la continuité de v et des $\bar{a}_k$, on voit que notre énoncé concernant les quantités (74) se réduit à un énoncé analogue relatif à la quantité

$$\frac{\partial v}{\partial T}. \tag{76}$$

Or en tenant compte de (75), et en posant, pour abréger, $\partial S / \partial a_k = S_k$, on tire de la relation $S(\mathbf{a}) = 0$

$$0 = v \cdot \frac{\partial S}{\partial T} = v \cdot \sum_k S_k \frac{\partial a_k}{\partial T} = v \cdot \sum S_k \frac{\partial \bar{a}_k}{\partial T} \tag{77}$$
$$= \frac{\partial v}{\partial T} \cdot \sum S_k \bar{a}_k + v^2 \cdot \sum S_k c_k = \frac{\partial v}{\partial T} \cdot M + v^2 \cdot \sum_k S_k c_k,$$

où l'on a posé

$$M = \sum_k S_k \bar{a}_k = \sum_k S_k (a_k - x_k) = (\mathbf{grad}\, S, \mathbf{a} - \mathbf{x}). \tag{78}$$

Aucun des rayons vecteurs issus du point $\mathbf{x}$ n'étant tangent à la surface S, on a

(79) $$M \approx 1,$$

ce qui entraîne la continuité de M^{-1} et par là l'existence et la continuité de (76). Puisque $v \approx 1$, il en résulte pour un exposant arbitraire γ l'existence et la continuité de

(80) $$\frac{\partial v^\gamma}{\partial T}.$$

On établit aussi, S étant toujours supposé suffisamment dérivable, l'existence et la continuité de

(81) $$\frac{\partial^p a_k}{\partial T^p} = \frac{\partial^p \bar{a}_k}{\partial T^p} \quad \text{et de} \quad \frac{\partial^p v^\gamma}{\partial T^p}.$$

En effet, en différentiant (75) $p-1$ fois, on voit que

$$v \cdot \partial^p \bar{a}_k / \partial T^p - \bar{a}_k \cdot \partial^p v / \partial T^p$$

s'exprime comme un polynome à coefficients constants en v et dans les dérivées successives de v et de $\bar{a}_k$, toutes ces dérivées étant d'ordre $\leqq p-1$. Dès lors, les faits énoncés étant supposés démontrés pour $p \leqq q-1$, la démonstration pour $p = q$ s'achève comme plus haut ($p = 1$).

29. Dérivabilité par rapport aux coordonnées x_j. — En ce qui concerne les dérivées des a_k et de v^γ par rapport aux x_j (et leurs dérivées mixtes par rapport aux x_j et à T), on peut, du point de vue de la méthode, se restreindre à celles du premier ordre. Montrons donc que les dérivées

(82) $$\frac{\partial a_k}{\partial x_j} \quad \text{ou} \quad \frac{\partial \bar{a}_k}{\partial x_j} = \frac{\partial (a_k - x_k)}{\partial x_j} = \frac{\partial a_k}{\partial x_j} - \delta_{jk} \quad \text{et} \quad \frac{\partial v^\gamma}{\partial x_j}$$

existent et sont continues. Ici encore, tout dépend de la relation $M \approx 1$. En effet, en différentiant par rapport à x_j la relation

$$S(\mathbf{a}) = S(\bar{\mathbf{a}} + \bar{\mathbf{x}}) = S(r \cdot \mathbf{z} + \mathbf{x}) = 0,$$

on obtient

(83) $$S_j + \sum_k S_k z_k \cdot \frac{\partial r}{\partial x_j} = 0.$$

En multipliant par $v = r \cdot t^{-1}$, et en observant que $r z_k = \bar{a}_k$ et $t^{-1} \cdot \partial r / \partial x_j = \partial v / \partial x_j$, on en tire

(84) $$S_j \cdot v + M \cdot \frac{\partial v}{\partial x_j} = 0.$$

D'autre part, en multipliant par $\bar{a}_l = r \cdot z_l$ et notant que

$$r\, z_l \cdot z_k \frac{\partial r}{\partial x_j} = r\, z_k \cdot z_l \frac{\partial r}{\partial x_j} = \bar{a}_k \cdot \frac{\partial \bar{a}_l}{\partial x_j},$$

on obtient

$$S_j \bar{a}_l + M \cdot \frac{\partial \bar{a}_l}{\partial x_j} = 0. \tag{85}$$

Il en résulte comme plus haut que toutes les dérivées $\partial v/\partial x_j$ ou $\partial v^\gamma/\partial x_j$ et $\partial \bar{a}_k/\partial x_j$ existent et sont continues.

En itérant les procédés ci-dessus et en les combinant avec ceux concernant les différentiations par rapport à T, on arrive au théorème suivant.

La surface S étant suffisamment dérivable, il en est de même des quantités a_k et v considérées comme fonctions des variables T et x_j.

30. Système de coordonnées pour D_S^P. — Afin de caractériser les points $\boldsymbol{\xi}$ du domaine D_S^P il faut compléter les variables T et x_j caractérisant les points $\mathbf{a}$ de S^P par une nouvelle variable σ. On posera simplement $\boldsymbol{\xi} - \mathbf{x} = \sigma(\mathbf{a} - \mathbf{x})$, $0 \leqq \sigma \leqq 1$, ce qui peut aussi s'écrire

$$\boldsymbol{\xi} = (1-\sigma)\mathbf{x} + \sigma \mathbf{a} \quad \text{ou} \quad \xi_k = (1-\sigma) x_k + \sigma a_k \qquad (0 \leqq \sigma \leqq 1). \tag{86}$$

On est par là conduit au théorème:

Si la surface S est suffisamment dérivable et s'il en est de même des fonctions $f(\boldsymbol{\xi})$ et $l(\mathbf{a})$ par rapport aux variables respectives ξ_k et a_k,[1] *ces fonctions seront des fonctions suffisamment dérivables, toutes les deux par rapport aux variables T et x_j et la première en outre par rapport à la variable σ.*

31. Transformation de l'intégrale I_*. — Nous appliquons maintenant ce qui précède à la transformation des intégrales figurant dans $I_*^\alpha \overline{f, g, h}(P)$, en commençant par l'intégrale de volume

$$I^\alpha f(P) = \frac{1}{H_m(\alpha)} \int_{D_S^P} f(Q)\, r_{PQ}^{\alpha-m}\, dQ, \tag{87}$$

où, d'après les formules (20) du n° **16** et (5) du n° **6**,

$$H_m(\alpha) = \pi^{\frac{m-2}{2}} \cdot 2^{\alpha-1} \cdot \Gamma\left(\frac{\alpha}{2}\right) \Gamma\left(\frac{\alpha+2-m}{2}\right) = \pi^{\frac{m-1}{2}}\, \Gamma\left(\frac{\alpha+2-m}{2}\right) \Gamma(\alpha) : \Gamma\left(\frac{\alpha+1}{2}\right). \tag{88}$$

---

[1] *Cf.* la note [1], p. 51.

Dans les considérations qui suivent on peut, sans restreindre la généralité, placer l'origine dans le point P. Soit $Q = \mathfrak{z}$ un point du domaine D_S^P et désignons par $\mathbf{a}$ et par $\mathbf{z}$ ses projections respectives sur la surface S et le demi-hyperboloïde H ($\mathbf{z}^2 = 1$, $z_1 < 0$), toutes les projections dont il sera question ici étant des projections centrales au centre P. En posant, pour abréger, $r_{PQ} = r_Q$ et $r_{P\mathbf{a}} = r_a$, on aura $r_Q = \sigma \cdot r_a$, avec $0 \leqq \sigma \leqq 1$. Pour calculer l'élément de volume dQ, nous envisageons encore le demi-hyperboloïde $\bar{H}$ passant par Q et composé des points $r_Q \cdot \mathbf{z}$ et nous désignons par $dr_Q = r_a \cdot d\sigma$ la différentielle de r_Q correspondant à un déplacement radial de Q. Les rayons vecteurs issus de P étant orthogonaux à $\bar{H}$, on a, en désignant respectivement par $d\bar{H}$ et dH l'élément de surface de $\bar{H}$ et sa projection sur H, l'expression suivante pour l'élément de volume

(89) $$dQ = d\bar{H} \cdot dr_Q = r_Q^{m-1}\, dH \cdot dr_Q = \sigma^{m-1} r_a^{m-1}\, dH \cdot r_a\, d\sigma = r_a^m\, dH \cdot \sigma^{m-1}\, d\sigma,$$

et

(90) $$r_Q^{\alpha-m}\, dQ = r_a^{\alpha-m} \sigma^{\alpha-m}\, dQ = r_a^\alpha\, dH \cdot \sigma^{\alpha-1}\, d\sigma.$$

Cela posé, il vient

(91) $$I^\alpha f(P) = \frac{1}{H_m(\alpha)} \int_H F(\mathbf{a}, \alpha)\, r_a^\alpha\, dH,$$

où

(92) $$F(\mathbf{a}, \alpha) = \int_0^1 f(\sigma \cdot \mathbf{a})\, \sigma^{\alpha-1}\, d\sigma.$$

Nous passons maintenant à nos intégrales de surface et considérons d'abord, pour des raisons qu'on va voir, la double couche

$$I_*^\alpha \overline{0, 0, h}\,(P) = -\frac{1}{H_m(\alpha)} \int_{S^P} h \cdot \frac{d\, r_a^{\alpha-m}}{dn_a} \cdot dS = -\frac{\alpha - m}{H_m(\alpha)} \int_{S^P} h \cdot r_a^{\alpha-m-1} \cdot \frac{dr_a}{dn_a} \cdot dS.$$

En analogie avec le cas euclidien, on doit avoir, si l'on désigne par dH la projection sur H de dS

(93) $$r_a^{1-m} \cdot \left|\frac{dr_a}{dn_a}\right| \cdot dS = dH,$$

cette dernière expression n'étant que l'angle solide (lorentzien) sous lequel l'élément dS est vu du point P. Cette formule est d'ailleurs facile à vérifier par le calcul suivant où, pour simplifier l'écriture, on supprime en général l'indice a.

Désignons par $\bar{\mathbf{n}} = \mathbf{a}/r$ l'une des deux normales unitaires au point $\mathbf{a}$ de l'hyperboloïde $\bar{H} = r_a \cdot H = r \cdot H$ qui passe par ledit point; on a évidemment pour deux éléments dS et $d\bar{H}$ qui se correspondent par projection $d\bar{H} = |(\mathbf{n}, \bar{\mathbf{n}})| \cdot dS$. D'autre part, d'après la formule (48) du n° **21**, on trouve

$$(94) \qquad r \cdot \frac{dr}{dn} = (\mathbf{a}, \mathbf{n}) \quad \text{ou} \quad \frac{dr}{dn} = (\mathbf{n}, \bar{\mathbf{n}}).$$

Par suite

$$\left|\frac{dr}{dn}\right| dS = d\bar{H} = r^{m-1}\, dH,$$

ce qui n'est autre chose que la formule (93). En observant encore que, pour la normale intérieure, dont nous nous servons en général, dr/dn est négatif, on en tire

$$(95) \qquad I_*^\alpha\, \overline{0, 0, h}\,(P) = \frac{\alpha - m}{H_m(\alpha)} \int_H h \cdot r_a^{\alpha-2}\, dH.$$

Transformons enfin l'intégrale de simple couche

$$I_*^\alpha\, \overline{0, g, 0}\,(P) = \frac{1}{H_m(\alpha)} \int_{S^P} g \cdot r_a^{\alpha-m}\, dS.$$

On a, d'après les formules (78), (93), (94) et la formule (37) du n° **21**,

$$(96) \qquad dS = \left|\frac{dr}{dn}\right|^{-1} r^{m-1}\, dH = |(\mathbf{a}, \mathbf{n})|^{-1} r^m\, dH = |M|^{-1} \cdot N \cdot r^m\, dH$$

et par là

$$(97) \qquad I_*^\alpha\, \overline{0, g, 0}\,(P) = \frac{1}{H_m(\alpha)} \int_H |M|^{-1} N \cdot g \cdot r_a^\alpha\, dH.$$

Nous avons maintenant à exprimer l'élément de surface dH par les paramètres T et φ_i considérés plus haut. En posant incidemment $t = e^{-\theta}$, le plan à $m - 1$ dimensions défini par $z_1 = -\cosh\theta$ coupe l'hyperboloïde suivant une sphère de caractère euclidien de rayon $\sinh\theta$. L'élément de cette sphère est donc $\sinh^{m-2}\theta\, d\omega_{m-1}$, le dernier élément étant celui d'une sphère unitaire dans un espace à $m - 1$ dimensions; pour la notation *cf.* le n° **6**, *Remarque*. Par là, l'élément de surface de H est $dH = \sinh^{m-2}\theta \cdot d\theta \cdot d\omega_{m-1}$.

Il faut maintenant exprimer cela à l'aide du paramètre $T = t^2$. On a $T = e^{-2\theta}$, $dT = -2e^{-2\theta} d\theta = -2T d\theta$ et par suite $|d\theta| = \frac{1}{2} T^{-1} |dT|$; d'autre part $\sinh\theta$

$= \frac{1}{2}(t^{-1} - t) = \frac{1}{2}t^{-1}(1 - T) = \frac{1}{2}T^{-\frac{1}{2}}(1 - T)$. Dès lors, il vient après quelques simplifications

$$dH = 2^{1-m}T^{-\frac{m}{2}}(1 - T)^{m-2}\cdot dT\cdot d\omega_{m-1}, \tag{99}$$

toutes les différentielles admettant le signe positif. Cela posé, et en reprenant la notation $r/t = v$ (formule (72) du n° 27), on trouve

$$r^\alpha \cdot dH = v^\alpha t^\alpha \cdot dH = 2^{1-m}\cdot v^\alpha \cdot T^{\frac{\alpha-m}{2}}(1 - T)^{m-2}\cdot dT\cdot d\omega_{m-1}. \tag{100}$$

Ce point acquis, nos trois intégrales peuvent s'écrire sous la forme commune suivante

$$I^\alpha = \frac{1}{H_m(\alpha)}\int_0^1 T^{\beta-1}\cdot K(T,\alpha)\cdot dT, \tag{101}$$

où l'on a dans les différents cas, en posant encore

$$2^{1-m}\cdot(1 - T)^{m-2} = k(T), \tag{102}$$

I. *Simple couche:*

$$\beta = \frac{\alpha + 2 - m}{2};\quad K(T,\alpha) = k(T)\int_{\omega_{m-1}} N\cdot|M|^{m-1}\cdot v^\alpha\cdot g(\mathbf{a})\cdot d\omega_{m-1}. \tag{103}$$

II. *Double couche:*

$$\beta = \frac{\alpha - m}{2};\quad K(T,\alpha) = 2\beta\cdot k(T)\int_{\omega_{m-1}} v^{\alpha-2}\cdot h(\mathbf{a})\cdot d\omega_{m-1}. \tag{104}$$

III. *Intégrale de volume:*

$$\beta = \frac{\alpha + 2 - m}{2};\quad K(T,\alpha) = k(T)\int_{\omega_{m-1}} v^\alpha\cdot F(\mathbf{a},\alpha)\cdot d\omega_{m-1}, \tag{105}$$

la fonction $F(\mathbf{a},\alpha)$ étant donnée par la formule (92).

Nous allons maintenant effectuer le prolongement analytique et montrer en particulier que l'on a, dans les cas I et II, pour $p = 0$ ou entier positif, $I^{-2p} = 0$ et, dans le cas III, $I^0 = f(P)$, $I^{-2p} = \Delta^p f(P)$, ou, en résumant les cas I—III,

$$I_*^0\overline{f, g, h}(P) = f(P),\quad I_*^{-2p}\overline{f, g, h}(P) = \Delta^p f(P)\qquad (p = 1, 2, \ldots). \tag{106}$$

Les cas I et II se traitent facilement par les méthodes du Chapitre I, en appliquant l'une des deux formules de prolongement (4) (n° **2**) ou (13) (n° **5**), par préférence la dernière.

32. Simple couche. — Avec la valeur que β admet dans (103) on peut écrire (*voir* (88) et (101))

$$I^\alpha = \frac{1}{\pi^{\frac{m-2}{2}} 2^{\alpha-1} \Gamma\left(\frac{\alpha}{2}\right)} \cdot \frac{1}{\Gamma(\beta)} \int_0^1 T^{\beta-1} K(T, \alpha)\, dT, \tag{107}$$

cette intégrale n'étant convergente en général que pour $\beta > 0$ ou $\alpha > m-2$. Nous avons montré plus haut que, la surface S et la fonction $g(\mathbf{a})$ étant suffisamment dérivables[1], la fonction $K(T, \alpha)$ est une fonction suffisamment dérivable de T. D'autre part, on voit par (103) que la fonction $K(T, \alpha)$ ainsi que ses dérivées sont des fonctions holomorphes de α pour toute valeur de α, cette variable n'entrant dans $K(T, \alpha)$ que par le facteur v^α. Cela étant, il résulte de la formule (13) du n° **5** que I^α, considéré comme fonction de α, est arbitrairement prolongeable (p. 51).

Il convient de préciser le dernier résultat. Nous dirons qu'une fonction est « *dérivable n fois* » par rapport à une ou à plusieurs variables si elle admet des dérivées continues de tout ordre $\leqq n$. On voit alors que si la fonction $K(T, \alpha)$ est dérivable n fois par rapport à T, n étant un nombre $> -\beta_0 = \frac{m-2-\alpha_0}{2}$, la fonction I^α pourra être prolongée analytiquement jusqu'à la valeur α_0. D'autre part, cette dernière valeur étant fixée, on pourra choisir $n = \left[\frac{m-2-\alpha_0}{2}\right] + 1 = \left[\frac{m-\alpha_0}{2}\right]$. Si l'on exige seulement que le prolongement soit possible jusqu'à toute valeur $\alpha_0 + \varepsilon$, $\varepsilon > 0$, et puisse ensuite s'opérer par un passage à la limite, $\varepsilon \to 0$, il suffira $n \geqq \frac{m-2-\alpha_0}{2}$. En particulier, pour $\alpha_0 = 0, -1, -2, \ldots$, on pourra alors poser $n = \left[\frac{m-1-\alpha_0}{2}\right]$. On voit, par les résultats généraux du n° **28** et par la formule (103) et les formules (78) du n° **28** et (37) du n° **21**, que $K(T, \alpha)$ est pour n arbitraire dérivable n fois, si $g(\mathbf{a})$ est dérivable n fois et $S(\mathbf{a})$ dérivable $n+1$ fois. En particulier, le pro-

[1] *Cf.* la note [1], p. 51.

longement sera possible jusqu'à $\alpha = 0$ ou, plus généralement, jusqu'à $\alpha = -2p$ (p entier) si $g(\mathrm{a})$ et $S(\mathrm{a})$ sont dérivables $\left[\frac{m-1}{2}\right] + p$ fois et $\left[\frac{m+1}{2}\right] + p$ fois respectivement.[1] Dans ces conditions, on aura

$$I^0 = 0, \quad I^{-2p} = 0, \tag{109}$$

puisque $\Gamma\left(\frac{\alpha}{2}\right)$ qui figure au dénominateur du premier facteur de l'expression (107) de I^α admet un pôle pour les valeurs de α en question.

33. Double couche. — On a, avec la valeur de β qui figure dans (104),

$$\frac{2\beta}{\Gamma\left(\frac{\alpha+2-m}{2}\right)} = \frac{2\beta}{\Gamma(\beta+1)} = \frac{2}{\Gamma(\beta)}. \tag{110}$$

Dès lors, le raisonnement ci-dessus s'applique. le seul changement étant que, pour arriver à une certaine valeur fixée de α_0, le degré de dérivabilité de $K(T, \alpha)$ devra être augmenté de 1. En particulier, on aura

$$I^0 = 0, \quad I^{-2p} = 0, \tag{111}$$

si $h(\mathrm{a})$ et $S(\mathrm{a})$ sont dérivables $\left[\frac{m+1}{2}\right] + p$ fois et $\left[\frac{m+3}{2}\right] + p$ fois, respectivement.[2]

34. Intégrale de volume. — Dans les deux cas précédents, l'intégration n'a eu lieu que par rapport à une seule variable essentielle, à savoir T. On voit par les formules (92), (101) et (105) que pour l'intégrale de volume il est question de deux variables d'intégration essentielles, à savoir T et σ, ce qui fait que les résultats du Chap. I ne suffisent pas pour éclaircir le cas actuel. Pour en faciliter la discussion. nous allons démontrer quelques lemmes.

35. Lemmes. Remarques. — Nous commençons par

Lemme I. *Toute fonction $G(u, v)$, définie pour $0 \leqq u \leqq a < \infty$, $0 \leqq v \leqq b < \infty$, et suffisamment dérivable, peut se mettre sous la forme*

$$G(u, v) = \pi(u, v) + \sum_{r=0}^{p-1} \frac{h_r(v)}{r!} u^r + \sum_{s=0}^{q-1} \frac{k_s(u)}{s!} v^s + l(u, v), \tag{112}$$

[1] *Cf.* FREMBERG **2**, p. 45.
[2] *Cf.* FREMBERG **2**, p. 51. *Cf.* aussi la note [1], p. 51.

où $\pi(u, v)$ est un polynome en u et en v tandis que les autres fonctions introduites satisfont aux inégalités

$$h_r(v) = O(v^q), \quad k_s(u) = O(u^p), \quad l(u, v) = O(u^p v^q). \tag{113}$$

La chose est évidente, si la fonction $G(u, v)$ peut être développée en série de Maclaurin, au moins pour u assez petit et $0 \leqq v \leqq b$ et, d'autre part, pour v assez petit et $0 \leqq u \leqq a$. Il s'agit pourtant d'établir cette formule de décomposition dans les conditions plus générales où nous nous trouvons ($G(u, v)$ est suffisamment dérivable). Il suffit de poser, à cet effet, en écrivant

$$\partial^{r+s} G(u, v)/\partial u^r \partial v^s = G^{(r s)}(u, v),$$

$$\pi(u, v) = \sum_{r=0}^{p-1} \sum_{s=0}^{q-1} \frac{G^{(r s)}(0, 0)}{r!\, s!} u^r v^s,$$

$$h_r(v) = G^{(r 0)}(0, v) - \sum_{s=0}^{q-1} \frac{G^{(r s)}(0, 0)\, v^s}{s!} = \frac{1}{(q-1)!} \int_0^v G^{(r q)}(0, \eta)(v - \eta)^{q-1}\, d\eta,$$

$$k_s(u) = G^{(0 s)}(u, 0) - \sum_{r=0}^{p-1} \frac{G^{(r s)}(0, 0)\, u^r}{r!} = \frac{1}{(p-1)!} \int_0^u G^{(p s)}(\xi, 0)(u - \xi)^{p-1}\, d\xi$$

et

$$l(u, v) = G(u, v) - \sum_{r=0}^{p-1} \frac{G^{(r 0)}(0, v)\, u^r}{r!} - \sum_{s=0}^{q-1} \frac{G^{(0 s)}(u, 0)\, v^s}{s!} + \pi(u, v)$$

$$= \frac{1}{(p-1)!\,(q-1)!} \int_0^u \int_0^v G^{(p q)}(\xi, \eta)(u - \xi)^{p-1}(v - \eta)^{q-1}\, d\eta\, d\xi.$$

L'identité entre les premiers et les seconds membres correspondants se vérifie moyennant des intégrations par parties. Les premiers membres mettent en évidence l'identité (112) et les seconds les inégalités (113).

Lemme II. [1] *Nous posons*

$$J^{\alpha\beta} G(0, 0) = J^{\alpha\beta} G = \frac{1}{\Gamma(\alpha)\Gamma(\beta)} \int_0^a \int_0^b G(u, v)\, u^{\alpha-1} v^{\beta-1}\, dv\, du. \tag{114}$$

Cette fonction est holomorphe en α et en β dans le domaine A: $\alpha > 0$, $\beta > 0$, si $G(u, v)$ est borné par exemple. Si $G(u, v)$ est suffisamment dérivable, la fonction $J^{\alpha\beta}$ admet un prolongement qui est holomorphe dans le domaine B: $\alpha \geqq \alpha_0$, $\beta \geqq \beta_0$ où α_0 et β_0 sont des nombres négatifs arbitraires ou zéro.

---

[1] Ce lemme est intimement lié à un lemme donné par M. Fremberg, **1**, p. 10.

Lemme III.[1]. *La valeur $J^{\alpha_0 \beta_0} G$, obtenue par prolongement analytique, est indépendante du chemin, situé dans B, suivant lequel le point (α, β) va au point (α_0, β_0).* En particulier, on peut effectuer d'abord le prolongement par rapport à α, en formant la valeur $J^{\alpha_0 \beta} G$ avec $\beta > 0$, pour en arriver à la valeur $J^{\alpha_0 \beta_0} G$ par le prolongement de $J^{\alpha_0 \beta} G$ par rapport à β.

Lemme IV. *Les lemmes II et III restent valides si $G(u, v)$ dépend aussi de α et de β de manière que $G(u, v)$ et les $G^{(r\,s)}(u, v)$, qui interviennent dans le calcul, sont des fonctions holomorphes de α et de β dans le domaine B.*

Soient p et q deux nombres entiers, $p > -\alpha_0, q > -\beta_0$. Écrivons $G(u, v)$ sous la forme donnée par le lemme I. Pour les monomes $u^r v^s$ on a — d'abord pour α et β assez grands —

$$J^{\alpha\beta} u^r v^s = \frac{a^{r+\alpha}}{(\alpha + r)\,\Gamma(\alpha)} \cdot \frac{b^{s+\beta}}{(\beta + s)\,\Gamma(\beta)},$$

par quoi le prolongement analytique de $J^{\alpha\beta}$ se trouve effectué pour ce monome et alors pour le polynome $\pi(u, v)$ du lemme I, et cela pour toute valeur de α et de β. Les pôles $\alpha = -r$, $\beta = -s$ ne sont qu'apparents, vu que $(\alpha + r)\,\Gamma(\alpha)$ et $(\beta + s)\,\Gamma(\beta)$ sont différents de zéro.

Ensuite on a

$$J^{\alpha\beta} h_r(v)\, u^r = \frac{a^{r+\alpha}}{(r + \alpha)\,\Gamma(\alpha)} \cdot \frac{1}{\Gamma(\beta)} \int_0^b h_r(v)\, v^{\beta-1}\, dv.$$

La dernière intégrale étant, en vertu de (113), convergente pour $\beta > -q$, où $-q < \beta_0$, le prolongement analytique se trouve effectué dans le domaine B entier.

Le raisonnement est analogue pour les fonctions $k_s(u)$. Enfin, pour $l(u, v)$, l'intégrale $J^{\alpha\beta}$ converge dans le domaine B entier, encore en vertu des inégalités (113). Ainsi, la démonstration du lemme II se trouve achevée.

La valeur obtenue pour $J^{\alpha_0 \beta_0} G$ étant manifestement indépendante du chemin suivi quand il s'agit des fonctions de forme particulière que nous venons de considérer, il en est de même de la fonction $G(u, v)$, ce qui établit les lemmes III et IV.

Remarque I. Dans nos applications, α et β ne sont pas indépendants l'un de l'autre. On aura p. ex. $\beta = \dfrac{\alpha + 2 - m}{2}$. Rien n'empêche cependant, d'après ce qui précède, de choisir le chemin allant dans le domaine B au point

[1] Ce lemme ainsi que le lemme IV sont en principe dus à M. Gårding.

$$\left(\alpha_0, \beta_0 = \frac{\alpha_0 + 2 - m}{2}\right)$$

d'une manière arbitraire (non nécessairement rectiligne).

Remarque II. Les conditions de dérivabilité imposées à $G(u, v)$ peuvent être telles que le prolongement analytique de $J^{\alpha\beta}$ ne peut s'effectuer que dans le domaine $\alpha > \alpha_0$, $\beta > \beta_0$ et que $J^{\alpha\beta_0}$ $(\alpha > \alpha_0)$, $J^{\alpha_0\beta}$ $(\beta > \beta_0)$ et $J^{\alpha_0\beta_0}$ s'obtiennent en ajoutant au prolongement un ou plusieurs passages à la limite. Ici encore, les valeurs finales, p. ex. celle de $J^{\alpha_0\beta_0}$, sont indépendantes du chemin suivi par (α, β). Nous nous contentons du cas simple que voici.

Lemme II'. Si $G(u, v)$ est continu dans le domaine $0 \leqq u \leqq a < \infty$, $0 \leqq v \leqq b < \infty$, l'intégrale $J^{\alpha\beta}$ convergente pour $\alpha > 0$, $\beta > 0$, tend vers $G(0, 0)$ quand $\alpha \to 0$ et $\beta \to 0$.

On arrive à une démonstration simple par l'observation suivante. La fonction

$$\begin{aligned} l(u, v) &= G(u, v) - G(0, v) - G(u, 0) + G(0, 0) \\ &= (G(u, v) - G(0, v)) - (G(u, 0) - G(0, 0)) = (G(u, v) - G(u, 0)) - (G(0, v) - G(0, 0)) \end{aligned}$$

tend uniformément vers zéro lorsque l'une au moins des variables tend vers 0.

Par cette propriété de $l(u, v)$, l'intégrale $J^{\alpha\beta}$ devient accessible à un procédé d'évaluation donné au Chap. I (n° **3**). On voit aussi que le passage à la limite $\lim\limits_{(\alpha, \beta) \to (0, 0)} I^{\alpha\beta}$ pourra être effectué en formant d'abord p. ex. $J^{0\beta} = \lim\limits_{\alpha \to 0} J^{\alpha\beta}$ et après, $\lim\limits_{\beta \to 0} J^{0\beta}$.

36. Intégrale de volume. (Suite). — Nos lemmes démontrés, reprenons maintenant l'intégrale $I^\alpha f(P)$, exprimée par les formules (101), (105), (92) et (102). En tenant compte du dernier membre de (88), on peut poser — *cf.* (114) — $\left(\text{avec } \dfrac{\alpha + 2 - m}{2} = \beta\right)$

$$(114^{bis}) \qquad I^\alpha f(P) = \frac{1}{\Gamma(\alpha)\,\Gamma(\beta)} \int_0^1\!\!\int_0^1 G(\sigma, T, \alpha) \cdot \sigma^{\alpha-1}\, T^{\beta-1}\, d\sigma\, dT = J^{\alpha\beta} G,$$

où

$$(115) \qquad G(\sigma, T, \alpha) = \frac{\Gamma\left(\dfrac{\alpha+1}{2}\right) k(T)}{\pi^{\frac{m-1}{2}}} \int_{\omega_{m-1}} f(\sigma \cdot \mathbf{a})\, v^\alpha\, d\,\omega_{m-1}.$$

Il découle du théorème du n° **30** que, $f(\mathfrak{x})$ et $S(\mathfrak{a})$ étant suffisamment dérivables, $G(\sigma, T)$ sera une fonction suffisamment dérivable des variables σ et T. Dès lors, il résulte de nos lemmes que $I^\alpha f(P)$ est une fonction arbitrairement prolongeable de α. Quant à la régularité de I^α, il faut jusqu'à nouvel ordre faire exception des valeurs négatives impaires de α où $\Gamma\left(\frac{\alpha+1}{2}\right)$ admet des pôles. Les pôles que I^α pourrait admettre pour les mêmes valeurs — qui d'ailleurs ne présentent aucun intérêt particulier — ne sont pourtant qu'apparents, comme on le verra tout à l'heure. Afin que le prolongement puisse être poussé jusqu'à une valeur fixée α_0, il suffit que $G(\sigma, T, \alpha)$ soit dérivable n fois, où n est $> -\alpha_0 - \beta_0 = -\alpha_0 + \frac{m-2-\alpha_0}{2}$ (le signe $>$ pouvant être remplacé par $\geqq$ si l'on se contente d'un passage à la limite $\alpha + \varepsilon \to \alpha$). Pour que $G(\sigma, T, \alpha)$ soit dérivable n fois, il suffit qu'il en soit ainsi de $f(\mathfrak{x})$ et de $S(\mathfrak{a})$. Pour arriver à I^0, on pourra prendre[1] (en se contentant d'un passage à la limite) $n = \left[\frac{m-1}{2}\right]$ et pour arriver à I^{-2p}, $n = 3p + \left[\frac{m-1}{2}\right]$.

Le fait que les dits pôles ne sont qu'apparents résulte immédiatement de la formule de Green (itérée)[2] ((60), n° **22** ou (61bis), n° **23**), et du fait que, dans les cas de simple ou de double couche, la fonction prolongée était holomorphe.

37. Les opérateurs I^0, I^{-p}, I^0_*, I^{-2p}_*. — Il est maintenant facile de démontrer que $I^0 f(P) = f(P)$, ce qui signifie que I^0 est *l'opérateur identique*. En effet, pour $\alpha = 0$, on a $\beta = \frac{2-m}{2}$, c'est-à-dire que pour calculer $I^0 f(P)$, nous n'avons qu'à calculer $J^{\alpha\beta} G$ pour $\alpha = 0$, $\beta = \frac{2-m}{2}$. D'après les lemmes III et IV, on peut y arriver en calculant d'abord $J^{0\beta} G$, β étant un nombre positif quelconque, et passer ensuite par prolongement à la valeur $\beta = \frac{2-m}{2}$. On a d'abord par (114bis) et par la formule fondamentale (6) du n° **3**, que, pour $\alpha = 0$,

$$(116) \qquad J^{0\beta} G = \frac{1}{\Gamma(\beta)} \int_0^1 T^{\beta-1}\, dT \left(\frac{1}{\Gamma(\alpha)} \int_0^1 G(\sigma, T, \alpha)\, \sigma^{\alpha-1}\, d\sigma\right)_{\alpha=0}$$

$$= \frac{1}{\Gamma(\beta)} \int_0^1 T^{\beta-1}\, G(0, T, 0)\, dT.$$

---

(1) *Cf.* Fremberg **2**, p. 40.

(2) En appliquant la formule une seule fois on voit que $\alpha = -1$ n'est pas de pôle, et ainsi de suite.

D'autre part, en vertu de (115),

$$(117) \qquad G(0, T, 0) = \frac{\Gamma\left(\frac{1}{2}\right) k(T)}{\pi^{\frac{m-2}{2}}} \int_{\omega_{m-1}} f(0 \cdot \mathbf{a}) v^0 d\omega_{m-1} = \frac{\Gamma\left(\frac{1}{2}\right) k(T)}{\pi^{\frac{m-1}{2}}} f(P) \omega_{m-1},$$

ou, grâce à la formule (102) et à la formule (8) du n° **6**,

$$(118) \qquad G(0, T, 0) = f(P) \cdot \frac{\Gamma\left(\frac{1}{2}\right) \cdot 2^{1-m} (1-T)^{m-2}}{\pi^{\frac{m-1}{2}}} \cdot \frac{2\pi^{\frac{m-1}{2}}}{\Gamma\left(\frac{m-1}{2}\right)}$$

$$= f(P) \cdot \frac{2^{2-m} \pi^{\frac{1}{2}}}{\Gamma\left(\frac{m-1}{2}\right)} \cdot (1-T)^{m-2}$$

En tenant encore compte de la relation

$$\int_0^1 T^{\beta-1}(1-T)^{m-2} dT = \frac{\Gamma(\beta)\Gamma(m-1)}{\Gamma(\beta+m-1)},$$

(116) peut s'écrire

$$J^{0\beta} G = f(P) \cdot \frac{2^{2-m} \cdot \pi^{\frac{1}{2}} \cdot \Gamma(m-1)}{\Gamma\left(\frac{m-1}{2}\right) \cdot \Gamma(\beta+m-1)},$$

et enfin, par prolongement analytique, pour $\beta = \frac{2-m}{2}$ (correspondant à $\alpha = 0$)

$$I^0 f(P) = J^{0, \frac{2-m}{2}} G = f(P) \cdot \frac{2^{2-m} \cdot \pi^{\frac{1}{2}} \cdot \Gamma(m-1)}{\Gamma\left(\frac{m-1}{2}\right) \Gamma\left(\frac{2-m}{2} + m - 1\right)}$$

$$= f(P) \cdot \frac{2^{2-m} \cdot \pi^{\frac{1}{2}} \Gamma(m-1)}{\Gamma\left(\frac{m-1}{2}\right) \cdot \Gamma\left(\frac{m}{2}\right)} = f(P),$$

la dernière réduction étant faite moyennant la formule (5) du n° **6**.

Ce point important acquis, nous allons démontrer que l'on a $I^{-2p} f(P) = \Delta^p f(P)$. En appliquant la formule de Green ((61bis) du n° **23**),

$$I^\alpha f(P) = I_*^{\alpha+2} \overline{\Delta f, \frac{df}{dn}, f}(P),$$

on trouve, grâce aux faits $I^0 f(P) = f(P)$ et $I_*^0 \overline{0, g, h}(P) = 0$ (n^{os} **32** et **33**), que $I^{-2} f(P) = \Delta f(P)$. La relation générale en résulte par itération.

En résumé, on a démontré que

$$(119)\qquad I_*^0\overline{f, g, h}(P) = f(P) \text{ et } I_*^{-2p}\overline{f, g, h}(P) = \Delta^p f(P) \qquad (p = 1, 2, \ldots).$$

En ce qui concerne les conditions de validité de ces formules, on consultera les n^os **32**, **33** et **36**.

38. Cône rétrograde entier. — Il faut encore revenir sur le cas particulier important, quoique très simple, traité déjà au n° **26**, où le domaine D_S^P se confond avec le cône entier D^P. Ce cône forme alors le domaine d'intégration pour l'intégrale de volume, et il n'y aura pas d'intégrales de surface. Or nous avons vu, au n° **26**, que dans le cas actuel, le prolongement analytique de $I^\alpha f(P)$ jusqu'à n'importe quelle limite s'obtient immédiatement par la formule de Green et ses itérées, bien entendu en admettant certaines conditions de régularité par rapport à $f(P)$ et à ses dérivées et spécialement à l'allure de ces fonctions à l'infini. Il y a pourtant une lacune dans la représentation du prolongement telle qu'elle est fournie par les formules dudit numéro. Ces formules ne mettent nullement en évidence les relations $I^0 f(P) = f(P)$ et $I^{-2p} f(P) = \Delta^p f(P)$. D'autre part, la méthode de prolongement, appliquée dans les n^os **27—37**, est basée sur l'emploi de certaines coordonnées curvilignes qu'on a introduites par l'intermédiaire d'une certaine surface S qui nous fait défaut ici. Il est pourtant facile de remédier à cet inconvénient et d'adapter la méthode au cas où nous nous trouvons.

Nous remplaçons provisoirement dans nos formules les points $\mathbf{a}$ de S par les points $\mathbf{z}$ de H et nous posons pour les points $\boldsymbol{\xi}$ du cône D^P (l'origine étant toujours placée en P) $\boldsymbol{\xi} = \sigma \mathbf{z}$, le paramètre σ variant cette fois de 0 à ∞. En observant que r_a devient $r_z = 1$ et que par conséquent $v = t^{-1} r_a$ devient t^{-1}, les formules (114bis) et (115) peuvent s'écrire

$$(114^{ter})\qquad I^\alpha f(P) = \frac{1}{\Gamma(\alpha)\,\Gamma(\beta)} \int_0^1 \left\{ \int_0^\infty G(\sigma, T, \alpha)\, \sigma^{\alpha-1}\, T^{\beta-1}\, d\sigma \right\} d\,T,$$

où

$$G(\sigma, T, \alpha) = \frac{\Gamma\left(\frac{\alpha+1}{2}\right) k(T)}{\pi^{\frac{m-1}{2}}} \cdot t^{-\alpha} \int_{\omega_{m-1}} f(\sigma \cdot \mathbf{z})\, d\,\omega_{m-1}.$$

Puisque $\mathbf{z}$ ne reste pas borné, nous le remplaçons par $\bar{\mathbf{z}} = t\mathbf{z}$, qui reste borné d'après (69). Posant encore $\bar{\sigma} = t^{-1}\sigma$, nous aurons

$$\boldsymbol{\xi} = \sigma \cdot \mathbf{z} = \bar{\sigma} \cdot \bar{\mathbf{z}}, \quad \sigma^{\alpha-1} d\sigma = t^{\alpha} \bar{\sigma}^{\alpha-1} d\bar{\sigma}.$$

Les $\bar{z}_k$ étant des fonctions suffisamment dérivables de T (leurs dérivées du premier ordre sont constants et celles d'ordre supérieur s'annulent d'après (70)), $f(\boldsymbol{\xi}) = f(\bar{\sigma} \cdot \bar{\mathbf{z}})$ sera une fonction suffisamment dérivable de T et de $\bar{\sigma}$. En posant encore

$$\bar{G}(\bar{\sigma}, T, \alpha) = t^{\alpha} G(\sigma, T, \alpha) = \frac{\Gamma\left(\frac{\alpha+1}{2}\right) k(T)}{\pi^{\frac{m-1}{2}}} \int\limits_{\omega_{m-1}} f(\bar{\sigma} \cdot \bar{\mathbf{z}})\, d\omega_{m-1}, \tag{120}$$

on aura

$$I^{\alpha} f(P) = \frac{1}{\Gamma(\alpha)\,\Gamma(\beta)} \int\limits_0^1 \left\{ \int\limits_0^{\infty} \bar{G}(\bar{\sigma}, T, \alpha)\, \bar{\sigma}^{\alpha-1}\, T^{\beta-1}\, d\sigma \right\} dT. \tag{121}$$

La seule différence entre la dernière formule et (114bis) est que l'un des intervalles d'intégration est infini. Ceci peut cependant être compensé par des conditions convenables concernant l'allure de f à l'infini. Cela admis, les procédés des numéros qui précèdent pourront être appliqués et on aura donc aussi dans ce cas particulier la relation $I^0 f(P) = f(P)$. Moyennant la formule (67) du n° **26**, il en résulte sur le champ $I^{-2p} f(P) = \Delta^p f(P)$.

IV. Différentiation sous le signe $\int$.

39. Différentiation de I^{α} pour α assez grand. — Terminons ce chapitre, qui est déjà devenu bien long, par l'examen du problème de la différentiation de nos intégrales I^{α}_{*} par rapport aux coordonnées du point $P(x_1, x_2, \ldots, x_m)$, considéré comme point variable. Nous allons montrer que ces différentiations (d'ordre quelconque) peuvent s'exécuter sous le signe $\int$, sans qu'on ait à prendre égard à la variation avec P de la frontière du domaine d'intégration. En particulier, l'opération Δ appliquée à nos expressions pourra être accomplie sous le signe d'intégration, ce qui nous donnera la formule importante (63) du n° **24**, dont nous avons ajourné la démonstration.

Admettons d'abord que α soit un nombre assez grand pour que $r_{PQ}^{\alpha-m}$ et ses dérivées, autant qu'elles interviennent, soient continues et par conséquent s'annulent sur la nappe C^P du cône rétrograde au sommet P. Définissons en outre dans un domaine $\bar{D}$, embrassant le domaine D_S^P, la fonction $K(P, Q) = r_{PQ}^{\alpha-m}$ autant que Q se trouve dans la portion d'espace commune aux domaines $\bar{D}$ et D^P, et $= 0$ ailleurs. Alors $K(P, Q)$ considéré comme fonction des deux points variables P et Q est continu et différentiable par rapport à P un certain nombre de fois. Dans l'expression $\overline{I_*^\alpha f, g, h}(P)$, la fonction $r_{PQ}^{\alpha-m}$ pourra évidemment être remplacée par la fonction $K(P, Q)$, c'est-à-dire qu'on aura des intégrales étendues respectivement au domaine fixe $\bar{D}$ et à la portion fixe $\bar{S}$ de la surface S qui se trouve dans ce domaine. (Pour que cette portion soit bien définie, on aura à choisir $\bar{D}$ dans un entourage assez étroit de D_S^P.) Les dernières intégrales peuvent manifestement être différentiées sous le signe $\int$ et alors il en sera de même des intégrales originales (identiques à celles-là).

40. Différentiation de l'intégrale prolongée dans un cas particulier. — Nous allons montrer que la règle énoncée reste valide pour toute valeur de α jusqu'à laquelle on peut poursuivre le prolongement de nos intégrales et de leurs dérivées en question. Considérons d'abord le cas particulièrement simple, traité aux n^{os} **26** et **38**, où le domaine d'intégration se confond avec le cône entier D^P.

Posons donc

$$I^\alpha f(P) = \frac{1}{H_m(\alpha)} \int_{D^P} f(Q) r_{PQ}^{\alpha-m} dQ.$$

En appliquant la règle de différentiation sous le signe $\int$, il vient

$$\frac{\partial}{\partial x_k} I^\alpha f(P) = \frac{1}{H_m(\alpha)} \int_{D^P} f(Q) \frac{\partial}{\partial x_k} r_{PQ}^{\alpha-m} dQ \tag{122}$$

et cette relation est exacte, d'après ce que nous venons de prouver, pour $\alpha > m + 2$, puisque, pour ces valeurs de α, $r_{PQ}^{\alpha-m}$ et $\partial r_{PQ}^{\alpha-m} / \partial x_k = (\alpha - m) r_{PQ}^{\alpha-2-m} (x_k - \xi_k)$ s'annulent sur la nappe C^P. En outre, le second membre de l'expression (122) est une fonction *holomorphe* de α, soit $A(\alpha)$, pour $\alpha > m + 2$.[1] Il en est donc de même du premier membre, c'est-à-dire de $\partial I^\alpha f(P) / \partial x_k$. Or nous montrerons à l'instant que, dans des conditions de régularité qu'il serait facile de spécifier, la dernière fonction est une fonction *holomorphe* de α, soit $B(\alpha)$, dans un intervalle plus

[1] En réalité, on voit immédiatement que le second membre de (122) est holomorphe, pour $\alpha > m$, mais l'existence du premier membre n'est démontrée pour le moment que pour $\alpha > m + 2$. Une remarque analogue s'applique à la formule (123) en y remplaçant m par $m - 2p$.

étendu que le précédent, savoir pour $\alpha > m + 2 - 2p$. Les deux fonctions $A(\alpha)$ et $B(\alpha)$ étant identiques pour $\alpha > m+2$, le seront encore pour tout $\alpha > m+2-2p$. *On obtient donc la dérivée* $\partial I^\alpha / \partial x_k$ pour les valeurs $m + 2 - 2p < \alpha \leqq m + 2$ *par le prolongement analytique du second membre de* (122), ce prolongement étant exécuté *d'une manière ou d'une autre*. Cela signifie dans notre langage que *la règle de différentiation sous le signe* $\int$ *reste valide pour* $\alpha > m + 2 - 2p$, *c'est-à-dire même pour des intégrales éventuellement divergentes.*

Pour achever cette démonstration, il suffit d'écrire, moyennant la formule (itérée) de Green du n° **26**,

$$I^\alpha f(P) = \frac{1}{H_m(\alpha + 2p)} \int_{D^P} \Delta^p f(Q)\, r_{PQ}^{\alpha+2p-m}\, dQ$$

et d'appliquer à cette intégrale les remarques faites au sujet de l'intégrale (107). On voit alors que

$$\frac{\partial I^\alpha f(P)}{\partial x_k} = \frac{1}{H_m(\alpha + 2p)} \int_{D^P} \Delta^p f(Q) \frac{\partial}{\partial x_k} r_{PQ}^{\alpha+2p-m}\, dQ \tag{123}$$

est bien une fonction holomorphe $(B(\alpha))$ pour tout $\alpha > m + 2 - 2p$.(¹)

Nous soulignons encore une fois, même au risque d'ennuyer le lecteur, que dans l'ordre d'idées que nous poursuivons, l'expression explicite pour $\partial I^\alpha / \partial x_k$ donnée par le second membre de la dernière formule est pour nous dépourvue de tout intérêt. Elle ne sert qu'à mettre en évidence que la dérivée en question *existe* pour les dites valeurs de la variable α et qu'elle en est une fonction *holomorphe*. Nous insistons sur ce point puisque, dans le cas général qui va suivre, nous ne serons pas à même de donner une expression explicite semblable.

Le raisonnement ci-dessus s'étend immédiatement aux dérivées d'ordre supérieur.

41. Différentiation dans le cas général. — Envisageons maintenant l'intégrale $I^\alpha f(P)$ sous la forme (114bis), qui s'écrit brièvement $I^\alpha f(P) = J^{\alpha\beta} G$. Or le prolongement de cette dernière expression est effectué par les formules explicites qui se retrouvent dans la démonstration du lemme II (n° **35**). Les données originales $f(\xi)$ et $S(\mathrm{a})$ étant suffisamment dérivables, les fonctions figurant dans les dites formules seront, d'après le théorème du n° **30**, suffisamment dérivables par rapport aux variables x_k. Pour autant qu'il s'agit d'intégrales, elles sont prises

(¹) *Voir* la note précédente.

entre des limites fixes, o et 1, et leur différentiation pourra par conséquent être exécutée sous le signe $\int$. Les fonctions obtenues seront encore des fonctions holomorphes de α jusqu'à une certaine valeur, soit α_0', qu'il serait facile de préciser. (Pour arriver à une valeur fixée α_0, il suffit d'augmenter les nombres n respectifs figurant au n° **36**, d'une unité pour chaque dérivation $\partial/\partial x_k$.)

Il en résulte, d'après ce qui a été dit aux n^os^ **39—40**, que $\partial I^\alpha f(P)/\partial x_k$ est identique à la fonction figurant au second membre de (122), n° **40**, interprétée, le cas échéant, par son prolongement analytique.

42. L'opérateur Δ. — La méthode esquissée ici pour l'intégrale de volume $I^\alpha f$ s'applique aisément aux deux intégrales de surface figurant dans $I_*^\alpha \overline{f, g, h}$. L'extension aux dérivées d'ordre supérieur n'implique non plus aucune difficulté. Il ne nous paraît pas nécessaire de nous arrêter sur les détails.

L'application des résultats obtenus à l'opérateur Δ nous fournit la formule importante (63) du n° **24** (*cf.* la formule (59) du n° **22**). Nous remarquons que cette formule peut aussi servir à effectuer le prolongement analytique de I_*^α. (*Cf.* les n^os^ **52** et **53**.)

Enfin, en posant dans la formule en question $\alpha = 2$, il vient (*cf.* (64), n° **24**)

$$\Delta I_*^2 \overline{f, g, h}(P) = I_*^0 \overline{f, g, h}(P) = f(P). \tag{124}$$

Cette relation remarquable, l'analogue de la formule de Poisson, relative au potentiel newtonien, va jouer un rôle important, quand il s'agira, au chapitre suivant, d'analyser notre solution du problème de Cauchy.

CHAPITRE IV.

Le problème de Cauchy pour l'équation des ondes.

43. Énoncé du problème. — Pour l'équation des ondes[1], le problème de Cauchy consiste — nous suivons M. Hadamard[2] — à trouver une solution u de l'équation $\Delta u = f$ quand on connaît les valeurs de u, soit $u = h$, et d'une de ses dérivées premières, soit $du/dn = g$, en chaque point d'une surface S, orientée dans l'espace[3], la dérivée étant prise dans une direction n qui n'est pas tangente à la surface. On peut, sans restreindre la généralité, admettre que la direction n est celle de la normale intérieure (lorentzienne) à la surface.

[1] Dans ce chapitre, nous admettons, en général, que les ε_k ont leurs valeurs originales, $\varepsilon_1 = 1$, $\varepsilon_2 = \cdots = \varepsilon_m = -1$.

[2] Hadamard **1**, p. 219.

[3] Pour d'autres orientations *voir* les n^os^ **61—64**.

La région de l'espace à m dimensions où l'on veut calculer la solution u est telle que, si l'on construit le cône rétrograde D^P qui a un des points P de la région comme sommet, la nappe C^P de ce cône découpera de la surface S une certaine portion S^P et formera avec S^P la frontière d'un domaine borné D_S^P de l'espace. (Pour les notations, voir le chapitre précédent.)

Quant à l'historique du problème et aux travaux qui s'y rapportent, nous sommes dispensés de les faire passer en revue ici, nous trouvant dans la situation agréable de pouvoir renvoyer le lecteur aux exposés de M. Hadamard, incorporés pour la plupart à son livre magistral sur le sujet.([1]) La place du problème de Cauchy parmi les divers problèmes aux limites, *correctement posés ou non* (« bien posés ou non », comme disait autrefois l'illustre auteur) s'y trouve aussi élucidée d'une manière incomparable. Nous tenons en outre à mentionner une conférence plus récente que le livre cité, faite par M. Hadamard à Genève en 1935 et imprimée dans l'*Enseignement Mathématique.*([2])

44. Solution générale. — En récrivant ici la dernière identité du n° **23**,

$$(1) \qquad I^\alpha u(P) = I_*^{\alpha+2}\,\overline{\Delta u,\ \frac{du}{dn},\ u(P)},$$

et en y posant $\alpha = 0$, on obtient, en raison de $I^0 u(P) = u(P)$, l'identité particulière

$$(2) \qquad u(P) = I_*^2\,\overline{\Delta u,\ \frac{du}{dn},\ u(P)},$$

c'est-à-dire que la valeur de u au point P peut se calculer par une intégrale spatiale portant sur Δu et étendue à un certain volume D_S^P, par des intégrales de simple couche et de double couche, portant respectivement sur $\dfrac{du}{dn}$ et u et étendues à une certaine portion S^P de la surface.

En ce qui concerne le problème de Cauchy, il vient de la dernière identité, en y portant les données du problème,

$$(3) \qquad u(P) = I_*^2\,\overline{f,\ g,\ h}(P).$$

Cela veut dire que si le problème de Cauchy admet une solution, ce qui reste à voir, cette solution est donnée par la dernière formule.

([1]) HADAMARD **1**. *Cf.* aussi diverses parties du *Cours d'Analyse* de GOURSAT, tome III.
([2]) HADAMARD **2**.

En réalité, la seule voie qu'on connaisse pour établir l'existence d'une solution du problème dans l'étendue[1] dont il s'agit ici, est de vérifier que la solution donnée par la formule ci-dessus satisfait aux conditions du problème, bien entendu dans des hypothèses convenables sur les données et sur la surface S. Une telle vérification trop sommaire sous plusieurs rapports sera faite plus loin (nº **60**). Néanmoins, pour ne pas alourdir notre exposé, nous y admettrons tacitement qu'une solution existe et qu'elle est, en conséquence, fournie par la formule (3).

Malheureusement, nous avons vu que les intégrales figurant dans cette solution sont en général divergentes, et il faut par suite leur donner un sens par prolongement analytique. La discussion de ce prolongement qui nous fournira l'expression I_*^2 (effectué d'ailleurs d'une manière ou d'une autre) forme l'objet principal du présent Chapitre. Nous y rencontrerons une différence essentielle entre les solutions se référant aux dimensions d'ordre pair et impair. Pour les dimensions paires on a le principe de Huygens[2], tandis que ce principe fait défaut pour les dimensions impaires. Dans ce dernier cas, les expressions sous forme finie, obtenues par le prolongement analytique des intégrales divergentes, sont essentiellement identiques à celles qu'on trouve par la méthode de la «*partie finie*», due à M. Hadamard. Or, au point où nous nous trouvons pour le moment, ce n'est pas la différence dépendant de la parité de l'espace qui nous intéresse en premier lieu. Tout au contraire, c'est que sous sa forme générale, donnée plus haut, *la solution est indépendante de la parité de la dimension de l'espace.* On sait qu'il n'en est pas ainsi des solutions dues à M. Hadamard qui sont de formes différentes suivant la parité de la dimension[3].

Avant d'entamer le cas général, nous traiterons deux cas particuliers simples, correspondant à $m = 1$ et à $m = 2$. Le cas $m = 2$ échappe dans un certain sens à la règle générale. Le principe de Huygens n'y intervient que partiellement.

45. Problème linéaire, $m = 1$. — Rappelons la formule (20) du nº **16**

$$H_m(\alpha) = \pi^{\frac{m-2}{2}} \cdot 2^{\alpha-1}\,\Gamma\left(\frac{\alpha}{2}\right)\Gamma\left(\frac{\alpha+2-m}{2}\right). \tag{4}$$

[1] Pour le théorème d'existence de Cauchy-Kowalewski et la critique de M. Hadamard, voir Hadamard **1**, p. 19, 14, 27.

[2] Pour le principe de Huygens voir préliminairement l'Introduction. Pour le problème de Cauchy ce principe se manifeste par le fait que seules les valeurs que les données admettent au voisinage infinitésimal de la nappe C^P du cône D^P influencent la valeur de la solution au point P (*cf.* plus loin, nº **55**).

[3] *Voir* Hadamard **1**, p. 227 (formule (39)) et p. 311 (formule (29)).

On trouve facilement que $H_1(\alpha) = \Gamma(\alpha)$. La surface S se réduit ici à un seul point, soit a, la dérivation d/dn se réduit à d/da et Δ à d^2/dx^2. La formule générale (60) du n° **22**, dont la formule (61bis) du n° **23** et la formule identique (1) du n° **44** offrent la traduction symbolique, se réduit donc à

$$\frac{1}{\Gamma(\alpha)}\int_a^x u(\xi)(x-\xi)^{\alpha-1}\,d\xi$$

$$= \frac{1}{\Gamma(\alpha+2)}\left\{\int_a^x u''(\xi)(x-\xi)^{\alpha+2-1}\,d\xi + u'(a)(x-a)^{\alpha+2-1} - u(a)\frac{d}{da}(x-a)^{\alpha+2-1}\right\}$$

$$= \frac{u(a)}{\Gamma(\alpha+1)}(x-a)^{\alpha} + \frac{u'(a)}{\Gamma(\alpha+2)}(x-a)^{\alpha+1} + \frac{1}{\Gamma(\alpha+2)}\int_a^x u''(\xi)(x-\xi)^{\alpha+1}\,d\xi,$$

identité qui n'est qu'un cas particulier de l'identité (4) du n° **2**. Les intégrales qui y figurent convergent pour $\alpha > 0$ et pour $\alpha > -1$ respectivement. En passant à la limite $\alpha \to +0$, on arrive à la formule classique, rencontrée incidemment au Chapitre II (n° **12**)

$$u(x) = u(a) + u'(a)(x-a) + \int_a^x u''(\xi)(x-\xi)\,d\xi,$$

qui fournit la solution du problème de Cauchy:

Déterminer $u(x)$ *pour* $a \leqq x < b$ $(b \leqq +\infty)$ *lorsqu'on connaît* $u''(x)$ *dans le même intervalle et, en outre,* $u(a)$ *et* $u'(a)$.

46. Problème dans le plan, $m = 2$. — Dans ce cas, Δ se réduit à l'opérateur des cordes vibrantes $\partial^2/\partial x_1^2 - \partial^2/\partial x_2^2$. La nappe C^P du cône se réduit à deux droites $x_1 - \xi_1 = \pm(x_2 - \xi_2)$ (les caractéristiques), issues du point $P(x_1, x_2)$, le cône rétrograde D^P à l'angle rétrograde D^P limité par ces droites et caractérisé par $r_{PQ}^2 = (x_1-\xi_1)^2 - (x_2-\xi_2)^2 \geqq 0$ et $x_1 - \xi_1 \geqq 0$, où $Q = Q(\xi_1, \xi_2)$. La surface S devient une courbe S à orientation d'espace avec $d\xi_1^2 - d\xi_2^2 < 0$ (fig. 5 et 5 a).

L'élément de surface dS, défini pour une dimension arbitraire par la formule (39) du n° **21**, se réduit ici à

$$dS = |d\xi_1^2 - d\xi_2^2|^{\frac{1}{2}} = (d\xi_2^2 - d\xi_1^2)^{\frac{1}{2}}.$$

La portion S^P de S interceptée par l'angle D^P limite avec les deux côtés de cet angle un domaine à deux dimensions D_S^P. Les points frontières de l'arc S^P seront désignés par B_1 et B_2.

De notre formule générale (1) nous déduirons dans ce qui suit l'identité bien connue

$$(5)\qquad u(P) = \frac{1}{2}\left\{\iint\limits_{D_S^P} \Delta u\, dQ + \int\limits_{S^P} \frac{du}{dn}\, dS + u(B_1) + u(B_2)\right\},$$

qui, sous une forme légèrement différente et avec une déduction tout à fait élémentaire se trouve p. ex. dans le Cours d'Analyse de Goursat.(1) Chez M. Hadamard on trouve la formule ci-dessus dans les notations utilisées ici, et cela même pour l'équation aux coefficients variables étudiée par Riemann.(2)

Pour nous, le problème consiste à obtenir (5) comme cas particulier de (2), dont on a donc à calculer le second membre, à savoir I_*^2, dans les conditions actuelles. Il vient de (4), pour $m = 2$, $H_2(\alpha) = 2^{\alpha-1}\Gamma^2\left(\frac{\alpha}{2}\right)$. En observant encore que $\alpha + 2 - m$ devient α pour $m = 2$, on aura

$$(6)\qquad I_*^{\alpha+2}\,\overline{\Delta u,\ \frac{du}{dn},\ u(P)}$$
$$= \frac{1}{2^{\alpha+1}\Gamma^2\left(\frac{\alpha+2}{2}\right)}\left\{\iint\limits_{D_S^P} \Delta u\cdot r^\alpha\cdot dQ + \int\limits_{S^P}\left(\frac{du}{dn}r^\alpha - u\frac{dr^\alpha}{dn}\right)dS\right\}.$$

47. Méthode erronée. — On est maintenant tenté de calculer I_*^2 en *posant* dans la dernière formule $\alpha = 0$. Il viendrait alors $r^\alpha = r^0 = 1$, $dr^\alpha/dn = dr^0/dn = 0$. C'est-à-dire, par ce mode de calcul qui, nous allons le voir, est plus ou moins illusoire, on arriverait à la formule (5), les termes $u(B_1)$ et $u(B_2)$ y étant supprimés. Or, sous cette forme l'identité (5) est fausse, puisqu'elle l'est pour $u = \text{const.} \neq 0$ par exemple.

Voici maintenant l'explication de ce fait paradoxal. Sur les deux droites PB_1 et PB_2 on n'a pas $r^0 = 1$, car on y a $r = 0$. Nous verrons tout à l'heure que cette circonstance, sans conséquences dangereuses en tant qu'il s'agit des deux premières intégrales, est responsable de l'erreur que nous venons de commettre au sujet de la troisième. La discordance provient du fait que le calcul de I_*^2 par la substitution $\alpha = 0$ dans $I_*^{\alpha+2}$ ne donne pas le prolongement analytique correct.

47*bis*. Méthode correcte. — Il faut donc reprendre l'étude du second membre de la formule (6) pour une valeur générale de $\alpha > 0$. Pour toutes ces valeurs,

(1) Tome III (1915) p. 117.
(2) Hadamard **1**, p. 314 et *suiv.*

nos intégrales convergent, et l'identité est exacte, sa validité ayant été démontrée au n° **24** pour $\alpha > m - 2$. Cela posé, notre problème se réduit à exécuter dans (6) le passage à la limite $\alpha \to 0$ d'une façon rigoureuse.

Le passage en question est immédiat pour les deux premières intégrales, puisque r^α tend uniformément vers 1, sauf au voisinage du sommet P et des lignes PB_1 et PB_2, où r^α reste borné. Le passage à la limite correct conduit donc, lui aussi, aux deux premiers termes, comme il fallait s'y attendre.

Il est plus délicat d'exécuter le passage à la limite pour la troisième intégrale dans (6). Désignons d'abord par C_1 et C_2 deux points de l'arc $B_1 B_2$, voisins de B_1 et de B_2 respectivement, et tels que r_{PQ} varie d'une manière croissante de 0 à δ lorsque Q varie de B_1 à C_1 ou de B_2 à C_2. Alors la convergence de dr^α/dn vers 0 étant uniforme sur l'arc $C_1 C_2$, la troisième intégrale, prise le long de cet arc, tendra effectivement vers 0, et on pourra se restreindre aux intégrales prises le long des arcs $B_1 C_1$ et $B_2 C_2$. D'autre part, en désignant par BC l'un quelconque de ces arcs et posant $r^2 = R$, on peut écrire[1]

$$(7) \qquad \int_B^C u \frac{dr^\alpha}{dn} dS = \int_B^C u \frac{dR^{\frac{\alpha}{2}}}{dn} dS = \frac{\alpha}{2} \int_0^{\delta^2} u R^{\frac{\alpha}{2}-1} \frac{dR}{dn} \frac{dS}{dR} dR.$$

Nous allons démontrer à l'instant que la valeur de $\frac{dR}{dn} \cdot \frac{dS}{dR}$ est $= -1$ aux points B, résultat qui sera présumé pour le moment. D'autre part, on voit par le mécanisme développé au Chap. I (n° **3**) que l'expression tend avec $\alpha \to 0$ vers la valeur de $u \cdot \frac{dR}{dn} \cdot \frac{dS}{dR}$ au point B, c'est-à-dire que, par le résultat présumé, elle tend vers $-u(B)$. On tire de là que

$$\lim_{\alpha \to 0} \left(- \frac{1}{2^{\alpha+1} \Gamma^2 \left(\frac{\alpha+2}{2} \right)} \int_B^C u \frac{dr^\alpha}{dn} dS \right) = \tfrac{1}{2} u(B).$$

Ainsi le passage à la limite, exécuté d'une façon correcte, nous donne la formule (5) comme conséquence de la formule (6). On voit que le principe de Huygens n'intervient dans la solution que dans les termes provenant de la double couche.

48. L'identité $dR/dn + dR/dS = 0$. — Il nous reste à démontrer qu'à l'un quelconque des points B_1 et B_2, désigné comme plus haut par B, on a

---

[1] Le lecteur voudra bien observer qu'on attribue partout une valeur positive à dS.

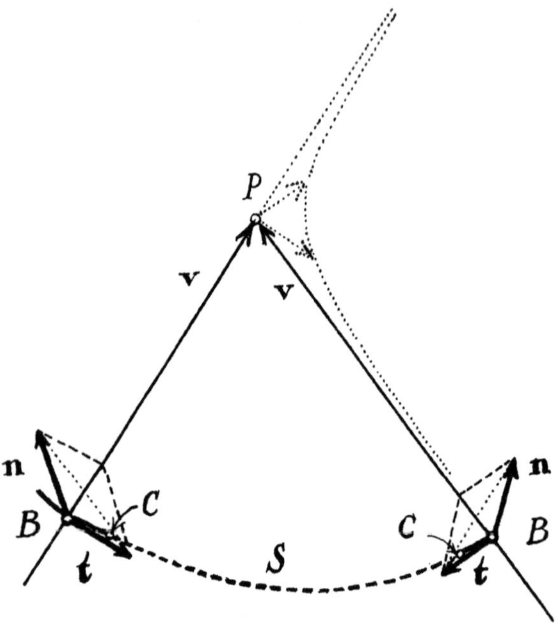

Fig. 5. C'est avec intention qu'on a modifié ici l'orientation et l'ouverture usuelles du « cône ». Les deux vecteurs tangents **t** de la figure, qui dans le texte sont supposés unitaires, ont la même « longueur », car il en est ainsi des deux vecteurs pointillés qui leur sont parallèles et qui aboutissent à une hyperbole admettant les deux génératrices comme asymptotes. Dans les deux parallélogrammes chacune des diagonales pointillées est parallèle à une génératrice, ces deux diagonales sont donc isotropes elles aussi. Par conséquent, les vecteurs **n** sont orthogonaux aux vecteurs **t** correspondants et ils ont la même « longueur ».

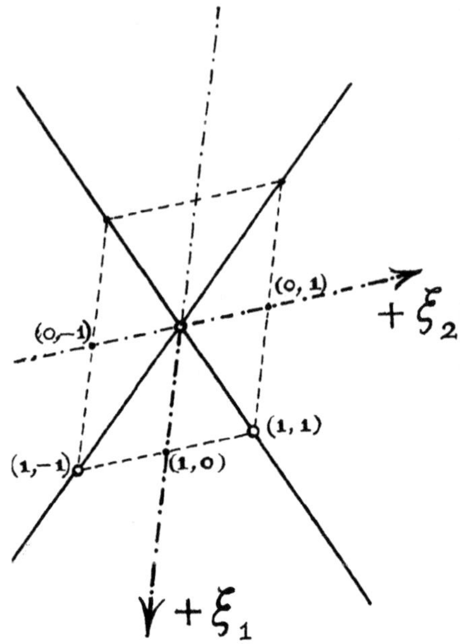

Fig. 5 a. Les systèmes lorentziens de coordonnées admissibles sont donnés par le postulat que la figure constituée des deux « génératrices » y admette l'équation $(\xi_1 - x_1)^2 - (\xi_2 - x_2)^2 = 0$. C'est un tel système qui est illustré par la figure. Les vecteurs **t** et **n** de la fig. 5 sont manifestement de « longueur » < 1 dans l'échelle de la fig. 5 a.

$$\frac{dR}{dn} \cdot \frac{dS}{dR} = -1$$

ou, ce qui revient au même,

$$\frac{dR}{dn} + \frac{dR}{dS} = 0.$$

En désignant le vecteur allant de B à P par $\mathbf{v}_{BP} = \mathbf{v}$, la normale intérieure[1] unitaire par **n**, et la tangente unitaire menée en B vers l'intérieur de l'angle D^P par **t**, on a, d'après la *Remarque* qui se trouve à la fin du n° **21**,

$$\frac{dR}{dn} + \frac{dR}{dS} = -2(\mathbf{v}, \mathbf{n} + \mathbf{t}).$$

[1] Évidemment, les normales « intérieures » menées aux points frontières B ne pénètrent pas dans l'intérieur du domaine D_S^P, mais elles peuvent être considérées comme cas limites de normales intérieures proprement dites.

(Il est à observer que $\mathbf{v}_{BP} = -\mathbf{v}_{PB}$.) Nous allons démontrer que

$$\mathbf{n} + \mathbf{t} = \lambda \cdot \mathbf{v},$$

d'où il résultera

$$\frac{dR}{dn} + \frac{dR}{dS} = -2\lambda\,\mathbf{v}^2 = 0,$$

puisque la droite reliant B à P est une « génératrice du cône D^P », ce qui fait que le vecteur $\mathbf{v}$ est isotrope: $\mathbf{v}^2 = 0$. Or, le vecteur $\mathbf{v}$ se trouve manifestement dans l'angle saillant formé par les vecteurs $\mathbf{n}$ et $\mathbf{t}$. On a par conséquent $\mathbf{v} = \nu\mathbf{n} + \tau\mathbf{t}$, ν et τ étant positifs tous les deux. Nous allons démontrer que $\nu = \tau$. En effet, $\mathbf{n}$ et $\mathbf{t}$ sont unitaires et orthogonaux l'un à l'autre, $\mathbf{n}$ est un vecteur de temps et $\mathbf{t}$ un vecteur d'espace. Il s'ensuit

$$0 = \mathbf{v}^2 = (\nu\,\mathbf{n} + \tau\,\mathbf{t})^2 = \nu^2\mathbf{n}^2 + 2\nu\tau(\mathbf{n}, \mathbf{t}) + \tau^2\mathbf{t}^2 = \nu^2 - \tau^2.$$

On a donc bien $\nu = \tau$, ou $\mathbf{n} + \mathbf{t} = \frac{\mathbf{v}}{\nu}$, ce qui achève la démonstration.

49. L'opérateur I_*^m. — Nous avons longuement insisté sur la résolution du problème des cordes vibrantes qui pourtant, nous l'avons dit, est facile à obtenir par les méthodes usuelles. La raison en est que nous tenions à faire rentrer la solution de ce problème dans le cadre de notre méthode générale, méthode dont l'application à une dimension particulière quelconque n'est en principe ni plus facile, ni plus difficile que l'application à une autre. Or nous allons voir à l'instant que nous avons obtenu beaucoup plus qu'il ne paraît au premier coup d'œil; nous avons, chemin faisant, trouvé une méthode générale qui fournira, pour toute dimension d'ordre *pair*, une solution explicite du problème de Cauchy qui est, au point de vue formel, d'une simplicité remarquable.

Cependant, la méthode que nous venons de développer ne conduira pas directement à l'expression I_*^2 qui nous intéresse en premier lieu, mais elle nous donnera presque immédiatement, pour toute dimension d'ordre *pair ou impair*, l'expression de I_*^m, identique, pour $m = 2$, à l'expression I_*^2.

Désignons par s^P le bord de la portion de surface S^P, c'est-à-dire la variété à $m - 2$ dimensions formée par l'intersection de la nappe C^P et de la surface S. Nous désignons encore par ds l'élément de « surface » de s^P.(¹)

Ces notations posées, nous trouverons l'identité suivante qui ne laisse rien à désirer en simplicité et en élégance:

(¹) C'est-à-dire élément de volume à $m - 2$ dimensions; *voir* le n° **21**^{bis}.

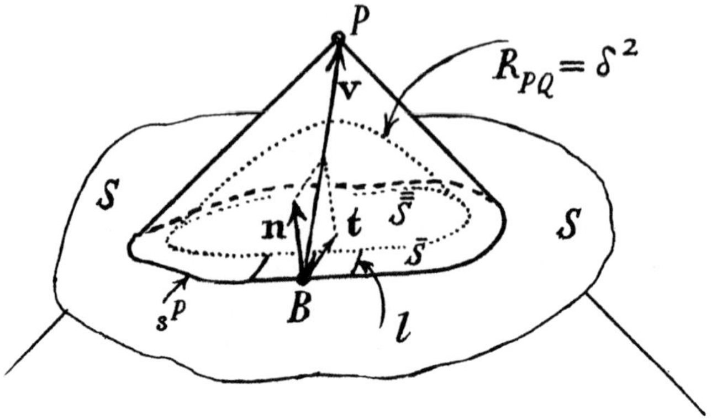

Fig. 6.

(8) $$I_*^m \overline{f, g, h}(P) = \frac{1}{H_m(m)} \left\{ \int_{D_S^P} f\, dQ + \int_{S^P} g\, dS + \int_{s^P} h\, ds \right\}$$

avec [1]

(9) $$H_m(m) = \pi^{\frac{m-2}{2}} 2^{m-1} \Gamma\left(\frac{m}{2}\right) = (m-2)!\, \omega_{m-1}.$$

La formule (8) s'obtient de $I_*^{\alpha+m}$ écrite pour $\alpha > 0$ par le passage à la limite $\alpha \to 0$. [2] On montre, tout comme au n° **47**, que, quant aux deux premiers termes, ce passage à la limite s'exécute par la substitution $\alpha = 0$ dans les deux premiers termes de $I_*^{\alpha+m}$. Pour le troisième terme il est nécessaire d'entrer dans les détails.

50. Terme provenant de la double couche. — Le troisième terme dans $I_*^{\alpha+m}\overline{f, g, h}(P)$ est, en tenant compte de $(\alpha + m) - m = \alpha$,

(10) $$-\frac{1}{H_m(\alpha+m)} \int_{S^P} h \frac{d r^\alpha}{dn} dS = -\frac{W}{H_m(\alpha+m)}.$$

En posant $r^2 = R$, on peut écrire

(11) $$W = \int_{S^P} h \frac{d R^{\frac{\alpha}{2}}}{dn} \cdot dS = \frac{\alpha}{2} \int_{S^P} h R^{\frac{\alpha}{2}-1} \cdot \frac{dR}{dn} \cdot dS.$$

En calquant notre raisonnement sur celui du n° **47**, nous divisons S^P par son intersection avec l'hyperboloïde

(1) *Cf.* formule (8) du n° **6**.

(2) *Voir* pour I_*^α la formule (61) du n° **23**.

$$r_{PQ}^2 = (\mathbf{x} - \boldsymbol{\xi})^2 = \delta^2$$

en deux parties (fig. 6). Nous désignons par $\bar{S}$ la partie annulaire extérieure et par $\bar{\bar{S}}$ la partie intérieure. Alors on voit, tout comme au numéro cité, que la contribution provenant de $\bar{\bar{S}}$ à la dernière expression de W dans (11) tend vers o avec α. Cela étant, on peut se restreindre à l'expression

$$\overline{W} = \beta \int\limits_{\bar{S}} h R^{\beta-1} \cdot \frac{dR}{dn} \cdot dS,$$

où l'on a posé $\alpha/2 = \beta$.

Faisons maintenant varier la valeur de R de o à δ^2 et désignons par s_R l'intersection de S avec l'hyperboloïde $(\mathbf{x} - \boldsymbol{\xi})^2 = R$. Considérons de plus les trajectoires orthogonales (bien entendu, au sens lorentzien) des variétés s_R, l'existence de telles trajectoires étant garantie par les théorèmes d'existence les plus simples des systèmes d'équations différentielles ordinaires, du moins pour δ assez petit. De chaque point B du bord s^P il part une trajectoire et une seule qui établit une correspondance biunivoque entre les points de s^P et ceux d'une variété s_R quelconque. Un tube de telles trajectoires issues des points d'un élément ds de s^P fait correspondre à cet élément un élément de s_R qui sera désigné, ainsi que sa mesure, par ds_R. L'élément ds se rétrécissant au point B, ds_R va se rétrécir au point de s_R correspondant à B et $ds_R : ds$ tendra vers une valeur limite qui est fonction du point B et de la valeur R. Cette limite sera encore désignée par $ds_R : ds$.

D'autre part, en définissant l'élément d'arc d'une trajectoire quelconque par $dl = (-\Sigma \varepsilon_k d\xi_k^2)^{\frac{1}{2}}$, on aura évidemment (formule (55), n° **21**^{bis}), pour l'élément de surface dS, l'expression $dS = ds_R \cdot dl$. Cela posé, il vient

$$\overline{W} = \beta \int\limits_{S^P} h R^{\beta-1} \cdot \frac{dR}{dn} \cdot dS = \beta \int\limits_{S^P} h R^{\beta-1} \cdot \frac{dR}{dn} \cdot ds_R \cdot dl$$

$$= \beta \int\limits_{s^P} ds \int\limits_0^{\delta^2} h R^{\beta-1} \cdot \frac{dR}{dn} \cdot \frac{ds_R}{ds} \cdot \frac{dl}{dR} dR.$$

Il est maintenant à peu près évident que ds_R/ds tend vers 1 lorsque R tend vers zéro. En présumant alors un instant que $\frac{dR}{dn} \cdot \frac{dl}{dR} = -1$ au point B du bord s^P on voit encore par la méthode du Chapitre I que la dernière expression de $\overline{W}$ tend pour $\beta = \alpha/2 \to 0$ vers

$$-\int_{s^P} h\,ds,$$

ce qui en raison de (10) donne le troisième terme dans (8).

Il faut encore vérifier la relation $\frac{dR}{dn}\cdot\frac{dl}{dR}=-1$ ou la relation équivalente

$$\frac{dR}{dn}+\frac{dR}{dl}=0, \tag{12}$$

cette relation ayant lieu aux points B du bord s^P. Or, pour y arriver, nous avons très peu à ajouter au raisonnement du n° **48**.

51. L'identité $dR/dn+dR/dl=0$. — Notons d'abord que l'identité $R_{PC}=0$ valable pour tout point C de la nappe C^P donne $\Sigma\,\varepsilon_k(x_k-c_k)\,dc_k=0$, les x_k et les c_k étant les coordonnées respectives de P et de C. Cela veut dire que le vecteur PC est orthogonal à tout déplacement infinitésimal du point C suivant la nappe. En particulier, B étant un point du bord s^P, le vecteur PB est orthogonal aux déplacements $\{dB\}$ de B suivant le bord. Or, ces déplacements forment une variété linéaire *à $m-2$ dimensions*. De leur côté, les vecteurs $\mathbf{n}$ et $\mathbf{t}$ sont aussi orthogonaux à cette variété ou, ce qui revient au même, orthogonaux au bord; la normale $\mathbf{n}$ puisqu'elle est orthogonale à la variété à $m-1$ dimensions des déplacements suivant S dont $\{dB\}$ fait partie, le vecteur $\mathbf{t}$ puisqu'il est, par définition, tangent à la trajectoire orthogonale issue de B. On en conclut que les trois vecteurs BP, $\mathbf{n}$ et $\mathbf{t}$ appartiennent à un plan à deux dimensions, engendré par les droites orthogonales au bord au point B. Il en résulte que le vecteur BP peut s'exprimer comme combinaison linéaire des vecteurs $\mathbf{n}$ et $\mathbf{t}$. En désignant donc, comme au n° **48**, le vecteur BP par $\mathbf{v}$, on voit comme là que $\mathbf{v}=\nu(\mathbf{n}+\mathbf{t})$, d'où, dans l'ordre d'idées du même numéro, il s'ensuit que

$$\frac{dR}{dn}+\frac{dR}{dl}=0.$$

52. Une solution explicite du problème de Cauchy pour $m=2l$. — Dans tout ce qui suit, nous supposerons que $m\geqq 3$ et alors, pour m pair, $m\geqq 4$. A partir de cette valeur, toutes les intégrales figurant dans la solution générale du n° **48** sont divergentes, et il faut, pour les obtenir sous forme finie, les interpréter par quelque méthode de prolongement analytique. L'expression explicite pour I_*^m déduite au n° **49** nous fournit un procédé effectif de prolongement. En effet, en y appliquant (l — 1)-fois la formule (63) du n° **24** (démontrée au n° **42**), on voit immédiatement que

$$(13)\quad u(P) = I_*^2 \overline{f, g, h}(P) = \Delta^{l-1} I_*^m \overline{f, g, h}(P) = \\ = \frac{1}{(m-2)!\,\omega_{m-1}} \Delta^{l-1} \left\{ \int_{D_S^P} f\,d\,Q + \int_{S^P} g\,d\,S + \int_{s^P} h\,d\,s \right\}.$$

Cette expression met en évidence le principe de Huygens[1] pour toute dimension d'ordre pair $\geqq 4$. En effet, il ressort immédiatement de la formule (8) que les dérivées de I_*^m ne dépendent que des valeurs que les données admettent au voisinage infinitésimal de la nappe. Il en est donc ainsi de $u = \Delta^{l-1} I_*^m$ pour $l \geqq 2$, c'est-à-dire $m \geqq 4$. Dans le n° **55**, nous déduirons le principe en question directement de la solution générale (3) (n° **44**).

53. Une solution explicite pour $m = 2\,l + 1$. — L'application (itérée) de l'opération Δ à I_*^m conduit à des I_* dont l'ordre est de la même parité que m. Pour un nombre impair m, elle ne conduira donc jamais à I_*^2. Or on a

$$(14)\quad I_*^{m+1} \overline{f, g, h}(P) = \frac{1}{H_m\,(m+1)} \left\{ \int_{D_S^P} f\,r\,d\,Q + \int_{S^P} \left(g\,r - h \frac{d\,r}{d\,n} \right) d\,S \right\},$$

toutes les intégrales étant manifestement convergentes. (Le facteur numérique peut aussi s'écrire $2/(m-1)!\,\omega_m$, où ω_m est la surface totale de la sphère unitaire dans l'espace euclidien à m dimensions; *cf.* n° **31**). On a alors pour la solution du problème de Cauchy

$$(15)\quad u(P) = I_*^2 \overline{f, g, h}(P) = \Delta^l I_*^{m+1} \overline{f, g, h}(P).$$

54. Caractère invariantif des solutions obtenues. — Les solutions que nous venons de déduire s'appliquent en particulier au deux cas classiques $m = 4$ et $m = 3$, traités par Kirchhoff et par Volterra. Dans sa critique instructive, adressée aux solutions de ces auteurs, M. Hadamard[2] leur reproche que les noyaux utilisés deviennent singuliers sur la ligne droite passant par le point P et parallèle à l'axe des ξ_1, singularité artificielle et étrangère au problème à résoudre. En particulier, le noyau de Volterra[3]

$$\log \left\{ \frac{x_1 - \xi_1 + \sqrt{(x_1-\xi_1)^2 - (x_2-\xi_2)^2 - (x_3-\xi_3)^2}}{\sqrt{(x_2-\xi_2)^2 + (x_3-\xi_3)^2}} \right\}$$

---

(1) *Cf.* pour l'énoncé du principe l'Introduction et, en particulier, le n° **55**.

(2) Hadamard **1**, p. 96 et *suiv.*

(3) *Cf.* p. ex. Goursat, *Cours d'Analyse* III (1915), p. 162.

est une fonction primitive par rapport à x_1 de la solution élémentaire $r^{-1} = \{(x_1 - \xi_1)^2 - (x_2 - \xi_2)^2 - (x_3 - \xi_3)^2\}^{-\frac{1}{2}}$, cette fonction primitive étant choisie de manière qu'elle s'annule sur la nappe du cône au sommet P. Ce noyau, qui présente une singularité logarithmique sur ladite ligne, donne lieu à trois intégrales convergentes, qu'il faut différentier par rapport à x_1, pour arriver à la solution définitive. Les reproches de M. Hadamard s'accentuent encore quand il devient question de la généralisation aux dimensions impaires d'ordre supérieur, due à M. Tedone. Là il s'agit des intégrations itérées de la solution élémentaire par rapport à x_1, suivies d'un nombre égal de différentiations, appliquées aux intégrales formées avec les noyaux correspondants.

Pour faire bien ressortir le point de vue de M. Hadamard, nous ne pouvons faire mieux que de reproduire textuellement le passage suivant, concernant la connexion éventuelle entre la ligne droite en question et le problème dont il s'agit:

« Les deux auteurs (Kirchhoff et Volterra) auraient évidemment vu que cette connexion était lointaine, si, à l'époque où ils ont composé leurs travaux, la Science avait été en possession de nos idées actuelles sur la Relativité. On sait maintenant qu'il n'y a pas de sens défini (ou, si on préfère, qu'il y a une infinité de sens) à parler d'un point fixe de l'espace considéré à des instants successifs, qu'il existe (comme on le savait même auparavant) une infinité de transformations linéaires en x, y, z, t (ou x, y, t) — formant le 'groupe de Lorentz — qui laissent notre équation aux dérivées partielles invariante, et que de telles transformations ne changent pas le cône caractéristique, mais peuvent transformer la droite en question en n'importe quelle autre droite, menée par le sommet et à l'intérieur du cône caractéristique. Il est donc clair que cette droite n'a aucun rôle spécial à jouer dans nos calculs. »

Les lecteurs qui connaissent la célèbre solution de Volterra, et même ceux qui ne la connaissent que par nos indications sommaires, ne pourront manquer d'observer que la méthode que nous venons d'exposer est très apparentée à celle de Volterra. Il y a pourtant une différence essentielle qui fait que notre méthode échappe entièrement aux reproches de M. Hadamard. Cela tient au fait que l'opération par laquelle on a passé des intégrales divergentes figurant dans I_*^2 aux intégrales convergentes dans I_*^m ou I_*^{m+1} (augmentation de l'exposant α dans $r^{\alpha-m}$) et les différentiations (Δ ou ses itérés) par lesquelles on retourne à I_*^2 sont des opérations invariantives qui, par conséquent, n'introduisent aucune singularité étrangère au problème.

Avant d'abandonner cette méthode de solution, il nous paraît utile d'observer que les différentiations([1]) faciles à écrire, qui y figurent, sont souvent difficiles

([1]) Nous entendons par là les opérations $\Delta^{\frac{m-2}{2}}$ et $\Delta^{\frac{m-1}{2}}$, respectivement.

à exécuter. Il ne sera donc pas superflu d'indiquer d'autres solutions explicites où il n'intervient pas de telles différentiations. On peut tirer de telles solutions, au moins en principe, des formules générales des n[os] **31—36**, où nous avons donné le prolongement analytique de I^α_* pour un indice α arbitraire. Or, ces formules étant adaptées au cas général, il n'y aura pas de simplifications notables pour $\alpha = 2$. Nous reviendrons sur la question de solutions explicites en donnant une solution de plus pour l'équation des ondes cylindriques au n° **56** et des solutions sous la forme de potentiels retardés pour les dimensions d'ordre pair au n° **57**.

55. Propriétés de la solution générale. Principe de Huygens. Diffusion des ondes. — En étudiant la solution générale

$$u(P) = I^2_* \overline{f, g, h}(P), \tag{3}$$

donnée au n° **44**, nous supposons dans la suite que $m \geqq 3$. Pour $m = 3$, ce n'est que l'intégrale provenant de la double couche qui diverge, tandis que, à partir de $m = 4$, il en est de même de toutes les intégrales figurant dans la solution. En effet, on a montré au n° **15**, que les deux premières intégrales en I^α_* divergent pour $\alpha \leqq m - 2$, et au n° **25**, que la troisième diverge même pour $\alpha < m$. (Aux n[os] **50—51** on a étudié l'allure singulière que cette intégrale présente pour $\alpha = m$.) Nous connaissons déjà le rôle que le principe du prolongement analytique joue pour mettre notre solution sous forme finie et nous avons déjà rencontré des méthodes effectives de prolongement. A présent nous nous proposons d'élucider l'allure générale de la solution sans effectuer aucun prolongement explicite.

On voit par la forme même de la solution que la valeur de $u(P)$ dépend uniquement des valeurs que les données admettent aux points Q situés à l'intérieur ou sur la frontière du cône D^P ou, pour parler avec M. Hadamard, points qui sont « *sous onde* » [1] par rapport à P. (Cela a d'ailleurs encore lieu pour $m = 1$ et $m = 2$.) En effet, notre solution est donnée par des intégrales étendues à un domaine et une portion de surface, tous les deux compris dans le cône rétrograde D^P.

On sait aussi, par les travaux de M. Hadamard, qu'on a là un trait général chez les solutions des équations que M. Hadamard appelle *hyperboliques normales*. [2] La valeur de $u(P)$ ne dépend que des valeurs des données admises aux points qui sont sous onde par rapport à P.

---

(1) HADAMARD **1**, p. 74.

(2) HADAMARD **1**, p. 47 et 219. *Cf.* aussi le n° **56**[bis] du présent travail.

Notre formule fait aussi nettement ressortir la différence fondamentale qui, pour l'équation particulière que nous traitons ici, a lieu entre les cas de m impair et de m pair. Cette différence, à laquelle nous avons déjà fait allusion plusieurs fois, consiste en ce que, pour les dimensions paires, entre en vigueur *le principe de Huygens*.

Remarquons d'abord que pour toute valeur paire de $m (\geqq 4)$, la fonction $H_m(\alpha)$ admet un pôle[1] au point $\alpha = 2$, ce pôle provenant du facteur $\Gamma\left(\frac{\alpha + 2 - m}{2}\right)$ (*cf.* pour $H_m(\alpha)$ le n° **16**). En effet, l'argument de cette fonction devient pour $\alpha = 2$ égal à $(4 - m)/2$, nombre égal à zéro ou à un entier négatif pour m pair et $\geqq 4$.

Dans ces circonstances, on voit, même sans effectuer aucun prolongement explicite, que la valeur de $u(P)$, c'est-à-dire la valeur de $I_*^\alpha \overline{f, g, h}(P)$ pour $\alpha = 2$ ne pourra être influencée par les valeurs que les données admettent à l'intérieur du cône D^P. Il suffit pour le voir d'adapter au cas actuel le raisonnement de caractère intuitif fait dans l'Introduction et relatif au cas linéaire, $m = 1$. Là on a montré que les opérateurs I^0 et I^{-n} sont de caractère local, grâce aux deux faits que l'intégrale $\int_a^{x-\delta} f(t) \cdot (x - t)^{\alpha - 1} dt$ $(\delta > 0)$ ne cesse jamais de converger et que le facteur $1/\Gamma(\alpha)$ s'annule pour α égal à zéro ou à un entier négatif. *Mutatis mutandis*, il en est de même ici. En effet, si, dans le second membre de l'expression de I_*^α (formule (61) du n° **23**), on n'étend l'intégration qu'à un domaine ou à une portion de surface arbitraires, *intérieurs au sens strict* au cône D^P, les intégrales correspondantes ne cesseront jamais de converger. Multipliées par $1/H_m(\alpha)$, facteur s'annulant pour $\alpha = 2$, elles n'apporteront donc aucune contribution à la valeur finale I_*^2 de I_*^α. Ce point admis, l'intégrale spatiale totale dans I_*^α devra dans I_*^2 se réduire à une intégrale étendue à la nappe du cône (ou, plus précisément, à une certaine portion de cette nappe) et les intégrales de surface à des intégrales étendues à l'intersection s^P de la nappe avec la surface S.[2]

Or, ce fait constitue précisément le principe de Huygens énoncé sous la forme suivante: La solution $u(P)$ ne dépend que des valeurs que les données admettent

---

[1] La fonction $H_2(\alpha) = 2\,\Gamma^2\left(\frac{\alpha}{2}\right)$ n'a pas de pôle pour $\alpha = 2$. Le principe de Huygens, auquel obéit le troisième terme de la solution pour $m = 2$, dépend uniquement du fait que le noyau $d r^{\alpha-2}/d n$ contient le facteur $\alpha - 2$, qui s'annule pour $\alpha \to 2$; (*cf.* les n^{os} **47—48** et pour I_m^* les n^{os} **50—51**).

[2] *Cf.* le n° **57**.

aux points de la nappe du cône D^P, points qui, suivant la terminologie de M. Hadamard([1]), sont « *juste en onde* » par rapport au point P.

On peut aussi dire, pour citer l'Introduction, que le principe de Huygens dérive du fait que l'opérateur I_*^2 présente, pour *m pair* $\geqq 4$, un caractère *quasi local*; il n'y intervient que de tels points qui sont à *distance zéro* du point P.([2])

Pour m impair, la fonction $H_m(\alpha)$ n'admet pas de pôle en $\alpha = 2$. En effet, il en est ainsi pour chacun des deux facteurs Γ qui y interviennent. Notre formule de solution devient en principe identique à celle de M. Hadamard([3]), la seule différence étant que nous interprétons les intégrales divergentes par prolongement analytique, tandis que M. Hadamard le fait en appliquant sa méthode de la partie finie. L'une ou l'autre des deux méthodes montre que, les données étant générales, la solution, écrite sous forme finie, contient des intégrales étendues au volume entier D_S^P et à la portion de surface entière S^P. On a là ce qu'on appelle *diffusion des ondes*([4]) ou encore l'apparition *d'intégrales résiduelles*.

56. Diffusion des ondes. (Suite). — La diffusion des ondes se présente en particulier pour l'équation des ondes cylindriques ($m = 3$) et tout à fait inattendue lors de sa découverte, elle fut mise en évidence par Volterra([5]) moyennant la méthode dont nous venons de donner un aperçu au n° **54**.

Dans la solution de cette équation, il n'y a pas lieu de faire de prolongement pour les intégrales spatiale et de simple couche qui sont convergentes et s'écrivent, en tenant compte de ce que $H_3(\alpha) = 2\pi\Gamma(\alpha - 1)$ et par suite $H_3(2) = 2\pi$,

$$I_*^2\overline{f, g, 0}(P) = \frac{1}{2\pi}\left\{\int\limits_{D_S^P} f r^{-1}\, dQ + \int\limits_{S^P} g r^{-1}\, dS\right\}.$$

On a là une intégrale triple et une intégrale double, toutes les deux « résiduelles ».

Pour obtenir le prolongement analytique de l'intégrale de double couche

---

([1]) Hadamard **1**, p. 74, 239, 324.

([2]) Il en est de même — les raisons sont analogues —, pour *m pair*, des opérateurs I_*^{2k}, k entier positif, $2k \leqq m - 2$, et pour *m impair*, des opérateurs I_*^{2k+1}, k entier, $2k + 1 \leqq m - 2$. I_*^{-2p} a, pour p entier $\geqq 0$, un caractère *strictement local*, indépendamment de la parité de m. En effet on a, d'après le n° **42**, $I_*^0\overline{f, g, h}(P) = f(P)$ et $I_*^{-2p}\overline{f, g, h}(P) = \Delta^p f(P)$.

([3]) Hadamard **1**, p. 227 (formule (39)).

([4]) Hadamard **1**, p. 238.

([5]) Volterra **1**.

sous une forme explicite, le plus commode sera de partir de l'intégrale (95) du n° **31** (au lieu de la formule (107) du n° **32**, modifiée selon les indications du n° **33**). Posons donc, dans les notations des n°s **27—33**,

$$I_*^\alpha \overline{0, 0, h}(P) = \frac{\alpha - m}{H_m(\alpha)} \int_{s^P} h(\mathbf{a})\, r_a^{\alpha-2}\, dH.$$

La sphère unitaire ω_{m-1} donnée par la formule (99) du n° **31** se réduit ici à une circonférence de rayon 1. En introduisant l'angle azimutal Φ, on peut exprimer les coordonnées des points $\mathbf{z}$ de H par les formules

$$z_1 = -\tfrac{1}{2}(t^{-1} + t),\quad z_2 = -\tfrac{1}{2}(t^{-1} - t)\cos\Phi,\quad z_3 = -\tfrac{1}{2}(t^{-1} - t)\sin\Phi.$$

Pour chaque valeur déterminée de Φ, le point $\mathbf{z}$ décrit un méridien[1] sur H. La projection (centrale au centre P) d'un tel méridien sera une certaine ligne l, qui coupe l'intersection s^P de S et de la nappe C^P du cône D^P en un certain point $\mathbf{b}$, point qu'on fait correspondre à tous les points $\mathbf{a}$ de la ligne l. Tandis que ces points $\mathbf{a}$ dépendent encore de t ou de $T = t^2$, $\mathbf{a} = \mathbf{a}(T, \Phi)$, le point $\mathbf{b}$ ne dépend que de Φ, $\mathbf{b} = \mathbf{b}(\Phi)$.

Spécialisons l'intégrale ci-dessus, en y posant $m = 3$, et notons qu'elle converge pour $\alpha > m = 3$. Écrivons aussi, d'après la formule (72) du n° **27**

$$r_a = \frac{r_a}{t} \cdot t = v_a \cdot t = v_a T^{\frac{1}{2}}.$$

Cela étant admis, il vient

$$(16)\qquad I_*^\alpha \overline{0, 0, h}(P) = \frac{\alpha - 3}{H_3(\alpha)} \int h(\mathbf{a})\, r_a^{\alpha-2}\, dH = \frac{\alpha - 3}{H_3(\alpha)} \int h(\mathbf{a})\, v_a^{\alpha-2}\, T^{\frac{\alpha-2}{2}}\, dH.$$

Écrivons

$$(17)\qquad h(\mathbf{a}) \cdot v_a^{\alpha-2} = (h(\mathbf{a})\, v_a^{\alpha-2} - h(\mathbf{b})\, v_b^{\alpha-2}) + h(\mathbf{b})\, v_b^{\alpha-2}$$

et, d'après la formule (99) du n° **31**, avec $m = 3$,

$$(18)\qquad T^{\frac{\alpha-2}{2}}\, dH = \tfrac{1}{4} \cdot T^{\frac{\alpha-5}{2}} \cdot (1 - T)\, dT\, d\Phi.$$

Le dernier terme de (17), indépendant de T, introduit dans (16), donne alors, d'abord pour $\alpha > 3$

---

[1] Ces méridiens dépendent évidemment du choix de l'axe des ξ_1.

$$\frac{\alpha-3}{4H_3(\alpha)}\int\limits_0^{2\pi} h(\mathbf{b})v_b^{\alpha-2}\,d\Phi\cdot\int\limits_0^1 T^{\frac{\alpha-5}{2}}(1-T)\,dT$$

$$=\frac{\alpha-3}{4H_3(\alpha)\cdot\frac{\alpha-3}{2}\cdot\frac{\alpha-1}{2}}\cdot\int\limits_0^{2\pi} h(\mathbf{b})v_b^{\alpha-2}\,d\Phi.$$

Vu que $H_3(2)=2\pi$, il en devient par prolongement analytiqne, pour $\alpha=2$,

$$\frac{1}{2\pi}\int\limits_0^{2\pi} h(\mathbf{b})\,d\Phi.$$

Il s'ensuit finalement

$$I_*^2\overline{0,0,h}(P)=-\frac{1}{2\pi}\int\limits_{S^P}(h(\mathbf{a})-h(\mathbf{b}))\,dH+\frac{1}{2\pi}\int\limits_0^{2\pi} h(\mathbf{b})\,d\Phi.$$

Le premier terme du second membre peut, d'après (18), aussi s'écrire

$$-\frac{1}{8\pi}\int\limits_0^{2\pi} d\Phi\int\limits_0^1 (h(\mathbf{a})-h(\mathbf{b}))\,T^{-\frac{3}{2}}(1-T)\,dT.$$

On voit, soit par la dernière formule, soit par les évaluations des n^{os} **32** et **33** qu'il suffit largement pour la validité des expressions obtenues que $h(\mathbf{a})$ et $S(\mathbf{a})$ admettent des dérivées partielles continues du premier ordre. [1]

L'intégrale double, figurant dans la solution, est encore une intégrale résiduelle. [2]

Les recherches de M. Hadamard, continuant celles de Volterra, ont montré clairement que, dans la théorie générale des équations linéaires hyperboliques normales du second ordre, à coefficients *constants* ou *variables*, c'est la *diffusion des ondes* qui fait la *règle* et son absence qui fait l'exception. [3]

---

[1] *Cf.* la note [1], p. 51.

[2] Il y a lieu de mentionner à cette occasion que MM. Baker et Copson ont appliqué notre méthode d'une manière explicite à l'équation des ondes cylindriques en élucidant les principes aussi pour des cas plus généraux. (Baker-Copson, *The Mathematical Theory of Huygens' Principle*, Oxford 1939, p. 54—67.) Je tiens à les remercier de leur intérêt, d'autant plus remarquable, que ce n'était qu'un résumé très précaire de la méthode (Riesz **1**) qu'ils avaient à leur disposition.

[3] Pour des coefficients constants, on trouve la démonstration dans le numéro suivant; pour des coefficients variables, voir le dernier Chapitre. M. Hadamard a émis l'hypothèse que les seules équations du *second ordre* obéissant au principe de Huygens sont celles qui contiennent un nombre pair de variables indépendantes et qui, par des transformations ponctuelles de ces variables et par la multiplication de la fonction inconnue avec une fonction convenable, peuvent se réduire à

Or, c'est précisement ce dernier cas, exceptionnel au point de vue de la théorie générale, qui a lieu pour l'équation des ondes dans tout espace — ou mieux, *dans tout univers — à une dimension d'ordre pair.* Cet état des choses est d'une importance capitale pour le monde physique, régi dans ses manifestations les plus nettes, telles les phénomènes électro-magnétiques — y compris la lumière — par l'équation des ondes sphèriques. (1)

56bis. Équation des ondes amorties. — Étant donnée une équation aux dérivées partielles, linéaire et du second ordre, nous dirons, avec M. Hadamard(2), qu'il s'agit d'une équation *hyperbolique normale,* si la forme quadratique formée avec les coefficients des dérivées du second ordre — transformée en somme de carrés — contient m carrés qui ont *tous, sauf un, le même signe.*

Il est facile de faire voir que toute équation hyperbolique normale à coefficients constants se réduit soit à l'équation des ondes $\Delta u = f$, soit à l'équation des ondes amorties $\Delta u + \mu u = f$, où $\mu \neq 0$ est constant. On montre ensuite que cette dernière équation conduit toujours à des intégrales résiduelles, et cela indépendamment de la parité de m. On aura ainsi pour le cas des coefficients constants confirmé le fait annoncé au numéro précédent.

L'équation générale (à coefficients constants) ayant été, par une substitution linéaire des variables indépendantes, mise sous la forme

$$\sum_1^m \varepsilon_k^{-1} \frac{\partial^2 u}{\partial x_k^2} + 2 \sum_1^m \varepsilon_k^{-1} B_k \frac{\partial u}{\partial x_k} + Cu = f,$$

on pose

$$e^{\sum\limits_1^m B_k x_k} = E,\ Eu = v,\ C - \sum_1^m \varepsilon_k^{-1} B_k^2 = \mu = \lambda^2,\ Ef = g$$

et on obtient l'équation

$$(\Delta + \lambda^2) v = g. \tag{19}$$

(3)

Comme nous l'avons fait au chapitre III, en y partant de l'opérateur Δ, nous définirons ici une intégration d'ordre fractionnaire rattachée à l'opérateur $\Delta + \lambda^2$. Le noyau correspondant W^α remplira, entre autres choses, la relation

$$(\Delta + \lambda^2) W^{\alpha+2} = W^\alpha.$$

Nous écrirons d'abord les formules définitives et indiqueront ensuite la voie qui conduit au noyau en question. Considérons, pour commencer, la fonction de Bessel d'ordre n

---

l'équation des ondes. Myron Mathisson, qu'une mort prématurée a empêché de poursuivre ses importantes recherches, a vérifié cette hypothèse dans le cas où les coefficients des termes du second ordre sont constants. Quant au principe de Huygens pour les équations hyperboliques d'ordre supérieur à deux, on pourra consulter les travaux de M. Petrowsky et de M. Gårding. (Pour les travaux des trois auteurs cités *voir* la Bibliographie.)

(1) Pour cette observation, remontant à Boussinesq, *cf.* Goursat, *Cours d'Analyse* III (1915), p. 109.

(2) Hadamard **1**, p. 47 et 219.

(3) *Cf.* p. ex. Hadamard, **1**, p. 148.

(20) $$J_n(z) = \sum_{p=0}^{\infty} \frac{(-1)^p z^{n+2p}}{2^{n+2p} \cdot p!\, \Gamma(n+p+1)}$$

et posons, avec les notations du n° **14**(¹),

(21) $$W^\alpha(P, Q) = \frac{\left(\lambda^{-1} r_{PQ}\right)^{\frac{\alpha-m}{2}}}{\pi^{\frac{m-2}{2}} \cdot 2^{\frac{\alpha+m-2}{2}}\, \Gamma\left(\frac{\alpha}{2}\right)} J_{\frac{\alpha-m}{2}}\left(\lambda r_{PQ}\right)$$

dans le cône rétrograde au sommet P et nul ailleurs. On trouve alors, en appliquant le développement en série (20) que pour $r \neq 0$

$$\left(\Delta_P + \lambda^2\right) W^{\alpha+2} = \left(\Delta_Q + \lambda^2\right) W^{\alpha+2} = W^\alpha$$

et en particulier

$$(\Delta + \lambda^2)\, W^2 = 0.$$

L'intégrale

$$I^\alpha_{(\lambda)} f(P) = \int_{D^P_S} f(Q)\, W^\alpha(P, Q)\, dQ$$

satisfait aux relations

$$I^\alpha_{(\lambda)} I^\beta_{(\lambda)} = I^{\alpha+\beta}_{(\lambda)} \quad \text{et} \quad \left(\Delta_P + \lambda^2\right) I^{\alpha+2}_{(\lambda)} = I^\alpha_{(\lambda)},$$

dont la première exige en général que S soit d'orientation d'espace ou que D^P_S se confonde avec le cône entier D^P.

La formule de Green sera de la même forme qu'au chapitre III. En posant donc

$$I^\alpha_{*(\lambda)} \overline{f, g, h}(P) = \int_{D^P_S} f(Q)\, W^\alpha(P, Q)\, dQ + \int_{S^P} \left\{ g(Q)\, W^\alpha(P, Q) - h(Q) \frac{d\, W^\alpha(P, Q)}{dn} \right\} dS,$$

on aura l'identité

$$I^\alpha_{(\lambda)} u(P) = I^{\alpha+2}_{*(\lambda)} \overline{(\Delta + \lambda^2)\, u, \frac{du}{dn}, u}(P).$$

On a aussi dans le cas actuel la relation importante

$$I^0_{*(\lambda)} \overline{f, g, h}(P) = f(P).$$

En rapprochant les deux dernières formules, on obtient l'identité

(22) $$u(P) = I^2_{*(\lambda)} \overline{(\Delta + \lambda^2)\, u, \frac{du}{dn}, u}(P),$$

qui, si l'on y porte les données, doit fournir la solution du problème de Cauchy correspondant.

On voit de suite par le développement (20) et par la formule (21) que les facteurs Γ figurant dans les dénominateurs des termes constituant W^α n'ont pas de pôles en $\alpha = 2$ pour m impair,

(¹) *Cf.* M. Riesz **2**, dernières formules de la Note additionnelle.

et que, pour m pair, ces termes n'admettent de pôles en $\alpha = 2$ que si l'on a $p \leqq \frac{m}{2} - 2$. Ce ne sont donc que ces termes-là, complétés, en ce qui concerne la double couche, et cela pour des raisons expliquées au n° **15**, par le terme correspondant à $p = \frac{m}{2} - 1$, qui donnent lieu au principe de Huygens. Tous les autres termes, en nombre infini, fournissent des intégrales résiduelles (convergentes) étendues à D_S^P ou à S^P.

Il nous reste à indiquer comment on arrive au noyau (19). On peut le faire par la théorie de l'intégrale de Fourier, en exigeant que l'opération $I_{(\lambda)}^{\alpha}$, étendue au cône rétrograde entier et appliquée à une fonction de la forme $e^{\Sigma b_k x_k}$, avec

$$b_k = b'_k + i b''_k, \text{ et } \Sigma \varepsilon_k^{-1} b'^2_k > 0,\ b'_1 > 0,$$

reproduise cette fonction multipliée par

$$(\Sigma \varepsilon_k^{-1} b_k^2 + \lambda^2)^{-\frac{\alpha}{2}}.$$

Il est pourtant plus commode de procéder comme il suit. On pose, d'une manière symbolique,

$$I_{(\lambda)}^{\alpha} = (\Delta + \lambda^2)^{-\frac{\alpha}{2}} = \Delta^{-\frac{\alpha}{2}} (1 + \lambda^2 \Delta^{-1})^{-\frac{\alpha}{2}} =$$
$$= \sum_{p=0}^{\infty} \binom{-\frac{\alpha}{2}}{p} \lambda^{2p} \Delta^{-\frac{\alpha}{2} - p} = \sum_{p=0}^{\infty} \binom{-\frac{\alpha}{2}}{p} \lambda^{2p} I^{\alpha+2p}.$$

L'opérateur I^{α} dérivant du noyau $r^{\alpha-m}/H_m(\alpha)$, l'opérateur $I_{(\lambda)}^{\alpha}$ doit dériver du noyau

$$W^{\alpha} = \sum_{p=0}^{\infty} \binom{-\frac{\alpha}{2}}{p} \lambda^{2p} \frac{r^{\alpha+2p-m}}{H_m(\alpha+2p)}, \tag{23}$$

qui est identique à celui donné par la formule (21).

57. Potentiels retardés. — Considérons *d'abord*, pour une *dimension m arbitraire*, l'intégrale convergente pour $\alpha > m - 2$

$$I^{\alpha} f(P) = \frac{1}{H_m(\alpha)} \int_{D_S^P} f(Q)\, r_{PQ}^{\alpha-m}\, d Q,$$

où, comme d'ordinaire, $P = \mathbf{x}$, $Q = \boldsymbol{\xi}$. Envisageons, dans l'ordre d'idées du n° **31**, le demi-hyperboloïde au centre $\mathbf{x}$ et de rayon r, c'est-à-dire $(\mathbf{x} - \boldsymbol{\xi})^2 = r^2$, $\xi_1 - x_1 < 0$, et posons la notation H_r pour la portion de cette surface qui est intérieure au domaine D_S^P. Eu égard à la relation évidente (*cf.* la formule (89) du n° **31**)

$$d\,Q = d\,H_r \cdot d\,r = \frac{1}{2}\frac{d\,H_r}{r} \cdot d\,R \qquad (R = r^2),$$

on peut écrire

$$I^\alpha f(P) = \frac{1}{2\,H_m(\alpha)} \int_0^{R_0} R^{\frac{\alpha-m}{2}}\,d\,R \int_{H_r} f(Q) \frac{d\,H_r}{r}, \tag{24}$$

où R_0 désigne le maximum de R en D_S^P. Si l'on pose encore

$$\int_{H_r} f(Q) \frac{d\,H_r}{r} = A(R) \text{ et } \pi(R) = \sum_{k=0}^{n-1} \frac{A^{(k)}(0)}{k!} \cdot R^k, \tag{25}$$

on aura, d'après la formule (13) du n° **5**,

$$I^\alpha f(P) = \frac{1}{2\,H_m(\alpha)} \int_0^{R_0} (A(R) - \pi(R))\, R^{\frac{\alpha-m}{2}}\,d\,R + \frac{1}{2\,H_m(\alpha)} \sum_{k=0}^{n-1} \frac{A^{(k)}(0)}{k!} \cdot \frac{R_0^{\frac{\alpha-m}{2}+k+1}}{\frac{\alpha-m}{2}+k+1}, \tag{26}$$

formule qui fournit le prolongement analytique de $I^\alpha f(P)$.

Nous ne nous sommes préoccupés ni de l'existence, ni de la continuité des dérivées $A^{(k)}(R)$, ni même de la signification de $A(0)$ et des $A^{(k)}(0)$. En ce qui concerne le premier point, on peut sans doute démontrer par les méthodes des n^{os} **27—29** que, dans nos conditions usuelles, les quantités en question existent et possèdent les propriétés désirées. Pour des surfaces S très simples, p. ex. un plan d'espace, la chose devient, pour ainsi dire, évidente, puisque l'intersection de H_r avec S devient très simple. D'autre part, dans le cas général, les calculs seraient certainement assez pénibles, surtout pour les dérivées d'ordre supérieur. Nous verrons à l'instant que, pour I^2 et même pour I^2_* et pour la dimension $m = 4$, les difficultés de calcul disparaissent entièrement, au moins en ce qui concerne l'intégrale de volume et la simple couche, dont nous allons nous occuper tout à l'heure.

Le second point auquel nous venons de faire allusion, à savoir la signification de $A(0)$ (et des $A^{(k)}(0)$), est de la plus grande importance et dépend des propriétés de la différentielle $d\,H_r/r$.

Considérons d'abord, dans l'espace euclidien à m dimensions, la sphère Σ_r, donnée par l'équation $z_1^2 + z_2^2 + \cdots + z_m^2 = r^2$. Les cosinus directeurs de la normale extérieure — *ou les composantes de la normale unitaire* — étant z_k/r, on a, pour l'élément $d\,\Sigma_r$ de cette sphère, la relation

$$|dz_2 \ldots dz_m| = \frac{|z_1|}{r} d\Sigma_r,$$

ou, d'une manière générale,

$$|dz_1 \ldots dz_{k-1} dz_{k+1} \ldots dz_m| = \frac{|z_k|}{r} d\Sigma_r.$$

Il est évident que des relations analogues ont lieu dans toute métrique quadratique et que l'on a en particulier pour H_r

$$\frac{d\xi_2 \ldots d\xi_m}{x_1 - \xi_1} = \frac{dH_r}{r}, \tag{27}$$

toutes les quantités étant positives. Cette formule met en évidence que le premier membre est invariant par toute transformation (lorentzienne) de coordonnées, il en étant ainsi pour le second membre. D'un autre côté, le premier membre garde sa signification aussi pour le cône caractéristique au sommet P, correspondant à $r = 0$, tandis que le second membre, qui devient $0:0$, la perd.

Nous terminons cette digression en indiquant pour le cône H_0, qui n'est que l'hyperboloïde dégénéré $(\mathbf{x} - \boldsymbol{\eta})^2 = 0$, une interprétation géométrique du premier membre de (27), qui met en évidence son caractère invariantif (d'ailleurs facile à établir par un calcul direct). Pour éviter toute confusion, nous désignons pour un instant la mesure d'un élément dH_r par $|dH_r|$, en gardant la première notation pour la variété géométrique (ensemble de points) en question — deux notions que nous désignons d'habitude par la même notation. Nous posons encore $\overline{dH_r}$ pour la projection de dH_r sur le plan $\xi_1 = 0$ dans la direction de l'axe des ξ_1 et $|\overline{dH_r}|$ pour la mesure de la projection. On aura alors, pour un élément dH_0 autour du point η du cône, $|dH_0| = 0$, tandis que $\dfrac{|\overline{dH_0}|}{x_1 - \eta_1}$ est la valeur limite de $|dH_r|/r$ lorsque la variété dH_r tend, point par point, vers la variété dH_0.[1]

Si l'on désigne par C_S^P la portion interceptée de la nappe par la surface S, par $\bar{C}_S^P$ la projection de C_S^P sur le plan $\xi_1 = 0$ et par $\boldsymbol{\eta}$ un point arbitraire de C_S^P, on pourra écrire par ce qui précède

$$A(0) = \int_{\bar{C}_S^P} f(\boldsymbol{\eta}) \frac{d\eta_2 \ldots d\eta_m}{x_1 - \eta_1}, \tag{28}$$

[1] Pour fixer les idées, on pourra établir une correspondance biunivoque entre les points des H_r et ceux de H_0 par des droites parallèles à une direction de temps déterminée.

la différentielle, répétons-le, étant invariantive. En posant encore

$$\left\{\sum_{i=2}^{m}(x_i-\eta_i)^2\right\}^{\frac{1}{2}}=\varrho,$$

on aura $x_1-\eta_1=\varrho$ ou $\eta_1=x_1-\varrho$, et $A(0)$ pourra s'écrire sous la forme

$$(28^{bis})\qquad A(0)=\int_{\bar{C}_S^P} f(x_1-\varrho,\eta_2,\ldots,\eta_m)\frac{d\eta_2\ldots d\eta_m}{\varrho}.$$

On voit que la continuité de f (et de S) suffit pour rendre légitime le passage à la limite ci-dessus. Ajoutons qu'il serait beaucoup plus compliqué de donner des formules explicites pour les $A^{(k)}(0)$, mais on comprend que, dans des conditions convenables, on pourra leur attribuer un sens bien défini.

Cela étant, retournons à la formule (26). En posant

$$(29)\qquad \frac{\alpha+2-m}{2}=\beta,\quad H_m(\alpha)=\pi^{\frac{m-2}{2}}\cdot 2^{\alpha-1}\cdot\Gamma\left(\frac{\alpha}{2}\right)\Gamma\left(\frac{\alpha+2-m}{2}\right)$$
$$=\pi^{\frac{m-2}{2}}\cdot 2^{\alpha-1}\cdot\Gamma\left(\frac{\alpha}{2}\right)\cdot\Gamma(\beta),$$

ladite formule s'écrit

$$(\mathrm{A})\qquad I^\alpha f(P)=\frac{1}{2^\alpha\pi^{\frac{m-2}{2}}\Gamma\left(\frac{\alpha}{2}\right)}\cdot\frac{1}{\Gamma(\beta)}\left\{\int_0^{R_0}(A(R)-\pi(R))R^{\beta-1}dR+\sum_{k=0}^{n-1}\frac{A^{(k)}(0)}{k!}\frac{R_0^{k+\beta}}{k+\beta}\right\}.$$

Admettant dans la suite que m est un nombre *pair* $\geqq 4$, nous allons calculer $I^2 f(P)$. Pour $\alpha=2$, la valeur $\beta=\frac{4-m}{2}$ sera égale à zéro ou à un entier négatif. En tenant compte des résultats du Chap. I, et surtout de la formule (13bis) du n° **5**(¹), on obtient donc

$$(30)\qquad I^2 f(P)=\frac{1}{4\pi^{\frac{m-2}{2}}}(-1)^{\frac{m-4}{2}}A^{\left(\frac{m-4}{2}\right)}(0)=\frac{(-1)^{\frac{m}{2}}A^{\left(\frac{m-4}{2}\right)}(0)}{4\pi^{\frac{m-2}{2}}}.$$

En particulier, pour $m=4$, eu égard à (28bis), on arrive au potentiel retardé bien connu

---

(¹) Pour $m=4$, on n'a pas besoin de la formule (13bis), car $A^{\left(\frac{m-4}{2}\right)}(0)$ devient $A(0)$ qui, comme nous venons de le voir, peut s'obtenir par un passage à la limite.

$$(30^{bis})\qquad I^2 f(P) = \frac{A(0)}{4\pi} = \frac{1}{4\pi}\int_{\overline{C}_S^P} f(x_1-\varrho, \eta_2, \eta_3, \eta_4)\cdot \frac{d\eta_2\, d\eta_3\, d\eta_4}{\varrho}.$$

Nous indiquons encore succinctement les relations correspondantes qui ont lieu pour nos intégrales de surface. Tout comme dans le n° **50**, nous divisons S^P, moyennant son intersection avec l'hyperboloïde $(\mathbf{x}-\mathfrak{x})^2 = \delta^2$ en deux parties et désignons par $\overline{S}$ la partie annulaire extérieure et par $\overline{\overline{S}}$ la partie intérieure. Notre but final étant de calculer I_*^2, nous pourrons nous restreindre à $\overline{S}$, puisque la contribution de $\overline{\overline{S}}$ à I_*^2 s'annule (vu que $1 : H_m(2) = 0$).

Il viendra pour la *simple couche* (*voir* pour les notations et pour la relation $dS = ds_R \cdot dl$ le n° **50**)

$$(\mathrm{B})\qquad \overline{I}_*^\alpha\, \overline{0, g, 0}(P) = \frac{1}{H_m(\alpha)}\int_{\overline{S}} g\, r^{\alpha-m}\, dS = \frac{1}{H_m(\alpha)}\int_0^{\delta^2} R^{\frac{\alpha-m}{2}}\, dR \int_{s_R} g\, \frac{dl}{dR}\, ds_R$$

$$= \frac{1}{H_m(\alpha)}\int_0^{\delta^2} R^{\beta-1} B(R)\, dR,$$

où $B(R)$ désigne l'intégrale étendue à s_R qui figure au troisième membre, et où $\beta = (\alpha + 2 - m)/2$ comme dans la formule (29). La dernière intégrale converge pour $\alpha > m-2$ ou $\beta > 0$ et son prolongement analytique s'obtient de la même manière que celui de (24). Il vient (puisque $I_*^2 = \overline{I}_*^2$)

$$(31)\qquad I_*^2\, \overline{0, g, 0}(P) = \frac{(-1)^{\frac{m}{2}}}{2\cdot \pi^{\frac{m-2}{2}}} \cdot B^{\left(\frac{m-4}{2}\right)}(0)$$

et, en particulier, pour $m = 4$,

$$(31^{bis})\qquad I_*^2\, \overline{0, g, 0}(P) = \frac{B(0)}{2\pi} = \frac{1}{2\pi}\int_{s^P} g\, \frac{dl}{dR}\, ds.$$

Pour la *double couche*, on aura d'abord

$$\frac{d r^{\alpha-m}}{dn} = (\alpha - m)\, r^{\alpha-m-1}\frac{dr}{dn} = \frac{\alpha-m}{2}\, r^{\alpha-m-2}\frac{dR}{dn}$$

et alors

$$(\mathrm{C})\qquad \overline{I}_*^\alpha\, \overline{0, 0, h}(P) = \frac{-1}{H_m(\alpha)}\int_{\overline{S}} h\, \frac{d r^{\alpha-m}}{dn}\, dS = \frac{-(\alpha-m)}{2\, H_m(\alpha)}\int_0^{\delta^2} R^{\frac{\alpha-m}{2}-1}\, dR \int_{s_R} h\, \frac{dR}{dn}\, \frac{dl}{dR}\, ds_R.$$

Désignons la dernière intégrale par $C(R)$, et posons cette fois-ci $\beta = \frac{\alpha - m}{2}$
On trouve d'abord

$$\Gamma\left(\frac{\alpha + 2 - m}{2}\right) = \frac{\alpha - m}{2} \cdot \Gamma\left(\frac{\alpha - m}{2}\right) = \beta\, \Gamma(\beta),$$

ce qui donne

$$(C^{bis}) \qquad \bar{I}_*^\alpha \overline{0, 0, h}(P) = \frac{-1}{\pi^{\frac{m-2}{2}} \cdot 2^{\alpha-1} \cdot \Gamma\left(\frac{\alpha}{2}\right)} \cdot \frac{1}{\Gamma(\beta)} \int_0^{\delta^2} R^{\beta-1}\, C(R)\, dR.$$

On en tire pour $\alpha = 2$ $\left(\beta = \frac{2 - m}{2}\right)$,

$$(32) \qquad I_*^2 \overline{0, 0, h}(P) = \frac{-(-1)^{\frac{m-2}{2}}}{2\pi^{\frac{m-2}{2}}} C^{\left(\frac{m-2}{2}\right)}(0) = \frac{(-1)^{\frac{m}{2}}}{2\pi^{\frac{m-2}{2}}} C^{\left(\frac{m-2}{2}\right)}(0)$$

et, en particulier, pour $m = 4$,

$$(32^{bis}) \qquad I_*^2 \overline{0, 0, h}(P) = \frac{1}{2\pi} C'(0).$$

On voit que $I_*^2 \overline{f, g, h}(P)$ s'exprime, d'une part, par les valeurs que f et ses dérivées d'ordre $\leqq \frac{m-4}{2}$ admettent sur C_S^P et, d'autre part, par celles que g et h et leurs dérivées d'ordre $\leqq \frac{m-4}{2}$ et d'ordre $\leqq \frac{m-2}{2}$, respectivement, admettent sur s^P. L'expression de I_*^2 obéit donc au principe de Huygens, chose évidente d'après ce qu'on a vu au n° **55**.

Le Chap. V est consacré à une interprétation géométrique des formules (31bis) et (32bis).

58. La solution élémentaire. — Dans tout ce qui précède, le lecteur chercherait en vain [1] un concept bien familier à tous les initiés, celui de la *solution élémentaire.* [2]

En s'exprimant d'une manière générale et assez vague, on entend par solution élémentaire d'une équation linéaire aux dérivées partielles une fonction V qui, dans une certaine région, satisfait à l'équation sans second membre et admet,

---

[1] En réalité, il faut faire exception pour deux mentions concernant les méthodes de Volterra et de M. Tedone (n° **54**).

[2] Cette expression, due à M. Hadamard, est à notre avis bien préférable à l'ancien terme de *solution fondamentale*; *cf.* HADAMARD **1**, p. XII.

dans un certain point une singularité d'un ordre d'infinitude convenable.[1] Pour fixer les idées, on a, dans la théorie de l'équation de Laplace, pour toute dimension $m \neq 2$, comme solution élémentaire le potentiel élémentaire newtonien

$$V = g_m \cdot r^{2-m}.$$

Le plus naturel nous paraît de choisir la constante numérique arbitraire g_m de façon que la formule de Poisson reproduise la densité spatiale avec le coefficient 1, c'est-à-dire que $\Delta \int f V dQ = f$. Par exemple, pour $m = 3$, la solution élémentaire ainsi normée est $-\frac{1}{4\pi r}$. (*Cf.* aussi le Chap. II du présent travail.)

Une pareille solution élémentaire n'intervenait pas directement dans nos considérations. Cette notion y a joué en quelque sorte le rôle d'un élément idéal; du moins pour $m \geqq 4$ (et en partie même pour $m = 2$ et 3). Nous entendions ainsi par $I^2 f$ — qui dans la théorie de l'opérateur de Laplace est (au signe près) identique au potentiel newtonien tel que nous l'avons défini — le prolongement analytique de $I^\alpha f$, représenté, pour les valeurs assez grandes de α, par des intégrales convergentes. Il s'ensuit que, si l'on veut parler d'une solution élémentaire dans la théorie actuelle, cela ne peut être que

$$(33) \qquad V = \begin{cases} \gamma_m \cdot r^{2-m} & \text{dans le cône rétrograde,} \\ 0 & \text{ailleurs,} \end{cases}$$

où

$$\gamma_m = \frac{1}{H_m(2)}.$$

Pour *m impair*, il n'y a en réalité rien à objecter à cette solution élémentaire, si ce n'est que les intégrales formées avec elle sont divergentes pour $m > 3$, et sont à interpréter par prolongement analytique — comme nous l'avons fait —, ou par la méthode de la partie finie de M. Hadamard. Le facteur numérique γ_m a, nous le savons, une valeur finie $\neq 0$. Pour sa valeur explicite, il vient par un calcul simple

$$(34) \qquad \gamma_m = \frac{1}{H_m(2)} = (-1)^{\frac{m+1}{2}} \cdot \frac{\Gamma\left(\frac{m-2}{2}\right)}{2\pi^{\frac{m}{2}}} = (-1)^{\frac{m+1}{2}} \cdot \frac{1}{\pi\,\omega_{m-2}},$$

où ω_{m-2} signifie, comme d'habitude, la surface totale de la sphère unitaire dans l'espace euclidien à $m-2$ dimensions (*cf.* la formule (8) du n° **6**).

[1] Pour des renseignements plus précis, *cf.* HADAMARD **1**, p. 115 et *suiv.*, en particulier p. 134.

Pour *m pair*, on a $\gamma_m = 0$, et par conséquent la solution élémentaire apparaît sous la forme paradoxale

$$V = 0 \cdot r^{2-m}. \tag{35}$$ [1]

On peut préciser cette forme en posant

$$V = \lim_{\varepsilon \to 0} \varepsilon\, \delta_m \cdot r^{2+\varepsilon-m}, \tag{36}$$

où

$$\delta_m = \lim_{\alpha \to 2} \frac{1}{(\alpha-2)\, H_m(\alpha)} = (-1)^{\frac{m}{2}} \cdot \frac{\Gamma\left(\frac{m-2}{2}\right)}{4\pi^{\frac{m-2}{2}}} = (-1)^{\frac{m}{2}} \cdot \frac{1}{2\,\omega_{m-2}}. \tag{37}$$

La forme « précisée » de la solution élémentaire pour m pair est très utile dans le cas spécial — fort important au point de vue de la Physique —, $m = 4$, où les intégrales spatiale et de simple couche convergent pour toute valeur de $\alpha > 2$ et où par conséquent on peut obtenir $I_*^2\overline{f, g, 0}(P)$ par un passage à la limite. On peut donc appliquer directement la formule (36) à l'évaluation de ladite expression et déduire par là d'une manière immédiate les potentiels retardés correspondants, sans faire le détour par la théorie générale. D'autre part, on peut aussi appliquer la formule en question à l'évaluation des intégrales de simples couches, ces dernières étalées sur des variétés à dimension 1 ou 2, en particulier sur une « ligne d'univers ». On arrive par là aisément au potentiel-quadrivecteur connu sous le nom de potentiel de Liénard-Wiechert. [2]

59. Examen critique de la solution élémentaire. — Il est clair que, dans la théorie du potentiel newtonien, relative à notre espace ordinaire à trois dimensions, par exemple, il est à peu près indifférent d'écrire le potentiel sous sa forme habituelle $\int f r^{-1}\, dQ$ ou d'accepter la « réforme » que nous avons l'air d'avoir préconisée, à savoir de le multiplier par $-1/4\pi$. Le facteur -4π figurerait dans le premier cas — comme il le fait, d'habitude — au second membre de la formule de Poisson, ce qui ne serait pas un malheur trop grave. Les circonstances, ne sont-elles pas analogues dans la théorie actuelle? Y a-t-il aucune différence à appliquer la formule (33) pour m impair et la formule (35) sous sa

---

(1) Une pareille forme paradoxale se présente, pour m pair, chez ceux des termes de W^2, défini par les formules (20) et (21) du n° **56**^{bis}, qui correspondent à $p \leqq \frac{m}{2} - 2$.

(2) *Cf.* FREMBERG **2**, Chapter 10 et le Chapitre VI du présent ouvrage.

forme précisée (36) pour m pair ou bien à poser simplement $V = r^{2-m}$ dans les deux cas?

Cela revient à poser la définition modifiée $I^2 f(P) = \int f(Q) r_{PQ}^{2-m} dQ$, cette dernière expression étant définie par un procédé de prolongement quelconque, p. ex. prolongement analytique de l'expression $\int f(Q) r_{PQ}^{\alpha-m} dQ$, ou prolongement par « partie finie ». Pour m impair, la différence serait minime, tout analogue à celle rencontrée dans le cas newtonien. Il est pour ainsi dire évident qu'il en serait tout autrement pour les dimensions d'ordre pair.

En effet, l'intégrale $\int f r^{\alpha-m} dQ$ est égale à $H_m(\alpha) \cdot I^\alpha f(P)$. D'autre part, nous avons vu que le prolongement analytique de $I^\alpha f(P)$ conduit, dans nos conditions, à des valeurs finies pour $I^2 f(P)$ qui évidemment ne sont pas nulles en général. Le facteur $H_m(\alpha)$ possédant, pour *m pair*, un pôle en $\alpha = 2$, le prolongement de l'expression modifiée ne saurait fournir pour $I^2 f(P)$ des valeurs finies.

Il en serait de même si au lieu de supprimer dans I^α le facteur $1/H_m(\alpha)$ on le remplaçait par un autre qui ne s'annule pas pour $\alpha = 2$.

Tel étant le cas pour le prolongement analytique, on peut se demander comment se présente la question de la solution élémentaire si l'on applique la méthode de la partie finie. Or, c'est précisément l'absence, dans les formules de M. Hadamard, du facteur $1/H_m(\alpha)$ — ou de facteurs qui se comportent d'une manière analogue au voisinage de $\alpha = 2$ — qui faisait que cette dernière méthode ne pouvait s'appliquer que dans le cas de m impair.(1)

Pour m pair, on trouve dans le livre de M. Hadamard (p. 143) une solution élémentaire qui n'est déterminée qu'à une solution régulière additive près, et qui, dans le cas général de coefficients variables, renferme un terme logarithmique. Nous allons montrer au dernier Chapitre de ce travail que cette solution élémentaire se réduit dans le cas actuel — à ladite indétermination près — à l'expression $V = \delta_m \cdot r^{2-m}$, δ_m étant précisement la valeur $\neq 0$ donnée par la formule (37) du n° **58**. La solution élémentaire en question ne pourra donc nous rendre aucun service.

M. Hadamard(2) lui-même a bien constaté (**1**, p. 287) que ladite solution élémentaire n'était pas utilisable dans la théorie générale et que par suite il lui

---

(1) *Cf.* HADAMARD **1**, p. 190.

(2) Pour une autre méthode d'intégration *voir* HADAMARD **1**, p. 316.

fallait, pour m pair, recourir à la méthode de descente. Cette méthode consiste à monter d'abord à une équation au nombre impair $m + 1$ de variables et d'en redescendre à l'équation originale.

60. Vérification de la solution. — En réalité, la formule (3) du n° **44**, appelée dans ce qui précède « formule de solution », n'est au point où nous nous trouvons qu'une transcription — consciemment illégitime — d'une formule de représentation, à savoir de l'identité (2) du même numéro. L'identité en question permet d'exprimer la fonction $u(P)$ par les valeurs que la fonction Δu admet au domaine D_S^P et que la fonction u elle-même et sa dérivée du/dn, prise suivant la normale intérieure, admettent sur la portion de surface S^P. La transcription consistait en ce que nous avons, dans le second membre de la formule, remplacé les quantités Δu, du/dn, u par les données f, g, h, Appeler la formule ainsi obtenue une formule de solution n'était pour nous qu'une façon commode de parler — comme nous l'avons souligné dès le début de ce chapitre.

Or, dans l'hypothèse que la surface S a une orientation d'espace, admise dans le cours de ce travail, sauf avis contraire, la formule dont il s'agit est une vraie formule de solution, ce qui resterait à voir à présent. Voici quelques indications sur ce sujet.

Chacun des trois termes dont le second membre de la formule (2) est composé apporte sa propre contribution à la solution demandée, en ce qu'il donne la solution complète pour l'une des données f, g, h et la solution nulle pour les deux autres. Ainsi $I_*^2\overline{f, 0, 0}$ résout complètement le problème avec les données $f, 0, 0$ ($\Delta u = f$, $du/dn = 0$, $u = 0$) et ainsi de suite.

Le problème le plus difficile, lorsqu'on applique les méthodes habituelles, différentes de celles de M. Hadamard et des nôtres, consiste à démontrer que l'équation $\Delta u = f$ elle-même est vérifiée, problème qui chez nous se trouve résolu par la formule (124) du n° **42**. Il ne nous reste donc que les conditions aux limites, à savoir la démonstration de ce que la fonction u, donnée par la formule (3) et sa dérivée du/dn admettent sur la surface S les valeurs limites respectives h et g. Or il en est certainement ainsi, grâce à notre hypothèse sur l'orientation de S, bien entendu dans des conditions de dérivabilité suffisantes. Cela tient au fait qu'en raison de ladite hypothèse, les domaines d'intégration D_S^P et S^P deviennent infinitésimaux lorsque le point P s'approche indéfiniment de la surface. Nous omettons ici la vérification détaillée des dites propriétés limites des fonctions u et du/dn, vu que ces propriétés ont fait l'objet d'une étude soigneuse dans la Thèse de M. Fremberg, à laquelle nous avons fait allusion plus haut. (1) Nous pouvons aussi renvoyer le lecteur au livre de M. Hadamard, tant de fois cité, où l'on trouve les vérifications nécessaires aux pages 249—250 et 397—405 pour le cas général des équations hyperboliques normales à coefficients variables.

61. Frontière orientée dans le temps. — Une surface (ou une portion de surface) est dite orientée dans le temps, si ses normales sont des droites d'espace Une condition caractéristique équivalente est que la forme différentielle quadratique $\Sigma \varepsilon_k d\xi_k^2$ soit indéfinie en tout point (régulier) de cette surface.

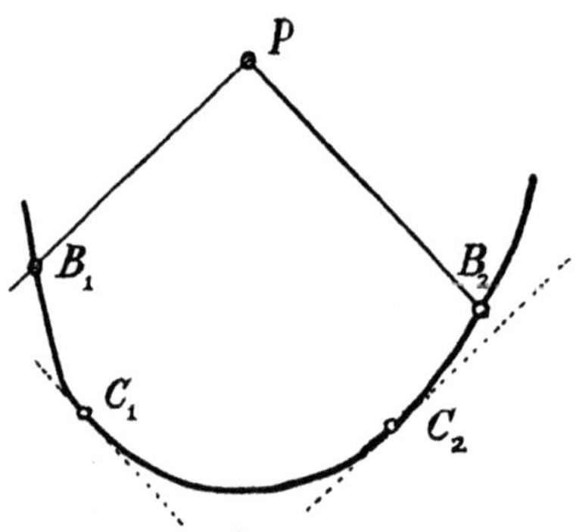

Fig. 7.

Dans un ordre d'idées général, considérons une frontière S (d'orientation arbitraire) pour laquelle il existe une région analogue à celle caractérisée dans le deuxième alinéa du n° **43**. Tout cône D^P dont le sommet se trouve dans la région en question devra délimiter avec la frontière S un domaine borné D_S^P. On supposera aussi que les lignes droites issues des points P et intérieures à D_S^P, les génératrices du cône y comprises, coupent S en un seul point sans lui être tangentes. Il est clair qu'une telle frontière ne pourra, dans toute son étendue, être orientée dans le temps si elle n'admet aucun point singulier (point sans plan tangent déterminé).

Cet état de choses est illustré par la fig. 7 ci-dessus, relative au cas de $m = 2$, où les arcs $B_1 C_1$ et $B_2 C_2$ sont de caractère de temps et l'arc $C_1 C_2$ de caractère d'espace. Les tangentes aux points C_1 et C_2 sont des caractéristiques (droites isotropes), parallèles aux génératrices PB_2 et PB_1 respectivement. De

(1) Fremberg **2**, p. 54.

telles frontières donnent lieu à des problèmes aux limites mixtes, auxquels M. Hadamard a voué des recherches assidues.[1] Comme on le verra dans la suite, il n'y a lieu de poser le problème de Cauchy — où l'on donne u et du/dn — que pour celle des parties d'une telle frontière qui est orientée dans l'espace, tandis que pour la partie orientée dans le temps, on aura à se contenter d'une seule de ces données.

Si l'on admet des singularités, p. ex. un ou plusieurs points coniques, ou, si l'on permet à la frontière de s'étendre à l'infini dans la direction des ξ_1 négatifs, la frontière peut être orientée entièrement dans le temps, comme il ressort des fig. 8 et 9 ci-jointes. Un exemple explicite est celui d'un cône direct

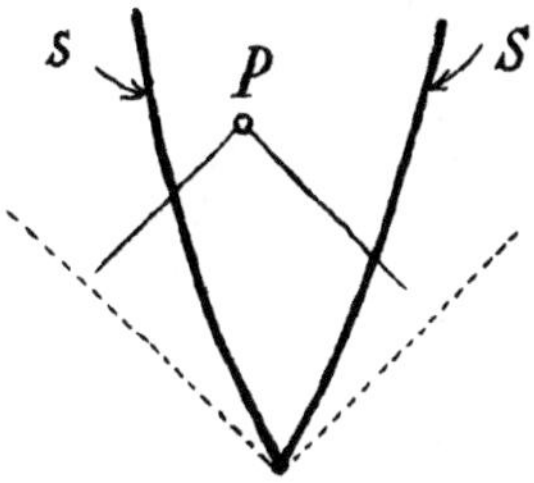

Fig. 8.

$\Sigma \eta_k x_k^2 = 0,\ x_1 > 0,\ \eta_1 = 1,\ \eta_i < -1\ (i = 2, 3, \ldots, m)$. On comprendra par la suite que pour une telle frontière il ne peut être question que d'un problème à une donnée aux limites unique, tout comme on pose, pour l'équation de Laplace, le problème de Dirichlet ou le problème de Neumann (problème hydrodynamique).

Avant d'aborder cette question assez délicate, nous allons faire voir que pour les frontières orientées dans le temps, telles qu'on vient de les décrire et assez régulières en général, l'identité (2) du n° **44** (notre point de départ pour la solution du problème de Cauchy) reste valide. Nous allons voir, en même temps, pourquoi cette identité ne conduit pas à une formule de solution pour de telles frontières. Ajoutons que les considérations qui suivent seraient faciles à étendre aux frontières de caractère mixte.

Remarquons d'abord que pour une frontière orientée dans le temps il faudra changer le signe de l'expression qui figure sous la racine carrée dans la définition de la quantité J (formule (34) du n° **21**) pour rendre cette expression positive. Ce changement fait, toutes les formules des n°s **21—25** dont on a besoin

[1] Hadamard **1**, p. 48—52, 337—347, 453—486; 2, p. 23—29.

pourront être conservées. Pour effectuer le prolongement analytique en présence d'un nombre fini de points singuliers on remplace les parties de S voisines des singularités par de petites surfaces telles que la surface S modifiée soit suffisamment dérivable. En appliquant les méthodes des n^os **27—37**, on peut former le prolongement des intégrales modifiées. Si l'on observe encore que nos intégrales étendues aux voisinages des points singuliers (supposés intérieurs à D_S^P) ne cessent jamais de converger, on voit que le prolongement des intégrales originales s'obtient par un passage à la limite de celui des intégrales modifiées. Par ces

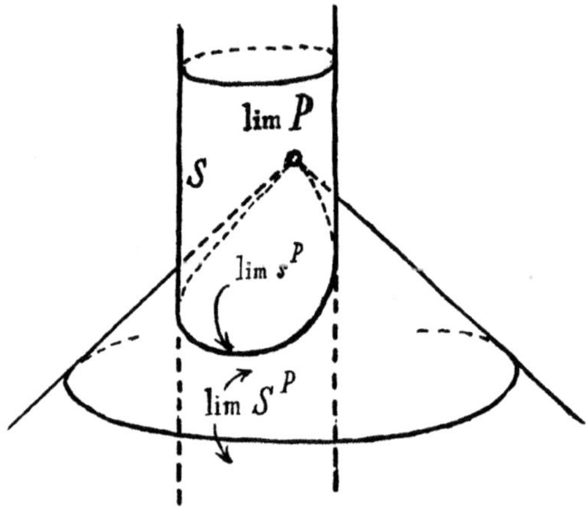

Fig. 9. La surface S de la figure est un cylindre portant (par derrière) le point $Q = \lim P$.

considérations on a en particulier démontré que l'identité (2) reste valide dans le cas actuel. [1]

Cette identité, conduit-elle à une solution, si l'on y porte les données f, g, h, c'est-à-dire qu'on passe à la formule (3)? Elle y conduit en ce qui concerne l'équation $\Delta u = f$, comme on le voit immédiatement. Supposons, pour simplifier, que $f = 0$, de sorte qu'on n'ait à considérer dans I_*^α que les deux intégrales de surface, étendues à S^P qui, pour $\alpha = 2$ et pour m pair, se réduisent à des intégrales suivant l'intersection s^P de la nappe C^P avec la surface S, complétées, pour m impair, par des intégrales étendues à S^P elle-même. Si P se rapproche d'un point Q de la frontière, S^P et s^P ne se rétréciront plus au point Q, mais tendront vers les variétés S^Q et s^Q, à dimensions $m - 1$ et $m - 2$ respectivement.

[1] Si le domaine s'étend à l'infini on prendra soin de faire des hypothèses convenables sur l'allure de la fonction.

Les intégrales correspondantes tendront vers des intégrales singulières, auxquelles on pourra en général donner un sens convenable. Postuler que l'expression limite ainsi obtenue soit égale à la valeur prescrite $h(Q)$ en tout point Q de la frontière revient à poser une infinité de conditions pour les fonctions g et h qu'on peut résumer par une équation intégrale de caractère singulier. Ces deux fonctions ne peuvent donc être données indépendamment l'une de l'autre.(1) On voit alors qu'il ne peut être question que de donner *a priori* une seule de ces fonctions et de chercher à tirer l'autre de l'équation intégrale à laquelle on paraît être arrivé. Or, on verra par les remarques qui vont suivre que l'esquisse ci-dessus, tout en mettant en évidence l'impossibilité du problème de Cauchy, ne pourra guère rendre de trop grands services quand il s'agit de résoudre le problème « correctement posé », conforme aux conditions actuelles, c'est-à-dire le problème où une seule des données est prescrite, soit les valeurs limites de la fonction u elle-même.

En effet, la situation est tout analogue à celle qui interviendrait si l'on voulait résoudre le problème de Dirichlet par la formule ordinaire de Green,

$$u(P) = \frac{1}{4\pi}\int\left(u\frac{d\frac{1}{r}}{dn} - \frac{1}{r}\frac{du}{dn}\right)dS.$$

Là aussi, on aboutirait à une équation intégrale entre les valeurs limites admissibles de u et de du/dn, les intégrales étant cette fois étendues à la surface totale S. Cette équation intégrale, qui présente des singularités assez faibles, est très facile à écrire. Néanmoins, il n'est pas trop facile d'en tirer soit la résolution effective du problème de Dirichlet, soit un théorème d'existence. Dans le cas actuel, l'équation à laquelle on arrive par l'aperçu de nature tout à fait théorique que nous venons de donner, est de caractère incomparablement plus singulier que dans le cas du problème de Dirichlet, et il paraît difficile de parvenir dans cette voie à une solution effective ou même à une démonstration d'existence.

Terminons cet exposé de caractère bien négatif en rappelant que Goursat(2) a résolu le problème dont il s'agit, dans des conditions très générales, pour le cas du plan ($m = 2$). La frontière considérée par Goursat devient identique à

(1) En conséquence, la formule célèbre de Kirchhoff ne peut être considérée comme une formule de solution. *Cf.* HADAMARD **1**, p. 332.

(2) GOURSAT III (1915), p. 123 et *suiv.*

celle représentée par notre fig. 8, si l'on remplace son opérateur $\frac{\partial^2}{\partial x_1 \partial x_2}$ par l'opérateur $\frac{\partial^2}{\partial x_1^2} - \frac{\partial^2}{\partial x_2^2}$. La structure de la solution très remarquable de Goursat fait pressentir les difficultés auxquelles on devra se heurter en attaquant le problème concernant les dimensions d'ordre pair > 2. Les difficultés ne sauraient être moindres pour les dimensions impaires.

62. Frontière caractéristique. — Considérons le cône caractéristique

$$\Sigma \varepsilon_k (\xi_k - a_k)^2 = 0.$$

Pour tout déplacement $(d\xi_k)$ tangent au cône au point (ξ_k), on a manifestement $\Sigma \varepsilon_k (\xi_k - a_k) d\xi_k = 0$. Ce fait implique que le plan tangent mené à un cône caractéristique suivant l'une de ses génératrices et qui, par conséquent, contient cette génératrice, l'admet aussi comme normale (et cela évidemment en tout point appartenant à la génératrice). Il en résulte que de tels plans tangents contiennent toutes leurs normales, tandis que, pour tout autre plan, la normale menée en un point n'a aucun autre point en commun avec le plan. N'oublions pas que les génératrices en question sont identiques aux droites que nous avons appelées *isotropes* (n° **14**).

On donne le nom de *plan caractéristique* aux susdits plans tangents. L'équation d'un tel plan étant $\Sigma \alpha_k \xi_k - \beta = 0$, on a $\Sigma \varepsilon_k^{-1} \alpha_k^2 = 0$ et inversement. Dès lors, en écrivant sous la forme

(38) $$S(\xi_1, \xi_2, \ldots, \xi_m) = 0$$

soit l'équation d'un cône caractéristique, soit celle d'un plan caractéristique, on voit que (38) entraîne l'équation aux dérivées partielles

(39) $$\sum_1^m \varepsilon_k^{-1} \left(\frac{\partial S}{\partial \xi_k}\right)^2 = 0.$$

Considérons maintenant, dans un ordre d'idées général, une surface S donnée par une équation de la forme (38). Si l'équation (39) est remplie en chaque point (régulier) de la surface, celle ci est dite *surface caractéristique*. Voici quelques indications sommaires sur les surfaces qu'on vient de définir.

La direction de la normale au point régulier (ξ_k) étant, en composantes covariantes, donnée par les quantités $\partial S / \partial \xi_k$ (*cf.* le n° **21**), la condition (39) exprime que les normales d'une surface caractéristiques sont isotropes. Il n'y a donc pas lieu de parler de normales unitaires. Nous allons faire voir que la surface admet ses normales comme génératrices.[1] Il suffit à cet effet, de montrer que toute courbe tracée sur la surface qui en chacun de ses points est tangente à la normale menée en ce point à la surface est elle-même une (portion de) ligne droite.

En rapportant la surface à $m-1$ paramètres, on établit facilement l'existence des courbes en question. Ce point admis et en posant pour abréger $\partial S/\partial \xi_k = S_k$, on aura à étudier les équations

$$\frac{\varepsilon_k\, d\xi_k}{S_k} = d\tau, \tag{40}$$

où $d\tau$, défini par les premiers membres, est indépendant de k. En observant que $\partial S_k/\partial \xi_j = \partial S_j/\partial \xi_k$ et en tenant compte de (40), on obtient

$$\frac{dS_k}{d\tau} = \sum_{j=1}^{m} \frac{\partial S_k}{\partial \xi_j}\frac{d\xi_j}{d\tau} = \sum_{j=1}^{m} \varepsilon_j^{-1}\frac{\partial S_j}{\partial \xi_k} S_j = \tfrac{1}{2}\frac{\partial T}{\partial \xi_k}, \tag{41}$$

où l'on a désigné par T le premier membre de (39). Or T étant par hypothèse identiquement nul sur la surface $S=0$, on a, pour tout déplacement $(d\xi_k)$ le long de cette surface, $\sum \frac{\partial T}{\partial \xi_k} d\xi_k = 0$. Cela implique que le gradient de T est orthogonal à la surface. Il est donc, à un facteur de proportionnalité près, identique au gradient de S, c'est-à-dire que

$$\frac{\partial T}{\partial \xi_k} = \lambda \frac{\partial S}{\partial \xi_k} = \lambda S_k \tag{42}$$

et alors d'après (41)

$$\frac{dS_k}{d\tau} = \frac{\lambda}{2} S_k, \tag{43}$$

λ étant indépendant de k. On en tire par des quadratures que, pour j et k arbitraires, le rapport $S_j : S_k$ est constant suivant la courbe; il en est donc de même de $d\xi_j : d\xi_k$, ce qui veut dire que la courbe est une ligne droite, évidemment

[1] L'équation (39) est équivalente au fait que la surface est, en chacun de ses points, tangente à un cône caractéristique qui admet le point comme sommet. Le fait énoncé ici revient à ce que les deux surfaces sont tangentes suivant une génératrice commune.

isotrope, qui en tous ses points est normale à la surface. Le plan tangent restant le même suivant cette droite, la surface peut être considérée comme une surface développable.

Une frontière caractéristique est formée d'une ou plusieurs portions de surfaces caractéristiques. Pour arriver à des frontières présentant le même caractère topologique que les frontières orientées dans le temps, considérées plus haut, on aura encore à admettre des singularités.(1) On prendra p. ex. un dièdre formé de deux surfaces caractéristiques régulières, un polyèdre à faces caractéristiques et de préférence un cône caractéristique direct.

Les problèmes aux limites conformes aux frontières caractéristiques sont en quelque sorte intermédiaires entre ceux qui conviennent aux deux espèces de frontières considérées plus haut. En effet, comme on le verra (après une légère modification de la définition de du/dn) l'une des deux données figurant d'habitude dans le problème de Cauchy détermine immédiatement l'autre (à une constante additive près quant il s'agira de déterminer la fonction moyennant sa dérivée).

Remarquons d'abord que, la normale étant isotrope, on ne pourra former la dérivée du/dn puisque $dn = 0$. Peu importe; on pourra toujours former la dérivée dans la *direction* de la normale (*cf.* les formules (43) ou (44) du n° **21**) et on le fera par rapport à une variable différente de la longueur identiquement nulle du segment correspondant de la normale.

Tel étant le cas, on voit que, la normale faisant partie de la surface S, les valeurs de u, une fois données sur S, déterminent la valeur de toute dérivée prise suivant la direction de cette normale. La réciproque est aussi vraie, à une constante additive près. Il n'y a donc lieu que de prescrire une seule des données habituelles.

En nous contentant pour le cas général de ces indications sommaires, nous allons, pour le cas particulier d'un *cône caractéristique direct* donner une solution et une vérification complètes pour l'équation sans second membre. Le cas d'un tel cône fera bien ressortir les traits essentiels du même problème posé pour des frontières caractéristiques plus générales. Ajoutons que la présence d'un second membre, tout en nécessitant des calculs supplémentaires, n'amènerait aucune difficulté essentielle.

---

(1) *Cf.* Hadamard **1**, p. 253.

63. Problème aux limites pour un cône caractéristique direct. [1] — Nous allons démontrer que l'équation $\Delta u = 0$ admet une solution et une seule, si l'on prescrit les valeurs de u — ces valeurs étant suffisamment dérivables — sur la surface S d'un cône caractéristique direct. [2] Plus précisement, si les valeurs sont prescrites sur une portion $\bar{S}$ de S, la valeur de u sera déterminée en tout point P tel que la portion S^P découpée de S par le cône rétrograde D^P est intérieure à $\bar{S}$.

La solution de notre problème reposait, dans ce qui précède, essentiellement sur la formule de Green (formule (60) du n° **22** ou formule (61bis) du n° **23**). Or il entre dans cette formule deux éléments géométriques qui ne peuvent nous servir dans le cas actuel; ce sont l'élément de surface dS dans le numérateur et l'élément de la normale dn dans le dénominateur, tous les deux étant nuls lorsque la surface est caractéristique. A l'aide de l'angle solide au sommet P, il est facile de donner à la formule de Green une forme qui est intéressante sous le point de vue général et qui est applicable aussi dans le cas où nous nous trouvons.

Considérons, dans nos notations usuelles, le domaine D^P_S, la surface S étant, jusqu'à nouvel ordre et pour fixer les idées, une surface d'espace. Les fonctions u et v étant suffisamment dérivables, dans D^P_S, et v s'annulant avec ses dérivées du premier ordre sur la portion de nappe C^P_S, on peut écrire la formule générale (41) du n° **21**

$$\int\limits_{D^P_S} (u\,\Delta v - v\,\Delta u)\,dQ = \int\limits_{S^P} \left(u\frac{dv}{dn} - v\frac{du}{dn}\right) dS, \tag{44}$$

la dérivation étant cette fois-ci faite suivant la normale extérieure. On peut maintenant introduire l'élément d'angle solide dH au sommet P et écrire, dans des notations légèrement modifiées, en vertu de la formule (93) du n° **31**,

$$dH = r^{1-m}\frac{dr}{dn}\cdot dS \quad \text{ou} \quad dS = r^{m-1}\left(\frac{dr}{dn}\right)^{-1} dH.$$

En portant cela dans la formule (44), on aura, avec la notation

$$\frac{dF}{dn}\cdot\left(\frac{dr}{dn}\right)^{-1} = \frac{dF}{dn}\cdot\frac{dn}{dr} = \frac{d_nF}{d_nr}, \tag{45}$$

---

[1] Nous admettons dans les n^{os} **63—64** que les ε_k ont leurs valeurs originales, $\varepsilon_1 = 1$, $\varepsilon_2 = \cdots = \varepsilon_m = -1$.

[2] *Cf.* pour ce problème d'Adhémar **1**.

$$\int_{D_S^P} (u\,\Delta v - v\,\Delta u)\,d\,Q = \int_{S^P} \left(u \frac{d_n v}{d_n r} - v \frac{d_n u}{d_n r}\right) r^{m-1}\,d\,H. \tag{46}$$

La quantité r, par rapport à laquelle on effectue la différentiation étant scalaire, on n'a pas besoin (dans nos conditions de dérivabilité usuelles) de distinguer ici entre les deux *sens* de la normale; la différentiation se fait simplement dans la *direction* de la normale.

La formule ci-dessus reste évidemment valide, si S signifie le cône direct considéré plus haut, comme on le voit par un passage à la limite facile[1]. Pour simplifier les calculs qui suivent, admettons que le sommet du cône direct S se trouve à l'origine O, ce qui ne restreint pas la généralité. Nous allons exprimer l'élément $d\,H$ relatif à S par d'autres éléments géométriques rattachés à ce cône. L'intersection[2] s^P (à $m-2$ dimensions) des cônes au sommet O et au sommet $P = \mathbf{x}$ est évidemment située dans un plan à $m-1$ dimensions, orthogonal au rayon-vecteur $\mathbf{x}$ et passant par le point $\mathbf{x}/2$. (s^P n'est qu'une sphère de caractère euclidien au centre $\mathbf{x}/2$, comme on le verra au prochain numéro.)

L'intersection s^P comprend ceux des points $\boldsymbol{\xi}$ de S qui satisfont à la relation $r_{P\xi} = r_{x\xi} = 0$. Nous considérons sur S les deux variétés σ et σ', homothétiques à s^P qui sont caractérisées[3] par les relations respectives $r_{x\xi} = r$ et $r_{x\xi} = r + d\,r$, ces dernières quantités étant fixées jusqu'à nouvel ordre. Nous allons maintenant calculer l'angle solide total sous lequel la zone de S délimitée par σ et σ' est vue du point P. On aurait donc à projeter (tout comme au n° **27**) la zone en question sur l'hyperboloïde $\mathbf{x} + H$. Il est pourtant plus commode d'évaluer la projection faite sur l'hyperboloïde $\mathbf{x} + r\,H$, dont σ fait partie. Cette projection-là qui, pour des raisons d'homothétie, est égale à l'angle total cherché, multiplié par r^{m-1}, peut être évaluée par les considérations suivantes, où l'on se servira, avec un grand avantage, de la fig. 10 ci-jointe.

Un segment de génératrice $A\,A'$, découpé par σ et σ', sera projeté dans un arc $A\,A''$, dont nous désignons la longueur par $d\tau$. Dès lors, en désignant encore par σ la mesure totale de la variété σ, la projection aura (en première approximation) la mesure $\sigma\,d\tau$. Nous allons démontrer que $d\tau = d\,r$ et obtenir ainsi, pour la valeur cherchée, l'expression $\sigma \cdot d\,r$.

---

(1) On peut arriver au même résultat en partant de la formule de Green écrite sous sa forme affine, formule (32) du n° **20**.

(2) Pour certains détails concernant les propriétés de s^P énumérées ici, *voir* le n° **64**.

(3) Il est évident que les variétés σ et σ' font partie de deux plans respectifs parallèles au plan qui contient s^P.

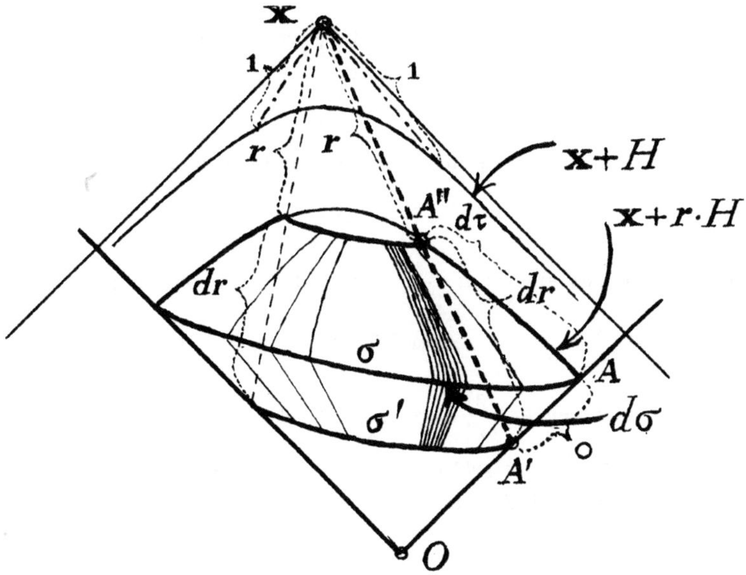

Fig. 10.

En effet, les points P, A'', A' se trouvant en ligne droite, la longueur du segment $A''A'$ sera la différence entre les longueurs de PA' et PA'', c'est-à-dire égale à $r + dr - r = dr$. Notons en outre, que $A''A'$, étant normal à l'hyperboloïde, est en même temps orthogonal à $A''A$ et de plus que $A''A'$, $A''A$, $A'A$ sont respectivement segment de temps, segment d'espace et segment isotrope (tous infinitésimaux). En exprimant que $A'A$ a la longueur o, on trouve $dr^2 - d\tau^2 = 0$, c'est-à-dire que $d\tau = dr$.

La projection d'un élément qui est infinitésimal dans tous les sens se trouve par le même raisonnement. Ainsi, la projection d'un élément à $m - 1$ dimensions, constitué par les segments de génératrices issus des points d'un élément infinitésimal $d\sigma$ de σ est $d\sigma \cdot dr$, ce qui donne enfin, en écrivant, pour préciser, $d\sigma_r$ au lieu de $d\sigma$

(47) $$r^{m-1}\, dH = d\sigma_r \cdot dr.$$

Observons encore que dans (46) la différentiation par rapport à r suivant la direction normale n'est, lorsqu'il s'agit d'un cône caractéristique, qu'une différentiation dans la *direction des génératrices* — opération que *nous désignerons désormais par* d/dr. Cette convention posée et portant (47) dans la formule (46), on trouve enfin

$$\int_{D_S^P} (u \Delta v - v \Delta u)\, dQ = \int_0^{r_0} \left\{ \int_{\sigma_r} \left(u \frac{dv}{dr} - v \frac{du}{dr} \right) d\sigma_r \right\} dr,$$

où l'on a désigné par r_0 la valeur que r admet au sommet O,

(48) $$r_0 = (\mathbf{x}^2)^{\frac{1}{2}}.$$

Nous introduisons encore $R = r^2$ à côté de r, posant en particulier

(48bis) $$R_0 = r_0^2 = \mathbf{x}^2.$$

On aura, pour commencer, $\frac{dv}{dr} dr = \frac{dv}{dR} dR$. En outre, R varie d'une manière linéaire sur toute génératrice, ce qui présente un avantage considérable. En effet, $\boldsymbol{\xi}$ variant sur une génératrice de S, on a

(49) $$R = (\mathbf{x} - \boldsymbol{\xi})^2 = \mathbf{x}^2 - 2\,\mathbf{x}\,\boldsymbol{\xi} = R_0 - 2\,\mathbf{x}\,\boldsymbol{\xi}.$$

Nous rapportons maintenant chaque point $\boldsymbol{\xi}$ de la portion de surface S^P au point $\mathbf{b}$ dans lequel la génératrice $O\xi$ coupe s^P. Nous avons là $m-2$ paramètres. Le dernier paramètre sera la valeur $R = R_{x\xi}$ (carré de la distance lorentzienne entre $\mathbf{x}$ et $\boldsymbol{\xi}$). En observant, de plus, que la quantité $1 - \frac{R}{R_0}$ varie aussi d'une manière linéaire sur toute génératrice, qu'elle s'annule au sommet O (où $R = R_0$) et qu'elle est $= 1$ aux points de s^P (où $R = 0$), il vient par homothétie

$$d\sigma_r = \left(1 - \frac{R}{R_0}\right)^{m-3} ds,$$

où ds désigne l'élément de s^P. On a donc, eu égard aux remarques qui précèdent,

(50) $$\int_{D_S^P} (u \Delta v - v \Delta u)\, dQ = \int_{s^P} ds \int_0^{R_0} \left(u \frac{dv}{dR} - v \frac{du}{dR} \right) \left(1 - \frac{R}{R_0}\right)^{m-3} dR.$$

Nous spécialisons maintenant la fonction v en posant

$$v = \frac{r^{\alpha+2-m}}{H_m(\alpha+2)} = \frac{R^{\frac{\alpha+2-m}{2}}}{H_m(\alpha+2)},$$

où $\alpha > m$ jusqu'à nouvel ordre. Il vient alors

$$(51)\quad I^{\alpha} u(P) = I^{\alpha+2} \Delta u(P) + \frac{1}{H_m(\alpha+2)} \int\limits_{s^P} ds \int\limits_0^{R_0} \left(u \cdot \frac{d}{dR} R^{\frac{\alpha+2-m}{2}} - \frac{du}{dR} \cdot R^{\frac{\alpha+2-m}{2}} \right) \left(1 - \frac{R}{R_0} \right)^{m-2} dR = I^{\alpha+2} \Delta u(P) + U^{\alpha+2}(P).$$

On voit facilement que les intégrales figurant dans la formule convergent toutes et que la formule reste valable jusqu'à $\alpha > m - 2$.

En supposant que $\Delta u = 0$, notre formule se réduit à

$$(52)\quad I^{\alpha} u(P) = U^{\alpha+2}(P),$$

et en particulier (par prolongement analytique) à

$$(53)\quad u(P) = U^{2}(P).$$

Le prolongement analytique du premier membre peut être effectué, tout comme dans le cas d'une frontière orientée dans l'espace (ou dans le temps) par la méthode développée plus haut (n^{os} **27—37** et p. 102). Il y aurait même des simplifications considérables, puisqu'on pourrait expliciter tous les calculs. En particulier on trouve $I^0 u(P) = u(P)$.

Il s'agit maintenant d'effectuer le prolongement analytique du second membre. Posons pour abréger

$$(54)\quad \frac{\alpha + 2 - m}{2} = \beta, \quad \pi^{\frac{m-2}{2}} 2^{\alpha+1} \Gamma\left(\frac{\alpha+2}{2}\right) = K_m(\alpha+2),$$

$$H_m(\alpha+2) = \pi^{\frac{m-2}{2}} 2^{\alpha+1} \Gamma\left(\frac{\alpha+2}{2}\right) \Gamma\left(\frac{\alpha+4-m}{2}\right) = K_m(\alpha+2)\, \Gamma(\beta+1).$$

Nous allons d'abord transformer l'intégrale, par rapport à R, qui figure dans l'expression (51) de $U^{\alpha+2}$. L'exposant de R dans cette expression étant β, d'après la première des formules (54), écrivons

$$\frac{du}{dR} R^{\beta} - u \frac{dR^{\beta}}{dR} = R^{2\beta} \frac{d}{dR}(u R^{-\beta}).$$

D'autre part, en exprimant α par β, on a $\alpha = 2\beta + m - 2$ et alors

$$\frac{d}{dR}\left\{ R^{2\beta} \left(1 - \frac{R}{R_0}\right)^{m-2} \right\} = R^{2\beta-1} \left(1 - \frac{R}{R_0}\right)^{m-3} \left\{ 2\beta \left(1 - \frac{R}{R_0}\right) - \frac{m-2}{R_0} R \right\}$$

$$= R^{2\beta-1} \left(1 - \frac{R}{R_0}\right)^{m-3} \left(2\beta - \alpha \frac{R}{R_0}\right) = 2 \left(1 - \frac{R}{R_0}\right)^{m-3} \left(\beta R^{2\beta-1} - \frac{\alpha}{2 R_0} R^{2\beta}\right).$$

Cela posé, on obtient en intégrant par parties, grâce au fait que le terme tout intégré s'annule pour $m > 2$[1] et β assez grand ($\beta > 0$), et dès lors, par prolongement analytique, pour toute valeur de β,

$$\int_0^{R_0}\left(u\frac{dR^\beta}{dR}-\frac{du}{dR}R^\beta\right)\left(1-\frac{R}{R_0}\right)^{m-2}dR=-\int_0^{R_0}R^{2\beta}\frac{d}{dR}(uR^{-\beta})\left(1-\frac{R}{R_0}\right)^{m-2}dR$$

$$=\int_0^{R_0}uR^{-\beta}\frac{d}{dR}\left\{R^{2\beta}\left(1-\frac{R}{R_0}\right)^{m-2}\right\}dR=2\int_0^{R_0}u\left(1-\frac{R}{R_0}\right)^{m-3}\left(\beta R^{\beta-1}-\frac{\alpha}{2R_0}R^\beta\right)dR.$$

Après p et $p-1$ intégrations par parties exécutées respectivement sur

$$\left\{u\cdot\left(1-\frac{R}{R_0}\right)^{m-3}\right\}R^{\beta-1}\quad\text{et}\quad\left\{u\cdot\left(1-\frac{R}{R_0}\right)^{m-3}\right\}R^\beta$$

et quelques simplifications évidentes, il vient, en tenant compte de (54),

$$(55)\quad U^{\alpha+2}(P)=\frac{2}{K_m(\alpha+2)}\int_{s^P}ds\left\{[C^p]_0^{R_0}+\frac{1}{\Gamma(\beta+p)}\int_0^{R_0}\left(A^p+\frac{\alpha}{2R_0}A^{p-1}\right)R^{\beta+p-1}dR\right\}$$

avec les notations

$$(56)\qquad A^p=(-1)^p\frac{d^p}{dR^p}\left(u\cdot\left(1-\frac{R}{R_0}\right)^{m-3}\right)=R_0^{-p}\left(1-\frac{R}{R_0}\right)^{m-3-p}\cdot D^p(\xi),$$

$$(57)\qquad D^p(\xi)=\sum_{h=0}^{p}B_h^p\frac{d^hu}{dR^h}\cdot(R-R_0)^h,\ \text{où}\ B_h^p=\binom{p}{h}\frac{(m-3)!}{(m-3-p+h)!},$$

$$(58)\qquad C^p=\sum_{i=0}^{p-1}\left(A^i+\frac{\alpha}{2R_0}A^{i-1}\right)\frac{R^{\beta+i}}{\Gamma(\beta+i+1)}.$$

Pour des applications ultérieures, observons déjà ici que la comparaison entre les deux expressions de A^p donne aussi

$$(57^{bis})\qquad D^p(\xi)=(R-R_0)^{-(m-3-p)}\frac{d^p}{dR^p}(u\cdot(R-R_0)^{m-3}).$$

L'intégrale par rapport à R dans (55) est convergente lorsque $\beta+p-1>-1$, ou formule (54)) $p>-\beta=\frac{1}{2}(m-2-\alpha)$, ce qu'on admettra dans la suite. D'autre

(1) Le cas de $m=2$, où l'équation $\Delta u=0$ admet la solution évidente $u=F(x_1+x_2)+G(x_1-x_2)$ (F et G arbitraires), sera définitivement laissé de côté.

part, le terme tout intégré s'annule pour $R = 0$, lorsque β est assez grand, et dès lors, par prolongement analytique, comme plus haut, pour toute valeur de β. Pour $R = R_0$, ce terme s'évanouit quand $p \leqq m - 3$, tandis que, quand $p > m - 3$, il contient la valeur de la fonction u et de certaines de ses dérivées au sommet O de notre cône direct (ainsi que certaines puissances de R_0). Nous donnerons tout à l'heure des formules explicites pour ce terme dans deux cas particuliers ($m = 3$ et 4).

C'est l'expression de $U^2(P)$ qui, en raison de la relation (53), nous intéresse en premier lieu. Nous le tirons de la formule (55) en y posant $\alpha = 0$. Or, justement pour cette valeur-là de α, il y aura une simplification considérable dans toutes nos formules, en tant que les termes qui contiennent α en facteur disparaîtront. — Dans l'examen détaillé de $U^2(P)$, nous aurons, comme d'habitude, à distinguer les cas de m impair et pair.

I. *m impair, $m = 2l + 1$.*

Nous aurons à poser dans la formule (55) $\alpha = 0$, $\beta = (2 - m)/2 = \frac{1}{2} - l$. Choisissons $p = l$, on aura $\beta + p - 1 = -\frac{1}{2}$ et l'intégrale suivant R dans (55) sera convergente. En admettant d'abord que $m > 3$ ou $l > 1$, on a $p = l \leqq m - 3 = 2l - 2$, ce qui fait que le terme tout intégré dans (55) s'annule (*cf.* (56) et (58)). En tenant compte de (54), on trouve $\frac{1}{2} K_m(2) = \pi^{l-\frac{1}{2}}$, $\Gamma(\beta + p) = \Gamma(\frac{1}{2} - l + l) = \Gamma(\frac{1}{2}) = \pi^{\frac{1}{2}}$, ce qui nous fournit (*cf.* (53))

$$u(P) = U^2(P) = \frac{1}{\pi^l} \int_{s^P} ds \int_0^{R_0} A^l R^{-\frac{1}{2}} \, dR, \tag{59}$$

A^l étant donné par les formules (56) et (57).

I'. *Le cas particulier $m = 3$.*

Ici, l est $= 1$ et $\beta = \frac{1}{2} - l = -\frac{1}{2}$. Tout se passe comme dans le cas général, sauf que le terme tout intégré dans (55) n'est pas nul. On a d'après (56) et (58) $C^p = C^1 = A^0 R^\beta / \Gamma(\beta + 1) = u \cdot (\pi R)^{-\frac{1}{2}}$, d'où en désignant la valeur de u au sommet O de notre cône direct par $u(O)$ (cette valeur correspond dans le terme tout intégré à $R = R_0$)

$$[C^p]^{R_0} = \frac{u(O)}{(\pi R_0)^{\frac{1}{2}}} \quad \text{et} \quad \frac{2}{K_3(2)} [C^p]^{R_0} = \frac{2}{2\pi^{\frac{1}{2}}} \frac{u(O)}{(\pi R_0)^{\frac{1}{2}}} = \frac{u(O)}{\pi R_0^{\frac{1}{2}}}.$$

La mesure totale de s^P étant $R_0^{\frac{1}{2}}\pi$, la contribution à $u(P)$ du terme tout intégré sera $u(O)$. D'autre part, dans l'intégrale figurant dans (59) on aura à poser, d'après (56), $A^l = A^1 = -du/dR$. Il s'ensuit la formule définitive

$$(59^{bis}) \qquad u(P) = U^2(P) = u(O) - \frac{1}{\pi}\int\limits_{s^P} ds \int\limits_0^{R_0} \frac{du}{dR} R^{-\frac{1}{2}}\, dR.$$

II. *m pair*, $m = 2l$.

On aura à poser dans les formules (55) et (56) $\alpha = 0$, $\beta = 1 - \frac{m}{2} = 1 - l$, $\frac{1}{2}K_m(2) = \pi^{l-1}$. En choisissant $p = l$, il vient $\beta + p = 1$, $\beta + p - 1 = 0$. L'intégration par rapport à R dans (55) pourra donc être exécutée sur le champ

$$(60) \qquad \int\limits_0^R A^l\, dR = -[A^{l-1}]_0^{R_0}.$$

Admettons d'abord que $m > 4$ ou $l > 2$; on a donc $l - 1 < m - 3 = 2l - 3$. Dans cette condition, le terme tout intégré dans la formule (60) s'annule pour $R = R_0$, comme on le voit immédiatement par (56). Il en sera de même du terme tout intégré dans la formule (55). Il ne nous reste donc que celui des termes tout intégrés qui dans la formule (60) correspond à $R = 0$, c'est-à-dire aux points du bord s^P. Le résultat final sera donc

$$(61) \qquad u(P) = U^2(P) = \frac{1}{\pi^{l-1}}\int\limits_{s^P} A^{l-1}\, ds,$$

où l'on pourrait, pour préciser, écrire $A^{l-1} = A^{l-1}_{(R=0)}$. En raison de (56), on a sur le bord

$$(62) \qquad A^{l-1}_{(R=0)} = (-1)^{l-1}\frac{d^{l-1}}{dR^{l-1}}\left(u\left(1 - \frac{R}{R_0}\right)^{2l-3}\right)_{(R=0)}$$
$$= R_0^{1-l}\sum_{h=0}^{l-1} B_h^{l-1}\left(\frac{d^h u}{dR^h}\right)_{(R=0)} \cdot (-R_0)^h.$$

II'. *Le cas particulier* $m = 4$.

Ici $l = 2$, $\beta = 1 - l = -1$ et les termes tout intégrés dans (60) et (55) sont différents de 0. Celui dans (60) est $-A'(R_0)$, tandis que dans (55) il faut poser, d'après (58),

$$C^2(R_0) = \sum_{i=0}^{1} A^i(R_0)\frac{R_0^{-1+i}}{\Gamma(i)} = A^1(R_0),$$

ce qui montre que les termes tout intégrés se détruisent. La formule (61) subsiste donc aussi pour $m = 4$. En explicitant (62), on obtient

$$A^1_{(R=0)} = -\frac{d}{dR}\left(u\left(1 - \frac{R}{R_0}\right)\right)_{(R=0)} = \left(\frac{u}{R_0} - \frac{du}{dR}\right)_{(R=0)},$$

ce qui nous fournit, en supprimant l'affixe $R = 0$,

(61*bis*) $$u(P) = U^2(P) = \frac{1}{\pi}\int_{s^P}\left(\frac{u}{R_0} - \frac{du}{dR}\right)ds. \text{ (1)}$$

Il suffiit d'un coup d'œil sur les formules générales (59) et (61) ou sur les formules particulières (59*bis*) et (61*bis*) pour voir qu'il y a ici la même différence fondamentale entre les solutions relatives aux dimensions d'ordre impair et pair que dans le cas d'une frontière orientée dans l'espace.

64. Vérification de la solution. — Jusqu'à nouvel ordre, les formules générales (59) et (61) ou leurs cas particuliers (59*bis*) et (61*bis*) — formules que, pour abréger, nous citerons dans la suite par (59—61*bis*) — ne sont que des identités relatives à une fonction arbitraire (suffisamment dérivable) u, qui satisfait à l'équation $\Delta u = 0$. Il faudra maintenant *vérifier* que, les valeurs limites (suffisamment dérivables) de u étant données sur (une portion $\bar{S}$ de) la surface S, la fonction $u(P)$ qu'on peut en former moyennant les formules citées satisfait à l'équation $\Delta u = 0$ et admet sur S les valeurs limites prescrites.

Soit maintenant t une quantité qui varie d'une manière linéaire sur une génératrice et s'annule au sommet O. Telles quantités sont p. ex. la coordonnée ξ_1, les quantités géométriques $R - R_0$ ou $1 - \frac{R}{R_0}$. Le point ξ étant considéré comme fonction de t, on pourra former les expressions

(63) $$t^h \cdot \frac{d^h u(\xi)}{dt^h} = t^h \cdot \frac{d^h u}{dt^h}.$$

(1) Cette formule qui est un cas particulier d'une formule générale, à laquelle est consacré le Chapitre V, s'ensuit facilement de la formule de Poisson, concernant le cas où la surface portant les données de Cauchy est un plan $\xi_1 = \text{const.}$ (*voir* le chapitre indiqué).

Les expressions (63) *sont indépendantes du choix de* t, c'est-à-dire qu'elles ne dépendent que de ξ. Le fait est évident; en effet, $t = \lambda \bar{t}$, où $\lambda =$ const., donne $t^h = \lambda^h \bar{t}^h$ et $d^h u/d t^h = \lambda^{-h} d^h u/d\bar{t}^h$. Il n'est pourtant pas inutile d'observer que, si la fonction u admet des dérivées partielles par rapport aux ξ_k dans un certain voisinage de S, on a, eu égard à $t \cdot d\xi_k/dt = \xi_k$, l'identité

$$(63^{bis}) \qquad t^h \cdot \frac{d^h u}{d t^h} = \left(\sum_k \xi_k \frac{\partial}{\partial \xi_k}\right)^h \cdot u,$$

où la dernière expression est manifestement indépendante de t. Notons que le premier membre fait voir que ladite expression ne dépend pas du système de coordonnées non plus.

Cela posé, l'expression $D^p(\xi)$ qui figure dans les formules (56), (57) et (57′) pourra aussi s'écrire

$$(64) \qquad D^p(\xi) = \sum_{h=0}^{p} B_h^p t^h \cdot \frac{d^h u}{d t^h} = t^{-(m-p-3)} \frac{d^p}{d t^p}(u \cdot t^{m-3}).$$

En reprenant l'ordre d'idées général, la fonction u sera considérée comme *suffisamment dérivable sur la portion* $\bar{S}$ *de* S *si les expressions* (63) *existent* jusqu'à une valeur assez élevée de h, qu'elles *sont continues* sur toute partie bornée et fermée de $\bar{S}$ et qu'elles *s'annulent au sommet* pour $h \geqq 1$. Évidemment, dans ces conditions-là, $D^p(\xi)$ est aussi continu sur toute partie bornée et fermée de $\bar{S}$. On voit par (63bis) que les dites conditions sont largement remplies si u admet un nombre suffisant de dérivées partielles continues au voisinage de $\bar{S}$.

Les fonctions $U^2(P)$ figurant dans les formules (59—61bis) seront donc dès maintenant à former avec des valeurs prescrites u suffisamment dérivables dans le sens qu'on vient d'adopter. Il est alors presque évident que ces fonctions satisfont à l'équation $\Delta U^2(P) = 0$. En effet, en imitant le raisonnement des n^{os} **39—42**, on voit facilement que $\Delta U^{\alpha+2} = U^\alpha$ et en particulier que $\Delta U^2 = U^0$. Or, U^α contient le facteur $\Gamma\left(\frac{\alpha}{2}\right)$ au dénominateur (*voir* la formule (55) qui est relative à $U^{\alpha+2}$ et où il figure au dénominateur $K_m(\alpha+2) = \pi^{\frac{m-2}{2}} \cdot 2^{\alpha+1} \cdot \Gamma\left(\frac{\alpha+2}{2}\right)$), ce qui entraîne que $U^0 = 0$.

La vérification du fait que $U^2(P)$ admet les valeurs limites prescrites u est beaucoup plus délicate et exige un examen attentif des formules explicites en question. Il figure dans toutes ces formules l'intersection s^P des deux cônes, variété que de temps en temps nous appellerons aussi *bord* de S^P, et que nous allons maintenant paramétriser.[1] Désignons par $\mathbf{b}$ les points de s^P et, en même temps, les rayons vecteurs qui de l'origine mènent à ces points. Les vecteurs $\mathbf{b}$ et $\mathbf{x}-\mathbf{b}$ étant isotropes, le dernier puisque le vecteur menant de $\mathbf{b}$ à $\mathbf{x}$ est un segment de génératrice du cône rétrograde, on a

$$\mathbf{b}^2 = 0, \quad (\mathbf{x}-\mathbf{b})^2 = 0. \tag{65}$$

Ces relations caractérisent les points $\mathbf{b}$ de s^P. En les écrivant sous la forme $(\mathbf{x}-\mathbf{b})^2 = 0$, $(\mathbf{x}-(\mathbf{x}-\mathbf{b}))^2$, on voit que deux points $\mathbf{b}$ et $\mathbf{x}-\mathbf{b}$ appartiennent à s^P simultanément. On tire de (65)

$$(\mathbf{x}, \mathbf{b}) = \frac{\mathbf{x}^2}{2} = \frac{R_0}{2} \tag{66}$$

(pour R_0 *voir* la formule (48bis)). La dernière relation étant linéaire en $\mathbf{b}$, on voit que s^P est situé dans un plan (d'espace) Π qui est orthogonal au vecteur $\mathbf{x}$ et qui passe par le point $\frac{\mathbf{x}}{2}$. En posant, d'après (48) et (48bis),

$$R_0^{\frac{1}{2}} = r_0, \quad \mathbf{b} - \frac{\mathbf{x}}{2} = \frac{r_0}{2}\cdot\mathbf{c}, \tag{67}$$

on aura

$$\frac{r_0^2}{4}\cdot\mathbf{c}^2 = \left(\mathbf{b}-\frac{\mathbf{x}}{2}\right)^2 = \mathbf{b}^2 - (\mathbf{b}, \mathbf{x}) + \frac{\mathbf{x}^2}{4} = 0 - \frac{R_0}{2} + \frac{R_0}{4} = -\frac{r_0^2}{4}.$$

De (66), (67) et de la dernière formule, on tire pour le vecteur $\mathbf{c}$ les relations

$$(\mathbf{x}, \mathbf{c}) = 0, \quad \mathbf{c}^2 = -1. \tag{68}$$

En résumé, on voit que le bord s^P (variété à $m-2$ dimensions) est une sphère (de caractère euclidien) de centre $\mathbf{x}/2$ et de rayon $r_0/2$, située dans le plan Π à $m-1$ dimensions. Les vecteurs $\mathbf{c}$ sont, de leur côté, des vecteurs unitaires; en plaçant l'une de leurs extrémités à l'origine, l'extrémité libre décrit la sphère unitaire du plan Π' mené par l'origine parallèlement à Π, engendré d'ailleurs par les droites menées par l'origine et orthogonales à $\mathbf{x}$. En paramétrisant cette dernière sphère, on en aura fait autant pour la sphère s^P. Pour

[1] Il ne s'agira que d'une paramétrisation légère, adaptée à nos besoins.

y arriver, nous nous servirons de la méthode par laquelle on construit un système de coordonnées polaires dans l'espace ordinaire.

Soit $\mathbf{b}^*$ un point particulier de s^P, $\mathbf{c}^*$ le vecteur unitaire correspondant. Nous considérons en outre l'ensemble des vecteurs unitaires (d'espace) $\mathbf{e}^*$, orthogonaux à $\mathbf{x}$ et à $\mathbf{c}^*$.[1] Si les vecteurs $\mathbf{e}^*$ sont issus de l'origine, leur extrémité décrit la sphère unitaire du plan π' à $m-2$ dimensions, qui fait partie du plan Π' et qui est engendré par les droites menées par l'origine et orthogonales à $\mathbf{x}$ et à $\mathbf{c}^*$. L'élément de cette dernière sphère sera, comme d'habitude, désigné par $d\omega_{m-2}$ et sa surface totale par ω_{m-2}.

Tout vecteur situé dans le plan à $m-1$ dimensions Π' pourra, d'une manière unique, être décomposé en deux vecteurs, l'un appartenant à π', c'est-à-dire de la forme $\lambda \cdot \mathbf{e}^*$ $(\lambda \geqq 0)$ et l'autre, extérieur à π', et de la forme $\mu \cdot \mathbf{c}^*$ $(|\mu| \geqq 0)$. En particulier, tout vecteur unitaire $\mathbf{c}$ de Π' peut, d'une manière unique, être décomposé d'après le schéma

$$\mathbf{c} = \cos\psi \cdot \mathbf{c}^* + \sin\psi \cdot \mathbf{e}^*, \qquad 0 \leqq \psi \leqq \pi. \tag{69}$$

Il s'ensuit, d'après ce qu'on a vu plus haut, la paramétrisation suivante de s^P

$$\mathbf{b} = \frac{\mathbf{x}}{2} + \frac{r_0}{2}(\cos\psi \cdot \mathbf{c}^* + \sin\psi \cdot \mathbf{e}^*), \qquad 0 \leqq \psi \leqq \pi. \tag{70}$$

En particulier, pour $\psi = 0$ et $\psi = \pi$, respectivement,

$$\mathbf{b}^* = \frac{\mathbf{x}}{2} + \frac{r_0}{2} \cdot \mathbf{c}^*, \quad \mathbf{x} - \mathbf{b}^* = \frac{\mathbf{x}}{2} - \frac{r_0}{2} \cdot \mathbf{c}^*. \tag{71}$$

Notons encore les relations, conséquences immédiates de (70) et de (71),

$$(\mathbf{b}, \mathbf{x} - \mathbf{b}^*) = \frac{\mathbf{x}^2}{4} - \frac{r_0^2}{4}\cos\psi \cdot \mathbf{c}^{*2} = \frac{r_0^2}{4}(1 + \cos\psi) = \frac{R_0}{4}(1 + \cos\psi), \tag{72}$$

et en particulier

$$(\mathbf{b}^*, \mathbf{x} - \mathbf{b}^*) = \frac{R_0}{2}. \tag{73}$$

De la formule de paramétrisation (70), on obtient facilement l'élément ds de s^P. Observons d'abord que les multiplicités où seule l'une des variables ψ et $\mathbf{e}^*$ varie sont orthogonales l'une à l'autre. On le voit par les relations $(\mathbf{c}^*, \mathbf{e}^*) = 0$ et $\mathbf{e}^{*2} = -1$, qui donnent encore $(\mathbf{c}^*, d\mathbf{e}^*) = 0$ et $(\mathbf{e}^*, d\mathbf{e}^*) = 0$. On peut donc appliquer la formule de décomposition (55) du n° **21**^{bis} et poser $ds = ds' \cdot ds''$.

[1] On voit par (67) qu'ils sont en même temps orthogonaux à $\mathbf{b}^*$ et à $\mathbf{x} - \mathbf{b}^*$.

En variant ψ, on a $d\mathbf{b}' = \frac{r_0}{2}(-\sin\psi\cdot\mathbf{c}^* + \cos\psi\cdot\mathbf{e}^*)\,d\psi$; par conséquent $ds' = |d\mathbf{b}'^2|^{\frac{1}{2}} = \frac{r_0}{2}\,d\psi$. D'autre part, quand $\mathbf{e}^*$ varie, on aura $d\mathbf{b}'' = \frac{r_0}{2}\sin\psi\cdot d\mathbf{e}^*$, et, puisque $\mathbf{e}^*$ décrit la sphère unitaire ω_{m-2}, il vient $ds'' = \left(\frac{r_0}{2}\sin\psi\right)^{m-3} d\omega_{m-2}$. On a donc la valeur finale

$$ds = \left(\frac{r_0}{2}\right)^{m-2} \sin^{m-3}\psi\cdot d\psi\, d\omega_{m-2}. \tag{74}$$

Dans le n° **63**, tout point $\mathfrak{x}$ de la portion de surface S^P fut rapporté 1°. au point $\mathbf{b}$, point d'intersection de la génératrice, menée par $\mathfrak{x}$, avec le bord s^P, et 2°. à la valeur que R admettait au point $\mathfrak{x}$. Pour arriver à une vérification de nos prétendues formules de solution (59—61*bis*), il convient de remplacer R par une autre variable τ, adaptée à ce but, ce qui se fera de la manière suivante.

Soit $\mathfrak{x}^* = \tau\cdot\mathbf{b}^*$, $0 \leqq \tau \leqq 1$, un point du segment de génératrice $\mathbf{b}^*$. Nous menons par $\mathfrak{x}^*$ un plan A_τ, parallèle au plan tangent à S le long de la génératrice qui porte le vecteur $\mathbf{x} - \mathbf{b}^*$. Le plan A_τ coupe S suivant un paraboloïde σ_τ et S^P suivant une calotte de paraboloïde σ_τ^P.[1] L'équation du plan A_τ qui admet le dernier vecteur comme normale et qui passe par $\mathfrak{x}^*$, est manifestement $(\mathfrak{x}, \mathbf{x} - \mathbf{b}^*) = (\mathfrak{x}^*, \mathbf{x} - \mathbf{b}^*)$. Le point d'intersection $\mathfrak{x} = \varkappa\cdot\mathbf{b}$ de la génératrice, portant un certain vecteur $\mathbf{b}$, avec le plan A_τ ou, ce qui revient au même, avec le paraboloïde σ_τ, sera donc fourni par l'équation $\varkappa\cdot(\mathbf{b}, \mathbf{x} - \mathbf{b}^*) = \tau\cdot(\mathbf{b}^*, \mathbf{x} - \mathbf{b}^*)$, qui, par (72) et (73), se réduit à $\varkappa\cdot\cos^2\frac{\psi}{2} = \tau$. En posant, pour simplifier l'écriture ici et plus loin, $\psi = 2\varphi$, on aura donc

$$\varkappa = \frac{\tau}{\cos^2\varphi}, \quad \varphi = \frac{\psi}{2}, \quad 0 \leqq \varphi \leqq \frac{\pi}{2}. \tag{75}$$

D'autre part, puisque $1 - R/R_0$ varie d'une manière linéaire sur toute génératrice, qu'il s'annule à l'origine et qu'il devient 1 en $\mathbf{b}$, on a au point d'intersection $\mathfrak{x} = \varkappa\,\mathbf{b}$

$$1 - \frac{R}{R_0} = \frac{\tau}{\cos^2\varphi}, \quad R = R_0\cdot\left(1 - \frac{\tau}{\cos^2\varphi}\right), \quad |dR| = \frac{R_0}{\cos^2\varphi}\,d\tau. \tag{76}$$

En résumé, les formules (76) expriment $1 - R/R_0$, R et $|dR|$ sur la géné-

(1) Dans le cas de trois dimensions (auquel se rapporte la fig. 11) le paraboloïde se réduit à une parabole.

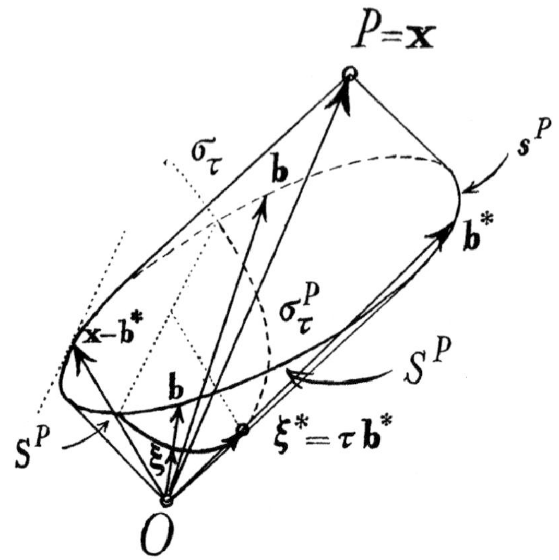

Fig. 11. La figure se rapporte au cas de $m = 3$ où la calotte de paraboloïde devient un segment de parabole, découpé de S^P par le plan Λ_τ, mené par le point $\xi^* = \tau\,\mathbf{b}^*$ parallèlement au plan tangent à S^P suivant le vecteur $\mathbf{x} - \mathbf{b}^*$. Quelques lignes auxiliaires pointillées font nettement ressortir la parallélité des deux plans.

ratrice portant le vecteur $\mathbf{b}$ en fonction de τ et de $d\tau$, respectivement. Les points $\boldsymbol{\xi}$ du paraboloïde σ_τ sont, de leur côté, selon (70) et (75) donnés par

$$(77)\qquad \boldsymbol{\xi} = \frac{\tau}{\cos^2\varphi}\cdot\mathbf{b} = \frac{\tau}{\cos^2\varphi}\left(\frac{\mathbf{x}}{2} + \frac{r_0}{2}(\cos 2\varphi\cdot\mathbf{c}^* + \sin 2\varphi\cdot\mathbf{e}^*)\right).$$

Puisque, comme nous venons de le rappeler, $1 - R/R_0$ varie de 0 à 1 sur tout rayon vecteur (segment de génératrice $\mathbf{b}$), ceux des points de σ_τ — les seuls qui nous intéressent — qui appartiennent en même temps à la calotte σ_τ^P, sont caractérisés par les inégalités

$$(78)\qquad 0 \leqq \frac{\tau}{\cos^2\varphi} \leqq 1,\ \text{ou}\ 0 \leqq \tau \leqq \cos^2\varphi,\ \text{ou}\ 0 \leqq \varphi \leqq \arccos\tau^{\frac{1}{2}}.$$

Ces calculs achevés, voici un premier pas vers le passage à la limite qu'il nous faudra exécuter. Soit $P = \mathbf{x}$ un point très rapproché de la surface S. Cette propriété de P sera ici exprimée par la condition que la valeur $\mathbf{x}^2 = R_0$ soit très petite, ce qui veut dire que le vecteur $\mathbf{x}$ sera sensiblement isotrope. Posons $\varrho = (x_2^2 + \cdots + x_m^2)^{\frac{1}{2}}$ et désignons par $\mathbf{v}$ le vecteur aux composantes $v_1 = 1$, $v_i = x_i/\varrho$, $i = 2, \ldots, m$. On a $\mathbf{v}^2 = 1 - \varrho^2/\varrho^2 = 0$, c'est-à-dire que $\mathbf{v}$ est isotrope.

De plus, $(\mathbf{x}, \mathbf{v}) = x_1 - \varrho^2/\varrho = x_1 - \varrho$ et $R_0 = \mathbf{x}^2 = x_1^2 - \varrho^2 = (x_1 - \varrho)(x_1 + \varrho)$. En posant $\mathbf{b}^* = \frac{1}{2}(x_1 + \varrho)\mathbf{v}$, on aura donc

$$\mathbf{b}^{*2} = 0, \quad (\mathbf{x}, \mathbf{b}^*) = \frac{R_0}{2}, \quad (\mathbf{x} - \mathbf{b}^*)^2 = 0,$$

ce qui veut dire, d'après ce que nous avons vu plus haut, que les points $\mathbf{b}^*$ et $\mathbf{x} - \mathbf{b}^*$ *appartiennent à* s^P.

On voit facilement que le point $\mathbf{x}$ et le point $\mathbf{b}^*$ sont très rapprochés l'un de l'autre, dans le sens que les valeurs numériques de toutes les composantes de $\mathbf{x} - \mathbf{b}^*$ sont très petites, étant toutes $\leqq r_0/2$. En effet, on a d'abord, à cause de $(x_1 - \varrho)(x_1 + \varrho) = R_0 = r_0^2$ et de $x_1 - \varrho \leqq x_1 + \varrho$, l'inégalité $x_1 - \varrho \leqq r_0$. Ensuite

$$|x_1 - b_1^*| = \left| x_1 - \frac{x_1 + \varrho}{2} \right| = \frac{x_1 - \varrho}{2} \leqq \frac{r_0}{2}.$$

Enfin, le vecteur $\mathbf{x} - \mathbf{b}^*$ étant isotrope, on a

$$(79) \qquad |x_k - b_k^*| \leqq |x_1 - b_1^*| \leqq \frac{r_0}{2}, \qquad k = 1, 2, \ldots, m.$$

Cela veut dire que toutes ces valeurs tendent vers 0 avec R_0.

On arrive à une limite supérieure pour l'oscillation des ξ_k sur la calotte σ_τ^P par les considérations suivantes. En posant, dans la formule (77), $\cos 2\varphi = 2\cos^2\varphi - 1$, et en tenant compte de (71), on obtient

$$\boldsymbol{\xi} = \frac{\tau}{\cos^2\varphi}\left(\mathbf{x} - \mathbf{b}^* + \frac{r_0}{2}\sin 2\varphi \cdot \mathbf{e}^*\right) + \tau r_0 \mathbf{c}^*$$

et en particulier $\boldsymbol{\xi}^* = \tau(\mathbf{x} - \mathbf{b}^*) + \tau r_0 \mathbf{c}^*$ et dès lors

$$\boldsymbol{\xi} - \boldsymbol{\xi}^* = \frac{\tau \sin^2\varphi}{\cos^2\varphi}(\mathbf{x} - \mathbf{b}^*) + \frac{\tau}{\cos^2\varphi} \cdot \frac{r_0}{2}\sin 2\varphi \cdot \mathbf{e}^*.$$

Tout vecteur unitaire $\mathbf{e}^*$ orthogonal à $\mathbf{x}$ et à $\mathbf{c}^*$ est aussi orthogonal à

$$\frac{\mathbf{x}}{2} + \frac{r_0}{2} \cdot \mathbf{c}^* = \mathbf{b}^* = \frac{x_1 + \varrho}{2} \cdot \mathbf{v}$$

et, par conséquent, aussi à $\mathbf{v}$ et à $\mathbf{x} - \varrho\mathbf{v}$. Ce dernier vecteur a les composantes $x_1 - \varrho, 0, \ldots, 0$, ce qui implique que $e_1^* = 0$; cela étant, $e_2^{*2} + \cdots + e_m^{*2} = 1$, ce qui donne $|e_i^*| \leqq 1$. Alors, vu que $\tau/\cos^2\varphi \leqq 1$ et en vertu de (79),

$$|\xi_k - \xi_k^*| \leqq |x_k - b_k| + \frac{r_0}{2}|e_k^*| \leqq \frac{r_0}{2} + \frac{r_0}{2} = r_0.$$

Cela veut dire que les ξ_k varient très peu sur σ_τ^P, et cela pour toute valeur de τ.

Il nous paraît convenable de *résumer les faits géométriques* qu'on vient d'établir.

Si l'on choisit un point $\mathbf{b}^*$ arbitraire sur s^P, on peut décomposer S^P en menant par les différents points $\boldsymbol{\xi}^* = \tau \mathbf{b}^*$, $0 \leqq \tau \leqq 1$, du rayon-vecteur $\mathbf{b}^*$ des plans de direction fixe qui coupent S^P suivant des calottes de paraboloïdes σ_τ^P. Lorsque $P = \mathbf{x}$ est très rapproché de la surface S, ce qui veut dire que $\mathbf{x}^2 = R_0$ est très petit, on peut choisir $\mathbf{b}^*$ de manière que les coordonnées de $\boldsymbol{\xi}$ variant sur la calotte σ_τ^P soit sensiblement égales à celles du point $\boldsymbol{\xi}^* = \tau \mathbf{b}^*$ correspondant.

Tel étant le cas pour les coordonnées, il en sera de même pour une fonction continue arbitraire $F(\boldsymbol{\xi})$ définie sur $\bar{S}$, c'est-à-dire que si $\boldsymbol{\xi}$ se trouve sur σ_τ^P, on aura sensiblement $F(\boldsymbol{\xi}) = F(\boldsymbol{\xi}^*)$. Or, en tenant compte de ce qu'on a dit au début de ce numéro sur les fonctions $D^P(\boldsymbol{\xi})$, on peut immédiatement appliquer ce résultat à ces dernières fonctions.

Si les valeurs limites prescrites u sont suffisamment dérivables, on aura, sensiblement, sur les susdites calottes, $D^p(\boldsymbol{\xi}) = D^p(\boldsymbol{\xi}^*)$.

Ces préliminaires posés, il est facile de donner des valeurs approchées convenables pour les différentes expressions $U^2(P)$. Nous commençons par le cas impair $m = 2l + 1 > 3$. La formule (59), relative à ce cas-là, s'écrit, si l'on y tient compte de la relation (56) (en posant dans cette dernière $p = l$)

$$(80) \qquad U^2(P) = (\pi R_0)^{-l} \int_{s^P} ds \int_0^{R_0} R^{-\frac{1}{2}} \left(1 - \frac{R}{R_0}\right)^{l-2} D^l(\boldsymbol{\xi})\, dR.$$

On a, d'après (74), (76), et en posant toujours $\psi = 2\varphi$,

$$\begin{aligned} R^{-\frac{1}{2}} &\left(1 - \frac{R}{R_0}\right)^{l-2} ds \cdot dR \\ &= R_0^{-\frac{1}{2}} \left(1 - \frac{\tau}{\cos^2\varphi}\right)^{-\frac{1}{2}} \left(\frac{\tau}{\cos^2\varphi}\right)^{l-2} r_0^{2l-1} \left(\frac{\sin 2\varphi}{2}\right)^{2l-2} d\varphi \cdot d\omega_{m-2} \cdot R_0 \frac{d\tau}{\cos^2\varphi} \\ &= R_0^l \left[(\cos^2\varphi - \tau)^{-\frac{1}{2}} \sin^{2l-2}\varphi \cos\varphi\right] d\varphi \cdot d\omega_{m-2} \cdot \tau^{l-2}\, d\tau \end{aligned}$$

et par suite

$$U^2(P) = \pi^{-l} \int_0^1 \tau^{l-2}\, d\tau \int_{\omega_{m-2}} d\omega_{m-2} \int_0^\theta [\,]\, D^l(\boldsymbol{\xi})\, d\varphi,$$

où la limite θ a la valeur $\arccos \tau^{\frac{1}{2}}$, qui s'ensuit de l'inégalité (78).

L'intégration par rapport à la différentielle $d\varphi \cdot d\omega_{m-2}$ s'étend — τ y étant constant — à la calotte σ_τ^P. La valeur de $D^l(\xi)$ qui y entre est donc sensiblement égale à $D^l(\xi^*) = D^l(\tau\,\mathbf{b}^*)$, ce qui fournit pour $U^2(P)$ la valeur *approchée*

(81) $$U^2(P) = \pi^{-l}\int_0^1 D^l(\xi^*)\,\tau^{l-2}\,d\tau \cdot \int_{\omega_{m-2}} d\,\omega_{m-2} \cdot \int_0^\theta [\;]\,d\varphi.$$

On a (*cf.* (8) du n° **6**)

$$\int_{\omega_{m-2}} d\,\omega_{m-2} = \omega_{m-2} = \frac{2\,\pi^{\frac{m-2}{2}}}{\Gamma\left(\frac{m-2}{2}\right)} = \frac{2\,\pi^{l-\frac12}}{\Gamma(l-\frac12)} \text{ et } \int_0^\theta [\;]\,d\varphi = \tfrac12(1-\tau)^{l-1}\frac{\Gamma(l-\frac12)\,\Gamma(\frac12)}{\Gamma(l)}$$

et alors, après des réductions évidentes,

$$\pi^{-l}\int_{\omega_{m-2}} d\,\omega_{m-2}\int_0^\theta [\;]\,d\varphi = \frac{(1-\tau)^{l-1}}{(l-1)!}.$$

D'autre part, on trouve par la formule (64)

$$D^l(\xi^*) = \tau^{-(l-2)}\frac{d^l}{d\tau^l}\left(u(\tau\,\mathbf{b}^*)\cdot\tau^{m-3}\right).$$

Tout cela étant admis, la valeur approchée (81) devient, après $l-1$ intégrations par parties,

$$U^2(P) = \frac{1}{(l-1)!}\int_0^1 (1-\tau)^{l-1}\frac{d^l}{d\tau^l}\left(u(\tau\,\mathbf{b}^*)\cdot\tau^{m-3}\right)d\tau = \left[u(\tau\,\mathbf{b}^*)\cdot\tau^{m-3}\right]_{\tau=0}^{\tau=1} = u(\mathbf{b}^*),$$

puisqu'on a, par hypothèse, $m > 3$.

Dans le cas de $m = 3$, le terme tout intégré devient $u(\mathbf{b}^*) - u(O)$. D'autre part, en se reportant à la déduction de la formule (59bis), on voit que, dans ce cas particulier, pour arriver à $U^2(P)$, il faut ajouter $u(O)$ au second membre de (80). On a donc, aussi dans ce dernier cas, la valeur approchée $u(\mathbf{b}^*)$ pour $U^2(P)$.

En résumé, nous avons montré, pour *m impair* que, le point P étant très rapproché de la surface S, il existe sur cette surface un point $\mathbf{b}^*$ très rapproché de P et tel que $U^2(P)$ est sensiblement égal à la valeur (prescrite) $u(\mathbf{b}^*)$. Lorsque P tend vers un point déterminé Q de la surface, il en sera de même de $\mathbf{b}^*$, et

$U^2(P)$ tendra par conséquent vers la valeur prescrite $u(Q)$. Dans les conditions de continuité posées, la convergence sera uniforme sur toute portion de surface bornée et fermée. Les formules (59) et (59bis) sont donc bien des *formules de solution.* (1)

Le cas de *m pair* est maintenant assez facile. L'intégrale, qui, en vertu de la formule (61)(2), fournit $U^2(P)$ s'étend au bord s^P seulement. On a

$$U^2(P) = \pi^{1-l} \int_{s^P} A^{l-1}\, ds. \tag{82}$$

Vu que $1 - R/R_0 = 1$ sur le bord, l'expression A^{l-1}, donnée encore par la formule (56), se réduit à

$$A^{l-1} = R_0^{1-l} D^{l-1}(\xi).$$

La valeur de ds est comme plus haut (formule (74), $m = 2l$ et $\psi = 2\varphi$)

$$ds = r_0^{2l-2} \left(\frac{\sin 2\varphi}{2}\right)^{2l-3} d\varphi\, d\omega_{m-2}.$$

La fonction $D^{l-1}(\xi)$ étant sensiblement égale à $D^{l-1}(\xi^*)$ sur σ_τ^P, elle le sera en particulier dans tous les points qui appartiennent à la fois à σ_τ^P et à s^P. Or, pour ces points-là, la valeur $\varkappa = \tau/\cos^2\varphi$ (formule (75)) devient égale à 1, ce qui donne $\cos^2\varphi = \tau$. (3) En variant τ, on en tire $|\sin 2\varphi \cdot d\varphi| = |d\tau|$ et ensuite, par un calcul simple,

$$ds = R_0^{\overline{\frac{l-1}{2}}} (\tau(1-\tau))^{l-2}\, d\tau\, d\omega_{m-2},$$

où toutes les différentielles admettent le signe positif. En portant cela dans la formule (82), on obtient pour $U^2(P)$ la valeur *approchée* (4)

---

(1) Pour dire un mot de l'aspect géométrique du passage à la limite qui vient d'avoir lieu, la portion de surface S^P tend, en se rétrécissant, vers le segment de génératrice OQ, tandis que chaque calotte σ_τ^P tend vers un point déterminé de ce segment. Les propriétés d'une frontière caractéristique sont donc bien intermédiaires entre celles d'une frontière d'espace et d'une frontière de temps; *cf.* en particulier la fig. 9.

(2) Cette formule renferme la formule (61bis) relative au cas particulier $m = 4$.

(3) Les points en question décrivent une sphère de centre $\frac{\mathbf{x}}{2} + \frac{r_0}{2} \cos 2\varphi\, \mathbf{c}^*$, de rayon $\frac{r_0}{2} \sin 2\varphi$ et située dans un plan orthogonal à $\mathbf{x}$ et à $\mathbf{c}^*$.

(4) L'erreur commise dans $U^2(P)$ tend vers 0 en même temps que celle relative à $D^l(\xi)$. En effet, si l'on remplace dans l'expression (80) la fonction $D^l(\xi)$ par 1, on aura une constante nu-

$$U^2(P) = \frac{\pi^{1-l}}{2}\,\omega_{m-2}\int_0^1 D^{l-1}(\xi^*)\cdot(\tau(1-\tau))^{l-2}\,d\tau.$$

Le facteur numérique devant l'intégrale est

$$\frac{\pi^{1-l}}{2}\cdot\frac{2\cdot\pi^{l-1}}{(l-2)!} = \frac{1}{(l-2)!}.$$

La formule (64) donne de son côté

$$D^{l-1}(\xi^*) = \tau^{-(l-2)}\frac{d^{l-1}}{d\tau^{l-1}}(u(\tau\,\mathbf{b}^*)\,\tau^{m-3}).$$

Eu égard aux deux dernières évaluations, la formule ci-dessus peut s'écrire

$$U^2(P) = \frac{1}{(l-2)!}\int_0^1 (1-\tau)^{l-2}\frac{d^{l-1}}{d\tau^{l-1}}(u(\tau\,\mathbf{b}^*)\,\tau^{m-3})\,d\tau = [u(\tau\,\mathbf{b}^*)\,\tau^{m-3}]_{\tau=0}^{\tau=1} = u(\mathbf{b}^*),$$

puisque $m \geqq 4$. Cette formule approchée étant démontrée, la vérification s'achève comme dans le cas *impair*. [1]

Nous donnons, pour terminer, quelques indications sur le degré de dérivabilité désirable de u. Un examen rapide montre que nos calculs sont légitimes, si l'on a soin de poser les conditions qui se trouvent au début de ce numéro pour toute valeur $h \leqq l+2$; les deux dernières valeurs $h = l+1$ et $h = l+2$ n'interviennent que dans les calculs concernant $\Delta\,U^2(P) = U^0(P) = 0$. Dans le cas pair, $m = 2\,l$, il ne figure dans l'expression finale de $U^2(P)$ que des dérivées d'ordre $\leqq l-1$ (formule (61)). Appliquant les méthodes du Chap. I, on pourrait arriver à cette expression en ne se servant que de ces dernières dérivées, c'est-à-dire en se passant de l'expression A^l, où entre la dérivée (63) d'ordre l. De même, dans la vérification de $\Delta\,U^2(P) = 0$, on pourrait se contenter des dérivées d'ordre $\leqq l+1$.

---

mérique, ne dépendant que de l, c'est-à-dire que de la dimension m. Cela est clair *a priori* par les calculs qui ont conduit à $U^2(P)$ dans le numéro précédent. On le voit presque sans calcul en observant que ds ne dépend de R_0 que par le facteur $r_0^{m-2} = R_0^{l-1}$, ce qui fait que l'expression qu'on vient de former est indépendante de R_0. (La constante numérique est d'ailleurs égale à $(l-2)!/(2\,l-2)!$.)

[1] Il est clair que les variétés S^P et s^P tendent toujours vers le même segment de génératrice, quand P tend vers un point de S.

CHAPITRE V.

Interprétation géométrique de la solution de l'équation des ondes sphériques.

65. L'opérateur I_*^{m-2}. — L'objet principal de ce chapitre est une étude approfondie de la solution de l'équation des ondes, relative au cas le plus important, celui des ondes sphériques. Il s'agit donc de l'équation posée pour l'espace-temps ordinaire — *trois* dimensions d'espace et, comme toujours, *une* dimension de temps. Par conséquent, il sera question de l'opérateur I_*^2 pour $m=4$, qui manifestement est un cas particulier de I_*^{m-2} pour m arbitraire. Vu que les considérations géométriques dont on aura besoin ont un intérêt en elles-mêmes et qu'elles sont indépendantes de la dimension, nous traiterons d'abord le cas général pour $m \geqq 3$. Nous déduirons ensuite de nos considérations générales la solution de caractère géométrique de l'équation des ondes sphériques, sans second membre, donnée par la formule (10) de l'Introduction.

On va appliquer les calculs du n° **57**, d'où sont tirées, sauf avis contraire, toutes les formules citées dans le présent numéro. Observons d'abord que $1/H_m(m-2)=0$; on peut donc remplacer la portion de surface S^P par sa partie annulaire extérieure $\bar{S}$, tout comme on l'a fait au susdit numéro en y calculant I_*^2 pour m pair. D'autre part, les nombres $\beta=\dfrac{\alpha+2-m}{2}$ figurant dans les formules (A) et (B) et $\beta=\dfrac{\alpha-m}{2}$ figurant dans les formules (C) et (C^{bis}) deviennent respectivement $\beta=0$ et $\beta=-1$ pour $\alpha=m-2$. Il s'ensuit donc des formules citées et de la formule (29) au moyen de la formule (13bis) du n° **5**

$$(1)\qquad I_*^{m-2}\overline{f,g,h}(P)=\frac{1}{\pi^{\frac{m-2}{2}}\,2^{m-3}\,\Gamma\left(\dfrac{m-2}{2}\right)}\bigl(2A(0)+B(0)+C'(0)\bigr).$$

Ici $A(0)$ est donné par la formule (28bis), tandis que $B(0)$ et $C'(0)$ sont les valeurs de $B(R)$ et de $dC(R)/dR$ pour $R=0$. Ces dernières expressions sont définies par les formules (B) et (C):

$$(2)\qquad B(R)=\int\limits_{s_R} g\frac{dl}{dR}ds_R \quad\text{et}\quad C(R)=\int\limits_{s_R} h\frac{dR}{dn}\frac{dl}{dR}ds_R.$$

En ce qui concerne la valeur $A(0)$, nous n'avons rien à ajouter à la formule (28bis). Pour calculer $B(0)$ et $C'(0)$, rappelons que la variété s_R devient pour $R=0$ le bord s^P dont l'élément de volume est désigné par ds. En tenant encore compte de la formule (12) du n° **50**, nous pouvons écrire

$$B(0) = -\int_{s^P} g \frac{dn}{dR} ds, \tag{3}$$

Le calcul explicite de $C'(0)$ est plus difficile. On peut écrire

$$C(R) = \int_{s^P} h K ds \quad \text{où} \quad K = \frac{dR}{dn} \cdot \frac{dl}{dR} \cdot \frac{ds_R}{ds}. \tag{4}$$

En rappelant la paramétrisation de $\bar{S}$ à l'aide des trajectoires issues des points B de s^P et de la quantité R que nous avons donnée au n° **50**, on voit que

$$C'(R) = \int_{s^P} \frac{d}{dR}(hK) ds, \tag{4^{bis}}$$

la différentiation par rapport à R étant faite suivant les trajectoires l. On a

$$\frac{d}{dR}(hK) = \frac{dh}{dR} K + h \frac{dK}{dR}. \tag{5}$$

Pour $R = 0$, on a évidemment $ds_R/ds = 1$, ce qui, rapproché de la formule (12) du n° **50**, montre que

$$K = -1 \text{ pour } R = 0. \tag{6}$$

Dès lors, pour $R = 0$, c'est-à-dire aux points du bord s^P, la formule (5) ci-dessus se réduit à

$$\frac{d}{dR}(hK)_{(R=0)} = \left(-\frac{dh}{dR} + h\frac{dK}{dR}\right)_{(R=0)} \tag{7}$$

Posons maintenant, pour abréger,

$$I_*^\alpha \overline{0, g, h}(P) = I_*^\alpha \overline{g, h}(P),$$

alors, par les formules (1), (3), (4bis) et (7) ci-dessus,

$$I_*^{m-2} \overline{g, h}(P) = \frac{1}{\pi^{\frac{m-2}{2}} 2^{m-3} \Gamma\left(\frac{m-2}{2}\right)} \int_{s^P} \left\{h\frac{dK}{dR} - g\frac{dn}{dR} - \frac{dh}{dR}\right\} ds, \tag{8}$$

où, soulignons-le, les expressions sont formées aux points du bord s^P et les différentiations par rapport à R dans dK/dR et dh/dR sont faites suivant les trajectoires l.

En posant maintenant $g = \frac{du}{dn}$ et $h = u$, il vient

$$(8^{bis}) \quad I_*^{m-2}\overline{\frac{du}{dn}}, u(P) = \frac{1}{\pi^{\frac{m-2}{2}}\, 2^{m-3}\, \Gamma\left(\frac{m-2}{2}\right)} \int_{s_P} \left\{u\frac{dK}{dR} - \frac{du}{dn}\frac{dn}{dR} - \frac{du}{dR}\right\} ds.$$

66. Différentiation par rapport à R dans différentes directions. Les droites ν. — On va encore simplifier la dernière formule en introduisant un symbole de différentiation approprié. Rappelons que nous avons déjà à plusieurs reprises différentié dans certaines directions par rapport à $R = R_{PQ}$ ou $r = r_{PQ}$ (*cf.* en particulier l'Introduction, formule (10) et les nos **63** et **65**) et même utilisé de temps en temps le symbole que nous allons maintenant introduire d'une manière systématique.

Désignons, dans un ordre d'idées général, par γ une direction déterminée. Soient d'autre part $U(Q)$ et $V(Q)$ des fonctions, continues avec leurs dérivées du premier ordre. On entend par $d_\gamma U$ et $d_\gamma V$ les accroissements infinitésimaux de ces fonctions correspondant au déplacement infinitésimal dQ dans la direction γ. Ceci posé, le rapport $d_\gamma U / d_\gamma V$ sera appelé *la dérivée de U par rapport à V dans la direction* γ(1). Cette dérivée est manifestement indépendante du sens de parcours positif de γ.

Dans cette notation, les deux derniers termes sous le signe $\int$ dans la formule (8bis) peuvent s'écrire

$$(9) \qquad \frac{du}{dn}\frac{dn}{dR} + \frac{du}{dR} = \frac{d_n u}{d_n R} + \frac{d_l u}{d_l R}.$$

Les coordonnées du point Q étant désignées, comme d'ordinaire, par ξ_k, on a manifestement

$$(10) \qquad \frac{d_\gamma u}{d_\gamma R} = \sum_{k=1}^{m} \frac{\partial u}{\partial \xi_k}\frac{d_\gamma \xi_k}{d_\gamma R}.$$

Alors, en posant

$$(11) \qquad \frac{d_\gamma \xi_k}{d_\gamma R} = \gamma'_k,$$

on peut aussi écrire

(1) On prendra évidemment soin de ne pas former la dérivée dans les directions tangentes à la surface $V = $ const. qui passe par Q.

(12) $$\frac{d_\gamma u}{d_\gamma R} = \sum_{k=1}^{m} \frac{\partial u}{\partial \xi_k} \gamma'_k.$$

D'autre part, en différentiant la relation $R = \sum_{k=1}^{m} \varepsilon_k (\xi_k - x_k)^2$ par rapport à R dans la direction γ, il vient

$$1 = 2 \sum_{1}^{m} \varepsilon_k (\xi_k - x_k) \frac{d_\gamma \xi_k}{d_\gamma R}$$

ou

(13) $$\sum_{1}^{m} \varepsilon_k (\xi_k - x_k) \gamma'_k = \frac{1}{2} \quad \text{ou encore} \quad (\boldsymbol{\xi} - \mathbf{x}, \boldsymbol{\gamma}') = \frac{1}{2}.$$

Le déplacement aux composantes $d_\gamma \xi_k$ ayant manifestement la direction γ, il en sera de même du vecteur $\boldsymbol{\gamma}'$ aux composantes γ'_k. Inversement, pour qu'un vecteur $\boldsymbol{\gamma}'$ aux composantes γ'_k puisse s'écrire sous la forme (11), il suffit qu'il ait la direction γ et qu'il satisfasse à la relation (13).

Ce point admis, on voit que si γ et δ sont des directions arbitraires, on pourra toujours écrire en vertu de (11)

(14) $$\frac{1}{2}\left(\gamma'_k + \delta'_k\right) = \nu'_k \quad \text{et} \quad \frac{1}{2}\left(\frac{d_\gamma \xi_k}{d_\gamma R} + \frac{d_\delta \xi_k}{d_\delta R}\right) = \frac{d_\nu \xi_k}{d_\nu R},$$

ν étant la direction du vecteur $\boldsymbol{\nu}'$, aux composantes ν'_k, qui de son côté est déterminé par les directions γ et δ. En effet, les relations (13)

$$(\boldsymbol{\xi} - \mathbf{x}, \boldsymbol{\gamma}') = \frac{1}{2} \quad \text{et} \quad (\boldsymbol{\xi} - \mathbf{x}, \boldsymbol{\delta}') = \frac{1}{2}$$

entraînent

(15) $$(\boldsymbol{\xi} - \mathbf{x}, \boldsymbol{\nu}') = \frac{1}{2}.$$

On aura en même temps, en vertu des formules (10) et (14),

(16) $$\frac{d_\gamma u}{d_\gamma R} + \frac{d_\delta u}{d_\delta R} = 2 \frac{d_\nu u}{d_\nu R}.$$

Cela étant, retournons à la formule (9) et tenons compte des faits que $\mathbf{n}$ est la normale «intérieure» [1] à la surface S en un certain point $\mathbf{b}$ du bord s^P, tandis que l est (la direction de) la trajectoire issue de $\mathbf{b}$. En désignant, comme au n° **51**, la tangente unitaire à l par $\mathbf{t}$, nous tirons du même numéro que

[1] *Cf.* la note p. 76.

(17) $$\mathbf{n} + \mathbf{t} = \lambda(\mathbf{x} - \mathbf{b}),$$

λ étant un nombre positif. Par conséquent, $(\mathbf{n} + \mathbf{t}, \mathbf{b} - \mathbf{x}) = -\lambda(\mathbf{b} - \mathbf{x})^2 = 0$. C'est-à-dire que

$$(-\mathbf{n}, \mathbf{b} - \mathbf{x}) = (\mathbf{t}, \mathbf{b} - \mathbf{x}) = \frac{1}{2\varkappa},$$

où $\varkappa$ est aussi un nombre positif. Cela donne

$$(-\varkappa\,\mathbf{n}, \mathbf{b} - \mathbf{x}) = (\varkappa\,\mathbf{t}, \mathbf{b} - \mathbf{x}) = \frac{1}{2}.$$

Par suite

(18) $$\mathbf{n}' = -\varkappa\,\mathbf{n},\ \mathbf{t}' = \varkappa\,\mathbf{t}$$

et le vecteur ν' défini par la relation $\nu' = \frac{1}{2}(\mathbf{n}' + \mathbf{t}')$ satisfait à

(19) $$\nu' = \frac{1}{2}(\mathbf{n}' + \mathbf{t}') = \frac{\varkappa}{2}(\mathbf{t} - \mathbf{n}),$$

ce qui entraîne

(19bis) $$(\nu', \nu') = \frac{\varkappa^2}{4}((\mathbf{t}, \mathbf{t}) - 2(\mathbf{t}, \mathbf{n}) + (\mathbf{n}, \mathbf{n})) = \frac{\varkappa^2}{4}(-1 + 1) = 0;$$

il satisfait en outre à la relation

(20) $$(\mathbf{b} - \mathbf{x}, \nu') = \frac{1}{2}.$$

Le vecteur ν' étant isotrope, la droite ν menée par le point $\mathbf{b}$ *et ayant la direction du vecteur ν' est une droite isotrope.* Les formules (14), (16) et (19) fournissent la transcription suivante de l'expression (9)

(21) $$\frac{du}{dn}\frac{dn}{dR} + \frac{du}{dR} = \frac{d_n u}{d_n R} + \frac{d_t u}{d_t R} = 2\frac{d_\nu u}{d_\nu R}.$$

Cela posé, la relation (8bis) peut s'écrire

(22) $$I_*^{m-2}\overline{\frac{du}{dn}}, u(P) = \frac{1}{\pi^{\frac{m-2}{2}}\, 2^{m-3}\, \Gamma\left(\frac{m-2}{2}\right)} \int\limits_{s^P} \left(\frac{d_t K}{d_t R}\cdot u - 2\frac{d_\nu u}{d_\nu R}\right) ds.$$

Il ne manque pas d'intérêt de remarquer que R varie d'une manière linéaire sur ν ou, plus généralement, sur toute droite isotrope[1]. Soit en effet C une telle droite, $\mathbf{c}$ un vecteur qui lui est parallèle, $(\mathbf{c}, \mathbf{c}) = 0$, $\boldsymbol{\eta}$ un point fixe et $\boldsymbol{\xi} = \boldsymbol{\eta} + \tau\mathbf{c}$ un point mobile sur C, τ étant un paramètre. Alors $R_{\mathbf{x}\xi} = (\boldsymbol{\xi} - \mathbf{x})^2 = R_{\mathbf{x}\eta} + 2\tau(\boldsymbol{\eta} - \mathbf{x}, \mathbf{c})$, expression linéaire en τ (ou dans les coordonnées de ξ).

(1) *Cf.* p. 110.

Observons encore que $d_\nu u/d_\nu R$ peut toujours être calculé par la formule (21) si l'on connaît sur S les données de Cauchy u et du/dn.

On vient de voir que par chaque point **b** du bord s^P il passe une droite ν, déterminée par le vecteur $\mathbf{v}^1$, qui de son côté est donné par la formule (19). Le vecteur **n** et le vecteur **t** étant respectivement orthogonal et tangent à la surface S, les formules (17) et (19) mettent en évidence que *la droite* ν *est la symétrique (au sens lorentzien) de la génératrice* $\mathbf{b}-\mathbf{x}$ indéfiniment prolongée *par rapport au plan tangent* à la surface S au point **b**.

Nous allons montrer qu'en réalité la droite ν et le vecteur $\mathbf{v}^1$ *dépendent beaucoup moins de la surface* S qu'il ne paraît au premier instant. En effet, *deux surfaces* S *qui admettent la même intersection* s^P *avec la nappe* C^P *du cône de sommet* P *admettent les mêmes droites* ν *et les mêmes vecteurs* $\mathbf{v}^1$ (¹).

Observons d'abord que la génératrice et la droite ν passant par un point **b** du bord s^P sont isotropes toutes les deux et, de plus, qu'elles sont orthogonales au bord, puisqu'il en est ainsi de **n** et de **t** (*cf.* les formules (17) et (19) et le n° **51**). Cela posé, considérons autour du point **b** un élément infinitésimal du bord. Cet élément, qui est à $m-2$ dimensions, est évidemment orienté dans l'espace puisqu'il en est de même du bord tout entier. Le plan à deux dimensions π mené par le point **b** et orthogonal à cet élément est donc orienté dans le temps, ce qui veut dire que la forme lorentzienne exprimant le carré de la distance de deux points arbitraires de ce plan est indéfinie. Dès lors, on peut introduire un nouveau système de coordonnées lorentziennes, l'axe des ξ_1 et celui des ξ_2 étant parallèles à π. De là il s'ensuit que *ce plan contient deux directions isotropes* (données par $\xi_1 \pm \xi_2 = 0$) qui, par suite, doivent être *celles de la génératrice et de la droite* ν, considérées tout à l'heure. On voit alors que la droite ν est entièrement déterminée par l'élément infinitésimal de s^P. Il en est de même du vecteur $\mathbf{v}^1$, puisque ce vecteur est déterminé par la direction de ν et la relation (20).

Les droites ν constituent une famille à $m-2$ paramètres et engendrent une surface réglée Σ^P à $m-1$ dimensions, dont nous allons montrer qu'elle est une *surface caractéristique* (*cf.* le n° **62**)(²).

Le point essentiel est que les ν sont des droites isotropes, orthogonales à une variété à $m-2$ dimensions, soit s. (Le caractère particulier de s^P d'ap-

(¹) Cet énoncé pourrait être remplacé par un énoncé plus précis, de caractère local; *cf.* la démonstration.

(²) Dans l'Introduction, la dernière surface s'obtient aussi comme enveloppe de cônes caractéristiques, construction rattachée à la méthode classique de Cauchy pour l'intégration des équations aux dérivées partielles du premier ordre.

partenir à la nappe d'un cône caractéristique n'intervient pas dans la démonstration.) Désignons par η les points de s, par ν' un vecteur parallèle à ν, $(\nu', \nu') = 0$ et $(\nu', d\eta) = 0$; le point $\xi = \eta + \tau\nu'$ décrit, quand τ varie, la droite ν et, quand η et τ varient tous les deux, la surface Σ. On a $d\xi = d\eta + d\tau \cdot \nu' + \tau \cdot d\nu'$. On en tire $(\nu', d\xi) = 0$, puisque $(\nu', d\eta) = 0$ et $(\nu', \nu') = 0$, par hypothèse, et $(\nu', d\nu') = 0$ s'ensuit de $(\nu', \nu') = 0$. On voit donc que les génératrices ν de Σ sont normales à Σ, ce qui met en évidence que Σ est une surface caractéristique (n° **62**).

67. Calcul explicite d'une dérivée. — Pour pousser plus loin l'étude de $\overline{I_*^{m-2} \frac{du}{dn}}, u(P)$, qui à présent est exprimé par la formule (8bis), reprenons la quantité K figurant dans la formule (4). Nous écrivons K sous la forme facile à comprendre

$$K = \frac{d_n R}{dn} \cdot \frac{dl}{d_l R} \cdot \frac{ds_R}{ds} = \frac{dS}{d_l R \cdot ds} \cdot \frac{d_n R}{dn}, \tag{23}$$

$dS = ds_R \cdot dl$ étant la mesure d'un élément infinitésimal de la surface S autour d'un point de S, soit $\mathbf{a}$. Observons maintenant que le produit $dS \cdot \frac{d_n R}{dn}$ peut, à un facteur constant près, être interprété comme le volume (à m dimensions) d'un cylindre infinitésimal qui admet dS comme base et dont les génératrices sont des segments équipollents au rayon vecteur $\mathbf{x} - \mathbf{a}$, mené du point $\mathbf{a}$ au sommet $P = \mathbf{x}$. La hauteur de ce cylindre est la valeur absolue de la projection orthogonale du vecteur $\mathbf{x} - \mathbf{a}$ sur la normale $\mathbf{n}$, c'est-à-dire égale à

$$|(\mathbf{x} - \mathbf{a}, \mathbf{n})| = -r\frac{d_n r}{dn} = -\frac{1}{2}\frac{d_n R}{dn}$$

(formule (48) du n° **21**). Le volume du cylindre est donc

$$dV = -\frac{1}{2}\frac{d_n R}{dn} \cdot dS. \tag{24}$$

Nous rapportons les points $\mathbf{b}$ du bord s^P et les trajectoires l issues de ces points à $m-2$ paramètres $\lambda^1, \lambda^2, \ldots, \lambda^{m-2}$. Ces paramètres et R fournissent une paramétrisation des points $\mathbf{a}$ de la portion de surface S^P, au moins *dans le voisinage du bord*, soit sur la partie annulaire extérieure $\tilde{S}$ (n° **50**). Enfin, pour simplifier l'écriture, nous plaçons *l'origine des coordonnées au sommet* $P = \mathbf{x}$.

Tout cela étant admis, nous formons les vecteurs

$$\frac{\partial \mathbf{a}}{\partial R} = \frac{d_l \mathbf{a}}{d_l R} \quad \text{et} \quad \frac{\partial \mathbf{a}}{\partial \lambda^j}, \quad j = 1, 2, \ldots, m-2, \tag{25}$$

au nombre total de $m-1$, et le déterminant d'ordre m

$$(26) \qquad D = \left| \mathbf{a} \frac{\partial \mathbf{a}}{\partial R} \frac{\partial \mathbf{a}}{\partial \lambda^1} \cdots \frac{\partial \mathbf{a}}{\partial \lambda^{m-2}} \right| = \left| \mathbf{a} \frac{\partial \mathbf{a}}{\partial R} \frac{\partial \mathbf{a}}{\partial \lambda^j} \right|, \quad j = 1, 2, \ldots, m-2,$$

dont les colonnes sont données par les composantes du rayon vecteur $\mathbf{a}$ et des vecteurs dérivés ci-dessus. D'autre part, on peut identifier l'élément infinitésimal dS au parallélépipède construit sur les vecteurs $\frac{\partial \mathbf{a}}{\partial R} dR$ et $\frac{\partial \mathbf{a}}{\partial \lambda^j} d\lambda^j$, $j=1, 2, \ldots, m-2$, ce qui donne

$$(27) \qquad dV = |D|\, dR \prod_1^{m-2} d\lambda^j,$$

toutes les différentielles étant positives. On tire de (24) et de (27), en observant que $d_n R/dn$ est négatif,

$$(28) \qquad \frac{d_n R}{dn} dS = -2|D|\, dR \prod d\lambda^j$$

et alors, de (23), puisque $d_l R = dR$,

$$K = -\frac{2|D| \prod d\lambda^j}{ds}.$$

Or s^P ayant été paramètrisé par les λ^j, on a d'après les règles du n° **21**[bis], $ds = p(\lambda^j) \prod d\lambda^j$, où $p(\lambda^j) > 0$ est une fonction des λ^j. On a donc

$$K = -\frac{2|D|}{p}.$$

Enfin, puisque p est indépendant de R,

$$(29) \qquad \frac{d_l K}{d_l R} = -\frac{2}{p} \frac{d_l |D|}{d_l R} = -\frac{2|D|}{p} \cdot \frac{1}{D} \frac{d_l D}{d_l R} = K \cdot \frac{1}{D} \frac{d_l D}{d_l R}.$$

C'est l'évaluation de la dernière dérivée aux points du bord dont on va s'occuper dans le reste du présent numéro.

On trouve immédiatement, grâce aux relations

$$\frac{d_l \mathbf{a}}{d_l R} = \frac{\partial \mathbf{a}}{\partial R}, \quad \frac{d_l}{d_l R} \frac{\partial \mathbf{a}}{\partial R} = \frac{\partial^2 \mathbf{a}}{\partial R^2}, \quad \frac{d_l}{d_l R} \frac{\partial \mathbf{a}}{\partial \lambda^j} = \frac{\partial^2 \mathbf{a}}{\partial R \partial \lambda^j},$$

que

$$(30) \qquad \frac{d_l D}{d_l R} = \left| \mathbf{a} \frac{\partial^2 \mathbf{a}}{\partial R^2} \frac{\partial \mathbf{a}}{\partial \lambda^1} \cdots \frac{\partial \mathbf{a}}{\partial \lambda^{m-2}} \right| + \sum_{j=1}^{m-2} D_j,$$

où D_j est identique à D sauf pour la colonne $\frac{\partial \mathbf{a}}{\partial \lambda^j}$, qui est remplacée par

$$\frac{\partial^2 \mathbf{a}}{\partial R \, \partial \lambda^j}.$$

Nous allons encore transformer les formules (29) et (30) en y introduisant un nouveau vecteur qui, à la limite, devient identique au vecteur isotrope $\mathbf{v}^1$, donné par les formules (19bis) et (20). Posons à cet effet (dans la région annulaire $\bar{S}$)

$$\mathbf{a}' = \frac{\partial \mathbf{a}}{\partial R} - N \mathbf{a}, \tag{31}$$

N étant un scalaire, dépendant de $\mathbf{a}$, qui sera spécialisé plus loin. On a d'abord

$$\frac{\partial^2 \mathbf{a}}{\partial R \, \partial \lambda^j} = \frac{\partial \mathbf{a}'}{\partial \lambda^j} + \frac{\partial N}{\partial \lambda^j} \mathbf{a} + N \frac{\partial \mathbf{a}}{\partial \lambda^j} \quad \text{et} \quad D_j = D'_j + N D,$$

où D'_j est identique à D sauf pour la colonne $\frac{\partial \mathbf{a}}{\partial \lambda^j}$, qui cette fois-ci est remplacée par $\frac{\partial \mathbf{a}'}{\partial \lambda^j}$. Il en résulte

$$\sum_1^{m-2} D_j = \sum_1^{m-2} D'_j + (m-2) N D. \tag{32}$$

On a en même temps

$$D = \left| \mathbf{a} \frac{\partial \mathbf{a}}{\partial R} \frac{\partial \mathbf{a}}{\partial \lambda^j} \right| = \left| \mathbf{a} \, \mathbf{a}' \frac{\partial \mathbf{a}}{\partial \lambda^j} \right|, \quad j = 1, 2, \ldots, m-2. \tag{33}$$

Observons d'autre part que, les vecteurs, au nombre de m, $\mathbf{a}$, $\frac{\partial \mathbf{a}}{\partial R}$, $\frac{\partial \mathbf{a}}{\partial \lambda^j}$ étant linéairement indépendants, on pourra poser

$$\frac{\partial^2 \mathbf{a}}{\partial R^2} = A \mathbf{a} + B \frac{\partial \mathbf{a}}{\partial R} + \sum_1^{m-2} C_j \frac{\partial \mathbf{a}}{\partial \lambda^j}, \tag{34}$$

les coefficients A, B, C_j étant des scalaires, dépendant de $\mathbf{a}$. On en tire pour le premier déterminant dans (30)

$$\left| \mathbf{a} \frac{\partial^2 \mathbf{a}}{\partial R^2} \frac{\partial \mathbf{a}}{\partial \lambda^j} \right| = B D. \tag{35}$$

Cela étant, nous faisons tendre $\mathbf{a}$ vers un point $\mathbf{b}$ du bord s^P. Toutes nos formules resteront valides, si nous y remplaçons $\mathbf{a}$ et $\mathbf{a}'$ par $\mathbf{b}$ et

(36) $$\mathbf{b}' = \frac{\partial \mathbf{b}}{\partial R} - N\,\mathbf{b}$$

respectivement, où $\frac{\partial \mathbf{b}}{\partial R}$ est une notation abrégée pour $\left(\frac{\partial \mathbf{a}}{\partial R}\right)_{(R=0)}$ Rappelons maintenant que $R = \mathbf{a}^2 = (\mathbf{a}, \mathbf{a})$, ce qui donne

(37) $$\left(\mathbf{a}, \frac{\partial \mathbf{a}}{\partial R}\right) = \frac{1}{2} \text{ et } \left(\mathbf{a}, \frac{\partial^2 \mathbf{a}}{\partial R^2}\right) + \left(\frac{\partial \mathbf{a}}{\partial R}, \frac{\partial \mathbf{a}}{\partial R}\right) = 0,$$

formules valables encore quand on y remplace **a** *par* **b**. Si l'on tient compte de (37) et de ce que le vecteur **b** (segment de génératrice du cône) est isotrope, c'est-à-dire que

(38) $$\mathbf{b}^2 = 0,$$

on trouve pour le vecteur $\mathbf{b}'$, donné par la formule (36),

$$\mathbf{b}'^2 = \left(\frac{\partial \mathbf{b}}{\partial R}\right)^2 - 2N\left(\mathbf{b}, \frac{\partial \mathbf{b}}{\partial R}\right) = \left(\frac{\partial \mathbf{b}}{\partial R}\right)^2 - N.$$

Cela veut dire que, si l'on choisit N de façon qu'au point **b** on a

(39) $$N = \left(\frac{\partial \mathbf{b}}{\partial R}\right)^2,$$

$\mathbf{b}'$ devient isotrope,

(40) $$\mathbf{b}'^2 = 0.$$

D'autre part, d'après (36), (37) et (38),

(41) $$(\mathbf{b}, \mathbf{b}') = \left(\mathbf{b}, \frac{\partial \mathbf{b}}{\partial R}\right) = \frac{1}{2}.$$

Cela posé, on voit (*cf.* p. 131) que $\mathbf{b}'$ *est identique au vecteur isotrope* ν^{I}, donné par les formules (19bis) et (20). En effet, les vecteurs isotropes $\mathbf{b}'$ est ν^{I} sont orthogonaux au bord tous les deux et ont même produit scalaire avec le vecteur **b**. En tenant compte de ce qu'on a dit au sujet du vecteur ν^{I} au n° **66** (p. 131), on voit que $\mathbf{b}'$ ne dépend que du bord s^P, dans le sens qu'on a prêté à ce mot à l'endroit indiqué. La même remarque s'applique aux dérivées $\partial \mathbf{b}/\partial \lambda^j$, $\partial \mathbf{b}'/\partial \lambda^j$ et dès lors aux quantités L_j^k définies par les formules (48).

Pour obtenir la valeur du coefficient B (formules (34) et (35)) au point **b**, nous notons qu'il s'ensuit de (38)

(42) $$\left(\mathbf{b}, \frac{\partial \mathbf{b}}{\partial \lambda^j}\right) = 0.$$

Il vient alors au moyen de (38), (37), (42) et (39)

$$\frac{1}{2} B = -N,$$

et le premier déterminant en (30) devient en raison de (35) égal à $-2ND$. En vertu de cette dernière évaluation et de (32), on obtient (30) sous la forme

$$\frac{d_l D}{d_l R} = (m-4) ND + \sum_1^{m-2} D'_j, \tag{43}$$

où N est donné par (39), D peut, d'après (33), s'écrire

$$D = \left| \mathbf{b}\,\mathbf{b}' \frac{\partial \mathbf{b}}{\partial \lambda^j} \right|, \quad j = 1, 2, \ldots, m-2, \tag{44}$$

et D'_j s'obtient de D en y remplaçant la colonne $\partial \mathbf{b}/\partial \lambda^j$ par $\partial \mathbf{b}'/\partial \lambda^j$.

Pour aller plus loin, exprimons le fait que les trajectoires l sont orthogonales au bord par les formules $\left(\frac{\partial \mathbf{b}}{\partial R}, \frac{\partial \mathbf{b}}{\partial \lambda^j}\right) = 0$, qui, en raison de (36) et de (42), donnent

$$\left(\mathbf{b}', \frac{\partial \mathbf{b}}{\partial \lambda^j}\right) = 0. \tag{45}$$

On en tire, eu égard à (41), que

$$\left(\mathbf{b}, \frac{\partial \mathbf{b}'}{\partial \lambda^j}\right) = 0. \tag{46}$$

Enfin, de (40) il vient

$$\left(\mathbf{b}', \frac{\partial \mathbf{b}'}{\partial \lambda^j}\right) = 0. \tag{47}$$

En résumé, les formules (38), (40)—(42), (45)—(47) disent que chacun des vecteurs $\mathbf{b}$ et $\mathbf{b}'$ est orthogonal à soi-même et aux vecteurs $\partial \mathbf{b}/\partial \lambda^j$ et $\partial \mathbf{b}'/\partial \lambda^j$, tandis que $(\mathbf{b}, \mathbf{b}') = \frac{1}{2}$.

Les vecteurs $\mathbf{b}$, $\mathbf{b}'$, $\partial \mathbf{b}/\partial \lambda^j$ sont linéairement indépendants et au nombre total de m en raison du fait analogue pour $\mathbf{b}$, $\partial \mathbf{b}/\partial R$, $\partial \mathbf{b}/\partial \lambda^j$ et de la formule (36); on pourra donc exprimer les $\partial \mathbf{b}'/\partial \lambda^j$ comme leurs combinaisons linéaires

$$\frac{\partial \mathbf{b}'}{\partial \lambda^j} = L_j \mathbf{b} + L'_j \mathbf{b}' - \sum_{k=1}^{m-2} L_j^k \frac{\partial \mathbf{b}}{\partial \lambda^k}.$$

Il résulte des formules énumérées que $L_j = L'_j = 0$. Il ne reste donc que

(48) $$\frac{\partial \mathbf{b}'}{\partial \lambda^j} = - \sum_{k=1}^{m-2} L_j^k \frac{\partial \mathbf{b}}{\partial \lambda^k}.$$

Si l'on porte les expressions (48) dans les D'_j, définis par les explications qui se rattachent à la formule (44), l'expression (43) devient

$$\frac{1}{D}\frac{d_l D}{d_l R} = (m-4) N - \sum_{j=1}^{m-2} L_j^j,$$

cette formule étant valable en tout point $\mathbf{b}$ du bord. On a en ces points aussi $K = -1$, d'après la formule (6) du n° **65**. Par conséquent, l'expression (29), dont l'évaluation explicite est l'objet du présent numéro, peut, en tout point $\mathbf{b}$, s'écrire

(49) $$\frac{d_l K}{d_l R} = (4-m) N + \sum_{j=1}^{m-2} L_j^j,$$

où N est donné par la formule (39), tandis que les L_j^j sont définis par les relations (48).

En portant la dernière expression dans (8bis), on obtient

(50) $$I_*^{m-2} \overline{\frac{du}{dn}}, u(P) = \frac{1}{\pi^{\frac{m-2}{2}} 2^{m-3} \Gamma\left(\frac{m-2}{2}\right)} \int_{s^P} \left\{\left((4-m) N + \sum_{j=1}^{m-2} L_j^j\right) u - 2 \frac{d_\nu u}{d_\nu R}\right\} ds$$

et, *en particulier*, *pour* $m = 4$

(51) $$I_*^2 \overline{\frac{du}{dn}}, u(P) = \frac{1}{2\pi} \int_{s^P} \left\{(L_1^1 + L_2^2) u - 2 \frac{d_\nu u}{d_\nu R}\right\} ds.$$

On voit que (51) est *indépendant* de la surface S dans le sens qu'on a prêté à ce mot plus haut (p. 131 et p. 135) et que (50) en dépend très légèrement, à savoir par le facteur N (formule (39)).

68. Interprétation géométrique. — Pour arriver à une telle interprétation, nous chercherons sur le bord s^P des courbes C telles que les droites isotropes ν appartenant aux points $\mathbf{b}$ de C (*cf.* le n° **65**) soit tangentes à une seconde courbe. Une telle courbe C sera appelée *ligne de courbure*[1] de s^P. Le rayon vecteur

[1] Les considérations qui suivent sont tout analogues à celles qu'on rencontre dans la géométrie différentielle des surfaces, relative à l'espace ordinaire à trois dimensions, le bord s^P joue ici le rôle de la surface et les droites isotropes ν, orthogonales au bord jouent le rôle des normales de la surface.

mené de l'origine (du sommet $P = \mathfrak{x}$) au point de contact aura la forme $\mathfrak{b} + \varrho\,\mathfrak{b}'$, ϱ étant un paramètre (¹). Il s'ensuit

$$(52) \qquad d(\mathfrak{b} + \varrho\,\mathfrak{b}') = \varepsilon\,\mathfrak{b}' \quad \text{ou} \quad d\,\mathfrak{b} + \varrho \cdot d\,\mathfrak{b}' + d\,\varrho \cdot \mathfrak{b}' = \varepsilon\,\mathfrak{b}',$$

ε étant une quantité infinitésimale. En formant le produit scalaire avec $\mathfrak{b}$, il vient d'après (41), (42) et (46)

$$0 = (d\,\varrho - \varepsilon)(\mathfrak{b}, \mathfrak{b}') = \frac{1}{2}(d\,\varrho - \varepsilon),$$

par quoi (52) se réduit à

$$(52^{bis}) \qquad d\,\mathfrak{b} + \varrho\, d\,\mathfrak{b}' = 0,$$

formule tout à fait analogue à celle d'Olinde Rodrigues (²).

Avant d'aller plus loin, remarquons que, pour $m = 3$, $m - 2$ se réduit à 1, le bord s^P devient donc une courbe. Il n'y aura qu'un seul paramètre soit λ^1. On voit de (48) que s^P satisfait à l'équation (52bis) avec $\varrho^{-1} = L_1^1$. La surface caractéristique Σ^P, engendrée par les droites ν, est donc une surface développable.

Reprenons le cas général ($m \geqq 3$). On peut donner pour ϱ l'interprétation géométrique suivante. Posons $\mathfrak{b} + \varrho\,\mathfrak{b}' = \mathfrak{f}$ pour le rayon vecteur menant de l'origine $\mathfrak{o}$ au point de contact. Formons $\mathfrak{f}^2 = \mathfrak{b}^2 + 2\varrho(\mathfrak{b}, \mathfrak{b}') + \varrho^2\mathfrak{b}'^2$, identité qui d'après (38), (40), (41) se réduit à $\varrho = \mathfrak{f}^2 = R_{\mathfrak{o}\mathfrak{f}}$. Dans le cas général où le sommet du cône est désigné par $P = \mathfrak{x}$, on aura

$$(53) \qquad \varrho = R_{\mathfrak{x}\mathfrak{f}}.$$

Si nous posons maintenant $\varrho = \varkappa^{-1}$, la relation (52bis) s'écrit

$$(54) \qquad d\,\mathfrak{b}' + \varkappa\, d\,\mathfrak{b} = 0,$$

et la formule (53) devient

$$(53^{bis}) \qquad \varkappa = \frac{1}{R_{\mathfrak{x}\mathfrak{f}}}.$$

Écrivons

$$(55) \qquad d\,\mathfrak{b} = \sum_{q=1}^{m-2} \frac{\partial\,\mathfrak{b}}{\partial\,\lambda^q}\, d\,\lambda^q \quad \text{et} \quad d\,\mathfrak{b}' = \sum_{q=1}^{m-2} \frac{\partial\,\mathfrak{b}'}{\partial\,\lambda^q}\, d\,\lambda^q,$$

ce qui, en raison de (54) et de (48), donne

(¹) On se rappellera que $\mathfrak{b}'$ est identique à v^1 (p. 135); $\mathfrak{b}'$ a par conséquent la direction ν.

(²) *Cf.* p. ex. Goursat, *Cours d'Analyse* I (1910), p. 611.

$$o = -(d\mathbf{b}' + \varkappa\, d\mathbf{b}) = \sum_{p,q=1}^{m-2} (L_q^p - \varkappa\, \delta_q^p) \frac{\partial \mathbf{b}}{\partial \lambda^p} d\lambda^q, \quad \text{où } \delta_q^p = \begin{cases} 1, & p = q \\ 0, & p \neq q \end{cases}.$$

Les vecteurs $\partial \mathbf{b}/\partial \lambda^p$ étant linéairement indépendants, on a pour $\varkappa$ et les $d\lambda^q$ le système d'équations

$$(56) \qquad \sum_{q=1}^{m-2} (L_q^p - \varkappa\, \delta_q^p)\, d\lambda^q = 0, \quad p = 1, 2, \ldots, m-2,$$

qui fournit pour $\varkappa$ l'équation

$$(57) \qquad |L_q^p - \varkappa\, \delta_q^p| = 0, \quad p, q = 1, 2, \ldots, m-2.$$

Nous verrons tout à l'heure que toutes les racines de cette équation sont réelles. Admettons ce point pour l'instant et, pour simplifier, admettons aussi, jusqu'à nouvel ordre, que les racines sont distinctes. A chaque racine $\varkappa$ correspondra par (56) un système de $d\lambda^q$ et les déplacements $d\mathbf{b}$ et $d\mathbf{b}'$ donnés par (55). D'autre part, on tire de (57), en numérotant les racines $\varkappa$,

$$(58) \qquad \sum_{A=1}^{m-2} \varkappa_A = \sum_{j=1}^{m-2} L_j^j.$$

Si l'on numérote aussi les points de contact $\mathbf{f}$ correspondants et pose pour abréger $R_{\mathbf{x}\mathbf{f}_A} = R_A$, on obtient des formules (53bis) et (58)

$$(59) \qquad \sum_{j=1}^{m-2} L_j^j = \sum_{A=1}^{m-2} \frac{1}{R_A}.$$

Alors, si l'on remplace dans (50) et (51) le premier membre de la dernière identité par le second, on parvient à une nouvelle forme de I_*^{m-2} et, en particulier, de I_*^2 pour $m = 4$. Nous bornant à ce cas particulier important, *nous écrivons la solution de l'équation des ondes sphériques, sans second membre,* $\Delta u = 0$, *sous la forme,* annoncée déjà dans l'Introduction,

$$(60) \qquad u(P) = I_*^2 \overline{\frac{du}{dn}, u(P)} = \frac{1}{\pi} \int_{s^P} \left\{ \frac{1}{2} \left(\frac{1}{R_1} + \frac{1}{R_2} \right) \cdot u - \frac{d_\nu u}{d_\nu R} \right\} ds.$$

Nous reviendrons sur cette expression, qui est l'*objet principal de ce chapitre,* au n° **71**. En attendant, il nout faut reprendre les points ajournés plus haut, en particulier la question des racines de l'équation (57). Nous introduisons d'abord la forme métrique de la variété s^P. Un déplacement infinitésimal $d\mathbf{b}$

s'écrit $d\mathbf{b} = \sum_1^{m-2} \frac{\partial b}{\partial \lambda^j} d\lambda^j$ et par suite

$$(61) \qquad d\mathbf{b}^2 = \sum_{j,k=1}^{m-2} \gamma_{jk}\, d\lambda^j\, d\lambda^k, \quad \text{où } \gamma_{jk} = \left(\frac{\partial \mathbf{b}}{\partial \lambda^j}, \frac{\partial \mathbf{b}}{\partial \lambda^k}\right).$$

La variété ayant une orientation d'espace (*cf.* p. 131), la forme métrique ci-dessus est *définie* négative.

Il sera utile de considérer en même temps que les L_j^k un second système de quantités L_{jk}. Nous posons par définition

$$(62) \qquad L_{jk} = -\left(\frac{\partial \mathbf{b}}{\partial \lambda^j}, \frac{\partial \mathbf{b}'}{\partial \lambda^k}\right).$$

La différention de (45) par rapport à λ^k donne

$$\left(\frac{\partial \mathbf{b}}{\partial \lambda^j}, \frac{\partial \mathbf{b}'}{\partial \lambda^k}\right) + \left(\mathbf{b}', \frac{\partial^2 \mathbf{b}}{\partial \lambda^k\, \partial \lambda^j}\right) = 0.$$

Le dernier terme étant *symétrique* en j et k, on voit que

$$(63) \qquad L_{jk} = L_{kj}.$$

Les L_{jk} s'expriment par les L_j^k de la manière suivante (*cf.* formules (48) et (61)).

$$(64) \quad L_{jk} = -\left(\frac{\partial \mathbf{b}}{\partial \lambda^j}, \frac{\partial \mathbf{b}'}{\partial \lambda^k}\right) = \left(\frac{\partial \mathbf{b}}{\partial \lambda^j}, \sum_{p=1}^{m-2} L_k^p \frac{\partial \mathbf{b}}{\partial \lambda^p}\right) = \sum_{p=1}^{m-2} \left(\frac{\partial \mathbf{b}}{\partial \lambda^j}, \frac{\partial \mathbf{b}}{\partial \lambda^p}\right) \cdot L_k^p = \sum_{p=1}^{m-2} \gamma_{jp} L_k^p,$$

ce qui met en évidence que les L_{jk} sont les composantes covariantes d'un tenseur symétrique du second ordre dont les L_j^k sont les composantes mixtes.

Nous allons maintenant démontrer que les racines de l'équation (57) sont réelles. En effet, en ajoutant les équations (56), multipliées par les facteurs respectifs γ_{jp}, on obtient, en vertu de (64), le système équivalent

$$(65) \qquad \sum_{k=1}^{m-2} (L_{jk} - \varkappa \gamma_{jk})\, \delta \lambda^k = 0, \quad j = 1, 2, \ldots, m-2.$$

De là on tire

$$(66) \qquad |L_{jk} - \varkappa \gamma_{jk}| = 0, \quad j, k = 1, 2, \ldots, m-2,$$

équation qui est équivalente à (57). Or cette dernière équation a toutes ses racines réelles, car la forme métrique (61) est une forme quadratique définie.

A chaque racine simple $\varkappa$ de (66) correspond par (56) un système de différentielles $d\lambda^1, \ldots, d\lambda^{m-2}$ à rapports déterminés, qui fournissent un déplacement infinitésimal $d\mathbf{b}$ de direction déterminée. Plus généralement, à chaque racine $\varkappa$ de multiplicité h correspond une variété linéaire de déplacements, dont h sont linéairement indépendants(¹).

Il est d'ailleurs clair que les équations (65) et (66) sont identiques à celles qui fournissent les valeurs extrêmes du rapport

$$\sum_{j,k} L_{jk}\, d\lambda^j\, d\lambda^k : \sum_{j,k} \gamma_{jk}\, d\lambda^j\, d\lambda^k.$$

Pour arriver aux lignes de courbure C, il faut intégrer les équations vectorielles identiques (52bis) ou (54) ou, plus précisément, le système d'équations (56) qui leur est équivalent. Par tout point de s^P où (57) admet des racines distinctes, passent $m - 2$ courbes distinctes, tandis que la coïncidence de certaines racines ou de toutes les racines amène une indétermination(²) partielle ou complète. Ce dernier fait a lieu, comme on le verra au nº **70**, dans le cas où la surface S est un plan ou un hyperboloïde $\mathbf{a}^2 = \text{const.}$ ou, à la limite, un cône caractéristique.

Voici encore quelques remarques complémentaires. Les déplacements $d\mathbf{b}_A$ et $d\mathbf{b}_B$ correspondant à deux racines différentes $\varkappa_A \neq \varkappa_B$ sont *orthogonaux*, $(d\mathbf{b}_A, d\mathbf{b}_B) = 0$. En effet, il s'ensuit de (65) et de la symétrie des L_{jk} et des γ_{jk}

$$(\varkappa_B - \varkappa_A) \sum_{j,k} \gamma_{jk}\, d\lambda_A^j\, d\lambda_B^k = 0$$

et, puisque $\varkappa_B - \varkappa_A \neq 0$,

$$(d\mathbf{b}_A, d\mathbf{b}_B) = \sum_{j,k} \gamma_{jk}\, d\lambda_A^j\, d\lambda_B^k = 0. \tag{67}$$

On peut encore tirer de (65) une seconde relation

$$(d\mathbf{b}_A, d\mathbf{b}'_B) = (d\mathbf{b}_B, d\mathbf{b}'_A) = -\sum_{j,k} L_{jk}\, d\lambda_A^j\, d\lambda_B^k = 0. \tag{68}$$

En nous inspirant de la terminologie usuelle de la théorie des surfaces, nous dirons que (68) exprime que les directions de $d\mathbf{b}_A$ et de $d\mathbf{b}_B$ — qui, comme nous venons de le voir, sont orthogonales — sont aussi *conjuguées*.

Les points de contact $\mathbf{f}_A$ $(A = 1, 2, \ldots, m - 2)$, distincts ou non, situés sur une droite de la famille des droites ν sont dits *foyers relatifs à la droite*. Ces

(¹) Pour tous ces faits bien connus, *voir* p. ex. COURANT-HILBERT **1**, I, p. 32—33.

(²) Un tel point peut être appelé *ombilic*.

points décrivent une variété à $m-2$ dimensions et à $m-2$ nappes, en général, que nous appelons la *surface focale* ou *caustique* de la famille de droites.

69. La plus courte distance. — La manière dont nous venons de caractériser les foyers et la surface focale a certains avantages sur celle qui se trouve dans l'Introduction. Ainsi l'équation (54), qui servait de base à toutes nos considérations, a été obtenue presque sans calcul. D'autre part, l'autre voie à laquelle nous venons de faire allusion est certainement à préférer sous certains rapports, notamment dans les cas de dégénérescence. Il suffira de l'indiquer succinctement.

On sait que dans l'espace à trois dimensions les surfaces développables se distinguent entre les surfaces réglées par les deux propriétés suivantes qui, en général, sont équivalentes: 1°. Les génératrices sont tangentes à une courbe (gauche). 2°. La plus courte distance de deux génératrices infiniment voisines est infiniment petite d'ordre supérieur au premier. On peut aussi exprimer cette propriété en disant que les génératrices voisines d'une surface développable se rencontrent. C'est la première propriété qui nous a servi de modèle plus haut, c'est la seconde qui va nous guider dans ce qui suit.

Considérons une ligne Γ, tracée sur le bord s^P et les droites ν menées par les points $\mathbf{b}$ de Γ. Soient[1]

$$(69)\qquad \mathbf{c} = \mathbf{b} + \tau\mathbf{b}' \quad\text{et}\quad \mathbf{c} + \Delta\mathbf{c} = \mathbf{b} + \Delta\mathbf{b} + (\tau + \Delta\tau)(\mathbf{b}' + \Delta\mathbf{b}')$$

les points qui sur deux droites voisines forment les extrémités du segment fournissant le plus courte distance de ces droites. On devra avoir

$$(70)\qquad (\mathbf{b}', \Delta\mathbf{c}) = 0 \quad\text{et}\quad (\mathbf{b}' + \Delta\mathbf{b}', \Delta\mathbf{c}) = 0$$

où, en vertu de (69),

$$(71)\qquad \Delta\mathbf{c} = \Delta\mathbf{b} + \tau\Delta\mathbf{b}' + \Delta\tau(\mathbf{b}' + \Delta\mathbf{b}').$$

On tire de la dernière des relations (70) et de (71)

$$(72)\quad \Delta\mathbf{c}^2 = (\Delta\mathbf{b} + \tau\Delta\mathbf{b}', \Delta\mathbf{c}) = (\Delta\mathbf{b} + \tau\Delta\mathbf{b}')^2 + \Delta\tau(\Delta\mathbf{b} + \tau\Delta\mathbf{b}', \mathbf{b}' + \Delta\mathbf{b}').$$

On cherche la condition dans laquelle $|\Delta\mathbf{c}^2|$ *est infiniment petit par rapport à* $|\Delta\mathbf{b}^2| = \varepsilon^2$, ce que nous exprimons par la notation $|\Delta\mathbf{c}^2| = o(\varepsilon^2)$. Écrivons $\Delta\mathbf{b} = d\mathbf{b} + \frac{1}{2}d^2\mathbf{b} + \cdots$ et pareillement pour $\Delta\mathbf{b}'$, les termes non écrits étant d'ordre supérieur au deuxième. Puisque $(\mathbf{b}', d\mathbf{b}) = 0$ et $(\mathbf{b}', d\mathbf{b}') = 0$ (formules (45)

[1] Nous faisons, de nouveau, coïncider l'origine avec le sommet $P = \mathbf{x}$.

et (47)), le dernier terme du second membre de (72) est $O(\varepsilon^3) = o(\varepsilon^2)$ [1]. De ce fait et de $|\Delta \mathbf{c}^2| = o(\varepsilon^2)$ on tire $(d\mathbf{b} + \tau\, d\mathbf{b}')^2 = o(\varepsilon^2)$. La quantité τ dépend des points $\mathbf{b}$ et $\mathbf{b} + \Delta\mathbf{b}$. Supposons encore que, si $\mathbf{b} + \Delta\mathbf{b}$ tend vers $\mathbf{b}$, τ tend vers une valeur limite τ^1, qui alors ne dépendra que de $\mathbf{b}$. Les valeurs τ et τ^1 ne différant que très peu, on tire de la dernière relation $(d\mathbf{b} + \tau^1 d\mathbf{b}')^2 = o(\varepsilon^2)$, ce qui veut dire que $(d\mathbf{b} + \tau^1 d\mathbf{b}')^2$ est infiniment petit d'ordre supérieur au deuxième. Or si la différentielle $d\mathbf{b} + \tau^1 d\mathbf{b}'$ n'est pas nulle, elle sera un vecteur d'espace [2] de direction constante, qui ne pourra satisfaire à une telle relation. Par conséquent

$$d\mathbf{b} + \tau^1 d\mathbf{b}' = 0. \tag{73}$$

Inversement, si (73) est rempli, on aura $(d\mathbf{b} + \tau\, d\mathbf{b}')^2 = o(\varepsilon^2)$ et $(\Delta\mathbf{b} + \tau\,\Delta\mathbf{b}')^2 = o(\varepsilon^2)$. Il en résulte aussi pour la plus courte distance $\Delta\mathbf{c}$ (formule (71)), grâce aux relations (40), (45) et (47), $\Delta\mathbf{c}^2 = o(\varepsilon^2)$.

On voit que dans (73) on est tombé de nouveau sur une équation de la forme (52bis); la ligne Γ est donc identique à une ligne de courbure C. Donc, au lieu de définir un foyer relatif à une droite ν comme point de contact, on peut le définir comme point limite de l'extrémité du segment qui fournit la plus courte distance de cette droite et d'une droite voisine, la dernière correspondant à un déplacement qui satisfait à (73).

Pour terminer cette digression, montrons que *la plus courte distance* $|\Delta\mathbf{c}|$ de deux droites telles qu'il a été dit *est du troisième ordre*, fait pareil à celui trouvé par Bouquet au sujet de génératrices voisines d'une surface développable (espace à trois dimensions) [3].

Considérons en effet sur les deux droites voisines les points respectifs $\mathbf{g} = \mathbf{b} + \left(\tau^1 + \frac{\Delta\tau^1}{2}\right)\mathbf{b}'$ et $\mathbf{g} + \Delta\mathbf{g} = \mathbf{b} + \Delta\mathbf{b} + \left(\tau^1 + \frac{\Delta\tau^1}{2}\right)(\mathbf{b}' + \Delta\mathbf{b}')$. On a

$$\Delta\mathbf{g} = \Delta\mathbf{b} + \left(\tau^1 + \frac{\Delta\tau^1}{2}\right)\Delta\mathbf{b}' = d\mathbf{b} + \tau^1 d\mathbf{b}' + \frac{1}{2}(d^2\mathbf{b} + \tau^1 d^2\mathbf{b}' + d\tau^1 d\mathbf{b}') + \cdots,$$

les termes non écrits étant d'ordre $\geqq 3$. Or $d\mathbf{b} + \tau^1 d\mathbf{b}' = 0$, par hypothèse, et de là, par différentiation, $d^2\mathbf{b} + \tau^1 d^2\mathbf{b}' + d\tau^1 d\mathbf{b}' = 0$. Par conséquent, $\Delta\mathbf{g} = O(\varepsilon^3)$, $|\Delta\mathbf{g}| = O(\varepsilon^3)$ et, à plus forte raison, la plus courte distance $|\Delta\mathbf{c}|$ est $O(\varepsilon^3)$.

---

(1) Pour la notation O, *cf.* la note (4), p. 50.

(2) En effet, le vecteur en question appartient, en raison de la formule (48), à l'élément infinitésimal du bord s^P situé autour du point $\mathbf{b}$ (*cf.* p. 131).

(3) *Cf.* Picard **1**, I, p. 323.

70. Trois cas particuliers. — Nous avons déjà dit dans l'Introduction que la surface Σ^P constituée par les droites ν, ses génératrices, peut être considérée comme l'image de la nappe C^P du cône caractéristique de sommet P par rapport à la surface S, tandis que la caustique de Σ^P peut s'interpréter comme l'image (diffuse) du point P. Nous allons envisager trois cas particuliers, celui d'un plan d'espace et ceux d'un hyperboloïde $\mathbf{a}^2 = \text{const.}$ et d'un cône caractéristique direct, où l'image du cône devient un second cône caractéristique $C^{P'}$. *La caustique se réduit à un point unique*, savoir au sommet P' du cône $C^{P'}$. Ce point peut donc s'interpréter comme l'image *distincte* du point P. Les lignes de courbure, au contraire, sont complètement indéterminées; tout point du bord est un ombilic parfait, toutes les racines de l'équation (57) sont identiques et même indépendantes du point considéré du bord.

Soit d'abord S un plan (à $m-1$ dimensions[1]), d'orientation d'espace. Soit de plus P' le point symétrique (au sens lorentzien) à P par rapport à S, ce qui veut dire que le segment PP' est orthogonal à S et divisé par S en deux segments égaux. Il résulte de la symétrie que l'intersection s^P du plan S et du cône C^P fait aussi partie du cône caractéristique $C^{P'}$ au sommet P'. Les images des génératrices de C^P seront les génératrices de $C^{P'}$, qui se rencontrent toutes au point P', auquel la caustique de $\Sigma^P = C^{P'}$ se réduit dans le cas actuel.

Le cas particulier qui précède peut être considéré comme cas limite de celui qui suit. Soit S l'une des nappes de l'hyperboloïde dont les points $\mathbf{a}$ satisfont à l'équation $\mathbf{a}^2 = A = \text{const.} > 0$. Le fait que le centre se trouve à l'origine ne restreint nullement la généralité. Soit de plus $P = \mathbf{x}$ un point intérieur à celui des demi-cônes caractéristiques $\boldsymbol{\xi}^2 = 0$ qui renferme S. Dès lors, le cône C^P coupe S en des points réels $\mathbf{b}$ à distance finie. Ces points satisfont aux équations $\mathbf{b}^2 = A$ et $(\mathbf{b}-\mathbf{x})^2 = 0$. D'autre part, il est clair que les points $\mathbf{b}$ qui satisfont à deux des équations $\mathbf{b}^2 = A$, $(\mathbf{b}-\mathbf{x})^2 = 0$ et $\mathbf{b}^2 - (\mathbf{b}-\mathbf{x})^2 = 2(\mathbf{b}, \mathbf{x}) - \mathbf{x}^2 = A$, satisfont à la troisième. Or la dernière étant linéaire en $\mathbf{b}$, elle définit un certain plan (à $m-1$ dimensions et orienté dans l'espace), soit Π. L'intersection du cône et de l'hyperboloïde, désignée, comme d'habitude, par s^P est en même temps l'intersection du cône et du plan Π. Par suite, en raison de ce que nous avons vu au n° **66**, l'image de C^P par rapport à S est identique à son image par rapport au plan Π. Alors, d'après nos remarques concernant le premier cas particulier, cette image est un cône caractéristique $C^{P'}$, le point $P' = \mathbf{x}'$ étant l'image de P par rapport à Π. On trouve d'ailleurs aisément

[1] Dans l'Introduction il n'est question que de $m = 4$.

$$\mathbf{x}' = \frac{A}{(\mathbf{x}, \mathbf{x})} \cdot \mathbf{x}\ ^{(1)}.$$

Les considérations précédentes s'étendent encore par un passage à la limite ou autrement au cas où l'hyperboloïde dégénère en un cône caractéristique, soit C. L'image de C^P par rapport au cône C se trouve identique à ce cône lui-même.

Les cônes C^P et $C^{P'}$ ci-dessus se correspondent aussi par *inversion* (2). La définition de cette transformation est dans l'espace lorentzien essentiellement la même que dans l'espace euclidien. Deux points $\mathbf{y}$ et $\mathbf{y}'$ se correspondent par une inversion s'ils sont en ligne droite avec un point fixe $\mathbf{o}$, le centre de l'inversion, et satisfont à une relation de la forme $(\mathbf{y} - \mathbf{o}, \mathbf{y}' - \mathbf{o}) = A = \text{const.}$ Remarquons d'abord que tout point $\mathbf{a}$ de l'hyperboloïde $(\mathbf{a} - \mathbf{o})^2 = A$ se transforme en lui-même. Cela posé, notre énoncé concernant deux cônes qui sont les images l'un de l'autre par rapport à cet hyperboloïde résulte immédiatement du fait qu'*une droite isotrope se transforme par une inversion en une droite isotrope.* Voici la démonstration de ce dernier fait. Plaçons de nouveau l'origine en $\mathbf{o}$. Soient $\mathbf{y}$ et $\mathbf{y}' = \lambda\mathbf{y}$ d'une part, $\mathbf{z}$ et $\mathbf{z}' = \mu\mathbf{z}$ de l'autre, des couples de points qui satisfont à $(\mathbf{y}, \mathbf{y}') = (\mathbf{z}, \mathbf{z}') = A$. On en tire $\lambda\mathbf{y}^2 = \mu\mathbf{z}^2$ et $(\mathbf{y}' - \mathbf{z}')^2 = \lambda^2\mathbf{y}^2 - 2\lambda\mu(\mathbf{y}, \mathbf{z}) + \mu^2\mathbf{z}^2 = \lambda\mu(\mathbf{z}^2 - 2(\mathbf{y}, \mathbf{z}) + \mathbf{y}^2) = \lambda\mu(\mathbf{y} - \mathbf{z})^2$, ce qui met en évidence que les deux carrés scalaires s'annulent simultanément.

71. Résumé. — Pour une vue d'ensemble nous renvoyons en premier lieu à l'Introduction, où l'on trouve un exposé assez complet des résultats du présent chapitre. Il ne nous reste ici qu'à amplifier certains détails.

Signalons d'abord un cas particulier de la formule (60) qui, dans le cas des ondes sphériques, fournit une solution de caractère géométrique du problème de Cauchy, posé pour l'équation $\Delta u = 0$ et pour une surface S d'orientation d'espace qui porte les données. Ce cas particulier est relatif aux trois types de surfaces, traités dans le numéro précédent, à savoir, plan d'espace, hyperboloïde $(\mathbf{a} - \mathbf{o})^2 = \text{const.} > 0$ et cône caractéristique (limite de surfaces d'espace). Dans le cas de telles surfaces, les deux foyers $\mathbf{f}_1$ et $\mathbf{f}_2$, distincts et variables en général, coïncident et restent identiques à un point fixe P', l'image de P par rapport à la surface considérée. En posant $R_{PP'} = R_0$, on tombe sur la *formule de solution*

(1) Pour éviter toute confusion, on a écrit au dénominateur $(\mathbf{x}, \mathbf{x})$ au lieu de $\mathbf{x}^2$.

(2) On peut choisir comme centre de l'inversion un point arbitraire de la droite PP', excepté les points P et P' eux-mêmes.

$$u(P) = \frac{1}{\pi} \int_{s^P} \left\{ \frac{u}{R_0} - \frac{d_\nu u}{d_\nu R} \right\} ds. \tag{74}$$

On voit donc que (60) comprend la formule (61bis) du n° **63**, relative au cas où S est un cône caractéristique (direct). En effet, cette dernière formule est exactement de la forme (74).

La formule (60) comprend aussi la solution de Poisson, relative au cas classique où la surface S est le plan $x_1 = 0$. Il est, en effet, aisé de vérifier que (74) peut dans ce dernier cas être transformé dans la formule de Poisson, écrite sous sa forme habituelle (¹) (ou réciproquement).

Une grande partie des notions géométriques que nous avons utilisées dans ce chapitre et qui rentrent dans la solution (60) — *image*, *foyer*, *caustique* — appartiennent à l'optique géométrique. Cependant il est ici question de l'*optique géométrique de l'espace-temps*, où les rayons de lumière sont fournies par les droites isotropes. Nous reviendrons ailleurs sur les liens qui existent entre les conceptions de cette dernière optique et celles de l'optique géométrique ordinaire et montrerons comment on arrive dans cette voie aux théorèmes classiques, tels la règle des sinus et le théorème de Malus.

CHAPITRE VI.

Applications à la théorie relativiste de l'électron.

72. Notations. — Les problèmes que nous allons traiter dans ce chapitre se rapportent — tout comme le problème principal du chapitre précédent — à l'espace lorentzien à *quatre* dimensions, *espace-temps ou univers de la relativité restreinte*. Nous adopterons ici le langage habituel de cette dernière théorie et les notations accoutumées du *calcul tensoriel*. Pour simplifier l'écriture, *nous supprimerons* aussi *les caractères gras* utilisés plus haut pour représenter les vecteurs.

La vitesse de la lumière est égalée à 1, les quatre coordonnées d'un point sont désignées par x^0, x^1, x^2, x^3; x^0 correspondant au temps, x^1, x^2, x^3 à l'espace à trois dimensions. Le carré scalaire (x, x) du rayon vecteur x mené de l'origine au point x est mesuré par la forme métrique

(¹) *Cf.* p. ex. Goursat III (1915), p. 103.

(1) $$(x, x) = x^{0 2} - x^{1 2} - x^{2 2} - x^{3 2} = \sum_{j,k=0}^{3} g_{jk} x^j x^k;$$

$$g_{00} = 1, \; g_{11} = g_{22} = g_{33} = -1; \; g_{jk} = 0, \; j \neq k.$$

En adoptant encore la convention due à Einstein, à savoir que, sauf indication contraire, *une sommation est à effectuer pour tout couple d'indices identiques (l'un supérieur, l'autre inférieur)*, on peut aussi écrire

(1bis) $$(x, x) = g_{jk} x^j x^k.$$

Les *composantes contravariantes* d'un vecteur U sont désignées par U^j et son *carré scalaire* (U, U) est mesuré par la forme analogue à (1)

(2) $$(U, U) = g_{jk} U^j U^k.$$

V étant un second vecteur aux composantes contravariantes V^j, le *produit scalaire* des vecteurs U et V est donné par

(3) $$(U, V) = g_{jk} U^j V^k.$$

Les *composantes covariantes* U_j du vecteur U sont définies par $U_j = g_{jk} U^k$ (c'est-à-dire que $U_0 = U^0$, $U_j = -U^j$, $j \neq 0$). Pour garder la symétrie, on pose en même temps $U^j = g^{jk} U_k$, où, dans le cas particulier actuel, $g^{jk} = g_{jk}$. Cela étant, le carré scalaire de U et le produit scalaire de U et de V peuvent s'écrire

(2bis) $$(U, U) = g_{jk} U^j U^k = g^{jk} U_j U_k = U^j U_j;$$

(3bis) $$(U, V) = g_{jk} U^j V^k = g^{jk} U_j V_k = U^j V_j = U_j V^j.$$

L'*opérateur des ondes* est désigné par $\square$, c'est-à-dire que

(4) $$\square = \frac{\partial^2}{\partial x^{0 2}} - \frac{\partial^2}{\partial x^{1 2}} - \frac{\partial^2}{\partial x^{2 2}} - \frac{\partial^2}{\partial x^{3 2}} = g^{jk} \frac{\partial^2}{\partial x^j \partial x^k}.$$

Les indices j et k figurant dans les opérations $\partial / \partial x^j$ et $\partial^2 / \partial x^j \partial x^k$ sont à considérer comme indices inférieurs (covariants).

Introduisons encore les coordonnées covariantes x_j du point x (composantes covariantes du rayon vecteur)

(5) $$x_j = g_{jk} x^k, \text{ et inversement } x^j = g^{jk} x_k.$$

On a

(6) $$\frac{\partial}{\partial x_j} = g^{jk} \frac{\partial}{\partial x^k}$$

et dès lors, l'opérateur des ondes peut s'écrire

$$\square = \frac{\partial^2}{\partial x_j \partial x^j}. \tag{4^{bis}}$$

L'indice j figurant dans l'opération $\partial / \partial x_j$ est à considérer comme indice supérieur (contravariant).

73. Les équations de Maxwell. — Le tenseur antisymétrique F qui représente le *champ électromagnétique* admet les composantes covariantes $F_{jk} = -F_{kj}$ et les composantes contravariantes $F^{jk} = g^{jr} g^{ks} F_{rs} = -F^{kj}$ (c'est-à-dire que $F^{0k} = -F_{0k}$ et $F^{jk} = F_{jk}$, j, $k \neq 0$). On introduit aussi les composantes contravariantes s^j du *vecteur-courant* s, vecteur-densité de charge et de courant électrique. Cela posé, on arrange les équations de Maxwell en deux systèmes, chacun à quatre équations,

$$\frac{\partial F^{jk}}{\partial x^k} = -s^j, \qquad j = 0, 1, 2, 3, \tag{7}$$

$$\bigcirc \frac{\partial F_{kl}}{\partial x^j} = \frac{\partial F_{kl}}{\partial x^j} + \frac{\partial F_{lj}}{\partial x^k} + \frac{\partial F_{jk}}{\partial x^l} = 0, \text{(1)} \tag{8}$$

où, dans les équations du dernier système, j, k, l forment une combinaison quelconque à trois éléments tirés de 0, 1, 2, 3. (L'interversion de deux indices, j et k par exemple, ne produit qu'un changement de signe dans chaque terme.) Le premier système donne, grâce à l'antisymétrie des F^{jk}, l'*équation de continuité*

$$\operatorname{div} s = \frac{\partial s^j}{\partial x^j} = 0. \tag{9}$$

à laquelle le vecteur-courant, supposé donné d'avance, doit satisfaire *a priori*.

Pour intégrer ces équations, on introduit d'habitude le *potentiel vecteur* A, aux composantes covariantes et contravariantes A_j et A^j respectivement. En posant

$$F_{jk} = \operatorname{rot}_{jk} A = \frac{\partial A_k}{\partial x^j} - \frac{\partial A_j}{\partial x^k}, \tag{10}$$

le système (8) se trouve satisfait. La formule (6) donne immédiatement

$$F^{jk} = \frac{\partial A^k}{\partial x_j} - \frac{\partial A^j}{\partial x_k}$$

---

(1) Le premier membre de (7) est connu sous le nom de *divergence tensorielle* du tenseur F. Nous nous permettons de proposer ici une notation abrégée pour la sommation cyclique figurant dans (8).

et par suite

(11) $$\frac{\partial F^{jk}}{\partial x^k} = \frac{\partial}{\partial x_j}\left(\frac{\partial A^k}{\partial x^k}\right) - \frac{\partial^2 A^j}{\partial x_k\, \partial x^k}.$$

En posant encore la condition supplémentaire

(12) $$\operatorname{div} A = \frac{\partial A^k}{\partial x^k} = 0,$$

il ne reste dans le second membre de (11) que le dernier terme qui, d'après (4bis), n'est autre chose que $-\Box A^j$. Alors le système (7) se réduit au système

(13) $$\Box A^j = s^j \quad \text{ou} \quad \Box A = s,$$

qui s'intègre par des potentiels retardés.

74. Les potentiels retardés. — Il est instructif de faire rentrer le problème d'intégration des équations (13) dans le cadre de la méthode développée au Chap. IV. Ceci se trouve tout fait dans les travaux de M. Fremberg(1) dont nous nous permettrons de reproduire, dans les n^{os} **74** et **75**, quelques parties essentielles. Nous nous bornerons comme lui à des domaines illimités tels qu'on les trouve caractérisés au début du n° **15**, ce qui fait qu'il n y aura pas de conditions aux limites. On admet que les composantes s^j, qui sont supposées données, comme nous venons de le dire, satisfont aux conditions indiquées aux n^{os} **15**, **26** et **40**, et en outre à l'équation (9). La solution des équations est, selon la formule (2) du n° **44**, donnée par $A(x) = I^2 s(x)$. Pour arriver à l'expression de $I^2 s(x)$, on forme d'abord

(14) $$A^\alpha(x) = I^\alpha s(x) = \frac{1}{H_4(\alpha)} \int_{D^x} s(y)\, r_{xy}^{\alpha-4}\, dy, \qquad dy = dy^0\, dy^1\, dy^2\, dy^3,$$

où (*cf.* formule (20) du n° **16**)

(15) $$H_4(\alpha) = 2^{\alpha-1} \cdot \pi \cdot \Gamma\left(\frac{\alpha}{2}\right) \cdot \Gamma\left(\frac{\alpha-2}{2}\right).$$

L'intégrale converge pour $\alpha > 2$ et est définie, pour $\alpha \leqq 2$, par prolongement analytique

En désignant l'origine par o, (14) peut s'écrire

$$A^\alpha(x) = \frac{1}{H_4(\alpha)} \int_{D^0} s(x+y)\, r_{oy}^{\alpha-4}\, dy,$$

(1) Fremberg **2** et **3**.

d'où découle, en raison de (9),

$$\operatorname{div} A^\alpha(x) = \frac{1}{H_4(\alpha)} \int_{D^0} \operatorname{div}_x s(x+y) r_{0y}^{\alpha-4}\, dy = 0.$$

Les composantes covariantes

$$F^\alpha_{jk} = \frac{\partial A^\alpha_k}{\partial x^j} - \frac{\partial A^\alpha_j}{\partial x^k} \tag{16}$$

du tenseur antisymétrique F^α et ses composantes contravariantes satisfont aux équations

$$\frac{\partial F^{\alpha jk}}{\partial x^k} = - s^{\alpha j}, \tag{17}$$

$$\bigcirc \frac{\partial F^\alpha_{kl}}{\partial x^j} = 0, \tag{18}$$

où

$$s^{\alpha j} = \Box A^{\alpha j}.$$

En passant des composantes aux vecteurs eux-mêmes, et en tenant compte du résultat établi au n° **42**, on trouve

$$s^\alpha(x) = \Box A^\alpha(x) = A^{\alpha-2}(x) = I^{\alpha-2} s(x). \tag{19}$$

Pour $\alpha = 2$, $I^{\alpha-2} s(x)$ devient $I^0 s(x) = s(x)$, tandis que $A^\alpha(x)$ devient la solution cherchée de (13), à savoir $A(x) = I^2 s(x)$, c'est-à-dire, d'après la formule (30bis) du n° **57**, le *potentiel retardé*

$$A(x) = \frac{1}{4\pi} \int_{\Omega_3} s(x^0 - \varrho, y^1, y^2, y^3) \frac{dy^1\, dy^2\, dy^3}{\varrho},$$

où

$$\varrho = \left(\sum_1^3 (x^k - y^k)^2 \right)^{\frac{1}{2}},$$

et l'intégrale est étendue à l'espace entier à trois dimensions.

75. Le champ produit par un électron. — Après ces préparatifs, considérons dans notre espace-temps une particule, soit un électron, représentée par un point matériel $z = (z^0, z^1, z^2, z^3)$ qui décrit une *ligne d'univers* L. Le déplacement infinitésimal dz sera, d'après les conceptions relativistes (toute vitesse mécanique est inférieure à la vitesse de la lumière), un vecteur de temps. Le carré de l'élément d'arc s'exprime par la relation $d\tau^2 = (dz, dz) > 0$. Le champ produit par un tel électron est donné par la rotation d'un potentiel vecteur, connu sous le nom du potentiel de Liénard-Wiechert. Les considérations qui suivent conduisent à de nouvelles expressions pour le potentiel et le champ en question.

Les coordonnées z^j des points de L pourront être exprimées comme fonctions d'un paramètre, p. ex. de l'arc τ de L, temps propre de l'électron, mesuré à partir d'un point fixe arbitraire et croissant avec z^0. On suppose que ces fonctions sont suffisamment dérivables (*cf.* le n° **27**). On suppose aussi que la ligne L s'étend à l'infini du côté des z^0 négatifs et y admet une asymptote qui est une droite de temps. Les valeurs correspondantes de z^0 et de τ sont égales à $-\infty$ toutes les deux, et on désignera encore par $-\infty$ le point infini lui-même. Pour des raisons de commodité, nous supposons que L s'étend à l'infini aussi du côté des z^0 positifs, mais nous ne faisons aucune hypothèse sur son allure à $+\infty$. Enfin on suppose que la charge portée par le point z est constamment égale à 1.

On voit que le vecteur-courant est nul partout, excepté sur la ligne L où il devient infini. Soit maintenant x un point qui n'est pas situé sur la ligne L, et formons, en analogie avec (14), le potentiel vecteur (retardé) d'ordre α

$$(20) \qquad A^\alpha(x) = \frac{1}{H_4(\alpha)} \int_{L^x} r_{xz}^{\alpha-4}\, dz, \text{ [1]}$$

où $r_{xz}^2 = (z-x,\ z-x)$ et L^x est la partie de L intérieure au cône rétrograde D^x au sommet x. L'arc L^x est donc formé de ceux des points de L qui sont «sous onde» par rapport à x (*cf.* le n° **55**). La limite inférieure de l'intégrale est le point $-\infty$, tandis que sa limite supérieure est le point d'intersection — unique, comme on le verra plus loin, — de la ligne L avec la nappe C^x du cône D^x.

On peut, par des formules analogues à (16) et (17), définir le champ et le courant d'ordre α et arriver à des équations de la forme (17) et (19). Une difficulté, particulière au cas actuel, provient du point $-\infty$. L'intégrale (20) converge pour $2 < \alpha < 3$, la limitation supérieure venant justement du point en question. La convergence peut, au moyen d'une intégration par parties, être étendue à l'intervalle $2 < \alpha < 5$, ce qui suffit à tous les besoins. [2]

---

[1] On ne confondra pas la notation dz qui figure dans l'intégrale (20) et qui signifie un vecteur, avec la notation $dy = dy^0\, dy^1\, dy^2\, dy^3$ figurant dans l'intégrale (14).

[2] *Voir* Fremberg **2**, p. 83.

On déduit du potentiel vecteur A^α le champ électromagnétique F^α et le vecteur-courant s^α par un mécanisme analogue à celui développé au numéro précédent. Nous renvoyons pour les détails aux travaux cités de M. Fremberg.

Nous y renvoyons également pour l'application de notre méthode à l'élimination de certaines quantités infinies qui interviennent dans la théorie classique de l'électron (telle son énergie propre), problème résolu antérieurement par MM. Wentzel et Dirac[1] en associant les potentiels avancés aux potentiels retardés. Le fait que cette dernière méthode et la nôtre sont entièrement équivalentes dans la théorie *classique* de l'électron a été établi par M. Ma.[2] Notre méthode a aussi été appliquée en théorie *quantique* par MM. Gustafson, Nilsson et d'autres.[3] Il ressort de leurs recherches que notre méthode conduit à l'élimination de certaines quantités infinies, mais non pas de toutes, du moins dans l'état actuel de l'Électrodynamique Quantique.

76. Potentiel de Liénard-Wiechert. — Nous avons dit plus haut que la ligne L et la nappe C^x du cône rétrograde ont un point d'intersection unique. Nous allons examiner cette question d'une façon plus détaillée. Posons, en effet, $R_{xz} = r^2_{xz} = (z - x, z - x)$. On a alors, en raison de la formule (23bis), où τ désigne le temps propre, $dR/d\tau < 0$. On voit par là que R est une fonction continue et décroissante de τ et qu'il en est donc de même de τ considéré comme fonction de R. Cela fait voir que toute nappe rétrograde d'hyperboloïde $z^0 - x^0 < 0$ et $(z - x, z - x) = R > 0$, où R est fixé, coupe L en un seul point $z(x, R)$. Il en est ainsi en particulier de la nappe C^x du cône rétrograde qui coupe L au point $z(x, 0)$ qui est «juste en onde» par rapport à x. Notons encore que R décroît de $+\infty$ jusqu'à 0, quand le point z va de $-\infty$ à $z(x, 0)$.

En résumé, *à tout point x et à toute valeur $\geqq 0$ de R correspond un point unique z de L qui pourra être désigné par $z(x, R)$* et dont les coordonnées peuvent être considérées comme fonctions de cinq variables, savoir des quatre coordonnées x^j de x et de R, et peuvent être différentiées par rapport à ces variables.

Fixons d'abord le point x. Le point $z(x, R)$ sera alors une fonction de R et l'intégrale peut s'écrire

$$(21) \qquad A^\alpha(x) = \frac{1}{H_4(\alpha)} \int_{+\infty}^{0} R^{\frac{\alpha-4}{2}} \frac{\partial z}{\partial R} dR = -\frac{1}{H_4(\alpha)} \int_{0}^{\infty} R^{\frac{\alpha-2}{2}-1} \frac{\partial z}{\partial R} dR.$$

---

(1) *Voir* p. ex. DIRAC **1**. (2) MA **1**. (3) Pour la Bibliographie *voir* NILSSON **1**.

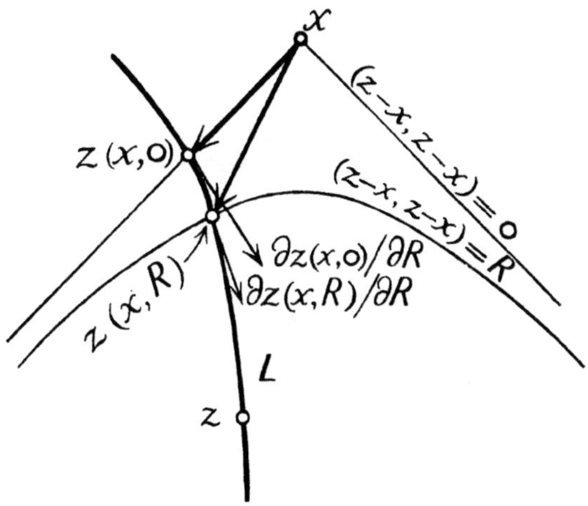

Fig. 12. La ligne d'univers L.

La valeur limite de cette intégrale pour $\alpha = 2$ *nous fournira le potentiel de Liénard-Wiechert.* On a $H_4(\alpha) = 2^{\alpha-1}\pi\,\Gamma\left(\frac{\alpha}{2}\right)\cdot\Gamma\left(\frac{\alpha-2}{2}\right)$. Pour $\alpha = 2$, le facteur $2^{\alpha-1}\pi\,\Gamma\left(\frac{\alpha}{2}\right)$ est $= 2\pi$, et alors on tire de la formule (13bis) du nº **5** que *le potentiel cherché devient*

$$A(x) = A^2(x) = -\frac{1}{2\pi}\left(\frac{\partial z}{\partial R}\right)_{(R=0)}. \tag{22}$$

La forme usuelle de ce potentiel [1] est, si l'on tient compte de notre façon de normer la charge,

$$\frac{1}{4\pi}\,\frac{\frac{dz}{d\tau}}{\left(x - z,\ \frac{dz}{d\tau}\right)}, \tag{23}$$

où τ signifie le temps propre de l'électron. L'identité des deux expressions ressort du calcul suivant. On a manifestement

$$\frac{\partial z}{\partial R} = \frac{dz}{d\tau} : \frac{dR}{d\tau}.$$

D'autre part, $R = (z - x,\ z - x)$ nous donne

[1] DIRAC **1**, p. 163.

$$(23^{bis})\qquad \frac{dR}{d\tau}=2\left(z-x,\ \frac{dz}{d\tau}\right)=-2\left(x-z,\ \frac{dz}{d\tau}\right),$$

par où l'identité des expressions (22) et (23) se trouve vérifiée. Il est clair que dans (23) τ peut être remplacé par un paramètre quelconque.

77. Le point $z(x, R)$ en fonction de x et de R. — On montre facilement que

$$(24)\qquad \frac{\partial z}{\partial x^j}=2(z_j-x_j)\frac{\partial z}{\partial R}.$$

En effet, on a d'abord l'équation vectorielle

$$(25)\qquad \frac{\partial z}{\partial x^j}=\lambda_j\cdot\frac{\partial z}{\partial R},$$

où λ_j est un facteur de proportionnalité. D'autre part, en tenant compte de l'identité $R=(z-x,\ z-x)$ et du fait que R et x sont considérés comme des variables indépendantes, on obtient

$$(26)\qquad 1=2\left(z-x,\ \frac{\partial(z-x)}{\partial R}\right)=2\left(z-x,\ \frac{\partial z}{\partial R}\right),$$

$$(27)\qquad 0=2\left(z-x,\ \frac{\partial(z-x)}{\partial x^j}\right)=2\left(z-x,\ \frac{\partial z}{\partial x^j}\right)-2\left(z-x,\ \frac{\partial x}{\partial x_j}\right)$$

ou

$$(28)\qquad \left(z-x,\ \frac{\partial z}{\partial x^j}\right)=\left(z-x,\ \frac{\partial x}{\partial x^j}\right)=(z_k-x_k)\cdot\frac{\partial x^k}{\partial x^j}$$

$$=(z_k-x_k)\cdot\delta_j^k=z_j-x_j;\quad \delta_j^k=\begin{cases}0, & j\neq k\\ 1, & j=k\end{cases}.$$

En formant le produit scalaire des deux membres de (25) avec $z-x$, on tire des formules (26) et (28) $\lambda_j=2(z_j-x_j)$. Il résulte de (24)

$$(29)\qquad \operatorname{div}_x z=\frac{\partial z^j}{\partial x^j}=2(z_j-x_j)\frac{\partial z^j}{\partial R}$$

$$=2(z_j-x_j)\frac{\partial(z^j-x^j)}{\partial R}=\frac{\partial}{\partial R}(z-x,\ z-x)=\frac{\partial R}{\partial R}=1,$$

ce qui nous donne la *formule importante*

$$(30)\qquad \operatorname{div}_x\frac{\partial z}{\partial R}=\frac{\partial}{\partial R}\operatorname{div}_x z=0.$$

On a donc en particulier pour le potentiel $A(x)$ de Liénard-Wiechert, à l'extérieur de la ligne d'univers,

$$\operatorname{div}_x A(x) = 0. \tag{31}$$

On peut adopter un point de vue plus général. Posons, pour abréger,

$$\frac{\partial z(x, R)}{\partial R} = v(x, R). \tag{32}$$

Les fonctions $g(R)$ et $G(R)$ étant définies dans l'intervalle $0 \leqq R < \infty$, on peut former les intégrales

$$V(x) = \int_0^\infty g(R)\, v(x, R)\, dR = -\int_{L^x} g(R)\, dz$$

et

$$W(x) = \int_0^\infty v(x, R)\, d\, G(R).$$

Dans des conditions assez générales, que nous n'avons pas l'intention de préciser, ces intégrales ont un sens, et on peut conclure de la relation $\operatorname{div}_x v(x, R) = 0$ que $\operatorname{div}_x V(x) = 0$ et $\operatorname{div}_x W(x) = 0$. La fonction $G(R) = 0$ pour $R = 0$ et $= -1/2\pi$ pour $R > 0$ fournit le potentiel de Liénard-Wiechert (formule (22)), tandis que $g(R) = -R^{\frac{\alpha-4}{2}} / H_4(\alpha)$ fournit le potentiel d'ordre α défini par les formules (20) ou (21).

Il s'ensuit de la formule (35) du prochain numéro que

$$\Box_x V(x) = 4 \int_0^\infty g(R)\, R\, \frac{\partial^2 v(x, R)}{\partial R^2}\, dR.$$

Si Rg et $d(Rg)/dR$ s'annulent pour $R = 0$ et ∞, on obtient, au moyen de deux intégrations par parties,

$$\Box_x V(x) = 4 \int_0^\infty \frac{d^2\{R\, g(R)\}}{dR^2}\, v(x, R)\, dR.$$

78. Le champ électromagnétique. — Nous obtenons

$$F_{jk} = \operatorname{rot}_{jk} A = \frac{\partial A_k}{\partial x^j} - \frac{\partial A_j}{\partial x^k} = -\frac{1}{2\pi}\left(\frac{\partial}{\partial x^j}\frac{\partial z_k}{\partial R} - \frac{\partial}{\partial x^k}\frac{\partial z_j}{\partial R}\right)$$

$$= -\frac{1}{2\pi}\frac{\partial}{\partial R}\left(\frac{\partial z_k}{\partial x^j} - \frac{\partial z_j}{\partial x^k}\right) = -\frac{1}{2\pi}\frac{\partial}{\partial R}\left\{(z_j - x_j)\frac{\partial z_k}{\partial R} - (z_k - x_k)\frac{\partial z_j}{\partial R}\right\}$$

$$= -\frac{1}{2\pi}\left\{(z_j - x_j)\frac{\partial^2 z_k}{\partial R^2} - (z_k - x_k)\frac{\partial^2 z_j}{\partial R^2}\right\} = -\frac{1}{2\pi}\left\{(z_j - x_j)\frac{\partial v_k}{\partial R} - (z_k - x_k)\frac{\partial v_j}{\partial R}\right\},$$

ou, en utilisant une notation usuelle (¹),

$$(33) \qquad F = \frac{1}{2\pi}\left[x - z, \frac{\partial v}{\partial R}\right] = \frac{1}{2\pi}\frac{\partial}{\partial R}[x - z, v] = \frac{\partial}{\partial R}[z - x, A].$$

On trouve facilement que pour tout point x (extérieur à la ligne L) F satisfait aux équations (7) et (8) de Maxwell, où dans (7) on a identiquement $s^j = 0$. D'abord il est clair que ce champ, qui est une rotation, satisfait au second système de Maxwell. En observant que, d'après (30), $\operatorname{div}_x v(x, 0) = 0$, on voit, dès qu'on a montré que $\Box_x v(x, 0) = 0$, que F, en raison de (11), satisfait aussi au premier système. Nous allons calculer, dans un ordre d'idées plus général, $\Box_x v(x, R)$ et $\Box_x z(x, R)$ et *établir les formules suivantes:*

$$(34) \qquad \Box_x z = 4\left(R\frac{\partial^2 z}{\partial R^2} - \frac{\partial z}{\partial R}\right) = 4\left(R\frac{\partial v}{\partial R} - v\right)$$

et

$$(35) \qquad \Box_x v = 4R\frac{\partial^2 v}{\partial R^2}.$$

De la formule (24) on tire

$$(36) \qquad \frac{\partial z}{\partial x_j} = 2(z^j - x^j)\frac{\partial z}{\partial R}.$$

Observons ensuite que, d'après la formule (21), $\operatorname{div}_x z = 1$ et que $\operatorname{div}_x x = 4$, ce qui donne $\operatorname{div}_x (z - x) = -3$. Il vient, en utilisant (36),

$$\Box_x z = \frac{\partial^2 z}{\partial x^j \partial x_j} = 2\frac{\partial}{\partial x^j}\left\{(z^j - x^j)\frac{\partial z}{\partial R}\right\} = 2\frac{\partial}{\partial x^j}(z^j - x^j)\frac{\partial z}{\partial R} + 2(z^j - x^j)\frac{\partial^2 z}{\partial x^j \partial R}$$

$$= -6\frac{\partial z}{\partial R} + 2(z^j - x^j)\frac{\partial}{\partial R}\frac{\partial z}{\partial x^j}.$$

(¹) Nous désignons par $[U, V]$ (produit vectoriel des vecteurs U et V) le bivecteur (tenseur antisymétrique du second ordre) dont les composantes covariantes et contravariantes sont respectivement $[U, V]_{jk} = U_j V_k - U_k V_j$ et $[U, V]^{jk} = U^j V^k - U^k V^j$.

De plus

$$(z^j - x^j)\frac{\partial}{\partial R}\frac{\partial z}{\partial x^j} = (z^j - x^j)\frac{\partial}{\partial R}\left\{2(z_j - x_j)\frac{\partial z}{\partial R}\right\} = 2(z^j - x^j)(z_j - x_j)\frac{\partial^2 z}{\partial R^2}$$

$$+ 2\frac{\partial z}{\partial R}(z^j - x^j)\frac{\partial}{\partial R}(z_j - x_j) = 2R\frac{\partial^2 z}{\partial R^2} + \frac{\partial z}{\partial R}\frac{\partial R}{\partial R} = 2R\frac{\partial^2 z}{\partial R^2} + \frac{\partial z}{\partial R}.$$

Cela, porté dans la formule précédente, fournit la formule (34) qu'on voulait démontrer. Ensuite on obtient

$$\Box_x \frac{\partial z}{\partial R} = \frac{\partial}{\partial R}\Box_x z = 4\frac{\partial}{\partial R}\left(R\frac{\partial^2 z}{\partial R^2} - \frac{\partial z}{\partial R}\right)$$

$$= 4\left(\frac{\partial^2 z}{\partial R^2} + R\frac{\partial^3 z}{\partial R^3} - \frac{\partial^2 z}{\partial R^2}\right) = 4R\frac{\partial^3 z}{\partial R^3},$$

ce qui n'est que la formule (35).

En particulier, si $R = 0$, $\Box_x v(x, 0) = 0$, ce qui établit le premier système de Maxwell.

79. Nouvelle expression du potentiel de Liénard-Wiechert. — Si l'on pose dans la formule (34) $R = 0$, on obtient

$$\left(\frac{\partial z(x, R)}{\partial R}\right)_{(R=0)} = v(x, 0) = -\frac{1}{4}\Box_x z(x, 0), \tag{37}$$

d'où il s'ensuit pour le potentiel de Liénard-Wiechert (formule (22))

$$A(x) = \frac{1}{8\pi}\Box_x z(x, 0). \tag{38}$$

Pour éliminer l'origine des coordonnées, on peut utiliser le fait que $\Box_x x = 0$, et on aura

$$A(x) = \frac{1}{8\pi}\Box_x (z - x), \tag{39}$$

où, pour abréger, on a désigné le point retardé $z(x, 0)$ par z.

En réalité, *je suis arrivé à ces formules assez remarquables par une voie plus intuitive*, qui se rattache à l'ordre d'idées du n° **52**. Là on a obtenu la solution du problème de Cauchy pour toute dimension paire $m = 2l$ par la formule

$$I_*^2 = \Delta^{l-1} I_*^m.$$

A cette formule correspondrait, dans le cas actuel où $m = 4$ et $l - 1 = 1$,

$$A(x) = A^2(x) = \Box_x A^4(x). \tag{40}$$

Or la formule (20) fournit (en raison de $H_4(4) = 8\pi$)

(41) $$A^4(x) = \frac{1}{8\pi}\int_{L^x} dz = \frac{1}{8\pi} z(x, 0),$$

ce qui, en vertu de (40), conduit de nouveau au résultat (38).

A vrai dire, dans le dernier terme de (41) on a supprimé un vecteur constant soustractif relatif au point $-\infty$, vecteur qui pour comble de malheur est infini lui-même. Or ce vecteur disparaît de la formule (40), puisqu'on y applique à $A^4(x)$ l'opération $\Box_x$. En réalité, il disparaît aussi de l'expression de $A^4(x)$, si on la calcule d'une manière rigoureuse, c'est-à-dire par prolongement analytique.

Pour le champ électromagnétique F, la nouvelle expression (37) de v donne

$$F = \frac{1}{2\pi}\left[x - z, \frac{\partial v}{\partial R}\right] = -\frac{1}{8\pi}\left[x - z, \frac{\partial}{\partial R}\Box z\right] = -\frac{1}{8\pi}\left[x - z, \Box\frac{\partial}{\partial R}z\right]$$
$$= -\frac{1}{8\pi}\left[x - z, \Box\left(-\frac{1}{4}\Box z\right)\right] = \frac{1}{32\pi}[x - z, \Box^2 z]$$

ou enfin, puisque $\Box^2 x = 0$,

$$F = -\frac{1}{32\pi}[z - x, \Box^2(z - x)].$$

Le bivecteur F est, comme on l'a vu, la rotation du potentiel vecteur A et satisfait aux équations de Maxwell (7) et (8), le vecteur-courant correspondant s étant nul sauf sur la ligne L. Formons aussi la rotation G du rayon vecteur $z - x$.

$$G_{jk} = \mathrm{rot}_{jk}(z - x) = \mathrm{rot}_{jk}\, z = \frac{\partial z_k}{\partial x^j} - \frac{\partial z_j}{\partial x^k} = 2(z_j - x_j)\frac{\partial z_k}{\partial R} - 2(z_k - x_k)\frac{\partial z_j}{\partial R}$$
$$= -\frac{1}{2}\{(z_j - x_j)\Box z_k - (z_k - x_k)\Box z_j\},$$

c'est-à-dire que G s'écrit

$$G = -\frac{1}{2}[z - x, \Box(z - x)].$$

Ce dernier bivecteur se trouve sur le même plan que la fonction de Hertz[1] donnée par le grand physicien à propos du problème d'un dipôle vibrant. Le potentiel vecteur A est, à un facteur constant près, la divergence tensorielle de G, c'est-à-dire le vecteur-courant correspondant au champ G, champ qui satisfait aux équations

$$\frac{\partial G^{jk}}{\partial x^k} = -\Box(z^j - x^j) = -8\pi A^j \quad \text{et} \quad \bigcirc\frac{\partial G_{kl}}{\partial x^j} = 0.$$

(1) *Voir* p. ex. ABRAHAM **1**, II, p. 62.

La relation entre le champ G et le champ électromagnétique F est simplement

$$\square\, G = -8\pi F.$$

En résumé, le rayon vecteur $z-x$, allant de x au point retardé z, définit en tout point x de l'espace-temps un *champ vectoriel*, constitué d'ailleurs par des vecteurs isotropes. Le champ s'annule sur L, qui en est une ligne singulière. Ce champ vectoriel se trouve à la tête d'une *hiérarchie de champs successifs*, alternativement vectoriels et bivectoriels, $z-x$, G, A, F, *tous engendrés par la ligne d'univers de l'électron* et qui se déduisent l'un de l'autre par les deux opérations fondamentales, rotation et divergence tensorielle, ou, en sautant un champ sur deux, par l'opérateur des ondes. Cette succession de champs s'arrête à la ligne L elle-même, qui représente le vecteur-courant dégénéré, inhérent au problème.

CHAPITRE VII.

L'équation des ondes dans les espaces de Riemann.

80. Énoncé du problème. — Le but de ce dernier chapitre est d'étendre la méthode d'intégration exposée au Chapitre IV aux équations linéaires du type hyperbolique normal *à coefficients variables.* Nous nous restreignons ici au cas où le premier membre de l'équation en question est fourni par le *paramètre différentiel du second ordre* $\Delta_2 u$ de Beltrami[1], attaché à une certaine forme différentielle quadratique. L'extension au cas général est immédiate[2].

Soit donc

$$ds^2 = g_{jk}\, dx^j\, dx^k, \text{[3]}$$

la forme métrique attachée à un espace de Riemann à m dimensions. On se trouve dans le cas de *signature lorentzienne* ou dans le cas hyperbolique normal, suivant la terminologie de M. Hadamard, si cette forme, transformée en somme de carrés, admet *un* carré positif et $m-1$ carrés négatifs. En désignant par g la valeur absolue du déterminant $|g_{jk}| \neq 0$ et par g^{jk} les coefficients de la forme réciproque, le paramètre différentiel du second ordre correspondant $\Delta_2 u$ s'écrit

---

[1] On trouve pourtant au n° **98** un aperçu succinct de l'équation légèrement plus générale dont le premier membre est donné par $\Delta_2 u + \lambda^2 u$.

[2] Un problème d'extension plus général et plus difficile, concernant certains *systèmes* particuliers d'équations, qui pourtant renferment le cas d'une *seule* équation de forme générale, se trouve résolu dans la Thèse de mon ancien élève, M. Malmheden; *voir* MALMHEDEN **1**, Chapt. VII—IX.

[3] Nous appliquons dans ce chapitre l'écriture tensorielle; *cf.* p. 147 et le n° **81**.

$$\Delta_2 u = \frac{1}{\sqrt{g}} \frac{\partial}{\partial x^j}\left(\sqrt{g}\, g^{jk} \frac{\partial u}{\partial x^k}\right).\ (^1)$$

C'est ce dernier opérateur Δ_2 qui dans le cas d'un espace de Riemann généralise *l'opérateur des ondes* de l'espace lorentzien.

Un problème difficile, résolu par M. Hadamard d'une manière ingénieuse, est celui de construire la solution élémentaire dans le cas de coefficients variables. Ces solutions sont pourtant de caractères très différents (²) suivant que la dimension est impaire ou paire et il en est de même des méthodes d'intégration. La méthode de la partie finie n'est applicable que pour les dimensions impaires, et pour les dimensions paires il faut recourir à d'autres méthodes, dont l'une est la méthode de descente. Nous allons montrer que l'introduction d'un paramètre convenable permet encore de donner la solution du problème de Cauchy sous une forme qui est indépendante de la parité de la dimension de l'espace. Chemin faisant, nous allons développer un procédé d'intégration fractionnaire tout analogue à celui donné plus haut.

81. Généralités. — Nous commençons par rappeler quelques notions du calcul tensoriel (³). Nous appelons tout système de valeurs admises par les coordonnées $x^1, \ldots, x^m$ *point*, désigné par x, d'une variété abstraite ou d'un *espace général* à m dimensions. Les considérations qui suivent visent l'espace entier ou, de préférence, une région de cet espace.

Soit $x^j = x^j(x^{1'}, \ldots, x^{m'}) = x^j(x^{k'})$ une transformation ponctuelle qui établit une correspondance biunivoque entre les systèmes de coordonnées $x^1, \ldots, x^m$ et $x^{1'}, \ldots, x^{m'}$, de sorte qu'on a aussi $x^{k'} = x^{k'}(x^j)$. On suppose de plus que les fonctions qui établissent la correspondance sont *suffisamment dérivables* (⁴) et que d'ailleurs il en est de même de toute fonction introduite dans la suite.

On a les relations équivalentes

$$(1)\qquad \frac{\partial x^j}{\partial x^{k'}} \frac{\partial x^{k'}}{\partial x^l} = \delta^j_l = \begin{cases} 1, & j = l \\ 0, & j \neq l \end{cases} \quad \text{et} \quad \frac{\partial x^{j'}}{\partial x^k} \frac{\partial x^k}{\partial x^{l'}} = \delta^{j'}_{l'}\ (^5),$$

qui expriment que les matrices fonctionnelles $(\partial x^j / \partial x^{k'})$ et $(\partial x^{j'} / \partial x^k)$ sont réciproques. Il en résulte pour les déterminants fonctionnels

---

(¹) Les détails nécessaires concernant les espaces de Riemann, l'expression $\Delta_2 u$ etc. seront donnés plus loin (n° **81**).

(²) *Cf.* en particulier le n° **90** du présent ouvrage.

(³) Pour un exposé plus complet, on pourra consulter p. ex. le livre excellent de M. L. Brillouin; Brillouin **1**.

(⁴) *Cf.* le n° **27**.

(⁵) Pour certaines conventions du calcul tensoriel *cf.* p. 147.

(2) $$\frac{d(x)}{d(x')} = \frac{d(x^1, \ldots, x^m)}{d(x^{1'}, \ldots, x^{m'})} \quad \text{et} \quad \frac{d(x')}{d(x)} = \frac{d(x^{1'}, \ldots, x^{m'})}{d(x^1, \ldots, x^m)}$$

que

(3) $$\frac{d(x)}{d(x')} \cdot \frac{d(x')}{d(x)} = 1 .$$

Pour les différentielles des coordonnées, il vient

(4) $$d x^j = \frac{\partial x^j}{\partial x^{k'}} d x^{k'} \quad \text{ou} \quad d x^{j'} = \frac{\partial x^{j'}}{\partial x^k} d x^k .$$

Pour les dérivées partielles d'un *scalaire* u, on a

(5) $$\frac{\partial u}{\partial x^j} = \frac{\partial u}{\partial x^{k'}} \frac{\partial x^{k'}}{\partial x^j} \quad \text{ou} \quad \frac{\partial u}{\partial x^{j'}} = \frac{\partial u}{\partial x^k} \frac{\partial x^k}{\partial x^{j'}} .$$

Les quantités $U^1, \ldots, U^m$ attachées au point x forment les composantes d'un *vecteur contravariant* s'ils obéissent à la même loi de transformation que les différentielles (4). Les composantes $V_1, \ldots, V_m$ d'un *vecteur covariant* auront par contre à obéir à la même loi que les dérivées partielles (5). Ou, d'une manière explicite,

(6) $$U^j = \frac{\partial x^j}{\partial x^{k'}} U^{k'} \quad \text{ou} \quad U^{j'} = \frac{\partial x^{j'}}{\partial x^k} U^k$$

et

(7) $$V_j = \frac{\partial x^{k'}}{\partial x^j} V_{k'} \quad \text{ou} \quad V_{j'} = \frac{\partial x^k}{\partial x^{j'}} V_k .$$

Le produit scalaire

(8) $$(U, V) = U^j V_j$$

est invariant par les transformations (6) et (7) qui sont donc *contragrédientes.*

En particulier, les différentielles $d x^j$ forment les composantes contravariantes du vecteur $d x$ qui exprime le *déplacement infinitésimal* du point x, tandis que les dérivées partielles $\frac{\partial u}{\partial x^j} = \text{grad}_j\, u$ forment les composantes covariantes du *gradient* de u. Le produit scalaire $\frac{\partial u}{\partial x^j} d x^j$ des susdits vecteurs est un invariant, la *différentielle totale* $d u$ de la fonction u.

Au point où nous sommes, les vecteurs contravariants et covariants sont des objets géométriques distincts. La situation change complètement au moment où, par l'introduction d'une métrique, l'espace général devient un *espace de Riemann.* Dans un tel espace, on pourra assigner à chaque vecteur des composantes contravariantes *et* covariantes.

On aura un *espace de Riemann*, dès qu'on aura défini le carré de la distance ds de deux points infiniments voisins x et $x + dx$ par une relation telle que

(9) $$ds^2 = g_{jk}\, dx^j\, dx^k, \quad g_{jk} = g_{kj}.$$

La *forme métrique* $g_{jk}\, dx^j\, dx^k$ est donc une forme différentielle quadratique, dont les coefficients dépendent en général des coordonnées du point x, $g_{jk} = g_{jk}(x)$. Cette forme peut jusqu'à nouvel ordre être définie ou indéfinie. On suppose pourtant que le déterminant $|g_{jk}| \neq 0$ et on désigne, comme plus haut, par g la valeur absolue de ce déterminant. Les coefficients g^{jk} de la forme réciproque sont symétriques eux aussi, $g^{jk} = g^{kj}$, et satisfont aux relations

(10) $$g_{jk}\, g^{kl} = g^{lk}\, g_{kj} = \delta^l_j.$$

Du caractère invariant de

$$ds^2 = g_{jk}\, dx^j\, dx^k = g_{j'k'}\, dx^{j'}\, dx^{k'},$$

il résulte que

(11) $$g_{jk} = g_{p'q'} \frac{\partial x^{p'}}{\partial x^j} \frac{\partial x^{q'}}{\partial x^k} \quad \text{ou} \quad g_{j'k'} = g_{pq} \frac{\partial x^p}{\partial x^{j'}} \frac{\partial x^q}{\partial x^{k'}}.$$

Ces dernières relations et (10) entraînent

(12) $$g^{jk} = g^{p'q'} \frac{\partial x^j}{\partial x^{p'}} \frac{\partial x^k}{\partial x^{q'}} \quad \text{ou} \quad g^{j'k'} = g^{pq} \frac{\partial x^{j'}}{\partial x^p} \frac{\partial x^{k'}}{\partial x^q}.$$

Le vecteur U de composantes contravariantes U^j aura, *par définition*, les composantes covariantes

(13) $$U_j = g_{jk}\, U^k \quad (\text{d'où } U^j = g^{jk}\, U_k),$$

tandis que le vecteur V de composantes covariantes V_j aura les composantes contravariantes

(14) $$V^j = g^{jk}\, V_k \quad (\text{d'où } V_j = g_{jk}\, V^k).$$

On pourra dès lors parler du *carré scalaire* d'un vecteur U quelconque

(15) $$(U, U) = g_{jk}\, U^j\, U^k = g^{jk}\, U_j\, U_k = U^j\, U_j, \text{[1]}$$

et du *produit scalaire* de deux vecteurs U et V quelconques

(16) $$(U, V) = g_{jk}\, U^j\, V^k = g^{jk}\, U_j\, V_k = U^j\, V_j = U_j\, V^j.$$

---

[1] Vecteur de temps, vecteur d'espace, vecteur isotrope et vecteur unitaire se définissent comme à la page 27.

Deux vecteurs U et V dont le produit scalaire $(U, V) = 0$ sont dits *orthogonaux.*

Les *composantes contravariantes du gradient* d'une fonction u sont

$$\text{grad}^j u = g^{jk} \,\text{grad}_k u = g^{jk} \frac{\partial u}{\partial x^k}. \tag{17}$$

Le carré scalaire du gradient de u et le produit scalaire des gradients de u et de v sont les *paramètres différentiels du premier ordre* de Beltrami

$$\Delta_1 u = g^{jk} \frac{\partial u}{\partial x^j} \frac{\partial u}{\partial x^k} \tag{18}$$

et

$$\Delta_1 (u, v) = \Delta_1 (v, u) = g^{jk} \frac{\partial u}{\partial x^j} \frac{\partial v}{\partial x^k}. \tag{19}$$

Les expressions (15)—(16) et (18)—(19) sont *invariantes* par rapport à une transformation de coordonnées arbitraire en vertu des relations données plus haut.

Le déterminant $|g_{jk}|$ se transforme en raison de (11) d'après les formules

$$|g_{jk}| = |g_{j'k'}| \left(\frac{d(x')}{d(x)}\right)^2 \quad \text{ou} \quad |g_{j'k'}| = |g_{jk}| \left(\frac{d(x)}{d(x')}\right)^2 \tag{20}$$

et la racine carrée de sa valeur absolue g d'après les formules

$$\sqrt{g} = \sqrt{g'} \left|\frac{d(x')}{d(x)}\right| \quad \text{ou} \quad \sqrt{g'} = \sqrt{g} \left|\frac{d(x)}{d(x')}\right|. \tag{20^{bis}}$$

Pour l'*élément de volume invariant* de l'espace de Riemann, la formule (56) du n° **21**^{bis} nous fournit l'expression

$$d\Omega = \sqrt{g} \begin{vmatrix} d_1 x^1, & \dots, & d_1 x^m \\ \cdot & \cdot\ \cdot\ \cdot & \cdot \\ d_m x^1, & \dots, & d_m x^m \end{vmatrix} \tag{21}$$

ou, en particulier,

$$d\Omega = \sqrt{g} \cdot dx^1 \dots dx^m, \tag{21^{bis}}$$

la première expression générale étant le volume du parallélépipède construit sur les déplacements infinitésimaux $d_1 x, d_2 x, \dots, d_m x$, tandis que dans le cas particulier (21bis) une seule des coordonnées varie à chaque déplacement.

Nous aurons plus loin besoin de la *divergence* d'un vecteur A. On arrive, soit par différentiation covariante, soit par l'application du théorème de Gauss (n° **92**), à la définition

$$\operatorname{div} A = \frac{1}{\sqrt{g}} \frac{\partial}{\partial x^j}(\sqrt{g}\, A^j). \tag{22}$$

Cette définition peut aussi être rendue légitime en observant qu'elle se réduit à la définition ordinaire, lorsque les coefficients de la forme métrique sont constants, et en vérifiant que (31) *est invariant par rapport à tout changement de coordonnées.* Ce dernier fait découle aisément du théorème suivant de Jacobi, théorème qui d'ailleurs sert de base à la théorie des multiplicateurs d'un système d'équations différentielles.

Les mineurs D_γ^j *relatifs aux éléments* $\partial f^\gamma / \partial x^j$ *d'un déterminant fonctionnel* $d(f^1, \ldots, f^\gamma, \ldots, f^m)/d(x^1, \ldots, x^j, \ldots, x^m)$ *satisfont à la relation*

$$\sum_{j=1}^{m} \frac{\partial D_\gamma^j}{\partial x^j} = 0.\ (^1)$$

Or les matrices $(\partial x^{k'}/\partial x^j)$ et $(\partial x^k/\partial x^{j'})$ étant réciproques, le mineur relatif à l'élément $\partial x^{k'}/\partial x^j$ du déterminant $d(x')/d(x)$ est égal à $\dfrac{d(x')}{d(x)} \cdot \dfrac{\partial x^j}{\partial x^{k'}}$. Par suite, le théorème de Jacobi peut aussi s'écrire

$$\frac{\partial}{\partial x^j}\left(\frac{d(x')}{d(x)} \frac{\partial x^j}{\partial x^{k'}}\right) = 0.$$

Eu égard à cette relation et aux relations (6) et (20bis), il vient

$$\frac{1}{\sqrt{g}} \frac{\partial}{\partial x^j}(\sqrt{g}\, A^j) = \frac{1}{\sqrt{g}} \frac{\partial}{\partial x^j}\left(\sqrt{g'} \cdot \left|\frac{d(x')}{d(x)}\right| \frac{\partial x^j}{\partial x^{k'}} \cdot A^{k'}\right)$$

$$= 0 + \frac{1}{\sqrt{g}} \left|\frac{d(x')}{d(x)}\right| \frac{\partial}{\partial x^j}(\sqrt{g'}\, A^{k'}) \frac{\partial x^j}{\partial x^{k'}} = \frac{1}{\sqrt{g'}} \frac{\partial}{\partial x^{k'}}(\sqrt{g'}\, A^{k'}).\ (^2)$$

En égalant les composantes A^j qui figurent dans (22) aux composantes contravariantes de grad u, données par la formule (17), on conclut *l'invariance du paramètre différentiel du second ordre*

---

(1) *Cf.* p. ex. GOURSAT II (1911), p. 310.

(2) La vérification ci-dessus rappelle beaucoup celle donnée par M. Brillouin (**1**, p. 71—72) qui pourtant ne s'appuie pas de manière explicite sur le théorème de Jacobi. Les deux vérifications s'appliquent aussi à l'expression plus générale $\tau^{-1} \partial(\sigma A^k)/\partial x^k$, où σ et τ sont des densités scalaires ($\sigma = \sigma' \cdot d(x')/d(x)$, $\tau = \tau' \cdot d(x')/d(x)$). Cette dernière expression ne contient aucun élément métrique, la propriété d'invariance envisagée appartient donc en réalité à la géométrie vectorielle affine, point de vue que M. Brillouin fait soigneusement ressortir.

(23) $$\Delta_2 u = \Delta u = \text{div grad } u = \frac{1}{\sqrt{g}} \frac{\partial}{\partial x^j}\left(\sqrt{g}\, g^{jk} \frac{\partial u}{\partial x^k}\right),$$

dont l'étude forme un des principaux objets du présent chapitre. Faisons remarquer que *c'est uniquement de la notation abrégée Δu que nous nous servirons dans la suite.*

Les vecteurs sont les *tenseurs* du premier ordre. Nous pouvons nous dispenser de parler longuement des tenseurs d'ordre supérieur au premier et nous contenter de signaler un tenseur du second ordre, qui d'ailleurs est d'une importance capitale, le *tenseur métrique* de l'espace. Ses *composantes covariantes* et *contravariantes* sont les quantités g_{jk} et g^{jk} respectivement, avec leurs formules de transformation (11) et (12), tandis que les quantités invariantes δ^j_k sont ses *composantes mixtes.*

82. Géodésiques. — Pour la commodité du lecteur, nous donnons ici un aperçu des géodésiques d'un espace de Riemann, lignes qui vont jouer un rôle fondamental dans la suite. Il nous faudra tenir compte du fait que *la forme métrique de l'espace est indéfinie dans les applications que nous avons en vue.*

Envisageons des courbes C, données sous la forme $x^j = x^j(\sigma)$, σ étant un paramètre, et le *problème de variation* relatif à l'intégrale

(24) $$\int_C L(x, \dot{x})\, d\sigma, \quad \text{où } \dot{x}^j = \frac{d x^j}{d\sigma} \quad \text{et} \quad L(x, \dot{x}) = g_{jk}(x)\, \dot{x}^j \dot{x}^k.$$

Les *extrémales* de ce problème sont données par les *équations d'Euler*

(25) $$\frac{\partial L}{\partial x^j} - \frac{d}{d\sigma}\frac{\partial L}{\partial \dot{x}^j} = 0, \quad j = 1, \ldots, m.$$

De ces équations et du fait que L est homogène et du second degré dans les $\dot{x}^k$, on tire

$$\frac{dL}{d\sigma} = \frac{\partial L}{\partial x^k}\dot{x}^k + \frac{\partial L}{\partial \dot{x}^k}\ddot{x}^k = \frac{d}{d\sigma}\frac{\partial L}{\partial \dot{x}^k}\dot{x}^k + \frac{\partial L}{\partial \dot{x}^k}\ddot{x}^k = \frac{d}{d\sigma}\left(\frac{\partial L}{\partial \dot{x}^k}\dot{x}^k\right) = \frac{d}{d\sigma}(2L) = 2\frac{dL}{d\sigma}.$$

C'est-à-dire que $dL/d\sigma = 0$, ou, en d'autres termes, *L est constant le long de toute extrémale C. Si cette valeur constante est différente de zéro, on peut diviser* (25) *par* $2|L|^{\frac{1}{2}}$, ce qui fournit les équations

(26) $$\frac{\partial |L|^{\frac{1}{2}}}{\partial x^j} - \frac{d}{d\sigma}\frac{\partial |L|^{\frac{1}{2}}}{\partial \dot{x}^j} = 0,$$

qui ne sont que les équations d'Euler relatives au problème de variation

$$\int\limits_C |L|^{\frac{1}{2}}\, d\sigma = \int\limits_C |g_{jk}\, dx^j\, dx^k|^{\frac{1}{2}}. \tag{27}$$

La dernière intégrale exprime la longueur de l'arc de la courbe à laquelle cette intégrale est étendue, les équations (26) fournissent donc les *géodésiques* de l'espace de Riemann envisagé. En conséquence, *les équations* (25) *fournissent, elles aussi, les géodésiques en question* (¹).

Or le système (25) admet aussi des solutions pour lesquelles la valeur constante de L est nulle. *Par extension*, ces lignes seront encore considérées comme des géodésiques, bien que le système (26) cesse d'avoir un sens pour une telle ligne.

Le fait que la valeur de L est constante sur toute courbe fournie par le système (25) implique que, sur une géodésique arbitraire, la forme métrique $g_{jk}\, dx^j\, dx^k$ garde son signe ou s'annule identiquement. Suivant que ce signe est *positif*, *négatif*, *nul*, on pourra parler de *géodésiques de temps*, *géodésiques d'espace*, *géodésiques isotropes* ou *de longueur nulle*. Ces dernières sont encore appelées *bicaractéristiques* suivant la terminologie de M. Hadamard.

Il y a une différence essentielle entre le rôle que le paramètre σ joue dans le système d'équations (26) d'un côté et dans le système (25) de l'autre. En effet, il est clair par la seconde forme de l'intégrale (27) que, dans la première forme de cette intégrale, on peut remplacer σ par un paramètre tout à fait arbitraire, soit τ, en posant $\dot{x}^j = dx^j/d\tau$. Cela étant, les équations (26) ne seront pas changées si on y introduit τ au lieu de σ, ce qui d'ailleurs est aisé à vérifier par un calcul direct. En d'autres termes, les équations (26) (et des conditions aux limites appropriées) déterminent la géodésique C, mais non σ, qui reste un paramètre arbitraire. *Il en est tout autrement des équations* (25), *où la coordination entre les points x de la courbe et le paramètre σ est déterminée à une substitution linéaire $\sigma \to a\sigma + b$ près* (¹), résultat extrêmement important pour la suite.

D'abord il est clair que les équations (25) restent valides si l'on y remplace σ par $a\sigma + b$. Réciproquement, si la relation $L = \text{const.} \neq 0$ a lieu en deux représentations paramétriques (²) relatives à deux paramètres σ et τ, elle implique que $d\tau/d\sigma$ doit être constant.

(¹) Quant aux rapports entre les équations (25) et (26) et les principes de la Dynamique, *voir* HADAMARD 1, p. 117—118.

(²) Évidemment, les valeurs constantes en question seront, en général, différentes.

Le raisonnement ci-dessus tombe en défaut pour une géodésique de longueur nulle, la valeur constante de L étant nulle sur une telle géodésique. Voici une démonstration qui embrasse tous les cas.

Admettons qu'une extrémale satisfait en même temps aux équations (25) relatives soit au paramètre σ, soit au paramètre τ. En posant $d\sigma/d\tau = A$, on a $d/d\tau = A \cdot d/d\sigma$, tandis que les quantités L, $\partial L/\partial x^j$, $\partial L/\partial \dot{x}^j$, relatives à σ, deviennent respectivement $A^2 L$, $A^2 \partial L/\partial x^j$, $A\,\partial L/\partial \dot{x}^j$, par où les équations relatives à τ s'écrivent

$$0 = A^2 \frac{\partial L}{\partial x^j} - A\frac{d}{d\sigma}\left(A\frac{\partial L}{\partial \dot{x}^j}\right) = A^2\left(\frac{\partial L}{\partial x^j} - \frac{d}{d\sigma}\frac{\partial L}{\partial \dot{x}^j}\right) - A\frac{dA}{d\sigma}\frac{\partial L}{\partial \dot{x}^j}.$$

La comparaison avec (25) donne que le dernier terme s'annule pour tout indice j. Il en résulte que $dA/d\sigma = 0$, car $\partial L/\partial \dot{x}^j = 0$, pour tout j, donnerait, en vertu de $|g_{jk}| \neq 0$, $\dot{x}^k = 0$, pour tout k. Par conséquent $d\sigma/d\tau =$ const. ou $d\tau/d\sigma$ = const.

Les paramètres dans lesquels les équations des lignes géodésiques admettent la forme (25) et qui, comme nous venons de le voir, sont déterminés, à une substitution linéaire près, seront appelés *paramètres admissibles. Seuls ces derniers paramètres seront utilisés dans la suite.*

83. Équations canoniques. — Le système de m équations du second ordre (25) à m inconnues x^j est équivalent à un système de $2m$ équations du premier ordre à $2m$ inconnues. Pour arriver aux *équations hamiltoniennes* ou *canoniques*, on introduit à côté des x^j comme nouvelles inconnues les quantités (composantes covariantes du vecteur tangent de composantes contravariantes $\dot{x}^j$)

$$p_j = g_{jk}\dot{x}^k = \frac{1}{2}\frac{\partial L}{\partial \dot{x}^j}, \tag{28}$$

relations équivalentes à

$$\dot{x}^k = \frac{1}{2} g^{kj}\frac{\partial L}{\partial \dot{x}^j} = g^{kj} p_j. \tag{29}$$

Introduisons encore la *fonction hamiltonienne*

$$H(x, p) = g^{jk} p_j p_k. \tag{30}$$

On tire des formules qui précèdent

$$\frac{1}{2}\frac{\partial H}{\partial p_j} = g^{jk} p_k = \dot{x}^j = \frac{dx^j}{d\sigma} \tag{31}$$

et

(32) $$H(x, p) = \frac{1}{2}\frac{\partial H}{\partial p_j} p_j = \dot{x}^j p_j = \frac{1}{2}\dot{x}^j \frac{\partial L}{\partial \dot{x}^j} = L(x, \dot{x}).$$

On a encore les relations

(33) $$\frac{\partial H(x, p)}{\partial x^j} = -\frac{\partial L(x, \dot{x})}{\partial x^j},$$

qui résultent des propriétés connues de la transformation de Legendre, si l'on observe que $L + H = 2\dot{x}^k p_k$, mais qui peuvent aussi se vérifier par un calcul direct. En effet

$$\frac{\partial L(x, \dot{x})}{\partial x^j} + \frac{\partial H(x, p)}{\partial x^j} = \frac{\partial g_{kl}}{\partial x^j}\dot{x}^k \dot{x}^l + \frac{\partial g^{kr}}{\partial x^j} p_k p_r = \frac{\partial g_{kl}}{\partial x^j} g^{kr} p_r \dot{x}^l + \frac{\partial g^{kr}}{\partial x^j} g_{kl} \dot{x}^l p_r$$

$$= \frac{\partial (g_{kl} g^{kr})}{\partial x^j}\dot{x}^l p_r = \frac{\partial \delta_l^r}{\partial x^j}\dot{x}^l p_r = 0.$$

Des relations (31), (25), (28) et (33), on tire les *équations canoniques des géodésiques*

(34) $$\frac{dx^j}{d\sigma} = \frac{1}{2}\frac{\partial H}{\partial p_j}, \quad \frac{dp_j}{d\sigma} = -\frac{1}{2}\frac{\partial H}{\partial x^j}.$$

Il résulte immédiatement de ces équations que $dH = 0$, c'est-à-dire que $H = \text{const.}$ le long de toute géodésique, en accord parfait avec le fait identique établi plus haut pour $L (= H)$.

84. La distance géodésique. — Dans la théorie habituelle des géodésiques, où la forme métrique est définie positive, c'est l'arc de la géodésique qui figure comme paramètre dans les deux systèmes équivalents (25) et (34). Cet arc, compté à partir d'un point arbitraire x_0 de la géodésique, se définit par l'intégrale, prise suivant la géodésique, $s = \int_{x_0}^{x} (g_{jk}\, dx^j\, dx^k)^{\frac{1}{2}}$. Avec s comme paramètre, on a évidemment $L(x, \dot{x}) = 1 = \text{const.}$, ce qui fait voir que s est un paramètre admissible. Dans le cas d'une forme indéfinie, on pourra faire de même pour les géodésiques de temps, sur lesquelles $g_{jk}\, dx^j\, dx^k > 0$; c'est précisément ce que nous ferons au n° **89**. L'emploi de l'arc comme paramètre est encore possible sur les géodésiques d'espace, où $g_{jk}\, dx^j\, dx^k < 0$, si l'on utilise l'expression $\int |g_{jk}\, dx^j\, dx^k|^{\frac{1}{2}}$, comme nous l'avons fait incidemment plus haut, ou si l'on admet des valeurs imaginaires (pures) pour l'arc. Cependant dans le cas des géodésiques isotropes, dont la longueur d'arc est identiquement nulle, on doit garder le paramètre sous sa forme générale.

Plus important que cette question de paramétrisation moyennant l'arc est le problème de la *distance géodésique* $s(x; y)$ de deux points x et y, donnée (dans certaines conditions qu'on va encore préciser) par l'arc de la géodésique qui joint ces points. Pour éviter des quantités imaginaires, nous préférons considérer le carré $s^2(x; y)$ de cette distance, qui aura toujours une valeur réelle, cette valeur étant, suivant les cas, positive, nulle ou négative.

Il ressort des théorèmes d'existence des systèmes d'équations différentielles que *les géodésiques menées en un point fixe* x_0 *sont parfaitement déterminées par les rapports mutuels entre les différentielles* dx_0^k. On peut aussi montrer que, le point x se trouvant dans un voisinage suffisamment restreint [1] du point x_0, il existe dans ce voisinage *une géodésique et une seule qui joint les deux points* et que cette géodésique varie d'une manière continue et même suffisamment dérivable avec ces points [2].

Tout cela étant bien compris, le carré $s^2(x; y)$ de l'arc géodésique, précisé tout à l'heure, joignant deux points suffisamment voisins x et y sera une quantité bien déterminée. Cette quantité, qui est d'ailleurs *symétrique* en x et en y, sera appelée *carré de la distance géodésique des points* x *et* y. Nous posons encore, avec M. Hadamard, $s^2(x; y) = \Gamma(x; y)$ et nous allons déduire *l'équation aux dérivées partielles du premier ordre* que $\Gamma(x; y)$ vérifie en fonction soit de x, soit de y.

On a vu plus haut que la fonction L admet une valeur constante le long d'une géodésique arbitraire et ainsi, en particulier, le long de la géodésique joignant x et y, valeur qui pourtant dépend du choix du paramètre σ, la substitution $\sigma \to A\sigma + B$ amenant le changement $L \to A^{-2} L$. Nous allons montrer que la valeur constante en question devient égale à $\Gamma(x; y)$ si l'on normalise le paramètre σ de façon qu'il admet les valeurs respectives 0 et 1 aux points x et y. Observons d'abord que le paramètre est *entièrement déterminé* par ces conditions; nous l'appellerons *paramètre normalisé relativement à* x *et* y. Désignons ensuite par z le point courant de la géodésique et par σ_z la valeur de σ qui,

---

(1) On entend ici par voisinage d'un certain point l'intérieur d'une surface simple fermée entourant ce point. On dira, dans les circonstances figurant au texte, que le point x est suffisamment voisin du point x_0 et, dans des circonstances analogues, que deux points x et y sont suffisamment voisins.

(2) Pour des énoncés plus précis *voir* Hadamard **1**, p. 120—123, 156—159, 408; Hadamard **8**, p. 106—123 (pour p. 121 *cf.* les *Errata*), 497—511; A. Szücs **1**, p. 380—391.

dans une paramétrisation arbitraire([1]), correspond à z. Admettons, pour commencer, qu'il s'agit d'une géodésique de temps. Si l'on pose d'abord σ_z égal à l'arc $s(x; z)$ (compté à partir de x) de cette géodésique, on obtient évidemment $L = 1$. Pour le paramètre normalisé $\tau_z = s(x; z) : s(x; y)$, il vient par conséquent $L = s^2(x; y) = \Gamma(x; y)$. Pour des géodésiques d'espace, le choix $\sigma_z = |s(x; z)|$ donne $L = -1$, et le paramètre normalisé $\tau_z = |s(x; z)| : |s(x; y)|$ fournit encore $L = -|s(x; y)|^2 = s^2(x; y) = \Gamma(x; y)$. Enfin, si la géodésique est isotrope, on a encore $L = \Gamma(x; y)$, puisque ces valeurs sont nulles, toutes les deux. *En résumé*

$$L(z, \dot{z}) = \Gamma(x; y) \tag{35}$$

où z est un point arbitraire de la géodésique qui joint x et y, $\dot{z}^k = dz^k/d\tau$ et τ est le paramètre normalisé admettant les valeurs respectives 0 *et* 1 *en x et en y.*

Puisque $L(z, \dot{z}) = \Gamma(x; y)$, on a aussi $\Gamma(x; y) = \int_0^1 L(z, \dot{z})\, d\tau$, l'intégrale étant prise le long de la géodésique C joignant x et y. Soient maintenant $x + \delta x$ et $y + \delta y$ deux points infiniment voisins, le premier de x, le second de y, et $C + \delta C$ la géodésique qui les joint. On a cette fois-ci $\Gamma(x + \delta x; y + \delta y) = \int_0^1 L(z, \dot{z})\, d\tau$, l'intégrale étant prise le long de $C + \delta C$ et le paramètre τ normalisé relativement à $x + \delta x$ et $y + \delta y$. Il vient dès lors, grâce à la *formule aux limites*([2]) du Calcul des Variations,

$$\delta \Gamma(x; y) = \Gamma(x + \delta x; y + \delta y) - \Gamma(x, y) = \left[\frac{\partial L(z, \dot{z})}{\partial \dot{z}^k}\, \delta z^k\right]_{(z=x)}^{(z=y)}, \tag{36}$$

où c'est le paramètre τ, normalisé relativement à x et y, qui figure dans les dérivées $\dot{z}^k = dz^k/d\tau$. Cette formule nous donne

$$\frac{\partial \Gamma(x; y)}{\partial x^k} = -\frac{\partial L(x, \dot{x})}{\partial \dot{x}^k} \quad \text{et} \quad \frac{\partial \Gamma(x, y)}{\partial y^k} = \frac{\partial L(y, \dot{y})}{\partial \dot{y}^k}, \tag{37}$$

et dès lors, en raison de ce qui précède et des formules (35), (32) et (28),

$$\Gamma(x; y) = L(x^j, \dot{x}^k) = H(x^j, p_k) = \frac{1}{4} H\left(x^j, \frac{\partial \Gamma(x; y)}{\partial x^k}\right), \tag{38}$$

ou, écrit d'une manière détaillée,

([1]) Il n'est évidemment jamais question que de paramètres *admissibles*; *voir* plus haut, p. 167.

([2]) *Voir* p. ex. HADAMARD **3**, p. 144 et 152; GOURSAT III (1915), p. 567.

$$(38^{bis}) \qquad \Delta_1 \Gamma(x;\, y) = g^{jk}(x) \frac{\partial \Gamma(x;\, y)}{\partial x^j} \frac{\partial \Gamma(x;\, y)}{\partial x^k} = 4\, \Gamma(x;\, y), \text{[1]}$$

équation d'importance capitale, qui subsiste évidemment si l'on y échange x et y.

Remarquons dans un autre ordre d'idées que, dans le cas où les géodésiques C et $C + \delta C$ sont *isotropes*, on peut encore tirer de la formule (36) la relation importante

$$(39) \qquad \frac{\partial L(y,\, \dot y)}{\partial \dot y^k} \delta y^k - \frac{\partial L(x,\, \dot x)}{\partial \dot x^k} \delta x^k = 0 \quad \text{ou} \quad g_{jk}(y)\, \dot y^j\, \delta y^k - g_{jk}(x)\, \dot x^j\, \delta x^k = 0,$$

valable pour un paramètre admissible *arbitraire*.

En effet, les égalités $\Gamma(x;\, y) = \Gamma(x + \delta x;\, y + \delta y) = 0$ entraînent $\delta \Gamma(x;\, y) = 0$, ce qui, de son côté, entraîne (39) pour le paramètre normalisé τ. Le passage au paramètre σ arbitraire est immédiat, car les termes de (39) se multiplient par le même facteur A^{-1} quand $\sigma = A\tau + B$.

En vue d'une application prochaine, nous transcrivons encore les formules (35) et (37), relatives au paramètre normalisé τ, pour le cas d'un paramètre arbitraire σ. Ce dernier paramètre admettant les valeurs respectives σ_x et σ_y aux points x et y, on aura $\tau = (\sigma - \sigma_x) : (\sigma_y - \sigma_x)$ et par conséquent

$$(35^{bis}) \qquad \Gamma(x;\, y) = (\sigma_y - \sigma_x)^2\, L(z,\, \dot z).$$

On aura de même

$$(37^{bis}) \qquad \frac{\partial \Gamma(x;\, y)}{\partial x^k} = (\sigma_x - \sigma_y) \frac{\partial L(x,\, \dot x)}{\partial \dot x^k} \quad \text{et} \quad \frac{\partial \Gamma(x;\, y)}{\partial y^k} = (\sigma_y - \sigma_x) \frac{\partial L(y,\, \dot y)}{\partial \dot y^k}.$$

85. Coordonnées normales. — Considérons une région $\mathcal{R}$, un point x_0 intérieur à $\mathcal{R}$ et supposons que tout autre point x intérieur à $\mathcal{R}$ peut être joint à x_0 par une géodésique parfaitement déterminée, située dans $\mathcal{R}$ et variant d'une manière continue et suffisamment dérivable avec x [2]. On reprend le paramètre σ des équations (25) et *on pose* $\sigma = 0$ *en* x_0, par où la paramétrisation de chaque géodésique est déterminée à une substitution $\sigma \to A\sigma$ près. Envisageons un point x de $\mathcal{R}$, la géodésique C qui le joint au point x_0, et formons les quantités

---

[1] On peut encore procéder de la manière suivante

$$\Gamma = L = \frac{1}{2} \dot x^k \frac{\partial L}{\partial \dot x^k} = \frac{1}{4} g^{jk} \frac{\partial L}{\partial \dot x^j} \frac{\partial L}{\partial \dot x^k} = \frac{1}{4} g^{jk} \frac{\partial \Gamma}{\partial x^j} \frac{\partial \Gamma}{\partial x^k}.$$

[2] *Voir* HADAMARD **1**, p. 156—159.

$$\xi^k = \sigma_x \left(\frac{d\,x^k}{d\,\sigma}\right)_0 = \sigma_x\,\dot{x}_0^k, \tag{40}$$

où les paramètres directeurs sont ceux de la géodésique C au point x_0, σ_x est la valeur que le paramètre σ admet au point x, le tout étant d'abord calculé au moyen d'une paramétrisation fixée. Or les quantités ξ^k, étant invariantes par toute substitution $\sigma \to A\,\sigma$, ne dépendent pas du choix du paramètre, c'est-à dire qu'*elles ne dépendent que du point* x ou, si l'on veut, des coordonnées x^k de ce point. Inversement, *les quantités* ξ^k *déterminent le point* x. Il suffit de montrer que deux points x différents fournissent deux systèmes de valeurs ξ^k différents. Supposons donc un instant, pour fixer les idées, que σ est normalisé sur chaque géodésique de façon qu'il admet la valeur 1 au point d'intersection de la géodésique avec une certaine surface fermée, intérieure à $\mathcal{R}$, entourant x_0 et coupée en un seul point par chaque géodésique issue de x_0. Cela étant, considérons deux points x différents. Si ces points appartiennent à des géodésiques tangentes en x_0 à des droites différentes, les rapports mutuels des ξ^k seront différents. Pour deux points situés sur deux arcs opposés admettant la même tangente, les signes des ξ^k seront opposés. Enfin pour deux points différents situés sur le même arc géodésique, on aura les mêmes paramètres directeurs, mais des valeurs différentes de σ.

En résumé, *le point* x *et le système des quantités* ξ^k *sont mutuellement déterminés* l'un par l'autre, ce qui fait que ces quantités pourront être introduites comme de nouvelles coordonnées de x. Ces coordonnées, inventées par Riemann, sont appelées *coordonnées normales* (¹) du point x (autour de x_0). Le point x_0, dont toutes les coordonnées deviennent nulles, est appelé l'*origine* du système de coordonnées en question.

Le système de coordonnées (40) est tangent en x_0 au système original des x^k. On a visiblement $\partial x^j / \partial \xi^k = \delta_k^j$ au point x_0, les coefficients g_{jk} de la forme métrique resteront donc inaltérés en ce point (*cf.* la formule (11)). Si l'on part d'un second système de coordonnées $x^{k'}$, le système de coordonnées normales $\xi^{k'}$, déduit de ce dernier système de la manière ci-dessus, est lié au système ξ^k par une transformation linéaire à coefficients constants

$$\xi^{j'} = \left(\frac{\partial x^{j'}}{\partial x^k}\right)_0 \xi^k. \tag{41}$$

(¹) Ces coordonnées sont des fonctions suffisamment dérivables des anciennes coordonnées et *vice versa*; *voir* la note précédente.

Les géodésiques issues du point x_0 sont dans un système de coordonnées normales, prises comme coordonnées affines, représentées par des lignes droites issues de l'origine de ces coordonnées.

Pour des géodésiques de temps, il est souvent avantageux de spécialiser σ et d'écrire, tout comme dans le cas d'une forme métrique définie positive,

$$\xi^k = s\left(\frac{dx^k}{ds}\right)_0 = s\,\dot{x}_0^k, \tag{42}$$

s désignant la longueur de l'arc géodésique (x_0, x).

L'emploi de coordonnées normales amène de grandes simplifications dans beaucoup de calculs. Pour commencer, nous modifions légèrement nos notations. Le point x_0, origine des coordonnées normales, sera désigné par o, le point x de coordonnées normales ξ^k par ξ; le point courant z de la géodésique sera ici ζ. Nous écrirons respectivement $g_{jk}(o)$, $g_{jk}(\xi)$, $g_{jk}(\zeta)$ au lieu de $g_{jk}(x_0)$, $g_{jk}(x)$, $g_{jk}(z)$ etc. L'expression $\Gamma(x_0; x)$, carré de la distance géodésique entre x_0 et x devient dans les nouvelles notations $\Gamma(o; \xi)$. Les formules (40), transcrites d'une manière convenable, deviennent à leur tour

$$\zeta^k = \sigma_\zeta\left(\frac{d\zeta^k}{d\sigma}\right)_0 = \sigma\,\dot{\zeta}_0^k. \tag{43}$$

Il en vient, le long de chaque géodésique,

$$\dot{\zeta}^k = \frac{d\zeta^k}{d\sigma} = \dot{\zeta}_0^k = u^k = \text{const.} \tag{44}$$

Dès lors, la relation $L(z, \dot{z}) = \text{const.}$ s'écrit dans les notations actuelles

$$L(\zeta, \dot{\zeta}) = L(\zeta, u) = g_{jk}(\zeta)\,u^j u^k = \text{const.} \tag{45}$$

On a, en particulier, $L(\xi, u) = L(o, u)$ et par suite

$$\sigma_\xi^2 L(\xi, u) = \sigma_\xi^2 L(o, u) \tag{46}$$

ou

$$g_{jk}(\xi)\,\xi^j \xi^k = g_{jk}(o)\,\xi^j \xi^k. \tag{47}$$

Le premier membre de (46) étant, en vertu de (35bis), égal à $\Gamma(o; \xi)$, on a l'*identité fondamentale*

$$\Gamma(o; \xi) = g_{jk}(o)\,\xi^j \xi^k. \tag{48}$$

Nous donnons encore quelques identités importantes, remontant à Riemann. On tire de (48)

$$\frac{\partial \Gamma(o;\zeta)}{\partial \zeta^j} = 2\, g_{jk}(o)\zeta^k = 2\,\sigma_\zeta\, g_{jk}(o)\, u^k,$$

ce qui montre que $\partial \Gamma/\partial \zeta^j$ est linéaire et homogène en σ_ζ. Il en résulte, d'après (37bis), que

(49) $$\frac{\partial L(\zeta, \dot\zeta)}{\partial \dot\zeta^j} = \frac{\partial L(\zeta, u)}{\partial u^j} = 2\, g_{jk}(\zeta) u^k = 2\, u_j = \text{const.}$$

le long de chaque géodésique(¹). On a donc, en particulier, $g_{jk}(\xi)u^k = g_{jk}(o)u^k$, ce qui, multiplié par σ_ξ, donne

(50) $$g_{jk}(\xi)\xi^k = g_{jk}(o)\xi^k. \text{(²)}$$

Ces formules impliquent évidemment (47).

86. Quelques formules auxiliaires. — Ces formules se rapportent aux paramètres Δ_1 et $\Delta_2 = \Delta$ (nº 81). La plus grande partie des calculs sera faite en coordonnées générales, et ce n'est que pour expliciter $\Delta(\Gamma)$ qu'on aura à recourir à des coordonnées normales.

Nous posons, pour simplifier l'écriture, $\Gamma(x_0; x) = \Gamma$, $g_{jk}(x) = g_{jk}$, $g^{jk}(x) = g^{jk}$. Nous désignons, comme d'habitude, par g la valeur absolue du déterminant $|g_{jk}|$ et par σ le paramètre usuel; $\sigma = o$ en x_0.

1°. *On a* pour une fonction U arbitraire

(51) $$\Delta_1(\Gamma, U) = g^{jk}\frac{\partial \Gamma}{\partial x^j}\frac{\partial U}{\partial x^k} = 2\,\sigma\frac{dU}{d\sigma},$$

la différentiation étant faite suivant la géodésique joignant x_0 et x.

En effet, il résulte de (37bis) et de (29)

(52) $$g^{jk}\frac{\partial \Gamma}{\partial x^j} = \sigma\, g^{jk}\frac{\partial L}{\partial \dot x^k} = 2\,\sigma\, \dot x^k,$$

par conséquent

(¹) On voit que, en coordonnées normales, non seulement les composantes contravariantes u^j de la «vitesse» sont constantes le long de la «trajectoire», mais qu'il en est de même de ses composantes covariantes u_j. On tire de (49) au moyen des équations (25), ou de (50) par différentiation

$$\frac{\partial g_{jk}}{\partial \xi^l}\xi^j\xi^k = o \quad\text{et}\quad \frac{\partial g_{kl}}{\partial \xi^j}\xi^j\xi^k = o \quad\text{ou}\quad \frac{\partial g_{jk}}{\partial \xi^l}u^j u^k = o \quad\text{et}\quad \frac{\partial g_{kl}}{\partial \xi^j}u^j u^k = o.$$

A l'origine des coordonnées normales, l'une quelconque des deux dernières identités est valable pour des u^j arbitraires, d'où l'on conclut que toutes les dérivées $\partial g_{jk}/\partial \xi^l$ s'annulent en ce point.

(²) J'emprunte la déduction élégante qui précède à la Thèse de M. Malmheden (**1**, p. 57). La méthode utilisée au numéro précédent aussi a été sensiblement influencée par celle de M. Malmheden.

$$\Delta_1(\Gamma, U) = 2\sigma \dot{x}^k \frac{\partial U}{\partial x^k} = 2\sigma \frac{\partial U}{\partial x^k}\frac{d x^k}{d\sigma} = 2\sigma \frac{d U}{d\sigma}.$$

On tire de (51) pour une fonction F arbitraire

(51bis) $$\Delta_1(F(\Gamma), U) = 2F'(\Gamma)\cdot\sigma\frac{d U}{d\sigma}.$$

2°. *On a* pour l'opérateur Δ, défini par la formule (23), U et V étant des fonctions arbitraires,

(53) $$\Delta(UV) = U\Delta V + 2\Delta_1(U, V) + V\Delta U.$$

En effet, en utilisant la notation (17), on peut écrire

$$\Delta(UV) = \frac{1}{\sqrt{g}}\frac{\partial}{\partial x^j}(\sqrt{g}\ \mathrm{grad}^j(UV)).$$

Dès lors, l'identité (53) découle immédiatement de

$$\sqrt{g}\ \mathrm{grad}^j(UV) = U(\sqrt{g}\ \mathrm{grad}^j V) + V(\sqrt{g}\ \mathrm{grad}^j U).$$

3°. *On a*

(54) $$\Delta(F(\Gamma)U) = U\Delta F(\Gamma) + 4F'(\Gamma)\cdot\sigma\frac{dU}{d\sigma} + F(\Gamma)\cdot\Delta U$$

et

(55) $$\Delta F(\Gamma) = F'(\Gamma)\cdot\Delta\Gamma + 4F''(\Gamma)\cdot\Gamma.$$

La première formule est une conséquence immédiate de (53) et de (51bis). Quant à la seconde,

$$\Delta(F(\Gamma)) = \frac{1}{\sqrt{g}}\frac{\partial}{\partial x^j}(\sqrt{g}\ \mathrm{grad}^j F(\Gamma)) = \frac{1}{\sqrt{g}}\frac{\partial}{\partial x^j}(F'(\Gamma)\sqrt{g}\ \mathrm{grad}^j \Gamma)$$
$$= F'(\Gamma)\cdot\Delta\Gamma + \mathrm{grad}^j\,\Gamma\cdot\frac{\partial}{\partial x^j}F'(\Gamma).$$

Le dernier terme est égal à $F''(\Gamma)\,\mathrm{grad}^j\,\Gamma\,\mathrm{grad}_j\,\Gamma$, ce qui, en vertu de (38bis) se réduit à $4F''(\Gamma)\cdot\Gamma$.

4°. Pour $F(\Gamma) = \Gamma^\beta$, *on tire* des formules (54) et (55)

(56) $$\Delta(\Gamma^\beta U) = \beta\Gamma^{\beta-1}\left[(4(\beta-1) + \Delta\Gamma)U + 4\sigma\frac{dU}{d\sigma}\right] + \Gamma^\beta\cdot\Delta U.$$

5°. *On a*

(57) $$\Delta\Gamma = 2\left(m + \sigma\frac{d\log\sqrt{g}}{d\sigma}\right).$$

où m désigne la dimension de l'espace de Riemann et g désigne, cette fois-ci, la valeur absolue du déterminant g_{jk}, non plus dans un système de coordonnées générales, mais *dans un système de coordonnées normales arbitraire qui admet x_0 comme origine.*

Vérifions d'abord que l'expression ci-dessus reste inaltérée si on passe d'un système de coordonnées normales à un autre. Or on a vu qu'un tel changement s'opère au moyen de coefficients constants (formule (41)), g sera donc multiplié par une constante (formule (20)) et $d \log \sqrt{g}/d\sigma$ ne sera pas changé. Bien entendu, ce dernier fait résulte aussi, quoique d'une manière implicite, du calcul qui suit.

On a d'abord en coordonnées générales (*cf.* les formules (23) et (52))

$$\Delta \Gamma = \frac{1}{\sqrt{g}} \frac{\partial}{\partial x^j}\left(\sqrt{g}\, g^{jk} \frac{\partial \Gamma}{\partial x^k}\right) = \frac{2}{\sqrt{g}} \frac{\partial}{\partial x^j}(\sqrt{g}\, \sigma \dot{x}^j)$$

$$= \frac{2\sigma}{\sqrt{g}} \frac{\partial \sqrt{g}}{\partial x^j} \dot{x}^j + 2 \frac{\partial}{\partial x^j}(\sigma \dot{x}^j) = 2\sigma \frac{d \log \sqrt{g}}{d\sigma} + 2 \frac{\partial}{\partial x^j}(\sigma \dot{x}^j).$$

Dès lors, si le système des x^k est un système de coordonnées normales admettant x_0 comme origine, on aura, d'après les formules (43) et (44), $\sigma \dot{x}^j = x^j$ et

$$\frac{\partial(\sigma \dot{x}^j)}{\partial x^j} = \frac{\partial x^j}{\partial x^j}\left(= \sum \frac{\partial x^j}{\partial x^j}\right) = m.$$

87. Le conoïde caractéristique. — Les *géodésiques de longueur nulle* ou *bicaractéristiques* issues d'un certain point x engendrent le *conoïde caractéristique*(1) de sommet x. Les points y de ce conoïde satisfont à l'équation $\Gamma(x; y) = 0$. La fonction $\Gamma(x; y)$ s'annule du second ordre au point x, point singulier du conoïde, où toutes les dérivées $\partial \Gamma / \partial y^k$ s'annulent. Par contre, tout autre point(2) du conoïde est un point régulier; $\Gamma(x; y)$ s'y annule du premier ordre, l'une au moins des susdites dérivées étant différente de zéro en un tel point. Tous ces faits découlent immédiatement des formules (37bis) et (29).

Dans le cas de *signature lorentzienne*, le seul qui intervient dans nos applications, le conoïde divise l'espace (voisin de x) en trois régions. Deux d'entre elles, où l'on a $\Gamma(x; y) > 0$, sont intérieures au conoïde et une, avec $\Gamma(x; y) < 0$, lui est extérieure. Les géodésiques, issues de x et menant à l'intérieur, sont des géodésiques de temps, tandis que celles menant à son extérieur sont des géodésiques d'espace.

(1) HADAMARD **1**, p. 47 et 124.

(2) Toutes nos considérations se rapportent à un voisinage suffisamment restreint de x.

Les faits énumérés se réduisent aux faits correspondants relatifs à l'espace lorentzien traité dans les chapitres précédents. En effet, en introduisant des coordonnées normales, avec leur origine en x, le conoïde devient un cône ordinaire $g_{jk}(0)\eta^j\eta^k = 0$. Par un choix convenable de ces dernières coordonnées on peut même s'arranger de manière à ce que $\Gamma(x; y)$ se réduise à la forme lorentzienne (1) du n° **14**

$$\Gamma(x; y) = \eta^{1^2} - \eta^{2^2} - \cdots - \eta^{m^2}. \tag{58}$$

88. Surfaces caractéristiques. — Une surface caractéristique est définie par les équations *simultanées*

$$S(y^1, y^2, \ldots, y^m) = S(y) = 0 \tag{59}$$

et

$$g^{jk}\frac{\partial S}{\partial y^j}\frac{\partial S}{\partial y^k} = 0. \tag{60}$$

Le conoïde du numéro précédent, appelé de prime abord conoïde caractéristique, est une surface caractéristique d'après la définition que nous venons de poser. En effet, en prenant $S(y) = \Gamma(x; y)$, le conoïde satisfait aux équations (59) et (60) (*cf.* l'identité (38*bis*)). L'équation (60) admet l'interprétation géométrique que voici. [1]

Posons, selon la formule (30),

$$H(y; \pi) = g^{jk}\pi_j\pi_k = \pi_j\pi^j, \tag{61}$$

où

$$\pi_j = \frac{\partial S}{\partial y^j} \quad \text{et} \quad \pi^j = g^{jk}\pi_k = \frac{1}{2}\frac{\partial H}{\partial \pi_j} \tag{62}$$

sont respectivement les composantes covariantes et contravariantes du gradient de S. La direction *normale* à la surface $S = 0$ étant, en chaque point régulier de S, donnée par celle du gradient de S, l'équation (60) exprime que *cette direction est isotrope.* Elle sera par suite orthogonale à elle-même et dès lors aussi *tangente* à la surface. On a là d'ailleurs la seule direction isotrope qui soit tangente, puisque deux directions isotropes différentes ne sont jamais orthogonales l'une à l'autre. [2]

---

[1] *Cf.* pour tout ce qui suit le n° **62**, où sont envisagées les surfaces caractéristiques de l'espace lorentzien.

[2] En effet, il est géométriquement évident que toute direction, orthogonale à une direction isotrope et différente de cette dernière, est une direction d'espace, tout plan tangent à un cône caractéristique étant extérieur à ce cône.

Tout cela étant admis, nous allons montrer[1] que *par tout point régulier d'une surface caractéristique il passe une et une seule bicaractéristique située sur la surface.* Il résultera aussi que *deux surfaces caractéristiques tangentes en un point sont tangentes le long d'une bicaractéristique.*

Considérons les lignes tracées sur la surface $S = 0$ qui en tous leurs points sont tangentes à la normale menée en ce point ou, en d'autres termes, au gradient. En rapportant la surface à $m - 1$ paramètres, l'existence de ces lignes s'établit aisément.

Ceci étant admis, une telle ligne satisfait au système d'équations différentielles[2]

$$\frac{d y^j}{d t} = \pi^j \quad \text{ou} \quad \frac{d y^j}{d t} = \frac{1}{2} \frac{\partial H}{\partial \pi_j}, \quad j = 1, 2, \ldots, m. \tag{63}$$

Le fait que la ligne est de longueur nulle est garanti par les dernières équations et l'équation (60). Tout revient donc à démontrer qu'elle est une géodésique.[3]

Observons maintenant que la fonction $S(y)$ et ses dérivées partielles π_j sont, comme fonctions du point y, manifestement définies dans un certain voisinage de la surface $S = 0$. Il en sera donc de même de la fonction $H(y, \pi)$. On a évidemment

$$\frac{\partial \pi_k}{\partial y^j} = \frac{\partial \pi_j}{\partial y^k} \left(= \frac{\partial^2 S}{\partial y^j \partial y^k} \right). \tag{64}$$

On pose $H(y, \pi(y)) = \bar{H}(y)$ et on trouve, en utilisant (64) et (63),

$$\frac{\partial \bar{H}}{\partial y^j} = \frac{\partial H}{\partial y^j} + \frac{\partial H}{\partial \pi_k} \cdot \frac{\partial \pi_k}{\partial y^j} = \frac{\partial H}{\partial y^j} + 2 \frac{\partial \pi_j}{\partial y^k} \cdot \frac{d y^k}{d t} = \frac{\partial H}{\partial y^j} + 2 \frac{d \pi_j}{d t}.$$

D'autre part, puisque $\bar{H}(y) = 0$ aux points réguliers de $S(y) = 0$, le gradient de $\bar{H}(y)$ aura en ces points la direction du gradient de $S(y)$, c'est-à-dire que

$$\frac{\partial \bar{H}(y)}{\partial y^j} = 2 \cdot \lambda(y) \cdot \frac{\partial S(y)}{\partial y^j},$$

la fonction $\lambda = \lambda(y)$ étant bien déterminée aux susdits points. Cela nous permet d'écrire

$$\frac{d \pi_j}{d t} - \lambda \cdot \pi_j = -\frac{1}{2} \frac{\partial H(y, \pi)}{\partial y^j}. \tag{65}$$

Il s'agit d'éliminer λ. Posons à cet effet, le long de la ligne considérée, $\pi_j = k(t) \cdot p_j$, la fonction $k(t)$ étant arbitraire jusqu'à nouvel ordre. On aura, en raison de l'homogénéité de H et de ses dérivées,

$$H(y, \pi) = k^2 H(y, p), \quad \frac{\partial H(y, \pi)}{\partial y^j} = k^2 \frac{\partial H(y, p)}{\partial y^j}, \quad \frac{\partial H(y, \pi)}{\partial \pi_j} = k \frac{\partial H(y, p)}{\partial p_j}.$$

Dès lors, les relations (63) et (65) deviennent

---

[1] Les remarques qui suivent, et qui ne sont qu'une mise au point de quelques faits connus, nous seront très utiles dans un ouvrage à venir. Aucune application n'en étant donnée dans le présent travail, le lecteur pourra passer sur le reste de ce numéro.

[2] M. Malmheden a étendu la méthode d'intégration qui suit au cas où H est en fonction des π_j une fonction homogène de degré arbitraire; Malmheden **1**, Appendice II, p. 110—112.

[3] Bien entendu, il s'agit de montrer qu'elle a le caractère géodésique par rapport à l'espace de Riemann envisagé et non seulement par rapport au sous-espace défini par $S = 0$.

$$\frac{d y^j}{d t} = \frac{1}{2} k \frac{\partial H(y, p)}{\partial p_j} \quad \text{et} \quad k \frac{d p_j}{d t} + (k' - \lambda k) p_j = -\frac{1}{2} k^2 \frac{\partial H(y, p)}{\partial y^j}.$$

En choisissant k de façon que $k' - \lambda k = 0$ et en posant $d\sigma = k \cdot dt$, on arrive aux *équations canoniques* (*cf.* formule (34))

$$\frac{d y^j}{d \sigma} = \frac{1}{2} \frac{\partial H(y, p)}{\partial p_j} \quad \text{et} \quad \frac{d p_j}{d \sigma} = -\frac{1}{2} \frac{\partial H(y, p)}{\partial y^j}.$$

La ligne est donc une *géodésique* et dès lors, par ce qui précède, une *bicaractéristique*.

Le dernier système d'équations est *indépendant de la surface caractéristique* envisagée. Il en ressort que *si deux surfaces de cette nature ont un élément de contact (point et plan tangent) en commun, il en émanera une bande caractéristique commune d'éléments de contact; en d'autres termes, les deux surfaces se toucheront le long d'une bicaractéristique commune.* En effet, les π_j, correspondant aux deux surfaces, auront les mêmes rapports mutuels au point commun. On pourra donc choisir les valeurs initiales des p_j égales, puisque les valeurs initiales des $k(t)$ sont arbitraires.

On voit aussi par ce qui précède que toute surface caractéristique peut être obtenue comme enveloppe de conoïdes caractéristiques.

Voici une dernière remarque qui appartient à un ordre d'idées légèrement différent. Nous avons vu que les bicaractéristiques tracées sur une surface caractéristique sont orthogonales à tout déplacement fait le long de la surface. Cela étant, prenons sur chaque bicaractéristique un point unique (ces points variant d'une manière continue et suffisamment dérivable). Il s'ensuit que *les points en question forment une variété d'espace*[1] *T à $m - 2$ dimensions, à laquelle chaque bicaractéristique est orthogonale.*

Réciproquement, envisageons une variété d'espace T arbitraire, à $m - 2$ dimensions. En chaque point de T on peut mener *deux bicaractéristiques orthogonales à* T[2], qui engendrent deux surfaces distinctes. Nous disons que *ces surfaces sont caractéristiques.* Soit Σ l'une de ces surfaces. Il suffit de montrer qu'une bicaractéristique arbitraire C appartenant à Σ est, en chacun de ses points y, orthogonale à un déplacement infinitésimal arbitraire δy fait le long de la surface. Désignons à cet effet par x et $x + \delta x$ les points d'intersection de la variété T et des bicaractéristiques C et $C + \delta C$ passant respectivement par y et $y + \delta y$. On aura, en vertu de la formule (39),

$$g_{jk}(y) \frac{d y^j}{d \sigma} \delta y^k = g_{jk}(x) \frac{d x^j}{d \sigma} \delta x^k.$$

Le second membre s'annulant par hypothèse, le premier membre s'annulera lui aussi.

On voit aussi que la variété initiale T ne joue aucun rôle particulier parmi les variétés d'espace (à $m - 2$ dimensions) situées sur Σ et qu'elle pourrait, dans la construction ci-dessus, être remplacée par n'importe laquelle de ces variétés.

89. Le noyau fractionnaire. — Nous voilà arrivés à un des problèmes principaux du présent chapitre, celui de construire le noyau qui, dans un espace de Riemann, correspond au noyau $r^{\alpha - m} / H_m(\alpha)$ servant de base à la méthode d'intégration fractionnaire, développée dans les chapitres précédents.

---

(1) Une variété d'espace se caractérise ici, en analogie avec le cas lorentzien, *cf.* p. 37, par l'inégalité $g_{jk}\, dx^j\, dx^k < 0$, vérifiée par tout déplacement infinitésimal le long de la variété.

(2) *Cf.* p. 131.

Soit x un point de l'espace de Riemann envisagé, et considérons en même temps les points y suffisamment voisins[1] de x et satisfaisant à l'inégalité $\Gamma(x; y) > 0$, c'est-à-dire appartenant à l'un quelconque des deux domaines intérieurs au conoïde caractéristique qui a son sommet en x (*cf.* le n° **87**).

Désignons comme plus haut par $s = s(x; y) = s(y; x)$ la distance géodésique des points x et y, c'est-à-dire que $s = \{\Gamma(x; y)\}^{1/2}$. Nous nous proposons de construire une fonction $V^\alpha(x; y)$, dépendant d'un paramètre α, qui satisfait à l'équation

$$\Delta_y V^{\alpha+2}(x; y) = V^\alpha(x; y) \tag{66}$$

et qui devient singulier sur le conoïde comme $s^{\alpha-m}$. A l'exemple de M. Hadamard nous admettons d'abord que *les coefficients* $g_{jk}(y)$ *sont des fonctions holomorphes des variables* y^j. Il en sera alors de même de s, sauf sur le conoïde, et de s^2 partout.[2] Nous posons

$$V^\alpha = \sum_{k=0}^{\infty} \frac{V_k \cdot s^{\alpha+2k-m}}{K_m(\alpha) \cdot L_m(\alpha + 2k)} = s^{\alpha-m} \sum_{k=0}^{\infty} \frac{V_k \cdot s^{2k}}{H_m(\alpha, k)}, \tag{67}$$

les coefficients V_k ne dépendant que de x et de y, tandis que $K_m(\alpha)$ et $L_m(\alpha)$ ne dépendent que de α (et de la dimension m). Introduisons maintenant un système de coordonnées normales, admettant le point x comme origine, en gardant la notation g pour la valeur absolue du déterminant des coefficients g_{jk} correspondants. Si nous remplaçons encore le paramètre σ, utilisé dans les numéros **82**—**86**, par s qui, nous le savons, est un paramètre *admissible* (*cf.* le n° **84**), les formules (56) et (57) nous permettent d'écrire

$$\Delta_y V^{\alpha+2} = \sum_{k=0}^{\infty} s^{\alpha+2k-m} \left\{ \frac{\alpha+2+2k-m}{K_m(\alpha+2) \cdot L_m(\alpha+2+2k)} \right. \tag{68}$$

$$\left. \times \left[\left(\alpha + 2k + s\frac{d \log \sqrt{g}}{ds}\right) V_k + 2s\frac{d V_k}{ds}\right] + \frac{\Delta V_{k-1}}{K_m(\alpha + 2) \cdot L_m(\alpha + 2k)} \right\},$$

la notation d/ds désignant une différentiation suivant la géodésique (de temps) qui va de x à y. On choisit maintenant $L_m(\alpha)$ de sorte qu'on ait

[1] Nous admettons en particulier que x et chaque y peuvent être joints par une géodésique d'une manière univoque (*cf.* p. 169).

[2] Les coordonnées normales qu'on va introduire tout à l'heure seront aussi des fonctions holomorphes des anciennes coordonnées et réciproquement. Ces faits résultent des théorèmes d'existence classiques relatifs aux systèmes d'équations différentielles ordinaires.

$$\frac{\alpha + 2 - m}{L_m(\alpha + 2)} = \frac{1}{L_m(\alpha)},$$

et on détermine les V_k par le système d'équations différentielles récurrentes

$$(69)\qquad \begin{cases} 2s\dfrac{dV_k}{ds} + \left(2k + s\dfrac{d\log\sqrt{g}}{ds}\right)V_k + \Delta V_{k-1} = 0; \\ V_{-1} \equiv 0, \quad V_0(x;x) = 1. \end{cases}$$

Comme chez M. Hadamard, la solution est uniquement déterminée par le postulat que les $V_k(x; y)$ sont des fonctions régulières (c'est-à-dire suffisamment dérivables) de y.[1] Notons qu'on trouve en particulier

$$V_0(x; y) = \left(\frac{g(y)}{g(x)}\right)^{-\frac{1}{4}}.$$

Les formules de récurrence posées étant satisfaites, il vient

$$\Delta V^{\alpha+2} = \alpha \sum_{k=0}^{\infty} \frac{V_k \cdot s^{\alpha+2k-m}}{K_m(\alpha+2)\cdot L_m(\alpha+2k)}.$$

En choisissant encore $K_m(\alpha)$ en sorte que

$$\frac{\alpha}{K_m(\alpha+2)} = \frac{1}{K_m(\alpha)},$$

l'équation (66) se trouvera résolue.

$K_m(\alpha)$ et $L_m(\alpha)$ auront les propriétés désirées si l'on pose

$$K_m(\alpha) = K\cdot 2^{\frac{\alpha}{2}}\cdot\Gamma\left(\frac{\alpha}{2}\right) \quad \text{et} \quad L_m(\alpha) = L\cdot 2^{\frac{\alpha}{2}}\cdot\Gamma\left(\frac{\alpha+2-m}{2}\right),$$

K et L étant indépendants de α. Nous écrivons encore

$$H_m(\alpha, k) = K_m(\alpha)\cdot L_m(\alpha + 2k)$$

---

[1] Pour $k = 0$, la solution est déterminée par la condition $V_0(x; x) = 1$. Pour $k > 0$, l'équation peut s'écrire

$$2\frac{d}{ds}(s^k V_k) + \frac{d}{ds}\log\sqrt{g}\cdot s^k V_k + s^{k-1}\cdot\Delta V_{k-1} = 0.$$

La solution est encore déterminée. En effet, V_k étant régulier pour $y = x$, c'est-à-dire pour $s = 0$, on aura pour cette valeur $s^k V_k = 0$, ce qui fait que $s^k V_k$ est déterminée, et il en est alors évidemment de même de V_k. On trouve par un calcul facile

$$V_0 = \left(\frac{g(y)}{g(x)}\right)^{-\frac{1}{4}} \quad \text{et} \quad V_k = -V_0 s^{-k}\int_0^s \frac{s^{k-1}}{2V_0}\cdot\Delta V_{k-1}\,ds.$$

et nous déterminons le produit KL en sorte qu'on ait

$$(70)\qquad H_m(\alpha, 0) = H_m(\alpha) = \pi^{\frac{m-2}{2}} \cdot 2^{\alpha-1} \cdot \Gamma\left(\frac{\alpha}{2}\right) \cdot \Gamma\left(\frac{\alpha + 2 - m}{2}\right),$$

la fonction $H_m(\alpha)$ étant le facteur spécial, relatif à l'espace lorentzien (formule (20) du n° **16**). Il vient en définitive

$$(71)\qquad H_m(\alpha, k) = K_m(\alpha) L_m(\alpha + 2k) = \pi^{\frac{m-2}{2}} \cdot 2^{\alpha+k-1} \cdot \Gamma\left(\frac{\alpha}{2}\right) \Gamma\left(\frac{\alpha + 2k + 2 - m}{2}\right).$$

Signalons encore la formule de récurrence

$$(72)\qquad H_m(\alpha, k) = (\alpha + 2k - m) \cdot H_m(\alpha, k - 1),$$

et observons que cette formule, associée à la formule (70), détermine entièrement la fonction $H_m(\alpha, k)$.[1]

Les questions de convergence de la série (67) seront étudiées au n° **91**.

90. Comparaison avec les solutions élémentaires de M. Hadamard. — Il y a lieu de comparer les calculs ci-dessus avec ceux de M. Hadamard.

Considérons à côté du système récurrent (69) le système analogue, dans lequel q désigne un nombre fixe arbitraire qui n'est pas un entier négatif,

$$(73)\qquad \begin{cases} 2s\dfrac{dT_k}{ds} + \left(2k + s\dfrac{d \log \sqrt{g}}{ds}\right) T_k + \dfrac{\Delta T_{k-1}}{2(q + k)} = 0, \\ T_{-1} \equiv 0, \quad T_0(x; x) = 1. \end{cases}$$

Il est clair qu'on a

$$(74)\qquad T_0 = V_0,\ T_k = \frac{V_k}{2^k (q + 1)(q + 2) \ldots (q + k)} = \frac{\Gamma(q + 1)}{2^k \Gamma(q + 1 + k)} \cdot V_k.$$

En particulier, si l'on prend $q = \dfrac{\alpha - m}{2}$, on trouve à l'aide des formules (70) et (71)

---

[1] Dans le cas elliptique, la fonction correspondante $H_m(\alpha, k)$ se détermine par la formule de récurrence (72) combinée avec la relation

$$H_m(\alpha, 0) = \frac{\pi^{\frac{m}{2}} \cdot 2^\alpha \cdot \Gamma\left(\frac{\alpha}{2}\right)}{\Gamma\left(\frac{m - \alpha}{2}\right)},$$

cette dernière fonction étant le facteur adapté à l'opérateur de Laplace (*cf.* la formule (7) du n° **6**).

(75) $$T_k = \frac{H_m(\alpha)}{H_m(\alpha, k)} \cdot V_k,$$

c'est-à-dire que

(76) $$T^\alpha = s^{\alpha-m} \sum_{k=0}^{\infty} T_k s^{2k} = H_m(\alpha) \cdot s^{\alpha-m} \sum_{k=0}^{\infty} \frac{V_k s^{2k}}{H_m(\alpha, k)} = H_m(\alpha) \cdot V^\alpha.$$

Le système d'équations (69), qui détermine les V_k, étant indépendant de q ou de α, il en est de même des V_k, tandis que les T_k dépendent de α de la façon mise en évidence par la formule (75).

Si $q = \frac{\alpha - m}{2}$ est un entier négatif, il y aura dégénérescence pour les T_k à partir de $k = -q$, puisque l'équation (73) cesse d'avoir un sens pour $k = -q$.

M. Hadamard ne considère que la valeur particulière $q = \frac{2-m}{2} = p$. Dans le cas impair, il en tire la solution élémentaire[1]

(77) $$U = s^{2-m} \sum_{k=0}^{\infty} U_k s^{2k},$$

où nous avons désigné par U_k les quantités T_k qui correspondent à la susdite valeur particulière. Pour les valeurs paires de m, p est un entier négatif; l'équation (73) cessera donc d'avoir un sens pour $k = \frac{m-2}{2}$ et il n'existera pas de solution élémentaire de la forme (77).

On voit que nous avons tourné cette difficulté en introduisant le noyau (ou — si l'on veut — la solution élémentaire) fractionnaire V^α. M. Hadamard de son côté[2] introduit pour m pair, dans la voie ouverte par Picard, une solution élémentaire renfermant un terme logarithmique. Elle a la forme

$$s^{2-m} \sum_{k=0}^{\frac{m}{2}-2} U_k \cdot s^{2k} + w - 2\,\mathcal{U} \log s,$$

où les U_k, w et $\mathcal{U}$ sont des fonctions régulières, w n'étant déterminé qu'à une solution régulière près de l'équation $\Delta u = 0$.[3] Il ne manque pas d'intérêt d'observer que cette solution élémentaire logarithmique est, à un facteur constant

---

[1] HADAMARD **1**, p. 136.

[2] HADAMARD **1**, p. 143; application à la résolution du problème de Cauchy p. 316. La remarque qui commence au bout de la page 98 du présent ouvrage est erronée et doit être supprimée.

[3] Bien entendu, M. Hadamard traite l'équation générale du second ordre au lieu de l'équation relative à Δ, traitée ici.

et à une fonction régulière additive près, identique à la fonction

$$(77^{bis})\qquad \left(\frac{\partial V^{\alpha}}{\partial \alpha}\right)_{(\alpha=2)}.\text{[1]}$$

91. Questions de convergence et de majoration. — Venons à la question de convergence de nos séries. M. Hadamard[2] a, par une application ingénieuse de la méthode des séries majorantes[3], montré que la série figurant dans la formule (77) converge absolument et uniformément tant que s est inférieur à une quantité fixe. La convergence est uniforme aussi par rapport à x quand x varie dans une région strictement intérieure à la région $\mathcal{R}$ considérée au nº **85**. Nous pouvons largement profiter de la susdite démonstration, en observant que nonobstant qu'on y suppose que le nombre q a la valeur particulière p et que m est impair, on n'y utilise réellement que le fait que q n'est pas un entier négatif et que, par conséquent, les équations (73) ne cessent jamais d'avoir un sens.

En résumé, lorsque les coefficients de la forme métrique sont des fonctions holomorphes des coordonnées, la fonction $\Gamma = s^2$ et les coefficients V_k figurant dans l'expression (67) de V^{α} seront des fonctions holomorphes[4] des coordonnées du point y et la série (67) convergera absolument et uniformément dès que s est assez petit (sauf peut-être — quant à l'uniformité — au voisinage du conoïde, où le facteur $s^{\alpha-m}$ peut devenir infini). Il résulte de l'holomorphisme en fonction de y que les séries obtenues de (67) par des différentiations terme à terme ont des propriétés analogues, d'où il résulte que nos calculs, qui reposaient sur de telles différentiations, étaient légitimes.

Occupons-nous maintenant de la convergence uniforme de la série (76) par rapport à α. Soit a un nombre réel qui n'est pas un entier négatif.

On sait par ce qui précède que la série $\sum\limits_{k=0}^{\infty} |V_k|\, s^{2k} : |H_m(a, k)|$ est convergente.

Cela admis, et $l \geqq 0$ étant un entier arbitraire, la série

---

(1) *Cf.* pour tout ce qui précède les nos **58** et **59**. L'expression (77bis) se réduit dans le cas lorentzien à $\delta_m r^{2-m}$, où δ_m est donné par la formule (37) du nº **59**.

(2) Hadamard **1**, p. 137—140 et p. 408—412.

(3) Rappelons qu'on se trouve toujours dans le cas *holomorphe*, ce qui veut dire que les coefficients g_{jk} de la forme métrique sont des fonctions holomorphes des coordonnées. Le domaine de convergence de la série (77) dépend des rayons de convergence des développements en séries entières des coefficients g_{jk} autour des points de la la région $\mathcal{R}$.

(4) *Voir* note (2) p. 180. L'holomorphisme en fonction des coordonnées de x résultera de la symétrie en x et y établie plus loin.

$$\sum_{k=l}^{\infty} \frac{|H_m(a, l)|}{|H_m(a, k)|} |V_k| s^{2k}$$

sera convergente, elle aussi. Pour $\Re(\alpha) \geqq a$ et $a - m + 2k \geqq 0$ on a aussi $\Re(\alpha - m + 2k) \geqq 0$, par conséquent $|\alpha - m + 2k| \geqq \Re(\alpha - m + 2k) \geqq a - m + 2k$. On a donc, en vertu de la formule (72), pour les mêmes valeurs de k

$$\frac{|H_m(\alpha, k)|}{|H_m(\alpha, k-1)|} \geqq \frac{|H_m(a, k)|}{|H_m(a, k-1)|}.$$

En désignant par l un quelconque de ces nombres k, il ressort que la série de factorielles

$$(78) \qquad F(\alpha, l) = \sum_{k=l}^{\infty} \frac{H_m(\alpha, l)}{H_m(\alpha, k)} \cdot V_k s^{2k}$$

$$= V_l s^{2l} + \frac{V_{l+1} s^{2(l+1)}}{\alpha - m + 2l + 2} + \cdots + \frac{V_{l+h} s^{2(l+h)}}{(\alpha - m + 2l + 2) \ldots (\alpha - m + 2l + 2h)} + \cdots$$

est absolument et *uniformément* convergente pour $\Re(\alpha) \geqq a$.

Avant d'aller plus loin, notons la relation

$$(78^{bis}) \qquad \lim_{|\alpha| \to \infty} F(\alpha, l) = V_l s^{2l}.$$

Les termes de la série uniformément convergente (78) étant des fonctions holomorphes de α, cette série représentera, elle aussi, une fonction holomorphe de α. En la multipliant par la fonction holomorphe $s^{\alpha-m}/H_m(\alpha, l)$ et en y ajoutant un nombre fini de termes holomorphes, on obtient la fonction V^α (*cf.* la formule (67)), qui est donc une fonction *holomorphe* de α.

Nous allons maintenant construire une majorante de la fonction V^α en nous restreignant à des valeurs de α dont la partie réelle est assez grande, soit $\Re(\alpha) \geqq m$. Prenons d'abord $\alpha = m$, ce qui donne $q = \frac{\alpha - m}{2} = 0$.

La série $\Sigma\, T_k s^{2k}$ figurant dans (76) devient d'après (74) $\Sigma\, V_k s^{2k}/2^k k!$. Nous savons que cette série converge absolument (et uniformément) tant que s est inférieur à une quantité fixe. C'est ce que nous supposons dans la suite et nous posons

$$(79) \qquad \sum_{k=0}^{\infty} \frac{|V_k| s^{2k}}{2^k \cdot k!} = C.$$

D'autre part, il vient des formules (72) et (70), pour $\Re(\alpha) > m$,

$$|H_m(\alpha, k)| = \left| H_m(\alpha) \cdot \prod_{p=1}^{k} (\alpha + 2p - m) \right| \geqq |H_m(\alpha)| \cdot \prod_{p=1}^{k} (2p) = 2^k \cdot k! \, |H_m(\alpha)| \cdot$$

Il s'ensuit

(80) $$|V^\alpha| = \left| s^{\alpha-m} \sum_{k=0}^{\infty} \frac{V_k s^{2k}}{H_m(\alpha, k)} \right| \leqq \frac{|s^{\alpha-m}|}{|H_m(\alpha)|} \cdot \sum_{k=0}^{\infty} \frac{|V_k| \, s^{2k}}{2^k \cdot k!} \leqq C \frac{|s^{\alpha-m}|}{|H_m(\alpha)|},$$

la valeur de C étant indépendante de α.

Il est facile de donner une majorante de forme très simple de la fonction

$$\frac{1}{H_m(\alpha)} = \frac{1}{\pi^{\frac{m-2}{2}} 2^{\alpha-1} \Gamma\left(\frac{\alpha}{2}\right) \Gamma\left(\frac{\alpha + 2 - m}{2}\right)}.$$

On tire de la formule de Stirling pour $\Re(z) \geqq 0$ et $|z| \to \infty$

$$\log \Gamma(z) = z (\log z - 1) + O(\log |z|).$$

Il s'ensuit pour $\Re(\alpha) \geqq m$ et $|\alpha| \to \infty$

$$\log H_m(\alpha) = \alpha (\log \alpha - 1) + O(\log |\alpha|).$$

On a donc, dans les mêmes conditions,

(81) $$\frac{1}{|H_m(\alpha)|} \leqq \left| \left(\frac{\alpha}{e}\right)^{-\alpha} \right| \cdot |\alpha|^p, \text{(1)}$$

l'exposant p étant *indépendant* de α. Cette inégalité se trouve établie pour $|\alpha|$ assez grand, soit $|\alpha| > A$, et pour $\Re(\alpha) \geqq m$. Nous disons que, en augmentant p (le cas échéant), l'inégalité subsiste dans tout le demi-plan $\Re(\alpha) \geqq m$. En effet, pour $\Re(\alpha) \geqq m$ et $|\alpha| \leqq A$, le premier membre est borné supérieurement, le second membre inférieurement, et on a de plus $|\alpha| \geqq m > 1$.

Soit maintenant β un nombre *fixe*, tel que $\Re(\beta) \geqq 0$. On montre aisément que la quantité $1 : |H_m(\alpha + \beta)|$ satisfait, elle aussi, à une inégalité de la forme (81), l'exposant p étant toujours indépendant de α. On a en effet, d'après ce qui précède

(82) $$\frac{1}{|H_m(\alpha + \beta)|} \leqq \left| \left(\frac{\alpha + \beta}{e}\right)^{-\alpha-\beta} \right| \cdot |\alpha + \beta|^p.$$

---

(1) En posant $\alpha = \xi + i\eta = \varrho e^{i\theta}$, on a évidemment $\left| \left(\frac{\alpha}{e}\right)^{-\alpha} \right| = \varrho^{-\xi} e^{\theta\eta + \xi}$.

Le rapport entre la dernière expression et le second membre de (81) est asymptotiquement égal à $(|\alpha| \cdot e^{-1})^{-\Re(\beta)}$, quand $|\alpha|$ tend vers l'infini. Vu que $\Re(\beta) \geqq 0$, le second membre de (82) se trouve asymptotiquement majoré par celui de (81). Ce point acquis, la démonstration s'achève comme plus haut.

En vertu de l'inégalité (80), on a en résumé:

Le nombre β étant fixe et tel que $\Re(\beta) \geqq 0$, *les noyaux* V^α et $V^{\alpha+\beta}$, fonctions holomorphes de α toutes les deux, *satisfont dans le demi-plan* $\Re(\alpha) \geqq m$ *aux inégalités*

$$(83) \quad |V^\alpha| \leqq C \cdot s^{\Re(\alpha)-m} \left| \left(\frac{\alpha}{e}\right)^{-\alpha} \right| \cdot |\alpha|^p \quad \text{et} \quad |V^{\alpha+\beta}| \leqq C s^{\Re(\alpha+\beta)-m} \left| \left(\frac{\alpha}{e}\right)^{-\alpha} \right| \cdot |\alpha|^p,$$

les valeurs de C et de p étant indépendantes(1) *de* α.

92. La formule de Green. — Reprenons le domaine D, limité par une ou plusieurs surfaces fermées T, considéré au n° **18.** Envisageons le champ des vecteurs A et écrivons, en nous reportant aux formules (21bis) et (22) du n° **81**, l'intégrale

$$\int_D \operatorname{div} A \, d\Omega = \int_D \frac{1}{\sqrt{g}} \frac{\partial}{\partial y^k} (\sqrt{g} A^k) \cdot \sqrt{g} \, dy^1 \ldots dy^m = \int_D \frac{\partial}{\partial y^k} (\sqrt{g} A^k) \, dy^1 \ldots dy^m.$$

En utilisant la formule (26) du n° **18**, on obtient

$$(84) \qquad \int_D \operatorname{div} A \, d\Omega = \int_T A^k J_k \cdot \sqrt{g} \cdot d\lambda^1 \ldots d\lambda^{m-1},$$

où les J_k sont les déterminants fonctionnels (*cf.* formule (25) du n° **18**)

$$J_k = (-1)^{k-1} \cdot \frac{d(y^1, y^2, \ldots, y^{k-1}, y^{k+1}, \ldots, y^m)}{d(\lambda^1, \lambda^2, \quad \ldots \quad, \lambda^{m-1})}.$$

La relation (84) exprime l'une des formes principales du *théorème de Gauss* relatif à un espace de Riemann.

Désignons maintenant par T' une portion de la surface T ayant une orientation d'espace.(2) Nous nous proposons de transformer l'intégrale figurant au second membre de (84), en tant qu'elle est étendue à T', dans une autre intégrale, ne renfermant que des éléments géométriques intrinsèques. Observons

(1) Une forme plus précise de la formule de Stirling que celle utilisée plus haut fournirait aisément $p = 1$.

(2) *Cf.* la note (1), p. 179.

à cet effet que l'expression $A^k J_k$ est égale au déterminant d'ordre m dont les lignes sont formées des composantes contravariantes du vecteur A et de celles des vecteurs $\partial y/\partial \lambda^p$ $(p = 1, 2, \ldots, m-1)$. En introduisant encore un nombre positif infinitésimal η, on voit, en vertu de la formule (21), que la quantité $\sqrt{g}\, A^k J_k \cdot \eta \prod d\lambda^p$ est, au signe près, égale au volume du parallélépipède infinitésimal construit sur les vecteurs $\eta \cdot A$ et $d\lambda^p \cdot \partial y/\partial \lambda^p$. D'autre part, ce parallélépipède peut être considéré comme ayant comme base l'élément de surface dT, parallélépipède à $m-1$ dimensions, construit sur les vecteurs $d\lambda^p \cdot \partial y/\partial \lambda^p$, et comme hauteur le segment $\eta \cdot A_{<n>}$, où $A_{<n>}$ est la projection orthogonale du vecteur A sur la normale extérieure; en formule

(85) $$A_{<n>} = A^k n_k,$$

les n_k étant les composantes covariantes de la normale extérieure unitaire n. Le volume en question s'exprime donc aussi par le produit[1] $dT \cdot \eta A_{<n>}$, au signe près. On a par conséquent, après avoir divisé par η, toujours au signe près,

(86) $$A^k J_k \cdot \sqrt{g} \prod d\lambda^p = A_{<n>}\, dT.$$

Le signe aussi est correct, si l'on attribue à dT et aux $d\lambda^p$ des valeurs positives. En effet, d'après la convention posée au n° **18**, relativement aux signes des J_k, l'expression $B^k J_k$ a, respectivement, le signe positif ou négatif pour tout vecteur B menant à l'extérieur ou à l'intérieur de la surface T, tandis que $B^k J_k$ s'annule quand le vecteur B est tangent à la surface. Les circonstances sont les mêmes quant au signe de $B^k n_k$, puisque n est une normale extérieure, et que l'on a $n^k n_k = 1 > 0$, vu que T' a une orientation d'espace et que, par conséquent, n est un vecteur de temps (unitaire).

L'identité (86) fournit pour la portion de surface T' l'identité analogue

(87) $$\int_{T'} A^k J_k \cdot \sqrt{g} \cdot d\lambda^1 \ldots d\lambda^m = \int_{T'} A_{<n>}\, dT.$$

Nous passons maintenant à la formule de Green. Établissons d'abord la formule

(88) $$u\,\Delta v - v\,\Delta u = \operatorname{div}(u \operatorname{grad} v - v \operatorname{grad} u),$$

u et v étant des scalaires. On a, en utilisant la formule (22)

[1] *Cf.* pour ces questions le n° **21**^{bis}, en particulier la formule (55).

$$\begin{aligned}\operatorname{div}(u \operatorname{grad} v) &= \frac{1}{\sqrt{g}} \frac{\partial}{\partial y^k}(\sqrt{g}\, u \operatorname{grad}^k v) \\ &= u \operatorname{div}(\operatorname{grad} v) + \frac{\partial u}{\partial y^k} \operatorname{grad}^k v = u \,\Delta v + \operatorname{grad}_k u \operatorname{grad}^k v \\ &= u\,\Delta v + (\operatorname{grad} u, \operatorname{grad} v),\end{aligned}$$

ce qui entraîne (88). Dès lors, la formule de Gauss (84) nous fournit la formule de Green

$$(89) \qquad \int_D (u\,\Delta v - v\,\Delta u)\, d\Omega = \int_T (u \operatorname{grad}^k v - v \operatorname{grad}^k u) J_k \cdot \sqrt{g}\, d\lambda^1 \ldots d\lambda^m.$$

Le long de la portion de surface T', la quantité figurant dans la dernière intégrale peut, d'après les formules (86) et (85), se mettre sous la forme

$$(u \operatorname{grad}^k v \cdot n_k - v \operatorname{grad}^k u \cdot n_k)\, dT.$$

Or $\operatorname{grad}^k v \cdot n_k = \operatorname{grad}_k v \cdot n^k = \frac{\partial v}{\partial y^k} n^k = \frac{dv}{dn}$, d'après la formule (42) du n° **21**. On a donc en définitive

$$(90) \qquad \int_{T'} (u \cdot \operatorname{grad}^k v - v \cdot \operatorname{grad}^k u) J_k \sqrt{g} \prod d\lambda^p = \int_{T'} \left(u \frac{dv}{dn} - v \frac{du}{dn}\right) dT.$$

Remarquons en passant que le long d'une portion de surface T'' orientée dans le temps l'égalité (86) aura lieu avec le signe opposé. Cela tient au fait que, la normale n étant dans ce dernier cas un vecteur d'espace, on aura $n^k n_k = -1 < 0$. Enfin, sur une portion caractéristique (ou en un point où le plan tangent est caractéristique) aucune transformation de la nature indiquée n'est possible.

Envisageons pour terminer un domaine D limité par une surface fermée T, celle-ci étant composée de certaines portions orientées respectivement dans l'espace et dans le temps et séparées par des variétés (arêtes) caractéristiques à dimensions $\leqq m-2$. On tire des considérations qui précèdent la formule de Gauss

$$\int_D \operatorname{div} A\, d\Omega = \int_T A_{<n>}\, \varepsilon\, dT$$

et la formule de Green

$$\int_D (u\,\Delta v - v\,\Delta u)\, d\Omega = \int_T \left(u \frac{dv}{dn} - v \frac{du}{dn}\right) \varepsilon\, dT,$$

la quantité ε ayant la valeur $+1$ le long des portions d'orientation d'espace, la valeur -1 le long des portions orientées dans le temps et la valeur 0 le long des variétés caractéristiques.[1]

Nulle application des deux dernières formules ne sera donnée dans le présent travail.

---

[1] Ce facteur ε doit évidemment entrer dans les intégrales dont il est question au n° **61**.

93. Les intégrales I^α et I_*^α. La formule fondamentale. — Nous avons vu au n° 87 que, dans le cas de signature lorentzienne, le seul cas qui nous intéresse dans la suite, le conoïde au sommet x et à l'équation $\Gamma(x; y) = 0$ divise l'espace voisin de x en trois régions. Certaines conventions faites, on pourra, comme dans le cas de l'espace lorentzien, parler sans ambiguïté de la nappe rétrograde et de la nappe directe d'un tel conoïde.[1]

Soit maintenant S une surface à $m - 1$ dimensions *d'orientation d'espace*, et admettons que les conoïdes rétrogrades par exemple ayant comme sommet un point arbitraire x situé d'un certain côté de S et assez voisins de S délimitent avec S certains domaines D_S^x. Nous admettons aussi, jusqu'à nouvel ordre, que nos séries (67) convergent dans ce domaine. Cela étant, on pose

$$(91) \qquad I^\alpha f(x) = \int_{D_S^x} f(y)\, V^\alpha(x; y)\, dy.\,[2]$$

Nous savons que $V^\alpha(x; y)$ devient singulier sur le conoïde comme le facteur $s(x; y)^{\alpha-m} = \Gamma(x; y)^{\frac{\alpha-m}{2}}$; plus précisement $V^\alpha(x; y) = s(x; y)^{\alpha-m} \cdot v^\alpha(x; y)$, où $v^\alpha(x; y)$ est régulier sur le conoïde et où l'on a $v^\alpha(x; x) = V_0(x; x) = 1$. (formules (67) et (69)). Vu encore que, dans un système de coordonnées normales approprié, $\Gamma(x; y)$ admet l'aspect lorentzien (58), l'étude de l'intégrale (91) se réduit immédiatement à celle de notre ancienne intégrale $I^\alpha f(P)$ relative à l'espace lorentzien (p. 47).[3] On pourra donc répéter presque mot à mot les faits démontrés plus haut au sujet de cette dernière intégrale. En particulier, la nouvelle intégrale converge et elle est une fonction holomorphe de α pour $\Re(\alpha) > m - 2$),[4] qui peut être prolongée vers la gauche[5] jusqu'à une valeur arbitraire de $\Re(\alpha)$, si la fonction f et la surface S sont suffisamment dérivables. On a en outre

$$(91^{bis}) \qquad I^0 f(x) = f(x), \quad I^{-2p} f(x) = \Delta^p f(x).\,[6]$$

Nous allons maintenant appliquer à nos intégrales la formule de Green (89). La marche sera tout à fait analogue à celle suivie dans le n° **22**. Pour commencer, nous posons dans la susdite formule $v = V^{\alpha+2}$. En désignant alors par

(1) *Cf.* Bureau **1**, p. 33.

(2) Nous posons ici, pour éviter toute ambiguïté, $dy = \sqrt{g}\, dy^1\, dy^2 \ldots dy^m$, quantité désignée auparavant par $d\Omega$ (*cf.* formule (21bis)).

(3) Notons que la quantité $\sqrt{g}$ devient $= 1$ au point x.

(4) *Cf.* n° **25**.

(5) *Cf.* n^{os} **27**—**36**, en particulier p. 51.

(6) *Cf.* n° **37**.

T la surface totale délimitant D_S^x, il vient pour d'assez grandes valeurs de $\Re(\alpha)$, qu'on va préciser,

$$\int_{D_S^x} (u(y)\,\Delta_y\, V^{\alpha+2}(x;y) - \Delta\, u(y)\, V^{\alpha+2}(x;y))\,dy$$
$$= \int_T (u \operatorname{grad}^k V^{\alpha+2} - V^{\alpha+2} \operatorname{grad}^k u)\, J_k \sqrt{g}\, d\lambda^1 \ldots d\lambda^{m-1}.$$

La surface T est composée d'une portion de la nappe du conoïde et de la portion de surface S^x découpée de S par le conoïde. Les intégrales de volume ci-dessus convergent certainement si l'on prend $\Re(\alpha) > m - 2$. Les circonstances sont moins favorables pour les intégrales de surface, puisque les dérivées premières de $V^{\alpha+2}$ deviennent infinies sur le conoïde tant que $\Re(\alpha) < m$. Par contre, pour $\Re(\alpha) > m$, ces dérivées restent finies et continues sur la surface T. Il y en a plus; pour les mêmes valeurs, la fonction $V^{\alpha+2}$ et ses dérivées en question s'annulent sur le conoïde (*cf.* le n° **22**). Il ne reste donc de l'intégrale de surface que la partie étendue à la portion de surface S^x; celle-ci ayant une orientation d'espace, l'intégrale correspondante pourra s'écrire sous la forme (90), savoir

$$-\int_{S^x} \left(u \frac{d\,V^{\alpha+2}}{d\,n} - \frac{d\,u}{d\,n} V^{\alpha+2}\right) d\,S,$$

le changement de signe étant dû à la nouvelle convention d'après laquelle on entendra désormais par n la normale unitaire *intérieure*. Observons encore que $\Delta_y V^{\alpha+2} = V_\alpha$ selon (66). Dès lors, nous sommes à même d'écrire la *formule fondamentale*

$$(92)\quad \int_{D_S^x} u(y)\, V^\alpha(x;y)\, dy = \int_{D_S^x} \Delta_y u\; V^{\alpha+2}(x;y)\, dy$$
$$+ \int_{S^x} \left[\frac{d\,u(y)}{d\,n} V^{\alpha+2}(x;y) - u(y) \frac{d\,V^{\alpha+2}(x;y)}{d\,n}\right] d\,S.$$

Pour mettre cette formule sous une forme plus condensée, nous posons

$$(93)\quad I_*^\alpha \overline{f, g, h}\,(x) = \int_{D_S^x} f(y)\, V^\alpha(x;y)\, dy + \int_{S^x} \left[g(y)\, V^\alpha(x;y) - h(y) \frac{d\,V^\alpha(x;y)}{d\,n}\right] d\,S,$$

par où la *formule fondamentale* s'écrit, comme dans le cas lorentzien (formule (61bis) du n° **23**),

$$I^{\alpha} u(x) = I_{*}^{\alpha+2} \overline{\Delta u, \frac{du}{dn}, u(x)}. \tag{94}$$

Les formules ci-dessus ne sont jusqu'ici démontrées que pour $\Re(\alpha) > m$. Or les intégrales figurant dans (92) restent convergentes pour $\Re(\alpha) > m - 2$ (*cf.* n° **25**), d'où l'on conclut, par les principes du prolongement analytique, que la formule reste valable pour toutes les valeurs vérifiant la dernière condition. Il résulte du même principe que les formules équivalentes (92) et (94) subsistent pour toute valeur de α, jusqu'à laquelle le prolongement est possible, en raison des hypothèses relatives à la fonction u et à la surface S. Comme plus haut, on aura en particulier pour $\alpha = 0$ la *formule d'importance capitale*(¹)

$$u(x) = I_{*}^{2} \overline{\Delta u, \frac{du}{dn}, u(x)}, \tag{95}$$

formule par laquelle se résout le problème de Cauchy, exposé succinctement au numéro qui suit.

Signalons encore les relations $\Delta I_{*}^{\alpha+2} = I_{*}^{\alpha}$ et $I^{\alpha} I_{*}^{\beta} = I_{*}^{\alpha+\beta}$(²) dont la première se vérifie comme le fait correspondant établi dans le cas lorentzien (n° **42**), tandis que la seconde exige une démonstration qu'on donnera plus loin (n° **95**).

94. Problème de Cauchy. Absence du principe de Huygens. — Pour éviter des redites continuelles, se faisant déjà pressentir au numéro précédent, nous nous bornerons, en ce qui concerne le problème de Cauchy, à quelques indications sommaires. Un tel procédé est d'autant plus légitime que le problème en question a été traité dans le cas lorentzien avec une profusion de détails. Nous renvoyons le lecteur au n° **43** pour l'énoncé du problème, à la formule (95), transcrite à la manière de la formule (3) du n° **44**, pour la solution sous sa forme générale indépendante de la parité de l'espace, enfin aux endroits cités au n° **60** pour la vérification de la solution.

On a vu aux n^os^ **55** et **56** que, dans le cas lorentzien, la solution obéit au principe de Huygens pour les dimensions paires, tandis qu'il y a des intégrales résiduelles (diffusion des ondes) pour les dimensions impaires. Pour les dernières dimensions, il n'y a aucune différence essentielle entre le cas actuel et le cas particulier. Pour les dimensions paires, il y a au contraire une différence fondamentale en ce que ce ne sont que certains termes de la série (67) qui donnent

(¹) *Cf.* le n° **44**.
(²) *Cf.* les n^os^ **42** et **24**.

lieu au principe de Huygens. Tout dépend encore des propriétés de la fonction *gamma*. En effet, les fonctions $H_m(\alpha, k)$, figurant dans la formule (67), chacune produit de deux fonctions *gamma*, n'admettent de pôles en $\alpha = 2$ que si l'on a $k \leqq \frac{1}{2}m - 2$; *cf.* la formule (71). Ce ne sont donc que ces termes-là, complétés, en ce qui concerne la double couche (pour des raisons expliquées au n° **15**) par le terme correspondant à $k = \frac{1}{2}m - 1$, qui obéissent au principe de Huygens. Tous les autres termes, en nombre infini, fournissent des intégrales résiduelles (convergentes), étendues à D^x_S ou à S^x.

Les dernières lignes ne sont qu'une reproduction presque littérale de ce que nous avons dit au n° **56**^{bis} au sujet de la solution de l'équation des ondes amorties. Cependant, dans ce cas particulier, on sait que les termes fournissant les intégrales résiduelles sont effectivement différentes de zéro, puisqu'on les connaît d'une façon tout à fait explicite. Dans le cas général qui nous occupe, ces termes sont définis comme solutions d'un système récurrent d'équations différentielles, et on est loin de les connaître effectivement. Cela explique la question posée par M. Hadamard[1] indiquée dans la note [3] à la page 87 du présent ouvrage et que nous y avons encore élucidée par des renvois à des résultats récents dus à différents auteurs.

95. Formules de composition. Propriétés d'échange. — Dans le cas lorentzien, la formule de composition

$$I^\alpha(I^\beta f(x)) = I^{\alpha+\beta} f(x) \tag{96}$$

et la formule légèrement plus générale

$$I^\alpha(I^\beta_* \overline{f, g, h}(x)) = I^{\alpha+\beta}_* \overline{f, g, h}(x) \tag{96^{bis}}$$

(*cf.* les n^{os} **23** et **24**) étaient des conséquences faciles d'une formule de composition relative au noyau respectif $r^{\alpha-m}/H_m(\alpha)$ (*cf.* le n° **5**), que nous écrivons ici sous la forme

$$\int_{D^x_y} \frac{r(x;z)^{\alpha-m}}{H_m(\alpha)} \cdot \frac{r(z;y)^{\beta-m}}{H_m(\beta)} \cdot dz = \frac{r(x;y)^{\alpha+\beta-m}}{H_m(\alpha+\beta)}. \tag{97}$$

---

[1] HADAMARD **1**, p. 325.

Les formules de composition de caractère général relatives au cas actuel sont du même aspect que les formules (96) *et* (96bis) *et dérivent de leur côté de la formule de composition relative au noyau* V^α

$$\int_{D_y^x} V^\alpha(x;\,z)\cdot V^\beta(z;\,y)\,dz = V^{\alpha+\beta}(x;\,y).\text{(}^1\text{)} \tag{98}$$

On suppose ici que les points x et y sont tels que le conoïde rétrograde appartenant à l'un d'eux et le conoïde direct appartenant à l'autre délimitent un certain double-conoïde borné D_y^x, domaine d'intégration dans la formule ci-dessus. (²)

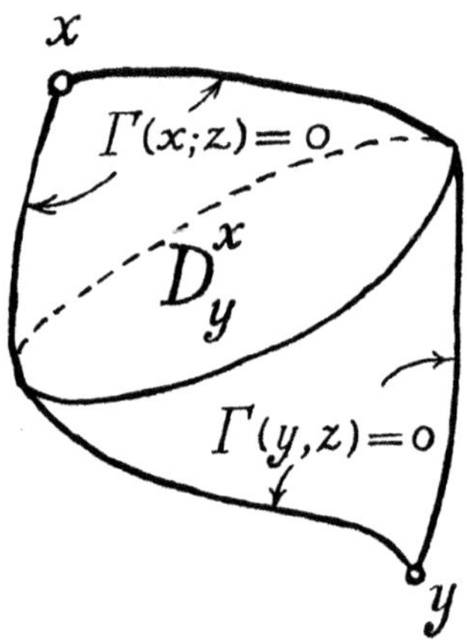

Fig. 13. Le double-conoïde D_y^x.

Si l'on savait déjà, ce qui ne sera démontré que plus loin, que le noyau $V^\alpha(x;\,y)$ est symétrique en fonction de x et de y, la formule (98) serait identique à la suivante, dont nous allons nous occuper en premier lieu,

$$\int_{D_y^x} V^\alpha(x;\,z)\cdot V^\beta(y;\,z)\,dz = V^{\alpha+\beta}(x;\,y). \tag{99}$$

La démonstration de la formule (97) reposait sur des calculs explicites, tandis que celle de (99) sera basée sur une application de la formule de Green, sur le fait que l'opérateur I^0 est l'opérateur identique et sur un théorème d'unicité très simple, emprunté à la théorie des fonctions analytiques.

(¹) Pour la signification de dz *cf.* la note(²) à la page 190.

(²) *Cf.* le nº 17 et en particulier la fig. 1, où les points P et Θ jouent les rôles de x et de y.

La formule (99), une fois démontrée, fournit sur le champ (comme on le verra à la fin de ce numéro) la propriété d'échange (ou de symétrie) de la fonction V^α, savoir $V^\alpha(x; y) = V^\alpha(y; x)$, d'où s'ensuit immédiatement d'une part la formule (98) et d'autre part la propriété d'échange des coefficients V_k, savoir $V_k(x; y) = V_k(y; x)$.[1]

Nous démontrons d'abord la relation

$$\int_{D^x_y} V^{\alpha+2n}(x; z) \cdot V^\beta(y; z)\, dz = \int_{D^x_y} V^\alpha(x; z) \cdot V^{\beta+2n}(y; z)\, dz, \tag{100}$$

qui, pour commencer, est certainement valable lorsque $\Re(\alpha)$ et $\Re(\beta)$ sont assez grands et pour $n = 0, 1, 2, \ldots$

Observons à cet effet 1°. que par exemple $\Delta_z V^{\alpha+2}(x; z) = V^\alpha(x; z)$ (formule (66)) et 2°. que les produits $V^\gamma(x; z) \cdot \partial V^\delta(y; z)/\partial z^j$ et $V^\delta(y; z) \cdot \partial V^\gamma(x; z)/\partial z^j$ s'annulent sur la surface du double-conoïde, dès que $\Re(\gamma)$ et $\Re(\delta)$ sont assez grands, fait dû à la présence des facteurs respectifs $s^{\gamma-m}$ et $s^{\delta-m}$ dans les fonctions V^γ et V^δ. Dès lors, une application réitérée de la formule de Green (89) d'après le schéma

$$\int V^{\alpha+2}(x; z) \cdot V^\beta(y; z)\, dz = \int V^{\alpha+2}(x; z) \cdot \Delta_z V^{\beta+2}(y; z)\, dz$$
$$= \int \Delta_z V^{\alpha+2}(x; z) \cdot V^{\beta+2}(y; z)\, dz = \int V^\alpha(x; z) \cdot V^{\beta+2}(y; z)\, dz$$

fournit la relation demandée. L'absence des intégrales de surface dans les formules ci-dessus repose sur l'observation 2°.

C'est de la relation (100) qu'on va déduire, à l'aide du théorème d'unicité auquel nous avons fait allusion, la formule (99) — notre objet principal — qui, comme nous l'avons dit, comprend pour ainsi dire tous les autres résultats indiqués. Cependant, pour des raisons méthodiques, il nous paraît préférable de faire précéder cet arrangement par une déduction directe et très simple de la propriété d'échange des coefficients $V_k(x; y)$.

La formule (100) n'est démontrée jusqu'ici que pour d'assez grandes valeurs de $\Re(\alpha)$ et de $\Re(\beta)$, par exemple pour $\Re(\alpha) > m$ et $\Re(\beta) > m$, mais on voit par prolongement analytique qu'elle reste valide par exemple pour $\alpha = \beta = 0$. Écrivons donc la formule

---

[1] M. Hadamard obtient cette propriété par la méthode de descente et en outre, mais seulement pour $k = 0$, par une étude approfondie du système d'équations des géodésiques; *cf.* Hadamard **1**, p. 366—377.

$$(101)\qquad \int\limits_{D^x_y} V^{2n}(x;z)\cdot V^0(y;z)\,dz = \int\limits_{D^x_y} V^0(x;z)\cdot V^{2n}(y;z)\,dz,$$

où, si l'on veut être rigoureux, on doit écrire au lieu du second membre, par exemple,

$$\underset{\gamma=0}{\text{prol. anal.}} \int\limits_{D^x_y} V^\gamma(x;z)\cdot V^{2n}(y;z)\,dz.$$

Si l'on identifie la nappe du conoïde de sommet y à la surface S figurant dans la formule de définition (91) de l'intégrale $I^\alpha f(x)$, la dernière formule peut s'écrire $I^0\, V^{2n}(y;x) = V^{2n}(y;x)$; on se rappelle que I^0 est l'opérateur identique. En d'autres termes, le second membre de la formule (101) se réduit a $V^{2n}(y;x)$; le premier membre devient donc $V^{2n}(x;y)$. Dès lors, la formule (101) se réduit à la relation d'échange

$$(102)\qquad V^{2n}(x;y) = V^{2n}(y;x).$$

En réalité, le fait que l'opérateur I^0 est l'opérateur identique n'a été énoncé jusqu'ici que pour des surfaces S ayant une orientation d'espace, tandis que la nappe en question est une surface caractéristique, admettant même un point singulier, à savoir son sommet. Cependant, les calculs donnés aux n^{os} **27—37** s'appliquent au cas actuel presque sans modifications, et la difficulté insignifiante provenant du point singulier s'élimine aisément à l'aide des indications données à la page 102.

On tire facilement de l'identité (102) la propriété d'échange des coefficients V_k, c'est-à-dire

$$(103)\qquad V_k(x;y) = V_k(y;x).$$

Il suffit de remplacer dans la formule (78) V_k par $V_k(x;y) - V_k(y;x)$, de faire tendre α vers l'infini, en parcourant les nombres pairs $2n$, et d'appliquer la formule (78bis), en y posant successivement $l = 0, 1, 2, \ldots$

En portant le résultat (103) dans la formule de définition (67) écrite pour $V^\alpha(x;y)$ et $V^\alpha(y;x)$ et en tenant compte de la symétrie de la distance géodésique $s(x;y)$, on obtient aussi la formule d'échange

$$(104)\qquad V^\alpha(x;y) = V^\alpha(y;x).$$

Il en résulte en particulier que V^α qui, comme nous le savons, satisfait à la relation (66) $\Delta_y V^{\alpha+2}(x;y) = V^\alpha(x;y)$, satisfait aussi à $\Delta_x V^{\alpha+2}(x;y) = V^\alpha(x;y)$.

Revenons maintenant à la démonstration de la formule (99). Fixons dans la formule (100) la valeur de β, avec $\Re(\beta) \geqq m$, et posons $\alpha = 0$. En s'appuyant toujours sur le fait que I^0 est l'opérateur identique, on obtient

$$\int_{D_y^x} V^{2n}(x;\, z)\, V^{\beta}(y;\, z)\, dz = V^{\beta+2n}(y;\, x). \tag{105}$$

Nous formons l'expression

$$G(\alpha) = \int_{D_y^x} V^{\alpha}(x;\, z)\, V^{\beta}(y;\, z)\, dz - V^{\alpha+\beta}(y;\, x), \tag{106}$$

où $\Re(\alpha) \geqq m$, $\Re(\beta) \geqq m$, ce qui fait que la quantité à intégrer est *bornée* et continue. La valeur de β étant fixée, $G(\alpha)$ est une fonction *holomorphe* de α qui, d'après la relation (105), s'annule pour $\alpha = 2n$ assez grand.(¹) Remarquons encore qu'on a toujours $s(x;\, z) \leqq s(x;\, y)$. Ceci posé, on voit par l'inégalité (83) que la fonction holomorphe $G(\alpha)$ satisfait dans le demi-plan $\Re(\alpha) \geqq m$ à une inégalité de la forme

$$|\, G(\alpha)\,| \leqq \left|\, \bar{C} \cdot \left(\frac{\alpha}{e}\right)^{-\alpha} \alpha^p \cdot s(x;\, y)^{\alpha} \right|, \tag{107}$$

les valeurs de $\bar{C}$ et de p étant indépendantes de α.(²)

Ceci étant admis, il résultera du théorème d'unicité qui suit que la fonction $G(\alpha)$ s'annule identiquement ou, en d'autres termes, *notre relation* (99) *se trouve établie pour* $\Re(\alpha) \geqq m$, $\Re(\beta) \geqq m$.

Voici le *théorème d'unicité* en question. Supposons que les fonctions $g(z)$ et $h(z)$ satisfont à l'inégalité $|g(z)| \leqq |h(z)|$ et qu'elles sont holomorphes dans le demi-plan $\Re(z) \geqq 0$, la fonction $h(z)$ y étant constamment différente de zéro, tandis que $g(z) = 0$ aux points $z_1, z_2, \ldots, z_n, \ldots$ avec $\Re(z_n) > 0$ et $|z_n| > 1$; dans ces hypothèses la divergence de la série $\sum \Re(1/z_n)$ entraîne que la fonction $g(z)$ s'annule identiquement.

Dans l'énoncé habituel de ce théorème, la condition de majoration ci-dessus est remplacée par la condition que $g(z)$ est borné.(³) Notre théorème apparemment plus général se réduit immédiatement au théorème particulier, si on applique ce dernier à la fonction $g(z) : h(z)$.

Appliquons le théorème d'unicité ci-dessus à la fonction $G(\alpha)$, en posant $z = \alpha - m$. Remplaçons $g(z)$ par $G(\alpha)$ et $h(z)$ par la fonction holomorphe de α,

(¹) P. ex. pour $2n \geqq m$.

(²) Bien entendu, les valeurs en question dépendent de x, de y et de β.

(³) *Cf.* p. ex. Pólya-Szegö 1, t. I, Abschnitt III, problème 298, p. 142 et 327.

dont la valeur absolue forme le second membre de la formule (107). Observons encore que les zéros connus de $G(\alpha)$ donnent lieu, après la substitution $z = \alpha - m$, à la série divergente $\sum_{2n>m} 1/(2n-m)$. On conclut que $G(\alpha)$ s'annule identiquement pour $\Re(\alpha) \geqq m$, ce qu'il fallait démontrer. *La formule importante* (99) *se trouve ainsi établie* pour $\Re(\alpha) \geqq m$, $\Re(\beta) \geqq m$. Elle s'étend par prolongement analytique à des valeurs arbitraires de α et de β.

A vrai dire, la formule en question se trouve établie avec $V^{\alpha+\beta}(y; x)$ figurant au second membre, tandis que dans l'énoncé on a $V^{\alpha+\beta}(x; y)$. Ceci dépend du fait que, dans la démonstration, on a fixé β et qu'on a fait varier α. Le chemin opposé aurait conduit à $V^{\alpha+\beta}(x; y)$.

Des dernières remarques il ressort de nouveau qu'on a pour une valeur arbitraire de γ la propriété d'échange $V^\gamma(x; y) = V^\gamma(y; x)$. En faisant tendre $|\gamma|$ vers l'infini, on retrouve la propriété d'échange des coefficients $V_k(x; y) = V_k(y; x)$, démontrée plus haut moyennant $\gamma = 2n \to \infty$. Mais ce qui importe davantage, c'est que, par la symétrie de V^γ, *la formule* (98) *se trouve démontrée en même temps que la formule* (99), *ce qui à son tour implique les formules de composition générales* (96) *et* (96bis).

96. Le cas non analytique. Aperçu de la méthode. (1) — Plaçons-nous maintenant dans le cas où les coefficients g_{jk} de la forme métrique ne sont plus nécessairement analytiques; nous supposons pourtant qu'ils sont *réguliers* suivant la terminologie de M. Hadamard, c'est-à-dire qu'ils sont *suffisamment dérivables.* (2) L'expérience acquise dans le cas lorentzien concernant l'allure appropriée des données nous apprend qu'une telle hypothèse est non seulement commode, mais pour ainsi dire indispensable. (3) Nous allons montrer que la méthode développée plus haut peut encore s'étendre au cas généralisé. Nous nous laisserons guider par quelques observations supplémentaires relatives au cas analytique.

Du point de vue de la résolution du problème de Cauchy, l'utilité du noyau fractionnaire consiste à permettre de calculer par prolongement analytique certaines intégrales qui deviennent divergentes si l'on utilise la solution élémentaire proprement dite. Le cas lorentzien fait ressortir ce point d'une manière très claire, la solution élémentaire y étant tout simplement r^{2-m} et le noyau fraction-

---

(1) Pour éviter des longueurs et des redites, nous nous contenterons en général dans la suite d'esquisser la marche du raisonnement sans trop nous arrêter sur les détails.

(2) *Voir* p. 51 et 160.

(3) *Cf.* Hadamard **1**, p. 415 et Courant-Hilbert **1**, II, note (1), p. 469.

naire $r^{\alpha-m} : H_m(\alpha)$. Dans le cas actuel, le noyau fractionnaire V^α est donné par la série (67) qui, en général, contient un nombre illimité de termes. En posant dans cette série $\alpha = 2$, on obtient la solution élémentaire proprement dite V^2, qui pourtant, comme nous le savons, ne peut être immédiatement substituée dans la formule (95). Or si l'on prend la série (67) terme à terme, en y posant $\alpha = 2$, on voit que ce n'est qu'un nombre limité de termes au début de la série qui donne lieu à des intégrales dangereuses, et le prolongement analytique n'est nécessaire que pour le compte de ces termes-ci. Quant aux autres termes, on pourrait les introduire dans la formule (95) en *substituant* dans leurs expressions $\alpha = 2$, en s'abstenant ainsi de tout prolongement analytique pour autant qu'il est question de ces derniers termes.

Conformément à cela posons

$$(108) \qquad V^\alpha(x; z) = \sum_{k=0}^{\infty} \frac{V_k s^{\alpha-m+2k}}{H_m(\alpha, k)} = \sum_{k=0}^{p} + \sum_{k=p+1}^{\infty} = V_{\mathrm{I}}^\alpha(x; z) + V_{\mathrm{II}}^\alpha(x; z).$$

Nous choisissons p assez grand pour que $V_{\mathrm{II}}^2(x; z)$ en fonction de z s'annule avec ses dérivées du premier ordre sur le conoïde de sommet x. Il suffira de prendre $2p > m - 2$.

On obtient aisément, en raison du système récurrent (69)

$$(109) \qquad \Delta_z V_{\mathrm{I}}^{\alpha+2}(x; z) = V_{\mathrm{I}}^\alpha(x; z) + s(x; z)^{\alpha+2-m+2p} \frac{\Delta_z V_p(x; z)}{H_m(\alpha+2, p)}$$

et

$$(109^{bis}) \qquad \Delta_z V_{\mathrm{II}}^{\alpha+2}(x; z) = V_{\mathrm{II}}^\alpha(x; z) - s(x; z)^{\alpha+2-m+2p} \frac{\Delta_z V_p(x; z)}{H_m(\alpha+2, p)}.$$

En posant $\alpha = 0$, on a $V_{\mathrm{I}}^0 = 0$, sauf sur le conoïde de sommet x, et $V_{\mathrm{II}}^0 = 0$ partout.[1] Il vient donc pour $s > 0$

$$(109^{ter}) \quad \Delta_z V_{\mathrm{I}}^2(x; z) = -\Delta_z V_{\mathrm{II}}^2(x; z) = s(x; z)^{2-m+2p} \frac{\Delta_z V_p(x; z)}{H_m(2, p)} = A(x; z),$$

pour abréger.

Nous allons maintenant utiliser la formule de Green à peu près de la même manière qu'au sujet de la formule (99). Soit $v(z)$ une fonction qui s'annule avec ses dérivées du premier ordre sur le conoïde de sommet x. Pour $\Re(\alpha)$ assez grand, $V_{\mathrm{I}}^{\alpha+2}(y; z)$ se comporte de même sur le conoïde de sommet y. Dès lors, on obtient par la formule de Green et la formule (109)

[1] Cela tient au facteur $\Gamma\left(\frac{\alpha}{2}\right)$ de $H_m(\alpha, k)$ (formule (71)).

$$(110) \quad \int\limits_{D_y^x} \Delta v(z) \cdot V_{\mathrm{I}}^{\alpha+2}(y; z)\, dz$$

$$= \int\limits_{D_y^x} v(z) \cdot \left\{ V_{\mathrm{I}}^{\alpha}(y; z) + s(y; z)^{\alpha+2-m+2p} \cdot \frac{\Delta_z V_p(y; z)}{H_m(\alpha + 2, p)} \right\} dz.$$

Par prolongement analytique, cette relation devient pour $\alpha = 0$, en vertu de la relation (109ter),

$$(111) \qquad F(x; y) = v(y) + \int\limits_{D_y^x} A(y; z)\, v(z)\, dz,$$

où $F(x; y)$, prolongement analytique du premier membre de (110), peut être considéré comme connu, dès que Δv est donné. Pour $v(z) = V_{\mathrm{II}}^2(x; z)$, on aura, d'après ($109^{ter}$), à substituer dans le premier membre de (110) $\Delta v(z) = -A(x; z)$.

On a donc, dans le cas *holomorphe*, obtenu pour la fonction $V_{\mathrm{II}}^2(x; z)$ une équation intégrale du type de Volterra à noyau borné et continu.

Passons au cas *non analytique*. Nous y avons déjà supposé que les coefficients g_{jk} sont réguliers. Naturellement on doit aussi supposer qu'il en est de même de la région considérée, surtout au point de vue du problème des géodésiques. On peut alors calculer la distance géodésique s et, à l'aide du système récurrent (69), les coefficients $V_0, V_1, \ldots, V_p$, toutes ces quantités étant des fonctions assez régulières. Cela étant, on peut encore d'une part calculer le noyau $A(y; z)$ et d'autre part le noyau fractionnaire tronqué $V_{\mathrm{I}}^{\alpha}(x; z)$ et dès lors la fonction $F(x; y)$. Enfin, *on définit* $v(z) = V_{\mathrm{II}}^2(x; z)$ *comme solution de l'équation de Volterra* (111). Cette solution, qui sera examinée au numéro suivant, a des propriétés de convergence excellentes, chose qu'on devine par la connaissance qu'on a de l'équation de Volterra dans le cas linéaire. On peut aussi montrer que $V_{\mathrm{II}}^2(x; z)$ a les propriétés requises: évanouissement avec les dérivées du premier ordre sur le conoïde de sommet x, $\Delta(V_{\mathrm{I}}^2 + V_{\mathrm{II}}^2) = 0$ etc.[1] Dès lors, on arrivera à la solution du problème de Cauchy si l'on porte dans la formule (95), transcrite à la manière de la formule (3) du nº **44**, le noyau $V^2 = V_{\mathrm{I}}^2 + V_{\mathrm{II}}^2$, les intégrales formées avec V_{I}^2 étant définies comme prolongements analytiques des intégrales correspondantes formées avec V_{I}^{α}.

---

[1] *Voir* les remarques faites à la page 212 au sujet des propriétés analogues de la fonction W_{II}^2.

Avant d'aller plus loin, rappelons que la première solution du problème de Cauchy dans le cas non analytique est encore due à M. Hadamard. (1) C'est en s'inspirant de la méthode utilisée par E. Elia Levi et Hilbert dans le cas elliptique, méthode appelée par Hilbert celle de la *«parametrix»* (solution élémentaire incomplète), que M. Hadamard a formé une «parametrix» adaptée au cas actuel. Tandis que, dans le cas elliptique, les auteurs cités arrivent à une équation intégrale de Fredholm, M. Hadamard réduit la résolution du problème de Cauchy à celle d'une équation intégrale de Volterra — le domaine d'intégration étant du type que nous avons désigné par D_S^x au n° **93** — pour remonter (2) ensuite de la solution du problème général à la solution élémentaire. Afin de traiter les dimensions impaires et paires, il faut encore recourir à des méthodes différentes. Le double-conoïde D_y^x, à l'aide duquel nous formons notre noyau fractionnaire (espèce de solution élémentaire généralisée), ne figure pas chez M. Hadamard.

97. Recours à une équation intégrale de Volterra. — Rappelons brièvement le mécanisme de l'équation de Volterra de seconde espèce dans le cas linéaire. (3) Écrivons l'équation

$$\varphi(x) - \mu \int_a^x K(x; y)\, \varphi(y)\, dy = f(x), \tag{112}$$

K et f étant des fonctions données et $\varphi(x)$ la fonction inconnue. Nous supposerons que la fonction $K(x; y)$, le noyau, est bornée dans le triangle limité par les droites $y = a$, $x = b$, $y = x$ $(b > a)$.

Formons les *noyaux itérés* successifs de $K(x; y) = K^{(1)}(x; y)$, définis de proche en proche par les relations de récurrence

$$K^{(n+1)}(x; y) = \int_y^x K(x; s)\, K^{(n)}(s; y)\, ds = \int_y^x K^{(n)}(x; s)\, K(s; y)\, ds;$$

ces noyaux sont tous bornés dans le même domaine que $K(x; y)$. Plus précisément, l'inégalité $|K(x; y)| \leqq M$ entraîne les inégalités

$$|K^{(n+1)}(x; y)| \leqq M^{n+1} \cdot \frac{(x-y)^n}{n!}.$$

(1) Hadamard **1**, p. 377—379, 415—426.
(2) Hadamard **1**, p. 421—424.
(3) *Cf.* Goursat III (1915), p. 324—329.

En effet, l'inégalité précédente étant supposée valable pour n au lieu de $n+1$, on obtient

$$|K^{(n+1)}(x;y)| \leqq M \cdot \frac{M^n}{(n-1)!}\int_y^x (x-s)^{n-1}\,ds = M^{n+1}\cdot\frac{(x-y)^n}{n!}.$$

Dès lors, la série

$$L(x;y\,|\,\mu) = K(x;y) + \mu\cdot K^{(2)}(x;y) + \cdots + \mu^{n-1}\cdot K^{(n)}(x;y) + \cdots \tag{113}$$

est absolument et uniformément convergente tant que μ reste borné. La fonction $L(x;y\,|\,\mu)$ est une fonction entière de μ, qui s'appelle le *noyau résolvant* relatif au noyau $K(x;y)$. A l'aide du noyau résolvant, la solution de l'équation (112) s'écrit

$$\varphi(x) = f(x) + \mu\int_a^x L(x;y\,|\,\mu) f(y)\,dy. \tag{114}$$

Nous appliquerons des principes analogues à l'équation intégrale (111). Soit dans un ordre d'idées général D_a^b un double-conoïde intérieur à la région $\mathcal{R}$ (nº 85) et $K(x;y)$ une fonction définie pour tout couple de points x et y tels que x appartient à D_a^b et y à D_a^x. Posons $K^{(1)}(x;y) = K(x;y)$ et

$$K^{(n+1)}(x;y) = \int_{D_y^x} K(x;z)K^{(n)}(z;y)\,dz = \int_{D_y^x} K^{(n)}(x;z)K(z;y)\,dz. \tag{115}$$

Nous allons établir l'existence d'une constante B, ne dépendant que de la région $\mathcal{R}$ et telle que l'inégalité

$$|A(x;y)| \leqq M \tag{116}$$

entraîne les inégalités

$$|K^{(n+1)}(x;y)| \leqq \Gamma(\tfrac{1}{2})\frac{M\cdot(MB)^n}{\Gamma\left(\frac{mn+1}{2}\right)}\cdot s(x;y)^{mn} \tag{117}$$

avec

$$B = \frac{\omega_{m-1}\,\Gamma\left(\frac{m+1}{2}\right)\sqrt{g_{\max}}}{2(m-1)},\text{[1]}$$

---

[1] Dans le cas lorentzien, on peut améliorer cette estimation en montrant que

$$|A^{(n+1)}(x;y)| \leqq \frac{M\cdot(MB)^n}{(mn)!}\cdot s(x;y)^{mn},$$

avec une certaine valeur de B. Cette majoration est facile à étendre au cas riemannien pour la dimension 2. L'extension aux dimensions supérieures semble exiger des calculs assez laborieux.

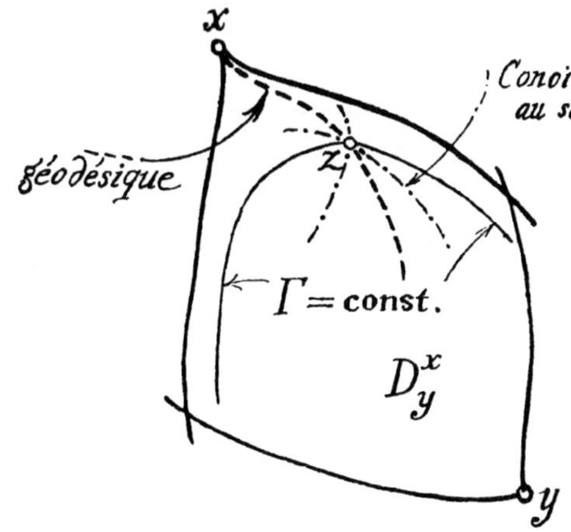

Fig. 14. Les deux nappes du conoïde au sommet z séparées par une surface Γ = const.

où m désigne comme toujours la dimension de l'espace, ω_{m-1} est donné par la formule (8) du n° **6** et $g_{\max}$ désigne le maximum de g dans le domaine D_y^x, la quantité g étant relative à certains systèmes de coordonnées normales autour de x.

Pour parvenir à cette limitation, nous aurons besoin de quelques considérations géométriques. Nous montrons d'abord que les surfaces $\Gamma(x; z) = \text{const.} > 0$, x étant fixé et z variable, sont coupées orthogonalement par les géodésiques issues de x. Cela résulte sur le champ de la formule aux limites (36)([1]), qui donne $g_{jk}\, dz^j\, \delta z^k = 0$, dz signifiant un déplacement infinitésimal le long de la géodésique et δz un déplacement infinitésimal arbitraire le long de la surface. Les géodésiques en question étant des lignes de temps, il ressort de l'orthogonalité établie tout à l'heure que les surfaces Γ = const. ont une orientation d'espace.([2]) On en conclut qu'une bicaractéristique (géodésique de longueur nulle) ne saura être tangente à aucune des surfaces en question, d'où il s'ensuit qu'une telle ligne rencontrera chaque surface en un seul point au plus. On voit par ceci que *les deux nappes du conoïde qui a un certain point z comme sommet seront séparées par la surface Γ =* const. *passant en z.*

Ces faits étant admis, nous introduisons un système de coordonnées normales autour du point x comme origine, ce système étant choisi de façon que les coef-

---

([1]) L'application de cette formule est légitime puisque, $\Gamma = s^2$ étant constant, le paramètre s qui remplace le paramètre normalisé τ, figurant dans la formule originale, ne subit aucun changement au cours de la variation.

([2]) *Cf.* la note ([1]) p. 179.

ficients au point x de la forme métrique se réduisent à ceux de la forme lorentzienne (*cf.* formule (58) du n° **87**). On aura dès lors, dans des notations faciles à comprendre, $\Gamma(x; z) = r_{xz}^2$ (*cf.* formule dernièrement citée, formule (48) du n° **85** et formule (1) du n° **14**); les surfaces $\Gamma(x; z) = \text{const.}$ deviennent des *hyperboloïdes* $r_{xz}^2 = \text{const.}$, telles qui figuraient au n° **57**, par exemple. En particulier, *le conoïde de sommet x devient le cône caractéristique de même sommet.*[1]

Ajoutons à ce qui précède que *les plans tangents* des hyperboloïdes en question *sépareront, eux aussi, les deux nappes du conoïde ayant comme sommet le point de contact.* Cela implique en particulier que *le double-conoïde D_y^x est tout entier compris dans le cône solide D* délimité par l'une des nappes du cône de sommet x (savoir celle qui renferme y) et par le plan Π tangent en y à l'hyperboloïde passant par ce dernier point.

En admettant alors que l'inégalité (117) est valable pour n au lieu de $n+1$ et pour z au lieu de y, z étant un point arbitraire intérieur à D_y^x, on trouve, en raison de (116),

$$(118) \qquad |K^{(n+1)}(x; y)| = \left| \int_{D_y^x} K^{(n)}(x; z) K(z; y)\, dz \right|$$

$$\leqq \int_{D_y^x} \frac{M \cdot (MB)^{n-1}}{\Gamma\left(\frac{m(n-1)+1}{2}\right)} \cdot r_{xz}^{m(n-1)} \cdot M \cdot dz = \frac{M \cdot (MB)^{n-1}}{\Gamma\left(\frac{m(n-1)+1}{2}\right)} \cdot M \int_{D_y^x} r_{xz}^{m(n-1)}\, dz.$$

Afin d'arriver à une majorante de la dernière intégrale, notons d'abord que

$$dz = \sqrt{g} \cdot dz^1 \ldots dz^m \leqq \sqrt{g_{\max}} \cdot dz^1 \ldots dz^m. \text{[2]}$$

D'autre part, en nous reportant à la formule qui précède la formule (24) du n° **57** et aux notations du même numéro, nous aurons

$$dz^1 \ldots dz^m = dH_r \cdot dr,$$

l'élément de surface de l'hyperboloïde H_r étant maintenant mesuré au sens lorentzien. Faisons observer que toutes les notions géométriques, telles mesure

---

(1) Dans les notations du n° **85**, on désignerait ce point par o.

(2) Dans un système de coordonnées lorentziennes dont l'axe de temps, soit l'axe des x^1, coïncide avec la droite xy, on a évidemment $r_{xz} \leqq |x^1 - z^1|$. Par cette remarque on arrive, moyennant des calculs très simples, à une majorante du type de (117), où pourtant le dénominateur $\Gamma\left(\frac{mn+1}{2}\right)$ est remplacé par $(n+1)!$. La dernière majorante, bien que plus grossière que celle du texte, suffit largement à tous les besoins.

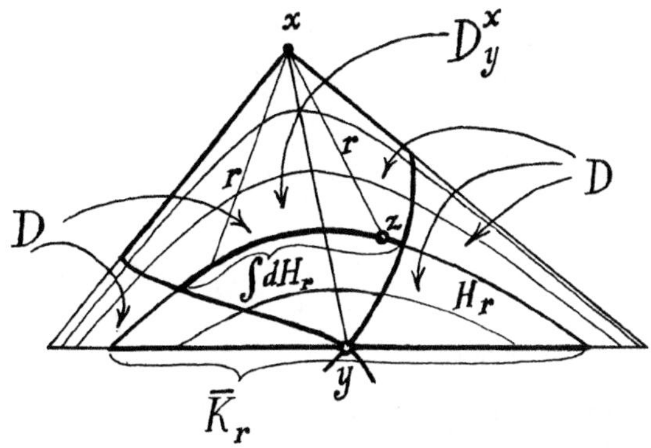

Fig. 15. Le double-conoïde D_y^x et le cône solide D.

de certains éléments, orthogonalité etc., seront à prendre dans le même sens au cours des présentes évaluations. On a donc, en notant que $r = r_{xz}$ varie de o à r_{xy},

(119)
$$\int_{D_y^x} r_{xz}^{km}\, dz \leqq \sqrt{g_{\max}} \int_0^{r_{xy}} \left(\int dH_r\right) r^{km}\, dr,$$

l'expression $\int dH_r$ signifiant la surface totale de la calotte d'hyperboloïde comprise dans le domaine D_y^x, qui, évidemment, *est inférieure à la surface totale, soit K_r, de la calotte comprise dans le cône solide D qui renferme D_y^x.*

En se reportant, d'autre part, à la formule (27) du n° **57**, où l'on a $x_1 - \xi_1 \geqq r$, on voit que (la mesure de) dH_r est inférieur(e) à (celle de) sa projection orthogonale sur un plan quelconque $\xi_1 = \text{const.}$ Or, par une transformation de Lorentz convenable, on peut faire jouer à tout plan d'orientation d'espace le rôle d'un de ces derniers plans, ce qui fait voir que dH_r est inférieur à sa projection orthogonale sur un plan d'espace quelconque. Il est manifestement de même de la surface totale K_r de la calotte de H_r comprise dans le solide D, qui est donc inférieure à sa projection orthogonale sur le plan Π limitant D. Or cette projection est identique au domaine intercepté sur Π par H_r. En remarquant d'autre part que la droite xy est orthogonale au plan Π, on voit que l'intersection de H_r et de Π n'est qu'une sphère dans Π, considéré comme un espace à $m - 1$ dimensions, sphère, dont le centre est y et dont le rayon ϱ est donné par $\varrho^2 = r_{xy}^2 - r^2$. Par suite, K_r est inférieur au volume, soit $\bar{K}_r$, de cette sphère ou, en raison de la formule (8) du n° **6** et de celle qui la précède,

$$K_r < \bar{K}_r = \varrho^{m-1} \cdot \frac{\omega_{m-1}}{m-1} = \left(r_{xy}^2 - r^2\right)^{\frac{m-1}{2}} \cdot \frac{\omega_{m-1}}{m-1}.$$

D'après ce qu'on a dit plus haut, la quantité $\int dH_r$ figurant dans la formule (119) admet la même majoration. On a donc

$$(120)\qquad \int\limits_{D_y^x} r_{xz}^{km}\,dz \leqq \frac{\omega_{m-1}}{m-1}\sqrt{g_{\max}}\int\limits_0^{r_{xy}} r^{km}(r_{xy}^2-r^2)^{\frac{m-1}{2}}\,dr$$

$$= \frac{\omega_{m-1}}{m-1}\sqrt{g_{\max}}\,r_{xy}^{(k+1)m}\int\limits_0^1 u^{km}(1-u^2)^{\frac{m-1}{2}}\,du.$$

La dernière intégrale est égale à

$$(121)\qquad \frac{1}{2}\cdot\frac{\Gamma\left(\frac{km+1}{2}\right)\Gamma\left(\frac{m+1}{2}\right)}{\Gamma\left(\frac{(k+1)m}{2}+1\right)} < \frac{1}{2}\cdot\frac{\Gamma\left(\frac{km+1}{2}\right)\Gamma\left(\frac{m+1}{2}\right)}{\Gamma\left(\frac{(k+1)m+1}{2}\right)}.$$

Les formules (119)—(121) avec $k=n-1$ et la formule (118) fournissent, après quelques simplifications, les inégalités (117) que nous avions à démontrer.

Les dernières inégalités une fois établies, la solution de l'équation du type de Volterra

$$(122)\qquad \varphi(x)-\mu\int\limits_{D_a^x} K(x;y)\,\varphi(y)\,dy = f(x),$$

rattachée au noyau $K(x;y)$, s'obtient en analogie parfaite avec le cas linéaire.

On passe de l'équation (122) à l'équation (111) en remplaçant a, x, y et $K(x;y)$ par x, y, z et $A(y;z)$ respectivement et en posant $\mu=-1$. Les noyaux itérés $A^{(n+1)}(y;z)$, correspondant au noyau $A(y;z)$, satisfont à des inégalités de la forme (117), excepté que l'arc $s(x;y)$ doit être remplacé par l'arc $s(y;z)$. Il en résulte que le noyau résolvant, formé en analogie avec (114),

$$L(y;z\,|\,\mu) = A(y;z)+\mu A^{(2)}(y;z)+\cdots+\mu^{n-1}A^{(n)}(y;z)+\cdots$$

a des propriétés de convergence éminentes et qu'il en est de même de la solution de l'équation (111), savoir

$$(123)\qquad V_{\mathrm{II}}^2(x;y) = v(y) = F(x;y)-\int\limits_{D_y^x} L(y;z\,|-1)\,F(x;z)\,dz.$$

98. L'équation des ondes amorties dans les espaces de Riemann. — On a vu que le noyau accessoire $V_{\mathrm{I}}^{\alpha}+V_{\mathrm{II}}^{2}$ suffit complètement à résoudre le problème de Cauchy dans le cas régulier non analytique. Cependant ce noyau considéré en soi ne possède aucune propriété intéressante. Il lui manque par exemple toute loi de composition. *Nous nous proposons de former aussi dans le cas non ana-*

lytique le noyau fractionnaire complet V^α *et de démontrer qu'il possède toutes les propriétés essentielles rencontrées dans le cas analytique.* En particulier, ce noyau vérifiera la formule (66) $\Delta V^{\alpha+2} = V^\alpha$, propriété qui, dans le cas holomorphe, servait de point de départ lors de la construction de V^α; il admettra les lois de composition (98) et (99). La première de ces lois permettrait évidemment de construire, aussi dans le cas actuel, un procédé d'intégration fractionnaire, possédant les propriétés fondamentales $I^\alpha I^\beta = I^{\alpha+\beta}$, $I^\alpha I^\beta_* = I^{\alpha+\beta}_*$.

Nous arriverons à V^α à l'aide de certains noyaux rattachés à l'équation des ondes amorties

(124) $$\Delta u(x) + \lambda^2 u(x) = f(x),\text{[1]}$$

relative à l'espace de Riemann considéré. A l'opérateur $\Delta + \lambda^2$ correspondent, dans le cas analytique, un noyau W^α et, dans le cas non analytique, un noyau $W^\alpha_{\mathrm{I}} + W^2_{\mathrm{II}}$, formés en analogie avec les noyaux rattachés à l'opérateur Δ. Les deux noyaux indiqués permettent, suivant les cas, de résoudre le problème de Cauchy posé pour l'équation (124). Mais il y en a plus, et voilà notre point principal, *le noyau* W^2_{II} *nous mettra en état d'accomplir la construction de* V^α.

Admettons encore une fois jusqu'à nouvel ordre que les coefficients g_{jk} sont holomorphes et posons, dans l'ordre d'idées du n° **56**^{bis},

(125) $$W^\alpha = \sum_{h=0}^{\infty} \binom{-\frac{\alpha}{2}}{h} \lambda^{2h} V^{\alpha+2h}$$
$$= \frac{1}{\pi^{\frac{m-2}{2}} \cdot 2^{\frac{\alpha+m-2}{2}} \cdot \Gamma\left(\frac{\alpha}{2}\right)} \sum_{k=0}^{\infty} V_k \cdot \left(\frac{s}{\lambda}\right)^{\frac{\alpha-m}{2}+k} \cdot J_{\frac{\alpha-m}{2}+k}(\lambda s),\text{[2]}$$

où les J sont les fonctions de Bessel, définies par la formule (20) du numéro cité. Il est manifeste que W^α devient singulier sur le conoïde de la même manière que V^α et on vérifie facilement la relation

$$(\Delta + \lambda^2)\, W^{\alpha+2} = W^\alpha.$$

On peut, tout comme dans le numéro cité, définir les intégrales $I^\alpha_{(\lambda)}$ et $I^\alpha_{*(\lambda)}$, arriver à une formule de l'aspect de la formule (22) du même numéro et donner, moyennant cette formule, la solution du problème de Cauchy.

---

[1] Bien entendu, l'opérateur Δ signifie ici, comme dans tout ce chapitre, l'opérateur du second ordre de Beltrami, défini par la formule (23).

[2] L'identité des deux expressions de W^α découle aisément des calculs qui ont conduit à la formule (23) du numéro cité.

Laissant de côté ce dernier point de vue, envisageons l'inégalité suivante, valable pour ν réel et $\geqq -\frac{1}{2}$,

$$(126) \qquad |J_\nu(t)| \leqq \frac{|\frac{1}{2}t|^\nu}{\Gamma(\nu+1)} \cdot e^{|\mathfrak{J}(t)|}, \text{ [1]}$$

t étant une variable complexe. Il résulte facilement de cette limitation et de la convergence de la série (79) que W^2 est une fonction entière de λ.

On a d'autre part [2]

$$\int_0^\infty \frac{J_\nu(t)\,dt}{t^{\nu+\alpha-1}} = \frac{\Gamma\left(1-\frac{\alpha}{2}\right)}{2^{\nu+\alpha-1}\cdot\Gamma\left(\nu+\frac{\alpha}{2}\right)},$$

formule valable pour $\frac{3}{2} - \mathfrak{R}(\nu) < \mathfrak{R}(\alpha) < 2$. En désignant par C un contour qui va de $-\infty$ à $+\infty$ le long de l'axe réel, évitant l'origine par une courbe tracée dans le demi-plan *inférieur*, et en observant que $J_\nu(t)/t^{\nu-1}$ est une fonction impaire, on obtient

$$\int_C \frac{J_\nu(t)\,dt}{t^{\nu+\alpha-1}} = -2ie^{\frac{i\pi\alpha}{2}}\cdot\sin\frac{\pi\alpha}{2}\cdot\frac{\Gamma\left(1-\frac{\alpha}{2}\right)}{2^{\nu+\alpha-1}\cdot\Gamma\left(\nu+\frac{\alpha}{2}\right)} = \frac{\pi e^{\frac{i\pi(\alpha-1)}{2}}}{2^{\nu+\alpha-2}\cdot\Gamma\left(\frac{\alpha}{2}\right)\cdot\Gamma\left(\nu+\frac{\alpha}{2}\right)}$$

ou encore

$$\frac{1}{\pi}\int_C \frac{J_\nu(t)\,dt}{t^\nu(it)^{\alpha-1}} = \frac{1}{2^{\nu+\alpha-2}\cdot\Gamma\left(\frac{\alpha}{2}\right)\Gamma\left(\nu+\frac{\alpha}{2}\right)},$$

avec $-\frac{\pi}{2} \leqq \arg(it) \leqq \frac{\pi}{2}$, les intégrales curvilignes étant convergentes et les formules étant valables pour $\mathfrak{R}(\alpha) > \frac{3}{2} - \mathfrak{R}(\nu)$. Dans nos applications, il sera question des valeurs $\nu = \frac{2-m}{2} + k$ $(k = 0, 1, 2, \ldots)$; la dernière formule sera donc valable, dès que $\mathfrak{R}(\alpha)$ sera assez grand $\left(\mathfrak{R}(\alpha) > \frac{m+1}{2}\right)$.

---

[1] WATSON **1**, p. 49. On entend par $\mathfrak{J}(t)$ la partie imaginaire de t.
[2] WATSON **1**, p. 391, formule (1), en y posant $\mu = 2 - \alpha$.

Posons $t = \lambda \cdot s$ où $s > 0$. On obtient

$$(127) \qquad \frac{1}{\pi} \int_C \left(\frac{s}{\lambda}\right)^{\nu} \cdot \frac{J_\nu(\lambda s) \cdot d\lambda}{(i\lambda)^{\alpha-1}} = \frac{s^{2\nu+\alpha-2}}{2^{\nu+\alpha-2} \cdot \Gamma\left(\frac{\alpha}{2}\right) \Gamma\left(\nu + \frac{\alpha}{2}\right)}.$$

En appliquant alors la seconde expression de la formule (125) qui donne W^α, on trouve, pour $\Re(\alpha)$ assez grand,[1]

$$(128) \qquad \frac{1}{\pi} \int_C \frac{W^2\, d\lambda}{(i\lambda)^{\alpha-1}} = \sum_{k=0}^{\infty} V_k \cdot \frac{s^{\alpha-m+2k}}{\pi^{\frac{m-2}{2}} \cdot 2^{\alpha+k-1} \cdot \Gamma\left(\frac{\alpha}{2}\right) \Gamma\left(\frac{\alpha+2-m+2k}{2}\right)} = V^\alpha.$$

C'est cette même formule qui, légèrement modifiée, nous fournira plus loin V^α dans le cas *non analytique*, dès que nous aurons, dans ce cas-là, réussi à calculer W^2 ou W^2_{II}, ce qui se fera à l'aide d'une nouvelle équation de Volterra.

En restant toujours dans le cas holomorphe, choisissons de nouveau un nombre entier p tel que $2p > m - 2$. Posons, en partant de la seconde expression (125),

$$(129) \qquad W^\alpha = \frac{1}{\pi^{\cdots} 2^{\cdots} \Gamma\left(\frac{\alpha}{2}\right)} \sum_{k=0}^{\infty} V_k \cdot \left(\frac{s}{\lambda}\right)^{\frac{\alpha-m}{2}+k} \cdot J_{\frac{\alpha-m}{2}+k}(\lambda s)$$

$$= \frac{1}{\pi^{\cdots} 2^{\cdots} \Gamma\left(\frac{\alpha}{2}\right)} \sum_{k=0}^{p} + \frac{1}{\pi^{\cdots} 2^{\cdots} \Gamma\left(\frac{\alpha}{2}\right)} \sum_{k=p+1}^{\infty} = W^\alpha_{\mathrm{I}} + W^\alpha_{\mathrm{II}}.$$

On trouve par un réarrangement analogue mais de sens opposé à celui appliqué dans la formule citée

$$(130) \qquad W^\alpha_{\mathrm{I}} = \sum_{h=0}^{\infty} \binom{-\frac{\alpha}{2}}{h} \lambda^{2h}\, V^{\alpha+2h}_{\mathrm{I}},$$

V^α_{I} étant défini par (108). Nous allons démontrer que

$$(131) \qquad (\Delta + \lambda^2)\, W^{\alpha+2}_{\mathrm{I}}$$

$$= W^\alpha_{\mathrm{I}} + \frac{1}{\pi^{\frac{m-2}{2}} \cdot 2^{\frac{\alpha+m}{2}} \cdot \Gamma\left(\frac{\alpha+2}{2}\right)} \left(\frac{s}{\lambda}\right)^{\frac{\alpha+2-m}{2}+p} \cdot J_{\frac{\alpha+2-m}{2}+p}(\lambda s) \cdot \Delta V_p.$$

---

[1] *Cf.* les formules (67) et (71).

On tire en effet de (130), moyennant (109),

$$(\Delta + \lambda^2)\, W_{\mathrm{I}}^{\alpha+2} = \sum_{h=0}^{\infty} \binom{-\frac{\alpha+2}{2}}{h} \lambda^{2h} (\Delta + \lambda^2)\, V_{\mathrm{I}}^{\alpha+2+2h}$$

$$= \sum_{h=0}^{\infty} \binom{-\frac{\alpha+2}{2}}{h} \lambda^{2h} \left\{ \left(V_{\mathrm{I}}^{\alpha+2h} + \lambda^2 V_{\mathrm{I}}^{\alpha+2+2h} \right) + s^{\alpha+2+2h+2p-m} \cdot \frac{\Delta V_p}{H_m(\alpha+2+2h,\, p)} \right\}.$$

La première partie de la série, savoir celle comprenant les expressions

$$\left(V_{\mathrm{I}}^{\alpha+2h} + \lambda^2 V_{\mathrm{I}}^{\alpha+2+2h} \right),$$

peut s'écrire

$$\sum_{h=0}^{\infty} \left[\binom{-\frac{\alpha+2}{2}}{h} + \binom{-\frac{\alpha+2}{2}}{h-1} \right] \lambda^{2h} V_{\mathrm{I}}^{\alpha+2h} = \sum_{h=0}^{\infty} \binom{-\frac{\alpha}{2}}{h} \lambda^{2h} V_{\mathrm{I}}^{\alpha+2h} = W_{\mathrm{I}}^{\alpha}.$$

La seconde partie de la même série est, d'après la formule (20) du n° **56**^{bis} et la formule (71) du n° **89**, égale au dernier terme de (131).

Vu que $W_{\mathrm{I}}^{0} = 0$, pour $s \neq 0$ [1], il s'ensuit de (131), si l'on y pose $\alpha = 0$ ($\alpha + 2 = 2$) (*cf.* (109ter)),

$$(132) \qquad (\Delta_y + \lambda^2)\, W_{\mathrm{I}}^{2}(x;\, y) = -(\Delta_y + \lambda^2)\, W_{\mathrm{II}}^{2}(x;\, y)$$

$$= \frac{1}{\pi^{\frac{m-2}{2}} \cdot 2^{\frac{m}{2}}} \left(\frac{s}{\lambda}\right)^{\frac{2-m}{2}+p} \cdot J_{\frac{2-m}{2}+p}(\lambda s) \cdot \Delta V_p = A_{(\lambda)}(x;\, y),$$

pour abréger.

Au moyen de la formule de Green, on obtient, en analogie avec (110), pour toute fonction $w(z)$ qui s'annule sur le conoïde de sommet x avec ses dérivées du premier ordre,

$$(133) \qquad \int_{D_y^x} (\Delta + \lambda^2)\, w(z) \cdot W_{\mathrm{I}}^{\alpha+2}(y;\, z)\, dz = \int_{D_y^x} w(z) (\Delta_z + \lambda^2)\, W_{\mathrm{I}}^{\alpha+2}(y;\, z)\, dz.$$

En particulier pour $w(z) = W_{\mathrm{II}}(x;\, z)$, on arrive, en analogie avec (111), à l'équation intégrale

$$(134) \qquad F_{(\lambda)}(x;\, y) = w(y) + \int_{D_y^x} w(z)\, A_{(\lambda)}(y;\, z)\, dz,$$

où $F_{(\lambda)}(x;\, y)$ est le prolongement analytique pour $\alpha = 0$ du premier membre de (133), avec $(\Delta + \lambda^2)\, w(z) = (\Delta_z + \lambda^2)\, W_{\mathrm{II}}^{2}(x;\, z) = -A_{(\lambda)}(x;\, z)$.

[1] *Cf.* la note [1], p. 199.

99. Le noyau V^α dans le cas non analytique. — Dans le cas non analytique mais assez régulier, on connaît W_{I}^2 et même W_{I}^α et par là $F_{(\lambda)}(x; y)$ et on *définit* W_{II}^2 par l'équation intégrale (134), d'aspect identique à celui de (111). Le noyau $A_{(\lambda)}$ satisfait, d'après les formules (132) et (126), à une inégalité de la forme

$$(135) \qquad |A_{(\lambda)}(y; z)| \leqq M e^{|\Im\{\lambda s(y; z)\}|},$$

M ne dépendant que de la région $\mathcal{R}$.

On forme maintenant les noyaux itérés $A_{(\lambda)}^{(n)}$ et on établit, en analogie avec l'inégalité (117), que

$$(136) \qquad \left|A_{(\lambda)}^{(n+1)}(y; z)\right| \leqq \Gamma\left(\frac{1}{2}\right) \frac{M\ (MB)^n}{\Gamma\left(\frac{mn+1}{2}\right)} s(y; z)^{mn} \cdot e^{|\Im\{\lambda s(y; z)\}|},$$

où B ne dépend que de la région $\mathcal{R}$ et m est la dimension de l'espace.

La seule chose exigeant des commentaires, qui, d'ailleurs, sont d'un assez grand intérêt, est le facteur exponentiel. Considérons p. ex. le noyau itéré d'ordre 2. Il est égal à

$$A_{(\lambda)}^{2}(y; z) = \int\limits_{D_z^y} A_{(\lambda)}(y; u)\, A_{(\lambda)}(u; z)\, du.$$

Dans la limitation de $A_{(\lambda)}^{2}$ entre, d'après (135), le produit

$$e^{|\Im\{\lambda s(y; u)\}|} \cdot e^{|\Im\{\lambda s(u; z)\}|} = e^{|\Im\{\lambda [s(y; u) + s(u; z)]\}|}.$$

Or, grâce au caractère lorentzien de l'espace, on a $s(y; u) + s(u; z) \leqq s(y; z)$; par conséquent, l'exponentielle en question est

$$\leqq e^{|\Im\{\lambda s(y; z)\}|}.$$

On observe que l'inégalité géométrique ci-dessus est de sens opposé à l'inégalité accoutumée relative aux triangles géodésiques. On appliquera une méthode de démonstration donnée dans le cas classique par Darboux [1], méthode basée sur une espèce de coordonnées polaires.

Menons du point y un faisceau d'arcs géodésiques intérieurs au domaine D_z^y et remplissant une certaine région $\mathcal{B}$ qui, de son côté, comprend le triangle géodésique yuz. Choisissons de plus une valeur positive a assez petite, en sorte que la surface $\Gamma(y; t) = a$, y étant fixé et t variable, coupe toutes les géodésiques du faisceau. Nous paramétrisons d'abord la portion de la surface $\Gamma = a$ com-

[1] DARBOUX **1**, II, p. 410—412.

prise dans la région $\mathcal{B}$ par $m-1$ paramètres $\lambda^2, \lambda^3, \ldots, \lambda^m$, et rapportons ensuite les points v de la région $\mathcal{B}$ à m paramètres, savoir à $\lambda^1 = s = s(y; v) =$ longueur de l'arc géodésique yv et aux $m-1$ paramètres $\lambda^2, \lambda^3, \ldots, \lambda^m$ correspondant au point d'intersection t de la géodésique yv avec la surface $\Gamma(y; t) = a$. Ceci posé, l'élément linéaire de l'espace, désigné incidemment par $d\sigma$, prend la forme $d\sigma^2 = \gamma_{pq} \cdot d\lambda^p \cdot d\lambda^q$. Nous disons que $\gamma_{11} = 1$, tandis que $\gamma_{1q} = 0$ pour $q \neq 1$. Afin de vérifier le premier énoncé, il suffit de remarquer que, pour $d\lambda^2 = \cdots = d\lambda^m = 0$, $d\sigma^2$ se réduit à $ds^2 (= d\lambda^1 \cdot d\lambda^1)$. Le second énoncé résulte de ce que les géodésiques issues du point y coupent les surfaces $\Gamma(y; v) = \text{const.}$ orthogonalement, fait établi au nº **97**. On en a conclu au même numéro que les dernières surfaces ont une orientation d'espace. Il s'ensuit que

$$d\sigma^2 = ds^2 + \sum_{p,\, q=2}^{m} \gamma_{pq} \cdot d\lambda^p \cdot d\lambda^q,$$

la dernière forme quadratique $\sum \ldots$ étant *définie négative.*

Ce point acquis, considérons un chemin de temps C quelconque, compris entièrement dans la région $\mathcal{B}$, réunissant les points y et z et coupant chaque surface $\Gamma = \text{const.}$ en un seul point. On va voir que la longueur de ce chemin, exprimée par l'intégrale $\int_C d\sigma$, est inférieure à $s(y; z)$. En effet

$$\int_C d\sigma = \int_C (ds^2 + \sum \ldots)^{\frac{1}{2}} \leqq \int_C ds = s(y; z).$$

Le dernier résultat comprend l'inégalité demandée $s(y; u) + s(u; z) \leqq s(y; z)$.

En résumé, on a d'une part, avec une constante convenable K,

$$(137) \qquad | W_{\mathrm{II}}^2(x; y) | = w(y) \leqq K \cdot e^{|\Im\{\lambda s(x;y)\}|}$$

et d'autre part, on voit facilement que $W_{\mathrm{II}}^2(x; y)$ s'annule avec ses dérivées du premier ordre sur le conoïde de sommet x, comme il fallait s'y attendre. On a aussi pour $s \neq 0$

$$(138) \qquad (\Delta_y + \lambda^2)\, W^2(x; y) = (\Delta_y + \lambda^2)\, \{ W_{\mathrm{I}}^2(x; y) + W_{\mathrm{II}}^2(x; y) \} = 0.$$

De plus, toutes les fonctions figurant dans l'équation intégrale (134) étant des fonctions entières de λ, il en sera de même de W_{II}^2. Les derniers faits se démontrent sans difficulté, grâce aux propriétés de convergence de la formule de solution de (134).

En nous laissant guider par la formule (128), valable dans le cas holomorphe, nous posons dans le cas non analytique, *par définition*, pour $\Re(\alpha) > 2$,

$$(139) \qquad V^\alpha = V_{\mathrm{I}}^\alpha + \frac{1}{\pi}\int\limits_C \frac{W_{\mathrm{II}}^2\, d\lambda}{(i\lambda)^{\alpha-1}} = V_{\mathrm{I}}^\alpha + V_{\mathrm{II}}^\alpha,$$

où C est la courbe figurant dans la formule (128) et où l'on choisit

$$-\frac{\pi}{2} \leqq \arg(i\lambda) \leqq \frac{\pi}{2}.$$

L'intégrale converge en raison de l'inégalité (137).

On a encore, d'après les formules (108) et (127)[1], pour $\Re(\alpha)$ assez grand ($\Re(\alpha) > m+1$),

$$(140) \qquad V^\alpha = \frac{1}{\pi}\int\limits_C \frac{W^2\, d\lambda}{(i\lambda)^{\alpha-1}}.$$

Il est à peu près évident que la première formule, dans laquelle V_{I}^α peut être considéré comme connu, forme le prolongement analytique de la seconde.

Le fait qui nous intéresse en premier lieu est que *la fonction* V^α définie de cette manière *possède toutes les propriétés essentielles que nous avons rencontrées dans le cas analytique*. Il est d'abord clair que les termes principaux sur le conoïde restent les mêmes, puisque V_{I}^α est conservé et que V_{II}^α devient singulier sur le conoïde d'un ordre moins élevé que les termes de V_{I}^α. Vérifions aussi la formule $\Delta V^{\alpha+2} = V^\alpha$. Il suffit évidemment de le faire pour $\Re(\alpha)$ assez grand au moyen de la formule (140). Il vient (*cf.* formule (138))

$$\Delta V^{\alpha+2} = \frac{1}{\pi}\int\limits_C \frac{\Delta W^2\, d\lambda}{(i\lambda)^{\alpha+1}} = \frac{1}{\pi}\int\limits_C \frac{-\lambda^2 W^2\, d\lambda}{(i\lambda)^{\alpha+1}} = \frac{1}{\pi}\int\limits_C \frac{W^2\, d\lambda}{(i a)^{\alpha-1}} = V^\alpha.$$

J'arrive au dernier point, *la validité des formules de composition* (98) *et* (99) *dans les conditions nouvelles*, question qui me tient particulièrement au cœur. La seule difficulté qui reste est de montrer que le noyau V^α admet une majoration du même genre que plus haut (formule (83)). Or ceci est clair quant à V_{I}^α. En ce qui concerne V_{II}^α, nous arriverons au fait demandé en examinant de plus près les fonctions données par des intégrales du type de (139).

Soit λ une variable complexe et supposons que la fonction $f(\lambda)$ est holo-

[1] *Cf.* aussi la formule (128).

morphe dans le demi-plan inférieur $\mathfrak{I}(\lambda) \leqq 0$ [1], sauf peut-être à l'origine, et qu'elle y satisfait à l'inégalité $|f(\lambda)| \leqq M e^{k|\mathfrak{I}(\lambda)|}$, avec $k \geqq 0$. Formons la fonction de $\alpha = \xi + i\eta$, holomorphe pour $\Re(\alpha) > 0$,

$$g(\alpha) = \int_C \frac{f(\lambda)}{(i\lambda)^{\alpha+1}} d\lambda, \tag{141}$$

le chemin d'intégration C étant le même que dans la formule (128), et les inégalités $-\frac{\pi}{2} \leqq \arg(i\lambda) \leqq \frac{\pi}{2}$ étant toujours remplies. Ceci posé, *on a l'inégalité*

$$|g(\alpha + b)| \leqq \bar{M} |\alpha^{-\alpha} e^{k\alpha}| \text{ avec } \bar{M} = M\left(\pi e^k + \frac{2}{b}\right), \tag{142}$$

valable pour $\Re(\alpha) > 0$ *et pour un nombre positif* b *arbitraire.*

Remarquons d'abord qu'on peut, sans changer la fonction $g(\alpha)$, remplacer le contour C par un autre contour, composé d'une part d'une demi-circonférence de rayon $\xi + 1$, où $\xi = \Re(\alpha)$, décrite autour de l'origine et située dans le demi-plan inférieur, et d'autre part des segments infinis adjacents de l'axe réel. En posant $i\lambda = \varrho e^{i\varphi}$, on a, avec $\eta = \mathfrak{I}(\alpha)$,

$$|(i\lambda)^{-(\alpha+1+b)}| = |\varrho^{-(\alpha+1+b)}| \cdot |e^{-i\varphi(\alpha+1+b)}| = \varrho^{-(\xi+1+b)} e^{\varphi\eta} \leqq \varrho^{-(\xi+1+b)} e^{\frac{\pi}{2}|\eta|}. \tag{143}$$

On obtient ensuite pour $\xi \geqq 0$

$$\int_{\xi+1}^{\infty} \varrho^{-(\xi+1+b)} d\varrho = \frac{(\xi+1)^{-(\xi+b)}}{\xi+b} = \frac{(\xi+1)^{-\xi}}{\xi+b} \cdot (\xi+1)^{-b} < \frac{\xi^{-\xi}}{b},$$

ce qui donne sur les deux segments infinis (*cf.* formule (143))

$$\left| \int_{\pm(\xi+1)}^{\pm\infty} f(\lambda)(i\lambda)^{-(\alpha+1+b)} d\lambda \right| \leqq M e^{\frac{\pi}{2}|\eta|} \int_{\xi+1}^{\infty} \varrho^{-(\xi+1+b)} d\varrho < M e^{\frac{\pi}{2}|\eta|} \cdot \frac{\xi^{-\xi}}{b}.$$

La valeur absolue de l'intégrale étendue à la demi-circonférence de rayon $\xi + 1$ est de son côté inférieure à

$$M e^{k(\xi+1)} \cdot e^{\frac{\pi}{2}|\eta|} \cdot \pi(\xi+1)(\xi+1)^{-(\xi+1+b)} < \pi M e^{k(\xi+1)+\frac{\pi}{2}|\eta|} \cdot \xi^{-\xi}.$$

---

[1] Puisque la fonction $f(\lambda)$ est bornée par hypothèse, il nous suffirait, grâce à un théorème fondamental de Fatou, de supposer que $f(\lambda)$ est holomorphe pour $\mathfrak{I}(\lambda) > 0$.

On a donc l'inégalité préliminaire

$$(144) \qquad |g(\alpha+b)| < M\left(\pi e^{k(\xi+1)} + \frac{2}{b}\right)\cdot \xi^{-\xi}\cdot e^{\frac{\pi}{2}|\eta|}$$

$$\leqq \overline{M} e^{k\xi}\cdot e^{\frac{\pi}{2}|\eta|}\cdot \xi^{-\xi} \text{ avec } \overline{M} = M\left(\pi e^{k} + \frac{2}{b}\right).$$

De la dernière inégalité on déduit aisément l'inégalité définitive (142). Remarquons en effet que la fonction $h(\alpha) = \alpha^{-\alpha} e^{k\alpha}$ est holomorphe dans le demi-plan $\Re(\alpha) \geqq 0$, à l'exception de l'origine, et que l'on a sur l'axe réel positif et sur l'axe imaginaire

$$(145) \qquad \left|\frac{g(\alpha+b)}{h(\alpha)}\right| \leqq \frac{\overline{M} e^{k\xi}\cdot e^{\frac{\pi}{2}|\eta|}\cdot \xi^{-\xi}}{|h(\alpha)|} = \overline{M}.\ (^1)$$

D'autre part le premier membre est dans le demi-plan $\Re(\alpha) \geqq 0$ manifestement inférieur à const. $\exp(|\alpha|^{2-\varepsilon})$ où $0 < \varepsilon < 1$. Il suffit donc, pour voir que l'inégalité $|g(\alpha+b) : h(\alpha)| \leqq \overline{M}$ est remplie dans tout le demi-plan $\Re(\alpha) \geqq 0$, d'appliquer aux deux domaines angulaires (d'ouverture $\pi/2$), limités chacun par l'axe réel positif et par l'un des demi-axes imaginaires, un théorème classique de Phragmén-Lindelöf.[2] L'inégalité (142) est identique à celle que nous venons de vérifier.

Nous sommes maintenant en mesure d'établir la formule de composition (99) aussi dans le cas non analytique. Remarquons à cet effet que la formule (105) se déduit exactement comme plus haut. Cela étant, on forme de nouveau la fonction $G(\alpha)$ définie par la formule (106), fonction qui, d'après la formule (105), s'annule pour tout nombre pair $\alpha = 2n$ assez grand. Pour arriver à la formule (99), il suffira donc, aussi dans le cas actuel, de donner *une majorante holomorphe convenable de* $G(\alpha)$, problème qui se réduit au même problème concernant V^{α}.

Nous commençons par appliquer l'inégalité (142) à la fonction

$$(146) \qquad V_{\mathrm{II}}^{\alpha} = \frac{1}{\pi}\int_{C} \frac{W_{\mathrm{II}}^{2}\, d\lambda}{(i\lambda)^{\alpha-1}}$$

(*cf.* la formule (139)). Nous notons que W_{II}^{2} est une fonction holomorphe de λ dans le demi-plan $\Im(\lambda) \leqq 0$[3] et que, d'après (137), cette fonction satisfait à

(1) On a $|h(\alpha)| = \varrho^{-\xi}\cdot e^{\theta\eta + k\xi}$.

(2) PHRAGMÉN-LINDELÖF 1, p. 385.

(3) Elle est même une fonction entière ; *cf.* p. 212.

une limitation de la forme $|f(\lambda)| \leqq e^{k|\mathfrak{I}(\lambda)|}$, posée plus haut au sujet de la fonction $f(\lambda)$ figurant dans la formule (141). On a dans le cas actuel $k = s(x; y)$. Il faut encore remarquer que dans (146) on a l'exposant $\alpha - 1$ (au dénominateur), tandis que pour $g(\alpha + b)$ c'est l'exposant $\alpha + 1 + b$ qui figure dans l'intégrale (141). Avec $b = 1$, le dernier exposant devient $\alpha + 2 = (\alpha - 1) + 3$. On a donc, en vertu de (142), pour d'assez grandes valeurs de $\Re(\alpha)$, les inégalités préliminaires

$$|V_{\mathrm{II}}^{\alpha+3}| \leqq |\bar{M} \cdot \alpha^{-\alpha} \cdot e^{\alpha s(x;y)}| \quad \text{ou} \quad |V_{\mathrm{II}}^{\alpha}| \leqq |\bar{M} \cdot (\alpha - 3)^{-(\alpha-3)} \cdot e^{(\alpha-3)s(x;y)}|,$$

ou encore (*cf.* le n° **91**), toujours pour $\Re(\alpha)$ assez grand,

$$|V_{\mathrm{II}}^{\alpha}| \leqq |\bar{M} \cdot \alpha^{-\alpha} \cdot e^{\alpha s} \cdot \alpha^{p}|,$$

majoration qui est essentiellement de la même forme que celle donnée dans la formule (83).

Quant à V_{I}^{α}, défini par la formule (108), il suffit de se reporter aux évaluations du n° **91** pour arriver à une majorante de la forme $|C \cdot \alpha^{-\alpha} \cdot e^{k\alpha} \cdot \alpha^{p}|$. Dès lors, $V^{\alpha} = V_{\mathrm{I}}^{\alpha} + V_{\mathrm{II}}^{\alpha}$ admet aussi une majorante de même forme, les valeurs de C, k, p étant indépendantes de α. Enfin on montre, comme au n° **95**, que la fonction $G(\alpha)$, définie par la formule (106), satisfait à une inégalité semblable. Donc $G(\alpha)$ s'annule identiquement en vertu du théorème d'unicité donné au numéro cité.

La formule de composition (99) étant démontrée, la relation d'échange $V^{\alpha}(x; y) = V^{\alpha}(y; x)$ en découle comme plus haut. Dès lors, la formule de composition (98) et, avec elle, les formules de composition de caractère général (96) et (96bis) se trouvent également établies.

Bibliographie.

M. ABRAHAM. 1. *Theorie der Elektrizität*, II, Leipzig 1905.

R. D'ADHÉMAR. 1. *Sur une équation aux dérivées partielles du type hyperbolique. Étude de l'intégrale près d'une frontière caractéristique*, Palermo Rend. 20 (1905).

B. B. BAKER et E. T. COPSON. 1. *The mathematical theory of Huygens principle*, Oxford 1939.

R. BALTZER. 1. *Theorie und Anwendung der Determinanten*, 5e éd., Leipzig 1881.

L. BRILLOUIN. 1. *Les tenseurs en mécanique et en élasticité*, Paris 1938.

F. BUREAU. 1. *L'intégration des équations linéaires aux dérivées partielles du second ordre et du type hyperbolique normal*, Mém. Soc. Roy. des Sc. Liége (4) 3 (1938).

H. CARTAN. 1. *Sur les fondements de la théorie du potentiel*, Bull. Soc. Math. de France 69 (1941).

2. *La théorie générale du potentiel dans les espaces homogènes*, Bull. des Sc. Math. (2) 66 (1942).

R. COURANT et D. HILBERT. 1. *Methoden der mathematischen Physik*, I, 2e éd., Berlin 1931, II, Berlin 1937.

G. DARBOUX. 1. *Leçons sur la théorie générale des surfaces et les applications géométriques du calcul infinitésimal*, II, 2e éd., Paris 1889.

P. A. M. DIRAC. 1. *Classical theory of radiating electrons*, Proc. Roy. Soc. A 167 (1938).

N. E. FREMBERG. 1. *Proof of a theorem of M. Riesz concerning a generalization of the Riemann-Liouville integral*, Kungl. Fysiogr. Sällsk. i Lund Förhandl. 15 (1945).

2. *A study of generalized hyperbolic potentials with some physical applications*, Thèse, Communic. Sémin. Math. de l'Univ. de Lund 7 (1946).

3. *Some applications of the Riesz potential to the theory of the electromagnetic field and the meson field*, Proc. Roy. Soc., A 188 (1946).

O. FROSTMAN. 1. *Potentiel d'équilibre et capacité des ensembles avec quelques applications à la théorie des fonctions*, Thèse, Communic. Sémin. Math. de l'Univ. de Lund 3 (1935).

2. *Sur le balayage des masses*, Acta Szeged 9 (1938) et Communic. Sémin. Math. de l'Univ. de Lund 4 (1939).

3. *Sur les fonctions surharmoniques d'ordre fractionnaire*, Arkiv för mat., astr. och fys. 26 A (1938) et Communic. Sémin. Math. de l'Univ. de Lund 4 (1939).

É. GOURSAT. 1. *Cours d'analyse mathématique*, I—III, 2e éd., Paris 1910—1915.

2. *Sur les invariants intégraux*, Journ. de Math. (6) 4 (1908).

L. GÅRDING. 1. *The solution of Cauchys problem for two totally hyperbolic linear differential equations by means of Riesz integrals*, Ann. Math. 48 (1947).

J. HADAMARD. 1. *Le problème de Cauchy et les équations aux dérivées partielles linéaires hyperboliques*, Paris 1932.

2. *Équations aux dérivées partielles. Les conditions définies en général. Le cas hyperbolique*, Enseign. Math. 35 (1936).

3. *Leçons sur le calcul des variations*, I, Paris 1910.

S. T. MA. 1. *Equivalence of the Riesz method and the λ-limiting process for the classical electromagnetic field of a point source*, Phys. Rev. 71 (1947).

H. MALMHEDEN. 1. *A class of hyperbolic systems of linear differential equations*, Thése, Communic. Sémin. Math. de l'Univ. de Lund 8 (1947).

M. MATHISSON. 1. *Le problème de M. Hadamard relatif à la diffusion des ondes*, Acta Math. 71 (1939).

2. *Eine neue Lösungsmethode für Differentialgleichungen von normalem hyperbolischen Typus*, Math. Ann. 107 (1932); une *correction* dans le même volume.

S. B. NILSSON. 1. *The e^4 approximation of the self-energy of the electron treated by analytic continuation*, Arkiv för mat., astr. och fys., 36 A (1948).

I. PETROWSKY. 1. *Sur la diffusion des ondes et les lacunes pour les systèmes d'équations hyperboliques*, Bull. Acad. Sci. USSR, Sér. Math. 8 (1944). En russe. Avec un résumé en français.

2. *On the diffusion of waves and the lacunas for hyperbolic equations*, Rec. Math. N. S. 17 (59) (1945).

E. PHRAGMÉN et E. LINDELÖF. 1. *Sur une extension d'un principe classique de l'Analyse etc.*, Acta Math. 31 (1908).

É. PICARD. 1. *Traité d'analyse*, I, 2e éd., Paris 1901.

H. POINCARÉ. 1. *Sur les résidus des intégrales doubles*, Acta Math. 9 (1887).

G. PÓLYA et G. SZEGÖ. 1. *Aufgaben und Lehrsätze aus der Analysis*, I, Berlin 1925.

M. RIESZ. 1. *Intégrale de Riemann-Liouville et solution invariantive du problème de Cauchy pour l'équation des ondes*, Comptes Rendus du Congrès internat. des math. Oslo 1936, II, Oslo 1937.

2. *L'intégrale de Riemann-Liouville et le problème de Cauchy pour l'équation des ondes*, Conférence à la Réunion internat. des math., Paris 1937, Soc. Math. de France 66 (1938) et Communic. Sémin. Math. de l'Univ. de Lund 4 (1939).

3. *Intégrales de Riemann-Liouville et potentiels*, Acta Szeged 9 (1938) et Communic. Sémin. Math. de l'Univ. de Lund 4 (1939).

V. VOLTERRA. 1. *Sur les vibrations de corps élastiques isotropes*, Acta Math. 18 (1894).

G. N. WATSON. 1. *A treatise on the theory of Bessel functions*, 1ère éd., Cambridge 1922, 2e éd., Cambridge 1944.

Table des Matières.

CHAPITRE IV.

Le problème de Cauchy pour l'équation des ondes.

CHAPITRE V.

Interprétation géométrique de la solution de l'équation des ondes sphériques.

CHAPITRE VI.

Applications à la théorie relativiste de l'électron.

CHAPITRE VII.

L'équation des ondes dans les espaces de Riemann.

Errata.

Page:	Ligne en descendant:	Ligne en remontant:	*Au lieu de:*	*Lire:*
60	13		$\left[\frac{m+3}{2}\right]+p$	$\left[\frac{m+1}{2}\right]+p$
80		3	**48**	**44**
87	6		s^P	S^P
98		4	*Supprimer* la phrase: M. Hadamard lui-même etc.; *cf.* la note (2), p. 183.	
			Au lieu de:	*Lire:*
101		1	2	**2**
110	9 (deux fois)		$\mathbf{x}\,\boldsymbol{\xi}$	$(\mathbf{x}, \boldsymbol{\xi})$
113	9		le tirons	la tirons

48.

Remarque sur les fonctions holomorphes

Acta Szeged Sect. Math. *12* (1950), 53–56

1. Dans un travail commun, paru en 1921, MM. L. FEJÉR et F. RIESZ[1]) ont publié quelques résultats qui sont devenus très connus. Ils ont démontré entre autres choses que lors de la représentation conforme d'un cercle sur un domaine limité par une courbe de Jordan rectifiable de longueur l l'image de chaque diamètre a une longueur $\leqq \frac{1}{2} l$. La constante $\frac{1}{2}$ ne peut pas être abaissée.

M. GABRIEL[2]) a considéré un problème plus général et montré que, la fonction $f(x)$ étant holomorphe à l'intérieur et sur le contour C d'un cercle et Γ étant une courbe convexe fermée intérieure à C, on a l'inégalité

$$\int_{\Gamma} |f(x)|\,|dx| \leqq 2\int_{C} |f(z)|\,|dz|. \tag{1}$$

La constante 2 ne peut pas être abaissée non plus.

On voit que le résultat de M. GABRIEL ne comprend pas celui de MM. FEJÉR et RIESZ. En effet un diamètre, parcouru deux fois, peut être considéré comme une courbe convexe fermée. Pour une telle courbe l'inégalité (1) est, d'après ce qu'on a dit au début, valable avec la constante 1. M. CARLSON a de son côté donné une inégalité d'une généralité remarquable qui comprend tous les résultats antérieurs. Voici le théorème de M. CARLSON[3]):

Soit $f(x)$ holomorphe à l'intérieur et sur le contour C d'un cercle et soit L une courbe rectifiable intérieure à C. On a l'inégalité

$$\int_{L} |f(x)|\,|dx| \leqq \frac{1}{\pi}\int |f(z)|\,V(z)\,|dz|, \tag{2}$$

---

1) L. FEJÉR und F. RIESZ, Über einige funktionaltheoretische Ungleichungen, *Math. Zeitschrift,* **11** (1921), p. 305–314.

2) R. M. GABRIEL, Some results concerning the integrals of moduli of regular functions along curves of certain types, *Proceedings London Math. Society,* **28** (1928), p. 121—127.

3) F. CARLSON, Quelques inégalités concernant les fonctions analytiques, *Arkiv för mat., astronomi och fysik,* **29 B,** n⁰ 11 (1943), p. 1—6.

où $V(z)$ est la borne supérieure de la somme des angles sous lesquels sont vus du point z les éléments d'arcs de L.[4])

En particulier, si l'inégalité $V(z) \leqq W$ est remplie pour tout point z de C, le second membre de (2) sera $\leqq \frac{W}{\pi} \int |f(z)|\, |dz|$. Si L est un diamètre de C, on a $W = \frac{\pi}{2}$, ce qui amène à l'inégalité de MM. FEJÉR et RIESZ. Le résultat de M. GABRIEL s'obtient, si l'on observe que, pour une courbe convexe fermée, on a toujours $W \leqq 2w \leqq 2\pi$, où w est l'angle le plus grand sous lequel on voit la courbe d'un point de C.

M. CARLSON déduit son théorème de l'inégalité suivante, dans laquelle l'intégrale est étendue au segment de droite AB joignant les points A et B intérieurs au cercle C,

$$\int_A^B |f(x)|\, |dx| \leqq \frac{1}{\pi} \int_C |f(z)|\, V(z)\, |dz|, \tag{3}$$

où $V(z)$ est l'angle sous lequel est vu le segment AB du point z. Cette inégalité, appliquée à des segments infinitésimaux, fournit le théorème plus haut.

Comme une contribution modeste au volume jubilaire dédié à MM. FEJÉR et RIESZ, nous déduirons les relations (équivalentes) (9) et (10), qui peuvent présenter un certain intérêt. De ces relations, on obtient immédiatement l'inégalité infinitésimale de M. CARLSON.

2. Après avoir introduit un système de coordonnés rectangulaires, désignons les points du plan par les nombres complexes correspondants et admettons que le centre du cercle C se trouve à l'origine. Soit $u(x)$ une fonction des coordonnés x' et x'' du point $x = x' + ix''$ qui est harmonique à l'intérieur de C et sur C. Désignons encore par z les points de C et posons

$$\arg z = \theta(z) = \theta \quad \text{et} \quad \arg (z - x) = \psi(z, x) = \psi. \tag{4}$$

Le potentiel de double couche

$$g(x) = \frac{1}{\pi} \int_C u(z)\, d\psi \tag{5}$$

(où $d\psi$ est une abréviation pour $d_z \psi(z, x) = \frac{d}{d\theta} \psi(z, x)\, d\theta$ ou $= \frac{\partial}{\partial z} \psi(z, x)\, dz$) est évidemment harmonique à l'intérieur de C, puisqu'il en est ainsi de $\psi(z, x)$. Quand x tend vers un point z_0 de C d'une façon arbitraire, $g(x)$ tend manifestement vers

$$u(z_0) + \frac{1}{\pi} \int_C u(z)\, d_z \psi(z, z_0). \tag{6}$$

[4]) M. GABRIEL et M. CARLSON montrent, au moyen d'un artifice utilisé dans un but analogue par MM. FEJÉR et RIESZ (*loc cit.*, p. 307), que leurs résultats subsistent, si l'on remplace $|f|$ par $|f|^k$, $k > 0$.

Pour le voir, décomposons le contour C en deux arcs C' et C'', l'arc C' entourant le point z_0 et étant assez petit, pour que l'oscillation de $u(z)$ sur cet arc soit très petite. On décompose en même temps l'intégrale figurant dans (5) en deux parties, l'une étendue à C' et l'autre à C''. L'angle sous lequel est vu l'arc C' d'un point x, assez rapproché du point z_0, étant sensiblement égal à π, la partie de l'intégrale étendue à C' sera sensiblement égale à $u(z_0)$, tandis que la partie étendue à C'' sera sensiblement égale à l'intégrale figurant dans (6). Dès lors, l'énoncé ci-dessus se trouve vérifié.

Or l'angle périphérique $d_z\psi(z, z_0)$ est indépendant de z_0 et égal à $\frac{1}{2}\,d\theta(z)$; le dernier terme de (6) aura donc la valeur

$$\frac{1}{2\pi}\int\limits_C u(z)\,d\theta(z) = u(0).$$

Il s'ensuit que la fonction $g(x) - u(0)$ tend vers $u(z_0)$, quand x tend vers z_0 d'une façon arbitraire. Une fonction harmonique étant déterminée par ses valeurs frontières, on conclut que $u(x) = g(x) - u(0)$, ou

$$u(x) = \frac{1}{2}\int\limits_C u(z)\,d\psi - u(0) = \frac{1}{\pi}\int\limits_C u(z)\left(d\psi - \frac{d\theta}{2}\right). \tag{7}$$ [5])

3. Appliquons maintenant cette formule à une fonction $F(x)$, holomorphe à l'intérieur de C et sur C. Nous aurons d'abord

$$F(x) = \frac{1}{\pi}\int\limits_C F(z)\,d\psi - F(0).$$

Désignons la dérivée $F'(x)$ de $F(x)$ par $f(x)$ et formons les dérivées par rapport à x *dans une direction déterminée.* Il vient

$$f(x) = \frac{1}{\pi}\int\limits_C F(z)\,d_z\left(\frac{\partial\psi}{\partial x}\right). \tag{8}$$

Le premier membre étant indépendant de la direction de différentiation, il en est de même du second membre. Bien entendu il n'en est ainsi que pour l'intégrale elle-même, tandis que les éléments infinitésimaux de l'intégrale dépendent de la direction en question. Une intégration par parties donne

$$f(x) = \frac{1}{\pi}\int\limits_C F(z)\frac{\partial}{\partial z}\frac{\partial\psi}{\partial x}\,dz = -\frac{1}{\pi}\int\limits_C f(z)\frac{\partial\psi}{\partial x}\,dz, \tag{9}$$

formule valable pour toute fonction $f(x)$ holomorphe à l'intérieur de C et sur C. Eu égard à (4), on a

$$-\frac{\partial\psi}{\partial x}\,dx = -\frac{\partial}{\partial x}\arg(z-x)\,dx = \frac{\partial}{\partial x}\arg(x-z)\,dx = d_x\arg(x-z).$$

[5]) *Cf.* É. Goursat, *Cours d'analyse mathématique*, vol. III, 2e édition (Paris, 1915), p. 185, formule (16).

Dès lors, on tire de (9)

$$f(x)\,dx = \frac{1}{\pi}\int_C f(z)\,d_x \arg(x-z)\,dz, \tag{10}$$

d'où il vient

$$|f(x)|\,|dx| \leqq \frac{1}{\pi}\int_C |f(z)|\,|d_x \arg(x-z)|\,|dz| \tag{11}$$

Or $|d_x \arg(x-z)|$ étant précisément l'angle sous lequel l'élément dx est vu du point z, la dernière inégalité est identique à la forme infinitésimale de l'inégalité (3) de M. Carlson.

(Reçu le 8 décembre 1949)

49.

Sur le potentiel de Liénard-Wiechert attaché à une ligne d'univers

C. R. Acad. Sci. Paris *234* (1952), 2159–2161

Dans l'espace-temps, muni de la métrique $ds^2 = (dz, dz) = g_{ik}\, dz^i\, dz^k$ avec $g_{00} = 1$, $g_{11} = g_{22} = g_{33} = -1$, $g_{ik} = 0$ pour $i \neq k$, nous considérons une ligne d'univers L, décrite par une particule de charge unité, représentée par un point matériel $z = (z^i)$. Soit x un point de l'espace-temps qui n'est pas situé sur L et formons le potentiel vecteur [1] retardé d'ordre α

$$A^\alpha(x) = \frac{1}{H(\alpha)} \int_{L^x} r_{xz}^{\alpha-4}\, dz, \tag{1}$$

([2]) P. Lejay, *Développements modernes de la gravimétrie*, Paris, Gauthier-Villars, 1947, p. 165-168.

([1]) Pour tous les détails, nous renvoyons à notre Mémoire : *L'intégrale de Riemann-Liouville et le problème de Cauchy* (*Acta Math.* **81**, 1949, p. 3-223).

où $H(\alpha) = 2^{\alpha-1}\pi\Gamma(\alpha/2)\Gamma((\alpha-2)/2)$, L^x est la partie de L intérieure au cône de lumière rétrograde de sommet x, r_{xz} est la distance lorentzienne des points x et z, le vecteur dz est le déplacement infinitésimal du point z sur L. On a

$$r_{xz}^2 = R_{xz} = (z - x,\ z - x).$$

L'intégrale ne converge que pour les indices $2 < \alpha < 3$, mais nous allons l'étendre par *prolongement analytique* aux indices $2 < \alpha < 5$. Pour $\alpha = 2$, on obtient par un passage à la limite *le potentiel de Liénard-Wiechert*, qui dans les notations de notre Mémoire cité (*cf.* p. 153) s'écrit

$$A(x) = A^2(x) = -\frac{1}{2\pi}\left(\frac{\partial z}{\partial R}\right)_{(R=0)}.$$

Dans le même Mémoire nous donnons également *la nouvelle expression*

$$A(x) = \frac{1}{8\pi}\square_x z_{ret} = \frac{1}{8\pi}\square_x(z_{ret} - x), \tag{2}$$

où $\square_x = g^{ik}\partial^2/\partial x^i\partial x^k$, et z_{ret} est le point d'intersection de la ligne L et du cône rétrograde de sommet x.

Après avoir déduit cette expression par un calcul explicite assez laborieux, nous avons indiqué une voie plus intuitive se rattachant à l'ordre d'idées du prolongement analytique de A^α par rapport à l'indice α et à la relation fondamentale $\square A^{\alpha+2} = A^\alpha$. En effet, il ressort de cette dernière relation que $A(x) = A^2(x) = \square A^4(x)$. Or par une considération de caractère formel on obtient $8\pi A^4(x) = \int_{L^x} dz$, ce qui à un vecteur constant près (relatif au point initial $-\infty$) est égal à z_{ret}. Ceci fournirait sur le champ la formule (2). Dès lors, le problème se pose de rendre cette méthode rigoureuse en calculant A^4 par prolongement analytique.

L'arc s étant compté à partir d'un point arbitraire de L, supposons que cette ligne admet une asymptote pour $s = -\infty$, avec une approximation telle qu'on a, avec des vecteurs constants B, C, D, $z = Bs + C + Ds^{-1} + o(s^{-2})$, $dz/ds = B - Ds^{-2} + o(s^{-3})$, $d^2z/ds^2 = 2Ds^{-3} + o(s^{-4})$. On en déduit, pour un point fixé x, des formules analogues, avec $r = r_{xz}$ au lieu de s. Dès lors, au moyen d'une intégration par parties on obtient, d'abord pour $2 < \alpha < 3$,

$$A^\alpha(x) = \frac{1}{H(\alpha)}\int_{L^x} r^{\alpha-4}\frac{dz}{dr}\,dr = -\frac{1}{H(\alpha)(\alpha-3)}\int_{L^x} r^{\alpha-3}\frac{d^2z}{dr^2}\,dr,$$

les termes tout intégrés étant nuls. En particulier celui relatif au point z_{ret} est nul, puisque $dz/dr = 2r\,dz/dR = 0$ et que dz/dR est fini. La dernière intégrale, qui converge pour $2 < \alpha < 5$, fournit, par prolongement analytique, $A^\alpha(x)$ pour toutes ces valeurs de α. Dès lors, pour $\alpha = 4$, l'intégrale divergente $\int_{L^x} dz/8\pi$ peut être remplacée par

$$A^4(x) = -\frac{1}{8\pi}\int_{L^x} r\frac{d^2z}{dr^2}\,dr.$$

Nous allons montrer que cette expression peut se mettre sous la forme

$$A^4(x) = \frac{1}{8\pi}(z_{\text{ret}} - b), \tag{3}$$

où le point b est la projection orthogonale du point x sur l'asymptote.

On obtient d'abord au moyen d'une intégration par parties

$$\int_z^{z_{\text{ret}}} r \frac{d^2 z}{dr^2} dr = z - r\frac{dz}{dr} - z_{\text{ret}}.$$

En posant ensuite $b_z = z - r(dz/dr)$, on constate que b_z est situé sur la tangente à L en z. D'autre part

$$(x - z, x - b_z) = \left(x - z, x - z + r\frac{dz}{dr}\right)$$
$$= r^2 - r\left(x - z, \frac{d(x-z)}{dr}\right) = r^2 - \frac{r}{2}\frac{d(r^2)}{dr} = 0.$$

Par conséquent $x - b_z$ est orthogonal à $x - z$. En faisant tendre z vers $-\infty$ et en désignant $\lim b_z$ par b, on voit que ce dernier point n'est que la projection orthogonale de x sur l'asymptote, comme nous l'avons dit tout à l'heure. Ainsi la formule (3) se trouve vérifiée. La tangente unitaire au point $-\infty$ étant désignée, comme plus haut, par B, on aura $b = (B, x)B + E$, E étant un vecteur constant. On en tire $\square_x b = 0$. Dès lors, (2) résulte de (3) par la relation de caractère général (*cf.* plus haut) $A(x) = A^2(x) = \square_x A^4(x)$. On a aussi $\text{rot}_x b = 0$, ce qui met en évidence que le bivecteur $G = \text{rot}_x(z_{\text{ret}} - x)$ figurant dans notre Mémoire (p. 158) est, à un facteur constant près, égal à $\text{rot}_x A^4(x)$. Ajoutons que $\text{div}_x z_{\text{ret}} = \text{div}_x b = 1$ et, par suite, $\text{div}_x(z_{\text{ret}} - b) = 0$.

PHYSIQUE THÉORIQUE. — *Les représentations du groupe des rotations et des retournements. Applications de la conservation de la parité en mécanique quantique* (1). II. *Annihilation d'une particule et d'une antiparticule de Dirac.* Note de M. **Louis Michel**, présentée par M. Louis de Broglie.

Le groupe R des rotations d'un espace euclidien à trois dimensions n'est pas simplement connexe; son revêtement universel $\mathcal{R}$ simplement connexe est holomorphe à R; à l'unité $1 \in R$ correspond $(1 + \omega)$, sous-groupe distingué de $\mathcal{R}$; ω est la rotation de 2π autour d'un axe. Toutes les représentations irréductibles, donc univalentes, de $\mathcal{R}$ sont connues; pour tout rang $n = 2J + 1$ il en existe une seule notée D_J; leur produit se décompose ainsi :

$$D_{J'} . D_{J''} = \Sigma_J D_J, \quad \text{avec} \quad |J' - J''| \leq J \leq J' + J''. \tag{1}$$

(1) L. Michel, *Comptes rendus*, **234**, 1952, p. 703.

50.

Sur le potentiel retardé attaché á un courant continu

C. R. Acad. Sci. Paris *234* (1952), 2260–2261

Nous voulons généraliser quelques résultats donnés dans une Note précédente ([1]) et relatifs au potentiel vecteur engendré par une particule ([2]). Soit $j(x)$ un vecteur-courant continu satisfaisant à l'équation de continuité $\mathrm{div}\, j = 0$. Formons les lignes d'univers L qui admettent les vecteurs j comme tangentes, intégrales du système d'équations différentielles ordinaires, $dz/d\sigma = j$, σ étant un paramètre. Nous supposons que ces lignes se comportent régulièrement et que dans l'infini négatif elles remplissent les conditions posées dans la Note précédente pour une ligne unique (existence d'une asymptote, etc.).

Soit D^x l'intérieur du cône caractéristique rétrograde de sommet x, H_r la nappe rétrograde de l'hyperboloïde $(x - z, x - z) = r^2$ et désignons par dH_r une portion infinitésimale de celui-ci, par $|dH_r|$ la mesure (à trois dimensions) de dH_r. Nous formons le potentiel vecteur ([3]) d'ordre α et utilisons la relation évidente $d^4 z = |dH_r|\, dr$.

$$A^\alpha(x) = \frac{1}{H(\alpha)} \int_{D^x} j(z) r^{\alpha-4} d^4 z = \frac{1}{H(\alpha)} \int_{D^x} j(z) r^{\alpha-4} |dH_r|\, dr. \tag{1}$$

En différentiant la relation $r^2 = (z - x, z - x)$ le long de la ligne d'univers passant par un point z, on obtient $r dr = (dz, z - x)$, $dr = (dz, n_r)$ où $n_r = (z - x)/r$ désigne la normale unitaire intérieure de H_r au point z. On peut donc écrire dans l'intégrale (1)

$$j(z)\, dr\, |dH_r| = j(z)(dz, n_r)\, |dH_r| = (j(z), n_r)\, |dH_r|\, dz. \tag{2}$$

Considérons maintenant un tube infiniment mince de lignes d'univers qui découpe de chaque H_r une portion infinitésimale dH_r. Puisque $\mathrm{div}\, j = 0$, par hypothèse, l'expression

$$(j(z), n_r)\, |dH_r| = (j(z), z - x) \frac{|dH_r|}{r} \tag{3a}$$

est en vertu du théorème de Gauss indépendante de r. Elle peut être considérée comme le nombre des lignes d'univers compris dans le tube, et elle est pour un

---

(*) Séance du 26 mai 1952.

([1]) *Voir*, aussi pour les notations, *Comptes rendus*, **234**, 1952, p. 2159. Nous renvoyons également à notre Mémoire cité, *Acta math.*, **81**, 1949, p. 3-223.

([2]) Ces deux Notes émanent d'entretiens que j'ai eus à Paris, en été 1949, avec M. Jean Valatin.

([3]) « Mémoire » p. 149 et 91.

point arbitraire z, situé sur une ligne d'univers déterminée, égale à la même expression prise au point retardé de cette ligne,

$$(3b)\qquad \left((j(z),\, z-x)\frac{|dH_r|}{r}\right)_0 = (j(z_{ret}),\, z_{ret}-x)\left(\frac{|dH_r|}{r}\right)_{(r=0)}$$

En désignant pour chaque ligne d'univers L la partie de cette ligne qui est intérieure au cône D^x par L^x, et par H_0 la nappe de ce cône, le potentiel vecteur (1) s'écrit

$$(4)\qquad A^\alpha(x) = \int_{H_0}(j(z_{ret}),\, z_{ret}-x)\left(\frac{|dH_r|}{r}\right)_0\left(\frac{1}{H(\alpha)}\int_{L^x} r^{\alpha-4}\,dz\right).$$

Ajoutons qu'en chaque point z de H_0 on a $(|dH_r|/r)_0 = dz^1\,dz^2\,dz^3/(x^0-z^0)$. L'intégrale relative à L^x dans (4) ne converge que pour $2 < \alpha < 3$, (*cf.* la Note précédente), mais peut être prolongée analytiquement aux valeurs $3 < \alpha < 5$. En utilisant les résultats de cette première Note, et en désignant par b les projections orthogonales de x sur les asymptotes de nos lignes d'univers, on trouve pour $\alpha = 4$ et $\alpha = 2$

$$(5a)\qquad A^4(x) = \frac{1}{8\pi}\int_{H_0}(z_{ret}-b)(j(z_{ret}),\, z_{ret}-x)\left(\frac{|dH_r|}{r}\right)_0,$$

$$(5b)\qquad A(x) = A^2(x) = \frac{1}{8\pi}\int_{H_0}\Box_x(z_{ret}-x)(j(z_{ret}),\, z_{ret}-x)\left(\frac{|dH_r|}{r}\right)_0.$$

$A(x)$ n'est que le potentiel vecteur retardé ordinaire, attaché au courant $j(x)$. Il est ici décomposé en des potentiels de Liénard-Wiechert relatifs à des lignes d'univers. La remarque qui suit la formule (3a) rend l'expression (5b) intuitive. De cette remarque résulte également qu'on a $\Box_x A^4(x) = A^2(x)$, vu que $\Box_x b = 0$.

Les calculs donnés au n° 78 de notre Mémoire mettent en évidence que la formule (5b) subsiste même dans le cas où les lignes d'univers L n'admettent pas d'asymptotes.

MAGNÉTISME. — *Étude magnétique d'un monocristal de Pyrrhotine aux basses températures.* Note (*) de M. **René Pauthenet**, présentée par M. Gaston Dupouy.

On sait, depuis les travaux de P. Weiss [1], que la pyrrhotine possède un plan de facile aimantation et une direction de très difficile aimantation perpendiculaire à ce plan, suivant laquelle la substance est pratiquement paramagnétique. Dans le plan magnétique, on dispose d'excellentes mesures

(*) Séance du 26 mai 1952.

(1) *J. Phys.*, 8, 1899, p. 542.

52.

Sur le lemme de Zolotareff et sur la loi de réciprocité des restes quadratiques

Math. Scand. *1* (1953), 159–169

Sommaire. Le but principal du présent travail est de donner une démonstration du lemme de Zolotareff qui ne sort pas du domaine des congruences quadratiques. On y arrive en écrivant la permutation considérée par Zolotareff comme produit de deux involutions (n°s 2—3). Aux n°s 4—5 on démontre que les symboles de Zolotareff et de Gauss sont identiques. Au n° 6 on donne une variante de la troisième démonstration de Gauss. Aux n°s 7—9 on montre de deux manières que, dans le cas d'un module composé impair, les symboles de Zolotareff et de Gauss sont identiques au symbole de Jacobi.

Introduction. Gauss, qui a été le premier à démontrer la loi de réciprocité des restes quadratiques, a donné en tout huit démonstrations de cette loi. La première de ces démonstrations, qui procède par induction, porte l'empreinte du génie de Gauss, mais est assez pénible à cause de la multitude des cas qu'il faut distinguer. Elle fut notablement simplifiée par Dirichlet [2] dans un mémoire paru en 1854, où Dirichlet insiste sur le caractère exceptionnel de cette démonstration dans les termes suivants (traduits de l'allemand).

« Parmi les nombreuses démonstrations de la loi de réciprocité, la première démonstration de Gauss, trouvée déjà en 1796 et insérée aux *Disquisitiones arithmeticae* (1801), m'a toujours paru particulièrement remarquable, non seulement à cause de l'idée tellement simple qui lui sert de base, mais encore parce que cette démonstration est la seule, que je sache, qui ne sort nulle part du domaine des congruences du second degré, auquel le théorème appartient lui-même, tandis que toutes les autres démonstrations connues reposent sur des principes qui paraissent plus ou moins étrangers au domaine en question. »

Kronecker [5] a appuyé sur le même point en 1876.

Un certain nombre des démonstrations auxquelles Dirichlet fait allusion, sont basées sur des méthodes transcendantes, ainsi deux démonstrations dues à Gauss et une démonstration très connue, due à Eisenstein. Cependant, la grande majorité des démonstrations en question reposent sur un théorème, appelé *lemme de Gauss*, primitivement point de départ de la démonstration la plus connue de Gauss, la troisième (1808). Gauss déduit ce lemme du *critère d'Euler*, qui, aussi simple qu'il soit, fait appel à une congruence d'ordre supérieur. Il en est de même de la démonstration, fondée sur la théorie des indices, du *lemme* extrêmement élégant *de Zolotareff* (1872) [8]. D'autre part, les deux lemmes résultent sur le champ l'un de l'autre, le passage ne faisant intervenir aucun élément étranger au problème.

Notre but primaire est de donner une démonstration du lemme de Zolotareff qui reste dans le cadre des congruences quadratiques. Aux n°s 2—3 on montre que le symbole de Zolotareff est identique au symbole de Legendre quand le module est un nombre premier (impair), aux n°s 4—5 on établit l'identité des symboles de Zolotareff et de Gauss, au n° 6 on donne une variante de la troisième démonstration de Gauss, conduisant à la loi de réciprocité par l'intermédiaire d'une très belle formule due à Kronecker [5] et répondant aux desiderata de Dirichlet.

Dans la plupart de ces considérations on envisage aussi le cas des modules composés (impairs). Aux n°s 7—9 on trouve deux démonstations du fait que, dans le cas généralisé, les symboles de Zolotareff et de Gauss sont identiques au symbole de Jacobi.

Ce qu'il y a de nouveau dans ce travail se restreint à la matière exposée aux n°s 2, 3, 8. Le reste n'est que compilation, qui pourtant, je l'espère, pourra rendre quelques services au lecteur.

1. *Le lemme de Zolotareff* consiste en ceci.

Soit p un nombre premier impair et D un nombre entier non divisible par p. Les plus petits restes positifs des nombres

$$D, 2D, \ldots, (p-1)D \tag{1}$$

par rapport à p forment évidemment une permutation des nombres

$$1, 2, \ldots, p-1\,. \tag{2}$$

Suivant que cette permutation est paire ou impaire, D est reste quadratique ou non-reste quadratique par rapport à p.

2. Considérons dans un ordre d'idées plus général m éléments $u_1, u_2, \ldots, u_m$ et une permutation $v_1, v_2, \ldots, v_m$ de ces éléments. Posons $v_k = Tu_k$ ou, plus brièvement, $v = Tu$ et $\operatorname{sgn} T = +1$ ou $\operatorname{sgn} T = -1$ suivant que T est une permutation paire ou impaire ou, en d'autres termes, suivant que la transformation $u \to Tu$ conserve ou ne conserve pas la classe des permutations auxquelles on l'applique. Il est très facile de alcculer sgn T, dans le cas où T est une involution, ce qui veut dire que $T^{-1} = T$. Chaque cycle d'une telle permutation est constitué par un élément fixe $w \to w$ ou par une transposition $u \longleftrightarrow v$ $(u \neq v)$. Soient α le nombre des éléments fixes et β le nombre des transpositions. On a évidemment $\alpha + 2\beta = m$ et

$$\operatorname{sgn} T = (-1)^\beta = (-1)^{\frac{m-\alpha}{2}}.$$

Soient maintenant T' et T'' deux involutions et $T'T''$ leur produit, ce qui veut dire que $(T'T'')u = T'(T''u)$. On a manifestement $\operatorname{sgn}(T'T'') = \operatorname{sgn} T' \cdot \operatorname{sgn} T''$ et alors, dans des notations évidentes,

$$\operatorname{sgn}(T'T'') = (-1)^{m - \frac{\alpha' + \alpha''}{2}}.$$

En admettant encore que m est un nombre *pair*, hypothèse toujours remplie dans les cas particuliers qui suivent, on aura

(3) $$\operatorname{sgn}(T'T'') = (-1)^{\frac{\alpha' + \alpha''}{2}}.$$

3. La transformation Z_D, qui consiste en ce que $u \to$ le plus petit reste positif de Du modulo p, n'est pas en général une involution (elle ne l'est que pour $D \equiv \pm 1$), mais elle peut toujours s'écrire comme le *produit de deux involutions* très simples.

u étant un élément quelconque de la suite (2), la congruence $uu' \equiv D$ (mod. p) a une solution et une seule u' qui appartient à (2) elle aussi. Nous désignons cette solution par D/u et nous posons $D/u = T_D u$. Dans le cas particulier où $D = 1$ cela devient $1/u = T_1 u$. Il est clair que $Z_D = T_D T_1$ (d'une manière symbolique $Du = D/(1/u)$). Le nombre des éléments fixes de T_D étant désigné par α_D, on a d'après (3)

(4) $$\operatorname{sgn} Z_D = (-1)^{\frac{\alpha_D + \alpha_1}{2}}.$$

Reste à calculer le dernier exposant, c'est-à-dire les nombres des éléments fixes de T_D et de T_1. Si D est reste quadratique mod. p, la congruence $x^2 \equiv D$ (mod. p) aura deux solutions (mod. p) soit x et $p - x \not\equiv x$, et on a $\alpha_D = 2$, tandis que $\alpha_D = 0$ dans le cas où D est non-reste. Le nombre $D = 1$ est toujours reste (les deux solutions de $x^2 \equiv 1$ étant 1 et $p-1$). On a donc $\alpha_1 = 2$. L'exposant sera par suite $= 2$ quand D est reste et $= 1$ quand D est non-reste. Moyennant le *symbole de Legendre* (D/p) qui, *par définition*, est $= 1$ quand D est reste et $= -1$ quand D est non-reste quadratique suivant p, notre résultat peut s'écrire

$$\operatorname{sgn} Z_D = \left(\frac{D}{p}\right).$$ C.Q.F.D.

En introduisant encore le symbole de Zolotareff $(D/p)_Z = \operatorname{sgn} Z_D$ et en désignant, pour plus de netteté, le symbole de Legendre par $(D/p)_L$, le résultat ci-dessus peut s'exprimer par la formule

$$(5) \qquad \left(\frac{D}{p}\right)_L = \left(\frac{D}{p}\right)_Z,$$

valable dans les conditions posées, savoir que p est un nombre premier impair et que le nombre entier D n'est pas divisible par p. (D'ailleurs le résultat subsiste pour $p = 2$, mais il devient trivial.)

Remarque I. Il résulte de la définition même du symbole de Zolotareff que ce symbole est un caractère du groupe multiplicatif mod. p c'est-à-dire que

$$(6) \qquad \left(\frac{D_1 D_2}{p}\right)_Z = \left(\frac{D_1}{p}\right)_Z \left(\frac{D_2}{p}\right)_Z.$$

Frobenius [4] a fortement insisté sur ce point au début d'une analyse pénétrante des diverses démonstrations de la loi de réciprocité.

Remarque II. Il est clair que la valeur du symbole de Zolotareff est conservée, si l'on remplace la suite (2) par une suite arbitraire $\{\alpha_k\}$ de $p-1$ nombres incongrus et $\not\equiv 0$ (mod. p), les plus petits restes positifs de la suite (1) suivant p étant remplacés en même temps par ceux des restes des nombres $\{D\alpha_k\}$ qui appartiennent à la suite $\{\alpha_k\}$.

On peut aussi opérer avec les classes de restes (classes d'équivalence) suivant p, former leurs produits, leurs permutations etc. Si l'on traduit les définitions et le résultat qui précèdent dans le langage de ces classes, on arrive à un énoncé très clair et très concis. Néanmoins la transcription en question sera laissée de côté dans ce qui suit.

4. La définition du symbole de Zolotareff s'étend facilement au cas où *le module p n'est plus nécessairement un nombre premier, mais D est premier avec p.* Pour arriver à des résultats intéressants, nous supposons en outre que *p est un nombre impair* (positif).

On a en effet dans ces conditions (*cf.* le premier alinéa de la *Remarque* II) que, les nombres $\{\alpha_k\}$, $k = 1, 2, \ldots, p-1$, étant incongrus entre eux suivant p et non divisibles par p, il en est de même des nombres $\{D\alpha_k\}$. Ceux des restes de ces nombres qui appartiennent à la suite $\{\alpha_k\}$ fournissent donc une permutation Z_D de ces derniers nombres. Le signe (signature) de cette permutation ne dépend que de p et de D. *On pose aussi dans les nouvelles conditions* $(D/p)_Z = \operatorname{sgn} Z_D$.

Ce qu'on perd dans ce cas généralisé, c'est le renseignement précis sur le caractère quadratique de D par rapport à p que le symbole fournissait dans le cas d'un module premier (*cf.* les n°s 7—9, où on montre que le symbole généralisé est identique au symbole de Jacobi; *voir*, en particulier, la fin du n° 7).

On va maintenant démontrer que les symboles de Zolotareff et de Gauss — tous les deux définis soit dans les conditions primitives soit dans les conditions généralisées — sont identiques.

5. Tandis que le symbole de Zolotareff se rattache à des systèmes complets de restes incongrus ($\not\equiv 0$), le symbole de Gauss se rattache à des demi-systèmes de restes, c'est-à-dire à des suites de $(p-1)/2$ nombres $\{x_j\}$ tels que les $p-1$ nombres $\{\pm x_j\}$ constituent un système complet. Un tel demi-système est fourni par l'ensemble

$$\{x\} = \left\{1,\, 2, \ldots, \frac{p-1}{2}\right\}. \tag{7}$$

Par le *reste minimum* (en valeur absolue) d'un nombre w modulo p, on entend le nombre $\overline{w}$, unique puisque p est impair, tel que $\overline{w} \equiv w$ (mod. p) et $|\overline{w}| \leqq (p-1)/2$. Le reste minimum d'un nombre quelconque non divisible par p appartient évidemment à l'ensemble $\{\pm x\}$. Nous notons encore la relation évidente $\overline{-w} = -\overline{w}$.

Le *symbole de Gauss* $(D/p)_G$ se définit de la manière suivante.

On fait parcourir à x la suite $\{x\}$, on désigne par μ *le nombre des restes minima des produits Dx qui sont négatifs* et on pose

$$\left(\frac{D}{p}\right)_G = (-1)^\mu\,. \tag{8}$$

Pour voir que ce symbole coïncide avec celui de Zolotareff, appliquons la transformation Z_D au système complet $\{x^*\} = \{\pm x\}$ écrit dans l'ordre naturel, c'est-à-dire que $x_1^* < x_2^*$ implique que x_1^* précède x_2^*. La transformation s'exprime par la formule $x^* \to \overline{Dx^*}$. On entend par une *inversion* une constellation telle que

$$u < v \text{ et } \overline{Du} > \overline{Dv}\,.$$

Si l'on pose $-u = u'$ et $-v = v'$, on aura en même temps l'inversion

$$v' < u' \text{ et } \overline{Dv'}(= -\overline{Dv}) > \overline{Du'}(= -\overline{Du})\,.$$

Si les couples (u, v) et (u', v') sont différents, les deux inversions se compensent en tant qu'il ne s'agit que de la classe de la permutation $x^* \to \overline{Dx^*}$. Dans le cas où les deux couples coïncident, les deux inversions simultanées se réduisent à une seule. On a évidemment dans ce cas $u = -v$ et

$v = -u$, $-v = u < v$ et $-\overline{Dv} = \overline{Du} > \overline{Dv}$, ce qui veut dire que v est positif et $\overline{Dv}$ négatif. Il y aura donc autant d'inversions non compensées qu'il y a des nombres $v > 0$ tels que $\overline{Dv} < 0$. Le nombre de ces inversions est donc égal au nombre μ introduit par Gauss et

$$(9) \qquad \left(\frac{D}{p}\right)_Z = \operatorname{sgn} Z_D = (-1)^\mu = \left(\frac{D}{p}\right)_G.$$

Remarque III. Le fait que la valeur du symbole de Zolotareff est indépendante du système complet auquel on applique la transformation Z_D entraîne que la valeur du symbole de Gauss est indépendante du demi-système moyennant laquel on calcule cette valeur. D'une manière plus précise: Les nombres z_j étant les éléments d'un demi-système arbitraire $\{z_j\}$ et μ le nombre des produits Dz_j qui admettent des restes appartenant au demi-système complémentaire $\{-z_j\}$, la parité du nombre μ est indépendante du demi-système $\{z_j\}$. (Pour une autre démonstration de ce fait *voir* Scholz [7], p. 75.)

Retournons un instant au cas particulier d'un *module premier*. L'identité (9) combinée avec l'identité (5) nous fournit $(D/p)_L = (D/p)_G$. Cette dernière identité constitue le *lemme de Gauss.*

6. En nous plaçant de nouveau dans les conditions généralisées, nous allons déduire la loi de réciprocité relative au symbole de Gauss. Les nombres positifs p et q étant impairs et premiers entre eux, écrivons les suites

$$(10) \qquad \{x\} = \left\{1, 2, \ldots, \frac{p-1}{2}\right\} \quad \text{et} \quad \{y\} = \left\{1, 2, \ldots, \frac{q-1}{2}\right\}.$$

Posons pour tout $x \in \{x\}$

$$(11) \qquad qx = g_x p + r_x,\ 0 < r_x < p\,,$$

c'est-à-dire que r_x est le plus petit reste positif de qx suivant p. On introduit aussi les restes minima (en valeur absolue) ϱ_x, en posant $\varrho_x = r_x$ ou $\varrho_x = r_x - p$ suivant que $r_x < p/2$ ou $r_x > p/2$. Les ϱ_x sont positifs dans le premier cas et négatifs dans le second. On a toujours $|\varrho_x| < p/2$ ou, ce qui revient au même, $|\varrho_x| \in \{x\}$.

Puisque p et q sont impairs, on obtient de (11), $x \equiv g_x + r_x \pmod{2}$, ce qui peut aussi s'écrire

$$(12) \qquad x \equiv g_x + \varrho_x \equiv g_x + |\varrho_x| \pmod{2} \quad \text{quand } \varrho_x > 0$$

et

$$(13) \qquad x \equiv g_x + p + \varrho_x \equiv g_x + |\varrho_x| + 1 \pmod{2} \quad \text{quand } \varrho_x < 0\,.$$

Les $|\varrho_x|$ sont différents entre eux. On a en effet $\varrho_k \pm \varrho_l \equiv q(k \pm l)$ (mod. p). Puisque q est premier avec p et que $k \pm l$ n'est pas divisible par p

si $k \neq l$ (vu que $k \in \{x\}$, $l \in \{x\}$), on a aussi $\varrho_k \pm \varrho_l \not\equiv 0 \pmod{p}$ et, à plus forte raison, $\varrho_k \pm \varrho_l \neq 0$. Il en résulte que l'ensemble des valeurs $|\varrho_x|$ est identique à l'ensemble $\{x\}$.

Ce point acquis, on voit que si x parcourt le système $\{x\}$, $|\varrho_x|$ en fera de même, ce qui donne $\Sigma_x |\varrho_x| = \Sigma_x x$. En désignant avec Gauss le nombre des ϱ_x négatifs par μ, on tire de (12) et de (13)

$$\sum_x g_x + \mu \equiv 0 \ (\mathrm{mod.}\ 2) \text{ ou } \mu \equiv \sum_x g_x (\mathrm{mod.}\ 2)\,,$$

et, par conséquent,

$$\left(\frac{q}{p}\right)_G = (-1)^\mu = (-1)^{\Sigma_x g_x}.$$

Il résulte de la formule (11) que $g_x = [qx/p]$. Cela étant, on va donner une seconde interprétation de $\Sigma_x g_x$. En effet, si l'on fixe le nombre $x \in \{x\}$, l'expression $qx-py$ est évidemment positive pour $y = 1, 2, \ldots, [qx/p]$, nombres qui appartiennent à $\{y\}$ ($qx/p < q \cdot \frac{1}{2}p/p = q/2$), tandis que l'expression est négative pour tout autre $y \in \{y\}$. La somme $\Sigma_x g_x$ est donc égale au nombre des couples $x \in \{x\}, y \in \{y\}$ pour lesquels $qx-py > 0$ ou, ce qui revient au même, $y/q-x/p < 0$. On a ainsi établi la formule de Kronecker

$$\text{(14a)} \qquad \left(\frac{q}{p}\right)_G = \operatorname{sgn} \prod_{x,y} \left(\frac{y}{q} - \frac{x}{p}\right), \qquad x \in \{x\}, y \in \{y\}\,.$$

Il vient par symétrie

$$\text{(14b)} \qquad \left(\frac{p}{q}\right)_G = \operatorname{sgn} \prod_{x,y} \left(\frac{x}{p} - \frac{y}{q}\right), \qquad x \in \{x\}, y \in \{y\}\,.$$

Chacun des produits contient $((p-1)/2)((q-1)/2)$ facteurs. Ces facteurs ayant des signes opposés dans les deux produits, on obtient la loi de réciprocité

$$\text{(15)} \qquad \left(\frac{q}{p}\right)_G \left(\frac{p}{q}\right)_G = (-1)^{\frac{p-1}{2} \cdot \frac{q-1}{2}}.$$

Dans les formules ci-dessus on peut évidement substituer le symbole de Zolotareff à celui de Gauss. Dans le cas particulier, où *p et q sont des nombres premiers* (impairs), nos formules subsistent aussi pour le *symbole de Legendre*. La formule (15) exprime, dans ce cas particulier, *la loi de réciprocité proprement dite, relative à deux nombres premiers impairs*. Cette loi se présente ici avec une démonstration qui répond aux desiderata de Dirichlet.

7. Nous allons montrer que, dans le cas généralisé, les symboles de

Zolotareff et de Gauss, identiques entre eux, sont aussi identiques au symbole de Jacobi.

Soit n un nombre impair positif et D un nombre entier premier avec n. Il est clair que la *Remarque* I peut encore s'appliquer dans les conditions généralisées, c'est-à-dire que le symbole (soit de Zolotareff, soit de Gauss) (D/n) possède un caractère multiplicatif par rapport à D, savoir $(D_1D_2/n) = (D_1/n)(D_2/n)$. Or le symbole possède un caractère multiplicatif aussi par rapport à n, savoir

$$\left(\frac{D}{n_1 n_2}\right) = \left(\frac{D}{n_1}\right)\left(\frac{D}{n_2}\right). \tag{16}$$

Pour le voir, posons $n_1 n_2 = n$ et admettons d'abord que D aussi est un nombre impair positif. On a alors, en vertu de la loi de réciprocité et de la *Remarque* I,

$$\left(\frac{D}{n}\right) = (-1)^{\frac{D-1}{2}\cdot\frac{n-1}{2}}\left(\frac{n}{D}\right) = (-1)^{\frac{D-1}{2}\cdot\frac{n-1}{2}}\left(\frac{n_1}{D}\right)\left(\frac{n_2}{D}\right) = (-1)^N\left(\frac{D}{n_1}\right)\left(\frac{D}{n_2}\right),$$

avec

$$N = \frac{D-1}{2}\left(\frac{n_1 n_2 - 1}{2} - \frac{n_1 - 1}{2} - \frac{n_2 - 1}{2}\right) \tag{17}$$

$$= \frac{D-1}{2}\,\frac{(n_1-1)(n_2-1)}{2} \equiv 0 \pmod{2},$$

ce qui entraîne (16) pour tout nombre impair positif D.

Pour aller plus loin, il nous faut calculer les valeurs explicites des symboles $(-1/n)$ et $(2/n)$. Quant au premier, on tire immédiatement de la définition du symbole de Gauss que

$$\left(\frac{-1}{n}\right) = (-1)^{\frac{n-1}{2}}.$$

Le symbole $(2/n)$ s'obtient facilement, si l'on observe que la transformation Z_2 appliquée à la suite $1, 2, \ldots, n-1$ produit $1+2+\ldots+(n-1)/2 = (n^2-1)/8$ inversions. On a donc

$$\left(\frac{2}{n}\right) = (-1)^{\frac{n^2-1}{8}}.$$

Le calcul exécuté dans la formule (17) met en évidence que la relation (16) est valide pour $D = -1$. Si l'on observe encore que le carré $(4k \pm 1)^2$ de tout nombre impair est $\equiv 1 \pmod{8}$, on voit par un calcul analogue que la relation (16) tient aussi pour $D = 2$. Enfin, en recourant de nouveau à la *Remarque* I, on voit que la formule (16) subsiste pour un nombre composé D arbitraire.

Ce point acquis, décomposons n en ses facteurs premiers $n = \Pi\, p$, les p étant différents ou non. La formule (16) nous donne immédiatement

$$\left(\frac{D}{n}\right) = \Pi \left(\frac{D}{p}\right), \tag{18}$$

où, d'après (5), on peut entendre par (D/p) le symbole de Legendre.

C'est par le dernier produit que Jacobi a défini la généralisation remarquable du symbole de Legendre qui porte son nom et que nous désignons par $(D/n)_J$. Nous avons donc, en fin de compte, démontré que

$$\left(\frac{D}{n}\right)_Z = \left(\frac{D}{n}\right)_G = \left(\frac{D}{n}\right)_J \tag{19}$$

Remarque IV. On sait depuis les travaux de Gauss que la condition nécessaire et suffisante pour que D soit reste quadratique par rapport à un nombre impair n est que D soit reste quadratique de tous les facteurs premiers de n. Il en résulte que, si D est reste quadratique de n, on a $(D/n) = 1$, mais la réciproque n'est pas vraie. Pour que (D/n) soit égal à 1 il faut et il suffit évidemment que les p suivant lesquels D est non-reste soient en nombre pair.

8. Pour montrer que la méthode utilisée aux n°s 2—3 dans le cas d'un module premier peut aussi s'appliquer dans le cas d'un module composé impair, nous indiquons ici une seconde démonstration du fait que dans ce dernier cas les symboles de Zolotareff et de Jacobi sont identiques.

Soit donc n un nombre composé impair. Nous répartissons les nombres $1, 2, \ldots, p-1$ en des systèmes partiels Σ_d, chaque tel système étant constitué par les nombres qui ont le même plus grand commun diviseur d avec n. En posant $n/d = n'$, le système Σ_d sera constitué par les nombres $d\xi$, ξ parcourant tous les nombres $< n'$ qui sont premiers avec n'. D étant un nombre premier avec n, la transformation Z_D transforme chaque système partiel Σ_d en lui-même. Cela étant, on pourra étudier l'effet de Z_D sur chaque Σ_d séparément et former ensuite le «produit» de ces effets.

La transformation $u \to Du$ (mod. n) avec $u = d\xi \in \Sigma_d$ est identique à la transformation $\xi \to D\xi$ (mod. n'). En appliquant le procédé donné au n° 3, on écrit cette transformation comme le produit de deux involutions et on obtient pour chaque système partiel une relation de la forme

$$\operatorname{sgn} Z_D = (-1)^{\frac{\alpha_D + \alpha_1}{2}},$$

α_D étant le nombre des solutions de la congruence $x^2 \equiv D$ (mod. n'). On voit facilement que $\alpha_D = \alpha_1$ quand D est reste quadratique, tandis que

$\alpha_D = 0$ quand D est non-reste quadratique suivant n'. L'exposant est donc égal à α_1 dans le premier cas et à $\frac{1}{2}\alpha_1$ dans le second. Vu que α_1 est un nombre pair, le symbole de Zolotareff relatif au système partiel est donc égal à 1 si D est reste quadratique suivant n' et égal à $(-1)^{\frac{1}{2}\alpha_1}$ quand D est non-reste.

Le nombre α_1 qui, par définition, est égal au nombre des solutions de la congruence $x^2 \equiv 1 \pmod{n'}$, ne dépend que de n'. En posant donc $\alpha_1 = \psi(n')$, on obtient pour le symbole de Zolotareff relatif à la suite entière $\{1, 2, \ldots, n-1\}$

$$\left(\frac{D}{n}\right)_Z = (-1)^{\frac{1}{2}\Sigma\psi(n')}, \tag{20}$$

la sommation étant étendue à tous les diviseurs n' de n suivant lesquels D est non-reste.

9. C'est la dernière expression à laquelle Schering a su réduire le symbole de Gauss dans le cas d'un module impair composé. Il a aussi montré que cette expression coïncide avec le symbole de Jacobi. Ce n'est que ce dernier point qui nous intéresse ici, et nous allons brièvement indiquer le raisonnement par lequel Schering [6] a établi ce point; *voir* aussi Bachmann [1], p. 193.

Nous citons d'abord deux propositions dues à Gauss; *cf.* Dirichlet [3] p. 81.

I. On a pour tout nombre impair n, $\psi(n) = 2^\varkappa$, $\varkappa$ étant le nombre des facteurs premiers *différents* de n.

II. Tout nombre D premier avec le nombre premier impair p est en même temps reste ou non-reste quadratique de toutes les puissances p^α $(\alpha \geqq 1)$.

On voit à l'aide de la proposition I que le second membre de (20) ne sera pas changé si l'on supprime tous les n' qui admettent au moins deux facteurs premiers différents. Il ne reste donc que les n' de la forme p^α où p^α divise n et D est non-reste par rapport à p^α. La dernière hypothèse revient, d'après la proposition II, à ce que D est non-reste par rapport à p. D'après la proposition I, chaque telle puissance p^α apporte à $\frac{1}{2}\Sigma\psi(n')$ une contribution égale à 1. Donc si $n = \Pi p_k^{\pi_k}$ les p_k étant différents entre eux, l'expression (20) se réduit à $(-1)^{\Sigma\pi_l}$, où les p_l sont ceux des facteurs premiers p_k de n suivant lesquels D est non-reste. Vu que $(D/p_l) = -1$, la dernière expression est égale à $\Pi(D/p_l)^{\pi_l}$. Pour les p_k autres que les p_l, on a $(D/p_k) = 1$. Le produit précédent pourra donc être étendu à tous les p_k, qui divisent n, sans que sa valeur soit changée. Dès lors, en posant $n = \Pi p$, les nombres premiers p étant différents ou non, on voit que le second membre de (20) est égal au symbole de Jacobi (*cf.* (18))

$$\Pi\left(\frac{D}{p}\right) = \left(\frac{D}{n}\right)_J,$$

résultat qui met en évidence que les symboles de Zolotareff et de Jacobi ont des valeurs identiques.

BIBLIOGRAPHIE

1. P. Bachmann, *Niedere Zahlentheorie* I, Leipzig, 1902.
2. P. G. Lejeune Dirchlet, *Über den ersten von Gauss gegebenen Beweis des Reciprocitätsgesetzes in der Theorie der quadratischen Reste*, J. reine angew. Math. 47 (1854), 139–150. (Werke II, Berlin, 1897, 123–138.)
3. P. G. Lejeune Dirichlet, *Vorlesungen über Zahlentheorie*, Vierte Auflage, Braunschweig, 1894.
4. G. Frobenius, *Über das quadratische Reciprocitätsgesetz*, S.-Ber. Preuss. Akad. Wiss. Berlin, math.-naturw. Kl. 1914 (I), 335–349, 484–488.
5. L. Kronecker, *Über das Reciprocitätsgesetz*, S.-Ber. Preuss. Akad. Wiss. Berlin, math.-naturw. Kl. 1876, 331–341. (Werke II, Leipzig, 1897, 13–23.)
6. E. Schering, *Zur Theorie der quadratischen Reste*, Acta math. 1 (1882), 153–170.
7. A. Scholz, *Einführung in die Zahlentheorie*, Sammlung Göschen 1131, Berlin, 1939.
8. Zolotareff, *Nouvelle démonstration de la loi de réciprocité de Legendre*, Nouvelles Annales de Math. (2e série) 11 (1872), 354–362.

UNIVERSITÉ DE LUND, SUÈDE

53.

L'équation de Dirac en relativité générale

12e Congr. des Math. Scand. Lund 1953, 241–259

Introduction. L'équation de Dirac posée pour une particule libre, de masse au repos m, s'écrit, dans des unités physiques convenables,

$$\left(\sum_{k=0}^{3}\gamma^k\frac{\partial}{\partial x^k}-i\,m\right)\psi=0.$$

Ici, les x^k sont les coordonnées d'un point x de l'espace-temps de la relativité restreinte, les γ^k des matrices d'ordre 4, satisfaisant à certaines relations — celles de Clifford — et ψ une colonne de 4 éléments. Lors d'une transformation de Lorentz $x^k \to x^{k'}$ (les γ^k restant inaltérés) la colonne ψ, appelée spineur, subit une certaine transformation de caractère différent de celles subies par les tenseurs.

Le but du présent travail est de donner une extension de la théorie de Dirac en relativité générale. L'espace-temps affine étant remplacé par un espace de Riemann de signature lorentzienne, on construit en chaque point de cet espace une algèbre de Clifford *locale*, basée sur les formules $e_ie_j+e_je_i=2g_{ij}$, où e_i est le vecteur de base correspondant à la coordonnée x^i et les g_{ij} sont les coefficients de la forme métrique. En chaque point, on introduit *tous* les repères lorentziens qui peuvent s'obtenir du repère riemannien formé par les e_i par déformation continue.

Les *spineurs* sont chez nous certains *collectifs* de nombres de Clifford qui dépendent des repères lorentziens d'une manière bien déterminée. La principale difficulté dans le problème d'extension qui nous occupe consiste dans la différentiation des spineurs. Autant que nous pouvons en juger, cette difficulté se résout ici d'une manière très naturelle et très simple.

C'est le point de vue géométrique qui domine notre exposé, où nous croyons avoir donné une synthèse assez satisfaisante entre tenseurs riemanniens et spineurs. Malheureusement, notre exposé n'est pas sans longueurs. En revanche, le caractère covariant de nos définitions et de nos procédés est manifeste dès le début.

Outre les considérations théoriques, qui prévalent dans notre travail, nous donnons au n° 10 un aperçu du calcul effectif. Le physicien y retrouvera ses outils habituels, matrices et colonnes.

Le problème traité ici a occupé plusieurs auteurs de grand mérite. Il convient de citer en particulier E. Cartan, qui le mentionne dans la préface et le traite à la fin de ses *Leçons sur la théorie des spineurs, I—II*, Actualités Sci. Ind. 643, 701, Paris 1938. On y trouve aussi une bibliographie, à laquelle il convient d'ajouter un travail de V. Bargmann, Sitzungsber. Akad. Berlin, 1932 et un travail de W. Pauli, Ann. der Physik (5), t. 18 (1933). Cartan, qui laisse les questions de covariance de côté, juge ses prédécesseurs assez sévèrement, en leur reprochant en particulier l'absence de points de vue géométriques. D'un autre côté, la plupart des auteurs en question se reprochent mutuellement que la covariance des procédés qu'ils utilisent est difficile à voir.

Malgré la divergence des points de vue, les résultats finals obtenus par des procédés différents semblent être identiques. Enfin, si l'on compare le n° 10 du présent travail et le n° 174 du fascicule II de Cartan, on trouvera un accord parfait.

Je tiens à remercier très chaleureusement mon ancien élève M. Karl Greger qui m'a prêté un précieux concours dans la préparation et la rédaction du présent travail.

1. Espace de Riemann. On considère un espace de Riemann (espace-temps de la relativité générale) avec une métrique $ds^2=g_{ij}\,dx^i\,dx^j$ $(i, j=0, 1, 2, 3)$ de *signature lorentzienne*, ce qui veut dire que, réduite à une somme de carrés, la forme métrique se compose d'un carré positif (temps) et de trois carrés négatifs (espace). On suppose $g_{00}>0$, $g_{11}<0$, $g_{22}<0$, $g_{33}<0$. Les vecteurs de base[1] e_i correspondant aux coordonnées x^i $(i=0, 1, 2, 3)$ admettent les composantes contrevariantes respectives $(1, 0, 0, 0)$, $(0, 1, 0, 0)$, ...

Le *produit scalaire* de deux vecteurs A et B de composantes contrevariantes A^i et B^i $(A=A^ie_i,\ B=B^ie_i)$ est défini par la formule

$$(A, B)=g_{ij}\,A^i\,B^j \tag{1}$$

et, en particulier,

$$(e_i, e_j)=g_{ij}. \tag{2}$$

[1] *Cf.* E. Cartan, Leçons sur la géométrie des espaces de Riemann, 2e éd., Paris 1946, p. 31 *et suiv.*

En posant, pour abréger, $\partial/\partial x^i = \partial_i$, on a

(3) $$\partial_i e_j = \partial_j e_i = \Gamma_{ij}{}^r e_r, \quad \Gamma_{ij}{}^r = \Gamma_{ji}{}^r,$$[1]

d'où l'on obtient, pour un vecteur $A = A^j e_j$ quelconque, la *différentielle covariante*,

(3bis) $$dA = dA^j e_j + A^j \Gamma_{ij}{}^r e_r dx^i.$$

Le *transport parallèle* infinitésimal est donné par $dA = 0$, c'est-à-dire par

(4) $$dA^j + \Gamma_{ir}{}^j A^r \, dx^i = 0.$$

On introduit aussi la *base duale* au moyen des formules

(5) $$(e^i, e_j) = \delta_j{}^i \text{ ou } e^i = g^{ij} \, e_j.$$

On tire de (2) et de (5)

(6) $$\partial_i e^j = -\Gamma_{ir}{}^j e^r.$$

Lors d'une transformation $x^i \to x^{i'}$ des coordonnées riemanniennes, les e_i et les e^i se transforment suivant les lois habituelles

(7) $$e_{i'} = \partial_{i'} x^j \cdot e_j, \quad e^{i'} = \partial_j x^{i'} \cdot e^j.$$

On n'admettra que de telles transformations de coordonnées qui 1) gardent les signes respectifs des g_{ii}, fixés au début, 2) gardent l'orientation (c'est-à-dire que le déterminant fonctionnel $|\partial_i x^{j'}| > 0$), 3) n'intervertissent pas le futur et le passé (c'est-à-dire que la dérivée partielle $\partial_0 x^{0'} > 0$). Les conditions posées sont symétriques et transitives.

Selon que le carré scalaire (A, A) d'un vecteur A, de composantes A^i réelles, a une valeur positive, nulle ou négative, A sera classifié comme *vecteur de temps, de lumière ou d'espace*. En raison de notre convention sur les signes des g_{ii}, e_0 est un vecteur de temps, tandis que e_1, e_2, e_3 sont des vecteurs d'espace. Un vecteur de temps ou de lumière A sera dit *positif* ou *négatif* selon que la composante A^0 est positive ou négative. En particulier e_0 sera un *vecteur de temps positif*. Grâce à la condition 3) qu'on vient de poser, le caractère positif ou négatif d'un vecteur est indépendant du système de coordonnées.

2. **Algèbres de Clifford locales.**[2] On forme en chaque point x de l'espace de Riemann une algèbre de Clifford *locale* C_x, définie sur

[1] *Cf.* Cartan *l.c.* p. 35.

[2] Pour la terminologie et pour différents faits concernant les nombres de Clifford, *voir* notre conférence faite au dixième congrès des mathématiciens scandinaves tenu à Copenhague 1946 (København 1947); citée dans la suite comme «*Conférence*». Notons pourtant que, dans cette conférence, les spineurs sont définis comme éléments de l'algèbre de Clifford, tandis qu'ici, ils sont définis comme collectifs de tels éléments.

le corps des nombres complexes. Dans une telle algèbre, on a une multiplication associative et distributive, dont les règles se déduisent des formules

$$e_i e_j + e_j e_i = 2(e_i, e_j) = 2g_{ij}. \tag{8}$$

De ces formules résulte, pour deux vecteurs $A = A^i e_i$ et $B = B^i e_i$ arbitraires,

$$AB + BA = 2(A, B). \tag{9}$$

Un élément général c de l'algèbre C_x peut s'écrire, d'une et d'une seule façon, comme somme d'un scalaire, d'un vecteur, d'un bivecteur, d'un trivecteur et d'un pseudoscalaire

$$c = c^S + c^i e_i + \frac{c^{ij}}{2!} e_i e_j + \frac{c^{ijk}}{3!} e_i e_j e_k + \frac{c^{ijkl}}{4!} e_i e_j e_k e_l, \tag{10}$$

les coefficients étant des nombres complexes et les c^{ij}, c^{ijk}, c^{ijkl} étant *antisymétriques.*

Le coefficient c^S est appelé la *partie scalaire* de c. Cette partie scalaire est *invariante* par rapport à toute transformation de coordonnées. En outre, la partie scalaire du produit d'un nombre arbitraire de nombres de Clifford est invariante par rapport à des permutations circulaires des facteurs:

$$(c_1 c_2 \ldots c_n)^S = (c_2 \ldots c_n c_1)^S. \tag{11}$$

Le nombre de Clifford c donné par la formule (10) sera considéré comme *réel*, si tous ses coefficients sont *réels*. Cette convention, qui rend toutes les questions de réalité concernant les résultats finals extrêmement simples, est entièrement différente des conventions usuelles.

En introduisant, d'une manière générale, le commutateur (ou produit extérieur) d'ordre p de p quantités arbitraires $U_1, U_2, \ldots, U_p$

$$[U_1, U_2, \ldots, U_p] = \varepsilon^{i_1 i_2 \ldots i_p} U_{i_1} U_{i_2} \ldots U_{i_p},$$

où $\varepsilon^{i_1 i_2 \ldots i_p}$ est le tenseur totalement alterné d'ordre p, on obtient 16 éléments linéairement indépendants qui fournissent une base de l'algèbre, savoir 1 scalaire $= 1$, 4 vecteurs $= e_i$, 6 bivecteurs $= [e_i, e_j]/2!$, $i < j$, 4 trivecteurs $= [e_i, e_j, e_k]/3!$, $i < j < k$, 1 pseudoscalaire $= [e_0, e_1, e_2, e_3]/4!$.

Les calculs se simplifient beaucoup, si l'on remplace les e_i par des combinaisons linéaires f_i qui forment un système *orthogonal* (non nécessairement normé), $(f_i, f_j) = 0$, $i \neq j$. On a dans ce cas, d'après (9),

$$f_i f_j + f_j f_i = 0, \; i \neq j, \tag{12}$$

ce qui exprime que les f_i *anticommutent.* Les 16 éléments de base peuvent dans ce cas s'écrire

$$(13) \qquad 1;\ f_i;\ f_i f_j,\ i<j;\ f_i f_j f_k,\ i<j<k;\ f_0 f_1 f_2 f_3.$$

Dans la vérification de nos énoncés ci-dessus, on peut avec avantage se servir de tels systèmes orthogonaux.

Soulignons encore que — si l'espace n'est pas plat[1] — *seules* les quantités cliffordiennes définies au *même point* ou en des *points infiniment voisins* peuvent être composées les unes avec les autres. Dans le dernier cas, *on convient d'identifier les quantités qui se déduisent l'une de l'autre par transport parallèle.*

Les algèbres C_x admettent deux antiautomorphismes, *renversement* et *renversement conjugué* (et un automorphisme, conjugaison) que nous allons maintenant envisager.[2]

Nous commençons par écrire l'expression (10) sous la forme condensée $c = \Sigma c^A e_A$.

1) **Renversement.** Cette opération consiste dans le renversement dans chaque produit e_A de l'ordre des facteurs e_i. Par exemple, $1 \to 1$, $e_0 \to e_0$, $e_2 e_0 e_3 \to e_3 e_0 e_2$, $e_2 e_1 e_0 e_3 \to e_3 e_0 e_1 e_2$. L'opération en question sera désignée par $e_A \to \tilde{e}_A$, $c = \Sigma c^A e_A \to \Sigma c^A \tilde{e}_A = \tilde{c}$.

2) **Conjugaison.** Chaque coefficient c^A est remplacé par son imaginaire conjugué $\bar{c}^A$, c'est-à-dire que $c = \Sigma c^A e_A \to \Sigma \bar{c}^A e_A = \bar{c}$.

3) **Renversement conjugué.** Les deux opérations ci-dessus sont commutatives. Leur produit consiste en $c \to \hat{c} = \Sigma \bar{c}^A \tilde{e}_A$.

Remarquons que toutes les opérations définies plus haut sont indépendantes du choix de vecteurs de base. Il suffit de le montrer pour la dernière, qui peut se caractériser de la manière suivante, où les e_i n'interviennent pas.

L'opération $c \to \hat{c}$ est un *antiautomorphisme* ($c' \to \hat{c}'$, $c'' \to \hat{c}''$ impliquent $c'c'' \to \hat{c}''\hat{c}'$) de l'algèbre de Clifford qui conserve les scalaires et les vecteurs réels et change le scalaire $i = \sqrt{-1}$ en $-i$.

Nous aurons besoin du théorème suivant dont la démonstration se trouve dans ma «Conférence» (p. 136).

U et V étant deux vecteurs de temps positifs (dans l'algèbre de Clifford) et c un nombre de Clifford $\neq 0$, *on a*

$$(14) \qquad (U\,\hat{c}\,V\,c)^S > 0.$$

[1] *Cf.* la note au n° 3.

[2] L'exposé donné ici est tiré de ma «Conférence». Une des notations se trouve légèrement changée.

Dans la suite, *l'opérateur différentiel, invariant par rapport à toute transformation de coordonnées*,

$$\nabla = e^j \partial_j \tag{15}$$

va jouer un rôle important. Moyennant la formule (6), on vérifie facilement la relation

$$\nabla^2 = \triangle = \frac{1}{\sqrt{|g|}} \partial_i (\sqrt{|g|}\, g^{ij} \partial_j), \text{ avec } g = |g_{ij}|. \tag{15bis}$$

C'est-à-dire que le carré de ∇ est égal à l'opérateur différentiel du second ordre de Beltrami.

3. **Repères lorentziens locaux.** Il est clair par la signature de la forme métrique qu'on peut en chaque point introduire, d'une infinité de manières, des repères lorentziens qui donnent lieu à la forme métrique

$$(A, A) = (A^0)^2 - (A^1)^2 - (A^2)^2 - (A^3)^2.$$

Cela posé, nous formons en chaque point tous les repères lorentziens Σ_x, $\dot{\Sigma}_x$, $\ddot{\Sigma}_x$, ... ou, plus brièvement, Σ, $\dot{\Sigma}$, $\ddot{\Sigma}$, ..., avec les vecteurs de base $\gamma_{k\Sigma}$, $\gamma_{k\dot{\Sigma}}$, $\gamma_{k\ddot{\Sigma}}$, ..., qui possèdent les deux propriétés suivantes. Les vecteurs $\gamma_{0\Sigma}$, $\gamma_{0\dot{\Sigma}}$, $\gamma_{0\ddot{\Sigma}}$, ... sont des vecteurs de temps positifs, les repères eux-mêmes ont la même orientation que le repère riemannien e_0, e_1, e_2, e_3.

Les vecteurs de base lorentziens $\gamma_{k\Sigma}$, $\gamma_{k\dot{\Sigma}}$, ... satisfont, d'après (9), aux relations

$$(\gamma_{k\Sigma}, \gamma_{l\Sigma}) = (\gamma_{k\dot{\Sigma}}, \gamma_{l\dot{\Sigma}}) = \ldots = \delta_{kl} \text{ où } (\delta_{kl}) = \begin{pmatrix} 1 & 0 & 0 & 0 \\ 0 & -1 & 0 & 0 \\ 0 & 0 & -1 & 0 \\ 0 & 0 & 0 & -1 \end{pmatrix}. \tag{16}$$

Ces vecteurs de base sont des combinaisons linéaires à coefficients réels des vecteurs e_i et inversement; considérés comme nombres de Clifford, ils satisfont, en vertu de (9) et de (16), aux relations

$$\gamma_{k\Sigma}\gamma_{l\Sigma} + \gamma_{l\Sigma}\gamma_{k\Sigma} = \ldots = 2\,\delta_{kl}. \tag{17}$$

Manifestement, tout nombre de Clifford c attaché au point x peut aussi s'exprimer à l'aide des $\gamma_{k\Sigma}$, $\gamma_{k\dot{\Sigma}}$, ... attachés au même point. (Grâce aux formules (17) et (13) l'expression (10) se trouvera beaucoup simplifiée.)

Observons que dans l'ordre d'idées de la géométrie riemannienne il n'existe aucun rapport (intégrable) entre deux repères lorentziens attachés à deux points non voisins[1] x et y. Ainsi, un des repères attachés au point x étant désigné par Σ_x, n'importe lequel des repères attachés au point y pourra être désigné par Σ_y. Ce n'est que dans le cas de repères attachés à des points infiniment voisins qu'on pourra établir une correspondance bien définie.

Soit Σ un repère arbitraire attaché au point x. (Nous supprimons ici les indices respectifs x et $x+dx$ pour ne pas encombrer nos formules.) Le repère attaché au point $x+dx$ qui se déduit de Σ par transport parallèle, sera désigné par Σ''. Les vecteurs de base de Σ étant $\gamma_{k\Sigma}$, ceux de Σ'' seront désignés par $\gamma_{k\Sigma''}$. Considérés comme nombres de Clifford, $\gamma_{k\Sigma}$ et $\gamma_{k\Sigma''}$ $(k=0, 1, 2, 3)$ sont à regarder comme *identiques*; *voir* une convention posée au n° 2.

4. Éléments stables et roteurs. Tout élément de l'algèbre C, formée en un point x, qui est indépendant des repères lorentziens attachés à ce point, sera appelé un *élément stable* de l'algèbre. Tels éléments sont, par exemple, le scalaire 1, les vecteurs de base riemanniens e_i, leurs commutateurs etc. Pour des raisons qu'on va voir un peu plus loin (n° 7), nous disons qu'un élément stable est *de l'espèce* $\{\,\}$. A côté des éléments stables, on considère aussi certains *collectifs* de nombres de Clifford (roteurs, spineurs) dépendant des repères en question. Comme exemple on peut citer les vecteurs de base $\gamma_{k\Sigma}, \gamma_{k\dot{\Sigma}}, \gamma_{k\ddot{\Sigma}}, \ldots$.

On considère d'abord un polynome »type«

$$\text{(18)} \qquad a(u_0, u_1, u_2, u_3)= \\ =a^S+\sum_k a^k u_k+\sum_{k<l} a^{kl} u_k u_l+\sum_{k<l<m} a^{klm} u_k u_l u_m+a^{0123} u_0 u_1 u_2 u_3,$$

où les u_k sont des indéterminées satisfaisant aux relations $u_k u_l+u_l u_k=2\delta_{kl}$ (analogues à (17)) et les coefficients $a^S, a^k, \ldots$ sont des nombres complexes arbitraires.

Si l'on substitue aux u_k les vecteurs de base des repères lorentziens $\Sigma, \dot{\Sigma}, \ddot{\Sigma}, \ldots$ attachés à un *point déterminé* x, on obtient des éléments de l'algèbre locale C_x, savoir

$$\text{(19)} \qquad a_\Sigma=a(\gamma_{0\Sigma}, \gamma_{1\Sigma}, \gamma_{2\Sigma}, \gamma_{3\Sigma}),\ a_{\dot{\Sigma}}=\ldots,\ a_{\ddot{\Sigma}}=\ldots, \ldots.$$

On dit que *ces éléments sont de structure identique*, par rapport aux différents repères. Le *collectif* constitué par ces éléments sera appelé *roteur* et désigné par a. On écrit

$$\text{(20)} \qquad a=\{\vdots a_\Sigma, a_{\dot{\Sigma}}, a_{\ddot{\Sigma}}, \ldots\vdots\}=\{\vdots a_\Sigma \vdots\}$$

[1] On suppose ici qu'il s'agit d'un espace de Riemann général et non pas d'un espace plat, comme l'espace-temps de la relativité restreinte.

et dit qu'un roteur est un collectif *de l'espèce* $\{\vdots\ \vdots\}$. Les notations ci-dessus s'expliqueront tout à l'heure (nº 7). Les nombres a_{Σ}, $a_{\dot{\Sigma}}$, $a_{\ddot{\Sigma}}$, ... sont appelés les *composantes* du roteur par rapport aux repères Σ, $\dot{\Sigma}$, $\ddot{\Sigma}$, ...

Toutes les composantes d'un roteur ont manifestement la même partie scalaire. Les scalaires, et les scalaires seulement, peuvent être considérés soit comme éléments stables, soit comme roteurs.

Voici encore une *extension* importante de l'une des définitions qu'on vient de poser. Si l'on substitue aux u_k les vecteurs de base de deux repères arbitraires attachés à deux points *différents* x et y, on obtient deux éléments des algèbres respectives C_x et C_y qu'on considère toujours comme étant *de structure identique*.

5. **Idéaux à gauche et à droite (minimaux).** On entend par un idéal à gauche I dans une algèbre C un ensemble d'éléments de C tel que $\eta_1 \in I$, $\eta_2 \in I$ entraînent $\eta_1 - \eta_2 \in I$ et que $\eta \in I$, $c \in C$ entraînent $c\eta \in I$. Définition analogue pour un idéal à droite. Un idéal à gauche (à droite) est dit *minimal*, s'il ne contient aucun idéal à gauche (à droite) sauf lui-même et l'idéal nul, consistant du seul élément 0. Jusqu'à nouvel ordre, il ne sera question que des idéaux à gauche; on reviendra aux idéaux à droite et à certaines correspondances entre idéaux à gauche et à droite plus loin (nº 11).

Aux repères Σ_x, $\dot{\Sigma}_x$, $\ddot{\Sigma}_x$, ... on associe les idéaux I_{Σ_x}, $I_{\dot{\Sigma}_x}$, $I_{\ddot{\Sigma}_x}$, ... de l'algèbre C_x. Ces idéaux doivent être de structure identique, par quoi on entend qu'ils sont constitués par des éléments de structure identique. En posant

$$I_{\Sigma} = \{\eta_{\Sigma}, \zeta_{\Sigma}, \tau_{\Sigma}, \ldots\},\ I_{\dot{\Sigma}} = \{\eta_{\dot{\Sigma}}, \ldots\},\ I_{\ddot{\Sigma}} = \{\eta_{\ddot{\Sigma}}, \ldots\}, \ldots, \tag{21}$$

où, pour simplifier l'écriture, on a supprimé l'indice x, les nombres η_{Σ}, $\eta_{\dot{\Sigma}}$, $\eta_{\ddot{\Sigma}}$, ..., par exemple, ne sont que les composantes du roteur η, c'est-à-dire des éléments de structure identique de C_x.

Les idéaux appartenant à des points différents x et y devront, eux aussi, être de structure identique par rapport aux repères respectifs; *cf.* l'extension de la définition de l'identité de structure à des éléments appartenant à des algèbres diffèrents C_x et C_y (fin du nº 4).

On construit tous ces idéaux à l'aide des indéterminées u_k, introduites au nº 4. Dans l'algèbre »type», engendrée par ces indéterminées, on forme un idéal à gauche (minimal). Les éléments de cet idéal seront donnés par certains polynomes $\eta(u_k)$, $\zeta(u_k)$, $\tau(u_k)$, ... (*cf.* la formule (18)). Enfin on substitue aux u_k, tout comme on l'a fait plus haut, les vecteurs de base des repères respectifs. Tous ces idéaux sont minimaux ou non en même temps.

Exemples triviaux d'idéaux: l'algèbre totale et l'idéal nul; pour des exemples non triviaux voir ma »Conférence« et la Note I du présent travail.

Tandis que l'algèbre totale a 16 éléments de base, un idéal minimal (non nul) n'en a que 4. L'inconnue ψ dans l'équation de Dirac sera chez nous, dans un certain sens, un élément d'idéal minimal. Elle pourra alors, comme chez Dirac, être donnée par 4 fonctions numériques (à valeurs complexes); *cf.* le n° 10 et la Note I.

6. **Le sous-groupe des rotations lorentziennes propres.** Nous allons étudier de plus près les transformations de Lorentz qui emmènent les repères lorentziens attachés à un *point déterminé* x l'un dans l'autre. Ces transformations forment évidemment un sous-groupe, désigné par L, du groupe des transformations de Lorentz opérant sur l'espace vectoriel local attaché au point x. Le sous-groupe L est constitué par ceux des éléments du groupe entier qui sont reliés à l'identité. On peut encore caractériser ce sous-groupe comme étant constitué par les rotations lorentziennes propres, qui ne sont que les transformations de Lorentz qui gardent l'orientation de l'espace ainsi que le passé et le futur.

Envisageons un élément arbitraire l du sous-groupe des rotations lorentziennes propres L. La rotation l ayant un sens clair pour tout vecteur de C_x, on peut, par les conventions $l(c'c'')=lc'\cdot lc''$ et $l(\lambda'c'+\lambda''c'')=$ $=\lambda'lc'+\lambda''lc''$, posées pour deux éléments arbitraires c' et c'' de l'algèbre C_x et deux nombres complexes λ' et λ'', également arbitraires, étendre l de façon qu'elle opère sur l'algèbre entière C_x. Les conventions ci-dessus mettent en évidence que l engendre un automorphisme de l'algèbre C_x.

On peut montrer[1] qu'à toute transformation l correspond un élément réel R_l de C_x, *déterminé au signe près*, tel que $R_l^{-1}=\tilde{R}_l$ (*voir*, pour cette notation et pour la notation $\hat{R}_l$ qui suit, le n° 2) et qu'on a, pour chaque vecteur $U \in C_x$, $lU=R_l^{-1}UR_l$. Il en résulte, pour un élément arbitraire c de C_x,

$$lc=R_l^{-1}cR_l. \tag{22}$$

Remarquons que R_l étant réel, on a $\tilde{R}_l=\hat{R}_l$, c'est-à-dire que

$$R_l^{-1}=\tilde{R}_l=\hat{R}_l. \tag{23}$$

Les R_l forment évidemment un groupe; c'est $R_{l'}R_{l''}$ qui correspond à la transformation $l''l'$.

---

[1] *Voir* Cartan, Spineurs I, n°s 10, 58 et II, n°s 96—98 et »Conférence«, p. 136—137.

La formule (22) met de nouveau en évidence le fait que toute transformation l fournit un automorphisme de l'algèbre C_x.

Ajoutons que pour tout nombre R_l existe un bivecteur réel[1] F_l (en réalité, il en existe en général une infinité) tel que

$$(24) \qquad R_l = e^{\frac{1}{2}F_l}.$$

Inversement, F étant un bivecteur réel arbitraire, $R=e^{\frac{1}{2}F}$ aura les propriétés 1) $R^{-1}=\tilde{R}$ (ce qui résulte de $-F=\tilde{F}$), 2) la transformation $U\to R^{-1}UR$ est une rotation lorentzienne propre.

Nous observons encore que dans le cas où l est une *rotation infinitésimale* — le cas le plus important dans ce qui suit — on peut se débarrasser de la bivalence de R_l et de la multivalence de F_l en se bornant à celle des deux valeurs de R_l qui est rapprochée de l'unité et à celle de F_l qui est infinitésimale. Cela posé, R_l et F_l seront entièrement déterminés par l.[2] *Voir* aussi la Note III.

7. **Retour aux repères lorentziens locaux.** Considérons deux repères arbitraires attachés à un point déterminé x et désignés par $\dot{\Sigma}$ et $\ddot{\Sigma}$. Il existe une rotation l bien déterminée qui emmène $\dot{\Sigma}$ en $\ddot{\Sigma}$ et qui sera désignée par $l_{\ddot{\Sigma}/\dot{\Sigma}}$. (Nous ne nous inquiétons pas du fait évident que cette rotation correspond à une infinité de couples de repères, dont l'un peut d'ailleurs être choisi arbitrairement.) Pour abréger, on écrira dans la suite l au lieu de $l_{\ddot{\Sigma}/\dot{\Sigma}}$.

Les formules

$$(25) \qquad \gamma_{k\ddot{\Sigma}} = l\gamma_{k\dot{\Sigma}}, \quad \ddot{\Sigma} = l\dot{\Sigma}$$

n'ont pas besoin d'être expliquées. La première de ces formules entraîne

$$(26) \qquad a_{\ddot{\Sigma}} = l\, a_{\dot{\Sigma}},$$

où $a_{\dot{\Sigma}}$ et $a_{\ddot{\Sigma}}$ sont les composantes respectives par rapport aux repères $\dot{\Sigma}$ et $\ddot{\Sigma}$ du roteur a.

Le nombre (bivalent) R_l qui correspond à l sera, de son côté, désigné par $R_{\ddot{\Sigma}/\dot{\Sigma}}$ ou plus brièvement par R. En vertu des formules (25) et (22), on a

[1] Un élément F de C_x est appelé bivecteur réel si $F=\frac{1}{2}F^{ij}e_ie_j$, $F^{ji}=-F^{ij}$ et F^{ij} réel; *cf.* le n° 2.

[2] Pour la démonstration de (24) dans le cas infinitésimal, *voir* Cartan, Spineurs I, n° 19. Dans le cas général, la démonstration est assez délicate; je pense y revenir dans un autre travail.

(27) $$\gamma_{k\ddot{\Sigma}}=R^{-1}\gamma_{k\dot{\Sigma}}R,$$

ou, sous forme condensée,

(27bis) $$\ddot{\Sigma}=R^{-1}\dot{\Sigma}R.$$

Enfin, en vertu des formules (26) et (22),

(28) $$a_{\ddot{\Sigma}}=R^{-1}a_{\dot{\Sigma}}R.$$

Ce point acquis, il est facile d'expliquer la notation $\{\vdots a_\Sigma \vdots\}$ (formule (20)) et la locution »de l'espèce $\{\vdots\ \vdots\}$». Les deux colonnes de points indiquent que le passage $a_{\dot{\Sigma}}\to a_{\ddot{\Sigma}}$, c'est-à-dire le passage entre une composante d'un roteur à une autre, peut s'obtenir par deux multiplications, l'une par R^{-1} à gauche et l'autre par R à droite. Nous écrivons de manière symbolique

(29) $$\{\vdots\ \vdots\}=\{R^{-1}\ldots R\}.$$

Cela posé, la locution que tout élément stable (c'est-à-dire tout élément indépendant des repères lorentziens (nº 4)) est de l'espèce $\{\ \}$ s'explique d'elle-même.

Il résulte de la formule (28), mais s'ensuit aussi de la définition d'un roteur que, si l'on définit le produit de deux roteurs $a=\{\vdots a_\Sigma\vdots\}$ et $\beta=$ $=\{\vdots\beta_\Sigma\vdots\}$ par $a\beta=\{\vdots(a_\Sigma\beta_\Sigma)\vdots\}$, on aura

(30) $$\text{roteur}\times\text{roteur}=\text{roteur ou } \{\vdots\ \vdots\}\times\{\vdots\ \vdots\}=\{\vdots\ \vdots\}.$$

Nous notons, en passant, que si l'on complète la définition ci-dessus par la définition $a-\beta=\{\vdots(a_\Sigma-\beta_\Sigma)\vdots\}$, on voit que les roteurs attachés à un point x constituent une algèbre Γ_x. Toute application $a\to a_\Sigma$ (où Σ est fixé) établit un isomorphisme entre les algèbres Γ_x et C_x.

En résumé, si $l\in L$, on a pour deux *éléments stables* c' et c'', $l(c'c'')=lc'\cdot lc''$, pour deux *roteurs* a' et a'', $l(a'a'')=la'\cdot la''$; par contre pour un *élément stable* c et un *roteur* a, $l(ca)\neq lc\cdot la$ $(=c\cdot la)$, excepté le cas où c ou a sont des scalaires. Cela veut dire que le produit élément stable$\times$roteur n'a pas de loi de transformation naturelle. Mais dans la suite nous aurons aussi besoin de collectifs ψ tels que $l(c\psi)=c\cdot l\psi$; c'est-à-dire qu'il s'agit d'un isomorphisme à opérateurs, suivant la terminologie algébrique usuelle. La raison est que l'opérateur différentiel invariant $\nabla=e^j\partial_j$ ne dépend pas des repères lorentziens.

8. Spineurs (à gauche). Comme plus haut, on attache des idéaux à gauche I_Σ, $I_{\dot{\Sigma}}$, $I_{\ddot{\Sigma}}$, ... aux repères lorentziens Σ, $\dot{\Sigma}$, $\ddot{\Sigma}$, ... (Dans les applications on se restreindra à des idéaux minimaux, pour des raisons discutées plus haut (fin du n° 5)). La formule (21) établit par son aspect même une correspondance par éléments entre les idéaux I, de sorte que les collectifs d'éléments correspondants sont des roteurs. Avec une interprétation facile de la formule (28) on peut écrire

$$(31) \qquad I_{\ddot{\Sigma}} = R^{-1}\, I_{\dot{\Sigma}} R, \text{ où } R = R_{\ddot{\Sigma}/\dot{\Sigma}}.$$

Tous les I étant des idéaux à gauche, on a aussi

$$(32) \qquad I_{\ddot{\Sigma}} = I_{\dot{\Sigma}} R, \text{ puisque } R^{-1} I_{\dot{\Sigma}} = I_{\dot{\Sigma}}.$$

Cette dernière formule définit une nouvelle correspondance par éléments entre les idéaux I_{Σ}, $I_{\dot{\Sigma}}$, $I_{\ddot{\Sigma}}$, ... Si l'on tire de chaque idéal un élément, ces éléments se correspondant de la nouvelle manière (formule (32)), on obtient un collectif d'éléments appelé *spineur* (*à gauche*).

D'une façon plus détaillée, la formule (32) peut s'écrire

$$I_{\Sigma} = \{\psi_{\Sigma}, \varphi_{\Sigma}, \ldots\},\ I_{\dot{\Sigma}} = \{\psi_{\dot{\Sigma}}, \varphi_{\dot{\Sigma}}, \ldots\},\ I_{\ddot{\Sigma}} = \{\psi_{\ddot{\Sigma}}, \varphi_{\ddot{\Sigma}}, \ldots\}, \ldots ,$$

où, par exemple,

$$(33) \qquad \psi_{\ddot{\Sigma}} = \psi_{\dot{\Sigma}}\, R,\ \varphi_{\ddot{\Sigma}} = \varphi_{\dot{\Sigma}}\, R, \ldots,\ R = R_{\ddot{\Sigma}/\dot{\Sigma}},$$

pour un couple arbitraire $\dot{\Sigma}$ et $\ddot{\Sigma}$. Les éléments ψ_{Σ}, $\psi_{\dot{\Sigma}}$, $\psi_{\ddot{\Sigma}}$, ... sont appelés *composantes* du spineur ψ par rapport aux repères Σ, $\dot{\Sigma}$, $\ddot{\Sigma}$, ... Nous écrivons

$$(34) \qquad \psi = \{\psi_{\Sigma}, \psi_{\dot{\Sigma}}, \psi_{\ddot{\Sigma}}, \ldots \vdots\} = \{\psi_{\Sigma} \vdots\}$$

et disons que le collectif ψ est *de l'espèce* $\{\ \vdots\}$, notation bien en accord avec les conventions posées plus haut.

On définit *l'addition* et la *soustraction* des spineurs par

$$(35) \qquad \psi \pm \varphi = \{(\psi_{\Sigma} \pm \varphi_{\Sigma}) \vdots\} = \text{spineur}.$$

L'élément c étant *stable* et ψ étant un *spineur* on pose $c\psi = \{(c\psi_{\Sigma}) \vdots\}$, c'est-à-dire que élément stable $\times$ spineur $=$ spineur ou symboliquement $\{\ \} \times \{\ \vdots\} = \{\ \vdots\}$.

ψ étant un spineur et a un roteur on pose $\psi a = \{(\psi_{\Sigma} a_{\Sigma}) \vdots\}$, c'est-à-dire que spineur $\times$ roteur $=$ spineur, symboliquement $\{\ \vdots\} \times \{\vdots\ \vdots\} = \{\ \vdots\}$.

Les produits roteur $\times$ spineur et spineur $\times$ spineur manquent d'intérêt.

Transformation des spineurs. Le résultat du présent alinéa ne sera appliqué nulle part. Soit $\psi = \{\psi_{\Sigma}, \psi_{\dot{\Sigma}}, \psi_{\ddot{\Sigma}}, \ldots \vdots\}$ un spineur et $l \in L$ un automorphisme de l'algèbre de Clifford C_x. Par $l\psi$ nous entendons l'ensemble $(l\psi_{\Sigma}, l\psi_{\dot{\Sigma}}, l\psi_{\ddot{\Sigma}}, \ldots)$, interprété d'une manière convenable. La transformation $(\psi_{\Sigma}) \rightarrow (l\psi_{\Sigma})$ peut s'écrire $(\psi_{\Sigma}) \rightarrow (R_l^{-1}\psi_{\Sigma}R_l)$ et peut s'effectuer en deux pas: $(\psi_{\Sigma}) \rightarrow (\psi_{\Sigma}R_l)$ et $(\psi_{\Sigma}R_l) \rightarrow (R_l^{-1}\psi_{\Sigma}R_l)$. Le premier pas $(\psi_{\Sigma}) \rightarrow (\psi_{\Sigma}R_l)$ transforme l'ensemble (ψ_{Σ}) en lui-même en permutant toutefois les ψ_{Σ}. Le second pas multiplie tous les éléments de l'ensemble par R_l^{-1} à gauche; tout cela peut s'écrire $(l\psi_{\Sigma}) = R_l^{-1}(\psi_{\Sigma})$. On conclut que les éléments de l'ensemble $(l\psi_{\Sigma})$ sont identiques aux composantes du spineur $R_l^{-1}\psi$. Nous écrivons symboliquement $l\psi = R_l^{-1}\psi$, ce qui se trouve en accord avec la loi de transformation des spineurs-colonnes.

La différentiation dans un champ de spineurs. En chaque point x de l'espace de Riemann on choisit parmi tous les repères lorentziens attachés à ce point un repère arbitraire, désigné par Σ_x. Les vecteurs de base de Σ_x s'expriment par les vecteurs riemanniens $e_j(x)$ à l'aide d'une matrice (a_{ij}). L'ensemble de ces repères est appelé un *champ de repères*, si la matrice (a_{ij}) est (suffisamment) différentiable.

Un spineur $\psi(x)$ étant attaché à chaque point x, on dira que ces spineurs constituent un champ de spineurs, si la composante $\psi_{\Sigma_x}(x)$, écrite sous la forme (10), admet des coefficients (suffisamment) différentiables, et cela pour un champ de repères Σ_x quelconque. En ce qui concerne l'existence et la construction de tels champs *cf.* le n° 10.

D'après la définition donnée au n° 3 on entend par Σ''_{x+dx}, ou plus brièvement Σ'', le repère provenant de Σ_x, ou plus brièvement de Σ, par transport parallèle $(x \rightarrow x+dx)$. On pose par définition

$$(36) \qquad d\psi_{\Sigma}(x) = \dot{\psi}_{\Sigma''}(x+dx) - \psi_{\Sigma}(x) \text{ et } d\psi(x) = \{d\psi_{\Sigma}(x) \vdots\}.$$

Les notations sont légitimes et $d\psi(x)$ est un spineur puisque $d\psi_{\Sigma}(x) \in I_{\Sigma}$ et $d\psi_{\ddot{\Sigma}}(x) = d\psi_{\dot{\Sigma}}(x) \cdot R_{\ddot{\Sigma}/\dot{\Sigma}}$.

9. **L'équation de Dirac.** Par la formule (36), les dérivées partielles $\partial_i\psi(x)$ se trouvent définies en même temps. L'équation de Dirac pour une particule de masse au repos m, de charge e et en présence d'un champ électromagnétique donné par le potentiel vecteur $A = A^j e_j = A_j e^j$ a la même apparence que chez Dirac, savoir

$$(37) \qquad (\nabla - ieA - im)\psi = 0,$$

ou, d'une manière explicite, $(e^j(\partial_j - ieA_j) - im)\psi = 0$.

Bien entendu, il faut interpréter l'équation entière et en particulier la différentiation du spineur ψ de la manière exposée plus haut.

10. **Calcul effectif.** En chaque point x de l'espace de Riemann on considère un repère lorentzien *particulier*, soit Σ_x, ces repères constituant un *champ de repères* d'après la définition qu'on vient de poser. On écrit, pour abréger, $\Sigma_x = \Sigma$ et $\Sigma_{x+dx} = \Sigma^*$. Lors d'un transport parallèle correspondant à $x \to x+dx$ on aura le passage $\Sigma \to \Sigma''$. Les repères Σ^* et Σ'' étant attachés au même point $x+dx$, on peut former $R = R_{\Sigma''/\Sigma^*}$ et on a $\Sigma'' = R^{-1}\,\Sigma^* R$.

On donne un spineur $\psi(x)$ par sa composante $\psi_{\Sigma_x}(x)$ on plus brièvement $\psi_\Sigma(x)$. On arrive ainsi à un champ de spineurs. On a $\psi_{\Sigma''}(x+dx) = \psi_{\Sigma^*}(x+dx) R$.

Bases d'idéaux. On introduit dans chaque idéal I_{Σ_x} des éléments de base $\sigma_1(x)$, $\sigma_2(x)$, ..., les éléments de base appartenant à des idéaux différents étant de structure identique. On peut écrire $\psi_{\Sigma_x}(x) = \Sigma a_p(x)\sigma_p(x)$, les $a_p(x)$ étant des nombres complexes. En vertu des conventions posées à la fin du n° 3, on identifie les éléments de base σ_p et σ_p'' relatifs aux idéaux attachés aux repères parallèles Σ et Σ''.

Soient $\psi_\Sigma = \Sigma\, a_p\, \sigma_p$ et $\psi_{\Sigma^*} = \Sigma\, b_p\, \sigma_p^*$, $a_p = a_p(x)$ et $b_p = a_p(x+dx)$. On veut calculer $\psi_{\Sigma''}$. On a $\psi_{\Sigma''} = \psi_{\Sigma^*} R = \Sigma\, b_p\, \sigma_p^* R = \Sigma\, b_p (R\, \sigma_p''\, R^{-1}) R = \Sigma\, b_p (R\, \sigma_p R^{-1}) R = \Sigma\, b_p R\, \sigma_p = R \Sigma\, b_p\, \sigma_p$.

Représentation des spineurs et des nombres cliffordiens. Soit $\psi_\Sigma \in I_\Sigma$ une composante du spineur ψ. σ_0, σ_1, ... étant une base de I_Σ, on peut écrire $\psi_\Sigma = \Sigma\, a_p\, \sigma_p$, ce qui donne une représentation de ψ_Σ par la colonne

$$\begin{pmatrix} \vdots \\ a_p \\ \vdots \end{pmatrix}_\Sigma .$$

Soit $c \in C_x$ un nombre cliffordien quelconque. De $c\sigma_k = \sum_j c_{jk}\, \sigma_j$ on obtient une représentation de c par la matrice $(c_{jk})_\Sigma$.

Le produit $c\psi_\Sigma$ se trouve ainsi représenté par

$$\begin{pmatrix} c_{11} & c_{12} & \cdots \\ c_{21} & c_{22} & \cdots \\ \vdots & & \end{pmatrix}_\Sigma \begin{pmatrix} \vdots \\ a_p \\ \vdots \end{pmatrix}_\Sigma .$$

Si l'idéal I_Σ est minimal, le nombre des éléments σ_p ainsi que l'ordre des matrices (c_{jk}) est égal à 4. Cela posé, les matrices à éléments complexes arbitraires fournissent une représentation fidèle de l'algèbre C_x.

Si

$$\psi_\Sigma(x) = \begin{pmatrix} \vdots \\ a_p(x) \\ \vdots \end{pmatrix}_\Sigma \quad \text{et} \quad \psi_{\Sigma^*}(x+dx) = \begin{pmatrix} \vdots \\ a_p(x+dx) \\ \vdots \end{pmatrix}_{\Sigma^*} ,$$

on a, moyennant le résultat obtenu tout à l'heure,

$$\psi_{\Sigma''} = R \begin{pmatrix} \vdots \\ a_p(x+dx) \\ \vdots \end{pmatrix}_\Sigma .$$

D'après les formules qui précèdent

$$d\psi_\Sigma(x)=\psi_{\Sigma'}(x+dx)-\psi_\Sigma(x)=R\begin{pmatrix}\vdots\\ a_p(x+dx)\\ \vdots\end{pmatrix}_\Sigma-\begin{pmatrix}\vdots\\ a_p(x)\\ \vdots\end{pmatrix}_\Sigma .$$

Nous écrivons d'après la formule (24) $R=e^{\frac{1}{2}F}=1+\frac{1}{2}F+\ldots$, F étant un bivecteur infinitésimal réel de la forme $F=\sum_{i=0}^{3}F_i\,dx^i$, où les F_i sont des bivecteurs réels dépendant de x.

Par conséquent

$$d\psi_{\Sigma x}(x)=\begin{pmatrix}\vdots\\ da_p(x)\\ \vdots\end{pmatrix}_{\Sigma x}+\frac{1}{2}F\begin{pmatrix}\vdots\\ a_p(x)\\ \vdots\end{pmatrix}_{\Sigma x} .$$

Dans les formules ci-dessus, R et F sont les matrices représentant les nombres cliffordiens auxquels elles correspondent.

11. **Spineurs à droite.** Un spineur à droite φ est un collectif de la forme

(38) $$\varphi=\{\vdots\varphi_\Sigma,\ \varphi_{\dot\Sigma},\ \varphi_{\ddot\Sigma},\ \ldots\}=\{\vdots\varphi_\Sigma\},$$

où $\varphi_{\ddot\Sigma}=R^{-1}_{\ddot\Sigma/\dot\Sigma}\varphi_{\dot\Sigma}$. Un tel spineur est un collectif de l'espèce $\{\vdots\ \}$. Au spineur à gauche $\psi=\{\psi_\Sigma\vdots\}$ correspond un spineur à droite $\hat\psi=\{\vdots\hat\psi_\Sigma\}$ (*voir* le n° 2) et *vice versa*.

On a manifestement

(39a) $$\{\vdots\ \ \}\times\{\ \ \ \}=\{\vdots\ \ \}$$
(39b) $$\{\vdots\ \vdots\}\times\{\vdots\ \ \}=\{\vdots\ \ \}$$
(39c) $$\{\ \ \vdots\}\times\{\vdots\ \ \}=\{\ \ \ \}$$
(39d) $$\{\vdots\ \ \}\times\{\ \ \vdots\}=\{\vdots\ \vdots\}$$ (*cf.* n° 8).

Outre $\hat\psi$, déduit de ψ, nous formons encore le spineur à droite (cf. (39b))

(40) $$\psi^\dagger=\gamma_0\,\hat\psi,$$

où $\gamma_0=\{\vdots\gamma_{0\Sigma}\vdots\}$ est un roteur. De l'équation (37) résulte, par un renversement conjugué (n° 2), $\hat\psi\big((\overleftarrow{\partial}_j+ieA_j)e^j+im\big)=0$ et d'ici

(41) $$\psi^\dagger\big((\overleftarrow{\partial}_j+ieA_j)e^j+im\big)=0.$$

Ce dernier passage est légitime, parce que $d\psi^\dagger=d(\gamma_0\hat\psi)=\gamma_0 d\hat\psi$, vu que $d\psi^\dagger{}_\Sigma=d(\gamma_{0\Sigma}\hat\psi_\Sigma)=\gamma_{0\Sigma'}\hat\psi_{\Sigma'}-\gamma_{0\Sigma}\hat\psi_\Sigma=\gamma_{0\Sigma}(\hat\psi_{\Sigma'}-\hat\psi_\Sigma)=\gamma_{0\Sigma}\,d\hat\psi_\Sigma$. On appelle l'équation (41) adjointe à (37).

12. **Le vecteur-courant de Dirac.** Nous formons le produit

$$\psi\psi^\dagger = \ldots + j^i e_i + \ldots, \tag{42}$$

où seulement la partie vectorielle est écrite d'une manière explicite. D'après (39c) ce produit est un élément stable, ce qui implique que les composantes contrevariantes j^i du vecteur-courant j sont indépendantes des repères lorentziens tandis qu'elles se transforment de la manière usuelle lors de transformations riemanniennes.

On obtient facilement[1]

$$j^i = (\psi\psi^\dagger e^i)^S = (\psi^\dagger e^i \psi)^S. \tag{43}$$

On tire de l'inégalité (14) que j^0 est réel et >0. Il en sera de même dans tout système riemannien admissible, ce qui entraîne que j est un vecteur de temps positif.

Enfin on voit aisément, que

$$\operatorname{div} j = \frac{1}{\sqrt{|g|}} \partial_i \left(\sqrt{|g|}\, j^i\right) = 0, \text{ où } g = |g_{ij}|. \tag{44}$$

Cela résulte des équations (37) et (41) complétées par la formule $\partial_i \left(\sqrt{|g|}\, e^i\right) = 0$. Celle-ci est une conséquence de (6) et de la formule connue $\Gamma_{ir}{}^i = \partial_r \left(\log \sqrt{|g|}\right)$.

Note I. Exemples d'idéaux minimaux. Le premier exemple correspond à la représentation donnée par v. d. Waerden et Élie Cartan.

Soient u_0, u_1, u_2, u_3 des indéterminées satisfaisant aux relations $u_j u_k + u_k u_j = 2\,\delta_{jk}$ (*cf.* (18)). Elles engendrent une algèbre cliffordienne C. Formons

$$\mu_1 = \frac{iu_1 + u_2}{2}, \quad \mu'_1 = \frac{iu_1 - u_2}{2}, \quad \mu_2 = \frac{u_0 + u_3}{2}, \quad \mu'_2 = \frac{u_0 - u_3}{2}.$$

Ces nouvelles quantités satisfont, de leur côté, aux relations

$$\mu_j \mu_k + \mu_k \mu_j = \mu'_j \mu'_k + \mu'_k \mu'_j = 0,$$
$$\mu_j \mu'_k + \mu'_k \mu_j = \varepsilon_{jk}, \text{ où } (\varepsilon_{jk}) = \begin{pmatrix} 1 & 0 \\ 0 & 1 \end{pmatrix}.$$

---

[1] Il est d'un certain intérêt de noter que $\psi^\dagger e^i \psi$ est un roteur (en vertu des formules (39)). D'après une remarque faite au n° 4, la partie scalaire d'un roteur, dans le cas actuel j^i, est indépendante des repères lorentziens.

Nous écrivons encore les relations suivantes, conséquences immédiates de celles qui précèdent.

$$\mu_j \mu'_k \mu_j = \varepsilon_{jk} \mu_j, \quad \mu'_j \mu_k \mu'_j = \varepsilon_{jk} \mu'_j.$$

Formons $N_1 = \mu_1 \mu'_1$, $N_2 = \mu_2 \mu'_2$ et $\sigma_0 = N_1 N_2 = N_2 N_1$. On a $N_1^2 = N_1$, $N_2^2 = N_2$, $\sigma_0^2 = \sigma_0$. De $(\sigma_0)^S = \frac{1}{4}$ suit que $\sigma_0 \neq 0$. Envisageons l'idéal à gauche engendré par σ_0, c'est-à-dire l'idéal $C\sigma_0$. Nous allons vérifier que cet idéal est minimal. Observons d'abord que, si c est un nombre cliffordien quelconque, $\sigma_0 c \sigma_0 = \lambda \sigma_0$, où λ est un scalaire. Cela est une conséquence immédiate du fait que $\sigma_0^2 = \sigma_0$ et que $\sigma_0 \pi \sigma_0 = 0$, quand π est un produit des μ_j et des μ'_j, sauf dans les cas où π est égal à $\pm N_1$ ou $\pm N_2$ ou $\pm N_1 N_2$, cas dans lesquels $\sigma_0 \pi \sigma_0 = \pm \sigma_0$.

Nous avons encore besoin du

Lemme. Soit $a\,\sigma_0 \neq 0$ un élément quelconque de l'idéal à gauche engendré par σ_0. Il existe alors un nombre cliffordien b tel que $ba\,\sigma_0 = \sigma_0$.

Observons (*cf.* l'inégalité (14)) que $(u_0 \hat{\sigma}_0 \hat{a} u_0 a \sigma_0)^S > 0$, ce qui donne, en notant que $u_0 \hat{\sigma}_0 u_0 = \sigma_0$, $0 \neq u_0 \hat{\sigma}_0 \hat{a} u_0 a \sigma_0 = (u_0 \hat{\sigma}_0 u_0)(u_0 \hat{a} u_0) a \sigma_0 = \sigma_0 (u_0 \hat{a} u_0) a \sigma_0 = \lambda \sigma_0$; le scalaire $\lambda \neq 0$. On peut poser $b = \frac{1}{\lambda} \sigma_0 u_0 \hat{a} u_0$. Ceci met en évidence que l'idéal peut être engendré par l'un quelconque de ses éléments $\neq 0$ et il est, par conséquent, un idéal minimal.

Les nombres σ_0, $\sigma_1 = \mu'_1\, \sigma_0$, $\sigma_2 = \mu'_2\, \sigma_0$, $\sigma_3 = \mu'_1\, \mu'_2\, \sigma_0$ et 0 se transforment les uns dans les autres, au signe près, par multiplication à gauche avec 1, μ_j ou μ'_j. Les éléments $\sigma_0, \ldots, \sigma_3$ sont linéairement indépendants et on conclut qu'ils forment une base de l'idéal $C\sigma_0$.

On peut se borner à démontrer l'indépendance linéaire des σ_j. Soit $\Lambda = \lambda_0 \sigma_0 + \lambda_1 \sigma_1 + \lambda_2 \sigma_1 + \lambda_3 \sigma_3 = 0$. Alors $\sigma_0 \Lambda = \lambda_0 \sigma_0 = 0$, $\mu_1 \mu_2 \Lambda = -\lambda_3 \sigma_0 = 0$, ce qui donne $\lambda_0 = \lambda_3 = 0$. Cela étant, $\mu_1 \Lambda = \lambda_1 \sigma_0 = 0$, $\mu_2 \Lambda = \lambda_2 \sigma_0 = 0$, ce qui donne $\lambda_1 = \lambda_2 = 0$.

Voici un second exemple d'idéal minimal qui correspond à la représentation donnée par Dirac. Comme éléments de base on choisit

$$\sigma_0 = \frac{(1 + u_0)(1 + i u_1 u_2)}{2}, \quad \sigma_1 = u_1 \sigma_0, \quad \sigma_2 = u_3 \sigma_0, \quad \sigma_3 = u_1 u_3 \sigma_0.$$

Pour certains détails *voir* notre «Conférence», p. 144—147 et A. Sommerfeld, Atombau und Spektrallinien, t. II, Braunschweig 1939, p. 217—263, en particulier, p. 249.

Note II. Quelques remarques concernant l'équation de Dirac. Dans un espace plat, l'espace-temps de la relativité restreinte, par exemple, notre procédé se réduit immédiatement à celui de Dirac. En chaque point, on choisit un seul repère lorentzien local (*cf.* n° 10), parallèle au repère lorentzien auquel l'espace est rapporté. Cela fait, on peut négliger les repères locaux. Après, on identifie toutes les algèbres cliffordiennes locales.

Voici une remarque concernant les équations de Maxwell et une confrontation de ces équations avec l'équation de Dirac. En chaque point x de l'espace-temps de la relativité générale on forme l'algèbre cliffordienne locale C_x. Dès lors, les équations de Maxwell peuvent être condensées dans une seule équation

$$\nabla F = -s,$$

où $\nabla = e^j \partial_j$, $F \in C_x$ est un bivecteur représentant le champ électromagnétique et $s \in C_x$ est un vecteur, le vecteur-courant donné d'avance, qui doit satisfaire à la condition div $s = 0$.

Pour simplifier, nous nous plaçons un instant dans le cas de la relativité restreinte. Lors d'une rotation lorentzienne $x \to x'$, $\nabla F = -s$ se transforme en $\nabla' F' = -s'$, où le passage suit les lois tensorielles. Les deux égalités sont correctes en même temps. Le passage en question peut aussi s'obtenir en multipliant la première équation par R^{-1} à gauche et par R à droite (avec un choix convenable de R). Si, au lieu de la multiplication bilatérale, on n'effectue qu'une multiplication unilatérale (par exemple en multipliant à gauche par R^{-1}), on arrive encore à une égalité correcte, mais celle-ci manque d'intérêt, puisque le caractère tensoriel de nos quantités se trouve détruit.

Chez Dirac, l'opérateur $\nabla - \ldots \to R^{-1}(\nabla - \ldots)R$ et $\psi \to R^{-1}\psi$, le passage étant correct. Il en serait de même si l'on appliquait à ψ une multiplication bilatérale. Cependant, le spineur ψ est un élément d'un idéal à gauche (minimal) et non pas un tenseur et il s'agit de rester dans le même idéal. Pour cette raison, c'est la première transformation et non pas le seconde qu'il faut choisir.

Nous passons maintenant à une comparaison entre notre schéma et celui de Dirac en ce qui concerne les lois de transformations. Dirac transforme les coordonnées, $x^k \to x^{k'}$, tandis que les γ^k restent inaltérés. Cela montre que chez lui $\nabla = \gamma^j \partial_j$ est un roteur. Chez nous, où $\nabla = e^j \partial_j$, le tout étant riemannien, il est naturel de ne pas faire tourner les e^j par exemple (ce qui pourtant serait possible), mais de considérer ∇ comme un élément stable. Tout de même, le problème de la transformation spécifique des spineurs se pose chez nous d'une manière assez naturelle; *cf.* le n° 8, *Transformation des spineurs.*

Note III. Sur la bivalence de R_l. On peut, dans une certaine mesure, éliminer la bivalence des R_l en se restreignant, en chaque point x, à un faisceau connexe de repères variant d'une manière continue avec x. On suppose que la déviation mutuelle des repères appartenant au même faisceau est assez petite pour que la partie scalaire de $R_{\ddot{\Sigma}/\dot{\Sigma}}$ ne s'annule pour aucun couple $\dot{\Sigma}$, $\ddot{\Sigma}$. Dans ces conditions on va désigner par $R_{\ddot{\Sigma}/\dot{\Sigma}}$ celle des deux valeurs possibles dont la partie scalaire est positive. Les R fixés de cette façon varient d'une manière continue avec $\dot{\Sigma}$ et $\ddot{\Sigma}$.

On peut arriver à une élimination complète de la bivalence en construisant un groupe de revêtement du groupe des rotations lorentziennes. Ce groupe de revêtement est isomorphe au groupe des R_l. Je reviendrai sur cette question à une autre occasion.

Note IV. Tenseurs, roteurs, spineurs. Les tenseurs attachés aux différents points d'un espace de Riemann mettent en évidence certaines propriétés de la structure de l'espace. Ainsi la possibilité de passer d'un système déterminé de coordonnées à un système »arbitraire« ou de composantes contrevariantes à des composantes covariantes et inversement, ou encore la possibilité de transport parallèle infinitésimal. Les collectifs introduits plus haut, savoir ceux des roteurs et des spineurs font davantage, puisqu'ils mettent en outre en évidence la propriété suivante des espaces de Riemann. Un tel espace peut, en chacun de ses points, être muni d'une famille de repères locaux, les repères attachés au même point ou à des points différents étant égaux entre eux. La famille des repères attachés au même point est invariante par rapport au groupe des rotations lorentziennes. Les familles attachées à des points voisins peuvent être emmenées en coïncidence par transport parallèle.

UNIVERSITY OF LUND, SWEDEN

55.

Problems related to characteristic surfaces*

Proc. Conf. Diff. Eq. Ded. to A. Weinstein, Univ. of Maryland (1956), 57–71

I. Introduction.

In 1949 I published in the *Acta Mathematica* (vol. 81) a voluminous paper: Intégrales de Riemann-Liouville et le problème de Cauchy. This paper will be quoted in the sequel as "Acta paper."

In this paper I introduce multiple integrals of the Riemann-Liouville type depending on a parameter α and converging for sufficiently large values of α. I give the solution of the Cauchy problem for the wave equation in a unique formula, the same for space-time of odd or even dimension, the formula implying an analytic continuation with respect to the parameter α. When this analytic continuation is carried out, it leads to final formulae of quite different types for odd or even dimensions, the one relative to even dimensions being ruled by the Huygens principle. We shall meet a similar situation in the present report.

The problem mentioned above is concerned with a space-like boundary. However, I also give in the Acta paper the solution of the boundary value problem relative to a special characteristic surface, namely, a characteristic half-cone (Section 63 and 64). Only one boundary value (preferably that of the unknown function) is required.

Characteristic surfaces enter also in the solution of the problem for space-like boundaries. The most intricate part of my paper is without any doubt Chapter V, where I give a geometric interpretation of the solution relative to ordinary space-time of four dimensions, i.e. the wave equation for ordinary spherical waves. The geometric elements entering in the solution all belong to geometric optics in space-time, and a great role is played by characteristic surfaces, which can be considered as images of characteristic half-cones with respect to the given space-like surface.

* This paper constitutes a technical report written in 1954 at the University of Chicago under research contract AF 18(600)-659, supported by the U. S. Air Force through the Office of Scientific Research.

These facts and the possible interpretation of ordinary geometric optics in space-time, through which Snell's and Malus' theorems become entirely intuitive, aroused my interest in characteristic surfaces.

The present research program is concerned with different problems connected with characteristic surfaces, of which I only quote the following.

We give the solution of the wave equation for a very general class of characteristic boundaries. The restrictions are of a topological nature. The class in question with its obvious extensions by limiting cases is the only one for which the boundary value problem has a good meaning.

In the three-dimensional case a specialization of the solution formula leads to an interesting case of *curvatura integra* for twisted curves. The differential geometrical implications of the formula obtained are then examined independently. The curves consisting of simple arcs which may form keels (to be defined later) of a characteristic surface are determined.

It is known that discontinuities of a solution of a partial differential equation can occur only on characteristic surfaces of the equation. According to an important theorem of Delassus the same is true for infinities of a solution.

In the last section we construct a sequence of special functions which are related to a given characteristic surface in the same way as different powers of the Lorentz distance are related to a characteristic cone. In particular we give an explicit solution of the homogeneous wave equation which becomes infinite on the characteristic surface in the same way as the classical elementary solution does on a characteristic cone. Thus we obtain an interesting complement to Delassus' theorem. Another of the above functions furnishes a formula of *curvatura integra* in the even-dimensional case.

II. On a Certain Class of Characteristic Surfaces.

By a light-like plane we mean a plane tangent to a characteristic cone. Its equation being written in the form $(u, x) = \text{const.}$, u is a light-like vector, i.e. $(u, u) = 0$ and conversely.

A characteristic surface, related to the wave operator

$$\Delta = \partial_0^2 - \partial_1^2 - \cdots - \partial_{m-1}^2,$$

can be defined by the postulate that all its tangent planes should be light-like. The tangent planes touch the surface along a generator. The generators of the surfaces are at the same time its normals.

We also mention that in point coordinates a characteristic surface is defined by two simultaneous equations

$$S=0, \left(\frac{\partial S}{\partial x^0}\right)^2-\left(\frac{\partial S}{\partial x^1}\right)^2-\cdots-\left(\frac{\partial S}{\partial x^{m-1}}\right)^2=0.$$

Our aim is to treat and solve the Cauchy problem relative to the wave equation $\Delta u=0$ for characteristic boundaries. For obvious reasons we look for those boundaries which enclose domains bounded in space-like directions.

Now it can be shown that a characteristic surface which for large positive (negative) values of the time-like coordinate x^0 behaves more or less like a characteristic cone encloses a convex domain denoted D^+ with a characteristic boundary denoted S^+. The most remarkable feature of such a surface—an $(m-1)$-dimensional variety—is that it intersects itself in an $(m-2)$-dimensional variety, which we call the *keel* of the surface. The strong singularities of the surface, the focal points (see below), form an $(m-2)$-dimensional and $(m-2)$-sheeted variety, called the focal surface. This surface lies in the exterior of S^+, except for an $(m-3)$-dimensional part which it has in common with the keel. Thus in three dimensions there are only isolated focal points on the keel.

Sufficient conditions for the existence and sufficiently regular behavior of the keel are 1) a one to one correspondence between parallel tangent planes of S and a characteristic cone, 2) differentiability conditions of suitable order.

The existence of the keel can be established by its property of being the locus of the vertices of characteristic positive half cones contained in D^+ which touch S^+ along two different generators, or have a multiple contact with S^+ along one generator. The points of the keel are also characterized as being the points of intersection between two generators or limits of such points.

More will be said about the keel in the three-dimensional case.

III. Notations, Definitions, Generalities.

The coordinates in m-dimensional space-time being denoted by $x^0, x^1, \cdots, x^{m-1}$, the metric form will be

(1) $$(x,x)=(x^0)^2-(x^1)^2-\cdots-(x^{m-1})^2=g_{ik}x^ix^k,$$

where the ordinary summation convention is used.

The square of the distance of two points x and y is given by

(2) $$R_{xy}=r_{xy}{}^2=(x-y,x-y)=g_{ik}(x^i-y^i)\cdot(x^k-y^k).$$

The scalar product (A,B) of two vectors A and B, with the components A^k and B^k respectively, is defined by

(3) $$(A,B)=g_{ik}A^iB^k.$$

According as the scalar square (A, A) of a vector A is positive, null or negative, the vector is said to be *time-like, light-like* or *space-like*. A time-like or a light-like vector A is called *positive* or *negative* according as its time component A^0 is positive or negative.

Time-like unit vectors U and *space-like unit vectors* V are defined by the relations $(U, U) = 1$ and $(V, V) = -1$ respectively.

Two vectors A and B whose scalar product vanishes are said to be orthogonal to each other. Orthogonality and normality are always meant in this sense.

Any vector orthogonal to a time-like vector is space-like. Any vector orthogonal to a light-like vector is either colinear to this vector or space like.

The *light cone,* or characteristic cone with vertex a is given by the equation $(x - a, x - a) = 0$. The positive and negative half-cones correspond to $x^0 - a^0 > 0$ or < 0 respectively. The generators of a light cone are at the same time its normals.

The tangent planes of a light cone are said to be light-like (or characteristic). The vectors issued from the vertex of a light cone and orthogonal to one of its generators lie in the tangent plane along this generator.

The p-dimensional volume $\Omega (p \leqq m)$ of the parallelepiped spanned by p finite or infinitesimal vectors $U, V, \cdots$ is given by $\Omega = | G |^{\frac{1}{2}}$, where

$$G = \begin{vmatrix} (U,U) & (U,V) \cdots \\ (V,U) & (V,V) \cdots \\ \cdots & \cdots \end{vmatrix}. \tag{4}$$

The angle ϕ between two unit vectors U and V lying in a space-like two-plane is defined in a multivalent way by the formula

$$\cos \phi = -(U, V). \tag{5}$$

The minus sign is due to the fact that $(U, U) = (V, V) = -1$. One has always $|(U, V)| \leqq 1$.

* * *

A surface is said to be space-like if its normal is time-like. Let S be a space-like surface or a characteristic surface. Suppose that the negative light cone with vertex x and the surface S enclose a bounded domain $D_S{}^x$. Let $f(x)$ be a continuous function. We form the integral

$$I^\alpha f(x) = \frac{1}{H_m(\alpha)} \int_{D_S{}^x} f(y) r_{xy}{}^{\alpha - m} \, dy \tag{6}$$

where

$$H_m(\alpha) = \pi^{(m-2)/2} 2^{\alpha - 1} \Gamma\left(\frac{\alpha}{2}\right) \Gamma\left(\frac{\alpha + 2 - m}{2}\right). \tag{7}$$

This integral converges for $\alpha > m-2$. Under suitable differentiability conditions upon f and S, the integral considered as a function of α has analytic continuation to values $< m-2$, down to some negative value of α say. Then $I^0(f,x) = f(x)$.

The above volume integral supplemented by two surface integrals along S furnishes the solution of a Cauchy problem for $\alpha = 2$; see Acta paper, p. 71.

* * *

Consider a light-like line L and a point x outside of it. The running point on the line being denoted by z, we form $R_{xz} = (x-z, x-z) = (x-z)^2$, for short. The intersection between the characteristic cone with vertex x and the line being y, we have $R_{xz} = (x-y+y-z)^2 = 2(x-y, y-z)$, because $x-y$ and $y-z$ are light-like. Thus R_{xz} *varies linearly* when z varies on the line.

Now, if a, b, c, d are points on the line, the relations

$$(8) \qquad d-c = \lambda(b-a) \text{ and } \lambda = \frac{R_{xd}-R_{xc}}{R_{xb}-R_{xa}}$$

are equivalent. We note that the above quotient is independent of x, if a, b, c, d are fixed points on the line.

IV. On the Measure of Infinitesimal Varieties Lying on a Characteristic Surface.

Consider two space-like curves $z_I(t)$ and $z_{II}(t)$ drawn on the positive light cone $(z,z) = 0$, $z^0 > 0$, in *three-dimensional* space-time. The parameter, t is chosen so that the points of the curves which lie on the same generator correspond to the same value of t. Then we can write $z_{II}(t) = \lambda(t) z_I(t)$, $\lambda(t)$ being a scalar. We shall prove that corresponding arc elements ds_I and ds_{II} satisfy the relation

$$(1) \qquad ds_{II}(t) = \lambda(t)\, ds_I(t),$$

or $ds_{II} = \lambda ds_I$, for short. Indeed, for any curve $z(t)$ drawn on the cone we have $(z(t), z(t)) = 0$ and thus $(z(t), \dot{z}(t)) = 0$. Furthermore $dz = \dot{z}dt$, $(dz, dz) = (\dot{z}, \dot{z})dt^2$, $ds = |dz| = |(\dot{z},\dot{z})|^{\frac{1}{2}} dt$. Now $\dot{z}_{II} = \lambda \dot{z}_I + \dot{\lambda} z_I$. Thus $(\dot{z}_{II}, \dot{z}_{II}) = \lambda^2 (\dot{z}_I, \dot{z}_I)$.

Another way of expressing (1) is to consider dz_I and dz_{II} as two displacements between two infinitesimally near generators,

$$(2) \qquad z_{II} = \lambda z_I \Longrightarrow |dz_{II}| = \lambda\, |dz_I| \text{ or } ds_{II} = \lambda ds_I.$$

If the cone has its vertex in a point c, λ will be defined by $z_{II} - c = \lambda(z_I - c)$.

For a *characteristic surface* in three dimensions there is a similar relation. Denote by f the focal point on a fixed generator g and define λ by $z_{II}-f=\lambda(z_I-f)$, where z_I and z_{II} are two regular points of the surface lying on g, on the same side of f. Then we have

$$(3) \qquad ds_{II}=\lambda ds_I.$$

The proof becomes analogous to that given below, if we note that two neighboring generators of any developable surfaces cut each other to the first approximation (as a matter of fact, to the second approximation).

Now consider a *characteristic surface* S in m dimensions. We shall have to compare the measures of corresponding infinitesimal elements of two $(m-2)$-dimensional space-like varieties A_I and A_{II} drawn on the surface.

By a slight generalization of the reasoning to be found in the Acta paper, Section 68, one can prove the following facts, analogous to and connected with those belonging to the curvature theory of $(m-2)$-dimensional surfaces in $(m-1)$-dimensional Euclidean space.

Through every generator g of S there pass *in general* $m-2$ different two-dimensional developable surfaces $D_1, D_2, \cdots, D_{m-2}$, whose generators are also generators of S. Vectors lying in the tangent bi-planes along g of these surfaces are mutually orthogonal to each other. The focal points $f_1, f_2, \cdots, f_{m-2}$ of these surfaces lying on g are all distinct from each other. They will be called focal points of the characteristic surface S. When g varies, these points describe the $(m-2)$-sheeted focal surface of S.

The lines $C_1, C_2, \cdots, C_{m-2}$ in which an $(m-2)$-dimensional space-like variety A traced on S is cut by $D_1, D_2, \cdots, D_{m-2}$ are called lines of curvature of A. They determine in any regular point of A the $m-2$ directions of (principal) curvature.

Denote by a_I and a_{II} the (regular) points of intersection of the generator g with two $(m-2)$-dimensional space-like varieties A_I and A_{II} traced on S. Denote further by C_I and C_{II} the lines of curvature in which one of the developable surfaces, D_k say, cuts A_I and A_{II} respectively. According to (3), the lengths of the infinitesimal arcs $d\sigma_I$ and $d\sigma_{II}$ between the points of intersection of g and a neighboring generator with C_I and C_{II} satisfy the relations $d\sigma_{II}=\lambda_k d\sigma_I$, where λ_k is determined by the relation $a_{II}-f_k=\lambda_k(a_I-f_k)$. (The fact that the developable surface D_k does not in general lie in a three dimensional linear space does not invalidate the arguments leading to formula (3)). The directions of curvature being orthogonal to each other, it follows that the $(m-2)$-dimensional areas ds_I and ds_{II}, which an infinitesimal bundle of generators, including g, cuts out from A_I and A_{II} satisfy the relation

$$(4) \qquad ds_{II} = \prod_{k=1}^{m-2} \lambda_k \cdot ds_I.$$

It is easy to show that the last formula subsists in the entirely umbilic case or partly umbilic case, in which all the f_k, or part of them, coincide, and the directions of curvature are entirely or partly undetermined. This happens for every charateristic cone $(z-c, z-c)=0$ $(m>3)$.

According to formula (III. 8) the formula (4) can be written

$$(5) \qquad \frac{ds^{II}}{ds^I} = \prod_{k=1}^{m-2} \frac{R_k - R_{II}}{R_k - R_I},$$

with $R_{xf_k} = R_k$, $R_{xa_I} = R_I$, $R_{xa_{II}} = R_{II}$, for shortness.

V. Solution of the Cauchy Problem for Characteristic Boundaries.

One can apply a technique very nearly related to that we used in the Acta paper (§ 63) in solving the more special problem when the boundary is a characteristic cone.

We aim to determine the solution of the homogeneous wave equation $\Delta u = 0$ in the convex domain D^+, when the boundary values of u on S^+ are given; for the notations cf. Section II.

Let x be a point of D^+. We form the Riemann-Liouville integral $I^\alpha u(x)$; cf. formulae (III. 6) and (III. 7). By D_{S^x} we mean here the domain bounded by the negative light cone with vertex x and the surface S^+. We have in analogy with formula (51) on p. 111 of the Acta paper

$$(1) \qquad I^\alpha u(x) = U^{\alpha+2}(x) = \frac{1}{H_m(\alpha+2)} \int_{s^x} ds \int_0^{R_q} \left(u \frac{dR^\beta}{dR} - \frac{du}{dR} R^\beta \right) F(R)\, dR$$

for $\alpha > m-2$. Here

$$\beta = \frac{\alpha+2-m}{2}, \qquad F(R) = \prod_{k=1}^{m-2} \left(1 - \frac{R}{R_k}\right),$$

The differentiations and integrations with respect to R are taken along a (variable) generator g. The final integration is extended to the intersection s^x of S^+ with the negative cone of vertex x. The point q is the intersection of the generator g with the keel; $R_q = R_{xq}$ (Cf. p. 111 of Acta paper and formula (V. 5) of the present paper. Observe also that $R_{xb} = 0$ for any point b on s^x.)

The solution of $\Delta u = 0$ is then given by $u(x) = I^0 u(x) = U^2(x)$, where

$U^2(x)$ is to be calculated by analytic continuation. As usual, this continuation leads to solution formulae of very different types for odd and even values of m.

I. *m odd,* $m = 2h+1$.

The formula of solution is

$$(3)\quad u(x) = \frac{1}{2\pi^{h-\frac{1}{2}}}\int_{s^x} ds \left\{ \left(\sum_{i=0}^{h-1} \frac{A^i R_q^{\frac{1}{2}-h+i}}{\Gamma(\frac{1}{2}-h+i+1)} \right) - \frac{F(R_q) R_q^{\frac{1}{2}-h}}{\Gamma(\frac{1}{2}-h+1)} \right\} u(q)$$
$$+ \frac{1}{\pi^h}\int_{s^x} ds \int_0^{R_q} A^h R^{-\frac{1}{2}}\, dR.$$

Here

$$(4)\qquad (-1)^i A^i = 2\frac{d^i}{dR^i}\left\{ u\cdot \left(F(R) - \frac{R}{m-2}F'(R)\right)\right\}.$$

The special case $m = 3$.

The solution is

$$(5)\qquad u(x) = \frac{1}{2\pi}\int_{s^x}\left(1 + \frac{R_{xq}}{R_{xf}}\right) R_{xq}^{-\frac{1}{2}}\cdot u(q)ds - \frac{1}{\pi}\int_{s^x} ds \int_0^{R_q} \frac{du}{dR} R^{-\frac{1}{2}}\, dR,$$

where f is the unique focal point on a variable generator.

II. *m even,* $m = 2h \geqq 4$.

The solution is

$$(6)\qquad u(x) = \frac{1}{2\pi^{h-1}}\int_{s^x} A^{h-1}_{(R=0)}\, ds.$$

The subscript implies that A^{h-1} is taken on s^x, more precisely, in the point of intersection of s^x with a variable generator.

The special case $m = 4$.

The solution is

$$(7)\qquad u(x) = \frac{1}{\pi}\int_{s^x}\left\{ \frac{1}{2}\left(\frac{1}{R_1} + \frac{1}{R_2}\right)\cdot u - \frac{du}{dR}\right\} ds.$$

This formula has the same form as formula (60) on p. 139 of the Acta paper, giving the solution for a space-like boundary and formula (61bis) p. 115 ibid., which gives the solution for the case when the boundary is a characteristic cone. Clearly $R_1 = R_{xf_1}$, $R_2 = R_{xf_2}$, f_1 and f_2 being the focal points on a variable generator.

A glance at the solution formulae (3) and (6) shows that they are of a very different character. For example, in the odd case the values which u

and certain derivatives of u assume on the keel play a very important role, while in the even case the Huygens principle rules entirely and only the data relative to the intersection s^x enter into the solution.

VI. Closer Study of the Three-Dimensional Case.

We see that the first integral in the solution formula (V. 5) only contains those values of u which are admitted in the points g of the keel Q. Consequently, it seems natural to transform this integral into another which is taken along the keel itself.

Observe that in general two generators g and $\bar{g}$ pass through every point q. Denote by f and $\bar{f}$ the focal points lying on g and $\bar{g}$ respectively. Denote by σ the arc length of the keel. The generators passing through $q=q(\sigma)$ cut s^x in two points $b(s)$ and $b(\bar{s})$ say, which we also denote by b and $\bar{b}$. To $d\sigma$ around q correspond two elements ds and $\overline{ds}$ on s^x. Because $R_{xb}=R_{x\bar{b}}=0$, we have

$$(1)\qquad ds=\frac{R_{xf}}{R_{xf}-R_{xq}}\,d\sigma \text{ and } d\bar{s}=\frac{R_{x\bar{f}}}{R_{x\bar{f}}-R_{xq}}\,d\sigma$$

by formula (IV. 5). Only those elements $d\sigma$ figure in (1) which are contained in the negative light cone C^x with vertex x. So, if we denote by Q^* that part of the keel Q which is contained in the cone C^x, the formula (V. 5) can be transcribed:

$$(2)\qquad u(x)=\frac{1}{2\pi}\int_{Q^*}\left(\frac{R_{xf}+R_{xq}}{R_{xf}-R_{xq}}+\frac{R_{x\bar{f}}+R_{xq}}{R_{x\bar{f}}-R_{xq}}\right)R_{xq}^{-\frac{1}{2}}\cdot u(q)\,d\sigma$$
$$-\frac{1}{\pi}\int_{s^x}ds\int_0^{R_q}\frac{du}{dR}\,R^{-\frac{1}{2}}\,dR.$$

A similar transcription can be made in any odd dimension, which does not concern us here.

If the given boundary value is constant on every generator, the last term of (2) vanishes. Owing to obvious continuity hypothesis, the constant value must be the same on every couple of generators g and $\bar{g}$ which meet in the point q of the keel. Finally, because the solution of the problem is unique and the function which is identically 1 is obviously a solution of the problem with boundary values 1, we obtain

$$(3)\qquad \int_{Q^*}K(x,q)\,d\sigma=2\pi,$$

where

$$(4)\qquad K(x,q)=\left(\frac{R_{xf}+R_{xq}}{R_{xf}-R_{xq}}+\frac{R_{x\bar{f}}+R_{xq}}{R_{x\bar{f}}-R_{xq}}\right)\cdot R_{xq}^{-\frac{1}{2}}.$$

The relation (3), in which only geometric elements enter, can be considered as a formula of *curvatura integra*. It is of a certain interest that *this geometric formula is obtained as a by-product of the theory of the wave equation.*

The following comments should be made. 1) The formula differs from the analogous formulae in any Euclidean space, because it contains an arbitrary point x. 2) No formula of *curvatura integra* is known for twisted curves in Euclidean space of three dimensions (or of higher dimension).

Behind any formula of *curvatura integra* in Euclidean space lies an angle of some kind. In the formula $\int \mathrm{K}\, ds = 2\pi$, relative to simple closed plane curves, $\mathrm{K}\, ds$ is an infinitesimal angle. Behind the Gauss-Bonnet formula and its generalization lies the spherical image. In the back of our present formula lies a certain double-valued mapping of the keel on a one-sheeted hyperboloid, neither the discovery of this nor its proof have been easy.

Here we define the angle lying behind the formula (3) without performing any mapping of the keel. Consider the straight line h joining the point x and a point q of the keel which lies inside the half-cone C^x. Denoting the generators through q by g and $\bar{g}$, we form the planes spanned by h and g, and by h and $\bar{g}$ respectively. The angles between these planes, i. e., between their normals, are defined by (III. 5). Any of these angles, ϕ say, varies monotonically. We have the relation

$$K(x, q)\, d\sigma = |\, d\phi \,|, \tag{5}$$

which makes the curvature formula (3) entirely intuitive.

The general solution formula (2) can now be written

$$u(x) = \frac{1}{2\pi} \int_{Q^*} u(q)\, |\, d\phi \,| - \frac{1}{\pi} \int_{s^*} ds \int_0^{R_q} \frac{du}{dR} R^{-\frac{1}{2}}\, dR. \tag{6}$$

The above formula of *curvatura integra* still has a drawback, since it is only concerned with curves which are given as keels of characteristic surfaces. Now such curves have the properties that they possess light-like end-tangents and (positive) time-like principal normals. In ramification points the plane spanned by the tangents of two successive branches are light-like. Conversely, any curve with these properties determines a characteristic surface having this curve as its keel. Furthermore, the curvature formula (3) can be directly established for any curve with the above properties. However, until quite recently I was unable to construct such a curve without the aid of a characteristic surface. A direct construction will be given in the following section.

VII. An Independent Construction of a Keel.

We shall construct a curve Q such that there exists a characteristic surface of the type considered above which has Q as a keel. We restrict ourselves here to simple arcs. The treatment of the ramified case is similar.

The problem consists of finding a finite curve with light-like end-tangents for which $\frac{d^2x}{ds^2} = \ddot{x}$ is positive time-like. In other words the principal normal $\ddot{x}/\sqrt{(\ddot{x},\ddot{x})}$ is positive time-like and so is the vector from the center of curvature to the point on the curve to which it belongs. From the time-likeness of the principal normal it follows that the unit tangent $u = \frac{dx}{ds} = \dot{x}$ of the curve is space-like. From this it again follows that, if the initial point of u is placed in the origin, its end-point describes a curve lying on the one-sheeted hyperboloid $(u,u) = -1$, being tangent to two asymptotes of this hyperboloid. It is easy to give such a curve in an explicit way. Let

$$u^0 = t, \quad u^1 = \sqrt{1+t^2}\cos\phi, \quad u^2 = \sqrt{1+t^2}\sin\phi, \quad \phi = \phi(t),$$

t being a parameter running from $-\infty$ to $+\infty$. The tangent vector du/dt to this curve is colinear with the principal normal of the corresponding (unknown) curve $x = x(t)$. The time-likeness of this vector is then expressed by $\left(\frac{du}{dt}, \frac{du}{dt}\right) > 0$, which after an easy calculation reduces to

$$\left|\frac{d\phi(t)}{dt}\right| < \frac{1}{1+t^2}.$$

From this it follows that ϕ tends to finite limits $\phi(\pm\infty)$ when $t \to \pm\infty$ and that the direction of u and du/dt also tend to limits, namely to those of the vectors $(1, \cos\phi(\pm\infty), \sin\phi(\pm\infty))$ which also give the directions of the asymptotes of the curve $u(t)$.

If we map the curve $u(t)$ in a one to one way on a finite interval $(-a, +a)$ say, let the variable s describe this interval and define the vector x (except for a constant additive vector) by $dx/ds = u$, we shall have

$$\left(\frac{dx}{ds}, \frac{dx}{ds}\right) = (u,u) = -1.$$

that is, s (or s plus a constant) is really the arc-length of the curve. On the other hand the positive time-likeness of dx^2/ds^2 will be assured. The directions of the tangents of the curve x tend to light-like directions, when $s \to \pm a$, namely those of the above asymptotes. Differentiability conditions follow from suitable conditions about the function $\phi(t)$ and the mapping $t \to s$. However, if we want these conditions to be fulfilled at the end points $s = \pm a$ as well, some special additional conditions will be necessary.

Let $G(v)$ be any function such that for small values of v and for some n, $G(v) = a_0 + a_1 v + \cdots + a_n v^n + 0(v^{n+1})$.

If we also require that the coordinates x^1 and x^2 of the keel should be differentiable a certain number of times with respect to the coordinate x^0 in the vicinity of and at the end-points, this requirement will be equivalent to the conditions

$$\frac{d\phi(t)}{dt} = t^{-2}G(t^{-2}), \quad \frac{ds(t)}{dt} = t^{-4}G(t^{-2}), \quad t \to \pm\infty,$$

with a suitable n, the constant term a_0 being different from zero in the two functions G corresponding to s.

The light-like normals of a curve constructed as above generate a characteristic surface which has the curve as its keel.

VIII. Special Functions Related to a Characteristic Surface.

We propose to form functions which are related to a characteristic surface in the same way as the powers of r_{xc} are related to the characteristic cone with vertex c. I originally found the functions in question by the following geometric considerations.

Suppose for example that m (the dimension of space-time) is four. Consider the intersection s^x of the fixed positive cone whose vertex is c with the variable negative cone whose vertex is x. If we draw the time-axis in the direction of the vector $x - c$, we immediately see that s^x is an ordinary sphere with the radius $\frac{1}{2}r_{cx}$. Thus, for its measure, also denoted by s^x, it follows that $s^x = \pi r_{cx}^2$. In m dimensions the corresponding formula is

$$s^x = \frac{2^{3-m}\pi^{m-\frac{1}{2}}}{\Gamma\left(\frac{m-1}{2}\right)} r_{cx}{}^{m-2}.$$

So the measure of the intersection s^x furnishes a certain power of r. Certain other powers can then be obtained by repeated application of the Δ operator on s^x. All that can be carried over to the case of a characteristic surface.

In the sequel we follow a slightly different way, making use of the operator I^α.

Consider the function identically equal to one and apply to it the operator I^α. According to (V.1) we have

$$I^\alpha \cdot 1 = \frac{\beta}{H_m(\alpha+2)} \int_{s^x} ds \int_0^{R_q} R^{\beta-1} F(R)\, dR.$$

Here $\beta = \dfrac{\alpha+2-m}{2}$ or $\alpha = 2\beta + m - 2$ and

$$F(R) = \prod_{k=1}^{m-2} \left(1 - \frac{R}{R_k}\right) = \sum_{j=0}^{m-2} (-1)^j C_j R^j, \tag{1}$$

with

$$C_0 = 1, \quad C_j = \sum \frac{1}{R_1 \cdots R_j} \tag{2}$$

= elementary symmetric function in

$1/R_1, \cdots, 1/R_{m-2}$ of degree j, $\qquad (j = 1, 2, \cdots, m-2)$.

We have according to formula (III.7)

$$\begin{aligned} J^\alpha &= \pi^{(m-2)/2} 2^{\alpha+1} \Gamma\left(\frac{\alpha+2}{2}\right) I^\alpha \cdot 1 \\ &= \frac{1}{\Gamma(\beta)} \cdot \int_{s^x} \sum_0^{m-2} (-1)^j C_j \frac{R_q^{\beta+j}}{\beta+j}\, ds. \end{aligned} \tag{3}$$

Because $\Gamma(\beta)$ has a pole with the residue $(-1)^p/p!$ for $\beta = -p$, that is $\alpha = m - 2 - 2p$, we have

$$J^{m-2-2p} = p! \int_{s^x} C_p\, ds = M_p. \tag{4}$$

We note that according to (2) the integrand only depends on the point x and the generator g_b passing through the point b of s^x. Thus the integral only depends on x, and we write $M_p = M_p(x)$. Thus denoting the total measure of s^x by the same sign s^x, we have

$$M_0(x) = s^x \tag{5}$$

and

$$M_p(x) = p! \int_{s^x} \sum \frac{1}{R_1 R_2 \cdots R_p}\, ds. \tag{6}$$

We also note the relations obtained from the combination of (3) and (4)

$$\pi^{m-1-2p} 2^{m-1-2p}\, \Gamma\left(\frac{m-2p}{2}\right) I^{m-2-2p} \cdot 1 = p!\, M_p(x). \tag{7}$$

Noting that $I^0 1 = 1$ we get

$$\frac{(h-1)!}{2\pi^{(m-2)/2}} \int_{s^x} \sum \frac{1}{R_1 R_2 \cdots R_{h-1}} \, ds = 1 \tag{8}$$

from (6) and (7) for *even* $m = 2h$, and especially for $m = 4$

$$\frac{1}{2\pi} \int_{s^x} \left(\frac{1}{R_1} + \frac{1}{R_2} \right) ds = 1. \tag{9}$$

The last two formulae can be looked upon as giving instances of *curvatura integra.*

It follows from formula (60) on page 139 of my Acta paper, that (9) is valid for space-like boundaries also.

The fundamental relation $\Delta I^{\alpha+2} = I^\alpha$ yields

$$\Delta M_{p-1}(x) = 2(m-2p) M_p(x), \tag{10}$$

when applied to (7), and

$$\Delta^p s^x = 2(m-2)(m-4) \cdots (m-2p) M_p(x). \tag{11}$$

We consider the function

$$M_{m-2}(x) = \int_{s^x} \frac{ds}{R_1 R_2 \cdots R_{m-2}} \tag{12}$$

which consists of a single term and which is the most interesting of all in certain respects. Because there are only $m-2$ values $1/R_k$, their elementary symmetric function of degree $m-1$ vanishes. Hence

$$\Delta M_{m-2}(x) = 0, \tag{13}$$

which could also be obtained by direct computation. That is $M_{m-2}(x)$ is a solution of the wave equation $\Delta u = 0$. This function can be considered as a generalization of the classical elementary solution r^{2-m}. Besides satisfying the wave equation it becomes infinite in all regular points of the characteristic surface to the same order as r_{cx}^{2-m} does in all regular points of a characteristic cone with vertex c, namely, $(m-2)/2$.

Delassus' theorem quoted in the Introduction only states that a function with the properties of $M_{m-2}(x)$ might exist. Here not only the existence of a function with the required properties is proved, but the function in question is given in an explicit form.

The preceding functions M_p, $0 \leqq p < m-2$, can also be considered as generalization of powers of r. Thus M_p behaves in the regular points of our characteristic surfaces as r^{m-2-2p} behaves in the regular points of a cone. The relation (10) generalizes the well-known relation $\Delta r^n = n(n+m-2)r^{n-2}$.

When m is odd, the numerical factor in (11) never vanishes. Thus the functions $M_p(x)$ form with respect to the Δ operator an uninterrupted sequence, $M_p(x) = c_p \Delta s^x$, where the c_p are numerical factors. When m is even, $\Delta M_{m-2/2}$ vanishes according to (10) or to (8). Thus for m even the sequence splits into two parts $0 \leqq p \leqq (m-2)/2$ and $m/2 \leqq p \leqq m-2$. The functions belonging to the first and second part can be obtained by repeated application of the Δ operator on s^x and on $M_{m/2}$ respectively.

We conclude by noting that the relation (10) which also can be obtained by direct computation, that is, without the use of the properties of the operator I^α, is in reality of a local character. It remains valid if the integration is not extended to the entire s^x but only to an arbitrary part of s^x cut out of it by any fixed bundle of generators.

56.

A special characteristic surface — a new relativistic model for a particle?[1]

Fund. Res. in Appl. Math. Dept. Math., Univ. of Maryland
Interim technical report *25* (1957)

Introduction

One finds a thorough discussion of the general problem of progressive waves in Courant-Hilbert, Methoden der Mathematischen Physik, II (1937), pp. 147–151, 448–449. There it is made clear that positive results only can be expected for very special equations.

The starting point of my own contributions to this problem is to be found in a very remarkable paper of F. G. Friedlander: Simple progressive solutions of the wave equation. *Proc. Cambridge Phil. Soc.* 43 (1947), 360–373. The object of this paper is to determine all the solutions of the wave equation

$$\frac{\partial^2 u}{\partial t^2}-\frac{\partial^2 u}{\partial x^2}-\frac{\partial^2 u}{\partial y^2}-\frac{\partial^2 u}{\partial z^2}=\frac{\partial^2 u}{\partial t^2}-\Delta u=0$$

which are of the simple form

$$u=v(x,y,z)\,F\{t-\phi(x,y,z)\},$$

where F denotes an arbitrary function.

We give in this Introduction a summary of Friedlander's paper. However, our own treatment can be fully understood without the study of this summary and, in particular, without any knowledge of the Dupin cyclides.

The arbitrariness of F leads to three equations

$$\left(\frac{\partial\phi}{\partial x}\right)^2+\left(\frac{\partial\phi}{\partial y}\right)^2+\left(\frac{\partial\phi}{\partial z}\right)^2=1,$$

$$2\left(\frac{\partial v}{\partial x}\frac{\partial\phi}{\partial x}+\frac{\partial v}{\partial y}\frac{\partial\phi}{\partial y}+\frac{\partial v}{\partial z}\frac{\partial\phi}{\partial z}\right)+v\Delta\phi=0,$$

$$\Delta v=0,$$

whose common solutions are to be found.

The first of these equations implies that the surfaces $\phi(x,y,z)=t=\text{const.}$ are the wave fronts of ordinary geometrical optics in vacuum, the velocity of light being $=1$. They are parallel surfaces orthogonal to a normal congruence of straight lines (the rays). The quantity $\phi(x,y,z)$ is the distance of the point

[1] This work was supported by the Office of Ordnance Research, U. S. Army, under Contract DA-36-034-ORD-1486 with the University of Maryland.

(x,y,z) from the surface $\phi=0$, and can be interpreted as the time t at which the wave front reaches the point in question.

For the solvability of these equations Friedlander finds the *necessary condition* that both sheets of the common focal surface (the locus of the principal centers of curvature) to the surfaces $\phi=$const. should degenerate into curves (or points). The surfaces possessing this property are the cyclides of Dupin. These are in general algebraic surfaces of the fourth order, but there are degenerate cases of lower order. They are the envelope of two one-parameter families of spheres; every such sphere is tangent to the surface along a circular line of curvature. To all the points of such a line there corresponds the center of the sphere as a common center of principal curvature. The focal lines are two conics, in the general case an ellipse and a hyperbola, which are related to each other as the focal conics of a quadric. The straight lines joining any point of one of these curves to any point of the other form a normal congruence and the corresponding Dupin cyclides can also be obtained as the surfaces orthogonal to the lines of this congruence.

An important special case is that of a torus or, more precisely, a family of parallel toruses, whose common normals all meet a circle C, the locus of the centers of the meridian circles, and a straight line A, the common axis of the toruses; C and A are the focal lines in this special case.

A degenerate case is that of parabolic focal lines when the corresponding cyclides are of the third order. Finally, cones, cylinders, spheres and planes can be considered as degenerate Dupin cyclides.

Friedlander shows that the above necessary condition also is *sufficient* and gives the corresponding explicit solutions both in the general case and the other cases. These solutions are of a rather unexpected character in several respects.

We are going to show that in a four-dimensional treatment the general Dupin cyclides of *different shapes* can be replaced by a *single characteristic surface of entirely determined shape*, with four limiting cases, of which two, namely those of a characteristic plane and a light cone, are obvious.

1. The geometry of the problem

If for any train of waves $\phi(x,y,z)=t$, we associate with every point (x,y,z) the value t as a fourth coordinate, the train will be represented in space-time by a characteristic surface, which, of course, will be a three-dimensional manifold.

We set $t=x^0$, $x=x^1$, $y=x^2$, $z=x^3$ and write x for the position vector of the space-time point with the coordinates x^k. We introduce a Lorentz metric, defining the scalar product of the two vectors x and y by

$$\langle x,y\rangle = x^0y^0 - x^1y^1 - x^2y^2 - x^3y^3. \tag{1}$$

Two vectors are *orthogonal* to each other, if this scalar product *vanishes*. Also in other respects we use the terminology of special relativity.

The covariant components of the gradient of an arbitrary $f(x)$ being $\partial f/\partial x^k$, we set

$$\langle \operatorname{grad} f, \operatorname{grad} g\rangle = \frac{\partial f}{\partial x^0}\frac{\partial g}{\partial x^0} - \frac{\partial f}{\partial x^1}\frac{\partial g}{\partial x^1} - \frac{\partial f}{\partial x^2}\frac{\partial g}{\partial x^2} - \frac{\partial f}{\partial x^3}\frac{\partial g}{\partial x^3}. \tag{2}$$

A characteristic surface S is defined by the simultaneous equations

$$S(x) = \text{const.}, \qquad \langle \operatorname{grad} S, \operatorname{grad} S\rangle = 0. \tag{3}$$

A train of waves can be written

$$\psi(x,y,z,t) = t - \phi(x,y,z) = 0.$$

Then

$$\left(\frac{\partial\phi}{\partial x}\right)^2 + \left(\frac{\partial\phi}{\partial y}\right)^2 + \left(\frac{\partial\phi}{\partial z}\right)^2 = 1$$

gives

$$\left(\frac{\partial\psi}{\partial t}\right)^2 - \left(\frac{\partial\psi}{\partial x}\right)^2 - \left(\frac{\partial\psi}{\partial y}\right)^2 - \left(\frac{\partial\psi}{\partial z}\right)^2 = 0. \tag{3'}$$

If $\psi(x,y,z,t) = 0$ is written in any form $S(x) = \text{const.}$, the ratios of the $\partial S/\partial x^k$ are the same as those of the $\partial\psi/\partial x^k$; thus (3) follows from (3′).

A characteristic surface is a ruled surface; from the second of the equations (3) follows that the *generators* of the surface S are at the same time its *normals*.

Now let us consider the family of parallel toruses in three-space whose meridians have their centers on the circle C

$$x^2 + y^2 = a^2, \qquad z = 0, \qquad t = 0.$$

The last relation means that we take this circle as the zero surface. The rays are the straight lines joining an arbitrary point of C to an arbitrary point of the z-axis. The time t counted from C is given in a point z of this axis by $t = \sqrt{a^2 + z^2}$. Hence the four-dimensional image of the axis is the hyperbola

$$x^1 = x^2 = 0, \qquad (x^0)^2 - (x^3)^2 = a^2.$$

This means that the images of both curves are Lorentz circles with the radius a, lying in two orthogonal two-planes S and T, the one space-like, the other time-like. The images of the rays are the straight lines joining arbitrary points of the two circles. These lines are, as easily seen, null lines; they are the generators of the characteristic surface which is the image of the family of parallel toruses.

The remarkable fact is that this seemingly very special case furnishes the essential features of the general case. The general figure, namely the solution of the geometrical problem we are going to formulate, can be obtained from the special one by a Lorentz motion, followed by a homothety if the radii a are different. This means that all these figures are of the same shape.

The direct space-time treatment of the above analytic problem consists in the equations (10)–(12) below, with $\tau(x)$ and $v(x)$ *both* unknown. From the subsequent equation (I) follows that the surfaces $\tau(x) = \text{const.}$ must be characteristic. Then equations (I)–(III) lead to the following geometric problem. Find the characteristic surfaces whose generators consist of the straight lines joining each point of a curve C' with each point of a curve C''. It will be even sufficient to require a little less, namely that all the lines in question should be null lines. It turns out that under this condition the lines are normal to the surface they generate, that is, this surface is characteristic.

Let a', b' and a'', b'' be arbitrary points of the curves C' and C''. The (skew) quadrangle obtained by joining both points a', b' with both points a'', b'' is a rhombus bent along either of its diagonals; consequently the diagonals $a'b'$ and $a''b''$ are orthogonal to each other.

[Analytic verification:

$$\left\langle a'' - \frac{a'+b'}{2}, b'-a' \right\rangle = \frac{1}{2}\langle (a''-a') + (a''-b'), (a''-a') - (a''-b') \rangle$$
$$= \frac{1}{2}(\langle a''-a', a''-a' \rangle - \langle a''-b', a''-b' \rangle) = 0,$$

since the two scalar squares are equal. In the same way

$$\left\langle b'' - \frac{a'+b'}{2}, b'-a' \right\rangle = 0,$$

and then by subtraction

$$\langle b''-a'', b'-a' \rangle = 0.]$$

This means that an arbitrary chord of C' is orthogonal to an arbitrary chord of C''. If one of the curves, C' for example, is a straight line or contains at least three points lying in a straight line, C'' can not have any finite point. Excluding this case for the moment, it follows that both curves are plane and that they lie in two-planes which are orthogonal to each other. Postponing the case that these planes are light-like or, what is the same thing, one of them is light-like, one of the curves will lie in a space-like two-plane S, the other in a time-like two-plane T, S and T being orthogonal to each other. Such planes have one and only one point in common; this point is finite and we denote it by o. Denoting now the curves by C_S and C_T, arbitrary points of C_S and C_T respectively by z_S and z_T, and writing here and from time to time in the sequel a scalar square as an ordinary square, we have

$$0 = (z_T - z_S)^2 = [(z_T - o) - (z_S - o)]^2 = (z_T - o)^2 + (z_S - o)^2 - 2\langle z_T - o, z_S - o \rangle.$$

The last term vanishes because of the orthogonality of the planes and there remains $(z_T - o)^2 + (z_S - o)^2 = 0$. Keeping z_T constant, we obtain $(z_S - o)^2 = \text{const.}$ Since $(z_S - o)$ is space-like, the last constant is negative and $= -a^2$, say. Then

$$(z_T - o)^2 = a^2, \quad (z_S - o)^2 = -a^2,$$

which shows that both curves are Lorentz circles, C_S being an ellipse and C_T being a hyperbola, from the affine point of view.

Choosing the point o as origin, we have

$$z_T^2 = a^2, \quad z_S^2 = -a^2, \quad \langle z_S, z_T \rangle = 0. \tag{4}$$

Any point y of the surface lying on the generator which joins z_T and z_S can be written

$$y = \frac{1+\mu}{2} z_T + \frac{1-\mu}{2} z_S. \tag{5}$$

An infinitesimal displacement of y is given by

$$dy = \frac{1+\mu}{2} dz_T + \frac{1-\mu}{2} dz_S + \frac{1}{2}(z_T - z_S)\,d\mu. \tag{6}$$

From $(z_T - z_S)^2 = 0$ follows that $\langle z_T - z_S, dy \rangle = 0$, which expresses the fact that the generator is normal to the surface. Consequently, the surface is characteristic.

The vector y can be decomposed according to the scheme $y = y_T + y_S$, where the vectors y_T and y_S are the projections of y onto the planes T and S. If we write

$$D(y) = \langle y, y \rangle^2 + 2a^2 \langle y_S, y_S \rangle - 2a^2 \langle y_T, y_T \rangle + a^4, \tag{7}$$

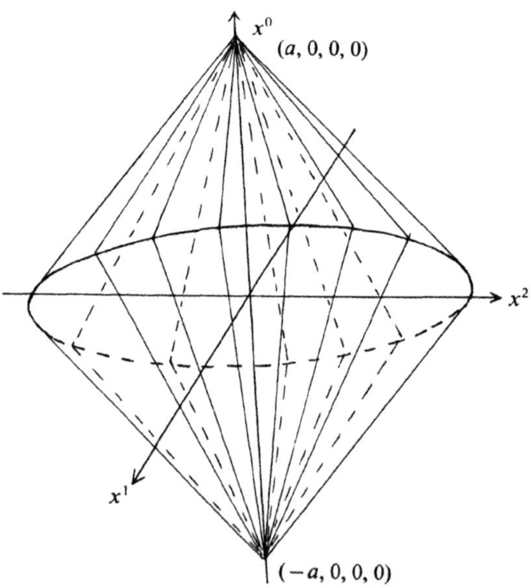

Fig. 1

the surface has the equation $D(y)=0$. In the sequel, this surface will be denoted by D.

The surface D may also be defined by either of the curves C_S and C_T alone. The (three-dimensional) normal plane at an arbitrary point z of the curve is time-like and contains a two-dimensional light cone with vertex at z. All these light cones pass through the other curve and together they constitute the surface D. This construction is illustrated by Figs. 1 and 2 giving two sections of the surface.

The construction above allows a simple description of the limiting case, when one of the curves is a space-like straight line. The two-dimensional light-cones with vertices on the line and lying in the normal planes intersect in a Lorentz circle at infinity and constitute a characteristic surface. This degenerate case will be designated as the cylindrical case. All these characteristic surfaces can obviously be carried over into each other by Lorentz motions.

The intermediate case, the parabolic one, is more difficult to describe. We consider two light-like two-planes L_1 and L_2, orthogonal to each other, having no finite point, but a light-like direction, that is a light-like point at infinity, in common. The Lorentz circles lying in the respective planes are parabolas, related to each other in such a way that the lines connecting arbitrary points of them are null lines. These lines generate the characteristic surface. However, it should be emphasized that the planes L_1 and L_2 in their entirety belong to the surface analytically as well as geometrically; they can be considered as generated by the lines connecting the finite points of either of the parabolas with the point at infinity of the other. All parabolic characteristic surfaces can be carried over into each other by Lorentz motions.

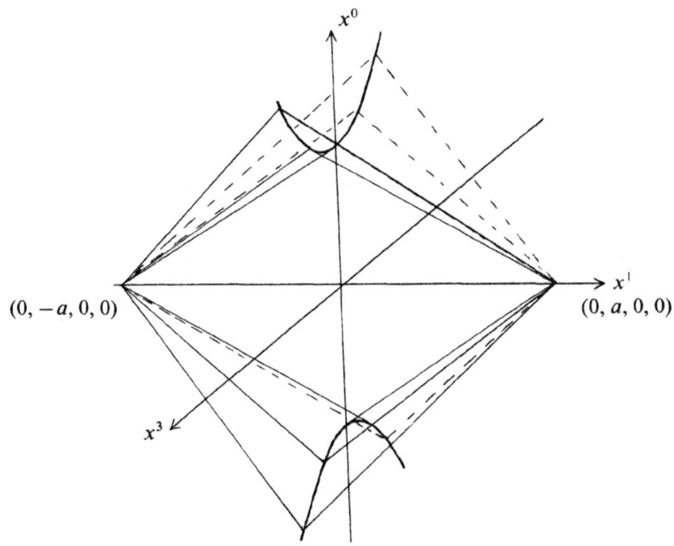

Fig. 2

Summing up, we have found three different types of surfaces: A) the general or principal case, that is $D(x-q)=0$, obtained from (7) by a translation carrying the origin into a point q; B) the parabolic case; C) the cylindrical case.

We shall study the effect of an *inversion* on these surfaces. The definition of this transformation is essentially the same in space-time as in Euclidean space. Two points x and x' correspond to each other by an inversion, if they are in a straight line with a fixed point o, the center of inversion, and if the scalar product $\langle x-o, x'-o\rangle = \text{const}$.

According as the center of inversion lies a) outside the surface, b) on a generator, but not on a singular curve (the Lorentz circles of different types), c) on the singular curves, the transformed surface will be of type A), B), C).

In fact, since null lines transform into null lines, any surface belonging to one of the above types transforms into some such surface. Now take for example a surface of type A); if we transform it according to b), the two Lorentz circles transform into two Lorentz circles having a light-like point at infinity in common, which means that we obtain a surface of type B).

2. The solution of the wave equation

In the present exposé we restrict ourselves to the general case, which means, characteristic surfaces of type (7).

Let p be a positive time-like unit vector, that is

$$\langle p,p\rangle = 1, \quad p^0 > 0, \tag{8}$$

let τ be the *proper time* parameter and, referring to (7), consider the translated surfaces D_τ, whose equation is

$$D(x-\tau p) = 0. \tag{9}$$

The Lorentz circles C_S and C_T describe two (two-dimensional) cylinders (of which the one described by C_T degenerates into the plane T, if p is parallel to this plane). For an arbitrary point x in space-time the above equation defines $\tau = \tau(x)$ as the root of an algebraic equation of the fourth degree in τ. If x does not lie on the cylinders just mentioned, the equation has distinct roots, and we may consider the surfaces D_τ corresponding to the different roots separately.

We write

$$\Box u = \left(\frac{\partial^2}{(\partial x^0)^2} - \frac{\partial^2}{(\partial x^1)^2} - \frac{\partial^2}{(\partial x^2)^2} - \frac{\partial^2}{(\partial x^3)^2}\right) u, \tag{10}$$

and propose to find a function $v(x)$ such that for an arbitrary function $F(\tau)$,

$$u(x) = v(x)\, F(\tau(x)) \tag{11}$$

satisfies the wave equation

$$\Box u(x) = 0. \tag{12}$$

Owing to the arbitrariness of F, this equation splits into the three equations

$$\text{(I) } \operatorname{grad}^2 \tau = 0, \quad \text{(II) } 2\langle \operatorname{grad} v, \operatorname{grad} \tau \rangle + v \Box \tau = 0, \quad \text{(III) } \Box v = 0.$$

The first equation is automatically fulfilled, because it expresses only that the surfaces $\tau = \text{constant}$, that is the surfaces D_τ, are characteristic.

We denote the respective arc lengths of the Lorentz circles C_S and C_T, measured from arbitrary initial points, by σ_S and σ_T and define the (elliptic and hyperbolic) angles α_S and α_T by $\alpha_S = \sigma_S / a$, $\alpha_T = \sigma_T / a$. The equations (II)–(III) have four linearly independent solutions of the form

$$v_k(x) = \frac{w_k(\alpha_T, \alpha_S)}{\rho}, \qquad k = 1, 2, 3, 4, \tag{13}$$

to which correspond four solutions of the wave equation (12)

$$u_k(x) = F_k(\tau)\, v_k(x), \tag{14}$$

with arbitrary $F_k(\tau)$.

We are going to explain these solutions with the aid of Fig. 3, and begin with the explanation of the denominator ρ. The left hand part of this figure illustrates the solution of the corresponding problem in the case of a point-shaped particle in uniform motion, that means moving along a time-like straight line. (In three-dimensional language we have the case of uniformly expanding spherical waves.) The surface D_τ is in this case the light cone with its vertex at the retarded point $z_{\text{ret}} = z(\tau)$. The corresponding solution of the equation (12) is, as well known,

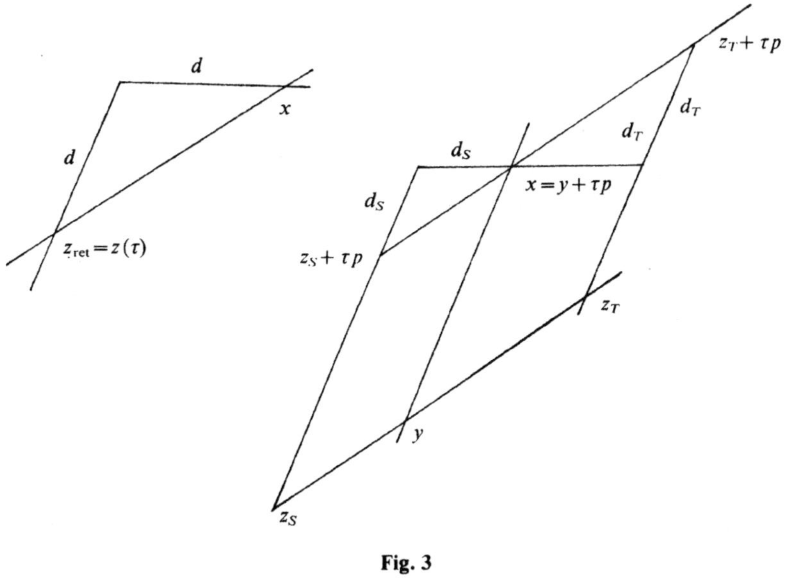

Fig. 3

$$u = \frac{F(\tau)}{d},$$

where d is the distance of the point x from the straight line or the length of the projection of the vector $x - z_{\text{ret}}$ onto this straight line. Written out explicitly, $d = \langle x - z_{\text{ret}}, p\rangle$, where p designates a positive unit vector along the line.

In our present problem the straight line is replaced by the two cylinders discussed above. The right hand part of Fig. 3 illustrates the motion of two particular points $z_S + \tau p$ and $z_T + \tau p$ describing two straight lines (generators of the cylinders). The point x lies on the generator of D_τ passing through the points $z_T + \tau p$ and $z_S + \tau p$. We denote by d_T and d_S the respective distances of the point x from the straight lines described by the latter points, and define ρ as the geometric mean of d_T and d_S, that is we set

$$\rho = \sqrt{d_T \cdot d_S}\,. \tag{15}$$

In the case, when the radius $a = 0$, we recover the classical solution, because $d_S = d_T = d$ and thus $\rho = d$.

The numerators w_k in (13) are given by

$$w_k = \exp\left(\pm\frac{\alpha_T}{2} \pm i\frac{\alpha_S}{2}\right), \qquad k = 1,2,3,4, \tag{16}$$

where the subscript k corresponds to the four possible combinations of the signs plus and minus, and i designates the imaginary unit. Obviously, the function

$$u(x) = \rho^{-1}\sum_{k=1}^{4} F_k(\tau)\exp\left(\pm\frac{\alpha_T}{2} \pm i\frac{\alpha_S}{2}\right), \tag{17}$$

obtained by superposition of the four particular solutions, also satisfies the wave equation (12).

The Fig. 4 refers to the case of an accelerated motion. Let the point $q(\tau)$ describe a world line with the arc length τ and the positive unit tangent vector $p(\tau) = dq(\tau)/d\tau$. The proper time τ corresponding to a space-time point x and the characteristic surface D_τ are now defined by the equation $D(x - q(\tau)) = 0$. The solutions of the equation (12) or of the equations (I)–(III) romain formally unchanged, if they are interpreted in the right way.

Let the space-time point x be situated on the generator connecting the points $z_T + q(\tau)$ and $z_S + q(\tau)$. We define d_T and d_S as the respective distances between the point x and the tangents at the points $z_T + q(\tau)$ and $z_S + q(\tau)$ of the world lines described by these points, and define again ρ as the geometric mean of d_T and d_S. With these definitions the above solutions remain valid.

4. The vector potential

By means of the scalar solution (17) of the wave equation we now form the vector

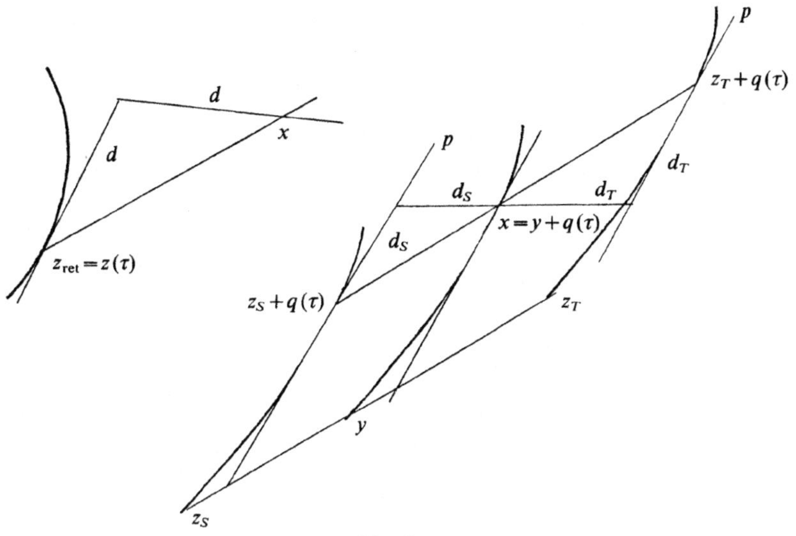

Fig. 4

$$A(x)=u(x)\cdot p(\tau), \tag{18}$$

where $p(\tau)$, as above, denotes the unit tangent vector of the world lines at the proper time τ which is the common velocity vector of all points of the characteristic surface D_τ. Since the components of $p(\tau)$ only depend on τ, the vector $A(x)$ satisfies the wave equation

$$\Box A(x)=0. \tag{19}$$

One readily finds

$$\operatorname{div} A=\rho^{-1}\sum_k \frac{d}{d\tau} F_k(\tau)\exp\left(\pm\frac{\alpha_T}{2}\pm i\frac{\alpha_S}{2}\right). \tag{20}$$

Because of this and because the exponentials are linearly independent of each other, div A vanishes identically, if and only if the coefficients F_k are constant.

5. Physical implications

We have already alluded to the fact that the Liénard-Wiechert potential and the corresponding electro-magnetic field are produced by a point-shaped particle describing a world line, the only acknowledged relativistic model of the electron. Since a point determines in a unique way the light cone whose vertex it is, and vice versa, one might also, for a given proper time τ, identify the particle with the light cone, that is the wave front along which it acts at the given proper time.

It is the characteristic surface D_τ which in our problem corresponds to the classical light cone, while the Lorentz circles $C_S+q(\tau)$ and $C_T+q(\tau)$ correspond to the moving point. The world line is replaced by two cylinders, or, in the generalized case, by two cylindroids described by the Lorentz circles.

If one could arrive at a satisfactory physical interpretation of the above mathematical results, one would have a particle with finite radius, polarized in two planes, S and T. There is a past and a future corresponding to the lower and higher branches of the hyperbola C_T. The surface D_τ has obviously the phase τ, moreover every generator has four complementary phases, namely $\pm\frac{\alpha_T}{2}\pm i\frac{\alpha_S}{2}$. It seems to me that the primary spin concept, as proposed by Uhlenbeck and Goudsmith, would be easier to interpret in the case of the more complicated model than in the case of the point-shaped one.

A sensible interpretation presents, of course, very great difficulties. There are, for example, the questions of the smearing out of the charge and of the behaviour in an external field. However, I should like to point out that even in the most general case treated here, the one given by the formula (18), the motion is a rather particular one, implying a high degree of stability. Namely, for a given proper time τ, all points of D_τ have the same velocity $p(\tau)$.

Obviously, it belongs to the physicists to find a natural interpretation of all these facts — if there is any.

I conclude by expressing my gratitude to Professor Karl Ludwig Stellmacher who aroused my interest in the subject.

58.

A geometric solution of the wave equation in space-time of even dimension

Comm. Pure Appl. Math. *13* (1960), 329–354

1. Introduction

In this paper I give a new geometric solution of the Cauchy problem for the wave equation in space-time (or Lorentz space) of an arbitrary even dimension $m \geqq 4$ and for an arbitrary space-like surface carrying the Cauchy data. The main feature of this solution is that, besides the data, only elements of purely geometrical, more precisely, of geometric optical character enter into the solution. By this I believe I have shown that geometric optics is not only an approximation to physical optics, but also forms its skeleton.

I found the present method during my stay at New York University in spring 1955. The method is based on a kind of Beltrami technique, which in the simplest case, namely the Poisson formula for the initial value problem for $m = 4$ is given in Courant-Hilbert II, pp. 370—371. The geometry of characteristic surfaces plays a great role in the presentation which follows.

The special case for $m = 4$ of the present solution figures already in Chapter V of my paper: *L'intégrale de Riemann-Liouville et le problème de Cauchy* (Acta Mathematica, Vol. 81, 1949, pp. 1—223). Only minor local references will be given to this paper, quoted in the sequel as *Acta paper*, and its knowledge is not required for the reading of the present article.

No verification of the solution and no existence proof are given. The identities (4.7) and (5.6) are nevertheless called solution formulae. Since it is known that, under suitable conditions, a solution actually exists, these formulae not only prove the uniqueness of the solution, but even represent it.

2. Preliminaries

We consider the space-time of m dimensions provided with the metric form

$$\text{(L)} \qquad (x, y) = x^0 y^0 - x^1 y^1 - \cdots - x^{m-1} y^{m-1} = l_{ik}\, x^i y^k$$

for the scalar product of two vectors

$$x = (x^0, x^1, \cdots, x^{m-1}) \quad \text{and} \quad y = (y^0, y^1, \cdots, y^{m-1}).$$

For the geometry of such spaces see Chapter III, in particular Section 3.6, of my lecture notes: *Clifford numbers and spinors*, Lecture Series, No. 38, Institute for Fluid Dynamics and Applied Mathematics, University of Maryland (1957—1958).

All metric notions, in particular orthogonality, arc length, measure of infinitesimal elements of arbitrary dimensions etc. *are to be interpreted according to the metric* (L) (cf. Acta paper, No. 21bis).

We suppose that all derivatives figuring explicitly or implicitly in our computations actually exist and are continuous.

A *characteristic surface* is defined as an $(m-1)$-dimensional ruled surface whose generators are *normal* to the surface. Since the generators are orthogonal to themselves, they must be light-like = isotropic (null lines).

Through a space-like manifold σ of $m-2$ dimensions there pass two characteristic surfaces. Indeed, the two-plane which in a fixed point of σ is orthogonal to σ is time-like (cf. quoted lecture notes, p. 109). Such a plane contains two light-like directions. In order to see this, we choose the x^0 and x^1 axes of a Lorentz frame in the two-plane, and find as light-like directions those of the lines $x^0 \pm x^1 = 0$. Thus we have proved that in every point of σ there are two light-like normals to σ. These normals form two distinct families. We shall show that each of these families generates a characteristic surface.

Denote by Σ the surface generated by one of these families, by g a variable generator of Σ, by x the point of intersection of g with the original manifold σ and by l a vector along g. An arbitrary point y of g can be written $y = x+\lambda l$. Now $(l, dx) = 0$,

$$(l, l) = 0 \Rightarrow (l, dl) = 0 \Rightarrow (l, dy) = (l, dx+d\lambda \cdot l+\lambda \cdot dl) = 0.$$

This means that the generator g is orthogonal to any infinitesimal displacement of its points along the surface Σ. Thus g is normal to the surface, which consequently is a characteristic surface.

The characteristic surfaces in question can also be obtained as the envelopes of a moving characteristic cone whose vertex describes σ.

3. The Wave Operator in Special Curvilinear Coordinates

Consider now an $(m-1)$-dimensional *space-like surface* S, which together with a negative *light-cone* or *characteristic cone* C^x with vertex x encloses a certain domain D_S^x (Figure 2 on page 38 of Acta paper). Denote the $(m-2)$-dimensional and obviously space-like intersection of S and C^x by s^x. One of the characteristic surfaces passing through s^x is obviously the light-cone C^x. Denote by Σ^x the other characteristic surface, which can be considered as the image of C^x with respect to S or s^x.

Our first aim is to parametrize that part of space-time between the cone and the surface S which lies sufficiently near to the cone by suitable parameters and to express the wave operator by these parameters.

We place the origin O in the vertex x of the cone C^x and until further notice drop all the superscripts x, thus writing C, s, Σ instead of C^x, s^x, Σ^x. The points of s and the corresponding position vectors are denoted by b. Such a vector satisfies the first equation of the subsequent formula (3.1), because it is a null vector. Next we consider along the generator of Σ through the point b a vector c. This vector is again a null vector and thus satisfies the second of the equations (3.1). Finally this vector is normalized by the third of the equations (3.1).

Thus we have

$$(3.1) \qquad (b, b) = 0, \qquad (c, c) = 0, \qquad (b, c) = \tfrac{1}{2}.$$

From here we deduce

$$(3.2) \qquad (b, db) = (c, dc) = (b, dc) = (c, db) = 0.$$

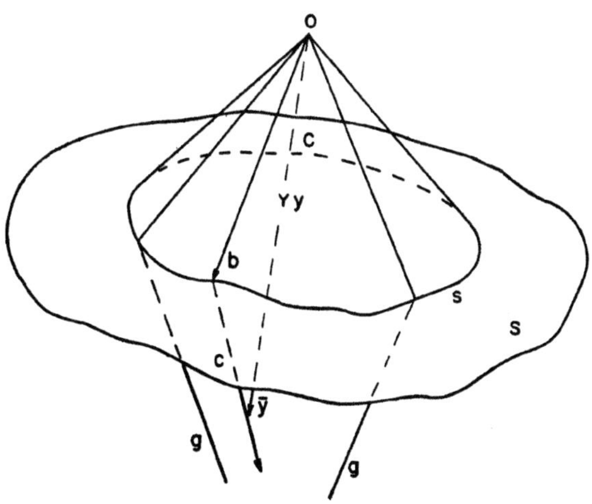

The first two equations of (3.2) follow immediately from (3.1). From (3.1) also follows that $(b, dc)+(c, db) = 0$. Now $(c, db) = 0$, because the displacement db lies along the characteristic surface. This completes the proof of (3.2).

To begin with, we parametrize the intersection s by setting $b = b(\lambda^1, \lambda^2, \cdots, \lambda^{m-2})$ for any point b of s, the $\{\lambda^k\}$ being suitable local parameters. The corresponding vector c along the generator g can then be written $c = c(\lambda^1, \lambda^2, \cdots, \lambda^{m-2})$.

In order to parametrize the characteristic surface Σ, we write an arbitrary point $\bar{y}$ of g as $\bar{y} = b+\tau c$. It is easy to see that for a sufficiently small $\tau_0 > 0$ and $0 \leqq \tau \leqq \tau_0$ a certain narrow portion of Σ is parametrized in a unique way by the $m-1$ parameters τ, $\{\lambda^k\}$. Finally by $y = \sigma\bar{y}$, with $0 \leqq \sigma \leqq 1$, we have accomplished the parametrization by the m parameters σ, τ, $\{\lambda^k\}$ of a certain m-dimensional domain limited by 1) the negative characteristic cone issuing from the origin, 2) the characteristic surface Σ, 3) a certain conical surface with vertex at O which lies inside the cone and is sufficiently near the cone. We write here the formulae

$$\bar{y} = b+\tau c, \qquad y = \sigma\bar{y} = \sigma(b+\tau c). \tag{3.3}$$

Writing from time to time the scalar square of a vector as an ordinary square, we obtain from (3.3) and (3.1)

$$\bar{y}^2 = \tau, \qquad 2(\bar{y}, d\bar{y}) = d\tau. \tag{3.4}$$

Since $d\bar{y} = db+d\tau \cdot c+\tau \cdot dc$, it follows from (3.1) and (3.2) that

$$d\bar{y}^2 = (db+\tau dc)^2. \tag{3.5}$$

Since $dy = d(\sigma\bar{y}) = d\sigma \cdot \bar{y}+\sigma \cdot d\bar{y}$, we obtain from (3.4) and (3.5)

$$\begin{aligned} dy^2 &= d\sigma^2 \bar{y}^2+2\sigma d\sigma(\bar{y}, d\bar{y})+\sigma^2 d\bar{y}^2 \\ &= \tau d\sigma^2+\sigma d\sigma d\tau+\sigma^2(db+\tau dc)^2. \end{aligned} \tag{3.6}$$

From

$$b = b(\lambda^1, \cdots, \lambda^{m-2}), \qquad c = c(\lambda^1, \cdots, \lambda^{m-2}) \tag{3.7}$$

we obtain

$$d\bar{y}^2 = (db+\tau dc)^2 = \bar{\gamma}_{ik}\, d\lambda^i d\lambda^k, \tag{3.8}$$

where

$$\bar{\gamma}_{ik} = \left(\frac{\partial(b+\tau c)}{\partial\lambda^i}, \frac{\partial(b+\tau c)}{\partial\lambda^k}\right). \tag{3.9}$$

Setting furthermore

$$\gamma_{ik} = \sigma^2 \bar{\gamma}_{ik}, \tag{3.10}$$

we can write (3.6) in the form

$$dy^2 = \tau d\sigma^2+\sigma d\sigma d\tau+\gamma_{ik} d\lambda^i d\lambda^k, \tag{3.11}$$

which is precisely the *metric form* of space-time expressed in the parameters $\sigma, \tau, \lambda^1, \cdots, \lambda^{m-2}$. If we moreover set $\{\alpha^r\}_{r=1,2,\cdots,m} = \{\sigma, \tau, \lambda^1, \cdots, \lambda^{m-2}\}$, we can write

$$dy^2 = g_{rs}\, d\alpha^r\, d\alpha^s.$$

The matrix (g_{rs}) splits into two quadratic blocks, namely

$$(3.12)\qquad \begin{pmatrix} g_{\sigma\sigma} & g_{\sigma\tau} \\ g_{\tau\sigma} & g_{\tau\tau} \end{pmatrix} = \begin{pmatrix} \tau & \frac{1}{2}\sigma \\ \frac{1}{2}\sigma & 0 \end{pmatrix} \quad \text{and} \quad (\gamma_{ik})_{i,k=1,\cdots,m-2}\,.$$

We recall that $\bar{\gamma}_{ik}$ and $\gamma_{ik} = \sigma^2 \bar{\gamma}_{ik}$ are given by (3.9) and (3.10) and introduce the determinants

$$(3.13)\qquad |\bar{\gamma}_{ik}| = \bar{\gamma} \quad \text{and} \quad |\gamma_{ik}| = \gamma = \sigma^{2(m-2)}\bar{\gamma}.$$

The determinant of the entire matrix (g_{rs}) has the value

$$g = -\tfrac{1}{4}\sigma^2\gamma = -\tfrac{1}{4}\sigma^{2m-2}\bar{\gamma}.$$

We shall see in (7.5) that $\bar{\gamma} > 0$ when m is even. Therefore we have

$$(3.14)\qquad \sqrt{|g|} = \tfrac{1}{2}\sigma\sqrt{\gamma} = \tfrac{1}{2}\sigma^{m-1}\sqrt{\bar{\gamma}}.$$

Besides the expression (3.11) of the arc element of space-time in our parameters α^r we also need the expression of the *wave operator*

$$(3.15)\qquad \varDelta = \frac{\partial^2}{(\partial x^0)^2} - \frac{\partial^2}{(\partial x^1)^2} - \cdots - \frac{\partial^2}{(\partial x^{m-1})^2}$$

in these parameters. To this end we introduce the inverse matrix $(g^{rs}) = (g_{rs})^{-1}$ and first of all note that this matrix too splits into two quadratic blocks, namely

$$(3.16)\qquad \begin{pmatrix} g^{\sigma\sigma} & g^{\sigma\tau} \\ g^{\tau\sigma} & g^{\tau\tau} \end{pmatrix} = \begin{pmatrix} 0 & \dfrac{2}{\sigma} \\ \dfrac{2}{\sigma} & -\dfrac{4\tau}{\sigma^2} \end{pmatrix} \quad \text{and} \quad (\gamma^{ik}) = (\gamma_{ik})^{-1}.$$

The wave operator (3.15) can also be written in the Beltrami form relative to the parameters α^r. This is

$$(3.17)\qquad \varDelta = \frac{1}{\sqrt{|g|}}\frac{\partial}{\partial\alpha^r}\left(\sqrt{|g|}\, g^{rs}\frac{\partial}{\partial\alpha^s}\right).$$

According to what precedes, this Beltrami operator splits into two parts, the first containing differentiations with respect to σ and τ only, the second with respect to the λ^i only. Since, according to (3.14), $\sqrt{|g|} = \frac{1}{2}\sigma\sqrt{\gamma}$, where only γ depends on the λ^i, the second part of the operator reduces to $\varDelta_{(\lambda)}$, by which we mean *the Beltrami operator with respect to the λ^i only:*

$$(3.18)\qquad \varDelta_{(\lambda)} = \frac{1}{\sqrt{\gamma}}\frac{\partial}{\partial\lambda^i}\left(\sqrt{\gamma}\,\gamma^{ik}\frac{\partial}{\partial\lambda^k}\right).$$

If we also note that, according to (3.14) and (3.16),

$$\sqrt{|g|}\,g^{\sigma\sigma} = 0,\ \sqrt{|g|}\,g^{\sigma\tau} = \sqrt{|g|}\,g^{\tau\sigma} = \sqrt{\gamma},\quad \sqrt{|g|}\,g^{\tau\tau} = -\sqrt{\gamma}\,\frac{2\tau}{\sigma},$$

we obtain

$$\Delta u = \frac{2}{\sigma\sqrt{\gamma}}\left\{\frac{\partial}{\partial\sigma}\left(\sqrt{\gamma}\,\frac{\partial u}{\partial\tau}\right) + \frac{\partial}{\partial\tau}\left[\sqrt{\gamma}\left(\frac{\partial u}{\partial\sigma} - \frac{2\tau}{\sigma}\frac{\partial u}{\partial\tau}\right)\right]\right\} + \Delta_{(\lambda)}u.$$

Since, according to (3.13), $\sqrt{\gamma} = \sigma^{m-2}\sqrt{\bar{\gamma}}$ and $\bar{\gamma}$ is independent of σ, we have $\partial\sqrt{\gamma}/\partial\sigma = (m-2)\sqrt{\gamma}/\sigma$. Hence the above relation can also be written

$$\Delta u = \frac{4}{\sigma}\,\frac{\partial^2 u}{\partial\sigma\partial\tau} + \frac{2}{\sigma\sqrt{\gamma}}\,\frac{\partial\sqrt{\gamma}}{\partial\tau}\,\frac{\partial u}{\partial\sigma} - \frac{4\tau}{\sigma^2\sqrt{\gamma}}\,\frac{\partial}{\partial\tau}\left(\sqrt{\gamma}\,\frac{\partial u}{\partial\tau}\right) + \frac{2(m-4)}{\sigma^2}\,\frac{\partial u}{\partial\tau} + \Delta_{(\lambda)}u. \tag{3.19}$$

The expression $\sqrt{\gamma} = \sqrt{\gamma(\sigma, \tau, \lambda^j)} = \sigma^{m-2}\sqrt{\bar{\gamma}(\tau, \lambda^j)}$ figures in two terms both in the numerator and the denominator. It is readily seen that (3.19) does not change if we replace $\sqrt{\gamma}$ by

$$\frac{\sqrt{\gamma(\sigma, \tau, \lambda^j)}}{\sqrt{\gamma(\sigma, 0, \lambda^j)}} = \frac{\sqrt{\bar{\gamma}(\tau, \lambda^j)}}{\sqrt{\bar{\gamma}(0, \lambda^j)}} = F(\tau, \lambda^j) = F. \tag{3.20}$$

Hence, denoting for brevity $\partial F/\partial\tau$ by F', we can write

$$\Delta u = \frac{4}{\sigma}\,\frac{\partial^2 u}{\partial\sigma\partial\tau} + 2\,\frac{F'}{\sigma F}\,\frac{\partial u}{\partial\sigma} - \frac{4\tau}{\sigma^2 F}\,\frac{\partial}{\partial\tau}\left(F\,\frac{\partial u}{\partial\tau}\right) + \frac{2(m-4)}{\sigma^2}\,\frac{\partial u}{\partial\tau} + \Delta_{(\lambda)}u. \tag{3.21}$$

It is seen from (3.9) and (3.13) that $\bar{\gamma}$ is a polynomial (of degree $\leqq 2(m-2)$) in τ. We shall prove in Section 7 that $\sqrt{\bar{\gamma}}$ also is a polynomial (of degree $\leqq m-2$) in τ and hence the same is true for F.

We now define s_σ as the manifold σs, homothetical to s. If we divide s_σ into two parts by means of an $(m-3)$-dimensional closed manifold and apply Green's theorem (cf. Acta paper, pp. 187—189) to each part, then we obtain

$$\int_{s_\sigma} \Delta_{(\lambda)}u(\sigma, 0, \lambda^i)\,ds_\sigma = 0.$$

Because $ds_\sigma = \sigma^{m-2}ds$, we also have

$$\int_s \Delta_{(\lambda)}u(\sigma, 0, \lambda^i)\,ds = 0. \tag{3.22}$$

This relation is entirely sufficient for the needs of Section 4, where only the case $m = 4$ is considered. In Section 5, however, where the case $m > 4$ is treated, we shall need a more general relation.

The manifold $s_\sigma = \sigma s$ is described by the point σb, where σ is fixed and b is the generic point of s. For every fixed pair σ and τ we designate by $s_{\sigma,\tau}$ the manifold described by the point $y = \sigma(b+\tau c)$ defined in formula (3.3). We find by the reasoning used above that

$$\int_{s_{\sigma,\tau}} \Delta_{(\lambda)} u(\sigma, \tau, \lambda^i) ds_{\sigma,\tau} = 0,$$

which also can be written as

$$\int_{s_\sigma} \Delta_{(\lambda)} u \frac{ds_{\sigma,\tau}}{ds_\sigma} ds_\sigma = 0. \tag{3.23}$$

From (3.9), (3.10), (3.20) we obtain that

$$\frac{ds_{\sigma,\tau}}{ds_\sigma} = \frac{\sqrt{\gamma(\sigma, \tau, \lambda^i)}}{\sqrt{\gamma(\sigma, 0, \lambda^i)}} = F(\tau, \lambda^i).$$

Moreover $ds_\sigma = \sigma^{m-2} ds$, and thus (3.23) can also be written as

$$\int_s F(\tau, \lambda^i) \Delta_{(\lambda)} u(\sigma, \tau, \lambda^i) ds = 0.$$

Since s and ds are independent of τ, it finally follows for an arbitrary integer $p \geqq 0$ that

$$\int_s \frac{\partial^p}{\partial \tau^p} (F(\tau, \lambda^i) \Delta_{(\lambda)} u(\sigma, \tau, \lambda^i)) ds = 0. \tag{3.24}$$

We will apply this formula only on the cone C, that is only for $\tau = 0$.

4. The Cauchy Problem for $m = 4$

We are going to solve the *Cauchy problem for the wave operator* Δ, defined in (3.15). This problem consists in finding a solution u of the equation $\Delta u = f$ if the *Cauchy data*, i.e. the values of u and of some non-tangential first derivative of u, are known on a space-like surface S. We begin with the case of dimension $m = 4$, which is much easier than the case of higher dimensions.

As we have seen in Section 3, Δu can be expressed by the right-hand side of equation (3.21). *In the case* $m = 4$, this equation reduces to

$$\Delta u = \frac{4}{\sigma} \frac{\partial^2 u}{\partial \sigma \partial \tau} + \frac{2F'}{\sigma F} \frac{\partial u}{\partial \sigma} - \frac{4\tau}{\sigma^2 F} \frac{\partial}{\partial \tau} \left(F \frac{\partial u}{\partial \tau} \right) + \Delta_{(\lambda)} u. \tag{4.1}$$

On the surface of the cone, where $\tau = 0$, the equation reduces still further; one more term cancels and $F(0, \lambda^j) = 1$. The equation becomes

$$(4.2) \qquad \Delta u = \frac{4}{\sigma}\frac{\partial^2 u}{\partial\sigma\partial\tau} + \frac{2}{\sigma} F' \frac{\partial u}{\partial\sigma} + \Delta_{(\lambda)} u.$$

Here

$$(4.3) \qquad u = u(\sigma, 0, \lambda^i), \quad \frac{\partial u}{\partial\tau} = \frac{\partial u(\sigma, \tau, \lambda^i)}{\partial\tau}_{(\tau=0)}, \quad F' = \frac{\partial F(\tau, \lambda^i)}{\partial\tau}_{(\tau=0)}.$$

We multiply (4.2) by $\sigma d\sigma ds/4\pi$ and integrate following the scheme

$$\int_0^1 d\sigma \int_s \cdots ds = \int_{C_s} \cdots d\sigma ds,$$

where C_s denotes that portion of the negative cone with vertex O which is bounded by s. According to (3.22) the integral corresponding to the term $\Delta_{(\lambda)} u$ vanishes. For the other terms of the right-hand side we perform first the integration with respect to σ; that is, we integrate along a generator segment of the cone joining the vertex O to a point b on s. We obtain

$$\frac{1}{\pi}\left[\frac{\partial u}{\partial\tau}\right]_{\sigma=0}^{\sigma=1} + \frac{F'}{2\pi}[u]_{\sigma=0}^{\sigma=1}.$$

At the vertex, that is for $\sigma = 0$, all derivatives $\partial^j u/\partial\tau^j$ vanish. Indeed, according to (3.3), we have

$$(4.4) \qquad \frac{\partial u}{\partial\tau} = \frac{\partial u}{\partial y^j}\frac{\partial y^j}{\partial\tau} = \sigma\frac{\partial u}{\partial y^j} c^j, \quad \frac{\partial^2 u}{\partial\tau^2} = \sigma^2 \frac{\partial^2 u}{\partial y^j \partial y^k} c^j c^k, \cdots.$$

Hence

$$(4.5) \qquad \left[\frac{\partial u}{\partial\tau}\right]_{\sigma=0}^{\sigma=1} = \frac{\partial u(y)}{\partial\tau}_{(y=b)} \quad \text{while} \quad [u]_{\sigma=0}^{\sigma=1} = u(b) - u(0).$$

We shall show that

$$(4.6) \qquad -\frac{1}{2\pi}\int_s F' ds = -\frac{1}{2\pi}\int_s \frac{\partial F(\tau, \lambda^i)}{\partial\tau}_{(\tau=0)} ds = 1.$$

By means of this we find the following *solution formula*

$$(4.7) \qquad u(0) = \frac{1}{4\pi}\int_{C_s} \sigma\Delta u_{(\sigma b)} ds db - \frac{1}{\pi}\int_s \left(\frac{F'}{2} u + \frac{\partial u}{\partial\tau}\right)_{(b)} ds.$$

Here $\Delta u = f$ is supposed to be known in space-time (or a suitable part of it) and in particular in the points σb, $0 \leqq \sigma \leqq 1$, of C, u and a non-tangential derivative du/dn (the Cauchy data) are supposed to be known on S and in particular in the infinitesimal neighborhood of s. From u and such a derivative, $\partial u/\partial \tau$ can be computed.

The first term on the right-hand side is equal to the well known retarded potential. In the case $\Delta u = 0$, this term vanishes.

Now we have to verify the relation (4.6). I should like to observe that in my original proof, based on fractional integration and analytic continuation, this relation follows automatically, expressing only that $I^0 1 = 1$. Here we verify it in the following way.

Let a be a fixed negative time-like unit vector, that is $(a, a) = 1$ and $a^0 < 0$. Set

$$\mu = (a, y), \qquad \nu = \log \mu = \log (a, y). \tag{4.8}$$

Since a is negative time-like, $(a, y) > 0$ inside and on the surface of the cone (except at the vertex), hence $\log (a, y)$ is regular in these points. One readily finds

$$\Delta \nu = -\frac{1}{\mu^2}, \quad \frac{\partial \nu}{\partial \sigma} = \frac{1}{\sigma}, \quad \frac{\partial^2 \nu}{\partial \sigma \partial \tau} = 0. \tag{4.9}$$

The first formula follows immediately if the time axis is chosen collinear with a, or by writing $\Delta \nu = \operatorname{div} \operatorname{grad} \nu$. The other two formulae follow from (3.3) and (4.8), which give

$$\mu = (a, y) = \sigma(a, b+\tau c) \text{ and } \nu = \log \mu = \log \sigma + \log (a, b+\tau c).$$

Now, by means of (4.2) we find on the cone

$$-\mu^{-2} = \frac{2}{\sigma^2} F' + \Delta_{(\lambda)} \nu$$

and, in particular, on s, where $\sigma = 1$,

$$-\mu^{-2} = 2F' + \Delta_{(\lambda)} \nu.$$

Thus, in view of (3.22),

$$-\frac{1}{2\pi} \int_s F' \, ds = \frac{1}{4\pi} \int_s \mu^{-2} ds. \tag{4.10}$$

We will prove in Section 6 that the last integral is independent of s, which means that it has identically the same value for the intersection of the cone C with any space-like surface S. If we compute it for the three-plane $\mu = \text{const.}$, we obtain

$$\frac{1}{4\pi} \int_s \mu^{-2} ds = \frac{\mu^{-2}}{4\pi} \int_s ds = \frac{1}{4\pi} \mu^{-2} 4\pi \mu^2 = 1,$$

from which the relation (4.6) follows. Except for the fact that the integral (4.10) is independent of s, the proof of the solution formula (4.7) is now complete.

5. The Cauchy Problem for $m = 2h > 4$

The success of our method in the preceding particular case depended largely on the fact that the relevant terms on the right-hand side of (4.2), when multiplied by σ, are derivatives with respect to σ. Thus we could explicitly carry out the integration in σ by cancelling the corresponding sign of differentiation.

In the case $m > 4$ the situation is less favorable. One of the bad terms on the right-hand side of (3.21) not containing any differentiation with respect to σ vanishes on the cone owing to the factor τ, but the term containing the factor $m-4$, which vanishes in the case $m = 4$, remains. This difficulty will be removed by means of a suitable number of differentiations with respect to τ.

So let us suppose $m = 2h > 4$. By an easy application of Leibniz' formula we obtain from (3.21)

$$(5.1)\qquad \begin{aligned} \frac{\partial^p}{\partial\tau^p}(F\Delta u) &= \frac{2}{\sigma}\frac{\partial^p}{\partial\tau^p}\left(2F\frac{\partial^2 u}{\partial\sigma\partial\tau} + F'\frac{\partial u}{\partial\sigma} + \frac{m-4-2p}{\sigma}F\frac{\partial u}{\partial\tau}\right) \\ &\quad - \frac{4\tau}{\sigma^2}\frac{\partial^{p+1}}{\partial\tau^{p+1}}\left(F\frac{\partial u}{\partial\tau}\right) + \frac{\partial^p}{\partial\tau^p}(F\Delta_{(\lambda)}u). \end{aligned}$$

In order to save space and to make some involved formulae more legible, we use in the sequel the following abbreviations:

$$(5.2)\qquad \begin{aligned} &\frac{\partial}{\partial\tau} = \partial_\tau, \quad \frac{\partial^p}{\partial\tau^p} = \partial_\tau^p, \quad \frac{\partial}{\partial\sigma} = \partial_\sigma, \quad \frac{\partial^2}{\partial\sigma\partial\tau} = \partial^2_{\sigma\tau}, \\ &\frac{\partial}{\partial\lambda^i} = \partial_i, \quad \frac{\partial^2}{\partial\lambda^i\partial\lambda^k} = \partial^2_{ik}. \end{aligned}$$

In this notation formula (5.1) becomes

$$(5.3)\qquad \begin{aligned} \partial_\tau^p(F\Delta u) &= \frac{2}{\sigma}\partial_\tau^p\left(2F\partial^2_{\sigma\tau}u + F'\partial_\sigma u + \frac{m-4-2p}{\sigma}F\partial_\tau u\right) \\ &\quad - \frac{4\tau}{\sigma^2}\partial_\tau^{p+1}(F\partial_\tau u) + \partial_\tau^p(F\Delta_{(\lambda)}u). \end{aligned}$$

In particular, for $p = h-2 = (m-4)/2$,

$$(5.4)\quad \begin{aligned}\partial_\tau^{h-2}(F\Delta u) &= \frac{2}{\sigma}\,\partial_\tau^{h-2}(2F\partial_{\sigma\tau}^2 u + F'\,\partial_\sigma u)\\ &\quad - \frac{4\tau}{\sigma^2}\,\partial_\tau^{h-1}(F\partial_\tau u) + \partial_\tau^{h-2}(F\Delta_{(\lambda)} u).\end{aligned}$$

On the cone this reduces to

$$\partial_\tau^{h-2}(F\Delta u)_{(\tau=0)} = \left\{\frac{2}{\sigma}\,\partial_\tau^{h-2}(2F\partial_{\sigma\tau}^2 u + F'\,\partial_\sigma u) + \partial_\tau^{h-2}(F\Delta_{(\lambda)} u)\right\}_{(\tau=0)}.$$

We now recall formula (3.24), according to which

$$\int_s \partial_\tau^{h-2}(F\Delta_{(\lambda)} u)ds = 0.$$

We also point out that all the derivatives $\partial_\tau^i u$ vanish at the origin, since formula (4.4) is valid in any dimension.

Moreover if we can show that

$$(5.5)\qquad \int_s (\partial_\tau^{h-1}F)_{(\tau=0)}\,ds = 2(-\pi)^{h-1},$$

the arguments used in order to establish formula (4.7) will give in the present case the *solution*

$$(5.6)\quad \begin{aligned}u(0) &= -\tfrac{1}{4}(-\pi)^{1-h}\int_{C_s}\partial_\tau^{h-2}(F\Delta u)_{(\sigma b)}\sigma d\sigma ds\\ &\quad +(-\pi)^{1-h}\int_s \partial_\tau^{h-2}\left(\frac{F'}{2}\,u + F\partial_\tau u\right)_{(b)} ds.\end{aligned}$$

From the known Cauchy data and the known Δu all the needed derivatives $\partial_\tau^p u$ on s can be computed.

We emphasize that the *differentiations* in (5.6) are all carried out *under the integral sign.*

In order to verify (5.5), we again consider a fixed time-like negative unit vector a and the functions μ and ν given in (4.8). We shall show that the left-hand side of (5.5) is, except for a certain numerical factor, equal to

$$\int_s \mu^{2-m}\,ds,$$

which, as will be proved in Section 6, is independent of s.

Using formula (5.4) with $u = \nu$ and formula (4.9) we find

$$(5.7)\qquad -\{\partial_\tau^{h-2}(F\mu^{-2})\}_{(\tau=0)} = \frac{2}{\sigma^2}\,(\partial_\tau^{h-1}F)_{(\tau=0)} + \{\partial_\tau^{h-2}(F\Delta_{(\lambda)}\nu)\}_{(\tau=0)}.$$

This formula is in a certain sense a degenerate case of a general recursion formula, which we now will deduce. We apply formula (5.3) to a function $u(y)$ which is homogeneous of some integer order r. Considered as a function of σ, τ and the λ^i, u will be homogeneous of order r in σ. Consequently $\partial_\sigma u = ru/\sigma$. Thus the first parenthesis on the right-hand side of (5.3) becomes equal to

$$\frac{m-4-2p+2r}{\sigma} F \cdot \partial_\tau u + \frac{r}{\sigma} \partial_\tau F \cdot u.$$

If furthermore $m-4-2p+2r = r$, or

$$r = 4-m+2p, \tag{5.8}$$

the last expression can be written $r\sigma^{-1}\partial_\tau(Fu)$, and the first part of the right-hand side of (5.3) reduces to $2r\sigma^{-2}\partial_\tau^{p+1}(Fu)$. If we set $u = \mu^r$, where $\mu = (a, y)$ as above, the last expression becomes $2r\sigma^{-2}\partial_\tau^{p+1}(F\mu^r)$. Since moreover $\Delta\mu^r = r(r-1)\mu^{r-2}$, the formula (5.3) becomes for $u = \mu^r$

$$\begin{aligned} r(r-1)\partial_\tau^p(F\mu^{r-2}) = {} & 2r\sigma^{-2}\partial_\tau^{p+1}(F\mu^r) \\ & -4\tau\sigma^{-2}\partial_\tau^{p+1}(F\partial_\tau\mu^r)+\partial_\tau^p(F\Delta_{(\lambda)}\mu^r), \end{aligned} \tag{5.9}$$

r and p being connected by (5.8). On the cone, that is for $\tau = 0$, this reduces to

$$r(r-1)\partial_\tau^p(F\mu^{r-2})_{(\tau=0)} = 2r\sigma^{-2}\partial_\tau^{p+1}(F\mu^r)_{(\tau=0)}+\partial_\tau^p(F\Delta_{(\lambda)}\mu^r)_{(\tau=0)}. \tag{5.10}$$

If we observe that $r+2 = 2(p+1)+4-m$, we find that $r+2$ and $p+1$ are connected in the same way as r and p and thus (5.9) and (5.10) are recursion formulae.

The last term of the right-hand side of (5.10) for any connected couple p and r, as well as any linear combination with constant coefficients of such terms will be denoted by (λ). Then formula (5.10), applied successively to the couples $(p; r) = (0; 4-m), (1; 6-m), \cdots, (h-3; -2,)$ and the relation $F(0, \lambda^i) = 1$ give, after some simplifications,

$$\partial_\tau^{h-2}(F\mu^{-2})_{(\tau=0)} = (-1)^{h-2}\frac{1\cdot 3\cdots(m-3)\sigma^{m-4}}{2^{h-2}}\mu^{2-m}{}_{(\tau=0)}+(\lambda),$$

and finally, according to (5.7), which is a substitute for the trivial relation $0 = 0$ into which (5.10) degenerates for $p = h-2$ and $r = 0$, we obtain

$$\partial_\tau^{h-1}F_{(\tau=0)} = (-1)^{h-1}\frac{1\cdot 3\cdots(m-3)}{2^{h-1}}\sigma^{m-2}\mu^{2-m}{}_{(\tau=0)}+(\lambda). \tag{5.11}$$

We shall use (5.11) on s, where $\sigma = 1$. We note that

$$\frac{1\cdot 3\cdots(m-3)}{2^{h-1}} = \frac{\Gamma(\frac{1}{2}(m-1))}{\Gamma(\frac{1}{2})} = \frac{\Gamma(\frac{1}{2}(m-1))}{\pi^{1/2}}$$

and that, in view of (3.24), $\int_s (\lambda) ds = 0$. We thus obtain

$$\int_s \partial_\tau^{h-1} F_{(\tau=0)}\, ds = (-1)^{h-1} \frac{\Gamma(\frac{1}{2}(m-1))}{\pi^{1/2}} \int_s \mu^{2-m}{}_{(\tau=0)}\, ds. \tag{5.12}$$

As already mentioned, the last integral is independent of s (see the following section). This being granted for a moment, we can choose s as the intersection of an $(m-1)$-plane $\mu =$ const. with the cone C, in which case the integral has the value

$$\mu^{2-m} \int ds = \mu^{2-m} \mu^{m-2} |S_{m-2}| = |S_{m-2}|,$$

where S_{m-2} is the $(m-2)$-dimensional unit sphere and $|S_{m-2}|$ its surface area. In explicit form we have

$$|S_{m-2}| = \frac{2\pi^{(m-1)/2}}{\Gamma(\frac{1}{2}(m-1))}. \tag{5.13}$$

The formulae (5.12) and (5.13) give the desired formula (5.5), from which the final formula (5.6) follows.

6. An Invariant Integral

It remains to prove that the integral

$$\int_s \mu^{2-m} ds \tag{6.1}$$

figuring in (5.12) and, for $m = 4$, in (4.10) is independent of s.

We consider a family of manifolds of dimension $p \leq m-2$ lying on the surface of the cone C. We suppose that each element of this family is intersected in at most one point by any generator of C and that all elements of the family meet exactly the same generators. Clearly the generators establish a one-to-one correspondence between the various elements of the family consisting in a *central projection* with the origin as center. We shall first prove the following facts.

(I) The central projection maps the manifolds conformally onto each other.

(II) Let a again be a fixed negative time-like unit vector, z a point of one of the manifolds and dz an infinitesimal displacement of z along the manifold. Then the expression

$$(a, z)^{-1}|dz| = \mu^{-1}|dz| \tag{6.2}$$

is invariant with respect to the central projection.

The vector $(a, z)^{-1} z$ is identically the same for all vectors z lying along a fixed generator. Consider now a curve $z = z(\tau)$ in one of the manifolds, set

$$\frac{dz}{dt} = \dot{z}(t) = \dot{z}, \qquad \frac{d\mu(t)}{dt} = \dot{\mu}(t) = \dot{\mu}.$$

Then $(\mu(t))^{-1} z(t)$ is invariant under central projection, hence its derivative with respect to t

$$\frac{d}{dt}(\mu^{-1} z) = \mu^{-1}\dot{z} - \mu^{-2}\dot{\mu}z \tag{6.3}$$

is also invariant, and the same is true for the scalar square of this vector. Since z is a null vector, we have

$$(z, z) = 0 \Rightarrow (z, \dot{z}) = 0. \tag{6.4}$$

By means of these relations the scalar square reduces to $\mu^{-2}(\dot{z}, \dot{z})$. This proves our statement (II).

To prove (I), let $u = u(\alpha)$, $v = v(\beta)$ be two curves in a manifold intersecting in a point $p = u(\alpha_0) = v(\beta_0)$. The angle ω between the tangent vectors at p is given by

$$\cos \omega = -\frac{(\dot{u}(\alpha_0), \dot{v})(\beta_0))}{\sqrt{(\dot{u}(\alpha_0), \dot{u}(\alpha_0)) \cdot (\dot{v}(\beta_0), \dot{v}(\beta_0))}}.$$

(The minus sign is inessential, it depends on the fact that formula (L) in Section 2 attributes a negative scalar square to space-like vectors.)

Set $(a, p) = \mu$. We already know by (II) that $\mu^{-2}(\dot{u}(\alpha_0), \dot{u}(\alpha_0))$ and $\mu^{-2}(\dot{v}(\beta_0), \dot{v}(\beta_0))$ are invariant under central projection. If we can prove that also $\mu^{-2}(\dot{u}(\alpha_0), \dot{v}(\beta_0))$ is invariant, the invariance of $\cos \omega$ follows. Now, according to (6.3), the vectors

$$\mu^{-1}\dot{u}(\alpha_0) - \mu^{-2}(a, \dot{u}(\alpha_0))u(\alpha_0) \quad \text{and} \quad \mu^{-1}\dot{v}(\beta_0) - \mu^{-2}(a, \dot{v}(\beta_0))v(\beta_0)$$

are invariant, so their scalar product is invariant also. But

$$u(\alpha_0) = v(\beta_0) = p, \quad (p, p) = 0, \quad (p, \dot{u}(\alpha_0)) = 0, \quad (p, \dot{v}(\beta_0)) = 0,$$

as is seen from (6.4). By means of this, the invariant scalar product reduces to $\mu^{-2}(\dot{u}(\alpha_0), \dot{v}(\beta_0))$. This completes the proof of (I).

The volume element dH of our p dimensional manifold H can be written as

$$dH = |dz_1| \cdots |dz_p|,$$

where the infinitesimal displacements are taken along p families of curves forming an orthogonal net. The combination of (I) and (II) then yields:

(III) The product $\mu^{-p} dH$ is invariant under central projection.

The invariance of the integral (6.1) is a corollary of (III) and with this the proof of the main results (4.7) and (5.6) is now complete.

7. Geometric Interpretation

We now pass to a geometric interpretation of the function $F(\tau, \lambda^j)$ figuring in our solution formulae (4.7) and (5.6).

We first prove that F is a polynomial of degree $\leqq m-2$ in τ. In view of (3.20), this is the same as to say that $\sqrt{\bar{\gamma}}$, where the determinant $\bar{\gamma} = |\bar{\gamma}_{ik}|$ is given by (3.9), is such a polynomial. The $\bar{\gamma}_{ik}$ and $\bar{\gamma}$ depend on τ and the λ^j but not on σ. We will show that $\bar{\gamma}$ can be written as the square of a determinant which is a polynomial of degree $\leqq m-2$ in τ. We recall the abbreviated notation (5.2) and calculate the product of two closely related determinants

$$
\begin{aligned}
&\bar{D}(\tau, \lambda^j) \cdot \underline{D}(\tau, \lambda^j) \\
&\quad = |b^k c^k \partial_i(b^k + \tau c^k)| \cdot |b_k c_k \partial_i(b_k + \tau c_k)|, \qquad i = 1, \cdots, m-2,
\end{aligned} \tag{7.1}
$$

where b^k and b_k, respectively, denote columns whose elements are the contravariant and the covariant components of the vector b and similarly for the other columns. According to (3.1) and (3.2), we have

$$(b, b) = (c, c) = (b, \partial_i(b+\tau c)) = (c, \partial_i(b+\tau c)) = 0, \quad (b, c) = \tfrac{1}{2}.$$

Hence, according to (3.9), the above product is equal to

$$
\begin{vmatrix} (b, b) & (b, c) \\ (c, b) & (c, c) \end{vmatrix} \cdot |\bar{\gamma}_{ik}(\tau, \lambda^j)| = -\tfrac{1}{4}\bar{\gamma}(\tau, \lambda^j). \tag{7.2}
$$

Now, for any vector y we have $y_0 = y^0$, $y_k = -y^k$, $k = 1, \cdots, m-1$. Hence

$$\bar{D}(\tau, \lambda^j) \cdot \underline{D}(\tau, \lambda^j) = (-1)^{m-1}(\bar{D}(\tau, \lambda^j))^2.$$

Since m is even, we obtain

$$\bar{\gamma}(\tau, \lambda^j) = 4(\bar{D}(\tau, \lambda^j))^2. \tag{7.3}$$

This shows that $\bar{\gamma} \geqq 0$. However, for $\tau = 0$ we have

$$\bar{\gamma}(0, \lambda^j) = |(\partial_i b, \partial_k b)|. \tag{7.4}$$

This determinant is the discriminant of the definite metric form of the space-like manifold s, thus it is $\neq 0$. Hence, for sufficiently small values of τ,

(7.5) $$\bar{\gamma}(\tau, \lambda^j) > 0.$$

This being the case, we obtain from (3.20) and (7.3)

(7.6) $$F(\tau, \lambda^j) = \frac{\bar{D}(\tau, \lambda^j)}{\bar{D}(0, \lambda^j)}.$$

This shows that $F(\tau, \lambda^j)$ is a polynomial of degree $\leqq m-2$ in τ and that $F(0, \lambda^j) = 1$.

The degree is $m-2$, in general. In order that the coefficient of τ^{m-2} in $F(\tau, \lambda^j)$ or, equivalently, in $\bar{D}(\tau, \lambda^j)$ shall vanish, it is necessary and sufficient that there be a linear relation between b, c and the $\partial c/\partial \lambda^k$. By means of (3.1) and (3.2) this relation reduces to a relation between the $\partial c/\partial \lambda^k$ alone. Such exceptional cases can actually occur. However, for convenience, we suppose in the following treatment that $F(\tau, \lambda^j)$ is of degree $m-2$ in τ, and consequently has $m-2$ finite zeros (distinct or not). In Section 8 we give some indications concerning the exceptional case.

In the sequel we mostly drop the parameters λ^j and write $\bar{\gamma}(\tau)$, $\bar{D}(\tau)$, $\underline{D}(\tau)$, $F(\tau)$ etc. If the zeros of $F(\tau)$ are denoted by τ_A, $A = 1, \cdots, m-2$, then we can write

(7.7) $$F(\tau) = \prod_{A=1}^{m-2} \left(1 - \frac{\tau}{\tau_A}\right).$$

We will show that all the zeros are real. Since $\bar{D}(0)\underline{D}(0) \neq 0$, $F(\tau)$ has, according to (7.6), the same zeros as $F(\tau)\bar{D}(0)\underline{D}(0) = \bar{D}(\tau)\underline{D}(0)$. By the same kind of computation as used above, we find

(7.8) $$\bar{D}(\tau)\underline{D}(0) = -\tfrac{1}{4}|\mu_{ik}+\tau\kappa_{ik}|,$$

where (cf. (7.4))

(7.9) $$\mu_{ik} = \bar{\gamma}_{ik}(0) = (\partial_i b, \partial_k b), \qquad \kappa_{ik} = (\partial_i b, \partial_k c).$$

We already know that the matrix $(\bar{\gamma}_{ik}(0))$, which obviously is symmetric, is also definite. We now show that the matrix (κ_{ik}) too is symmetric. We know from (3.2) that $(\partial_i b, c) = 0$. Differentiating this with respect to λ^k we find

(7.10) $$\kappa_{ik} = (\partial_i b, \partial_k c) = -(\partial^2_{ik} b, c).$$

Hence $\kappa_{ik} = \kappa_{ki}$. It is well known (see Courant-Hilbert I, English edition (1953), pp. 37—41) that under these conditions the zeros τ_A of the determinant $|\mu_{ik}+\tau\kappa_{ik}|$ are real and that to every zero τ_A of multiplicity p there corresponds a p-dimensional linear variety of solutions $\{\omega^k_A\}$ of the system of homogeneous equations

(7.11) $$\sum_{k=1}^{m-2} (\mu_{ik}+\tau_A \kappa_{ik})\omega_A^k = 0, \qquad i = 1, \cdots, m-2.$$

The total number of linearly independent solutions for all the τ_A is $m-2$.

The last equations can also be written as

(7.12) $$(\partial_i b, \sum_{k=1}^{m-2} (\partial_k b+\tau_A \partial_k c)\omega_A^k) = 0.$$

The vectors $\sum_{k=1}^{m-2} (\partial_k b+\tau_A \partial_k c)\omega_A^k$ are orthogonal to b and c according to formula (3.2) and to the $m-2$ vectors $\partial_i b$ according to equation (7.12). Thus they are orthogonal to m linearly independent vectors and hence they vanish, i.e.

(7.13) $$\sum_{k=1}^{m-2} (\partial_k b+\tau_A \partial_k c)\omega_A^k = 0.$$

The converse is obvious, (7.13) implies (7.12). Hence the systems (7.12) and (7.13) are equivalent.

Choosing infinitesimal displacements $d\lambda_A^k$ proportional to the ω_A^k, we obtain

(7.14) $$\sum_{k=1}^{m-2} (\partial_k b+\tau_A \partial_k c)d\lambda_A^k = 0,$$

or

(7.15) $$d_A b+\tau_A d_A c = 0,$$

a formula which has the same aspect as the formula of Olinde Rodrigues (cf. for example Struik, Lectures on Classical Differential Geometry, 1950, p. 94). The latter formula occurs in the theory of curvature of surfaces in three-dimensional Euclidean space and is valid along the lines of curvature of a surface.

Guided by this analogy, we will call a curve L_A traced on the manifold of intersection s and given by the parametric representation $\lambda_A^k = \lambda_A^k(t)$, along which the equation (7.15) or the equation (7.14) is satisfied, a *line of curvature* of the manifold s. Such a curve is obtained by integrating the system of differential equations $d\lambda_A^k/dt = \omega_A^k$.

We should like to observe that the following considerations about lines of curvature and related items are of entirely local character.

In the ordinary theory of curvature the lines of curvature can be characterized by the property that the normals of the surface along such a line generate a developable surface. We will show that in the present case the curves L defined above can be characterized in an analogous way, namely by the property that *the generators of the characteristic surface through the points of such a line constitute a two-dimensional developable surface.*

To this end we have to show that the two-dimensional ruled surface just defined has a constant (two-dimensional) tangent plane along every generator. For brevity, we drop in the following discussion the subscript A. An arbitrary point z on the generator g through the point b of s can be written $z = b+\mu c$. Thus, according to (7.15),

$$dz = db+d\mu \cdot c+\mu \cdot dc = \left(1 - \frac{\mu}{\tau}\right) db+d\mu \cdot c.$$

This shows that at any point z the tangent plane is spanned by the same vectors c and db, consequently it is constant.

Conversely, consider a two-dimensional ruled surface formed by those generators of Σ which pass through the points of a curve L^1 traced on s and suppose that this surface has a constant tangent plane along every generator. At the point b the tangent plane is spanned by the vectors c and db, thus it must be spanned by the same vectors at any point $b+\mu c$ of this generator. This means that $dz = db+d\mu \cdot c+\mu \cdot dc$ must be a linear combination of the vectors c and db. But then $\mu \cdot dc$ is also a linear combination of the same vectors, $\mu \cdot dc = \varepsilon c+\nu \cdot db$, say ($\varepsilon$ is infinitesimal). Taking the scalar product with b we find that, in view of (3.1) and (3.2), the factor ε equals 0. Thus the vectors db and dc are collinear, and we have an equation $db+\tau^1 dc=0$ analogous to (7.15). From here follows, in view of the equivalence of the equations (7.12) and (7.13), that τ^1 satisfies the equation $|\mu_{ik}+\tau^1\kappa_{ik}| = 0$ and that consequently L^1 is a line of curvature, as defined above.

Two displacements $d_A b$ and $d_B b$ corresponding to two different zeros τ_A and τ_B are *orthogonal* to each other. Indeed, we have the relations

$$d_A b+\tau_A d_A c = 0, \qquad d_B b+\tau_B d_B c = 0. \tag{7.16}$$

We observe that in view of the symmetry of κ_{ik} established by the formula (7.10), we have $(d_A b, d_B c) = (d_B b, d_A c)$. By means of this we obtain from (7.16) that

$$(\tau_B-\tau_A)(d_A b, d_B b) = 0,$$

hence also

$$(d_A b, d_B b) = 0,$$

since $\tau_B-\tau_A \neq 0$.

In an analogous way we find that under the same condition

$$(d_A b, d_B c) = (d_B b, d_A c) = 0.$$

Inspired by the usual terminology of the theory of surfaces, we may say that the directions $d_A b$ and $d_B b$—which, as we just have seen, are orthogonal—are also *conjugate*.

As a consequence of what precedes we note that through a point b at which all the τ_A are different from each other there pass $m-2$ orthogonal and conjugate lines of curvature. As in the usual theory of curvature, no generally valid simple statement can be made about the behavior of the lines of curvature at or around an *umbilical point*, i.e. a point at which the τ_A entirely or partly coincide.

Consider now a point b of s, the generator g of Σ which passes through b and the point

$$f_A = b+\tau_A c \tag{7.17}$$

lying on this generator. If b describes a line of curvature L_A corresponding to the zero τ_A, f_A will describe a curve Γ_A, which might degenerate. We show that in the non-degenerate case the generator g is tangent to Γ_A at the point f_A. Indeed,

$$df_A = db+\tau_A dc+d\tau_A \cdot c = d\tau_A \cdot c,$$

according to (7.15). Thus the curve Γ_A is the *edge of regression* of the two-dimensional developable surface through L_A constructed in the preceding section.

The points f_A distinct or not lying on the same generator g are called *focal points* relative to g. If g varies, the points f_A describe the *caustic* or the *focal surface* of the characteristic surface Σ. In a case which is not degenerate in any respect (the exact formulation is left to the reader) the caustic is a variety of $m-2$ dimensions, which consists of $m-2$ sheets. There are also important degenerate cases of which we shall give some examples at the end of this paper.

8. Exceptional Cases and Various Remarks

This section contains some indications concerning the exceptional case when the degree of the polynomial $F(\tau)$ is $< m-2$. We start by some general remarks of various kinds.

Except at the beginning of Section 3, the vertex of our negative characteristic cone was fixed at the origin. At the place just quoted we have defined for an arbitrary point x the negative cone C^x, its intersection s^x with the given surface S, the characteristic surface Σ^x, which all depend on x. So do also the vectors corresponding to the vectors b and c used above.

If the surface S, instead of being space-like, is a characteristic surface itself, then the characteristic surface Σ^x passing through the intersection s^x of C^x and S will necessarily coincide with S, since, by Section 2, there are exactly two characteristic surfaces passing through s^x, one of which is C^x.

Thus, in this case, $\Sigma^x = S$ is independent of x. From the converse part of our digression about two-dimensional developable surfaces in Section 7, it is apparent that these surfaces are also independent of x. Hence so is also the caustic of S constituted by the edges of regression of these surfaces.

As a matter of fact, the arguments used at the place just quoted show that, for a characteristic surface S given *a priori*, the caustic can be defined as the $(m-2)$-sheeted, $(m-2)$-dimensional surface constituted by the edges of regression of *all* two-dimensional developable surfaces formed by the generators of S. If we again denote the intersection of S with a characteristic cone C^x by s^x, the connection with the wording in Section 7 can be reestablished.

Incidentally I would like to mention that the Cauchy problem posed for characteristic boundaries is treated in my paper: *Problems related to characteristic surfaces*, Proceedings of the Conference on Differential Equations, held at the University of Maryland, March 17, 18, 19, 1955.

Now we turn to the exceptional case mentioned at the beginning of this section. The caustic of a characteristic $(m-1)$-plane P obviously lies at infinity in the light-like direction within P. Consider the intersection s of P with the characteristic cone C whose vertex lies at the origin and the corresponding vectors b and c (cf. Section 3). The vectors c clearly lie in the light-like direction within P, and it is easy to prove that $c = \text{const.}$. Consequently, by virtue of the formulae (7.1) and (7.6), $F(\tau) = \text{const.}\ (= 1)$. Thus, in all points of s, $F(\tau)$ exhibits the most exceptional behavior possible. Clearly, the situation is unchanged if O, C, s are replaced by x, C^x, s^x. It is also easy to construct characteristic surfaces of more general shape which exhibit the exceptional case. By means of these facts I am able to construct some rather artificial examples of space-like surfaces, for which the exceptional case occurs in a certain part of some s^x. I refrain from giving the details.

Let now the polynomial $F(\tau)$ be of degree $m-2-p$ and thus possess $m-2-p$ zeros instead of $m-2$. We show that several of the previous results remain valid if the lacking zeros are assumed to lie at infinity.

1) The product formula (7.7) remains valid.

2) The equation (7.15) has to be replaced by the equation $dc = 0$. More precisely, if the degree of $F(\tau)$ is exactly $m-2-p$, there exists a p-dimensional linear variety of infinitesimal displacements $db = \sum \partial_k b \cdot d\lambda^k$ such that the corresponding $dc = \sum \partial_k c \cdot \partial\lambda^k = 0$ (cf. again Courant-Hilbert I, loc. cit.).

3) Any displacement db of the above type is orthogonal and conjugate to any displacement $d_A b$ satisfying an equation of the type $d_A b + \tau_A d_A c = 0$.

The conjugate character follows from $(db, d_A c) = (d_A b, dc) = 0$, since $dc = 0$. This symmetry relation follows from the symmetry of the κ_{ik}, established in (7.10). The orthogonality follows from

$$0 = (db, d_A b) + \tau_A (db, d_A c) = (db, d_A b).$$

4) Our statements in Section 7 about two-dimensional developable surfaces remain valid in a rather trivial way in the exceptional case.

9. Geometric Form of the Solution

In this section we rewrite the solution formulae (4.7) and (5.6) in geometric form, but, for the sake of simplicity, only for the homogeneous wave equation, that is for $\Delta u = 0$. To this end we first complete the results of Section 7 by the fact that on the generators g our main parameter τ has a clearcut geometric meaning. Indeed, let $z = b + \tau c$ be an arbitrary point of the generator g through the point b. Then, by virtue of (3.1), we find $(z, z) = \tau$. Thus τ is *the square of the Lorentz distance* between the origin (the vertex) and the point z on the generator g. We also set $(z, z) = R_z = R$ and thus have $\tau = R_z = R$. In particular, for the focal points f_A on g we have $\tau_A = R_{f_A} = R_A$. We replace $F(\tau)$ on the generators by $F(R)$ and thus obtain

$$F(R) = \prod_{A=1}^{m-2} \left(1 - \frac{R}{R_A}\right) = \sum_{k=0}^{m-2} (-1)^k C_k R^k, \tag{9.1}$$

where the C_k are the elementary symmetric functions of the $1/R_A$. The differentiation $\partial/\partial\tau$ along the generator g can be written as

$$\frac{\partial}{\partial \tau} = \frac{d_g}{d_g R}, \tag{9.2}$$

which means differentiation with respect to R in the direction of g. With these conventions, formula (5.6) for *a solution u of the equation $\Delta u = 0$* becomes

$$u(0) = (-\pi)^{1-h} \int_s \frac{d_g^{h-2}}{d_g R^{h-2}} \left(\frac{1}{2} \frac{d_g F}{d_g R} u + F \frac{d_g u}{d_g R}\right)_{(R=0)} ds, \tag{9.3}$$

where, as before, $h = m/2$.

In the case $m = 4$ we have $m-2 = 2$ and thus the subscript A only takes the values 1 and 2. Hence

$$F(R) = \left(1 - \frac{R}{R_1}\right)\left(1 - \frac{R}{R_2}\right)$$

and

$$\frac{d_g F}{d_g R}_{(R=0)} = -\left(\frac{1}{R_1} + \frac{1}{R_2}\right), \quad F(0) = 1.$$

Moreover $h-2 = 0$. Hence we have for a *solution u of the equation* $\Delta u = 0$ *in the case* $m = 4$

$$u(0) = \frac{1}{\pi}\int_s \left\{\frac{1}{2}\left(\frac{1}{R_1}+\frac{1}{R_2}\right)u - \frac{d_g u}{d_g R}\right\}_{(R=0)} ds, \tag{9.4}$$

which is identical with the solution (60) given on page 139 of my Acta paper and which there is obtained by the method of fractional integration.

It is easy to rewrite formulae (9.3) and (9.4) for an arbitrary point x by changing $u(0)$ into $u(x)$, s into s^x and setting

$$R = (z-x, z-x), \qquad R_A = (f_A-x, f_A-x).$$

Formulae (9.3) and (9.4) exhibit an accentuated Huygens principle. Indeed, this principle in its original form affirms that the value of a solution u of the equation $\Delta u = 0$ at a point x depends only on the values which u assumes in the infinitesimal neighborhood of s^x, a manifold which belongs to the light-like past of x (i.e. the cone C^x). In our formulae all the differentiations are taken in light-like directions, i.e. only the infinitesimal light-like past of u at the points of s^x enters into the expression of $u(x)$. These directions lie along the generators of the characteristic surface Σ^x— enveloped by the characteristic cones with vertices on s^x.

10. Three Special Cases

In the case of a general initial surface S the characteristic surface Σ^x can be considered as the image of the characteristic cone C^x with respect to the surface S, while the caustic can be interpreted as a *diffuse* image of the point x. In this section we will treat three special cases in which the characteristic surface Σ^x reduces to a positive characteristic cone C^{x^1}. The caustic reduces to the vertex x^1 of this cone, which can be considered as the *distinct* image of the point x. The special cases are those of a *space-like plane*, a *Lorentz hyperboloid*, a *positive characteristic cone*. More precise descriptions will follow in a moment.

To begin with, let P be a space-like $(m-1)$-plane. Further let x^1 be the point symmetric to x (in the Lorentz sense) with respect to P, which means that the segment xx^1 is orthogonal to P and is divided into two equal parts by P.

It follows from this symmetry that the intersection s^x of the cone C^x with the plane P belongs also to the positive cone C^{x^1} whose vertex is x^1. All the generators of C^{x^1} meet in the vertex x^1. Consequently the caustic of $\Sigma^x = C^{x^1}$ consists of the single point x^1.

The case of a plane surface is a limiting case of the one which now follows. Let o be a fixed point, u a variable point and consider the positive sheet H of the hyperboloid $(u-o, u-o) = A = a^2 > 0$. Let the point x be interior to the (convex) domain enclosed by H and consider the negative characteristic cone $(u-x, u-x) = 0$. The intersection s^x of these two surfaces satisfies the equation

$$(u-o, u-o)-(u-x, u-x) = A,$$

which is linear in u and can be written $(u, x-o) = \text{const.}$. Since the vector $x-o$ is time-like, the last equation yields a space-like plane P. The characteristic surface Σ^x is again the cone C^{x^1}, where x^1 is the symmetric image of x with respect to the plane P, and the caustic reduces to the point x^1.

Let finally C^{x^1} be a positive characteristic cone with a *fixed* vertex x^1. Such a cone can also be considered as a limiting case of a hyperboloid of the above type. Take an arbitrary point x interior to C^{x^1} and consider the negative characteristic cone C^x. Both surfaces being characteristic, it is clear that Σ^x (the image of C^x in C^{x^1}) will coincide with C^{x^1} and that the caustic again reduces to the point x^1.

Set $(x-x^1, x-x^1) = R_0$. Then

$$F(R) = \left(1-\frac{R}{R_0}\right)^{m-2}.$$

The formula (9.3), in which the origin is replaced by x, becomes in all three cases

$$u(x) = (-\pi)^{1-h}\int_{s^x}\frac{d_g^{h-2}}{d_gR^{h-2}}\left\{\frac{1-h}{R_0}\left(1-\frac{R}{R_0}\right)^{m-3}u+\left(1-\frac{R}{R_0}\right)^{m-2}\frac{d_gu}{d_gR}\right\}_{(R=0)}ds^x.$$

In particular for $m = 4$ we obtain

$$u(x) = \frac{1}{\pi}\int_{s^x}\left(\frac{u}{R_0}-\frac{d_gu}{d_gR}\right)_{(R=0)}ds^x.$$

Received January, 1960.

59.

The analytical continuation of the Riemann-Liouville integral in the hyperbolic case

Canad. J. Math. *13* (1961), 37–47

Introduction. In 1949 I published in the *Acta Mathematica* (vol. 81) a rather long paper: "L'intégrale de Riemann-Liouville et le problème de Cauchy." This work will be quoted in the sequel as *Acta* paper. Only minor local references to this paper will be made here, and knowledge of it is not required for the reading of the present article. The notations used here are slightly different from those used in my former paper.

In the *Acta* paper I introduce multiple integrals I^α and I^α_* of the Riemann-Liouville type depending on a parameter α and converging for sufficiently large values of α. I give the solution of the Cauchy problem for the wave equation in a unique formula, the same for space-time of odd or even dimensions, implying an analytic continuation with respect to the parameter α. When this analytic continuation is carried out, it leads to final formulae of quite different types for odd or even dimensions, the one relative to even dimensions obeying the Huygens principle.

The main difficulty concerning the analytic continuation was to prove that I^0 is the identity operator. My way of doing this was neither simple nor elegant. The principal aim of the present paper is to give a more satisfactory proof.

I hope that the present approach will be useful in other connections as well. Indeed, this method of analytic continuation has found unexpected applications in other fields. Here I only make reference to results of Gelfand and Grajew.[1]

1. Preliminaries. If the co-ordinates of a point x in m-dimensional space-time or Lorentz-space are denoted by $x^0, x^1, \ldots, x^{m-1}$, the metric form will be

$$(x, x) = (x^0)^2 - (x^1)^2 - \ldots - (x^{m-1})^2 = l_{ik}x^i x^k, \tag{1.1}$$

where the ordinary summation convention is used. The square of the distance of two points x and y is given by

$$R_{xy} = r^2_{xy} = (x - y, x - y) = l_{ik}(x^i - y^i)(x^k - y^k). \tag{1.2}$$

The scalar product (a, b) of two vectors a and b, with the respective components a^k and b^k, is defined by

Received October 19, 1959.

[1]See Appendix III of the book by I. M. Gelfand and M. A. Neumark, *Unitäre Darstellungen der klassischen Gruppen* (Berlin: Akademie-Verlag, 1957), also A.M.S. translations, Series 2, vol. 9, pp. 123–154.

(1.3) $$(a, b) = l_{ik}a^i b^k.$$

Two vectors whose scalar product vanishes are said to be *orthogonal* to each other. In what follows, *orthogonality and normality are always meant in this sense.*

According as the scalar square (a, a) of a vector a is (1) positive, (2) zero, (3) negative, the vector is said to be (1) *time-like*, (2) *light-like* or a *null vector*, (3) *space-like*. A time-like or a light-like vector a is called *positive* or *negative* according as its time component a^0 is positive or negative.

Time-like unit vectors u and *space-like unit vectors* v are defined by the relations $(u, u) = 1$ and $(v, v) = -1$ respectively.

The *light cone* or characteristic cone with vertex a is given by the equation $(x - a, x - a) = 0$. The *positive* and *negative* half-cones correspond to $x^0 - a^0 > 0$ or < 0 respectively. These half-cones will be called *positive and negative light cones* in the sequel.

Consider now a p-dimensional (curved) variety S whose points y are referred to p parameters $\lambda^1, \lambda^2, \ldots, \lambda^p$. The p-dimensional *volume element* dS of S, or alternately *surface element* if $1 < p < m$, can be defined in the following way (cf. *Acta* paper pp. 44–45). Let $ds^2 = (dy, dy) = \sum_{i,k} \gamma_{ik} d\lambda^i d\lambda^k$ be the square of the arc element in S. Form the determinant $\gamma = |\gamma_{ik}|$. Then

(1.4) $$dS = \sqrt{|\gamma|}\, d\lambda^1 d\lambda^2 \ldots d\lambda^p.$$

An $(m-1)$-dimensional surface is said to be space-like if its normal is time-like. Let S be a space-like surface. Suppose that the negative light cone C^x with vertex x and the surface S enclose a bounded domain $D_S{}^x$. We shall consider functions defined in domains including $D_S{}^x$ and make the blanket hypothesis that the functions and all their derivatives with respect to the Cartesian co-ordinates which explicitly or implicitly enter into our computations exist and are continuous. We express this by saying that the functions are *well behaved*. The same phrase will be used in an appropriate sense in connection with the surface S and functions defined on S.

We form the *volume potential*

(1.5) $$I^\alpha f(x) = \frac{1}{H_m(\alpha)} \int_{D_s{}^x} f(y) r_{xy}{}^{\alpha - m} dV,$$

where $dV = dy^0 dy^1 \ldots dy^{m-1}$ is the volume element of m-space and

(1.6) $$H_m(\alpha) = \pi^{\frac{1}{2}(m-2)} 2^{\alpha-1} \Gamma(\tfrac{1}{2}\alpha) \Gamma(\tfrac{1}{2}(\alpha + 2 - m)).$$

The integral in (1.5) converges for $\alpha > m-2$ (cf. *Acta* paper, p. 31), or more generally for $Re\, \alpha > m$-2, if we admit complex values of α. Similarly, our subsequent assertions about convergence of integrals or analytic continuation of I^α or $I^{*\alpha}$ (see below) remain valid for complex α, if we replace all inequalities of the type $\alpha > \alpha_0$ by $Re\, \alpha > Re\, \alpha_0$.

Besides the volume potentials we also consider potentials of a *simple layer* and of a *double layer.*

Let S^x be that part of the surface S which is interior to the cone C^x. Denote further by n the positive unit normal to S and let g and h be two functions defined on S. We write

$$(1.7)\quad I_*^\alpha\overline{f, g, h}(x) = \frac{1}{H_m(\alpha)}\int_{D_{S^x}} f(y) r_{xy}^{\alpha-m} dV + \frac{1}{H_m(\alpha)}\int_{S^x}\left\{g(y) r_{xy}^{\alpha-m} - h(y)\frac{d}{dn} r_{xy}^{\alpha-m}\right\} dS,$$

where dS is the surface element of S (cf. formula (1.4)).

The simple layer converges for $\alpha > m-2$, while the double layer, whose kernel has a stronger singularity, converges only for $\alpha > m$ (cf. *Acta* paper, pp. 48–49, and §4 of the present paper).

We will show that by virtue of our hypotheses about the behaviour of the surface S and the functions f, g, h the integral I_*^α can be continued analytically down to an arbitrary value $\alpha_0 \gtreqless 0$. Moreover, if $\alpha_0 < 0$, then

$$I_*^0\overline{f, g, h}(x) = I^0 f(x) = f(x).$$

For a specification of the derivatives needed for different purposes cf. *Acta* paper, pp. 59–60, 64, 223.

Some simple facts concerning the analytic continuation of the ordinary Riemann-Liouville integral in *one dimension* will be needed in the sequel (cf. *Acta* paper, pp. 14–16).

Set

$$(1.9)\quad J^\alpha f(0) = \frac{1}{\Gamma(\alpha)}\int_0^b f(t) t^{\alpha-1} dt.$$

If $f(t)$ is continuous in the closed interval $[0, b]$, this integral is convergent for $\alpha > 0$. If for $k \leqq n$ the derivatives $f^{(k)}(t)$ exist and are continuous in $[0, b]$, then $J^\alpha f(0)$ *has a holomorphic continuation to all* $\alpha > -n$. *Moreover, if* p *is an integer* $0 \leqq p < n$, *then*

$$(1.10)\quad J^{-p} f(0) = (-1)^p f^{(p)}(0).$$

(As a matter of fact, only the case $p = 0$ will be used explicitly in the sequel.)

To prove this, set

$$P(t) = \sum_{k=0}^{n-1} \frac{f^{(k)}(0)}{k!} t^k.$$

Then we have for $\alpha > 0$, to begin with, and subsequently by analytic continuation, for all $\alpha > -n$

$$(1.11)\quad J^\alpha f(0) = \frac{1}{\Gamma(\alpha)}\int_0^b [f(t) - P(t)] t^{\alpha-1} dt + \frac{1}{\Gamma(\alpha)}\sum_{k=0}^{n-1}\frac{f^{(k)}(0)}{k!(k+\alpha)} b^{k+\alpha}.$$

Indeed, the last integral is convergent and the whole expression (1.11) is holomorphic for $\alpha > -n$. $J^\alpha f(0)$ reduces to $(-1)^p f^{(p)}(0)$ at $\alpha = -p$, since $\Gamma(\alpha)$ has a simple pole with residue $(-1)^p/p!$ at this point.

The following extension will also be needed, and the corresponding result will be quoted as *the extended one-dimensional case.* Its verification is left to the reader.

Let $f(t)$ also depend on α. If $f(t)$ and its derivatives with respect to t up to the order n are continuous in the closed interval $[0, b]$ and moreover are holomorphic in α for $\alpha > -n$, then our above statement and its proof remain valid, except for some slight changes in the notations.

2. A co-ordinate system. We place the origin O at the point x and will eventually refer the domain $D_S{}^O$ to co-ordinates which are to be introduced here. We denote a *fixed* negative time-like unit vector by a and a *variable* space-like unit vector orthogonal to a by v. In a suitable Lorentz frame a and v can be written $a = (-1, 0, \ldots, 0)$ and $v = (0, v^1, \ldots, v^{m-1})$ with $\sum (v^k)^2 = 1$. If the vector v issues from the origin, its endpoint describes the unit sphere S_{m-2} lying in the $(m-1)$-plane orthogonal to a. We write out explicitly that

$$(a, a) = 1,\ (v, v) = -1,\ (a, v) = 0. \tag{2.1}$$

An arbitrary position vector y can be written

$$y = ta + \rho v,\ \rho \geqq 0. \tag{2.2}$$

We always suppose that also $t \geqq 0$. This inequality is obviously satisfied in the domain $D_S{}^O$.

The relation (2.2) can also be written

$$y = \tfrac{1}{2}(t + \rho)(a + v) + \tfrac{1}{2}(t - \rho)(a - v).$$

Furthermore, if we set

$$b = \tfrac{1}{2}(a + v),\ c = \tfrac{1}{2}(a - v), \tag{2.3}$$

then

$$y = (t + \rho)b + (t - \rho)c = (t + \rho)\left(b + \frac{t - \rho}{t + \rho}c\right).$$

Setting now

$$\tau = \frac{t - \rho}{t + \rho},\quad \sigma = t + \rho, \tag{2.4}$$

we obtain

$$y = \sigma(b + \tau c). \tag{2.5}$$

The inverted formulae (2.4) are

$$\rho = \tfrac{1}{2}\sigma(1 - \tau),\ t = \tfrac{1}{2}\sigma(1 + \tau). \tag{2.6}$$

It follows from (2.1) and (2.3) that

$$(b, b) = 0,\ (c, c) = 0,\ (b, c) = \tfrac{1}{2}. \tag{2.7}$$

Hence b and c are (negative) *null vectors.*

The variables τ and σ and the angular variable v which varies on the sphere S_{m-2} and determines the vectors b and c will be our new co-ordinates. Here are the principal merits of τ and σ. The square of the Lorentz distance of a point from the vertex can be expressed and "separated" in τ and σ. The vertex of the cone C^0 is given by the single equation $\sigma = 0$, while the cone apart from the vertex is given by the equation $\tau = 0$. The derivatives $\partial^p f/\partial\tau^p$ of an arbitrary function $f(y)$ vanish at the vertex since they contain the factor σ^p.

We prove these assertions and complete them in certain respects. The square r^2 of the Lorentz distance is according to (2.2) and (2.1)

$$(y, y) = (ta + \rho v, ta + \rho v) = t^2 - \rho^2 = (t + \rho)^2 \frac{t - \rho}{t + \rho}.$$

Hence

$$r^2 = (y, y) = \sigma^2\tau. \tag{2.8}$$

The same relation also follows from (2.5) and (2.7), since (2.7) gives $(b + \tau c, b + \tau c) = \tau$. The equation of the cone C^0 is $(y, y) = 0$. At the vertex $\sigma = 0$, while τ is indeterminate. On the cone, except at the vertex, $\tau = 0, \sigma > 0$. It follows from (2.4) that $0 < \tau \leqq 1$ inside the cone and that, in particular, $\tau = 1$ on the axis $y = \sigma a$ (or $\rho = 0$) and only there.

We always have $\sigma \geqq 0$, according to (2.4) and the inequalities $t \geqq 0$, $\rho \geqq 0$. The equation $\sigma = \text{const.} = \gamma > 0$, which, in view of (2.4), is equivalent to $t + \rho = \gamma$ is the equation of a *positive light cone* $C_{\gamma a}$, with the vertex γa. It is clear that the inequalities $0 \leqq \tau \leqq 1$ and $0 \leqq \sigma \leqq \gamma$ characterize the interior and the boundary of a double cone $D_{\gamma a}{}^0$ limited by the negative light cone C^0 and the positive light cone $C_{\gamma a}$.

From now on we make ample use of our hypothesis that the function $f(y)$ is well behaved (cf. p. 38). We have

$$\frac{\partial y}{\partial t} = \frac{\partial}{\partial \tau}[\sigma(b + \tau c)] = \sigma c, \quad \frac{\partial^p y}{\partial \tau^p} = 0, \quad p = 2, 3, \ldots. \tag{2.9}$$

From this it follows for any function $f(y)$

$$\frac{\partial f}{\partial \tau} = \sigma\Big(\sum c^k \partial_k\Big) f, \quad \text{where } \partial_k = \frac{\partial}{\partial y^k},$$

and, more generally, for any positive integer p,

$$\frac{\partial^p f}{\partial \tau^p} = \sigma^p\Big(\sum c^k \partial_k\Big)^p f. \tag{2.10}$$

This proves our assertion about the behaviour of the derivatives with respect to τ at the vertex.

3. The volume potential. If the surface element of the sphere S_{m-2} is denoted by dS_{m-2}, the volume element dV of the m-space can be written

$dV = \rho^{m-2} d\rho dt dS_{m-2}$. From (2.6) we have that $\rho = \frac{1}{2}\sigma(1-\tau)$ and that the Jacobian

$$\frac{d(\rho, t)}{d(\sigma, \tau)} = \frac{1}{2}\sigma.$$

Making use of (2.5) and (2.8) we obtain after some simplifications

$$\frac{1}{H_m(\alpha)} f(y) r^{\alpha-m} dV \tag{3.1}$$
$$= \frac{2^{1-m}}{H_m(\alpha)} f[\sigma(b + \tau c)] \sigma^{\alpha-1} \tau^{\frac{1}{2}(\alpha-m)} (1-\tau)^{m-2} d\tau d\sigma dS_{m-2}.$$

In order to get $I^\alpha f(O)$, we have to integrate this expression over the domain $D_S{}^O$. However, it will be convenient to divide this domain into two parts and treat these parts separately. First we choose γ small enough, so that the double-cone $D_{\gamma a}{}^O$ should be contained in $D_S{}^O$. Then we divide the latter domain into $D_{\gamma a}{}^O$ and $D_S{}^O - D_{\gamma a}{}^O$ and show by rather different methods that the corresponding parts of the integral I^α, denoted incidentally by $I_I{}^\alpha$ and $I_{II}{}^\alpha$, are holomorphic for $\alpha > -1$ and that $I_I{}^0 = f(O)$, $I_{II}{}^0 = 0$, which gives that the original $I^0 f(O) = f(O)$. By this our *main objective* will be attained. The more difficult part of the proof, the one concerning I_I, will be carried out in the present section. The easy part I_{II} can be treated by a method similar to that used in §4 for a simple layer. Therefore it is postponed to §5. In the same section we apply our results concerning the analytic continuation of a simple and a double layer to carry out the "unlimited" analytic continuation of the volume potentials.

The integral of the right-hand side of (3.1) extended over the double-cone $D_{\gamma a}{}^O$ gives us the functional $I^\alpha f(O)$ relative to this special domain. A very great simplification arises here from the fact that the limits of integration with respect to τ and σ are fixed, τ varying between 0 and 1 and σ between 0 and γ. Thus we have in the present case

$$I^\alpha f(O) = \int_{S_{m-1}} dS_{m-2} \int_0^\gamma d\sigma \int_0^1 \ldots d\tau,$$

where the dots stand for the integrand given in the right-hand side of (3.1)

Besides the formula (1.6) for $H_m(\alpha)$ we shall need the relations

$$\int_0^1 t^{r-1}(1-t)^{s-1} dt = \frac{\Gamma(r)\Gamma(s)}{\Gamma(r+s)}, \tag{3.2}$$

$$\Gamma(r) = \pi^{-\frac{1}{2}} 2^{r-1} \Gamma(\tfrac{1}{2}r) \Gamma(\tfrac{1}{2}r + \tfrac{1}{2}) \tag{3.3}$$

and the explicit expression for the total surface $|S_{m-2}|$ of the sphere S_{m-2},

$$|S_{m-2}| = \frac{2\pi^{\frac{1}{2}m-\frac{1}{2}}}{\Gamma(\frac{1}{2}m - \frac{1}{2})}. \tag{3.4}$$

We develop $f[\sigma(b + \tau c)]$ in a finite power series in τ with a remainder term. We have

(3.5) $$f(y) = f[\sigma(b + \tau c)] = \sum_{p=0}^{N-1} \frac{\tau^p}{p!} \Phi_p(\sigma, v) + R_N(\tau),$$

where $\Phi_p(\sigma, v) = \left\{ \frac{\partial^p}{\partial \tau^p} f[\sigma(b + \tau c)] \right\}_{\tau=0}$, and

(3.6) $$R_N(\tau) = \frac{1}{(N-1)!} \int_0^\tau \frac{\partial^N}{\partial \bar{\tau}^N} f[\sigma(b + \bar{\tau} c)](\tau - \bar{\tau})^{N-1} d\bar{\tau}.$$

N is here a sufficiently large integer, to be specified later. Obviously $R_N(\tau) = O(\tau^N)$.

We first compute the expression

$$A(\alpha) = \frac{2^{1-m}}{H_m(\alpha)} \int_0^1 f[\sigma(b + \tau c)] \tau^{\frac{1}{2}(\alpha - m)} (1 - \tau)^{m-2} d\tau.$$

This integral and all the integrals which follow are convergent for $\alpha > m - 2$. On account of (3.5) we have

(3.7) $$A(\alpha) = \sum_{p=0}^{N-1} A_p(\alpha)\, \Phi_p(\sigma, v) + \frac{2^{1-m}}{H_m(\alpha)} \int_0^1 R_N(\tau) \tau^{\frac{1}{2}(\alpha-m)} (1 - \tau)^{m-2} d\tau,$$

where by means of (3.2) with $r = \frac{1}{2}(\alpha + 2 - m) + p$, $s = m - 1$

(3.8) $$A_p(\alpha) = \frac{2^{1-m}\Gamma(m-1)\Gamma(\frac{1}{2}(\alpha + 2 - m) + p)}{p! H_m(\alpha) \Gamma(\frac{1}{2}(\alpha + m) + p)}.$$

The most important term in (3.7) is $A_0(\alpha)\Phi_0(\sigma, v) = A_0(\alpha) f(\sigma b)$. In view of the expression (1.6) of $H_m(\alpha)$ we find

(3.9) $$A_0(\alpha) = \frac{2^{1-m}\Gamma(m-1)}{\pi^{\frac{1}{2}(m-2)} 2^{\alpha-1} \Gamma(\frac{1}{2}\alpha) \Gamma(\frac{1}{2}(\alpha + m))}.$$

Expressing $2^{\alpha-1}\Gamma(\frac{1}{2}\alpha)$ by means of (3.3), with $r = \alpha$, we find after some simplifications

(3.10) $$A_0(\alpha) = K_0(\alpha) \cdot \frac{1}{\Gamma(\alpha)}, \text{ where } K_0(\alpha) = \frac{2^{1-m}\Gamma(m-1)\Gamma(\frac{1}{2}\alpha + \frac{1}{2})}{\pi^{\frac{1}{2}(m-1)} \Gamma(\frac{1}{2}(\alpha + m))}.$$

Since $\Gamma(\frac{1}{2}) = \pi^{\frac{1}{2}}$, we have

(3.11) $$K_0(0) = \frac{2^{1-m}\Gamma(m-1)}{\pi^{\frac{1}{2}(m-2)} \Gamma(\frac{1}{2}m)}.$$

According to (3.3), with $r = m - 1$, and to (3.4)

(3.12) $$K_0(0) = \frac{\Gamma(\frac{1}{2}m - \frac{1}{2})}{2\pi^{(\frac{1}{2}m - \frac{1}{2})}} = \frac{1}{|S_{m-2}|}.$$

Our next step is to carry out the analytic continuation of the expression

(3.13) $$A_0(\alpha). \int_0^\gamma \Phi_0(\sigma, v) \sigma^{\alpha-1} d\sigma = K_0(\alpha) \cdot \frac{1}{\Gamma(\alpha)} \int_0^\gamma f(\sigma b) \sigma^{\alpha-1} d\sigma.$$

The integral converges for $\alpha > 0$ and, according to what we know about the *one-dimensional case* (cf. p. 39), the analytic continuation of (3.13) is holomorphic for $\alpha > -1$. (For $\alpha = -1$ the function $\Gamma(\frac{1}{2}(\alpha + 1))$ has a pole.) For

$\alpha = 0$ the expression (3.13) becomes $K_0(0)\, f(O) = f(O)/|S_{m-2}|$. The integral of this constant value with respect to the angular variable is clearly $f(O)$. Hence, and this contains virtually our main result, the term corresponding to $p = 0$ in the I^α relative to the double-cone $D_{\gamma a}{}^0$ *yields exactly* $f(O)$ *for* $\alpha = 0$.

The terms in (3.7) with $p > 0$ are easy to handle. In analogy with (3.13) we have to consider the expression

$$A_p(\alpha) \,.\, \int_0^\gamma \Phi_p(\sigma, v)\sigma^{\alpha-1} d\sigma, \quad p \geqq 1, \tag{3.14}$$

where $A_p(\alpha)$ is given in (3.8). We first note that

$$\Gamma(\tfrac{1}{2}(\alpha + 2 - m) + p) = \Gamma(\tfrac{1}{2}(\alpha + 2 - m))P_p(\alpha),$$

where $P_p(\alpha)$ is a polynomial of degree p in α. Hence, after the same simplifications as those performed for $A_0(\alpha)$ we obtain

$$A_p(\alpha) = K_p(\alpha) \,.\, \frac{1}{\Gamma(\alpha)}, \text{ where } K_p(\alpha) = \frac{2^{1-m}\Gamma(m-1)\Gamma(\frac{1}{2}(\alpha+1))}{p!\pi^{(\frac{1}{2}m-\frac{1}{2})}\Gamma(\frac{1}{2}(\alpha+m)+p)} \,.\, P_p(\alpha).$$

According to (2.10) $\partial^p/\partial\tau^p$ contains the factor σ^p, $p \geqq 1$. Hence all the integrals of the type given in (3.14) converge if $\alpha > -1$. Moreover, $K_p(0)$ is finite, $1/\Gamma(0) = 0$, consequently all expressions (3.14) vanish for $\alpha = 0$.

Since $R_N = O(\tau^N\sigma^N)$, the remainder term can be treated in an analogous way. It is clear that for the present purposes N may be any integer such that $-\frac{1}{2}(1 + m) + N \geqq -1$ or $2N \geqq m - 1$.

Summing up, it is now shown that the integral $I^\alpha f(O)$ extended to the double cone can be analytically continued to all values $\alpha > -1$ and that for these values it is a holomorphic function of α. Moreover $I^0 f(O) = f(O)$, that is I^0 *is the identity operator in the case of the double cone.*

We could have gone a bit farther and established the possibility of the analytic continuation down to arbitrary negative values of α. However, one difficulty would have remained, the possible occurrence of (simple) poles at the negative odd integers = the poles of $\Gamma(\frac{1}{2}(\alpha + 1))$. As a matter of fact, none of these poles actually occurs. Their disappearance must be the effect of the integration with respect to the angular variable, considered here in a very summary way. On p. 64 of my *Acta* paper I indicate how the holomorphic character of the unlimited analytic continuation can be established by an indirect method. This will be carried out here in §5.

4. Simple layer and double layer. We now pass to the simple layer

$$\frac{1}{H_m(\alpha)} \int_{S^O} g(y) r^{\alpha-m} dS \tag{4.1}$$

considered in formula (1.7), where now the vertex coincides with the origin and r is written instead of r_{xy}. The integral converges for $\alpha > m - 2$ and has to be continued analytically for $\alpha \leqq m - 2$. The portion S^O of the surface S can be parametrized by the variables τ and v in the following way. Through every point of S passes a unique ray issuing from the origin. On such a ray τ and v are constant, hence also the vectors b and c corresponding to v are constant, while σ varies. If we write the point of intersection of the ray with the surface S in the form $y = \sigma_S(b + \tau c)$, the equations $\sigma = \sigma_S(\tau, v)$ or $y = \sigma_S(\tau, v)(b + \tau c)$ and the additional condition $0 \leqq \tau \leqq 1$ yield the required parametrization of S^O, because b and c depend only on v.

Since v is indeterminate on the axis $y = \sigma a$, where $\tau = 1$, we divide S^O into two parts $S_A{}^O$ and $S_B{}^O$ in the following way. With an arbitrary δ such that $0 < \delta < 1$ the first part will be given by $0 \leqq \tau \leqq \delta$ and the second by $\delta < \tau \leqq 1$.

That part of the simple layer which relates to $S_B{}^O$ is an entire function of α vanishing for all even integers $\leqq 0$. Indeed the corresponding part of the integral in (4.1) never ceases to converge and $H_m(\alpha)$ has poles at these integers owing to the factor $\Gamma(\frac{1}{2}\alpha)$ (cf. formula (1.6)).

In order to treat that part of (4.1) which is taken over $S_A{}^O$, we have to express the surface element dS in a convenient way. The angular variable v on the sphere S_{m-2} can be expressed by $m - 2$ local parameters $\phi^1, \phi^2, \ldots, \phi^{m-2}$. Thus, according to (1.4), we can write in summary notations

$$dS = G(\tau, v) \,.\, d\tau \,.\, \prod d\phi^i, \quad dS_{m-2} = \Theta(v) \,.\, \prod d\phi^i,$$

hence $dS = H(\tau, v) d\tau dS_{m-2}$.

We set $\frac{1}{2}(\alpha + 2 - m) = \beta$ and can then write according to (1.6)

$$H_m(\alpha) = L_m(\alpha)\Gamma(\beta) \text{ where } L_m(\alpha) = \pi^{\frac{1}{2}m-1}2^{\alpha-1}\Gamma(\tfrac{1}{2}\alpha). \tag{4.2}$$

We also set $g(y) = g(\tau, v)$ and recall the relation $r^2 = \sigma^2\tau$ given in (2.8). Then we write that part of (4.1) which corresponds to $S_A{}^O$ in the form

$$U(\alpha) = \frac{1}{L_m(\alpha)} \int_{S_{m-2}} dS_{m-2} \frac{1}{\Gamma(\beta)} \int_0^\delta g(\tau, v) H(\tau, v) [\sigma_S(\tau, v)]^{\alpha-m} \tau^{\beta-1} d\tau. \tag{4.3}$$

Here $H(\tau, v)$, $\sigma_S(\tau, v)$, $g(\tau, v)$ are well-behaved even in τ and v by virtue of our hypothesis concerning the surface S and the function $g(y)$. Moreover σ_S is bounded away from 0. Hence we can apply our statement concerning the *extended one-dimensional case* (cf. p. 40), which gives that $(1/(\Gamma(\beta)) \int_0^\delta \ldots$ is a holomorphic function of β, hence also of α, and this is then also true for $U(\alpha)$. Owing to the presence of the factor $\Gamma(\frac{1}{2}\alpha)$ in $L_m(\alpha)$, the function $U(\alpha)$ vanishes for $\alpha = 0$ and $\alpha =$ a negative even integer.

There is very little to change in the case of a double layer. It is easily seen that $dr/dn = r^{-1}(y, n)$ hence

$$\frac{dr^{\alpha-m}}{dn} = (\alpha - m) r^{\alpha-m-1} \frac{dr}{dn} = (\alpha - m) r^{\alpha-m-2}(y, n).$$

The scalar product (y, n) (cf. (1.3)) is a well behaved function in τ and v. Owing to the lowered exponent the double layer integral in (1.7) converges only for $\alpha > m$, thus the need of continuation begins already at m.

The part relative to $S_B{}^O$ is again an entire function of α and vanishes for all even integers $\leqq 0$. On the other hand, with $\beta' = \frac{1}{2}(\alpha - m) = \beta - 1$,

$$\frac{\alpha - m}{\Gamma(\beta)} r^{\alpha-m} = \frac{2(\beta - 1)}{\Gamma(\beta)} r^{\beta-1} = \frac{2}{\Gamma(\beta - 1)} r^{\beta-1} = \frac{2}{\Gamma(\beta')} r^{\beta'}.$$

Thus, when treating the part relative to $S_A{}^O$, we obtain a formula of the same type as (4.3), $\beta' = \beta - 1$ playing for the double layer the same role as β played for the simple layer, and the results are essentially the same.

Our findings can be summed up as follows. *Both the simple layer and the double layer potentials can be continued analytically to arbitrary values of* α. *They are holomorphic functions of* α *which vanish for all even integers* $\leqq 0$.

5. The volume potential (continued). Now we return to the volume potential and clarify the properties of the part which relates to the domain $D_S{}^O - D_{\gamma_a}{}^O$ (cf. the beginning of §3). We want to prove that this part of I^α, when continued analytically, is holomorphic for $\alpha > -1$, and vanishes for $\alpha = 0$.

In the same way as we did in the previous section with the portion of surface S^O, we now divide the domain $D_S{}^O - D_{\gamma a}{}^O$ into two parts according as $0 \leqq \tau \leqq \delta$ or $\delta < \tau \leqq 1$. The volume potential relative to the second part is again an entire function vanishing for all even integers $\leqq 0$. Thus we only have to investigate the first part. This can be written in the form (cf. (3.1.))

$$\frac{2^{1-m}}{H_m(\alpha)} \int_{S_{m-2}} dS_{m-2} \int_0^\delta \tau^{\frac{1}{2}(\alpha-m)}(1 - \tau)^{m-2} d\tau \int_\gamma^{\sigma_S} f[\sigma(b + \tau c)]\sigma^{\alpha-1} d\sigma,$$

where σ_S, defined in the previous section, depends on τ and v, $\sigma_S = \sigma_S(\tau, v)$. We set

$$\int_\gamma^{\sigma_S} f[\sigma(b + \tau c)]\sigma^{\alpha-1} d\sigma = F(\tau, v, \alpha),$$

where F is well behaved in τ (and v) and holomorphic in α, since σ is bounded away from 0. Writing, as in the case of a simple layer, $H_m(\alpha) = L_m(\alpha)\Gamma(\beta)$ we see by virtue of our findings in the extended one-dimensional case that

$$\frac{1}{L_m(\alpha)} \frac{1}{\Gamma(\beta)} \int_0^\delta F(\tau, v, \alpha)\tau^{\beta-1}(1 - \tau)^{m-2} d\tau$$

can be continued analytically as a holomorphic function of α to any $\alpha > \alpha_0$, where α_0 is arbitrary. Thus it is a holomorphic function for $\alpha > -1$ any way, and again, owing to the presence of the factor $\Gamma(\alpha/2)$ in $L_m(\alpha)$, it vanishes for $\alpha = 0$.

This completes the proof of the fact stated in §3, that $I^0 f(O) = f(O)$, if I^0 is relative to the *original domain* $D_S{}^O$. This can clearly be expressed by the more inspiring formula

$$I^0 f(x) = f(x), \tag{5.1}$$

or by the statement that I^0 *is the identity operator*. This is *our principal result.* The passage from (5.1) to the relation

$$I_*^0 \overline{f, g, h}\,(x) = f(x) \tag{5.2}$$

follows from the properties of simple and double layers established in the previous section.

We conclude this paper by two additional remarks, the first concerning the unlimited analytic continuation of the volume potential, the second concerning the case of the infinite cone.

As indicated on p. 64 of our *Acta* paper, the unlimited holomorphic continuation of the volume potential can be reduced to that of simple layer and double layer potentials.

We consider the wave operator

$$\Delta = \frac{\partial^2}{(\partial x^0)^2} - \frac{\partial^2}{(\partial x^1)^2} - \dots - \frac{\partial^2}{(\partial x^{m-1})^2}.$$

Then, by virtue of Green's formula, we have in the notation (1.7) for $\alpha > m-2$

$$I^\alpha f(x) = I_*^{\alpha+2} \overline{\Delta f, \frac{df}{dn}, f(x)} \tag{5.3}$$

(see *Acta* paper, pp. 46–47). The left-hand side converges for $\alpha > m - 2$. On the right-hand side the volume potential and the simple layer converge for $\alpha + 2 > m - 2$, that is for $\alpha > m - 4$, while the double layer converges only for $\alpha + 2 > m$, that is only for $\alpha > m - 2$. Thus, seemingly nothing is gained as far as the analytic continuation of $I^\alpha f(x)$ is concerned. But if we take into account our results about the unlimited holomorphic continuation of the simple and the double layer and the fact that $I^{\alpha+2}\Delta$ is holomorphic for $\alpha > m - 4$, the possibility of the holomorphic continuation of $I^\alpha f(x)$ down to $m - 4$ is established. The iteration of this procedure, that is the application of formula (5.3) to $I^{\alpha+2}\Delta f$, $I^{\alpha+4}\Delta^2 f, \dots$, establishes the possibility of an unlimited holomorphic continuation of $I^\alpha f(x)$.

In the case of an infinite cone we suppose that $f(x)$ and its derivatives are not only well behaved, but also decrease rapidly enough at infinity. In this case formula (5.3) reduces to

$$I^\alpha f(x) = I^{\alpha+2}\Delta f(x), \tag{5.4}$$

and the iteration of this formula gives immediately the possibility of an unlimited holomorphic continuation. However, in order to establish the main relation $I^0 f(x) = f(x)$, we still have to go back to §3 and use the double-cone $D_{\gamma a}{}^O$. The treatment of the complementary expression $I_{II}{}^\alpha$ is in the infinite case still simpler than in the finite case.

University of Lund
University of Maryland

Bibliography

Most of the articles have been reproduced from offprints. In a few cases the page numbering is different from that in the original publications as given in the Bibliography. This is due to the fact that at that time, many journals changed the page make-up to produce the offprints.

The works of Marcel Riesz in chronological order

1. Representation of the analytical continuation of a given power series I, II (Hungarian)
   Math. és Phys. Lapok *16* (1907), 1–25, *17,* (1908), 96–108
2. Summable trigonometric series and summable power series
   Math. és Phys. Lapok *19* (1910), 1–56
3. The summability of a power series on the boundary of the circle of convergence (Hungarian)
   Hung. Acad. Sc. Math. és Természettud. értisitö *26* (1908), 221–229
4. Sur les séries trigonométriques
   C. R. Acad. Sci. Paris *145* (1907), 583–586
5. Sur les séries de Dirichlet
   C. R. Acad. Sci. Paris *148* (1909), 1658–1660
6. Sur la sommation des séries de Dirichlet
   C. R. Acad. Sci. Paris *149* (1909), 18–21
7. Sur les séries de Dirichlet et les séries entières
   C. R. Acad. Sci. Paris *149* (1909), 909–912
8. Sur un problème d'Abel
   Rend. Circ. Mat. Pal. *30* (1910), 339–345
9. Une méthode de sommation équivalente à la méthode des moyennes arithmétiques
   C. R. Acad. Sci. Paris *152* (1911), 1651–1654
10. Über einen Satz des Herrn Fatou
    J. reine angew. Math. *140* (1911), 89–99
11. Über summierbare trigonometrische Reihen
    Math. Ann. *71* (1911), 54–75
12. Sur la représentation analytique des fonctions définies par des séries de Dirichlet
    Acta. Math. *35* (1911), 253–270
13. Formule d'interpolation pour la dérivée d'un polynôme trigonométrique
    C. R. Acad. Sci. Paris *158* (1914), 1152–1154

14. Eine trigonometrische Interpolationsformel und einige Ungleichungen für Polynome
Jahresber. Deutsch. Math. Ver. *23* (1914), 354–368
15. A survey of the theory of trigonometric series (Swedish)
3 Skand. Mat. Kongr. 1913, printed 1915, 107–127
16. (with G.H. Hardy) The general theory of Dirichlet's series
Camb. Tracts in Math. *18* (1915), 1–78. Reprinted 1952
17. Neuer Beweis des Fatouschen Satzes
Gött. Nachr. (1916), 62–65
18. Sätze über Potenzreihen
Ark. Mat. Astr. Fys. *11* no. 12 (1916), 1–16
19. Sur l'hypothèse de Riemann
Acta Math. *40* (1916), 185–190
20. Über einen Satz des Herrn Serge Bernstein
Acta Math. *40* (1916), 337–347
21. Ein Konvergenzsatz für Dirichletsche Reihen
Acta Math. *40* (1916), 349–361
22. (with F. Riesz) Über die Randwerte einer analytischen Funktion
4e Congr. des Math. Scand. Stockholm 1916 (1920), 27–44
23. Sur le principe de Phragmén-Lindelöf
Proc. Camb. Phil. Soc. *20* (1920–21), 205–207; *21* (1922–23), 6
24. Sur le problème des moments
Première Note: Ark. Math. Astr. Fys. *16*, no. 12 (1921), 1–23. Deuxième Note: ibid *16*, no. 19 (1922), 1–21. Troisième Note: ibid *17*, no. 16 (1923), 1–52
25. Sur la sommation des séries de Fourier
Acta Szeged Sect. Math. *1* (1923), 104–113
26. Sur un théorème de la moyenne et ses applications
Acta Szeged Sect. Math. *1* (1923), 114–126
27. Sur le problème des moments et le théorème de Parseval correspondent
Acta Szeged Sect. Math. *1* (1923), 209–225
28. Sur l'équivalence de certaines méthodes de sommation
Proc. London Math. Soc. (2)*22* (1924), 412–419
29. Les fonctions conjuguées et les séries de Fourier
C. R. Acad. Sci. Paris *178* (1924), 1464–1467
30. Über die Summierbarkeit durch typische Mittel
Acta Szeged Sect. Math. *2* (1924), 18–31
31. (with E. Hilb) Neuere Untersuchungen über trigonometrische Reihen
Enc. d. Math. Wiss. *II C 10* (1924), 1189–1228
32. Sur les maxima des formes bilinéaires et sur les fonctionnelles linéaires
Acta Math. *49* (1927), 465–497
33. Sur les fonctions conjuguées
Math. Zeitschrift *27* (1927), 218–244
34. Sur certaines inégalités dans la théorie des fonctions avec quelques remarques sur les géometries non-euclidiennes
Proc. Roy. Physiogr. Soc. Lund *1* (1931), 1–21
35. Sur les ensembles compacts de fonctions sommables
Acta Szeged Sect. Math. *6* (1933), 136–142

36. Zum Eindeutigkeitssatz der fastperiodischen Funktionen
Proc. Roy. Physiogr. Soc. Lund *3*, no. 10 (1933), 1–9
37. Eine Bemerkung über den Eindeutigkeitssatz der Theorie der fastperiodischen Funktionen
Mat. Tidskr. B *1* (1934), 11–13
38. Intégrale de Riemann-Liouville et solution invariantive du problème de Cauchy pour l'équation des ondes
C. R. Congrès. Int. des Math. Oslo (1937) II, 44–45
39. Modules réciproques
C. R. Congrès Int. des Math. Oslo (1937) II, 36–37
40. Potentiels de divers ordres et leurs fonctions de Green
C. R. Congrès Int. des Math. Oslo (1937) II, 62–63
41. Volumes mixtes et facteurs invariants dans la théorie des modules
C. R. Congrès Int. des Math. Oslo (1937) II, 16
42. Intégrales de Riemann-Liouville et potentiels
Acta Szeged Sect. Math. *9* (1937–38), 1–42
Rectification ..., ibid, 116–118
43. L'intégrale de Riemann-Liouville et le problème de Cauchy pour l'équation des ondes
Soc. Math. de France. Conf. à la Réunion Int. des Math. Paris, Juillet 1937, 153–170
44. An intuitive picture of non-euclidean geometry. Geometric excursions into relativity theory (Swedish)
Acta Physiogr. Soc. Lund, NF, vol. 53, no. 9 (1943), 1–7
45. Sur certaines notions fondamentales en théorie quantique relativiste
10e Congr. des Math. Scand. Copenhagen 1946, pp. 123–148
46. Eléments de probabilité en théorie quantique relativiste
Försäkr. mat. studier till. Filip Lundberg. Stockholm, 1946
47. L'intégrale de Riemann-Liouville et le problème de Cauchy
Acta Math. *81* (1949), 1–223
48. Remarque sur les fonctions holomorphes
Acta Szeged Sect. Math. *12* (1950), 53–56
49. Sur le potentiel de Liénard-Wiechert attaché à une ligne d'univers
C. R. Acad. Sci. Paris *234* (1952), 2159–2161
50. Sur le potentiel retardé attaché à un courant continu
C. R. Acad. Sci. Paris *234* (1952), 2260–2261
51. Court exposé des propriétés principales de la mesure de Lebesgue
Ann. Soc. Pol. Math. *25* (1952), 298–308
52. Sur le lemme de Zolotareff et sur la loi de réciprocité des restes quadratiques
Math. Scand. *1* (1953), 159–169
53. L'équation de Dirac en relativité générale
12e Congr. des Math. Scand. Lund 1953, pp. 241–259
54. (with A.E. Livingstone) A short proof of a classical theorem in the theory of the Fourier integrals
Amer. Math. Monthly *62* (1955), 434–437

55. Problems related to characteristics surfaces
Proc. Conf. Diff. Eq. Ded. to A. Weinstein, Univ. of Maryland (1956), 57–71
56. A special characteristic surface – a new relativistic model for a particle?
Fund. Res. in Appl. Math., Dept. Math., Univ. of Maryland.
Interim technical report *25* (1957), pp. 1–18
57. Clifford numbers and spinors, Chaps. 1–4
Lecture notes, Inst. Fluid Dyn. Appl. Math., Univ. of Maryland (1957–1958), 1–193
58. A geometric solution of the wave equation in space-time of even dimension
Comm. Pure Appl. Math. *13* (1960), 329–354
59. The analytical continuation of the Riemann-Liouville integral in the hyperbolic case
Canad. J. Math. *13* (1961), 37–47

Acknowledgements

The editors would like to thank the publishers and copyright holders of Marcel Riesz's papers for granting permission to reprint them here.

The numbers following each source correspond to the numbering of the articles in the Bibliography.

Acta Scientiarium Mathematicarum: 25, 26, 27, 30, 35, 42, 48
Akademie der Wissenschaften zu Göttingen: 17
B. G. Teubner Verlag: 14
Cambridge University Press: 23
Canadian Mathematical Congress Montreal: 59
Clarendon Press: 28
CWK Gleerup Publishers: 34
Éditions Gauthier-Villars: 4, 5, 6, 7, 9, 13, 29, 49, 50
Institut Mittag-Leffler: 12, 19, 20, 21, 32, 47
Institute for Fluid Dynamics and Applied Mathematics: 55
John Wiley & Sons, New York: 58
Jul. Gjellerups Forlag a–s: 45
Matematisk Forening: 36
Mathematica Scandinavica: 52
Matematisk Institutt Oslo, Universitetsbiblioteket: 38, 39, 40, 41
Circolo Matematico di Palermo: 8
Royal Swedish Academy of Sciences: 18, 24
Société Mathématique de France: 43
Walter de Gruyter & Co.: 10

GPSR Compliance

The European Union's (EU) General Product Safety Regulation (GPSR) is a set of rules that requires consumer products to be safe and our obligations to ensure this.

If you have any concerns about our products, you can contact us on ProductSafety@springernature.com

In case Publisher is established outside the EU, the EU authorized representative is:

Springer Nature Customer Service Center GmbH
Europaplatz 3
69115 Heidelberg, Germany

Batch number: 10682269

Printed by Printforce, the Netherlands